The IGF System

CONTEMPORARY ENDOCRINOLOGY

P. Michael Conn, SERIES EDITOR

19. ***Human Growth Hormone:*** *Basics and Clinical Research,* edited by *ROY G. SMITH AND MICHAEL O. THORNER,* 1999
18. ***Menopause: Endocrinology and Management,*** edited by *DAVID B. SEIFER AND ELIZABETH A. KENNARD,* 1999
17. ***The IGF System:*** *Molecular Biology, Physiology, and Clinical Applications,* edited by *RON G. ROSENFELD AND CHARLES T. ROBERTS JR.,* 1999
16. ***Neurosteroids:*** *A New Regulatory Function in the Nervous System,* edited by *ETIENNE-EMILE BAULIEU, MICHAEL SCHUMACHER, AND PAUL ROBEL,* 1999
15. ***Autoimmune Endocrinopathies,*** edited by *ROBERT VOLPÉ,* 1999
14. ***Hormone Resistance Syndromes,*** edited by *J. LARRY JAMESON,* 1999
13. ***Hormone Replacement Therapy,*** edited by *A. WAYNE MEIKLE,* 1999
12. ***Insulin Resistance:*** *The Metabolic Syndrome X,* edited by *GERALD M. REAVEN AND AMI LAWS,* 1999
11. ***Endocrinology of Breast Cancer,*** edited by *ANDREA MANNI,* 1999
10. ***Molecular and Cellular Pediatric Endocrinology,*** edited by *STUART HANDWERGER,* 1999
9. ***The Endocrinology of Pregnancy,*** edited by *FULLER W. BAZER,* 1998
8. ***Gastrointestinal Endocrinology,*** edited by *GEORGE H. GREELEY, JR.,* 1999
7. ***Clinical Management of Diabetic Neuropathy,*** edited by *ARISTIDIS VEVES,* 1998
6. ***G Proteins, Receptors, and Disease,*** edited by *ALLEN M. SPIEGEL,* 1998
5. ***Natriuretic Peptides in Health and Disease,*** edited by *WILLIS K. SAMSON AND ELLIS R. LEVIN,* 1997
4. ***Endocrinology of Critical Diseases,*** edited by *K. PATRICK OBER,* 1997
3. ***Diseases of the Pituitary:*** *Diagnosis and Treatment,* edited by *MARGARET E. WIERMAN,* 1997
2. ***Diseases of the Thyroid,*** edited by *LEWIS E. BRAVERMAN,* 1997
1. ***Endocrinology of the Vasculature,*** edited by *JAMES R. SOWERS,* 1996

The IGF System

Molecular Biology, Physiology, and Clinical Applications

Edited by

Ron G. Rosenfeld, MD

Charles T. Roberts, Jr., PhD

Oregon Health Sciences University

Portland, OR

Humana Press

Totowa, New Jersey

999 Riverview Drive, Suite 208
Totowa, New Jersey 07512

For additional copies, pricing for bulk purchases, and/or information about other Humana titles, contact Humana at the above address or at any of the following numbers:
Tel: 973-256-1699; Fax: 973-256-8341; E-mail: humana@humanapr.com or visit our website at http://humanapress.com

Cover design by Patricia F. Cleary.
Cover art: Fig. 19-8; *see* complete legend on p. 442.

This publication is printed on acid-free paper. ∞
ANSI Z39.48-1984 (American National Standards Institute)
Permanence of Paper for Printed Library Materials.

Printed in the United States of America. 10 9 8 7 6 5 4 3 2 1

Library of Congress Cataloging-in-Publication Data

The IGF system : molecular biology, physiology, and clinical applications/edited by Ron G. Rosenfeld, Charles T. Roberts, Jr.
p. cm. -- (Contemporary endocrinology ; 17)
Includes index.
ISBN 0-89603-692-8 (alk. paper)
1. Somatomedin--Physiological effects. 2. Somatomedin--Pathophysiology. 3. Insulin-like grow factor-binding proteins. I. Roberts, Charles T. II. Rosenfeld, Ron G. III. Series: Contemporary endocrinology (Totowa, N.J.)
[DNLM: 1. Insulin-Like Growth Factor I--physiology. 2. Molecular Biology. 3. Insulin-Like Growth Factor I--therapeutic use. 4. Insulin-Like Growth Factor II--physiology. 5. Insulin-Like Growth Factor II--therapeutic use. 6. Insulin-Like Growth-Factor-Binding Proteins--physiology. QU 107 I24 1999]
QP552.S65I33 1999
612'.015756--dc21
DNLM/DLC
for Library of Congress 98-55264
CIP

PREFACE

It has been over 40 years since the original report by Salmon and Daughaday demonstrating that the ability of GH to stimulate sulfation of cartilage was mediated by a "sulfation factor." In the ensuing decades, it has become apparent that this "sulfation factor activity" encompasses a complex system of ligands (IGFs), receptors, and carrier proteins that are, in turn, responsible for a wide array of cellular actions. The IGF system has been demonstrated to be critically involved in both intrauterine and postnatal growth, and to have important implications in cancer biology as well, owing to the ability of the IGFs to function in endocrine, paracrine, and autocrine modes and given the wide distribution of IGFs in virtually every organ system.

The contributions to *The IGF System* reflect the wide span of interest in the IGF system and its implications for normal and abnormal growth and metabolism. The chapters have been divided into four broad sections: I. Molecular biology of the IGF system; II. Biological actions of the IGFs; III. IGF physiology; and IV. Clinical aspects of the IGFs. We have made every effort to highlight the major contemporary themes in IGF biology, but as is inevitable in such a fast-moving field, perspectives will continue to change as new information is accumulated. Most importantly, we have remained sensitive to the many remaining gaps in our knowledge of the role(s) of the IGF peptides, receptors, and binding proteins in cell growth and replication, and are always aware of the admonition that "it is better to have unanswered questions than unquestioned answers."

Ron G. Rosenfeld, *MD*
Charles T. Roberts, *JR., PhD*

CONTENTS

CONTRIBUTORS

Renato Baserga, MD, *Department of Microbiology and Immunology, Kimmel Cancer Center, Thomas Jefferson University, Philadelphia, PA*

Robert C. Baxter, PhD, DSc, *Kolling Institute of Medical Research, Royal North Shore Hospital, St. Leonards, Australia*

David J. Baylink, MD, *Mineral Metabolism, Jerry L. Pettis VA Medical Center; Departments of Medicine and Biochemistry, Loma Linda University, Loma Linda, CA*

Michel Binoux, MD, *Institut National de la Santé et de la Recherche Médicale, Hôpital Saint Antoine, Paris, France*

Vicky A. Blakesley, MD, PhD, *Diabetes Branch, National Institute of Diabetes and Digestive and Kidney Diseases, National Institutes of Health, Bethesda, MD*

Yves R. Boisclair, PhD, *Department of Animal Science, Cornell University, Ithaca, NY*

Christine P. Burren, MD, *Department of Pediatrics, School of Medicine, Oregon Health Sciences University, Portland, OR*

Andrew A. Butler, PhD, *Diabetes Branch, National Institute of Diabetes and Digestive and Kidney Diseases, National Institutes of Health, Bethesda, MD*

Won Kyu Choi, MD, *Department of Pediatrics, School of Medicine, Oregon Health Sciences University, Portland, OR*

David R. Clemmons, MD, *Department of Medicine, University of North Carolina at Chapel Hill, Chapel Hill, NC*

Pinchas Cohen, MD, *Department of Pediatrics, University of California at Los Angeles, Los Angeles, CA*

Cheryl A. Conover, PhD, *Endocrine Research Unit, Mayo Clinic and Mayo Foundation, Rochester, MN*

Anna L. Crown, MD, *University of Bristol Division of Surgery, Department of Hospital Medicine, Bristol Royal Infirmary, Bristol, United Kingdom*

William H. Daughaday, MD, *Balboa Island, CA*

Marlys R. Drange, MD, PhD, *Division of Endocrinology and Metabolism, Cedars-Sinai Medical Center, Los Angeles, CA*

A. Joseph D'Ercole, MD, *Department of Pediatrics, University of North Carolina at Chapel Hill, Chapel Hill, NC*

Eva L. Feldman, MD, PhD, *Department of Neurology, University of Michigan, Ann Arbor, MI*

E. Rudolph Froesch, MD, *Department of Internal Medicine, Division of Endocrinology and Diabetes, University Hospital, Zurich, Switzerland*

Robert A. Gabbay, MD, PhD, *Section of Endocrinology, Diabetes, and Metabolism, Hershey Medical Center, Penn State Geisinger Health System, Hershey, PA*

Linda C. Giudice, MD, PhD, *Division of Reproductive Endocrinology and Fertility, Department of Obstetrics and Gynecology, Stanford University Medical Center, Stanford, CA*

Donna L. Graham, *Department of Pediatrics, School of Medicine, Oregon Health Sciences University, Portland, OR*

Vivian Hwa, PhD, *Department of Pediatrics, School of Medicine, Oregon Health Sciences University, Portland, OR*
Jeff M. P. Holly, PhD, *University of Bristol Division of Surgery, Department of Hospital Medicine, Bristol Royal Infirmary, Bristol, United Kingdom*
P. Elly Holthuizen, PhD, *Laboratory for Physiological Chemistry, Utrecht University, Utrecht, The Netherlands*
Atsushi Hongo, PhD, *Department of Microbiology and Immunology, Kimmel Cancer Center, Thomas Jefferson University, Philadelphia, PA*
Angela Ingermann, *Department of Pediatrics, School of Medicine, Oregon Health Sciences University, Portland, OR*
Wieland Kiess, MD, *Children's Hospital, University of Leipzig, Leipzig, Germany*
Ho-Seong Kim, MD, *Department of Pediatrics, School of Medicine, Oregon Health Sciences University, Portland, OR*
Christian A. Koch, MD, *National Institute of Child Health and Human Development, National Institutes of Health, Bethesda, MD*
Anatolii P. Koval, PhD, *Diabetes Branch, National Institute of Diabetes and Digestive and Kidney Diseases, National Institutes of Health, Bethesda, MD*
Claude Lalou, PhD, *Institut National de la Santé et de la Recherche Médicale, Hôpital Saint Antoine, Paris, France*
Derek LeRoith, MD, PhD, *Diabetes Branch, National Institute of Diabetes and Digestive and Kidney Diseases, National Institutes of Health, Bethesda, MD*
Phillip S. Leventhal, PhD, *Department of Neurology, University of Michigan, Ann Arbor, MI*
Abel Lopez-Bermejo, MD, *Department of Pediatrics, School of Medicine, Oregon Health Sciences University, Portland, OR*
P. Kay Lund, PhD, *Departments of Physiology and Pediatrics and Center for Gastrointestinal Biology and Disease, University of North Carolina at Chapel Hill, Chapel Hill, NC*
Laura A. Maile, PhD, *University of Bristol Division of Surgery, Department of Hospital Medicine, Bristol Royal Infirmary, Bristol, United Kingdom*
Janet L. Martin, PhD, *Kolling Institute of Medical Research, Royal North Shore Hospital, St. Leonards, Australia*
Shlomo Melmed, MD, *Research Institute, UCLA School of Medicine, Cedars-Sinai Medical Center, Los Angeles, CA*
Giuseppe Minniti, MD, *Department of Pediatrics, School of Medicine, Oregon Health Sciences University, Portland, OR*
Subburaman Mohan, PhD, *Mineral Metabolism, Jerry L. Pettis VA Medical Center; Departments of Medicine, Biochemistry, and Physiology, Loma Linda University, Loma Linda, CA*
Sarah Mohseni-Zadeh, PhD, *Institut National de la Santé et de la Recherche Médicale, Hôpital Saint Antoine, Paris, France*
Alan C. Moses, MD, *General Clinical Research Center and Division of Endocrinology, Beth Israel Deaconess Medical Center, Harvard Medical Center, Boston, MA*
Srinivasa R. Nagalla, MD, *Department of Pediatrics, School of Medicine, Oregon Health Sciences University, Portland, OR*

S. Peter Nissley, MD, *National Cancer Institute, National Institutes of Health, Bethesda, MD*
Youngman Oh, PhD, *Department of Pediatrics, School of Medicine, Oregon Health Sciences University, Portland, OR*
Yumiko Okubo, MD, *Diabetes Branch, National Institute of Diabetes and Digestive and Kidney Diseases, National Institutes of Health, Bethesda, MD*
Guck T. Ooi, PhD, *Prince Henry's Institute of Medical Research, Clayton, Victoria, Australia*
Kisoo Pai, MD, *Department of Pediatrics, School of Medicine, Oregon Health Sciences University, Portland, OR*
Marco Prisco, PhD, *Department of Microbiology and Immunology, Kimmel Cancer Center, Thomas Jefferson University, Philadelphia, PA*
Arlan L. Rosenbloom, MD, *Department of Pediatrics, University of Florida College of Medicine, Children's Medical Services Center, Gainesville, FL*
Ron G. Rosenfeld, MD, *Department of Pediatrics, School of Medicine, Oregon Health Sciences University, Portland, OR*
Stephen M. Rosenthal, MD, *Department of Pediatrics, University of California–San Francisco, San Francisco, CA*
Jesse Roth, MD, FACP, *Department of Medicine, Johns Hopkins University School of Medicine, Baltimore, MD*
Kristina I. Rother, MD, *National Institute of Child Health and Human Development, National Institutes of Health, Bethesda, MD*
Peter S. Rotwein, MD, *Department of Medicine, Division of Molecular Medicine, Oregon Health Sciences University, Portland, OR*
James W. Russell, MD, *Department of Neurology, University of Michigan, Ann Arbor, MI*
William D. Salmon, Jr., MD
Martin O. Savage, MA, MD, FRCP, *Paediatric Endocrinology, St. Batholomew's Hospital, London, United Kingdom*
Christoph Schmid, MD, *Department of Internal Medicine, Division of Endocrinology and Diabetes, University Hospital, Zurich, Switzerland*
Kenneth Siddle, PhD, *Department of Clinical Biochemistry,University of Cambridge, Addenbrooke's Hospital, Cambridge, United Kingdom*
Maria A. Soos, PhD, *Department of Clinical Biochemistry,University of Cambridge, Addenbrooke's Hospital, Cambridge, United Kingdom*
Anna Spagnoli, MD, *Department of Pediatrics, School of Medicine, Oregon Health Sciences University, Portland, OR*
Paul H. Steenbergh, PhD, *Laboratory for Physiological Chemistry, Utrecht University, Utrecht, The Netherlands*
John S. Sussenbach, PhD, *Laboratory for Physiological Chemistry, Utrecht University, Utrecht, The Netherlands*
Louis E. Underwood, MD, *Division of Pediatric Endocrinology, School of Medicine, University of North Carolina at Chapel Hill, Chapel Hill, NC*
Peter Vorwerk, MD, *Department of Pediatrics, School of Medicine, Oregon Health Sciences University, Portland, OR*

DON L. V. WANEK, *Department of Pediatrics, School of Medicine, Oregon Health Sciences University, Portland, OR*
STUART A. WEINZIMER, MD, *Department of Pediatrics, Division of Endocrinology/ Diabetes, Children's Hospital of Philadelphia, University of Pennsylvania School of Medicine, Philadelphia, PA*
HAIM WERNER, PHD, *Department of Clinical Biochemistry, Sackler School of Medicine, Tel Aviv University, Tel Aviv, Israel*
ELIZABETH M. WILSON, *Department of Pediatrics, School of Medicine, Oregon Health Sciences University, Portland, OR*
KATIE A. WOODS, MD, MRCP, *University Department of Paediatrics, John Radcliffe Hospital, Oxford, United Kingdom*
YOSHITAKA YAMANAKA, MD, *Department of Pediatrics, School of Medicine, Oregon Health Sciences University, Portland, OR*
DOO HYUN YANG, MD, *Department of Pediatrics, School of Medicine, Oregon Health Sciences University, Portland, OR*
JÜRGEN ZAPF, MD, *Department of Internal Medicine, Division of Endocrinology and Diabetes, University Hospital, Zurich, Switzerland*

1 Historical Perspective

The Origins and Development of the Somatomedin Hypothesis

William H. Daughaday, MD and William D. Salmon, Jr, MD

Contents

RECOGNIZING THE ROLE OF THE PITUITARY IN SKELETAL GROWTH

The initial stimulus for the study of the role of the anterior pituitary gland in the regulation of growth arose from clinical observations. Marie is generally given the credit for characterizing and naming the syndrome of acromegaly in his 1886 paper *(1)*. In a review only 3 yr later *(2)* he described one characteristic of the pathologic anatomy as "hypertrophy of the pituitary body with enormous dilatation of the sella turcica." However, pituitary tumor and sellar enlargement were not always present, and there were various theories on the possible relationship of pituitary abnormalities to the growth disturbance. Harvey Cushing of Johns Hopkins Medical School introduced the terms

From: *Contemporary Endocrinology: The IGF System*
Edited by: R. Rosenfeld and C. Roberts Jr © Humana Press, Inc., Totowa, NJ

hyperpituitarism and *hypopituitarism*, and in 1912 he championed the "hyperpituitarism conception of acromegaly" in his monograph on the pituitary *(3)*. The argument was based on a review of the literature and his own extensive clinical, pathological, and experimental observations. He concluded, "Certainly most of the circumstantial evidence in our possession points in the direction of an oversecretion, whether normal or pathological. . . ." He added, "One point at least is now generally accepted, namely, that the skeletal changes in gigantism and acromegaly are expressions of the same morbid influence. . . ."

Cushing became the director of the experimental surgery division of the Hunterian Laboratory, which opened in 1905. His course in comparative surgery was one of the most popular in the medical school and attracted many future leaders in academic surgery and other fields. In 1907 one of the small group of select students participating in this course was Herbert M. Evans, who must have been affected by Cushing's intense interest in diseases of the pituitary. Evans was to assume a major role in research into the pituitary regulation of growth, and Melvin Grumbach *(4)*, in his Presidential Address at the 64th Annual Meeting of the Endocrine Society, presented a biography of this pioneer in 20th century endocrinology and gave us some insight into his dominating personality.

After graduation from Johns Hopkins Medical School in 1908 Evans joined the Anatomy Department of Franklin Mall at the same institution. Seven years later at the age of 33 he accepted the chairmanship of the Department Anatomy at the University of California in Berkeley. Isolation of the hormones of the anterior pituitary gland and characterization of their functions became the focus of his department. In 1921 Evans and Long *(5)* reported growth stimulation in normal rats by injections of bovine anterior pituitary extracts for up to 2 mo, and the following year *(6)* they characterized the results of extended treatment as "producing constantly a certain degree of true gigantism." In 1929 Putnam et al. *(7)* at Harvard Medical School and Peter Bent Brigham Hospital reported that chronic injections of bovine anterior pituitary extracts caused "experimental acromegaly" in an English bulldog. The evidence was now convincing that excess of an anterior pituitary growth hormone (GH) was the "morbid influence" in gigantism and acromegaly.

However, progress in isolating GH required a suitable assay animal. When Evans arrived at Berkeley, one of the two faculty members retained from the old Anatomy Department was Philip E. Smith, a technically gifted and independent investigator. He made an enormous contribution to pituitary research by the development of a relatively simple method for hypophysectomy of rats through a parapharyngeal approach. This made it possible to obtain long-term survival of a mammal after hypophysectomy without the effects of hypothalamic injury, which had complicated previous pituitary ablation procedures. He observed marked impairment of growth in hypophysectomized rats, which was corrected by injections of bovine anterior pituitary extracts. Smith left the University of California for Stanford in 1926, and the following year he accepted an invitation to become Professor of Anatomy at the College of Physicians and Surgeons of Columbia University. A preliminary report of his work on hypophysectomized rats was published in 1927, but the detailed description of the operative procedure and the results of replacement therapy appeared in his classic paper of 1930 *(8)*.

The ability of anterior pituitary extracts to cause a dose-related widening of the proximal tibial epiphyseal plate of hypophysectomized rats was the basis of a sensitive assay for GH *(9)*. After the full report in 1945 by Li et al. *(10)* of a method for isolation of bovine GH (bGH), a standardized bioassay procedure ("the tibia test") was proposed by Greenspan et al. *(11)*. A minimal effective dose in this assay was considered to be a total of 5 μg of purified bGH administered by daily injection over 4 d. Kinsell et al. *(12)* reported in 1948 that injections of plasma from a patient with acromegaly into hypophysectomized rats widened the tibial epiphyseal plate more than injections of normal human plasma, which had little effect. Gemzell et al. *(13)* reported in 1955 that plasma fractions from a patient with acromegaly and from another with gigantism appeared to contain high levels of GH as determined by the tibia test, but fractions from normal human plasma were inactive. In view of the assay sensitivity, the amount of plasma or plasma fractions injected, and the levels of GH determined by radioimmunoassay (RIA) in plasma of patients with acromegaly and gigantism, we know now that GH alone could not account for these bioassay results. Because insulin-like growth factors (IGF-I, IGF-II) are known to be present in plasma fractions similar to those used by Gemzell, it seems likely that they contributed to the tibial epiphyseal stimulation.

PRELIMINARY STUDIES ON SULFATE INCORPORATION AS A PARAMETER OF GH ACTION ON CARTILAGE

The early results of bioassay of human plasma for GH appeared promising, but it was clear that greater sensitivity was necessary to detect GH in normal subjects. In the early 1950s one of us (W. H. D.) was beginning a career in academic clinical endocrinology and was attracted to the possibility that measuring [^{35}S]sulfate incorporation into tibial epiphyseal cartilage might provide a more sensitive and less laborious GH assay. Radioactive sulfur became generally available to investigators in the immediate post-World War II years, and it had been shown that [^{35}S]sulfate was incorporated by cartilage into matrix mucopolysaccharides (glycosaminoglycans). Subsequently it was learned that the sulfated glycosaminoglycans of cartilage (chondroitin sulfate, keratan sulfate), along with certain oligosaccharides, are linked covalently to protein, and these proteoglycans form giant aggregates in the extracellular matrix with hyaluronan (hyaluronic acid) and a stabilizing (link) glycoprotein *(14)*. The proteoglycans play an important role not only in hydration of the matrix but also in the regulation of cellular function by their ability to bind certain growth factors *(15)* and IGF-binding proteins (IGFBPs) *(16)*.

Ellis et al. *(17)* reported in 1953 that hypophysectomy of young rats resulted in decreased [^{35}S]sulfate incorporation into costal cartilage in vivo, and partial restoration followed the injection of bGH daily for 3 d. Denko and Bergenstal *(18)* examined the effects of hypophysectomy and GH treatment in rats on [^{35}S]sulfate incorporation into cartilage from multiple sites including the proximal tibia. They used a schedule of 8 daily injections of [^{35}S]sulfate with or without GH over a period of 9 d. Decreased sulfate incorporation in cartilage of hypophysectomized rats compared to normal rats was found only in costal cartilage, but GH treatment increased sulfate incorporation into cartilage from all sites tested. These observations suggested that measurement of [^{35}S]sulfate incorporation into tibial epiphyseal cartilage might provide an assay for GH with improved sensitivity. In initial studies W. H. D. with Murphy and Hartnett *(19)* observed a progressive fall of sulfate incorporation in the proximal tibia (including epiphyseal cartilage) of rats during the first 2 wk after hypophysectomy. After this postoperative interval bGH administered 24 h before and at the time of injection of [^{35}S]sulfate resulted in a dose-related increase in proximal tibial sulfate incorporation (Fig. 1). A significant increase was detected with bGH in a total dose of 10 μg, which was similar to the sensitivity of the conventional tibial assay with a shortening of the treatment period from 4 d to 2 d. At this time a postdoctoral clinical fellow (W. D. S.) joined the laboratory, and we undertook a study of the

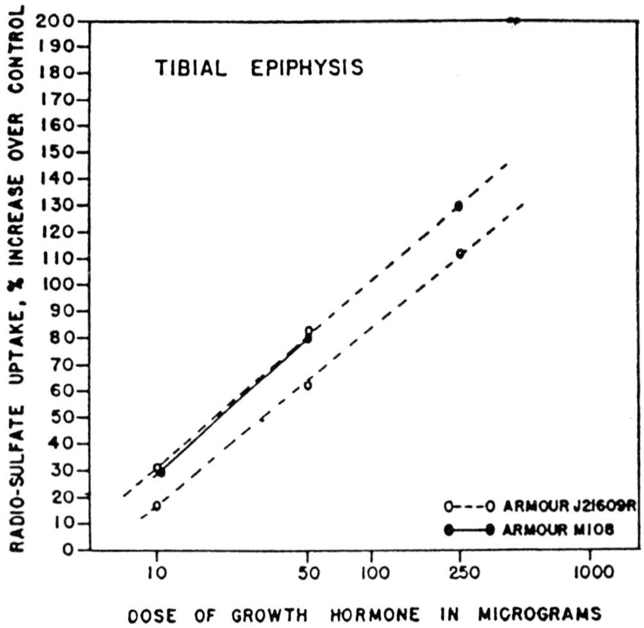

Fig. 1. Effects of treatment of hypophysectomized rats with different doses of bGH (one half administered 24 h before and one half at the time of injection of [^{35}S]sulfate) on proximal tibial sulfate incorporation (measured 24 h after the second injection). (Reprinted with permission from W. R. Murphy, W. H. Daughaday, and C. Hartnett [1956] J Lab Clin Med 47:715–722, Mosby-Year Book.)

ability of GH to increase [^{35}S]sulfate incorporation into isolated cartilage. It was demonstrated that sulfate incorporation in vitro by costal, nasal, and xiphoid cartilage of hypophysectomized rats was decreased and could be restored by bGH treatment of the animals *(20)*. Costal cartilage was the obvious choice for a future bioassay because multiple samples were easily obtained from each rat. It was disappointing to find that concentrations of bGH up to 50 µg/mL caused little or no increase of sulfate incorporation by cartilage incubated in a simple phosphate-buffered medium containing glucose as the only nutrient. Negative results were obtained when 140 ng/mL of bGH was added to an incubation medium containing hypophysectomized rat plasma.

BIRTH OF THE SOMATOMEDIN HYPOTHESIS

While in the laboratory of Carl Cori as a research fellow, W. H. D. with Charles Park et al. *(21)* had investigated the ability of GH to correct the insulin hypersensitivity of the diaphragm of hypophysectomized rats. This was easy to accomplish with GH administration 24 h before the removal of the diaphragm, but direct addition of GH to the diaphragm in vitro led to an insulin-like action rather than an anti-insulin action (a paradoxical response that has intrigued many investigators). The possibility of an intermediary serum factor responsible for the anti-insulin action of GH in vivo was discussed and later investigated by Bornstein and Park *(22)*. These observations of GH action on the diaphragm suggested the possibility that GH action on cartilage might also involve a circulating intermediary factor.

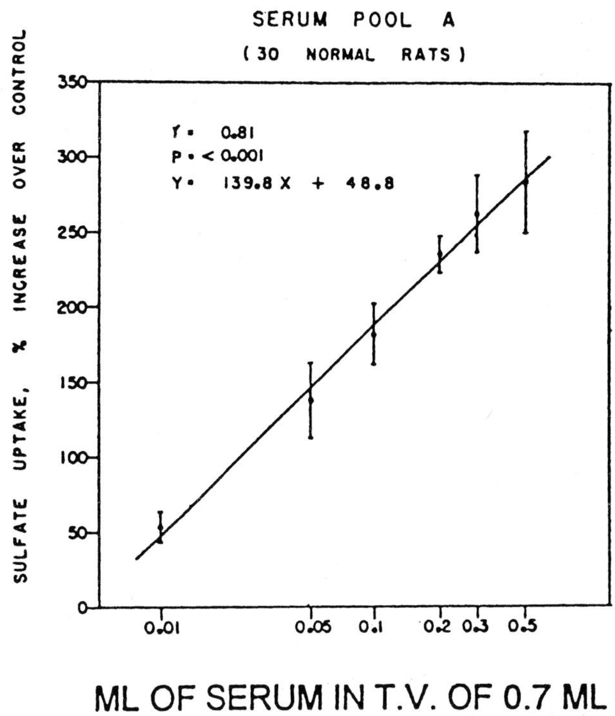

Fig. 2. Effects of different amounts of normal rat serum in the incubation medium on sulfate incorporation by costal cartilage from hypophysectomized rats. Indicated volumes of serum were added to KPS medium (Krebs' phosphate-buffered saline + glucose) containing [^{35}S]sulfate to yield a total volume (T.V.) of 0.7 mL for each cartilage sample; incubation period 24 h. (Reprinted with permission from W. D. Salmon, Jr. and W. H. Daughaday [1957] J Lab Clin Med 49:825–836, Mosby-Year Book.)

In experiments to test this hypothesis the addition of normal rat plasma to the tissue incubation medium restored [^{35}S]sulfate incorporation by costal cartilage from hypophysectomized rats to the level of cartilage from normal rats, but the addition of hypophysectomized rat plasma did not *(20)*. Sulfation-promoting activity was present in serum as well as plasma from normal rats, and a dose-related cartilage response was demonstrated at serum concentrations from 1.4% to 71% (Fig. 2). Insulin stimulated cartilage sulfate incorporation, but the high concentrations necessary for this action eliminated that hormone from serious consideration as the cartilage stimulator in normal rat plasma or serum. Thyroxine and triiodothyronine were ineffective. Cartilage stimulatory activity was detected in plasma of hypophysectomized rats 6 h after the intraperitoneal injection of 500 µg of bGH and reached a peak at 12 h (Fig. 3). There was a slight delay in the increase of sulfate incorporation by isolated cartilage from the GH-treated rats in comparison to the rise of sulfation-promoting activity in plasma from the same rats. Considerable stimulatory activity was retained in normal rat serum after 24 h of dialysis, which suggested that "an active component was either protein in nature or highly protein-bound" *(20)*. The operational term "sulfation factor" was given to this GH-dependent serum component.

It became apparent that amino acids were essential for optimal activity of the sulfation factor on rat cartilage in vitro, and subsequent studies were performed using a tissue

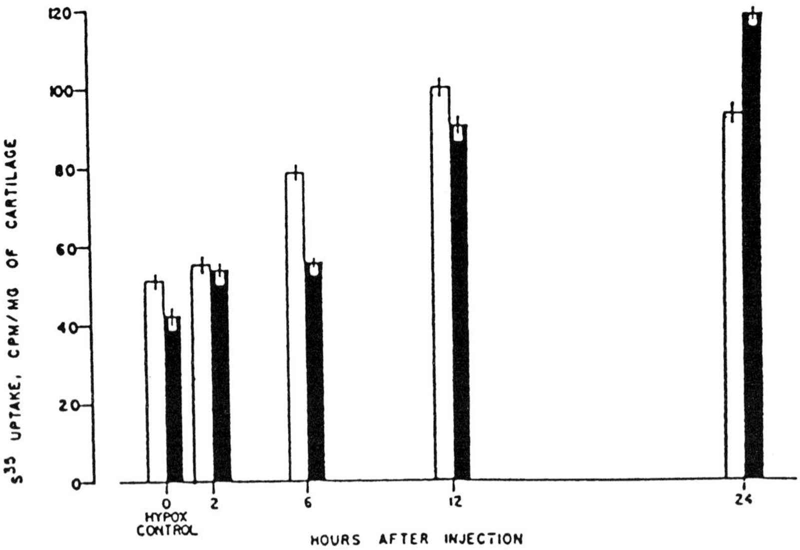

Fig. 3. Time course of the effects of bGH treatment of hypophysectomized rats on costal cartilage sulfate incorporation and on serum sulfation-promoting activity. Cartilage samples from control rats and from rats 2, 6, 12, and 24 h after the injection of bGH (500 μg) were incubated for 24 h in KPS medium containing [^{35}S]sulfate (*solid columns*). In separate incubations cartilage samples from untreated hypophysectomized rats were incubated individually for 24 h in 0.5 mL of pooled plasma from GH-treated or control rats + 0.2 mL of KPS medium containing [^{35}S]sulfate (*open columns*). (Reprinted with permission from W. D. Salmon, Jr. and W. H. Daughaday [1957] J Lab Clin Med 49:825–836, Mosby-Year Book.)

incubation medium containing amino acids. In 1966 W. H. D. and Reeder *(23)* reported that treating hypophysectomized rats with injections of bGH in a daily dose of 200 μg led to as much as a 26-fold increase in [^{3}H]thymidine incorporation by costal cartilage (Fig. 4). The stimulatory effect of bGH was evident 24 h after hormonal injections were started and reached a maximum at 48 h. A decline followed, but elevation of [^{3}H]thymidine incorporation to seveal times that of cartilage from untreated hypophysectomized rats was still present 6 d after the institution of GH treatment. [^{3}H]thymidine incorporation into costal cartilage from hypophysectomized rats was stimulated by the addition of normal rat serum to the incubation medium with a significant increase of incorporation at a serum concentration of 1.7% (Fig. 5). Serum from hypophysectomized rats was somewhat inhibitory. The presence in the incubation medium of 0.83–83 μg/mL of bGH for 24 h before the addition of [^{3}H]thymidine and continuation of the incubation for either 4 or 24 h did not affect cartilage thymidine incorporation. It was concluded that "GH does not act directly on cartilage cells but stimulates cell division indirectly by a circulating serum component which may be identical to sulfation factor" *(23)*.

The apparent inability of GH to directly stimulate sulfate incorporation into glycosaminoglycans (a differentiated response) or thymidine incorporation into DNA (a mitogenic response) of isolated costal cartilage and the ability of normal rat serum to stimulate both responses formed the basis of the somatomedin hypothesis, which postulated an endocrine mode of action of the factor now called IGF-I. The experimental conditions in these studies were fortunate because they permitted a clear separation of IGF-I and GH actions

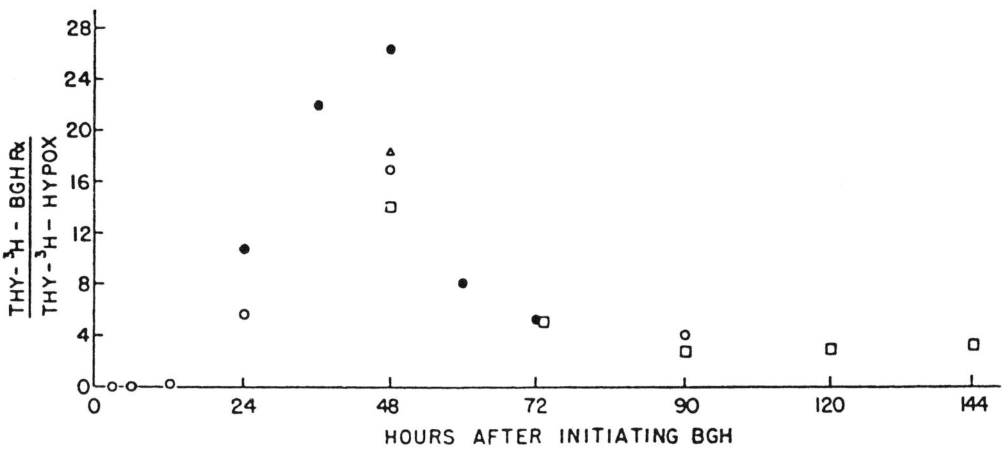

Fig. 4. Effects of bGH treatment (200 μg daily) of hypophysectomized rats on thymidine incorporation by costal cartilage. After the indicated times costal cartilage was removed from treated rats and incubated for 4 h in KPS medium containing amino acids and [^{3}H]thymidine. Results of four independent experiments are shown. (Reprinted with permission from W. H. Daughaday and C. Reeder [1966] J Lab Clin Med 68:357–368, Mosby-Year Book.)

on cartilage. The use of rat plasma or serum was obviously directed by the study design without knowledge of their advantages over human serum. The mean plasma IGF-I level determined by homologous RIA in normal young rats *(24)* is about 7 times the mean level of IGF-I in normal adult human serum *(25)*. Also, serum from normal young rats contains very low levels of IGF-II *(26)*, whereas the levels of IGF-II in normal human serum are 2–3 times greater than the levels of IGF-I and considerable IGF-II remains in serum of hypopituitary patients *(25)*. Finally, the net stimulating effect of hypophysectomized rat plasma or serum on hypophysectomized rat cartilage is low or nil, despite measurable levels of IGF-I and IGF-II *(26)* and the presence of pituitary-independent growth factors. These circumstances made a comparison of the cartilage-stimulating activity of plasma or serum from normal rats with the activity of plasma or serum from hypophysectomized rats relatively specific for IGF-I. As we shall see later, the conditions were also unfavorable for the demonstration of a direct action of GH on cartilage.

CLINICAL APPLICATION OF CARTILAGE BIOASSAYS

Hypophysectomized rat cartilage bioassays of human serum showed that sulfation factor activity was low in hypopituitarism and frequently was elevated in acromegaly *(27)*. The pituitary dependence of the sulfation factor activity of human serum was confirmed by studies on hypophysectomized rat cartilage by Almqvist *(28)*, on embryonic chick cartilage by Hall *(29)*, on pig cartilage by Van den Brande and Du Caju *(30)*, and on rabbit cartilage by Bala et al. *(31)*. These bioassays were capricious and susceptible to misleading results because of serum inhibitors, but they often provided useful information in the assessment of disorders of the GH/IGF-I axis. A noteworthy example other than cases of GH excess or deficiency was the finding in Laron-type dwarfism. In this condition the low serum sulfation factor activity determined by the hypophysectomized rat cartilage bioassay and the failure of that activity to rise after human GH (hGH) treatment clearly pointed to GH resistance *(32)*.

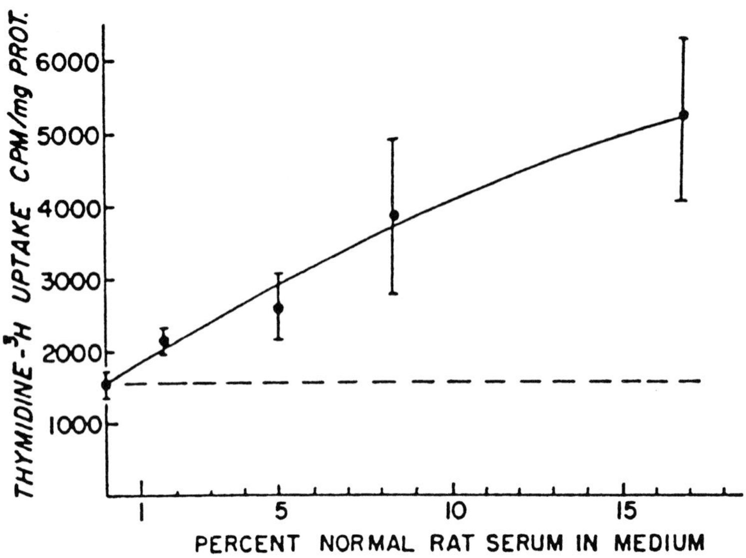

Fig. 5. Effects of different concentrations of normal rat serum on thymidine incorporation by costal cartilage from hypophysectomized rats. Cartilage samples were incubated without or with serum in KPS medium containing amino acids for 24 h, then incubated for another 24 h after addition of [^{3}H]thymidine. (Reprinted with permission from W. H. Daughaday and C. Reeder [1966] J Lab Clin Med 68:357–368, Mosby-Year Book.)

OTHER EFFECTS OF THE CARTILAGE-STIMULATING FACTOR: A CHANGE OF TERMINOLOGY AND INTRODUCTION OF RIA

Further studies of whole normal rat serum or of rat and human serum fractions showed other effects attributed to the sulfation factor. Effects demonstrated in hypophysectomized rat cartilage included stimulation of incorporation of proline into hydroxyproline of collagen, of leucine into protein and specifically into proteoglycans, and of uridine into RNA. Insulin-like effects that were not inhibited by antiinsulin serum were demonstrated in muscle and adipose tissue, and mitogenic action was shown in HeLa cells. The terms *sulfation factor* and *thymidine factor* had outlived their usefulness, and somatomedin was selected as a replacement *(33)*. After partial purification of somatomedin-C from human plasma fractions in Van Wyk's laboratory, the development in the same laboratory of RIA for this factor by Furlanetto et al. *(34)* was a major advance. The methodology and initial results of clinical testing with this RIA were reported in 1977, and elevated levels in acromegaly and low levels in hypopituitarism were demonstrated. Radioimmunoassay of somatomedin C in human serum made possible extensive clinical studies of changes in concentration in health and disease.

MERGING OF DIFFERENT LINES OF INVESTIGATION

An independent line of investigation had been directed at the nature of plasma insulin-like activity (ILA). In 1954 Randle reported that plasma ILA determined by glucose uptake into the diaphragm of fasting normal rats was elevated in patients with acromegaly *(35)* and decreased in patients with hypopituitarism *(36)*. Froesch et al. *(37)* reported in

1963 that most of the human serum ILA determined by adipose tissue assay could not be inhibited by anti-insulin serum. Two major components of the antibody-nonsuppressible ILA (NSILA) were described by Jakob et al. *(38)*, one precipitable by acid-ethanol and the other acid-ethanol-soluble (NSILA-S). In 1973 Zingg and Froesch *(39)* reported that partially purified NSILA-S stimulated sulfate incorporation into embryonic chick cartilage and into cartilage from starved pituitary-intact rats, which essentially confirmed the earlier findings of Hall and Uthne *(40)* with their sulfation factor preparation from human plasma. The extensive investigations of NSILA-S in Zurich, Switzerland culminated in a final purification and sequence determination by Rinderknecht and Humbel of two growth factors from human plasma fractions, IGF-I *(41)* and IGF-II *(42)*. The reports of these findings in 1978 were followed by the demonstration that somatomedin-C and IGF-I were identical *(43)*. Subsequently, the somatomedin terminology was abandoned, and the insulin-like growth factors were ready for introduction to the era of molecular endocrinology.

AUTOCRINE/PARACRINE ACTION OF IGF-I: A CHALLENGE TO THE SOMATOMEDIN HYPOTHESIS

D'Ercole et al. *(44)* reported in 1980 that explants of multiple tissues from the fetal mouse produced IGF-I, as determined by the RIA for somatomedin-C. The investigators suggested "the possibility that the primary biological actions of this hormone are exerted locally at the site of origin." In 1982 Isaksson et al. *(45)* in Göteborg, Sweden reported the first in a series of experiments that challenged the endocrine role of IGF-I in skeletal growth. Injections of hGH into the proximal tibial epiphysis stimulated tibial growth as determined by an oxytetracycline labeling/fluorescence microscopy technique. These results were confirmed by Russell and Spencer *(46)* using measurements of tibial epiphyseal plate width, and they also observed that tibial epiphyseal growth was increased by direct injections of IGF-I. Schlechter et al. *(47)* demonstrated that intraarterial infusions of rat GH into a hindlimb of hypophysectomized rats stimulated ipsilateral tibial epiphyseal growth, and the action of GH was inhibited by simultaneous infusion of anti-IGF-I serum. They concluded that the direct action of GH on cartilage growth was mediated by local production of IGF-I. Nilsson et al. *(48)* showed that GH administered to hypophysectomized rats, either systemically or by local injections into the proximal tibial epiphysis, increased the number of growth plate chondrocytes containing IGF-I as determined by an immunohistochemical method. Isgaard et al. *(49)* found that systemic GH treatment of hypophysectomized rats also increased the IGF-I messenger RNA content of costal cartilage growth plate chondrocytes. These observations were convincing evidence that IGF-I can act as an autocrine/paracrine mediator of GH stimulation of cartilage growth.

ENDOCRINE ACTION OF IGF-I: EVIDENCE FROM IN VIVO STUDIES

Direct evidence that IGF-I can also act in a hormonal fashion was provided by the results of systemic treatment with recombinant IGF-I, which stimulates growth in hypophysectomized rats *(50)* and in children with GH resistance of the Laron syndrome *(51)*. A comparison of the results of IGF-I treatment in children with the Laron syndrome and of hGH treatment in GH-deficient children indicated a greater skeletal growth response in the latter group over a 2-yr period of study *(51)*. The possibility of a specific requirement for the action of GH on skeletal growth independent of IGF-I was raised by the findings of Isaksson et al. *(52)*, which suggested that GH stimulates mitosis of chondroprogenitor cells and that chondrocyte responsiveness to IGF-I develops later. However, Hunziker et al. *(53)* demonstrated that infusions of either IGF-I or GH into hypophysectomized rats stimulated growth plate chondrocytes at all stages of differentiation. Also, normal linear growth and weight gain were observed in genetically engineered mice that expressed IGF-I but had few or no GH-producing cells in the pituitary *(54)*. Nevertheless, chondroprogenitor cells may be more responsive to GH than to IGF-I.

In separate studies with hypophysectomized rats *(55)* and with spontaneously GH-deficient dwarf rats *(56)* it was demonstrated that twice daily injections of GH led to growth without a detectable increase of total serum IGF-I levels. It should not be inferred that IGF-I of hepatic origin (the liver is the major source of serum IGF-I) was unavailable to nonhepatic tissues in these GH-treated rats. In the normal rat and human the major part of serum IGF-I is bound in a large ternary complex with IGFBP-3 and an acid-labile subunit (ALS), and this complex is relatively confined to the circulation. The serum level of ALS in rats is GH dependent *(57)*, and restoration from a very low level in GH-deficient rats occurs over many days in response to GH treatment. Free IGF-I and IGF-I associated with binding proteins smaller than albumin cross the capillary barrier, and this serum pool is predominant until ALS levels increase sufficiently. Therefore, bioavailable IGF-I may increase in serum with or without a detectable increase in total serum IGF-I levels in the early stage of GH treatment in GH-deficient rats. In the two studies cited previously an examination of the serum IGF-binding pattern after GH injections showed no evidence of reconstitution of the ternary complex.

ENDOCRINE ACTION OF IGF-I: EVIDENCE FROM IN VITRO STUDIES

In the face of the increasing evidence that GH has a direct action on rat cartilage, W. D. S. and Burkhalter *(58)* reexamined the effects of the hormone on the incorporation of sulfate and thymidine by hypophysectomized rat cartilage. The assay conditions were modified from those of the original studies by substitution of a *N*-2-hydroxyethylpiperazineethanesulfonic acid (HEPES)-buffered medium for a phosphate-buffered medium, addition to the medium of bovine serum albumin, inclusion of amino acids for studies of sulfate incorporation as well as thymidine incorporation, and the use of highly purified or recombinant preparations of GH. Measurements of [^{35}S]sulfate incorporation after hormonal effects were established, rather than as a cumulative total for the entire incubation period, increased the magnitude of differences from the control in experiments involving that parameter of hormonal action, but the timing of measurements of [^{3}H]thymidine incorporation was not a significant difference in the early and recent studies.

In contrast with the original findings, a direct stimulation of both sulfate and thymidine incorporation was observed with bGH in concentrations of 100 ng/mL or lower. Stimulation of sulfate incorporation by bGH and by IGF-I reached a maximum after 24 h. At this time the cartilage responses to both bGH and IGF-I were dose dependent, but the slope of the dose–response regression was steeper with IGF-I (Fig. 6). Stimulation of thymidine incorporation by bGH reached a maximum after 36–48 h (Fig. 7), but the peak effects of bGH and IGF-I were not synchronous (Fig. 7A). Coincubation with anti-IGF-I immunoglobulin inhibited the cartilage responses to both bGH and IGF-I, indicating that

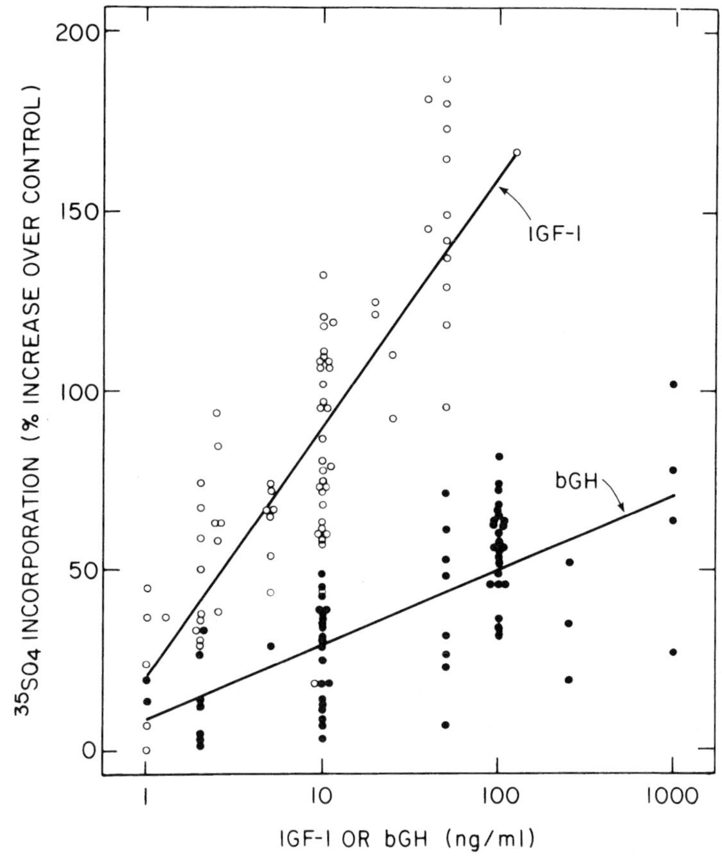

Fig. 6. Pooled results of 42 independent experiments in which one or more concentrations of bGH and IGF-I were tested for stimulation of sulfate incorporation into costal cartilage from hypophysectomized rats. Cartilage samples were incubated without or with bGH or IGF-I for 24 h in HEPES medium (HEPES-buffered saline + glucose, amino acids, and bovine serum albumin), then incubated for 2 h in HEPES medium containing [^{35}S]sulfate. (Reprinted with permission from W. D. Salmon, Jr. and V. J. Burkhalter [1997] J Lab Clin Med 129:430–438, Mosby-Year Book.)

most, if not all, of the effects of bGH on sulfate and thymidine incorporation were mediated by locally produced IGF-I. This is consistent with the earlier finding of Schlechter et al. *(47)* that the local action of rat GH infusion on tibial epiphyseal growth was inhibited by coinfusion of anti-IGF-I serum.

It should be emphasized that the direct action of physiological concentrations of GH on costal cartilage from hypophysectomized rats only partially restored sulfate incorporation toward the level of costal cartilage from normal rats. Full restoration can be achieved by the action of GH in vivo or by the action of either normal rat serum or IGF-I in vitro. Apparently the response of cartilage to GH in vivo represents both the hormonal action and the autocrine/paracrine action of IGF-I.

THE EXPLOSION OF IGF RESEARCH

There was no way to anticipate that simple experiments designed to understand the action of GH on cartilage would lead to the flourishing field of IGF research that forms

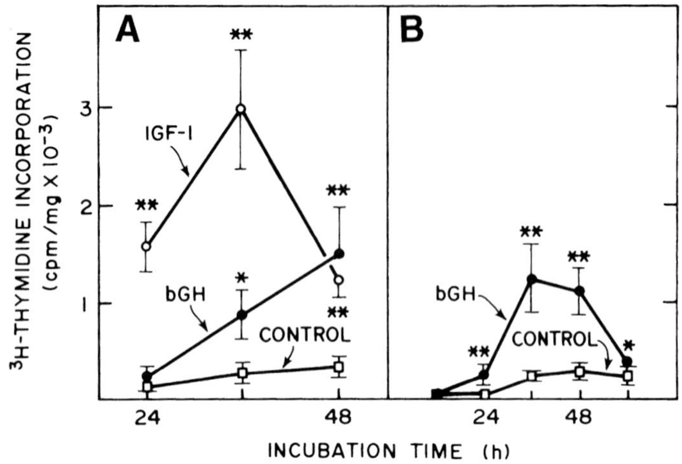

Fig. 7. Thymidine incorporation by costal cartilage from hypophysectomized rats after 24 to 48 h of incubation (Exp A) with bGH (100 ng/mL) or IGF-I (10 ng/mL), or after 12–60 h of incubation with bGH (Exp B). Cartilage was incubated without or with bGH or IGF-I in HEPES medium for the indicated times, then incubated for 2 h in HEPES medium containing [^{3}H]thymidine. (Reprinted with permission from W. D. Salmon, Jr. and V. J. Burkhalter [1997] J Lab Clin Med 129:430–438, Mosby-Year Book.)

the body of this book. Important milestones on the path from early studies were the determination of the structure of not one but two IGFs and the recognition of the structural homology of IGF-I and IGF-II to proinsulin. Later it was established that the expression of the IGF-I gene and synthesis of the peptide were not limited to the liver but were present in many, if not all, tissues. This information provided evidence for the autocrine/paracrine actions of IGF-I. GH was recognized early as an important regulator of IGF-I production and secretion, but subsequent research showed that in certain tissues other hormones and growth factors may assume a regulatory role. Both IGF-I *(59)* and IGF-II *(60)* have been shown to be essential for normal embryonic and fetal growth, and an unidentified pituitary-independent factor must regulate IGF-I gene expression until GH regulation appears late in fetal life and becomes dominant postnatally. Specific binding of IGF peptides by serum proteins was demonstrated many years ago, but the multiplicity of distinct binding proteins has been surprising. The regulation of their separate secretion, their IGF binding characteristics, their unique actions in the modulation of IGF action, and the role of IGFBP proteolysis as a factor in the modulation of IGF action have occupied literally hundreds of investigators for varying periods of time.

CONCLUSIONS

The studies summarized are consistent with both autocrine/paracrine and endocrine actions of IGF-I on cartilage growth, which are regulated by GH. The hypophysectomized rat costal cartilage assay system has provided evidence on the role of IGF-I in cartilage growth generally, but it offers a limited view of the processes involved in epiphyseal cartilage growth specifically. Isaksson and his co-workers have focused attention on the action of GH on chondroprogenitor cells, which form a distinct zone in the tibial epiphyseal growth plate. These cells may be more responsive to GH than to

IGF-I, but the mechanism underlying the quantitative difference between the actions of GH and IGF-I on growth plate chondrocytes remains undefined. Even after 40 years we await new insight into the pituitary control of skeletal growth.

REFERENCES

1. Marie P. Sur deux cas d'acromégalie. Hypertrophie singulière non congénitale des extrémités supérieures, inférieures et céphalique. Rev Méd (Paris) 1886; 6:297–333.
2. Marie P. Acromegaly (translated by Dudley, W.). Brain 1889; 12:59–81.
3. Cushing H. The Pituitary Body and Its Disorders. Clinical States Produced by Disorders of the Hypophysis Cerebri. Lippincott, Philadelphia.
4. Grumbach MM. Herbert McLean Evans, revolutionary in modern endocrinology: a tale of great expectations. J Clin Endocrinol Metab 1982; 55:1240–1247.
5. Evans HM, Long JA. The effect of the anterior lobe administered intraperitoneally upon growth, maturity, and oestrus cycles of the rat. Anat Rec 1921; 21:62–63.
6. Evans HM, Long JA. Characteristic effects upon growth, oestrus and ovulation induced by the intraperitoneal administration of fresh anterior hypophyseal substance. Proc Natl Acad Sci USA 8:38–39.
7. Putnam TJ, Benedict EB, Teel HM. Studies in acromegaly. VIII. Experimental canine acromegaly produced by injection of anterior lobe pituitary extract. Arch Surg 1929; 18:1708–1736.
8. Smith PE. Hypophysectomy and a replacement therapy in the rat. Am J Anat 45:205–273.
9. Evans HM, Simpson ME, Marx W, Kibrick E. Bioassay of the pituitary growth hormone. Width of the proximal epiphysial cartilage of the tibia in hypophysectomized rats. Endocrinology 1943; 32:13–16.
10. Li CH, Evans HM, Simpson ME. Isolation and properties of the anterior hypophyseal growth hormone. J Biol Chem 1945; 159:353–366.
11. Greenspan FS, Li CH, Simpson ME, Evans HM. Bioassay of hypophyseal growth hormone: the tibia test. Endocrinology 1949; 45:455–463.
12. Kinsell LW, Michaels GD, Li CH, Larsen WE. Studies in growth. I. Interrelationship between pituitary growth factor and growth-promoting androgens in acromegaly and gigantism. II. Quantitative evaluation of bone and soft tissue growth in acromegaly and gigantism. J Clin Endocrinol Metab 1948; 8:1013–1036.
13. Gemzell CA, Heijkenskjöld F, Ström L. A method for demonstrating growth hormone activity in human plasma. J Clin Endocrinol Metab 1955; 15:537–546.
14. Hardingham TE, Fosang AJ. Proteoglycans: many forms and many functions. FASEB J 1992; 6:861–870.
15. Ruoslahti E. Proteoglycans in cell regulation. J Biol Chem 1989; 264:13369–13372.
16. Hodgkinson SC, Napier JR, Spencer GSG, Bass JJ. Glycosaminoglycan binding characteristics of the insulin-like growth factor-binding proteins. J Mol Endocrinol 1994; 13:105–112.
17. Ellis S, Hublé J, and Simpson ME. Influence of hypophysectomy and growth hormone on cartilage sulfate metabolism. Proc Soc Exp Biol Med 1953; 84:603–605.
18. Denko CW, Bergenstal DM. The effect of hypophysectomy and growth hormone on S^{35} fixation in cartilage. Endocrinology 1955; 57:76–86.
19. Murphy WR, Daughaday WH, Hartnett C. The effect of hypophysectomy and growth hormone on the incorporation of labeled sulfate into tibial epiphyseal and nasal cartilage of the rat. J Lab Clin Med 1956; 47:715–722.
20. Salmon WD, Jr., Daughaday WH. A hormonally controlled serum factor which stimulates sulfate incorporation by cartilage in vitro. J Lab Clin Med 1957; 49:825–836.
21. Park CR, Brown DH, Cornblath M, Daughaday, WH, Krahl ME. The effect of growth hormone on glucose uptake by the isolated rat diaphragm. J Biol Chem 1952; 197:151–166.
22. Bornstein J, Park CR. Inhibition of glucose uptake by the serum of diabetic rats. J Biol Chem 1953; 205:503–511.
23. Daughaday WH, Reeder C. Synchronous activation of DNA synthesis in hypophysectomized rat cartilage by growth hormone. J Lab Clin Med 1966; 68:357–368.
24. Upton Z, Webb H, Tomas FM, Ballard FJ, Francis GL. Characterization of serum-derived and recombinant rat IGF-1 and their use for measuring true concentrations of IGF-1 in rat plasma. J Endocrinol 1996;149:379–387.
25. Zapf J, Walter H, Froesch ER. Radioimmunological determination of insulinlike growth factors I and II in normal subjects and in patients with growth disorders and extrapancreatic tumor hypoglycemia. J Clin Invest 1981; 68:1321–1330.

26. Glasscock GF, Gin KKL, Kim JD, Hintz RL, Rosenfeld RG. Ontogeny of pituitary regulation of growth in the developing rat: comparison of effects of hypophysectomy and hormone replacement on somatic and organ growth, serum insulin-like growth factor-I (IGF-1) and IGF-II levels, and IGF-binding protein levels in the neonatal and juvenile rat. Endocrinology 1991; 128:1036–1047.
27. Daughaday WH, Salmon WD Jr., Alexander F. Sulfation factor activity of sera from patients with pituitary disorders. J Clin Endocrinol Metab 1959;19:743–758.
28. Almqvist S. Studies on sulfation factor (SF) activity of human serum. Evaluation of an improved method of the in vitro bioassay of SF and the effects of glutamine and human growth hormone in the system. Acta Endocrinol (Copenh) 1961; 36:31–50.
29. Hall K. Quantitative determination of the sulphation factor activity in human serum. Acta Endocrinol (Copenh) 1970; 63:338–350.
30. Van den Brande JL, Du Caju MVL. An improved technique for measuring somatomedin activity in vitro. Acta Endocrinol (Copenh) 1974; 75:233–242.
31. Bala RM, Hankins C, Smith GR. A somatomedin assay using normal rabbit cartilage in clinical studies. Can J Physiol Pharmacol 1975; 53:403–409.
32. Daughaday WH, Laron Z, Pertzelan A, Heins JN. Defective sulfation factor generation: a possible etiological link in dwarfism. Trans Assoc Am Physicians 1969; 82:129–140.
33. Daughaday WH, Hall K, Raben MS, Salmon WD Jr, Van den Brande JL, Van Wyk JJ. Somatomedin: proposed designation for sulphation factor. Nature 1972; 235:107.
34. Furlanetto RW, Underwood LE, Van Wyk JJ, D'Ercole AJ. Estimation of somatomedin-C levels in normals and patients with pituitary disease by radioimmunoassay. J Clin Invest 1977; 60:648–657.
35. Randle PJ. Plasma-insulin activity in acromegaly. Assayed by the rat-diaphragm method. Lancet 1954; i:441–444.
36. Randle PJ. Plasma-insulin activity in hypopituitarism. Assayed by the rat-diaphragm method. Lancet, 1954; i:809–810.
37. Froesch ER, Bürgi H, Ramseier EB, Bally P, Labhart A. Antibody-suppressible and nonsuppressible insulin-like activities in human serum and their physiologic significance. An insulin assay with adipose tissue of increased precision and specificity. J Clin Invest 1963; 42:1816–1834.
38. Jakob A, Hauri C, Froesch ER. Nonsuppressible insulin-like activity in human serum. III. Differentiation of two distinct molecules with nonsuppressible ILA. J Clin Invest 1968; 47:2678–2688.
39. Zingg AE, Froesch ER. Effects of partially purified preparations with nonsuppressible insulin-like activity (NSILA-S) on sulfate incorporation into rat and chicken cartilage. Diabetologia 1973; 9:472–476.
40. Hall K, Uthne K. Some biological properties of purified sulfation factor (SF) from human plasma. Acta Med Scand 1971; 190:137–143.
41. Rinderknecht E, Humbel RE. The amino acid sequence of human insulin-like growth factor I and its structural homology with proinsulin. J Biol Chem 1978; 253:2769–2776.
42. Rinderknecht E, Humbel RE. Primary structure of human insulin-like growth factor II. FEBS Lett 1978; 89:283–286.
43. Klapper DG, Svoboda ME, Van Wyk JJ. Sequence analysis of somatomedin-C: confirmation of identity with insulin-like growth factor I. Endocrinology 1983; 112:2215–2217.
44. D'Ercole AJ, Applewhite GT, Underwood LE. Evidence that somatomedin is synthesized by multiple tissues in the fetus. Dev Biol 1980; 75:315.
45. Isaksson OGP, Jansson J-O, Gause IAM. Growth hormone stimulates longitudinal bone growth directly. Science 1982; 216:1237–1239.
46. Russell SM, Spencer EM. Local injections of human or rat growth hormone or of purified human somatomedin-C stimulate unilateral tibial epiphyseal growth in hypophysectomized rats. Endocrinology 1985; 116:2563–2567.
47. Schlechter NL, Russell SM, Spencer EM, Nicoll CS. Evidence suggesting that the direct growth-promoting effect of growth hormone on cartilage in vivo is mediated by local production of somatomedin. Proc Natl Acad Sci USA 1986; 83:7932–7934.
48. Nilsson A, Isgaard J, Lindahl A, Dahlström A, Skottner A, Isaksson OGP. Regulation by growth hormone of number of chondrocytes containing IGF-1 in rat growth plate. Science 1986; 233:571–574.
49. Isgaard J, Möller C, Isaksson OGP, Nilsson A, Mathews LS, Norstedt G. Regulation of insulin-like growth factor messenger ribonucleic acid in rat growth plate by growth hormone. Endocrinology 1988; 122:1515–1520.
50. Guler HP, Zapf J, Scheiwiller E, Froesch ER. Recombinant human insulin-like growth factor I stimulates growth and has distinct effects on organ size in hypophysectomized rats. Proc Natl Acad Sci USA 1988 85:4889–4893.

51. Guevara-Aguirre J, Rosenbloom AL, Vasconez O, Martinez V, Gargosky SE, Allen L, Rosenfeld RG. Two-year treatment of growth hormone (GH) receptor deficiency with recombinant insulin-like growth factor 1 in 22 children: comparison of two dosage levels and to GH-treated GH deficiency. J Clin Endocrinol Metab 1997; 82:629–633.
52. Isaksson OGP, Lindahl A, Nilsson A, Isgaard J. Mechanism of the stimulatory effect of growth hormone on longitudinal bone growth. Endocr Rev 1987; 8:426–438.
53. Hunziker EB, Wagner J, Zapf J. Differential effects of insulin-like growth factor I and growth hormone on developmental stages of rat growth plate chondrocytes in vivo. J Clin Invest 1994; 93:1078–1086.
54. Behringer RR, Lewin TM, Quaife CJ, Palmiter RD, Brinster RL, D'Ercole AJ. Expression of insulin-like growth factor I stimulates normal somatic growth in growth hormone-deficient transgenic mice. Endocrinology 1990; 127:1033–1040.
55. Orlowski CC, Chernausek SD. Discordance of serum and tissue somatomedin levels in growth hormone-stimulated growth in the rat. Endocrinology 1988; 122:44–49.
56. Gargosky SE, Tapanainen P, Rosenfeld RG. Administration of growth hormone (GH), but not insulin-like growth factor-I (IGF-I), by continuous infusion can induce the formation of the 150-kilodalton IGF-binding protein-3 complex in GH-deficient rats. Endocrinology 1994, 134:2267–2276.
57. Fielder PJ, Mortensen DL, Mallet P, Carlsson B, Baxter RC, Clark RG. Differential long-term effects of insulin-like growth factor-I (IGF-I), growth hormone (GH), and IGF-I plus GH on body growth and IGF binding proteins in hypophysectomized rats. Endocrinology 1996; 137:1913–1920.
58. Salmon WD Jr., Burkhalter VJ. Stimulation of sulfate and thymidine incorporation into hypophysectomized rat cartilage by growth hormone and insulin-like growth factor-I in vitro: the somatomedin hypothesis revisited. J Lab Clin Med 1997; 129:430–438.
59. Liu J-P, Baker J, Perkins AS, Robertson EJ, Efstratiadis A. Mice carrying null mutations of the genes encoding insulin-like growth factor I (*Igf-1*) and type 1 IGF receptor (*Igf1r*). Cell 1993; 75: 59–72.
60. DeChiara TM, Efstratiadis A, Robertson EJ. A growth-deficiency phenotype in heterozygous mice carrying an insulin-like growth factor II gene disrupted by targeting. Nature 1990; 345:78–80.

I

Molecular Biology of the IGF System

2 Molecular Biology of IGF-I and IGF-II

Peter Rotwein, MD

CONTENTS

INTRODUCTION AND OVERVIEW

The molecular era for the insulin-like growth factors (IGFs) began in 1978 with the purification and amino acid sequencing of human IGF-I and IGF-II *(1,2)*. During the past 20 yr, IGF proteins have been characterized from species as diverse as humans, chickens, frogs, and sharks, with the bulk of the information being obtained during the last decade through cDNA cloning. Although in most species IGF-I and IGF-II are small single-chain peptides of 70 and 67 amino acids, respectively, molecular cloning has revealed a remarkable complexity and diversity in both IGFs and in their precursor proteins, mRNAs, and genes. This diversity is all the more surprising because the genes and mRNAs encoding insulin, the other major member of the insulin–IGF gene family, are relatively simple *(3)*. The focus of this chapter is to summarize basic information on structural aspects of IGF-I and IGF-II and their genes. Readers are referred to other chapters for discussions of IGF gene regulation, and to previous reviews of IGF protein and gene structure and function *(4–6)*.

IGF-I

Introduction

The initial sequencing of human IGF-I indicated that it was a 70-residue single-chain basic peptide with a linear organization consisting of four contiguous domains, termed B, C, A, and D (Fig. 1). This structure is similar to proinsulin, which contains B, C, and A regions (the D domain is unique to the IGFs). During posttranslational processing of proinsulin to generate insulin, the C peptide is cleaved from the remainder of the molecule, and the separated B and A chains remain joined to one another by disulfide bonds *(3)*. No analogous processing step accompanies the biosynthesis of IGF-I or IGF-II, and amino acid similarity between insulin and the IGFs is restricted to the B and A portions

From: *Contemporary Endocrinology: The IGF System*
Edited by: R. Rosenfeld and C. Roberts Jr © Humana Press, Inc., Totowa, NJ

```
A  Mammals
                 1       10        20         30       40        50        60         70
   HUMAN         GPETLCGAELVDALQFVCGDRGFYFNKPT  GYGSSSRRAPQT  GIVDECCFRSCDLRRLEMYCA  PLKPAKSA
   COW           -----------------------------  ------------  ---------------------  --------
   HORSE         -----------------------------  ------------  ---------------------  --------
   PIG           -----------------------------  ------------  ---------------------  --------
   DOG           -----------------------------  ------------  ---------------------  --------
   GUINEA PIG    -----------------------------  ------------  ---------------------  --------
   SHEEP         -----------------------------  ------------  ---------------------  ---A----
   DEER          -----------------------------  ------------  ---------------------  ----T-A-
   GOAT          -----------------------------  ------------  ---------------------  ----T---
   RAT           -------------------P---------  -----I------  ---------------------  ----T---
   MOUSE         -------------------P---------  -----I------  ---------------------  ----T-A-

   Non-mammalian vertebrates

   CHICKEN       -------------------------S---  --------LHHK  --------Q------------  -I--P---
   XENOPUS 1     ------------T------------S---  ----NN--SHHR  --------Q---F--------  -A------
   XENOPUS 2                                        NN--AHHR  --------Q---F--------  -Q------
   SALMON        ------------T------E-----S---  ---P----SHNR  --------Q--E---------  -V-SG-A-
   TROUT         ------------T------E-----S---  ---P----SHNR  --------Q--E---------  -V-SG-A-
   CARP          ------------T------------S---  ---P----SHNR  --------Q--E---------  -V-PG-TP
   CATFISH       ------------T------E-----S---  ---P----SHNR  --------Q--E---------  -V-SG-A-
   CATFISH brain --Y---R-----S--P---Q-----SR-M  -----TCWTHNR  --L-K---Q--Q--WI-----  AFE-SLVT
   TILAPIA       ------------T------E---------  ---P-A--SR..  --------Q--E-Q-------  -V-TP-IS
   SEABREAM      S-----------T------------S--.  ---PNA--SR..  --------Q--E---------  -A-TS-A-
   SHARK         S-----------A------------S---  --RS-V--PH.R  ---N----Q----KL------ K-QRADGP-

   Domain                    B                       C                 A               D
```

Fig. 1. Amino acid sequence of IGF-I. The aligned amino acid sequences of mature IGF-I are presented in (**A**) in single-letter code. *Dashed lines* represent amino acids that are identical to human IGF-I. *Dots* indicate residues that are missing when compared with human IGF-I. The B, C, A, and D domains are indicated. The amino acid sequences of the C-terminal E region are shown in (**B**) and are divided into segments that are shared (common) and those that are unique to IGF-IA or IGF-IB. *Vertical arrows* indicate the locations of exon/intron junctions involved in alternative RNA splicing. The two frog IGF-I sequences are represented by *Xenopus* 1 and 2. Because the amino acid sequences of mature IGF-I are identical for different species of salmon, these are represented only once in (**A**). Only the E domains for chinook salmon are shown in (**B**), as these have been analyzed more completely than in other species. The two different carp IGF-I E domains are indicated as carp a and carp b. The mature proteins are identical to one another. The brain specific and more ubiquitous catfish IGF-I molecules are indicated.

```
B  Mammals
                         Common                 EA
   HUMAN        RSVRAQRHTDMPKTQK       EVHLKNASRGSAGNKNYRM
   COW          -------------A--       ------T------------
   PIG          -------------A--       ------T----S-------
   DOG          -------------A--       ---------------T-
   GUINEA PIG   ----------------       -------------------
   SHEEP        -------------A--       ------T------------
   GOAT         -------------A--       ------T------------
   RAT          --I-------------       ------T--------T---
   MOUSE        --I-------------       ------T--------T---
                               ⇑      ⇑
                                                EB
   HUMAN                              YQPPSTNKNTKSQRRKGWPKTHPGGEQKEGTEASLQIRGKKKEQRREIGSRNAECRGNKKGK
   RAT                                S--L--H-KR-L---RKGSTLEEHL
   MOUSE                              SPSL----K--L---RKGSTFEEHK

Non-mammalian vertebrates

                         Common                 EA
   CHICKEN      -------------A--       ------T---NT--R----
   XENOPUS 1    -------------A--       ---P--T---NT-SRGF--
   XENOPUS 2    ----T--
   SALMON (chin) ------------R-P-      ---Q--S---NT-GR----
   TROUT        ------------R-P-
   CARP a       --L-------S-R-P-
   CARP b       ----------S-R-A-
   CATFISH      ------------R-L-
   CATFISH brain -------V---CR-L-
   TILAPIA      ----S-------RAP-
   SEABREAM     ------------RAP-
   SHARK        -I   -L--EKGQREN       VWRNP-RTNA-SV-R---I

                               ⇑      ⇑
                                                EB
   SALMON (chin)                      VSTAVQNVDRGTERRTAQHPDKTKTKKKPLSGNSHTSCKEVHQKNSSRGNTGGRNYRM
                                      VSTAVQNVDRGTERRTAQHPDKTKTKK            EVHQKNSSRGNTGGRNYRM
                                                                 KPLSGNSHTSCKEVHQKNSSRGNTGGRNYRM
                                                                 ⇑          ⇑
   TROUT                              VSTAVQSVDRGTERRTAQHPDKTKPKK            EVHQKNSSRGNTGGRNYRM
   CARP a                                                        KPISGHSHSSCKEVHQKNSSRGNTGGRNYRM
   CARP b                                                        KPLPGQSHSSYKEVHQKNSSRGNTGGRNYRI
   CATFISH                                      VSTAVQNVDRGTERTAQHPDKKEVHNKNSSRGNTGGRNYRM
   CATFISH brain                      KSAVQNVERRTEIWTAQHPDKTKPKKEVYQQNTLRGNTWGGYYGMYKLTLAKGYEDSGYNTTGP
   TILAPIA                                VSSRANKGTERRTAPQPDKTKNKKRPSPGHSSSFKEVHQKNSSRGSSGGRNYRM
   SEABREAM                               VSTAGHKVDKGTERRTAQQPDKTKNKKRPLPGHSHSSPKEVHPKNSSRGNGGRN
```

Fig. 1. *(continued)*

of the molecules. The first report of a human IGF-I cDNA appeared in 1983 *(7)*, with studies on the organization of the human IGF-I gene following soon thereafter *(8)*. IGF-I cDNAs now have been cloned from many mammalian and nonmammalian vertebrate species, and IGF-I genes have been characterized from several species in sufficient detail to allow a fairly complete analysis of their structures and of the mechanisms governing IGF-I mRNA production.

IGF-I Peptides

The sequence of IGF-I is known for at least 11 mammalian and 9 nonmammalian vertebrates *(1,9–31)* (Fig. 1). In all species except shark; tilapia; and the gilthead seabream, *Sparus aurata (29–31)*, IGF-I is a 70-residue single-chain molecule containing an amino (N)-terminal B region of 29 amino acids, a central C domain of 12 residues, an A domain of 21 amino acids, and a carboxy (C)-terminal D region of 8 residues. In 70 amino acid shark IGF-I *(31)*, the penultimate residue of the C region is missing and an extra lysine has been inserted at the N-terminus of the D domain. In 68-residue tilapia IGF-I, the last two amino acids of the C region are absent *(29)*, while in 67-amino-acid seabream IGF-I, the final residue in the B domain and the last two amino acids of the C region are missing *(30)*. As also seen in Fig. 1, IGF-I has been highly conserved among mammals. In the 11 sequences pictured, 65/70 amino acids are invariant, with three substitutions being located in the D domain, and one each in the B and C regions.

By contrast with the high degree of amino acid conservation seen in mammalian IGF-I, substantially more variability is found in IGF-I molecules from nonmammalian vertebrates. In salmonides and *Xenopus laevis*, two nonallelic IGF-I genes have been identified *(23–25,32,33)*. This reflects the relatively recent tetraploidization of their genomes and the subsequent divergence of each pair of genes. In *Xenopus*, one consequence of gene divergence has been the appearance of distinct IGF-I proteins *(22,23)*, while in salmon and trout the sequences of mature IGF-I have remained identical *(24–26,32)*. A pair of IGF-I sequences also has been reported in carp, but this may represent allelic variation because a pure strain was not used as the source of RNA for cDNA cloning *(27)*. In catfish, both a brain-specific and a ubiquitous form of IGF-I were identified by cDNA cloning *(28)*, although the corresponding genes have not been characterized and the proteins have not been sequenced directly or tested for biological function. The two catfish IGF-I molecules differ from each other at 25/70 residues, with the ubiquitous form being identical to the proteins found in salmon and trout. Overall, only 34/70 amino acids are identical among the IGF-I molecules shown in Fig. 1, with 20/29 residues being conserved in the B domain, 2/12 in the C region, 12/21 in the A domain, and 0/8 in the D region. All 6 cysteines have been conserved, and because these residues are involved in formation of three intraprotein disulfide bonds, it is likely that all IGF-I molecules exhibit a spatially similar three-dimensional topography. Also conserved are 3 tyrosines, at positions analogous to B24, C31, and A60 in mammalian IGF-I. These residues have been shown to be involved in high-affinity binding of human IGF-I to the IGF-I receptor *(34)*, and it is thus likely that the same amino acids play a similar role in regulating ligand–receptor interactions in other species. By contrast, the tyrosines at B24 and C31 are not conserved in the single IGF protein identified in the primitive vertebrate, the Atlantic hagfish *(35)*, making it unlikely that this protein binds to a typical, mammalian-like IGF-I receptor.

As a result of alternative splicing of the primary gene transcript *(see below)*, IGF-I mRNAs are produced that encode protein precursors with different C-termini, termed

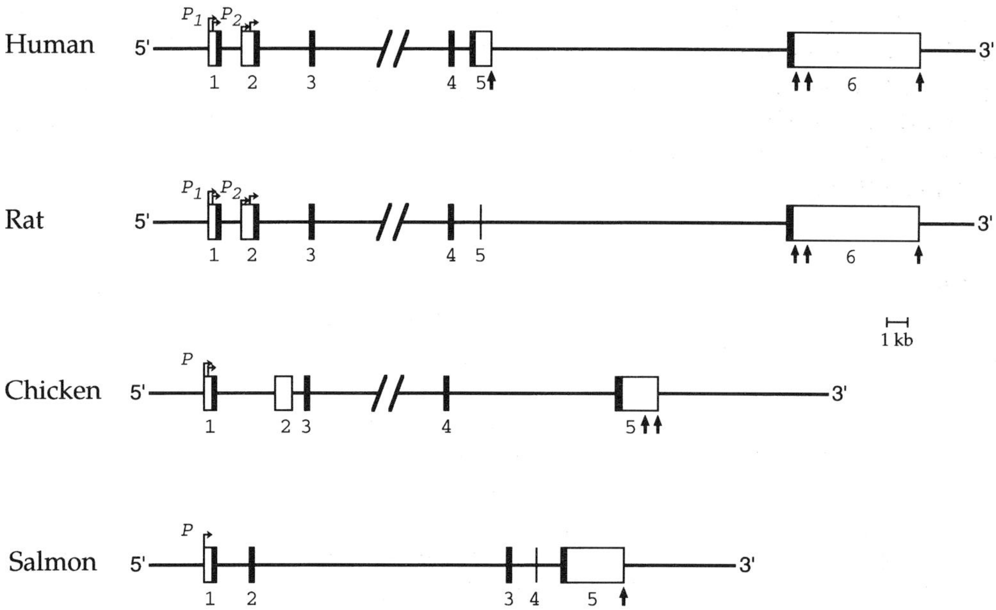

Fig. 2. Comparative anatomy of IGF-I genes. The structures of human, rat, chicken, and one of the two chum salmon IGF-I genes are shown. Exons are numbered and are indicated by *boxes,* with coding regions in *black* and noncoding segments in *white.* Promoters are delineated with the letter "P," transcription start sites by *bent arrows,* and polyadenylation sites by *vertical arrows.* The large central intron in the human, rat, and chicken IGF-I genes has been truncated, as indicated.

E domains. The human IGF-IA protein contains a 35-residue E domain, and human IGF-IB contains a 77-amino-acid C-terminal extension. The initial 16 residues of the two E domains are identical because they are encoded by the same IGF-I exon *(see below).* This common proximal E region has been fairly well conserved among mammals as shown in Fig. 1 (14/16 identical residues), but is not as conserved in birds, frogs, or fish (4/16 identical amino acids). The distal segments of IGF-I E domains are more divergent, although in mammals the IGF-IA-specific region has been conserved (16/19 identical residues). In fish, a variety of IGF-IB sequences have been identified through cDNA cloning *(24–30,32).* In salmon these arise as a consequence of alternative RNA splicing *(see below).* To date no functions have been reported for the larger IGF-I molecules with C-terminal extensions although preliminary evidence suggests that at least in mammals these proteins act as precursors for 70-residue IGF-I. The mechansisms of posttranslational cleavage have not been defined, although the subtilisin-related prohormone convertases may be involved *(36).*

IGF-I Genes and mRNAs

The complete or nearly complete structures of IGF-I genes have been established from several species (Fig. 2). In rats and in humans, the single-copy IGF-I gene consists of six exons and five introns, and spans more than 80 kb of chromosomal DNA *(5,6,8,37,38).* It is likely that IGF-I genes in other mammals are organized similarly. The human gene has been mapped to chromosome 12 *(39).* The chicken IGF-I gene contains five exons and four introns, and is 50 kb in length *(21,40),* whereas salmon IGF-I genes are smaller. The two nonallelic chum salmon IGF-I genes are approx 17 kb in length, and probably

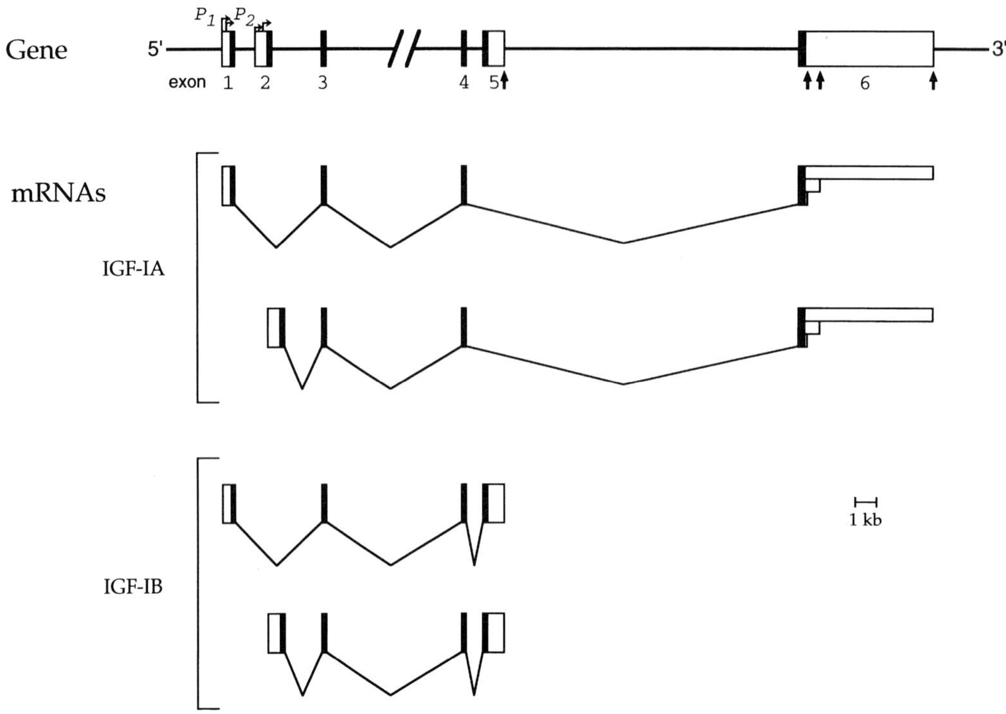

Fig. 3. Structure and expression of the human IGF-I gene. Structures of different human IGF-IA and IGF-IB transcripts are displayed below a map of the gene. Sites of pre-mRNA splicing are indicated by the *thin lines* and locations of differential polyadenylation at the 3′ end by *arrows* in the gene and by boxes of different lengths in the mRNAs.

contain five exons and four introns, although this has not been determined conclusively *(32,33)*. Based on current evidence, it is likely that IGF-I genes exhibit a similar organization in other salmonides *(24,25)*.

The major structural features of IGF-I genes are schematized in Fig. 2, and mechanisms of IGF-I mRNA production are outlined for the human gene in Fig. 3, and summarized for several genes in Table 1. In all species, multiple IGF-I mRNAs are produced through a combination of steps that include differential promoter use (mammals), variable transcription initiation sites (mammals, chickens), differential RNA splicing (all species), and alternative RNA polyadenylation (mammals, chickens). In mammals distinct promoters reside 5′ to exons 1 and 2 (promoters 1 and 2, respectively), whereas in other species the single promoter is adjacent to exon 1. Promoter 1 is the major promoter in mammals, and is active in all tissues in which IGF-I is expressed *(38,41–43)*. Promoter 1, which regulates IGF-I transcripts containing exon 1, is structurally related to the single chicken promoter *(40)*. The nucleotide sequences of rat and human exon 1 also are very similar to each other and to chicken and salmon exon 1 *(32,38,40)*. Exon 1 contains the 5′ untranslated region of IGF-I mRNA and the initial 21 codons of the IGF-I signal peptide. Mammalian IGF-I promoter 2 governs transcripts containing exon 2, and is active primarily in the liver, where it is responsible for approx 25% of IGF-I mRNAs under normal physiological conditions *(38,41–43)*. Mammalian exon 2 contains 5′ untranslated sequences and the intial six codons for an alternative signal peptide *(38,44,45)*. Both promoters 1 and 2 regulate transcription over dispersed start sites, and

Table 1
Characteristics of IGF-I Genes

Species	*Promoters*	*Variable tx initiation*	*Leader exons*	*Alternative RNA splicing*	*Differential RNA polyadenylation*
Human	P1, P2	E1, E2	E1, E2	E4→E5 E4→E6	E6
Rat	P1, P2	E1, E2	E1, E2	intra E1 E4→E5→E6 E4→E6	E6
Sheep	Pw?, P1, P2	E1, E2	Ew?, E1, E2	?	?
Chicken	P1	E1	E1	E1→E2→E3 E1→E3	E5
Salmon	P1	?	E1	E3→E4→E5 E3→E5 E3′→E4→E5 E3′→E5	?

?, information is unknown; Ew and Pw, putative 5′ promoter and exon identified in sheep IGF-I gene; E′ represents a site of alternative RNA splicing that occurs from the extended 3′ terminus of salmon IGF-I exon 3.

it is likely that heterogeneous transcription initiation reflects the absence of typical proximal promoter elements, such as TATA or CCAAT boxes *(38,46–51)*. A discussion of IGF-I promoter function can be found in another chapter. In sheep, another putative exon has been identified 5′ to exon 1, raising the possibility of a third promoter in this species *(49,50)*. There is no homologue of IGF-I promoter 2 or exon 2 in nonmammalian vertebrates. In the chicken genome, a unique exon 2 has been characterized that contains untranslated sequences and is expressed as part of an alternatively spliced IGF-I mRNA *(21)*. In mammals IGF-I transcripts containing either exons 1 or 2 are spliced onto exon 3 *(42–44,46)*, while in chickens, either exon 1 or a composite of exons 1 and 2 is spliced onto exon 3 *(21)*. In all three species, exon 3 encodes the distal common 27 amino acids of the signal peptide and most of the B domain of mature IGF-I. The remainder of 70-residue IGF-I is located in exon 4, which also contains the proximal shared 16 amino acids of the E domain. In salmonids, IGF-I exons 2 and 3 correspond to mammalian and chicken exons 3 and 4 *(32,33)*.

Alternative RNA splicing occurs at the 5′ end of the rat IGF-I gene, near the 5′ end of the chicken gene as noted previously, and at the 3′ ends of the human, rat, and salmon genes, but has not been detected in this latter part of the chicken gene. In the rat IGF-I gene, part of the 5′ untranslated region of exon 1 is differentially expressed in IGF-I mRNAs *(16)*. In the human gene, exon 5 encodes IGF-IB-specific protein sequences and contains DNA for both the 3′ untranslated region of IGF-IB mRNA and for an IGF-IB-specific polyadenylation signal and site *(52)* (Fig. 3). Exon 6 encodes the corresponding IGF-IA-specific region, as well as unique 3′ untranslated sequences and several alternatively used polyadenylation signals and sites *(53)*. In the rat, nucleotide changes within exon 5 have created a splice donor site for RNA processing, such that IGF-IB mRNAs contain both exons 5 and 6 *(54)*. The situation is more complicated in the salmon genome. Nucleotide alterations during speciation have created two differentially used splice donor sites at the 3′ end of exon 3 *(25,32,33)*. RNA processing at the more distal site adds 81 nucleotides to the resultant IGF-I mRNAs, leading to generation of an open reading frame containing 27 extra amino acids within the E domain. Because subsequent alternative splicing can include both exons 4 (12 codons) and 5 (19 codons), or just exon 5, four distinct types of IGF-I mRNAs encoding proteins with different lengths of E peptides can be generated from each of the two nonallelic salmon IGF-I genes *(24,25,32,33)*. In Fig. 1, three of these sequences have been classified as IGF-IB and one as IGF-IA. Thus at least eight different IGF-I mRNAs and proteins can be produced in salmon tissues. This appears to be the case in chinook salmon but is not true for chum salmon. In this latter salmon species only six classes of IGF-I transcripts are generated because a point mutation in one of the nonallelic IGF-I genes has destroyed the proximal splice site in exon 3 *(33)*. The mechanisms regulating production of these multiple alternative transcripts and encoded peptides have not been elucidated. The steps responsible for production of distinct types of IGF-I mRNAs in different species are summarized in Table 1.

IGF-II

Introduction

The initial amino acid sequencing of human IGF-II showed that it was a 67-residue single-chain peptide with a linear organization similar to that of IGF-I, and amino acid homology with both IGF-I and insulin *(2)*. The first publication of IGF-II cDNAs appeared in 1984, with analysis of the IGF-II gene following soon thereafter *(55–57)*. IGF-II cDNAs now have been cloned from several mammalian and nonmammalian vertebrates, and the gene has been characterized in detail from at least four different mammalian species *(57–66)*.

IGF-II Peptides

The sequence of IGF-II is known for at least 10 mammalian and 6 nonmammalian vertebrates *(2,15,26,30,31,56,67–76)* (Fig. 4). In 9 species, IGF-II is a 67-residue single-chain molecule containing an N-terminal B domain of 28 amino acids, a C region of 12 residues, an A domain of 21 amino acids, and a 6-residue C-terminal D segment. Several variant human IGF-II proteins have been characterized. A 70-amino-acid IGF-II arises through alternative use of a 5′ splice acceptor site in exon 9 of the IGF-II gene, resulting in the substitution of the tetrapeptide arginine–leucine–proline–glycine for serine 29 *(77,78)*. This variant molecule comprises ~25% of serum IGF-II and has about one third of the potency of 67-residue IGF-II for the IGF-I receptor *(78)*. A 69-amino-acid human IGF-II also was sequenced in which the tripeptide cysteine–glycine–aspartate was substituted for serine 33 *(79)*. Because the DNA sequence alteration responsible for these new amino acids does not occur near an intron–exon junction, this observation suggests the existence of allelic variation in the IGF-II gene. The biological significance of this variant protein is unknown. In 68-amino-acid mink IGF-II, an extra residue is present in the C domain *(72)*. In 67-residue chicken IGF-II, the first amino acid relative to human IGF-II is missing and an asparagine has been added near the C-terminus of the C domain *(74)*, while in 70-residue trout, salmon, seabream, and tilapia IGF-II, the C region has increased in length from 12 to 15 amino acids *(26,30,75,76)*. As indicated in Fig. 4, IGF-II has been highly conserved among mammals, with 61/67 residues being invariant. Four substitutions are located in the C region, and one each in the B and D domains. An even higher degree of conservation is seen among IGF-II sequences

```
A

Mammals
                1        10        20           30        40           50        60          67
HUMAN           AYRPSETLCGGELVDTLQFVCGDRGFYF   SRPASRVSRRSR   GIVEECCFRSCDLALLETYCA   TPAKSE
HUMAN Var 1     ---------------------------- RLPG----------   ---------------------   ------
HUMAN Var 2     ----------------------------   ----CGDR------ ---------------------   ------
COW             ----------------------------   ---S--SN----   ---------------------   ------
HORSE           ----------------------------   ------IN----   ---------------------   ------
PIG             ----------------------------   -------N----   ---------------------   ------
GUINEA PIG      ----------------------------   ------------   ---------------------   ------
SHEEP           ----------------------------   ---S--IN----   ---------------------   A-----
DEER            ----------------------------   ---S--IN----   ---------------------   ------
MINK            ----------------------------   -----------SR  ---------------------   ------
RAT             ---------------------S------   ---S--AN----   ---------------------   ------
MOUSE           ---------------------S------   ---S--AN----   ---------------------   ------

Non-mammalian vertebrates

CHICKEN         .-GTA-----------------------   ---VG-NN--INR  ---------------------   KSV---
TROUT           EVASA----------A-----E------   ---T--SNS-RSQNR -------------N---Q---  K-----
SALMON          EVASA----------A-----E------   ---T--SNS-RSQNR -------------N---Q---  K-----
SEA BREAM       EVASA----------A-----E------   ---T--GNN-RPQNR -------------N---Q---  K-----
TILAPIA         EVASA----------A-----E------   ---T--GNN-RPQTR -------------N---Q---  K-----
SHARK           EA-LE-----S--------I-AE-----   VSKVVGRRS-QNR  -------------LI------   V-PEAA

Domain                    B                         C                 A                 D
```

Fig. 4. Amino acid sequence of IGF-II. The aligned amino acid sequences of mature IGF-II are presented in (**A**). *Dashed lines* represent amino acids that are identical to human IGF-II. The B, C, A, and D domains are indicated. The amino acid sequences of the C-terminal E region are shown in (**B**). The two human variant nature IGF-II proteins are indicated as Var 1 and Var 2, respectively.

B

Mammals

```
HUMAN       RDVSTPPTVLPDNFPRYPVGKFFQYDTWKQSTQRLRRGLPALLRARRGHVLAKELEAFREAKRHRPLIALPTQDPAHGGAPPEMASNRK
COW         ----AST-----DVTA----------I--------------F------RT-------L----S-------------IH-GASSK--SD
PIG         -----------------------R-------A----------------RT-------V---------T-R--R---AH-GASPE--GHRK
SHEEP       ----ASTT----D-TA--------S----------------F------RT-------L----S-------------TH-GASSE--SD
RAT         -----SQA----D----------KF---R--AG---------------RM-----KE-----------V--PK-------SS--S--HQ
MOUSE       -----SQA----D---------------R--AG---------------RM------------------V--PK-------SS--S--HQ
```

Non-mammalian vertebrates

```
CHICKEN     --L-ATSLAG--ALNKESFQKPSHAKYS-YNVWQKKSSQRLQREVPGILRARRYRWQAEGLQAAEEARAMHRPLISLPSQR-PAPRASPEATGPQE
TROUT       ----ATSLQIIPMV-TIKQDVPRKHV-V-Y-KYEAWQRKA-QRLR-GVPAILRARKFR-Q-VKIKAQEQAMFHRPLITLPSKLPPVLPPTDNYVSHN
SALMON      ----ATSLQIIPMV-TIKQDVPRKHV-V-Y-KYEAWQRKA-QRLR-GVPAILRARKFR-Q-VKIKAQEQAMFHRPLITLPSKLPPVLPPTDNYVSHN
SEA BREAM   ----ATSLQVLPVM-PLKQEVSRKQHVTVKYSKYEVWQRK-AQ-L---VPAILRAKKY-RQAEKIKAQEQAIFHRPLISLGSKLPPVLLATDNYVNHK
TILAPIA     ----ATSLQVIPVM-ALKQEVPKKQHVTVKYSKYEVWQRK-AQ-L---VPAILRARKYKRHAEKIKAKEQAIFHRPLISLPSKLPPVLLTTDNFVSHK
SHARK       -STPSPAQQRTLMHRIVGKRSLGIEQERLTRPGGHLQQYLNNASPASSPKAEILDWPGEGEASPPIGDLTWRARYPPRHWVWLS-ALPSPSTS
```

Fig. 4. *(continued)*

identified in bony fishes, with 66/70 identical residues among trout, salmon, tilapia, and seabream proteins *(26,30,75,76)*. All four amino acid differences are located in the C domain. The 68 residue shark IGF-II molecule is more divergent, with 28 amino acid substitutions compared with bony fishes *(31)*. Overall, 37/67 residues are identical among the IGF-II molecules listed in Fig. 4, with 18/28 residues being conserved in the B domain, 1/12 in the C region, 18/21 in the A domain, and 0/6 in the D segment. All six cysteines have been conserved. Because these amino acids are involved in the formation of three intraprotein disulfide bonds, it is likely that the overall three-dimensional organization of IGF-II also has been conserved.

Like IGF-I, IGF-II is synthesized as a precursor with an extended C-terminal E domain, although unlike IGF-I only a single class of E proteins has been characterized. The IGF-II E region has been less conserved than mature IGF-II among different species, as indicated in Fig. 4, and its biological function has not been defined, although IGF-II molecules containing C-terminal extensions are secreted by tumors and have been shown to exert potent hypoglycemic effects in vivo *(80,81)*. The mechanisms of posttranslational cleavage of IGF-II from its large precursor protein have not been characterized.

IGF-II Genes and mRNAs

Extensive structural information is available for the IGF-II gene and the IGF-II chromosomal locus from several mammalian species *(57–66)*. In humans, the single-copy 10 exon IGF-II gene has been mapped to chromosome 11p15.5, just 3′ to the insulin gene and less than 200 kb 5′ to the H19 gene *(82,83)*. All three genes are in the same transcriptional polarity, as is the tyrosine hydroxylase gene, located just 1.5 kb 5′ to insulin *(84)*. This conserved linkage group has been mapped in mice to chromosome 7 *(83)*. IGF-II and insulin genes also have been shown to be contiguous in sheep and rats *(63,66)*.

The major features of IGF-II genes are outlined in Fig. 5, and mechanisms of IGF-II mRNA production are described for the human gene in Fig.6 and summarized for several genes in Table 2. In mammalian species, IGF-II transcripts are produced by an interplay of mechanisms that include differential promoter use, alternative transcription initiation sites, variable RNA polyadenylation, and endonucleolytic cleavage of mature IGF-II mRNAs. Four different promoters have been identified in the human IGF-II gene *(85)*, and three have been described in the rat and mouse genes *(64,86–87)*. To date a single promoter has been mapped in a salmon IGF-II gene *(75)*, but no promoter has been identified in the recently cloned chicken gene *(74)*. Each promoter governs expression of unique 5′ noncoding exons. The most 5′ human promoter and its three associated exons (human exons 1–3) are not found in the rodent genes, although two nonfunctional "pseudo-exons," corresponding in location and DNA sequence to human exons 2 and 3, have been mapped to mouse *igf2 (64)*. Human IGF-II promoters 2–4 are homologous to rodent promoters 1–3 and the DNA sequences of the adjacent 5′ noncoding exons also have been conserved *(59,62–64,87)*. The human locus additionally contains an alternatively spliced noncoding exon 5 that is associated with promoter 2 and is not found in other IGF-II genes, although a nonfunctional "pseudo-exon" has been identified in the same relative location in the rat gene *(58)*. In humans, IGF-II is expressed at relatively high levels throughout life, whereas in rodents IGF-II production becomes attenuated in most tissues soon after birth *(60,88)*. It has been proposed that the continued activity of promoter 1 in the liver is responsible for the maintenance of IGF-II expression during postnatal life in humans *(6)*, and that the absence of this promoter explains the decline in

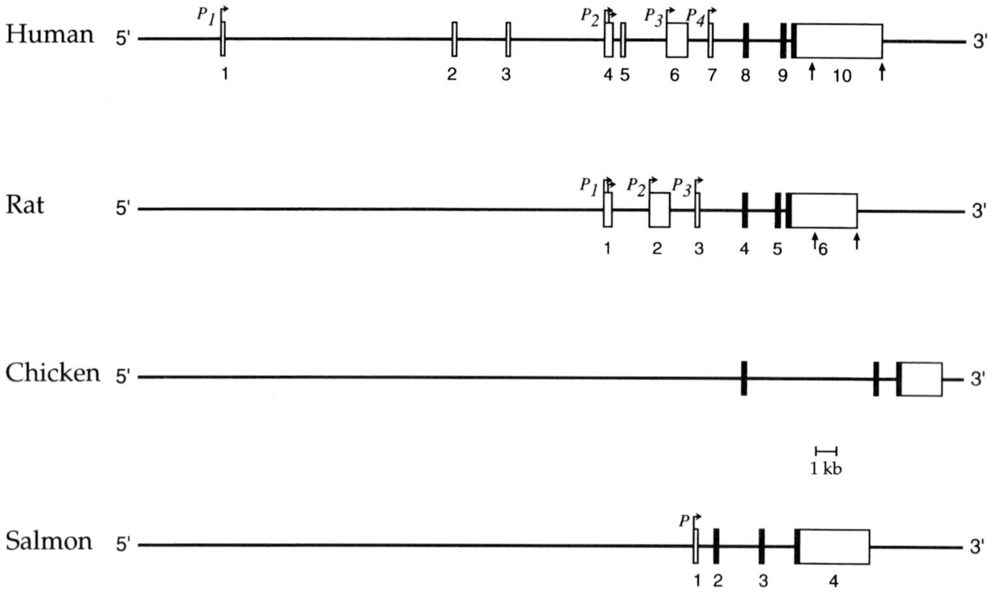

Fig. 5. Comparative anatomy of IGF-II genes. The structures of human, rat, chicken, and salmon IGF-II genes are shown. Exons are numbered (except for the partially characterized chicken gene) and are indicated by *boxes,* with coding regions in *black* and noncoding sequences in *white.* Where known, promoters are delineated by the letter "P," transcription start sites by *bent arrows,* and polyadenylation sites by *vertical arrows.*

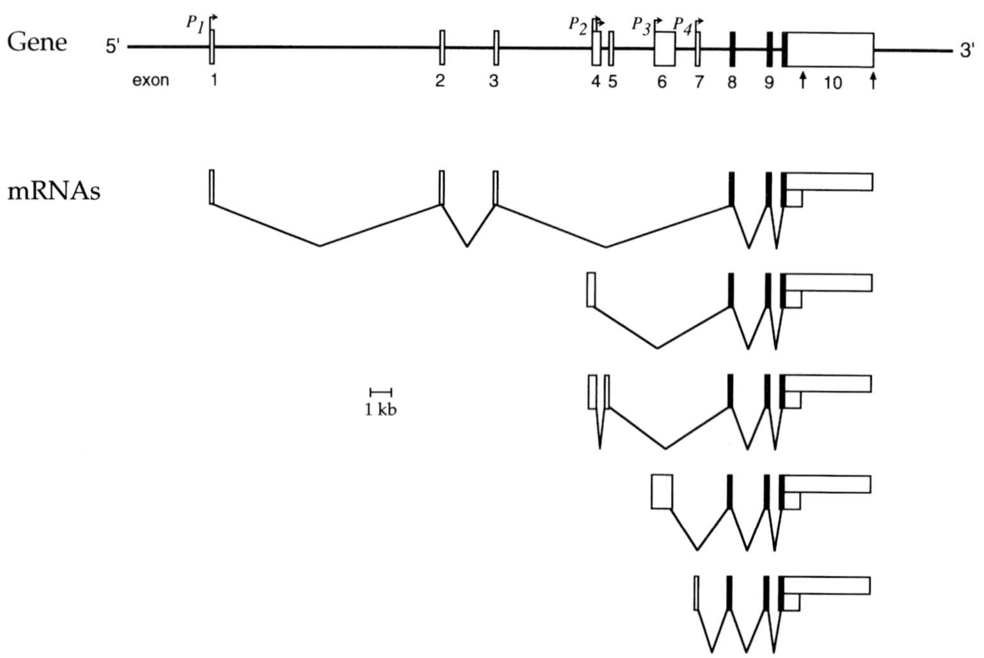

Fig. 6. Structure and expression of the human IGF-II gene. Structures of different human IGF-II transcripts are displayed below a map of the gene. Sites of pre-mRNA splicing are indicated by the *thin lines* and locations of differential polyadenylation at the 3′ end by *arrows* in the gene and by boxes of different lengths in the mRNAs.

Table 2
Characteristics of IGF-II Genes

Species	*Promoters*	*Variable tx initiation*	*Leader exons*	*Alternative RNA splicing*	*Differential RNA polyadenylation*
Human	P1, P2, P3, P4	E4	E1, E4, E6, E7	E4→E5→E6 E4→E6 intra E9	E10
Rat	P1, P2, P3	E1	E1, E2, E3	No	E6
Mouse	P1, P2, P3	E1	E1, E2, E3	No	E6
Sheep	P1, P2, P3, P4?	?	E1, E4, E6, E7	?	E10
Chicken	?	?	?	?	?
Salmon	P1	?	E1	?	?

?, information is unknown.

IGF-II production seen in rats and mice. A discussion of IGF-II promoter function can be found in a later chapter.

In all species in which IGF-II genes have been characterized to date, the coding region is distributed over three exons (exons 8–10 in humans, exons 4–6 in rodents). In mammalian genes, the first two of these coding exons contains the mature IGF-II molecule, as well as a 24-amino-acid signal peptide (human exon 8, rodent exon 4), and the initial 11 residues of the E domain (human exon 9, rodent exon 5). The reminder of the E domain is encoded by the last exon (human exon 10, rodent exon 6), which also contains the 3′ untranslated region of IGF-II mRNA.

In mammals, IGF-II mRNAs also differ in the length of the 3′ untranslated region secondary to alternative RNA polyadenylation *(64,89)*, as indicated in Figs. 6 and 8. Another class of IGF-II mRNAs arises through endonucleolytic cleavage within the 3′ noncoding region, generating an ~1.8 kb polyadenlylated transcript that does not encode a protein *(90,91)*. The physiological significance of this mRNA species is not known; to date it has been identified only in mammals. The mechanisms responsible for production of this transcript are described in a later chapter.

REFERENCES

1. Rinderknecht E, Humbel RE. The amino acid sequence of human insulin-like growth factor I and its structural homology with proinsulin. J Biol Chem 1978; 253:2769–2776.
2. Rinderknecht E, Humbel RE. Primary structure of human insulin-like growth factor II. FEBS Lett 1978; 89:283–286.
3. Chan SJ, Nagamatsu S, Cao Q-P, Steiner DF. Structure and evolution of insulin and insulin-like growth factors in chordates. Prog Brain Res 1992; 92:15–24.
4. Daughaday WH, Rotwein P. Insulin-like growth factors I and II. Peptide, messenger ribonucleic acid and gene structures, serum and tissue concentrations. Endocr Rev 1989; 10:68–91.
5. Rotwein P. Structure, evolution, expression and regulation of insulin-like growth factors I and II. *Growth Factors* 1991; 5:3–18.
6. Sussenbach JS, Steenbergh PH, Holthuizen P. Structure and expression of the human insulin-like growth factor genes. Growth Regul 1992; 2:1–9.
7. Jansen M, Van Schaik FMA, Ricker AT, Bullock B, Woods DE, Gabbay KH, Nussbaum AL, Sussenbach JS, Van Den Brande JL. Sequence of cDNA encoding human insulin-like growth factor I precursor. Nature 1983; 306:609–611.

8. Rotwein P, Pollack KM, Didier DK, Krivi GG. Organization and sequence of the human insulin-like growth factor-I gene. J Biol Chem 1986; 261:4828–4832.
9. Fotsis T, Murphy C, Gannon F. Nucleotide sequence of the bovine insulin-like growth factor I (IGF-I) and its IGF-IA precursor. Nucleic Acids Res 1989; 18:676.
10. Otte K, Rozell B, Gessbo A, Engstrom W. Cloning and sequencing of an equine insulin-like growth factor I cDNA and its expression in fetal and adult tissues. Gen Comp Endocrinol 1996; 102:11–15.
11. Tavakkol A, Simmen FA, Simmen RC. Porcine insulin-like growth factor-I (pIGF-I): complementary deoxyribonucleic acid cloning and uterine expression of messenger ribonucleic acid encoding evolutionarily conserved IGF-I peptides. Mol Endocrinol 1988; 2:674–681.
12. Delafontaine P, Lou H, Harrison DG, Bernstein KE. Sequence of a cDNA encoding dog insulin-like growth factor I. Gene 1993; 130:305–306.
13. Bell GI, Stempien MM, Fong NM, Seino S. Sequence of a cDNA encoding guinea pig IGF-I. Nucleic Acids Res 1990; 18:4275.
14. Wong EA, Ohlsen SM, Godfredson JA, Dean DM, Wheaton JE. Cloning of ovine insulin-like growth factor-I cDNAs: heterogeneity on the mRNA population. DNA 1989; 9:649–657.
15. Moore LG, Jones D, Lymburn MA, Hodgkinson SC, Davis SR, Suttie JM, Sadighi M, Carne A. Isolation and sequencing of deer and sheep insulin-like growth factors-I and -II. Gen Comp Endocrinol 1993; 92:302–310.
16. Shimatsu A, Rotwein P. Sequence of two rat insulin-like growth factor I mRNAs differing within the 5′ untranslated region. Nucleic Acids Res 1987; 15:7196–7190.
17. Casella SJ, Smith EP, Van Wyk JJ, Joseph DR, Hynes MA, Hoyt EC, Lund PK. Isolation of rat testis cDNAs encoding an insulin-like growth factor I precursor. DNA 1987; 4:325–330.
18. Murphy LJ, Bell GI, Duckworth ML, Friesen HG. Identification, characterization, and regulation of a rat complementary deoxyribonucleic acid which encodes insulin-like growth factor-I. Endocrinol 1987; 121:684–691.
19. Bell GI, Stempien MM, Fong NM, Rall LB. Sequences of liver cDNAs encoding two different mouse insulin-like growth factor-I precursors. Nucleic Acids Res 1986; 14:7873–7882.
20. Kajimoto Y, Rotwein P. Structure and expression of a chicken insulin-like growth factor I precursor. Mol Endocrinol 1989; 3:1907–1913.
21. Fawcett DH, Bulfield G. Molecular cloning, sequence analysis and expression of putative chicken insulin-like growth factor-I cDNAs. J Mol Endocrinol 1990; 4:201–211.
22. Kajimoto Y, Rotwein P. Evolution of insulin-like growth factor I (IGF-I): structure and expression of an IGF-I precursor from *Xenopus laevis.* Mol Endocrinol 1990; 4:217–226.
23. Shuldiner AR, Nirula AJ, Scott LA, Roth J. Evidence that *Xenopus laevis* contains two different nonallelic insulin-like growth factor-I genes. Biochem Biophys Res Commun 1990; 166:223–230.
24. Duguay SJ, Park LK, Samadpour M, Dickhoff WW. Nucleotide sequence and tissue distribution of three insulin-like growth factor I prohormones in salmon. Mol Endocrinol 1992; 8:1202–1210.
25. Wallis AE, Devlin RH. Duplicate insulin-like growth factor-I genes in salmon display alternative splicing pathways. Mol Endocrinol 1993; 8:409–422.
26. Shamblott MJ, Chen TT. Identification of a second insulin-like growth factor in a fish species. Proc Natl Acad Sci USA 1992; 89:8913–8917.
27. Hashimoto H, Mikawa S, Takayama E, Yokoyama Y, Toyohara H, Sakaguchi M. Molecular cloning and growth hormone-regulated gene expression of carp insulin-like growth factor-I. Biochem Mol Biol Intl 1997; 41:877–886.
28. McRory JE, Sherwood NM. Catfish express two forms of insulin-like growth factor-I (IGF-I) in the brain. J Biol Chem 1994; 269:18588-18592.
29. Reinecke M, Schmid A, Ermatinger R, Loffing-Cueni D. Insulin-like growth factor I in the teleost *Oreochromis mossambicus,* the tilapia: gene sequence, tissue expression, and cellular localization. Endocrinology 1997; 138:3613–3619.
30. Duguay SJ, Zhang JL, Steiner DF, Funkenstein B, Chan SJ. Developmental and tissue-regulated expression of IGF-I and IGF-II mRNAs in *Sparus aurata.* Mol Endocrinol 1996;16:123–132.
31. Duguay SJ, Chan SJ, Mommsen TP, Steiner DF. Divergence of insulin-like growth factors I and II in the elasmobranch, *Squalus acanthias.* FEBS Lett 1995; 371:69–72.
32. Kavsan VM, Koval AP, Grebenjuk VA, Chan SJ, Steiner DF, Roberts CT Jr, LeRoith D. Structure of the chum salmon insulin-like growth factor I gene. DNA Cell Biol 1993; 12:729–737.
33. Kavsan VM, Grebenjuk VA, Koval AP, Skorokhod AS, Roberts CT Jr, LeRoith D. Isolation of a second nonallelic insulin-like growth factor I gene from the salmon genome. DNA Cell Biol 1994; 13:555–559.

34. Bayne ML, Applebaum J, Chicchi GG, Miller RE, Cascieri MA. The roles of tyrosine 24, 31, and 60 in the high affinity binding of insulin-like growth factor-I to the type 1 insulin-like growth factor receptor. J Biol Chem 1990; 265:15648–15652.
35. Nagamatsu S, Chan SJ, Falkmer S, Steiner DF. Evolution of the insulin gene superfamily. J Biol Chem 1991; 266:2397–2402.
36. Duguay SJ, Milewski WM, Young BD, Nakayama K, Steiner DF. Processing of wild-type and mutant proinsulin-like growth factor-IA by subtilisin-related proprotein convertases. J Biol Chem 1997; 272:6663–6670.
37. Shimatsu A, Rotwein P. Mosaic evolution of the insulin-like growth factors. Organization, sequence and expression of the rat insulin-like growth factor I gene. J Biol Chem 1987; 262:7894–7900.
38. Hall LJ, Kajimoto Y, Bichell DP, Kim SW, James PL, Counts D, Nixon L, Tobin G, Rotwein P. Functional analysis of the rat insulin-like growth factor I gene and identification of an IGF-I gene promoter. DNA Cell Biol 1992; 11:301–313.
39. Brissenden JE, Ullrich A, Francke U. Human chromosomal mapping of genes for insulin-like growth factors I and II and epidermal growth factor. Nature 1984;310:781–784.
40. Kajimoto Y, Rotwein P. Structure of the chicken insulin-like growth factor I gene reveals conserved promoter elements. J Biol Chem 1991; 266:9724–9731.
41. Adamo ML, Ben-Hur H, Roberts CT Jr, LeRoith D. Regulation of start site usage in the leader exons of the rat insulin-like growth factor-I gene by development, fasting, and diabetes. Mol Endocrinol 1991;5:1677–1686.
42. Adamo ML, Lowe J, LeRoith D, Roberts CT Jr. Insulin-like growth factor I messenger ribonucleic acids with alternative 5′-untranslated regions are differentially expressed during development of the rat. Endocrinology 1989; 124:2737–2744.
43. Hoyt EC, Van Wyk JJ, Lund PK. Tissue and development specific regulation of a complex family of rat insulin-like growth factor I messenger ribonucleic acids. Mol Endocrinol 1988; 2:1077–1086.
44. Tobin G, Yee D, Brunner N, Rotwein P. A novel human insulin-like growth factor I messenger RNA is expressed in normal and tumor cells. Mol Endocrinol 1990; 4:1914–1920.
45. Bucci C, Mallucci P, Roberts CT Jr, Frunzio R, Bruni CB. Nucleotide sequence of a genomic fragment of the rat IGF-I gene spanning an alternative 5′ non coding exon. Nucleic Acids Res 1987; 17:3596–3590.
46. Kim SW, Lajara R, Rotwein P. Structure and function of a human insulin-like growth factor-I gene promoter. Mol Endocrinol 1991; 5:1964–1972.
47. Jansen E, Steenbergh PH, LeRoith D, Roberts CT Jr, Sussenbach JS. Identification of multiple transcription start sites in the human insulin-like growth factor-I gene. Mol Cell Endocrinol 1991; 78:115–125.
48. Adamo ML, Ben-Hur H, LeRoith D, Roberts CT Jr. Transcription initiation in the two leader exons of the rat IGF-I gene occurs from disperse versus localized sites. Biochem Biophys Res Commun 1991;176:887–893.
49. Ohlsen SM, Dean DM, Wong EA. Characterization of multiple transcription initiation sites of the ovine insulin-like growth factor-I gene and expression profiles of three alternatively spliced transcripts. DNA Cell Biol 1993;12:243–251.
50. Dickson MC, Saunders JC, Gilmour RS. The ovine insulin-like growth factor-I gene: characterization, expression and identification of a putative promoter. J Mol Endocrinol 1991; 6:17–31.
51. Weller PA, Dickson MC, Huskisson NS, Dauncey MJ, Buttery PJ, Gilmour RS. The porcine insulin-like growth factor-I gene: characterization and expression of alternate transcription sites. J Mol Endocrinol 1993; 11:201–211.
52. Rotwein P. Two insulin-like growth factor I messenger RNAs are expressed in human liver. Proc Natl Acad Sci USA 1986; 83:77–81.
53. Steenbergh PH, Koonen-Reemst AMCB, Cleutjens CBJM, Sussenbach JS. Complete nucleotide sequence of the high molecular weight human IGF-I mRNA. Biochem Biophys Res Commun 1991; 175:507–514.
54. Lowe J, Lasky SR, LeRoith D, Roberts CT Jr. Distribution and regulation of rat insulin-like growth factor I messenger ribonucleic acids encoding alternative carboxyterminal E-peptides: evidence for differential processing and regulation in liver. Mol Endocrinol 1988; 2:528–535.
55. Bell GI, Merryweather JP, Sanchez-Pescador R, Stempien MM, Priestley L, Scott J, Rall LB. Sequence of a cDNA clone encoding human preproinsulin-like growth factor II. Nature 1984; 310:775–777.
56. Whitfield HJ, Bruni CB, Frunzio R, Terrell JE, Nissley SP, Rechler MM. Isolation of a cDNA clone encoding rat insulin-like growth factor-II precursor. Nature 1984; 312:277–280.
57. Dull TJ, Gray A, Hayflick JS, Ullrich A. Insulin-like growth factor II precursor gene organization in relation to insulin gene family. Nature 1984; 310:777–780.

58. Ikejiri K, Wasada T, Haruki K, Hizuka N, Hirata Y, Yamamoto M. Identification of a novel transcription unit in the human insulin-like growth factor-II gene. Biochem J 1991; 280:439–444.
59. Holthuizen P, van der Lee FM, Ikejiri K, Yamamoto M, Sussenbach JS. Identification and initial characterization of a fourth leader exon and promoter of the human IGF-II gene. Biochim Biophys Acta 1990; 1087:341–342.
60. Gray A, Tam AW, Dull TJ, Hayflick J, Pintar J, Cavenee WK, Koufos A, Ullrich A. Tissue-specific and developmentally regulated transcription of the insulin-like growth factor 2 gene. DNA 1987; 6:283–295.
61. Soares MB, Turken A, Ishii D, Mills L, Episkopou V, Cotter S, Zeitlin S, Efstratiadis A. Rat insulin-like growth factor II gene: a single gene with two promoters expressing a multitranscript family. J Mol Biol 1986; 192:737–752.
62. Frunzio R, Chiariotti L, Brown AL, Graham DE, Rechler MM, Bruni C. Structure and expression of the rat insulin-like growth factor II (rIGF-II) gene. J Biol Chem 1986; 261:17138–17149.
63. Ueno T, Takahashi K, Matsuguchi T, Endo H, Yamamoto M. A new leader exon identified in the rat insulin-like growth factor II gene. Biochem Biophys Res Commun 1987; 148:344–349.
64. Rotwein P, Hall LJ. Evolution of insulin-like growth factor II: characterization of the mouse IGF-II gene and identification of two pseudo-exons. DNA Cell Biol 1990; 9:725–735.
65. Sasaki H, Shimozaki K, Zubair M, Aoki N, Ohta K, Hatano N, Moore T, Feil R, Constancia M, Reik W, Rotwein P. Nucleotide sequence of a 28-kb mouse genomic region comprising the imprinted *Igf2* gene. DNA Res 1996; 3:331–335.
66. Ohlsen SM, Lugenbeel KA, Wong EA. Characterization of the linked ovine insulin and insulin-like growth factor-II genes. DNA Cell Biol 1994; 13:377–388.
67. Brown WM, Dziegielewska KM, Foreman RC, Saunders NR. The nucleotide and deduced amino acid sequences of insulin-like growth factor II cDNAs from adult bovine and fetal sheep liver. Nucleic Acids Res 1990; 18:4614.
68. Otte K, Engstrom W. Insulin-like growth factor II in the horse: determination of a cDNA nucleotide sequence and expression in fetal and adult tissue. Gen Comp Endocrinol 1994; 96:270–275.
69. Catchpole IR, Engstrom W. Nucleotide sequence of a porcine insulin-like growth factor II cDNA. Nucleic Acids Res 1990; 18:6430.
70. Levinovitz A, Norstedt G, van den Berg S, Robinson IC, Ekstrom TJ. Isolation of an insulin-like growth factor II cDNA from guinea pig liver: expression and developmental regulation. Mol Cell Endocrinol 1992; 89:105–110.
71. O'Mahoney JV, Adams TE. Nucleotide sequence of an ovine insulin-like growth factor-II cDNA. Nucleic Acids Res 1989; 17:5392.
72. Ekstrom TJ, Backlin BM, Lindqvist Y, Engstrom W. Insulin-like growth factor II in the mink (*Mustela vison*): determination of a cDNA nucleotide sequence and developmental regulation of its expression. Gen Comp Endocrinol 1993; 90:243–250.
73. Stempien MM, Fong NM, Rall LB, Bell GI. Sequence of a placental cDNA encoding the mouse insulin-like growth factor II precursor. DNA 1986; 5:357–361.
74. Darling DC, Brickell PM. Nucleotide sequence and genomic structure of the chicken insulin-like growth factor-II (IGF-II) coding region. Gen Comp Endocrinol 1996; 102:286–287.
75. Palamarchuk AY, Holthuizen PE, Muller WEG, Sussenbach J, Kavsan VM. Organization and expression of the chum salmon insulin-like growth factor II gene. FEBS Lett 1997; 416:344–348.
76. Chen J, Chang C, Chen J, Shen S, Wu J. Production of biologically active recombinant tilapia insulin-like growth factor II polypeptides in *Escherichia coli* cells and characterization of the genomic structure of the coding region. DNA Cell Biol 1997; 16:883–892.
77. Jansen M, van Schaik FMA, van Tol H, Van Den Brande JL, Sussenbach JS. Nucleotide sequences of cDNA encoding precursors of human insulin-like growth factor II (IGF-II) and an IGF-II variant. FEBS Lett 1985; 179:243–246.
78. Hampton B, Burgess WH, Marshak DR, Cullen KJ, Perdue JF. Purification and characterization of an insulin-like growth factor II variant from human plasma. J Biol Chem 1989; 264:19155–19160.
79. Zumstein PP, Luthi C, Humbel RE. Amino acid sequence of a variant pro-form of insulin-like growth factor II. Proc Natl Acad Sci USA 1985; 82:3169–3172.
80. Daughaday WH, Emanuele MA, Brooks MH, Barbato AL, Kapadia M, Rotwein P. Synthesis and secretion of insulin-like growth factor II by a leiomyosarcoma with associated hypoglycemia. N Engl J Med 1988; 317:1434–1440.
81. Daughaday WH, Trivedi B, Baxter RC. Serum "big insulin-like growth factor II" from patients with tumor hypoglycemia lacks normal E-domain O-linked glycosylation, a possible determinant of normal propeptide processing. Proc Natl Acad Sci USA 1993; 90:5823–5827.

82. Bell GI, Gerhard DS, Fong NM, Sanchez-Pescador R, Rall LB. Isolation of the human insulin-like growth factor genes: insulin-like growth factor II and insulin genes are contiguous. Proc Natl Acad Sci USA 1985; 82:6450–6454.
83. Zemel S, Bartolomei MS, Tilghman SM. Physical linkage of two mammalian imprinted genes, H19 and insulin-like growth factor 2. Nature Genet 1992; 2:61–65.
84. O'Malley KL, Rotwein P. Human tyrosine hydroxylase and insulin genes are contiguous on chromosome II. Nucleic Acids Res 1988; 16:4437–4446.
85. van Dijk MA, Van Schaik FMA, Bootsma HJ, Holthuizen P, Sussenbach JS. Initial characterization of the four promoters of the human insulin-like growth factor II gene. Mol Cell Endocrinol 1991; 81:81–94.
86. Matsuguchi T, Takahashi K, Ikejiri K, Ueno T, Endo H, Yamamoto M. Functional analysis of multiple promoters of the rat insulin-like growth factor II gene. Biochim Biophys Acta 1990; 1048:165–170.
87. Ikejiri K, Furuichi M, Ueno T, Matsuguchi T, Takahashi K, Endo H, Yamamoto M. The presence and active transcription of three independent leader exons in the mouse insulin-like growth factor II gene. Biochim Biophys Acta 1991; 1089:77–82.
88. Soares MB, Ishii DN, Efstratiadis A. Developmental and tissue-specific expression of a family of transcripts related to rat insulin-like growth factor II mRNA. Nucleic Acids Res 1985; 13:1119–1133.
89. Ueno T, Takahashi K, Matsuguchi T, Ikejiri K, Endo H, Yamamoto M. Multiple polyadenylation sites in a large 3′-most exon of the rat insulin-like growth factor II gene. Biochim Biophys Acta 1989; 1009:27–34.
90. Nielson FC, Christiansen J. Endonucleolysis in the turnover of insulin-like growth factor II mRNA. J Biol Chem 1992; 267:19404–19411.
91. Meinsma D, Scheper W, Holthuizen PE, Van Den Brande JL, Sussenbach JS. Site-specific cleavage of IGF-II mRNAs requires sequence elements from two distinct regions of the IGF-II gene. Nucleic Acids Res 1992; 20:5003–5009.

3 Regulation of IGF Gene Expression

P. Elly Holthuizen, PhD, *Paul H. Steenbergh,* PhD
and John S. Sussenbach, PhD

Contents

INTRODUCTION

The insulin-like growth factors (IGFs) are potent mitogens for many different cell types and play a central role in growth and development. The significance of the IGFs in these processes has recently been shown by the dramatic effects of mutations in IGF genes that result in the absence of these growth factors (for details, *see* Chapter 23). In addition to their function in general growth, locally produced IGFs have important roles in tissue regeneration and wound healing. All of these processes involve mitogenic stimulation and cell division and need to be controlled accurately. A multitude of regulatory principles modulating the biological effects of the IGFs can be discerned.

First, the number of receptors expressed on target cells determines the strength of the IGF signal. Both IGF-I and IGF-II act through the type 1 IGF receptor. A second receptor, the type 2 IGF receptor, may be involved only in degradation of IGF-II and not in IGF signaling, thereby directly regulating IGF-II levels (for details, *see* Chapters 4 and 5).

Second, the biological activities of the IGFs are controlled by their interaction with larger soluble proteins, the IGF binding proteins, of which six are known to date. The binding proteins serve to protect the IGFs against degradation, facilitate their transport to distinct body compartments, and interfere both positively and negatively with the interaction between the IGFs and their cell surface receptors (for details, *see* Chapter 6).

Third, and the main topic of this chapter, the bioavailability of IGFs is regulated by the amounts of these growth factors produced and secreted by the cells capable of their synthesis. Serum concentrations of IGFs vary widely as a result of their individual expression profiles during prenatal and postnatal growth and development. In human, serum levels of IGF-I are relatively low in the fetal stage, ranging from 20 ng/mL in week 30 to well over 100 ng/mL at term. The IGF-II levels amount to 100 ng/mL at 30 weeks to almost 300 ng/mL at term. After birth, the serum levels of IGF-I increase up to 200

From: *Contemporary Endocrinology: The IGF System*
Edited by: R. Rosenfeld and C. Roberts Jr © Humana Press, Inc., Totowa, NJ

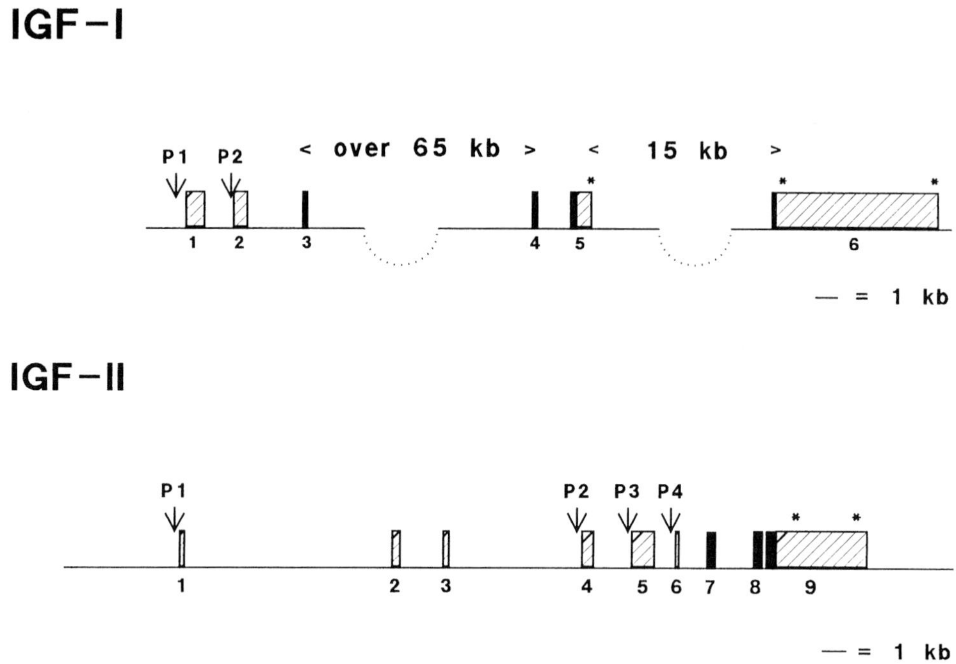

Fig. 1. Structure of the human IGF-I and IGF-II genes. *Boxes* denote the exons of the IGF genes. Coding exons are *filled*; untranslated regions are *striped*. The positions of the two IGF-I and the four IGF-II promoters are indicated by *arrows*. *Asterisks* denote alternative polyadenylation signals.

ng/mL and the IGF-II levels rise to 700 ng/mL *(1)*. Analogous profiles are seen in other mammals with the exception of rodents, where IGF-II levels decline very rapidly after birth and become virtually undetectable in the adult stages.

In addition to the developmental stage-dependent synthesis of endocrine IGF, originating mainly from hepatocytes, IGF with local paracrine and autocrine functions is synthesized in a number of other cell types. Numerous factors are involved in both types of IGF synthesis. Many of these factors act by ultimately influencing the activity of the multiple promoters of the IGF genes in their target cells, often via complex intracellular signaling routes. The human IGF-I gene, consisting of six exons including two leader exons, has two promoters and the human IGF-II gene, comprising nine exons, has six leader exons preceded by four promoters (Fig. 1; for details on IGF gene structures, *see* Chapter 2). The aim of this chapter is to summarize what is presently known about the regulation of the expression levels of the IGF genes by reviewing the mechanisms and factors that play a role in this process.

REGULATION OF IGF-I EXPRESSION

The expression levels of the IGF-I gene are quite different in the consecutive stages of development. Increasingly sensitive techniques have allowed the detection of IGF-I protein and IGF-I-encoding mRNA in very early stages of development. The IGF-I transcript has been shown to be present already in the preimplantation mouse embryo. Increasing mRNA levels have been observed in embryos between the eight-cell and blastocyst stages *(2)*. Numerous reports have described the presence and significance of low levels of IGF-I protein and mRNA during later stages of embryogenesis. The IGF-I transcript is then found in many different cell types and tissues, the liver not being a particularly high producer. The mode of action of IGF-I in the embryonic stages is generally regarded as autocrine and paracrine. The importance of IGF-I for embryonic growth and development is best illustrated by the effect of an IGF-I null mutation. IGF-I knockout mice are infertile dwarfs and at birth their size is approx 60% of the size of wild-type littermates *(3)*.

Hepatocytic IGF-I mRNA levels increase markedly around birth when the liver becomes the main producer of endocrine IGF-I. At the peak of postnatal growth, the levels of circulating IGF-I reach their maximum. A slow decrease in hepatic IGF-I production with age is generally observed. In humans, a second peak of IGF-I production coincides in time with the so-called growth spurt during puberty. In mice, the null mutation ultimately results in an adult-stage size of only 30% of wild-type littermates *(3)*.

Factors Determining Bioactive IGF-I Levels

In view of the developmental stage-dependent and tissue-specific expression of the IGF-I gene, it is not surprising that the agents involved in the regulation of expression of this growth factor have been the subject of many investigations. A vast array of hormones and other signaling molecules, compiled in Table 1, have been implicated in the regulation of the levels of bioactive IGF-I. These agents may be involved in direct regulation of the transcription rate of the IGF-I gene, as has been demonstrated for some of the factors, but may also affect posttranscriptional processes governing formation, stability, and translation of IGF-I mRNA, and posttranslational processes such as proteolytic cleavage and glycosylation. All of these processes together determine the actual level of bioactive IGF-I protein.

Transcription Factors Involved in the Regulation of IGF-I Expression

As discussed in Chapter 2, the known mammalian IGF-I genes contain two promoters *(4)*. Both promoters lack TATA and CAAT elements and show dispersed initiation of transcription. A number of transcription factors regulating the expression of the IGF-I gene have been identified to date. All of them influence the activity of promoter P1, the promoter located upstream of exon 1. No factors have been described that directly regulate the activity of the second promoter, P2, located upstream of exon 2. P1 is the more active promoter in most of the mammalian cells expressing IGF-I, yielding approx 80% of the IGF-I mRNA. The single promoters identified in the IGF-I genes of nonmammalian species (chicken, salmon) are homologous to P1.

The liver is the main source of endocrine IGF-I. This may be explained by the fact that a number of liver-specific transcription factors are able to bind to and enhance the activity of P1. Six binding sites for three different hepatocyte-enriched transcription factors, C/EBP, HNF-1, and HNF-3, have been identified in P1 of the human IGF-I gene, and these sites seem to be conserved quite well in the IGF-I genes of other species. In addition, a number of more ubiquitously expressed transcription factors have been shown to interact with P1 and regulate its activity. Binding sites for members of the AP-1 and E2F families of transcription factors have been described, and a cAMP-responsive site as well as a glucocorticoid-responsive region have been established in P1. These factors are probably also involved in the regulation of P1 expression in numerous nonliver tissues that produce IGF-I with local para- and autocrine functions.

Table 1
Compilation of Factors Stimulating (+) or Inhibiting (–) IGF-I Expression

Factor	*Tissue / Cells*	*Effect on*	*+ / –*	*Reference*
ACTH	Adrenal fasciculata cells	Protein	+	*(5)*
Angiotensin II	Adrenal fasciculata cells	Protein	+	*(5)*
	Vascular smooth muscle cells	Transcription	+	*(6)*
Calcitonin	MC3T3-E1 osteoblast cells	mRNA	+	*(7)*
Cortisol	(pre)Osteoblast cells	mRNA	–	*(8)*
Dexamethasone	Osteoblast-like cells	Protein	–	*(9)*
		mRNA	–	*(10)*
EGF	Renal collecting duct cells	mRNA	+	*(11)*
Estrogen	Osteoblast cells	Transcription	+	*(12)*
	Uterus (in vivo)	mRNA	+	*(13)*
	HepG2 hepatoma cells	Transcription	+	*(14)*
FGF	Aortic smooth muscle cells	mRNA	–	*(15)*
	Osteoblast cells	mRNA	–	*(16)*
FSH	Granulosa cells	Protein	+	*(17)*
Glucagon	Primary hepatocytes	mRNA	+	*(18)*
Glucose	C6 glioma cells	mRNA	+	*(19)*
Growth hormone	Primary hepatocytes	mRNA	+	*(20)*
	Ob1771 adipocytes	mRNA	+	*(21)*
	Liver (in vivo)	Transcription	+	*(22)*
IGF-I	Aortic smooth muscle cells	mRNA	+	*(15)*
IGF-II	MC3T3-E1 osteoblast cells	Protein	+	*(23)*
Insulin	Aortic smooth muscle cells	mRNA	+	*(15)*
	Primary hepatocytes	mRNA	+	*(24)*
	Liver (in vivo)	Transcription	+	*(25)*
Interferon-γ	Macrophages	mRNA	–	*(26)*
Interleukin-1β	Leydig cells	mRNA	–	*(27)*
LH	Testis (in vivo)	mRNA	+	*(28)*
PDGF	Aortic smooth muscle cells	mRNA	–	*(15)*
	Vascular smooth muscle cells	mRNA	–	*(29)*
	Osteoblast cells	mRNA	–	*(16)*
Progesterone	Uterus (in vivo)	mRNA	+	*(13)*
Prostaglandin E_2	Osteoblast cells	Transcription	+	*(30)*
		Transcription	+	*(31)*

(*continued*)

The transcription factor binding sites identified in promoter P1 of the IGF-I gene are discussed in detail in the following paragraphs and are summarized in Fig. 2.

Liver-Specific Transcription Factors

Members of different families of liver-enriched transcription factors have been shown to stimulate the activity of the human promoter P1 in transient transfection assays. Of the CAAT-box/enhancer binding protein (C/EBP) family, both C/EBPα and C/EBPβ enhance the activity of the human promoter P1. The transactivation is dependent on a specific binding site around position –115 relative to the transcription start site at +1 *(42)*. A second binding site around position –147 has a much lower binding affinity and only marginally contributes to the activation of P1 by C/EBP. The nucleotide sequences of these

Table 1 (*continued*)

Factor	*Tissue / Cells*	*Effect on*	+ / –	*Reference*
PTH	(pre)Osteoblast cells	mRNA	+	*(32)*
PTH-related protein	Fetal dermal fibroblasts	Transcription	+	*(33)*
Relaxin	Uterus (in vivo)	Protein	+	*(34)*
Somatostatin	Serum (in vivo)	Protein	–	*(24)*
Testosterone	Uterus (in vivo)	mRNA	+	*(35)*
T_3	GH3 pituitary cells	mRNA	+	*(36)*
	Liver (in vivo)	mRNA	+	*(37)*
	Primary hepatocytes	mRNA	+	*(38)*
	Heart (in vivo)	mRNA	+	*(39)*
TGF-β	MC3T3-E1 osteoblast cells	Protein	–	*(23)*
	Thyroid follicular cells	Protein	–	*(40)*
	Osteoblast cells	mRNA	–	*(16)*
Thyrotropine	Thyroid follicles (ex vivo)	mRNA	+	*(41)*
Vitamin D_3	Osteoblast-like cells	Protein	+	*(9)*

If experimental proof (e.g., run-on assay or reporter gene expression in transient transfections) has been obtained for a direct effect on the transcription rate of the IGF-I gene, "transcription" is indicated. Effects on steady-state levels of IGF-I mRNA are indicated by "mRNA," and "protein" stands for all techniques showing altered IGF-I levels after administration of the listed factor. More general factors known to affect IGF-I levels (e.g., nutritional status, injury, mechanical strain) have not been included in this table. *Abbreviations used:* ACTH, adrenocorticotrophic hormone; EGF, epidermal growth factor; FGF, fibroblast growth factor; FSH, follicle-stimulating hormone; LH, luteinizing hormone; PDGF, platelet-derived growth factor; PTH, parathyroid hormone; T_3, triiodothyronine; TGF-β, transforming growth factor-β.

binding sites in the human IGF-I gene and the homologous sequences in the IGF-I genes of other species are shown below. The lower, noncoding strand sequence is given for the –115 site, whereas the upper strand sequence is shown for the –147 site. Both adhere well to the consensus C/EBP binding sequence 5′- T–G/T–N–N–G–C/T–A–A–G/T –3′ *(43)*.

	–111 –119	–151 –143
Human *(44)*	5′– TTGAGTAAG –3′	5′– TGATGTTAT –3′
Pig *(45)*	TCGAGTAAG	TGATGTTAT
Sheep *(46)*	TCGAGTAAG	TGATGTTAT
Rat *(47)*	TCGAGTAAG	TGATGTTAT
Chicken *(48)*	CCGAGTAAG	TGATGTTAT
Salmon *(49)*	TTCATTAAG	TGACGTTAT

Of the hepatocyte nuclear factor 1 (HNF-1) family of transcription factors, HNF-1α has been shown to enhance significantly the activity of promoter P1. Of the two binding sites for this factor in the proximal promoter region, one completely overlaps the C/EBP recognition sequence around position –115. The upstream binding site is located around position –275 and has about two-fold higher affinity for HNF-1α than the downstream HNF-1 binding site. Nevertheless, the downstream binding site is of more importance to the activating effect of HNF-1α on P1 than the upstream one *(50)*. Both of the HNF-1 binding sites are well conserved during evolution as can be seen from the comparison below. In the salmon sequence no site homologous to the upstream binding site is found. The downstream sequence has been shown to be a functional HNF-1 binding site also in the

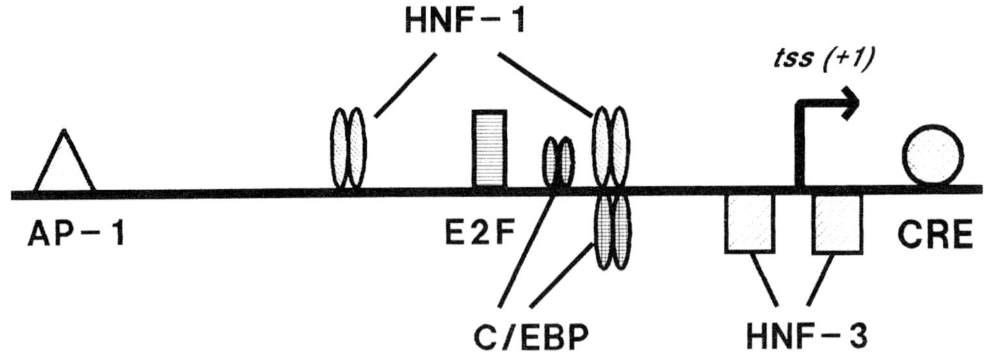

Fig. 2. Schematic presentation of elements and factors that participate in the activation of the human IGF-I promoter P1. The position of the major start site of transcription in the human IGF-I gene is indicated (+1). It should be noted that data obtained for different species are compiled in this figure.

salmon IGF-I promoter, although it contains residues that differ from the human sequence at three positions *(51)*. The suggested consensus sequence for HNF-1 binding sequence elements is the palindromic sequence 5′- G–T–T–A–A–T– N–A–T–T–A–A–C -3′ *(52)*, but many HNF-1 binding sites deviating from the consensus sequence in several residues have been described.

	–119 –107	–282 –270
Human	5′– CTTACTCAATAAC –3′	5′– GTTAATCAGAAAA –3′
Pig	CTTACTCGATAAC	GTTAATCAGAGGG
Sheep	CTTACTCGATAAC	GTTAATCAGAGGA
Rat	CTTACTCGATAAC	GTTAATCAGAGAA
Chicken	CTTACTCGGTAAC	GTTAACCAGAATA
Salmon	CTTAATGAATAAT	Not conserved

The third family of liver-enriched transcription factors involved in the regulation of P1 is the HNF-3 family. Of the three isoforms, HNF-3β is the strongest stimulator of IGF-I transcription. Again, two binding sites are found in the proximal promoter region of the human P1 promoter around positions –29 and +10 *(53)*. The two sites contribute equally to the transactivation by HNF-3β. The sequence of the upper, coding strand is indicated below for the –29 site, whereas the lower, noncoding strand sequence is shown for the +10 site. Both adhere well to the consensus HNF-3 binding sequence A/T–A/G–A/G–A/G–C/T–A/C–A–A–C/T–A *(54)*.

	–33 –24	+14 +5
Human	5′– TAGATAAATG –3′	5′– TTAGCAAACA –3′
Pig	TAGATAAATG	TTAGCAAACA
Sheep	TAGATAAAAT	TTAGCAAACA
Rat	TAGATAAATG	TTAGCAAGCA
Chicken	TAGATAAATG	TTAGCAAACA
Salmon	TAGATACATG	TTAGCGAACT

With six binding sites for three liver-enriched transcription factors in its proximal region, the architecture of P1 resembles that of a gene predominantly expressed in the liver (e.g., the albumin gene). All three factors are simultaneously expressed at high levels after birth and their concerted action may well account for the observed high levels of IGF-I expression in the liver at this stage of development. Because each of these transcription factors is also expressed in a number of nonhepatic tissues, albeit at much lower levels, IGF expression may also be regulated by liver-enriched transcription factors in other tissues. However, ubiquitously expressed transcription factors are likely to play a more significant role in nonliver IGF-I-expressing tissues.

Ubiquitous Transcription Factors

Members of the adaptor protein (AP)-1 family of transcription factors have been reported to enhance the activity of the chicken IGF-I promoter by binding to a site located around position –425 in the published sequence of the chicken IGF-I gene *(55)*. The phorbol ester TPA induces binding of the AP-1 factors to the response element, suggesting a direct role for protein kinase C (PKC)-mediated signaling pathways in the activation of the promoter. Furthermore, stimulation of the activity of the chicken promoter by estrogen has been shown to depend on enhanced *fos–jun* binding to the AP-1 binding site, which is directly or indirectly mediated by the estrogen receptor *(14)*. The AP-1 element in the chicken promoter, 5′- TTAGTCAG -3′, is located in a region of the chicken IGF-I gene that does not show significant homology to the IGF-I genes of the mammalian species of which the nucleotide sequences have been published.

Cyclic AMP has been shown to induce IGF-I synthesis via a responsive element (CRE) located within the untranslated region of exon 1 *(56)*. Transcriptional activation of the rat IGF-I gene via protein kinase A (PKA) and cAMP has been demonstrated in osteoblasts *(31)*. Parathyroid hormone *(32)*, parathyroid hormone-related protein *(33)*, and prostaglandin E_2 (PGE_2) *(31)* have been reported to enhance IGF-I expression via this pathway, whereas 17β-estradiol reportedly potently suppresses cAMP-induced IGF-I gene activation in primary osteoblast cultures *(57)*. The reported CRE within the rat exon 1 sequence (positions 202–209) is rather different from the consensus CRE sequence. Comparison to the homologous sequences in the IGF-I genes of other species shows that the C-residue at position 7 of the response sequence element is an insertion specific for the rat sequence within this otherwise strongly conserved region.

Human	(+55)	5′– CGCAAT-G(G) –3′	(+62)
Pig		CGCAAT-G(G)	
Sheep		CGCAAT-G(G)	
Rat	(+202)	CGCAATCG(G)	(+210)
Chicken		CGCAAT-G(G)	
Salmon		CGCAATGG(G)	

An E2F binding sequence has been reported to negatively regulate the rat P1 promoter of the IGF-I gene *(58)*. Complexes of proteins, including E2F, cyclin A, and p107 bind to the E2F site. Disaggregation of the complexes of cell-cycle-regulated proteins upon stimulation by growth factors, e.g., PDGF or EGF, or by serum, is suggested to increase expression from P1. This mechanism would provide a link between the regulation of cell proliferation by growth factors and the E2F-associated proteins. The nucleotide sequence reported to constitute the E2F site deviates at one position from the published rat promoter sequence *(47)*. The homologous sequences in the promoters of IGF-I genes in different species are shown below.

Reported rat E2F site *(58)*		5′– GCGGGGAAA –3′	
Human	(–198)	TAAGGGAAA	(–190)
Pig		TAAGGGAAA	
Sheep		TAAGGGAAA	
Rat	(–51)	GGGGGGAAA	(–43)
Chicken		AAGGGGAAA	
Salmon		AGCCTTAAA	

Finally, regulation of the rat IGF-I promoter P1 by glucocorticoids has been documented, but the nature and exact location of the cis-acting element is still unknown. Cortisol was shown to down-regulate IGF-I expression in osteoblasts, and transient transfection experiments have indicated that the +34 to +192 region of the rat promoter (homologous to the –120 to + 40 region of the human promoter) harbors the element involved in this process *(59)*.

Regulation of IGF-I Expression by Growth Hormone

The growth promoting effects of growth hormone (GH) are known to be mediated to a large extent by IGF-I. Since this concept was first formulated more than 40 years ago *(60)*, it has been shown that GH stimulates specific cells to produce IGF-I which in turn acts as a mitogenic agent on cells of many different lineages. From more recent work, it has become clear that transcription of the IGF-I gene, e.g., in liver tissue in vivo *(22)*, and in primary hepatocytes in vitro *(20)*, is strongly enhanced by the action of GH. The signal transduction pathway leading from the GH receptor (GHR) to the promoters of the IGF-I gene has not fully been unraveled at present, although much is known about GH signaling from recent work. After the binding of GH, the GHR transiently binds to and activates the tyrosine kinase JAK2 *(61)*. Tyrosine residues in both the GHR and JAK2 are rapidly phosphorylated. Like the tyrosine kinase surface receptors, e.g., the insulin and EGF receptors, the activated GHR-JAK2 complex is subsequently able to bind various SH2 domain-containing proteins and to transmit the activating signal to the PI-3 kinase and ras–raf–MAP kinase pathways *(62,63)*. In addition, the active complex is able to directly activate members of the family of signal transducers and activators of transcription, the STATs. Growth hormone has been shown to activate STAT-1, -3, and 5 *(64)*. Stimulation of c-*fos* gene transcription by GH involves the binding of STAT-1 and -3 to the SIE (*sis*-inducible element) in the c-*fos* promoter *(65)*. STAT-5 has been shown to be involved in GH-induced up-regulation of transcription of specific genes, e.g., the insulin gene *(66)*, by interaction with the GLE sequence (gamma-interferon-activated sequence-like element). Elements comparable in function to the SIE or GLE elements have not yet been detected in the promoter of an IGF-I gene and the mechanism by which GH stimulates the transcription of this gene awaits further elucidation.

REGULATION OF IGF-II EXPRESSION

The IGF-II gene is a complex transcription unit with many interesting regulatory aspects. Expression of IGF-II is controlled at multiple levels ranging from tissue-specific and developmental stage-dependent transcription initiation, alternative splicing, usage of multiple polyadenylation sites, site-specific endonucleolytic cleavage of IGF-II mRNAs, and translational control *(4)*. The structures of the IGF-II gene and mRNA species have been discussed in Chapter 2. Here, a summary is presented of what is currently known about the regulation of IGF-II expression, with a detailed description of the factors influencing the activity of the four promoters of the human IGF-II gene.

Expression of the IGF-II Gene During Fetal Development and After Birth

IGF-II gene expression has been detected in the mouse embryo from the two-cell stage onwards *(67)*. The fact that IGF-II has an essential growth-promoting function during embryogenesis became clear from targeted disruption of the IGF-II gene in the mouse. From embryonal day E11 onwards, mouse embryos with an IGF-II null mutation exhibit proportionate growth retardation and they are born as viable, fertile, proportionate dwarfs with a body weight that is 60% that of wild-type littermates *(68)*. Furthermore, it was found that homozygous IGF-II mutants and heterozygous progeny carrying a paternally derived mutated IGF-II gene are phenotypically indistinguishable as a consequence of parental imprinting. This phenomenon results in nonequivalence of the paternal and maternal IGF-II alleles. It was demonstrated that the paternal allele is transcriptionally active during mouse embryonal development, while the maternal allele is transcriptionally silent *(69)*. IGF-II mRNA and protein levels are high in most fetal tissues, notably the liver, and are expressed in all species examined *(70)*. After birth, IGF-II expression remains high in humans, whereas it is markedly decreased in all rodent tissues with the exception of the leptomeninges and choroid plexus of the brain, where IGF-II expression persists throughout adult life *(71,72)*. In human, IGF-II gene expression not only persists in these neuronal tissues after birth, but also in the liver due to the activation of the adult stage-specific promoter, P1 *(73)*. In addition, the IGF-II gene continues to be expressed postnatally from promoters P2, P3, and P4 in tissues such as heart, brain, kidney, muscle, skin, and the reproductive organs. However, the levels of the P2–P4 promoter-specific mRNAs in the latter tissues are much lower than that of the P1-derived mRNA in the liver *(74)*.

Factors Influencing IGF-II Expression

Hormonal regulation of IGF-II gene expression has been demonstrated in several instances. Beck et al. *(75)*, using Northern blot analysis and *in situ* hybridization, showed that cortisone acetate rapidly extinguishes IGF-II mRNA expression in the neonatal rat liver, but not in skeletal muscle and choroid plexus. It was postulated that this tissue-specific down-regulation of IGF-II expression by glucocorticoids could be effected by putative glucocorticoid-responsive elements in the rat IGF-II gene. Down-regulation of IGF-II mRNA expression has also been observed when the effects of glucocorticoid and thyroid hormone on rat hepatic IGF-II levels were investigated in 8-day-old pups *(76)*. In contrast, dexamethasone treatment of cultured human pheochromocytoma cells resulted in a three- to fivefold increase in IGF-II mRNA *(77)*. The highly differentiated rat insulin-producing β-cell INS-1 cell line has been used to study the regulation of IGF-II expression by glucose. IGF-II mRNA levels were increased three-fold due to enhanced transcription from rat promoter P3 (homologous to human promoter P4) and the effect was confirmed in transient transfection assays using IGF-II P3 promoter constructs *(78)*. In cultured human ovarian granulosa cells IGF-II mRNA synthesis is stimulated by FSH, chorionic gonadotropin, and dibutyryl cAMP, whereas in cultured human fetal adrenal cells IGF-II synthesis is stimulated by ACTH and dibutyryl cAMP *(79)*. In contrast,

ACTH and cortisol decreased IGF-II gene expression in ovine fetal adrenal cells *(80)*. Yee et al. *(81)* have found that IGF-II mRNA synthesis is increased in the T47D human breast tumor cell line after estrogen treatment. Furthermore, the effects of certain growth factors on IGF-II expression have been studied in bone cell cultures *(82)*. IGF-II mRNA and protein concentrations were measured in osteoblast-enriched (Ob) cell cultures from 22-day-old fetal rat calvariae. A continuous 48-h treatment with bFGF, TGFβ1, and PDGF-BB caused a dose-dependent decrease in steady-state levels of IGF-II mRNA, and TGFβ1 and PDGF-BB also decreased IGF-II polypeptide levels. The same cells were used in a study where the effect of retinoic acid on IGF-II expression was examined and it was demonstrated that both the IGF-II mRNA levels and the polypeptide levels were elevated after retinoic acid treatment of the cells *(83)*. Enhanced IGF-II mRNA expression and increased IGF-II polypeptide levels as a result of retinoic acid treatment were also described for several neuroblastoma cell lines such as Lan-1-15N cells *(84)*, BE*(2)*-M17 cells *(85)*, and SK-N-SH cells *(86)*, although the effect of retinoic acid on cell proliferation differed in these cell types. The mechanisms by which these hormones act or the identification of specific responsive elements that may be involved in regulation are still elusive.

Transcription of the IGF-II Gene

With the elucidation of the complete IGF-II gene structure, attention has shifted towards the characterization of the transcription factors involved in activation of the multiple IGF-II promoters. A variety of approaches have been applied to identify *cis*-acting elements and their cognate binding proteins involved in IGF-II gene regulation.

Regulation of the Human Promoter P1

Expression of IGF-II in human adult liver tissue is derived from promoter P1, giving rise to a 5.3-kb mRNA, containing the transcript of leader exons 1, 2, and 3 *(73)*. This promoter, located immediately downstream of the insulin gene, comprises approximately 1 kb. It is the main activator of IGF-II transcription in the adult liver, which functions as the source of circulating endocrine IGF-II. In rodents, a homologue of this promoter is absent, and thus the expression of IGF-II in rodents is rapidly shut off after birth. In other mammals such as horse, cow, sheep, and pig, expression of a specific IGF-II mRNA species in adult liver tissues has been detected by Northern blotting, suggesting the presence of an adult promoter as in the human gene *(87–90)*. This was confirmed for the ovine IGF-II gene, for which the gene structure as well as the nucleotide sequence of the adult promoter P1 has been determined *(91)*.

Regulation of the human P1 promoter has been studied in detail and both activating and repressing elements have been identified (Fig. 3). Based on transient transfection assays the promoter can be subdivided into two regions. The first 175 nt upstream of the transcription start site contain several regions to which nuclear proteins can bind, resulting in activation of transcription *(92–94)*. The distal promoter region (positions –900 to –175) contains two homologous 67 nucleotide (nt)-long inverted repeat elements (IR1 and IR2). In transient transfection experiments using P1 deletion constructs that lack one or both of the elements, it has been shown that these two IR elements act as cell-type-dependent suppressors of P1 activity. Similar to previously identified silencer elements, the IR elements affect promoter activity in an orientation- and position-independent manner *(95)*.

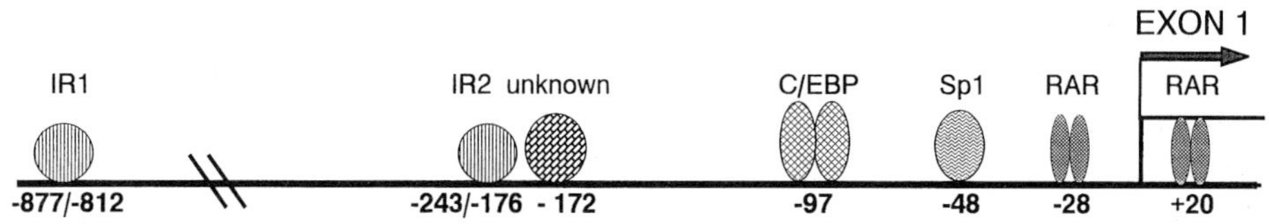

Fig. 3. Schematic presentation of elements and factors that participate in the activation of the human IGF-II promoter P1. The transcription start site, indicated by an *arrow*, is set to +1 and the positions of the elements in the promoter region are numbered accordingly. The various *cis*-acting elements and their *trans*-acting factors were identified by DNase I footprinting analysis, electrophoretic mobility shift assays, and supershifting with specific antibodies *(92,95,96)*. RAR represents two nuclear hormone responsive elements (RARE) that can be bound by retinoic acid receptor RAR/RXR heterodimers. (Rodenburg and Holthuizen, *unpublished results*.)

A single binding site for the ubiquitous transcription factor Sp1 is located around position –50. Mutations in the Sp1 binding site result in an 85% decrease in P1 promoter activity in transient transfections assays. Examination of P1 mutants in which the spacing between the Sp1 binding site and the transcription start site was increased showed that the role of the Sp1 binding site in regulation of P1 is position dependent and is absolutely required for basal P1 activity. This suggests that Sp1 may be involved in recruitment of the transcriptional machinery to this TATA-less promoter *(96)*. In addition, the effects of various liver-specific transcription factors on P1 expression were examined using transient cotransfection assays. An important activator element was identified around position –100, where a functional C/EBP binding site is located *(92,94)*. Two of the C/EBP family members are able to activate P1, C/EBPα (6-fold) and C/EBPβ (15-fold). DBP (D-element binding protein), another member of the C/EBP family, does not activate P1 and mutational analysis indicated that DBP binding is prevented by a specific G–C basepair in the P1 element *(92)*. C/EBP itself is expressed in a tightly controlled manner, the concentration being minimal in fetal liver, increasing around birth, and reaching maximum levels in adult liver tissue *(97,98)*. The 10-fold activation of P1 by C/EBP and the matching expression patterns of IGF-II and C/EBP suggest that these transcription factors are major contributors to the postnatal liver-specific activation of the human IGF-II promoter P1. The role of the hepatocyte nuclear factors (HNFs) in IGF-II P1 expression was studied in cotransfection experiments, electrophoretic mobility shift assays (EMSA), and mutational analysis. P1 activity is stimulated 10-fold by the presence of HNF-3β, whereas HNF-3α and HNF-3γ and HNF-1 hardly affect P1 promoter activity. Interestingly, HNF-4, a factor belonging to the family of steroid/thyroid hormone nuclear receptors, strongly suppresses IGF-II P1 activity. This suppression was observed in liver-derived Hep3B cells but not in kidney-derived 293 cells. The negative effect of HNF-4 on P1 activity was completely abolished by coexpression of C/EBP transcription factors, but not by HNF-3β. The IGF-II promoter P1 contains at least two tandem binding sites for nuclear hormone receptors in the proximal promoter region. Although these two sites deviate slightly from the consensus binding site, it was shown that the minimal P1 promoter construct encompassing the region from –57 to +52, containing both responsive elements, can be activated 15-fold when cotransfected with the retinoic acid receptor (RAR) and the retinoic X receptor (RXR) (Rodenburg and Holthuizen, *unpublished results*). In addition, it was shown that HNF-4 inhibits this activation of P1 by RAR/RXR. This may be explained by the fact that HNF-4 itself is a member of the nuclear hormone receptor family and can compete for binding to these sites. The identification of the specific nuclear hormone receptor and its ligand that are involved in the P1 regulation needs further investigation.

The observed effects of liver-specific transcription factors influencing P1 in combination with what is known about the expression of the HNFs during liver development have led to the following model for P1 expression. During the early phase of fetal liver development, HNF-4 is the dominant transcription factor and its presence suppresses P1 activation via a still unknown mechanism. Although the stimulatory factor HNF-3β is also expressed, this does not counteract the suppressing effect of HNF-4. Around the time of birth, the expression of C/EBPα and C/EBPβ is initiated and these factors do have the ability to counteract the suppression of P1 by HNF-4. After birth, expression of HNF-4 decreases, while the expression of HNF-3β and the C/EBPs increases, resulting in maximal expression of P1 in adult liver tissue. The effects of retinoic acid and its receptors in this context still need to be examined further.

Regulation of Promoter P2

Transcription from the human IGF-II promoter P2 results in a 5.0-kb mRNA, containing the transcript of exon 4 as the leader exon *(99)*. P2 transcripts have been detected at very low levels in fetal and adult tissues and may be elevated in some human tumor tissues *(100,101)*. The P2 promoter is a very weak promoter that lacks TATA- and CAAT-boxes and transcription starting at multiple initiation sites is observed. No Sp1 recognition sequences were found *(99)*, and no other enhancer elements have been described to date. Promoters homologous to the human P2 were detected and characterized in sheep and in rodents, where this promoter is designated P1, as rodents lack a homologue of the human promoter P1. Based on Northern blotting experiments with rat and mouse tissues, it can be concluded that in these species the homologous mRNA also represents a minor transcript *(102)*.

Regulation of Promoter P3

The human IGF-II promoter P3 is active in many fetal and nonhepatic adult tissues and in most IGF-II-expressing cell lines. The human P3 transcripts, containing exon 5-derived leader sequences, give rise to an abundant 6.0-kb mRNA and a minor 2.2-kb transcript that contains a shorter 3′ untranslated region (UTR) due to usage of an internal polyadenylation signal *(73)*. P3 contains a functional TATA-box and CAAT-box and the overall sequence of the promoter region is highly GC rich. Transient transfection experiments using truncated promoter P3 constructs have revealed that the distal promoter region (–1300/–289) is responsible for cell-type-specific IGF-II expression. The proximal region (–289/+140) exhibits basal levels of expression that are similar in several different cell lines, including HeLa and Hep3B *(103,104)*. It has been shown in EMSA experiments that in the upstream P3 region (–1231/–1063) multiple cell-type-specific DNA–protein complexes are formed *(105)*, but the exact nature of these binding proteins has not yet been identified.

In the proximal region a number of elements that are recognized by nuclear proteins were identified by DNase I footprint analysis, EMSA, and in vitro transcription (Fig. 4). TBP (TATA-binding protein) can bind to the TATA motif around position –25. In addition, p53 can interfere with the binding of TBP to the TATA motif, thereby inhibiting IGF-II transcription *(106)*. P3 can be bound and activated by the general transcription factors

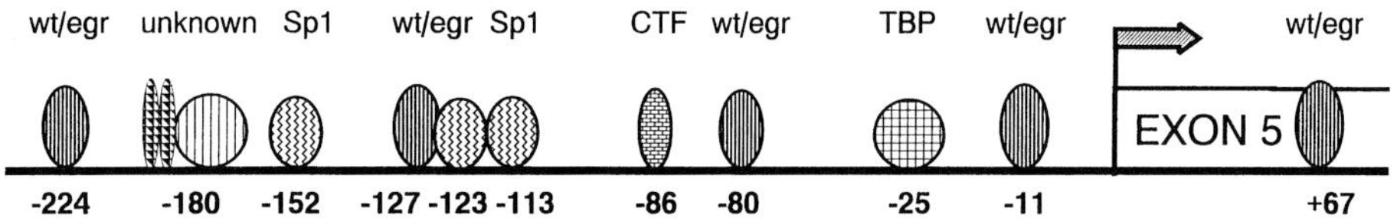

Fig. 4. Schematic presentation of elements and factors that participate in the activation of the human IGF-II promoter P3. The transcription start site, indicated by an *arrow*, is set to +1 and the position of the elements in the promoter region are numbered accordingly. The various *cis*-acting elements and their accompanying *trans*-acting factors were identified by DNase I footprinting analysis, electrophoretic mobility shift assays, and competition assays *(93,104,107–109,154)*. TBP represents the TATA-box binding protein, CTF the CAAT transcription factor, also known as NF1. The binding of Sp1 and wt/egr to the element around position –120 is mutually exclusive.

CTF (CAAT transcription factor), also known as nuclear factor 1, and there are multiple Sp1 sites. The early growth response proteins Egr-1 and Egr-2 have multiple recognition sites in promoter P3. The activation of P3 by the Egr zinc-finger transcription factors has received much attention, because the specific DNA motif that can be bound by the zinc-fingers of Egr-1 and Egr-2 is also recognized by the Wilms' tumor WT1 gene product. Binding studies have revealed that both Egr and WT1 proteins are able to recognize and bind to multiple sites in P3, albeit with different affinities *(107,108)*. Of the four isotypes of WT1, the WT1(-KTS) form, which is a minor component in vivo, was shown to bind to all Egr/WT1 sites of P3, while WT1(+KTS) could bind with high affinity only to the site located within exon 5 at position +63/+71. Using transient transfection experiments it was shown that Egr-1 has a strong stimulatory effect on P3 activity and a high level of WT1 expression of the –KTS type represses IGF-II P3 activity *(108)*. Based on these and other results it has been postulated that the Egr proteins may play a role in stimulating expression of the IGF-II gene resulting in autocrine growth stimulation of specific tumors, whereas WT1 may act as a repressor of P3 activity *(109)*.

An element located around position –180 and identified by DNase I footprinting analysis was shown to be quite important for promoter P3 transcription, because truncation of P3 at position –180 leads to a severe loss of promoter activity both in transient transfections and in vitro transcription assays *(103)*. Recently, this key regulatory element P3–4, located between positions –192 and –172 relative to the transcription start site, has been further examined. Using oligonucleotides containing different mutations in the P3–4 element in EMSA competition experiments and differential copper–phenanthroline footprinting it was shown that two distinct regions within element P3–4 are involved in the specific binding of three proteins. Box A (–193 to –188) can be bound by two highly related proteins and promoter activity decreased fourfold when protein binding to the box is inhibited by specific mutations. The binding of these proteins to box A occurs in a Zn^{2+} ion-dependent manner. Box B (–183 to –172) is bound by a third protein. Although the proteins bind to the P3–4 element in close proximity of each other, they can bind simultaneously without steric hindrance *(104)*.

The results obtained so far indicate that promoter P3 is a complex promoter and activation involves binding of multiple *trans*-acting factors that act in concert in an intricate mechanism of regulation. It is not known whether some of the above mentioned factors compete for the same binding sites or that they cooperate in transactivation. Moreover, not all of these factors are present in each cell type.

Regulation of Promoter P4

Transcription of the human promoter P4, preceding leader exon 6, results in an mRNA species of 4.8 kb. P4 contains a TATA-box and expression is moderate in most fetal and nonhepatic adult human tissues examined. The general transcription factor Sp1 is the major regulator of P4. Four binding sites for Sp1 are located within the first 125 nt upstream of the transcription start site *(93,110)*. Two of the sites fit the consensus sequence of Sp1 perfectly; the remaining two sites contain a few mismatches, suggesting that the contribution of the latter to transcriptional activation is less significant *(111)*. The fact that Sp1 is a major regulator of P4 activity was demonstrated when transient transfection experiments with a P4 promoter-CAT construct were performed in insect cells that do not produce Sp1 endogenously. No residual P4 activity could be detected in the absence of Sp1 *(112)*. Flanking the Sp1 site between positions –67 and –58, two retinoblastoma control elements (RCEs) were identified, and it was shown that the retinoblastoma gene product can act as a positive regulator on Sp1-mediated transcription of P4 *(112)*. In addition, the human P4 promoter can also be activated by Egr-1 *(113)*, and inhibited by WT1 *(114)*. Transfection experiments further demonstrated that an Egr-1 consensus element in P4 mediates induction of IGF-II gene expression after okadaic acid treatment *(113)*.

The homologous promoter in rat and mouse, P3, was shown to be the most active promoters in these species. Transcripts of 3.6 kb derived from this promoter are in general 10-fold more abundant in rat and mouse tissues than the 4.6-kb P2 transcripts from the promoter homologous to the prominent human P3 promoter. Interestingly, in the rat *(115,116)* and the mouse *(117)* P3 promoters four Sp1 binding sites were also identified, and all four sites were shown to bind Sp1 (compiled in *[111]*). Furthermore, it was shown that the mouse IGF-II P3 contains two noncanonical AP-1 binding sites that can be bound by recombinant c-jun protein. In a transient expression system using CP-1 embryonic stem cells it was demonstrated that the mouse P3 promoter is activated by c-jun and c-fos through the AP-1 binding sites *(118)*.

Differential Promoter Activity in Normal Tissues vs Tumor Tissues

A distinct example of developmental regulation of IGF-II transcription is found in the human liver. Before birth, IGF-II transcription is directed by the promoters P2, P3, and P4, of which P3, yielding an mRNA transcript of 6.0 kb, is the most active one *(73)*. After birth, these promoters are down-regulated and within the first postnatal year they are switched off *(119,120)*. Simultaneously, promoter P1 is activated and subsequently the 5.3-kb IGF-II mRNA becomes the only IGF-II-encoding transcript in the adult liver.

The pattern of IGF-II mRNA expression in human liver and liver cancers has been studied extensively *(121,122)*. IGF-II expression in benign liver tumors and in cirrhotic liver shows the same pattern of promoter activation as that of normal adult liver tissue. However, primary hepatic cancers and surrounding cirrhotic tissues exhibit a fetal pattern of IGF-II expression. These results suggest a differential regulation of IGF-II mRNA expression related to the pathological state of the liver *(121)*. Similar results were obtained when other groups examined hepatocellular carcinoma samples *(123–126)*, and showed that in 80% of the examined samples a switch in IGF-II promoter activity has taken place, the adult P1 promoter being shut off and the fetal promoters P2, P3, and P4 being preferentially active, with P3 activity most notably up-regulated. Similar observations have been made in a variety of human tumor tissues where a silenced IGF-II allele in normal tissue becomes activated in tumor tissues such as rhabdomyosarcoma *(127)*, lung carcinoma *(128)*, and choriocarcinoma *(129)*.

Also in human, modifications affecting the IGF-II gene locus have led to abnormalities in growth of the newborn. Uniparental paternal disomy of chromosome 11p results in an increased IGF-II expression due to biallelic expression of the normally monoallelically expressed IGF-II. Such enhanced fetal expression of IGF-II, as seen in Beckwith–Wiedemann syndrome, is characterized by neonatal gigantism, exomphalos, macroglossia, and an increased risk to develop embryonal tumors, especially Wilms' tumors *(130,131)*. Overproduction of IGF-II mRNA and protein in Wilms' tumors has been described, supporting the role of IGF-II as an embryonal growth factor.

Genomic Imprinting of IGF-II Promoters

Genomic imprinting of the IGF-II gene has been clearly demonstrated in human, rat, and mouse. Here we briefly address the issue of the existence of the promoter-specific imprinting status in humans. In most tissues the IGF-II gene is actively transcribed from the paternal allele, which is highly methylated in the IGF-II coding region in spite of its activity *(132)*. The maternal allele is transcriptionally inactive in all of the tissues that have been examined, and this seems to be the normal pattern of IGF-II expression *(133)*. Again adult human liver tissue is the exception to this rule, as it expresses IGF-II transcripts from both the maternal and the paternal allele, whereas human fetal liver exhibits monoallelic expression *(130, 134–137)*. In addition, biallelic IGF-II expression is detected in the choroid plexus and the leptomeninges of the central nervous system *(72)*. Monoallelic IGF-II expression from the paternal allele, corresponding to transcription from the fetal promoters P2, P3, and P4, is observed in most fetal tissues and more than 80% of the transcripts are derived from P3 *(136)*. In adult liver, however, where the promoter switch occurs from promoters P2–P4 to P1, IGF-II expression is biallelic and solely derived from the adult promoter P1. These results clearly indicate that promoter P1 is not imprinted, whereas the IGF-II promoters P2–P4 are imprinted. In addition, imprinting of the human IGF-II gene is developmentally regulated and tissue-specific and distinct imprinting patterns occur in a promoter-dependent manner *(135,137)*.

Site-Specific Endonucleolytic Processing of IGF-II mRNAs

Expression of the IGF-II gene is regulated not only at the level of initiation of transcription, but also at the level of posttranscriptional mRNA processing. In human, rat, and mouse a non-IGF-II-encoding RNA of 1.8 kb, derived from the IGF-II gene but not corresponding to any of the promoters, was detected on Northern blots when a 3′-UTR-specific IGF-II probe was used *(111,138)*. The 1.8-kb mRNA was subsequently identified as a product generated by a site-specific endonucleolytic cleavage reaction in the 3′-UTR of full-length IGF-II mRNAs *(139)*. Endonucleolytic cleavage of IGF-II mRNAs yields a rapidly degraded 5′-cleavage product that contains a cap structure, but no poly(A) tail. The 3′-cleavage product of 1.8-kb RNA lacks a cap structure but is polyadenylated and relatively stable *(139)*. The 1.8-kb uncapped RNA is associated with the 30S ribosomal particle and it is not actively engaged in protein synthesis *(140)*. Although the 3′-terminal 1.8-kb cleavage product lacks a cap structure, it is an extremely stable RNA that can be detected in IGF-II-expressing fetal and postnatal tissues and in cell lines expressing IGF-II. The endonucleolytic cleavage of IGF-II mRNAs was shown to occur in all types of full length IGF-II transcripts, irrespective of the promoter from which the transcripts are derived. In human and rat IGF-II mRNAs, the cleavage site has been mapped to a single nucleotide resolution by S1 mapping and primer extension experiments *(138,140)*.

Employing mutant IGF-II genes in an in vivo system, it was demonstrated that two widely separated sequence elements (designated I and II) in the 3′-UTR of IGF-II mRNAs are essential for the site-specific endonucleolytic cleavage reaction to occur *(141)* (Fig. 5). It was shown that a 350-nt region (element II) surrounding the cleavage site is necessary but not sufficient for cleavage, and that an additional 150 nt element (element I) located 2 kb upstream of element II is also required to confer cleavage. The presence of both elements I and II in the 3′-UTR of a mRNA is sufficient for cleavage, as was demonstrated when these two elements were transferred to the 3′-UTR of the heterologous β-globin gene *(141)*.

Based on these results a model was tested suggesting that the two structural elements, which are distant in the primary sequence, are brought into proximity in higher order RNA structures, thereby forming the recognition determinant for endonucleolytic cleavage. Structural analysis using RNA folding algorithms demonstrates that the cleavage site in element II is situated between two highly structured domains, a region immediately upstream of the cleavage site that can form two stem-loop structures and a region downstream of the cleavage site that is very G-rich and that may form intramolecular structures. A stable 80 nt-long double-stranded (ds) RNA stem structure ($\Delta G = -100$ kcal/mol) was identified that can be formed between the C-rich element I and the G-rich region in element II *(142)*. Furthermore, using the in vivo IGF-II minigene expression system, it was shown that the presence of this stable stem-structure as well as its sequence are important for proper processing of IGF-II mRNAs *(142)*.

An alternative explanation was presented by Christiansen et al. *(143)*. It is proposed that the G-rich stretch directly downstream of the cleavage site adopts an intramolecular quadruplex structure in a metal-ion sensitive manner, thereby exposing the cleavage site to the endonuclease. Support for this model was obtained by structural probing of synthetic RNA with RNase T1 and reverse transcriptase. However, formation of the double-stranded RNA structure between elements I and II, and the formation of the quadruplex structure, are mutually exclusive. The same G residues are involved either in the formation of the stem-structure or in the quadruplex structure, but not simultaneously in both. Given the requirement of both elements I and II for the cleavage reaction to occur, we favor the possibility that the stem-structure is the predominant structure in full-length IGF-II mRNA, while subsequently, after cleavage of full-length IGF-II RNA has taken place, the quadruplex structure may form at the 5′ end of the 1.8-kb RNA. The intramolecular guanosine quadruplex structure may possibly protect the 1.8-kb RNA from rapid 5′-exonucleolytic degradation which would account for the unusual stability of the 3′-terminal cleavage product.

Now that the regions required for endonucleolytic cleavage have been identified, the search has shifted towards the identification of the proteins participating in the cleavage reaction as well as the physiological meaning of this process. Examples of complex RNA structures interacting with *trans*-acting factors to regulate turnover of eukaryotic mRNAs are evolving rapidly *(144,145)*. Because both the exceptionally large stem-structure of IGF-II mRNA as well as its sequence and flanking structures are important for cleavage, it is probable that elements I and II cooperate in the binding of *trans*-acting factors involved in cleavage of IGF-II mRNAs. The activity of the specific endonuclease

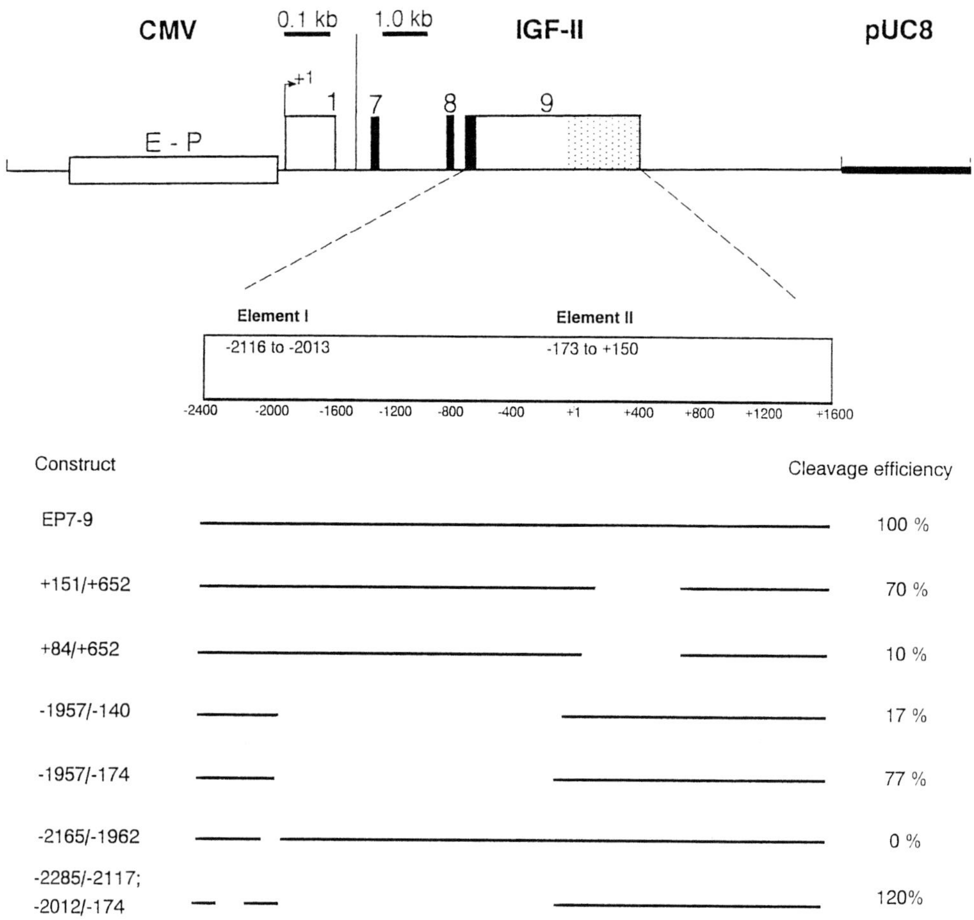

Fig. 5. Localization of elements I and II involved in site-specific endonucleolytic cleavage of IGF-II mRNAs. Schematic representation of the CMV-IGF-II minigene expression constructs. The wild-type construct, EP7-9, consists of the enhancer-promoter region of cytomegalovirus (CMV) linked to exons 7, 8, and 9 of the human IGF-II gene. The IGF-II coding regions are indicated in *black*. Exon 9 contains the end of the open reading frame of preprolGF-II and 4 kb of 3′-UTR. The cleavage site is numbered +1. The region in exon 9 that represents the 1.8-kb RNA is *stippled*. The cleavage efficiency observed for the wild-type construct EP7-9 was set to 100%. The mutant constructs are depicted with numbers indicating the 5′ and 3′ positions of the deleted fragments *(139,141,142)*.

may be modulated by additional *trans*-acting factors binding to these flanking sequences. Identification of proteins binding to the structured RNA elements required for cleavage may provide insight into the mechanism underlying the specific endonucleolytic cleavage.

The biological function of the cleavage reaction is still elusive. It may be a first step in the degradation of IGF-II mRNAs, as has been suggested for other endonucleolytic cleavage reactions of mRNA molecules *(146,147)*. Alternatively, the IGF-II 1.8-kb 3′-UTR RNA, albeit untranslated, may have intrinsic biological activity. Interestingly, it has been reported that 3′-UTR sequences of muscle-specific mRNAs (e.g., α-tropomyosin) have stimulatory effects on myogenic differentiation, suppression of tumorigenesis in myogenic cells, and inhibition of fibroblast proliferation *(148)*.

Translational Regulation of IGF-II Expression

Regulation of translation of mRNAs is an important regulatory step in gene expression and a number of mechanisms by which translation of specific mRNAs can be regulated are presently known *(144,145)*. Although translation takes place in the cytoplasm, compartmentalization can occur by interaction of the translational machinery with the cytoskeleton or by selective mobilization of specific mRNAs.

Multiple IGF-II mRNAs are generated by four different promoters, P1–P4, resulting in four different types of 5′-UTR leader sequences each linked to identical preproIGF-II coding regions and 3′-UTR sequences *(149)*. The leaders 1–4 are numbered according to the corresponding promoters. Three of the four IGF-II leaders contain much longer than average 5′-UTRs (leader 1, exons 1–3, 586 nt; leader 2, exon 4, 408 nt; leader 3, exon 5, 1171 nt) and can be expected to be translated less efficiently than leader 4 (exon 6, 109 nt). However, not only the length of the 5′-UTR leader sequence determines the efficiency of translation. A unique feature of the IGF-II leader 1 mRNA is the presence of an internal ribosomal entry site (IRES) located in the 5′-UTR, allowing translation initiation to occur by internal initiation as well. The function of this IRES as well as the factors influencing internal ribosomal entry are still unknown, but could provide for an additional regulatory step in translation enabling cell-stage-dependent expression of IGF-II *(150)*.

Studies on the polysomal distribution of the four types of mRNA in different cell lines and in fetal liver have revealed that the most abundant human IGF-II mRNA of 6.0 kb carrying leader 3 is hardly found in the polyribosome fraction, indicating that this mRNA species is poorly translated, whereas mRNAs with leader 4 sequences are associated with the membrane-bound polysomes and are actively engaged in IGF-II protein synthesis *(151–153)*. It was further demonstrated that translation of IGF-II mRNA in the rhabdomyosarcoma cell line RD is considerably affected by the growing state (e.g., cell density) of the cells *(151)*. P3-derived 6.0-kb mRNA (leader 3) is not translated in quiescent cells, but is selectively mobilized and translated in exponentially growing cells. Activation of translation is inhibited by rapamycin and mimicked by anisomycin, suggesting that translation of leader 3 mRNAs is regulated by the $p70^{s6k}$ signaling pathway. In contrast, the translation of P4-derived 4.8-kb mRNA is not dependent on the growth state of the cells. This suggests that leader 4 mRNAs are constitutively translated, whereas translation of leader 3 mRNAs is correlated with cellular proliferation and is regulated by the S6 kinase signaling pathway. Because it is known that the activation of the S6 kinase is controlled by growth factors, administration of IGF-I or IGF-II to IGF-II-producing cells may lead to enhanced translation of the leader 3 IGF-II mRNAs via stimulation of the PI-3 kinase pathway and the activation of S6 kinase *(151)*.

REFERENCES

1. Bennett AL, Wilson DM, Liu F. Levels of insulin-like growth factors I and II in human cord blood. J Clin Endocrinol Metab 1983; 57:609–612.
2. Doherty AS, Temeles GL, Schultz RM. Temporal pattern of IGF-1 expression during mouse implantation embryogenesis. Mol Reprod Dev 1994; 37:21–26.
3. Baker J., Hardy MP, Zhou J, Bondy C, Lupu F, Bellvé AR, Efstratiadis A. Effects of an *Igf1* gene null mutation on mouse reproduction. Mol Endocrinol 1996; 10:903–918.

4. Sussenbach JS. The gene structure of the insulin-like growth factor family. Prog Growth Factor Res 1989; 1:33–48.
5. Penhoat A, Leduque P, Jaillard C, Chatelain PG, Dubois PM, Saez JM. ACTH and angiotensin II regulation of insulin-like growth factor-1 and its binding proteins in cultured bovine adrenal cells. J Mol Endocrinol 1991; 7:223–232.
6. Delafontaine P, Lou H. Angiotensin II regulates insulin-like growth factor I gene expression in vascular smooth muscle cells. J Biol Chem 1993; 268:16866–16870.
7. Kobayashi T, Sugimoto T, Saijoh K, Fukase M, Chihara K. Calcitonin directly acts on mouse osteoblastic MC3T3-E1 cells to stimulate mRNA expression of c-fos, insulin-like growth factor-1 and osteoblastic phenotypes (type 1 collagen and osteocalcin). Biochem Biophys Res Commun 1994; 199:876–880.
8. McCarthy TL, Centrella M, Canalis E. Cortisol inhibits the synthesis of insulin-like growth factor-I in skeletal cells. Endocrinology 1990; 126:1569–1575.
9. Chen TL, Mallory JB, Hintz RL. Dexamethasone and $1,25(OH)_2$ vitamin D_3 modulate the synthesis of insulin-like growth factor-I in osteoblast-like cells. Calcif Tissue Int. 1991; 48:278–282.
10. Conover CA, Rosenfeld RG, Hintz RL. Serum glucocorticoids have persistent and controlling effects on insulin-like growth factor I action under serum-free assay conditions in cultured human fibroblasts. In Vitro Cell Dev Biol 1989; 25:521–527.
11. Rogers SA, Miller SB, Hammerman MR. Insulin-like growth factor-I gene expression in isolated rat renal collecting duct is stimulated by epidermal growth factor. J Clin Invest 1991; 87:347–351.
12. Ernst M, Rodan GA. Estradiol regulation of insulin-like growth factor-I expression in osteoblastic cells: evidence for transcriptional control. Mol Endocrinol 1991; 5:1081–1089.
13. Kapur S, Tamada H, Dey SK, Andrews GK. Expression of insulin-like growth factor-I (IGF-I) and its receptor in the peri-implantation mouse uterus, and cell-specific regulation of IGF-I gene expression by estradiol and progesterone. Biol Reprod 1992; 46:208–219.
14. Umayahara Y, Kawamori R, Watada H, Imano E, Iwama N, Morishima T, Yamasaki Y, Kajimoto Y, Kamada T. Estrogen regulation of the insulin-like growth factor I gene transcription involves an AP-1 enhancer. J Biol Chem 1994; 269:16433–16442.
15. Bornfeldt KE, Arnqvist HJ, Norstedt G. Regulation of insulin-like growth factor-I gene expression by growth factors in cultured vascular smooth muscle cells. J Endocrinol 1990; 125:381–386.
16. Canalis E, Pash J, Gabbitas B, Rydziel S, Varghese S. Growth factors regulate the synthesis of insulin-like growth factor-I in bone cell cultures. Endocrinology 1993; 133:33–38.
17. Hsu CJ, Hammond JM. Gonadotropins and estradiol stimulate immunoreactive insulin-like growth factor I and progesterone production by porcine granulosa cells. Endocrinology 1987; 120:198–207.
18. Kachra Z, Barash I, Yannopoulos C, Khan MN, Guyda HJ, Posner BI. The differential regulation by glucagon and growth hormone of insulin-like growth factor (IGF)-I and IGF binding proteins in cultured rat hepatocytes. Endocrinology 1991; 128:1723–1730.
19. Straus DS, Burke EJ. Glucose stimulates IGF-I gene expression in C6 glioma cells. Endocrinology 1995; 136:365–368.
20. Norstedt G, Moeller C. Growth hormone induction of insulin-like growth factor I messenger RNA in primary cultures of rat liver cells. J Endocrinol 1987; 115:135–139.
21. Doglio A, Dani C, Fredrikson G, Grimaldi P, Aihaud G. Acute regulation of insulin-like growth factor-I gene expression by growth hormone during adipose cell differentiation. EMBO J 1987; 6:4011–4016.
22. Bichell DP, Kikuchi K, Rotwein P. Growth hormone rapidly activates insulin-like growth factor I gene transcription *in vivo*. Mol Endocrinol 1992; 6:1899–1908.
23. Tremollieres FA, Strong DD, Baylink DJ, Mohan S. Insulin-like growth factor II and transforming growth factor β1 regulate insulin-like growth factor I secretion in mouse bone cells. Acta Endocrinol (Copenh) 1991; 125:538–546.
24. Pollak MN, Polychronakos C, Guyda H. Somatostatin analogue SMS 201–995 reduces serum IGF-I levels in patients with neoplasms potentially dependent on IGF-I. Anticancer Res 1989; 9:889–892.
25. Pao CI, Farmer PK, Begovic S, Goldstein S, Wu GJ, Phillips LS. Expression of hepatic insulin-like growth factor-I and insulin-like growth factor-binding protein-1 genes is transcriptionally regulated in streptozotocin-diabetic rats. Mol Endocrinol 1992; 6:969–977.
26. Arkins S, Rebeiz N, Brunke-Reese DL, Biragyn A, Kelley KW. Interferon-gamma inhibits macrophage insulin-like growth factor-I synthesis at the transcriptional level. Mol Endocrinol 1995; 9:350–360.
27. Lin T, Wang D, Nagpal ML, Chang W, Calkins JH. Down-regulation of Leydig cell insulin-like growth factor-I gene expression by interleukin-1. Endocrinology 1992; 130:1217–1224.

28. Closset J, Gothot A, Sente B, Scippo ML, Igout A, Vandenbroeck M, Dombrowicz D, Hennen G. Pituitary hormones dependent expression of the insulin-like growth factors I and II in the immature hypophysectomized rat testis. Mol Endocrinol 1989; 3:1125–1131.
29. Giannella-Neto D, Kamyar A, Sharifi B, Pivola CJ, Kupfer J, Rosenfeld RG, Forrester JS, Fagin JA. Platelet-derived growth factor isoforms decrease insulin-like growth factor I gene expression in rat vascular smooth muscle cells and selectively stimulate the biosynthesis of insulin-like growth factor binding protein 4. Circ Res 1992; 71:646–656.
30. Pash JM, Delany AM, Adamo ML, Roberts CT Jr., LeRoith D, Canalis E. Regulation of insulin-like growth factor I transcription by prostaglandin E_2 in osteoblast cells. Endocrinology 1995; 136:33–38.
31. Thomas MJ, Umayahara Y, Shu H, Centrella M, Rotwein P, McCarthy TL. Identification of the cAMP response element that controls transcriptional activation of the insulin-like growth factor-I gene by prostaglandin E_2 in osteoblasts. J Biol Chem 1996; 271:21835–21841.
32. McCarthy TL, Centrella M, Canalis E. Parathyroid hormone enhances the transcript and polypeptide levels of insulin-like growth factor I in osteoblast-enriched cultures from fetal rat bone. Endocrinology 1989; 124:1247–1253.
33. Shin JH, Ji CH, Casinghino S, McCarthy TL, Centrella M. Parathyroid hormone-related protein enhances insulin-like growth factor-I expression by fetal rat dermal fibroblasts. J Biol Chem 1997; 272:23498–23502.
34. Ohleth KM, Lenhart JA, Ryan PL, Radecki SV, Bagnell CA. Relaxin increases insulin-like growth factors (IGFs) and IGF-binding proteins of the pig uterus *in vivo*. Endocrinology 1997; 138:3652–3658.
35. Sahlin L, Norstedt G, Eriksson H. Androgen regulation of the insulin-like growth factor-I and the estrogen receptor in rat uterus and liver. J Steroid Biochem Mol Biol 1994; 51:57–66.
36. Fagin JA, Fernandez-Mejia C, Melmed S. Pituitary insulin-like growth factor-I gene expression: regulation by triiodothyronine and growth hormone. Endocrinology 1989; 125:2385–2391.
37. Wolf M, Ingbar SH, Moses AC. Thyroid hormone and growth hormone interact to regulate insulin-like growth factor-I messenger ribonucleic acid and circulating levels in the rat. Endocrinology 1989; 125:2905–2914.
38. Tollet P, Enberg B, Mode A. Growth hormone (GH) regulation of cytochrome P450IIC12, insulin-like growth factor-I (IGF-I), and GH receptor messenger RNA expression in primary rat hepatocytes: a hormonal interplay with insulin, IGF-I, and thyroid hormone. Mol Endocrinol 1990; 4:1934–1942.
39. Kupfer JM, Rubin SA. Differential regulation of insulin-like growth factor I by growth hormone and thyroid hormone in the heart of juvenile hypophysectomized rats. J Mol Cell Cardiol 1992; 24:631–639.
40. Beere HM, Soden J, Tomlinson S, Bidey SP. Insulin-like growth factor I production and action in porcine thyroid follicular cells in monolayer: regulation by transforming growth factor-beta. J Endocrinol 1991; 130:3–9.
41. Hofbauer LC, Rafferzeder M, Janssen OE, Gartner R. Insulin-like growth factor I messenger ribonucleic acid expression in porcine thyroid follicles is regulated by thyrotropin and iodine. Eur J Endocrinol 1995; 132:601–602.
42. Nolten LA, Van Schaik FMA, Steenbergh PH, Sussenbach JS. Expression of the insulin-like growth factor I gene is stimulated by the liver-enriched transcription factors C/EBP α and LAP. Mol Endocrinol 1994; 8:1636–1645.
43. De Simone V, Cortese R. Transcription factors and liver-specific genes. Biochim Biophys Acta 1992; 1132:119–126.
44. Rotwein P, Pollock KM, Didier DK, Krivi GG. Organization and sequence of the human insulin-like growth factor I gene. J Biol Chem 1986; 261:4828–4832.
45. Müller M, Brem G. Nucleotide sequence of porcine insulin-like growth factor I: 5′ untranslated region, exons 1 and 2 and mRNA. Nucleic Acids Res 1990; 18:364.
46. Wong EA, Ohlsen SM, Godfredson JA, Dean DM, Wheaton JE. Cloning of ovine insulin-like growth factor-I cDNAs: heterogeneity in the mRNA population. DNA 1989; 8:649–657.
47. Shimatsu A, Rotwein P. Mosaic evolution of the insulin-like growth factors. Organization, sequence and expression of the rat insulin-like growth factor I gene. J Biol Chem 1987; 262:7894–7890.
48. Kajimoto Y, Rotwein P. Structure of the chicken insulin-like growth factor I gene reveals conserved promoter elements. J Biol Chem 1991; 266:9724–9731.
49. Kavsan VM, Koval AP, Grebenjuk VA, Chan SJ, Steiner DF, Roberts CT Jr, LeRoith D. Structure of the chum salmon insulin-like growth factor I gene. DNA Cell Biol 1993; 12:729–737.

50. Nolten LA, Steenbergh PH, Sussenbach JS. Hepatocyte nuclear factor 1α activates promoter 1 of the human insulin-like growth factor gene via two distinct binding sites. Mol Endocrinol 1995; 9:1488–1499.
51. Kulik VP, Kavsan VM, Van Schaik FMA, Nolten LA, Steenbergh PH, Sussenbach JS. The promoter of the salmon insulin-like growth factor I gene is activated by hepatocyte nuclear factor 1. J Biol Chem 1995; 270:1068–1073.
52. Courtois G, Baumhueter S, Crabtree GR. Purified hepatocyte nuclear factor 1 interacts with a family of hepatocyte-specific promoters. Proc Natl Acad Sci USA 1988; 85:7937–7941.
53. Nolten LA, Steenbergh PH, Sussenbach JS. The hepatocyte nuclear factor 3β stimulates the transcription of the human insulin-like growth factor I gene in a direct and indirect manner. J Biol Chem 1996; 271:31846–31854.
54. Roux J, Pictet R, Grange T. Hepatocyte nuclear factor 3 determines the amplitude of the glucocorticoid responses of the rat tyrosine aminotransferase gene. DNA Cell Biol 1995; 14:385–396.
55. Kajimoto Y, Kawamori R, Umayahara Y, Iwama N, Imano E, Morishima T, Yamasaki Y, Kamada T. An AP-1 enhancer mediates TPA-induced transcriptional activation of the chicken insulin-like growth factor I gene. Biochem Biophys Res Commun 1993;190:767–773.
56. McCarthy TL, Thomas MJ, Centrella M, Rotwein P. Regulation of insulin-like growth factor I transcription by cyclic adenosine 3′,5′-monophosphate (cAMP) in fetal rat bone cells through an element within exon 1: protein kinase A-dependent control without a consensus AMP response element. Endocrinology 1995; 136:3901–3908.
57. McCarthy TL, Ji CH, Shu H, Casinghino S, Crothers K, Rotwein P, Centrella M. 17β-Estradiol potently suppresses cAMP-induced insulin-like growth factor-I gene activation in primary rat osteoblast cultures. J Biol Chem 1997; 272:18132–18139.
58. Porcu P, Grana X, Li S, Swantek J, De Luca A, Giordano G, Baserga R. An E2F binding sequence negatively regulates the response of the insulin-like growth factor 1 (IGF-I) promoter to simian virus 40T antigen and to serum. Oncogene 1994; 9:2125–2134.
59. Delany AM, Canalis E. Transcriptional repression of insulin-like growth factor I by glucocorticoids in rat bone cells. Endocrinology 1995; 136:4776–4781.
60. Salmon WD, Daughaday WH. A hormonally controlled serum factor which stimulates sulfate incorporation by cartilage in vitro. J Lab Clin Med 1957; 49:825.
61. Argetsinger LS, Campbell GS, Yang X, Witthuhn BA, Silvennoinen O, Ihle JN, Carter-Su C. Identification of JAK-2 as a growth hormone receptor-associated tyrosine kinase. Cell 1993; 74:237–244.
62. Kilgour E, Gout I, Anderson NG. Requirement for phosphoinositide 3-OH kinase in growth hormone signalling to the mitogen-activated protein kinase and $p70^{s6k}$ pathways. Biochem J 1996; 315:517–522.
63. VanderKuur JA, Butch ER, Waters SB, Pessin JE, Guan KL, Carter-Su C. Signaling molecules involved in coupling growth hormone receptor to mitogen-activated protein kinase activation. Endocrinology 1997; 138:4301–4307.
64. Smit LS, Meyer DJ, Billestrup N, Norstedt G, Schwartz J, Carter-Su C. The role of the growth hormone (GH) receptor and JAK1 and JAK2 kinases in the activation of Stats 1, 3, and 5 by GH. Mol Endocrinol 1996; 10:519–533.
65. Campbell GS, Meyer DJ, Raz R, Levy DE, Schwartz J, Carter-Su C. Activation of acute phase response factor (APRF)/Stat3 transcription factor by growth hormone. J Biol Chem 1995; 270:3974–3979.
66. Galsgaard ED, Gouilleux F, Groner B, Serup P, Nielsen JH, Billestrup N. Identification of a growth hormone-responsive STAT5-binding element in the rat insulin 1 gene. Mol Endocrinol 1996; 10:652–660.
67. Rappolee DA, Sturm KS, Behrendtsen O, Schultz GA, Pedersen RA, Werb Z. Insulin-like growth factor II acts through an endogenous growth pathway regulated by imprinting in early mouse embryos. Genes Dev 1992; 6:939–952.
68. DeChiara TM, Efstratiadis A, Robertson EJ. A growth-deficiency phenotype in heterozygous mice carrying an insulin-like growth factor II gene disrupted by targeting. Nature 1990; 345:78–80.
69. DeChiara TM, Robertson EJ, Efstratiadis A. Parental imprinting of the mouse insulin-like growth factor II gene. Cell 1991; 64:849–859.
70. Han VKM, Lund PK, Lee DC, D'Ercole AJ. Expression of somatomedin/insulin-like growth factor messenger ribonucleic acids in the human fetus: identification, characterization and tissue distribution. J Clin Endocrinol Metab 1988; 66:422–429.
71. Hynes MA, Brooks PJ, Van Wyk JJ, Lund PK. Insulin-like growth factor II messenger ribonucleic acids are synthesized in the choroid plexus of the rat brain. Mol Endocrinol 1988; 2:47–54.

72. Stylianopoulou F, Herbert J, Soares MB, Efstratiadis A. Expression of the insulin-like growth factor II gene in the choroid plexus and the leptomeninges of the adult rat central nervous system. Proc Natl Acad Sci USA 1988; 85:141–145.
73. De Pagter-Holthuizen P, Jansen M, Van Schaik FMA, van der Kammen RA, Oosterwijk C, Van den Brande JL, Sussenbach JS. The human insulin-like growth factor II contains two development-specific promoters. FEBS Lett 1987; 214:259–264.
74. Chin E, Bondy C. Insulin-like growth factor system gene expression in the human kidney. J Clin Endocrinol Metab 1992; 75:962–968.
75. Beck F, Samani NJ, Senior P, Byrne S, Morgan K, Gebhard R, Brammar WJ. Control of IGF-II mRNA by glucocorticoids in the neonatal rat. J Mol Endocrinol 1988; 1:5–8.
76. Kitraki E, Philippidis H, Stylianopoulou F. Hormonal control of insulin-like growth factor-II gene expression in the rat liver. J Mol Endocrinol 1992; 9:131–136.
77. Liu J, Kahri AI, Heikkilä P, Blum WF, Voutilainen R. Glucocorticoids increase insulin-like growth factor-II mRNA accumulation in cultured human phaeochromocytoma cells. J Endocrinol 1994; 142:29–35.
78. Asfari M, De W, Noël M, Holthuizen PE, Czernichow P. Insulin-like growth factor-II gene expression in a rat insulin-producing beta-cell line (INS-1) is regulated by glucose. Diabetologia 1995; 38:927–935.
79. Voutilainen R, Miller WL. Coordinate tropic hormone regulation of mRNAs for insulin-like growth factor II and the cholesterol side-chain-cleavage enzyme P450ssc, in human steroidogenic tissues. Proc Natl Acad Sci USA 1987; 84:1590–1594.
80. Lu F, Han VKM, Milne WK, Fraser M, Carter AM, Berdusco ET, Challis JRG. Regulation of insulin-like growth factor-II gene expression in the ovine fetal adrenal gland by adrenocorticotropic hormone and cortisol. Endocrinology 1994; 134:2628–2635.
81. Yee D, Cullen KJ, Paik S, Perdue JF, Hampton B, Schwartz A, Lippman ME, Rosen N. (1988) Insulin-like growth factor II m RNA expression in human breast cancer. Cancer Res 1988; 48:6691–6696.
82. Gabbitas B, Pash J, Canalis E. Regulation of insulin-like growth factor-II synthesis in bone cell cultures by skeletal growth factors. Endocrinology 1994; 135:284–289.
83. Gabbitas B, Canalis E. Retinoic acid regulates the expression of insulin-like growth factors I and II in osteoblasts. J Cell Physiol 1997; 172:253–264.
84. Matsumoto K, Gaetano C, Daughaday WH, Thiele CJ. Retinoic acid regulates insulin-like growth factor II expression in a neuroblastoma cell line. Endocrinology 1992; 130:3669–3676.
85. Melino G, Stephanou A, Annicchiarico-Petruzzelli M, Knight RA, Finazzi-Agró A, Lightman SL. Modulation of IGF-2 expression during growth and differentiation of human neuroblastoma cells: retinoic acid may induce IGF-2. Neurosci Lett 1993; 151:187–191.
86. Babajko S, Binoux M. Modulation of retinoic acid of insulin-like growth factor (IGF) and IGF binding protein expression in human SK-N-SH neuroblastoma cells. Eur J Endocrinol 1996; 134:474–480.
87. Otte K, Engström W. Insulin-like growth factor II in the horse: determination of a cDNA nucleotide sequence and expression in fetal and adult tissue. Gen Comp Endocrinol 1994; 96:270–275.
88. Brown WM, Dziegielewska KM, Foreman RC, Saunders NR. The nucleotide and deduced amino acid sequences of insulin-like growth factor II cDNAs from adult bovine and fetal sheep liver. Nucleic Acids Res 1990; 18:4614.
89. O'Mahoney JV, Adams TE. Nucleotide sequence of an ovine insulin-like growth factor-II cDNA. Nucleic Acids Res 1989; 17:5392.
90. Catchpole IR, Engström W. Nucleotide sequence of a porcine insulin-like growth factor II cDNA. Nucleic Acids Res 1990; 18:6430.
91. Ohlsen SM, Lugenbeel KA, Wong EA. Characterization of the linked ovine insulin and insulin-like growth factor-II genes. DNA Cell Biol 1994; 13:377–388.
92. Rodenburg RJT, Teertstra W, Holthuizen PE, Sussenbach JS. Postnatal liver-specific expression of human insulin-like growth factor-II is highly stimulated by the transcriptional activators liver-enriched activating protein and CCAAT/enhancer binding protein-α. Mol Endocrinol 1995; 9:424–434.
93. van Dijk MA, Van Schaik FMA, Bootsma HJ, Holthuizen P, Sussenbach J.S. Initial characterization of the four promoters of the human IGF-II gene. Mol Cell Endocrinol 1991; 81:81–94.
94. van Dijk MA, Rodenburg RJT, Holthuizen P, Sussenbach JS. The liver-specific promoter of the human insulin-like growth factor II gene is activated by CCAAT/enhancer binding protein (C/EBP). Nucleic Acids Res 1992; 20:3099–3104.

95. Rodenburg RJT, Krijger JJT, Holthuizen PE, Sussenbach JS. The liver-specific promoter of the human insulin-like growth factor-II gene contains two negative regulatory elements. FEBS Lett 1996; 394:25–30.
96. Rodenburg RJT, Holthuizen PE, Sussenbach JS. A functional Sp1 binding site is essential for the activity of the adult liver-specific human insulin-like growth factor II promoter. Mol Endocrinol 1997; 11:237–250.
97. Birkenmeier EH, Gwynn B, Howard S, Jerry J, Gordon JI, Landschulz WH, McKnight SL. Tissue-specific expression, developmental regulation, and genetic mapping of the gene encoding CCAAT/enhancer binding protein. Genes Dev 1989; 3:1146–1156.
98. Descombes P, Chojkier M, Lichtsteiner S, Falvey E, Schibler U. LAP, a novel member of the C/EBP gene family, encodes a liver-enriched transcriptional activator protein. Genes Dev 1990; 4:1541–1551.
99. Holthuizen P, van der Lee FM, Ikejiri K, Yamamoto M, Sussenbach JS. Identification and initial characterization of a fourth leader exon and promoter of the human IGF-II gene. Biochim Biophys Acta 1990; 1087:341–343.
100. Gloudemans T, Prinsen I, Van Unnik JAM, Lips CJM, Den Otter W, Sussenbach JS. Insulin-like growth factor gene expression in human smooth muscle tumors. Cancer Res 1990; 50:6689–6695.
101. Ikejiri K, Furuichi M, Ueno T, Matsuguchi T, Takahashi K, Endo H, Yamamoto M. The presence and active transcription of three independent leader exons in the mouse insulin-like growth factor II gene. Biochim Biophys Acta 1991; 1089:77–82.
102. Kou K, Rotwein P. Transcriptional activation of the insulin-like growth factor II gene during myoblast differentiation. Mol Endocrinol 1993; 7:291–302.
103. van Dijk MA, Holthuizen P, Sussenbach JS. Elements required for activation of the major promoter of the human insulin-like growth factor II gene. Mol Cell Endocrinol 1992; 88:175–185.
104. Rietveld LEG, Holthuizen PE, Sussenbach JS. Identification of a key regulatory element for the basal activity of the human insulin-like growth factor II gene promoter P3. Biochem J 1997; 327:689–697.
105. Schneid H, Holthuizen PE, Sussenbach JS. Differential promoter activation in two human insulin-like growth factor-II-producing tumor cell lines. Endocrinology 1993; 132:1145–1150.
106. Zhang LJ, Kashanchi F, Zhan QM, Zhan SL, Brady JN, Fornace AJ, Seth P, Helman LJ. Regulation of insulin-like growth factor II P3 promoter by p53: a potential mechanism for tumorigenesis. Cancer Res 1996; 56:1367–1373.
107. Rauscher FJ III. Tumor suppressor genes which encode transcriptional repressors: studies on the EGR and Wilms' tumor (WT1) gene products. Adv Exp Med Biol 1993; 348:23–29.
108. Drummond IA, Madden SL, Rohwer-Nutter P, Bell GI, Sukhatme VP, Rauscher FJ III. Repression of the insulin-like growth factor II gene by the Wilms tumor suppressor WT1. Science 1992; 257:674–678.
109. Madden SL, Rauscher FJ III. Positive and negative regulation of transcription and cell growth mediated by the EGR family of zinc-finger gene products. Ann NY Acad Sci 1993; 684:75–84.
110. Hyun SW, Kim S-J, Park K, Rho HM, Lee YI. Characterization of the P4 promoter region of the human insulin-like growth factor II gene. FEBS Lett 1993; 332:153–158.
111. Holthuizen PE, Cleutjens CBJM, Veenstra GJC, van der Lee FM, Koonen-Reemst AMCB, Sussenbach JS. Differential expression of the human, mouse and rat IGF-II genes. Regul Pept 1993; 48:77–89.
112. Kim S-J, Wagner S, Liu F, O'Reilly MA, Robbins PD, Green MR. Retinoblastoma gene product activates expression of the human *TGF*-β2 gene through transcription factor ATF-2. Nature 1992; 358:331–334.
113. Hyun SW, Park K, Lee YS, Lee YI, Kim S-J. Inhibition of protein phosphatases activates P4 promoter of the human insulin-like growth factor II gene through the specific promoter element. J Biol Chem 1994; 269:364–368.
114. Lee YI, Kim SJ. Transcriptional repression of human insulin-like growth factor-II P4 promoter by Wilms' tumor suppressor WT1. DNA Cell Biol 1996; 15:99–104.
115. Evans T, De Chiara T, Efstratiadis A. A promoter of the rat insulin-like growth factor II gene consists of minimal control elements. J Mol Biol 1988; 199:61–81.
116. Matsuguchi T, Takahashi K, Ikejiri K, Ueno T, Endo H, Yamamoto M. Functional analysis of multiple promoters of the rat insulin-like growth factor II gene. Biochim Biophys Acta 1990; 1048:165–170.
117. Rotwein P, Hall LJ. Evolution of insulin-like growth factor II: characterization of the mouse IGF-II gene and identification of two pseudo-exons. DNA Cell Biol 1990; 9:725–735.
118. Caricasole A, Ward A. Transactivation of mouse insulin-like growth factor II (IGF-II) gene promoters by the AP-1 complex. Nucleic Acids Res 1993; 21:1873–1879.

119. Davies SM. Developmental regulation of genomic imprinting of the IGF2 gene in human liver. Cancer Res 1994; 54:2560–2562.
120. Li X, Cui H, Sandstedt B, Nordlinder H, Larsson E, Ekström TJ. Expression levels of the insulin-like growth factor-II gene (IGF2) in the human liver: developmental relationships of the four promoters. J Endocrinol 1996; 149:117–124.
121. Cariani E, Lasserre C, Seurin D, Hamelin B, Kemeny F, Franco D, Czech MP, Ullrich A, Brechot C. Differential expression of insulin-like growth factor II mRNA in human primary liver cancers, benign liver tumors, and liver cirrhosis. Cancer Res 1988; 48:6844–6649.
122. Lasserre C, Cariani E, Zindy F, Lamas E, Brechot C. Insulin-like growth factor II and human primary liver cancer. In: Brechot C, ed. Primary Liver Cancer. Etiological and Progression Factors. CRC Press, Boca Raton, FL, 1994, pp. 283–297.
123. Takeda S, Kondo M, Kumada T, Koshikawa T, Ueda R, Nishio M, Osada H, Suzuki H, Nagatake M, Washimi O, Takagi K, Takahashi T, Nakao A. Allelic-expression imbalance of the insulin-like growth factor 2 gene in hepatocellular carcinomas and underlying disease. Oncogene 1996; 12:1589–1592.
124. Uchida K, Kondo M, Takeda S, Osada H, Takahashi T, Nakao A. Altered transcriptional regulation of the insulin-like growth factor 2 gene in human hepatocellular carcinoma. Mol Carcinog 1997; 18:193–198.
125. Li XR, Nong Z, Ekström C, Larsson E, Nordlinder H, Hofmann WJ, Trautwein C, Odenthal M, Dienes HP, Ekström TJ, Schirmacher P. Disrupted IGF2 promoter control by silencing of promoter P1 in human hepatocellular carcinoma. Cancer Res 1997; 57:2048–2054.
126. Nardone G, Romano M, Calabro A, Pedone PV, De Sio I, Persico M, Budillon G, Bruni CB, Riccio A, Zarrilli R. Activation of fetal promoters of insulinlike growth factor II gene in hepatitis C virus-related chronic hepatitis, cirrhosis, and hepatocellular carcinoma. Hepatology 1996; 23:1304–1312.
127. Zhan SL, Shapiro DN, Helman LJ. Loss of imprinting of IGF2 in Ewing's sarcoma. Oncogene 1995; 11:2503–2507.
128. Suzuki H, Ueda R, Takahashi T. Altered imprinting in lung cancer. Nature Genet 1994; 6:332–333.
129. Hashimoto K, Azuma C, Koyama M, Ohashi K, Kamiura S, Nobunaga T, Kimura T, Tokugawa Y, Kanai T, Saji F. Loss of imprinting in choriocarcinoma. Nature Genet 1995; 9:109–110.
130. Weksberg R, Shen DR, Fei YL, Song QL, Squire J. Disruption of insulin-like growth factor 2 imprinting in Beckwith–Wiedemann syndrome. Nature Genet 1993; 5:143–150.
131. Ogawa O, Eccles MR, Szeto J, McNoe LA, Yun K, Maw MA, Smith PJ, Reeve AE. Relaxation of insulin-like growth factor II gene imprinting implicated in Wilms' tumor. Nature 1993; 362:749–751.
132. Feil R, Handel MA, Allen ND, Reik W. Chromatin structure and imprinting: developmental control of DNase-1 sensitivity in the mouse insulin-like growth factor 2 gene. Dev Genet 1995; 17:240–252.
133. Giannoukakis N, Deal C, Paquette J, Goodyer CG, Polychronakos C. Parental genomic imprinting of the human IGF2 gene. Nature Genet 1993; 4:98–101.
134. Vu TH, Hoffman AR. Promoter-specific imprinting of the human insulin-like growth factor-II gene. Nature 1994; 371:714–717.
135. Ekström TJ, Cui H, Li X, Ohlsson R. Promoter-specific IGF2 imprinting status and its plasticity during human liver development. Development 1995; 121:309–316.
136. Vu TH, Hoffman AR. Alterations in the promoter-specific imprinting of the insulin-like growth factor-II gene in Wilms' tumor. J Biol Chem 1996; 271:9014–9023.
137. Wu HK, Squire JA, Song QL, Weksberg R. Promoter-dependent tissue-specific expressive nature of imprinting gene, insulin-like growth factor II, in human tissues. Biochim Biophys Res Commun 1997; 233:221–226.
138. De Pagter-Holthuizen P, Jansen M, van der Kammen RA, Van Schaik FMA, Sussenbach JS. Differential expression of the human insulin-like growth factor II gene. Characterization of the IGF-II mRNAs and an mRNA encoding a putative IGF-II-associated protein. Biochim Biophys Acta 1988; 950:282–295.
139. Meinsma D, Holthuizen P, Van den Brande JL, Sussenbach J.S. Specific endonucleolytic cleavage of IGF-II mRNAs. Biochim Biophys Res Commun 1991; 179:1509–1516.
140. Nielsen FC, Christiansen J. Endonucleolysis in the turnover of insulin-like growth factor II mRNA. J Biol Chem 1992; 267:19404–19411.
141. Meinsma D, Scheper W, Holthuizen P, Van den Brande JL, Sussenbach JS. Site-specific cleavage of IGF-II mRNAs requires sequence elements from two distinct regions of the IGF-II gene. Nucleic Acids Res 1992; 20:5003–5009.

142. Scheper W, Meinsma D, Holthuizen PE, Sussenbach JS. Long-range RNA interaction of two sequence elements required for endonucleolytic cleavage of human insulin-like growth factor II mRNAs. Mol Cell Biol 1995; 15:235–245.
143. Christiansen J, Kofod M, Nielsen FC. A guanosine quadruplex and two stable hairpins flank a major cleavage site in insulin-like growth factor II mRNA. Nucleic Acids Res 1994; 22:5709–5716.
144. Sachs AB, Sarnow P, Hentze MW. Starting at the beginning, middle, and end: translation initiation in eukaryotes. Cell 1997; 89:831–838.
145. Jacobson A, Peltz SW. Interrelationships of the pathways of mRNA decay and translation in eukaryotic cells. Annu Rev Biochim 1996; 65:693–739.
146. Beelman CA, Parker R. Degradation of mRNA in eukaryotes. Cell 1995; 81:179–183.
147. Binder R, Horowitz JA, Basilion JP, Koeller DM, Klausner RD, Harford JB. Evidence that the pathway of transferrin receptor mRNA degradation involves an endonucleolytic cleavage within the 3′ UTR and does not involve poly(A) tail shortening. EMBO J 1994; 13:1969–1980.
148. Rastinejad F, Conboy MJ, Rando TA, Blau HM. Tumor suppression by RNA from the 3′ untranslated region of α-tropomyosin. Cell 1993; 75:1107–1117.
149. Sussenbach JS, Rodenburg RJT, Scheper W, Holthuizen P. Transcriptional and post-transcriptional regulation of the human IGF-II gene expression. In: LeRoith D and Raizada MK, eds. Current Directions in Growth Factor Research. Plenum Press, New York, 1993, pp. 63–71.
150. Teerink H, Voorma HO, Thomas AAM. The human insulin-like growth factor II leader 1 contains an internal ribosomal entry site. Biochim Biophys Acta Gene Struct Expression 1995; 1264:403–408.
151. Nielsen FC, Ostergaard L, Nielsen J, Christiansen J. Growth-dependent translation of IGF-II mRNA by a rapamycin-sensitive pathway. Nature 1995; 377:358–362.
152. de Moor CH, Jansen M, Sussenbach JS, Van den Brande JL. Differential polysomal localization of human insulin-like growth factor 2 mRNAs, both in cell lines and foetal liver. Eur J Biochem 1994; 222:1017–1022.
153. de Moor CH, Jansen M, Bonte E, Thomas AAM, Sussenbach JS, Van den Brande JL. Influence of the four leaders of the human insulin-like growth factor mRNAs on the expression of reporter genes. Eur J Biochem 1994; 226:1039–1047.
154. Raizis AM, Eccles MR, Reeve AE. Structural analysis of the human insulin-like growth factor-II P3 promoter. Biochem J 1993; 289:133–139.

4 Molecular Biology of the Type 1 IGF Receptor

Haim Werner, PHD

CONTENTS

INTRODUCTION

The biological actions of the insulin-like growth factors (IGFs), as described in this book, are mediated by a family of cell-surface receptors that includes the insulin, the type 1, and the type 2 IGF receptors *(1–6)*. Although there is a certain degree of cross-interaction between the various ligands and their receptors, current evidence supports the view that most of the effects of both IGF-I and IGF-II on growth and differentiation are elicited by the ligand-dependent activation of the type 1 receptor. The type 2 receptor, which is identical to the cation-independent mannose-6-phosphate receptor, is mainly involved in the recycling of lysosomal enzymes containing mannose-6-phosphate residues and its role in the transduction of IGF-mediated biological signals has not yet been elucidated. This chapter focuses on the type 1 IGF receptor and discusses the molecular organization of its gene, mRNA, and protein. Functional aspects of the IGF receptor are examined in light of its specific physical features.

From: *Contemporary Endocrinology: The IGF System*
Edited by: R. Rosenfeld and C. Roberts Jr © Humana Press, Inc., Totowa, NJ

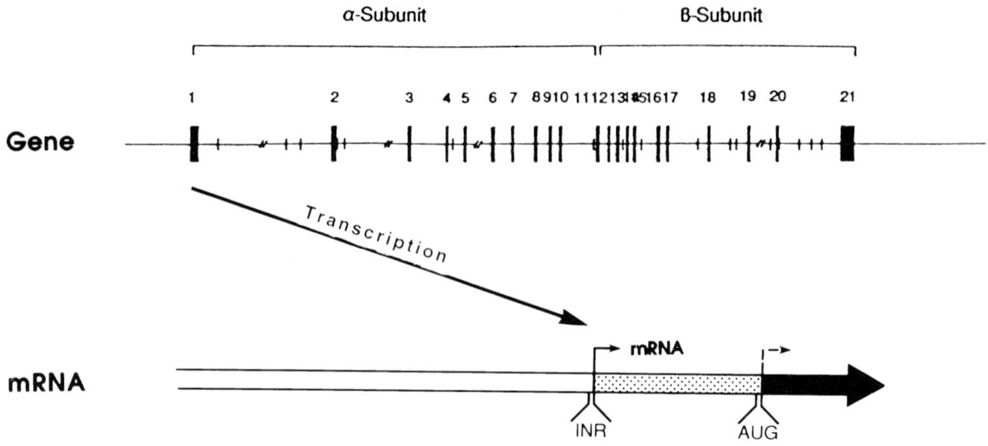

Fig. 1. Type 1 IGF receptor gene and mRNA. The type 1 IGF receptor gene consists of 21 exons (*closed blocks*). Exons 1–10 encode the 5′-untranslated region (*dotted area*), the signal peptide, and the α-subunit (*black area*). Exon 11 encodes the proteolytic cleavage site, and exons 12–21 encode the β-subunit. mRNA transcription is initiated from a unique site at the 5′ end of exon 1, contained within an initiator (INR) motif.

GENE STRUCTURE

The human type 1 IGF receptor is the product of a single-copy gene that spans more than 100 kilobases (kb) of genomic DNA at the distal end of chromosome 15 (bands q25→26) *(7–9)*. The gene contains 21 exons and its exon–intron organization is similar to that of the structurally related insulin receptor gene (Fig. 1, top panel) *(10–12)*.

The organization of the IGF receptor into functional domains is a reflection of its exonic architecture (Fig. 2A). Thus, the long 5′-untranslated (UTR) region, the signal peptide, and the amino (N)-terminal non-cysteine-rich and the cysteine-rich domains of the α-subunit are encoded by exons 1–3. Exons 4–10 encode the rest of the α-subunit, while the proteolytic cleavage site (Arg–Lys–Arg–Arg) that generates mature α- and β-subunits from the polypeptide precursor is encoded by exon 11. The receptor β-subunit is encoded by exons 12–21, with exon 14 encoding the transmembrane domain and exons 16–20 encoding the tyrosine kinase domain. This last domain exhibits the highest amino acid identity (80–95%) between the type 1 and insulin receptors.

Alternative splicing at the 5′ end of exon 14 of the type 1 IGF receptor gene results in two transcripts that differ in the absence or presence of three nucleotides (CAG), and encode either an arginine or a threonine and glycine in the extracellular portion of the β-subunit *(13,14)*.

Overexpression of both forms of the receptor in mammalian cells indicated that the CAG-containing form showed decreased ligand-mediated endocytosis. As a result of this reduction in receptor internalization, autophosphorylation of the receptor β-subunit and phosphorylation of insulin receptor substrate-1 (IRS-1) are augmented. The alternatively spliced exon 11 of the insulin receptor gene is absent in the type 1 IGF receptor gene.

MESSENGER RNA

Type 1 IGF receptor mRNA in human tissues appears in Northern blots as an ~11-kb transcript, with a minor band of ~7 kb in some tissues. In rodents, only the ~11 kb mRNA species is detected (Fig. 3) *(8,15)*.

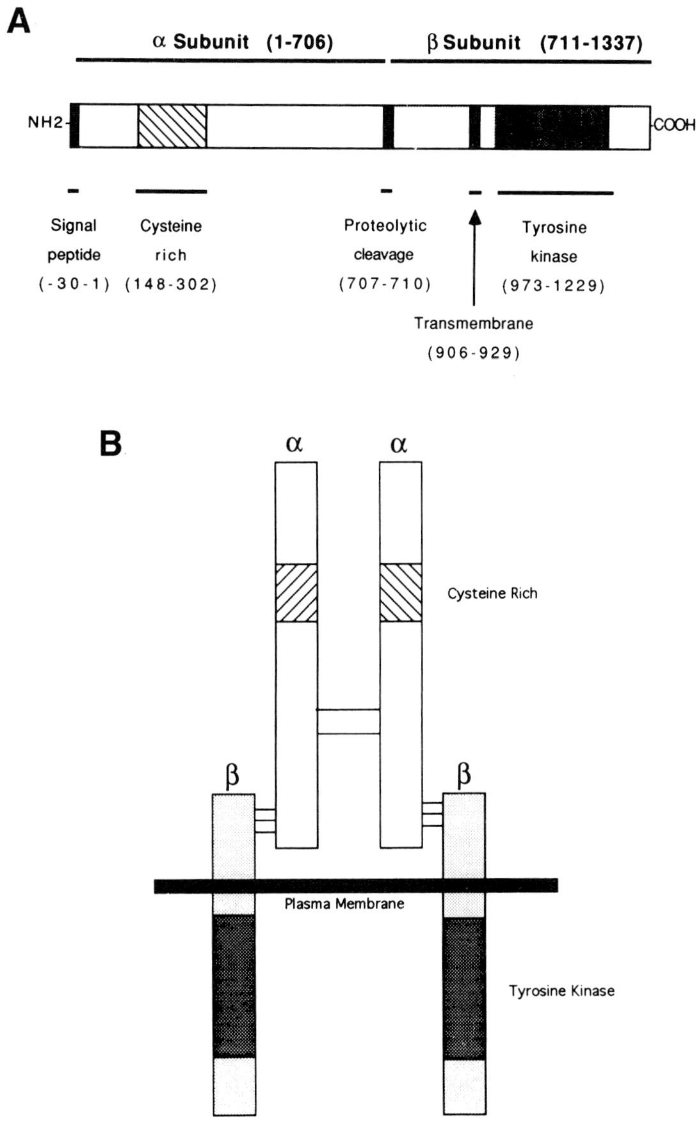

Fig. 2. Type 1 IGF receptor precursor and mature molecules. The 1367-amino-acid IGF receptor precursor (**A**) contains a 30-amino-acid signal peptide responsible for the transfer of the nascent polypeptide chain into the endoplasmic reticulum. Endopeptidase digestion at the proteolytic cleavage site (Arg–Lys–Arg–Arg) located between residues 707 and 710 releases α- and β-subunits, which bind by means of disulfide bonds to give the configuration of the mature heterotetrameric receptor (**B**).

The major transcript includes ~1 kb of 5′-UTR, ~5.1 kb of coding region, and ~5 kb of 3′-UTR.

The type 1 IGF receptor gene is expressed by virtually every tissue and cell type, consistent with the role of IGF-I as a progression factor through the cell cycle *(16–21)*.

Usually, high levels of expression are seen at embryonic stages, which are characterized by a large proportion of actively proliferating cells. Levels of receptor mRNA decrease postnatally, reaching lowest values in the postmitotic fully differentiated adult cells *(22)*.

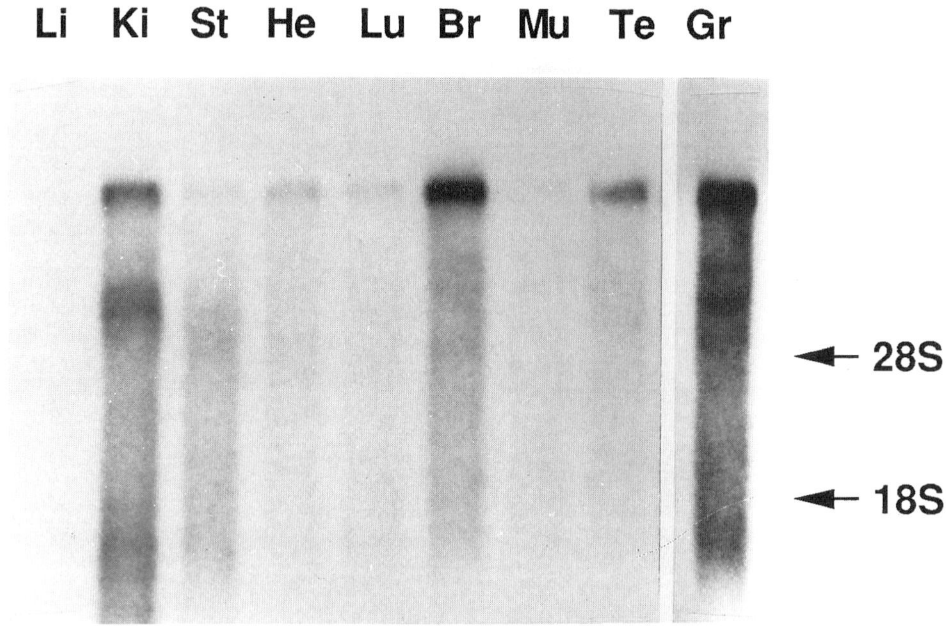

Fig. 3. Northern blot hybridization of rat type 1 IGF receptor mRNA. Four micrograms of poly (A^+)-enriched RNA from adult rat tissues were electrophoresed through a 1.5% agarose/formaldehyde gel, transferred to a nylon membrane, and hybridized to a ^{32}P-labeled rat type 1 IGF receptor cDNA probe. The positions of the 28S and 18S rRNA bands are indicated. Li, liver; Ki, kidney; St, stomach; He, heart; Lu, lung; Br, brain; Mu, skeletal muscle; Te, testes; Gr, SV40-transformed granulosa cell line.

In the adult rat, highest levels of expression are seen in the central nervous system (Fig. 3) *(23–28)*.

Other organs such as kidney, stomach, testes, lung, and heart show intermediate levels of mRNA. Liver, the organ with the highest levels of IGF-I ligand expression, exhibits almost undetectable levels of receptor mRNA, and this pattern of expression may be partially explained by "down-regulation" of the receptor gene by locally produced IGF-I. Dedifferentiated states, such as those associated with malignancy, are typically characterized by high levels of type 1 IGF receptor mRNA and protein *(2,29,30)*.

This phenomenon is illustrated by the extremely high concentrations of IGF receptor transcripts in the simian virus (SV)40-transformed rat granulosa cell line depicted in Fig. 3, right lane *(22)*.

PRIMARY STRUCTURE OF THE TYPE 1 IGF RECEPTOR

Early biosynthetic studies using specific inhibitors that retard the translocation of newly synthesized proteins through the Golgi apparatus to the plasma membrane suggested that the type 1 IGF receptor, like the insulin receptor, is synthesized as a single polypeptide chain that is processed to yield a glycopeptide of M_r 180 kDa (Fig. 2A) *(31)*.

Precursor processing includes core glycosylation and removal of the signal peptide. Partially processed proreceptors form disulfide-linked dimers which are subsequently terminally glycosylated and proteolitically cleaved to yield mature α- and β-subunits. The mature heterotetramer has a β-α-α-β conformation (Fig. 2B).

Deduction of the amino acid sequence of the type 1 IGF receptor from the nucleotide sequence of cloned cDNAs confirmed that α- and β-subunits are synthesized colinearly from a single precursor molecule composed of 1367 amino acids *(8)*.

Each precursor chain contains at its N-terminal domain a 30-amino-acid signal peptide rich in polar residues that is involved in the transfer of the nascent protein into the endoplasmic reticulum. Proteolytic cleavage at the basic tetrapeptide Arg–Lys–Arg–Arg (residues 707–710) releases α- and β-subunits (Fig. 2A).

The α-subunit of the mature receptor resides entirely extracellularly and includes a cysteine-rich region (containing 24 Cys residues) and 11 potential N-linked glycosylation sites (Asn–X–Ser/Thr motifs). The cysteine-rich domain of the type 1 IGF receptor is extremely important for high-affinity IGF-I binding, unlike the insulin receptor in which regions N- and C-terminal to the cysteine-rich domain are critical for insulin binding *(32)*.

Furthermore, all of the cysteines present in this domain are conserved in human and rat type 1 receptors.

The β-subunit features a unique 24-amino-acid hydrophobic sequence located between residues 906 and 929 that constitutes the transmembrane domain. The region located between residues 711 and 905 represents the extracellular portion of the β-subunit and it includes five potential N-linked glycosylation sites. Immediately C-terminal to the membrane-spanning region there is a stretch of basic amino acids that most probably help anchor the receptor to the plasma membrane. The juxtamembrane domain contains an Asn–Pro–X–Tyr (NPXY) motif (residues 947–950) that is important for receptor internalization and biological functioning of both insulin and type 1 IGF receptors *(33–35)*.

Finally, the cytoplasmic portion of the β-subunit contains a tyrosine kinase enzymatic domain located between residues 973 and 1229. Inside this catalytic region there is a glycine-rich conserved element (Gly–X–Gly–X–X–Gly; residues 976–981) that participates in the transfer of the phosphate moiety of ATP to specific substrates *(36–38)*.

Another functionally important segment in the tyrosine kinase domain is the catalytic loop of seven amino acid residues that lies between two β-pleated sheets (residues 1104-1110) *(37,38)*.

COMPARISON WITH INSULIN RECEPTOR

There is a striking similarity in the genomic organization of the insulin and type 1 IGF receptors *(7,12,39)*.

Twelve exons of the type 1 receptor gene are identical in size with the homologous exons of the insulin receptor. Furthermore, with the exception of the first and last exons, differences in size between the remaining homologous exons do not exceed 15 nucleotides. The impressive similarity in overall structure between these two genes undoubtedly suggests a close evolutionary origin.

As expected from their conserved gene arrangements, the primary structures of the insulin and type 1 IGF receptor proteins also share a high degree of similarity. The length of the insulin receptor precursor is 1370 amino acids (including a 27-amino-acid signal peptide), compared to 1367 amino acids (including a 30-amino-acid leader peptide) for the type 1 IGF receptor (Table 1). The highest similarity (84%) between both molecules is found in the tyrosine kinase domain of the β-subunit. Relatively high similarity (64–67%) is also seen between the extracellular α-subunit regions flanking the cysteine-rich domains. The cysteine-rich domains themselves exhibit a degree of homology of 48%, despite conservation in the type 1 receptor of 24 out of 26 cysteines that are present in the insulin receptor.

Table 1
Comparison of Insulin and Type 1 IGF Receptors

	Insulin Receptor	*Type 1 IGF Receptor*
Precursor size	1370 Amino acids	1367 Amino acids
Signal peptide size	27 Amino acids	30 Amino acids
Predicted M_r of precursor	152,000	151,869
Predicted M_r of α-subunit	82,400	80,423
Predicted M_r of β-subunit	69,700	70,866
Location of Arg–Lys–Arg–Arg sequence (cleavage site)	At position 720	At position 707
α-Subunit size	719 Amino acids	706 Amino acids
β-Subunit size	620 Amino acids	627 Amino acids
Transmembrane domain	Residues 915 (or 918)–940	Residues 906–929

Homology between Insulin and Type 1 IGF Receptors

β-Subunit tyrosine kinase domain	84%
α-Subunit regions flanking the cysteine-rich domain	64–67%
α-Subunit cysteine-rich domain	48%
β-Subunit transmembrane domain	27%

In addition, 15 out of 16 putative N-linked glycosylation sites (Asn–X–Thr/Ser) in the extracellular region of the type 1 receptor are located at almost identical positions in the insulin receptor. Lowest amino acid identity (27%) is found in exon 14 of the IGF receptor, which contains sequences encoding the membrane-spanning domain.

Despite this generally high level of similarity between insulin and type 1 IGF receptors, each ligand demonstrates high-affinity binding to its own receptor *(40–44)*.

Scatchard analysis of IGF-I and IGF-II binding to the type 1 receptor generates a linear plot, suggesting a single class of receptors with an affinity of ~10^{-10} *M*. Insulin binds to the IGF receptor with lower affinity (10^{-8} *M*). On the other hand, insulin receptors yield curvilinear plots that classically suggest the presence of high- (~10^{-10} *M*) and low-affinity binding sites. IGF-II binds the insulin receptor with an affinity 10- to 50-fold lower than insulin, and IGF-I with an affinity 100- to 500-fold lower.

TYPE 1 IGF RECEPTOR PROMOTER

Many physiological and pathological factors that affect the expression of the type 1 IGF receptor gene have been shown to act at the level of transcription. Characterization of the regulatory region of this gene is thus important to understand the molecular mechanisms that control its expression. The promoter region of the type 1 IGF receptor gene has a number of features that are shared by a family of genes which are constitutively expressed by virtually every cell and that are collectively referred to as "housekeeping genes." The promoter is extremely rich in G and C nucleotides and lacks TATA or CAAT motifs, two control elements that are usually required for accurate positioning of the transcription machinery and for efficient transcription initiation of most eukaryotic genes *(45–48)*.

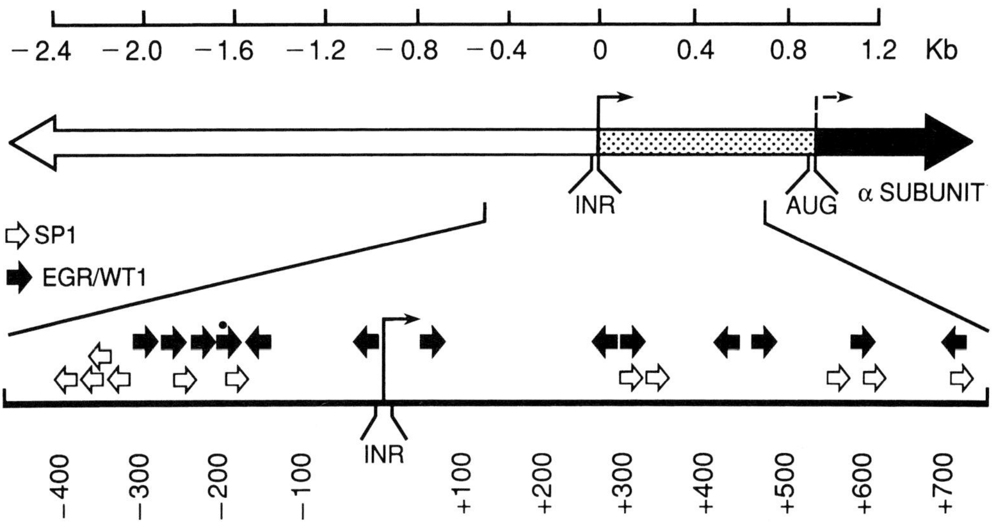

Fig. 4. Structure of the type 1 IGF receptor promoter. The IGF receptor gene contains at its 5′ end a TATA-less, CAAT-less, GC-rich promoter. The 5′-untranslated region (*dotted area*) is extremely long (~1 kb) and has the capacity to form significant secondary structures that may be involved in translational regulation. The coding region is shown in *black*. Both the 5′-flanking (*open bar*) and 5′-untranslated regions contain GC boxes of two major types: 5′-GGGCGG-3′, which are binding sites for transcription factor Sp1, and 5′-GCGGGGGCG-3′ (or related sequences), which are binding sites for members of the early growth response (EGR)/WT1 family of *trans*-activators. The EGR/WT1 site with a dot is a perfectly conserved consensus site, whereas the others differ by one or two nucleotides from this sequence. The orientation of the *arrows* denotes whether the consensus sequences are in the sense or in the antisense DNA strand.

In spite of the absence of these two important *cis*-elements, however, primer extension assays showed that transcription of the type 1 IGF receptor gene starts from a unique site (Fig. 1, bottom panel), whereas transcription of the insulin receptor gene, which also lacks TATA and CAAT boxes, starts from multiple sites *(49–52)*.

The single start site of the IGF receptor gene is contained within an "initiator" motif, a discrete promoter element that can direct specific transcription initiation from an internal position in the absence of a TATA element *(53)*.

Initiator elements are present in the promoters of genes that, like the type 1 IGF receptor, are highly regulated during differentiation and development. Thus, coexistence of control elements classically found in constitutive genes with those found in regulatable genes may provide the physical basis for the universal expression of the IGF receptor gene by virtually every cell, and for its potential to respond to specific, local needs by stimulating or suppressing transcription.

The type 1 IGF receptor promoter contains multiple GC boxes, which are potential binding sites for members of two families of transcription factors: the Sp1 and the early growth response (EGR) families (Fig. 4) *(54,55)*.

Sp1 binds preferentially to the 5′-GGGCGG-3′ sequence, whereas proteins of the EGR gene family (including WT1) recognize sequences related to the 5′-GCGGGGGCG-3′ consensus motif. Regulation of transcription by Sp1 and WT1 is described below. The 5′-flanking region of the type 1 IGF receptor gene contains, in addition, potential binding sites for at least three other known transcription factors: GC factor (GCF), adaptor protein 2 (AP-2) and ETF *(45,47,48)*.

GCF has been shown to repress the expression of the type 1 IGF receptor, probably via interaction with its specific binding element *(56)*.

The AP-2 element has been demonstrated to mediate both cyclic AMP (cAMP) and phorbol ester control of transcription in a number of genes *(57)*.

The presence of a putative AP-2 site in the receptor promoter may explain early studies that indicated that the expression of the type 1 IGF receptor gene is controlled by both cAMP and phorbol esters *(58,59)*.

The participation of the epidermal growth factor receptor transcription factor (ETF) element in promoter activity has not yet been established *(60)*.

The transcription initiation site of the type 1 IGF receptor gene is located ~1000 bp upstream of the ATG translation start codon in the human gene and 940 bp upstream in the rat (Fig. 1, bottom panel). This exceptionally long 5′-UTR is one of the longest described to date in eukaryotic genes. Albeit the function of this long 5′-UTR has not yet been elucidated, it has been shown that it can potentially form secondary structures, including hairpin motifs *(48)*.

These type of configurations have been reported in genes involved in the regulation of cellular proliferation, such as growth factors, their receptors, and oncogenes *(61)*.

Another interesting feature that is found in the 5′-UTR of the IGF receptor gene is an 84-bp open reading frame that begins and ends upstream of the translation start site for the ORF encoding the type I IGF receptor precursor.

The activity of the type 1 IGF receptor gene promoter was measured by subcloning genomic fragments containing different portions of 5′-flanking and 5′-UTR upstream of a luciferase reporter gene in the promoterless vector p0LUC *(46)*.

These constructs were transiently transfected into Chinese hamster ovary (CHO) and buffalo rat liver (BRL3A) cells, and the levels of luciferase activity were measured in cellular extracts. The IGF receptor gene displays a high basal promoter activity, reaching levels between 10% and 50% of the activity elicited by the potent SV40 enhancer/promoter control. Basal promoter activity was mainly mapped to the proximal ~400–500 bp of 5′-flanking region. Sequences in the 5′-UTR had no major contribution to basal levels of transcriptional activity. Using an heterologous reporter gene system, however, 5′-UTR fragments of the IGF receptor gene were shown to enhance transcription elicited by the SV40 promoter and, concomitantly, these same fragments were able to inhibit message translatability *(62)*.

The initiator element, by itself, was devoid of any promoter activity, although its orientation in the sense direction was crucial for efficient and accurate initiation.

TRANS-ACTIVATORS OF THE TYPE 1 IGF RECEPTOR PROMOTER

The levels of expression of the type 1 IGF receptor gene in any given tissue at each developmental stage are determined by the finely tuned interplay between a number of transcription factors and other regulatory proteins (Fig. 5). Some of these *trans*-activators can stimulate transcription of the receptor gene whereas others negatively regulate its expression.

One of the best characterized activators of the type 1 IGF receptor gene is Sp1. Sp1 is an ubiquitous zinc-finger nuclear protein that stimulates transcription from a specific group of RNA polymerase II-dependent promoters in vertebrates *(54,63)*.

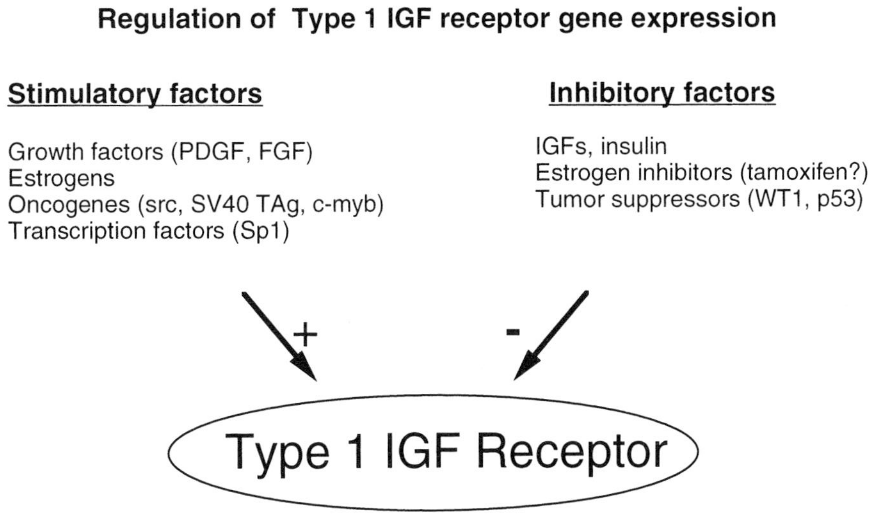

Fig 5. Regulation of type 1 IGF receptor gene expression. The level of expression of the IGF receptor gene is determined by the interplay between a number of positive and negative factors. Disruption of the finely tuned balance between stimulatory and inhibitory pathways often results in aberrant gene expression.

This family of promoters, which usually lack TATA and CAAT elements, includes among others the epidermal growth factor receptor, insulin receptor, and nerve growth factor receptor, in addition to the type 1 IGF receptor *(12,64,65)*.

Sp1 binding to GC boxes appears to be critical to achieve significant levels of transcription from TATA-less, CAAT-less promoters, although such interactions do not appear to determine the precise site of transcription initiation *(66–68)*.

Sp1 binds DNA by means of three zinc finger motifs and activates transcription via a glutamine-rich domain *(63)*.

Physical and functional interactions of transcription factor Sp1 with the type 1 IGF receptor promoter were analyzed in *Drosophila* Schneider cells which lack endogenous Sp1 and therefore provide a "null" background for these type of studies *(46,69)*.

Basal promoter activity was very low in Schneider cells, compared to the high activity displayed by this promoter in all mammalian cell lines tested. Transient cotransfection of Schneider cells with a "proximal" fragment of the type 1 IGF receptor promoter containing 476 bp of 5′-flanking region and 640 bp of 5′-UTR fused to a luciferase reporter gene, together with increasing amounts of an Sp1 expression vector, resulted in as much as an 500-fold activation of the type 1 receptor promoter. Thus, type 1 IGF receptor gene expression is extremely dependent upon Sp1 action. Electrophoretic mobility-shift and DNase I footprinting assays using purified Sp1 protein and crude nuclear extracts localized a number of Sp1 binding sites in the 5′-flanking region (Fig. 6). A cluster of four sites around nucleotide position –400 is apparently responsible for most of the effect of Sp1, although other sites in the 5′-flanking region may also contribute to this activity (Fig. 4). Thus, the capacity of Sp1 to *trans*-activate the IGF receptor promoter is consistent with its ability to bind specifically to consensus sites present in this region.

Although ubiquitously expressed, developmental studies have shown substantial variations in the expression of the Sp1 gene during ontogeny. Fully differentiated cells with specialized functions exhibited the lowest levels of Sp1 mRNA and protein *(70)*.

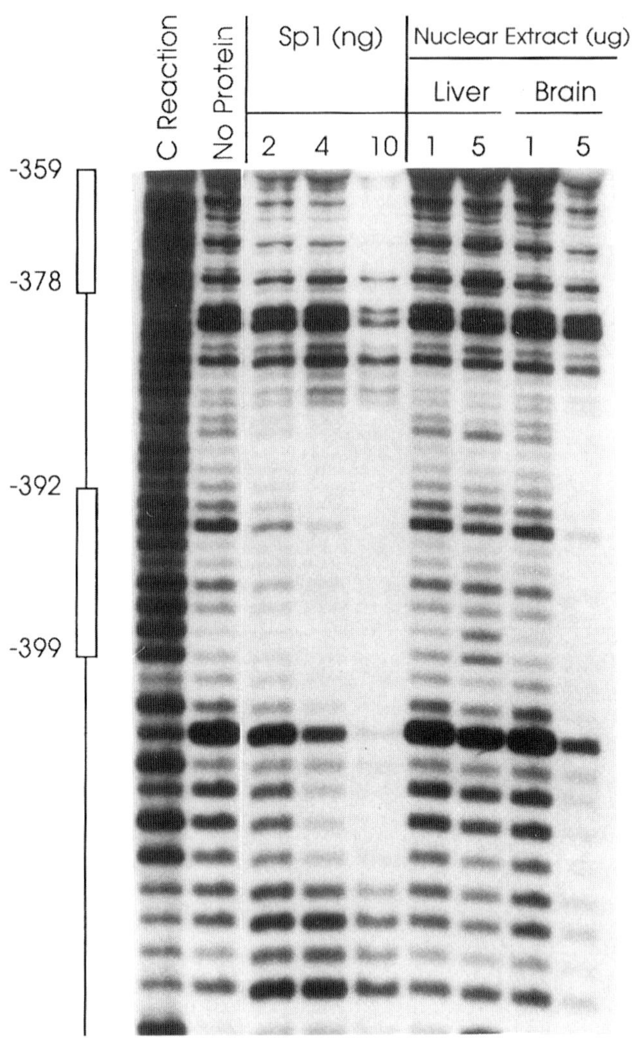

Fig. 6. Sp1 binding to the type 1 IGF receptor promoter. DNase I footprinting analysis was employed to map the binding sites of Sp1 to the 5′-flanking region of the receptor promoter. A ^{32}P-labeled probe extending from –416 to –135 was incubated with increasing amounts of purified Sp1 protein, or with crude nuclear extracts from adult rat liver or neonate brain. The footprints generated by neonate brain are similar to those obtained with purified Sp1, and correlate with higher levels of expression of type 1 IGF receptor mRNA in this organ in comparison to adult liver. *Open boxes* denote consensus Sp1 elements.

Furthermore, the DNA-binding efficiency of Sp1, as revealed by mobility-shift assays, is greatly reduced in tissues of aged animals *(71)*.

The expression of the type 1 IGF receptor mRNA and protein during development is similarly regulated, with highest levels in embryonic and early postnatal stages and lowest levels in postmitotic adult tissues *(22) (See below.)*

When combined with the functional and in vitro results described previously on Sp1 action at the IGF receptor gene promoter, it is evident that Sp1 is an important regulator of type 1 IGF receptor gene expression.

REGULATION OF THE TYPE 1 IGF RECEPTOR GENE BY ONCOGENES AND TUMOR SUPPRESSORS

In addition to its basic requirement for cell cycle progression, the type 1 IGF receptor gene is implicated in malignant transformation, as shown by its high level of expression in most tumors and cancer cell lines *(19,29,72)*.

The importance of the receptor in the tumoral process was empirically demonstrated by experiments that showed that artificial overexpression of functional receptors in NIH3T3 fibroblasts resulted in a ligand-dependent, highly transformed phenotype that included the formation of tumors in nude mice *(73)*.

On the other hand, Rat-1 fibroblasts expressing a truncated IGF receptor β-subunit (952 STOP) did not undergo transformation *(74)*.

Furthermore, it was demonstrated that a fibroblast cell line established from mouse embryos homozygous for a targeted disruption of the type 1 IGF receptor gene, R- cells, cannot be transformed by any of a number of oncogenes, including the SV40 large tumor antigen, bovine papillomavirus, and Ha-*ras* *(75–77)*.

Only when these cells were first transfected to express normal human IGF receptors were the oncogenes capable of inducing the transformed phenotype. These findings suggested that the receptor gene can be the target of multiple oncogenic agents, and constituted the conceptual frame for studies in which the direct effects of oncogenes and tumor suppressors on the expression of the type 1 IGF receptor gene were evaluated.

Oncogenes

The protooncogene c-*myb*, the cellular equivalent of the viral transforming oncogene v-*myb*, is usually expressed in cells of the hematopoietic lineage. Constitutive overexpression of this protooncogene in Balb/c3T3 cells has been shown to abrogate the requirement for IGF-I in the culture media *(78)*.

This effect of c-*myb* was associated with an increase in the levels of both type 1 IGF receptor and IGF-I mRNAs *(79)*.

Run-on assays indicated that this increase in mRNA was due, at least in part, to augmented transcriptional rates. It seems, therefore, that a potential mechanism by which c-*myb* affects cellular proliferation is by increasing the expression of the type 1 IGF receptor and of its specific ligand. Likewise, the hepatitis B virus X (HBx) gene product has been shown to stimulate IGF receptor promoter activity, mRNA levels, and binding by two- to threefold in HepG2 cells transfected with an HBx expression vector. In hepatocellular carcinoma-derived cell lines containing HBx protein, the endogenous levels of type 1 IGF receptor mRNA were fivefold higher than in cells lacking HBx transcripts. These results suggest that HBx may play a role in the process of hepatocellular carcinoma by stimulating the expression of the type 1 IGF receptor gene *(80)*

In addition to modulating transcriptional activity, oncogenes can affect the phosphorylation status of the type 1 IGF receptor. Transformation of human cells by $pp60^{src}$, the protein encoded by the *src* oncogene of Rous sarcoma virus, resulted in the constitutive tyrosine phosphorylation of the type 1 receptor β-subunit *(81)*.

Between 10% and 50% of the receptors are phosphorylated in the unstimulated *src*-transformed cells, and IGF-I stimulation synergistically increased phosphorylation levels *(82)*.

These findings raise the possibility that $pp60^{src}$ alters growth regulation by rendering the cells constitutively subject to a mitogenic signal, i.e., the elevated level of phosphorylated type 1 IGF receptor.

Tumor Suppressors

Tumor suppressors constitute an heterogeneous, rapidly growing family of negative growth regulators *(83,84)*.

Because tumor suppressors are mutated in a large proportion of human cancers in which the type 1 IGF receptor is overexpressed, it has been postulated that the expression of the receptor gene in most adult, terminally differentiated cells may be constitutively inhibited by wild-type tumor suppressors *(85,86)*.

Loss of tumor suppressor activity, in broad terms, may result from constitutional deletions or missense or nonsense mutations. All, or some, of these mechanisms may lead to transcriptional *de*repression of the type 1 IGF receptor gene and of one of the IGF ligand genes (mostly IGF-II). Activation of the receptor by locally produced IGFs may elicit a mitogenic event which may be a key mechanism for the progression of the tumors.

The WT1 Wilms' Tumor Gene Product

The promoter region of the type 1 IGF receptor gene contains, in addition to Sp1 binding sites, several putative binding sites for members of the EGR family of transcription factors (Fig. 4). This family of nuclear proteins, whose transcription is rapidly induced by a variety of cell surface stimuli, includes EGR1 (also known as NGFI-A, Zif268, and Krox24), EGR2, and EGR3 *(55,87,88)*.

EGR proteins bind and activate promoters containing the consensus sequence 5′-GCGGGGGCG-3′ by means of a zinc-finger domain at the C-terminus *(89)*.

In addition, the EGR family includes WT1, a tumor suppressor gene product whose inactivation was postulated to be a key event in the etiology of Wilms' tumor, a pediatric kidney cancer *(90–93)*.

The WT1 gene was cloned by several groups and shown to encode a 52–54-kDa protein with four zinc-fingers of the Kruppel C2-H2 class at its C-terminus *(94–96)*.

The N-terminus of WT1 is rich in proline, glutamine, serine, and glycine residues, similar to the transactivation domain of many transcription factors *(97,98)*.

Cotransfection of CHO cells with type 1 IGF receptor promoter/ luciferase reporter constructs together with a WT1 expression vector resulted in a dose-dependent repression of promoter activity, suggesting that this promoter is a target for the inhibitory action of WT1 *(99)*.

Using electrophoretic mobility-shift and DNase I footprinting assays, multiple WT1 binding sites were mapped both in the 5′-flanking and 5′-UTR regions *(100)*.

Furthermore, the initiator element itself constituted a weak binding site for WT1 (Fig. 4). In addition to its inhibitory effect on transfected promoter constructs, WT1 has been shown to control the expression of the endogenous type 1 IGF receptor gene. Thus, the levels of IGF receptor mRNA and the activity of the receptor promoter were found to be significantly higher in G401 cells (a cell line derived from a rhabdoid tumor of the kidney, lacking detectable levels of WT1 mRNA) than in 293 cells (a human embryonic kidney cell line which expresses significant levels of WT1 mRNA) (Fig. 7) *(86)*.

Reintroduction of a functional WT1 gene by stable transfection of G401 cells resulted in a significant decrease in the rate of cellular proliferation, which was associated with a reduction in the levels of type 1 IGF receptor mRNA, promoter activity, and ligand binding, and with a reduction in IGF-I-stimulated cellular proliferation, thymidine incorporation, and anchorage-independent growth. These data suggest that a major aspect of WT1 action is the repression of IGF receptor gene expression. Because many cases of

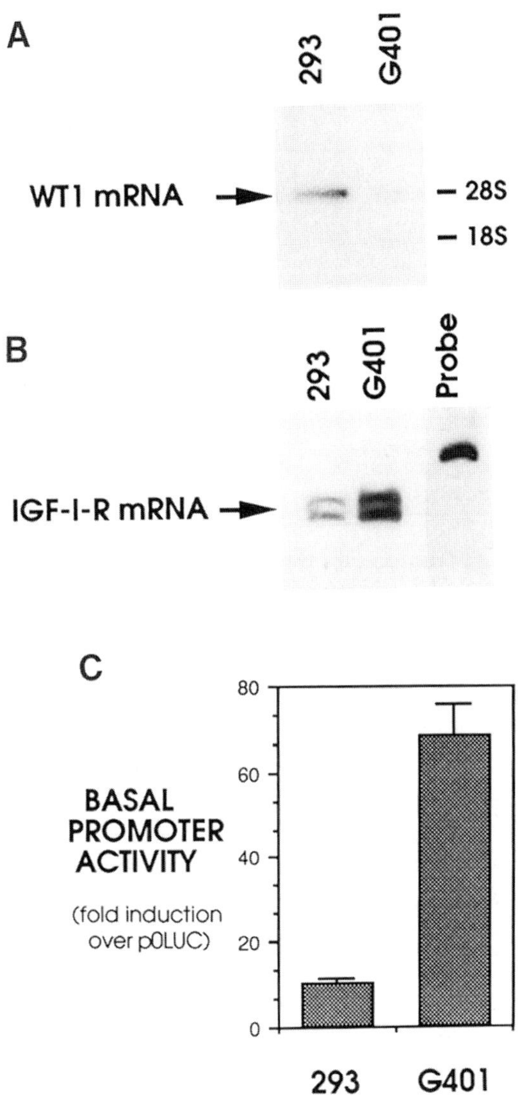

Fig. 7. Regulation of type 1 IGF receptor gene expression by WT1. The regulation of the endogenous IGF receptor gene by WT1 was studied in two kidney-derived human cell lines: 293, an embryonic line that expresses significant levels of WT1 mRNA (**A**), and G401, a rhabdoid-derived cell line lacking detectable levels of WT1 mRNA. Solution hybridization/RNase protection assay using a ^{32}P-labeled IGF receptor antisense RNA probe showed that the levels of type 1 IGF receptor mRNA were ~2.5-fold higher in G401 than in 293 cells (**B**). In addition, basal promoter activity was several-fold higher in G401 than in 293 cells (**C**). It seems, therefore, that the endogenous type 1 gene is constitutively inhibited in 293 cells by the high levels of WT1, with ensuing decreases in the levels of receptor mRNA.

Wilms' tumor and related nephropathies, such as Denys-Drash syndrome, are associated with mutations in the WT1 locus *(91)*, it is possible that lack of repression of the receptor gene (and of IGF-II, which also contains WT1 sites) by mutant WT1 may be an important step in the etiology of Wilms' tumor.

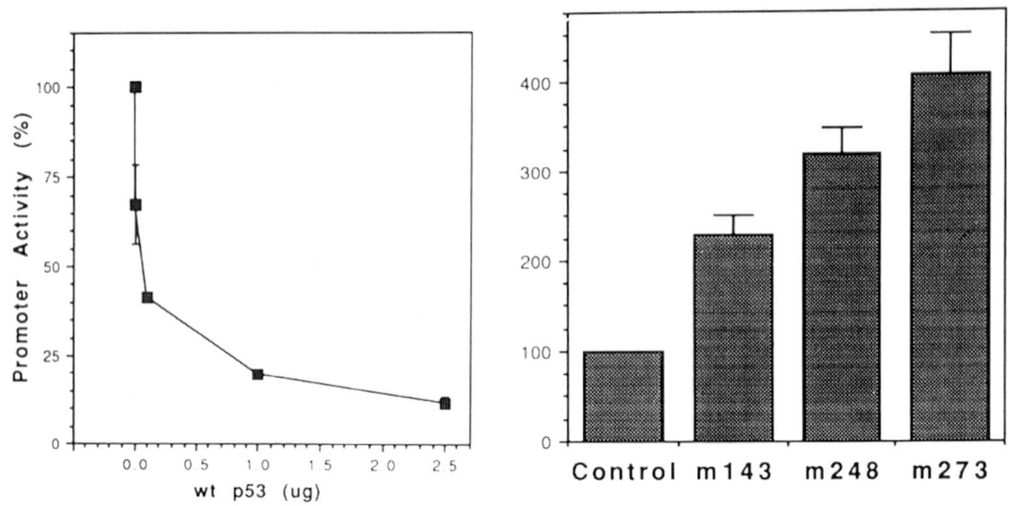

Fig. 8. Regulation of type 1 IGF receptor promoter by wild-type and mutant p53. The reporter plasmid p(–2350/+640)LUC (10 μg) was cotransfected into the osteosarcoma cell line Saos-2 with increasing amounts of a wild-type p53 expression vector (**left**), or with 2.5 μg of the expression vectors pC53–SCX3, pC53–248W, and pC53–273H, which encode mutant versions of p53 harboring mutations at codons 143, 248, and 273, respectively (**right**). The values of luciferase activity are expressed as a percentage of the levels seen in the absence of p53.

The p53 Tumor Suppressor

Wild-type p53 is a tumor suppressor that, in its hyperphosphorylated state, blocks progression of cells through the cell cycle *(101–103)*.

p53 is localized to the nucleus, where it functions as a DNA sequence-specific transcription factor. Mutations in the *p53* gene are the most frequent mutation in human cancer, with most mutations occuring in the central domain of p53, which is the region involved in DNA binding *(104)*.

Transient coexpression studies of type 1 IGF receptor promoter fragments fused to a luciferase reporter gene, together with increasing amounts of a wild-type p53 expression vector, resulted in a dose-dependent suppression of promoter activity (Fig. 8, left) *(85)*.

On the other hand, transfection of tumor-derived, mutant versions of p53 (encoding point mutations at codons 143, 248, and 273) stimulated promoter activity to 227%, 319%, and 406% of control values, respectively (Fig. 8, right).

In vitro transcription assays using an IGF receptor promoter template and extracts of HeLa cells showed that addition of exogenous purified p53 protein (produced as a glutathione-*S*-transferase [GST], fusion protein) abolished transcription of the IGF receptor gene in a dose-dependent fashion (Fig. 9). Although the mechanism for transcriptional suppression by p53 is not entirely clear, results of mobility shift assays indicated that TBP, the TATA box binding subunit of the general initiation factor TFIID, binds specifically to the "initiator" element of the type 1 IGF receptor promoter *(85)*.

p53 precludes binding of TBP to the promoter region, most probably through protein–protein interaction and, as a result, TBP is no longer able to assemble a functional transcription initiation complex. p53 binds, in addition, directly to promoter DNA, although this is most probably a nonspecific type of interaction.

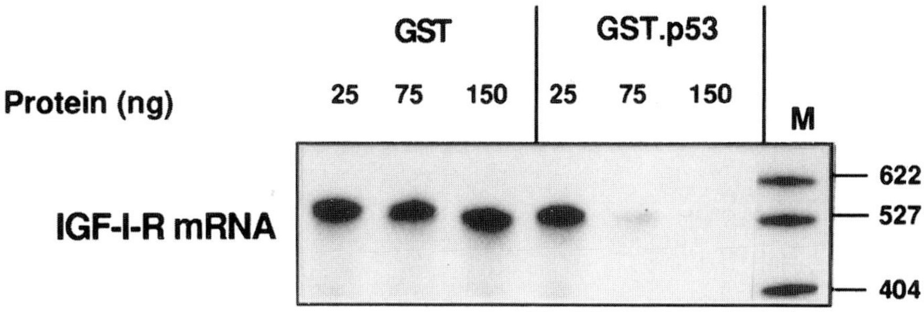

Fig. 9. Transcriptional control of type 1 IGF receptor gene expression by p53. In vitro transcription reactions were performed using HeLa cell extracts and a purified IGF receptor promoter template extending from –476 to +640. Addition of exogenous p53 (produced as a GST fusion protein, GST.p53), but not of the control GST protein, abolished transcription in a dose-dependent fashion.

The levels of wild-type p53 in normal cells are usually increased following DNA damage (for instance, as a result of ultraviolet radiation) *(105)*.

A number of target genes for p53 have been described, including the *p21/WAF* gene that inhibits the formation of cyclin/cdk complexes and thus induces cell cycle arrest *(101,106,107)*.

p53 is, in addition, a potent inducer of apoptosis *(108,109)*.

It is, therefore, possible that the effect of p53 on apoptosis and cell-cycle progression is mediated, at least partially, through suppression of the type 1 IGF receptor promoter. Lack of inhibition of the receptor gene by mutant p53 in malignant states may accelerate tumor growth and reduce apoptosis, thus conferring enhanced survival to a tumorigenic population (Fig. 10).

Disrupted Transcription Factors

Disrupted transcription factors constitute a particular category of aberrant proteins that, in many cases, are the product of chromosomal translocations usually resulting in the fusion of genes. Some human malignancies are characterized by this type of recurrent translocations *(110)*.

One of the best studied examples is the fusion of the *bcr* and c-*abl* genes on the Philadelphia chromosome in chronic myelogenous leukemia, with generation of an oncogenic fusion protein *(111)*.

A relatively new clinical entity shown to be associated with a recurrent chromosomal translocation is desmoplastic small round cell tumor (DSRCT) *(112–114)*.

DSRCT is characterized by the translocation t(11;22) (p13;q12). A genomic DNA fragment containing an Ewings sarcoma *(EWS)* and *WT1* fusion gene has been isolated from these tumors *(115–117)*.

The *EWS* gene encodes an ubiquitous 656-amino-acid of unknown function that comprises an N-terminal domain homologous to eukaryotic RNA polymerase II and a C-terminal domain homologous to RNA-binding domains. Analyses of chimeric transcripts showed fusion of RNAs encoding the N-terminal domain of EWS to the last three zinc-fingers of WT1 (i.e., the DNA-binding domain of WT1), suggesting that the fusion protein can bind and activate target promoters of WT1. This hypothesis was tested by means of transient transfection assays performed in Saos-2 and G401 cells *(118)*.

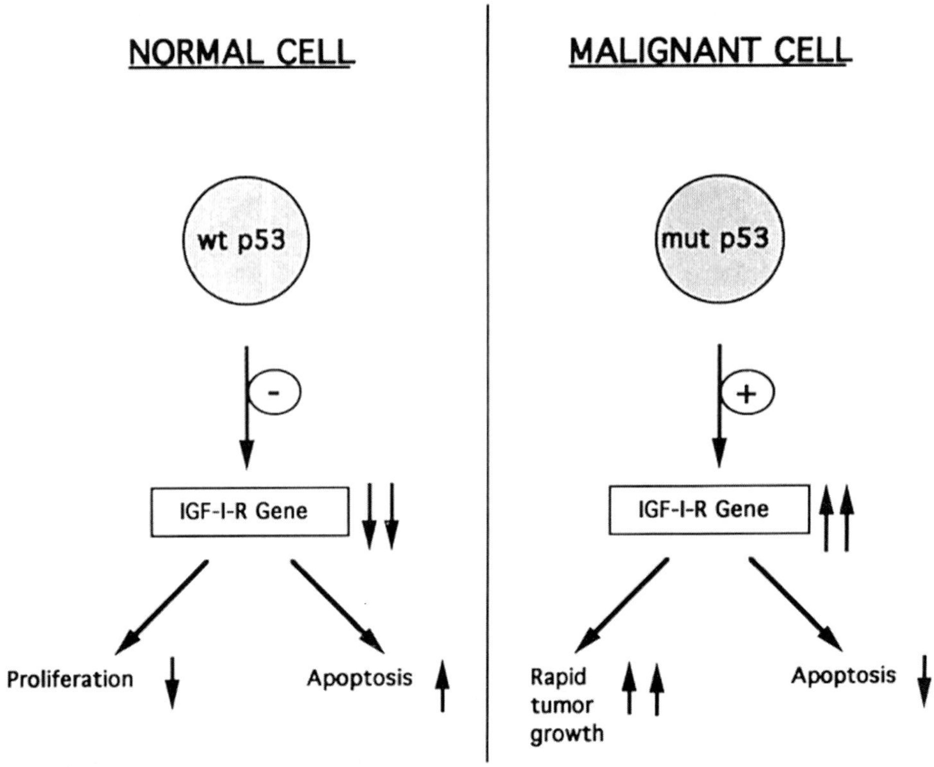

Fig. 10. Proposed model for p53 regulation of type 1 IGF receptor gene expression. The levels of p53 in normal cells (**left**) are usually increased following DNA damage, such as UV radiation. Wild-type p53 suppresses type 1 IGF receptor gene expression, with ensuing decrease in IGF-stimulated cellular proliferation and increase in apoptosis. Malignant cells (**right**) contain mutant p53 which may enhance the expression of the type 1 receptor gene. As a result, IGF-mediated mitogenic stimuli are enhanced, with an increase in tumor growth rate and a decrease in apoptosis.

These experiments showed that expression vectors encoding the EWS-WT1 chimeric protein activated the type 1 IGF receptor promoter ~160–340%, whereas a WT1 expression vector significantly suppressed promoter activity. EWS-WT1 expression vectors that contain a three-amino-acid insert (Lys–Thr–Ser) between zinc-fingers 3 and 4, which impairs DNA binding, did not affect promoter activity. Finally, mobility- shifts and DNase I footprinting assays showed that the chimeric EWS-WT1 protein binds WT1 sites in the promoter region of the type 1 IGF receptor gene with an affinity similar to that of native WT1. Thus, pathological fusion of *EWS* to *WT1* abrogates the tumor suppressor effect of WT1 and generates an oncogenic chimera which is able to bind and activate the IGF receptor promoter (and probably other WT1 target genes).

HORMONAL REGULATION OF TYPE 1 IGF RECEPTOR GENE EXPRESSION

The number of cell-surface IGF-I binding sites in any given tissue, as well as the expression of the type 1 IGF receptor gene, are tightly controlled by the local and circulating levels of a number of hormones and growth factors. Likewise, the level of IGF receptors in cultured cells is clearly affected by the hormonal milieu. IGF-I itself can modulate receptor levels, with increased IGF-I concentrations generally causing a decrease in receptor number and vice versa. In vivo, decreased levels of circulating IGF-I in growth disorders such as Laron-type dwarfism caused a several-fold increase in the levels of type 1 IGF receptor mRNA and binding in mononuclear cells *(119)*.

Caloric deprivation in rats was similarly shown to increase the specific binding of [^{125}]IGF-I in lung, testes, stomach, and kidney. Increased binding was accompanied by a ~1.6- to 2.5-fold increase in type 1 IGF receptor mRNA abundance *(15)*.

Because the local and circulating levels of IGF-I are reduced in these animals, it is possible that the increases in type 1 IGF receptor gene expression and binding are secondary to the decrease in IGF-I. Similar effects have been reported in a number of cultured cells, including IM-9 lymphoid cells, endothelial cells, FRTL-5 thyroid cells, and others *(120–122)*.

Regulation of the type 1 IGF receptor by IGF-I involves translocation of cell-surface receptors to an intracellular pool; removal of IGF-I from the medium reverses this effect *(123)*.

In addition, IGF-I down-regulated IGF receptor mRNA levels in the SH-SY5Y neuroblastoma and C2C12 muscle cell lines *(124)*.

This effect of IGF-I is due to a decrease in transcription activity rate with no change in mRNA stability. Thus, IGF-I decreased pre-mRNA and mature mRNA levels to a similar extent. Although the type 1 IGF receptor promoter contains a sequence between nucleotides –583 and –555 which is similar to a putative IGF-I response element described in the elastin gene that mediates activation of the elastin promoter by IGF-I, this specific sequence did not mediate the response to IGF-I in the IGF receptor promoter *(125)*.

The region responsible for this effect in the receptor promoter has yet to be identified.

Other growth factors, in general, stimulate the expression of the type 1 IGF receptor gene. Platelet-derived growth factor (PDGF), for example, has been shown to increase the expression of the receptor gene and the number of IGF-I binding sites in cells in culture *(126)*.

This effect is mediated at the level of transcription, with PDGF increasing the activity of the type 1 IGF receptor promoter *(127)*.

Specifically, a ~100-bp promoter fragment located immediately upstream of the initiator element contains all the information required for the response to PDGF. Because this DNA region contains a canonical c-*myc* binding site, and because PDGF induces c-*myc*, it has been suggested that the effect of PDGF on type 1 IGF receptor gene expression may be mediated by c-*myc (128)*.

Basic fibroblast growth factor (bFGF) has been also shown to increase IGF-I binding and type 1 IGF receptor mRNA and protein levels in the C2C12 and BC3H-1 muscle cell lines *(124,129)*.

bFGF stimulated the activity of the IGF receptor promoter by 90%. Using different fragments of the type 1 IGF receptor promoter in transient transfection assays, the major bFGF-responsive element has been mapped to a region located between nucleotides –476 and –188 of the 5′-flanking region. Because bFGF and PDGF are competence factors that stimulate entry of cells into the G_1 phase of the cell cycle it has been postulated that a potential mechanism by which these growth factors achieve this goal is by way of their ability to increase the expression of the type 1 IGF receptor gene *(2,30)*.

High levels of cell-surface IGF receptors can enhance the progression factor activity of IGF-I, thus allowing the cells to proceed through the remaining phases of the cycle. Transcriptional activation of the type 1 receptor by bFGF in rat aortic smooth muscle cells has been recently shown to be mediated by a protein kinase C-dependent pathway *(130)*.

The expression of the type 1 IGF receptor in the ovaries is tightly controlled by circulating gonadotropin levels *(59,131)*.

Hypophysectomy of immature rats led to significant decrements in the abundance of transcripts corresponding to the type 1 receptor in whole ovaries as well as in isolated granulosa cells *(132)*.

Follicle-stimulating hormone treatment produced a fourfold increase in the levels of receptor mRNA in both ovaries and granulosa cells. Similar increases in the abundance of receptor mRNA were seen following diethylstilbestrol infusion, indicating that the expression of this gene is positively regulated by estrogens.

In a number of breast carcinoma cell lines, as well as in Ishikawa endometrial cancer cells, estradiol caused an increase in the levels of type 1 IGF receptor mRNA, suggesting that estradiol sensitizes these cells to the mitogenic effects of the IGFs by augmenting receptor concentrations *(133–135)*.

Progestins, on the other hand, reduced IGF-I binding and receptor mRNA in estrogen-responsive breast carcinoma cell lines *(136,137)*.

This effect seems to be mediated by IGF-II, whose secretion is stimulated by progestins and that can, in turn, down-regulate the expression of the receptor gene.

DEVELOPMENTAL REGULATION OF TYPE 1 IGF RECEPTOR GENE EXPRESSION

The IGF system plays an important role in embryonic development *(138,139)*.

Targeted disruption of the type 1 IGF receptor gene resulted in growth-retarded mouse embryos *(140,141)*.

These mice, weighing 45% of controls, exhibit generalized hypoplasia, delayed bone development, impaired skin formation, and abnormal central nervous system morphology. They invariably die at birth. On the other hand, disruption of the IGF-II gene results in viable mice weighing 60% of controls at birth. These mice develop into fertile proportionate dwarfs *(142)*.

The phenotype displayed by null mutants for the IGF-I gene seems to depend on the genetic background of the animal. Some of them die shortly after birth, while others survive and reach adulthood, although they are infertile and show delays in the ossification process, underdeveloped muscle tissue, and poorly organized lungs.

In preimplantation mouse embryos type 1 IGF receptors (and insulin receptors) are first detected at the eight-cell stage. Because insulin and IGF-I are not expressed at any time during preimplantation stages, it seems that IGF-II, which is already detectable at the two-cell stage, is the predominant ligand at these early stages. The type 1 IGF receptor gene is widely expressed after implantation *(16,17)*, consistent with the fact that most cells require IGF-I (or large concentrations of insulin) to grow in culture. IGF receptor mRNA levels are especially high in the developing nervous system and muscle. Given the widespread distribution of the type 1 IGF receptor, it appears that IGF-II, which is expressed in developing muscle, cartilage, vascular tissue, liver, and pituitary, may play an endocrine role in embryogenesis, circulating both in bloodstream and in cerebrospinal fluid.

The high levels of type 1 IGF receptor mRNA and binding at perinatal stages suggest that this receptor may be involved in the processes of differentiation and growth at this stage *(22)*.

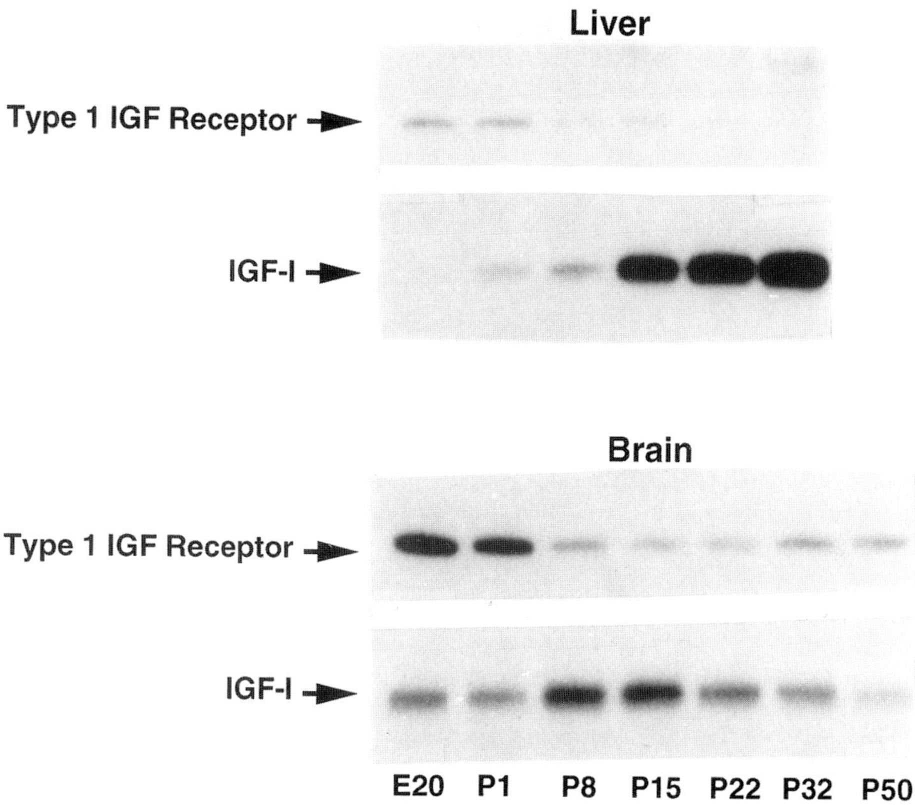

Fig. 11. Developmental regulation of IGF-I and type 1 IGF receptor mRNAs in rat liver and brain. Total RNA was prepared from rat tissues at various developmental stages, and the levels of both transcripts were measured by solution hybridization/RNase protection assay. The levels of receptor mRNA are maximal at embryonic (E) and early postnatal (P) stages, decreasing to very low levels in the adult animal. The developmental trend of IGF-I ligand mRNA is usually opposite to that of the receptor.

During postnatal development the levels of receptor mRNA decrease dramatically in most tissues, a finding that is consistent with the attenuation in the proliferative capacity of most tissues *(143)*.

In some of these tissues, the decrease in the expression of the IGF receptor gene is correlated with an increase in the levels of IGF-I mRNA and peptide (Fig. 11). This finding suggests that the increment in local and circulating levels of IGF-I may downregulate the expression of the receptor gene.

CONCLUSIONS

The wealth of information that has accumulated over the last several years demonstrates that the type 1 IGF receptor fulfills a central role in normal development and differentiation processes, as well as in the establishment and maintenance of the malignant phenotype. This chapter focused on the molecular aspects of the IGF receptor gene, mRNA, and protein. Those aspects were described from structural and functional standpoints, with the clear understanding that clarification of these basic aspects will help us to better understand the functioning of the receptor at the cellular and organismal levels. Special emphasis was placed on the characterization of the regulatory region of the receptor gene. Further studies of the molecular mechanisms responsible for type 1 IGF receptor function will undoubtely contribute to unravel new basic and clinical aspects of this receptor.

ACKNOWLEDGMENTS

H. W. is the recipient of a Guastalla Fellowship, The Rashi Foundation, Israel.

REFERENCES

1. Nissley SP, Lopaczynski W. Insulin-like growth factor receptors. Growth Factors 1991; 5:29–49.
2. Rubin R, Baserga R. Insulin-like growth factor-I receptor. Lab Invest 1995; 73:311–331.
3. Werner H, Woloschak M, Stannard B, Shen-Orr Z, Roberts CT, Jr, LeRoith D. Insulin-like growth factor receptor: Molecular Biology, Heterogeneity, and Regulation. In: LeRoith D, ed. Insulin-Like Growth Factors: Molecular and Cellular Aspects. CRC Press, Boca Raton, FL, 1991, pp. 17–47.
4. Cohick WS, Clemmons DR. The insulin-like growth factors. Ann Rev Physiol 1993; 55:131–153.
5. LeRoith D, Werner H, Beitner-Johnson D, Roberts CT Jr. Molecular and cellular aspects of the insulin-like growth factor I receptor. Endocr Rev 1995: 16:143–163.
6. Werner H, Adamo M, Roberts CT Jr, LeRoith D. Molecular and cellular aspects of Insulin-like growth factor action. In: Litwack G, ed. Vitamins and Hormones, Vol. 48. Academic Press, San Diego, CA, 1994, pp. 1–58.
7. Abbott AM, Bueno R, Pedrini MT, Murray JM, Smith RJ. Insulin-like growth factor I receptor gene structure. J Biol Chem 1992; 267:10759–10763.
8. Ullrich A, Gray A, Tam AW, Yang-Feng T, Tsubowka M, Collins C, Henzel W, Lebon T, Kathuria S, Chen F, Jacobs S, Francke U, Ramachandran J, Fujita-Yamaguchi Y. Insulin-like growth factor I receptor primary structure: comparison with insulin receptor suggests determinants that define functional specificity. EMBO J 1986; 5:2503–2512.
9. Sundaresan S, Francke U. Insulin-like growth factor I receptor gene is concordant with c-fes protooncogene and mouse chromosome 7 in somatic cell hybrids. Som Cell Mol Gen 15:373–378.
10. Ebina Y, Ellis L, Jarnagin K, Edery M, Graf L, Clauser E, Qu J, Masiarz F, Kan YW, Goldfine ID, Roth RA, Rutter WJ. The human insulin receptor cDNA: the structural basis for hormone-activated transmembrane signaling. Cell 1985; 40:747–758.
11. Ullrich A, Bell JR, Chen EY, Herrera R, Petruzzeli LM, Dull TJ, Gray A, Coussens L, Liao Y-C, Tsubokawa M, Mason A, Seeburg PH, Grunfeld C, Rosen OM, Ramachandran J. Human insulin receptor and its relationship to the tyrosine kinase family of oncogenes. Nature 1985; 313:756–761.
12. Seino S, Seino M, Nishi S, Bell GI. Structure of the human insulin receptor gene and characterization of its promoter. Proc Natl Acad Sci USA 1989; 86:114–118.
13. Condorelli G, Bueno R, Smith R. Two alternatively spliced forms of the human insulin-like growth factor I receptor have distinct biological activities and internalization kinetics. J Biol Chem 1994; 269:8510–8516.
14. Pedrini MT, Giorgino F, Smith RJ. cDNA cloning of the rat IGF-I receptor: structural analysis of rat and human IGF-I and insulin receptors reveals differences in alternative splicing and receptor-specific domain conservation. Biochem Biophys Res Commun 1994; 202:1038–1046.
15. Lowe WL Jr, Adamo M, Werner H, Roberts CT Jr, LeRoith D. Regulation by fasting of rat insulin-like growth factor I and its receptor. Effects on gene expression and binding. J Clin Invest 1989; 84:619–626.
16. Bondy CA, Werner H, Roberts CT Jr, LeRoith D. Cellular pattern of insulin-like growth factor I (IGF-I) and type I IGF receptor gene expression in early organogenesis: comparison with IGF-II gene expression. Mol Endocrinol 1990; 4:1386–1398.
17. Bondy CA, Werner H, Roberts CT Jr, LeRoith D. Cellular pattern of Type I insulin-like growth factor receptor gene expression during maturation of the rat brain: comparison with insulin-like growth factors I and II. Neuroscience 1992; 46:909–923.
18. Baserga R. The double life of the IGI-I receptor. Receptor 1992; 2:261–266.
19. Baserga R, Rubin R. Cell cycle and growth control. Crit Rev Euk Gene Exp 1993; 3:47–61.

20. Baserga R, Sell C, Porcu P, Rubini M. The role of the IGF-I receptor in the growth and transformation of mammalian cells. Cell Proliferat 1994; 27:63–71.
21. Lowe WL Jr. Biological actions of the insulin-like growth factors. In LeRoith D, ed. Insulin-Like Growth Factors: Molecular and Cellular Aspects. CRC Press, Boca Raton, FL, 1991, pp. 49–85.
22. Werner H, Woloschak M, Adamo M, Shen-Orr Z, Roberts CT Jr, LeRoith D. Developmental regulation of the rat insulin-like growth factor I receptor gene. Proc Natl Acad Sci USA 1989; 86:7451–7455.
23. Pomerance M, Gavaret J-M, Jacquemin C, Matricon C, Toru-Delbauffe D, Pierre M. Insulin and insulin-like growth factor I receptors during postnatal development of rat brain. Dev Brain Res 1988; 470:77–83.
24. Bohannon NJ, Corp ES, Wilcox BJ, Figlewicz DP, Dorsa DM, Baskin DG. Localization of binding sites for insulin-like growth factor I (IGF-I) in the rat brain by quantitative autoradiography. Brain Res 1988; 444:205–213.
25. Lesniak MA, Hill JM, Kiess W, Rojeski M, Pert CB, Roth J. Receptors for insulin-like growth factor I and II; autoradiographic localization in rat brain and comparison to receptors for insulin. Endocrinology 1988; 123:2089–2099.
26. Werther GA, Hogg A, Oldfield BJ, McKinley MJ, Figdor R, Mendelsohn D. Localization and characterization of IGF-I receptors in rat brain and pituitary gland using in vitro autoradiography and computerized densitometry: distinct distribution from insulin receptors. J Endocrinol 1989; 1:369–377.
27. Matsuo K, Niwa M, Kurihara M, Shigematsu K, Yamashita S, Ozaki M, Nagataki S. Receptor autoradiographic analysis of IGF-I binding in the rat forebrain and pituitary gland. Cell Mol Neurobiol 1989; 9:357–367.
28. Marks JL, Porte D, Baskin DG. Localization of type I IGF receptor mRNA in the adult rat brain by in situ hybridization. Mol Endocrinol 5:1158–1168.
29. Werner H, LeRoith D. The role of the insulin-like growth factor system in human cancer. Adv Cancer Res 1996; 68:183–223.
30. Baserga R. The insulin-like growth factor I receptor. Cancer Res 1995; 55:249–252.
31. Jacobs S, Kull FC Jr, Cuatrecasas P. Monensin blocks the maturation of receptors for insulin and somatomedin C: identification of receptor precursors. Proc Natl Acad Sci USA 1983; 80:1228–1232.
32. Kjeldsen T, Andersen AS, Wiberg FC, Rasmussen JS, Schaffer L, Balschmidt P, Moller KB, Moller NPH. The ligand specificities of the insulin receptor and the insulin-like growth factor I receptor reside in different regions of a common binding site. Proc Natl Acad Sci USA 1991; 88:4404–4408.
33. Backer JM, Kahn CR, Cahill DA, Ullrich A, White MF. Receptor-mediated internalization of insulin requires a 12-amino acid sequence in the juxtamembrane region of the insulin receptor beta-subunit. J Biol Chem 1990; 265:16450–16454.
34. Hsu D, Knudson PE, Zapf A, Rolband GC, Olefsky JM. NPXY motif in the insulin-like growth factor-I receptor is required for efficient ligand-mediated receptor internalization and biological signaling. Endocrinology 1994; 134:744–750.
35. Prager D, Li H-L, Yamasaki H, Melmed S. Human insulin-like growth factor I receptor internalization. J Biol Chem 1994; 269:11934–11937.
36. Hanks SK, Quinn AM, Hunter T. The protein kinase family: conserved features and deduced phylogeny of the catalytic domains. Science 1988; 241:42–52.
37. Knighton DR, Zheng J, Ten Eyck LF, Ashford VA, Xuong N-H., Taylor SS, Sowadski JM. Crystal structure of the catalytic subunit of cyclic adenosine monophosphate-dependent protein kinase. Science 1991; 253; 407–414.
38. Knighton DR, Zheng J, Ten Eyck LF, Xuong N-H, Taylor SS, Sowadski JM. Structure of a peptide inhibition bound to the catalytic subunit of cyclic adenosine monophosphate-dependent protein kinase. Science 1991; 253:414–420.
39. Seino S, Seino M, Bell GI. Human insulin receptor gene. Diabetes 1990; 39:129–133.
40. Hintz RL, Clemmons DR, Underwood LE, Van Wyk JJ. Competitive binding of somatomedin to the insulin receptors of adipocytes, chondrocytes, and liver membranes. Proc Natl Acad Sci USA 1972; 69:2351–2355.
41. Marshall RN, Underwood LE, Voina SJ, Foushee DB, Van Wyk JJ. Characterization of the insulin and somatomedin-C receptors in human placental cell membranes. J Clin Endocrinol Metab 1974; 39:283–289.
42. Rechler M, Zapf J, Nissley S, Froesch E, Moses A, Podskalny J, Schilling E, Humbel R. Interactions of insulin-like growth factors I and II and multiplication-stimulating activity with receptors and serum carrier proteins. Endocrinology 1980; 107:1451–1458.

43. Bayne ML, Applebaum J, Underwood D, Chicchi GG, Green BG, Hayes NS, Cascieri MA. The C region of the human insulin-like growth factor (IGF) I is required for high affinity binding to the type 1 IGF receptor. J Biol Chem 1989; 264:11004–11011.
44. Tollefsen SE, Thompson K. The structural basis for insulin-like growth factor I receptor high affinity binding. J Biol Chem 1988; 263:16267–16273.
45. Werner H, Stannard B, Bach MA, LeRoith D, Roberts CT Jr. Cloning and characterization of the proximal promoter region of the rat insulin-like growth factor I (IGF-I) receptor gene. Biochem Biophys Res Commun 1990; 169:1021–1027.
46. Werner H, Bach MA, Stannard B, Roberts CT, Jr, LeRoith D. Structural and functional analysis of the insulin-like growth factor I receptor gene promoter. Mol Endocrinol 1992; 6:1545–1558.
47. Mamula PW, Goldfine ID. Cloning and characterization of the human insulin-like growth factor-I receptor gene 5′-flanking region. DNA Cell Biol 1992; 11:43–50.
48. Cooke DW, Bankert LA, Roberts CT Jr, LeRoith D, Casella SJ. Analysis of the human type I insulin-like growth factor receptor promoter region. Biochem Biophys Res Commun 1991; 177:1113–1120.
49. McKeon C, Moncada V, Pham T, Salvatore P, Kadowaki T, Accili D, Taylor SI. Structural and functional analysis of the insulin receptor promoter. Mol Endocrinol 1990; 4:647–656.
50. Mamula PW, Wong W-Y, Maddux BA, McDonald AR, Goldfine ID. Sequence and analysis of promoter region of the human insulin receptor gene. Diabetes 1988; 37:1241–1246.
51. Araki E, Shimada F, Uzawa H, Mori M, Ebina Y. Characterization of the promoter region of the human insulin receptor gene. J Biol Chem 1987; 262:16186–16191.
52. Sibley E, Kastelic T, Kelly TJ, Lane MD. Characterization of the mouse insulin receptor gene promoter. Proc Natl Acad Sci USA 1989; 86:9732–9736.
53. Smale ST, Baltimore D. The "initiator" as a transcription control element. Cell 1989; 57:103–113.
54. Courey AJ, Tjian R. Mechanisms of transcriptional control as revealed by studies of human transcription factor Sp1. In: McKnight S and Yamamoto K, eds. Transcriptional Regulation, Cold Spring Harbor Laboratory Press, Cold Spring Harbor, New York, 1992, pp. 743–769.
55. Sukhatme VP. The EGR transcription factor family: from signal transduction to kidney differentiation. Kidney Int. 1992; 41:550–553.
56. Morrione A, DeAngelis T, Baserga R. The GC factor regulates the expression of the insulin-like growth factor-I receptor. Cell Proliferat 1995; 28:659–671.
57. Roesler WJ, Vandenbark GR, Hanson RW. Cyclic AMP and the induction of eukaryotic gene transcription. J Biol Chem 1988; 263:9063–9066.
58. Ota A, Shen-Orr Z, Robert CT Jr, LeRoith D. TPA-induced neurite formation in a neuroblastoma cell line (SH-SY5Y) is associated with increased IGF-I receptor mRNA and binding. Mol Brain Res 1989; 6:69–76.
59. Adashi EY, Resnick CE, Svoboda ME, Van Wyk JJ. Follicle-stimulating hormone enhances somatomedin-C binding to cultured rat granulosa cells: evidence for cAMP-dependence. J Biol Chem 1986; 261:3923–3926.
60. Kageyama R, Merlino GT, Pastan I. Nuclear factor ETF specifically stimulates transcription from promoters without a TATA box. J Biol Chem 1989; 264:15508–15514.
61. Kozak M. An analysis of vertebrate mRNA sequences: intimations of translational control. J Cell Biol 1991; 115:887–903.
62. Cooke DW, Casella SJ. The 5′-untranslated region of the IGF-I receptor gene modulates reporter gene expression by both pre- and post-transcriptional mechanisms. Mol Cell Endocrinol 1994; 101:77–84.
63. Courey AJ, Tijian R. Analysis of Spl in vivo reveals multiple transcriptional domains, including a novel glutamine-rich activation motif. Cell 1988; 55:887–898.
64. Ishii S, Xu YH, Stratton RH, Row BA, Merlino GT. Characterization and sequence of the promoter region of the human epidermal growth factor receptor gene. Proc Natl Acad Sci USA 1985; 82:4920–4924.
65. Sehgal A, Patil N, Chao M. A constitutive promoter directs expression of the nerve growth factor receptor gene. Mol Cell Biol 1988; 8:3160–3167.
66. Blake MC, Jambou RC, Swick AG, Kahn JW, Azizkhan JC. Transcriptional initiation is controlled by upstream GC-box interactions in a TATA-less promoter. Mol Cell Biol 1990; 10:6632–6641.
67. O'Shea-Greenfield A, Smale ST. Roles of TATA and initiator elements in determining the start site location and direction of RNA polymerase II transcription. J Biol Chem 1992; 267:1391–1402.
68. Lu J, Lee W, Jiang C, Keller EB. Start site selection by Sp1 in the TATA-less human Ha-ras promoter. J Biol Chem 1994; 269:5391–5402.

69. Beitner-Johnson D, Werner H, Roberts CT Jr, LeRoith D. Regulation of insulin-like growth factor I receptor gene expression by Sp1: Physical and functional interactions of Sp1 at GC boxes and at a CT element. Mol Endocrinol 1995; 9:1147–1156.
70. Saffer JD, Jackson SP, Annarella MB. Developmental expression of Sp1 in the mouse. Mol Cell Biol 1991; 11:2189–2199.
71. Ammendola R, Mesuraca M, Russo T, Cimino F. Sp1 DNA binding efficiency is highly reduced in nuclear extracts from aged rat tissues. J Biol Chem 1992; 267:17944–17948.
72. Macaulay VM. Insulin-like growth factors and cancer. Br J Cancer 1992; 65:311–320.
73. Kaleko M, Rutter WJ, Miller AD. Overexpression of the human insulin-like growth factor I receptor promotes ligand-dependent neoplastic transformation. Mol Cell Biol 1990; 10:464–473.
74. Prager D, Li H-L, Asa S, Melmed S. Dominant negative inhibition of tumorigenesis in vivo by human insulin-like growth factor I receptor mutant. Proc Natl Acad Sci USA 1994; 91:2181–2185.
75. Sell C, Dumenil G, Deveaud C, Miura M, Coppola D, DeAngelis T, Rubin R, Efstratiadis A, Baserga R. Effect of a null mutation of the insulin-like growth factor I receptor gene on growth and transformation of mouse embryo fibroblasts. Mol Cell Biol 1994; 14:3604–3612.
76. Sell C, Rubini M, Rubin R, Liu J-P, Efstratiadis A, Baserga R. Simian virus 40 large tumor antigen is unable to transform mouse embryonic fibroblasts lacking type 1 insulin-like growth factor receptor. Proc Natl Acad Sci USA 1993; 90:11217–11221.
77. Morrione A, DeAngelis T, Baserga R. Failure of the bovine papillomavirus to transform mouse embryo fibroblasts with a targeted disruption of the insulin-like growth factor I receptor gene. J Virol 1995; 69:5300–5303.
78. Travali S, Reiss K, Ferber A, Petralia S, Mercer WE, Calabreta B, Baserga R. Constitutively expressed c-*myb* abrogates the requirement for insulinlike growth factor I in 3T3 fibroblasts. Mol Cell Biol 1991; 11:731–736.
79. Reiss K, Ferber A, Travali S, Porcu P, Phillips PD, Baserga R. The protooncogene c-*myb* increases the expression of insulin-like growth factor I and insulin-like growth factor I receptor messenger RNAs by a transcriptional mechanism. Cancer Res 1991; 51:5997–6000.
80. Kim SO, Park JG, Lee YI. Increased expression of the insulin-like growth factor I (IGF-I) receptor gene in hepatocellular carcinoma cell lines: implications of IGF-I receptor gene activation by hepatitis B virus X gene product. Cancer Res 1996; 56:3831–3836.
81. Kozma LM, Weber MJ. Constitutive phosphorylation of the receptor for insulinlike growth factor I in cells transformed by the src oncogene. Mol Cell Biol 1990; 10:3626–3634.
82. Peterson JE, Jelinek T, Kaleko M, Siddle K, Weber MJ. C phosphorylation and activation of the IGF-I receptor in src-transformed cells. J Biol Chem 1994; 269:27315–27321.
83. Marshall CJ. Tumor suppressor genes. Cell 1991; 64:313–326.
84. Knudson AG. Antioncogenes and human cancer. Proc Natl Acad Sci USA 1993; 90:10914–10921.
85. Werner H, Karnieli E, Rauscher FJ III, LeRoith D. Wild type and mutant p53 differentially regulate transcription of the insulin-like growth factor I receptor gene. Proc Natl Acad Sci USA 1996; 93:8318–8323.
86. Werner H, Shen-Orr Z, Rauscher FJ III, Morris JF, Roberts CT Jr, LeRoith D. Inhibition of cellular proliferation by the Wilms' tumor suppressor WT1 is associated with suppression of insulin-like growth factor I receptor gene expression. Mol Cell Biol 1995; 15:3516–3522.
87. Lemaire P, Vesque C, Schmitt J, Stunnenberg H, Frank R, Charnay P. The serum-inducible mouse gene *Krox*-24 encodes a sequence-specific transcriptional activator. Mol Cell Biol 1990; 10:3456–3467.
88. Milbrandt J. A nerve growth factor-induced gene encodes a possible transcriptional regulatory factor. Science 1987; 238:797–799.
89. Gashler AL, Swaminathan S, Sukhatme VP. A novel repression module, an extensive activation domain, and a bipartite nuclear localization signal defined in the immediate-early transcription factor Egr-1. Mol Cell Biol 1993; 13:4556–4571.
90. Haber DA, Buckler AJ. WT1: a novel tumor suppressor gene inactivated in Wilms' tumor. The New Biologist 1992; 4:97–106.
91. Coppes MJ, Campbell CE, Williams BRG. The role of WT1 in Wilms tumorigenesis. FASEB J 1993; 7:886–895.
92. Rauscher FJ III. The WT1 Wilms tumor gene product: a developmentally regulated transcription factor in the kidney that functions as a tumor suppressor. FASEB J 1993; 7:896–903.
93. Hastie N. The genetics of Wilms' tumor—A case of disrupted development. Annu Rev Genet 1994; 28:523–558.

94. Call KM, Glaser T, Ito CY, Buckler AJ, Pelletier J, Haber DA, Rose EA, Kral A, Yeger H, Lewis WH, Jones C, Housman DE. Isolation and characterization of a zinc finger polypeptide gene at the human chromosome 11 Wilms' tumor locus. Cell 1990; 60:509–520.
95. Gessler M, Poustka A, Cavence W, Neve RL, Orkin SH, Bruns GAP. Homozygous deletion in Wilms' tumors of a zinc-finger gene identified by chromosome jumping. Nature 1990; 343: 774–778.
96. Rose EA, Glaser T, Jones C, Smith CL, Lewis WH, Call KM, Minden M, Champagne E, Bonetta L, Yeger H, Housman DE. Complete physical map of the WAGR region of 11 p13 localizes a candidate Wilms' tumor gene. Cell 1990; 60: 495–508.
97. Madden SL, Cook DM, Morris JF, Gashler A, Sukhatme VP, Rauscher FJ III. Transcriptional repression mediated by the WT1 Wilms' tumor gene product. Science 1991; 253:1550–1553.
98. Morris JF, Madden SL, Tournay OE, Cook DM, Sukhatme VP, Rauscher FJ III. Characterization of the zinc finger protein encoded by the WT1 Wilms' tumor locus. Oncogene 1991; 6:2339–2348.
99. Werner H, Re GG, Drummond IA, Sukhatme VP, Rauscher FJ III, Sens DA, Garvin AJ, LeRoith D, Roberts CT Jr. Increased expression of the insulin-like growth factor-I receptor gene, IGFIR, in Wilms' tumor is correlated with modulation of IGFIR promoter activity by the WT1 Wilms' tumor gene product. Proc Natl Acad Sci USA 1993; 90:5828–5832.
100. Werner H, Rauscher FJ III, Sukhatme VP, Drummond IA, Roberts CT Jr, LeRoith D. Transcriptional repression of the insulin-like growth factor I receptor (IGF-I-R) gene by the tumor suppressor WT1 involves binding to sequences both upstream and downstream of the IGF-I-R gene transcription start site. J Biol Chem 1994; 269:12577–12582.
101. El-Deiry WS, Tokino T, Velculescu VE, Levy DB, Parsons R, Trent JM, Lin D, Mercer WE, Kinsler KW, Vogelstein B. WAF1, a potential mediator of p53 tumor suppression. Cell 1993; 75:817–825.
102. Kern SE, Kinzler KW, Bruskin A, Jarosz D, Friedman P, Prives C, Vogelstein B. Identification of p53 as a sequence-specific DNA-binding protein. Science 1991; 252:1708–1711.
103. Oren M. p53: The ultimate tumor suppressor gene? FASEB J 1992; 6:3169–3176.
104. Hollstein M, Sidransky D, Vogelstein B, Harris CC. p53 mutations in human cancers. Science 1991; 253:49–53.
105. Kuerbitz SJ, Plunkett BS, Walsh WV, Kastan MB. Wild-type p53 is a cell cycle checkpoint determinant following irradiation. Proc Natl Acad Sci USA 1992; 89:7491–7495.
106. Kastan MB, Zhan Q, El-Deiry W, Carrier F, Jacks T, Walsh WV, Plunkett BS, Fornace AJ Jr. A mammalian cell cycle checkpoint pathway utilizing p53 and GADD45 is defective in ataxia-telangiectasia. Cell 1992; 71:587–597.
107. Morris GF, Bischoff JR, Mathews MB. Transcriptional activation of the human proliferating-cell nuclear antigen promoter by p53. Proc Natl Acad Sci USA 1996; 93:895–899.
108. Vogelstein B, Kinzler K. p53 function and dysfunction. Cell 1992; 70:523–526.
109. Selivanova G, Wilman KG. p53: a cell cycle regulator activated by DNA damage. Adv Cancer Res 1995; 66:143–180.
110. Rabbitts TH. Chromosomal translocations in human cancer. Nature 1994; 372:143–149.
111. deKlein A, van Kessel AG, Grosveld G, Bartram CR, Hagemeijer A, Bootsma D, Spurr NK, Heiterkamp N, Groffen J. A cellular oncogene is translocated to the Philadelphia chromosome in chronic myelocytic leukemia. Nature 1982; 300:765–767.
112. Biegel JA, Conard K, Brooks JJ. Translocation (11;22)(p13;q12): Primary changes in intra-abdominal desmoplastic small round cell tumor. Genes Chromosomes Cancer 1993; 7:119–121.
113. Shen WP, Towne B, Zadeh TM. Cytogenetic abnormalities in an intra-abdominal desmoplastic small cell tumor. Cancer Genet Cytogenet 1992; 64:189–191.
114. Sawyer JR, Tryka AF, Lewis JM. A novel reciprocal chromosome translocation t(11;22)(p13;q12) in an intraabdominal desmoplastic small round cell tumor. Am J Surg Pathol 1992; 16:411–416.
115. Gerald WL, Rosai J, Ladanyi M. Characterization of the genomic breakpoint and chimeric transcripts in the EWS-WT1 gene fusion of desmoplastic small round cell tumor. Proc Natl Acad Sci USA 1995; 92:1028–1032.
116. Ladanyi M, Gerald W. Fusion of the EWS and WT1 genes in the desmoplastic small round cell tumor. Cancer Res 1994; 54:2837–2840.
117. Rauscher FJ III, Benjamin LE, Fredericks WJ, Morris JF. Novel oncogenic mutations in the WT1 Wilms' tumor suppressor gene: a t(11;22) fuses the Ewings' sarcoma gene, EWS1, to WT1 in desmoplastic small round cell tumor. Cold Spring Harbor Symp Quant Biol 1994; 59:137–146.
118. Karnieli E, Werner H, Rauscher FJ III, Benjamin LE, LeRoith D. The IGF-I receptor gene promoter is a molecular target for the Ewings' sarcoma-Wilms' tumor 1 fusion protein. J Biol Chem 1996; 271:19304–19309.

119. Eshet R, Werner H, Klinger B, Silbergeld A, Laron Z, LeRoith D, Roberts CT Jr. Up-regulation of insulin-like growth factor-I (IGF-I) receptor gene expression in patients with reduced serum IGF-I levels. J Mol Endocrinol 1993; 10:115–120.
120. Rosenfeld RG, Hintz RL. Characterization of a specific receptor for somatomedin C (Sm-C) on cultured human lymphoblasts: evidence that Sm-C modulates homologous receptor concentration. Endocrinology 1980; 107:1841–1848.
121. Rosenfeld RG, Dollar LA. Characterization of the somatomedin-c/insulin-like growth factor I (SM-C/IGF-I) receptor on cultured human fibroblast monolayers: regulation of receptor concentration by SM-C/IGF-I and insulin. J. Clin Endocrinol Metab 1982; 55:434–440.
122. Yang Y, Hoeflich A, Butenandt O, Kliess W. Opposite regulation of IGF-I and IGF-I receptor mRNA and concomitant changes of GH receptor and IGF-II/M6P receptor mRNA in human IM-9 lymphoblasts. Biochim Biophys Acta 1996; 1310:317–324.
123. Yamamoto H, Prager D, Yamasaki H, Melmed S. Rat pituitary GC cell insulin-like growth factor I receptor regulation. Endocrinology 1993; 133:1420–1425.
124. Hernandez-Sanchez C, Werner H, Roberts CT Jr, Woo EJ, Hun DW, Rosenthal SM, LeRoith D. Differential regulation of IGF-I receptor gene expression by IGF-I and basic fibroblast growth factor. J Biol Chem 1997; 272:4663–4670.
125. Wolfe BL, Rich CB, Goud HD, Terpstra AJ, Bashir M, Rosenbloom J, Sonenshein GE, Foster JA. Insulin-like growth factor-I regulates transcription of the elastin gene. J Biol Chem 1993; 268:12418–12426.
126. Clemmons DR, Shaw DS. Variables controlling somatomedin production by cultured human fibroblasts. J Cell Physiol 1983; 115:139–142.
127. Rubini M, Werner H, Gandini E, Roberts CT Jr, LeRoith D, Baserga R. Platelet-derived growth factor increases the activity of the promoter of the IGF-I receptor gene. Exp Cell Res 1994; 211:374–379.
128. Halazonetis TD, Kandil AN. Determination of the c-myc DNA-binding site. Proc Natl Acad Sci USA 1991; 88:6162–6166.
129. Rosenthal SM, Brown EJ, Brunetti A, Goldfine ID. Fibroblast growth factor inhibits insulin-like growth factor II (IGF-II) gene expression and increases IGF-I receptor abundance in BC3H-1 muscle cells. Mol Endocrinol 1991; 5:678–684.
130. Du J, Meng XP, Delafontaine P. Transcriptional regulation of the insulin-like growth factor-I receptor gene: evidence for protein kinase C-dependent and -independent pathways. Endocrinology 1996; 137:1378–1384.
131. Adashi EY, Resnick CE, Hernandez ER, Svoboda ME, Van Wyk JJ. In vivo regulation of granulosa cell somatomedin-C/insulin-like growth factor I receptors. Endocrinology 1988; 122:1383–1388.
132. Hernandez E, Hurwitz A, Botero L, Ricciarelli E, Werner H, Roberts CT Jr, LeRoith D, Adashi E. Insulin-like growth factor receptor gene expression in the rat ovary: divergent regulation of distinct receptor species. Mol Endocrinol 1991; 5:1799–1805.
133. Kleinman D, Karas M, Roberts CT Jr, LeRoith D, Phillip M, Segev Y, Levy J, Sharoni Y. Modulation of insulin-like growth factor I (IGF-I) receptors and membrane-associated IGF-binding proteins in endometrial cancer cells by estradiol. Endocrinology 1995; 136:2531–2537.
134. Stewart AJ, Johnson MD, May FEB, Westley BR. Role of insulin-like growth factors and the type I insulin-like growth factor receptor in the estrogen stimulated proliferation of human breast cancer cells. J Biol Chem 1990; 265:21172–21178.
135. Clarke RB, Howell A, Anderson E. Type 1 insulin-like growth factor receptor gene expression in normal human breast tissue treated with oestrogen and progesterone. Br J Cancer 1997; 75:251–257.
136. Papa V, Hartmann KKP, Rosenthal SM, Maddux BA, Siiteri PK, Goldfine ID. Progestins induce down-regulation of insulin-like growth factor (IGF-I) receptors in human breast cancer cells; potential autocrine role of IGF-II. Mol Endocrinol 1991; 5:709–717.
137. Goldfine ID, Papa V, Vigneri R, Siiteri P, Rosenthal SM. Progestin regulation of insulin and insulin-like growth factor I receptors in cultured human breast cancer cells. Mol Endocrinol 1992; 6:1665–1672.
138. Bassas L, DePablo F, Lesniak MA, Roth J. Ontogeny of receptors for insulin-like peptides in chick embryo tissues: early dominance of insulin-like growth factor over insulin receptors in brain. Endocrinology 1985; 117:2321–2329.
139. Bassas L, Girbau M, Lesniak MA, Roth J, DePablo F. Development of receptors for insulin-like growth factor-I in head and brain of chick embryos: autoradiographic localization. Endocrinology 1989; 125:2320–2326.

140. Baker J, Liu J-P, Robertson EJ, Efstratiadis A. Role of insulin-like growth factors in embryonic and postnatal growth. Cell 1993; 75:73–82.
141. Liu J-P, Baker J, Perkins AS, Robertson EJ, Estratiadis A. Mice carrying null mutations of the genes encoding insulin-like growth factor I (*Igf-1*) and type 1 IGF receptor (*Igf1r*). Cell 1993; 75:59–72.
142. DeChiara TM, Efstratiadis A, Robertson EJ. A growth-deficiency phenotype in heterozygous mice carrying an insulin-like growth factor II gene disrupted by targeting. Nature 1990; 345:78–80.
143. Cheng W, Reiss K, Kajstura J, Kowal K, Quaini F, Anversa P. Down-regulation of the IGF-I system parallels the attenuation in the proliferative capacity of rat ventricular myocytes during postnatal development. Lab Invest 1995; 72: 646–655.

5 Molecular Biology of the IGF-II/Mannose-6-Phosphate Receptor

Wieland Kiess, MD

Contents

INTRODUCTION

The insulin-like growth factor-II/Man-6-P (IGF-II/Man-6-P) receptor is a bifunctional binding protein that binds lysosomal enzymes bearing the Man-6-P recognition marker and IGF-II at distinct binding sites *(1–15)*. In addition, transforming growth factor (TGF)-β precursor, thyroglobulin, and proliferin, a protein that is expressed in rapidly proliferating cells, are also recognized by this receptor (Table 1) *(1–13)*. In avian and amphibian cells the receptor lacks the binding site for IGF-II but serves as a binding protein for Man-6-P-bearing ligands *(16–18)*. Almost all mammalian cells described to date express IGF-II/Man-6-P receptors that bind both classes of ligands, namely Man-6-P-containing glycoproteins and IGF-II *(1–13)*.

The receptor functions to target lysosomal enzymes bearing the Man-6-P recognition marker to lysosomes by binding Man-6-P-bearing ligands in the Golgi network and delivering them to a prelysosomal compartment with acidic pH. In the acidic prelysosomal compartment the ligands dissociate from the receptor and are transported to the lysosome via capillary movement *(1,3–4,7,11)*.

The majority of cellular IGF-II/Man-6-P receptors are localized intracellularly whereas only about 10% of the total receptor pool is cell-surface expressed. Cell-surface receptors serve to internalize extracellular lysosomal enzymes *(1,3,4,7,11,19–23)*.

From: *Contemporary Endocrinology: The IGF System*
Edited by: R. Rosenfeld and C. Roberts Jr © Humana Press, Inc., Totowa, NJ

Table 1
Ligands of the IGF-II/Man-6-P Receptor

1. IGF-II binding site
Insulin-like growth factor-II
(insulin-like growth factor-I)
2. Man-6-P-binding sites
Proliferin
Transforming growth factor-β1 precursor
Thyroglobulin
Unteroferrin
Prorenin
Acid hydrolases (approx 50 lysosomal enzymes bearing the Man-6-P recognition marker)

CHROMOSOMAL LOCATION

The genes that code for the IGF-II/Man-6-P receptor map to the centromeric third of chromosome 17 in the mouse *(24)*, and to the long arm of human chromosome 6, region 6q25–q27 *(25)*.

GENE AND RECEPTOR PROTEIN STRUCTURE

Overview, Homologies and Structural Repeats

The extracellular region of the IGF-II/Man-6-P receptor is composed of 15 homologous repeating domains, each of which is approx 147 residues in length. The extracellular part of the receptor contains two high-affinity Man-6-P recognition sites and one IGF-II specific binding site *(1–15,26–29)*.

There is a high degree of homology between the sequence of the human, bovine, mouse, and rat receptor. The aligned sequences of the full-length human and bovine receptors that are derived from the cDNA sequences are 80% identical. Homology between the rat and the bovine amino acid sequences also infers to approx 80% *(12–15,26)*. A 9-kb RNA transcript of the IGF-II/Man-6-P receptor is routinely detected in Northern blotting experiments using RNA from cells or tissues that express this receptor *(5,15,30)*.

Consensus sequences exist in the amino (N)-terminal sequence of the IGF-II/Man-6-P receptor, fibronectin, and a cell–cell adhesion regulator that has been recently purified and characterized from porcine seminal vesicle fluid *(31)*.

Sequence and Gene Structure

The mammalian IGF-II/Man-6-P receptor comprises a large extracellular domain that consists of 15 repeats with an average size of approx 147 amino acids *(1–15)*. These repeats share between 16% and 38% homology and are also homologous to the extracellular domain of the so-called cation-dependent Man-6-P receptor *(26)*. The cation-dependent Man-6-P receptor is a 46-kDa glycoprotein that also binds Man-6-P-containing ligands but does not recognize IGF-II *(1–4,7)*. Sixteen potential N-linked glycosylation sites have been identified upon examining the cDNA sequence of the IGF-II/Man-6-P receptor *(14,15)*. Differential glycosylation of the receptor in different tissues has been described *(32,33)*. The function of the carbohydrate chains of the receptor molecule remains unclear. However, it has become evident that glycosylation of the receptor is not necessary for binding of IGF-II or Man-6-P-containing ligands *(32,34)*.

Szebenyi and Rotwein reported on the cloning and characterization of the mouse receptor gene *(24)*. The gene was found to be 93 kb long and composed of 48 exons. The mouse gene codes for 2482 amino acids. The extracellular portion of the receptor is encoded by exons 1–46, with each of the 15 related motifs being determined by parts of three to five exons. A single fibronectin type II-like element is encoded by exon 39. The transmembrane portion of the receptor is alos encoded by exon 46, while exons 46–48 encode the cytoplasmic region. The positions of exon–intron splice junctions are conserved between several of the repeats in the IGF-II/Man-6-P receptor and the homologous extracellular region of the gene for the other known lysosomal enzyme sorting receptor, the 46-kDa cation-dependent Man-6-P receptor *(24)*.

Intracellular Domain

The cytoplasmic tail of the IGF-II/Man-6-P receptor has at least two signals for lysosomal enzyme sorting in the Golgi: deletion of the carboxy(c)-terminal Leu–Leu–His-Val residues of the 163-amino-acid cytoplasmic tail of the bovine IGF-II/Man-6-P receptor partially impairs the transport of lysosomal enzymes to a prelysosomal compartment *(35–40)*. This impairment results in the diversion of a portion of the receptor–ligand complexes to the cell surface from where they are endocytosed. When 134 residues of the cytoplasmic domain were deleted from the c-terminus, a similar effect on lysosomal enzyme transport was observed. Lastly, disrupture of the Tyr^{24}–Lys–Tyr–Ser–Lys–Val^{29} plasma membrane internalization signal had little effect on sorting of lysosomal enzmyes in the Golgi *(35)*. When this disruption was combined with either of the above mentioned deletions, a complete loss of lysosomal enzyme sorting resulted. The mutant receptors retained their ability to recycle to the Golgi and bind acid hydrolases. It is concluded from these data that the intracellular domain of the IGF-II/Man-6-P receptor contains at least two signals that contribute to Golgi sorting. It is hypothesized that this effect is mediated through interactions with clathrin-coated pit adaptor proteins in the Golgi apparatus *(39)*.

Fusion of the cytoplasmic and transmembrane domains of the bovine IGF-II/Man-6-P receptor to lysozyme, a monomeric secretory protein that is thought to be devoid of sorting information for the lysosomal system, was used as another approach to study the relative importance of various parts of the intracellular tail of the receptor for lysosomal enzyme sorting *(38)*. When the resulting chimera was transiently expressed in COS cells or stably expressed in CV1 cells, it had a predominantly intracellular distribution in the *trans*-Golgi region, with less than 10% present on the cell surface *(38)*. This distribution pattern is similar but not identical to the one of the natural form of the receptor *(7)*. The signal to specify the intracellular localization of the chimeric molecule must be located in the membrane-proximal third of the tail, as deletion of the last 120 residues of the 163-residue tail had no obvious effect on the distribution of the receptor chimera. However, immunofluorescence labeling studies suggested to the authors that the lumenal domain of the receptor may in fact partially determine the steady-state distribution of the receptor itself *(38)*.

Truncated Forms (Proteolytic Sites and Posttranslational Modification)

A truncated form of the receptor lacking the intracellular domain is present in the serum of the rat, sheep, monkey, and human *(41–48)*. In the sheep the serum form of the IGF-II/Man-6-P receptor seems to be an important carrier of IGF-II during fetal life *(42)*. The function of the truncated IGF-II/Man-6-P receptor form has not been elucidated. However, it has been suggested that release of the extracellular part of the receptor represents the major degradative pathway of the IGF-II/Man-6-P receptor *(45,46,48)*. Insulin injection increases the levels of serum receptors in intact cells *(45)*. The presence of truncated forms of the IGF-II/Man-6-P receptor is due to proteolytic cleavage and does not arise from differential splicing of receptor mRNA *(46)*.

The receptor is synthesized as a high molecular weight precursor protein that is glycosylated via N-linked sugars. The half-life of the receptor has been found to vary in different cell lines between approx 24 and 36 h *(32)*. The functional role of receptor glycosylation has not been elucidated. Glycosylation of the receptor protein is not required for binding of ligand to the receptor *(32,34)*.

IGF-II Binding Site

The mammalian IGF-II/Man-6-P receptor binds IGF-II with high affinity. Equilibrium binding experiments have revealed that the IGF-II/Man-6-P receptor binds 1 mol of IGF-II. IGF-II binds to the receptor with high affinity (K_d 2×10^{-10} *M*) *(27)*. The receptor seems to be rather specific for IGF-II. Insulin, IGF-I, and IGF-II variants do not recognize the receptor at all or with much lower affinity *(7,10,14,15,27–29,47,49–55)* (Fig. 1 and Tables 1 and 2). It had been suggested that IGF-II binds to a sequence that localizes close to repeat no. 13, which is close to the transmembrane spanning region of the receptor and that shares some homology with fibronectin *(12,14,15,56)*. Recently, the IGF-II binding site was found to reside within repeat 11 of the extracellular part of the receptor. To localize the IGF-II binding site, purified human IGF-II/Man-6-P receptor was digested with thermolysin, and the fragments were analyzed for their ability to bind radioiodinated IGF-II in a crosslinking assay *(57)*. Two IGF-II-binding receptor fragments were purified and sequenced. These fragments consist of disulfide connected peptides comprising amino acids 1331–1566 and 1331–1697. After further analysis using the expression of truncated forms of the receptor, it was concluded that peptide 1508–1566 within repeat 11 forms the IGF-II binding site of the human IGF-II/Man-6-P receptor *(57)*. Dahms et al. concluded from their initial work to clarify the exact location of the IGF-II binding site of the receptor, that this site actually resided in domains 5–11 *(58)*: truncated forms of the receptor were transiently expressed in COS-1 cells and assayed for their ability to bind iodinated human recombinant IGF-II. A construct encoding domains 5–11 bound 0.9 mol of IGF-II per mole of receptor, whereas a construct encoding domains 5–10 did not bind IGF-II at all. These data indicated that in fact the IGF-II binding site resided within domains 5–11 and that residues in domain 11 play a pivotal role in IGF-II binding *(58)*. In respect to repeat 11, a point mutation substituting threonine for isoleucine at residue 1572, located in the N-terminal half of repeat 11, completely abolished IGF-II binding *(47)*. It was concluded that repeat 11 of the receptor contains the minimal elements required for IGF-II binding (Table 3) *(2,47,57,58)*.

Whereas the mammalian IGF-II/Man-6-P receptor binds IGF-II with high affinity, the receptor protein in both birds and amphibians does not bind IGF-II at all *(16–18)*. The characteristics of IGF-II binding to the IGF-II/Man-6-P binding site of the receptor have also been delineated with the use of IGF analogs with high affinity for IGF-II/Man-6-P receptors but markedly reduced affinity for insulin and IGF-I receptors *(49,52,55)*.

Man-6-P Binding Sites

Equilibrium dialysis experiments revealed that the receptor binds 2 mol of Man-6-P or 1 mol of β-galactosidase or equivalent lysosomal enzymes via their Man-6-P-residues *(3,7,27)*. The process whereupon Man-6-P residues are attached to lysosomal enzymes

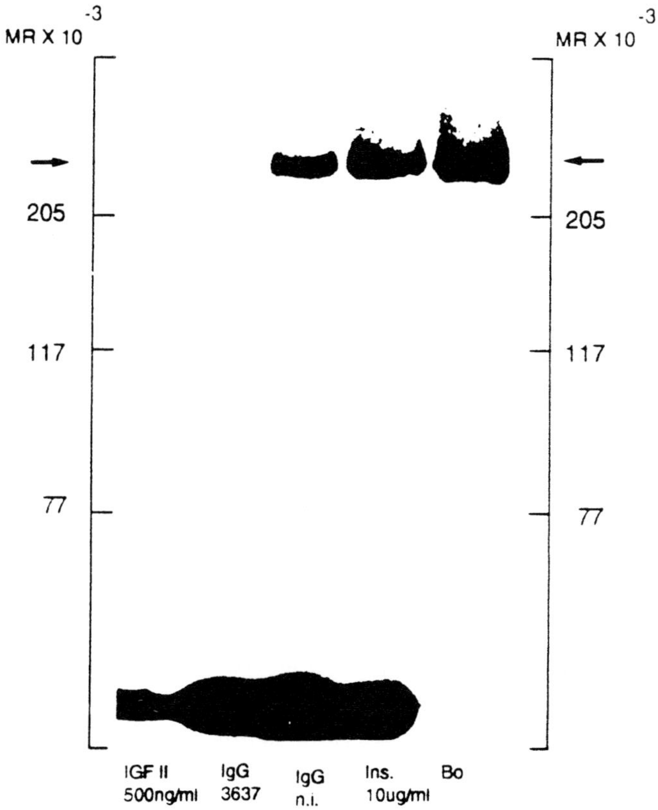

Fig. 1. Binding of [^{125}I]IGF-II to IGF-II/Man-6-P receptors in rat kidney embryonal cell (NRKE) membranes as shown by affinity crosslinking. The binding experiment was performed in the absence or presence of 500 ng/mL recombinant human IGF-II, 180 µg/mL of anti-IGF-II/Man-6-P receptor immunoglobulin, IgG 3637, 180 µg/mL of nonimmune control immunoglobulin, or 10 µg/mL of insulin. Affinity crosslinking was performed using disuccinimidyl suberate as the crosslinking agent. Radiolabeled bands were analyzed by SDS-PAGE and autoradiography (Kiess W, Siebler T, Nissley SP, *unpublished data*).

in the Golgi involves two enzymes, *N*-acetylglucosamine-1-phosphotransferase and α-*N*-acetylglucosaminyl-phosphodiesterase *(3,7,11,27)*. Only processed lysosomal enzymes and ligands bearing the Man-6-P recognition marker bind to the Man-6-P binding sites of the receptor and are targeted to lysosomes. The Man-6-P binding sites of the IGF-II/Man-6-P receptor localize to repeats 1–3 and 7–11 of the extracytoplasmic region *(59)*. The exact location of the Man-6-P binding sites has been elucidated by different approaches: first, bovine receptor was subjected to partial proteolysis with subtilisin followed by affinity chromatography on pentamannosyl phosphate-agarose *(59)*. Eleven proteolytic fragments were identified and characterized in respect to their ability to bind Man-6-P. The two fragments containing the actual Man-6-P binding domains were found to encompass domains 1–3 and 7–11 of the cytoplasmic region *(40,59)*. When truncated receptor forms were expressed in *Xenopus laevis* oocytes and assayed for their ability to bind phosphomannosyl residues, it again became clear that domain 3 was a necessary component of one of the Man-6-P binding sites of the receptor *(59)*. Second, to further localize the two Man-6-P binding domains and to determine the role of specific arginine

Table 2
Stoichiometry of Ligand Binding to the IGF-II/Man-6-P Receptor

Ligand	K_d	*Mol/mol*
Mannose-6-phosphate	7×10^{-6} *M*	2.17
β-galactosidase	2×10^{-8} *M*	0.9
Insulin-like growth factor-II	2×10^{-10} *M*	0.95

After data from Tong et al. (1988), ref. *27*.

Table 3
Amino Acid Sequences and Repeats of the IGF-II/Man-6-P Receptor Protein that Are Critical for Ligand Binding

Man-6-P binding domains in domains 1–3 and 7–11
Arginine 435 in domain 3
Arginine 1334 in domain 9
IGF-II binding site in repeat 11
Amino acids 1508–1566
Amino acid 1571

residues in Man-6-P binding, site-directed mutagenesis was utilized to create soluble receptor forms in conjunction with either conservative (Lys) or nonconservative (Ala) replacement of arginine residues *(60)*. Substitution of Arg 435 in domain 3 of the N-terminal binding site and Arg 1334 in domain 9 of the second Man-6-P binding site resulted in a dramatic loss of ligand binding activity. This finding suggests that Arg 435 and Arg 1334 are indeed essential components of the high-affinity Man-6-P binding sites of the IGF-II/Man-6-P receptor *(2,60)* (Table 3).

Internalization and Targeting Sequences

It is important to note that in many cell types a distinctive pattern of subcellular distribution is found: in gastrointestinal cells IGF-II/Man-6-P receptors are predominantly found close to the apical membranes. In proximal tubule cells of rat kidney apical localization of the IGF-II/Man-6-P receptor has also been described *(61)*. There is constant recycling of receptors from the cell surface to intracellular compartments *(7,19,23,39,62–67)*.

Deletion of the c-terminal Leu–Leu–His–Val residues of the 163-amino-acid cytoplasmic tail of the bovine IGF-II/Man-6-P receptor partially impairs the cycling of the receptor between the Golgi, endosomes, and the plasma membrane *(39)*. When 134 residues of the cytoplasmic tail were deleted from the c-terminus a portion of receptor–ligand complexes was diverted to the cell surface, where they were endocytosed. Disruption of the Tyr^{24}–Lys–Tyr–Ser–Lys–Val^{29} plasma membrane internalization signal alone had very little effect on Golgi sorting. However, the key sequence for receptor internalization seems to be the Tyr–Lys–Tyr–Ser–Lys–Val signal *(36)*. The crucial elements of this sequence are in fact present in the cytoplasmic tails of a number of other membrane receptors and proteins known to undergo rapid internalization *(36)*. Jadot et al. have investigated further the effect of a series of mutations of this internalization signal sequence. Their data showed that the essential elements of the signal are an aromatic residue, especially a Tyr in the first position, separated from a large hydrophobic residue in the last position by two amino acids. The residues in positions 2 and 3 of the signal, Tyr–Ser–Lys–Val, may have an additional modulating effect on its activity *(35)*.

AP-1 Golgi Assembly Proteins

Clathrin-coated vesicles mediate the transport of IGF-II/Man-6-P receptors and many other proteins from the *trans*-Golgi network and from the plasma membrane to endosomes. The major coat proteins of these carrier proteins are clathrin and assembly proteins also referred to as adaptor proteins (APs) *(68)*. Binding of AP-1 Golgi adaptors to membranes requires phosphorylated cytoplasmic domains of the IGF-II/Man-6-P receptor.

Phosphorylation of serines 2421 and 2492 is critical in inhibiting adaptor binding of the receptor. The recruitment of Golgi-specific adaptors has been shown to be cytosol and energy dependent and sensitive to GTP-γ-S and Brefeldin A *(68)*.

Putative Signaling Interaction Sites

G-Protein Interaction

Nishimoto and co-workers have extensively reported on a putative interrelation between the IGF-II/Man-6-P receptor and the G-protein signaling cascade. These workers have delineated the molecular basis of such an interaction as follows: a 14-amino-acid peptide, Arg^{2410}–Lys^{2423}, of the receptor directly activated G_i and deletion of C-terminal 4 residues from the peptide nullified this activity. In addition, single-residue substitutions led to modified G-protein interaction, suggesting that different regions of $G_{i\alpha}$ alpha might be capable of interacting with the IGF-II/Man-6-P receptor protein *(69,70)*. However, experiments using mutated receptors that lack the putative G-protein interaction site do not support the concept of IGF-II signaling through an IGF-II/Man-6-P receptor-mediated pathway *(71)* (*see* Chapter 8 by Nissley, *this* volume).

A number of kinases and phosphatases are being recognized to participate in the molecular mechanisms that lead to IGF-II/Man-6-P receptor translocation and redistribution *(62,65,66,72–75)*. In the *trans*-Golgi network cytoplasmic domains of the receptor molecules are being recognized by the Golgi-specific adaptor complex HA-2 whereupon specific kinases associate with the receptor *(4,7)*. Phosphorylation of the receptor leads to translocation to different cellular compartments *(64)*. The receptor exits from the *trans*-Golgi network via clathrin-coated pits. The degree of phosphorylation of receptor molecules predetermines their localization within the cell *(4,7,64)*.

Casein Kinase II Phosphorylation Site and PI3 Kinase Interaction

Analysis of a series of truncation and alanine scanning mutants revealed that a casein kinase II site followed by two leucines near the c-terminus was important for receptor function in sorting of lysosomal enzymes *(37)*. In a recent article, Nakajima and Pfeffer reported that wortmannin, an inhibitor of phosphatidylinositol 3-kinase, interfered with the delivery of newly synthesized lysosomal enzymes to lysosomes. However, receptor endocytosis and transport to the *trans*-Golgi network were not inhibited by wortmannin. In addition, transport of IGF-II/Man-6-P receptors from late endosomes to the *trans*-Golgi network was not affected by the drug. These workers concluded from their results that a wortmannin-sensitive process, i.e., phosphatidylinositol 3-kinase activity, is required for the export of the receptor from the *trans*-Golgi network *(74)*.

In a more recent article, Rosorius et al. reported on the interaction of a 91-kDa and a 35-kDa membrane-bound protein with the cytoplasmic tail of the IGF-II/Man-6-P receptor *(76)*. The formation of receptor-protein crosslinks appeared to be developmentally and regionally regulated in the brain and inhibited upon ATP hydrolysis *(76)*. It is hoped that the nature and function of these interactions will be elucidated in the near future.

PROMOTOR STRUCTURE AND FUNCTION

The structure and function of the mouse IGF-II/Man-6-P receptor gene promoter was elucidated by Liu et al. *(77)*. A series of chimeric plasmids linking different segments of IGF-II/Man-6-P receptor 5′ flanking DNA to luciferase as the reporter gene were transiently transfected. Promoter activity was orientation specific and maximal with a plasmid containing 266 basepairs (bp) of IGF-II/Man-6-P receptor DNA. DNA–protein binding studies by in vitro DNase I footprinting revealed an extended 54-bp footprint within the proximal promoter that contained two E boxes and potential binding sites for transcription factors Sp1, NGF-IA, and related proteins. Mutational analysis revealed that each E box contributed to more than half of the enhancer's activity. The authors conclude that a strong minimal IGF-II/Man-6-P receptor promoter of no more than 266 bp exists. In addition, they have identified a 54-bp enhancer within this promoter fragment. The additional molecular components that control the developmental, epigenetic, and tissue-specific regulation of IGF-II/Man-6-P receptor gene expression are still not known *(77)*.

GENE EXPRESSION AND REGULATION OF RECEPTOR PROTEIN LEVELS

The expression of the IGF-II/Man-6-P receptor varies considerably from tissue to tissue: in the rat and human, heart, kidney, and thymus are among the organs that contain high levels of the receptor. Brain expresses only small amounts of IGF-II/Man-6-P receptor *(5,6,33,61,78–83)*. There is a strong developmental pattern of receptor expression in the rat and sheep, with high concentrations of IGF-II/Man-6-P receptor found during fetal life. After birth, receptor concentrations in all tissues decline dramatically *(30,81,82)*. In the human, the developmental regulation of receptor levels is less evident *(33)*. It is unclear whether the high levels of receptor protein and mRNA expression during fetal life reflects the involvement of this receptor in growth processes. Alternatively, the receptor could function in tissue remodeling or play a role in tissue differentiation *(80,84,85)*. For example, it has been reported that both IGF-II and IGF-II/Man-6-P receptor levels increase during skeletal muscle differentiation *(86,87)* and during differentiation of colon cells *(88,89)* (Fig. 2). Compensatory growth of hepatic, thyroid, or renal tissue is accompanied by increased levels of receptors in such tissues *(90–92)*. In addition, a number of pathologic states such as the onset of diabetes *(93–95)* or fetal starvation *(42)* seem to modulate IGF-II/Man-6-P receptor expression. In addition, a number of hormones are thought to be capable of regulating IGF-II/Man-6-P receptor levels at least in some tissues *(45,72,96–104)*. The physiologic role of such differential expression of the receptor has not been elucidated. The differential and regulated expression of the IGF-II/Man-6-P receptor is described in more detail in the following paragraphs.

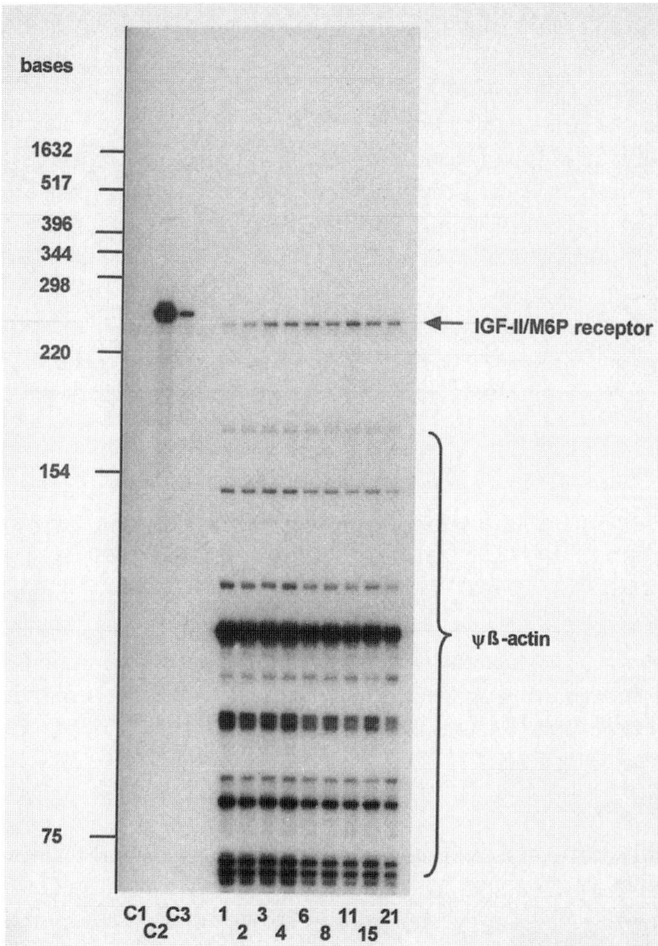

Fig. 2. Expression of IGF-II/Man-6-P receptor mRNA and of β-actin pseudogene during human colon carcinoma (Caco-2) cell differentiation as shown by RNase protection/solution hybridization assay. RNA was isolated from Caco-2 cells after 1, 2, 3, 4, 6, 8, 11, 15, and 21 d in culture as indicated in the figure. Cells show a differentiated phenotype after 4–6 d in culture. RNA was isolated from the cells and ^{32}P-labeled riboprobes specific for human IGF-II/Man-6-P receptor and β-actin RNA employed in the hybridisation step. Protected bands were visualized by autoradiography (Hoeflich A, Kiess W, *unpublished data*). C1, digested probes; C2 and C3, undigested probes.

Developmental Pattern

The first indication that the IGF-II/Man-6-P receptor is developmentally regulated came from studies of the truncated serum form of the receptor in the rat: the circulating form of the receptor was high in sera from fetal and neonatal rats and declined thereafter *(41)*. Later, it was found that receptor levels were high in fetal rat tissues and in most tissues declined dramatically in late gestation or after birth *(80)*. When the tissue expression of the receptor mRNA was measured using Northern blotting and solution hybridization/RNase protection assays, the concentration of receptor mRNA was significantly lower in postnatal rat tissue than in fetal tissues *(30,82)*. Similar results have been found in other species including sheep, cattle, rhesus monkey, mouse, and human; e.g., receptor mRNA levels were fourfold higher in fetal bovine lung and liver than in adult tissues *(33,42,81,84,102,105–110)*.

Embryogenesis

In fact, both IGF-II/Man-6-P receptor protein and mRNA are expressed in the mouse embryo soon after fertilization. Preimplantation parthenogenetic mouse embryos already express mRNA for the IGF-II/Man-6-P receptor *(107)*. IGF-II/Man-6-P receptor mRNA has been found in human fetal ovary and fetal uterus as early as at 10, 15, 19, and 22 weeks' gestation employing a reverse transcription-polymerase chain reaction (RT-PCR) technique *(108)*. In rabbit brain, a constant level of IGF-II/Man-6-P receptor mRNA was measured during fetal and postnatal life, whereas the amount of receptor protein declined with maturation from high levels in 23–30 days' gestation to the lowest levels in adult life *(105)*.

Tissue Distribution

In general, the IGF-II/Man-6-P receptor is ubiquitiously expressed in many tissues and in a large variety of different cell types *(5,8,9,33,81,82,85)*. Sklar et al. used a semiquantitative immunoblotting technique to measure receptor content of rat tissues: the rank order of receptor expression was heart > placenta > lung ≥ intestine > muscle = kidney > liver > brain. In heart, the receptor was 1.7% of total protein in extracted tissue *(82)*. Using Northern blotting and solution hybridization/RNase protection assays it was established that receptor mRNA concentration among tissues agreed with the rank order of receptor protein *(30)*.

Immune System

Normal freshly collected human lymphocytes express IGF-II/Man-6-P receptor mRNA as do transformed human IM-9 lymphoblasts *(104,111)*. In IM-9 lymphoblasts, the expression of IGF-II/Man-6-P receptor mRNA is modulated by the availability of serum factors *(104)*.

Central and Peripheral Nervous System

The IGF-II/Man-6-P receptor has been reported to be widely distributed in brains from adult rats and rabbits: in *in situ* hybridization studies receptor mRNA was detected in neuron-rich areas such as the hippocampus and to a lesser extent in the choroid plexus and the meninges *(105,112)*. In autoradiography and immunocytochemical studies receptor protein was visualized in neurons throughout the forebrain, the highest signal intensity being seen in the pyramidal cell, the polymorphic layers of the hippocampus, and the granule layer of the dentate gyrus *(113,114)*. In view of these data, a possible neurotrophic and/or neuromodulatory action of the IGF-II/Man-6-P receptor has been suggested *(114)*. In rabbits, a constant level of IGF-II/Man-6-P receptor mRNA was measured in the brain during fetal and postnatal life, while the amount of receptor protein declined with maturation *(105)*.

Reproductive System

When RNA was extracted from human fetal reproductive tissues and RT-PCR was performed, IGF-II/Man-6-P receptor mRNA was detected in human fetal ovary and fetal uterus *(108,115)*. In addition, *in situ* hybridization studies revealed the presence of IGF-II/Man-6-P receptor mRNA in testis from healthy young men who had died of trauma. IGF-II/Man-6-P receptor mRNA was most abundant in the germinal epithelium of the testis *(116)*. Furthermore, the receptor mRNA is colocalized with insulin degrading enzyme and IGF-I receptor mRNA in spermatocytes, suggesting an involvement with processes that degrade IGF-II in the testis *(78)*.

Regulation of IGF-II/Man-6-P Receptor Expression at the Transcriptional and Translational Level

Hormonal Regulation

Hormonal control of IGF-II/Man-6-P receptor expression has been reported in in vivo and in vitro situations *(45,72,96–104)*. For example, estradiol was found to specifically decrease the steady-state level of the IGF-II/Man-6-P receptor in three different breast cancer cell lines at both the transcriptional and translational levels *(97)*. Further evidence for a role of sex steroids in regulating IGF-II/Man-6-P receptor expression stems from studies using binding and affinity crosslinking experiments in the rat uterus: in mature cycling rats, the proestrus uterus had the lowest level of IGF-II binding activity when compared to other stages of the estrus cycle *(115)*.

A rather complex regulatory scheme connecting the GHRH–GH–IGF axis in the rat to the regulated expression of the IGF-II/Man-6-P receptor has been described by Spatola et al.: In rats, administration of antisera against GHRH during gestation led to an up-regulation of IGF-II/Man-6-P receptor protein in the fetus as was shown by Western blotting analysis *(102)*. In contrast, GH administration increased ^{125}I-IGF-II binding in regenerating rat skeletal muscle *(100)*. An increase of IGF-II/Man-6-P receptor binding was also seen after GH treatment of rat adipocytes *(96)*. In fact, in adipocytes GH had both an acute and a long-term stimulatory effect upon IGF-II binding in vitro. Paradoxically, IGF-II binding to adipocytes from GH-deficient, hypophysectomized animals was significantly higher than in control animals *(96)*. In two studies the effect of IGF-II upon IGF-II/Man-6-P receptor expression was investigated: Claussen et al. showed that incubation of NIH 3T3 cells with IGF-I, IGF-II, and IGF analogs increased both IGF binding protein-6 (IGFBP-6) and IGF-II/Man-6-P receptor expression. In addition, IGF-II overexpressing mouse fibroblasts transfected with the human IGF-II cDNA showed increased IGF-II binding and IGF-II/Man-6-P receptor RNA expression *(98)*. In contrast, Hoeflich et al. used human kidney embryonal fibroblasts that were also transfected with the full-length IGF-II cDNA but oversecreted 16 kDa proIGF-II rather than mature 7.5-kDa IGF-II and failed to show any effect of the proIGF-II oversecretion upon IGF-II/Man-6-P receptor expression *(99)* (Fig. 3).

Diabetes

Insulin generally is thought to stimulate ^{125}I-IGF-II binding to cells in culture. This effect is presumably caused by an increase in cell surface IGF-II/Man-6-P receptors through translocation of intracellular receptors to the cell surface *(23,67,96)*. In insulin-deficient streptozotocin-induced diabetes mellitus in rats, however, IGF-II/Man-6-P receptor mRNA expression and IGF-II binding to kidney membranes was significantly increased and returned to normal when the animals were treated with exogenous insulin *(93–95)*.

Tissue Remodeling (Bone, Liver, Kidney, Thyroid, Skin)

Acute injury induces the expression of selective growth factors and growth factor receptors (such as fibroblast growth factor [FGF], epidermal growth factor [EGF], and

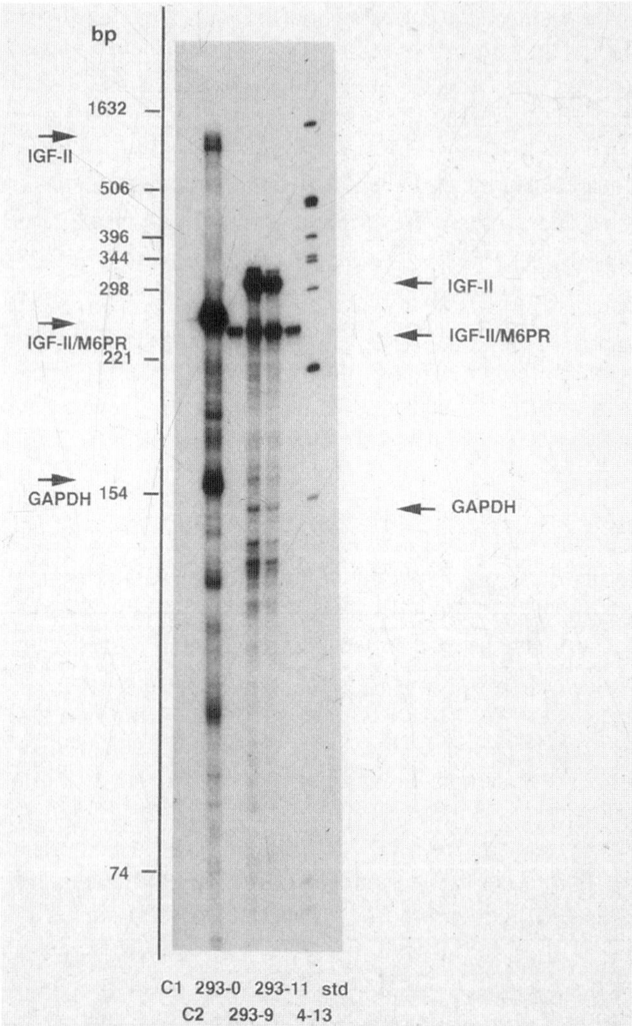

Fig. 3. Expression of IGF-II/Man-6-P receptor, IGF-II, and GAPDH mRNA by human embryonal kidney fibroblasts (293 cells) as shown by RNase protection/solution hybridization assay. RNA was isolated from wild-type 293-0 cells, from IGF-II transfected cell clones 293-9 and 293-11, and from 4–13 control cells as indicated in the figure. The transfected cell clones overexpress and secrete proIGF-II. RNA was isolated from the cells and ^{32}P-labeled riboprobes specific for human IGF-II/Man-6-P receptor, human IGF-II, and GAPDH RNA employed in the hybridization step. Protected bands were visualized by autoradiography (Kiess W, Koepf G, Wolf E, *unpublished data*). C1, digested probes; C2, undigested probes.

their receptors) in the epithelial cells of the wounded tissue: whereas FGF and EGF receptors are induced in wounded tissues, IGF-II/Man-6-P receptor mRNA is uniformly present both in normal skin epithelial cells and in cells from wounded tissue *(117)*. IGF-II/Man-6-P receptor mRNA and IGF-I receptor mRNAs are both selectively concentrated in developing chondrocytes and osteoblasts in developing murine skeleton from embryonic day 15 through postnatal day 25 *(83)*. Along with the expression patterns of the IGF ligands and their binding proteins, these data refer to a dynamic cell-specific pattern of gene expression during chondro- and osteogenesis *(118,119)*, supporting the view that the IGF-II/Man-6-P receptor has a complex and integral role within the developing and remodeling skeleton as do the other components of the IGF system. High levels of expression of the IGF-II/Man-6-P receptor are correlated with the process of liver regeneration after partial hepatectomy. Whether the receptor is required to bind some insulin-related growth factor, or alternatively is mainly involved in binding lysosomal enzymes that are involved in the remodeling process is still unclear *(90)*. An increase of IGF-II/Man-6-P receptor protein has also been observed after partial nephrectomy and partial thyroidectomy in the remaining kidney or thyroid tissue *(91,92)*.

Differentiation

During terminal differentiation of the myogenic C2 cell line, there is an increase in cell-surface expression of the IGF-II/Man-6-P receptor in parallel with a rise in secretion of IGF-II *(87)*. Similarly, receptor mRNA increases by more than 10-fold during the initial 48 h of C2 muscle differentiation. Comparable levels of IGF-II/Man-6-P receptor mRNA are expressed in differentiated C2 myotubes and primary cultures of fetal muscle *(86)*. A role for IGF-II and the IGF-II/Man-6-P receptor in muscle cell differentiation has been also suggested by data from Rosenthal et al. *(55)*. In human colon carcinoma (CaC0-2) cells, functional and morphological differentiation is also associated with an increase in IGF-II/Man-6-P receptor protein and mRNA expression (Fig. 2). This increase in receptor expression again parallels the concomitant increase in IGF-II secretion *(88,89)*.

Cancer

IGF-II/Man-6-P receptor-related IGF-II binding activity was found to be increased in both human thyroid cancer *(121)* and human adrenal tumors *(121)* when compared to normal adjacent tissues. In addition, in colonic cancer cells, overexpression of the IGF-II/Man-6-P receptor gene has been described *(88,89)*.

Chronic exposure of rats to phenobarbital can induce liver tumors and alter the ability of normal hepatocytes to respond to growth stimuli. In fact, phenobarbital exposure was found to increase transforming growth factor-β 1 (TGF-β) and IGF-II/Man-6-P receptor concentration in hepatocytes, with a concomitant up-regulation in gene expression *(122)*. Because the receptor facilitates the proteolytic activation of TGF-β, changes in receptor concentrations might be important in respect to hepatocyte growth and tumor progression. In contrast, in preneoplastic lesions IGF-II/Man-6-P receptors are not up-regulated by phenobarbital, suggesting that a reduction in receptor concentration may provide the tumor cells with an important selective growth advantage *(122)*. In human hepatocellular carcinomas the IGF-II/Man-6-P receptor gene is mutated and heterzygosity is lost. The observed mutations give rise to truncated receptor protein and significant receptor substitutions *(123,124)*. These data provide direct evidence that the IGF-II/Man-6-P receptor gene functions as a tumor suppressor in human liver carcinogenesis.

In contrast, no amplification of the receptor gene was observed in specimens of invasive breast carcinoma from 51 patients by Southern blotting techniques. In fact, IGF-II/Man-6-P gene expression was stable in nontumor breast tissue and tumor tissues irrespective of a concomitant amplification of c-*erbB2* or *int-2* genes in several of the tumors *(125)*. These findings imply that the receptor gene is stable in breast carcinoma and that the receptor involvement in breast cancer if at all may result from mechanisms other than gene amplification or gene copy numbers *(125–127)*.

IMPRINTING

The IGF-II/Man-6-P receptor gene is imprinted in rodents but not in normal human tissues *(128–135)*. However, The IGF-II/Man-6-P receptor gene is maternally imprinted, whereas the IGF-II gene is paternally imprinted *(128,133)*. It has been suggested that the differential imprinting of the IGF-II and the IGF-II/Man-6-P receptor genes has important functional consequence for fetal growth and development *(130,133)*. Imprinted genes are thought to influence development during preimplantation stages in mouse embryos *(107)*. In fact, reduction of IGF-II expression by antisense IGF-II oligonucleotides decreased the rate of progression to the blastocyst stage and decreased the cell number in blastocysts. In addition, preimplantation parthenogenetic mouse embryos expressed only IGF-II/Man-6-P receptor mRNA but failed to express IGF-II or IGF-I receptor mRNA. These findings indicate that the latter genes are not expressed when inherited maternally. Some growth factors and their receptors, regulated by genomic imprinting, may control cell proliferation from the earliest stages of embryonic development *(107)*.

Loss of imprinting has been described in Wilms' tumor and in several other malignancies. In contrast, in uterine smooth-muscle tumors, the IGF-II/Man-6-P receptor gene was expressed from both alleles in all tissues examined. These data confirm the finding that the IGF-II/Man-6-P receptor is not imprinted in humans *(136)*. Lau et al. reported that murine embryos that inherit a nonfunctional IGF-II/Man-6-P receptor gene from their fathers are viable and develop normally into adults. However, the majority of the mice inheriting the same mutated allele from their mothers die around birth. These mice are approx 30% larger than their normal siblings *(137)*. Importantly, it was reported that inactive and biallelic modes of expression of imprinted genes are potential mechanisms for the death of gynogenones and androgenones at the preimplantaion stage *(109)*.

PROTEIN BIOSYNTHESIS AND TRANSCRIPTIONAL MODIFICATION

The receptor is synthesized as a precursor protein and subsequently glycosylated via N-linked sugars. The half-life of the receptor has been found to be approx 24–36 hours *(32,34)*. Glycosylation of the receptor protein is not required for binding of IGF-II to the receptor *(32)*. Glycosylation could be important for the stability of the receptor protein or alternatively may play an important role in receptor trafficking. Phosphorylation of the receptor protein has been thought to be important for trafficking of the receptor through the Golgi network and also for internalization of the receptor protein *(62)*. Phosphorylation of the receptor determines its subcellular distribution and localization *(64)*.

CELLULAR TRANSLOCATION, RECEPTOR RECYCLING, AND TRAFFICKING

A number of chemical compounds, enzymes, and hormones have been found to modulate recycling and routing of the IGF-II/Man-6-P receptors. Some of these compounds, such as Brefeldin A, phorbol esters, and polymyxin B, have been useful tools for the elucidation of transport pathways of the lysosomal system *(62,65,66,72)*. Others such as the IGFs, insulin, and EGF might well play a physiologic role in modulating receptor localization and receptor function in vivo *(19)*. It is known that insulin stimulation of fat cells induces an increased surface expression of IGF-II/Man-6-P receptors *(23,64)*. In addition, insulin is capable of enhacing the shedding of truncated forms of the IGF-II/Man-6-P receptor *(45)*.

77. Liu Z, Mittanck DW, Kim S, Rotwein P. Control of IGF-II/Man6P receptor gene transcription by proximal promoter elements. Mol Endocrinol 1995; 9:1477–1487.
78. Bondy C, Zhou J, Chin E, Reinhardt RR, Ding L, Roth RA. Cellular distribution of insulin-degrading enzyme gene expression. Comparison with insulin and IGF receptors. J Clin Invest 1994; 93:966–973.
79. Guse AH, Kiess W, Funk B, Kessler U, Berg I, Gercken G. Identification and characterization of IGF receptors on adult rat cardiac myocytes: linkage to inositol 1,4,5,-trisphosphate formation. Endocrinology 1992; 130:145–151.
80. Nissley P, Kiess W, Sklar MM. The developmental expression of the IGF-II/M6P receptor. Mol Reprod Dev 1993; 35:408–413.
81. Pfuender M, Sauerwein H, Funk B, Kessler U, Barenton B, Schwarz HP, Hoeflich A, Kiess W. IGF-II/M6P receptors are expressed in bovine fetal tissues throughout gestation. Domest Anim Endocrinol 1995; 12:317–324.
82. Sklar MM, Kiess W, Thomas CL, Nissley SP. Developmental expression of the tissue IGF-II/Man6P receptor in the rat. J Biol Chem 1989; 264:16733–16738.
83. Wang E, Wang J, Chin E, Zhou J, Bondy CA. Cellular patterns of IGF system gene expression in murine chondrogenesis and osteogenesis. Endocrinology 1995; 136:2741–2751.
84. Harvey MB, Kaye PL. IGF-II receptors are first expressed at the 2-cell stage of mouse development. Development 1991; 111:1057–1060.
85. Senior PV, Byrne S, Brammar WJ, Beck F. Expression of the IGF-II/M6P receptor mRNA and protein in the developing rat. Development 1990; 109:67–73.
86. Szebenyi G, Rotwein P. Differential regulation of Man6P receptors and their ligands during the myogenic development of C2 cells. J Biol Chem 1991; 266:5534–5539.
87. Tollefsen SE, Sadow JL, Rotwein P. Coordinate expression of IGF-II and its receptor during muscle differentiation. Proc Natl Acad Sci USA 1989; 86:1543–1547.
88. Hoeflich AT, Koepf G, Kessler U, Brem G, Rascher W, Blum W, Kiess W. Coordinate expression of IGF-II and IGF-II/Man-6-P receptor mRNA and stable expression of IGF-I receptor mRNA during differentiation of human colon carcinoma cells. Eur J Endocrinol 1996; 135:49–59.
89. Hoeflich AT, Koepf G, Kessler U, Brem G, Rascher W, Blum W, Kiess W. Expression of IGFBP-2, -3, and -4 mRNA during differentiation of Caco-2 colon epithelial cells. Am J Physiol 1996; 271:E922–E931.
90. Burguera B, Werner H, Sklar M, Shen-Orr Z, Stannard B, Roberts CT, Jr, Nissley SP, Vore SJ, Caro JF, LeRoith D. Liver regeneration is associated with increased expression of the IGF-II/M6P receptor. Mol Endocrinol 1990; 4:1539–1545.
91. Polychronakos C, Guyda HJ, Posner BI. Increase in the type 2 IGF receptors in the rat kidney during compensatory growth. Biochem Biophys Res Commun 1985; 132:418–423.
92. Polychronakos C, Guyda HJ, Patel B, Posner BI. Increase in the number of type II IGF receptors during propylthiouracil-induced hyperplasia in the rat thyroid. Endocrinology 1986; 119:1204–1207.
93. Werner H, Shen-Orr Z, Stannard B, Burguera B, Roberts CT, Jr, LeRoith D. Experimental diabetes increases IGF-I and -II receptor concentration and gene expression in kidney. Diabetes 1990; 39:1490–1497.
94. Kiess W, Hoeflich A, Yang Y, Gronbaek H, Flyvbjerg A. Streptozotocin induction of diabetes in rats leads to increased IGF-II/Man-6-P receptor mRNA expression in kidney but not in lung or heart. Growth Regul 1996; 6:66–72.
95. Flyvbjerg A, Kessler U, Funk B, Dorka B, Orskov H, Kiess W. Transient increase of IGF-II/M6P receptor protein content in diabetic rat kidney and liver tissue. Growth Regul 1993;
96. Lönnroth P, Assmundsson K, Eden S, Enberg G, Gause I, Hall K, Smith U. Regulation of IGF-II receptors by growth hormone and insulin in rat adipocytes. Proc Natl Acad Sci USA 1987; 84:3619–3622.
97. Mathieu M, Vignon F, Capony F, Rochefort H. Estradiol down-regulates the Man6P/IGF-II receptor gene and induces cathepsin-D in breast cancer cells: a receptor saturation mechanism to increase the secretion of lysosomal proenzymes. Mol Endocrinol 1991; 5:815–822.
98. Claussen M, Buergisser D, Schuller AGP, Matzner U, Braulke T. Regulation of IGFBP-6 and Man6P/IGF-II receptor expression in IGF-II overexpressing NIH 3T3 cells. Mol Endocrinol 1995; 9:902–912.
99. Hoeflich A, Wolf E, Braulke T, Koepf G, Kessler U, Brem G, Rascher W, Blum W, Kiess W. Does the overexpression of pro-IGF-II in transfected human embryonic kidney fibroblasts increase the secretion of lysosomal enzymes? Eur J Biochem 1995; 232:172–178.
100. Kirk SP, Whittle MA, Oldham JM, Dobbie PM, Bass JJ. GH regulation of the type 2 IGF receptor in regenerating skeletal muscle of rats. J Endocrinol 1996; 149:81–91.

101. Rydziel S, Canalis E. Cortisol represses IGF-II receptor transcription in skeletal cell cultures. Endocrinology 1995; 136:4254–4260.
102. Spatola E, Pescovitz OH, Marsh K, Johnson NB, Berry SA, Gelato MC. Interaction of GHRH with the insulin-like growth factors during prenatal development in the rat. Endocrinology 1991; 129:1193–1200.
103. Taaffe DR, Jin IH, Hoffman AR, Marcus R. Lack of effect of recombinant hGH on muscle morphology and GH-IGF expression in resistance trained elderly men. J Clin Endocrinol Metab 1996; 81:421–425.
104. Yang Y, Hoeflich A, Butenandt O, Kiess W. Opposite regulation of IGF-I and IGF-I receptor mRNA and concommitant changes of GH receptor and IGF-II/Man6P receptor mRNA in human IM-9 lymphoblasts. Biochim Biophys Acta 1996; 1310:317–324.
105. Devaskar SU, Sadiq HF, Holtzclaw L, George M. The development pattern of rabbit brain insulin and insulin-like growth factor receptor expression. Brain Res 1993; 605:101–109.
106. Ho Y, Wigglesworth K, Eppig JJ, Schultz RM. Preimplantation development of mouse embryos in KSOM: augmentation by amino acids and analysis of gene expression. Mol Reprod Dev 1995; 41:232–238.
107. Rappolee DA, Sturm KS, Behrendtsen O, Schultz GA, Pedersen RA, Werb Z. IGF-II acts through an endogenous pathway regulated by imprinting in early mouse embryos. Genes Dev 1992; 6:939–952.
108. Shifren JL, Osathanoondh R, and Yeh J. Human fetal ovaries and uteri: developmental expression of genes encoding the insulin, IGF-I, and IGF-II receptors. Fertil Steril 1993; 59:1036–1040.
109. Szabo PE, Mann JR. Maternal and paternal genomes function independently in mouse ova in establishing expression of the imprinted genes Snrpn and Igf2r: no evidence for allelic trans-sensing and counting mechanisms. EMBO J 1996; 15:6018–6025.
110. Wang ZQ, Fung MR, Barlow DP, Wagner EF. Regulation of embryonic growth and lysosomal targeting by the imprinted IGF-II/Man6P receptor gene. Nature 1994; 372:464–467.
111. Nyman T, Pekonen F. The expression of IGFs and their binding proteins in normal human lymphocytes. Acta Endocrinol 1993; 128:168–172.
112. Stephenson DT, Rash K, Clemens JA. Increase in IGF-II receptor within ischemic neurons following focal cerebral infarction. J Cereb Blood Flow Metab 1995; 15:1022–1031.
113. Lesniak MA, Hill JM, Kiess W, Rojeski M, Pert CB, Roth J. Receptors for IGFs I and II: autoradiographic localization in rat brain and comparison to receptors for insulin. Endocrinology 1988; 123:2089–2099.
114. Couce ME, Weatherington AJ, McGinty JF. Expression of IGF-II and IGF-II/Man6P receptor in the rat hippocampus: an in situ hybridization and immunocytochemical study. Endocrinology 1992; 131:1636–1642.
115. Chandraskhar Y, Narayan S, Singh P, Nagamani M. Receptors for IGF-II in the rat uterus: characterization and variation throughout the estrus cycle. Acta Endocrinol 1991; 124:434–441.
116. Zhou J, Bondy C. Anatomy of the insulin-like growth factor system in the human testis. Fertil Steril 1993; 60:897–904.
117. Antoniades HN, Galanopoulos T, Neville-Golden J, Kiritsy CP, Lynch SE. Expression of growth factor and receptor mRNAs in skin epithelial cells following acute cutaneous injury. Am J Pathol 1993; 142:1099–1110.
118. De Ceninck F, Poiraudeau S, Pagano M, Tsagris L, Blachard O, Willeput J, Corvol M. Inhibition of chondrocyte cathepsin B and L activities by IGF-II and its Ser29 variant in vitro: possible role of the Man6P/IGF-II receptor. Mol Cell Endocrinol 1995; 113:205–213.
119. Fitzsimmons RJ, Ryaby JT, Magee FP, Baylink DJ. IGF-II receptor number is increased in TE-85 osteosarcoma cells by combined magnetic fields. J Bone Miner Res 1995; 10:812–819.
120. Yashiro T, Tsushima T, Murakami H, Obara T, Fujimoto Y, Shizume K, Ito K. IGF-II/Man6P receptors are increased in primary human thyroid neoplasms. Eur J Cancer 1991; 27:699–703.
121. Weber MM, Auernhammer CJ, Kiess W, Engelhardt D. IGF receptors in normal and tumorous adult human adrenocortical glands. Eur J Clin Endocrinol 1997; 136:296–303.
122. Jirtle RL, Hankins GR, Reisenbichler H, Boyer IJ. Regulation of Man6P/IGF-II receptors and TGFbeta during liver tumor promotion with phenobarbital. Carcinogenesis 1994; 15:1473–1478.
123. De Souza AT, Hankins GR, Washington MK, Orton TC, Jirtle RL. Man6P/IGF-IIR gene is mutated in human hepatocellular carcinomas with loss of heterozygosity. Nature Genet 1995; 11:447–449.
124. De Souza AT, Hankins GR, Washington MK, Fine RL, Orton TC, Jirtle RL. Frequent loss of heterozygosity on 6q at the Man6P/IGF-IIR receptor locus in human hepatocellular tumors. Oncogene 1995; 10:1725–1729.

125. Hebert E, Herbelin C, Bougnoux P. Analysis of the IGF-II receptor gene copy number in breast carcinoma. Br J Cancer 1994; 69:120–124.
126. Vignon F, Rochefort H. Interactions of pro-cathepsin D and IGF-II on the Man6P/IGF-IIR receptor. Breast Cancer Res Treat 1992; 22:47–57.
127. Zhoa Y, Escot C, Maudelonde T, Puech C, Rouanet P, Rochefort H. (1993) Correlation between Man6P/IGF-IIR receptor and cathepsin D RNA levels by in situ hybridization in benign and malignant mammary tumors. Cancer Res 1993; 53:2901–2905.
128. Barlow DP, Stöger R, Hermann BG, Saito K, Schweifer N. The mouse IGF-type-2 receptor is imprinted and closely linked to the Tme locus. Nature 1991; 349:84–87.
129. Deal CL. Parental genomic imprinting. Curr Opin Pediatr 1995; 7:445–458.
130. Haig D, Graham C. Genomic imprinting and the strange case of the IGF-II receptor. Cell 1991; 64:1045–1046.
131. Kalscheuer VM, Mariman EC, Schepens MT, Rehder H, Ropers HH. The IGF type 2 receptor is imprinted in the mouse but not in humans. Nature Genet 1993; 5:74–78.
132. Ogawa O, McNoe LA, Eccles MR, Morison IM, Reeve AE. Human IGF type I and type II receptors are not imprinted. Hum Mol Genet 1993; 2:2163–2165.
133. Willison K. Opposite imprinting of the mouse IGF-II and IGF-II receptor genes. Trends Genet 1991; 7:107–108.
134. Zemel S, Bartolomei MS, Tilghman SM. Physical linkage of two mammalian imprinted genes, H19 and IGF-II. Nature Genet 1992; 2:61–65.
135. Xu Y, Goodyer CG, Deal C, Polychronakos C. Functional polymorphism in the parental imprinting of the human IGF2R gene. Biochem Biophys Res Commun 1993; 197:747–754.
136. Vu TH, Yballe C, Boonyanit S, Hoffman AR. IGF-II in uterine smooth-muscle tumors: maintenance of genomic imprinting in leiomyomata and loss of imprinting in leiomyosarcomata. J Clin Endocrinol Metab 1995; 80:1670–1676.
137. Lau MMH, Stewart CEH, Liu Z, Bhatt H, Rotwein P, Stewart CL. Loss of the imprinted IGF-II/cation-independent Man6P receptor results in fetal overgrowth and perinatal lethality. Genes Dev 1994; 8:2953–2963.
138. Stagsted J, Olsson L, Holman GD, Cushman SW, Satoh S. Inhibition of internalization of glucose transporters and IGF-II receptors. Mechanism of action of MHC class I-derived peptides which augment the insulin response in rat adipose tissue. J Biol Chem 1993; 268:22809–22813.

6 Molecular Biology of the IGF Binding Proteins

Guck T. Ooi, PhD *and Yves R. Boisclair*, PhD

CONTENTS

INTRODUCTION

In adult animals, 80-85% of insulin-like growth factor-I and -II (IGFs) circulate as 150-kDa ternary complexes. These ternary complexes are comprised of one molecule each of IGF, IGF binding protein-3 (IGFBP-3), the predominant IGFBP in serum, and an acid-labile subunit (ALS) *(1)*. Serum also contains lower molecular weight complexes that are not fully saturated with IGFs. These complexes are comprised of the other smaller binding proteins (IGFBP-1, -2, -4, -5, and -6). These six IGFBPs are structurally related to each other (Fig. 1) and bind IGF-I and IGF-II with high affinity *(2,3)*, but share no homology with the IGF receptors. These proteins, however, have different biochemical and biological characteristics that are due to differences in their primary structure. The present chapter summarizes current molecular information on the six IGFBPs and ALS (regulation of gene expression, cDNA, and gene organization). This summary, however, does not attempt to be encyclopedic, but rather to provide a general outline on the molecular aspects of the six high-affinity IGF-binding proteins, underlying their similarities and differences.

IGFBP-1

General

IGFBP-1 is present at very low concentrations in adult serum, but is the predominant IGFBP found in human amniotic fluid *(4)*. In serum, IGFBP-1 showed dynamic regula-

From: *Contemporary Endocrinology: The IGF System*
Edited by: R. Rosenfeld and C. Roberts Jr © Humana Press, Inc., Totowa, NJ

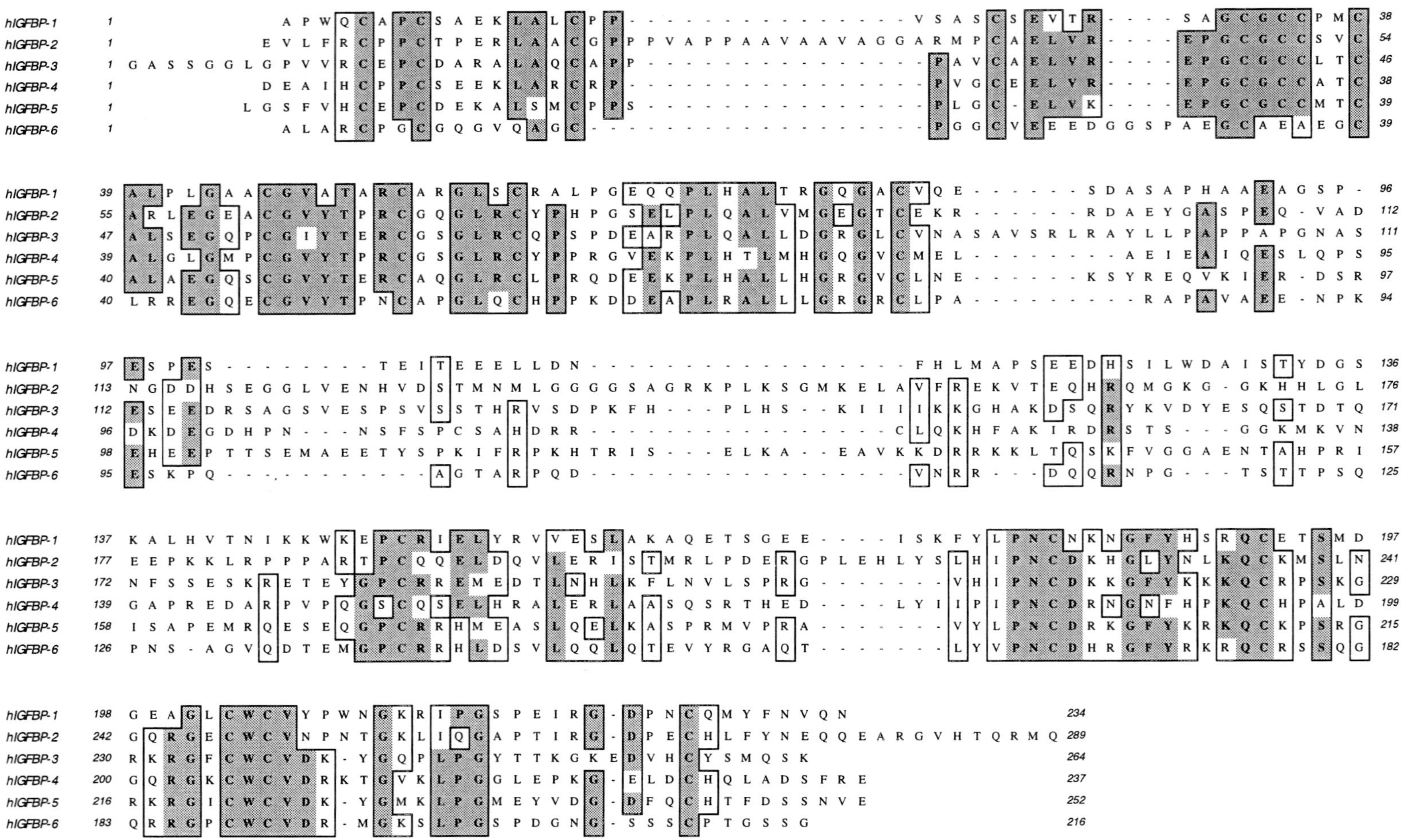

Fig. 1. Amino acid sequences of mature human IGFBP-1 to -6 proteins deduced from nucleotide sequences (*14,216,157,174,193,218*) were aligned using a ClustalW algorithm analysis (MacVector 6.0, Oxford Molecular Group, Beaverton, OR, USA). Conserved residues are indicated by *dark shadings,* whereas *light shadings* indicate residues with similar classification groups.

tion, with insulin being the main regulator *(see below)*. IGFBP-1 has equal affinity for IGF-I and IGF-II, and like IGFBP-3 and IGFBP-5, is known to be serine phosphorylated in its central domains *(5)*. IGFBP-1 from different tissues and body fluids is serine phosphorylated to varying extents *(6–8)*. Dephosphorylated forms that have a lower affinity for IGF-I are thought to potentiate IGF actions, whereas phosphorylated forms may inhibit the mitogenic actions of IGF-I *(7)*.

The deduced molecular mass of human IGFBP-1 is 25.3 kDa, although it migrated with an apparent molecular mass of 25–31 kDa on nonreducing sodium dodecyl sulfate-polyacrylamide gel electrophoresis (SDS-PAGE), and as a 32–34-kDa band on reducing SDS-PAGE *(7,60)*.

Regulation of Expression In Vivo

IGFBP-1 mRNA is developmentally regulated in the rat, being highest in the liver of fetal and term-gestation rats *(9,10)*. In mid-gestational (embryonic day 14 [e14]) rat liver, IGFBP-1 mRNA was detected by *in situ* hybridization in discrete groups of cells that appear to be hepatocytes, and then stably expressed by the liver throughout the fetal and postnatal stages *(11)*. At gestational age prior to e18, no extrahepatic IGFBP-1 mRNA was seen *(11)*. High expression of IGFBP-1 mRNA occurs in the liver of late fetal (e21) rats, with much lower levels occurring in the stomach, heart, lung, and kidneys *(9)*. These high levels of IGFBP-1 mRNA persist at 1 day of age, but are greatly reduced postnatally, with low levels seen in livers from 21-day and adult rats *(9,10,12,13)*. In adult rats, the major site of IGFBP-1 mRNA expression is in the liver *(9,11)*, although IGFBP-1 mRNA has been detected in the adult kidney and pseudopregnant rat uterus but not in the brain, diaphragm, heart, and lung *(13)*. In human fetal tissues, IGFBP-1 mRNA is also detected in the liver, but not in kidney, heart, lung, brain, adrenal, muscle, jejunum, tongue, spleen, thymus, and pancreas *(14,15)*. In human adult, major IGFBP-1 expression is seen only in the liver and the uterine decidua from late pregnancy, but not in placental trophoblasts *(15,16)*. Expression of IGFBP-1 mRNA also varies during the menstrual cycle, with high expression seen in the endometrium during the late secretory phase of the cycle *(17)*.

In humans, IGFBP-1 levels show a similar pattern of developmental regulation. IGFBP-1 levels are high in fetal serum, 10 times lower in term cord plasma, and decrease after birth to the low steady state levels seen during puberty *(18,19)*. Immunostaining of IGFBP-1 was seen in 14- to 16-week-gestation human fetuses in multiple tissues but not in cerebral cortex, spleen, and thyroid *(20)*. During pregnancy, serum IGFBP-1 levels are increased fivefold *(18,21)*, with maximal levels seen at 22–23 weeks postconception that then decrease before term *(22)*. IGFBP-1 levels are 1000-fold higher in human amniotic fluid than in serum, with peak levels seen during midterm pregnancy, followed by a gradual decrease at term to one half of the maximal levels *(22–25)*.

In human serum, IGFBP-1 is highly regulated by metabolic status, which reflects changes in insulin levels *(26,27)*. Glucose intake, however, may have an independent effect *(28)*. Metabolic conditions associated with decreased insulin levels such as fasting *(29)*, diabetes *(30)*, and prolonged exercise *(31)* increased serum IGFBP-1 levels. The increases seen after acute fasting and in diabetes can be rapidly reversed by refeeding *(29,32)* and insulin treatment *(26,30)*, respectively. IGFBP-1 levels in human plasma are elevated at night, although this apparent diurnal variation probably reflects the concomitant nocturnal fasting period during the night *(19,33)*.

In addition to insulin, growth hormone (GH) may suppress IGFBP-1 levels in serum, as elevated IGFBP-1 levels are observed in children and adults with GH deficiency *(29,34)*.

Regulation of Expression In Vitro

IGFBP-1 is the predominant IGFBP produced by a number of liver-derived cell lines *(35–39)*, breast cancer *(40)*, and endometrial adenocarcinoma cell lines *(41)*.

In the H4-II-E rat hepatoma cell model system, steady-state IGFBP-1 mRNA abundance was increased ~10-fold by physiologically appropriate concentrations of dexamethasone, but not by progesterone and testosterone *(42)*. This increase in IGFBP-1 mRNA is solely at the level of gene transcription *(36,43)*, and dexamethasone did not affect the half-life of IGFBP-1 mRNA *(42)*. Cyclic AMP (cAMP) *(44)* as well as phorbol esters *(45)* also can stimulate IGFBP-1 mRNA in H4-II-E cells. In contrast, insulin inhibited IGFBP-1 and IGFBP-1 mRNA by ~90% in H4-II-E rat hepatoma cells, whereas IGF-I, IGF-II, and GH had no effects *(36)*. Similar inhibition of IGFBP-1 protein and IGFBP-1 mRNA was also seen in rat hepatocytes *(46)*, human fetal liver explants *(47,48)*, and HepG2 human hepatocarcinoma cells *(49–51)*.

Although IGFBP-1 regulation is primarily regarded to bc independent of GH, Seneviratne et al. *(34)* and others *(52)* have shown that IGFBP-1 transcription may be down-regulated by GH. Both hepatic IGFBP-1 mRNA and serum IGFBP-1 levels are increased in hypophysectomized rats, and were reduced following a single injection of GH *(34)*. This decrease following GH treatment is at the level of transcription as demonstrated by nuclear run-on analysis *(34)*. Similarly, Grownowski and Rotwein *(52)* have observed that nuclear IGFBP-1 mRNA levels declined to 60% of baseline at 30 min following GH injection into hypophysectomized rats.

In human decidual stromal cultures, IGFBP-1 synthesis is stimulated fivefold by cAMP AMP, and 20-fold by relaxin *(53)*. Incubation of endometrial stromal cells with either medroxyprogesterone or relaxin increased IGFBP-1 mRNA *(54,55)*.

cDNA and Protein

Human IGFBP-1 cDNA have been cloned from libraries of human decidua *(15,56,57)*, placenta *(14)*, liver *(14)*, and HepG2 hepatocarcinoma cells *(58)*. The cDNA codes for a 259-amino-acid precursor, with a 25-residue prepeptide and a 234-residue mature protein. Northern blots analysis indicated the presence of a single mRNA species with a size of 1.5 kb. The IGFBP-1 mRNA contains 56–164 bp of 5′ untranslated region, and 564–612 bp of 3′ untranslated region culminating in a consensus poly(A) signal (AATAAA) and a poly(A) tail *(7,58,59)*.

Human IGFBP-1 contains 18 cysteine residues, clustered at the amino (N)-terminal and carboxy (C)-terminal ends. All 18 cysteines residues in human IGFBP-1 appear to be used in intrachain disulfide linkages *(61)*. Human IGFBP-1 lacks potential sites of *N*-glycosylation, but has an Arg-Gly-Asp (RGD) sequence at the C-terminus, a motif found in fibronectin and other cell matrix adhesion molecules that is recognized by cell surface integrins *(62,63)*. Human IGFBP-1 also contains a region rich in a Pro–Glu–Ser–Thr (PEST) sequence near the C-terminal domain *(15)*, the presence of which is common to rapidly degraded proteins *(64)*.

Frost and Tseng *(6)* have shown that IGFBP-1 from human decidual cultures was phosphorylated to varying extents. Similarly, Jones et al. *(8,65)* have shown that nonphosphorylated IGFBP-1 has lower affinity for IGF-I compared with phosphorylated IGFBP-1, and that phosphorylation at serine 101 is important in regulating the affinity of IGFBP-I binding to IGF-1. Interestingly, however, the serine 101 residue is not present in either rat or mouse IGFBP-1 *(66)*.

Rat IGFBP-1 cDNA clones have been isolated from H4-II-E rat hepatoma cells (G. T. Ooi, *unpublished results*), adult rat liver, and rat decidua *(13)* as well as from cycloheximide-treated regenerating rat liver *(67)*. The rat IGFBP-1 precursor has a 25-amino-acid signal peptide, and a mature peptide of 247 amino acid residues that is highly homologous with human IGFBP-1. The molecular mass deduced from the cDNA sequence is 26.9 kDa, although on nonreducing SDS-PAGE, rat IGFBP-1 migrated as a 30–32-kDa band *(35,37)*. Like human IGFBP-1, rat IGFBP-1 has no *N*-glycosylation sites, although it has several regions rich in serine and threonine that might be *O*-glycosylated.

Mouse IGFBP-1 has also been cloned, and is similar to rat IGFBP-1 *(68)*. The mouse IGFBP-1 gene has been localized to chromosome 11 *(66)*.

Gene Structure and Promoter Regulation

IGFBP-1 is encoded by a single-copy gene located on human chromosome 7p14–p12 where it is contiguous in a tail-to-tail configuration with the IGFBP-3 gene, and would therefore be transcribed in the opposite direction *(69,70)*. The gene spans 5.2 kb and has four exons *(69–71)*. There is a direct repeat of 20–24 bp from the splice junction between intron 3 and exon 4 *(70)*, with the 3′ acceptor splice site repeated 93 bp upstream of exon 4 *(70)*. As is the case in all IGFBPs, intron 1 is the largest of all the introns.

In humans, transcription of the IGFBP-1 gene is directed by a TATA sequence present 28 nt upstream of the transcription initiation site that is identified by primer extension analysis to be an adenine nucleotide beginning at 165 bp upstream from the first nucleotide of the A+1TG translation start codon *(39)*. However, S1 nuclease protection analysis suggested the transcription start site to be at nt –155 from the ATG *(71)*.

Suwanichkul et al. *(39)* have sequenced 1205 bp upstream from the transcription initiation site (at nt –165) and found a number of *cis*-regulatory elements. Besides a TATA box at nt –28 (relative to the adenine transcription start site), a CCAAT sequence (through which the transcription factor C/EBP may bind) was found at nt –72. This fragment showed orientation-specific promoter activity when transiently transfected into HepG2 cells. Similar promoter activity also was obtained when a longer promoter fragment (3.6 kb) was used *(39)*. Progressive 5′-deletion analysis of the promoter indicated that basal promoter activity was mediated by a *cis*-element that binds to hepatic nuclear factor 1 (HNF-1) *(39,72,73)*. This HNF-1 site is highly conserved in the rat and mouse promoters, and mutations in this site that abolished HNF-1 binding decreased basal activity *(39,74,75)*. Conversely, mutations in the HNF-1 site that increased HNF-1 binding enhanced IGFBP-1 promoter activity *(39)*.

In addition to HNF-1 which is important for basal promoter activity, extensive studies by Powell et al. *(39,76–78)* have indicated the presence of a number of *cis*-regulatory sequences that mediate the multihormonal regulation of the human IGFBP-1 promoter. The regulation of the IGFBP-1 gene by various hormonal agents is similar to that of the gene for the cytosolic form of phosphoenolpyruvate carboxykinase (PEPCK), a key enzyme in gluconeogenesis *(79)*, and is therefore consistent with the proposed role of IGFBP-1 in glucose homeostasis *(80)*.

A CGTCA motif, which mediates the effects of cAMP on a number of other gene promoters *(76)*, was present between nt –263 and –259 of the human IGFBP-1 promoter; mutation of this CGTCA motif to TAGCA caused a 51% decrease in the ability of dibutyryl cAMP and theophylline to stimulate IGFBP-1 promoter activity above control levels *(81)*. Interestingly, however, the CGTCA motif is not present in the rat *(75)* and mouse *(66)* IGFBP-1 promoters, although cAMP does stimulate IGFBP-1 mRNA *(44)* and promoter activity *(74)* in H4-II-E cells. The rat IGFBP-1 promoter, however, has a potential adaptor protein-2 (AP-2) site that might mediate the stimulation by both cAMP and phorbol esters *(74,82,83)*, indicating that the CGTCA motif may not be the only *cis*-regulatory element conferring the effects of cyclic AMP on the IGFBP-1 promoter.

Stimulation of IGFBP-1 mRNA *(42,84,85)* and transcription *(42)* is attributed to the presence of two glucocorticoid response elements (GREs) *(86–88)* located within the first 200 bp of the human IGFBP-1 promoter *(77)*. These two GREs bind to the human glucocorticoid receptor with low affinity, probably owing to their weak consensus with the canonical GRE motif *(86)*, although the loss of either one of the GREs resulted in a complete loss of glucocorticoid activation *(77)*. This is different from the rat IGFBP-1 promoter in which only a single GRE is present that is very similar to the canonical GRE *(74,82)*.

Like PEPCK, IGFBP-1 is dynamically regulated by insulin *(see above)*, and the similarities in their regulation can be attributed to the presence of a similar insulin response element (IRE) found in the proximal regions of both gene promoters *(78)*. The insulin response sequence was first identified in the PEPCK promoter *(89,90)* and is comprised of a 10-bp sequence located between nt –416 and –407. This sequence was present as a conserved palindrome (CAAAACAAACTTATTTTG) in both the human and rat IGFBP-1 promoters *(74,78,82)*. Although mutation of either half site decreases insulin inhibition of promoter activity slightly, insulin inhibition was abolished when both sites were mutated together *(78)*. Block scanning mutations *(74)* or internal deletion *(82)* of the IRE in the rat IGFBP-1 promoter also abolished the ability of insulin to inhibit promoter activity. Although transcription factors binding to this element that are responsible for mediating the inhibitory effects of insulin on the IGFBP-1 promoter are not known, the albumin D-site-binding protein (DBP), hepatic nuclear protein-3β (HNF-3β), and the high-mobility group I protein (HMGI) each protected the human IGFBP-1 IRE from DNase digestion *(91,92)*.

IGFBP-2

General

In human fetal serum, IGFBP-2 is the most abundant IGFBP at mid-gestation *(93,94)*. At birth, the concentration of IGFBP-2 is approximately fourfold higher than during adulthood, but is less than the concentration of IGFBP-3 *(93,95,96)*. In rodents, IGFBP-2 is the dominant IGFBP in serum during fetal and early postnatal life, but disappears almost completely in the adult animal *(97,98)*. On nonreducing SDS-PAGE, IGFBP-2 migrates as a nonglycosylated protein of ~ 34 kDa *(99)*.

Regulation of Expression In Vivo

In the rat, IGFBP-2 is expressed as early as e7 throughout the epiblast. During early embryonic life (until e13), expression localized to many rapidly growing tissues such as the neuroepithelium or those directing proliferation of neighboring cells such as the notochord *(100)*. By mid-gestation (e13.5–e15), IGFBP-2 is expressed in numerous ectodermal and endodermal derivatives *(101)*. Throughout these periods, the expression of IGFBP-2 is generally distinct and complementary to the expression of IGF-II *(100,101)*. A similar pattern of IGFBP-2 expression occurs in the mouse embryo *(102)*. At the end of fetal life, IGFBP-2 is detected by Northern analysis in liver, kidney, intestines, and lungs *(9)*. After birth, the rat IGFBP-2 gene becomes silent except in a few tissues such as the choroid plexus, testis, and kidney *(103–105)*. In liver, however, the levels of IGFBP-2 mRNA are reinduced by fasting, hypophysectomy, and diabetes *(9)*. At least during development and fasting, these changes in IGFBP-2 mRNA are due to changes in the rate of transcription of the gene *(98,106)*.

Expression of IGFBP-2 is ubiquitous in the ovine fetus until e80, but subsists at high levels at e120 and in adult sheep only in liver, kidney, and choroid plexus *(107)*. Throughout ovine development, serum IGFBP-2 correlates best with hepatic expression, suggesting that liver synthesizes most of circulating IGFBP-2. In lactating dairy cows, circulating levels of IGFBP-2 are decreased after 4 days of GH administration or of hyperinsulinemic euglycemia *(108,109)*. In adult human, circulating levels of IGFBP-2 are increased by fasting, GH deficiency, IGF-I infusion, and by tumor-induced hypoglycemia *(96,110–112)*.

Regulation of Expression in Vitro

Many cell models synthesize IGFBP-2, but few display regulated expression *(7,113)*. Examples of regulated expression include the decreased expression of IGFBP-2 in C2C12 mouse myoblasts and turkey satellite cells upon differentiation *(114,115)*, the insulin inhibition of basal and retinoic acid (RA)-stimulated expression of IGFBP-2 in primary rat hepatocyte *(116,117)*, and the human chorionic gonadotropin inhibition of IGFBP-2 gene transcription and mRNA abundance in rat Leydig cells *(118)*.

cDNAs and Protein

The human cDNA has an open reading frame of 328 amino acids corresponding to a signal peptide of 39 amino acids, and a mature protein of 289 amino acids *(119)*. The mature protein contains 18 cysteine residues and an Arg–Gly–Asp (RGD) sequence near the end of the C-terminus, but is devoid of *N*-glycosylation sites. The size of the corresponding mRNA is ~ 1.5 kb by Northern analysis *(119)*. Bovine and rat cDNAs encode IGFBP-2 proteins with near identical features *(104,105,120,121)*.

Gene Structure and Promoter Regulation

The human IGFBP-2 gene maps to region q33–q34 of chromosome 2 *(122,123)*. The gene contains four protein coding exons and covers ~ 32 kb of chromosomal DNA, with the first intron alone contributing 27 kb *(122,124)*. Transcription initiation occurs from a single site located 113 bp from the ATG in a region that is devoid of TATA or initiator elements *(124)*. The region between nt –521 and nt +112 (relative to the cap site) exhibited promoter activity when assayed in K16 human Jurkat cells *(124)*. The organization of the IGFBP-2 gene is similar in rat, mouse, and pig *(125–129)*.

Cis-elements required for transcription of the IGFBP-2 gene were studied in the rat promoter *(127,130)*. Binding of the transcription factor Sp1 to a cluster of three independent GC boxes located between nt –89 and nt –41 (relative to the cap site identified in *[127,130]*) is essential for efficient activation of the promoter in BRL-3A rat liver and 293 human embryonic kidney cell lines *(127,130)*. A fourth GC box, between nt –139 and nt –130, is also important in BRL-3A cells *(130)*. These GC boxes are well conserved in the human, mouse, and pig promoters, suggesting that they are equally important for the expression of the gene in these species *(124,128,129)*. In addition, the regions located between nt –493 and nt –423, between nt –291 and nt –188 also contribute to the activity of the rat promoter *(130)*. Location and identity of the *cis*-elements responsible for the developmental and metabolic regulation of the IGFBP-2 gene in liver have not been characterized.

IGFBP-3

General

IGFBP-3 is the most abundant circulating IGFBP in postnatal animal *(131)*. At birth, levels of IGFBP-3 are approximately one third of adult levels in humans, but are almost undetectable in rodents *(93,97)*. Serum levels increase with postnatal age to peak at puberty *(95,97,132)*. On nonreducing SDS-PAGE gels, IGFBP-3 migrates as a doublet of 40–45 kDa, reflecting variation in the degree of *N*-glycosylation *(7,99)*. IGFBP-3 secreted by CHO cells is phosphorylated on serine 111 and serine 113 *(133)*. IGFBP-3 has the unique ability, when occupied by a molecule of IGF-I or -II, to form a ternary complex of 150 kDa in plasma by binding to the acid labile subunit (ALS) (*see below* and *[7,131,134]*).

Regulation of Expression In Vivo

IGFBP-3 expression is undetectable at e12 in the mouse embryo, but localized to liver and vertebrae by e18 *(102)*. In the rat embryo, hybridization is detected in the developing urogenital system and in several muscles at e14 *(11)*, and in liver, kidney, muscle, developing cerebral cortex, and in periosteal condensations of mesenchyme at e20 *(135)*. In juvenile and adult rats, IGFBP-3 mRNA is most abundant in liver, kidney, uterus, placenta, ovary, stomach, and heart *(136–138)*. Expression of IGFBP-3 is particularly abundant in endothelial cells of many tissues (rat liver, placenta, and white adipose tissue [WAT], human testis) *(135,139–141)*.

After birth, serum IGFBP-3 is decreased by GH insufficiency, diabetes, and undernutrition *(132,142,143)*. IGF-I has been suggested as a direct regulator based on its ability to increase the serum levels of IGFBP-3 in various rodent models of GH deficiency *(143–145)*, but this effect of IGF-I is not seen in humans *(111,146–148)*. In the rat, regulation of IGFBP-3 synthesis after birth or hypophysectomy appears to occur primarily via translational/and or postranslational mechanisms as very few changes occur in the abundance of liver and kidney IGFBP-3 mRNA despite significant changes in serum IGFBP-3 *(135,136)*.

Regulation of Expression In Vitro

Villafuerte and colleagues have developed a cell model that bears direct relevance to the regulation of IGFBP-3 in rat liver: as seen in vivo *(135)*, only Kupffer and sinusoidal endothelial cells of liver express IGFBP-3 when cocultured with hepatocytes *(149)*. In these cells, insulin increases and dexamethasone decreases the abundance of IGFBP-3 mRNA via changes in transcription *(150,151)*. In contrast, IGF-I increases IGFBP-3 mRNA by extending its half-life *(152)*. In most other IGFBP-3 producing cell lines, the mechanism underlying the changes in IGFBP-3 mRNA remains undefined (reviewed in *[7,153]*). However, evidence has been obtained that osteogenic protein-1 in SaOS-2 human osteosarcoma cells, GH in SKHEP-1 human hepatocarcinoma cells, and IGF-I in bovine endothelial cells increase the levels of IGFBP-3 mRNA by stimulating transcription *(154–156)*.

cDNAs and Protein

The human cDNA predicts a prepeptide of 27 amino acids and a mature protein of 264 amino acids *(157)*. Examination of the primary sequence of the mature protein indicates the presence of three potential *N*-glycosylation sites, 18 cysteines, and the absence of an RGD sequence. By Northern analysis, a single mRNA species of ~ 2.5 kb is detected. Almost identical features can be deduced from bovine, rat, and porcine IGFBP-3 cDNAs *(137,158–160)*.

Gene Structure and Promoter Regulation

The single human IGFBP-3 gene maps to chromosome 7, 20 kb from the IGFBP-1 gene, in a tail-to-tail organization *(161)*. The gene spans 8.9 kb and contains four coding exons *(162)*. Unlike the other IGFBP genes, a fifth exon contributing most of the 3′ untranslated region of the mRNA and the polyadenylation signal is also present. Transcription initiation occurs from a single adenosine residue located 132 nt upstream of the ATG. A DNA fragment corresponding to nt –431 to nt +72 (relative to the cap site) was shown to contain the IGFBP-3 promoter *(162)*. Elements present in the proximal region of this promoter include a TATA box at nt –30 and a GC box containing overlapping Sp1 and AP-2 elements at nt –97. These elements are conserved in the rat and bovine IGFBP-3 promoters *(159,163)*.

Using nonparenchymal cells cocultured with rat hepatocytes, an insulin response element (IRE) was mapped between nt –1150 and nt –1134 of the rat IGFBP3 promoter *(164)*. This element bears little homology to other known IREs, including the IGFBP-1 IRE, and recognized a ~ 90-kDa nuclear protein whose abundance/activity is increased by insulin in extracts of nonparenchymal liver cells and diabetic liver. Also, the IGFBP-3 gene is induced by the tumor suppressor p53 in EB1 colon carcinoma cells *(165)*. Two elements, capable of binding p53, are present in intron 1 and intron 2 of the human IGFBP-3 gene, and confer p53 inducibility to a heterologous promoter *(165)*. In bovine epithelial cells, the activation of the bovine promoter by IGF-I was lost when the proximal GC box was deleted *(159)*.

IGFBP-4

General

IGFBP-4 is synthesized locally by normal human bone cells *(166,167)*, and is a potent inhibitor of IGF actions in bone *(167)*. IGFBP-4 has been isolated from human osteosarcoma cells *(167)*, the PC3 human prostatic tumor cell line *(168)*, human serum *(169)*, rat serum *(170)*, the B104 rat neuroblastoma cell line *(171,172)*, ovine plasma *(173)*, and from porcine ovarian follicular fluid *(60)*. In humans, serum concentrations of intact IGFBP-4 are low, although in adult rats, IGFBP-4 is the second most abundant IGFBP *(170)*.

The deduced molecular mass of human IGFBP-4 is 26 kDA *(174)*, and IGFBP-4 occurs both as in *N*-glycosylated and nonglycosylated forms. On nonreducing SDS-PAGE electrophoresis, glycosylated IGFBP-4 migrated with an apparent molecular mass of 28K whereas nonglycosylated forms migrated as a 24-kDa band.

Regulation of Expression In Vivo

IGFBP-4 is predominantly expressed in the liver, which represents the largest single source of IGFBP-4 in the adult rat *(11)* but is also present at lower levels in other tissues such as brain cortex, hypothalamus, stomach, kidney, lung, heart spleen, testis, and adrenal *(174)*. IGFBP-4 mRNA is one of the most abundant IGFBP transcripts in the mid- and late gestational embryo, particularly in mesodermally derived tissues such as smooth and striated muscles, lungs, liver, and to a lesser degree the heart *(11)*. IGFBP-4 mRNA is also highly expressed in the sinus precartilage layers, tongue, the muscularis of the esophagus, the mesenchymal periphery of the larynx, as well as the connective tissues around the follicles of the vibrissa and the thymic capsule of the e14 rat embryo *(11)*.

In late pregnancy in the rat, hepatic IGFBP-4 mRNA in liver is increased ~10-fold; these levels persist until postpartum *(175)*.

Regulation of Expression In Vitro

IGFBP-4 is one of the major IGFBPs produced by osteoblasts in vitro: it is a potent inhibitor of osteoblast proliferation *(167)* and is regulated by several key osteotropic agents such as parathyroid hormone, 1,25-dihydroxyvitamin D_3, and bone morphogenetic protein-7 *(176,177)*. In human chondrocytes, prostaglandin E_2 and calcium ionophore (ionomycin) stimulated IGFBP-4 protein and mRNA by three- to fourfold, and this stimulation is reversed by the calcium channel blocker Verapamil as well as the calcium/calmodulin inhibitor W7 *(178)*. Phorbol esters, which activate protein kinase C, did not affect IGFBP-4 production, whereas cAMP mimetics, which activate protein kinase A, inhibited IGFBP-4 mRNA *(178)*. In cultured human osteoblast-like SaOS-2 cells, parathyroid hormone but not 17β-estradiol increased IGFBP-4 mRNA *(179)*. In primary rat hepatocytes, triiodothyronine increases IGFBP-4 mRNA as well as protein levels *(180)*. Thyroid hormone and retinoic acid induce the synthesis of IGFBP-4 in mouse osteoblastic cells *(181)*.

cDNA and Protein

IGFBP-4 cDNA has been isolated from rat liver and human placenta *(174)*, TE89 human osteosarcoma *(176)*, bovine pulmonary artery *(182)*, and murine TC-1 bone marrow stromal cell line *(183)*. Human IGFBP-4 cDNA has ~400 bp of 5′-untranslated sequence, and >900 bp of 3′-untranslated sequence *(176)*, with the open reading frame coding for a 20–21-amino-acid signal peptide and 237-amino-acid mature peptide *(174,176)*. IGFBP-4 contains two extra cysteine residues in the exon 2 region, compared to the other IGFBPs; however the alignment of the remaining 18 cysteines is conserved in all the IGFBPs. IGFBP-4 does not contain an RGD sequence, but has a single N-linked glycosylation site.

Rat IGFBP-4 has also been cloned *(174)*, and is very similar to human IGFBP-4. Rat IGFBP-4 cDNA has ~185 bp 5′-untranslated sequence, and an ~318 bp 3′-untranslated sequence, with the open reading frame coding for a 21-amino-acid signal peptide, and a 233-amino-acid mature peptide *(174)*.

Mouse IGFBP-4 cDNA has been cloned from TC-1 bone marrow stromal cell line *(183)*. Mouse IGFBP-4 has a deduced molecular mass of 25.7 kDa and differs from rat IGFBP-4 by three amino acids *(183)*.

Gene Structure and Promoter Regulation

Although the chromosomal localization of human IGFBP-4 has been determined (chromosome 17q12–21) *(70,184,185)*, and the 5′-flanking sequence known *(186,187)*, full nucleotide sequence of the gene has not been determined. The 5′-flanking sequence, however, has been determined *(186,187)*. In human IGFBP-4, the proximal 1.3 kb promoter, exon 1, and some of intron 1 have a high GC content with a high density of CpG dinucleotides indicative of the presence of a CpG islands *(186)*. It contains a typical TATA box located at nt –299 from the ATG translation codon *(187)*, with the cap site located 14 bases downstream from the TATA box. A number of potential ATF/CREB,

AP-1, and Egr-1 (early growth response factor-1) sites as well as GC boxes are present *(187)*. The 1254-bp 5′ flanking sequence showed promoter activity when transfected into CaCo2 human colon cancer cell line *(187)*.

The rat IGFBP-4 gene spans at least 12 kb, with four exons separated by three introns of 6.4, 0.6, and 2.7 kb. A single transcriptional start site is located at 249 nt (guanine) upstream of the ATG codon. Like human IGFBP-4, the rat IGFBP-4 gene has a TATA box and a CAAT box, as well as multiple *cis*-regulatory elements including 3 cAMP responsive elements, three AP-1 sites, and one progesterone receptor binding site *(188)*.

IGFBP-5

General

In healthy men and women, IGFBP-5 circulates at ~ 15% of IGFBP-3 levels and is also the major IGFBP stored in bone *(189,190)*. Serum and bone levels decrease from prepuberty through adulthood *(189,190)*. IGFBP-5 migrates at 29–31 kDa on nonreducing SDS-PAGE gels, and is *O*-glycosylated *(99,191)*.

Regulation of Expression In Vivo

In the rat embryo, IGFBP-5 expression is first detected at e10.5 *(192)*. Expression is present in muscle precursors soon after myotone formation and remains prominent in skeletal muscles throughout fetal life. Other sites of expression in early embryonic life include the notochord, the floor plate, regions of the surface ectoderm, and axial regions of the neuroepithelium *(192)*. In late fetal and early postnatal life, expression also localizes to several regions of the developing brain *(192)*. In e18 mouse embryo, expression is detected in meninges, vertebrae, lung, kidney, and intestine *(102)*. In adult rodents, IGFBP-5 mRNA is most abundant in kidney and WAT; is easily detected in skeletal muscle, brain, stomach, lung, heart, adrenal, and intestine; and is low in liver, spleen, and testis *(193–195)*. After hypophysectomy, the abundance of IGFBP-5 in the rat is decreased in skeletal muscle and WAT but is unchanged in kidneys; levels of IGFBP-5 mRNA in skeletal muscle and WAT are restored partially by IGF-I, and completely by GH therapy *(140,194)*.

Regulation of Expression In Vitro

The expression of the IGFBP-5 gene is under dynamic regulation in osteoblasts prepared from fetal rat parietal cells (ob cells). In these cells, abundance of IGFBP-5 mRNA is increased by RA, IGF-I, prostaglandin E_2 (PGE_2), and by interleukin-6 in the presence of its soluble receptor *(196–199)*, and is decreased by transforming growth factor b-1 (TGF-b-1), fibroblast growth factor-2 (FGF-2), platelet-derived growth factor-BB (PDGF-BB), and cortisol *(200,201)*. Changes in the transcription of the IGFBP-5 gene account for most or all of these effects *(196–201)*. Transcription of the IGFBP-5 gene is also induced by IGF-I in porcine aortic smooth muscle cells *(202)* and by differentiation in the C2 and C21 mouse myoblast cell lines *(203,204)*. Other cell systems display regulation of IGFBP-5 mRNA, but the underlying mechanisms have not been described. For instance, increases in IGFBP-5 mRNA were reported after PTH treatment of UMR-106-01 rat osteosarcoma cells *(205)*, after IGF-I treatment of FRTL-5 rat thyroid follicular cells *(206)*, and after 1,25-dihydroxyvitamin D treatment of MCF-7 human breast cancer cells *(207)*; a decrease in IGFBP-5 mRNA was observed after RA treatment of T47D human breast cancer cells *(208)*.

cDNAs and Protein

The human cDNA encodes a prepeptide of 20 amino acids and a mature protein of 252 amino acids *(193,209)*. The mature protein is characterized by the presence of the 18 cysteine residues conserved among IGFBP-1 to -4 and lacks an *N*-glycosylation site or RGD sequence. Mature IGFBP-5 contains a stretch of mostly basic amino acid residues (Arg^{201}–Arg^{218}) that may mediate the binding of IGFBP-5 to extracellular matrix and glycosaminoglycans, and play a role in the heparin-induced reduction of binding affinity for IGF *(210,211)*. The amino acid identity between rat, mouse, and human IGFBP-5 is greater than 97%, making it the most conserved of all IGFBPs *(193,203,209)*. The ~6.0 kb IGFBP-5 mRNA is the longest of all IGFBP mRNAs because of a long 3′ untranslated region (4419 bp) *(212)*.

Gene Structure and Promoter Regulation

The human IGFBP-5 gene maps to region 2q33–34 of chromosome 2, 20–40 kb from the IGFBP-2 gene, in a tail-to-tail orientation *(212)*. This single gene is composed of four coding exons and covers 33 kb of DNA with intron 1 accounting for ~ 25 kb *(212)*. Transcription initiation occurs at a single adenosine residue located 722 bp upstream of the ATG. Elements identified in near vicinity of the start site include a TATA element at nt –33 and a CAAT box at nt –73. A fragment extending from nt –461 to nt +23 directed expression of a reporter gene in an orientation-specific manner in MDA-MB-468 human breast cancer cells *(212)*. The IGFBP-5 gene in the rat and mouse have similar structural features *(195,213)*. In particular, the nucleotide sequence of the proximal 200 bp of the promoter is 90% identical between mouse, rat, and human, including complete conservation of the TATA, CAAT, and potential AP-2 elements *(195,212,213)*.

Forskolin, an inducer of intracellular cAMP levels, increases the abundance of IGFBP-5 mRNA and the activity of a reporter gene driven by the human IGFBP-5 promoter in GM-10 human fetal dermal fibroblasts *(214,215)*. The authors proposed that cAMP mediates these effects by activating the AP-2 transcription factor. Consistent with this hypothesis, expression of AP-2 in GM-10 and in the HepG2 liver cells activates the human IGFBP-5 promoter by binding to an AP-2 element located between nt –47 and nt -39 *(215)*. In ob cells, a mouse promoter fragment (nt –51 to nt +120) retaining only this AP-2 element is activated by PGE_2, which acts partly by increasing cAMP *(198)*. However, deletion of this element in the context of a promoter fragment extending to nt –1004 only attenuates the stimulation mediated by PGE_2, suggesting the involvement of other AP-2 sites or alternative cAMP response elements *(198)*. In the myogenic cell line C21, the proximal region of the mouse promoter (nt –156 to nt +120) contributes more than 70% of the maximum activity, and mediates a portion of the stimulation of gene transcription occurring during differentiation *(204)*. In ob cells, the inhibition of the activity of the nt –70 to +120 mouse promoter fragment by cortisol is lost when a E box site between nt –56 and nt –51 is mutated *(201)*. This suggests that c-Myb or E box binding proteins are important in mediating the inhibitory effects of cortisol on the transcription of the IGFBP-5 gene.

IGFBP-6

General

IGFBP-6 is present in adult human *(216)* and rat *(193)* serum and human cerebrospinal fluid *(217)*, porcine follicular fluid *(218)*, and has been isolated from conditioned media of either SV-40-transformed *(219)* or nontransformed *(220)* human lung fibroblasts, as well as from U-2 human osteosarcoma cells *(221)*. IGFBP-6 has a marked preferential affinity for IGF-II compared to IGF-I *(216,217,219,220,222)*. In human serum, levels of IGFBP-6 are low, and studies on its regulation are limited.

On nonreducing SDS-PAGE, human IGFBP-6 migrates with an apparent molecular mass of 28–34 kDa.

Regulation of Expression In Vivo

Northern blot analysis indicated that the single 1.3-kb rat IGFBP-6 mRNA is found in all tissues, including testis, intestine, adrenal, kidney, stomach, spleen, heart, lung, brain, and liver, with the highest expression in the lung, indicating that IGFBP-6 is ubiquitous *(218)*. IGFBP-6 mRNA is detected in rat embryo as early as e14 in the liver cells, nasal epithelia *(11)* as well as in the peripheral cells of the perichondrial layers of most cartilage *(11)*.

In human, follicle-stimulating hormone and estrogens decrease the ovarian expression of IGFBP-6 *(223)*.

Regulation of Expression In Vitro

Gabbitas et al. *(64,224,225)* have used cultures of osteoblast-enriched cells from 22-day-old fetal rat calvariae (Ob cells) to study the in vitro regulation of IGFBP-6 expression. Cortisol and retinoic acid enhances the transcription of rat IGFBP-6 in fetal rat calvarial cells *(224,225)*, whereas TGF-b as well as agents that increase the intracellular levels of cAMP decreased IGFBP-6 transcription *(64,226)*. FGF-2, PDGF-BB, IGF-I, and IGF-II have no effects *(64)*. By far, retinoic acid appears to be the main regulator of IGFBP-6 expression *(225–227)*.

cDNA and Protein

cDNA clones have been isolated from pregnant mare serum gonadotrophins (PMSG)-stimulated rat ovary and human placenta libraries *(218)*, as well as from a human osteosarcoma library *(169)*. In the rat, as in the pig, the N-terminal residue of the mature protein is alanine *(218)*, although purified human IGFBP-6 isolated from serum *(216)*, cerebrospinal fluid *(217)*, and lung fibroblast cell lines *(219,220)* starts at an arginine residue. However, the deduced primary protein sequence for human IGFBP-6 predicted from its cDNA sequence isolated by Shimasaki et al. *(218)* contains an alanine sequence three amino acids upstream of the arginine residue.

Assuming that human IGFBP-6 starts at alanine at the N-terminal, the mature protein contains 216 amino acids, with a deduced molecular mass of 22.8 kDa *(218)*. Empirically, on nonreducing SDS-PAGE, human IGFBP-6 migrated with an apparent molecular mass of 28–34 kDa, due in part to the *O*-glycosylation *(228)*. The potential *N*-glycosylation sites present near the C-terminal of human IGFBP-6 are not utilized *(169,218,228)*, whereas similar sites are not present in rat IGFBP-6. A distinctive feature of rat and human IGFBP-6 is that they contain 14 and 16 cysteines respectively, instead of the usual 18 cysteine residues present in the other five binding proteins. This results in the absence of the invariant Gly–Cys–Gly–Cys–Cys motif present in the N-terminal region of the other IGFBPs *(218)* (Figs. 1 and 2).

Gene Structure and Promoter Regulation

The complete nucleotide sequence for human IGFBP-6 has not been determined, although it is known that the single-copy gene is localized to chromosome 12 *(218)*.

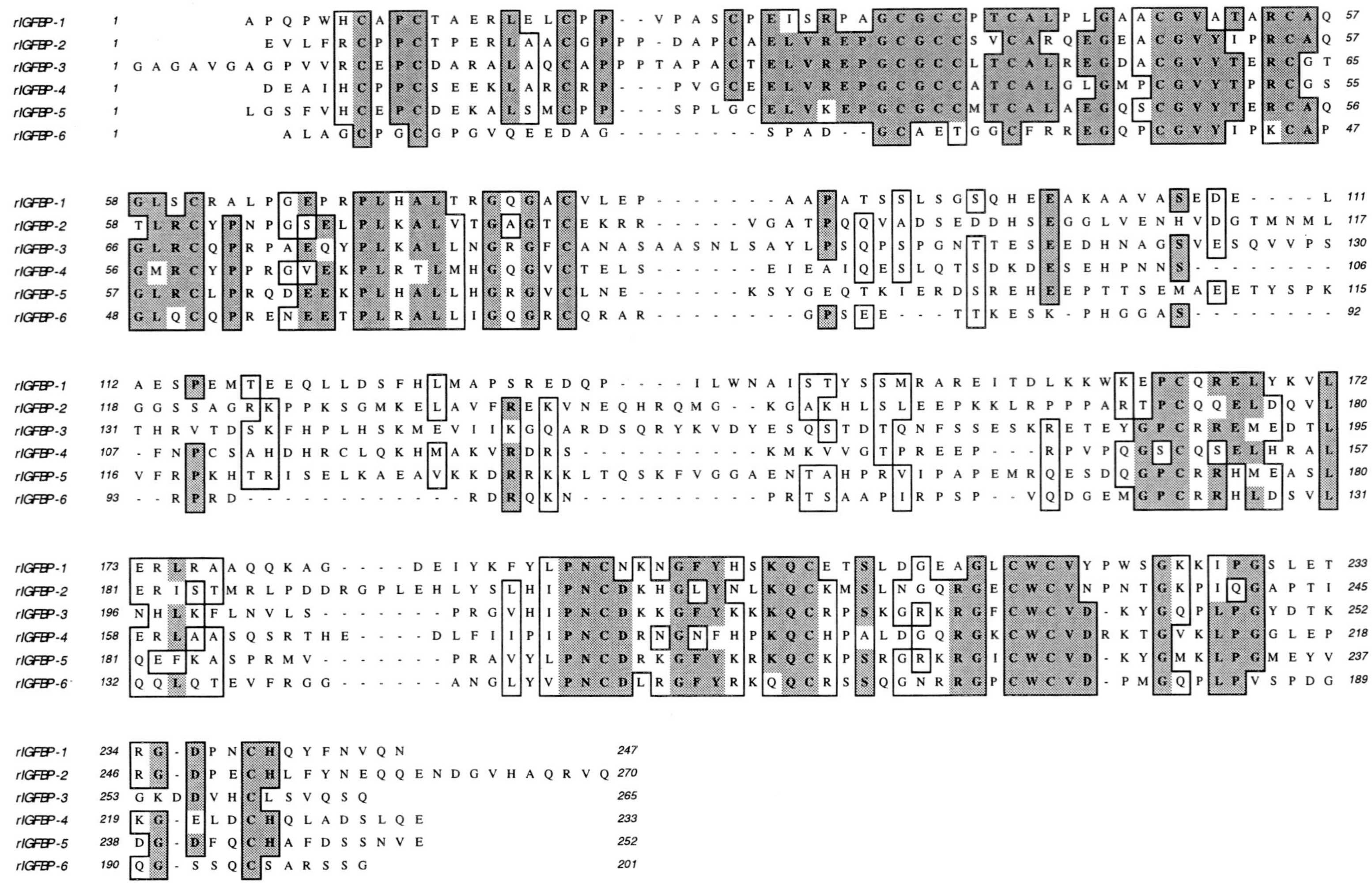

Fig. 2. Amino acid sequences of rat IGFBP-1 to -6 proteins deduced from nucleotide sequences (*104,137,163,174,193,218*) (IGFBP-1 from GenBank Accession M89791) were aligned using a ClustalW algorithm analysis (MacVector 6.0, Oxford Molecular Group, Beaverton, OR, USA). Conserved residues are indicated by *dark shadings,* whereas *light shadings* indicate residues with similar classification groups.

Recently, Strong et al. *(229)* have cloned and characterized the 5′-flanking sequence of the human IGFBP-6 gene. The human IGFBP-6 promoter contains an imperfect TATA box, a CAAT box, multiple CACC box elements, an AP-2 site, as well as at least three complex retinoic acid response elements (RAREs) distributed over a 1.5-kb length of the 5′-flanking region *(229)*.

The full sequence of the rat IGFBP-6 gene, however, has been cloned *(230)*. It spans 5.1 kb and contains four exons interrupted by three introns approximately 2.4, 0.2, and 1.2 kb in length, respectively. Primer extension analysis and ribonuclease protection assays using RNA from lung tissues demonstrated two transcriptional start sites located 85 (adenine) and 82 (thymidine) nucleotides upstream of the ATG translational initiation codon. The rat IGFBP-6 gene lacks TATA or CAAT consensus motifs *(230)* upstream of the transcription start sites. The upstream region is not GC rich and contains only one putative GC box in the proximal 700 bp upstream of the transcription start sites *(230)*. Putative RAREs as well as estrogen responsive element (EREs) were present at ~400 nucleotides upstream of the cap sites, although their functional significance has not been determined.

ACID-LABILE SUBUNIT

General

After birth, most of IGF-I and IGF-II circulate as a ternary complex formed by association of the binary IGFBP-3:IGF complex with ALS (reviewed in *[7,131,134]*). Unlike IGFBPs, ALS is found almost exclusively in serum *(231)*. In humans, ALS is undetectable in fetal serum at 27 wk of gestation, but is present at the end of fetal life, and increases fivefold from birth to puberty with little change during adulthood *(231,232)*. In adult humans and rat, ALS circulates at 2–3 times the molar concentration of IGFBP-3 and IGFs, driving 80–85% of IGF into ternary complexes *(231,233)*. Recruitment of binary IGFBP-3:IGF complexes into ternary complexes extends the half-life of IGFs from 30–90 min to 11–12 h *(234,235)*. On SDS-PAGE, ALS migrates at 84–86 kDa, and after *N*-glycanase treatment, at 63–66 kDa *(233,236)*.

Regulation of Expression In Vivo

In the rat embryo, ALS is expressed weakly at e20 in liver, kidney, and forming cartilaginous and membranous bone *(135)*. After birth, ALS expression localized only to hepatocytes and to the epithelial cells of the kidney proximal tubule *(135)*. However, hepatocytes are the principal site of synthesis as ALS mRNA is detected only in liver by Northern analysis *(237)*. In rat, hepatic expression increases rapidly during the first few weeks of postnatal life, and growth hormone is the most potent hormonal inducer of ALS in the circulation *(135,187,233,238)*. The decrease in the levels of ALS mRNA in liver of hypophysectomized rats results from a reduction of ALS gene transcription *(187)*. In adult human, circulating levels of ALS are decreased in patients suffering from GH deficiency or defect in the GH receptor, and by chronic IGF-I infusion, probably via inhibition of GH secretion *(146,147,231)*.

Regulation of Expression In Vitro

The liver cell lines studied thus far, including the rat liver cell lines H4-II-E, H35, HTC, 5123TC, FAO, and BRL-3A and the human Hep G2 liver cell line, do not express ALS mRNA or secrete ALS (Ooi and Boisclair, *unpublished results* and *[239]*). This is thought to reflect the silencing of the ALS gene in culture by modifications of chromatin structure such as methylation of CpG islands *(240)*. In contrast, primary rat hepatocytes retain basal and GH-regulated expression of the ALS gene *(239,241)*.

cDNAs and Protein

The human cDNA has an open reading frame of 605 amino acids encoding a signal peptide of 27 amino acids, and a mature protein of 578 amino acids *(242)*. The mature protein is organized into 18–20 leucine-rich domains of 24 amino acids thought to be involved in protein–protein interaction. The mature protein also has 13 cysteine residues and 7 potential *N*-glycosylation sites *(242)*. The size of the corresponding mRNA is ~ 2.0 kb by Northern analysis *(242)*. The sequences ofcDNAs from mouse, rat, baboon, and sheep suggest almost complete conservation of these features (*[243-245]* and Y. R. Boisclair, *unpublished* results) (Fig. 3).

Gene Structure and Promoter Regulation

The structure of the human ALS gene has not yet been reported. In the mouse, the ALS gene spans ~ 3.3 kb on chromosome 17 and is composed of 2 exons *(243)*. Exon 2 encodes the entire precursor protein with the exception of the first five amino acids of the signal peptide. Transcription initiation is dispersed over a 140-bp region immediately upstream of the ATG in a region that is devoid of TATA or initiator elements *(243)*. The region between nt –2001 and nt –49 (relative to the ATG) directs basal and GH-regulated expression of a reporter gene in H4-II-E cells and in primary hepatocytes, indicating promoter activity *(187)*. The organization of the rat and sheep ALS genes is very similar (*[246]* and Y. R. Boisclair, *unpublished results*).

In H4-II-E rat hepatoma liver cells and in rat primary hepatocytes, a single γ-interferon activated sequence (GAS) located between nt –633 to nt –625 is required for GH stimulation of the mouse promoter activity *(241,247)*. This stimulation is conferred by the binding of signal transducer and activator of transcription 5a (STAT5a) and STAT5b to this single GAS element upon GH treatment *(241,247)*.

CONCLUSION

In summary, molecular studies have contributed to our understanding of the functional and biochemical properties of IGFBP-1 to -6 and ALS. Isolation of cDNAs and genes encoding IGFBP-1 to -6 has provided the definitive proof that they belong to a family of related genes. Expression of each member, however, is subject to precise spatial, developmental, and hormonal regulation, suggesting that each IGFBP has unique roles throughout development. A number of mechanistic studies have also been performed such as determination of the structural basis for properties of specific IGFBPs and identification of the molecular mechanisms by which various hormones regulate transcription. The interest for these two research areas (structure–function relationship and promoter function) will likely remain high in the next few years.

An unresolved issue that needs to be addressed is the functional significance of IGFBP-1 to -6 and of ALS. Although IGFBPs have been shown to alter various cellular functions in vitro, elucidation of their roles in vivo has been difficult. Investigations of this issue should be advanced by the creation of transgenic mouse models in which expression of various IGFBPs or ALS is altered (overexpression and knockout models).

Fig. 3. Alignment of primate and rodent ALS amino acid sequences by ClustalW algorithm analysis (MacVector 6.0, Oxford Molecular Group, Beaverton, OR, USA). Conserved residues are indicated by *dark shadings*, whereas *light shadings* indicate residues with similar classification groups. Sequences are deduced from nucleotide sequences (*242–245*).

These models will yield invaluable insights on the functions of these proteins in vivo, and will directly test the hypothesis that they each play distinctive roles.

REFERENCES

1. Baxter RC. Martin JL. Structure of the Mr 140,000 growth hormone-dependent insulin-like growth factor binding protein complex: determination by reconstitution and affinity-labeling. Proc Natl Acad Sci USA 1989; 86:6898–6902.
2. Bach LA, Hsieh S, Sakano K, Fujiwara H, Perdue JF, Rechler MM. Binding of mutants of human insulin-like growth factor II to insulin-like growth factor binding proteins 1–6. J Biol Chem 1993; 268:9246–9254.
3. Bach LA, Hsieh S, Sakano K, Fujiwara H, Perdue JF, Rechler MM. Towards identification of a binding site on insulin-like growth factor-II for IGF-binding proteins. In: Raizada MK, Leroith D (eds.). Molecular Biology and Physiology of Insulin and Insulin-Like Growth Factors 2. Plenum Press, New York, 1993, pp. 55–61.
4. Povoa G, Enberg G, Jornvall H, Hall K. Isolation and characterization of a somatomedin-binding protein from mid-term human amniotic fluid. Eur J Biochem 1984; 144:199–2045.
5. Coverley JA, Baxter RC. Phosphorylation of insulin-like growth factor binding proteins. Mol Cell Endocrinol 1997; 128:1–5.
6. Frost RA, Tseng L. Insulin like growth factor binding protein-1 is phosphorylated by cultured human endometrial stromal cells and multiple protein kinases in vitro. J Biol Chem 1991; 266:18082–18088.
7. Rechler MM. Insulin-like growth factor binding proteins. Vitam Horm 1993; 47:1–114.
8. Jones JI, Busby WH Jr, Wright G, Smith CE, Kimack NM, Clemmons DR. Identification of the sites of phosphorylation in insulin-like growth factor binding protein-1. Regulation of its affinity by phosphorylation of serine 101. J Biol Chem 1993; 268:1125–1131.
9. Ooi GT, Orlowski CC, Brown AL, Becker RE, Unterman TG, Rechler MM. Different tissue distribution and hormonal regulation of messenger RNAs encoding rat insulin-like growth factor binding proteins-1 and -2. Mol Endocrinol 1990; 4:321–328.
10. Ooi GT, Tseng LY-H, Rechler MM. Transcriptional regulation of the rat IGFBP-1 and IGFBP-2 genes. Growth Reg 1993; 3:14–17.
11. Cerro JA, Grewal A, Wood TL, Pintar JE. Tissue-specific expression of the insulin-like growth factor binding protein (IGFBP) mRNAs in mouse and rat development. Regul Pept 1993;48:189–198.
12. Liu F, Powell DR, Styne DM, Hintz RL. Insulin-like growth factors (IGFs) and IGF-binding proteins in the developing rhesus monkey. J Clin Endocrinol Metab 1991; 72:905–911.
13. Murphy LJ, Seneviratne C, Ballejo G, Croze F, Kennedy TG. Identification and characterization of a rat decidual insulin-like growth factor-binding protein complementary DNA. Mol Endocrinol 1990; 4:329–336.
14. Brinkman A, Groffen C, Kortleve DJ, van Kessel AG, Drop SLS. Isolation and characterization of a cDNA encoding the low molecular weight insulin-like growth factor binding protein (IGFBP-1). EMBO J 1988; 7:2417–2423.
15. Julkunen M, Koistinen R, Aalto-Setala K, Seppälä M, Jänne OA, Kontula K. Primary structure of human insulin-like growth factor-binding protein/placental protein 12 and tissue-specific expression of its mRNA. FEBS Lett 1988; 236:295–302.
16. Rutanen EM, Koistinen R, Wahlström T, Bohn H, Ranta T, Seppälä M. Synthesis of placental protein 12 by human decidua. Endocrinology 1985; 116:1304–1309.
17. Julkunen M, Koistinen R, Suikkari A-M, Seppälä M, Jänne OA. Identification by hybridization histochemistry of human endometrial cells expressing RNAs encoding a uterine beta-lactoglobulin homologue and insulin-like growth factor-binding protein-1. Mol Endocrinol 1990; 4:700–707.
18. Drop SLS, Kortleve DJ, Guyda HJ, Posner BI. Immunoassay of a somatomedin-binding protein from human amniotic fluid: levels in fetal, neonatal, and adult sera. J Clin Endocrinol Metab 1984; 59:908–915.
19. Hall K, Lundin G, Povoa G. Serum levels of the low molecular weight form of insulin-like growth factor binding protein in healthy subjects and patients with growth hormone deficiency, acromegaly and anorexia nervosa. Acta Endocrinol 1988; 118:321–326.
20. Hill DJ, Clemmons DR, Wilson S, Han VKM, Strain AJ, Milner RDG. Immunological distribution of one form of insulin-like growth factor (IGF)-binding protein and IGF peptides in human fetal tissues. J Mol Endocrinol 1989; 2:31–38.

21. Hall K, Hansson U, Lundin G, Luthman M, Persson B, Povoa G, Stangenberg M, Ofverholm U. Serum levels of somatomedins and somatomedin-binding protein in pregnant women with type I or gestational diabetes and their infants. J Clin Endocrinol Metab 1986; 63:1300–1306.
22. Rutanen EM, Bohn H, Seppälä M. Radioimmunoassay of placental protein 12: levels of amniotic fluid, cord blood, and serum of healthy adults, pregnant women, and patients with trophoblastic disease. Am J Obstet Gynecol 1982; 144:460–463.
23. Baxter RC, Martin JL, Wood MJ. Two immunoreactive binding proteins for insulin-like growth factors in human amniotic fluid: relationship to fetal maturity. J Clin Endocrinol Metab 1987; 65:423–431.
24. Drop SLS, Kortleve DJ, Guyda HJ. Isolation of a somatomedin-binding protein from preterm amniotic fluid. Development of a radioimmunoassay. J Clin Endocrinol Metab 1984; 59:899–907.
25. Povoa G, Roovete A, Hall K. Cross-reaction of serum somatomedin-binding protein in a radioimmunoassay developed for somatomedin-binding protein isolated from human amniotic fluid. Acta Endocrinol 1984; 107:563–570.
26. Suikkari AM, Koivisto VA, Rutanen EM, Yki-Jarvinen H, Karonen SL, Seppälä M. Insulin regulates the serum levels of low molecular weight insulin-like growth factor-binding protein. J Clin Endocrinol Metab 1988; 66:266–272.
27. Suikkari A-M, Koivisto VA, Koistinen R, Seppälä M, Yki-Jarvinen H. Dose-response characteristics for suppression of low molecular weight plasma insulin-like growth factor-binding protein by insulin. J Clin Endocrinol Metab 1989; 68:135–140.
28. Snyder DK, Clemmons DR. Insulin dependent regulation of IGFBP-1. J Clin Endocrinol Metab 1990; 71:1632–1636.
29. Busby WH, Snyder DK, Clemmons DR. Radioimmunoassay of a 26,000-dalton plasma insulin-like growth factor-binding protein: control by nutritional variables. J Clin Endocrinol Metab 1988; 67:1225–1230.
30. Brismar K, Gutniak M, Povoa G, Werner S, Hall K. Insulin regulates the 35 kDa IGF binding protein in patients with diabetes mellitus. J Endocrinol Invest 1988; 11:599–602.
31. Suikkari A-M, Sane T, Seppälä M, Hannele Y-J, Karonen S-L, Koivisto VA. Prolonged exercise increases serum insulin-like growth factor-binding protein concentrations. J Clin Endocrinol Metab 1989; 68:141–144.
32. Cotterill AM, Cowell CT, Baxter RC, McNeil D, Silinik M. Regulation of the growth hormone-independent growth factor-binding protein in children. J Clin Endocrinol Metab 1988; 67:882–887.
33. Baxter RC, Cowell CT. Diurnal rhythm of growth hormone-independent binding protein for insulin-like growth factors in human plasma. J Clin Endocrinol Metab 1987; 65:432–440.
34. Seneviratne C, Jiangming L, Murphy LJ. Transcriptional regulation of rat insulin-like growth factor-binding protein-1 expression by growth hormone. Mol Endocrinol 1990; 4:1199–1204.
35. Yang YW-H, Brown AL, Orlowski CC, Graham DE, Tseng LY-H, Romanus JA, Rechler MM. Identification of rat cell lines that preferentially express insulin-like growth factor binding proteins rIGFBP-1, 2, or 3. Mol Endocrinol 1990; 4:29–38.
36. Orlowski CC, Ooi GT, Brown DR, Yang YW-H, Tseng LY-H, Rechler MM. Insulin rapidly inhibits insulin-like growth factor binding protein-1 gene expression in H4-II-E rat hepatoma cells. Mol Endocrinol 1991; 5:1180–1187.
37. Unterman TG, Oehler DT, Gotway MB, Morris PW. Production of the rat type 1 insulin-like growth factor-binding protein by well differentiated H4EIIC3 hepatoma cells: identification, purification, and N-terminal amino acid analysis. Endocrinology 1990; 127:789–797.
38. Babajko S, Tronche F, Groyer A. Liver-specific expression of human insulin-like growth factor-binding protein 1: functional role of transcription factor HNF1 in vivo. Proc Natl Acad Sci USA 1993; 90:272–276.
39. Suwanichkul A, Cubbage ML, Powell DR. The promoter of the human gene for insulin-like growth factor-binding protein-1: basal promoter activity in Hep G2 cells depends on liver factor B1. J Biol Chem 1990; 265:21185–21193.
40. Yee D, Favoni RE, Lupu R, Cullen KJ, Lebovic GS, Huff KK, Lee PDK, Lee YL, Powell DR, Dickson RB, Rosen N, Lippman ME. The insulin-like growth factor-binding protein BP-25 is expressed by human breast cancer cells. Biochem Biophys Res Commun 1989; 158:38–44.
41. Pekonen F, Nyman T, Rutanen E-M. Human endometrial adenocarcinoma cell lines HEC 1B and KLE secrete insulin-like growth factor-binding protein-1 and contain IGF-I receptors. Mol Cell Endocrinol 1991; 75:81–87.
42. Orlowski CC, Ooi GT, Rechler MM. Dexamethasone stimulates transcription of the insulin-like growth factor-binding protein-1 (IGFBP-1) gene in H4-II-E rat hepatoma cells. Mol Endocrinol 1990; 4:1592–1599.

43. Ooi GT, Brown DR, Suh D-S, Tseng LY-H, Rechler MM. Cycloheximide stabilizes insulin-like growth factor-binding protein-1 (IGFBP-1) mRNA and inhibits IGFBP-1 transcription in H4-II-E rat hepatoma cells. J Biol Chem 1993; 268:16664–16672.
44. Unterman TG, Oehler DT, Murphy LJ, Lacson RG. Multihormonal regulation of insulin-like growth factor-binding protein-1 in rat H4IIE hepatoma cells: the dominant role of insulin. Endocrinology 1991; 128:2693–2701.
45. Unterman TG, Lacson RG, Jentel JJ, Oehler DT. Divergent effects of phorbol esters and insulin on insulin-like growth factor-binding protein-1 (IGFBP-1) production and mRNA in rat H4IIE hepatoma cells. Biochem Biophys Res Commun 1992; 182:262–268.
46. Vilafuerte BC, Goldstein S, Murphy LJ, Phillips LS. Nutrition and somatomedin. XXV. Regulation of insulinlike growth factor binding protein 1 in primary cultures of normal rat hepatocytes. Diabetes 1991; 40:837–841.
47. Lewitt MS, Baxter RC. Regulation of growth hormone-independent insulin-like growth factor-binding protein (BP-28) in cultured human fetal liver explants. J Clin Endocrinol Metab 1989; 69:246–252.
48. Lewitt MS, Baxter RC. Inhibitors of glucose uptake stimulate the production of insulin-like growth factor-binding protein (IGFBP-1) by human fetal liver. Endocrinology 1990; 126:1527–1533.
49. Lewitt MS, Baxter RC. Cytochalasin B insulin-like growth factor-binding protein-1 production by Hep G2 cells. Mol Cell Endocrinol 1991; 77:149–157.
50. Conover CA, Lee PDK. Insulin regulation of insulin-like growth factor-binding protein production in cultured HepG2 cells. J Clin Endocrinol Metab 1990; 70:1062–1067.
51. Powell DR, Suwanichkul A, Cubbage ML, DePaolis LA, Snuggs MB, Lee PDK. Insulin inhibits transcription of the human gene for insulin-like growth factor-binding protein-1. J Biol Chem 1991; 266:18868–18876.
52. Gronowski AM, Rotwein P. Rapid changes in gene expression after in vivo growth hormone treatment. Endocrinology 1995; 136:4741–4748.
53. Clemmons DR, Thrailkill KM, Handwerger S, Busby WH, Jr. Three distinct forms of insulin-like growth factor-binding proteins are released by decidual cells in culture. Endocrinology 1990; 127:643–650.
54. Bell SC, Jackson JA, Ashmore J, Zhu HH, Tseng L. Regulation of insulin-like growth factor-binding protein-1 synthesis and secretion by progestin and relaxin in long term cultures of human endometrial stromal cells. J Clin Endocrinol Metab 1991; 72:1014–1024.
55. Tseng L, Gao JG, Chen R, Zhu HH, Mazella J, Powell DR. Effect of progestin, antiprogestin and relaxin on endometrial stromal cells. Biol Reprod 1994; 13:829–837.
56. Brewer MT, Stetler GL, Squires CH, Thompson RC, Busby WH, Clemmons DR. Cloning, characterization, and expression of a human insulin-like growth factor-binding protein. Biochem Biophys Res Commun 1988; 152:1289–1297.
57. Brewer MT, Stetler GL, Squires CH, Thompson RC, Busby WH, Clemmons DR. Cloning, characterization, and expression of a human insulin-like growth factor-binding protein (erratum). Biochem Biophys Res Commun 1988; 155:1485.
58. Lee Y-L, Hintz RL, James PM, Lee PDK, Shively JE, Powell DR. Insulin-like growth factor (IGF) binding protein complementary deoxyribonucleic acid from human HEP G2 hepatoma cells: predicted protein sequence suggests an IGF binding domain different from those of the IGF-I and IGF-II receptors. Mol Endocrinol 1988; 2:404–411.
59. Cubbage ML, Suwanichkul A, Powell DR. Structure of the human chromosomal gene for the 25 kilodalton insulin-like growth factor-binding protein. Mol Endocrinol 1989; 3:846–851.
60. Shimasaki S, Ling N. Identification and molecular characterization of insulin-like growth factor-binding proteins (IGFBP-1, -2, -3, -4, -5, and -6). Prog Growth Factor Res 1991; 3:243–266.
61. Brinkman A, Kortleve DJ, Schuller AGP, Zwarthoff EC, Drop SLS. Site-directed mutagenesis of the N-terminal region of IGF binding protein 1; analysis of IGF binding capability. FEBS Lett 1991; 291:264–268.
62. Ruoslahti E, Pierschbacher MD. New perspectives in cell adhesion: RGD and integrins. Science 1987; 238:491–497.
63. Ruoslahti E, Yamaguchi Y. Proteoglycans as modulators of growth factor activity. Cell 1991; 64:867–869.
64. Rogers S, Wells R, Rechsteiner M. Amino acid sequences common to rapidly degraded proteins: The PEST hypothesis. Science 1986; 234:364–368.

65. Jones JI, D'Ercole AJ, Camacho-Hubner C, Clemmons DR. Phosphorylation of insulin-like growth factor (IGF)-binding protein 1 in cell culture and in vivo: effects on affinity for IGF-1. Proc Natl Acad Sci USA 1991; 88:7481–7485.
66. Lee J, Greenbaum L, Haber BA, Nagle D, Lee V, Miles V, Mohn KL, Bucan M, Taub R. Structure and localization of the IGFBP-1 gene and its expression during liver regeneration. Hepatology 1994; 19:656–665.
67. Mohn KL, Melby AE, Tewari DS, Laz TM, Taub R. The gene encoding rat insulinlike growth factor-binding protein 1 is rapidly and highly induced in regenerating liver. Mol Cell Biol 1991; 11:1393–1401.
68. Schuller AGP, Groffen C, Van Neck JW, Zwarthoff EC, Drop SLS. cDNA cloning and mRNA-expression of the six mouse insulin-like growth factor binding proteins. Mol Cell Endocrinol 1994; 104:57–66.
69. Ekstrand J, Ehrenborg E, Stern I, Stellan B, Zech L, Luthman H. The gene for insulin-like growth factor-binding protein-1 is localized to human chromosomal region 7p14-p12. Genomics 1990; 6:413–418.
70. Allander SV, Bajalica S, Larsson C, Luthman H, Powell DR, Stern I, Weber G, Zazzi H, Ehrenborg E. Structure and chromosomal localization of human insulin-like growth factor-binding protein genes. Growth Regul 1993; 3:3–5.
71. Brinkman A, Groffen CAH, Kortleve DJ, Drop SLS. Organization of the gene encoding the insulin-like growth factor-binding protein IBP-1. Biochem Biophys Res Commun 1988; 157:898–907.
72. Mendel DB, Crabtree GR. HNF-1, a member of a novel class of dimerizing homeodomain proteins. J Biol Chem 1991; 266:677–680.
73. Mendel DB, Hansen LP, Graves MK, Conley PB, Crabtree GR. HNF-1a and HNF-1b (vHNF-1) share dimerization and homeo domains, but not activation domains, and form heterodimers in vitro. Genes Dev 1991; 5:1042–1056.
74. Suh D-S, Ooi GT, Rechler MM. Identification of cis-elements mediating the stimulation of rat insulin-like growth factor binding protein-1 promoter activity by dexamethasone, cyclic AMP and phorbol esters, and the inhibition by insulin. Mol Endocrinol 1994; 8:794–805.
75. Unterman TG, Lacson RG, McGary E, Whalen C, Purple C, Goswami RG. Cloning of the rat insulin-like growth factor binding protein-1 gene and analysis of its 5′ promoter region. Biochem Biophys Res Commun 1992; 185:993–999.
76. Montminy MR, Gonzalez GA, Yamamoto R. Characteristics of the cAMP response unit. Rec Prog Horm Res 1990; 46:219–230.
77. Suwanichkul A, Allander SV, Morris SL, Powell DR. Glucocorticoids and insulin regulate expression of the human gene for insulin-like growth factor-binding protein-1 through proximal promoter elements. J Biol Chem 1994; 269:30835–30841.
78. Suwanickul A, Morris SL, Powell DR. Identification of an insulin-responsive element in the promoter of the human gene for insulin-like growth factor binding protein-1. J Biol Chem 1993; 268:17063–17068.
79. O'Brien RM, Granner DK. PEPCK gene as model of inhibitory effects of insulin on gene transcription. Diabetes Care 1990; 13:327–339.
80. Lewitt MS, Denyer GS, Cooney GJ, Baxter RC. Insulin-like growth factor-binding protein-1 modulates blood glucose levels. Endocrinology 1991; 129:2254–2256.
81. Suwanichkul A, DePaolis LA, Lee PDK, Powell DR. Identification of a promoter element which participates in cAMP-stimulated expression of human insulin-like growth factor-binding protein-1. J Biol Chem 1993; 268:9730–9736.
82. Goswami R, Lacson R, Yang E, Sam R, Unterman T. Functional analysis of glucocorticoid and insulin response sequences in the rat insulin-like growth factor-binding protein-1 promoter. Endocrinology 1994; 134:736–743.
83. Roesler WJ, Vandenbark GR, Hanson RW. Cyclic AMP and the induction of eukaryotic gene transcription. J Biol Chem 1988; 263:9063–9066.
84. Luo J, Reid RE, Murphy LJ. Dexamethasone increases hepatic insulin-like growth factor-binding protein-1 (IGFBP-1) mRNA and serum IGFBP-1 concentrations in the rat. Endocrinology 1990; 127:1456–1462.
85. Powell D, Lee PDK, DePaolis LA, Morris SL, Suwanichkul A. Dexamethasone stimulates expression of insulin-like growth factor-binding protein-1 in HEP G2 human hepatoma cells. Growth Regul 1993; 3:11–13.
86. Beato M. Gene regulation by steroid hormones. Cell 1989; 56:335–344.

87. Yamamoto KR. Steroid receptor regulated transcription of specific genes and gene networks. Annu Rev Genet 1985; 19:209–252.
88. Yamamoto KR, Alberts BM. Steroid receptors: elements for modulation of eukaryotic transcription. Annu Rev Biochem 1976; 45:721–746.
89. O'Brien RM, Lucas PC, Forest CD, Magnuson MA, Granner DK. Identification of a sequence in the PEPCK gene that mediates a negative effect of insulin on transcription. Science 1990; 249:533–537.
90. O'Brien RM, Bonovich MT, Forest CD, Granner DK. Signal transduction convergence: phorbol esters and insulin inhibit phosphoenolpyruvate carboxykinase gene transcription through the same 10-base-pair sequence. Proc Natl Acad Sci USA 1991; 88:6580–6584.
91. Allander SV, Durham SK, Scheimann AO, Wasserman RM, Suwanichkul A, Powell DR. Hepatic nuclear factor 3 and high mobility group I/Y proteins bind the insulin response element of the insulin-like growth factor-binding protein-1 promoter. Endocrinology 1997; 138:4291–4300.
92. Powell DR, Allander SV, Scheimann AO, Wasserman RM, Durham SK, Suwanichkul A. Multiple proteins bind the insulin response element in the human IGFBP-1 promoter. Prog Growth Factor Res 1995; 6:93–101.
93. Bang P, Westgren M, Schwander J, Blum WF, Rosenfeld RG, Stangenberg M. Ontogeny of insulin-like growth factor-binding protein-1, -2, and -3: quantitative measurements by radioimmunoassay in human fetal serum. Pediatr Res 1994; 36:528–536.
94. Lassarre C, Hardouin S, Daffos F, Forestier F, Frankenne F, Binoux M. Serum insulin-like growth factors and insulin-like growth factor binding proteins in the human fetus. Relationships with growth in normal subjects and in subjects with intrauterine growth retardation. Pediatr Res 1991; 29:219–225.
95. Blum WF, Ranke MB, Kietzmann K, Gauggel E, Zeisel HJ, Bierich JR. A specific radioimmunoassay for the growth hormone (GH)-dependent somatomedin-binding protein: its use for diagnosis of GH deficiency. J Clin Endocrinol Metab 1990; 70:1292–1298.
96. Clemmons DR, Snyder DK, Busby WH, Jr. Variables controlling the secretion of insulin-like growth factor binding protein-2 in normal human subjects. J Clin Endocrinol Metab 1991; 73:727–733.
97. Donovan SM, Oh Y, Pham H, Rosenfeld RG. Ontogeny of serum insulin-like growth factor binding proteins in the rat. Endocrinology 1989; 125:2621–2627.
98. Babajko S, Hardouin S, Segovia B, Groyer A, Binoux M. Expression of insulin-like growth factor binding protein-1 and -2 genes through the perinatal period in the rat. Endocrinology 1993; 132:2586–2592.
99. Yang YW-H, Rechler MM. Identification and characterization of insulin-like growth factor-binding proteins. In: dePablo F, Scanes CG, Weintraub BD (eds). Techniques in Endocrine Research. Academic Press, New York, 1993, pp. 181–204.
100. Wood TL, Streck RD, Pintar JE. Expression of the IGFBP-2 gene in post-implantation rat embryos. Development 1992; 114:59–66.
101. Wood TL, Brown AL, Rechler MM, Pintar JE. The expression pattern of an insulin-like growth factor binding protein gene is distinct from IGF-II in the midgestational rat embryo. Mol Endocrinol 1990; 4:1257–1263.
102. Schuller AGP, Zwarthoff EC, Drop SLS. Gene expression of the six insulin-like growth factor binding proteins in the mouse conceptus during mid- and late gestation. Endocrinology 1993; 132:2544–2550.
103. Tseng LY-H, Brown AL, Yang YW-H, Romanus JA, Orlowski CC, Taylor T, Rechler MM. The fetal rat binding protein for insulin-like growth factors is expressed in the choroid plexus and cerebrospinal fluid of adult rats. Mol Endocrinol 1989; 3:1559–1568.
104. Brown AL, Chiariotti L, Orlowski CC, Mehlman T, Burgess WH, Ackerman EJ, Bruni CB, Rechler MM. Nucleotide sequence and expression of a cDNA clone encoding a fetal rat binding protein for insulin-like growth factors. J Biol Chem 1989; 264:5148–5154.
105. Margot JB, Binkert C, Mary J-L, Landwehr J, Heinrich G, Schwander J. A low molecular weight insulin-like growth factor binding protein from rat: cDNA cloning and tissue distribution of its messenger RNA. Mol Endocrinol 1989; 3:1053–1060.
106. Tseng LY-H, Ooi GT, Brown AL, Straus DS, Rechler MM. Transcription of the insulin-like growth factor binding protein-2 gene is increased in neonatal and fasted adult rat liver. Mol Endocrinol 1992; 6:1195–1201.
107. Delhanty PJD, Han VKM. The expression of insulin-like growth factor (IGF)-binding protein-2 and IGF-II genes in the tissue of the developing ovine fetus. Endocrinology 1993; 132:41–52.
108. Cohick WS, McGuire MA, Clemmons DR, Bauman DE. Regulation of insulin-like growth factor-binding proteins in serum and lymph of lactating cows by somatotropin. Endocrinology 1992; 130:1508–1514.

109. McGuire MA, Dwyer DA, Harrell RJ, Bauman DE. Insulin regulates circulating insulin-like growth factors and some of their binding proteins in lactating cows. Am J Physiol Endocrinol Metab 1995; 269:E723–E730.
110. Young SCJ, Underwood LE, Celniker A, Clemmons DR. Effects of recombinant insulin-like growth factor-I (IGF-I) and growth hormone on serum IGF-binding proteins in calorically restricted adults. J Clin Endocrinol Metab 1992; 75:603–608.
111. Young SCJ, Clemmons DR. Changes in insulin-like growth factor (IGF)-binding proteins after IGF-I injections in noninsulin-dependent diabetics. J Clin Endocrinol Metab 1994; 78:609–614.
112. Zapf J, Schmid C, Guler HP, Waldvogel M, Hauri C, Futo E, Hossenlopp P, Binoux M, Froesch ER. Regulation of binding proteins for insulin-like growth factor (IGF) in humans. J Clin Invest 1990; 86:952–961.
113. Boisclair YR, Yang YW-H, Stewart JM, Rechler MM. Insulin-like growth factor-I and insulin stimulate the synthesis of IGF- binding protein-2 in a human embryonic kidney cell line. Growth Regul 1994; 4:136–146.
114. Ernst CW, McCusker RH, White ME. Gene expression and secretion of insulin-like growth factor-binding proteins during myoblast differentiation. Endocrinology 1992; 130:607–615.
115. Ernst CW, McFarland DC, White ME. Expression of insulin-like growth factor II (IGF-II), IGF binding protein-2 and myogenin during differentiation of myogenic satellite cells derived from the turkey. Differentiation 1996; 61:25–33.
116. Böni-Schnetzler M, Schmid C, Mary J-L, Zimmerli B, Meier PJ, Zapf J, Schwander J, Froesch ER. Insulin regulates the expression of the insulin-like growth factor binding protein 2 mRNA in rat hepatocytes. Mol Endocrinol 1990; 4:1320–1326.
117. Schmid C, Schläpfer I, Waldvogel M, Meier PJ, Schwander J, Böni-Schnetzler-M, Zapf J, Froesch ER. Differential regulation of insulin-like growth factor binding protein (IGFBP)-2 mRNA in liver and bone cells by insulin and retinoic acid in vitro. FEBS Lett 1992; 303:205–209.
118. Wang D, Nagpal ML, Lin T, Shimasaki S, Ling N. Insulin-like growth factor-binding protein-2: The effect of human chorionic gonadotropin on its gene regulation and protein secretion and its biological effects in rat leydig cells. Mol Endocrinol 1994; 8:69–76.
119. Binkert C, Landwehr J, Mary J-L, Schwander J, Heinrich G. Cloning, sequence analysis and expression of a cDNA encoding a novel insulin-like growth factor binding protein (IGFBP-2). EMBO J 1989; 8:2497–2502.
120. Upton FZ, Szabo L, Wallace JC, Ballard FJ. Characterization and cloning of a bovine insulin-like growth factor-binding protein. J Mol Endocrinol 1990; 5:77–84.
121. Bourner MJ, Busby WH, Jr., Siegel NR, Krivi GG, McCusker RH, Clemmons DR. Cloning and sequence determination of bovine insulin-like growth factor binding protein-2 (IGFBP-2): Comparison of its structural and functional properties with IGFBP-1. J Cell Biochem 1992; 48:215–226.
122. Ehrenborg E, Wilhelmsdotter S, Bajalica S, Larsson C, Stern I, Koch J, Brondum-Nielsen K, Luthman H. Structure and localization of the human insulin-like growth factor-binding protein 2 gene. Biochem Biophys Res Commun 1991; 176:1250–1255.
123. Agarwal N, Hsieh C-L, Sills D, Swaroop M, Desai B, Francke U, Swaroop A. Sequence analysis, expression and chromosomal localization of a gene, isolated from a subtracted human retina cDNA library, that encodes an insulin-like growth factor binding protein (IGFBP2). Exp Eye Res 1991; 52:549–561.
124. Binkert C, Margot JB, Landwehr J, Heinrich G, Schwander J. Structure of the human insulin-like growth factor binding protein-2 gene. Mol Endocrinol 1992; 6:826–836.
125. Brown AL, Rechler MM. Cloning of the rat insulin-like growth factor binding protein-2 gene and identification of a functional promoter lacking a TATA box. Mol Endocrinol 1990; 4:2039–2051.
126. Boisclair YR, Brown AL. Use of reverse ligation-pcr to identify transcriptional start sites in GC-rich TATA-less genes: application to the rat IGFBP-2 gene. DNA Cell Biol 1995; 14:731–739.
127. Kutoh E, Margot JB, Schwander J. Genomic structure and regulation of the promoter of the rat insulin-like growth factor binding protein-2 gene. Mol Endocrinol 1993; 7:1205–1216.
128. Landwehr J, Kaupmann K, Heinrich G, Schwander J. Cloning and characterization of the gene encoding murine insulin-like growth factor-binding protein-2, mIGFBP-2. Gene 1993; 124:281–286.
129. Song S, Lee CY, Green ML, Chung CS, Simmen RC, Simmen FA. The unique endometrial expression and genomic organization of the porcine IGFBP-2 gene. Mol Cell Endocrinol 1996; 120:193–202.
130. Boisclair YR, Brown AL, Casola S, Rechler MM. Three clustered Sp1 sites are required for efficient transcription of the TATA-less promoter of the gene for insulin-like growth factor binding protein-2 from the rat. J Biol Chem 1993; 268:24892–24901.

131. Baxter RC. Insulin-like growth factor binding proteins in the human circulation: a review. Horm Res 1994; 42:140–144.
132. Baxter RC, Martin JL. Radioimmunoassay of growth hormone-dependent insulinlike growth factor binding protein in human plasma. J Clin Invest 1986; 78:1504–1512.
133. Hoeck WG, Mukku VR. Identification of the major sites of phosphorylation in IGF binding protein-3. J Cell Biochem 1994; 56:262–273.
134. Jones JI, Clemmons DR. Insulin-like growth factors and their binding proteins: Biological actions. Endocr Rev 1995; 16:3–34.
135. Chin E, Zhou J, Dai J, Baxter RC, Bondy CA. Cellular localization and regulation of gene expression for components of the insulin-like growth factor ternary binding protein complex. Endocrinology 1994; 134:2498–2504.
136. Albiston AL, Herington AC. Tissue distribution and regulation of insulin-like growth factor (IGF)-binding protein-3 messenger ribonucleic acid (mRNA) in the rat: comparison with IGF-I mRNA expression. Endocrinology 1992; 130:497–502.
137. Shimasaki S, Koba A, Mercado M, Shimonaka M, Ling N. Complementary DNA structure of the high molecular weight rat insulin-like growth factor binding protein (IGF-BP3) and tissue distribution of its mRNA. Biochem Biophys Res Commun 1989; 165:907–912.
138. Takenaka A, Miura Y, Mori M, Hirosawa M, Kato H, Noguchi T. Distribution of messenger RNAs of insulin-like growth factor (IGF)-binding proteins-1 and -3 between parenchymal and nonparenchymal cells in rat liver. Agric Biol Chem 1991; 55:1191–1193.
139. Zhou J, Bondy. Insulin-like growth factor-II and its binding proteins in placental development. Endocrinology 1992; 131:1230–1240.
140. Peter MA, Winterhalter KH, Böni-Schnetzler M, Froesch ER, Zapf J. Regulation of insulin-like growth factor-I (IGF-I) and IGF-binding proteins by growth hormone in rat white adipose tissue. Endocrinology 1993; 133:2624–2631.
141. Zhou J, Bondy C 1993. Anatomy of the insulin-like growth factor system in the human testis. Fertil Steril 1993; 60:897–904.
142. Thissen J-P, Ketelslegers J-M, Underwood LE. Nutritional regulation of the insulin-like growth factors. Endocr Rev 1994; 15:80–101.
143. Zapf J, Hauri C, Waldvogel M, Futo E, Hasler H, Binz K, Guler HP, Schmid C, Froesch ER. Recombinant human insulin-like growth factor I induces its own specific carrier protein in hypophysectomized and diabetic rats. Proc Natl Acad Sci USA 1989; 86:3813–3817.
144. Clemmons DR, Thissen JP, Maes M, Ketelslegers JM, Underwood LE. Insulin-like growth factor-I (IGF-I) infusion into hypophysectomized or protein-deprived rats induces specific IGF-binding proteins in serum. Endocrinology 1989; 125:2967–2972.
145. Camacho-Hubner C, Clemmons DR, D'Ercole AJ. Regulation of insulin-like growth factor (IGF) binding proteins in transgenic mice with altered expression of growth hormone and IGF-I. Endocrinology 1991; 129:1201–1206.
146. Kupfer SR, Underwood LE, Baxter RC, Clemmons DR. Enhancement of the anabolic effects of growth hormone and insulin-like growth factor I by use of both agents simultaneously. J Clin Invest 1993; 91:391–396.
147. Gargosky SE, Wilson KF, Fielder PJ, Vaccarello MA, Guevara-Aguirre J, Diamond FB, Baxter RC, Rosenbloom AL, Rosenfeld RG. The composition and distribution of insulin-like growth factors (IGFs) and IGF-binding proteins (IGFBPs) in the serum of growth hormone receptor-deficient patients: effects of IGF-I therapy on IGFBP-3. J Clin Endocrinol Metab 1993; 77:1683–1689.
148. Laron Z, Klinger B, Blum WF, Silbergeld A, Ranke MB. IGF binding protein 3 in patients with Laron type dwarfism: effect of exogenous rIGF-I. Clin Endocrinol (Oxf) 1992; 36:301–304.
149. Villafuerte BC, Koop BL, Pao C-I, Gu L, Birdsong GG, Phillips LS. Coculture of primary rat hepatocytes and nonparenchymal cells permits expression of insulin-like growth factor binding protein-3 in vitro. Endocrinology 1994; 134:2044–2050.
150. Villafuerte BC, Koop BL, Pao C-I, Phillips LS. Glucocorticoid regulation of insulin-like growth factor-binding protein-3. Endocrinology 1995; 136:1928–1933.
151. Villafuerte BC, Zhang W-N, Phillips LW. Insulin and insulin-like growth factor-I regulate hepatic insulin-like growth factor binding protein-3 by different mechanisms. Mol Endocrinol 1996; 10:622–630.
152. Suh D-S, Rechler MM. Hepatocyte nuclear factor 1 (HNF-1) and the glucocorticoid receptor synergistically activate transcription of the rat insulin-like growth factor binding protein-1 (IGFBP-1) gene. Mol Endocrinol 1997; 11:1822–1831.

153. Bach LA, Rechler MM. Insulin-like growth factor binding proteins. Diabetes Rev 1995; 3:38–61.
154. Hayden JM, Strong DD, Baylink DJ, Powell DR, Sampath TK, Mohan S. Osteogenic protein-1 stimulates production of insulin-like growth factor binding protein-3 nuclear transcripts in human osteosarcoma cells. Endocrinology 1997; 138:4240–4247.
155. Gucev ZS, Oh Y, Kelley KM, Labarta JI, Vorwerk P, Rosenfeld RG. Evidence for insulin-like growth factor (IGF)-independent transcriptional regulation of IGF binding protein-3 by growth hormone in SKHEP-1 human hepatocarcinoma cells. Endocrinology 1997; 138:1464–1470.
156. Erondu NE, Dake BL, Moser DR, Lin M, Boes M, Bar RS. Regulation of endothelial IFGBP-3 synthesis and secretion by IGF-I and TGF-b. Growth Regul 1996; 6:1–9.
157. Wood WI, Cachianes G, Henzel WJ, Winslow GA, Spencer SA, Hellmiss R, Martin JL, Baxter RC. Cloning and expression of the growth hormone-dependent insulin-like growth factor-binding protein. Mol Endocrinol 1988; 2:1176–1185.
158. Spratt SK, Tatsuno GP, Sommer A. Cloning and characterization of bovine insulin-like growth factor binding protein-3 (bIGFBP-3). Biochem Biophys Res Commun 1991; 177:1025–1032.
159. Erondu NE, Toland B, Boes M, Dake B, Moser DR, Bar RS. Bovine insulin-like growth factor binding protein-3: organization of the chromosomal gene and functional analysis of its promoter. Endocrinology 1997; 138:2856–2862.
160. Shimasaki S, Shimonaka M, Ui M, Inouye S, Shibata F, Ling N. Structural characterization of a follicle-stimulating hormone action inhibitor in porcine ovarian follicular fluid. Its identification as the insulin-like growth factor-binding protein. J Biol Chem 1990; 265:2198–2202.
161. Ehrenborg E, Larsson C, Stern I, Janson M, Powell DR, Luthman H. Contiguous localization of the genes encoding human insulin-like growth factor binding proteins 1 (IGBP1) and 3(IGBP3) on chromosome 7. Genomics 1992; 12:497–502.
162. Cubbage ML, Suwanichkul A, Powell DR. Insulin-like growth factor binding protein-3. Organization of the human chromosomal gene and demonstration of promoter activity. J Biol Chem 1990; 265:12642–12649.
163. Albiston AL, Saffery R, Herington AC. Cloning and characterization of the promoter for the rat insulin-like growth factor-binding protein-3 gene. Endocrinology 1995; 136:696–704.
164. Villafuerte BC, Zhao W, Herington AC, Saffery R, Phillips LS. Identification of an insulin-responsive element in the rat insulin-like growth factor-binding protein-3 gene. J Biol Chem 1997; 272:5024–5030.
165. Buckbinder L, Talbott R, Velasco-Miguel S, Takenaka I, Faha B, Seizinger BR, Kley N. Induction of the growth inhibitor IGF-binding protein 3 by p53. Nature 1995; 377:646–649.
166. Campbell PG, Novak JF. Insulin-like growth factor binding protein (IGFBP) inhibits IGF action on human osteosarcoma cells. J Cell Physiol 1991; 149:293–300.
167. Mohan S, Bautista CM, Wergedal J, Baylink DJ. Isolation of an inhibitory insulin-like growth factor (IGF) binding protein from bone cell-conditioned medium: a potential local regulator of IGF action. Proc Natl Acad Sci USA 1989; 86:8338–8342.
168. Perkel VS, Mohan S, Baylink DJ, Linkhart TA. An inhibitory insulin-like growth factor binding protein (IN-IGFBP) from human prostatic cell conditioned medium reveals N-terminal sequence identity with bone derived IN-IGFBP. J Clin Endocrinol Metab 1990; 71:533–535.
169. Kiefer MC, Masiarz FR, Bauer DM, Zapf J. Identification and molecular cloning of two new 30-kDa insulin-like growth factor binding proteins isolated from adult human serum. J Biol Chem 1991; 266:9043–9049.
170. Shimonaka M, Schroeder R, Shimasaki S, Ling N. Identification of a novel binding protein for insulin-like growth factors in adult rat serum. Biochem Biophys Res Commun 1989; 165:189–195.
171. Ceda GP, Fielder PJ, Henzel WJ, Louie A, Donovan SM, Hoffman AR, Rosenfeld RG. Differential effects of insulin-like growth factor (IGF)-I and IGF-II on the expression of IGF binding proteins (IGFBPs) in a rat neuroblastoma cell line: isolation and characterization of two forms of IGFBP-4. Endocrinology 1991; 128:2815–2824.
172. Cheung PT, Smith EP, Shimasaki S, Ling N, Chernausek SD. Characterization of an insulin-like growth factor binding protein (IGFBP-4) produced by the B104 rat neuronal cell line: chemical and biological properties and differential synthesis by sublines. Endocrinology 1991; 129:1006–1015.
173. Walton PE, Grant P, Owens JA, Wallace JC, Ballard FJ. Isolation and characterization of ovine plasma IGF binding proteins 1, 2 and a 24 kDa species. In: Program of the 72nd Annual Meeting of the Endocrine Society Atlanta, GA. 1990, p. 69.
174. Shimasaki S, Uchiyama F, Shimonaka M, Ling N. Molecular cloning of the cDNAs encoding a novel insulin-like growth factor-binding protein from rat and human. Mol Endocrinol 1990; 4:1451–1458.

175. Donovan SM, Giudice LC, Murphy LJ, Hintz RL, Rosenfeld RG. Maternal insulin-like growth factor-binding protein messenger ribonucleic acid during rat pregnancy. Endocrinology 1991; 129:3359–3366.
176. La Tour D, Mohan S, Linkhart TA, Baylink DJ, Strong DD. Inhibitory insulin-like growth factor binding protein: cloning, complete sequence, and physiological regulation. Mol Endocrinol 1990; 4:1806–1814.
177. Rosen CJ, Donahue LR, Hunter SJ. Insulin-like growth factors and bone: the osteoporosis connection. Proc Soc Exp Biol Med 1994; 206:83–102.
178. Di Battista JA, Dore S, Morin S, He Y, Pelletier J-P, Martel-Pelletier J. Prostaglandin E2 stimulates insulin-like growth factor binding protein-4 expression and synthesis in cultured human articular chondrocytes: possible mediation by Ca++-calmodulin regulated processes. J Cell Biochem 1997; 65:408–419.
179. Kudo Y, Iwashita M, Iguchi T, Takeda Y, Hizuka N, Takano K, Muraki T. Estrogen and parathyroid hormone regulate insulin-like growth factor binding protein-4 in SaOS-2 cells. Life Sci 1997; 61:165–170.
180. Demori I, Bottazzi C, Voci A, Gallo G, Scharf J-G, Fugassa E. Tri-iodothyronine increases insulin-like growth factor binding protein-4 expression in rat hepatocytes. J Endocrinol 1997; 154:155–165.
181. Glantschnig H, Varga F, Klaushofer K. Thyroid hormone and retinoic acid induce the synthesis of insulin-like growth factor-binding protein-4 in mouse osteoblastic cells. Endocrinology 1996; 137:281–286.
182. Moser DR, Lowe WL, Jr., Dake BL, Booth BA, Boes M, Clemmons DR, Bar RS. Endothelial cells express insulin-like growth factor-binding proteins 2 to 6. Mol Endocrinol 1992; 6:1805–1814.
183. Bethel CR, Vitullo JC, Miller RE, Aron DC. Molecular cloning of mouse insulin-like growth factor binding protein 4 (IGFBP4) cDNA and expression of a fusion protein with IGF-binding activity. Biochem Mol Biol Int 1994; 34:385–392.
184. Bajalica S, Allander SV, Ehrenborg E, Brondum-Nielsen K, Luthman H, Larsson C. Localization of the human insulin-like growth-factor-binding protein 4 gene to chromosomal region 17q12-21.1. Hum Genet 1992; 89:234–236.
185. Tonin P, Ehrenborg E, Lenoir G, Feunteun J, Lynch H, Morgan K, Zazzi H, Vivier A, Pollak M, Huynh H, Luthman H, Larsson C, Narod S. The human insulin-like growth factor-binding protein 4 gene maps to chromosome region 17q12-q21.1 and is close to the gene for hereditary breast-ovarian cancer. Genomics 1993; 18:414–417.
186. Qin X, Morales S, Lee K-W, Boonyaratanakornkit V, Baylink DJ, Mohan S, Strong DD. Structure and functional analysis of the 5′-flanking region of the human insulin-like growth factor binding protein (IGFBP)-4 gene. Biochim Biophys Acta 1997; 1350:136–140.
187. Dai B, Widen SG, Mifflin R, Singh P. Cloning of the functional promoter for human insulin-like growth factor binding protein-4 gene: endogenous regulation. Endocrinology 1997; 138:332–343.
188. Gao L, Ling N, Shimasaki S. Structure of the rat insulin-like growth factor binding protein-4 gene. Biochem Biophys Res Commun 1993; 190:1053–1059.
189. Mohan S, Farley JR, Baylink DJ. Age-related changes in IGFBP-4 and IGFBP-5 levels in human serum and bone: implications for bone loss with aging. Prog Growth Factor Res 1995; 6:465–473.
190. Mohan S, Libanati C, Dony C, Lang K, Srinivasan N, Baylink DJ. Development, validation, and application of a radioimmunoassay for insulin-like growth factor binding protein-5 in human serum and other biological fluids. J Clin Endocrinol Metab 1995; 80:2638–2645.
191. Conover CA, Kiefer MC. Regulation and biological effect of endogenous insulin-like growth factor binding protein-5 in human osteoblastic cells. J Clin Endocrinol Metab 1993; 76:1153–1159.
192. Green BN, Jones SB, Streck RD, Wood TL, Rotwein P, Pintar JE. Distinct expression patterns of insulin-like growth factor binding proteins 2 and 5 during fetal and postnatal development. Endocrinology 1994; 134:954–962.
193. Shimasaki S, Shimonaka M, Zhang H-P, Ling N. Identification of five different insulin-like growth factor binding proteins (IGFBPs) from adult rat serum and molecular cloning of a novel IGFBP-5 in rat and human. J Biol Chem 1991; 266:10646–10653.
194. Gosteli-Peter MA, Winterhalter KH, Schmid C, Froesch ER, Zapf J. Expression and regulation of insulin-like growth factor-I (IGF-I) and IGF-binding protein messenger ribonucleic acid levels in tissues of hypophysectomized rats infused with IGF-I and growth hormone. Endocrinology 1994; 135:2558–2567.
195. Kou K, Jenkins NA, Gilbert DJ, Copeland NG, Rotwein P. Organization, expression, and chromosomal location of the mouse insulin-like growth factor binding protein 5 gene. Genomics 1994; 20:412–418.

196. Dong Y, Canalis E. Insulin-like growth factor (IGF) I and retinoic acid induce the synthesis of IGF-binding protein 5 in rat osteoblastic cells. Endocrinology 1995; 136:2000–2006.
197. Pash JM, Canalis E. Transcriptional regulation of insulin-like growth factor-binding protein-5 by prostaglandin E2 in osteoblast cells. Endocrinology 1996; 137:2375–2382.
198. McCarthy TL, Casinghino S, Mittanck DW, Ji C-H, Centrella M, Rotwein P. Promoter-dependent and -independent activation of insulin-like growth factor binding protein-5 gene expression by prostaglandin E2 in primary rat osteoblasts. J Biol Chem 1996; 271:6666–6671.
199. Franchimont N, Durant D, Canalis E. Interleukin-6 and its soluble receptor regulate the expression of insulin-like growth factor binding protein-5 in osteoblast cultures. Endocrinology 1997; 138:3380–3386.
200. Canalis E, Gabbitas B. Skeletal growth factors regulate the synthesis of insulin-like growth factor binding protein-5 in bone cell cultures. J Biol Chem 1995; 270:10771–10776.
201. Gabbitas B, Pash JM, Delany AM, Canalis E. Cortisol inhibits the synthesis of insulin-like growth factor-binding protein-5 in bone cell cultures by transcriptional mechanisms. J Biol Chem 1996; 271:9033–9038.
202. Duan C, Hawes SB, Prevette T, Clemmons DR. Insulin-like growth factor-I (IGF-I) regulates IGF-binding protein-5 synthesis through transcriptional activation of the gene in aortic smooth muscle cells. J Biol Chem 1996; 271:4280–4288.
203. James PL, Jones SB, Busby WH, Jr., Clemmons DR, Rotwein P. A highly conserved insulin-like growth factor-binding protein (IGFBP-5) is expressed during myoblast differentiation. J Biol Chem 1993; 268:22305–22312.
204. Rotwein P. James PL, Kou K. Rapid activation of insulin-like growth factor binding protein-5 gene transcription during myoblast differentiation. Mol Endocrinol 1995; 9:913–923.
205. Conover CA, Bale LK, Clarkson JT, Torring O. Regulation of insulin-like growth factor binding protein-5 messenger ribonucleic acid expression and protein availability in rat osteoblast-like cells. Endocrinology 1993; 132:2525–2530.
206. Backeljauw PF, Dai Z, Clemmons DR, D'Ercole AJ. Synthesis and regulation of insulin-like growth factor binding protein-5 in FRTL-5 cells. Endocrinology 1993; 132:1677–1681.
207. Rozen F, Yang X-F, Huynh H, Pollak M. Antiproliferative action of vitamin D-related compounds and insulin-like growth factor-binding protein 5 accumulation. J Natl Cancer Inst 1997; 89:652–656.
208. Shemer J, Yaron A, Werner H, Shao Z-M, Sheikh MS, Fontana JA, LeRoith D, Roberts CT, Jr. Regulation of insulin-like growth factor (IGF) binding protein-5 in the T47D human breast carcinoma cell line by IGF-I and retinoic acid. J Clin Endocrinol Metab 1993; 77:1246–1250.
209. Kiefer MC, Ioh RS, Bauer DM, Zapf J. Molecular cloning of a new human insulin-like growth factor binding protein. Biochem Biophys Res Commun 1991; 176:219–225.
210. Arai T, Parker A, Busby W, Jr., Clemmons DR. Heparin, heparan sulfate, and dermatan sulfate regulate formation of the insulin-like growth factor-I and insulin-like growth factor-binding protein complexes. J Biol Chem 1994; 269:20388–20393.
211. Arai T, Clarke J, Parker A, Busby W, Jr., Nam T, Clemmons DR. Substitution of specific amino acids in insulin-like growth factor (IGF) binding protein 5 alters heparin binding and its change in affinity for IGF-I in response to heparin. J Biol Chem 1995; 271:6099–6106.
212. Allander SV, Larsson C, Ehrenborg E, Suwanichkul A, Weber G, Morris SL, Bajalica S, Kiefer MC, Luthman H, Powell DR. Characterization of the chromosomal gene and promoter for human insulin-like growth factor binding protein-5. J Biol Chem 1994; 269:10891–10898.
213. Zhu X, Ling N, Shimasaki S. Cloning of the rat insulin-like growth factor binding protein-5 gene and DNA sequence analysis of its promoter region. Biochem Biophys Res Commun 1993; 190:1045–1052.
214. Camacho-Hubner C, Busby WH Jr, McCusker RH, Wright G, Clemmons DR. Identification of the forms of insulin-like growth factor binding proteins produced by human fibroblasts and the mechanisms that regulate their secretion. J Biol Chem 1992; 267:11949–11956.
215. Duan CM, Clemmons DR. Transcription factor AP-2 regulates human insulin-like growth factor binding protein-5 gene expression. J Biol Chem 1995; 270:24844–24851.
216. Zapf J, Kiefer M, Merryweather J, Masiarz F, Bauer D, Born W, Fischer JA, Froesch ER. Isolation from adult human serum of four insulin-like growth factor (IGF) binding proteins and molecular cloning of one of them that is increased by IGF I administration and in extrapancreatic tumor hypoglycemia. J Biol Chem 1990; 265:14892–14898.
217. Roghani M, Lassarre C, Zapf J, Povoa G, Binoux M. Two insulin-like growth factor (IGF)-binding proteins are responsible for the selective affinity for IGF-II of cerebrospinal fluid binding proteins. J Clin Endocrinol Metab 1991; 73:658–666.

218. Shimasaki S, Gao L, Shimonaka M, Ling N. Isolation and molecular cloning of insulin-like growth factor-binding protein-6. Mol Endocrinol 1991; 5:938–948.
219. Martin JL, Willetts KE, Baxter RC. Purification and properties of a novel insulin-like growth factor-II binding protein from transformed human fibroblasts. J Biol Chem 1990; 265:4124–4130.
220. Forbes B, Ballard FJ, Wallace JC. An insulin-like growth factor-binding protein purified from medium conditioned by a human lung fibroblast cell line (HE[39]L) has a novel N-terminal sequence. J Endocrinol 1990; 126:497–506.
221. Andress DL, Birnbaum RS. A novel human insulin-like growth factor binding protein secreted by osteoblast-like cells. Biochem Biophys Res Commun 1991; 176:213–218.
222. Roghani M, Hossenlopp P, Lepage P, Balland A, Binoux M. Isolation from human cerebrospinal fluid of a new insulin-like growth factor-binding protein with a selective affinity for IGF-II. FEBS Lett 1989; 255:253–258.
223. Rohan RM, Ricciarelli E, Kiefer MC, Resnick CE, Adashi EY. Rat ovarian insulin-like growth factor binding protein-6: a hormonally regulated theca-interstitial-selective species with limited anti-gonadotropic activity. Endocrinology 1992; 132:2507–2512.
224. Gabbitas B, Canalis E. Cortisol enhances the transcription of insulin-like growth factor-binding protein-6 in cultured osteoblasts. Endocrinology 1996; 137:1687–1692.
225. Gabbitas B, Canalis E. Retinoic acid stimulates the transcription of insulin-like growth factor binding protein-6 in skeletal cells. J Cell Physiol 1996; 169:15–22.
226. Matsumoto K, Gaetano C, Daughaday WH, Thiele CJ. Retinoic acid regulates insulin-like growth factor II expression in a neuroblastoma cell line. Endocrinology 1992; 130:3669–3676.
227. Martin JL, Coverley JA, Baxter RC. Regulation of immunoreactive insulin-like growth factor binding protein-6 in normal and transformed human fibroblasts. J Biol Chem 1994; 269:11470–11477.
228. Bach LA, Thotakura NR, Rechler MM. Human insulin-like growth factor binding protein-6 is O-glycosylated. Biochem Biophys Res Commun 1992; 186:301–307.
229. Strong DD, Morales S, Baylink DJ, Mohan S. Structural and functional characterization of the human IGFBP-6 promoter. In: Program of the 4th International Symposium on Insulin-Like Growth Factors, 1997, Tokyo, Japan.
230. Zhu X, Ling N, Shimasaki S. Structural characterization of the rat insulin-like growth factor binding protein-6 gene. Biochem Biophys Res Commun 1993; 191:1237–1243.
231. Baxter RC. Circulating levels and molecular distribution of the acid-labile a subunit of the high molecular weight insulin-like growth factor-binding protein complex. J Clin Endocrinol Metab 1990; 70:1347–1353.
232. Lewitt MS, Scott FP, Clarke NM, Baxter RC. Developmental regulation of circulating insulin-like growth factor-binding proteins in normal pregnancies and in pre-eclampsia. Prog Growth Factor Res 1995; 6:475–480.
233. Baxter RC, Dai J. Purification and characterization of the acid-labile subunit of rat serum insulin-like growth factor binding protein complex. Endocrinology 1994; 134:848–852.
234. Guler H-P, Zapf J, Schmid C, Froesch ER. Insulin-like growth factors I and II in healthy man. Estimations of half-lives and production rates. Acta Endocrinol (Copenh) 1989; 121:753–758.
235. Zapf J, Hauri C, Futo E, Hussain M, Rutishauser J, Maack CA, Froesch ER. Intravenously injected insulin-like growth factor (IGF) I/IGF binding protein-3 complex exerts insulin-like effects in hypophysectomized, but not in normal rats. J Clin Invest 1995; 95:179–186.
236. Baxter RC, Martin JL, Beniac VA. High molecular weight insulin-like growth factor binding protein complex. J Biol Chem 1989; 264:11843–11848.
237. Dai J, Baxter RC. Regulation in vivo of the acid-labile subunit of the rat serum insulin-like growth factor-binding protein complex. Endocrinology 1994; 135:2335–2341.
238. Gargosky SE, Tapanainen P, Rosenfeld RG. Administration of growth hormone (GH), but not insulin-like growth factor-I (IGF-I), by continuous infusion can induce the formation of the 150-kilodalton IGF-binding protein-3 complex in GH-deficient rats. Endocrinology 1994; 134:2267–2276.
239. Dai J, Scott CD, Baxter RC. Regulation of the acid-labile subunit of the insulin-like growth factor complex in cultured rat hepatocytes. Endocrinology 1994; 135:1066–1072.
240. Antequera F, Boyes J, Bird A. High levels of de novo methylation and altered chromatin structure at CpG islands in cell lines. Cell 1990; 62:503–514.
241. Boisclair YR, Bassal S, Rechler MM, Ooi GT. Regulation of ALS gene expression by growth hormone. In: Takano K, Hizuka N, Takahashi S (eds). Molecular mechanisms to regulate the activities of insulin-like growth factors. Elsevier Science BV, Amsterdam, 1998, 49–56.

242. Leong SR, Baxter RC, Camerato T, Dai J, Wood WI. Structure and functional expression of the acid-labile subunit of the insulin-like growth factor binding protein complex. Mol Endocrinol 1992; 6:870–876.
243. Boisclair YR, Seto D, Hsieh S, Hurst KR, Ooi GT. Organization and chromosomal localization of the gene encoding the mouse acid labile subunit of the insulin-like growth factor binding complex. Proc Natl Acad Sci USA 1996; 93:10028–10033.
244. Dai J, Baxter RC. Molecular cloning of the acid-labile subunit of the rat insulin-like growth factor binding protein complex. Biochem Biophys Res Commun 1992; 188:304–309.
245. Delhanty P, Baxter RC. The cloning and expression of the baboon acid-labile subunit of the insulin-like growth factor binding protein complex. Biochem Biophys Res Commun 1997; 227:897–902.
246. Delhanty PJD, Baxter RC. Characterisation of the rat acid-labile subunit gene. Prog Growth Factor Res 1995; 6:141–149.
247. Ooi GT, Hurst KR, Poy MN, Rechler MM, Boisclair YR. Binding of Stat5a and Stat5b to a single element resembling a gamma-interferon activated sequence (GAS) mediates the growth hormone induction of the mouse acid-labile subunit promoter in liver cells. Mol Endocrinol 1998; 12:675–687.

II

Biological Actions of the IGFs

7 IGF-I Receptor Function

Transducing the IGF-I Signal into Intracellular Events

Vicky A. Blakesley, MD, PhD, *Andrew A. Butler*, PhD, *Anatolii P. Koval*, PhD, *Yumiko Okubo*, MD, *and Derek LeRoith*, MD, PhD

CONTENTS

INTRODUCTION

The insulin-like growth factor (IGF)-I receptor, or type 1 IGF receptor, is a transmembrane tyrosine kinase receptor that mediates the majority of the biological actions of IGF-I and IGF-II *(1,2)*. The ligands, IGF-I or IGF-II, bind to the extracellular domain of the receptor and initiate a conformational change that is transmitted to the intracellular domain. The receptor is then autophosphorylated on several intracellular tyrosine residues. The tyrosine-phosphorylated receptor is then fully active as a tyrosine kinase toward endogenous substrates. Both IGF-I and IGF-II circulate bound to IGF binding proteins (IGFBPs) *(3)*. The IGF–IGFBP complexes either enhance or inhibit IGF action in a tissue-specific manner. The known roles of the IGFBPs are presented in other chapters. Whereas the IGFBPs regulate the activation of the receptor by the binding of its cognate ligands, IGF-I receptor activation constitutes the ultimate requisite for the transduction of IGF-mediated signals.

Activation of the IGF-I receptor triggers intracellular events that, in the main, result in cellular growth and differentiation. The IGF-I receptor is structurally similar to the

From: *Contemporary Endocrinology: The IGF System*
Edited by: R. Rosenfeld and C. Roberts Jr © Humana Press, Inc., Totowa, NJ

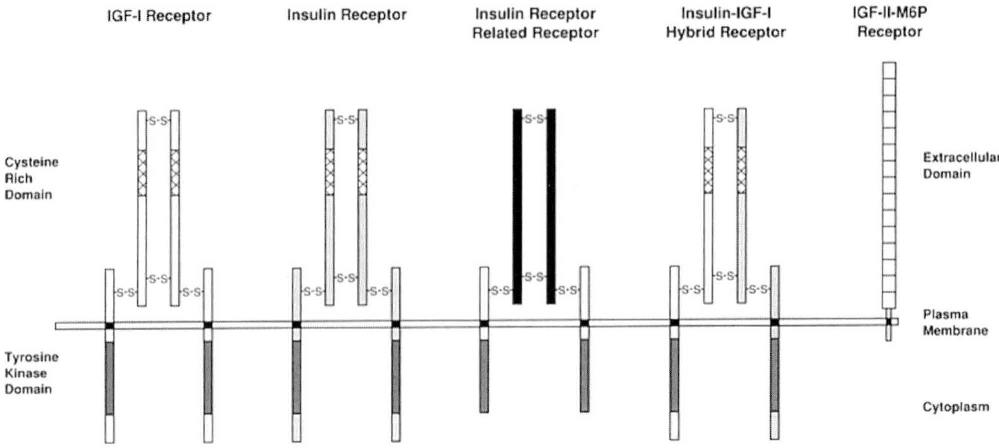

Fig. 1. Schematic representation of the IGF-I receptor. The IGF-I receptor is a heterotetrameric transmembrane tyrosine kinase receptor. Each hemireceptor is comprised of an α- and β-subunit, covalently bound by disulfide bonds. A doublet of hemireceptors constitutes the active IGF-I receptor. The α-subunit is entirely extracellular and contains the ligand (IGF-I or -II) binding domain. The β-subunit spans the membrane and is divided into the extracellular, transmembrane, juxtamembrane, tyrosine kinase, and C-terminal domains. PM, plasma membrane.The essential ATP-binding site is at lysine 1003 within the tyrosine kinase domain. Also within the catalytic domain is a triple tyrosine cluster (tyrosines 1131, 1135, and 1136).

insulin receptor, which is mostly involved in the transduction of metabolic signals. The IGF-II receptor (or type 2 IGF receptor) is structurally dissimilar to the IGF-I and insulin receptors. The IGF-II receptor is identical to the mannose-6-phosphate (Man-6-P) receptor and binds only IGF-II with high affinity. The major biological roles of the IGF-II receptor appear to be the recycling of lysosomal enzymes containing Man-6-P residues and clearance of IGF-II.

In the present chapter we describe the structure of the IGF-I receptor and provide examples of the known signal transduction pathways that mediate IGF effects.

STRUCTURE OF THE IGF-I RECEPTOR

Processing to Form the Mature Receptor

The heterotetrameric IGF-I receptor is composed of two α- and two β-subunits (Fig. 1) *(4)*. The ligand binding domain, a cysteine-rich region, is within the extracellular α-subunits (each 706 amino acids). Disulfide bonds between one α-subunit and one β-subunit and between the two α-subunits maintain the secondary structure of the mature, functional receptor. The β-subunits (each 627 amino acids) are transmembrane chains that each contain the tyrosine kinase domain (residues 973-1229) in their cytoplasmic portions (numbering system of Ullrich et al.) *(5)*. Activation of this enzymatic domain is crucial for the propagation of IGF effects, as discussed below. Other domains of each β-subunit are the carboxy-(c)-terminal domain, a juxtamembrane domain proximal to the tyrosine kinase domain, a 24-residue hydrophobic transmembrane domain, and a short extracellular region.

The mature IGF-I receptor is obtained by the processing of a larger precursor protein. The α- and β-subunits are synthesized colinearly as a preproreceptor composed of 1367 amino acids *(5)*. The polar residue-rich leader or signal peptide, the amino-(N)-terminal 30 amino acids, is important in the targeting of the proreceptor to the endoplasmic reticulum where the protein is further processed. The α- and β-subunits are released by proteolytic cleavage at the basic tetrapeptide Arg–Lys–Arg–Arg (residues 707 and 710). The subunits are subsequently assembled into their $\alpha_2\beta_2$ mature configuration *(6)*. Finally, multiple potential sites for N-linked glycosylation (Asn–X–Ser/Thr) are found in the receptor molecule. The actual glycosylation pattern varies between cell type and developmental stage. Primary structure thus dictates the functional domains and specifies the glycosylation sites that may afford additional regulation of the activity of the IGF-I receptor.

Heterogeneity of the IGF-I Receptor

The IGF-I receptor is not a unique molecular entity as noted by differences in the migration of the protein through electrophoretic gels. The receptor displays, in fact, a large degree of heterogeneity that can be explained by a number of mechanisms. These include potential primary structure variation, differential glycosylation, and hybrid formation with insulin receptors. The variation in primary structure might reflect a second IGF-I receptor gene, a possibility that has not been excluded.

Differences in Primary Structure: The Known Gene Transcript

Two IGF-I receptor mRNA species, that differ in the region encoding the extracellular domain of the β-subunit, have been detected in solution hybridization/RNase protection assays using antisense IGF-I receptor RNAs as probes *(7)*. These mRNAs are splice variants resulting from the alternative usage of 3′ splice acceptor sites at the 5′ end of exon 14 of the IGF-I receptor gene. When the upstream acceptor site is employed, an internal CAG sequence is added to the mature RNA. As a result, an arginine residue is replaced by two residues (i.e., threonine and glycine) in the extracellular domain of the β-subunit. Functional assays of both forms of the transfected receptor in Chinese hamster ovary (CHO) cells have indicated that the variant including the CAG insert displays decreased ligand-mediated endocytosis, increased receptor autophosphlorylation, and increased phosphorylation of insulin receptor substrate-1 (IRS-1) *(8,9)*. Both transcripts are expressed at similar levels in a number of human tissues and cell lines. The physiological significance of the individual receptors is not known.

Additional Structural Heterogeneity: Glycosylation and Variant Receptors

Multiple forms of the α- and β-subunits of the IGF-I receptor have been described in various tissues and cell lines. In the vast majority of cases this variability cannot be easily explained by differences in primary structure resulting from various transcripts from the known IGF-I receptor gene or by hybrid IGF-I/insulin receptors *(see next section)*.

Heterogeneity of the β-subunit has been found in skeletal muscle and brain and in a number of cell lines *(10)*. For example, adult rat muscle contains only a 95-kDa β-subunit, whereas fetal rat muscle expresses the 95- and a 105-kDa β-subunit *(11–13)*. In humans, the placenta exhibits both β-subunit variants, whereas adult skeletal muscle displays only the 95-kDa species. Differential glycosylation might explain this difference; however, a similar variability has been seen in a number of cell lines that contain both species of β-subunits. It appears that primary structure differences contribute to this heterogeneity because enzymatic deglycosylation reduced, but did not eliminate, the differences in size. Additional evidence for a higher molecular weight β-subunit due to a difference in primary structure has been described in HL 60 cells, where again the size of the 105-kDa β-subunit was not reduced by deglycosylation *(14)*.

The different β-subunits (i.e., 95- and 105-kDa) can be characterized further by immunoprecipitation with antibody αIR-3, which is directed against a specific epitope in the α-subunit. From fetal rat tissue, αIR3 immunoprecipitates the 105-kDa β-subunit but not the 95-kDa β-subunit found in both fetal and adult tissue *(13)*. These results suggest that the 95-kDa β-subunit is associated with a variant α-subunit that is not recognized by αIR-3. Thus, there appear to be at least two separate primary structures of both the α- and β-subunits. These results would be consistent with the presence of a second IGF-I receptor gene, although such a gene has not been identified.

Atypical IGF-I receptors containing α-subunits that are not recognized by αIR-3 have also been described in cell lines. The KB human epidermoid carcinoma cell line expresses two species of the β-subunit (95- and 102-kDa) *(15)*. Receptors with only the 102-kDa β-subunit isoform could be immunoprecipitated with αIR-3; however, atypical receptors with both the 95- and 102-kDa β-subunits could not. These results imply that there are at least two isoforms of the α-subunit, with each forming a hemireceptor with a particular isoform of the β-subunit. Stimulation of autophosphorylation of the hybrid receptors (both isoforms of the β-subunit) was several-fold greater with IGF-I than with IGF-II. Cells expressing only the 102-kDa β-subunit could be stimulated equally by IGF-I or IGF-II. This suggests that the different isoforms of the α-subunit have different affinities for binding IGF-I and IGF-II. Similar isoforms of the α- and β-subunits of the IGF-I receptor have been described in an inducible subclone of C2 myoblasts *(16)*. These cells contain a 140-kDa α-subunit that preferentially binds IGF-I. This α-subunit is lost following differentiation of C2 myoblasts into myotubes *(17)*. It thus appears that the magnitude of activation of the IGF-I receptor signal transduction system might be regulated at the level of ligand binding to one or more α-subunits. The differential binding affinities of the variant α-subunits may act as one more regulatory mechanism. Expression of this variant may be controlled in development and differentiation.

Hybrid IGF-I/Insulin Receptors

Hybrid receptors are heterotetramers consisting of an IGF-I α–β hemireceptor linked by disulfide bonds to an insulin α–β hemireceptor. Hybrid receptors can also be formed using the insulin receptor-related receptor (IRR) α–β hemireceptor *(18–20)*.

In those tissues and cell lines that express both the IGF-I receptor and the insulin receptor genes, hybrid receptors have been characterized *(18–20)*. Hybrid receptors can also be produced in vitro *(21–24)*. Hybrid receptors were overexpressed in HeLa cells using a T7 expression system *(25)*. The hybrids and "pure" IGF-I or insulin receptors were distinguished by using constructs that encoded a full-length IGF-I receptor and a C-terminally truncated insulin receptor *(19,25)*. Results of binding experiments using this approach, as well as results of studies of naturally occurring hybrid receptors, suggest that hybrid receptors bind IGF-I with an affinity similar to that of "pure" IGF-I receptors, whereas they bind insulin with lower affinity than "pure" insulin receptors *(20)*. It has not yet been determined if IGF-II binds these hybrid receptors with an affinity similar to that with which it binds the "pure" IGF-I receptor. Furthermore, it is not clear what, if any, is the physiologic function of naturally occurring hybrid receptors.

FUNCTION OF THE IGF-I RECEPTOR

The Receptor and Signaling

THE α-SUBUNIT

Together with the insulin receptor and the IRR, the IGF-I receptor is a member of the classic type 2 group of the tyrosine kinase membrane receptor family *(26–29)* (Fig. 1). The specificity of the IGF-I receptor for the IGFs as compared to insulin can be explained, in part, by the 10- to 100-fold higher affinity of the IGF-I receptor for its cognate ligands *(30)*. Ligand binding specificities were characterized in studies using chimeric receptors in which corresponding regions of the IGF-I and insulin receptors were exchanged *(31–33)*. High-affinity IGF-I binding was diminished when the cysteine-rich domain of the IGF-I receptor, corresponding to amino acids 131–315, was replaced by the equivalent region of the insulin receptor. Conversely, when this region was introduced into the insulin receptor, IGF-I binding to the insulin receptor was markedly increased. The abolition of IGF-I binding by antibodies αIR-3 and No. 24-60, which recognize amino acids 223–274 and 184–283, respectively, provides further evidence for the importance of the cysteine-rich domain of the IGF-I receptor in the binding of the cognate ligand *(24,34)*. These results have allowed investigators to map the cysteine-rich domain of the IGF-I receptor as being important for high-affinity IGF-I binding. Mapping of the insulin binding site of the insulin receptor by similar techniques has revealed that sequences N- and C-terminal to the cysteine-rich domain are essential for high-affinity insulin binding *(35,36)*. Finally, whereas amino acid residue phenylalanine 38 in the α-subunit of the insulin receptor is critical for the high-affinity insulin binding, the homologous phenylalanine residues are not important in IGF-I receptor/IGF-I binding *(37)*. Another crucial phenylalanine residue for binding of the cognate ligand has been identified in the IGF-I receptor at position 701 and at the homologous position (phenylalanine 714) in the insulin receptor *(38)*. Other amino acids have been found to be important for IGF-I binding to the IGF-I receptor, but their counterparts in the insulin receptor do not appear to be important for insulin binding (i.e., Asp^{8}, Asn^{11}, Phe^{58}, Phe^{692}, Glu^{693}, His^{697}, and Asn^{698}) *(38)*. Elucidation of the individual amino acids within the α-subunit important for binding the cognate ligand illustrates the level of specificity inherent in the ligand–receptor interactions and thus the impact on transduction of the extracellular signal to the specific intracellular signaling pathways.

THE β-SUBUNIT EXTRACELLULAR DOMAIN

IGF-I binding to the extracellular domain of the IGF-I receptor induces autophosphorylation of the β-subunit by an as yet unidentified mechanism. It has been postulated that ligand binding releases the inhibitory effect of the extracellular domain on the cytoplasmic domains that predominates in the unstimulated state. This potential mechanism has been best sudied for the analogous insulin receptor, where trypsinization of the α-subunit resulted in uninhibited functioning of the receptor *(39)*. Further evidence of this hypothesis was gained by replacement of the extracellular domain of the IGF-I receptor with the avian sarcoma virus UR2 gag p19 and subsequent enhancement of the transforming ability of the fusion receptor *(40,41)*. In addition, the IGF-I receptor can be constitutively activated by a mutation in the transmembrane domain (valine 922 to glutamate or isoleucine) *(42)*.

It has been postulated that ligand binding to its receptor induces a conformational change that it transmitted to the intracellular domains, thereby increasing the intrinsic tyrosine kinase activity. The studies noted above provide support for the proposition that disruptions of the primary structure of the either the α-subunit or extracellular β-subunit affect the intramolecular signaling.

The β-Subunit Intracellular Domain

Mutational analyses have been used to evaluate the function of the β-subunit of the IGF-I receptor. The initial event induced by ligand binding is autophosphorylation of three tyrosine residues (tyrosines 1131, 1135, and 1136) of the β-subunit located within the highly conserved tyrosine kinase domain *(5)*. This tyrosine phosphorylation greatly enhances the intrinsic tyrosine kinase activity of the IGF-I receptor toward intracellular substrates *(43)*. The essential role of ligand-stimulated autophosphorylation was demonstrated by substituting the lysine residue of the ATP-binding site (lysine 1003) in the glycine-rich domain with either alanine or arginine. The intrinsic tyrosine kinase activity of the lysine-mutated IGF-I receptor could not be activated by IGF-I binding to the receptor. All IGF-I-induced actions, including endogenous substrate phosphorylation and activation of signaling pathways, mitogenic activity, and the ability of the IGF-I receptor to transform cells, were blocked by lysine 1003 substitutions *(44)*. Replacement of any or all tyrosines within the tyrosine kinase domain (tyrosines 1131, 1135, and 1136) also demonstrates the importance of the activation of the intrinsic kinase activity. Single substitutions of the three tyrosines with phenylalanine, double substitutions of varying combinations, or a triple tyrosine substitution produced a reduction in the function of the IGF-I receptor, with the triple tyrosine substitution being the most severe *(45–50)*. Specifically, while substitution of tyrosine 1131 or 1135 alone had small effects on autophosphorylation and function, substitution of 1136 alone dramatically reduced substrate phosphorylation *(50)* and transforming ability *(47)*. The double substitutions 1131/1135, 1131/1136, and 1135/1136 caused significant reductions in the tyrosine phosphorylation of endogenous substrates, inhibition of the signaling pathways, and loss of tumor formation when cells with these mutant IGF-I receptors were injected into nude mice *(48)*. The most severe change in phenotype occurred with triple tyrosine substitutions, which resulted in the total absence of several IGF-I-stimulated biological functions *(45,46)*.

The transfer of the phosphate from ATP to endogenous substrates requires an intact catalytic loop domain, amino acids 1104–1110 (RDLAARN), of the IGF-I receptor. Replacement of alanine 1108 with glycine renders the IGF-I receptor temperature sensitive with respect to transforming ability, confirming the suspected importance of this domain to IGF-I receptor function *(51)*. The structures of the ATP-binding site and the catalytic loop with its substrate binding groove are maintained within a complex α-helix tertiary structure. The integrity of the surrounding α-helices appears to be important for proper functioning of the IGF-I receptor. For example, the substitution of tryptophan 1173, which lies distal to the catalytic loop, inhibits autophosphorylation and some biological functions, presumably by affecting receptor autophosphorylation *(52)*.

Other tyrosine residues within the intracellular domain are potential phosphorylation sites. Tyrosine residues at positions 943, 950, and 957 are located within the juxtamembrane domain just proximal to the tyrosine kinase domain. Tyrosine 950 lies in an NPXY motif previously shown to be important for internalization of some receptors and, in the case of the insulin receptor, for substrate interactions *(53)*. Deletion of the NPXY motif affected IGF-I receptor internalization *(54)*. The single substitution of tyrosine 950 resulted in decreased IGF-I receptor autophosphorylation and abrogation of IGF-I-induced post-receptor signaling *(55)*. The C-terminal domain of the β-subunit, which is distal to the tyrosine kinase domain, contains three tyrosines at positions 1250, 1251, and 1316. Deletion of the 44 most distal amino acids, with loss of tyrosine 1316, or deletion of 108 amino acids of this domain with loss of all three tyrosines, results in the loss of transforming ability *(56,57)*. Transforming ability is also lost in cells expressing IGF-I receptors with substitution of the individual tyrosines, particularly substitution of tyrosine 1251 *(58)*. Substitution of tyrosine residue 1316 has a lesser effect on mitogenesis, but abolishes tumor formation by cells expressing these receptors *(59)*.

To date, the mapping of important residues involved in the tyrosine kinase activity has focused on tyrosine residues in the β-subunit, because they are obvious targets for phosphorylation. Furthermore, it is known that phosphorylation of tyrosines within specific motfs increases the interaction of the receptor with endogenous substrates. Recently, non-tyrosine motifs have been shown to play an important role in signaling. Substitution of serines at positions 1280–1283 in the C-terminal domain with alanine residues results in loss of transforming ability without loss of IGF-I-induced mitogenicity *(60)*. Interestingly, there is evidence that the insulin receptor has a tightly associated serine kinase that may be intrinsic to the protein *(61)*. These results suggest that in the closely related IGF-I receptor the four serines at positions 1280–1283 may become phosphorylated upon ligand binding and play an important role in signaling.

Other proteins that interact with and affect the IGF-I receptor activity include various protein phosphatases and v-*Src*. Down-regulation of the IGF-I receptor tyrosine kinase occurs by dephosphorylation catalyzed by intracellular tyrosine phosphatases *(62–64)*. Conversely, activation of the IGF-I receptor tyrosine kinase by v-*Src* involves the increase in tyrosine phosphorylation of the IGF-I receptor *(65)*, an effect not readily seen with the analogous insulin receptor.

Signaling Substrates of the IGF-I Receptor

Members of at least five major families of protein substrates, all identified as important for propagation of intracellular signaling mediated by various transmembrane tyrosine kinase receptors, interact directly with the cytoplasmic domain of the IGF-I receptor (Fig. 2). These families include (1) the GRB (growth factor receptor-bound protein) family, adaptor proteins containing both Src-homology (SH) 2 and SH3 domains; (2) the SHC family, adaptor proteins with SH2 domains; (3) the IRS (insulin receptor substrate) family, adaptor proteins with a PTB (phosphotyrosine binding) domain and multiple tyrosine residue-containing motifs that are docking sites for other downstream proteins; (4) the CRK family, adaptor proteins containing SH2 and SH3 domains; and (5) p85, the regulatory subunit of phosphatidylinositol 3-kinase (PI-3K), an enzyme that phosphorylates intracellular lipids. The specificity of signaling from the IGF-I receptor is dictated at each downstream step in the signal cascade by interactions that are just now being elucidated.

Tyrosine phosphorylation of the IGF-I receptor and various downstream adaptor proteins creates binding sites for SH2 domain-containing signaling proteins. Some examples of these interactions include: YXXM motifs as binding sites for PI-3K; YVNM motifs as binding sites for Grb2; YIEV motifs as binding sites for protein-tyrosine phosphatase SHP-2 and phospholipase C_γ, and NPXY motifs as binding sites for IRS-1, SHC, and GTPase-activating protein (GAP). The specific binding sites for each of these protein families are being mapped. Docking of these proteins in turn activates specific downstream signal transduction pathways. Further details are noted below in the descriptions of each substrate family.

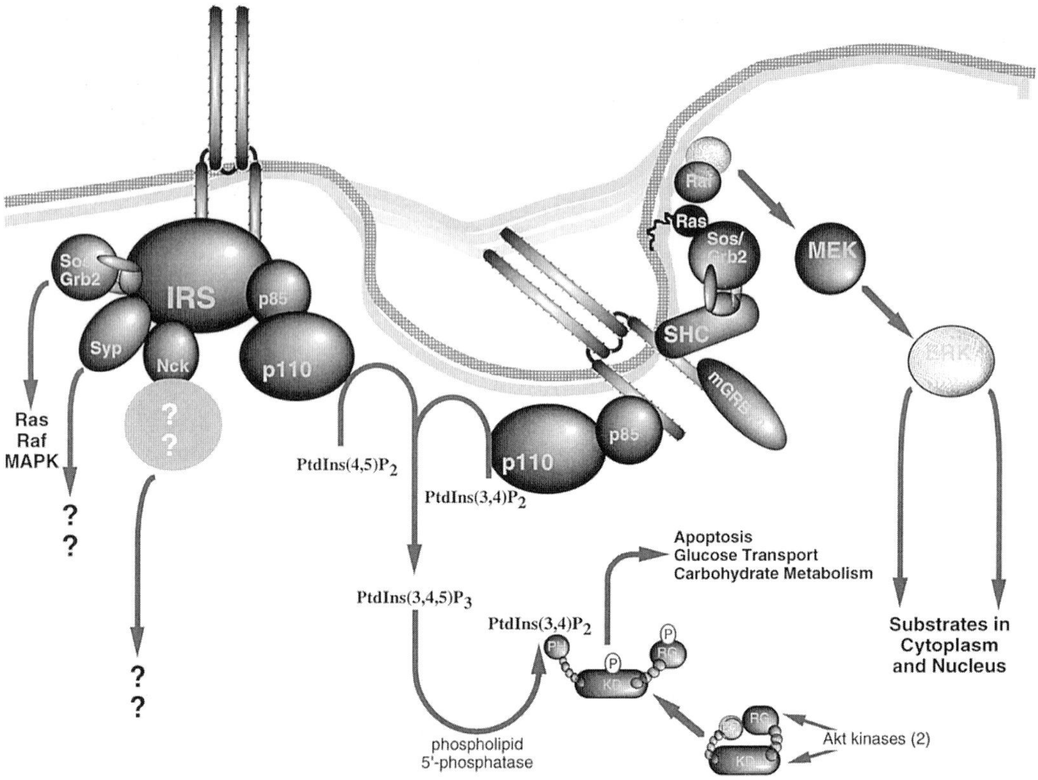

Fig. 2. Schematic representation of intracellular signaling pathways emanating from the IGF-I receptor. The activated IGF-I receptor undergoes autophosphorylation upon ligand binding. Thereafter several endogenous intracellular adaptor proteins bind specifically to the activated receptor and are phosphorylated on tyrosine residues. Tyrosyl-phosphorylated adaptor proteints are capable of binding the next substrates in the signaling cascade. The IGF-I receptor phosphorylates a number of adaptor proteins including IRS proteins, SHC, and Crk-II. Tyrosyl-phosphorylated SHC binds the Grb2/SOS complex. This complex associates with Ras, resulting in membrane association and activation of Raf. Thereafter, dual tyrosine/threonine kinases phosphorylate subsequent kinases such as MEK and MAP kinases. This pathway is active during IGF-I-mediated cellular proliferation. Tyrosine phosphorylation of p85, the regulatory subunit of PI3-K, induces activation of the catalytic p110 catalytic subunit. Activated PI3-K catalyzes phosphorylation of the phosphoinositol bisphosphonate. Subsequent dephosphorylation of the trisphosphonate with phosphorylation of Akt mediates signals associated with apoptosis, glucose transport, and carbohydrate metabolism. Tyrosyl-phosphorylated IRS proteins act as docking proteins for several downstream adaptor proteins, including p85 and Grb2/SOS. IRS proteins also associate with the phosphatase Syp and the kinase Nck. The specific pathways mediated by these enzymes are not fully understood.

Various splice variants of Grb10 and its homologues interact directly with the IGF-I receptor *(66–70)*. The mGrb10α isoform inhibits IGF-I-mediated growth, and appears to occupy a binding site in the C-terminus of the IGF-I receptor *(70,71)*. This isoform, the orginally described Grb10, also binds the C-terminal domain of the insulin receptor but does not affect insulin-mediated cell growth *(71,72)*. It is unclear exactly where on the receptor that mGrb10α1 binds. If it binds to the IGF-I receptor in an analogous manner to that of its binding the insulin receptor, it requires that the receptor be capable of autophosphorylation. However, it does not bind to the tyrosine-containing motifs of the C-terminal domain *(70)*. Microinjection of a peptide of a second isoform, mGrb10γ1, inhibits both IGF-I- and insulin-mediated DNA synthesis, suggesting that mGrb10γ1 mediates IGF-I- and insulin-induced cellular proliferation *(69)*. Analyses of the early results of studies of these two isoforms of Grb10 suggest that different isoforms may have different physiologic functions. The mechanism by which these molecules transmit signals to downstream elements have not, as yet, been identified.

There are three known protein isoforms of the SHC family, i.e., the p46-, p52-, and p66-kDa SHC proteins. SHC proteins each contain a SH2 domain, a SAIN (SHC and IRS NPXY) binding domain, and various tyrosine residues in motifs that act as binding sites for other SH2 domain-containing proteins. An intact NPXY motif in the receptor is essential for the interaction with SHC. SHC is phosphorylated on tyrosine residues by the activated IGF-I receptor which results in its association with Grb2 (a SH2 and SH3 domain-containing 24 kDa-adaptor protein) (Fig. 2) *(53,73,74)*. SHC and Grb2 activate the Ras/Raf/MAP kinase pathway via the association of Grb2 with mSos, a nucleotide-exchange protein, that loads GTP onto the Ras molecule *(75–77)*. In this manner the IGF-I receptor activates the Ras/Raf/MAP kinase, PI3′-kinase, and presumably other signaling pathways *(73,78–80)*. The receptor/SHC/Grb2/mSOS/Ras activation cascade has been well characterized for the epidermal growth factor (EGF) and platelet-derived growth factor (PDGF) receptors, and is considered the primary pathway for activating the mitogenic Ras/Raf/MAP kinase pathway. The IGF-I and insulin receptors can also activate this pathway via IRS–protein interactions with the Grb2–mSOS complex (Fig. 2).

The IRS proteins are related by the presence of common domains including a pleckstrin homology (PH) domain, a PTB domain, and the C-terminal domain that contain many tyrosine-containing motifs capable of binding several of the known SH2 domain-containing proteins important in the signaling pathways of the IGF-I receptor. To date four members of the IRS family have been described (i.e., IRS-1, IRS-2 or 4PS [IL-4-protein substrate], IRS-3, and IRS-4) *(81–87)*. Both IRS-1 and IRS-2 directly engage the stimulated IGF-I receptor and rapidly become phosphorylated on multiple tyrosine residues *(88–91)*. Interaction of these adaptor molecules then occurs with the SH2 domain-containing proteins including Grb2, SHP-2 (also known as Syp or PTP1D), p85 (the regulatory subunit of PI 3-K), Crk-II, and Nck (Fig. 2) *(88,92–95)*. The PH domain does not appear to interact directly with the receptor (i.e., insulin receptor); however, it is necessary for efficient tyrosine phosphorylation of the IRS proteins (i.e., IRS-1) *(96–98)*.

The N-terminal domain (amino acids 160–516), which includes the PTB motif (amino acids 160–174) of IRS-1, interacts with the IGF-I receptor. The more distal SAIN binding domain also plays a role in binding of IRS-1 to the IGF-I receptor *(91,99)*. More recently it has been shown that the motif around tyrosine 895 of IRS-1, now termed the RBD1 (receptor binding domain 1), is the preferred binding site for the IGF-I receptor. This is in contrast to the regions surrounding tyrosines 987 and 727 which are the preferred binding sites for the insulin receptor. An intact NPXY motif in the IGF-I receptor, with phosphorylation of the tyrosine residue, is essential for the interaction of IRS-1. Using the yeast two-hybrid system, IRS-1 has also been shown to interact with the major tyrosine phosphorylation site, tyrosines 1131, 1135, and 1136, of the IGF-I receptor *(99)*.

IRS-2 apparently binds via two domains to two separate regions of the IGF-I receptor: the N-terminal RBD1 of IRS-2 (amino acids 188–591) that contains a PTB domain and the RBD2 domain (amino acids 591–733) *(100)*. The NPXY motif containing tyrosine 950 of the IGF-I receptor has been shown to be essential for the binding of RBD1 of

IRS-2, but not for binding the RBD2 of IRS-2. Binding of the RBD2 of IRS-2 requires phosphorylation of the receptor on residues other than tyrosine residues, as yet undefined.

Several SH2 domain-containing downstream proteins are capable of binding tyrosine phosphorylated IRS-1 and IRS-2. The best characterized of this is p85, the regulatory subunit of PI3-K. The p85 subunit contains two SH2 domains that bind to tyrosine-phosphorylated YXXM motifs in IRS-1 *(83,101,102)*. PI3-K appears to preferentially phosphorylate phosphatidylinositol-4,5-bisphosphate to phosphatidylinositol-3,4,5-trisphosphate (PIP_3), which is a crucial, perhaps ubiquitous, second messenger in many cells. It has been suggested that phospholipase Cβ1 (the calcium- and diacylglycerol-independent isoform of phospholipases) is a target of this PIP_3 *(103)*. This isoform of phospholipase C is essential for IGF-I-induced cellular proliferation *(104)*. A second important pathway is activated following induction of PI3-K activity in response to IGF-I. The ribosomal protein S6 kinase ($p70^{S6K}$) becomes serine phosphorylated and activated in response to IGF-I and these actions can be abrogated by complete inhibition of PI3-K activity *(105–107)*.

Two newly identified members of the IRS family have been shown to associate with the insulin receptor. IRS-3 is tyrosine phosphorylated in response to insulin and then associates with PI3-K *(82,86)*. IRS-4 is a 160-kDa protein that is tyrosine phosphorylated in response to insulin and is immunologically distinct from IRS-1 *(87)*. Both IRS-3 and IRS-4 contain the defining domains of the IRS family of adaptor proteins. Each has tyrosine residues in the C-terminal region that, when phosphorylated, act as docking sites for SH2 domain-containing proteins, thus propagating the signaling further down the cascade. The potential interactions of IRS-3 and IRS-4 with the IGF-I receptor are under investigation.

The CRK family of SH2-SH3 domain-containing adaptor proteins have been identified as substrates for the IGF-I receptor. This family includes Crk-I, Crk-II, and Crk-L *(108,109)*. Tyrosine phosphorylation of Crk-II occurs rapidly in response to IGF-I *(110,111)*. Phosphorylated Crk-II can then interact with C3G and mSos, two well-characterized nucleotide-exchange proteins, thereby activateing the Ras/Raf/MAP kinase pathway (Fig. 2). Interaction of Crk-II with IRS-1 and IRS-2 can potentially mediate the mitogenic response of cells to IGF-I. Recently it has been shown that Crk-II directly binds the IGF-I receptor and preliminary mapping has been done *(112)*. The SH2 domain of Crk-II interacts with phosphorylated tyrosines 943 and 950 in the juxtamembrane domain and tyrosine 1250 in the C-terminal domain of the IGF-I receptor. In addition, Crk-II interacts with the IGF-I receptor on phosphorylated non-tyrosine residues (presumably serines). The exact motif of Crk-II responsible for this interaction has not yet been determined *(112)*.

Other substrates for the IGF-I receptor have been described. For example, substrates that bind primarily to IRS-1 and IRS-2 also may bind the IGF-I receptor directly. The regulatory subunit (p85) of PI3-K and the phosphatase SHP-2 (also known as Syp) bind the IGF-I receptor at tyrosine 1316 in the C-terminal domain. GAP interacts with tyrosine 950 in the juxtamembrane domain of the IGF-I receptor. Evidence that IGF-I-induced activation of the Ras/Raf/MAP kinase pathways via adaptor proteins other than SHC and IRS1/2 has been shown. Such activation of the MAP kinase pathway can occur via activation of $p21^{Ras}$ by the pertussis toxin-sensitive G protein subunit complex βγ *(113)*. IGF-I receptor activation in hematopoeitic cells results in rapid tyrosine phosphorylation of the *vav* proto-oncogene, an effect apparently not mediated by IRS-1 *(114)*. p95vav is an SH2–SH3–SH3-containing protein with motifs typical of guanine exchange factors, such as mSOS and C3G. Two isoforms of the 14–3–3 family of proteins, 14–3–3 β and ζ have been shown to interact directly with the IGF-I receptor, but not with the insulin receptor *(115)*, suggesting a specific response to IGF-I stimulation *(116)*. The IGF-I receptor may also transphosphorylate the EGF receptor, thereby enhancing its function *(117)*. The majority of IGF-I-stimulated pathways involve phosphorylation of substrates. However, IGF-I receptor signaling also involves dephosphorylation of certain substrates. Paxillin, a protein involved in focal adhesions, is rapidly dephosphorylated in response to IGF-I receptor activation *(118–120)*. IGF-I stimulation induced the association of paxillin with Crk-II in parallel with a dissociation of $p130^{Ras}$–Crk-II complex *(119,120)*. IGF-I receptor signaling also results in phosphorylation of phospholipase Cγ *(121)*, selective translocation of PKC to the nucleus in Swiss 3T3 fibroblasts *(122)*, as well as induction of tyrosine phosphorylation of a number of nuclear proteins *(123)*. The breadth of protein modifications in response to IGF-I activation is wide and as yet incompletely defined.

THE IGF-I RECEPTOR AND APOPTOSIS

The regulation of apoptosis, or programmed cell death, is essential to normal development and in particular disease states. During development of the nervous system, apoptosis is an important process leading to the death of large numbers of excess neurons, thereby allowing surviving neurons to establish correct synaptic connections *(124)*. In some disease states, e.g., viral infections and emerging cancers, apoptosis constitutes an important defense mechanism. The disregulation of apoptsis is intimately involved in the pathogenesis of many disease states, such as degenerative neuronal diseases and progressive cancers. The exact mechanisms involved in apoptosis are currently under study and a number of recent studies have elucidated some of the mediators of apoptosis.

Studies on the effects of the IGFs on cellular viability in culture have demonstrated that IGF-I and -II are very powerful survival factors. For instance, cerebellar granule cells are protected from subphysiologic potassium-induced apoptosis by a low concentration of IGF-I *(125)*. The exact mechanism of the action of IGF-I is unknown. IGF-I has been shown to protect SH-SY5Y neuroblastoma cells from hyperosmotic-induced apoptosis *(126)*. In an in vivo model of cell death produced by unilateral cerebral ischemia with hypoxia, intracerebral infusion of rhIGF-I reduced neuronal cell death *(127)*. Such findings are not restricted to neuronal cells. IGF-I markedly decreases apoptosis in erythroid colony-forming cells devoid of essential growth factors *(128)*, in COS cells deprived of serum but overexpressing interleukin-1β-converting enzyme *(129)*, and in fibroblasts induced to undergo apoptosis by the c-myc protein *(130)*. IGF-I inhibits apoptosis in atretic ovarian follicules, an effect associated with increased bcl-2 expression *(131)*. IGF-I inhibits apopotosis during mammary gland involution in transgenic mice *(132)*. Studies using mice with deficiencies of the IGF-I receptor or IGF ligands (IGF-I or -II) confirm that the IGF-I receptor signaling system plays an essential role in normal embryonic and postnatal development *(133,134)*. Normal development is dependent on appropriately regulated apoptosis, and presumably alteration in the signaling through the IGF-I receptor system impacts negatively on appropriate apoptosis and mitogenesis.

The results of several studies suggest a role for the IGF/IGF-I receptor system in controlling tumor cell survival. It has been hypothesized that the extent of apoptosis may affect the aggressiveness of the tumor *(135)*. Apoptosis induced by N-myc is inhibited by IGF-II, compatible with the overexpression of IGF-II in hepatocellular carcinoma

(136). In an in vivo system, it has been demonstrated that a decrease in the number of IGF-I receptors results in increased apoptosis of tumor cells. Conversely, when IGF-I receptor expression was increased, the level of apoptosis was markedly inhibited *(137,138)*. Most human cancers, including lung, breast, colon, thyroid, brain, and Ewing's sarcoma, express higher levels of IGF-I receptors than normal nontransformed tissues *(139–147)*. In breast cancer cell lines, chemotherapeutic agents, including puromycin, 5-fluorouracil, cisplatin, actinomycin D, adriamycin, methotrexate, tamoxifen, and camptothecin, induced apoptosis which was specifically and effectively inhibited by IGF-I *(148,149)*. IGF-I did not change the levels of bcl-2 protein or bax mRNA levels; both of these proteins are known to be involved in anti-apoptotic pathways *(149)*. The cytostatic action of tamoxifen on breast cancer cells was associated with a down-regulation of IGF-I-induced autophosphorylation of the IGF-I receptor, up-regulation of SHC tyrosine phosphorylation and SHC-Grb2 association, and inhibition of the IRS-1/PI3-K signaling pathways *(150)*. Exactly how these changes in IGF-I signaling affect of the balance of cell proliferation and apoptosis has yet to be determined. Based of studies such as these, it has been postulated that the resistance of some cancers to chemotherapeutic agents may in part be due to the anti-apoptotic effects of particular growth factors.

Although the effects of IGF-I on inhibiting apoptosis are well described, the intracellular signaling pathways involved are less clear. In rat pheochromocytoma PC12 cells undergoing apoptosis following serum and nerve growth factor (NGF) depletion, IGF-I inhibits apoptosis via at least two post-receptor signaling pathways, the PI3-K and MAP kinase pathways *(151)*. Okadaic acid (OKA) is a potent inhibitor of both serine/threonine protein phosphatase-1 and 2A. OKA induces apotosis in a variety of cell lines *(57,152)*. Stimulation of the IGF-I receptor protects cells from OKA-induced apoptosis. This protective effect is proportional to the density of functional IGF-I receptors on the cell *(138)*. Conversely, disruption of the function of the IGF-I receptor can reverse the transformed phenotype, inhibit tumorigenesis, and induce loss of the metastatic phenotype *(59,137,153–160)*. This evidence suggests that the IGF-I receptor signaling cascades are important for induction and/or maintenance of tumor growth. More recent studies have shown that the IGF-I receptor is not essential for this transformed phenotype. Expression of either v-Src or the G-protein $G_{\alpha 12}$ can confer a transformed phenotype to cells that lack the IGF-I receptor *(161)*. Some regions of the IGF-I receptor that are important for mediating the anti-apoptotic signal have been mapped. Important regions include the ATP-binding site (lysine 1003) and two residues in the C-terminal domain (tyrosine 1251 and histidine 1293) *(161,162)*. The triple tyrosine cluster in the tyrosine kinase domain is not essential to protect against apoptosis although it is essential for IGF-I to stimulate cellular proliferation.

THE IGF-I RECEPTOR AND CELLULAR MOTILITY

The effect of IGF-I on tumor cells may, in part, be the impact of IGF-I-mediated signaling on cellular motility. Tumor cell interactions with adhesion proteins and the extracellular matrix (ECM) are influenced by growth factors and this interaction likely contributes to the tumor's ability to spread locally and metastasize. Adhesion receptors of the integrin family of proteins promote cell attachment to the ECM and potentiate cellular migration and invasion *(163–165)*. Human tumors frequently express the integrin $\alpha v\beta 5$, but these cells require a growth factor (e.g., IGF-I)-mediated signal event for mobility on the ECM *(166)*. IGF-I can be secreted by tumor cells and this may lead to autocrine stimulation of tumor cell growth and motility *(154, 167–169)*. Interaction of the IGF-I receptor and αvβ5 resulted in pulmonary metastases of several tumor cell types in two in vivo models *(170)*. It was shown that IGF-I promoted actin polymerization at the leading edge of a migrating cell *(171)*. IGF-I-induced motility required the formation of new adhesions with involvement of tyrosine phosphorylated paxillin and focal adhesion kinase *(171)*. Similar to tumor cells, vascular smooth muscle cells (VSMCs) express IGF-I receptors and migrate in response to IGF-I and IGF-II via activation of the IGF-I receptor *(172,173)*. An antibody directed against a specific integrin (i.e., αvβ3) abrogated IGF-I-induced migration of VSMC *(174)*. VSMC migration is considered one of the most important components in the formation of atherosclerotic plaques *(175)*. IGF-I-mediated signals thus play an important role in cellular migration as illustrated by tumor cells and VSMCs.

CONCLUSIONS

Studies of the IGF-I receptor have yielded a substantial amount of information about the critical functions that are mediated by the receptor. Significantly, the pathways that mediate progression through the cell cycle resulting in cellular proliferation are being elucidated. Balancing mitogenesis, the function of the IGF-I receptor in modulating apoptosis is being defined. The crucial information that is being gathered indicates the level of complexity of the IGF-I receptor signaling pathways. Importantly the interrelationships between the the signaling pathways of the IGF-I receptor, a transmembrane receptor tyrosine kinase, and other receptor signaling systems are being investigated.

REFERENCES

1. LeRoith D, Adamo M, Werner H, Roberts CT Jr. Insulin-like growth factors and their receptors as growth regulators in normal physiology and pathological states. Trends Endocrinol Metab 1991; 2:134–139.
2. Nissley P, Lopaczynski W. Insulin-like growth factor receptors. Growth Factors 1991; 5:29–49.
3. Jones JI, Clemmons DR. Insulin-like growth factors and their binding proteins: biological actions. Endocr Rev 1995; 16:3–24.
4. LeRoith D, Werner H, Beitner-Johnson D, Roberts CT Jr. Molecular and cellular aspects of the insulin-like growth factor I receptor. Endocr Rev 1995; 16:143–163.
5. Ullrich A, Gray A, Ram AW, Yang-Feng T, Tsubokawa M, Collins C, Henzel W, Le Bon T, Kathuria S, Chen E, Jacobs S, Francke U, Ramachandran J, Fujita-Yamaguchi Y. Insulin-like growth factor I receptor primary structure: comparison with insulin receptor suggests structural determinants that define functional specificity. EMBO J 1986; 5:2503–2512.
6. Jacobs S, Kull FC Jr, Cuatrecasas P. Monensin blocks the maturation of receptors for insulin and somatomedin C: identification of receptor precursors. PNAS 1983; 80:1228–1231.
7. Yee D, Lebovic GS, Marcus RR, Rosen N. Identification of an alternate type I insulin-like growth factor receptor *β* subunits in mRNA transcript. J Biol Chem 1989; 264:21439–21441.
8. Condorelli G, Smith RJ. Two alternatively spliced forms of the IGF-I receptor have distinct biological activities and hormone-induced internalization rates. Exp Clin Endocrinol 1993; 101:95–97.
9. Condorelli G, Bueno R, Smith RJ. Two alternatively spliced forms of the human insulin-like growth factor I receptor have distinct biological activities and internalization kinetics. J Biol Chem 1994; 269:8510–8516.
10. Moss AM, Livingston JN. Distinct β-subunits are present in hybrid insulin-like-growth-factor-I receptors in the central nervous system. Biochem J 1993; 294:685–692.
11. Alexandrides TK, Moses AC, Smith RJ. Developmental expression of receptors for insulin, insulin-like growth factor I (IGF-I), and IGF-II in rat skeletal muscle. Endocrinology 1989; 124:1064–1076.

12. Alexandrides TK, Smith RJ. A novel fetal insulin-like growth factor (IGF) I receptor. Mechanism for increased IGF-I and insulin-stimulated tyrosine kinase activity in fetal muscle. J Biol Chem 1989; 264:12922–12930.
13. Alexandrides TK, Chen J-H, Bueno R, Giorgino F, Smith RJ. Evidence for two insulin-like growth factor I receptors with distinct primary structure that are differentially expressed during development. Regul Pept 1993; 48:279–290.
14. Kellerer M, Obernmaier-Kusser B, Ermal B, Wallner U, Haring H-U, Petrides P. An altered IGF-I receptor is present in human leukemic cells. J Biol Chem 1990; 265:9340–9345.
15. Garofalo RS, Barenton B. Functional and immunological distinction between insulin-like growth factor I receptor subtypes in KB cells. J Biol Chem 1992; 276:1140–11475.
16. Barenton B, Domeyne A, Garandel V, Garofalo RS. A developmentally regulated form of insulin-like growth factor receptor β-subunit in C2 myoblasts exhibiting altered requirements for differentiation. Endocrinology 1993; 133:651–660.
17. Domeyne A, Pinset C, Montarras D, Garandel V, Rosenfeld RG, Barenton B. Preferential binding of insulin-like growth factor-II (IGF-II) to a putative $\alpha 2\beta 2$IGF-II receptor type C2 myoblasts. Eur J Biochem 1992; 208:273–279.
18. Soos MA, Siddle K. Immunological relationships betwen receptors for insulin and insulin-like growth factor I. Evidence for structural heterogeneity of insulin-like growth factor I receptors involving hybrids with insulin receptors. Biochem J 1989; 263:553–563.
19. Soos MS, Whittaker J, Lammers R, Ullrich A, Siddle K. Receptors for insulin and insulin-like growth factor-I can form hybrid dimers. Characterization of hybrid receptors in transfected cells. Biochem J 1990; 270:383–390.
20. Soos MA, Field CE, Siddle K. Purified hybrid insulin/insulin-like growth factor-I receptors bind insulin-like growth factor-I, but not insulin, with high affinity. Biochem J 1993; 290:419–425.
21. Kasuya J, Paz B, Maddux AB, Goldfine ID, Hefta SA, Fujita-Yamaguchi Y. Characterization of human placental insulin-like growth factor-I/insulin hybrid receptors by protein microsequencing and purification. Biochemistry 1993; 32:13531–13536.
22. Moxham CP, Duronio V, Jacobs S. Insulin-like growth factor I receptor β subunit heterogeneity. Evidence for hybrid tetramers composed of insulin-like growth factor I and insulin receptor heterodimers. J Biol Chem 1989; 264:13238–13244.
23. Moxham CP, Jacobs S. Insulin/IGF-I receptor hybrids: a mechanism for increasing receptor diversity. J Cell Biochem 1992; 48:136–140.
24. Moxham CP, Jacobs S. Insulin-like growth factor receptors. In: Schofield PN, ed. The Insulin-Like Growth Factors. Structure and Biological Functions. Oxford University Press, Oxford, 1992, pp. 80–109.
25. Treadway JL, Morrison BD, Goldfine ID, Pessin JE. Assembly of insulin/insulin-like growth factor-I hybrid receptors in vitro. J Biol Chem 1989; 264:21450–21453.
26. Ullrich A, Bell JR, Chen EY, Herrera R, Petruzzelli LM, Dull TJ, Gray A, Coussen L, Liao Y-C, Tsubokawa M, Mason A, Seeburg PH, Grunfeld C, Rosen OM, Ramachandran J. Human insulin receptor and its relationship to the tyrosine kinase family of oncogenes. Nature 1985; 313:756–761.
27. Shier P, Watt VM. Primary structure of a putative receptor for a ligand of the insulin family. J Biol Chem 1989; 264:14605–14608.
28. Ullrich A, Schlessinger J. Signal transduction by receptors with tyrosine kinase activity. Cell 1990; 61:203–212.
29. Shier P, Watt VM. Tissue-specific expression of the rat insulin receptor-related gene. Mol Endocrinol 1992; 6:723–729.
30. Tollefson SE, Thompson K. The structural basis for insulin-like growth factor-I receptor high affinity binding. J Biol Chem 1988; 263:16267–16273.
31. Andersen AS, Kjeldsen T, Wiberg FC, Christensen PM, Rasmussen JS, Norris K, Moller KB, Moller NPH. Changing the insulin receptor to possess insulin-like growth factor I ligand specificity. Biochemistry 1990; 29:7363–7366.
32. Kjeldsen T, Andersen AS, Wiberg FC, Rasmussen JS, Schaffer L, Balschmidt P, Moller KB, Moller NPH. The ligand specificities of the insulin receptor and the insulin-like growth factor I receptor reside in different regions of a common binding site. PNAS 1991; 88:4404–4408.
33. Schumacher R, Mosthaf L, Schlessinger J, Brandenburg D, Ullrich A. Insulin and insulin-like growth factor-I binding specificity is determined by distinct regions of their cognate receptors. J Biol Chem 1991; 266:19288–19295.

34. Soos MA, Field CE, Lammers R, Ullrich A, Zhang B, Roth RA, Andersen AS, Kjeldsen T, Siddle K. A panel of monoclonal antibodies for the type I insulin-like growth factor receptor. J Biol Chem 1992; 267:12955–12963.
35. Andersen AS, Kjeldsen T, Wiberg FC. Identification of determinants that confer ligand specificity on the insulin receptor. J Biol Chem 1992; 267:13581–13686.
36. Zhang B, Roth RA. A region of the insulin receptor important for ligand binding (residues 450–601) is recognised by patients autoimmune antibodies and inhibitory monoclonal antibodies. PNAS 1991; 88:9858–9862.
37. Kjeldsen T, Wiberg FC, Andersen AS. Chimeric receptors indicate that phenylalanine 39 is a major contributor to insulin specificity of the insulin receptor. J Biol Chem 1994; 269:32942–32946.
38. Mynarcik DC, Williams PF, Schaffer L, Yu GQ, Whittaker J. Identification of common ligand binding determinants of the insulin and insulin-like growth factor 1 receptors. Insights into mechanisms of ligand binding. J Biol Chem 1997; 272:18650–18655.
39. Shoelson SE, White MF, Kahn CR. Tryptic activation of the insulin receptor. J Biol Chem 1988; 263:4852–4860.
40. Liu D, Rutter WJ, Wang L-H. Enhancement of transforming potential of human insulin-like growth factor I receptor by N-terminal truncation and fusion to avian sarcoma virus UR2 gag sequence. J Virol 1992; 66:374–385.
41. Liu D, Rutter WJ, Wang L-H. Modulating effects of the extracellular sequence of the human insulinlike growth factor I receptor on its transforming and tumorigenic potential. J Virol 1993; 67:9–18.
42. Takahashi K, Yonezawa K, Nishimoto I. Insulin-like growth factor-I receptor activated by a transmembrane mutation. J Biol Chem 1995; 270:19041–19045.
43. Pilch PF. The insulin-like growth factor-I receptor kinase. In: Moudgil VK, ed. Receptor Phosphorylation. CRC Press, Boca Raton, FL, 1989, pp. 135–150.
44. Kato H, Faria TN, Stannard B, Roberts CT Jr, LeRoith D. Role of tyrosine kinase activity in signal transduction by the insulin-like growth factor-I (IGF-I) receptor: characterization of kinase-deficient IGF-I receptors and the action of an IGF-I-mimetic antibody (αIR-3). J Biol Chem 1993; 268:2655–2661.
45. Grönborg M, Wulff BS, Rasmussen JS, Kjeldsen T, Gammeltoft S. Structure–function relationship of the insulin-like growth factor I receptor tyrosine kinase. J Biol Chem 1993; 268:23435–23440.
46. Kato H, Faria TN, Stannard B, Roberts CT Jr, LeRoith D. Essential role of tyrosine residues 1131, 1135, and 1136 of the insulin-like growth factor-I (IGF-I) receptor in IGF-I action. Mol Endocrinol 1994; 8:40–50.
47. Li S, Ferber A, Miura M, Baserga R. Mitogenicity and transforming activity of the insulin-like growth factor-I receptor with mutations in the tyrosine kinase domain. J Biol Chem 1994; 269:32558–32564.
48. Hernández-Sánchez C, Blakesley VA, Kalebic T, Helman L, LeRoith D. The role of the tyrosine kinase domain of the insulin-like growth factor-I receptor in intracellular signaling, cellular proliferation, and tumorigenesis. J Biol Chem 1995; 270:29176–29181.
49. Stannard B, Blakesley VA, Kato H, Roberts CT Jr, LeRoith D. Single tyrosine substitution in the insulin-like growth factor-I receptor inhibits biological activity and ligand-induced receptor internalization. Endocrinology 1995; 136:4918–4924.
50. Jiang Y, Chan JLK, Zong CS, Wang L. Effect of tyrosine mutations on the kinase activity of transforming potential of an oncogenic human insulin-like growth factor I receptor. J Biol Chem 1996; 271:160–167.
51. Chen J, Hanafusa T, Wang L-H. Ala-Gly mutation in the putative catalytic loop confers temperature sensitivity on Ros, insulin receptor, and insulin-like growth factor I receptor protein-tyrosine kinases. Proc Natl Acad Sci USA 1994; 91:321–325.
52. Blakesley VA, Kato H, Roberts CT Jr, LeRoith D. Mutation of a conserved amino acid residue (tryptophan 1173) in the tyrosine kinase domain of the IGF-I receptor abolishes autophosphorylation but does not eliminate biologic function. J Biol Chem 1995; 270:2764–2769.
53. White MF, Kahn CR. The insulin signaling system. J Biol Chem 1994; 269:1–4.
54. Hsu D, Knudson PE, Zapf A, Rolband GC, Olefsky JM. NPXY motif in the insulin-like growth factor-I receptor is required for efficient ligand-mediated receptor internalization and biological signaling. Endocrinology 1994; 134:744–750.
55. Prager D, Li H-L, Yamasaki H, Melmed S. Human insulin-like growth factor I receptor internalization. Role of the juxtamembrane domain. J Biol Chem 1994; 269:11934–11937.
56. Liu D, Zong CS, Wang L-H. Distinctive effects of the carboxyl-terminal sequence of the insulin-like growth factor I receptor on its signaling functions. J Virol 1993; 67:6835–6840.

57. Surmacz E, Sell C, Swantek J, Kato H, Roberts CT Jr, LeRoith D, Baserga R. Dissociation of mitogenesis and transforming activity by C-terminal truncation of the insulin-like growth factor I receptor. Exp Cell Res 1995; 218:370–380.
58. Miura M, Surmacz E, Burgaud J-L, Baserga R. Different effects on mitogenesis and transformation of a mutation at tyrosine 1251 of the insulin-like growth factor I receptor. J Biol Chem 1995; 270:22639–22644.
59. Blakesley VA, Kalebic T, Helman LJ, Stannard B, Faria TN, Roberts CT Jr, LeRoith D. Tumorigenic and mitogenic capacities are reduced in transfected fibroblasts expressing mutant insulin-like growth factor (IGF)-I receptors. The role of tyrosine residues 1250, 1251, and 1316 in the carboxy-terminus of the IGF-I receptor. Endocrinology 1996; 137:410–417.
60. Li S, Resnicoff M, Baserga R. Effect of mutations at serines 1280–1283 on the mitogenic and transforming activities of the insulin-like growth factor-I receptor. J Biol Chem 1996; 271:12254–12260.
61. Lewis RE, Wu GP, Macdonald RG, Czech M. Insulin sensitive phosphorylation of serine 1293/1294 on the human insulin receptor by a tightly associated serine kinase. J Biol Chem 1990; 265:947–954.
62. Peraldi P, Hauguel-de Mouzon S, Alengrin F, Van Obberghen E. Dephosphorylation of human insulin-like growth factor I (IGF-I) receptors by membrane-associated tyrosine phosphatases. Biochem J 1992; 285:71–78.
63. van der Geer P, Hunter T, Lindberg RA. Receptor protein-tyrosine kinases and their signal transduction pathways. Annu Rev Cell Biol 1994; 10:251–337.
64. Way BA, Mooney RA. Activation of phosphatidylinositol-3-kinase by platelet-derived growth factor and insulin-like growth factor-I is inhibited by a transmembrane phosphotyrosine phosphatase. J Biol Chem 1993; 269:26409–26415.
65. Peterson JE, Jelinek T, Kaleko M, Siddle K, Weber MJ. Phosphorylation and activation of the IGF-I receptor in src-transformed cells. J Biol Chem 1994; 269:27315–27321.
66. Ooi J, Yajnik V, Immanuel D, Gordon M, Moskow JJ, Buchberg AM, Margolis B. The cloning of Grb10 reveals a new family of SH2 domain proteins. Oncogene 1995; 10:1621–1630.
67. Dey BR, Frick K, Lopaczynski W, Nissley SP, Furlanetto RW. Evidence for the direct interaction of the insulin-like growth factor I receptor with IRS-1, Shc, and Grb10. Mol Endocrinol 1996; 10:631–641.
68. Morrione A, Valentinis B, Li S, Ooi JYT, Margolis B, Baserga R. Grb10: a new substrate of the insulin-like growth factor I receptor. Cancer Res 1996; 56:3165–3167.
69. O'Neill TJ, Rose DW, Pillay TS, Hotta K, Olefsky JM, Gustafson TA. Interaction of a GRB-IR splice variant (a human GRB10 homolog) with the insulin and insulin-like growth factor I receptors. Evidence for a role in mitogenic signaling. J Biol Chem 1996; 271:22506–22513.
70. Dong LQ, Farris S, Christal J, Liu F. Site-directed mutagenesis and yeast two-hybrid studies of the insulin and insulin-like growth factor-1 receptors: the Src homology-2 domain-containing protein hGrb10 binds to the autophosphorylated tyrosine residues in the kinase domain of the insulin receptor. Mol Endocrinol 1997; 11:1757–1765.
71. Morrione A, Valentinis B, Resnicoff M, Xu S-Q, Baserga, R. The role of mGrb10α in insulin-like growth factor I-mediated growth. J Biol Chem 1997; 272:26382–26387.
72. Hansen H, Svensson U, Zhu J, Laviola L, Giorgino F, Wolf G, Smith RJ, Riedel H. Interaction between the Grb10 SH2 domain and the insulin receptor carboxyl terminus. J Biol Chem 1996; 271:8882–8886.
73. Skolnik EY, Batzer A, Li N, Lee C-H, Lowenstein E, Mohammadi M, Margolis B, Schlessinger J. The function of GRB2 in linking the insulin receptor to Ras signaling pathways. Science 1993; 260:1953–1955.
74. Sasaoka T, Rose DW, Juhn BH, Saltiel AR, Draznin B, Olefsky JM. Evidence for a functional role of Shc proteins in mitogenic signaling induced by insulin, insulin-like growth factor-I, and epidermal growth factor. J Biol Chem 1994; 269:13689–13694.
75. Gale NW, Kaplan S, Lowenstein EJ, Schlessinger J, Bar-Sagi D. Grb2 mediates the EGF-dependent activation of guanine nucleotide exchange on Ras. Nature (Lond) 1993; 363:88–92.
76. Lowenstein EJ, Daly RJ, Batzer AG, Li W, Margolis B, Lammers R, Ullrich A, Skolnik EY, Bar-Sagi D, Schlessinger J. The SH2 and SH3 domain-containing protein GRB2 links receptor tyrosine kinases to ras signaling. Cell 1992; 70:431–442.
77. Rozakis-Adcock M, Fernley R, Wade J, Bowtell D. The SH2 and SH3 domains of mammalian Grb2 couple the EGF receptor to the Ras activator mSos1. Nature 1993; 363:83–85.
78. Webster J, Prager D, Melmed S. Insulin-like growth factor-I activation of extracellular signal-related kinase-1 and -2 in growth hormone-secreting cells. Mol Endocrinol 1994; 8:539–544.
79. Hansson A, Thorén M. Activation of MAP kinase in Swiss 3T3 fibroblasts by insulin-like growth factor-I. Growth Regul 1995; 5:92–100.

80. Lamothe B, Bucchini D, Jami J, Joshi RL. Interaction of p85 subunit of PI 3-kinase with insulin and IGF-1 receptors analysed by using the two-hybrid system. FEBS Lett 1995; 373:51–55.
81. Shemer J, Adamo MM, Wilson GL, Heffez D, Zick Y, LeRoith D. Insulin and insulin-like growth factor-I stimulate a common endogenous phosphoprotein substrate (pp185) in intact neuroblastoma cells. J Biol Chem 1987; 262:15476–15482.
82. Lavan BE, Lienhard GE. The insulin-elicited 60-kDa phosphotyrosine protein in rat adipocytes is associated with phosphatidylinositol 3-kinase. J Biol Chem 1993; 268:5921–5928.
83. Myers MG Jr, Sun XJ, Cheatham B, Jachna BR, Glasheen EM, Bsacker JM, White MF. IRS-1 is a common element in insulin and insulin-like growth factor-I signaling to the phosphatidylinositol 3′-kinase. Endocrinology 1993; 132:1421–1430.
84. Cheatham B, Kahn CR. Insulin action and the insulin signaling network. Endocr Rev 1995; 16:117–142.
85. Sun XJ, Wang LM, Zhang Y, Yenush L, Myers MGJ, Glasheen E, Lane WS, White M. Role of IRS-2 in insulin and cytokine signaling. Nature 1995; 377:173–177.
86. Lavan BE, Lane WS, Lienhard GE. The 60-kDa phosphotyrosine protein in insulin-treated adipocytes is a new member of the insulin receptor substrate family. J Biol Chem 1997; 272:11439–11443.
87. Lavan BE, Fantin VR, Chang ET, Lane WS, Keller SR, Lienhard GE. A novel 160-kDa phosphotyrosine protein in insulin-treated embryonic kidney cells is a new member of the insulin receptor substrate family. J Biol Chem 1997; 272:21403–21407.
88. Myers MG Jr, Backer JM, Sun X-J, Shoelson SE, Hu P, Schlessinger J, Yoakin M, Schaffhausen B, White MF. IRS-1 activates the phosphatidylinositol 3′-kinase by associating with the src homology 2 domains of p85. PNAS 1992; 89:10350–10354.
89. Araki E, Lipes MA, Patti ME, Bruning JC, Haag B, Johnson RS, Kahn CR. Alternative pathway of insulin signaling in mice with targeted disruption of the IRS-1 gene. Nature 1994; 372:186–190.
90. Keller S, Lienhard G. Insulin signalling: the role of insulin receptor substrate 1. Trends Cell Biol 1994; 4:115–119.
91. Craparo A, O'Neill TJ, Gustafson TA. Non-SH2 domains within insulin receptor substrate-1 and SHC mediate their phosphotyrosine-dependent interaction with the NPEY motif of the insulin-like growth factor I receptor. J Biol Chem 1995; 270:15639–15643.
92. Kuhné MR, Pawson T, Lienhard GE, Feng G-S. The insulin receptor substrate 1 associates with the SH2-containing phosphotyrosine phosphatase Syp. J Biol Chem 1993; 268:11479–11481.
93. Pawson T, Schlessinger J. SH2 and SH3 domains. Curr Biol 1993; 3:434–442.
94. Kotani K, Yonezawa K, Hara K, Ueda H, Kitamura Y, Sakaue H, Ando A, Chavanieu A, Calas B, Grigorescu F, Nishiyama M, Waterfield MD, Kasuga M. Involvement of phosphoinositide 3-kinase in insulin- or IGF-I-induced membrane ruffling. EMBO J 1994; 13:2313–2321.
95. Kooijman R, Lauf JJ, Kappers AC, Rijkers GT. Insulin-like growth factor induces phosphorylation of immunoreactive insulin receptor substrate and its association with phosphatidylinositol 3-kinase in human thymocytes. J Exp Med 1995; 182:593–597.
96. Myers MG Jr, Grammer TC, Brooks J, Glasheen EM, Wang L-M, Sun XJ, Blenis J, Pierce JH, White MF. The pleckstrin homology domain in insulin receptor substrate-1 sensitizes insulin signaling. J Biol Chem 1995; 270:11715–11718.
97. Voliovitch H, Schlindler DG, Hadari YR, Taylor SI, Accili D, Zick Y. Tyrosine phosphorylation of insulin receptor substrate-1 *in vivo* depends on the presence of its plekstrin homology region. J Biol Chem 1995; 270:18083–18087.
98. Yenush L, Makati KJ, Smith-Hall J, Ishibashi O, Myers MG Jr, White MF. The pleckstrin homology domain is the principal link between the insulin receptor and IRS-1. J Biol Chem 1996; 271:24300–24306.
99. Tartare-Deckert S, Sawka-Verhelle D, Murdaca J, Van Obberghen E. Evidence for a differential interaction of SHC and insulin receptor substrate-1 (IRS-1) with the insulin-like growth factor-I (IGF-I) receptor in the yeast two-hybrid system. J Biol Chem 1995; 270:23456–23460.
100. He W, Craparo A, Zhu Y, O'Niell TJ, Wang L-M, Pierce JH, Gustafson TA. Interaction of insulin receptor substrate-2 (IRS-2) with the insulin and insulin-like growth factor I receptors: evidence for two distinct phosphotyrosine-dependent interaction domains within IRS-2. J Biol Chem 1996; 271:11641–11645.
101. Backer JM, Myers MG Jr, Shoelson SE, Chin DJ, Sun X-J, Miralpeix M, Hu P, Margolis B, Skolnik EY, Schlessinger J, White MF. Phosphatidylinositol 3′-kinase is activated by association with IRS-1 during insulin stimulation. EMBO J 1992; 11:3469–3479.

102. Giorgetti S, Ballotti R, Kowalski-Chaubel A, Tartare S, Van Obberghen E. The insulin and insulin-like growth factor-I receptor substrate IRS-1 associates with and activates phosphatidylinositol 3-kinase *in vitro*. J Biol Chem 1993; 268:7358–7364.
103. Nakanishi H, Brewer KA, Exton JH. Activation of the zeta isozyme of protein kinase C by phosphatidylinositol 3,4,5-trisphosphate. J Biol Chem 1993; 268:13–16.
104. Manzoli L, Billi AM, Rubbini S, Bavelloni A, Faenza I, Gilmour RS, Rhee S-G, Cocco L. Essential role for nuclear phospholipase C β_1 in insulin-like growth factor I-induced mitogenesis. Cancer Res 1997; 57:2137–2139.
105. Cheatham B, Vlahos CJ, Cheatham L, Wang L, Blenis J, Kahn CR. Phosphatidylinositol 3-kinase activation is required for insulin stimulation of pp70 S6 kinase, DNA synthesis, and glucose transporter translocation. Mol Cell Biol 1994; 14:4902–4911.
106. Chung J, Grammer TC, Lemon KP, Kazlauskas A, Blenis J. PDGF- and insulin-dependent pp70S6k activation mediated by phosphatidylinositol-3-OH kinase. Nature 1994; 370:71–75.
107. Carter AN, Downest CP. Molecular mechanisms of regulation of phosphoinositide 3-kinase by insulin, insulin-like growth factor I and nerve growth factor in PC12 cells. Biochem Soc Trans 1995; 23:148–52.
108. Matsuda M, Tanaka S, Nagata S, Kojima A, Kurata T, Shibuya M. Two species of human Crk cDNA encode proteins with distinct biological activities. Mol Cell Biol 1992; 12:3482–3489.
109. Reichman CT, Mayer BJ, Keshav S, Hanafusa H. The product of the cellular crk gene consists primarily of SH2 and SH3 regions. Cell Growth Differ 1992; 3:451–460.
110. Beitner-Johnson D, LeRoith D. Insulin-like growth factor-I stimulates tyrosine phosphorylation of endogenous c-Crk. J Biol Chem 1995; 270:5187–5190.
111. Beitner-Johnson D, Blakesley VA, Shen-Orr Z, Spiegel S, LeRoith D. The proto-oncogene product c-Crk associates with insulin receptor substrate-1 and 4PS: modulation by insulin growth factor-I (IGF-I) and enhanced IGF-I signaling. J Biol Chem 1996; 271:9287–9290.
112. Koval AP, Blakesley VA, Roberts CT Jr, Zick Y, LeRoith D. Direct interaction of the protein of the c-CrkII proto-oncogene with the insulin-like growth factor-I receptor. Biochem J 1998; 330:923–932.
113. Luttrell LM, van Biesen T, Hawes BE, Koch WJ, Touhara K, Lefkowitz RJ. $G_{\beta\gamma}$ subunits mediate mitogen-activated protein kinase activation by the tyrosine kinase insulin-like growth factor-I receptor. J Biol Chem 1995; 270:16495–16498.
114. Uddin S, Yetter A, Katzav S, Hoffman C, White MF, Platanias LS. Insulin-like growth factor-I induces rapid tyrosine phosphorylation of the vav proto-oncogene product. Exp Hematol 1996; 24:622–627.
115. Furlanetto RW, Day BR, Lopaczynski W, Nissley SP. 14-3-3 proteins interact with the IGF-I receptor, but not the insulin receptor. Biochem J 1997; 327:765–771.
116. Xu B, Bird VG, Miller WT. Substrate specificities of the insulin and insulin-like growth factor I receptor tyrosine kinase catalytic domains. J Biol Chem 1995; 270:29825–29830.
117. Krane JF, Murphy DP, Carter DM, Krueger JG. Synergistic effects of epidermal growth factor (EGF) and insulin-like growth factor I/somatomedin C (IGF-I) on keratinocyte proliferation may be mediated by IGF-I transmodulation of the EGF receptor. J Invest Dermatol 1991; 96:419–424.
118. Konstantopoulos N, Clark S. Insulin and insulin-like growth factor-1 stimulate dephosphorylation of paxillin in parallel with focal adhesion kinase. Biochem J 1996; 314:387–390.
119. Butler AA, Blakesley VA, Koval A, deJong R, Groffen J, LeRoith D. *In vivo* regulation of CrkII and CrkL proto-oncogenes in the uterus by insulin-like growth factor-I. Differential effects on tyrosine phosphorylation and association with paxillin. J Biol Chem 1997; 272:27660–27664.
120. Ishiki M, Sasaoka T, Ishihara H, Imamura T, Usui I, Takata Y, Kobayashi M. Evidence for functional roles of Crk-II in insulin and epidermal growth factor signaling in Rat-1 fibroblasts overexpressing insulin receptors. Endocrinology 1997; 138:4950–4958.
121. Foncea R, Andersson M, Ketterman A, Blakesley V, Sapag-Hagar M, Sugden PH, LeRoith D, Lavandero S. Insulin-like growth factor-I rapidly activates multiple signal transduction pathways in cultured rat cardiac myocytes. J Biol Chem 1997; 272:19115–19124.
122. Neri LM, Billi AM, Manzoli L, Rubbini S, Gilmour RS, Cocco L, Martelli AM. Selective nuclear translocation of protein kinase Ca in Swiss 3T3 cells treated with IGF-I, PDGF and EGF. FEBS Lett 1994; 347:63–68.
123. Oemar BS, Law NM, Rosensweig SA. Insulin-like growth factor-I induces tyrosyl phosphorylation of nuclear proteins. J Biol Chem 1991; 266:24241–24244.
124. Oppenhein RW. Cell death during development of the nervous system. Annu Rev Neurosci 1991; 14:453–501.

125. Galli C, Meucci O, Scorziello A, Werge TM, Calissano P, Schettini G. Apoptosis in cerebellar granule cells is blocked by high KCl forskolin and IGF-1 through distinct mechanisms of action: the involvement of intracellular calcium and RNA synthesis. J Neurol 1995; 15:1172–1179.
126. Matthews CC, Feldman EL. Insulin-like growth factor-1 rescues SH-SY5Y human neuroblastoma cells from hyperosmotic induced programmed cell death. J Cell Physiol 1996; 166:323–331.
127. Johnston BM, Mallard EC, Williams CE, Gluckman PD. Insulin-like growth factor-1 is a potent neuronal rescue agent after hypoxic-ischemic injury in fetal lambs. J Clin Invest 1996; 97:300–308.
128. Muta K, Krantz SB. Apoptosis of human erythroid colony-forming cells is decreased by stem cell factor and insulin-like growth factor-I as well as erythropoietin. J Cell Physiol 1993; 156:264–271.
129. Jung Y-K, Miura M, Yuan J. Suppression of interleukin-1 β-converting enzyme-mediated cell death by insulin-like growth factor. J Biol Chem 1996; 271:5112–5117.
130. Harrington EA, Bennett MR, Fanidi A, Evan GI. c-Myc-induced apoptosis in fibroblasts is inhibited by specific cytokines. EMBO J 1994; 13:3286–3295.
131. Chung S-Y, Billig H, Tilly JL, Furata I, Tsafriri A, Hsueh AJW. Gonadotrophin suppression of apoptosis in cultured preovulatory follicles: mediatory role of endogenous insulin-like growth factor-I. Endocrinology 1994; 135:1845–1853.
132. Neuenschwander S, Scwartz A, Wood TL, Roberts CT Jr, Heninghausen L, LeRoith D. Involution of the lactating mammary gland is inhibited by the IGF system in a transgenic mouse model. J Clin Invest 1996; 97:2225–2232.
133. Baker J, Liu J-P, Robertson EJ, Efstratiadis A. Role of insulin-like growth factors in embryonic and postnatal growth. Cell 1993; 75:73–82.
134. Liu J-P, Baker J, Perkins AS, Robertson EJ, Estratiadis A. Mice carrying null mutations of the genes encoding insulin-like growth factor I (*Igf-1*) and type 1 IGF receptor (*Igf1r*). Cell 1993; 75:59–72.
135. Fisher DE. Apoptosis in cancer therapy: crossing the threshold. Cell 1994; 78:539–542.
136. Ueda K, Ganem D. Apoptosis is induced by N-myc expression in hepatocytes, a frequent event in hepadnavirus oncogenesis, and is blocked by insulin-like growth factor II. J Virol 1996; 70:1375–1383.
137. Resnicoff M, Abraham D, Yutanawibboonchai W, Rotman HL, Kajstura J, Rubin R, Zoltick P, Baserga R. The insulin-like growth factor I receptor protects tumor cells from apoptosis *in vivo*. Cancer Res 1995; 55:2463–2469.
138. D'Ambrosio C, Valentinis B, Prisco M, Reiss K, Rubini M, Baserga R. Protective effect of the insulin-like growth factor I receptor on apoptosis induced by okadaic acid. Cancer Res 1997; 57:3264–3271.
139. Lee PDK, Rosenfeld RG, Hintz RL, Smith SD. Characterization of insulin, insulin-like growth factors I and II and growth hormone receptors on human leukemic lymphoblasts. J Clin Endo and Metab 1986; 62:28–36.
140. DeLeon DD, Bakker B, Wilson DM, Hintz RL, Rosenfeld RG. Demonstration of insulin-like growth factor (IGF-I and -II) receptors and binding protein in human breast cancer cell lines. Biochem Biophys Res Curr 1988; 152:398–406.
141. Macauley VM, Teale J, Everard MJ, Joshi GP, Smith IE, Millar J. Somatomedin-C/insulin-like growth factor I is a mitogen for human small cell lung cancer. Br J Cancer 1988; 57:91–98.
142. Peyrat JP, Bonneterre J, Beuscart R, Dijane J, Demaille A. Insulin-like growth factor I receptors in human breast cancer and their relation to estradiol and progesterone receptors. Cancer Res 1988; 48:6429–6436.
143. Pollak MN, Polychronakos C, Yousefi S, Richard M. Characterization of insulin-like growth factor I (IGF-I) receptors of human breast cancer cells. BBRC 1988; 154:326–332.
144. Kurihara M, Tokunaga Y, TsuTsumi K. Characterization of insulin-like growth factor I and epidermal growth factor receptors in meningioma. J Neurosurg 1989; 71:538–546.
145. Cullen KI, Yee D, Sly WS, Perdue J, Hampton B, Lippman ME, Rosen N. Insulin-like growth factor receptor expression and function in human breast cancer. Cancer Res 1990; 50:48–53.
146. Kamio T, Shigematsu K, Kawai K, Tsuchiyama H. Immunoreactivity and receptor expression of insulin-like growth factor I and insulin in human adrenal tumors. An immunohistochemical study of 94 cases. Am J Pathol 1991; 138:83–91.
147. Antoniades HN, Galanopoulos T, Neville-Golden J, Maxwell M. Expression of insulin-like growth factors I and II and their receptor mRNAs in primary human astrocytomas and meningiomas; in vivo studies using in situ hybridization and immunocytochemistry. Int J Cancer 1992; 50:215–222.
148. Geier A, Beery R, Haimsohn M, Karasik A. Insulin-like growth factor-I inhibits cell death induced by anticancer drugs in the MCF-7 cells: involvement of growth factors in drug resistance. Cancer Invest 1995; 13:480–486.

149. Dunn SE, Hardman RA, Kari FW, Barrett JC. Insulin-like growth factor 1 (IGF-1) alters drug sensitivity of HBL100 human breast cancer cells by inhibition of apoptosis induced by diverse anticancer drugs. Cancer Res 1997; 57:2687–2693.
150. Guvakova MA, Surmacz E. Tamoxifen interferes with the insulin-like growth factor I receptor (IGF-IR) signaling pathway in breast cancer cells. Cancer Res 1997; 57:2606–2610.
151. Parrizas M, Saltiel AR, LeRoith D. Insulin-like growth factor 1 inhibits apoptosis using the phosphatidyl 3′-kinase and mitogen-activated protein pathways. J Biol Chem 1997; 272:154–161.
152. Christophori G, Naik P, Hanahan D. A second signal supplied by insulin-like growth factor II in oncogene-induced tumorigenesis. Nature 1994; 369:414–418.
153. Arteaga CL, Osborne CK. Growth inhibition of human breast cancer cells *in vitro* with an antibody against the type I somatomedin receptor. Cancer Res 1989; 49:6237–6241.
154. Kalebic T, Tsokos M, Helman LJ. *In vivo* treatment with antibody against IGF-1 receptor suppresses growth of human rhabdomyosarcoma and down-regulates $p34^{cdc2}$. Cancer Res 1994; 54:5531–5534.
155. Prager D, Li H-L, Asa S, Melmed S. Dominant negative inhibition of tumorigenesis *in vivo* by human insulin-like growth factor I receptor mutant. Proc Natl Acad Sci 1994; 91:2181–2185.
156. Resnicoff M, Sell C, Rubini M, Coppola D, Ambrose D, Baserga R, Rubin R. Rat glioblastoma cells expressing an antisense RNA to the insulin-like growth factor-1 (IGF-1) receptor are nontumorigenic and induce regression of wild-type tumors. Cancer Res 1994; 54:2218–2222.
157. Resnicoff M, Coppola D, Sell C, Rubin R, Ferrone S, Baserga R. Growth inhibition of human melanoma cells in nude mice by antisense strategies to the type 1 insulin-like growth factor receptor. Cancer Res 1994; 54:4848–4850.
158. Sell C, Dumenil G, Deveaud C, Miura M, Coppola D, DeAngelis T, Rubin R, Efstratiadis A, Baserga R. Effect of a null mutation of the insulin-like growth factor I receptor gene on growth and transformation of mouse embryo fibroblasts. Mol Cell Biol 1994; 14:3604–3612.
159. Shapiro DN, Jones BG, Shapiro LH, Dias P, Houghton PJ. Antisense-mediated reduction in insulin-like growth factor-I receptor expression suppresses the malignant phenotype of a human alveolar rhabdomyosarcoma. J Clin Invest 1994; 94:1235–1242.
160. D'Ambrosio C, Ferber A, Resnicoff M, Baserga R. A soluble insulin-like growth factor I receptor that induces apoptosis of tumor cells *in vivo* and inhibits tumorigenesis. Cancer Res 1996; 56:4013–4020.
161. Liu LJ, Blakesley VA, Gutkind JS, LeRoith D. The constitutively active mutant Gα13 transforms mouse fibroblast cells deficient in insulin-like growth factor-I receptor. J Biol Chem 1997; 272:29438–29441.
162. O'Connor R, Kauffman-Zeh A, Liu Y, Lehar S, Evan GI, Baserga R, Blättler WA. Identification of domains of the insulin-like growth factor I receptor that are required for protection from apoptosis. Mol Cell Biol 1997; 17:427–435.
163. Wayner EA, Orlando RA, Cheresh DA. Integrins αvβ3 and αvβ5 contribute to cell attachment to vitronectin but differentially distribute on the cell surface. J Cell Biol 1991; 113:919–929.
164. Felding-Habermann B, Mueller BM, Romerdahl CA, Cheresh DA. Involvement of integrin αv gene expression in human melanoma tumorigenicity. J Clin Invest 1992; 89:2018–2022.
165. Akiyama SK, Olden K, Yamada KM. Fibronectin and integrins in invasion and metastasis. Cancer Metastas Rev 1995; 14:173–189.
166. Klemke RL, Yebra Y, Bayna EM, Cheresh DA. Receptor tyrosine kinase signaling is required for integrin αvβ5-directed cell motility but not adhesion on vitronectin. J Cell Biol 1994; 127:850–866.
167. Rodeck U, Heryln M. Growth factors in melanoma. Cancer Metastas Rev 1991; 10:89–101.
168. Rodeck U, Melbor K, Kath R, Menssen H-D, Varello M, Atkinson B, Herlyn M. Constitutive expression of multiple growth factor genes by melanoma cells but not normal melanocytes. J Invest Dermatol 1991; 97:20–26.
169. Matsumoto K, Ziober BL, Yao C-C, Kramer RH. Growth factor regulation of integrin-mediated cell motility. Cancer Metastas 1995; 14:205–217.
170. Brooks PC, Klemke RL, Schön S, Lewis JM, Schwartz MA, Cheresh DA. Insulin-like growth factor receptor cooperates with integrin αVβ5 to promote tumor cell dissemination *in vivo*. J Clin Invest 1997; 99:1390–1398.
171. Leventhal PS, Feldman EL. Insulin-like growth factors as regulators of cell motility. Signaling mechanisms. Trends in Endosc and Metab 1997; 8:1–6.
172. Pfeifle B, Boeder H, Ditschuneit H. Interaction of receptors for insulin-like growth factor I, platelet-derived growth factor, and fibroblast growth factor in rat aortic cells. Endocrinology 1987; 120:2251–2258.

173. Gockerman A, Prevette T, Jones JI, Clemmons DR. Insulin-like growth factor (IGF)-binding proteins inhibit the smooth muscle cell migration responses to IGF-I and IGF-II. Endocrinology 1995; 136:4168–4173.
174. Jones JI, Prevette T, Gockerman A, Clemmons DR. Ligand occupancy of the αVβ3 integrin is necessary for smooth muscle cells to migrate in response to insulin-like growth factor I. Proc Natl Acad Sci 1996; 93:2482–2487.
175. Jackson CL, Reidy MA. The role of plasminogen activation in smooth muscle cell migration after arterial injury. Ann NY Acad Sci 1992; 667:141–150.

8 Type 2 IGF Receptor-Mediated Events

S. Peter Nissley, MD

CONTENTS

INTRODUCTION

In the insulin-like growth factor (IGF) field, receptors were identified by the classic approach of competitive binding and affinity crosslinking of radiolabeled IGF-I and IGF-II *(1)*. A receptor that preferred IGF-I over IGF-II, bound insulin with low affinity, and had a binding subunit of approx 130 kDa became known as the IGF-I receptor or the type 1 IGF receptor. A 250 kDa receptor that bound IGF-II with much higher affinity than IGF-I and did not recognize insulin was designated the IGF-II receptor or the type 2 IGF receptor. Meanwhile, in the lysosomal enzyme field, Hickman and Neufeld *(2)* had proposed a specific and saturable cellular uptake mechanism for lysosomal enzymes that would recognize a common feature of a long list of lysosomal enzymes. This prediction was fulfilled by the identification of a 215-kDa receptor that recognized mannose-6-phosphate (Man-6-P) residues on lysosomal enzymes *(3)*. These two research areas briefly converged when the molecular cloning of the cation-independent Man-6-P receptor and the IGF-II receptor revealed an identical molecule *(4)*. The two fields have again diverged somewhat with the demonstration that the frog and bird cation-independent Man-6-P receptors do not have binding sites for IGF-II, suggesting that the multifunctional nature of the receptor may be confined to mammals *(5,6)*. This chapter focuses on the multiple functions of the IGF-II/Man-6-P receptor which include well documented roles in lysosomal enzyme trafficking and IGF-II degradation and a more controversial role in signaling various intracellular events in response to ligand binding.

LIGAND BINDING TO THE IGF-II/MAN-6-P RECEPTOR

Lysosomal Enzymes

Attention was focused on mannose residues when treatment of the lysosomal enzyme β-galactosidase with mannosidase resulted in a 97% loss in the rate of uptake by cells and

From: *Contemporary Endocrinology: The IGF System*
Edited by: R. Rosenfeld and C. Roberts Jr © Humana Press, Inc., Totowa, NJ

100 m*M* mannose inhibited the uptake of β-galactosidase by 21% *(7)*. Sly and his colleagues made the important discovery that yeast mannans containing phosphate and Man-6-P itself potently inhibited the uptake of β-glucuronidase by human skin fibroblasts *(3)*. Man-1-P was 100-fold less effective than Man-6-P. Man-6-P also inhibited the binding of β-galactosidase to purified receptor from bovine liver *(8)*. Next, Man-6-P residues were identified in the acid hydrolysates of the lysosomal enzymes *(9,10)*. Treatment of lysosomal enzymes with endoglycosidase H inhibited their ability to be taken up by cells and released oligosaccarides containing the Man-6-P *(9)*. Taken together, these reports indicate that the IGF-II/Man-6-P receptor recognizes uncovered Man-6-P residues present in N-linked high-mannose oligosaccharides. An additional important determinant of binding of lysosomal enzymes to the IGF-II/Man-6-P receptor is the number of uncovered Man-6-P residues present in the lysosomal enzyme. Oligosaccharides generated from lysosomal enzymes that contain two uncovered Man-6-P residues bind to receptor immobilized on solid supports and are eluted with Man-6-P, whereas oligosaccharides containing only one uncovered Man-6-P are only retarded on the receptor affinity column and do not require Man-6-P for elution *(11,12)*. Tong et al. *(11)* provided direct experimental support for multivalent binding by measuring stoichiometry of binding of a variety of phosphorylated ligands to purified receptor using equilibrium dialysis of immobilized receptors. The dissociation constants for Man-6-P and pentamannose phosphate were 7×10^{-6} *M* and 6×10^{-6} *M*, whereas the values for β-galactosidase and a high-mannose oligosaccharide containing two phosphomonoesters were 2×10^{-8} *M* and 2×10^{-9} *M*. The stoichiometry values (moles of ligand bound/mole of receptor monomer) were 2.17 and 1.85 for Man-6-P and pentamannose phosphate, whereas the values were 0.9 and 1.0 for the two multivalent ligands. These results suggested that only 2 of the 15 repeats in the extracellular domain of the receptor are utilized for binding of Man-6-P residues, a result that has been confirmed by biochemical and molecular biology approaches (discussed in Chapter 5).

Other Ligands that Bind to the Man-6-Phosphate Recognition Site

Proliferin

Linzer and Nathans *(13)* cloned a cDNA for a mRNA that appeared in mouse 3T3 cells within a few hours of stimulating the growth-arrested cells with serum or platelet-derived growth factor. The deduced amino acid sequence showed 32% homology with mouse prolactin. The protein was named proliferin because of its relation to prolactin and growth hormone and its association with growing cells. Proliferin was found to be secreted by a number of mouse cell lines during growth and by mouse placenta in mid- and late gestation *(14)*. Proliferin was discovered to bind to the cation-independent Man-6-P receptor and binding was inhibited half-maximally with 10 μ*M* Man-6-P *(15)*. A separate proliferin receptor has also been described in uterine membrane preparations from pregnant mice *(16)*. The dissociation constant was 6×10^{-10} *M*. Growth hormone competed weakly for binding at high concentrations. Man-6-P did not compete for ^{125}I-proliferin binding, providing evidence that the binding site was not the IGF-II/Man-6-P receptor *(16)*. Thus, there is evidence for a second receptor for proliferin.

Transforming Growth Factor-β1

The precursor for transforming growth factor-β1 (TGF-β1), secreted by Chinese hamster ovary (CHO) cells transfected with TGF-β1 cDNA, was found to contain Man-6-P residues in oligosaccharides from two of the three N-linked glycosylation sites *(17)*. The

TGF-β1 precursor was shown to bind to purified IGF-II/Man-6-P receptor that had been immobilized on microtiter wells by interaction with receptor antibody–Protein A complexes. Binding of radiolabeled TGF-β1 precursor to the receptor was competed by Man-6-P but not by Man-1-P. ^{125}I-TGF-β1 precursor was also reported to bind to CHO cells overexpressing the IGF-II/Man-6-P receptor *(18)*. Natural TGF-β1 precursor that had been purified from human platelets competed for the binding of recombinant ^{125}I-TGF-β1 precursor to immobilized receptor.

Thyroglobulin

Thyroglobulin produced by thyroid follicle cells is ultimately conveyed to lysosomes where proteolysis leads to the release of thyroid hormones, triiodothyronine (T_3) and thyroxine (T_4). The pathway of delivery of thyroglobulin to lysosomes includes secretion, storage in follicles, and recapture by the follicle cells. Herzog et al. *(19)* reported that Man-6-P residues are found in thyroglobulin in monoester form and that ^{125}I-thyroglobulin binds to the Man-6-P receptor in CHO cells. It was proposed that binding of thyroglobulin to the IGF-II/Man-6-P receptor on the apical surface of follicle cells might be important for recapture of thyroglobulin from the follicles. However, a later article from the same group provided evidence that the recapture of thyroglobulin from the follicle is not mediated by the IGF-II/Man-6-P receptor, but rather occurs via low-affinity binding sites on the apical membrane of thyrocytes *(20)*. Nonetheless, thyroglobulin is endocytosed by human skin fibroblasts via the IGF-II/Man-6-P receptor. The authors proposed that the Man-6-P recognition marker is a signal by which cells other than thyrocytes are able to internalize thyroglobulin.

IGF-II

Measurements of binding affinity of IGF-II for highly purified IGF-II/Man-6-P receptor have ranged from K_d values of 0.7 n*M* to 0.017 n*M*. *(21–24)*. These values approximate K_d values obtained by many investigators using membranes and solubilized receptors. Binding stoichiometry values of IGF-II for the IGF-II/Man-6-P receptor have been reported to range from 0.2 to 0.95 mol/mol receptor monomer *(21–24)*. A value of 0.95 was obtained from experiments in which receptor concentration was accurately measured by amino acid composition, and all of the radioligand was shown to be capable of binding to the receptor *(24)*.

There is general agreement that IGF-I binds to the IGF-II/Man-6-P receptor with very low affinity such that physiologic concentrations of IGF-I would not interact with the receptor, and that insulin does not bind at all. Earlier reports that IGF-I bound to the IGF-II/Man-6-P receptor with 10-fold lower affinity than IGF-II can probably be explained by the use of natural IGF-I preparations that had a low level of contamination with IGF-II. Tong et al. *(24)* determined the binding affinity of IGF-I for the bovine liver IGF-II/Man-6-P receptor by Scatchard analysis of binding data obtained by varying the amount of highly purified receptor that was incubated with a fixed concentration of ^{125}I-IGF-I. The K_d value of 0.4 m*M* was 1000-fold higher than the K_d value for IGF-II binding to the same receptor preparation.

Haskell and Tucker *(25)* reported different results in a study of the binding of IGF-I and IGF-II in rat embryonic heart using ligand crosslinking. ^{125}I-IGF-I was crosslinked in a 260-kDa complex and, surprisingly, this binding was fully competed by insulin. ^{125}I-IGF-I binding was also competed by IGF-II. Evidence that the 260-kDa complex represented binding of ^{125}I-IGF-I to the IGF-II/Man-6-P receptor was provided by competition with an antibody to the IGF-II/Man-6-P receptor. The authors proposed two alternative explanations of the data. The first possibility was that ^{125}I-IGF-I binding to the IGF-II/Man-6-P receptor is readily detected because of the very high concentration of this receptor in embryonic heart and the ability of insulin to compete for ^{125}I-IGF-I binding but not to compete for ^{125}I-IGF-II binding might be explained by the relative binding affinities of IGF-I and IGF-II for the receptor. The authors' alternative explanation was that there is an independent binding site for IGF-I on the IGF-II/Man-6-P receptor and insulin competes for binding to this site. Although a second high-affinity binding site for IGF-I would have been missed in most studies which have routinely assessed IGF-I binding affinity by competitive binding experiments using ^{125}I-IGF-II, the method employed by Tong et al. *(24)* to measure binding of IGF-I to the bovine IGF-II/Man-6-P receptor would have detected a high-affinity binding site for IGF-I. They used a fixed concentration of ^{125}I-IGF-I and varied the amount of highly purified IGF-II/Man-6-P receptor. Although Tong et al. *(24)* did not test insulin for ability to compete for ^{125}I-IGF-I binding, their data tell us that if there is a second binding site for IGF-I on the bovine IGF-II/Man-6-P receptor, it is a very low-affinity site. Further studies on the binding of ^{125}I-IGF-I to the rat IGF-II/Man-6-P receptor need to be performed to determine which of the models proposed by Haskell and Tucker *(25)* is correct.

Retinoic Acid

Recently Kang et al. *(25a)* reported that the IGF-II/Man-6-P receptor is also a receptor for retinoic acid. Using photoaffinity labeling they found that retinoic acid bound specifically to a 300-kDa protein found in membrane preparations from rat cardiac myocytes and in neonatal rat serum. (Most laboratories have reported that the receptor is 260 kDa on sodium dodecyl sulfate-polyacrylamide gel electrophoresis [SDS/PAGE] under reducing conditions.) The 300-kDa band almost completely disappeared in a neonatal rat serum sample that had been immunodepleted of the IGF-II/Man-6-P receptor, and retinoic acid was bound to the 300-kDa species in the immunoprecipitate. The serum IGF-II/Man-6-P receptor was purified by pentamannosyl-6-phosphate affinity chromatography and affinity photolabeled with [^{3}H]retinoic acid. Coomassie blue staining of this preparation showed a major 300-kDa band and a minor 260-kDa band, but only the 300-kDa species was photoaffinity labeled with retinoic acid. Immunoblotting of the 300-kDa species from rat serum showed that it was recognized by an IGF-II/Man-6-P receptor antiserum directed toward a peptide in the cytoplasmic domain of the receptor. However, MacDonald et al. *(25b)* had reported earlier that the rat serum receptor was not recognized by this same antiserum, providing evidence that the serum receptor is truncated. Kang reported that Scatchard analysis of retinoic acid binding to partially purified rat serum IGF-II/Man-6-P receptor demonstrated high-affinity binding (K_d = 2.5 nM). Retinoic acid binding stoichiometry was not reported. IGF-II or Man-6-P did not inhibit the binding of retinoic acid to the 300-kDa protein; Man-6-P and β-glucuronidase actually increased the binding of retinoic acid to the 300-kDa protein. Conversely, in neonatal rat cardiac myocytes retinoic acid increased the endocytosis of IGF-II and β-glucuronidase.

Reciprocal Modulation of Binding of Lysosomal Enzymes and IGF-II to the IGF-II/Man-6-P Receptor

There is evidence for separate binding sites for IGF-II and lysosomal enzymes on the IGF-II/Man-6-P receptor. Receptor antibodies were reported to differentially inhibit the binding of IGF-II and pentamannosyl phosphate-substituted bovine serum albumin

(PMP-BSA) to the receptor *(26)*. Pentamannose-6-phosphate/trilysine substituted aprotinin and IGF-II were crosslinked to different tryptic fragments of the receptor *(27)*. Also, when the IGF-II/Man-6-P receptor was incubated with ^{125}I-IGF-II and pentamannose 6-phosphate substituted BSA, ^{125}I-IGF-II-receptor complexes could be immunoprecipitated with antibodies to PMP-BSA, providing evidence that the receptor can bind both ligands simultaneously. The binding sites for IGF-II and for phosphomannosyl residues have been localized by utilizing receptor mutants (discussed in Chapter 5).

Inhibition of IGF-II Binding by Lysosomal Enzymes

Kiess et al. *(28)* reported that β-galactosidase inhibited the binding of ^{125}I-IGF-II as measured by affinity crosslinking. Co-incubation with Man-6-P prevented the inhibition by the lysosomal enzyme, demonstrating that the inhibition by β-galactosidase required binding to the Man-6-P recognition site on the receptor. In a solution assay, β-galactosidase inhibited the binding of ^{125}I-IGF-II to pure receptor by 80% *(29)*. The concentration of β-galactosidase required for half-maximal inhibition of ^{125}I-IGF-II binding was 25 n*M*, which agreed with published K_d values for β-galactosidase binding to purified receptor. Scatchard analysis of the binding of IGF-II to the receptor in the presence and absence of β-galactosidase demonstrated that the lysosomal enzyme decreased the binding affinity of IGF-II for the receptor. It is not clear whether the lysosomal enzymes are sterically inhibiting the binding of IGF-II or changing the conformation of the receptor by bivalent binding to the receptor. β-Glucuronidase and cathepsin D have also been reported to inhibit the binding of IGF-II to the IGF-II/Man-6-P receptor (30,*31)*.

Earlier reports of the ability of Man-6-P to increase the binding of ^{125}I-IGF-II to the receptor can be explained by the removal of inhibitory lysosomal enzymes from the receptor by Man-6-P. Thus, crude receptor preparations or receptor that had been purified by IGF-II affinity chromatography exhibited the positive effect of Man-6-P on ^{125}I-IGF-II binding, whereas receptor purified on a lysosomal enzyme affinity column did not show the positive effect *(29)*. Similarly, the experiments of Polychronakos et al. *(32)* and MacDonald *(33)* showed that extensive washing of membranes with Man-6-P eliminated the positive effect of Man-6-P on ^{125}I-IGF-II binding.

Although the majority of newly synthesized lysosomal enzymes are targeted directly to lysosomes, some escape this intracellular pathway and are secreted. If present in sufficient concentration extracellularly, lysosomal enzymes would be expected to block the binding of IGF-II to the IGF-II/Man-6-P receptor. Because extracellular IGF-II is degraded in a pathway initiated by IGF-II/Man-6-P receptor-mediated internalization, inhibition of binding of IGF-II to the receptor by lysosomal enzymes would result in increased concentration of extracellular IGF-II. This would favor increased signaling via the IGF-I receptor. Thus, the IGF-II/Man-6-P receptor and extracellular lysosomal enzymes provide a potential regulatory system for IGF-I receptor signaling.

Inhibition of Binding of Lysosomal Enzymes to the IGF-II/Man-6-P Receptor by IGF-II

Kiess et al. *(34)* reported that IGF-II inhibited the uptake of ^{125}I-β-galactosidase by monolayer cultures of BRL 3A2 rat liver cells and C6 rat glial cells. The most likely explanation for the inhibition of uptake was that IGF-II inhibited the binding of the lysosomal enzyme to the IGF-II/Man-6-P receptor and this was confirmed by showing that the binding of ^{125}I-β-galactosidase to a purified receptor preparation was inhibited by IGF-II. IGF-II has also been shown to inhibit the cellular uptake or receptor binding of β-hexosaminidase and arylsulfatase *(35,36)*.

These observations suggest that overexpression of IGF-II (a property of many cancer cells) would lead to misrouting of lysosomal enzymes such that more lysosomal enzyme would be found extracellularly and less in lysosomes. This redistribution would be expected if intracellular IGF-II blocked targeting of lysosomal enzymes to lysosomes and/or extracellular IGF-II prevented reuptake of lysosomal enzymes by cell-surface IGF-II/Man-6-P receptors. This hypothesis has been tested by overexpressing IGF-II in MCF-7 breast cancer cells *(37)*. Breast cancer cells are an attractive model because they synthesize large quantitites of a particular lysosomal enzyme, cathepsin D, making lysosomal enzyme measurement easier. MCF-7 breast cancer cells were transfected with two IGF-II cDNAs, one encoding mature IGF-II and the other encoding a precursor for IGF-II. Two different protocols were utilized to assess cathepsin D trafficking in these transfected cells. With both protocols, the ratio of intracellular to extracellular cathepsin D was significantly lower in the cells overexpressing the precursor IGF-II than in vector control cells. These findings were consistent with IGF-II acting intracellularly and extracellularly to inhibit the binding of cathepsin D to the IGF-II/Man-6-P receptor. This resulted in misrouting of cathepsin D such that relatively more cathepsin D was found extracellularly and relatively less was found in lysomes. Earlier, Hoeflich et al. *(38)* had overexpressed IGF-II precursor in human embryonal kidney cells (293 cells) and examined the trafficking of β-hexosaminadase and cathepsin D. They found that the amount of β-hexosaminidase was approximately twofold increased in the media from pro-IGF-II-overexpressing cells compared to control cells as measured by immunoblotting. In addition, in metabolic labeling experiments the percentage of total β-hexosaminidase and cathepsin D found in the media of IGF-II overexpressing cells was increased compared to control cells. The authors proposed that in cells synthesizing high amounts of pro-IGF-II, IGF-II may modulate the targeting of lysosomal enzymes by enhancing the secretion of newly synthesized enzymes and inhibiting reuptake from the media. In contrast with these reports, Braulke et al. *(39)* found that IGF-II overexpression did not affect sorting of lysosomal enzymes in NIH-3T3 cells. The positive effects on lysosomal enzyme trafficking observed by Hoeflich et al. *(38)* and by De Leon et al. *(37)* were obtained with overexpression of IGF-II precursor whereas Braulke et al. *(39)* overexpressed the mature form of IGF-II. Another difference was the higher concentration of IGF-II in the media of the overexpressing cells in the studies of Hoeflich et al. *(38)* and De Leon et al. *(37)*.

In related experiments, De Ceuninck et al. *(40)* observed that addition of IGF-II to cultures of postnatal rabbit chondrocytes resulted in a decline in intracellular cathepsin B and L activities. Addition of IGF-I did not cause this decline. The authors proposed that internalized IGF-II could occupy the intracellular IGF-II/Man-6-P receptor binding sites required for the targeting of cathepsins B and L to lysosomes. However, the expected corresponding increase in the amount of cathepsin B and L in the media after addition of IGF-II was not observed. Thus, the binding of IGF-II to the mammalian IGF-II/Man-6-P receptor potentially provides an additional point of regulation of lysosomal enzyme trafficking by the receptor based on the ability of IGF-II to block the binding of lysosomal enzymes.

IGF-II/MAN-6-P RECEPTOR FUNCTIONS

Targeting of Lysosomal Enzymes to Lysosomes

Mouse L-cells or mouse macrophages lacking the IGF-II/Man-6-P receptor secrete a large proportion of their newly synthesized lysosomal enzymes *(41)* and this phenotype can be corrected by expressing the IGF-II/Man-6-P receptor in these cells *(42,43)*. Intracellular sorting of lysosomal enzymes was examined in primary embryo fibroblasts isolated from embryonic day 14.5 (e14.5) embryos from IGF-II/Man-6-P receptor knockout mice *(44)*. Steady-state levels of four out of five lysosomal enzymes were reduced by 30–80% in the embyro fibroblasts lacking the IGF-II/Man-6-P receptor. Also, human β-glucuronidase was not internalized in the IGF-II/Man-6-P receptor knockout fibroblasts. These results provide further in vivo evidence for the important role of the IGF-II/Man-6-P receptor in the trafficking of lysosomal enzymes.

Targeting of lysosomal enzymes to lysosomes can be divided into biosynthetic and endocytic arms *(45,46)*. In the major, intracellular biosynthetic arm, newly synthesized lysosomal enzymes are directed to lysosomes by first binding to IGF-II/Man-6-P receptors in the Golgi or *trans*-Golgi network, then traveling in coated vesicles to an acidic late endosome compartment where the acid hydrolases dissociate from the receptor and are incorporated into lysosomes. The unoccupied receptor then cycles back to the Golgi to bind newly synthesized lysosomal enzymes. The second minor arm is an endocytic one in which extracellular lysosomal enzymes bind to a cell-surface IGF-II/Man-6-P receptor and move to the same late endosome compartment via coated pits, clathrin-coated vesicles, and early endosomes.

Biosynthetic Pathway

Lysosomal enzymes are synthesized in the rough endoplasmic reticulum (RER) and enter the lumen of the RER by means of an amino (N)-terminal signal peptide. Cotranslational glycosylation occurs on selected asparagine residues by transfer of preformed oligosaccharides rich in mannose residues. The newly synthesized lysosomal enzymes move by vesicular transport to the Golgi stack. Phosphorylation of selected mannose residues on the 6 position is performed by two enzymes *(47,48)*. First, a phosphotransferase transfers *N*-acetylglucosamine 1-phosphate from UDP-GlcNAc to mannose residues on the lysosomal enzymes, resulting in a phosphodiester intermediate. Next, *N*-acetylglucosamine-1-phosphodiester α-*N*-acetylglucosaminidase removes *N*-acetylglucosamine, yielding a mannose phosphomonoester that can bind to the receptor. As soon as lysosomal enzymes have acquired their complement of phosphomonoester residues in the *cis*-Golgi, they are capable of binding to the IGF-II/Man-6-P receptor that is also being directed through the Golgi stack. If present in the membrane of a cisterna that contains lysosomal enzyme in the lumen, binding of ligand to the receptor could occur. Using double-labeling immunoelectron microscopy, Geuze et al. *(49)* reported that cathepsin D in the Golgi cisternae could be found in close association with the IGF-II/Man-6-P receptor. Sorting of lysosomal enzymes from membrane proteins and secreted proteins is thought to occur in the *trans*-Golgi network adjacent to the *trans*-Golgi cisternae, where coated vesicles are seen to bud from the tubular structures of the *trans*-Golgi network *(50)*. Visicles from the *trans*-Golgi network discharge lysosomal enzyme: receptor complexes into an acidic late endosome compartment where lysosomal enzymes dissociate from receptors *(51)*. The lysosomal enzymes then become incorporated into lysosomes and the unoccupied IGF-II/Man-6-P receptors recycle to the *trans*-Golgi network to bind newly synthesized ligand *(52)*; the receptor itself is absent from lysosomes.

Endocytic Pathway

A minor population of lysosomal enzymes escapes the intracellular pathway and is secreted by constitutive bulk flow. Also, about 10% of total cellular IGF-II/Man-6-P receptors are found on the cell surface. The cell surface receptors are in equilibrium with the intracellular pool of receptors and stay on the cell surface with a half-life of only a few minutes *(53–55)*. Extracellular lysosomal enzymes bind to the cell surface receptors that gather in coated pits and are internalized in clathrin-coated vesicles *(56)*. This endocytic pathway that leads to delivery of lysosomal enzymes to lysosomes utilizes the same late endosome/prelysosomal vesicle that is utilized in the intracellular biosynthetic pathway *(57)*.

Cation-Dependent Man-6-P Receptor

Hoflack and Kornfeld *(58)* identified a second receptor for phosphomannosyl residues in cells that lacked the IGF-II/Man-6-P receptor. The second receptor is only 46 kDa in size and binding of lysosomal enzymes to this receptor is increased by the presence of divalent cations. A striking feature of the cation-dependent Man-6-P receptor is that the extracytoplasmic domain is similar to each of the repeating units of the IGF-II/Man-6-P receptor; the sequence identity ranges from 14% to 28% *(59,60)*. The 46-kDa receptor does not bind IGF-II *(24,28)*. The 46-kDa receptor is also widely distributed and targets lysosomal enzymes to lysosomes along the intracellular pathway *(45)*. Although the receptor is also found on the cell surface, the cation-dependent Man-6-P receptor does not function to endocytose extracellular lysosomal enzymes *(61)*.

Comparison of the IGF-II/Man-6-P Receptor and the Cation-Dependent Man-6-P Receptor

Sleat et al. *(62)* compared the binding specificities of the IGF-II/Man-6-P receptor and the cation-dependent Man-6-P receptor by using an in vitro approach. Glycoproteins containing Man-6-P were affinity purified from metabolically labeled secretions of mutant mouse fibroblasts lacking both Man-6-P receptors. They were incubated with purified Man-6-P receptors and the ligand–ligand complexes were immunoprecipitated with specific receptor antibodies coupled to agarose beads. Bound ligands were eluted with Man-6-P and analyzed by SDS-PAGE and autoradiography. Saturating concentrations of the IGF-II/Man-6-P receptor completely depleted Man-6-P glycoproteins from the preparation whereas saturating concentrations of the cation-dependent Man-6-P receptor removed only 50% of total Man-6-P glycoproteins. However, ligands bound by the two different Man-6-P receptors appeared to be identical in terms of mobility on SDS-PAGE. The IGF-II/Man-6-P receptor displayed higher binding affinity for Man-6-P glycoproteins than the cation-dependent Man-6-P receptor, and the binding affinities of the two receptors for the different ligands were not correlated. The results suggest that the cation-dependent Man-6-P receptor binds a subpopulation of the Man-6-P glycoproteins recognized by the IGF-II/Man-6-P receptor. The authors suggest that the molecular basis for the differences in ligand binding by the two receptors may be that lysosomal enzymes exist in a number of mannose 6-phoshorylation isoforms (multiple oligosaccharide chains with zero to two Man-6-P residues per chain) that are bound preferentially by the two receptors. This in turn may be related to differences in the geometry of bivalent binding for the two receptors: binding to two binding sites within a receptor monomer (IGF-II/Man-6-P receptor) vs binding to two sites provided by a receptor dimer (cation-dependent Man-6-P receptor) *(63)*.

Munier-Lehmann et al. *(64)* compared lysosomal enzyme targeting by the IGF-II/Man-6-P receptor and the cation-dependent Man-6-P receptor by using an in vivo approach. Using two-dimensional PAGE to analyze metabolically labeled Man-6-P-containing glycoproteins that were secreted by fibroblasts lacking the IGF-II/Man-6-P receptor or the cation-dependent Man-6-P receptor, they found that the two cell types secreted different phosphorylated proteins. To investigate sorting of lysosomal enzymes as a function of receptor expression, receptor-negative fibroblasts were transfected with either the IGF-II/Man-6-P receptor or the cation-dependent Man-6-P receptor. Expression of the IGF-II/Man-6-P receptor or the cation-dependent Man-6-P receptor corrected the secretion of lysosomal enzymes in a similar manner; however, one receptor could not fully compensate for the loss of the other. In addition, reexpression of each of the receptors in the receptor-negative fibroblasts supported the notion that each receptor preferentially recognizes in vivo two different pools of phosphorylated lysosomal enzymes. Kasper et al. *(65)* also expressed the IGF-II/Man-6-P receptor and the cation-dependent Man-6-P receptor in receptor-deficient fibroblasts and examined the sorting of lysosomal enzymes. Only partial correction of missorting was observed after overexpression of the cation-dependent receptor even at levels five times the levels in wild-type cells. In contrast, twofold overexpression of the IGF-II/Man-6-P receptor corrected missorting of lysosomal enzymes. However, in contrast to wild-type cells in which the secretion-recapture pathway is of minor importance, at least one fourth of the lysosomal enzymes were transported along a secretion-recapture pathway in fibroblasts overexpressing the IGF-II/Man-6-P receptor. Therefore, neither receptor was able to restore normal targeting of lysosomal enzymes along intracellular routes.

Pohlmann et al. *(66)* and Ludwig et al. *(67)* also examined targeting of lysosomal enzymes in fibroblasts from mice lacking the IGF-II/Man-6-P receptor, the cation-dependent Man-6-P receptor or both receptors. They also concluded that neither receptor was able to completely substitute for the loss of the other receptor.

Taken together, these reports suggest that the complementary activities of these two closely related Man-6-P receptors are required for targeting of the diverse array of lysosomal enzymes to lysosomes.

Degradation of IGF-II

Cell Culture

It was first demonstrated in cell culture experiments with rat adipocytes that the IGF-II/Man-6-P receptor mediated the degradation of extracellular IGF-II. Oka et al. *(68)* demonstrated that rat adipocyte IGF-II/Man-6-P receptors that had been covalently attached to a photoreactive IGF-II analogue were rapidly internalized from the cell surface with a half-life of 3.5 min. Insulin increased the number of cell-surface IGF-II receptors and thereby increased the number of ligand-bound receptors that were internalized per unit time and increased the cellular degradation of ^{125}I-IGF-II. The insulin-stimulated increase in the cellular degradation of ^{125}I-IGF-II could be blocked by an antibody to the IGF-II/Man-6-P receptor.

In L6 rat myoblasts, Kiess et al. *(69)* demonstrated that a rabbit IgG that blocked the binding of IGF-II to the IGF-II/Man-6-P receptor totally blocked the degradation of IGF-II added to the medium. During an 8-h incubation of ^{125}I-IGF-II (45 ng/mL) with L6 myoblasts, 34% of the IGF-II was degraded and addition of the receptor antibody decreased degradation to only 4%. In rat C6 glial cells the same IGF-II/Man-6-P receptor antibody inhibited the degradation of ^{125}I-IGF-II in the medium *(70)*.

Nolan et al. *(31)* transfected IGF-II/Man-6-P-deficient mouse L-cells with a plasmid containing the cDNA for the human IGF-II/Man-6-P receptor. Compared to untransfected cells, the IGF-II/Man-6-P receptor-expressing cells were capable of degrading ^{125}I-IGF-II into trichloroacetic acid (TCA)-soluble material. The degradation occurred at least partly in an acidic compartment because it could be partially inhibited by NH_4Cl, a lysosomotrophic agent that is known to neutralize intracellular acidic compartments. Auletta et al. *(71)* compared the receptor-mediated endocytosis of IGF-I and IGF-II in primary cultures of neonatal rat astrocytes. Surface-bound IGF-II was rapidly internalized and 80% of cell-associated radioactivity was located intracellularly after 30 min. IGF-I was internalized at a slower rate and only 40% of cell-associated radioactivity was inside the cell after 30 min. A pulse-chase experiment showed that 55% and 70% of internalized IGF-I and IGF-II, respectively, was degraded to free amino acids after 3 h. Whereas IGF-I endocytosis led to a degradative endosomal pathway and a nondegradative retroendocytotic pathway, IGF-II endocytosis led only to a major degradative pathway. Thus, extracellular IGF-II was rapidly and efficienty removed by the cells via binding to the IGF-II/Man-6-P receptor. Earlier experiments comparing the processing of IGF-I, IGF-II, and insulin had also provided evidence for distinct pathways for processing each of the family members, with relatively greater channeling of IGF-II into a degradative pathway *(72,73)*.

The results of these cell culture experiments predict that the IGF-II/Man-6-P receptor would indirectly control IGF-I receptor signaling in response to IGF-II by controlling the extracellular concentration of IGF-II. Ellis et al. *(74)* tested this hypothesis by transfecting MCF-7 breast cancer cells with retroviruses containing IGF-II analogs with altered affinity for the IGF-II/Man-6-P receptor and then assessing cellular proliferation mediated by the IGF-I receptor. Unfortunately, an IGF-II analog (A 54 R, L 55 R, T 58 M), which was expected by the authors to exhibit greatly diminished binding to the IGF-II receptor, turned out to be only moderately impaired compared to IGF-II (K_d 2.7 nM vs 0.94 nM). Nonetheless, cells that expressed IGF-I or the IGF-II mutant proliferated at 1.7 times the rate of cells expressing the wild-type IGF-II, supporting the hypothesis. Perhaps a more dramatic effect would have been seen if the authors had employed the A 54 R, L 55 R IGF-II analog, which others had reported not to bind to the IGF-II/Man-6-P receptor.

In Vivo

The most compelling evidence for an important role of the IGF-II/Man-6-P receptor in the degradation of IGF-II comes from gene knockout experiments in mice *(44,75,76)*. Disruption of the IGF-II/Man-6-P receptor gene resulted in a phenotype in which the mice died in the perinatal period and were larger than normal. The excessive growth was postulated to be secondary to elevated IGF-II, which was to be expected if the IGF-II/Man-6-P receptor played a role in the degradation of IGF-II as had been demonstrated in cell culture experiments. This hypothesis was shown to be correct when serum and tissue IGF-II levels were found to be increased in the IGF-II/Man-6-P receptor-deficient embryos. Lau et al. *(75)* found that serum IGF-II levels in the IGF-II/Man-6-P receptor knockout embryos were 2–2.7-fold increased compared to the wild-type siblings, providing direct evidence for the proposed role of the IGF-II/Man-6-P in IGF-II degradation. IGF-II mRNA expression in the IGF-II/Man-6-P receptor knockout embryos was normal. Ludwig et al. *(76)* found that the serum level of IGF-II was 4.4-fold higher in the IGF-II/Man-6-P knockout embryos than in wild-type embryos at e18.5. They also examined the tissue levels of IGF-II by immunoblotting and found that levels were 2-fold higher in the mutants than in the wild-type embryos at e12.5 and declined to 1.4-fold 1 d later. Tissues levels could not be measured in older embryos because of the inadequate sensitivity of the immunoblotting assay. The overgrowth phenotype of these IGF-II/Man-6-P knockout mice also provide in vivo evidence supporting the hypothesis that the IGF-II/Man-6-P receptor indirectly controls signaling by the IGF-I receptor in response to IGF-II by providing a degradative pathway for IGF-II.

Activation of Latent TGF-β1

High molecular weight latent TGF-β1 binds to the IGF-II/Man-6-P receptor at Man-6-P recognition sites, leading to the activation of latent TGF-β1 *(77–79)*. The latent TGF-β1 consists of three subunits: active TGF-β1, its latency-associated peptide (LAP), and the latent TGF-β1 binding protein (LTBP). LAP and active mature TGF-β1 are synthesized as a 390-amino-acid preproprotein. The preproprotein is cleaved intracellularly at an Arg–Arg site, producing the mature TGF-β1, which is a disulfide-bonded homodimer. LAP is also a disulfide-bonded homodimer. After intracellular cleavage, LAP continues to be associated with mature TGF-β1 through noncovalent interactions. LAP is glycosylated at three N-linked sites and Man-6-P residues are present on at least two of these carbohydrate side chains. LTBP is synthesized separately from the TGF-β1 preproprotein, but becomes disulfide-bonded to LAP before being secreted from the cell. LTBP does not contain Man-6-P residues, so the binding of the latent TGF-β1 to the IGF-II/Man-6-P receptor is only through LAP. LTBP is not associated with mature TGF-β1.

After binding of latent TGF-β1 to the IGF-II/Man-6-P receptor, activation of the latent TGF-β1 involves proteolytic cleavage of LAP by extracellular plasmin. Presumably the proteolytic cleavage of LAP leads to a conformational change in LAP and the dissociation of LAP from mature TGF-β1. LTBP may play a role in the initial association of latent TGF-β1 with the cell surface. Thus, in this model of TGF-β1 activation, the proteolytic cleavage occurs at the cell surface rather than by a mechanism involving receptor-mediated endocytosis of the latent TGF-β1, proteolytic cleavage in an endosomal compartment, and secretion of active TGFβ-1.

Although the activation of latent TGF-β1 by a mechanism of initial binding to the IGF-II/Man-6-P receptor and subsequent cleavage of LAP by plasmin has been demonstrated in a cell culture model, it is not clear to what extent this mechanism operates in vivo.

Signaling of Biologic Responses to Ligands

This section discusses evidence that there are intracellular signaling pathways that are triggered by binding of a ligand to the IGF-II/Man-6-P receptor. Although most of these reports are concerned with signaling in response to IGF-II, other reports provide evidence for IGF-II/Man-6-P receptor signaling in response to proliferin, TGF-β1 precursor, or lysosomal enzymes.

Negative Evidence from IGF-II/Man-6-P Receptor Gene Knockout Mice

Wang et al. *(44)* generated mice lacking the IGF-II/Man-6-P receptor gene. Maternal inheritance of an IGF-II/Man-6-P receptor null allele as well as homozygosity for the null allele was generally lethal at birth and the offstpring were 30% larger. The mice were rescued from the lethal phenotype by introduction of a paternal IGF-II null allele, providing evidence that the lethal phenotype was caused by elevated IGF-II levels. Lau et al. *(75)* also generated IGF-II/Man-6-P receptor gene knockout mice. Mouse fetuses inheriting the nonfunctioning IGF-II/Man-6-P receptor gene from their mothers died perinatally and were 25–30% larger than normal littermates. A significant increase in weight was first noted at e17. As also observed by Wang et al. *(44)* the hearts of the *m–/p+* were nearly three times larger than those of the *m+/p+* siblings. Histological analysis revealed hyperplasia of the mycardium which was supported by increase in total DNA content. The authors proposed that these cardiac abnormalities were the cause of the perinatal lethality. During normal development the heart has the highest expression of the IGF-II/Man-6-P receptor of any fetal tissue and IGF-II is also abundant in the heart. Perhaps the heart pathophysiology is explained by increased local IGF-II in the absence of the IGF-II/Man-6-P receptor, leading to a proliferative response to IGF-II mediated by the IGF-I receptor. More recently, Efstratiadis and his collaborators *(76)* generated IGF-II/Man-6-P receptor knockout mice in IGF-II and IGF-I receptor null backgrounds to examine the cause of the perinatal lethality in the IGF-II/Man-6-P receptor knockout mice. IGF-II/Man-6-P receptor-deficient mice were 135% of normal birth weight. Careful analysis of the growth curves of the receptor-deficient embryos revealed that the increased growth rate could detected as early as e13.5. Perinatal lethality in the IGF-II/Man-6-P receptor-deficient mice could be reversed by generating mice that were deficient both in the expression of the IGF-II/Man-6-P receptor and IGF-II. The IGF-II/Man-6-P receptor-deficient mice could also be rescued by generating mice that were also deficient in IGF-I receptor expression. These findings of Ludwig et al. *(76)* tell us that the perinatal lethality noted in the IGF-II/Man-6-P receptor embryos is caused by elevated IGF-II levels and that the IGF-I receptor is signaling the deleterious responses to IGF-II. In addition, these double receptor knockout mice exhibited normal embryonic growth, consistent with the conclusion that overstimulation of the IGF-I receptor by elevated levels of IGF-II accounts for the excessive growth of the IGF-II/Man-6-P receptor-deficient embryos. Normal embryonic growth of these IGF-I receptor-deficient embryos was explained by stimulation of an unknown receptor by elevated levels of IGF-II. This receptor X has recently been identified as being the insulin receptor *(80)*.

These gene knockout experiments would seem to rule out the existence of major intracellular growth stimulatory pathways that directly emanate from the IGF-II/Man-6-P receptor.

IGF-II Analogs as Tools to Evaluate IGF-II/Man-6-P Receptor Signaling

Sakano et al. *(81)* designed and expressed mutants of IGF-II that were specific for either the IGF-I receptor or the IGF-II/Man-6-P receptor. [Leu^{27}]IGF-II and [Leu^{43}]IGF-II (class I mutants) bound specifically and with undiminished affinities to the IGF-II/Man-6-P receptor and had 80- to 220-fold decreased affinities respectively for the IGF-I and insulin receptors. Conversely, [Thr^{48},Ser^{49},Ile^{50}]IGF-II and [Arg^{54},Arg^{55}] (class II mutants) bound specifically and with undiminished affinities to the IGF-I receptor but had little or no affinity for the IGF-II/Man-6-P receptor. The receptor preparations used for these studies were partially or highly purified receptors from rat or human placental membranes. Thus, there would be less liklihood of errors being introduced by crossreactivity with heterologous receptors or IGF binding proteins in experiments to determine binding affinities of the analogs. Burgisser et al. *(82)* also prepared the [Leu^{27}]IGF-II mutant and found that this mutant exhibited a 25-fold decrease in affinity for the IGF-I receptor and a 3.4-fold decrease in affinity for the IGF-II/Man-6-P receptor. This is less discrimination between the two receptors for [Leu^{27}]IGF-II than reported by Sakano et al. *(81)* but Burgisser et al. *(82)* measured binding affinities using cells overexpressing IGF-I receptors or insulin receptors whereas Sakano et al. *(81)* had used purified or enriched solubilized receptor preparations. Beukers et al. *(83)* reported that the affinity of [Leu^{27}]IGF-II for the IGF-II/Man-6-P receptor in rat placental membranes was the same as for IGF-II and that its affinity for the human placental IGF-I receptor was 100-fold reduced.

The IGF-II analogs described by these three groups are useful reagents to determine whether the IGF-I receptor or the IGF-II/Man-6-P receptor is signaling a particular biologic response. They are employed most effectively if both class I and class II mutants are tested.

Mouse 3T3 Fibroblasts

Calcium Influx. Low concentrations of IGF-II (1 n*M*) were reported to cause a rapid increase in cytoplasmic free calcium as measured by aequorin *(84)* or fura-2 loading *(85)*, unidirectional calcium influx rate using ^{45}Ca *(85–87)*, and activity of a cation channel with patch-clamp technique *(88)*. These experiments were carried out in BALB/c 3T3 fibroblasts that were initially treated with platelet-derived growth factor and then with epidermal growth factor (primed, competent cells). Dose–response curves for the stimulation of Ca^{2+} influx by IGF-II *(87)* corresponded to binding isotherms for IGF-II *(85,86)*. An IgG fraction from an antiserum against the IGF-II/Man-6-P receptor also stimulated Ca^{2+} influx *(87)* and cation channel activity *(88)*. These results suggested that IGF-II stimulates Ca^{2+} influx in primed, competent 3T3 fibroblasts by activating a calcium-permeable cation channel via the IGF-II/Man-6-P receptor.

[^{3}H]Thymidine Incorporation into DNA. Evidence was presented using the primed, competent BALB/c 3T3 fibroblast model that the stimulation of Ca^{2+} influx by IGF-II was a trigger for DNA synthesis *(84,86,87)*. Thus, the stimulation of Ca^{2+} influx and the stimulation of DNA synthesis by low concentrations of IGF-II were both dependent upon the presence of extracellular Ca^{2+} *(84,86)* and inhibitors of the calcium-permeable cation channel also blocked IGF-II-stimulated [^{3}H]thymidine incorporation into DNA *(86)*. Pertussis toxin inhibited both IGF-II-stimulated Ca^{2+} influx and [^{3}H]thymidine incorporation into DNA. Because IGF-I also stimulates Ca^{2+} influx in 3T3 fibroblasts *(85)* and this effect is blocked by pertussis toxin, the above results do not necessarily imply signaling of DNA synthesis by the IGF-II/Man-6-P receptor. However, anti-IGF-II/Man-6-P receptor IgG was reported to stimulate [^{3}H]thymidine incorporation into DNA *(87)*. Using IGF-II and IGF-II analogs, Sakano et al. *(81)* measured the incorporation of [^{3}H]thymidine into DNA in Balb/c 3T3 cells. Mutants of IGF-II that bind to the IGF-I receptor but not to the IGF-II receptor were equipotent to IGF-II in stimulating DNA synthesis, whereas the mutants that bound normally to the IGF-II receptor but with diminished affintity to the IGF-I receptor were least effective as stimulators of DNA synthesis. Similar findings were reported by Burgisser et al. *(82)* and Beukers et al. *(83)*. These results point to the paramount involvement of the IGF-I receptor in the stimulation of DNA synthesis in Balb/c 3T3 cells.

Human Erythroleukemia Cell Line (K562)

Tally et al. *(89)* used a clonal growth assay in semisolid agar to measure the growth effects of IGFs and insulin on a parental K562 erythroleukemia cell line and a subclone designated K562cl.1. The authors were unable to demonstrate the presence of IGF-I receptors on the parental line and the subclone had twice as much ^{125}I-IGF-II binding and one fifth the binding of insulin compared to the parental line. In the parental line, IGF-II and insulin were equipotent in stimulating clonal growth in semisolid agar and IGF-I was

100-fold less potent. In the subclone, insulin was less effective than IGF-II and IGF-I was not active.

Tally et al. *(90)* crosslinked ^{125}I-IGF-II to K562 erythroleukemia cells and in addition to the IGF-II/Man-6-P complex, they noted a major affinity-labeled band at 70 kDa under reducing conditions. Zhang et al. *(91)* have now partially purified this lower molecular weight species from K562 cells. The partially purified receptor/binding protein exhibited high-affinity binding to ^{125}I-IGF-II with no competition of binding by the highest concentration of IGF-I tested. An antibody raised against the partially purified protein recognized a 68-kDa protein on immunoblotting of the purified receptor-binding protein, but did not recognize the IGF-II/Man-6-P receptor in a preparation of solubilized human placental membranes, providing evidence that the 68-kDa protein is not simply a fragment of the IGF-II/Man-6-P receptor. Furthermore, the ligand binding characteristics of the 68-kDa protein are unlike the ligand binding profile of any known IGF binding protein. It is not known whether the IGF-II-stimulated clonal growth of K562 cells in soft agar is mediated by the IGF-II/Man-6-P receptor or by the putative 68-kDa receptor. Based on an analysis of the growth curves for the various gene knockout mice, Efstratiadis and his colleagues *(76)* proposed that there are two additional receptors for IGF-II. One of these receptors has now been identified as the insulin receptor *(80)*. The most interesting possibility is that the 68-kDa protein from K562 cells represents a new receptor for IGF-II.

Blanchard et al. *(92)* also found that the binding of ^{125}I-IGF-II to k562 cells was much greater than the binding of ^{125}I-IGF-I, and in affinity crosslinking experiments ^{125}I-IGF-I appeared to only bind to the IGF-II/Man-6-P receptor. However, plasma-derived IGF-I was equipotent to IGF-II in stimulating [^{3}H]thymidine incorporation into DNA, and recombinant IGF-I was more potent than either of the plasma-derived IGFs. Hizuka et al. *(93)* readily demonstrated ^{125}I-IGF-I binding to an IGF-I receptor on K562 cells, and IGF-I stimulated [^{3}H]thymidine incorporation into DNA at 0.1 ng/mL. IGF-I at 10 ng/mL also stimulated an increase in cell number. Thus, it seems likely that K562 cells in use in different laboratories are not uniform in their complement of IGF-I receptors and IGF-II/Man-6-P receptors and consequently exhibit different relative growth responses to IGF-I and IGF-II.

Canine Kidney Proximal Tubules

Phosphorylation of Basolateral Membranes. IGF-II caused a rapid, concentration-dependent phosphorylation of a 135-kDa component in the membrane preparation *(94)*. The maximal increase in phosphorylation of the 135-kDa species was only 40% over basal. IGF-I was not active at the highest concentration tested (10^{-9} *M*).

Alkalinization of Proximal Tubular Cells. A very low concentration of IGF-II (10^{-11} *M*) caused a rapid alkalinization of cells within proximal tubular segments as measured by the uptake of a radiolabeled weak acid *(95)*. Alkalinization presumably occurred by activation of an amiloride-sensitive Na^+–H^+ exchanger in the brush border membrane of proximal tubular cells. Both insulin and IGF-I at 10^{-9} or 10^{-6} *M* were ineffective.

Production of Inositol Trisphosphate and Diacylglycerol in Basolateral Membranes. At a concentration of 10^{-10} *M*, IGF-II stimulated an increase in levels of inositol trisphosphate and diacylglycerol in basolateral membrane preparations *(96)*. IGF-I at 10^{-8} *M* was not effective in stimulating an increase in the level of inositol trisphosphate. Man-6-P produced a twofold potentiation of the effect of IGF-II on the generation of inositol trisphosphate in the basolateral membrane preparation *(97)*. Man-6-P potentiated the effect of 10^{-10}, 10^{-9}, and 10^{-8} *M* IGF-II by twofold but increased the binding of 10^{-10} *M* ^{125}I-IGF-II by only 26% and did not increase the binding of 10^{-9} and 10^{-8} *M* IGF-II, leading the authors to conclude that the potentiating effect of Man-6-P on signaling could not be explained entirely by increased binding of IGF-II to the IGF-II/Man-6-P receptor.

Rogers et al. *(98)* also reported that the Man-6-P-containing proteins TGF-βI precursor and proliferin also increased phospholipase C activity in proximal tubular basolateral membranes from canine kidney. Maximal activation of phospholipase C was achieved with very low concentrations of TGF-βI precursor and proliferin (10^{-13} *M*). This concentration is much less than the K_d value for binding of proliferin to the IGF-II/Man-6-P receptor. The lysosomal enzyme β-galactosidase did not activate phospholipase C in this membrane preparation. Man-6-P inhibited the effect of a low concentration (10^{-14} *M*) but actually enhanced the effect on phospholipase C at a higher concentration (10^{-11} *M*). An antibody to the IGF-II/Man-6-P receptor inhibited the effect of proliferin and TGF-βI precursor. These results suggest involvement of the IGF-II/Man-6-P receptor in the activation of phospholipase C by proliferin and TGF-βI precursor but raise questions about mechanism because of the very low concentrations of ligand that were required and the dual effects of Man-6-P.

Cartilage

Mouse Embryonic Limb Buds. Limb buds from 11-day-old embryos undergo differentiation in organ culture from the blastema stage to cartilagenous anlagen. ^{125}I-IGF-I binding declined during the time of differentiation whereas ^{125}I-IGF-II binding increased *(99)*. IGF-I stimulated sulfate and thymidine incorporation in differentiated limb bud cultures but not in undifferentiated limb buds. In contrast, IGF-II stimulated thymidine incorporation in undifferentiated cultures but had no effect on either sulfate or thymidine incorporation in differentiated cultures. IGF-II, but not IGF-I, stimulated glucose uptake, glycogen synthesis, and glycogen synthetase activity in both undifferentiated and differentiated limb buds *(100)*. Rather high concentrations of IGF-II were required for these effects but this might potentially be explained by high rates of degradation and problems of diffusion into tissue.

Bhaumick and Bala *(101)* developed a method for culturing differentiating cartilage cells from the mouse embryonic limb buds. IGF-I and IGF-II both stimulated glucose uptake in undifferentiated and differentiated cultures. However, when compared with receptor occupancy curves for their respective receptors, IGF-II was much more effective than IGF-I in stimulating glucose uptake, suggesting signaling by the IGF-II/Man-6-P receptor.

Human Fetal Chondrocytes. Vetter et al. *(102)* compared the response to IGF-I and IGF-II for the clonal growth of human fetal and adult chondrocytes. IGF-II was twice as effective as IGF-I in stimulating the clonal growth of the fetal chondrocytes, whereas the clonal growth of the adult chondrocytes was stimulated more potently by IGF-I.

Human Chondrosarcoma. The chondrosarcoma-derived cell line HCS 2/8 exhibits properties of chondrocytes, synthesizing aggrecan, collagen types II, IX, and XI, as well as integrins found in chondrocytes. Takigawa et al. *(103)* provided evidence that IGF-I and IGF-II stimulate proteoglycan synthesis by utilizing both IGF-I receptors and IGF-II/Man-6-P receptors. HCS 2/8 cells produce IGF-I and IGF-II. Addition of the monoclonal IGF-I receptor antibody αIR-3 inhibited 80% of autocrine stimulation of proteoglycan synthesis and an antibody to IGF-II blocked 40% of the autocrine stimulation. These results point to an important role of the IGF-I receptor in the autocrine stimulation of proteoglycan synthesis. However, addition of Man-6-P potentiated the autocrine stimulation of proteoglycan synthesis, consistent with removal of IGF-II/Man-6-P receptor-bound lysosomal enzymes, thus allowing the binding of endogenous IGF-II. Added IGF-II was more potent than IGF-I in stimulating proteoglycan synthesis, and IGF-II stimulated proteoglycan synthesis even in the presence of αIR-3 at a concentration that inhibited endogenous synthesis by 80%. [Leu^{27}]IGF-II, which binds with high affinity only to the IGF-II receptor, was less potent than IGF-II but more potent than IGF-I in stimulating proteoglycan synthesis. The authors also measured the effects of IGF-I and IGF-II on intracellular Ca^{2+} accumulation. Interestingly, only IGF-II was capable of stimulating Ca^{2+} accumulation. These results are consistent with IGF-II/Man-6-P receptor involvement in the stimulation of proteoglycan synthesis and Ca^{2+} accumulation by IGF-II in this chondrosarcoma cell line.

Rabbit Articular Chondrocytes. Poiraudeau et al. *(104)* examined IGF-I and IGF-II stimulation of intracellular Ca^{2+} accumulation in cultures of rabbit articular chondrocytes. Although both IGF-I and IGF-II stimulated Ca^{2+} accumulation, there were two mechanisms involved. There was evidence for stimulation of Ca^{2+} influx through voltage-gated calcium channels by IGF-I and pharmacologic concentrations of insulin, but not by IGF-II. IGF-I, IGF-II, and insulin also mobilized Ca^{2+} from endoplasmic reticulum; phospholipase C inhibitors partially blocked Ca^{2+} accumulation stimulated by IGF-I and insulin, but totally blocked the stimulation by IGF-II. Stimulation of Ca^{2+} mobilization by IGF-II was totally blocked by pertussis toxin and an IGF-I receptor antibody, which partially blocked the effect of IGF-I on Ca^{2+} accumulation but did not inhibit the effect of IGF-II. These results provide evidence that the effect of IGF-II on Ca^{2+} mobilization in these rabbit articular chondrocytes was signaled by a receptor that was not the IGF-I receptor, suggesting a role for the IGF-II/Man-6-P receptor. The putative IGF-II/Man-6-P receptor pathway involved G_i and phospholipase C.

Muscle

BC3H-1 Cells. IGF-II is the principal IGF expressed by differentiating muscle cells. [Leu^{27}]IGF-II stimulated differentiation of BC3H-1 muscle cells as measured by the expression of myogenin mRNA and the binding of α-bungarotoxin to the nicotinic acetylcholine receptor *(105)*. [Leu^{27}] displayed markedly reduced affinity for the IGF-I receptor in these muscle cells and was almost as potent as IGF-II in stimulating differentiation.

L6 Cells. Bach et al. *(106)* reported that [Leu^{27}]IGF-II was considerably less potent than IGF-II in stimulating L6A1 myoblast proliferation and differentiation, pointing to the importance of the IGF-I receptor in signaling these reponses to IGF-II. It is not clear whether the response to [Leu^{27}]IGF-II resulted from a weak interaction with the IGF-I receptor or reflected activation of the IGF-II/Man-6-P receptor.

In L6A1 muscle cells, although IGF-I stimulated myogenesis at lower concentrations than IGF-II, IGF-II treatment resulted in more extensive differentiation, suggesting a role for the IGF-II receptor in stimulating myogenesis *(107)*. However, the authors found that the greater stimulation by IGF-II was simply a consequence of a delay in IGF-I-stimulated myogenesis as a result of its greater mitogenic action. By using IGF-I and IGF-II analogs with altered affinity for the IGF-I receptor and the IGF-II/Man-6-P receptor, the authors provided evidence that the stimulation of myogenesis was mediated by the IGF-I receptor rather than the IGF-II/Man-6-P receptor. [YQ]IGF-I, an analog with sevenfold increased affinity for the IGF-II/Man-6-P receptor, had activity equal to native IGF-I and [Leu^{27}]IGF-II, an analog with normal affinity for the IGF-II/Man-6-P receptor, and 100-fold reduced affinity for the IGF-I receptor, showed a pronounced decrease in activity compared to IGF-I.

Thus, although some studies possibly suggest that the IGF-II/Man-6-P receptor is able to signal differentiation in muscle, most studies, including the IGF-I receptor gene knock-out experiments, point to the paramount role of the IGF-I receptor in mediating differentiation and proliferation in muscle *(108,109)*.

Rhabdomyosarcoma. Rhabdomysosarcoma tumors express high levels of IGF-II mRNA and in culture IGF-II provides autocrine growth stimulation *(110)*. The autocrine growth stimulation is signaled by the IGF-I receptor, as the receptor antibody αIR-3 blocks growth. Exogenous IGF-II stimulated cell motility, but this action was not inhibited by αIR-3. The mechanism of IGF-II action was chemokinetic rather than chemotactic. [Leu^{27}]IGF-II also stimulated motility of rhabdomyosarcoma cells at concentrations as low as 10^{-10} *M (111)*. At a concentration of 20 n*M* [Leu^{27}]IGF-II did not inhibit the binding of ^{125}I-IGF-I to the IGF-I receptor in these cells. An antibody that blocked the binding of IGF-II to the IGF-II/Man-6-P receptor blocked IGF-II-stimulated motility. These results point to signaling by the IGF-II/Man-6-P receptor.

Capillary Endothelial Cells

Proliferin or mitogen-regulated protein is a member of the PRL/GH family. In vivo, proliferin is expressed specifically in placental trophoblast giant cells. Proliferin was reported to stimulate capillary endothelial cell migration in culture and neovascularization in vivo *(112)*. This activity was a property of both recombinant proliferin from stably transfected CHO cells and proliferin from medium conditioned by placental tissue harvested at mid-gestation. It was proposed that proliferin stimulates endothelial cells from maternal blood vessels in the uterus to migrate toward the implantation site. Volpert et al. *(112)* reported that Man-6-P inhibited the binding of ^{125}I-proliferin to endothelial cells in placental tissue sections. Man-6-P also blocked the stimulation of the migration of capillary endothelial cells by proliferin and inhibited the action of proliferin to induce neovascularization in an in vivo rat cornea model. Interestingly, IGF-II also stimulated endothelial cell migration in a dose-dependent manner from 3 to 100 ng/mL *(112)*. Because IGF-I also stimulated cell migration at similar concentrations, it was not clear whether the stimulation by IGF-II utilized the IGF-II/Man-6-P receptor. However, endothelial cells responded equally well to [Leu^{27}]IGF-II, consistent with involvement of the IGF-II/Man-6-P receptor. [Leu^{27}]IGF-II was also effective in inducing neovascularization in the rat cornea model, and this action was completely blocked by Man-6-P.

Pertussis toxin, which inactivates G_i proteins, inhibited proliferin-induced chemotaxis of bovine capillary epithelial cells *(113)*. Proliferin stimulated mitogen-activated protein kinase (MAPK). This activity was blocked by pertussis toxin and by a specific inhibitor of MAPK kinase (PD098059). Finally, the MAPK kinase inhibitor also blocked the stimulation of chemotaxis by proliferin. Thus, activation of MAPK was required for the stimulation of chemotaxis by proliferin and this activation was coupled to a G_i protein. Similarly, the stimulation of chemotaxis by [Leu^{27}]IGF-II was also blocked by pertussis toxin and by the MAPK kinase inhibitor, providing further evidence that the IGF-II/Man-6-P receptor signaled the stimulation of chemotaxis in these capillary endothelial cells.

Human Skin Fibroblasts

Palka et al. *(114)* showed that the chemotaxis of human skin fibroblasts in response to platelet-derived growth factor (PDGF) was enhanced by IGF-II and Man-6-P. One proposed mechanism was that, as an inhibitor of the binding of lysosomal enzymes to the IGF-II/Man-6-P receptor, Man-6-P would raise the extracellular concentration of lysosomal enzymes in the extracellular matrix. In support of this proposal, the authors found that pepstatin, an inhibitor of cathepsin D, inhibited chemotaxis induced by PDGF and blocked enhancement by Man-6-P. It is not clear from these experiments whether IGF-II enhanced chemotaxis by inhibiting lysosomal enzyme binding or by direct signaling through the IGF-II/Man-6-P receptor or the IGF-I receptor.

Placental Cytotrophoblasts

Nestler *(115)* reported that IGF-II potently inhibited the conversion of androstenedione to estrogen (aromatase activity) in cultures of cytotrophoblasts. The effect of IGF-II was seen only after 20 h of culture, but by 24 h of culture the inhibition by IGF-II was greater than 95%. IGF-II at 20 ng/mL was much more effective than IGF-I at 20 ng/mL. A monoclonal antibody, αIR-3, which blocks activation of the IGF-I receptor by IGF-I and IGF-II, only partially inhibited the effect of IGF-II on aromatase, but totally blocked the effect of IGF-I. Interestingly, Man-6-P increased the inhibition of aromatase in the absence of IGF-II and potentiated the effect of low doses of IGF-II. The effect of Man-6-P in the absence of IGF-II could be explained by the stripping of lysosomal enzymes from the IGF-II/Man-6-P receptor, allowing endogenous IGF-II to inhibit aromatase. Taken together, these results suggest that at least part of the effect of IGF-II on aromatase activity is mediated by the IGF-II/Man-6-P receptor. Of course, because Man-6-P and IGF-II can increase the concentration of extracellular lysosomal enzymes, the inhibition of aromatase could be a secondary effect of extracellular lysosomal enzymes rather than direct activation of the IGF-II/Man-6-P receptor by IGF-II.

Testis

In a mouse model for Sertoli cell–spermatogenic cell interaction it was shown that Sertoli cell conditioned medium stimulated c-*fos* gene expression and increased the level of 18S RNA in spermatogenic cells *(116)*. These effects were totally blocked by addition of Man-6-P. By using IGF-II/Man-6-P receptor affinity chromatography the authors have been able to isolate at least 10 Man-6-P-glycoproteins from Sertoli cell conditioned medium. In addition, Sertoli cells express IGF-II mRNA and IGF-II stimulates an increase in c-*fos* mRNA levels in spermatogenic cells. The results suggest that interactions between Sertoli cells and spermatogenic cells are at least in part mediated by the IGF-II/Man-6-P receptor on the surface of spermatogenic cells. Presumably these interactions are triggered by Man-6-P-glycoproteins coming from Sertoli cells. Although IGF-II mRNA was found in the Sertoli cells, there was no evidence presented that IGF-II was found in the conditioned medium. It also was not proven that the stimulatory effect of added IGF-II on c-*fos* mRNA levels in spermatogenic cells was mediated by the IGF-II/Man-6-P receptor.

Myeloma Cells

The interaction between myeloma cells and stromal cells was investigated by using a human myeloma cell line (OMP-2) and human umbilical vein endothelial cells (HUVECs) *(117)*. Biotinylated glycoproteins at the plasma membrane of HUVECs specifically bound to the IGF-II/Man-6-P receptor of OMP-2 cells, and adhesion of OPM-2 cells to HUVECs was inhibited by Man-6-P. Man-6-P was also found to suppress the adhesion of human myeloma cell lines to bone marrow stromal cells established from patients with multiple myeloma. In addition, IGF-II stimulated the proliferation of OMP-2 cells and the effect of IGF-II was not blocked by an antibody to the IGF-I receptor (αIR-3). The authors concluded that the interaction of multiple myeloma cells with stromal cells or human vein endothelial cells involved interaction of cell surface glycoproteins with the multiple myeloma cell IGF-II/Man-6-P receptor and that, moreover, the IGF-II/Man-6-P receptor signaled a proliferative response to IGF-II.

Nerve

Fex Svenningsen et al. *(118)* provided evidence that IGF-I analogues, IGF-II, and insulin stimulate [^{3}H]thymidine incorporation into rat sciatic nerve segments in culture by acting through their respective receptors. The stimulation by the IGF-I analogues was partially blocked by the JB1 peptide which had been shown to have IGF-I receptor antagonistic properties. However, the stimulation by IGF-II was not inhibited by the JB1 peptide but was blocked completely by Man-6-P. The mechanism of the inhibitory action of Man-6-P is not clear as Man-6-P does not block the binding of IGF-II to the IGF-II/Man-6-P receptor.

Man-6-P stimulated [^{3}H]thymidine incorporation into DNA in neuronal precursor cells from developing rat brain and the effect was blocked by an antibody to the IGF-II/Man-6-P receptor *(119)*. IGF-II also stimulated DNA synthesis in these cells but the receptor antibody did not inhibit. Thus, the IGF-II/Man-6-P receptor appears to be involved in stimulation of DNA synthesis by Man-6-P but IGF-II is presumably utilizing the IGF-I receptor.

In SH-SY5Y human neuroblastoma cells Man-6-P alone was found to increase cell number and neurite outgrowth and to potentiate the effect the effect of IGF-II on both parameters *(120)*. Further evidence supporting mediation of the effect of IGF-II by the IGF-II/Man-6-P receptor was not presented. For example, the synergistic effect of Man-6-P could have resulted from interaction of postreceptor pathways emanating from the IGF-I receptor.

Pancreatic β Cells

Zhang et al. *(121)* reported that IGF-II promoted the exocytosis of insulin from pancreatic β cells from the *ob/ob* mouse and provided evidence that the IGF-II/Man-6-P receptor was signaling this response. IGF-I did not stimulate exocytosis of insulin from the β cells and there was no evidence for the presence of insulin receptors or IGF-I receptors on these cells either by ligand crosslinking or metabolic labeling. The stimulation of insulin exocytosis by IGF-II was blocked by pertussis toxin or by down-regulation of protein kinase C by phorbol ester treatment of the cells. The authors reported that an antiserum to the IGF-II receptor blocked binding of ^{125}I-IGF-II to the β cells and blocked IGF-II induced exocytosis of insulin. This is in disagreement with a previous report that the blocking antibody population in this rabbit polyclonal antibody did not block the binding of ^{125}I-IGF-II to the mouse IGF-II/Man-6-P receptor *(69)*. Nonetheless, the lack of response to IGF-I and absence of the IGF-I receptor in these *ob/ob* mouse pancreatic β cells indirectly support signaling of insulin exocytosis by the IGF-II/Man-6-P receptor.

Osteoblasts

The ability of IGF-II to evoke Ca^{2+} transients in rat calvaria osteoblasts was evaluated *(122)*. Osteoblasts were pretreated with IGF-I until no further response to IGF-I was observed. The osteoblasts were still capable of responding to IGF-II. A polyclonal antibody to the IGF-II/Man-6-P receptor was able to stimulate a Ca^{2+} response when tested alone and blocked the response to IGF-II. The Ca^{2+} transient appeared to be caused primarily by mobilization of intracellular Ca^{2+} stores.

Breast Cancer Cells

Cathepsin D inhibited the stimulation of cell growth of MCF-7 breast cancer cells by low concentrations of IGF-II (0.1 n*M*), but cathepsin D had no effect on stimulation by high concentrations of IGF-II (1 n*M*) *(30)*. Conversely, the αIR-3 antibody, which blocks the binding of IGF-I to the IGF-I receptor, did not inhibit stimulation of cell growth by low concentrations of IGF-II but did inhibit by 38% the effect of high concentrations of IGF-II. These results suggest that, in MCF-7 cells, low concentrations of IGF-II stimulate cell growth by acting through the IGF-II/Man-6-P receptor whereas at higher concentrations of IGF-II stimulation of cell growth is mediated by the IGF-I receptor.

De Leon et al. *(123)* also observed that the IGF-I receptor antibody αIR-3 did not block stimulation of growth of MCF-7 cells by low concentrations of IGF-II, whereas αIR-3 did inhibit the stimulation of cell growth by IGF-I. The lysosomal enzyme β-galactosidase caused a small inhibition of the effect of IGF-II on cell growth but the effect was not statistically significant.

Hepatoma Cell

Insulin, IGF-I, and IGF-II stimulate glycogen synthesis in the human hepatoma cell line HEP G2. Hari et al. *(124)* found that receptor antibodies that blocked binding of insulin and IGF-I to their respective receptors only partially blocked the stimulation of glycogen synthesis by IGF-II. In addition, IgG purified from a polyclonal antiserum for the IGF-II receptor mimicked the effect of IGF-II on glycogen synthesis. However, results from more recent experiments with IGF-II and IGF-II analogs that show preferential binding to either the IGF-I receptor or the IGF-II/Man-6-P receptor suggest that it is the IGF-I receptor that is signaling the stimulation of glycogen synthesis in these cells *(81)*.

Evidence for Coupling of the IGF-II/Man-6-P Receptor to a Heterotrimeric G Protein of the G_i Subclass

Pertussis toxin, which causes ADP ribosylation of the α-subunit of heterotrimeric G-proteins of the G_i subclass and thereby blocks receptor–effector interactions, was found to inhibit IGF-II-stimulated increases in cytoplasmic free Ca^{2+} and rate of Ca^{2+} influx in primed, competent 3T3 cells *(85)*. Pertussis toxin stimulated ADP ribosylation of a 41-kDa protein in 3T3 cell membranes. GTPγS inhibited ^{125}I-IGF-II binding to 3T3 cell membranes in a concentration-dependent manner. These results suggested that the IGF-II/Man-6-P is coupled to a calcium channel through a pertussis-toxin-sensitive G-protein. Over the past decade, Nishimoto and his colleagues have investigated the molecular basis of the coupling of the IGF-II/Man-6-P receptor to G_i. They have proposed that a segment of the cytoplasmic domain of the activated IGF-II/Man-6-P receptor interacts directly with the α-subunit of G_i, resulting in exchange of GTP for GDP on the α-subunit *(125)*.

In initial experiments broken cell membranes from 3T3 fibroblasts were incubated with IGF-II and GTP, and than ADP ribosylated with pertussis toxin *(126)*. Preincubation with 10 n*M* IGF-II reduced ADP ribosylation of a 40-kDa protein by 70–80%. Significant reduction in ADP ribosylation was observed in 5 min and the EC_{50} for IGF-II was 0.7 n*M*. Anti-IGF-II/Man-6-P receptor IgG also caused attenuation of ADP ribosylation of the 40-kDa substrate in 3T3 cell membranes. Since pertussis toxin-stimulated ADP ribosylation takes place only on heterotrimeric G-proteins, these results are consistent with dissociation of G-protein into α and $\beta\gamma$ in the presence of IGF-II and GTP. 3T3 cell membranes were pretreated with pertussis toxin to ADP ribosylate endogenous substrate, and the membranes were then incubated with IGF-II, GTP, and preparations of $G_{i\text{-}1}$, $G_{i\text{-}2}$, and G_o purified from brain. The subsequent ADP ribosylation of $G_{i\text{-}2}$ was attenuated by IGF-II, suggesting that it is $G_{i\text{-}2}$ or a related G-protein that is coupled to the IGF-II/Man-6-P receptor in 3T3 cell membranes.

IGF-II was unable to inhibit the ADP-ribosylationof a 40-kDa substrate in membranes from quiescent BALB/c 3T3 cells unless the cells had been treated with PDGF and EGF (primed, competent cells) *(127)*. In BALB/c 3T3 cells transfected with temperature-sensitive v-Ki-*ras*, IGF-II was able to inhibit the ADP-ribosylation of a 40-kDa substrate only in membranes from cells maintained at the permissive temperature. There was no change in IGF-II/Man-6-P receptor and $G_{i\text{-}2\alpha}$ expression when cells were treated with PDGF and EGF or when the v-Ki-*ras* transfected cells were maintained at the permissive temperature. The authors proposed that in quiescent cells or in the v-Ki-*ras* transfected cells maintained at the nonpermissive temperature, the activated receptor is uncoupled from $G_{i\text{-}2}$ and that addition of EGF and PDGF or activation of v-Ki-*ras* leads to coupling of the receptor with $G_{i\text{-}2}$.

To study a more defined experimental system, phospholipid vesicles were constructed that contained highly purified IGF-II/Man-6-P receptor from rat placenta and $G_{i\text{-}2}$ from porcine brain *(126)*. GTPγS inhibited the binding of IGF-II to the IGF-II/Man-6-P receptor, changing the dissociation constant from 8.9×10^{-10} to 7.0×10^{-9}. IGF-II also increased the initial rate of binding of GTPγS to $G_{i\text{-}2}$ by approximately twofold. Moreover, the rate of GTPγS binding to $G_{i\text{-}2}$ in the presence of IGF-II was inhibited by 70% in reconstituted vesicles that had been pretreated with pertussis toxin. These results obtained with purified components in reconstituted vesicles are consistent with direct coupling of the IGF-II/Man-6-P receptor to the transducing protein $G_{i\text{-}2}$.

In reconstituted phospholipid vesicles containing purified recombinant IGF-II/Man-6-P receptor and $G_{i\text{-}2}$, IGF-II stimulation of the rate of binding of GTPγS to the G protein was inhibited by Man-6-P or more potently by the lysosomal enzyme β-glucuronidase *(128)*. Man-6-P and β-glucuronidase had no effect when tested in the absence of IGF-II. Because Man-6-P does not inhibit the binding of IGF-II to the receptor, the authors postulated that the binding of Man-6-P modulates post-receptor signals after IGF-II binding. (It would be of interest to test whether or not Man-6-P also inhibits the IGF-II stimulation of Ca^{2+} influx in BALBc/3T3 cells that have first been washed with Man-6-P to remove lysosomal enzymes.)

The domain of the cytoplasmic domain of the IGF-II/Man-6-P receptor that interacts with the G protein $G_{i\text{-}2}$ was identified *(129)*. The cytoplasmic domain of the receptor was searched for sequences with similarities to mastoparan, a peptide present in wasp venom known to activate G-proteins. Based on this search, a synthetic peptide corresponding to residues 2410–2423 of the receptor was used for further studies. This peptide (peptide 14) increased the binding rate of GTPγS to $G_{i\text{-}2}$ by approximately threefold. Peptide 14 also stimulated the release of GDP from $G_{i\text{-}2}$ and stimulated the GTP hydrolysis activity of $G_{i\text{-}2}$. The action of peptide 14 was inhibited when $G_{i\text{-}2}$ was pretreated with pertussis toxin. Peptide 14 was more effective in stimulating binding of GTPγS to $G_{i\text{-}2}$ when peptide 14 was coupled to the transmembrane domain of the receptor and the reaction was performed in reconstituted phospholipid vesicles. Finally, an antibody raised against peptide 14 was shown to inhibit the effect of IGF-II on the binding rate of GTPγS to $G_{i\text{-}2}$ in reconstituted phospholipid vesicles containing pure IGF-II/Man-6-P receptor. The results of these in vitro experiments suggest that residues 2410–2423 are involved in the interaction of the receptor with $G_{i\text{-}2}$ upon activation of the receptor by the binding of IGF-II.

The IGF-II/Man-6-P receptor peptide Arg^{2410}–Lys^{2423} was shown to increase significantly the rate of binding of GTPγS to the α-subunit of $G_{i\text{-}2}$ *(130)*. This effect was potentiated by addition of βγ-subunits in stoichiometric amounts. An antibody to the C-terminal decapeptide of $G\alpha_{i\text{-}2}$ blocked peptide-stimulated GDP release from $G_{i\text{-}2}$, suggesting that the IGF-II/Man-6-P receptor interacts with the C-terminal portion of the α-subunit.

The authors were able to change the G-protein specificity of the IGF-II/Man-6-P receptor by exchanging a short sequence from the β-adrenergic receptor for the sequence that had been shown to interact with $G_{i\text{-}2}$ *(131)*. The 15-amino-acid peptide βIII-2 from the β-adrenergic receptor was substituted for Arg^{2410}–Lys^{2423} in the IGF-II/Man-6-P receptor and Cos-7 cells were transiently transfected with this receptor construct. Addition of IGF-II to a membrane preparation from these cells resulted in an increase in adenylate cyclase activity, consistent with activation of G_s.

To demonstrate direct activation of G_i by the IGF-II/Man-6-P receptor in vivo, Ikezu et al. *(132)* transfected COS cells with receptor mutants and measured the ability of IGF-II to inhibit adenylyl cyclase activity. They first demonstrated a G_i-responsive adenylyl cyclase in COS cells by showing that transfection of COS cells with constitutively activated $G\alpha_{i\text{-}2}$ decreased cylic AMP accumulation stimulated by cholera toxin. In cells transiently transfected with IGF-II/Man-6-P receptor cDNA, addition of IGF-II resulted in inhibition of cyclic AMP accumulation promoted by cholera toxin or forskolin. IGF-II failed to inhibit cholera toxin-stimulated cyclic AMP accumulation in cells that had been transfected with IGF-II/Man-6-P receptor cDNA constructs in which amino acids 2410–2423 had been deleted or with receptor constructs lacking the C-terminal 41 residues after Arg^{2410}, consistent with their earlier demonstration that the Arg^{2410}–Lys^{2423} peptide increased the rate of binding of GTPγS to $G_{i\text{-}2.}$ Surprisingly, in COS cells that had been transfected with a receptor cDNA encoding a receptor in which the C-terminal 28 residues after Ser^{2424} were deleted, IGF-II further enhanced cholera toxin-stimulated cyclic AMP accumulation. This enhancement of cyclic AMP accumulation by IGF-II was blocked by pertussis toxin. The authors proposed that the enhancement of adenylyl cyclase activity was mediated by βγ-subunit stimulation of another adenylyl cyclase subtype, the βγ-subunit being released by activation of G_i by the IGF-II/Man-6-P receptor lacking the C-terminal 28 residues. Activation by βγ-subunits was supported by the ability of transfected $G\alpha_t$ to prevent the IGF-II stimulated enhancement of cholera toxin-stimulated cyclic AMP accumulation. ($G\alpha_t$ would be expected to bind free βγ-subunits.) The authors further proposed that the C-terminal 28 residues of the IGF-II/Man-6-P residues functions to sequester free βγ-subunits although they had no direct evidence for binding of βγ-subunits to the C-terminus of the receptor. Thus, the IGF-II/Man-6-P receptor could signal biologic responses to IGF-II both by activating $G\alpha_i$ and sequestering βγ-subunits.

Korner et al. *(133)* attempted to confirm the experiments of Nishimoto and his colleagues that pointed to activation of G_{i-2} by a 14-amino-acid peptide found within the cytoplasmic fragment of the IGF-II/Man-6-P receptor. They transfected mouse L-cells with either wild-type receptor or mutant receptors that lacked various portions of the cytoplasmic tail including residues 2451–2465, the 14-amino-acid sequence described as being critical for interaction with G_{i-2} by Nishimoto and his colleagues. After addition of IGF-II to the cells, membranes were prepared and ADP-ribosylation of a 40-kDa protein by pertussis toxin was measured. In contrast to the results of Okamoto et al. *(126)* there was no decrease in the ADP-ribosylation of the 40-kDa substrate following addition of IGF-II. IGF-II was also added to membranes that had been prepared from the L-cells transfected with the various IGF-II/Man-6-P receptor constructs and the ADP-ribosylation of the 40-kDa sustrate by pertussis toxin was assessed. Again, ADP-ribosylation was not attenuated following addition of IGF-II. It should be pointed out that the experiments of Nishimoto and his colleagues were carried out with BALB/c 3T3 cells in which pertussis toxin had previously been shown to block an IGF-II-stimulated biologic function (Ca^{2+} influx) whereas Korner et al. *(133)* studied mouse L-cells. Experiments by Korner et al. *(133)* that more closely resembled the experiments of Nishimoto's group were those in which purified components of the system (IGF-II/Man-6-P receptor and G-proteins) were reconstituted in phospholipid vesicles and the effect of IGF-II on the rate of GTPγS binding to G-proteins was examined. There was a small 1.2- to 1.3-fold increase in GTPγS binding to both G_{i-2} in vesicles containing wild-type receptor after 5–20 min incubation with IGF-II. However, the increase in GTPγS binding was also 1.2- to 1.3-fold both in vesicles containing G_{o-1} and in vesicles containing an IGF-II receptor that had a C-terminal deletion that included peptide 14. As a positive control, there was a 2.2-fold increase in GTPγS binding when mastaparin was substituted for the receptor in the vesicles. Receptor-mediated stimulation of G_{i-2} GTPase acitivity was also examined in the reconstituted phospholipid vesicles. IGF-II failed to stimulate G_{i-2} GTPase activity in vesicles containing wild-type receptor or the mutant receptor with the C-terminal deletion. In contrast, mastaparin increased GTPase activity by 7- to 11-fold. Thus, these latter experiments with reconstituted phospholipid vesicles failed to duplicate the results of Nishimoto's group using the same system. It should be pointed out that these experiments of Korner et al. *(133)* examined the molecular mechanism whereby the IGF-II /Man-6-P receptor might interact with a G-protein but did not examine the more general question of G-protein involvement in signaling by the IGF-II/Man-6-P receptor.

The IGF-II/Man-6-P Receptor as a Tumor Suppressor

Recent genetic evidence that the IGF-II/Man-6-P receptor is a tumor suppressor points to an important role of the receptor in regulating the growth of cancer cells. TGF-β1 is a growth inhibitor for most epithelial cells. IGF-II overexpression has been reported in a number of human cancers *(134)*. Rogler et al. *(135)* produced two lines of transgenic mice in which IGF-II expression was driven by the major urinary protein promoter. This promoter is normally silent until turned on by male hormones at the time of puberty in the liver and perputial (scent) glands. These transgenic lines exhibit 20–30 times the level of serum IGF-II found in normal mice. Mice from both lines develop a diverse spectrum of tumors at higher frequency than in controls (22/101 vs 4/59) after 18 months of age. Because the IGF-II/Man-6-P receptor plays a role in the activation of TGF-β1 and mediates the degradation of IGF-II by receptor-mediated internalization, the receptor is a candidate for being a tumor suppressor gene. An inactivating mutation in one allele of a tumor suppressor gene is recessive, but becomes manifest when the other allele is lost. Loss of the other allele is detected by loss of heterozygosity at polymorphic DNA marker loci in or near the gene. De Souza et al. *(136)* used two polymorphisms in the 3′ untranslated region of the IGF-II/Man-6-P receptor gene to screen noncirrhotic, hepatitis virus-negative patients with hepatocellular tumors for loss of heterozygosity. Twenty-two of 36 patients were informative and 14/22 (64%) of liver tumors had loss of heterozygosity (11/16 carcinomas, 1/3 fibrolamellar tumors, and 2/3 adenomas). Using a coarse screen, De Souza et al. *(137)* identified point mutations in 25% of the hepatocellular carcinomas with loss of heterozygosity. These mutations gave rise to a truncated receptor protein and significant amino acid substitutions. The results provide evidence that the IGF-II/Man-6-P receptor functions as a tumor suppressor gene in hepatocellular carcinoma. Yamada et al. *(138)* also examined dysplastic liver lesions and hepatocellular carcinomas associated with the high risk factors of hepatitis virus infection and liver cirrhosis for loss of heterozygosity of the IGF-II/Man-6-P receptor. Loss of heterozygosity occurred in 5/8 (63%) of dysplastic liver lesions and 11/18 (61%) of hepatocellular carcinomas. Mutations in the remaining allele were found in 6/11 (55%) of hepatocellular carcinomas. Loss of heterozygosity was also found in cirrhotic tissue of clonal origin adjacent to the dysplastic liver lesions and hepatocellular carcinomas. The authors conclude that IGF-II/Man-6-P gene inactivation occurs early in liver carcinogenesis associated with hepatitis virus infection and cirrhosis.

Evidence for role for the IGF-II/Man-6-P receptor as a negative growth regulator in breast cancer was provided by a rat model in which the monoterpene *d*-limonene induced the regression of advanced mammary adenocarcinomas *(139)*. The tumors that regressed had an increased cellular concentration of both the IGF-II/Man-6-P receptor and TGF-β1. Hankins et al. *(140)* utilized two polymorphisms in the 3′ untranslated region of the IGF-II/Man-6-P receptor to screen breast tumors for loss of heterozygosity. Forty of 62 patients were informative and 12/40 (30%) of breast tumors had loss of heterozygosity (5/19 carcinomas *in situ* and 7/21 invasive carcinomas). The carcinomas *in situ* were screened for mutations in the remaining allele and 40% were found to have missense mutations that gave rise to significant amino acid substitutions. The presense of loss of heterozygosity and mutation in the remaining allele in the carcinomas *in situ* provide evidence that loss of IGF-II/Man-6-P receptor function may be an early event in the development of breast cancer. In breast cancer the IGF-II/Man-6-P receptor could be acting as a tumor suppressor in three different ways. The receptor lowers the concentration of extracellular IGF-II by receptor-mediated internalization. (IGF-II has been shown to be produced by breast cancer cell lines.) Second, the receptor activates latent TGF-β1. (TGF-β1 is a growth inhibitor for breast cancer cells.) Third, the receptor internalizes extracellular lysosomal enzymes, including cathepsin D. (Increased cathepsin D levels are found in breast tumors and high cathepsin D levels increase the liklihood of lymph node metastases *[141]*.)

Microsatellites are oligonucleotide repeat sequences present throughout the human genome. Disruption of the DNA mismatch repair system plays a major role in human carcinogenesis and is characterized by microsatellite instability. The mutations within the microsatellites consist of additions or deletions of repeat units. Most microsatellite instability has been described in noncoding DNA, within introns or intergenic regions. The first example of microsatellite instability within a coding region was identified in the

TGF-β1 type II receptor *(142)*. The mutations within the microsatellite of TGF-β1 type II receptor yielded nonsense mutations downstream that would result in truncated, non-functional proteins. A search of databases for other genes containing coding region microsatellites found the IGF-II/Man-6-P receptor *(143)*. The authors screened a total of 92 tumors positive for microsatellite instability for mutations within the microsatellites of the IGF-II/Man-6-P receptor coding region. Twelve of the 92 tumors (13%) contained IGF-II/Man-6-P receptor mutations. These included 7 of 29 gastric adenocarcinomas, 1 of 8 HNPCC tumors, 3 of 35 sporadic colorectal cancers, and 1 of 8 ulcerative colitis-associated colorectal cancers. No mutations were found in the microsatellites of the IGF-II/Man-6-P receptor in tumors that did not exhibit microsatellite instability. Of particular interest was the complementarity of mutations in the TGF-β1 type II receptor and the IGF-II/Man-6-P receptor. That is to say, of 31 tumors with mutations in the IGF-II/Man-6-P receptor or the TGF-β1 type II receptor, only 3 tumors were mutated in both receptors. This is consistent with the known function of the IGF-II/Man-6-P receptor to play a role in the activation of latent TGF-β1. Arguing teleologically, if a gastrointestinal tumor has acquired a mutation in the TGF-β1 type II receptor which is important in signaling the growth inhibitory response to TGF-β1, there is no further growth advantage to be provided to the tumor by acquiring a mutation in the IGF-II/Man-6-P receptor which activates latent TGF-β1.

Ouyang et al. *(144)* examined the IGF-II/Man-6-P receptor for mutations at repetitive sequences within its coding region in 405 human tumors of several different organs. Although five regions of repetitive sequence were examined within the IGF-II/Man-6-P receptor, mutations were observed only in the poly$(G)_8$ region. Mutations were detected in 4 of 26 microsatellite-positive endometrial cancers, 3 of 12 microsatellite-positive gastric cancers, 1 of 18 microsatellite-positive colorectal cancers, and 0 of 7 microsatellite-positive pancreatic cancers. No mutations were seen in cancers without microsatellite instability. Two of the gastric cancers and one colon cancer contained mutations in both the IGF-II/Man-6-P receptor and the TGF-β1 type II receptor, a higher coincidence than observed by Souza et al. *(143)*.

Wang *(145)* used an immunohistochemical approach to determine whether the IGF-II/Man-6-P receptor mutation in microsatellite-positive cancers affected the expression of IGF-II and the relative expression of active and latent forms of TGF-β1. Twenty gastrointestinal tumors with or without known IGF-II/Man-6-P receptor microsatellite mutations were examined along with matching normal tissues. There was decreased active TGF-β1 protein expression in mutant IGF-II/Man-6-P receptor tumors vs matching normal tissues or wild-type IGF-II/Man-6-P receptor tumors. There was increased latent TGF-β1 protein expression in mutant IGF-II/Man-6-P receptor tumors vs matching normal tissues or wild-type IGF-II/Man-6-P receptor tumors. The tumors containing mutant IGF-II/Man-6-P receptors had a higher level of expression of IGF-II than matching normal tissue or tumors containing wild-type receptor. However, in tumors containing mutant receptor or wild-type receptor the level of IGF-II was higher than in the matching normal tissue, suggesting that IGF-II expression in tumors was not entirely dependent upon the functional status of the IGF-II/Man-6-P receptor. Nonetheless, the overall results of this study tell us that the microsatellite mutations of the IGF-II/Man-6-P led to decreased ability of the receptor to down-regulate IGF-II levels and to activate latent TGF-β1.

Although it is not clear in individual cancers whether the tumor suppressor function of the IGF-II/Man-6-P is related to its role in IGF-II degradation or TGF-β1 activation or lysosomal enzyme trafficking, these recent reports do point to an important role of the receptor in the regulation of tumor growth.

CONCLUSION

Recent work on comparing the lysosomal enzyme trafficking function of the IGF-II/Man-6-P receptor and the cation-dependent Man-6-P receptor has provided evidence that although the function of these two receptors overlap they are both required to deliver the diverse array of lysosomal enzymes to lysosomes. The acquisition of an IGF-II binding site by the cation-independent Man-6-P receptor during evolution has equipped this widely distributed receptor to rapidly deliver extracellular IGF-II to lysosomes for degradation. This role in regulating IGF-II levels has now been clearly documented by elegant gene knockout experiments in mice. A role for the IGF-II/Man-6-P receptor in the activation of latent TGF-β1 has been demonstrated in cell culture experiments. An in vivo role for the receptor in the activation of TGF-β1 is supported by some of the recent reports on the IGF-II/Man-6-P receptor as a tumor suppressor. True receptor signaling functions for the IGF-II/Man-6-P receptor remain controversial. Because the IGF-II binding site is a mammalian feature of the cation-independent Man-6-P receptor, it might be expected that signaling roles of the receptor would be confined to more specialized functions. This is supported by the receptor gene knockout experiments in mice that demonstrate that the IGF-II/Man-6-P receptor is not directly involved in signaling growth responses to IGF-II. Nonetheless, there are now a number of reports in which some of the requirements for possible IGF-II/Man-6-P receptor signaling have been satisfied. Common themes have emerged such as G-protein involvement, stimulation of cell motility, and regulation of intracellular calcium levels. However, more work is required to identify the immediate post-receptor events in these pathways.

REFERENCES

1. Rechler MM, Nissley SP. The nature and regulation of the receptors for insulin-like growth factors. Annu Rev Physiol 1985; 47:425–442.
2. Hickman S, Neufeld EF. A hypothesis for I-cell disease: defective hydrolases that do not enter lysosomes. Biochem Biophys Res Commun 1972: 49:992–999.
3. Kaplan A, Achord DT, Sly WS. Phosphomannosyl components of a lysosomal enzyme are recognized by pinocytosis receptors on human fibroblasts. Proc Natl Acad Sci USA 1977; 74:2026–2030.
4. Morgan DO, Edman JC, Standring DN, Fried VA, Smith MC, Roth RA, Rutter WJ. Insulin-like growth factor II receptor as a multifunctional binding protein. Nature 1987; 329:301–307.
5. Canfield WM, Kornfeld S. The chicken liver cation-independent mannose 6-phosphate receptor lacks the high affinity binding site for insulin-like growth factor II. J Biol Chem 1989; 264:7100–7103.
6. Clairmont KB, Czech MP. Chicken and *Xenopus* mannose 6-phosphate receptors fail to bind insulin-like growth factor II. J Biol Chem 1989; 264:16390–16392.
7. Hieber V, Distler J, Myerowitz R, Schmickel RD, Jourdian GW. The role of glycosidically bound mannose in the assimilation of β-galactosidase by generalized gangliosidosis fibroblasts. Biochem Biophys Res Commun 1976; 73:710–717.
8. Sahagian GG, Distler J, Jourdian GW. Characterization of a membrane-associated receptor from bovine liver that binds phosphomannosyl residues of bovine testicular β-galactosidase. Proc Natl Acad Sci USA 1981; 78:4289–4293.
9. Natowicz MR, Chi MM-Y, Lowry OH, Sly WS. Enzymatic identification of mannose 6-phosphate on the recognition marker for receptor-mediated pinocytosis of β-glucuronidase by human fibroblasts. Proc Natl Acad Sci USA 1979; 76:4322–4326.
10. Hasilik A, Neufeld EF. Biosynthesis of lysosomal enzymes in fibroblasts. Phosphorylation of mannose residues. J Biol Chem 1980; 255:4946–4950.

11. Tong PY, Gregory W, Kornfeld S. Ligand interactions of the cation-independent mannose 6-phosphate receptor. The stoichiometry of mannose 6-phosphate binding. J Biol Chem 1989; 264:7962–7969.
12. Fischer HD, Creek KE, Sly WS. Binding of phosphorylated oligosaccharides to immobilized phosphomannosyl receptors. J Biol Chem 1982; 257:9938–9943.
13. Linzer DIH, Nathans D. Nucleotide sequence of a growth-related mRNA encoding a member of the prolactin-growth hormone family. Proc Natl Acad Sci USA 1984; 81:4255–4259.
14. Lee S-J, Nathans D. Secretion of proliferin. Endocrinology 1987; 120:208–213.
15. Lee S-J, Nathans D. Proliferin secreted by cultured cells binds to mannose 6-phosphate receptors. J Biol Chem 1988; 263:3521–3527.
16. Nelson JT, Rosenzweig N and Nilsen-Hamilton M. Characterization of the mitogen-regulated protein (proliferin) receptor. Endocrinology 1995; 136:283–288.
17. Purchio AF, Cooper JA, Brunner AM, Lioubin MN, Gentry LE, Kovacina KS, Roth RA, Marquardt H. Identification of mannose 6-phosphate in two asparagine-linked sugar chains of recombinant transforming growth factor-β1 precursor. J Biol Chem 1988; 263:14211–14215.
18. Kovacina KS, Steele-Perkins G, Purchio AF, Lioubin M, Miyazono K, Heldin C-H, Roth RA. Interactions of recombinant and platelet transforming growth factor-β1 precursor with the insulin-like growth factor II/mannose 6-phosphate receptor. Biochem Biophys Res Commun 1989; 160:393–403.
19. Herzog V, Neumuller W, Holzmann B. Thyroglobulin, the major and obligatory exportable protein of thyroid follicle cells, carries the lysosomal recognition marker mannose 6-phosphate. EMBO J 1987; 6:555–560.
20. Lemansky P, Herzog V. Endocytosis of thyroglobulin is not mediated by mannose-6-phosphate receptors in thyrocytes. Evidence for low-affinity-binding sites operating in the uptake of thyroglobulin. Eur J Biochem 1992; 209:111–119.
21. Oppenheimer CL, Czech MP. Purification of the type II insulin-like growth factor from rat placenta. J Biol Chem 1983; 258:8539–8542.
22. August GP, Nissley SP, Kasuga M, Lee L, Greenstein L, Rechler MM. Purification of an insulin-like growth factor II receptor from rat chondrosarcoma cells. J Biol Chem 1983; 258:9033–9036.
23. Scott CD, Baxter RC. Purification and immunological characterization of the rat liver insulin-like growth factor-II receptor. Endocrinology 1987; 120:1–9.
24. Tong PY, Tollefsen SE, Kornfeld S. The cation-independent mannose 6-phosphate receptor binds insulin-like growth factor II. J Biol Chem 1988; 263:2585–2588.
25. Haskell JF, Tucker DC. Binding of insulin-like growth factors (IGF-I and IGF-II) to the IGF-II/mannose 6-phosphate receptor in fetal rat myocardium. Endocrinology 1994; 135:231–239.
25a. Kang JX, Li Y, Leaf A. Mannose-6-phosphate/insulin-like growth factor-II receptor is a receptor for retinoic acid. Proc Natl Acad Sci USA 1997; 95:13671–13676.
25b. MacDonald RG, Tepper MA, Clairmont KB, Perregaux SB, Czech MP. Serum form of the rat insulin-like growth factorII/mannose 6-phosphate receptor is truncated in the carboxyl-terminal domain. J Biol Chem 1989; 264:3256–3261.
26. Braulke T, Causin C, Waheed A, Junghans U, Hasilik A, Maly P, Humbel RE, von Figura K. Mannose 6-phosphate/insulin-like growth factor II receptor: distinct binding sites for mannose 6-phosphate and insulin-like growth factor II. Biochem Biophys Res Commun 1988; 150:1287–1293.
27. Waheed A, Braulke T, Junghans U, von Figura K. Mannose 6-phosphate/insulin like growth factor II receptor: the two types of ligands bind simultaneously to one receptor at different sites. Biochem Biophys Res Commun 1988; 152:1248–1254.
28. Kiess W, Blickenstaff GD, Sklar MM, Thomas CL, Nissley SP, Sahagian GG. Biochemical evidence that the type II insulin-like growth factor receptor is identical to the cation-independent mannose 6-phosphate receptor. J Biol Chem 1988; 263:9339–9344.
29. Kiess W, Thomas CL, Sklar MM, Nissley SP. Beta-galactosidase decreases the binding affinity of the insulin-like growth facto-II/mannose 6-phosphate receptor for insulin-like growth factor II. Eur J Biochem 1990; 190:71–77.
30. Mathieu M, Rochefort H, Barenton B, Prebois C, Vignon F. Interactions of cathepsin-D and insulin-like growth factor-II (IGF-II) on the IGF-II/mannose-6-phosphate receptor in human breast cancer cells and possible consequences on mitogenic activity of IGF-II. Mol Endocrinol 1990; 4:1327–1335.
31. Nolan CM, Kyle JW, Watanabe H, Sly WS. Binding of insulin-like growth factor II (IGF-II) by human cation-independent mannose 6-phosphate/IGF-II receptor expressed in receptor-deficient mouse L cells. Cell Regul 1990; 1:197–213.

32. Polychronakos C, Guyda HJ, Posner BI. Mannose 6-phosphate increases the affinity of its cation-independent receptor for insulin-like growth factor II by displaying inhibitory endogenous ligands. Biochem Biophys Res Commun 1988; 157:632–638.
33. MacDonald RG. Mannose-6-phosphate enhances cross-linking efficiency between insulin-like growth factor-II (IGF-II) and IGF-II/mannose-6-phosphate receptors in membranes. Endocrinology 1991; 128:413–421.
34. Kiess W, Thomas CL, Greenstein LA, Lee L, Sklar MM, Rechler MM, Sahagian GG, Nissley SP. Insulin-like growth factor-II (IGF-II) inhibits both the cellular uptake of β-galactosidase and the binding of β-galactosidase to purified IGF-II/mannose 6-phosphate receptor. J Biol Chem 1989; 264:4710–4714.
35. Schmitz F, Bresciani R, Hartmann H, Braulke T. Effect of insulin-like growth factor II on uptake of arylsulfatase A by cultured rat hepatocytes and Kupffer cells. J Hepatol 1995; 22:356–363.
36. Kessler U, Aumeier S, Funk B, Kiess W. Biosynthetic labeling of beta-hexosaminidase B: inhibition of the cellular uptake of ^{3}H hexosaminidase B by IGF-II in rat C6 glial cells. Mol Cell Endocrinol 1992; 90:147–153.
37. De Leon DD, Terry C, Asmerom Y, Nissley SP. Insulin-like growth factor II modulates the routing of cathepsin D in MCF-7 breast cancer cells. Endocrinology 1996; 137:1851–1859.
38. Hoeflich A, Wolf E, Braulke T, Koepf G, Kessler U, Brem G, Rasher W, Blum W, Kiess W. Does the overexpression of pro-insulin-like growth factor-II in transfected human embryonic kidney fibroblasts increase the secretion of lysosomal enzymes? Eur J Biochem 1995; 232:172–178.
39. Braulke T, Bresciani R, Buergisser DM, von Figura K. Insulin-like growth factor II overexpression does not affect sorting of lysosomal enzymes in NIH-3T3 cells. Biochem Biophys Res Commun 1991; 179:108–115.
40. De Ceuninck F, Poiraudeau S, Pagano M, Tsagris L, Blanchard O, Willeput J, Corvol M. Inhibition of chondrocyte cathepsin B and L activities by insulin-like growth factor-II (IGF-II) and its Ser^{29} variant in vitro: possible role of the mannose 6-phosphate/IGF-II receptor. Mol Cell Endocrinol 1995; 113:205–213.
41. Gabel CA, Goldberg DE, Kornfeld S. Identification and characterization of cells deficient in the mannose 6-phosphate receptor: evidence for an alternative pathway for lysosomal enzyme targeting. Proc Natl Acad Sci USA 1983; 81:775–779.
42. Kyle JW, Nolan CM, Oshima A, Sly WS. Expression of human cation-independent mannose 6-phosphate receptor cDNA in receptor-negative mouse $P388D_1$ cells following gene transfer. J Biol Chem 1988; 263:16230–16235.
43. Lobel P, Dahms N, Kornfeld S. Cloning and sequence analysis of the cation-independent mannose 6-phosphate receptor. J Biol Chem 1988; 263:2563–2570.
44. Wang Z-Q, Fung MR, Barlow DP, Wagner EF. Regulation of embryonic growth and lysosomal enzyme targeting by the imprinted Igf2/Mpr gene. Nature 1994; 372:464–467.
45. Dahms NM, Lobel P, Kornfeld SA. Mannose 6-phosphate receptors and lysosomal enzyme targeting. J Biol Chem 1989; 264:12115–12118.
46. Kornfeld S. Structure and function of the mannose 6-phosphate/insulin-like growth factor II receptors. Annu Rev Biochem 1992; 61:307–330.
47. Nolan CM, Sly W. Intracellular traffic of the mannose 6-phosphate receptor and its ligands. Adv Exp Med Biol 1987; 225:199–212.
48. Kornfeld S. Trafficking of lysosomal enzymes in normal and disease states. J Clin Invest 1986; 77:1–6.
49. Geuze HJ, Slot JW, Strous GJAM, Hasilik A, von Figura K. Ultrastructural localization of the mannose 6-phosphate receptor in rat liver. J Cell Biol 1984; 98:2047–2054.
50. Griffiths G, Simons K. The trans Golgi network: sorting at the exit site of the Golgi complex. Science 1986; 234:438–443.
51. Griffiths G, Hoflack B, Simons K, Mellman I, Kornfeld S. The mannose 6-phosphate receptor and the biogenesis of lysosomes. Cell 1988; 52:329–341.
52. Duncan JR, Kornfeld S. Intracellular movement of two mannose 6-phosphate receptors: return to the Golgi apparatus. J Cell Biol 1988; 106:617–628.
53. Sahagian GG. The mannose 6-phosphate receptor: function, biosynthesis and translocation. Biol Cell 1984; 51:207–214.
54. Gartung C, Braulke T, Hasilik A, von Figura K. Internalization of blocking antibodies against mannose-6-phosphate specific receptors. EMBO J 1985; 4:1725–1730.
55. Rome LH, Weissmann B, Neufeld EF. Direct demonstration of binding of a lysosomal enzyme, α-L-iduronidase, to receptors on cultured fibroblasts. Proc Natl Acad Sci USA 1979; 76:2331–2334.

56. Willingham MC, Pastan IH, Sahagian GG. Ultrastructural immunocytochemical localization of the phosphomannosyl receptor in Chinese hamster ovary cells. J Histochem Cytochem 1983; 31:1–11.
57. Geuze HJ, Stoorvogel W, Strous GJ, Slot JW, Bleekemolen JE, Mellman I. Sorting of mannose 6-phosphate receptors and lysosomal membrane proteins in endocytic vesicles. J Cell Biol 1988; 107:2491–2501.
58. Hoflack B, Kornfeld S. Purification and characterization of a cation-dependent mannose 6-phosphate receptor from murine P388D_1 macrophages and bovine liver. J Biol Chem 1985; 260:12008–12014.
59. Dahms NM, Lobel P, Breitmeyer J, Chirgwin JM, Kornfeld S. 46 Kd mannose 6-phosphate receptor: cloning, expression, and homology to the 215 kd mannose 6-phosphate receptor. Cell 1987; 50:181–192.
60. Pohlmann R, Nagel G, Schmidt B, Stein M, Lorkowski G, Krentler C, Cully J, Meyer HE, Grzeschik K-H, Mersmann G, Hasilik A, von Figura K. Cloning of a cDNA encoding the human cation-dependent mannose 6-phosphate-specific receptor. Proc Natl Acad Sci USA 1987; 84:5575–5579.
61. Stein M, Zijderhand-Bleekemolen JE, Geuze H, Hasilik A, von Figura K. M_r 46 000 mannose 6-phosphate specific receptor: its role in targeting of lysosomal enzymes. EMBO J 1987; 6:2677–2681.
62. Sleat DE, Lobel P. Ligand binding specificities of the two mannose 6-phosphate receptors. J Biol Chem 1997; 272:731–738.
63. Munier-Lehmann H, Mauxion F, Hoflack B. Carbohydrate recognition proteins. Biochem Soc Trans 1996; 24:133–136.
64. Munier-Lehmann H, Mauxion F, Bauer U, Lobel P, Hoflack B. Re-expression of the mannose 6-phosphate receptors in receptor-deficient fibroblasts. Complementary function of the two mannose 6-phosphate receptors in lysosomal enzyme targeting. J Biol Chem 1996; 271:15166–15174.
65. Kasper D, Dittmer F, von Figura K, Pohlmann R. Neither type of mannose 6-phosphate receptor is sufficient for targeting of lysosomal enzymes along intracellular routes. J Cell Biol 1996; 34:615–623.
66. Pohlmann R, Boeker MWC, von Figura K. The two mannose 6-phosphate receptors transport distinct complements of lysosomal proteins. J Biol Chem 1995; 270:27311–27318.
67. Ludwig T, Munier-Lehmann H, Bauer U, Hollinshead M, Ovitt C, Lobel P, Hoflack B. Differential sorting of lysosomal enzymes in mannose 6-phosphate receptor-deficient fibroblasts. EMBO J 1994; 13:3430–3437.
68. Oka Y, Rozek LM, Czech MP. Direct demonstration of rapid insulin-like growth factor II receptor internalization and recycling in rat adipocytes. Insulin stimulates ^{125}I-insulin-like growth factor II degradation by modulating the IGF-II receptor recycling process. J Biol Chem 1985; 260:9435–9442.
69. Kiess W, Haskell JF, Lee L, Greenstein LA, Miller BE, Aarons AL, Rechler MM, Nissley SP. An antibody that blocks insulin-like growth factor (IGF) binding to the type II IGF receptor is neither an agonist nor an inhibitor of IGF-stimulated biologic responses in L6 myoblasts. J Biol Chem 1987; 262:12745–12751.
70. Kiess W, Lee L, Graham DE, Greenstein L, Tseng LY-H, Rechler MM, Nissley SP. Rat C6 glial cells synthesize insulin-like growth factor-I (IGF-I) and express IGF-I receptors and IGF-II/mannose 6-phosphate receptors. Endocrinology 1989; 124:1727–1736.
71. Auletta M, Nielsen FC, Gammeltoft S. Receptor-mediated endocytosis and degradation of insulin-like growth factor I and II in neonatal rat astrocytes. J Neurosci Res 1992; 31:14–20.
72. Hachiya HL, Carpentier J-L, King GL. Comparative studies on insulin-like growth factor II and insulin processing by vascular endothelial cells. Diabetes 1986; 35:1065–1072.
73. Bar RS, Boes M, Yorek M. Processing of insulin-like growth factors I and II by capillary and large vessel endothelial cells. Endocrinology 1986; 118:1072–1080.
74. Ellis MJC, Leav BA, Yang Z, Rasmussen A, Pearce A, Zweibel JA, Lippman ME, Cullen KJ. Affinity for the insulin-like growth factor-II (IGF-II) receptor inhibits autocrine IGF-II activity in MCF-7 breast cancer cells. Mol Endocrinol 1996; 10:286–297.
75. Lau MMH, Stewart CEH, Liu Z, Bhatt H, Rotwein P, Stewart CL. Loss of the imprinted IGF2/cation-independent mannose 6-phosphate receptor results in fetal overgrowth and perinatal lethality. Genes Dev 1994; 8:2953–2963.
76. Ludwig T, Eggenschwiler J, Fisher P, D'Ercole J, Davenport ML, Efstratiadis A. Mouse mutants lacking the type 2 IGF receptor (IGF2R) are rescued from perinatal lethality in *Igf2* and *Igf1r* null backgrounds. Dev Biol 1996; 177:517–535.
77. Dennis PA, Rifkin DB. Cellular activation of latent transforming growth factor β requires binding to the cation-independent mannose 6-phosphate/insulin-like growth factor type II receptor. Proc Natl Acad Sci USA 1991; 88:580–584.

78. Odekon LE, Blasi F, Rifkin DB. Requirement for receptor-bound urokinase in plasmin-dependent cellular conversion of latent TGF-β to TGF-β. J Cell Physiol 1994; 158:398–407.
79. Flaumenhaft R, Kojima S, Abe M, Rifkin DB. Activation of latent transforming growth factor β. Adv Pharmacol 1993; 24:51–76.
80. Louvi A, Accili D, Efstratiadis A. Growth-promoting interaction of IGF-II with the insulin receptor during mouse embryonic development. Dev Biol 1997; 189:33–48.
81. Sakano K, Enjoh T, Numata F, Fujiwara H, Marumoto Y, Higashihashi N, Sato Y, Perdue JF, Fujita-Yamaguchi Y. The design, expression, and characterization of human insulin-like growth factor II (IGF-II) mutants specific for either the ITF-II/cation-independent mannose 6-phosphate receptor or IGF-I receptor. J Biol Chem 1991; 266:20626–20635.
82. Burgisser DM, Roth BV, Giger R, Luthi C, Weigl S, Zarn J, Humbel RE. Mutants of human insulin-like growth factor II with altered affinities for the type 1 and type 2 insulin-like growth factor receptor. J Biol Chem 1991; 266:1029–1033.
83. Beukers MW, Oh Y, Zhang H, Ling N, Rosenfeld RG. [Leu^{27}]insulin-like growth factor II is highly selective for the type-II IGF receptor in binding, cross-linking and thymidine incorporation experiments. Endocrinology 1991; 128:1201–1203.
84. Nishimoto I, Ohkuni Y, Ogata E, Kojima I. Insulin-like growth factor II increases cytoplasmic free calcium in competent Balb/c 3T3 cells treated with epidermal growth factor. Biochem Biophys Res Commun 1987; 142:275–286.
85. Kojima I, Matsunaga H, Kurokawa K, Ogata E, Nishimoto I. Calcium influx: an intracellular message of the mitogenic action of insulin-like growth factor-I. J Biol Chem 1988; 263:16561–16567.
86. Nishimoto I, Hata Y, Ogata E, Kojima I. Insulin-like growth factor II stimulates calcium influx in competent Balb/c 3T3 cells primed with epidermal growth factor. Characteristics of calcium influx and involvement of GTP-binding protein. J Biol Chem 1987; 262:12120–12126.
87. Kojima I, Nishimoto I, Iiri T, Ogata E, Rosenfeld R. Evidence that type II insulin-like growth factor receptor is coupled to calcium gating system. Biochem Biophys Res Commun 1988; 154:9–19.
88. Matsunaga H, Nishimoto I, Kojima I, Yamashita N, Kurokawa K, Ogata E. Activation of a calcium-permeable cation channel by insulin-like growth factor II in Balb/c 3T3 cells. Am J Physiol 1988; 255:C442–C446.
89. Tally M, Li CH, Hall K. IGF-2 stimulated growth mediated by the somatomedin type 2 receptor. Biochem Biophys Res Commun 1987; 148:811–816.
90. Tally M, Engberg G, Li CH, Hall K. The specificity of the human IGF-2 receptor. Biochem Biophys Res Commun 1987; 147:1206–1212.
91. Zhang Z, Hall K, Tally M. Partial purification of an IGF-II receptor/binding protein from the erythroleukemia cell line K562. Mol Cell Biochem 1996; 154:47–54.
92. Blanchard MM, Barenton B, Sullivan A, Foster B, Guyda HJ, Posner BI. Characterization of the insulin-like growth factor (IGF) receptor in K562 erythroleukemia cells; evidence for a biological function for the type II IGF receptor. Mol Cell Endocrinol 1988; 56:235–244.
93. Hizuka N, Sukegawa I, Takano K, Asakawa K, Horikawa R, Tsushima T, Shizume K. Characterization of insulin-like growth factor I receptors on human erythroleukemia cell line (K-562 cells). Endocrinol Japon 1987; 34:81–88.
94. Hammerman MR, Gavin JR. Binding of insulin-like growth factor II and multiplication-stimulating activity-stimulated phosphorylation in basolateral membranes from dog kidney. J Biol Chem 1984; 259:13511–13517.
95. Mellas J, Gavin JR, Hammerman MR. Multiplication-stimulating activity-induced alkalinization of canine renal proximal tubular cells. J Biol Chem 1986; 261:14437–14442.
96. Rogers SA, Hammerman MR. Insulin-like growth factor II stimulates production of inositol trisphosphate in proximal tubular basolateral membranes from canine kidney. Proc Natl Acad Sci USA 1988; 85:4037–4041.
97. Rogers SA, Hammerman MR. Mannose 6-phosphate potentiates insulin-like growth factor II-stimulated inositol trisphosphate production in proximal tubular basolateral membranes. J Biol Chem 1989; 264:4273–4276.
98. Rogers SA, Purchio AF, Hammerman MR. Mannose 6-phosphate-containing peptides activate phospholipase C in proximal tubular basolateral membranes from canine kidney. J Biol Chem 1990; 265:9722–9727.
99. Bhaumick B, Bala RM. Receptors for insulin-like growth factors I and II in developing embryonic mouse limb bud. Biochim Biophys Acta 1987; 927:117–128.

100. Bhaumick B, Bala RM. Parallel effects of insulin-like growth factor-II and insulin on glucose metabolism of develooping mouse embryoonic limb buds in culture. Biochem Biophys Res Commun 1988; 152:359–367.
101. Bhaumick B, Bala MR. Differential effects of insulin-like growth factors I and II on growth, differentiation and glucogregulation in differentiating chondrocyte cells in culture. Acta Endocrinol 1991; 125:201–211.
102. Vetter U, Zapf J, Heit W, Helbing G, Heinze E, Froesch ER, Teller WM. Human fetal and adult chondrocytes. Effect of insulinlike growth factors I and II, insulin, and growth hormone on clonal growth. J Clin Invest 1986; 77:1903–1908.
103. Takigawa M, Okawa T, Pan H-O, Aoki C, Takahashi K, Zue J-D, Suzuki F, Kinoshita A. Insulin-like growth factors I and II are autocrine factors in stimulating proteoglycan synthesis, a marker of differentiated chondrocytes, acting through their respective receptors on a clonal human chondrosarcoma-derived chondrocyte cell line, HCS-2/8. Endocrinology 1997; 138:4390–4400.
104. Poiraudeau S, Lieberherr MA, Kergosie N, Corvol M-T. Different mechanisms are involved in intracellular calcium increase by insulin-like growth factors 1 and 2 in articular chondrocytes: voltage-gated calcium channels, and/or phospholipase C coupled to a pertussis-sensitive G-protein. J Cell Biochem 1997; 64:414–422.
105. Rosenthal SM, Hsiao D, Silverman LA. An insulin-like growth factor-II (IGF-II) analog with highly selective affinity for IGF-II receptors stimulates differentiation, but not IGF-I receptor down-regulation in muscle cells. Endocrinology 1994; 35:38–44.
106. Bach LA, Salemi R, Leeding KS. Roles of insulin-like growth factor (IGF) receptors and IGF-binding proteins in IGF-II-induced proliferation and differentiation of L6A1 rat myoblasts. Endocrinology 1995; 136:5061–5069.
107. Ewton DZ, Roof SL, Magri KA, McWade FJ, Florini JR. IGF-II is more active than IGF-I in stimulating L6A1 myogenesis: greater mitogenic actions of IGF-I delay differentiation. J Cell Physiol 1994; 16:277–284.
108. Liu J-L, Baker J, Perkins AS, Robertson EJ, Efstratiadis A. Mice carrying null mutations of the genes encoding insulin-like growth-factors I (Igf1) and type I IGF receptor (Igf1r). Cell 1993; 75:59–72.
109. Florini JR, Ewton DZ, Coolican SA. Growth hormone and the insulin-like growth factor system in myogenesis. Endocrin Rev 1996; 17:481–517.
110. El-Badry OM, Minniti C, Kohn EC, Houghton PJ, Daughaday WH, Helman LJ. Insulin-like growth factor II acts as an autocrine growth and motility factor in human rhabdomyosarcoma tumors. Cell Growth and Different 1990; 1:325–331.
111. Minniti CP, Kohn EC, Grubb JH, Sly WS, Oh Y, Muller HL, Rosenfeld RG, Helman L. The insulin-like growth factor II(IGF-II)/mannose 6-phosphate receptor mediates IGF-II-induced motility in human rhabdomyosarcoma cells. J Biol Chem 1992; 267:9000–9004.
112. Volpert O, Jackson D, Bouck N, Linzer DIH. The insulin-like growth factor II/mannose 6-phosphate receptor is required for proliferin-induced angiogenesis. Endocrinology 1996; 137:3871–3876.
113. Groskopf JC, Syu L-J, Saltiel AR, Linzer DIH. Proliferin induces endothelial cell chemotaxis through a G protein-coupled, mitogen-activated protein kinase-dependent pathway. Endocrinology 1997; 138:2835–2840.
114. Palka JA, Karna E, Miltyk W. Fibroblast chemotaxis and prolidase activity modulation by insulin-like growth factor II and mannose 6-phosphate. Mol Cell Biochem 1997; 168:177–183.
115. Nestler JE. Insulin-like growth factor II is a potent inhibitor of the aromatase activity of human placental cytotrophoblasts. Endocrinology 1990; 127:2064–2070.
116. Tsuruta JK, O'Brien DA. Sertoli cell-spermatogenic cell interaction: the insulin-like growth factor-II/cation-independent mannose 6-phosphate receptor mediates changes in spermatogenic cell gene expression in mice. Biol Reprod 1995; 53:1454–1464.
117. Nishiura T, Karasuno T, Yoshida H, Nakao H, Ogawa M, Horikawa Y, Yoshimura M, Okajima Y, Kanakura Y, Kanayama Y, Matsuzawa Y. Functional role of cation-independent mannose 6-phosphate/insulin-like growth factor II receptor in cell adhesion and proliferation of a human myeloma cell line OPM-2. Blood 1996; 88:3546–3554.
118. Fex Svenningsen AF, Kanje M. Insulin and the insulin-like growth factors I and II are mitogenic to cultured rat sciatic nerve segments and stimulate [^{3}H]thymidine incorporation through their respective receptors. Glia 1996; 18:68–72.

119. Nielsen F, Gammeltoft S. Mannose-6-phosphate stimulates proliferation of neuronal precursor cells. FEBS Lett 1990; 262:142–144.
120. Feldman EL, Randolph AE. Mannose 6-phosphate potentiates insulin-like growth factor II effects in cultured human neuroblastoma cells. Brain Res 1991; 562:111–116.
121. Zhang Q, Tally M, Larsson O, Kennedy RT, Huang L, Hall K, Berggren P-O. Insulin-like growth factor II signaling through the insulin-like growth factor II/mannose-6-phosphate receptor promotes exocytosis in insulin-secreting cells. Proc Natl Acad Sci USA 1997; 94:6232–6237.
122. Martinez DA, Zuscik MJ, Ishibe M, Rosier RN, Romano PR, Cushing JE, Puzas JE. Identification of functional insulin-like growth factor-II/mannose-6-phosphate receptors in isolated bone cells. J Cell Biochem 1995; 59:246–257.
123. De Leon DD, Wilson DM, Powers M, Rosenfeld RG. Effects of insulin-like growth factors (IGFs) and IGF receptor antibodies on the proliferation of human breast cancer cells. Growth Factors 1992; 6:327–336.
124. Hari J, Pierce SB, Morgan DO, Sara V, Smith MC, Roth RA. The receptor for insulin-like growth factor II mediates an insulin-like response. EMBO J 1987; 6:3367–3371.
125. Nishimoto I. The IGF-II receptor system: a G protein-linked mechanism. Mol Reprod Dev 1993; 35:398–407.
126. Nishimoto I, Murayama Y, Katada T, Ui M, Ogata E. Possible direct linkage of insulin-like growth factor-II receptor with guanine nucleotide-binding proteins. J Biol Chem 1989; 264:14029–14038.
127. Okamoto T, Asano T, Harada S, Ogata E, Nishimoto I. Regulation of transmembrane signal transduction of insulin-like growth factor II by competence type growth factors or viral *ras* p21. J Biol Chem 1991; 266:1085–1091.
128. Murayama Y, Okamoto T, Ogata E, Asano T, Iiri T, Katada T, Ui M, Grubb JH, Sly WS, Nishimoto I. Distinctive regulation of the functional linkage between the human cation-independent mannose 6-phosphate receptor and GTP-binding proteins by insulin-like growth factor II and mannose 6-phosphate. J Biol Chem 1990; 265:17456–17462.
129. Okamoto T, Katada T, Murayama Y, Ui M, Ogata E, Nishimoto I. A simple structure encodes G protein-activating function of the IGF-II/mannose 6-phosphate receptor. Cell 1990; 62:709–717.
130. Okamoto T, Nishimoto I. Analysis of stimulation-G protein subunit coupling by using active insulin-like growth factor II receptor peptide. Proc Natl Acad Sci USA 1991; 88:8020–8023.
131. Takahashi K, Murayama Y, Okamoto T, Yokota T, Ikezu T, Takahashi S, Giamarella U, Ogata E, Nishimoto I. Conversion of G-protein specificity of insulin-like growth factor II/mannose 6-phosphate receptor by exchanging of a short region with β-adrenergic receptor. Proc Natl Acad Sci USA 1993; 90:11772–11776.
132. Ikezu T, Okamoto T, Giambarella U, Yokota T, Nishimoto I. In vivo coupling of insulin-like growth factor II/mannose 6-phosphate receptor to heteromeric G proteins. Distinct roles of cytoplastmic domains and signal sequestration by the receptor. J Biol Chem 1995; 270:29224–29228.
133. Korner C, Nurnberg B, Uhde M, Braulke T. Mannose 6-phosphate/insulin-like growth factor II receptor fails to interact with G-proteins. Analysis of mutant cytoplasmic receptor domains. J Biol Chem 1995; 270:287–295.
134. Toretsky JA, Helman ILJ. Involvement of IGF-II in human cancer. J Endocrinology 1996; 149:367–372.
135. Rogler CE, Yang D, Rossetti L, Donohoe J, Alt E, Chang CJ, Rosenfeld R, Neely K, Hintz R. Altered body composition and increased frequency of diverse malignancies in insulin-like growth factor-II transgenic mice. J Biol Chem 1994; 269:13779–13784.
136. De Souza AT, Hankins GR, Washington MK, Fine RL, Orton TC, Jirtle R. Frequent loss of heterozygosity on 6q at the mannose 6-phosphate/insulin-like growth factor II receptor locus in human hepatocellular tumors. Oncogene 1995; 10:1725–1729.
137. De Souza AT, Hankins GR, Washington MK, Orton TC, Jirtle RL. M6P/IGF2R gene is mutated in human hepatocellular carcinomas with loss of heterozygosity. Nature Genet 1995; 11:447–449.
138. Yamada T, De Souza AT, Finkelstein S, Jirtle RL. Loss of the gene encoding mannose 6-phosphate/insulin-like growth factor II receptor is an early event in liver carcinogenesis. Proc Natl Acad Sci USA 1997; 94:10351–10355.
139. Jirtle RL, Haag JD, Ariazi EA, Gould MN. Increased mannose 6-phosphate/insulin-like growth factor II receptor and transforming growth factor β1 levels during monoterpene-induced regression of mammary tumors. Cancer Res 1993; 53:3849–3852.
140. Hankins GR, De Souza AT, Bentley RC, Patel MR, Marks JR, Iglehart JD, Jirtle RL. M6P/IGF2 receptor: a candidate breast tumor suppressor gene. Oncogene 1996; 12:2003–2009.

141. Tandon AK, Clark GM, Chamness GC, Chirgwin JM, McGuire WL. Cathepsin D and prognosis in breast cancer. N Engl J Med 1990; 322:297–302.
142. Markowitz S, Wang J, Myeroff L, Parsons R, Sun L, Lutterbaugh J, Fan RS, Zborowska E, Kinzler KW, Vogelstein B, Brattain M, Willson JKV. Inactivation of the type II TGF-β receptor in colon cancer cells with microsatellite instability. Science 1995; 268:1336–1338.
143. Souza RF, Appel R, Yin J, Wang S, Smolinski KN, Abraham JM, Zou T-T, Shi Y-Q, Lei J, Cottrell J, Cymes K, Biden K, Simms L, Leggett B, Lynch PM, Frazier M, Powell SM, Harpaz N, Sugimura H, Young J, Meltzer SJ. Microsatellite instability in the insulin-like growth factor II receptor gene in gastrointestinal tumours. Nature Genet 1996; 14:255–257.
144. Ouyang H, Shiwaku HO, Hagiwara H, Miura K, Abe T, Kato Y, Ohtani H, Shiiba K, Souza RF, Meltzer SJ, Horii A. The insulin-like growth factor II receptor gene is mutated in genetically unstable cancers of the endometrium, stomach, and colorectum. Cancer Res 1997; 57:1851–1854.
145. Wang S, Souza RF, Kong D, Yin J, Smolinski KN, Zou T-T, Frank T, Young J, Flanders KC, Sugimura H, Abraham JM, Meltzer SJ. Deficient transforming growth factor-β1 activation and excessive insulin-like growth factor II (IGFII) expression in IGFII receptor-mutant tumors. Cancer Res 1997; 57:2543–2545.

9 Alternative IGF-Related Receptors

Kenneth Siddle, PhD and Maria A. Soos, PhD

CONTENTS

INTRODUCTION AND SCOPE

A substantial number of reports over more than a decade have described variant forms of receptors for insulin-like growth factors (IGFs) and insulin, which differ in structure or ligand specificity from classic type I IGF receptors and insulin receptors as characterized in most mammalian tissues or by expression of their cloned cDNAs. As an introduction to receptor subtypes, we first describe briefly the structure and function of the insulin receptor (IR), type I IGF receptor (IGFR), and insulin-receptor-related receptor (IRR), as reflected in the products of their cloned genes. We then review the properties of known splice variants of these receptors, which represent an obvious potential source of functional heterogeneity. Next, we discuss "atypical" insulin and IGF receptors that apparently bind both insulin and IGFs with high affinity, and attempt to rationalize anomalous observations and assess the structural basis and functional significance of these receptors. Finally we consider the properties of insulin/IGF hybrid receptors, which combine the structures and functions of both receptors and may account for much of the reported heterogeneity in receptor subtypes.

OVERVIEW OF THE IGF-I AND INSULIN RECEPTORS

The type I IGF receptor is a member of a subfamily of receptor tyrosine kinases, which also includes the insulin receptor and the insulin receptor-related receptor (Fig. 1). These

From: *Contemporary Endocrinology: The IGF System*
Edited by: R. Rosenfeld and C. Roberts Jr © Humana Press, Inc., Totowa, NJ

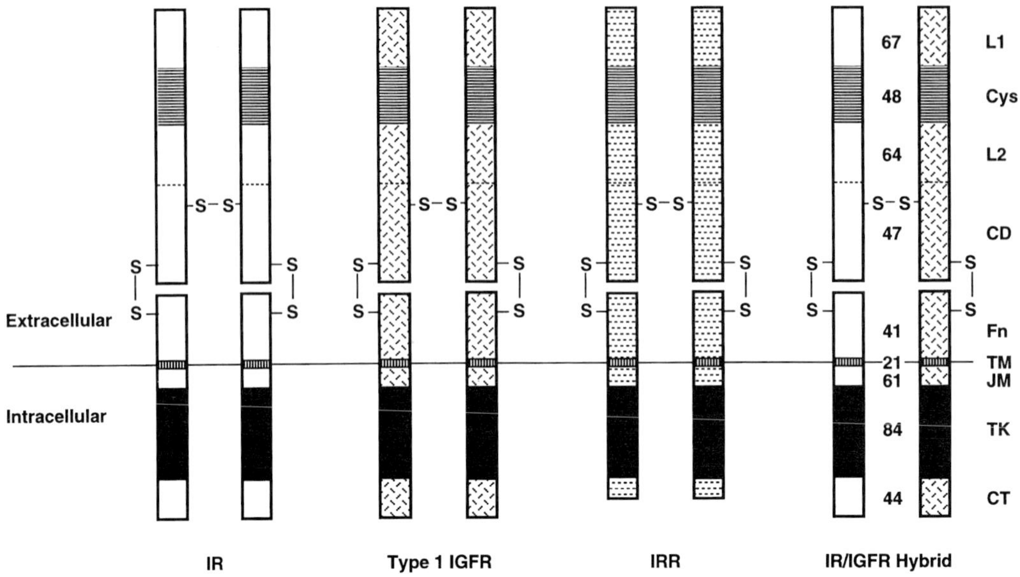

Fig. 1. The insulin/IGF-I receptor family. Receptor domains are schematically represented, approximately to scale in terms of primary sequence. In order from the N-terminus, structural domains are L1, cysteine-rich domain (Cys), L2, connecting domain (CD), fibronectin type III domains (Fn), transmembrane domain (TM), juxtamembrane domain (JM), and C-terminal domain (CT) (*5,25,154*). Numerical values shown with hybrid receptors are the percentage of residues within each domain that are identical in IR and IGFR (*3*).

receptors share substantial homology of primary sequence *(1–4)* and similarity of function *(5–8)*. The distinctive structural features of this group of receptors are that they contain two different subunits, α and β, generated by proteolysis from a single proreceptor polypeptide, and the mature receptors contain two (αβ) units disulfide-linked in a β–α–α–β structure. The α-subunits are wholly extracellular and contain the ligand binding site(s). The β-subunits span the membrane with a single hydrophobic α-helix and contain an intracellular tyrosine-specific protein kinase domain. Ligand binding activates the receptors via an intramolecular autophosphorylation reaction that is believed to occur reciprocally and in *trans* between the two β-subunits. This in turn activates the tyrosine kinase toward intracellular substrates of the IRS and Shc families, whose phosphorylation triggers multiple signaling cascades. Receptor autophosphorylation may in addition create binding sites for signaling molecules to associate directly with the receptors.

The specificity of the IGFR and IR for their respective ligands, as classically described, is such that each receptor would be expected to respond only to its own ligand under physiological conditions *(9)*. The IGFR binds IGF-I and IGF-II with 100- to 1000-fold higher affinity than insulin, whereas the IR binds insulin with 100-1000-fold higher affinity than IGF-I. No ligand has yet been identified for the IRR. The regions of IR and IGFR structure that contribute to ligand binding have been probed by ligand crosslinking *(10–12)*, by construction of receptor chimeras *(13–17)*, and by site-directed mutagenesis *(18–20)*. These studies have revealed that determinants of high-affinity insulin binding reside primarily in the amino (N)-terminal region of the IR, while high-affinity IGF-I binding depends on residues in the cysteine-rich region of the IGFR. However, other regions of the receptors on the carboxy (C)-terminal side of the cysteine-rich domain may also contribute to ligand binding. These findings suggest a model in which the ligand binding sites of the IR and IGFR are formed by folding of disparate regions of primary sequence, and in which the facets of the binding site contributing interactions critical to ligand specificity differ between the two receptors. Based on measurements of binding stoichiometry and negative cooperativity, it has been proposed that the two α-subunits within a native receptor each contribute to the high-affinity binding site, and that a single molecule of bound ligand effectively crosslinks the α-subunits in an asymmetric fashion *(21,22)*.

Notwithstanding the obvious differences in the physiological roles of IGF-I and insulin, it is clear that their respective receptors have fundamentally similar signaling potential when examined in the same cell background. Numerous studies have shown that both receptors will mediate both acute "metabolic" responses and longer term "mitogenic" effects. The specificty of action of IGF-I and insulin in vivo therefore arises in large part from differences in tissue distribution of receptors rather than their intrinsic activity. Attempts to compare the intrinsic activities of endogenous IR and IGFR in cultured cells are complicated by differences in levels of receptor expression and affinity for ligand. To overcome these problems, studies have been carried out with chimeric receptors, in which the same ligand may be used to activate either the IGFR or IR tyrosine kinase. One such study in transfected NIH3T3 fibroblasts concluded that a mitogenic response was more efficiently transduced by the IGFR than by the IR, with no detectable difference in signaling to glucose uptake *(23)*. Our own work with TrkC chimeras in transfected 3T3-L1 preadipocytes suggested that stimulation of glycogen synthesis was more effectively mediated by the IR than the IGFR, while the receptors were very similar in their capacity to mediate stimulation of DNA synthesis *(24)*. Studies with chimeras and site-directed mutants have further suggested that the relative efficiency of coupling to metabolic and mitogenic responses is influenced by the IR and IGFR C-terminal domains *(25–28)*. The molecular basis of these differences in signaling efficiency is unclear. There may be subtle differences in the specificity or kinetics of phosphorylation of intracellular substrates by the respective receptor tyrosine kinases *(29)*, which in turn could influence the activation of downstream signaling cascades involving phosphatidylinositol-3-kinase and protein kinase B (PKB) or Ras and mitogen-activated protein (MAP) kinase. Alternatively, differences in the capacity of the receptors to interact directly with other proteins such as Grb10 *(30)*, MAD2 *(31)*, or 14–3–3 *(32,33)* may influence signaling, although the function of these proteins and physiological significance of their interactions with the receptors remains to be established.

THE INSULIN RECEPTOR-RELATED RECEPTOR

The IRR was first identified by screening genomic DNA libraries for sequences that hybridized at low stringency with cDNA probes derived from exon 2 (corresponding to the N-terminus) of the IR *(4)*. Unlike the IR and IGFR, the IRR shows a very restricted pattern of expression in mammalian tissues. The consensus of several studies involving Northern blotting or reversed transcription-polymerase chain reaction (RT-PCR) is that highest levels of mRNA are found in kidney, with lower levels in thymus, brain, stomach, and heart and still lower levels in muscle, fat, liver, and placenta *(34–37)*. More detailed examination by *in situ* hybridization revealed that expression was restricted to discrete cellular populations, namely distal tubules in the kidney, neuroendocrine cells in the stomach, and cholinergic, nerve growth factor (NGF)-sensitive neurons in the forebrain and spinal cord *(37–42)*. Expression in peripheral neurons may account for the low level of IRR mRNA detected in other tissues. Among a large variety of cultured cell lines, IRR protein expression was detected only in certain neuroblastomas *(43)*. These findings imply a more tissue-specific role for the IRR than the IR or IGFR.

The amino acid sequence of the IRR is as closely related to those of the IR and IGFR as those two sequences are to each other, the level of overall sequence identity being approx 55% between any two members of the receptor family. The most prominent difference in sequence alignment is that the IRR has a considerably shorter and more divergent C-terminal domain, downstream of the tyrosine kinase domain, compared to the IR and IGFR. The gene structure is also well conserved, with the same pattern of exons in the IRR as in the IR and IGFR, although the intronic sequences are very much shorter in the IRR *(4)*.

Initial speculation that the IRR might represent a novel form of IGF receptor, perhaps with particular affinity for IGF-II, has not been borne out by subsequent experiments. The IRR does not detectably bind or respond to IGF-I, IGF-II, insulin, proinsulin, or relaxin when expressed in fibroblasts *(35,44)*, and IGF-I, insulin, and NGF fail to activate the tyrosine kinase of endogenous IRR in neuroblastoma cells *(43)*. The IRR therefore remains for the present an orphan receptor with no known ligand. That the IRR is potentially functional is shown by using insulin to stimulate a chimera consisting of the extracellular domain of the IR fused to the intracellular domain of the IRR *(35)*, or by using an agonistic monoclonal antibody to stimulate endogenous IRR in neuroblastoma cells *(43)*. These studies suggest that, in spite of substantial sequence divergence in the C-terminal domain, the IRR tyrosine kinase is similar to the IR and IGFR kinases in both the spectrum of substrates phosphorylated and the biological responses induced.

SPLICE VARIANTS OF THE IR, IGFR, AND IRR

Isoforms of the IR, IGFR, and IRR have been described that arise from alternative splicing of the respective primary mRNA transcripts. However, only in the case of the IR is there convincing evidence for a functional difference between alternative transcripts that is likely to be physiologically significant.

Alternative Splicing of the IR

Splice variants of the IR reflect the inclusion or exclusion of a discrete mini-exon, exon 11, that encodes a sequence of 12 amino acids toward the C-terminus of the α-subunit *(45)*. There is no evidence for additional heterogeneity in the coding region of insulin receptor mRNA other than that arising from alternative splicing of exon 11 *(46,47)*. The relative expression of exon 11+ and 11– isoforms varies between different tissues. Liver expresses predominantly the Ex11+ form, brain and haematopoietic tissues largely the Ex11– form, while placenta and skeletal muscle express both forms in comparable amounts *(48–50)*. As studied in transfected fibroblasts, the Ex11– (A) isoform displays approx twofold higher equilibrium binding affinity for insulin *(51,52)* and higher rates of endocytosis and recycling *(52,53)* compared to the Ex11+ (B) isoform. The two isoforms have very similar signaling properties when allowance is made for the effect of affinity differences on receptor occupancy *(52,54)*, although it has been reported that the Ex11+ isoform displays somewhat greater insulin-stimulated autophosphorylation and tyrosine kinase activity *(55)*.

The most marked difference between the IR isoforms yet determined is in their relative affinity for IGF-I, the IC_{50} for inhibition of ^{125}I-insulin binding by IGF-I being 10-fold lower for Ex11– compared to Ex11+ receptors *(56,57)*. Thus the affinity of the Ex11– isoform for IGF-I (IC_{50} approx 20–40 n*M*, compared to 1 n*M* for insulin) is such that this receptor species might, in sufficiently concentrated cell extracts, bind IGF-I even at tracer concentrations, and that it might in vivo be significantly responsive to IGFs at pathophysiological concentrations. The relative affinity of IR isoforms for IGF-II has not been documented. However, it has been shown that low concentrations of IGF-II, but not IGF-I, exert a significant mitogenic effect via the Ex11– IR isoform in transfected $IGFR^-$ cells *(58)*. Further, genetic evidence from mice with targeted disruptions of IGFR and IR genes suggests that the growth-promoting function of IGF-II during mouse development is mediated in part by signaling through the IR *(59)*.

Alternative Splicing of the IGFR

The cloned human and rat IGFR cDNAs lack the sequence corresponding to Ex11 of the IR *(3,60,61)*, and it seems that the IGFR is not subject to alternative splicing comparable to the IR. However, the human IGFR does undergo a different form of alternative splicing, as revealed by the presence of two mRNA transcripts in several tissues and cultured cell lines *(62)*. These two transcripts derive from use of alternative 3′ splicing acceptor sites, separated by three bases, at the 5′ end of exon 14, with no change in the 5′ donor site *(60)*. The predicted effect on amino acid sequence is an alternative Lys–Arg–Tyr or Lys–Thr–Gly–Tyr sequence in the extracellular portion of the receptor α-subunit close to the transmembrane domain. The two alternatively spliced forms of the human IGFR exhibit similar affinity for IGF-I but may differ somewhat in their responsiveness to IGF-I and in their internalization kinetics *(63)*. However, the rat IGFR gene sequence lacks the alternative splice acceptor site *(61)*, calling into question the physiological significance of alternative splicing of the human IGFR.

Alternative Splicing of the IRR

In contrast to the IGFR, both the human and guinea pig IRRs include the sequence corresponding to Ex11, with no evidence that this is alternatively spliced *(4,44)*. Like the IGFR, the human IRR apparently undergoes alternative splicing involving two splice acceptor sites, in this case in the 3′ portion of intron 13, which potentially introduces an additional sequence of eight amino acids in the IRR *(44)*. However, mRNA with the 24-base insert forms at best a small proportion of total IRR mRNA in human cells. Moreover the corresponding mouse alternative transcript is longer by only 23 bases and would not produce a functional protein. Therefore the physiological importance of this alternative splicing of the IRR, as with the IGFR, is questionable. Two further alternatively spliced rat IRR mRNA transcripts have been characterized, in which the predicted coding sequence terminates at the 3′ end of exon 5 and additional coding sequence corresponding to intron 4 is either present or absent *(64)*. The corresponding truncated proteins were detected in human tissues, and would presumably be secreted as soluble proteins, lacking a transmembrane domain but containing the domains implicated in ligand binding. The function of such potential binding proteins is no more clear than that of the IRR itself. Comparable truncated forms of the IR or IGFR have not been reported.

ATYPICAL RECEPTORS

Studies in various cell types have described receptors that apparently bind both insulin and IGFs with relatively high affinity, unlike classic receptors that show specificity for only one of these ligands at tracer concentrations (Fig. 2). These apparent variants may be classified as either atypical insulin receptors or atypical IGF receptors, according to their immunoreactivity and the ligand bound with greatest affinity.

Atypical IR

Atypical insulin receptors or type III IGF receptors have been described in placenta and IM-9 cells *(65–70)*. The defining properties of atypical IR are that: (1) they bind ^{125}I-IGF-II (MSA) as a tracer ligand; (2) they have a higher affinity for insulin than IGFs, as indicated by IC_{50} values for displacement of ^{125}I-IGF-II (in a ratio of approx 1:5:30 for insulin, IGF-II, and IGF-I) (Fig. 2); (3) they copurify with classic IR; (4) they react with anti-IR but not anti-IGFR antibodies. In these respects atypical IR are clearly distinct from classic IGFR, classic IR (as exemplified by the hepatic Ex11+ isoform), and insulin/IGF-I hybrid receptors *(see below)*. However, the properties are very similar to those that would be expected of the Ex11– (A) IR isoform *(57)*, and it seems very likely that the high levels of this IR isoform expressed in placenta and IM-9 cells account for the observed "atypical" binding. This hypothesis would be confirmed if it were shown that the Ex11– isoform had even higher affinity for IGF-II than for IGF-I. Against the hypothesis is a report that atypical insulin receptors are present in transfected Chinese hamster ovary (CHO) cells expressing the cloned Ex11+ IR isoform *(71)*. However, in this study the number of ^{125}I-IGF-II binding sites adsorbed to anti-IR antibodies was very small compared to the number of ^{125}I-insulin binding sites, and reactivity with anti-IGFR antibodies was not assessed. It therefore seems possible that the phenomena observed in CHO cells were only superficially similar to those studied in more detail in placenta, and that hybrids between transfected human IR and endogenous rodent IGFR *(see below)* were responsible for the observed insulin-inhibitable ^{125}I-IGF binding in transfected CHO cells. It is unclear what other structural heterogeneity could account for atypical insulin receptors, if not the recognized receptor isoforms and/or hybrid receptors.

Atypical IGFR

The converse phenomenon of atypical IGF-I receptors with unexpectedly high affinity for insulin has been reported for rat L6 myocytes, bovine neural retina, and at least some lines of human MCF-7 breast cancer cells *(72–74)*. The defining properties of atypical IGFR are that: (1) they bind ^{125}I-insulin as tracer ligand to a much higher level than can be accounted for by classic IGFR; (2) they have a higher affinity for IGF-I than insulin as indicated by IC_{50} values for displacement of ^{125}I-insulin (in a ratio of approx 1:10) (Fig. 2); (3) the sizes of both α- and β-subunits are more similar to those of IGFR than IR; (4) they react with anti-IGFR antibodies but not with most anti-IR antibodies (the exceptions being a subset of antibodies to the IR kinase domain). It is to be expected that such anomalous binding of ^{125}I-insulin would be apparent only when the level of classic IR is very low. Indeed, a small fraction of ^{125}I-insulin binding was also detected in these cells which behaved in the manner expected for classic IR, in terms of displacement only by high concentrations of IGF-I and reaction with anti-IR antibodies *(73,74)*. Paradoxically, in receptor preparations that displayed anomalous high-affinity binding of ^{125}I-insulin to sites that also had high affinity for IGF-I, binding of ^{125}I-IGF-I was displaced by insulin only at very high concentrations, as expected for classic IGFR. It therefore seemed likely that atypical IGFR represented a small subfraction of total IGFR that was apparent only in cells with a very low level of classic IR. Support for this hypothesis was provided by the reported partial purification of atypical IGFR, and

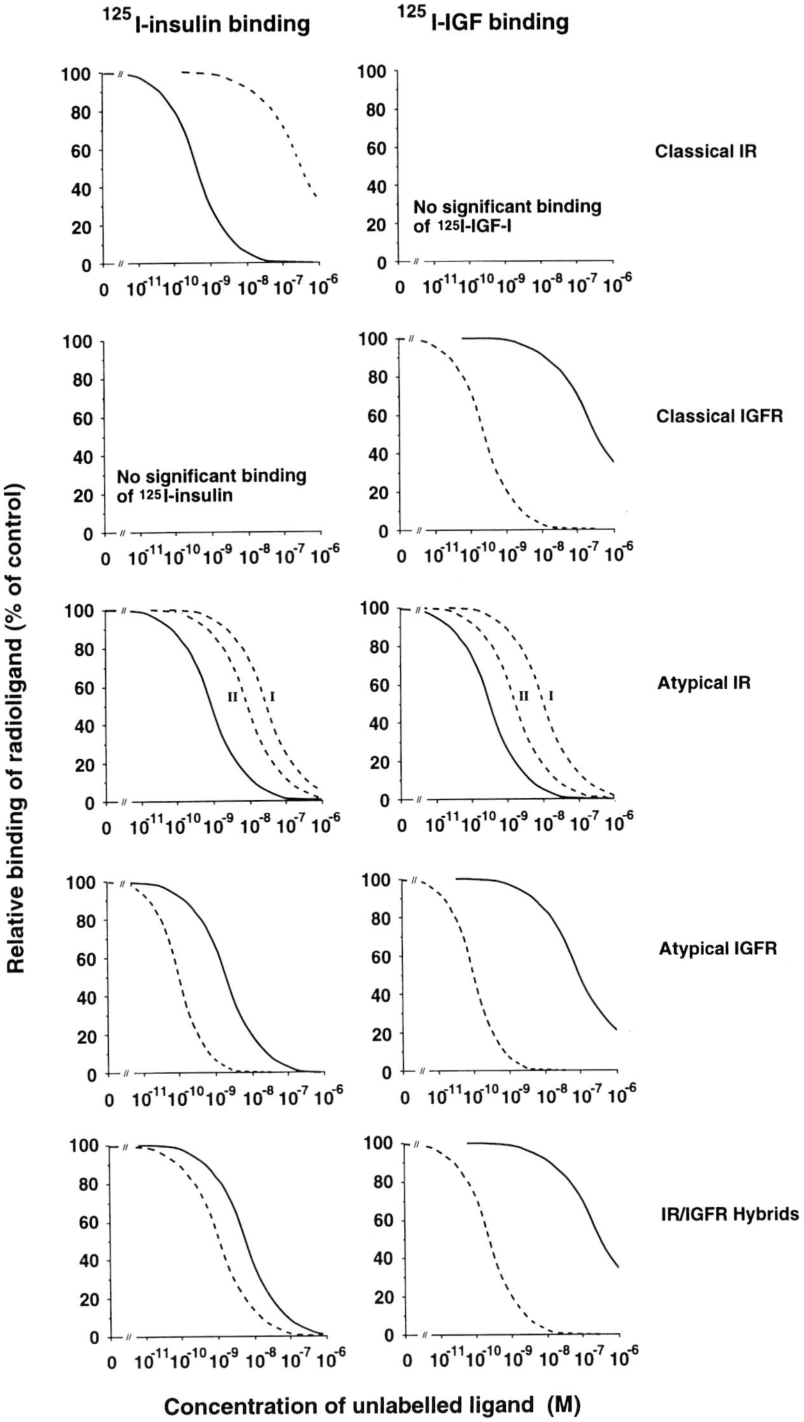

Fig. 2. Ligand specificities of receptor subtypes. Idealized curves are shown to illustrate the binding properties of classic, atypical, and hybrid receptors, as derived from published data for atypical IR (*69,70*), atypical IGFR (*74,77*), and hybrid receptors (*95,96*). The *left-hand panels* represent binding of ^{125}I-insulin, and the *right-hand panels* binding of ^{125}I-IGF-I. *Solid lines* indicate radioligand competition by unlabeled insulin, and *dotted lines* competition by unlabeled IGF-I (except for atypical IGFR where both IGF-I and IGF-II are shown). In each case, radioligand binding is normalized to 100% in the absence of competing unlabeled ligand, but it should be noted that absolute levels of radioligand binding will differ for insulin and IGF-I and for different receptor subtypes.

Table 1
Ligand Affinities of Atypical and Classic IGF Receptors

		IC_{50} (K_d) Values (nM) for ^{125}I-Ligand Binding			
		Solubilized (Atypical)		*Intact Cells (Classic)*	
^{125}I-ligand	*Competing Ligand*	*a*	*b*	*a*	*b*
IGF-I	IGF-I	0.1 (K_d 0.01)	0.05	0.2 (K_d 0.1)	1
	Insulin	200	100	400	300
Insulin	IGF-I	0.1	0.1	—	
	Insulin	2 (K_d 1)	1	—	

Results of radioligand binding assays on solubilized receptors or intact cells are summarized. The figures are IC_{50} values from competition assays with homologous or heterologous ligand, as idealized in Fig. 2, with dissociation constants calculated from Scatchard plots shown in parentheses. Data for IGF-1R/3T3 cells (a) and MCF-7 cells (b) were adapted from information in Soos et al. *(77)* and Milazzo et al. *(74)*, respectively.

separation from classic IGFR and classic IR, by sequential affinity chromatography of detergent extracts of MCF-7 cells on immobilized insulin and anti-IR antibody *(74)*. The atypical IGFR was presumed to be closely related in structure to classic IGFR, although the structural basis of the different binding specificities was not elucidated.

Data from our own laboratory support an alternative hypothesis, that atypical binding properties are a reflection of a high-affinity state of IGFR induced by detergent solubilization, rather than a distinct structure. It has been noted by others that IGF-I binds with particularly high affinity to solubilized receptors compared to intact cells *(75)* (also P. DeMeyts, *personal communication*). In detergent lysates of MCF-7 breast cancer cells and transfected IGF-1R/3T3 fibroblasts *(23)*, we have consistently observed a very high affinity of IGF-I binding together with substantial binding of ^{125}I-insulin to sites with all the features of atypical IGFR as defined previously. In particular, IGF-I displaces tracer binding with greater potency than insulin (Table 1 and Fig. 2), and the receptors are precipitable by anti-IGFR but not anti-IR monoclonal antibodies. In contrast, we found that intact cells or membrane fractions, with the same ^{125}I-IGF-I binding capacity as solubilized preparations, displayed only very low levels of ^{125}I-insulin binding that had properties similar to those expected of classic IR *(76,77)*. Others have similarly observed a dramatic increase in the level and affinity of atypical IGFR following detergent solubilization, while reporting some atypical insulin binding to intact membranes *(73,74)*. We showed that trypsinization of intact cells caused parallel loss of ^{125}I-insulin and ^{125}I-IGF-I binding capacity in subsequently prepared detergent lysates, ruling out the possibility that the different binding capacities of cells and lysates reflected intracellular sequestration of atypical receptors *(77)*.

The demonstration of atypical insulin binding to receptors produced from the cloned IGFR cDNA in IGF-1R/3T3 cells rules out the possibility that atypical IGFR are the product of a distinct gene or alternative splicing. Nevertheless, heterogeneity might be introduced by some other form of posttranslational modification. We attempted to separate atypical and classic IGFR in extracts of MCF-7 and IGF-1R/3T3 cells by affinity chromatography on insulin-Sepharose. However, we were unable to reproduce previous results *(74)*, and obtained little if any enrichment of ^{125}I-insulin binding relative to ^{125}I-IGF-I binding in specifically bound and eluted material compared to the flow-through of unbound material *(77)*. Moreover, Scatchard analysis performed on extracts of IGF-1R/3T3 cells indicated that the numbers of high-affinity binding sites for IGF-I (K_d approx 0.01 n*M*) and insulin (K_d approx 1 n*M*) were very similar, ruling out the possibility that insulin was binding to a minor subpopulation of IGFR. Scatchard analysis of IGF-I binding to intact cells and membranes indicated a significantly lower affinity for IGF-I (K_d approx 0.1 n*M*), and competition experiments suggested a much lower affinity for insulin (IC_{50} approx 450 n*M*). Because of the very high affinity of solubilised receptors for IGF-I, IC_{50} values for displacement of ^{125}I-IGF-I by insulin and IGF-I do not provide a valid estimate of the respective K_d values under conditions normally employed in binding assays. However, there is a clear anomaly in the IC_{50} concentrations of insulin determined with ^{125}I-insulin and ^{125}I-IGF-I (approx 1 n*M* and 100 n*M*, respectively). Thus the profile for inhibition of ^{125}I-IGF-I binding by insulin is actually quite similar in intact cells and lysates, and the IC_{50} value in lysates is not consistent with the K_d for insulin of approx 1 n*M* as determined by Scatchard analysis. We have no explanation at present for this anomaly, but all other data are consistent with the hypothesis that a high-affinity classic IGFR, with approx 10-fold greater affinity for IGF-I and 100-fold greater affinity for insulin compared to the receptor in intact cells, is responsible for atypical binding of ^{125}I-insulin. The high-affinity state exhibited in detergent lysates compared to cells or membranes might reflect the activity of a detergent-sensitive affinity regulator as originally suggested *(73)* or an effect of solubilization on the properties of the receptor *per se*. It should be noted that the IR does not exhibit a comparable change in properties following detergent solubilization.

The question arises as to whether atypical receptors have any physiological importance, or whether they are simply a feature of very artificial experimental conditions. In this context, the concept of an affinity regulator, whose activity might be modulated by physiological circumstances, is obviously attractive. Although there is no direct evidence for such a molecule, it is interesting to note that certain anti-IGFR monoclonal antibodies, most notably IGFR 26–3 *(78)*, dramatically increase the affinity of IGFR in intact cells for insulin, and to a lesser extent IGF-I, so that the properties become very similar to those of "atypical" detergent-solubilized receptors *(76)*. The enhancement of insulin binding by such antibodies is paralleled by an equally dramatic potentiation of insulin action on DNA synthesis *(78)*. Other antibodies differentially affect binding of ^{125}I-insulin and ^{125}I-IGF-I to solubilized receptors; for instance IGFR 24-60 inhibits binding of IGF-I but not insulin, whereas IGFR 17–69 has the converse effect (M. A. Soos and K. Siddle, *unpublished observations*). Thus it is clear that both the affinity and specificity of the IGFR can be influenced by interaction with other proteins. It remains to be determined whether there are endogenous proteins in tissues or in serum with the potential to exert similar effects, and the ability to render the type I IGFR more or less susceptible to stimulation by insulin. Certainly there are circumstances in which "specificity spillover" does apparently occur *(79,80)*. In states of insulin resistance, in which insulin action via its own receptor in skeletal muscle and adipose tissue is markedly impaired, paradoxical hyperresponsiveness is nevertheless observed in some other tissues, resulting in ovarian hyperandrogenism, the skin condition acanthosis nigricans, and cardiac hypertrophy. It seems very unlikely that these effects are mediated by the IR unless insulin resistance is highly tissue specific. The alternative possibility is clearly that insulin is acting via the type I IGFR, although even at the elevated concentrations that prevail it is questionable whether insulin would significantly activate the classic IGFR in the absence of mechanisms to sensitise the receptor. However, any suggestion that modulation of IGFR affinity for insulin does indeed contribute to the phenomenon of specificity spillover remains highly speculative. In fact, it remains doubtful whether IGFR in vivo can be induced to display "atypical" binding and anomalously high affinity for insulin.

INSULIN/IGF-I HYBRID RECEPTORS

It has long been apparent that the IR and IGFR are widely expressed in mammalian tissues and frequently coexpressed in the same cells. Even when it became clear that the receptors were functionally dimeric, and that dimerisation (preexisting or ligand-induced) was a prerequisite for activation of the superfamily of receptors with intrinsic tyrosine kinase activity *(81,82)*, IR and IGFR were considered as separate homodimeric proteins, each specifically binding and responding to its preferred ligand. However, three lines of evidence came together almost simultaneously that showed that these receptors can also undergo heterodimerisation to form insulin/IGF hybrid receptors.

First *(56,83)*, a substantial fraction of high-affinity IGF-I receptors in human placenta, HepG2 cells, IM-9 cells, and IR-transfected fibroblasts was recognized by a panel of IR-specific monoclonal antibodies, as demonstrated by inhibition of ligand binding or coprecipitation of ligand–receptor complexes. This immunoreactivity was lost, although overall IGF-I binding remained the same, when receptors were treated with dithiothreitol (DTT) to reduce intramolecular (α–α) disulfides and cause dissociation into half-receptors. It was concluded that the apparent reaction of IR antibodies with IGF-I receptors was due to the physical association of IGF half-receptors with insulin half-receptors in heteromeric hybrid receptors. Immunological heterogeneity of IGFR had in fact been noted previously, without interpretation in terms of hybrid structures *(84,85)*.

Second *(86)*, low concentrations of IGF-I were shown to stimulate the autophosphorylation of two β-subunits with slightly different electrophoretic mobilities in NIH3T3 and HepG2 cells. Both subunits of native receptors were precipitated with both IR-specific and IGFR-specific antibodies, but after dissociation into αβ half-receptors by reduction with DTT only the smaller subunit reacted with IR-specific antibody and only the larger with IGFR-specific antibody. Moreover, the two β-subunits had distinct tryptic phosphopeptide maps, that of the smaller subunit being similar to maps of the IR β-subunit. Again the deduction was made that a substantial fraction of IGF-I receptors in these cells existed as hybrids composed of disulfide-linked IGFR and IR αβ half-receptors. A very similar study of IGF receptors in the brain failed to make the deduction that apparent heterogeneity might be explained by formation of hybrids with insulin receptors *(87)*. Indeed, a doublet of autophosphorylated receptor β-subunits following stimulation with IGF-I had been noted previously in many studies without mechanistic interpretation *(88–92)*. However, as discussed below, it has not been universally accepted that hybrid structures account fully for this phenomenon in all cell types.

Third *(93)*, it was shown that hybrids could be assembled in vitro by the simultaneous treatment of pooled insulin and IGF-I half receptors with a combination of insulin and IGF-I, following earlier demonstrations that each ligand could induce homodimerization of its cognate half-receptors. The hybrids generated in this way bound both insulin and IGF-I as shown by immunoprecipitation with both IR-specific and IGFR-specific antibodies.

Table 2
Ligand Affinities of Hybrid and Classic Receptors

		IC_{50} values (nM) for Inhibition of ^{125}I-Ligand Binding					
		Hybrids		*IGFR*		*IR*	
^{125}I-ligand	*Competing Ligand*	*a*	*b*	*a*	*b*	*a*	*b*
IGF-I	IGF-I	0.15	1	0.1	1	3	10
	Insulin	100	1000	300	1500	0.2	4
Insulin	IGF-I	1	3	0.05	3	50	100
	Insulin	4	3	4	3	0.4	0.3

Results of radioligand binding assays on immunopurified receptors are summarized. The figures are IC_{50} values from competition assays with homologous or heterologous ligand, as idealized in Fig. 2. Data were adapted from information in Soos et al. *(95)* (a) and Kasuya et al. *(96)* (b).

More recently, it has been shown, again by using receptor-specific antibodies, that both the IR and IGFR can form hybrids with the IRR *(43,94)*.

Ligand Binding to IR/IGFR Hybrids

The fact that hybrids necessarily occur together with classic IR and IGFR presents some difficulties when it comes to analyzing their properties in intact cells. Two approaches have been used to circumvent this problem, namely purification and overexpression by controlled cotransfection. Hybrids have been purified from detergent-solubilized human placental membranes by combinations of immunodepletion and immunoaffinity chromatography with receptor-specific antibodies. We made use of an IGFR-specific monoclonal antibody (IGFR 24-55) that fortuitously showed very poor reactivity toward hybrids. This antibody was used to deplete extracts of classic IGFR, and hybrids were then isolated from the residual material with a specific antipeptide antibody (IGFR 1–2) directed against the C-terminus of the IGFR β-subunit *(95)*. This protocol had the advantage that only a single elution step was required, involving a competing peptide rather than potentially denaturing conditions. Other workers purified hybrids by sequential immunoaffinity chromatography on IGFR-specific and IR-specific antibodies, necessitating two elution steps *(96)*. Both studies went on to document ligand binding by purified hybrids, with similar results. It was found that hybrids bound 20–25 times as much IGF-I as insulin at tracer concentrations of radioligand. As judged by IC_{50} values in homologous binding competition assays, hybrids bound IGF-I with affinity similar to classic IGFR, but bound insulin with approx 10-fold lower affinity than classic IR (Table 2 and Fig. 2). It is to be expected therefore that hybrid receptors would be poorly responsive to insulin at physiological concentrations, and, indeed, that incorporation of a significant fraction of insulin receptors into hybrids would lead to insulin resistance.

Somewhat paradoxical results were obtained in heterologous competition assays designed to investigate interactions between the two potential ligands *(95)*. The IC_{50} concentration of insulin determined with ^{125}I-insulin as radioligand was considerably lower than that determined with ^{125}I-IGF-I, consistent with a model in which insulin does indeed bind to the insulin half-receptor within a hybrid, albeit with relatively low affinity, without inhibiting binding of IGF-I to the other half-receptor. On the other hand, although the IC_{50} concentration of IGF-I determined with ^{125}I-insulin as radioligand was somewhat higher than that determined with ^{125}I-IGF-I, IGF-I was significantly more effective at inhibiting ^{125}I-insulin binding than insulin itself, suggesting that occupancy of the IGF half-receptor did result in inhibition of insulin binding to its half-receptor. Similar results were obtained for hybrids generated by cotransfection *(97)*. It is difficult to reconcile these data based on a model in which each ligand binds primarily to its cognate halfreceptor in a mutually exclusive fashion. An alternative model, in which insulin in fact binds in an "atypical" manner to the IGF half-receptor, and the insulin half-receptor is non-functional as regards ligand binding, is similarly unable to account fully for the data. Crosslinking of insulin analogues with photoactivatable groups substituted at different positions might help to identify definitively which α-subunit within hybrids is primarily responsible for insulin binding.

As mentioned previously, it has been proposed that in spite of its functionally dimeric structure, the insulin receptor binds only a single molecule of ligand with high affinity, in a manner that involves asymmetric contacts with both α-subunits. Specifically, it has been suggested that the "dimer" face of insulin contacts the receptor in a region encoded by exons 2/3 of one α-subunit, while the "hexamer" face of ligand contacts a second region encoded by exons 6/7 of the other α-subunit *(21,22)*. The former interaction is believed to make the major contribution to binding. This model has many attractive features and considerable experimental support. In particular it provides a framework for understanding the phenomenon of negative cooperativity of insulin binding *(21,98)*. However, there are data that indicate that the binding mechanisms of insulin and IGF-I to their respective receptors may not be identical. Consistent with the model, reduction of insulin receptor disulfides to generate half-receptors results in a substantial decrease in insulin binding affinity and linearization of Scatchard plots *(99–101)*. However, although one study obtained similar findings with IGF-I receptors *(102)*, others have reported that IGFR disulfide reduction does not decrease IGF-I binding *(95,103,104)*, suggesting that native receptors may bind two molecules of IGF with high affinity. Moreover, the negative cooperativity exhibited by the the IGFR differs in detail from that of the IR *(105,106)*. We found that reduction of intersubunit disulfides in hybrid receptors actually increased insulin binding and slightly decreased IGF-I binding *(95)*. Predictably, dissociation of bound ^{125}I-IGF-I from hybrids was not influenced by unlabeled IGF-I, but it was also unaffected by insulin, even though it was accelerated by anti-IR antibodies *(76)*. These data suggest that the IGFR α-subunit within hybrids actually exerts a negative effect on insulin binding, and that insulin is unable to crosslink the α-subunits of hybrid receptors. Binding of IGF-I, on the other hand, appears relatively indifferent to the presence of the IR α-subunit, suggesting that if crosslinking of α-subunits does occur with IGF-I it makes little contribution to high-affinity binding and involves residues that are conserved between the two receptors. These differences between insulin and IGF-I binding may be related to the different receptor domains contributing to high-affinity binding of the two ligands (N-terminus of IR, cysteine-rich domain of IGFR), as described previously. Further studies of hybrid receptors may shed light on the mechanism of ligand binding to insulin and IGF receptors and the relationship between crosslinking of α-subunits and receptor activation.

Activation and Signaling Capacity of IR/IGFR Hybrids

Studies of the autophosphorylation of hybrid receptors are generally consistent with analyses of ligand binding, in suggesting that the receptors are more responsive to IGF-I than to insulin. All the work published to date has utilized detergent-solubilized receptor preparations, and the more relevant issue of hybrid receptor phosphorylation in situ in intact cells has so far been neglected, as has a detailed assessment of autophosphorylation sites in hybrid compared to classic receptors.

In extracts of mouse NIH3T3 fibroblasts, IGF-I at low concentrations (10-25 ng/mL, or approx 1.5–3 n*M*) stimulated the autophosphorylation of both β-subunits within hybrids, whereas insulin had little effect even at 10-fold higher concentrations *(86)*. Tryptic phosphopeptide mapping indicated that the major sites of IGF-I-induced autophosphorylation of insulin receptor β-subunits within hybrids were the same as for insulin-induced phosphorylation of classic IR. In hybrids purified from human placenta, IGF-I induced a somewhat greater stimulation of autophosphorylation and tyrosine kinase activity than insulin *(96)*, although the extent of ligand stimulation in this study was quite low, apparently due to partial inactivation during the extensive purification. We similarly could not demonstrate substantial ligand-stimulated autophosphorylation in hybrids purified from placenta, but found that IGF-I was more potent than insulin in stimulating autophosphorylation of hybrids in extracts of human fibroblasts (K. Dib, M.A. Soos, and K. Siddle, *unpublished observations*).

Others have studied hybrid receptors formed during transient high-level coexpression of IR and IGFR in fibroblasts. It was shown, by generating hybrids that incorporated mutant, kinase-inactive receptor halves, that β-subunit *trans*-autophosphorylation occurred independently of which β-subunit was ligand occupied *(97)*. In this system, no significant differences were observed in either the ligand sensitivity or extent of insulin- and IGF-I-stimulated β-subunit autophosphorylation within hybrids. Other work with stably transfected cell lines also showed that *trans*-autophosphorylation occurred in hybrids with a kinase-defective insulin receptor half, although the amount of phosphorylation was diminished compared to receptors in which both halves were kinase competent *(107)*. Overall, the results of these limited studies suggest that both insulin and IGF-I stimulate the reciprocal *trans*-autophosphorylation of both IR and IGFR β-subunits within hybrids. Taken together with the binding studies discussed earlier, it would be expected that IGF-I would be much the more effective ligand in stimulating hybrids, and that hybrids would be poorly responsive to insulin in vivo.

The obvious question that arises is whether the intracellular signalling properties of hybrid receptors are distinct from those of classic IGFR or IR. The answer is as yet unknown but will depend very much on the extent of signaling differences between the IR and IGFR. As discussed briefly in the introduction, there is a certain amount of evidence that the receptors differ in the efficiency with which they mediate different biological responses, and in their interactions with potential signaling intermediates, but this is all somewhat tenuous and no clear differences in intrinsic signaling capacity and consequent biological responses have been unambiguously demonstrated. It is nevertheless very tempting to speculate that hybrid receptors can mediate a broader range, or even a distinct set, of responses compared to either of the individual receptors. However, hard evidence is lacking, for want of systems in which an appropriate range of responses to IR, IGFR, and hybrids can be independently examined in the same cell background.

With regard to ligand internalization and degradation, hybrids appear to behave like IGFR rather than IR, in that the bulk of IGF-I bound to hybrids undergo retroendocytosis rather than degradation *(107)*. However, this might reflect the binding properties of the respective ligands as much as the trafficking or signaling properties of the receptors, and the relevance of these observations to other aspects of hybrid receptor function is unclear.

The availability of tissues or cell lines from animals with targeted disruptions of the IR and IGFR genes will undoubtedly facilitate studies designed to compare the activities of these receptors and to assess the properties of hybrids. Clearly, knockout of either gene precludes formation of hybrids. Mice lacking the IGFR exhibit severe growth deficiency, with generalized organ hypoplasia and developmental delays, and die at birth *(108)*. In contrast, mice lacking the IR appear to develop normally *in utero* and are born at normal weight, although they survive only a few days in the face of postnatal hyperglycaemia and ketoacidosis *(109,110)*. This phenotype indicates that neither the IR itself nor IR/IGFR hybrids are necessary for fetal growth and development in mice, and presumably classic IGFR alone are able to mediate these effects of IGFs. However, mutations that result in complete lack of functional insulin receptors in humans underly Donohue's syndrome (leprechaunism), in which fetal growth and development are markedly impaired *(111,112)*. The basis of the different phenotypes resulting from absence of IR in humans and rodents is unclear, but the possibility cannot be ruled out that lack of IR/IGFR hybrid receptors compromises the action of IGFs in human fetal development, because classic IGFR are not able to mediate the same magnitude or range of responses as hybrids.

There are precedents in the epidermal growth factor (EGF) receptor and platelet-derived growth factor (PDGF) receptor tyrosine kinase families which lend support to the hypothesis that hybrid structures may have unique signaling potential as well as specific ligand binding properties. Ligand-induced formation of heterodimers involving p185c-neu (*erb*B2), *erb*B3, and the EGF receptor itself has been shown to occur and to be necessary for certain biological responses *(113–119)*. In this case, the synergy between different family members likely reflects the creation of novel binding sites for specific ligands in heterodimers, together with the presence of unique phosphorylation sites on different family members that are coupled to distinct signaling pathways. The PDGF α- and β-receptors, which activate both unique and common signal transduction pathways *(120)*, have also been shown to form heterodimers when PDGF-AB is used as ligand *(121)*. In this situation it would be expected that signaling by receptor heterodimers, induced specifically by heterodimeric ligand, would be at least the sum of the signaling capacities of the individual receptor isoforms.

Distribution and Assembly of Insulin/IGF Hybrid Receptors

Although IR/IGFR hybrids were initially demonstrated and characterized largely in cultured cell lines, they have now been shown to be widely distributed in mammalian tissues, and to be responsible for the majority of IGF-I binding in some tissues such as heart and skeletal muscle *(122,123)*. The fraction of insulin receptors incorporated into hybrids is harder to estimate, because of the low affinity of insulin binding to hybrids compared to classic receptors. By using radiolabeled antibody as ligand, and by quantitative immunoblotting, we estimated that hybrids in human placenta contained 70–80% of IGFR but only 10-20% of IR, consistent with the overall four- to fivefold excess of IR over IGFR in this tissue *(95,122)*. Others have reported considerably higher fractions of IR as hybrids in placenta and skeletal muscle, as determined by two-site immunoassays employing IR-specific and IGFR-specific monoclonal antibodies *(123)*. However, the calculations of hybrid fractions based on these immunoassays may be unreliable, as they depend on an assumption that two different iodinated antibodies bound with equal affinity to their respective receptors. Notwithstanding the problems with precise quantification, the widespread occurrence of hybrids raises questions regarding the interpretation of earlier experiments on signaling specificity, in which insulin and IGF were assumed to be acting only through their classic receptors *(124,125)*.

Several studies have addressed the issue of efficiency of assembly of hybrids compared to classic receptors. When hybrids were generated experimentally by ligand-induced dimerisation of isolated half-receptors in vitro *(93)* or by coexpression of IR and IGFR in transfected fibroblasts *(97)*, it was observed that the extent of hybrid formation reflected the molar ratios of IR and IGFR, implying that heterodimerization to form hybrids and homodimerization to produce classic receptors occurred with equal efficiency. The relative levels of hybrid and classic receptors in placenta were similarly consistent with a process of random and unbiased assembly of half-receptors in vivo *(122)*. However, in contrast to the situation with IR/IGFR hybrids, it appears that the efficiency of formation of IR/IRR hybrids may be low *(94)*. The limited evidence available suggests that the two IR isoforms do not differ in their ability to form hybrids with IGFR *(57)*. On the other hand, it was reported that only one of the two forms of endogenous IGFR detected in transfected fibroblasts formed hybrids with overexpressed IR *(107)*, although the structural basis of the IGFR heterogeneity in these cells is unknown and the significance of the observation is therefore difficult to judge.

More generally in mammalian tissues, the fraction of hybrids is not a simple function of the relative levels of IR and IGFR. This might simply reflect heterogeneity in receptor distribution among different cell populations of whole tissues. Moreover, even if hybrids are initially assembled during biosynthesis with the same efficiency as classic receptors, the relative levels of expression at steady state may be affected by differential rates of degradation, especially as a consequence of ligand-induced down-regulation. However, the possibility cannot yet be ruled out that mechanisms exist for the regulation of hybrid assembly or turnover in vivo.

The dimerization and disulfide-linkage of insulin proreceptors is an early posttranslational event that appears to take place within the endoplasmic reticulum *(126)*. The dimerization process is preceded by a conformational transition that may reflect rearrangement of intermolecular disulfides and that is associated with acquisition of ligand binding activity. N-linked glycosylation, which occurs cotranslationally in the endoplasmic reticulum, is required for both the initial conformational transition and for dimerization. Proreceptor dimers are transferred to the Golgi apparatus, where proteolytic cleavage produces the separate α-and β-subunits and oligosaccharide maturation occurs. The mature receptor is then transferred to the plasma membrane. As far as is known, no disulfide exchange takes place subsequent to exit of receptors from the endoplasmic reticulum. It is not clear that this possibility has been tested experimentally, although it would be assumed that other subcellular compartments lack the necessary redox environment and catalytic activity to make such exchange possible. Comparable studies of IGFR biosynthesis have not been reported, but there is no reason to suppose this would differ significantly from IR biosynthesis in terms of the major steps in posttranslational processing. It would be expected therefore that assembly of both homotypic and heterotypic proreceptor dimers, destined to be processed to classic and hybrid receptors, respectively, should be completed in the endoplasmic reticulum. Any influences on the efficiency of hybrid assembly would have to operate at this stage, unless as yet unknown mechanisms exist that permit rearrangement of half-receptors in other compartments. It is possible in principle that the extent of hybrid formation might be governed not only by the relative levels of insulin and IGF proreceptors in the endoplasmic reticulum, and by the relative efficiency of their homo- and heterodimerization, but also by temporal and spatial differences in translation of IR and IGFR mRNA transcripts (involving different phases of the cell cycle and different regions of the endoplasmic reticulum, respectively). As alluded to previously, the relative levels of classic and hybrid receptors available at the cell surface will depend on rates of degradation as well as biosynthesis, and this process may also be susceptible to differential regulation for different receptor species. Thus there are a number of ways in which the level of hybrid receptors might in principle be regulated, but for the present these are no more than theoretical possibilities without experimental support.

The ability of the IR and IGFR half-receptors to form functional heterodimers as readily as homodimers implies high conservation of structural features in the dimerization interface of the two receptors. It is well established that the two halves of native receptors are linked through their α-subunits, but conflicting data have been presented regarding the precise number and location of inter-subunit disulfides *(127–131)*. Cysteine 524 has been identified as one of the residues involved in α–α disulfide linkage in the IR *(132)*, although probably it is not the only such residue *(131,133)*. As the IR extracellular region contains an odd number of cysteines (41), but reportedly no free thiol groups *(129)*, it follows that there must be an odd number of α–α inter-subunit disulfides. Therefore there may be at least two residues in addition to cysteine 524 that contribute α–α disulfide links. The IGFR extracellular region contains 39 cysteine residues, of which 38 are homologous to residues in the insulin receptor. The extracellular IR therefore contains three cysteines with no counterpart in the IGFR, and the extracellular IGFR contains one cysteine with no counterpart in the IR. Interestingly, the extracellular IRR contains only the 38 cysteines that are common to both the IR and IGFR, suggesting either that this must lack an inter-subunit disulfide relative to the IR and IGFR or that it must contain a free thiol group. Certainly it cannot be assumed that the pattern of inter-subunit disulfides is the same in all three receptors, although given the apparent ease of formation of hybrid receptors it would be surprising if this were not so at least for the IR and IGFR. Overall, the IR and IGFR share 55% sequence identity in their α-subunits. The region flanking cysteine 524 of the IR is identical to the corresponding region of IGFR over a continuous sequence of 11 amino acids, which is the longest uninterrupted identity in the extracellular portion of the two receptors. Thus there may very high conservation of sequence and structure in the region that provides the interface between α-subunits. However, it is notable that following mild reduction of IR, IGFR, or hybrid receptors, the two receptor halves dissociate quite readily, suggesting that, at least in the absence of ligand, noncovalent forces do not make a substantial contribution to maintaining oligomeric structure. It may be that the association between proreceptor polypeptides is stronger than that between mature receptor halves, and that the α–α disulfide linkage is necessary to maintain a dimeric receptor structure following proteolytic processing of proreceptors to generate α- and β-subunits, just as the disulfides formed in proinsulin help to maintain the structure of insulin following proteolytic removal of C-peptide.

Hybrid Receptors in Pathological States

Several studies have reported changes in the fraction of hybrid receptors in human tissues under pathophysiological conditions, which do not seem to be a simple consequence of changes in the relative expression of IR and IGFR. These observations lend support to the ideas that mechanisms may exist for the specific modulation of hybrid assembly or turnover, and that hybrids may play a specific pathophysiological role.

The first observation of altered expression of insulin/IGF hybrid receptors was in the placenta of insulin-resistant women with gestational hypertension *(134)*. The fraction of IGFR in hybrids was reportedly increased from 68% in controls to 85% in hypertensives. The hybrid fraction was positively correlated with overall IGF-I binding, and negatively correlated with insulin binding. Interpretation was complicated by the fact that changes in the relative proportions of IR isoforms were also observed. However, taking the results at face value it would appear that the increased fraction of IGFR in hybrids could not be accounted for by an increase in the ratio of IR to IGFR, but rather would require some change in the efficiency of hybrid assembly. The same laboratory has reported that the expression of hybrid receptors is similarly increased in the skeletal muscle of non-insulin-dependent (type 2) diabetes mellitus subjects *(135)*. It was suggested that the increased incorporation of IR into hybrids with a low affinity for insulin might contribute to decreased sensitivity to insulin and to increased sensitivity to IGF-I. However, receptors were quantified only in terms the fraction of IGF-I receptors incorporated into hybrids (estimated at 44% in controls and 74% in diabetics in terms of IGF-I binding, although higher values were obtained by immunoblotting). The fraction of IR in hybrids was not determined but is likely to be small. In our hands, the overall level of IR in muscle appears to be considerably greater than that of IGFR, as determined by immunoblotting with carefully standardized antibodies (M.A. Soos, E.M. Bailyes, and K. Siddle, *unpublished observations*). Moreover, we were unable to detect a significant difference between control and diabetic muscle in terms of the fraction of IGFR in hybrids (M.P. Coghlan, H.U. Häring, and K. Siddle, *unpublished observations*). Therefore we consider it unlikely that increased incorporation of IR into hybrids in diabetic muscle makes a significant contribution to changes in insulin sensitivity.

Other studies have examined the expression of hybrid receptors in tumors. Hybrid and classic receptors were determined in 39 breast cancer specimens and 8 breast cancer cell lines *(136)*. The relative and absolute levels of expression of IR and IGFR varied widely in different samples, with substantial up-regulation of one or both of the receptors compared to normal breast tissue. Nevertheless, the measured proportion of hybrids was in all cases very close to that which would be expected from random assembly of half-receptors according to their respective mole fractions. However, in a smaller series of ovary and colon cancer specimens, the levels of hybrids substantially exceeded what would be expected for random assembly of half-receptors, suggesting that in these tumors hybrids were assembled or retained preferentially relative to classic receptors *(136)*. In all types of tumor examined, the levels of hybrids were at least comparable to those of classic receptors, suggesting that they would play a major role in autocrine stimulation by IGFs. Up-regulation of hybrids, from 20% to 55% of total IGFR, was also observed during differentiation of HT29-D4 human colonic carcinoma cells *(137)*. In this case the increase in hybrids was paralleled by a twofold increase in IR number, with no change in the number of IGF-I binding sites. The increase in IR:IGFR ratio in the differentiated cells therefore appears sufficient to account for the increased fraction of IGFR in hybrids, with no requirement for changes in the efficiency of hybrid assembly.

Mutations of the IR gene have been shown to underly some conditions of severe insulin resistance, including Donohue's syndrome, Rabson–Mendenhall syndrome, and type A insulin resistance *(138,139)*. Some of these mutations, most especially those in the tyrosine kinase domain, act in a dominant-negative fashion as evidenced by the severity of phenotype in heterozygotes, or by expression of mutant receptors in transfected cell lines. It has been shown that mutant IR can similarly exert a dominant-negative influence when incorporated into hybrids with wild-type IGFR *(140)*. It is therefore possible that the phenotype of patients with IR mutations reflects in part impairment in the action of IGF as well as insulin, particularly in those tissues in which a substantial fraction of IGFR occurs as hybrids.

As mentioned earlier in the context of atypical IGFR, a phenomenon of specificity spillover has been noted in conditions of severe insulin resistance *(79,80)*. This is manifest as, for example, the skin lesion acanthosis nigricans or ovarian hyperandrogenism, apparently reflecting excessive insulin stimulation of these tissues even though insulin action on glucose uptake and metabolism in muscle and fat is clearly impaired. The molecular basis of insulin resistance is in most cases unclear, and it may be that the mechanisms are in part tissue specific. However, an alternative possibility is that under conditions of hyperinsulinaemia, insulin is able to act via hybrid receptors in certain tissues such as skin and ovary, activating the IGF receptor half and bypassing inhibitory mechanisms that are specific for the IR. However, this depends on the assumption that the IR and IGFR have signaling and regulatory mechanisms that are at least in part distinct. As discussed previously, there is still rather little direct experimental evidence in support of this, however attractive an idea it might be in theory.

IS THERE ANOTHER IGF RECEPTOR?

Heterogeneity of IGFR β-subunits, in terms of size or immunoreactivity, has been described in numerous studies *(87, 141–149)*. Various proposals, which are not necessarily mutually exclusive, have been made as to the mechanistic or structural basis of such apparent heterogeneity, including transphosphorylation between distinct insulin and IGF-I receptors *(141,150)*, differences in glycosylation *(141,145)*, or differences in primary sequence arising from alternative splicing or distinct gene products *(148,149)*, as well as the formation of receptor hybrids *(146,147)*. The most recent studies specifically addressed, and apparently ruled out, the possibility that heterogeneity was simply a reflection of hybrid formation between the known IR and IGFR. Some encouragement for the search for additional receptors was provided by the results of genetic experiments involving targeted disruption of the genes for IGFs and IGFRs, which show that growth retardation in embryos lacking both IGF-II and the type I IGFR is more severe than that manifested in mice with either of the individual deletions *(108)*. It was inferred from these experiments that IGF-II must act in part through an unidentified receptor (XR), although more recent studies suggest that the IR may fulfill this role *(59)*.

The β-subunits of the IGFR and the IR are commonly observed to differ in mobility on sodium dodecyl sulfate-polyacrylamide gel electrophoresis (SDS-PAGE), with reported molecular masses of 97–105 kDa, and 92–95 kDa, respectively. The predicted peptide masses based on primary sequence are 70.9kDa for the IGFR *(3)* and 69.7kDa for the IR *(1)*. There are six potential N-linked glycosylation sites in the extracellular portion of the IGFR β-subunit compared to only four in the IR, so that the difference in size of the mature glycosylated β-subunits is readily explained. However, glycosylation of IGFR and IR may differ in detail between tissues and between species *(47,144,151,152)*. Such glycosylation differences may account for high-M_r forms of IGFR β-subunit observed in human leukemic HL-60 cells *(143)* and murine myogenic C2 cells *(145,147)*. The high-M_r form of IGFR observed in C2 cells was developmentally regulated and appeared to possess substantially higher affinity for IGF-II than the classic type I IGF receptor.

Conversely, it has been reported that human IGFR have low affinity for IGF-II compared to IGF-I when expressed in murine NIH-3T3 fibroblasts *(75)*. It remains to be determined whether similar functionally distinct forms of IGFR occur in normal tissues.

Several laboratories have presented evidence that the doublet of β-subunits observed following IGF-I-stimulated autophosphorylation cannot be wholly accounted for by the formation of hybrids between the cloned IR and IGFR. Studies in skeletal muscle *(142,148)*, brain *(149)*, and cultured cell lines *(146,147)* have suggested that there exist distinct bona fide IGFR β-subunits of 92–95 and 99–105 kDa that can occur together in hybrid as well as homomeric structures. The smaller of these subunits would comigrate on SDS-PAGE with the IR β-subunit. The two forms of IGFR differed, and were distinguished from the IR, by criteria of immunoreactivity and tryptic phosphopeptide mapping. In particular, whereas the larger form reacted with antipeptide antibodies directed against deduced sequences in the β-subunit of the cloned IGFR, the smaller form appeared unreactive with at least some such antibodies. Moreover, in human tissues the IGFR-specific antibody αIR-3, which recognizes an α-subunit epitope, selectively coimmunoprecipitated the larger β-subunit and not the smaller, suggesting that the two forms differed in primary sequence in both subunits. In muscle the two forms were differentially expressed during development, the larger form being selectively expressed in fetal and early postnatal life, whereas the smaller form was detected in both fetal and adult tissues. It was thus concluded that the cloned and sequenced IGFR was in fact the fetal form, and that the primary structure of the adult type receptor remained to be established *(148,149)*. The obvious possibilities are that the novel "adult" receptor might arise by alternative splicing of the known IGFR gene, or be the product of a distinct gene. In view of the evidence for sequence differences affecting both the α-subunit and the C-terminal portion of the β-subunit, the latter might be considered the more likely. However, five years after the last publication on this topic a second IGFR gene has not been identified, nor has alternative splicing of the known gene been demonstrated, and the structural basis of the putative "adult" IGFR remains obscure. The very existence of an alternative receptor must therefore be questioned, and the evidence that led to the proposal reexamined.

The question that arises is whether the smaller β-subunit that is phosphorylated in response to IGF is in fact that of the IR rather than a novel IGFR, so that the data reported are after all a manifestation of IR/IGFR hybrids, rather than hybrids between two different forms of IGFR. The apparent developmental regulation of the novel receptor would then be interpreted in terms of changes in the relative levels of IGFR and IR. The critical analyses defining the nature of the putative novel receptor are those relating to phosphopeptide mapping and immunoreactivity. Taken at face value these are certainly highly suggestive, but interpretation of the relevant data is not necessarily straightforward. Moreover, these studies lack control data with *bona fide* IR/IGFR hybrids, to show that these behave as expected and differently from the putative novel "adult" IGFR.

The phosphopeptide maps of the IR and IGFR would be expected to be very similar, as the sequence containing the major autophosphorylation sites in the tyrosine kinase domain (involving Y1146/1150/1151 in the Ex11– IR, and Y1131/1135/1136 in the IGFR) is completely conserved between the two receptors. However, the maps should also exhibit some unique components, as the receptors differ in other autophosphorylation sites and/or their flanking sequences in the C-terminal and juxtamembrane regions, and in potential sites of serine/threonine phosphorylation. All the reported analyses *(142,147,149)* describe unique features of the phosphopeptide maps of the putative 92–

95-kDa novel IGFR compared to the 99–105-kDa species and the IR, against a background of substantial similarities. However, the differences between the maps are largely in minor components, the identity and significance of which are unclear. Moreover, questions arise as to the reproducibility of the different maps, the purity of the analyzed material as excised from gels, and the validity of comparisons between receptors obtained from different tissues. In addition, the possibility that phosphorylation patterns may differ when a given β-subunit is incorporated into hybrid rather than classic receptors cannot be excluded. Thus the available mapping data do not provide strong evidence for the existence of a novel IGFR.

Care must also be taken when interpreting immunoreactivity data, and in deducing the presence or absence of epitopes based on differential immunoprecipitation. At the very least, hybrid structures will necessarily contain only single copies of receptor-specific peptide epitopes, whereas homomeric receptors will possess two copies of corresponding epitopes. Further, the conformation and possibly even the very existence of specific epitopes may differ when polypeptides are incorporated into hybrid rather than classic receptors. Because of these factors, antibodies that react quite well with classic receptors may react poorly with the corresponding structures when incorporated into hybrids. This may result in differential precipitation of classic and hybrid receptors, particularly under suboptimal conditions where precipitation is in any case incomplete. For instance, we have found that a relatively low-affinity IR-specific antipeptide monoclonal antibody CT-1 *(153)* precipitates hybrid receptors very poorly under conditions in which precipitation of classic IR is nearly quantitative, whereas a high-affinity monoclonal antibody 83–14 precipitates both hybrid and classic receptors efficiently. Conformation-specific monoclonal antibodies such as IGFR 24-55 and αIR-3 have similarly been shown to react poorly with hybrids compared to classic IGFR *(95,136)*. In light of these observations, caution must be exercised in interpreting relative lack of immunoreactivity as evidence of novel structure, especially as it does appear in some of the published data relating to the putative "adult" form of IGFR that immunoprecipitation was far from complete *(148,149)*. The most convincing evidence for a novel receptor is that obtained in rat brain synaptosome extracts that had been depleted of insulin receptors by using a polyclonal antipeptide antibody *(149)*, but even in this study evidence that the antibody was capable of efficiently depleting *bona fide* IR/IGFR hybrids was lacking. Although the jury is still out on the existence of distinct forms of the type I IGFR, it is increasingly hard to believe that a novel gene or splice variant could have escaped identification for so long in a field of intense activity. It therefore seems simpler to believe that the observed heterogeneity results from posttranslational modification or hybrid formation among the established members of the IR/IGFR family.

CONCLUSIONS

In spite of numerous publications reporting receptors with anomalous properties, the world of IGF receptors may not after all be too complicated. It is our view that the evidence for physiologically significant forms of the IGFR other than the classic type I receptor and IR/IGFR hybrids is unconvincing. The IRR does not bind IGFs and can be discounted as a mediator of IGF action. Splice variants of the cloned IGFR appear to be restricted to certain species and do not differ significantly in their properties. Atypical IGFR are most likely a reflection of the effects of detergent solubilization on classic IGFR rather than a distinct structure. IR/IGFR hybrids on the other hand are widespread and in many tissues may represent the major form of IGFR. It remains to be determined whether hybrids and classic IGFR differ in their signaling capacity, and whether regulation of the assembly and activity of hybrid receptors has a significant impact on IGF action in vivo.

ACKNOWLDGMENTS

The financial support of the Wellcome Trust, British Diabetic Association and Medical Research Council is gratefully acknowledged. We also thank many colleagues and collaborators for their contributions to our work and for stimulating discussions, with particular thanks to Dr Pierre DeMeyts (Gentofte) and Dr Riccardo Vigneri (Catania).

REFERENCES

1. Ullrich A, Bell JR, Chen EY, Herrera R, Petruzzelli LM, Dull TJ, Gray A, Coussens L, Liao YC, Tsubokawa M, Mason A, Seeburg PH, Grunfeld C, Rosen OM, Ramachandran J. Human insulin receptor and its relationship to the tyrosine kinase family of oncogenes. Nature 1985; 313:756–761.
2. Ebina Y, Ellis L, Jarnagin K, Edery M, Graf L, Clauser E, Ou JH, Masiarz F, Kan YW, Goldfine ID, Roth RA, Rutter WJ. The human insulin receptor cDNA: the structural basis for hormone-activated transmembrane signalling. Cell 1985; 40:747–758.
3. Ullrich A, Gray A, Tam AW, Yang Feng T, Tsubokawa M, Collins C, Henzel W, Le Bon T, Kathuria S, Chen E, Jacobs S, Francke U, Ramachandran J, Fujita-Yamaguchi Y. Insulin-like growth factor I receptor primary structure: comparison with insulin receptor suggests structural determinants that define functional specificity. EMBO J 1986; 5:2503–2512.
4. Shier P, Watt VM. Primary structure of a putative receptor for a ligand of the insulin family. J Biol Chem 1989; 264:14605–14608.
5. Siddle K. The insulin receptor and type I IGF receptor: comparison of structure and function. Prog Growth Factor Res 1992; 4:301–320.
6. De Meyts P, Wallach B, Christofferson CT, Ursø B, Grønskov K, Latus LJ, Yakushiji F, Ilondo MM, Shymko RM. The insulin-like growth factor-I receptor. Structure, ligand-binding mechanism and signal transduction. Horm Res 1994; 42:152–169.
7. Cheatham B, Kahn CR. Insulin action and the insulin signaling network. Endocr Rev 1996; 16:117–142.
8. Saltiel AR. Diverse signaling pathways in the cellular actions of insulin. Am J Physiol 1996; 270:E375–E385.
9. Ballard FJ, Walton PE, Dunshea FR, Francis GL, Tomas FM. Does IGF-I ever act through the insulin receptor? In Baxter RC, Gluckman PD, Rosenfeld RG, eds. The Insulin-Like Growth Factors and Their Regulatory Proteins. Elsevier, Amsterdam, 1994, pp. 131–138.
10. Wedekind F, Baer-Pontzen K, Bala-Mohan S, Choli D, Zahn H, Brandenburg D. Hormone binding site of the insulin receptor: analysis using photoaffinity-mediated avidin complexing. Biol Chem Hoppe Seyler 1989; 370:251–258.
11. Yip CC, Grunfeld C, Goldfine ID. Identification and characterization of the ligand-binding domain of insulin receptor by use of an anti-peptide antiserum against amino acid sequence 241–251 of the α subunit. Biochemistry 1991; 30:695–701.
12. Fabry M, Schaefer E, Ellis L, Kojro E, Fahrenholz F, Brandenburg D. Detection of a new hormone contact site within the insulin receptor ectodomain by the use of a novel photoreactive insulin. J Biol Chem 1992; 267:8950–8956.
13. Zhang B, Roth RA. Binding properties of chimeric insulin receptors containing the cysteine-rich domain of either the insulin-like growth factor I receptor or the insulin receptor related receptor. Biochemistry 1991; 30:5113–5117.
14. Schumacher R, Mosthaf L, Schlessinger J, Brandenburg D, Ullrich A. Insulin and insulin-like growth factor-1 binding specificity is determined by distinct regions of their cognate receptors. J Biol Chem 1991; 266:19288–19295.
15. Andersen AS, Kjeldsen T, Wiberg FC, Vissing H, Schaffer L, Rasmussen JS, De Meyts P, Moller NP. Identification of determinants that confer ligand specificity on the insulin receptor. J Biol Chem 1992; 267:13681–13686.

16. Schumacher R, Soos MA, Schlessinger J, Brandenburg D, Siddle K, Ullrich A. Signaling-competent receptor chimeras allow mapping of major insulin receptor binding domain determinants. J Biol Chem 1993; 268:1087–1094.
17. Kjeldsen T, Wiberg FC, Andersen AS. Chimeric receptors indicate that phenylalanine 39 is a major contributor to insulin specificity of the insulin receptor. J Biol Chem 1994; 269:32942–32946.
18. De Meyts P, Gu JL, Shymko RM, Kaplan BE, Bell GI, Whittaker J. Identification of a ligand-binding region of the human insulin receptor encoded by the second exon of the gene. Mol Endocrinol 1990; 4:409–416.
19. Williams PF, Mynarcik DC, Yu GQ, Whittaker J. Mapping of an NH_2-terminal ligand binding site of the insulin receptor by alanine scanning mutagenesis. J Biol Chem 1995; 270:3012–3016.
20. Mynarcik DC, Yu GQ, Whittaker J. Alanine-scanning mutagenesis of a C-terminal ligand binding domain of the insulin receptor α-subunit. J Biol Chem 1996; 271:2439–2442.
21. De Meyts P. The structural basis of insulin and insulin-like growth factor-I receptor binding and negative co-operativity, and its relevance to mitogenic versus metabolic signalling. Diabetologia 1994; 37 Suppl 2, S135–148.
22. Schäffer L. A model for insulin binding to the insulin receptor. Eur J Biochem 1994; 221:1127–1132.
23. Lammers R, Gray A, Schlessinger J, Ullrich A. Differential signalling potential of insulin- and IGF-1 receptor cytoplasmic domains. EMBO J 1989; 8:1369–1375.
24. Kalloo-Hosein HE, Whitehead JP, Soos M, Tavaré JM, Siddle K, O'Rahilly S. Differential signaling to glycogen synthesis by the intracellular domain of the insulin versus the insulin-like growth factor-1 receptor. Evidence from studies of TrkC-chimeras. J Biol Chem 1997; 272:24325–24332.
25. Tavaré JM, Siddle K. Mutational analysis of insulin receptor function: consensus and controversy. Biochim Biophys Acta 1993; 1178:21–39.
26. Tartare S, Mothe I, Kowalski-Chauvel A, Breittmayer JP, Ballotti R, Van Obberghen E. Signal transduction by a chimeric insulin-like growth factor-1 (IGF-1) receptor having the carboxyl-terminal domain of the insulin receptor. J Biol Chem 1994; 269:11449–11455.
27. Faria TN, Blakesley VA, Kato H, Stannard B, LeRoith D, Roberts CT Jr. Role of the carboxyl-terminal domains of the insulin and insulin-like growth factor I receptors in receptor function. J Biol Chem 1994; 269:13922–13928.
28. Esposito DL, Blakesley VA, Koval AP, Scrimgeour AG, LeRoith D. Tyrosine residues in the C-terminal domain of the insulin-like growth factor-I receptor mediate mitogenic and tumorigenic signals. Endocrinology 1997; 138:2979–2988.
29. Xu B, Bird VG, Miller WT. Substrate specificities of the insulin and insulin-like growth factor 1 receptor tyrosine kinase catalytic domains. J Biol Chem 1995; 270:29825–29830.
30. Laviola L, Giorgino F, Chow JC, Baquero JA, Hansen H, Ooi J, Zhu J, Riedel H, Smith RJ. The adapter protein Grb10 associates preferentially with the insulin receptor as compared with the IGF-I receptor in mouse fibroblasts. J Clin Invest 1997; 99:830–837.
31. O'Neill TJ, Zhu Y, Gustafson TA. Interaction of MAD2 with the carboxyl terminus of the insulin receptor but not with the IGFIR. Evidence for release from the insulin receptor after activation. J Biol Chem 1997; 272:10035–10040.
32. Craparo A, Freund R, Gustafson TA. 14-3-3 (ε) interacts with the insulin-like growth factor I receptor and insulin receptor substrate I in a phosphoserine-dependent manner. J Biol Chem 1997; 272:11663–11669.
33. Furlanetto RW, Dey BR, Lopaczynski W, Nissley SP. 14-3-3 proteins interact with the insulin-like growth factor receptor but not the insulin receptor. Biochem J 1997; 327:765–771.
34. Shier P, Watt VM. Tissue-specific expression of the rat insulin receptor-related receptor gene. Mol Endocrinol 1992; 6:723–729.
35. Zhang B, Roth RA. The insulin receptor-related receptor. Tissue expression, ligand binding specificity, and signaling capabilities. J Biol Chem 1992; 267:18320–18328.
36. Kurachi H, Jobo K, Ohta M, Kawasaki T, Itoh N. A new member of the insulin receptor family, insulin receptor-related receptor, is expressed preferentially in the kidney. Biochem Biophys Res Commun 1992; 187: 934–939.
37. Mathi SK, Chan J, Watt VM. Insulin receptor-related receptor messenger ribonucleic acid: quantitative distribution and localization to subpopulations of epithelial cells in stomach and kidney. Endocrinology 1995; 136:4125–4132.
38. Reinhardt RR, Chin E, Zhang B, Roth RA, Bondy CA. Insulin receptor-related receptor messenger ribonucleic acid is focally expressed in sympathetic and sensory neurons and renal distal tubule cells. Endocrinology 1993; 133:3–10.

39. Reinhardt RR, Chin E, Zhang B, Roth RA, Bondy CA. Selective coexpression of insulin receptor-related receptor (IRR) and TRK in NGF-sensitive neurons. J Neurosci 1994; 14:4674–4683.
40. Tsujimoto K, Tsuji N, Ozaki K, Ohta M, Itoh N. Insulin receptor-related receptor messenger ribonucleic acid in the stomach is focally expressed in the enterochromaffin-like cells. Endocrinology 1995; 136:558–561.
41. Tsujimoto K, Tsuji N, Ozaki K, Minami M, Satoh M, Itoh N. Expression of insulin receptor-related receptor mRNA in the rat brain is highly restricted to forebrain cholinergic neurons. Neurosci Lett 1995; 188:105–108.
42. Tsuji N, Tsujimoto K, Takada N, Ozaki K, Ohta M, Itoh N. Expression of insulin receptor-related receptor in the rat brain examined by in situ hybridization and immunohistochemistry. Brain Res Mol Brain Res 1996; 41:250–258.
43. Kovacina KS, Roth RA. Characterization of the endogenous insulin receptor-related receptor in neuroblastomas. J Biol Chem 1995; 270:1881–1887.
44. Jui HY, Suzuki Y, Accili D, Taylor SI. Expression of a cDNA encoding the human insulin receptor-related receptor. J Biol Chem 1994; 269:22446–22452.
45. Seino S, Bell GI. Alternative splicing of human insulin receptor messenger RNA. Biochem Biophys Res Commun 1989; 159:312–316.
46. Goldstein BJ, Dudley AL. Heterogeneity of messenger RNA that encodes the rat insulin receptor is limited to the domain of exon 11. Analysis by RNA heteroduplex mapping, amplification of cDNA, and in vitro translation. Diabetes 1992; 41:1293–1300.
47. Kenner KA, Kusari J, Heidenreich KA. cDNA sequence analysis of the human brain insulin receptor. Biochem Biophys Res Commun 1995; 217:304–312.
48. Moller DE, Yokota A, Caro JF, Flier JS. Tissue-specific expression of two alternatively spliced insulin receptor mRNAs in man. Mol Endocrinol 1989; 3:1263–1269.
49. Goldstein BJ, Dudley AI. The rat insulin receptor: primary structure and conservation of tissue-specific alternative messenger RNA splicing. Mol Endocrinol 1990; 4:235–244.
50. Benecke H, Flier JS, Moller DE. Alternatively spliced variants of the insulin receptor protein. Expression in normal and diabetic human issues. J Clin Invest 1992; 89:2066–2070.
51. Mosthaf L, Grako K, Dull TJ, Coussens L, Ullrich A, McClain DA. Functionally distinct insulin receptors generated by tissue-specific alternative splicing. EMBO J 1990; 9:2409–2413.
52. Yamaguchi Y, Flier JS, Yokota A, Benecke H, Backer JM, Moller DE. Functional properties of two naturally occurring isoforms of the human insulin receptor in Chinese hamster ovary cells. Endocrinology 1991; 129:2058–2066.
53. Vogt B, Carrascosa JM, Ermel B, Ullrich A, Häring HU. The two isotypes of the human insulin receptor (HIR-A and HIR-B) follow different internalization kinetics. Biochem Biophys Res Commun 1991; 177:1013–1018.
54. McClain DA. Different ligand affinities of the two human insulin receptor splice variants are reflected in parallel changes in sensitivity for insulin action. Mol Endocrinol 1991; 5:734–739.
55. Kellerer M, Lammers R, Ermel B, Tippmer S, Vogt B, Obermaier-Kusser B, Ullrich A, Häring HU. Distinct α-subunit structures of human insulin receptor A and B variants determine differences in tyrosine kinase activities. Biochemistry 1992; 31:4588–4596.
56. Soos MA, Whittaker J, Lammers R, Ullrich A, Siddle K. Receptors for insulin and insulin-like growth factor-I can form hybrid dimers. Characterisation of hybrid receptors in transfected cells. Biochem J 1990; 270:383–390.
57. Yamaguchi Y, Flier JS, Benecke H, Ransil BJ, Moller DE. Ligand-binding properties of the two isoforms of the human insulin receptor. Endocrinology 1993; 132:1132–1138.
58. Morrione A, Valentinis B, Xu SQ, Yumet G, Louvi A, Efstratiadis A, Baserga R. Insulin-like growth factor II stimulates cell proliferation through the insulin receptor. Proc Natl Acad Sci USA 1997; 94:3777–3782.
59. Louvi A, Accili D, Efstratiadis A. Growth-promoting interaction of IGF-II with the insulin receptor during mouse embryonic development. Dev Biol 1997; 189:33–48.
60. Abbott AM, Bueno R, Pedrini MT, Murray JM, Smith RJ. Insulin-like growth factor I receptor gene structure. J Biol Chem 1992; 267:10759–10763.
61. Pedrini MT, Giorgino F, Smith RJ. cDNA cloning of the rat IGF I receptors: structural analysis of rat and human IGF I and insulin receptors reveals differences in alternative splicing and receptor-domain conservation. Biochem Biophys Res Commun 1994; 202:1038–1046.
62. Yee D, Lebovic GS, Marcus RR, Rosen N. Identification of an alternate type I insulin-like growth factor receptor β subunit mRNA transcript. J Biol Chem 1989; 264:21439–21441.

63. Condorelli G, Bueno R, Smith RJ. Two alternatively spliced forms of the human insulin-like growth factor I receptor have distinct biological activities and internalization kinetics. J Biol Chem 1994; 269:8510–8516.
64. Itoh N, Jobo K, Tsujimoto K, Ohta M, Kawasaki T. Two truncated forms of rat insulin receptor-related receptor. J Biol Chem 1993; 268:17983–17986.
65. Hintz RL, Thorsson AV, Enberg G, Hall K. IGF-II binding on human lymphoid cells: demonstration of a common high affinity receptor for insulin like peptides. Biochem Biophys Res Commun 1984; 118:774–782.
66. Misra P, Hintz RL, Rosenfeld RG. Structural and immunological characterization of insulin-like growth factor II binding to IM-9 cells. J Clin Endocrinol Metab 1986; 63:1400–1405.
67. Jonas HA, Newman JD, Harrison LC. An atypical insulin receptor with high affinity for insulin-like growth factors copurified with placental insulin receptors. Proc Natl Acad Sci USA 1986; 83:4124–4128.
68. Tollefsen SE, Thompson K, Petersen DJ. Separation of the high affinity insulin-like growth factor I receptor from low affinity binding sites by affinity chromatography. J Biol Chem 1987; 262:16461–16469.
69. Jonas HA, Cox AJ, Harrison LC. Delineation of atypical insulin receptors from classical insulin and type I insulin-like growth factor receptors in human placenta. Biochem J 1989; 257:101–107.
70. Jonas HA, Cox AJ. Insulin-like growth factor binding to the atypical insulin receptors of a human lymphoid-derived cell line (IM-9). Biochem J 1990; 266:737–742.
71. Jonas HA, Eckardt GS, Clark S. Expression of atypical and classical insulin receptors in Chinese hamster ovary cells transfected with cloned cDNA for the human insulin receptor. Endocrinology 1990; 127:1301–1309.
72. Burant CF, Treutelaar MK, Allen KD, Sens DA, Buse MG. Comparison of insulin and insulin-like growth factor I receptors from rat skeletal muscle and L-6 myocytes. Biochem Biophys Res Commun 1987; 147:100–107.
73. Waldbillig RJ, Chader GJ. Anomalous insulin-binding activity in the bovine neural retina: a possible mechanism for regulation of receptor binding specificity. Biochem Biophys Res Commun 1988; 151:1105–1112.
74. Milazzo G, Yip CC, Maddux BA, Vigneri R, Goldfine ID. High-affinity insulin binding to an atypical insulin-like growth factor-I receptor in human breast cancer cells. J Clin Invest 1992; 89:899–908.
75. Germain-Lee EL, Janicot M, Lammers R, Ullrich A, Casella SJ. Expression of a type I insulin-like growth factor receptor with low affinity for insulin-like growth factor II. Biochem J 1992; 281:413–417.
76. Soos MA, Field CE, Navé BT, Siddle K. Hybrid and atypical insulin-like growth factor receptors. In: Baxter RC, Gluckman PD, Rosenfeld RG, eds. The Insulin-Like Growth Factors and Their Regulatory Proteins. Elsevier, Amsterdam, 1994, pp. 95–106.
77. Soos MA, Siddle K. Atypical ligand binding to IGF-I receptors reflects a high affinity state of solubilised receptors, in preparation.
78. Soos MA, Field CE, Lammers R, Ullrich A, Zhang B, Roth RA, Andersen AS, Kjeldsen T, Siddle K. A panel of monoclonal antibodies for the type I insulin-like growth factor receptor. Epitope mapping, effects on ligand binding, and biological activity. J Biol Chem 1992; 267:12955–12963.
79. Poretsky L, Kalin MF. The gonadotropic function of insulin. Endocr Rev 1987; 8:132–141.
80. Fradkin JE, Eastman RC, Lesniak MA, Roth J. Specificity spillover at the hormone receptor—exploring its role in human disease. N Engl J Med 1989; 320:640–645.
81. Schlessinger J. Signal transduction by allosteric receptor oligomerization. Trends Biochem Sci 1988; 13:443–447.
82. Heldin CH. Dimerization of cell surface receptors in signal transduction. Cell 1995; 80:213–223.
83. Soos MA, Siddle K. Immunological relationships between receptors for insulin and insulin-like growth factor I. Evidence for structural heterogeneity of insulin-like growth factor I receptors involving hybrids with insulin receptors. Biochem J 1989; 263:553–563.
84. Jonas HA, Harrison LC. The human placenta contains two distinct binding and immunoreactive species of insulin-like growth factor-I receptors. J Biol Chem 1985; 260:2288–2294.
85. Morgan DO, Roth RA. Identification of a monoclonal antibody which can distinguish between two distinct species of the type I receptor for insulin-like growth factor. Biochem Biophys Res Commun 1986; 138:1341–1347.

86. Moxham CP, Duronio V, Jacobs S. Insulin-like growth factor I receptor β-subunit heterogeneity. Evidence for hybrid tetramers composed of insulin-like growth factor I and insulin receptor heterodimers. J Biol Chem 1989; 264:13238–13244.
87. Garofalo RS, Rosen OM. Insulin and insulinlike growth factor-1 (IGF-1) receptors during central nervous system development: expression of two immunologically distinct IGF-1 receptor β subunits. Mol Cell Biol 1989; 9:2806–2817.
88. Jacobs S, Kull FC, Earp HS, Svoboda ME, Van Wyk JJ, Cuatrecasas P. Somatomedin-C stimulates the phosphorylation of the β-subunit of its own receptor. J Biol Chem 1983; 258:9581–9584.
89. Yu KT, Peters MA, Czech MP. Similar control mechanisms regulate the insulin and type I insulin-like growth factor receptor kinases. Affinity-purified insulin-like growth factor I receptor kinase is activated by tyrosine phosphorylation of its β subunit. J Biol Chem 1986; 261:11341–11349.
90. Morgan DO, Jarnagin K, Roth RA. Purification and characterization of the receptor for insulin-like growth factor I. Biochemistry 1986; 25:5560–5564.
91. Kadowaki T, Koyasu S, Nishida E, Tobe K, Izumi T, Takaku F, Sakai H, Yahara I, Kasuga M. Tyrosine phosphorylation of common and specific sets of cellular proteins rapidly induced by insulin, insulin-like growth factor I, and epidermal growth factor in an intact cell. J Biol Chem 1987; 262:7342–7350.
92. Shemer J, Adamo M, Wilson GL, Heffez D, Zick Y, LeRoith D. Insulin and insulin-like growth factor-I stimulate a common endogenous phosphoprotein substrate (pp185) in intact neuroblastoma cells. J Biol Chem 1987; 262:15476–15482.
93. Treadway JL, Morrison BD, Goldfine ID, Pessin JE. Assembly of insulin/insulin-like growth factor-1 hybrid receptors in vitro. J Biol Chem 1989; 264:21450–21453.
94. Jui HY, Accili D, Taylor SI. Characterization of a hybrid receptor formed by dimerization of the insulin receptor-related (IRR) with the insulin receptor (IR): coexpression of cDNAs encoding human IRR and human IR in NIH-3T3 cells. Biochemistry 1996; 35:14326–14330.
95. Soos MA, Field CE, Siddle K. Purified hybrid insulin/insulin-like growth factor-I receptors bind insulin-like growth factor-I, but not insulin, with high affinity. Biochem J 1993; 290:419–426.
96. Kasuya J, Paz IB, Maddux BA, Goldfine ID, Hefta SA, Fujita-Yamaguchi Y. Characterization of human placental insulin-like growth factor-I/insulin hybrid receptors by protein microsequencing and purification. Biochemistry 1993; 32:13531–13536.
97. Frattali AL, Pessin JE. Relationship between α-subunit ligand occupancy and β-subunit autophosphorylation in insulin/insulin-like growth factor-1 hybrid receptors. J Biol Chem 1993; 268:7393–7400.
98. Hammond BJ, Tikerpae J, Smith GD. An evaluation of the cross-linking model for the interaction of insulin with its receptor. Am J Physiol 1997; 272:E1136–E1144.
99. Böni-Schnetzler M, Rubin JB, Pilch PF. Structural requirements for the transmembrane activation of the insulin receptor kinase. J Biol Chem 1986; 261:15281–15287.
100. Böni-Schnetzler M, Scott W, Waugh SM, DiBella E, Pilch PF. The insulin receptor. Structural basis for high affinity ligand binding. J Biol Chem 1987; 262:8395–8401.
101. Sweet LJ, Morrison BD, Pessin JE. Isolation of functional αβ heterodimers from the purified human placental alpha $\alpha_2\beta_2$ heterotetrameric insulin receptor complex. A structural basis for insulin binding heterogeneity. J Biol Chem 1987; 262:6939–6942.
102. Tollefsen SE, Thompson K. The structural basis for insulin-like growth factor I receptor high affinity binding. J Biol Chem 1988; 263:16267–16273.
103. Jonas HA, Harrison LC. Disulphide reduction alters the immunoreactivity and increases the affinity of insulin-like growth-factor-I receptors in human placenta. Biochem J 1986; 236:417–423.
104. Feltz SM, Swanson ML, Wemmie JA, Pessin JE. Functional properties of an isolated αβ heterodimeric human placenta insulin-like growth factor 1 receptor complex. Biochemistry 1988; 27:3234–3242.
105. Zhong P, Cara JF, Tager HS. Importance of receptor occupancy, concentration differences, and ligand exchange in the insulin-like growth factor I receptor system. Proc Natl Acad Sci USA 1993; 90:11451–11455.
106. Christoffersen CT, Bornfeldt KE, Rotella CM, Gonzales N, Vissing H, Shymko RM, ten Hoeve J, Groffen J, Heisterkamp N, De Meyts P. Negative cooperativity in the insulin-like growth factor-I receptor and a chimeric IGF-I/insulin receptor. Endocrinology 1994; 135:472–475.
107. Seely BL, Reichart DR, Takata Y, Yip C, Olefsky JM. A functional assessment of insulin/insulin-like growth factor-I hybrid receptors. Endocrinology 1995; 136:1635–1641.
108. Baker J, Liu JP, Robertson EJ, Efstratiadis A. Role of insulin-like growth factors in embryonic and postnatal growth. Cell 1993; 75:73–82.

109. Joshi RL, Lamothe B, Cordonnier N, Mesbah K, Monthioux E, Jami J, Bucchini D. Targeted disruption of the insulin receptor gene in the mouse results in neonatal lethality. EMBO J 1996; 15:1542–1547.
110. Accili D, Drago J, Lee EJ, Johnson MD, Cool MH, Salvatore P, Asico D, José PA, Taylor SI, Westphal H. Early neonatal death in mice homozygous for a null allele of the insulin receptor gene. Nature Genet 1996; 12:106–109.
111. Krook A, Brueton L, O'Rahilly S. Homozygous nonsense mutation in the insulin receptor gene in an infant with leprechaunism. Lancet 1993; 342:277–278.
112. Wertheimer E, Lu SP, Backeljauw PF, Davenport ML, Taylor SI. Homozygous deletion of the human insulin receptor gene results in leprechaunism. Nature Genet 1993; 5:71–73.
113. Wada T, Qian XL, Greene MI. Intermolecular association of the p185neu protein and EGF receptor modulates EGF receptor function. Cell 1990; 61:1339–1347.
114. Qian XL, Decker SJ, Greene MI. p185c-neu and epidermal growth factor receptor associate into a structure composed of activated kinases. Proc Natl Acad Sci USA 1992; 89:1330–1334.
115. Sliwkowski MX, Schaefer G, Akita RW, Lofgren JA, Fitzpatrick VD, Nuijens A, Fendly BM, Cerione RA, Vandlen RL, Carraway KL. Coexpression of erbB2 and erbB3 proteins reconstitutes a high affinity receptor for heregulin. J Biol Chem 1994; 269:14661–14665.
116. Soltoff SP, Carraway KL, Prigent SA, Gullick WG, Cantley LC. ErbB3 is involved in activation of phosphatidylinositol 3-kinase by epidermal growth factor. Mol Cell Biol 1994; 14:3550–3558.
117. Alimandi M, Romano A, Curia MC, Muraro R, Fedi P, Aaronson SA, Di Fiore PP, Kraus MH. Cooperative signaling of ErbB3 and ErbB2 in neoplastic transformation and human mammary carcinomas. Oncogene 1995; 10:1813–1821.
118. Wallasch C, Weiss FU, Niederfellner G, Jallal B, Issing W, Ullrich A. Heregulin-dependent regulation of HER2/neu oncogenic signaling by heterodimerization with HER3. EMBO J 1995; 14:4267–4275.
119. Riese DJ, van Raaij TM, Plowman GD, Andrews GC, Stern DF. The cellular response to neuregulins is governed by complex interactions of the erbB receptor family. Mol Cell Biol 1995; 15:5770–5776.
120. Eriksson A, Siegbahn A, Westermark B, Heldin CH, Claesson-Welsh I. PDGF α- and β-receptors activate unique and common signal transduction pathways. EMBO J 1992; 11:543–550.
121. Heidaran MA, Pierce JH, Yu JC, Lombardi D, Artrip JE, Fleming TP, Thomason A, Aaronson SA. Role of αβ receptor heterodimer formation in β platelet-derived growth factor (PDGF) receptor activation by PDGF-AB. J Biol Chem 1991; 266:20232–20237.
122. Bailyes EM, Nave BT, Soos MA, Orr SR, Hayward AC, Siddle K. Insulin receptor/IGF-I receptor hybrids are widely distributed in mammalian tissues: quantification of individual receptor species by selective immunoprecipitation and immunoblotting. Biochem J 1997; 327:209–215.
123. Federici M, Porzio O, Zucaro L, Fusco A, Borboni P, Lauro D, Sesti G. Distribution of insulin/insulin-like growth factor-I hybrid receptors in human tissues. Mol Cell Endocrinol 1997; 129:121–126.
124. Weiland M, Bahr F, Hohne M, Schurmann A, Zeihm D, Joost HG. The signaling potential of the receptors for insulin and insulin-like growth factor I (IGF-I) in 3T3-L1 adipocytes: comparison of glucose transport activity, induction of oncogene c-fos, glucose transporter mRNA, and DNA-synthesis. J Cell Physiol 1991; 149:428–435.
125. Adamo M, Roberts CT Jr, LeRoith D. How distinct are the insulin and insulin-like growth factor I signalling systems? Biofactors 1992; 3:151–157.
126. Olson TS, Bamberger MJ, Lane MD. Post-translational changes in tertiary and quaternary structure of the insulin proreceptor. Correlation with acquisition of function. J Biol Chem 1988; 263:7342–7351.
127. Waugh SM, DiBella EE, Pilch PF. Isolation of a proteolytically derived domain of the insulin receptor containing the major site of cross-linking/binding. Biochemistry 1989; 28:3448–3455.
128. Xu QY, Paxton RJ, Fujita-Yamaguchi Y. Substructural analysis of the insulin receptor by microsequence analyses of limited tryptic fragments isolated by sodium dodecyl sulfate-polyacrylamide gel electrophoresis in the absence or presence of dithiothreitol. J Biol Chem 1990; 265:18673–18681.
129. Finn FM, Ridge KD, Hofmann K. Labile disulfide bonds in human placental insulin receptor. Proc Natl Acad Sci USA 1990; 87:419–423.
130. Chiacchia KB. Quantitation of the class I disulfides of the insulin receptor. Biochem Biophys Res Commun 1991; 176:1178–1182.
131. Sparrow LG, McKern NM, Gorman JJ, Strike PM, Robinson CP, Bentley JD, Ward CW. The disulphide bonds in the C-terminal domains of the human insulin receptor ectodomain. J Biol Chem 1997; 272:29460–29467.
132. Schäffer L, Ljungqvist L. Identification of a disulfide bridge connecting the β-subunits of the extracellular domain of the insulin receptor. Biochem Biophys Res Commun 1992; 189:650–653.

133. Macaulay SL, Polites M, Hewish DR, Ward CW. Cysteine-524 is not the only residue involved in the formation of disulphide-bonded dimers of the insulin receptor. Biochem J 1994; 303:575–581.
134. Valensise H, Liu YY, Federici M, Lauro D, Dell'anna D, Romanini C, Sesti G. Increased expression of low-affinity insulin receptor isoform and insulin/insulin-like growth factor-I hybrid receptors in term placenta from insulin-resistant women with gestational hypertension. Diabetologia 1996; 39:952–960.
135. Federici M, Zucaro L, Porzio O, Massoud R, Borboni P, Lauro D, Sesti G. Increased expression of insulin/insulin-like growth factor-I hybrid receptors in skeletal muscle of noninsulin-dependent diabetes mellitus subjects. J Clin Invest 1996; 98:2887–2893.
136. Pandini G, Vigneri R, Costantino A, Ippolito A, Fujita-Yamaguchi Y, Siddle K, Goldfine ID, Belfiore A. Insulin/IGF-I hybrid receptors play a major role in IGF-I signalling in breast cancer. Clin Cancer Res, in press.
137. Garrouste FL, Remacle Bonnet MM, Lehmann MM, Marvaldi JL, Pommier GJ. Up-regulation of insulin/insulin-like growth factor-I hybrid receptors during differentiation of HT29-D4 human colonic carcinoma cells. Endocrinology 1997; 138:2021–2032.
138. Taylor SI, Cama A, Accili D, Barbetti F, Quon MJ, de la Luz Sierra M, Suzuki Y, Koller E, Levy Toledano R, Wertheimer E, Moncada VY, Kadowaki H, Kadowaki T. Mutations in the insulin receptor gene. Endocr Rev 1992; 13:566–595.
139. Accili D. Molecular defects of the insulin receptor gene. Diabetes Metab Rev 1995; 11:47–62.
140. Treadway JL, Morrison BD, Soos MA, Siddle K, Olefsky J, Ullrich A, McClain DA, Pessin JE. Transdominant inhibition of tyrosine kinase activity in mutant insulin/insulin-like growth factor I hybrid receptors. Proc Natl Acad Sci USA 1991; 88:214–218.
141. Ota A, Wilson G-L, LeRoith D. Insulin-like growth factor I receptors on mouse neuroblastoma cells. Two β subunits are derived from differences in glycosylation. Eur J Biochem 1988; 174:521–530.
142. Alexandrides TK, Smith RJ. A novel fetal insulin-like growth factor (IGF) I receptor. Mechanism for increased IGF I- and insulin-stimulated tyrosine kinase activity in fetal muscle. J Biol Chem 1989; 264:12922–12930.
143. Kellerer M, Obermaier-Kusser B, Ermel B, Wallner U, Häring HU, Petrides PE. An altered IGF-I receptor is present in human leukemic cells. J Biol Chem 1990; 265:9340–9345.
144. Hainaut P, Kowalski A, Giorgetti S, Baron V, Van Obberghen E. Insulin and insulin-like growth-factor-I (IGF-I) receptors in *Xenopus laevis* oocytes. Comparison with insulin receptors from liver and muscle. Biochem J 1991; 273:673–678.
145. Domeyne A, Pinset C, Montarras D, Garandel V, Rosenfeld RG, Barenton B. Preferential binding of insulin-like growth factor-II (IGF-II) to a putative $\alpha_2\beta_2$ IGF-II receptor type in C2 myoblasts. Eur J Biochem 1992; 208:273–279.
146. Garofalo RS, Barenton B. Functional and immunological distinction between insulin-like growth factor I receptor subtypes in KB cells. J Biol Chem 1992; 267:11470–11475.
147. Barenton B, Domeyne A, Garandel V, Garofalo RS. A developmentally regulated form of insulin-like growth factor receptor α-subunit in C2 myoblasts exhibiting altered requirements for differentiation. Endocrinology 1993; 133:651–660.
148. Alexandrides TK, Chen JH, Bueno R, Giorgino F, Smith RJ. Evidence for two insulin-like growth factor I receptors with distinct primary structure that are differentially expressed during development. Regul Pept 1993; 48:279–290.
149. Moss AM, Livingston JN. Distinct β-subunits are present in hybrid insulin-like-growth-factor-1 receptors in the central nervous system. Biochem J 1993; 294:685–692.
150. Beguinot F, Smith RJ, Kahn CR, Maron R, Moses AC, White MF. Phosphorylation of insulin-like growth factor I receptor by insulin receptor tyrosine kinase in intact cultured skeletal muscle cells. Biochemistry 1988; 27:3222–3228.
151. Heidenreich KA, Brandenburg D. Oligosaccharide heterogeneity of insulin receptors. Comparison of N-linked glycosylation of insulin receptors in adipocytes and brain. Endocrinology 1986; 118:1835.
152. Whittaker J, Okamoto AK, Thys R, Bell GI, Steiner DF, Hofmann CA. High-level expression of human insulin receptor cDNA in mouse NIH 3T3 cells. Proc Natl Acad Sci USA 1987; 84:5237–5241.
153. Ganderton RH, Stanley KK, Field CE, Coghlan MP, Soos MA, Siddle K. A monoclonal anti-peptide antibody reacting with the insulin receptor β-subunit. Characterization of the antibody and its epitope and use in immunoaffinity purification of intact receptors. Biochem J 1992; 288:195–205.
154. Bajaj M, Waterfield MD, Schlessinger J, Taylor WR, Blundell T. On the tertiary structure of the extracellular domains of the epidermal growth factor and insulin receptors. Biochim Biophys Acta 1987; 916:220–226.

10 IGF Binding Proteins as Modulators of IGF Action

Janet L. Martin, PhD
and Robert C. Baxter, PhD, DSc

Contents

INTRODUCTION

The insulin-like growth factors (IGFs) are able to modulate an extraordinarily wide variety of biological actions. First described as stimulators of proteoglycan biosynthesis in cartilage *(1,2)*, IGF peptides (then known as somatomedins or nonsuppressible insulin-like activity) were soon recognized as having insulin-like stimulatory effects on the cellular uptake of glucose and other substrates *(3)*, and were shown to increase protein, RNA, and DNA synthesis *(1,4–6)* in a variety of tissues. Subsequently IGFs have been found to act as positive regulators of an extensive range of differentiated functions in numerous tissues, often acting in synergy with other trophic or stimulatory factors, both hormones (e.g., adrenocorticotropic hormone [ACTH], follicle-stimulating hormone [FSH]) and growth factors (e.g.,platelet-derived growth factor, fibroblast growth factor) *(7)*. Not all IGF actions, however, are stimulatory, and their ability to block lipolysis in adipocytes *(8)* and to inhibit adenylate cyclase activity *(9,10)* was recognized more than two decades ago. Among other important inhibitory effects are the attenuation of pituitary growth hormone (GH) secretion *(11)* and the prevention of apoptosis *(12)*, a significant component of the "survival factor" role of the IGFs.

It was soon realized that IGF binding proteins (IGFBPs), originally identified as serum transport proteins for the IGFs *(13)*, act as modulators of IGF activity *(14,15)*. Six such proteins have been studied extensively, and have been the subject of many recent reviews *(7,16,17)*. These proteins were classified by international consensus as a single family—the IGFBPs—based on their extensive sequence homology, which includes well-

From: *Contemporary Endocrinology: The IGF System*
Edited by: R. Rosenfeld and C. Roberts Jr © Humana Press, Inc., Totowa, NJ

conserved amino (N)- and carboxy (C)-terminal domains with extensive conservation of cysteine residues in each of these domains *(18)*. The core proteins range in size from 22.8 kDa for human (h) IGFBP-6 to 31.3 kDa for hIGFBP-2, but N- and O-linked glycosylation can increase their size considerably—in the case of IGFBP-3, from 28.7 kDa to approx 45 kDa for the fully glycosylated protein. The structural similarities among these proteins reflect their highly conserved genes; IGFBPs -1 to -6 are each encoded by four exons in the IGFBP genes *(17)*. As discussed elsewhere in this volume, several other proteins with distinct sequence homology to regions of the IGFBPs have recently been identified *(18)*. However, these proteins are encoded by genes that do not appear to share more than a single exon of the conserved IGFBP gene structure, and although weak IGF-binding activity can be demonstrated in some cases, there has been little study to date of their possible physiological roles as modulators of IGF activity.

BINDING PROPERTIES

Affinity and Specificity for IGFs

IGFBPs -1 to -6 are characterized by high affinity for both IGF-I and IGF-II. This can be demonstrated by equilibrium binding in solution or by methods involving affinity labeling (i.e., covalent crosslinking) or immobilizing one of the ligands, as in the BIAcore™ biosensor (Pharmacia, Uppsala, Sweden). Such different techniques may not give entirely consistent results; for example, the affinities of human IGFBP-3 for IGF-I and IGF-II were originally determined by equilibrium binding to be 2×10^{10} and 3×10^{10} L/mol, respectively, for serum-derived IGFBP-3 *(19)*, and by BIAcore to be 0.4×10^{10} and 0.8×10^{10} L/mol, respectively, for recombinant human IGFBP-3 *(20)*, probably a reflection of the different methodologies used rather than the different source of the proteins. Table 1 summarizes some reported affinities for the binding of IGF-I and IGF-II to IGFBPs -1 to -6. With the exception of IGFBP-1, which shows a slight preferential affinity for IGF-I over IGF-II, each of the IGFBPs binds IGF-II with higher affinity than IGF-I. The difference in affinity ranges from 2-fold or less, in the case of IGFBP-3, to up to 100-fold in the case of IGFBP-6, which binds IGF-II with an K_a of approx 3×10^{11} L/mol *(21)*.

The effects of pH and ionic strength have not been studied in detail for most of the IGFBPs. In the case of hIGFBP-3 binding to IGF-I or IGF-II, optimal binding is seen around pH 5, with the apparent affinity decreasing markedly as the pH increases to near-physiological values *(22)*. High ionic strength appears to favor the interaction between IGFs and IGFBP-3 *(22)*; indeed, IGFBP-3 was first purified by affinity chromatography on a column of immobilized IGF-II in the presence of 2 *M* sodium acetate *(19)*. Like IGFBP-3, hIGFBP-4 appears to have a much higher affinity for both IGF-I and IGF-II when measured at pH 6.0 than at pH 7.4. There are also marked effects of pH on the specificity of IGFBP-4 for some IGF analogs; for example, the naturally truncated IGF-I derivative des(1–3) IGF-I binds with only 3% of the affinity of IGF-I at pH 6.0, but 25% of the relative affinity at pH 7.4 *(23)*. Human IGFBP-2 has an unusually broad pH optimum, extending from approximately pH 4 to above 9 *(24)*. Interestingly, a naturally occurring C-terminal fragment of IGFBP-2 retains IGF-II binding at neutral pH, but has lost the pH 4 binding seen for the intact protein *(24)*.

The use of IGF-I and IGF-II mutants has allowed the IGF domains involved in IGFBP interactions to be determined. N-terminal residues (in the IGF B-domain) are of key importance for both IGF-I and IGF-II. Des(1–3)IGF-I, mentioned previously, is

Table 1
Association Constants for the Interaction Between IGF Binding Proteins and IGF-I and -II

			Association Constant (K_a), L/mol		
Binding Protein	*Source/Species*	*Modification*	*IGI-I*	*IGF-II*	*Ref.*
IGFBP-1	h Amniotic fluid	Phosphorylated	6.6×10^9	3.2×10^9	*(204)*
	h Plasma	Phosphorylated	2.3×10^{10}	3.6×10^9	*(45)*
		Dephosphorylated	2.5×10^9	1.8×10^9	
IGFBP-2	r Liver cells	Intact	1.5×10^9		*(205)*
		Proteolyzed	1.0×10^8		
	h Recombinant			5.8×10^{10}	*(24)*
	h Milk	Proteolyzed		4.5×10^9	*(24)*
	h CSF		0.5×10^{10}	1.7×10^{10}	*(28)*
IGFBP-3	h Plasma	Glycosylated	2×10^{10}	3×10^{10}	*(19)*
	h Plasma	Proteolyzed	2×10^9	1.9×10^{10}	*(58)*
IGFBP-4	h Glioblastoma		0.6×10^{10}	1.1×10^{10}	*(28)*
IGFBP-5	h Glioblastoma	Soluble	1.4×10^{10}		*(206)*
		Matrix-bound	2.1×10^9		*(206)*
	h Rec (CHO cell)		2.2×10^9	1.6×10^{10}	*(28)*
IGFBP-6	h Fibroblasts	*O*-Glycosylated	2.6×10^9	2.9×10^{11}	*(21)*
	h CSF	*O*-Glycosylated	3×10^9	3×10^{10}	*(207)*

h, human; r, rat; b, bovine; rec, recombinant.

well recognized as having a somewhat reduced affinity for IGFBP-3 and a markedly reduced affinity for the other IGFBPs—although this is clearly dependent on the assay conditions used *(25)*. Glu^3 of IGF-I appears to be particularly important, and a substitution of this residue to Arg is sufficient to greatly decrease IGF binding to IGFBPs. Other notable IGF-I derivatives with low binding affinity for IGFBPs include [$Gln^3Ala^4Tyr^{15}Leu^{16}$]IGF-I, known as QAYL IGF-I, and N-terminally extended forms such as "long" [Arg^3]IGF-I *(26,27)*. An IGF-I derivative of exceptionally low binding affinity is the "B-chain" mutant which contains the first 16 residues of insulin in place of the first 16 IGF-I residues *(23,26)*. The rationale for this substitution is the well-accepted view that insulin has unmeasurably low affinity for the IGFBPs. Interestingly, a recent BIAcore study of the binding of insulin to human IGFBP-3 concluded that some interaction exists, with a K_a approx 1000-fold lower than that for IGF-I *(20)*.

N-terminal residues of IGF-II are similarly involved in IGFBP binding, with Glu^6 of IGF-II apparently as important as Glu^3 of IGF-I. Accordingly, both [Arg^6]IGF-II and des(1–6)IGF-II have been shown to bind IGFBPs with greatly reduced affinity *(27)*. This region of IGF-II appears to interact with C-terminal IGFBP residues, since a carboxy-terminal fragment of IGFBP-2 binds IGF-II with moderate affinity but shows no measurable binding of des(1–6)IGF-II *(24)*. For both IGF-I and IGF-II, A-domain residues also have an involvement in IGFBP interactions, with mutations of A-domain residues 49–51 of IGF-I, or 48–50 of IGF-II, showing a modest decrease in binding to IGFBP-3 and a much greater loss of binding to other IGFBPs *(23,26,28)*.

Affinity for ALS

The existence of an acid-labile component in the high molecular weight circulating complexes between IGFBPs and IGFs was first postulated in 1980 *(29,30)*, and confirmed several years later when the 85-kDa glycoprotein known as the acid-labile subunit (ALS) was purified from human serum *(31)*. ALS appears unable to bind IGF-I or IGF-II alone, but forms moderately high-affinity IGF-containing complexes in the presence of IGFBP-3 *(31)*. More recently, IGFBP-5 has also been shown to form ternary complexes with IGFs and ALS with similar high affinity *(32)*. Conversely, whether studied by gel chromatography *(33)*, immunoprecipitation with either IGFBP-3 or ALS antibodies *(17,31)*, or sodium dodecyl sulfate-polyacrylamide gel electrophoresis (SDS-PAGE) after affinity labeling *(34)*, human plasma IGFBP-3 consistently shows unmeasurably low binding to human ALS unless either IGF-I or IGF-II is present, leading to the concept that binary IGF–IGFBP-3 complexes, rather than free IGFBP-3, form the binding site for ALS. This view has been challenged on the basis of studies using rat ALS and a partially proteolyzed form of IGFBP-3 with reduced affinity for IGF-I *(35)*, and another study using nonglycosylated *(E. coli)* human IGFBP-3 *(36)*. Although these studies indicate that a weak interaction between IGFBP-3 and ALS can be visualized under some experimental conditions, the physiological significance of such complexes, if they exist in vivo, remains to be demonstrated.

The association constant for ALS binding to IGFBP-3 depends on the type of IGF bound to the IGFBP-3, and is also very dependent upon the pH and ionic strength. Although IGF-II has a slightly higher affinity than IGF-I for IGFBP-3, complexes containing IGF-I bind ALS with higher affinity than those containing IGF-II *(31)*. While affinity constants of approx 10^9 L/mol were originally determined for ternary complex formation by ALS under conditions of low ionic strength, slightly acidic pH, and room temperature *(31)*, measurement under near-physiological conditions of salt, pH, and temperature indicates considerably lower affinities for ALS binding to IGFBP-3 complexes: 6×10^7 L/mol in the presence of IGF-II, and 2.5×10^8 L/mol in the presence of IGF-I *(22)*. Because IGFs are thought to be confined to the vascular space when complexed with IGFBP-3 and ALS, dissociation of ALS from these complexes may be an important step in the passage of circulating IGFs to the tissues.

Functional Domains of the IGFBPs Responsible for Binding Properties

In the absence of X-ray crystallographic or nuclear magnetic resonance structures for any IGFBP, inferences about important structural domains of the proteins have generally been based on natural variants (e.g., partially proteolyzed forms) and mutagenesis and expression studies. Huhtala et al. *(37)* described a naturally occurring N-terminal 21-kDa fragment of human IGFBP-1 isolated from placenta, showed that it had IGF-I-binding activity, and concluded that the IGF binding domain must be in this part of the protein. In contrast, as noted previously, a naturally occurring, ~12-kDa C-terminal fragment of IGFBP-2 from human milk retains considerable affinity for IGF-II, with an absolute requirement for the N-terminal IGF-II hexapeptide *(24)*, while mutagenesis studies of bovine IGFBP-2 have indicated that residues of both the N- and C-terminal regions of IGFBP-2 are involved in ligand binding *(38,39)*.

Mutagenesis studies on IGFBP-1 confirm this conclusion, as a deletion of 60 N-terminal residues, or of 20 C-terminal residues, both abolish IGF binding *(40,41)*. The N-terminal IGFBP-3 fragment, IGFBP-3[1–88], expressed in Chinese hamster ovary (CHO) cells, has measurable IGF-binding activity, although greatly reduced compared to the intact protein, and removal of a C-terminal negative charge in IGFBP-3 ($Lys^{253}Glu^{254}$ to ArgGly) reduces both IGF-I and IGF-II affinity by 75–80% *(42)*. Modifications in the C-terminal region also affect ternary complex formation between ALS,

IGFs, and either IGFBP-3 or IGFBP-5 *(42,43)*. These and similar studies all support the conclusion that both domains of IGFBPs are important in maintaining the integrity of the IGF binding site, whether by direct contribution of key residues that interact with IGFs, or by maintaining the conformation of the protein.

Effects of Posttranslational Modifications on Binding Properties

Phosphorylation

As recently reveiewed *(44)*, various posttranslational modifications of IGFBPs can affect their binding of IGFs, although the effects differ for different proteins, and perhaps also different species of origin. For example, phosphorylated forms of IGFBP-1 from human serum and other sources have up to 10-fold higher affinity for IGF-I than the fully dephospho-form, although the affinities of the two forms for IGF-II do not differ significantly *(45,46)* (Table 1). By contrast, however, rat hepatoma cell-derived IGFBP-1 does not show any change in affinity for IGF-I after dephosphorylation *(47)*. IGFBP-1 phosphorylation appears to occur predominantly at serine residues amenable to casein kinase II phosphorylation *(48,49)*, and at least four isoforms of human IGFBP-1 of different charge are identifiable on native polyacrylamide gel electrophoresis and ion-exchange chromatography *(45)*. As discussed later, high-affinity phosphoforms of IGFBP-1 are invariably inhibitory to IGF action, whereas the nonphosphorylated protein has been reported to potentiate IGF-I activity *(46)*. IGFBP-3 is also secreted as a phosphoprotein, but in contrast to IGFBP-1, phosphorylation appears to have little or no effect on the binding affinity of IGFBP-3 *(50,51)*.

Glycosylation

Whereas IGFBP-1 and IGFBP-2 have little or no glycosylation, the other IGFBPs are all documented to have glycosylated isoforms. IGFBP-3 from natural sources always exists in two glycoforms, resulting from occupancy of either two or three of the three possible *N*-glycosylation sites by carbohydrate *(52)*. Comparison of IGFBP-3 expressed in *E. coli* with the protein of mammalian origin suggests that glycosylation has little influence in determining the affinity of IGF binding *(53)*, and IGFBP-3 variants, expressed in CHO cells, in which potential *N*-glycosylation sites have been removed by site-directed mutagenesis are also unchanged in their reactivity toward ALS, again indicating no effect on IGF binding *(52)*. IGFBP-4 is often found in both nonglycosylated (~24 kDa) and *N*-glycosylated (~28 kDa) forms *(54)*, but the effects of glycosylation on IGF binding affinity have not been reported; similarly, glycosylation variants of IGFBP-5 have not been studied separately. Finally, O-linked glycosylation of IGFBP-6 has been shown to have no influence on its ability to bind IGFs *(55)*. There is no evidence, therefore, that would support the idea that glycosylation of IGFBPs can modulate the effects of these proteins on IGF activity.

Limited Proteolysis

Limited proteolysis is increasingly recognized as a mechanism for modulating the activity of IGFBPs in the cellular environment. In general, proteolyzed forms of IGFBPs tend to have greatly reduced binding activity compared to the intact proteins, as demonstrated by the ability of protease resistant forms of IGFBP-4 and -5 to retain IGF-inhibitory activity under conditions where protease-sensitive IGFBPs are inactive *(56,57)*. There are, however, some documented examples of IGFBP fragments that retain considerable activity. As noted previously, a C-terminal fragment of human IGFBP-2 still binds

IGF-II with an affinity of 4.5×10^9 L/mol, about 10-fold lower than the value for intact IGFBP-2, although it retains little IGF-I binding activity; a corresponding naturally occurring fragment of rat IGFBP-2 has only a 10-fold reduced affinity for IGF-I (Table 1).

IGFBP-3 proteolysis can also occur in the circulation in a variety of conditions, notably pregnancy and catabolic states *(58,59)*. In human pregnancy, serum IGFBP-3 appears exclusively in a form of 30 kDa when analyzed by SDS-PAGE *(60)*, although its size appears normal by gel chromatographic analysis *(61)*. When stripped of its IGFs and analyzed for IGF binding affinity, the pregnancy serum protein is reduced in affinity for IGF-I by eightfold, and for IGF-II by twofold *(58)*. Paradoxically, the circulating high molecular weight IGFBP-3 complexes in pregnancy carry a normal or elevated concentration of IGF-I and IGF-II, implying that the binding affinity is not decreased. A possible explanation lies in the observation that the presence of ALS is able to increase the affinity of proteolyzed IGFBP-3 for IGF-I by several-fold, although it has no effect on the affinity of intact IGFBP-3 *(62)*; that is, ALS stabilizes ternary complexes containing proteolyzed IGFBP-3. Proteolysis of IGFBP-4 and IGFBP-5 by metalloproteinases in pregnancy serum has also been demonstrated *(63)*, although the effects on ligand binding affinity was not reported.

A selective loss of IGF-I binding, compared to IGF-II, has been described for other forms of proteolyzed IGFBP-3. For example, a 22-25 kDa plasmin fragment of nonglycosylated IGFBP-3 was reduced 50-fold in IGF-I affinity compared to a 20-fold loss of IGF-II affinity *(64)*, and a 30-kDa form of rat IGFBP-3 has been found to have a 50-fold preferential affinity for IGF-II *(35)*. Many other examples of effects of IGFBP proteolysis on IGF binding or action have been documented *(17)*, and this topic is also dealt with elsewhere in this volume. As a generalization, it may be said that, whereas the effect of proteolysis of circulating IGFBP-3 on its activity to transport IGFs is still controversial, there is little doubt that proteolysis of IGFBPs at the tissue level can in some circumstances release IGFs from inhibition by the IGFBPs, thereby increasing local IGF bioavailability.

INHIBITORY EFFECTS OF IGFBPS ON IGF ACTIONS

Endocrine Effects

The Ternary Complex as an Inhibitor of Insulin-Like Effects

IGF-I and IGF-II circulate in healthy adults at a total concentration of 700–800 ng/mL, or approx 100 n*M*. With a hypoglycemic potential estimated at approx 5–10% that of insulin on a molar basis *(65)*, circulating IGFs have the potential to vastly exceed the activity of insulin itself. It is the sequestration of IGFs into IGFBP complexes with ALS that is believed to prevent the expression of this activity. A small proportion of total IGFs are predicted (from simple equilibrium considerations) to be free from ternary complexes, but other IGFBPs might be expected to capture some or all of the "free" peptides. As recently reviewed *(66)*, estimates of free IGFs vary from 10% of total, measured by gel permeation chromatography, to approx 0.5% of total measured by methods such as ultrafiltration *(67)*, designed to minimize dissociation. If an average of 0.5% is free, and it is 10% as potent as insulin, activity equivalent to 50 p*M* insulin could be exerted by free IGFs, a significant concentration given typical normal fasting insulin values of up to 100 p*M*.

Ternary complexed IGFs are believed to have restricted bioavailability due to the inability of the complexes, of approx 150 kDa, to traverse the capillary barrier. Even though the ALS gene is expressed in kidney and embryonic bone as well as liver *(68)*, there is little evidence of ALS protein production in extrahepatic tissues, but high molecular weight IGF complexes have been identified in a number of extravascular fluids, such as skin interstitial fluid *(69)*, mammary lymph *(70)*, and wound fluid *(71)*, suggesting that some egress of ternary complexes from the circulation may be possible. It is assumed, however, that for IGFs to leave the circulation there would normally have to be dissociation of ternary complexes. Binary complexes between IGFs and IGFBPs appear to leave the circulation rapidly, but ALS would clearly have to dissociate from ternary complexes, as it is ALS that determines the long circulating half-life of complexed IGFs *(72)*.

The low association constant for ALS binding to IGFBP-3—around 10^8 L/mol under near-physiological conditions *(22)*—suggests that dissociation may occur quite readily, although the considerable concentration of free ALS in the circulation would tend to oppose it. It has also been suggested that proteoglycans might promote dissociation, as heparin has been shown to lower the affinity of ALS for IGFBP-3 *(73)*; there is, however, no evidence that this mechanism is important in vivo. Limited proteolysis of IGFBP-3, as seen in pregnancy and other conditions, might also be expected to decrease the binding affinity for ALS, but measurement of ALS binding to partially purified IGFBP-3 from nonpregnancy and pregnancy serum did not reveal any difference in affinity *(74)*.

There is evidence from a variety of sources that the insulin-like activity of IGFs in ternary complexes is blocked. Administration of potentially lethally hypoglycemic doses of IGF-I together with IGFBP-3 markedly ablates the hypoglycemic effect *(75)*. While this might appear to be evidence that binary IGF-IGFBP-3 complexes have reduced activity, it has been shown in other experiments that IGFBP-3 administered alone or with IGF-I has substantially formed into ternary complexes within a few minutes *(72,76)*, so it is likely to be these complexes that block IGF activity.

An interesting illustration of the inhibitory effect of ternary complexing on IGF activity comes from the clinical condition of non-islet-cell tumor hypoglycemia (NICTH), in which patients with IGF-II-secreting tumors, usually of mesenchymal origin, experience severe hypoglycemia. Although total IGF-II may be only modestly elevated, there is a large increase in the proportion of free IGFs *(77)*. GH secretion is markedly impaired, presumably due to feedback inhibition by IGF-II at the pituitary level *(11)*; as a consequence, ALS levels are low, and IGFBP-3 tends to be present mainly in binary complexes, which are assumed to be able to leave the circulation. Glucocorticoid treatment or GH administration has been shown to alleviate the hypoglycemia in some patients. However, whereas glucocorticoids can inhibit the secretion of tumor IGF-II, GH appears to block the hypoglycemia without decreasing IGF-II levels. A common mechanism nevertheless appears responsible for the effects of both agents, as they both markedly increase serum IGFBP-3 and ALS levels, and much of the IGFBP-3 once again forms normal ternary complexes, which lock the IGFs in the circulation.

Acute Modulation of Endocrine IGF Action by IGFBP-1

As noted previously, at least 0.5% of IGFs in serum samples can be separated by ultrafiltration under conditions that suggest that they are not associated with IGFBPs *(67)*, and there is some evidence that the "free" IGFs exert a tonic hypoglycemic activity that may be a significant contributor to the complex regulation of blood sugar levels. The first direct evidence that this may be so was the demonstration that a bolus injection of natural human IGFBP-1 in rats causes a significant transient hyperglycemia *(78)*. Other studies have shown that transgenic rats expressing the rat IGFBP-1 gene have fasting hyperglycemia *(79)*, and that recombinant human IGFBP-1 injection in rats causes a pulse of insulin secretion, suggestive of a transient hyperglycemia *(80)*. These experiments are all consistent with the hypothesis that free IGFs tonically suppress blood glucose, and that acutely blocking the IGFs with IGFBP-1 reverses this suppression.

Although these studies involve artificial concentrations of human IGFBP-1 in rats, several clinical reports suggest that endogenous IGFBP-1 has the same effect. The natural regulation of serum IGFBP-1 levels, falling after glucose ingestion and rising in response to hypoglycemia, is certainly consistent with a counterregulatory role for this protein in acutely inhibiting the activity of IGFs not bound in ternary complexes *(81)*. More explicit are experimental studies showing an inverse relationship between serum IGFBP-1 and free IGF-I levels *(82,83)*, notably an investigation of the response of healthy subject to oral glucose, demonstrating that the changes in IGFBP-1 and free IGF-I are significantly negatively correlated *(83)*. A strong positive correlation ($r = 0.94$) between peak IGFBP-1 levels and subsequent peak glucose levels in healthy subjects receiving a single IGF-I injection also supports the concept that endogenous IGFBP-1 may play a role in glucoregulation *(84)*. Thus, although there needs to be further investigation of this area, available data are consistent with the view that the insulin-like activity of IGFs is regulated acutely by changes in IGFBP-1.

Possible Endocrine Effects of Other IGFBPs

Owing to the relatively limited availability of binding proteins other than IGFBP-1 and -3 for physiological studies, there is little knowledge of their possible endocrine effects, and in the case of IGFBP-4 and -6, essentially no evidence to support a major serum transport function. IGFBP-2, as previously reviewed *(85)*, is observed to increase in situations where there is insufficient IGFBP-3 to carry the available IGFs; for example, during IGF-I infusion, in the presence of an IGF-II secreting tumor, or in states of relative IGFBP-3 deficiency. This has led to the assumption that IGFBP-2 might act as an alternative IGF carrier in these situations, but this has not been explicitly demonstrated, and its quite rapid half-life in serum *(86)* suggests that rather than acting as a medium-term serum transport protein for IGFs, IGFBP-2 might be involved in facilitating the egress of IGFs from the circulation.

IGFBP-5 is also emerging as a possible circulating regulator of IGF availability. Previously thought to have its major role at the tissue level, IGFBP-5 has recently been shown to circulate at quite high levels, second only to IGFBP-3 *(87)*. Its serum levels, like those of IGFBP-3, show a strong correlation with IGF-I and IGF-II concentrations, and respond like IGFBP-3 to GH treatment in GH-deficient subjects *(87,88)*. This is consistent with recent observations that IGFBP-5 can form an alternative ternary complex with IGFs and ALS *(32)*. These data suggest a transport function for IGFBP-5 similar to that of IGFBP-3, but its much lower concentration than IGFBP-3 implies that either it plays a quite subsidiary role to that of IGFBP-3, or else it has a more specialized function in IGF transport or delivery that has not yet been elucidated.

Effects at the Cellular Level

The use of cells in culture to study IGF action and its modulation by IGFBPs came about through the development of assays designed to measure the bioactivity of IGFs in serum. With the emergence of the concept that IGFs have paracrine and/or autocrine activity, in vitro techniques were further refined to examine the mechanisms by which the IGFs stimulate cell proliferation and metabolism, and the factors involved in modulating their bioactivity. Such studies have revealed the participation of the IGFs in a wide variety of cellular processes: as mitogens involved in cell cycle progression and cell proliferation; as inhibitors of apoptosis; as growth factors involved in cellular differentiation and migration; and as regulators of cell function via their effects on the production or action of other proteins and factors, such as hormones, receptors, cytokines, and extracellular matrix (ECM) components.

A growing body of data now supports the concept that the IGFBPs are critical regulatory factors in many of these processes; however, the original supposition that the IGFBPs universally act at the cellular level as they do in the circulation—that is, to bind and inhibit IGF-I and -II—now appears somewhat simplistic. Indeed, studies demonstrating that IGF–IGFBP complexes may have increased mitogenic activity compared with IGF alone *(89–91)*, that IGF action may be either enhanced or inhibited by a particular IGFBP *(92)*, and that the nature or degree of modulation may vary with the specific IGFBP or cell type *(93)*, together indicate that regulation of IGF action by IGFBPs is a complex and poorly understood phenomenon. Feedback regulation of IGF bioactivity must also be factored into the equation, with IGFs regulating IGFBP synthesis, secretion, and/or posttranslational modification, and thus altering their potential as inhibitors or enhancers of IGF action.

In the majority of studies, the actions of specific IGFBPs have been assessed either by adding them in purified form from an exogenous source, increasing endogenous production by transfection with recombinant cDNAs, or ablating their expression using antisense oligodeoxynucleotides. An alternative protocol analyzes differences in sensitivity to wild-type IGFs and synthetic analogs of IGF-I or -II with reduced capacity to bind IGFBPs; if an IGFBP in the system under investigation is inhibitory, IGFs will exhibit reduced mitogenic potency when compared with analogs having decreased affinity for IGFBPs but normal receptor binding. In addition, in some studies IGFBP modulation of IGF action has been inferred by correlating changes in IGFBP synthesis with changes in cell proliferation either stimulated by exogenous IGFs, or possibly involving the action of endogenous IGFs. Such analyses will not necessarily provide information regarding the action of a specific IGFBP as the production or action of more than one protein may be modified, and the final effect on IGF action may reflect the combined inhibitory and potentiating effects of a number of IGFBPs. As discussed elsewhere in this volume, similar protocols have been used to identify IGFBP effects on cell proliferation that apparently occur independently of the action of IGFs, being carried out in the absence of exogenous IGFs under conditions where production of IGF-I or IGF-II by the cell cannot be demonstrated, or in cells lacking the type I IGF receptor.

Inhibitory Effects in Normal Cell Types

Effects on IGF-Stimulated Cell Proliferation. Undoubtedly the most frequently studied actions of IGFBPs pertain to their inhibitory effects on IGF-stimulated DNA synthesis and cell proliferation. Mitogenic effects of IGFs have been observed in cells of, *inter alia*, liver, gut, breast, thyroid, reproductive organs, bone, muscle, and skin, and inhibition of these effects by IGFBPs has similarly been well documented. Inhibition of IGF action in vitro by a pure IGFBP preparation was first demonstrated in fibroblasts with the 34-kDa IGFBP-1 from amniotic fluid *(94)*, and since that time, inhibition of the mitogenic activity of IGF-I, determined either by measuring incorporation of [^{3}H]thymidine into DNA over 24 h, or cellular proliferation over a period of days, has been demonstrated for each of the IGFBPs in a wide variety of normal cell types.

IGFBP-1 effects on IGF-stimulated cell division have been documented in porcine *(95)* and chick embryo fibroblasts *(96)*, FRTL thyroid cells *(97)*, porcine and rat aortic smooth muscle cells *(95,98)*, chick embryo pelvic cartilage *(99)*, human *(100)* and bovine granulosa cells *(101)*, and bone marrow stromal cells *(102)*. As noted earlier, IGFBP-1 occurs in variably phosphorylated forms according to its source *(46,103)*, with the less phosphorylated forms exhibiting decreased affinity for IGFs *(46)*. Although the phosphorylation state of the IGFBP-1 used in inhibitory studies has not been determined in the majority of cases, the possibility that the different forms may have different biological activity has some support *(95)*. In endometrial stromal cells, phosphorylated and nonphosphorylated IGFBP-1 were reported to inhibit IGF-stimulated DNA synthesis equipotently, but progestin-stimulated DNA synthesis, which could be reversed by high concentrations of IGF-II, was markedly more sensitive to inhibition by the fully phosphorylated form of the protein *(104)*.

Mouse IGFBP-3 was purified from 3T3 fibroblast conditioned medium on the basis of its inhibitory effects on fibroblast proliferation *(105)*, and much of the work investigating the cell proliferative effects of IGFBP-3 has been conducted in skin fibroblasts derived from humans or other species. The conditions under which experiments are carried out—such as pre- or coincubation with the IGFBP, and the molar ratio of IGFBP to IGF—appear critical to whether IGFBP-3 is inhibitory or stimulatory to IGF action, and it is clear that there are complex factors involved in the regulation of IGF actions on cellular proliferation by IGFBP-3. Although IGFBP-3 is secreted by fibroblasts as a glycosylated *(106)* and phosphorylated *(51)* protein, these modifications probably do not directly affect its inhibition of IGF-stimulated cell proliferation *(53)*.

IGF-inhibitory effects of pure IGFBP-3 were first reported in human fibroblasts *(92)*, where coincubation of IGF-I and an equimolar concentration of IGFBP-3 reversed the growth stimulatory effects of IGF-I. By contrast, a large (10- to 20-fold) molar excess of IGFBP-3 was required for inhibition of IGF-stimulated DNA synthesis in porcine skin fibroblasts *(107)*. Constitutive overexpression of IGFBP-3 results in inhibition of cell proliferation in IGFBP-3 cDNA-transfected BALB/c 3T3 fibroblasts in the presence of serum *(108)*. Of interest in this study, it was observed that whereas insulin could override the effects of exogenous IGFBP-3 on inhibited BALB/c cell growth, it was unable to do so in the IGFBP-3 transfectants, implying different actions of endogenous and exogenous IGFBP-3 in this system. While a dual mechanism for IGFBP-3 action was postulated, incorporating both IGF-dependent and -independent effects of IGFBP-3 on cell proliferation, no data were presented to support this. The concept that paracrine/autocrine IGFBP-3 may be a regulator of cell proliferation finds further experimental support in the observation that in vitro senescence and decreased proliferative potential is associated with accumulation of IGFBP-3 in medium conditioned by human skin fibroblasts *(109)*. Inhibition of cell proliferation by IGFBP-3 has also been reported in rat *(110)* and mouse *(111)* osteoblasts and a number of transformed cell types, as discussed later.

The role of IGFBP-2 as an inhibitor of IGF-stimulated cell proliferation is somewhat less well documented than for either IGFBP-1 or IGFBP-3. Inhibition of IGF-I binding to type 1 IGF receptors and DNA synthesis was demonstrated for pure BRL-3A-derived rat IGFBP-2 in chick embryo fibroblasts *(112)*, and inhibitory effects of IGFBP-2 have been shown in smooth muscle cells *(113,114)* and fetal rat calvaria, though that study used a recombinant form of IGFBP-2 with an extra cysteine residue at position 281 *(115)*. A marked increase in IGFBP-2 production by a bovine mammary epithelial cell line correlated with decreased cell proliferation *(116)*. Similarly, a study in lung alveolar epithelial cells has demonstrated that glucocorticoid-induced growth arrest is associated with elevated production of IGFBP-2 *(117)*, and in mesangial cells, decreased IGFBP-2 production in response to elevated glucose is associated with increased sensitivity to IGF-I *(118)*. Addition of pure IGFBP-2 to the intestinal epithelial cell line IEC-6 attenuated basal and IGF-stimulated DNA synthesis *(119)*, while in the same cells ablation of IGFBP-2 expression by transfection with an antisense cDNA construct resulted in increased cell proliferation *(120)*, consistent with inhibitory influences of endogenous IGFBP-2. In addition, transforming growth factor-β (TGF-β)-induced inhibition of proliferation of these cells, which correlated with up-regulation of IGFBP-2 mRNA, could be partially reversed by anti-IGFBP-2 antibodies, implying IGFBP-2 mediation of the growth inhibitory effects of TGF-β *(121)*.

IGFBP-4, like IGFBP-3, has been purified from a variety of sources on the basis of its ability to inhibit cell growth, and its abundance in medium conditioned by bone and neural cells has prompted a number of groups to investigate the role of this protein in these tissues. Osteosarcoma-derived IGFBP-4 inhibits basal and IGF-II-stimulated DNA synthesis in human bone cells *(122)* chicken calvarial cells *(123)*, and mouse osteoblast-like MC3T3-E1 cells *(124)*, effects that correlated with reduced binding of IGF-I or -II to receptors *(122)*. IGFBP-4 purified from a subline of the rat neuronal B104 cell line inhibit IGF-I receptor binding and action in the same cells *(125)*, and increased synthesis of IGFBP-4 correlates with decreased cell proliferation in astroglial cells transfected with the *connexin43* gene *(126)*. The inhibitory effects of IGFBP-4 on IGF-stimulated DNA synthesis appear exquisitely sensitive to, and reversible by, the action of specific proteases in bone, fibroblasts, and neural cells, lending credence to the proposition that this is a mechanism by which cells are released from tonic growth inhibitory effects of IGFBP-4.

Although IGFBP-5 appears to predominate as a potentiator of IGF action in bone, reports of its inhibitory effects in the same cells are comparable to the dichotomous effects of IGFBP-3 on IGF action in fibroblasts. Recombinant human IGFBP-5 expressed in yeast inhibits IGF-I and IGF-II stimulation of DNA synthesis in osteoblastic cells *(127)*, and in the U-2 osteosarcoma cell line inhibitory effects of endogenous IGFBP-5 were inferred from the increased potency of des(1–3)IGF-I (which binds IGF receptors but not IGFBP-5) relative to IGF-I *(128)*. In SV-40 transformed human fibroblasts, increased cell proliferation was associated with decreased expression of IGFBP-5 *(129)*, suggestive of a role for endogenous IGFBP-5 in the regulation of autocrine IGF action in these cells.

Given the markedly higher affinity of IGFBP-6 for IGF-II compared with IGF-I (Table 1), it was initially thought that this may be reflected in greater inhibition of IGF-II-stimulated events compared with IGF-I, and there is now some evidence to support this. The recombinant protein purified from yeast transfectants completely inhibited IGF-II-stimulated DNA synthesis in rat osteoblastic cells, but only partially inhibited IGF-I *(127)*. When present in molar excess, rh IGFBP-6 (expressed in COS-7 cells) inhibited IGF-II stimulation of myoblast proliferation *(130)*. In the absence of exogenous IGFBP-6, the potency of an IGF-II analog with decreased affinity for IGFBP-6 was increased relative to wild-type IGF-II, consistent with inhibition by endogenous IGFBPs, although as these cells secrete IGFBP-4, -5, and -6, this could arise from the combined effects of a number of proteins *(130)*. In another study, a protein purified from keratinocyte cell conditioned medium on the basis of its cell growth inhibitory effects proved to be IGFBP-6 *(131)*; when added to MvLu1 (mink lung), BALB/MK (mouse keratinocyte) or HaCat (human keratinocyte), cells, this pure IGFBP-6 inhibited cell proliferation. As this inhibition could be overcome by high concentrations of insulin, it was deemed an IGF-dependent inhibitory effect. The observation that the IGFBP-6 was secreted only by confluent cultures of keratinocytes provides further support for IGFBP-6 constituting an important autocrine or paracrine growth regulator, via its inhibition of IGF-II, in keratinocyte proliferation *(131)*. The recent observation that neuroblastoma cells overexpressing IGFBP-6 show decreased proliferation in vitro and reduced tumorigenicity in vivo *(132)* suggests an important role for this protein in the regulation of tumor cell proliferation.

Effects on IGF Modulation of Apoptosis. Although the anti-apoptotic effects of the IGFs have been documented in a number of different cell types, studies more recently addressing the role of the IGFBPs in modulating this action have implicated IGFBP-2, -3, and -5 as potential players in the regulation of programmed cell death. Buckbinder and co-workers *(133)* reported induction of expression of IGFBP-3 by the tumor suppressor p53, and postulated that IGFBP-3 may mediate at least some of apoptotic effects of p53. This received further support with the observation that loss of apoptotic activity was associated with the inability of p53 mutants to induce IGFBP-3 transcription *(134,135)*.

One of the first studies in cells examining this aspect of IGF–IGFBP physiology demonstrated an inhibitory effect of exogenous IGFBP-3 on the suppression of apoptosis induced by IGF-I in ovarian follicles *(136)*, and follicular atresia is probably the most commonly studied cell model demonstrating IGFBP action in apoptosis. The anti-apoptotic effects of human chorionic gonadotropin (hCG), FSH *(136)* and vasoactive intestinal peptide *(137)* can be partially antagonized by IGFBP-3 in follicles, and although it has been assumed that this effect is due to inhibition of endogenous IGF action, reversal by high concentrations of mitogens (insulin or IGFs) was not examined in these studies, and the possibility of intrinsic bioactivity of IGFBP-3 cannot be ruled out. A correlation between secretion of IGFBP-2 by atretic cells and follicular apoptosis has also been presented as evidence for the autocrine involvement of this protein in follicular atresia *(138)*.

There is also evidence that IGFBPs are involved in mammary gland involution, a process involving apoptotic pathways; IGFBP-3 and -5 have both been implicated, albeit in opposite ways. Transgenic mice overexpressing IGFBP-3 exhibited delayed mammary involution (i.e., a reduced level of apoptosis) compared with control animals *(139)*, whereas induction of IGFBP-5, and to a lesser extent IGFBP-4, expression correlated with increased apoptosis in involuting mammary tissue in rats *(140)*. The possibility that these differences stem from specific actions of the different IGFBPs remains to be investigated. A possible role for IGFBP-5 in apoptosis in the prostate has also been postulated, with induction of IGFBP-5 expression apparent after removal of androgens (which promote cell survival by suppressing apoptosis) in the rat ventral prostate *(141)*.

Effects on IGFs in Cell Migration and Invasion. To date, little work has been carried out on IGFBP-mediated effects on cell migration, an important event in embryogenesis, wound healing and cancer cell invasion and metastasis, despite good evidence that IGFs promote the migration of a number of cell types. IGF-stimulated migration of porcine smooth muscle cells in a cell culture model of wound healing could be inhibited by IGFBP-1, albeit with reduced potency compared with IGFBP-2 *(142)*. The discrepancy between the two appeared due to intrinsic migration-stimulatory activity of IGFBP-1 as a result of its $\alpha_5\beta_1$ integrin binding, because mutation of the recognition sequence of IGFBP-1 which abolished its capacity to bind this integrin resulted in increased IGF-inhibitory activity *(142)*. These findings imply that the distribution of IGFBP-1 between ECM-bound and soluble forms is a critical determinant of its function as an inhibitor of IGFs in cell migration. A recent study indicates that IGFBP-5 may also inhibit mesangial cell migration, but only when present as a carboxy-truncated protein; the intact protein has IGF-independent stimulatory activity *(143)*.

Effects on IGF Action in Cell Differentiation. Direct effects of IGFBPs in modulating IGF action in cellular differentiation have been shown in a number of cell systems, and much has been inferred from studies identifying altered expression of an IGFBP correlated with increased differentiated function or morphology. In L6A1 myoblasts, recombinant human IGFBP-6 inhibited IGF-II-induced, but not IGF-I-induced, creatine kinase activity, a finding consistent with the preferential affinity IGFBP-6 for IGF-II compared with IGF-I *(144)*. Use of IGF analogs with altered affinities for the IGF receptors and IGFBPs showed inhibition by endogenous IGFBPs on IGF-mediated differentiation in these same cells *(130,145)*. Curiously, non-IGFBP-binding mutants used in these studies suggested differences in the degree to which IGFBPs modulate the processes of proliferation and differentation. Whereas an IGF-II mutant stimulated myoblast proliferation and differentiation with fivefold and twofold greater potency, respectively, than IGF-II, similar use of an IGF-I mutant showed greater potency in the stimulation of differentation (100-fold) than proliferation (10-fold) compared with wild-type IGF-I *(130,145)*. Whether these results reflect the relative contributions of different inhibitory IGFBPs in the processes of cellular proliferation and differentation remains to be determined. In another study in the C21 myoblast line, a non-IGFBP-binding analog showed enhanced potency for inducing myogenic differentiation, consistent with a bypass of the inhibitory effects of IGFBPs—in this case IGFBP-5—in this process *(146)*. Transfection of these cells with an IGFBP-5 transgene results in accumulation of IGFBP-5 in the extracellular matrix coincident with failure to differentiate normally *(147)*, an event that was overcome by high concentrations of IGF-I. These findings are consistent with IGFBP-5 inhibition of endogenous IGFs acting to stimulate muscle cell differentiation *(148)*.

Effects on IGF Action in Cell Function. The antigonadotrophic effects of IGFBPs have been recognized since an inhibitor of FSH-stimulated steroidogenesis purified from porcine follicular fluid was identified as IGFBP-3 *(149)*. FSH-stimulated estrogen and progesterone secretion by rat granulosa cells is inhibited by IGFBP-3 and IGFBP-2; that this effect can be mimicked by anti-IGF-I antiserum and overcome by exogenous IGF-I provides support for inhibition of IGF-mediated effects of FSH, rather than IGF-independent effects of the IGFBPs *(149,150)*. In human granulosa cells, IGFBP-1 and IGFBP-3 inhibited IGF-I or FSH-stimulated estradiol secretion, while progesterone production in response to IGF-I, but not FSH, was inhibited by both IGFBPs as well *(151,152)*. The possibility of species differences in the effects of the IGFBPs in ovarian function has been demonstrated using perfused rabbit ovaries capable of undergoing oocyte maturation, ovulation, and steroidogenesis in vitro *(153)*; inclusion of IGFBP-3 but not IGFBP-1 in the perfusate resulted in prevention of ovulation, and inhibition of estradiol and plasminogen activator secretion by perfused ovaries *(153)*.

Fewer data are available regarding IGFBP action in male reproductive tissue, but here again there appear to be specific actions of different IGFBPs. In rat Leydig cells, testosterone production in response to the combined effects of hCG and IGF-I, but not hCG alone, was inhibited to a greater degree by IGFBP-3 than IGFBP-4, while IGFBP-2 was without effect *(93)*. As each of the IGFBPs was present in concentrations sufficient to saturate the total (exogenous and endogenous) IGF-I, different effects due to their slight differences in IGF affinity are unlikely to explain these observations, and it appears that IGFBP specificity will prove to be of greater significance than originally thought.

The actions of IGFs in other cell functions are also subject to regulation by IGFBPs. The glycogenic effects of IGF-II in fetal rat hepatocytes are negated by the inclusion of either IGFBP-1 or IGFBP-3 *(154)*. A study examining glucose consumption by BALB/c-3T3 mouse fibroblasts indicated a complex mechanism of regulation by IGFBPs in this system *(155)*. When coincubated with either IGF-I or IGF-II, IGFBP-1 had greater inhibitory activity on IGF-I effects, while IGFBP-3 had greater IGF-II inhibitory activity, consistent with its slightly higher affinity for this ligand. By contrast, preincubation of cells with IGFBP-3 with subsequent addition of IGF-I or -II in combination with IGFBP-3 resulted in greater inhibitory activity for IGF-I than IGF-II. Inhibition of IGF-stimulated glycogen synthesis in rat osteoblasts by IGFBP-3 *(110)*, and human osteoblasts by IGFBP-4, -5, and -6 *(127)* has been demonstrated, while glucose oxidation and lipogenesis in porcine adipocytes has been reported for IGFBP-3 *(156)*.

IGF-stimulated collagen and proteoglycan synthesis by fibroblasts is also modulated by IGFBPs. Serum from vitamin C-deficient or scorbutic guinea pigs that contained elevated levels of endogenous IGFBP-1 and IGFBP-2 *(157)* inhibited IGF-I-stimulated collagen synthesis in 3T3 fibroblasts; immunodepletion of these IGFBPs from the serum resulted in loss of inhibitory activity. Inhibition of collagen synthesis by rhIGFBP-2 has also been reported in fetal rat calvaria *(115)*.

Inhibitory Effects in Cancer Cells

The possibility that the regulatory influences exerted by the IGFBPs are aberrant in disease states is currently the focus of much research. It must be noted, however, that as the majority of these studies use commercially available cell lines established many years ago from primary tumors or metastatic lesions, difficulties may arise from the way the cells are handled in different laboratories *(158)*. It is not uncommon to find in the literature disparate reports regarding growth behavior, sensitivity to growth factors, and patterns of expression of IGFs, IGF receptors, or IGFBPs, depending on where the work has been carried out and the precise conditions under which the cell line has been maintained.

Many of the studies examining the IGF axis in transformed cells in general, and breast cancer cells in particular, have documented actions of IGFBPs in the absence of added IGFs, and the possibility that this represents IGF-independent effects of the IGFBPs must be considered. Cell proliferative effects of estradiol or charcoal-stripped serum can be inhibited by IGFBP-1 *(159)* and IGFBP-3 *(160)* in MCF-7 cells that do not secrete produce IGF-I or -II *(161)*, whereas in the Hs578T cell line the growth inhibitory effects of TGF-β appear to be mediated by IGFBP-3. As this can be reversed by IGFBP-binding, but not receptor-binding, analogs of IGF-II *(162)* or IGFBP-3 antisense oligodeoxynucleotides in the MDA-MB-231 cell line *(163)*, it is possible that these findings reflect intrinsic IGFBP bioactivity, an aspect of IGFBP physiology that is dealt with in greater detail elsewhere in this volume.

Secretion of IGFBPs by T47D breast cancer cells in vitro was identified more than 15 years ago *(164)* but it was not until fairly recently that bioactivity of these proteins in breast cell lines was identified. The first of these reports documented the abolition of IGF-stimulated proliferation of MCF-7 breast cancer cells concomitant with increased levels of IGFBPs in response to retinoic acid treatment of cultures *(165)*. In the same cell line, a combination of retinoic acid and forskolin increased IGFBP-3 and IGFBP-6 levels 6- and 12-fold, respectively, and this was accompanied by a decrease in IGF-stimulated DNA synthesis under conditions in which sensitivity to an IGF analog with reduced affinity for IGFBPs, [QAYL]IGF-I, remained unchanged *(166)*. Addition of purified IGFBPs has also been used to demonstrate regulation of IGF action in MCF-7 cells: rh IGFBP-1 at high concentrations (40–80 n*M*, ~1–2 µg/mL) inhibited proliferation and anchorage-independent growth induced by IGF-I *(159)*. Overexpression of IGFBP-1 as a consequence of transfection with IGFBP-1 cDNA results in decreased IGF-I-stimulated type 1 receptor phosphorylation and cell proliferation *(167)*.

The possibility that the development of prostate disease may also be associated with alterations in IGFBP regulation of IGF-stimulated cell proliferation has some experimental support *(168,169)*. Most studies in vitro examining IGF and IGFBP action in this disease have used three cell lines: the androgen receptor negative Du145 and PC-3 cell lines, and the androgen receptor positive LNCaP, derived from metastatic lesions (Du145 and LNCaP) or a primary adenocarcinoma (PC-3). IGF-II is secreted by cells of each of these lines *(170)*, and the variable degree of responsiveness to exogenous IGFs *(171)* may be due either to this, or to the effects of endogenous IGFBPs secreted in differing amounts by the different lines. In the authors' laboratory LNCaP cells secrete predominantly IGFBP-2 with approximately 20-fold lower concentrations of IGFBP-3, Du145 cells have similar levels of IGFBP-3 and IGFBP-6 and lower levels of IGFBP-2 and -4, while PC-3 cells secrete predominantly IGFBP-6 and -4, with smaller concentrations of IGFBP-2, -3, and -5 *(see also ref. 170)*. Both IGFBP-4 *(172)* and IGFBP-6 *(173)* have been purified from PC-3 cell conditioned medium on the basis of their cell growth inhibitory activity. Given that these cells do secrete IGFs and IGFBPs, there is potential for autocrine regulation of IGF bioactivity and this has been demonstrated. In PC-3 cells, fragments of IGFBP-3 with low IGF-binding affinity, and thus reduced IGF-inhibitory activity, are generated by the action of serine proteases *(174)*. Prevention of IGFBP-3 proteolysis using a serine protease inhibitor results in an increase in the amount of intact IGFBP-3 in the medium, and this is accompanied by decreased cell proliferation. IGF-II-stimulated monolayer growth of Du145 cells is inhibited by rhIGFBP-1, which also inhibited both basal and IGF-II-stimulated anchorage independent growth of these cells *(175)*.

The IGFBPs are probably important regulators in many other IGF-responsive malignant cell types. Virtually all the IGF-II secreted by the HT-29 colon carcinoma cell line is associated with IGFBP *(176,177)* and antisense ablation of IGFBP-4 secretion by the cells resulted in increased cell proliferation *(178)*. A similar role for IGFBP-4 in the rat neuroblastoma B104 cell line has also been demonstrated, where it prevented IGF-I tracer binding to the cell, and IGF-stimulation of DNA synthesis *(125)*. Human gastric cancer cell responsiveness to IGF-I is also ablated by the presence of endogenous IGFBPs *(179)*, and in lung tumors both extracellular and membrane-associated IGFBP-2 appear to exert inhibitory effects on the proliferative action of IGF-II *(180)*. Membrane-associated IGFBP-3 has also been demonstrated to attenuate IGF-I-induced type 1 receptor signaling in the Ishikawa cell model of endometrial cancer *(181)*.

Modulation of Inhibitory Effects by Limited Proteolysis

Limited proteolysis of IGFBPs is recognized as a significant mechanism by which their IGF-regulatory effects may be modified. As noted earlier, proteolytic cleavage of IGFBPs in general reduces their affinity for IGFs, and the biological significance of this change in function has been demonstrated in vitro in a wide variety of cell systems. One of the first demonstrations of this phenomenon was in bone cells, where addition of intact but not truncated IGFBP-3 inhibited IGF-stimulated DNA and glycogen synthesis *(110)*; a later study showed that IGF-II-induced IGFBP-4 proteolysis in osteoblastic cells enhanced subsequent IGF-I stimulation of DNA synthesis *(182)*, indicating that proteolytic modification of endogenous IGFBPs is a likely mechanism by which IGF bioavailability is regulated in vivo. Similar observations have been made in other cell types: for example, in PC-3 prostate cancer cells, extracellular IGFBP-3 occurs predominantly as fragments with reduced affinity for IGFs, but if IGFBP-3 protease activity is inhibited, the presence of intact (high-affinity) IGFBP-3 correlates with decreased cell proliferation *(174)*. This general principle of increased IGFBP proteolysis releasing IGFs from inhibition may have important implications in malignancy, a disease state believed to be associated with increased activity of proteases. The actions of IGFBP fragments is discussed in greater detail in Chapter 15.

STIMULATORY EFFECTS OF IGFBPS ON IGF ACTION

Stimulatory Effects In Vivo

Relationship Between the Ternary Complex and Growth Effects of IGFs

Although there is good evidence that the complexing of IGFs with IGFBP-3 and ALS in the circulation is inhibitory to their insulin-like activities, there are numerous situations in which the concentration of the ternary complex correlates positively with growth status, implying that the pool of complexed IGFs is somehow able to exert an overarching positive control on growth-stimulatory processes. This is seen both in clinical conditions of GH deficiency and excess, where growth rates correlate broadly with the concentrations of ternary complex components, and in situations of pharmacological administration of GH, where GH treatment increases the ternary complex components concomitantly with a positive growth response.

These observations, from a great many clinical and animal studies, are consistent with the circulating IGF complexes acting as a large reservoir from which IGFs are made available for growth stimulation as required. Although in this context, IGFBP-3 in ternary complexes can be seen to have a positive relationship to IGF activity, studies with IGF analogues that complex poorly with IGFBP-3 nevertheless support the view that release from IGFBP-3 must be an obligatory step for endocrine IGFs to exert growth effects. For example, in both GH-deficient dwarf mice and streptozotocin-diabetic rats, des(1–3)IGF-I, which shows several-fold lower binding to IGFBP-3 than IGF-I *(25)*, is at least as effective, and in some cases more so, than IGF-I in normalizing growth parameters *(183,184)*, implying that in these pharmacological situations, the ability to form a tight complex with IGFBP-3 is not necessary for the maximum expression of IGF bioactivity.

Therapeutic Effects of Complexed IGFs

More explicit examples of IGFBP-bound IGFs being potentiated, rather than inhibited, in their activity in vivo come from studies of wound healing and bone growth. Sommer et al. *(75)* have described rat and pig wound healing models in which administration of IGF-I–IGFBP-3 as a 1:1 complex is more effective than IGF-I alone. For example, in the Schilling–Hunt wound model in rats, 17 d treatment with the complex resulted in about twice as much collagen deposition as IGF-I alone. Similarly, in steroid-treated pigs, the IGF-I–IGFBP-3 complex, applied topically, was more effective than IGF-I in stimulating wound recovery. In rats made relatively osteoporotic by ovariectomy, IGF-I–IGFBP-3 complex, administered subcutaneously, was effective in increasing bone formation rates and improving mineralization, and the use of the complex allowed IGF-I doses to be administered that would have cause severe hypoglycemia if given in an uncomplexed form. Administration of IGF-I–IGFBP complexes into rats may also affect tissue or cellular localization of IGF-I, with more IGF-I targeted to the glomerular components of the kidney if it was infused as a binary complex with IGFBP-3, than if given with another IGFBP or alone *(185)*. The observation that bone turnover may also be increased in response to a complex of IGF-I and IGFBP-3 has led to the suggestion that administration of such complexes may be of benefit in the treatment of bone loss induced by microgravity *(186)*. Thus the complexing of IGF-I to IGFBP-3 may still allow expression of its activity in some situations, and may effectively potentiate IGF-I activity by providing a more stable form of presentation to the tissues.

Stimulatory Effects at the Cellular Level

Although IGF-potentiating effects of IGFBPs in vitro have been recognized for more than a decade, the precise mechanisms by which such enhancement occurs remain unclear. Current evidence indicates that posttranslational modification of the IGFBP—via limited proteolysis, phosphorylation, or cell-surface or ECM association—results in reduced affinity for IGFs, and underlies binding protein potentiation of IGF bioactivity. It is apparent that the processes involved differ depending on the cell type and the specific IGFBP involved.

Effects in Fibroblasts

The potentiating effects of IGFBP-3 in fibroblasts were first described by De Mellow and Baxter, who showed that DNA synthesis in response to IGF-I was enhanced if cells were preincubated with IGFBP-3 for 4–48 h prior to IGF-I exposure *(92)*. Similarly, *E. coli* (nonglycosylated) derived IGFBP-3 enhanced aminoisobutyrate uptake in bovine fibroblasts, again under conditions of pre- but not coincubation with the binding protein *(187)*. This enhancement appears to correlate with an increase in the level of processed IGFBP-3 in association with cell surfaces *(188)*, possibly due to the action of membrane-associated IGFBP-3 proteases *(189)*. As with other IGFBP fragments, these appear to have diminished capacity to bind IGFs *(53)* and the apparent potentiation may reflect slow release of IGFs for receptor activation as the IGFBPs become proteolyzed. This concept of IGFBP-3 serving as a reservoir facilitating continued IGF receptor activation was also proposed by Blum and co-workers, who showed that IGF-I in binary complex with IGFBP-3 retained enhanced mitogenicity in human and baby hamster kidney fibroblasts compared with IGF-I alone *(90)*. Alternatively, preincubation with IGFBP-3 has been shown to block IGF-induced receptor down-regulation *(190)*; as this process normally desensitizes cells to subsequent IGF stimulation, what appears to be enhanced IGF action may reflect decreased refractoriness to IGF stimulation.

IGFBP-1 potentiation of IGF action has been reported in human, mouse, and chick embryo fibroblasts *(191,192)*. As with IGFBP-3, a change in ligand affinity appears to be associated with potentiation of IGF action by IGFBP-1; however, in this case it is brought about by dephosphorylation of the binding protein *(46)*. Thus the phosphorylation state of IGFBP-1 secreted by a cell or derived from extracellular sources will significantly affect its ability to modulate IGF action, with nonphosphorylated IGFBP-1 often potentiating, and phosphorylated IGFBP-1 inhibiting, IGF action. The requirement for a low concentration of platelet poor plasma for maximal enhancement in some cell systems suggests the involvement of a cofactor in the potentiating effect *(191)*, and although this has been characterized to the extent of determining that it is heat- and acid-stable and nondialyzable, its identity remains unknown.

Association with the ECM reduces the affinity of IGFBP-5 affinity for IGFs concomitant with a twofold increase in IGF-I-stimulated fibroblast proliferation *(193)*. In contrast to IGFBP-3 in which cell association and subsequent processing has been suggested to enhance the actions of IGFs, proteolysis of IGFBP-5 was prevented by matrix association, and this was interpreted as being advantageous to the potentiating effects of the binding protein *(193)*. Thus IGFBP-5 in association with fibroblast ECM undergoes reduction in affinity to the extent that it is brought closer to that of IGF-I for the type 1 IGF receptor; by localizing IGF-I to the ECM and therefore proximal to cell-surface IGF receptors, IGFBP-5 may serve as a reservoir for sustained presentation of IGFs to their receptors *(194)*. In support of this, a recent study in porcine smooth muscle cells showed that constitutive expression of mutant IGFBP-5 with weak binding to ECM correlated with reduced responsiveness to IGF-I *(195)*.

Effects in Bone

The possible potentiation of IGF action by an IGFBP in bone was proposed in 1990 by Ernst and Rodan, who noted increased IGF stimulation of DNA synthesis, cell proliferation, and collagen gene expression in osteoblasts in the presence of elevated levels of IGFBP-3 *(196)*. Similarly to the proposed mechanism of action requiring proteolysis of IGFBP-3 in fibroblasts, this appears to be an important factor in this effect in bone, with potentiation by C-terminally truncated forms of this protein *(110)* and also IGFBP-5 *(111)* under conditions where the intact proteins inhibit IGF action. Again, as in fibroblasts, bone ECM has an important role via its binding of IGFBP-5 *(111,122)*, but the surface of the osteoblasts themselves are also able to bind IGFBP-5 with higher affinity than that of ECM for IGFBP-5 *(197)*. The nature of the interaction between the ECM or cell surface and IGFBP-5 is not yet clear, although it does not appear to involve glycosaminoglycan interaction *(197)*, and the possibility of specific IGFBP-5 receptors on osteoblast membranes capable of internalizing IGFBP-5 has been raised *(122,198)*. Regardless, IGFBP-5 increases IGF-I tracer binding to bone cells *(122)*, and thus may serve to concentrate IGFs, and regulate their receptor-signaling potential by controlled release of peptide. IGF-independent stimulation of bone cell growth and function by IGFBP-5, perhaps by signaling through the putative receptor, has also been proposed *(111,122)*.

Effects in Breast and Prostate Cancer

Although early studies into the actions of IGFs and IGFBPs in breast cancer indicated predominantly inhibitory effects, recent data indicate that potentiation of IGF-stimulated cell proliferation by IGFBPs may also occur, although disparate results from different laboratories may derive from subtle differences in the experimental protocols employed. Chen et al. found that recombinant human IGFBP-2 and -3 enhanced IGF-stimulated DNA synthesis in MCF-7 cells under conditions where IGFBP-4 and IGFBP-5 had no effect *(199)*. Cells transfected with IGFBP-3 cDNA exhibited increased IGF-I binding to cells, accompanied by increased IGF-stimulated DNA synthesis and cell proliferation.

The proposed mechanism of action was that IGFBP-3 was preventing IGF-mediated down-regulation of its receptor.

A biphasic effect of rhIGFBP-3 on endogenous IGF-II-stimulated PC-3 prostate carcinoma cell proliferation has also been identified, in which low concentrations of exogenous IGFBP-3 enhanced cell proliferation, while high concentrations abolished IGF-II mediated cell proliferation *(200)*. As in many other situations, proteolysis of IGFBP-3 appears to be a critical factor in this case, with the relative proportion of IGFBP-3 fragments in the medium correlating with the degree of enhancement of cell proliferation, and inhibition of IGFBP-3 proteolysis blocking the potentiating effect of the protein.

Other Systems

In an in vitro model of fibroblast contraction, a process important in wound healing, a combination of IGF-I and IGFBP-1 promoted fibroblast-embedded collagen gel contraction *(201)*; this effect was partially reversed by an endothelin receptor antagonist, and increased by coincubation with endothelin-1, indicating the involvement of this vasoconstrictor peptide in the process of IGF-I–IGFBP-1 action by an unspecified mechanism. A modest stimulatory effect of IGFBP-2 on IGF-I-stimulated glucose transport and aminoisobutyrate uptake in microvascular endothelial cells has also been documented *(202)* and apparent bioactivity of IGFBP-2 in the absence of IGFs is this system was also noted *(202)*. In porcine aortic smooth muscle cells DNA synthesis stimulated by IGF-I can also be enhanced by IGFBP-2 *(203)*. In this system, however, there is an absolute requirement for factors(s) present in platelet-poor plasma: without their inclusion inhibition of IGF action was observed *(203)*. This is reminiscent of the effects observed for IGFBP-1 stimulation of DNA synthesis in these cells *(191)*, and purification and identification of these required cofactors will undoubtedly prove an important step in understanding the mechanisms by which potentiation of IGF action by IGFBP-1 and -2 occurs.

CONCLUDING COMMENTS

This chapter has presented an overview of the factors affecting IGF–IGFBP interactions, and the ways in which these interactions can modulate IGF bioactivity in positive and negative ways, both in vivo and in vitro. The great diversity of examples presented, and the seemingly contradictory examples of stimulatory and inhibitory effects of IGFBPs, serve to illustrate the daunting complexity of these regulatory systems.

Although there now seems little doubt that some IGFBP actions may occur independently of a functional type I IGF receptor, it remains to be proven whether any are truly IGF independent, because when IGFBPs are not acting as regulators of IGF action (as reviewed in this chapter), IGFs may serve as regulators of IGFBP action. Whatever emerges over the next few years about this form of cell regulation, it is clear from the many examples discussed here, as well as many more that could not be included, that a fundamental role of the IGFBP family is to regulate IGF activity.

In this context it is still unclear to what degree there is functional redundancy among the IGFBPs, and how much the effects demonstrated in vitro are truly representative of events in vivo. As our knowledge increases about the importance of posttranslational protein processing, and subcellular or extracellular distribution of the proteins, it is becoming increasingly clear that simple information about regulation of IGFBP gene expression will never provide an adequate explanation of the ways in which IGFBPs act, and that studying the proteins themselves is essential to understanding the importance of IGFBP effects on IGF action.

REFERENCES

1. Salmon WD, DuVall MR. A serum fraction with "sulfation factor activity" stimulates in vitro incorporation of leucine and sulfate into protein-polysaccharide complexes, uridine into RNA and thymidine into DNA of costal cartilage from hypophysectomized rats. Endocrinology 1970; 86:721–727.
2. Daughaday WH, Hall K, Raben MS, Salmon WD, Van den Brande JL, Van Wyk JJ. Somatomedin: proposed designation for sulphation factor. Nature 1972; 235:107.
3. Meuli C. Effects of insulin and NSILA-S on the perfused heart: glucose uptake, lacate production and efflux of 3-*O*-methyl glucose. Eur J Clin Invest 1975; 5:93–99.
4. Rechler MM, Podskalny JM, Goldfine ID, Wells CA. DNA synthesis in human fibroblasts: stimulation by insulin and by nonsuppressible insulin-like activity (NSILA-S). J Clin Endocrinol Metab 1974; 39:512–521.
5. Smith GL, Temin HM. Purified multiplication-stimulating activity from rat liver cell conditioned medium: comparison of biological activities with calf serum, insulin, and somatomedin. J Cell Physiol 1994; 84:181–192.
6. Uthne K, Reagan CR, Gimpel LP, Kostyo JL. Effects of human somatomedin preparations on membrane transport and protein synthesis in the isolated rat diaphragm. J Clin Endocrinol Metab 1974; 39:548–554.
7. Jones JI, Clemmons DR. Insulin-like growth factors and their binding proteins: biological actions. Endocr Rev 1995; 16:3–34.
8. Underwood LE, Hintz RL, Voina SJ, Van Wyk JJ. Human somatomedin, the growth hormone dependent sulfation factor, is antilipolytic. J Clin Endocrinol Metab 1972; 35:194–198.
9. Renner R, Hepp KD, Humbel RE, Froesch ER. Mechanism of the antilipolytic action of NSILA-a: inhibition of adenylate cyclase activity in lipocyte ghosts. Horm Metab Res 1973; 5:56–57.
10. Tell GP, Cuatrecasas P, Van Wyk JJ, Hintz RL. Somatomedin: inhibition of adenylate cyclase activity in subcellular membranes of various tissues. Science 1973; 180:312–315.
11. Tannenbaum GS, Guyda HJ, Posner BI. Insulin-like growth factors: a role in growth hormone negative feedback and body weight regulation in brain. Science 1983; 220:77–79.
12. Jung Y-K, Miura M, Yuan J. Suppression of interleukin-1β converting enzyme-mediated cell death by insulin-like growth factor. J Biol Chem 1996; 271:5112–5117.
13. Zapf J, Waldvogel M, Froesch ER. Binding of non-suppressible insulin-like activity to human serum. Evidence for a carrier protein. Arch Biochem Biophys 1975; 168:638–645.
14. Drop SLS, Valiquette G, Guyda HJ, Corvol MT, Posner BI. Partial purification and characterization of a binding protein for insulin-like activity in human amniotic fluid: a possible inhibitor of insulin-like activity. Acta Endocrinol 1979; 90:505–518.
15. Zapf J, Schoenle E, Jagars G, Sand I, Grunwald J, Froesch ER. Inhibition of the activity of nonsuppressible insulin-like activity on isolated rat fat cells by binding to its carrier protein. J Clin Invest 1979; 63:1077–1084.
16. Zapf J. Physiological roles of the insulin-like growth factor binding proteins. Eur J Endocrinol 1995; 132:645–654.
17. Baxter RC. Molecular aspects of insulin-like growth factor binding proteins. In: D. LeRoith ed., Advances in Molecular and Cellular Endocrinology. JAI Press, Greenwich, 1997, pp. 123–159.
18. Baxter RC, Binoux MA, Clemmons DR, Conover CA, Drop SLS, Holly JMP, Mohan S, Oh Y, Rosenfeld RG. Recommendations for nomenclature of the insulin-like growth factor binding protein superfamily. Endocrinology 1998; 139:4036.
19. Martin JL, Baxter RC. Insulin-like growth factor-binding protein from human plasma. Purification and characterization. J Biol Chem 1986; 261:8754–8760.
20. Heding A, Gill R, Ogawa Y, De Meyts P, Shymko RM. Biosensor measurement of the binding of insulin-like growth factors I and II and their analogues to the insulin-like growth factor-binding protein-3. J Biol Chem 1996; 271:13948–13952.
21. Martin JL, Willetts KE, Baxter RC. Purification and properties of a novel insulin-like growth factor-II binding protein from transformed human fibroblasts. J Biol Chem 1990; 265:4124–4130.
22. Holman SR, Baxter RC. Insulin-like growth factor binding protein-3: factors affecting binary and ternary complex formation. Growth Regul 1996; 6:42–47.

23. Clemmons DR, Dehoff ML, Busby WH, Bayne ML, Cascieri MA. Competition for binding to insulin-like growth factor (IGF) binding protein-2, 3, 4, and 5 by the IGFs and IGF analogs. Endocrinology 1992; 131:890–895.
24. Ho PJ, Baxter RC. Characterization of truncated insulin-like growth factor-binding protein-2 in human milk. Endocrinology 1997; 138:3811–3818.
25. Forbes B, Szabo L, Baxter RC, Ballard FJ, Wallace JC. Classification of the insulin-like growth factor binding proteins into three distinct categories according to their binding specificities. Biochem Biophys Res Commun 1988; 157:196–202.
26. Baxter RC, Bayne ML, Cascieri MA. Structural determinants for binary and ternary complex formation between insulin-like growth factor-I (IGF-I) and IGF binding protein-3. J Biol Chem 1992; 267:60–65.
27. Francis GL, Aplin SE, McNeil KA, Milner SJ, Forbes BE, Ross M, Ballard FJ, Wallace JC. Molecular interactions of IGF-I and IGF-II with their binding proteins and receptors. In: Baxter RC, Gluckman PD, Rosenfeld RG, eds. The Insulin-Like Growth Factors and Their Regulatory Proteins. Elsevier, Amsterdam, 1994, pp. 57–66.
28. Bach LA, Hsieh S, Sakano K, Fujiwara H, Perdue J, Rechler MM. Binding of mutants of human insulin-like growth factor II to insulin-like growth factor binding proteins 1–6. J Biol Chem 1993; 268:9246–9254.
29. Furlanetto RW. The somatomedin C binding protein: evidence for a heterologous subunit structure. J Clin Endocrinol Metab 1980; 51:12–19.
30. Hintz RL, Liu F. Somatomedin plasma binding proteins. In: Pecile A, Müller EE, eds. Growth Hormone and other Biologically Active Peptides. Excerpta Medica, Amsterdam, 1980, pp. 133–143.
31. Baxter RC, Martin JL, Beniac VA. High molecular weight insulin-like growth factor binding protein complex. Purification and properties of the acid-labile subunit from human serum. J Biol Chem 1989; 264:11843–11848.
32. Twigg SM, Baxter RC. Insulin-like growth factor (IGF)-binding protein 5 forms an alternative ternary complex with IGFs and the acid-labile subunit. J Biol Chem 1998; 273:6074–6079.
33. Baxter RC. Characterization of the acid-labile subunit of the growth hormone-dependent insulin-like growth factor binding protein complex. J Clin Endocrinol Metab 1988; 67:265–272.
34. Baxter RC, Martin JL. Structure of the Mr 140,000 growth hormone-dependent insulin-like growth factor binding protein complex: determination by reconstitution and affinity labeling. Proc Natl Acad Sci USA 1989; 86:6898–6902.
35. Lee CY, Rechler MM. A major portion of the 150-kilodalton insulin-like growth factor-binding protein (IGFBP) complex in adult rat serum contains unoccupied, proteolytically nicked IGFBP-3 that binds IGF-II preferentially. Endocrinology 1995; 136:668–678.
36. Barreca A, Ponzani P, Arvigo M, Giordano G, Minuto F. Effect of the acid-labile subunit on the binding of insulin-like growth factor (IGF)-binding protein-3 to [^{125}I]IGF-I. J Clin Endocrinol Metab 1995; 80:1318–1324.
37. Huhtala M-L, Koistinen R, Palomäki P, Partanen P, Bohn H, Seppälä M. Biologically active domain in somatomedin-binding protein. Biochem Biophys Res Commun 1986; 141:263–270.
38. Hobba GD, Lothgren A, Holmberg E, Forbes BE, Francis GL, Wallace JC. Alanine screening mutagenesis establishes tyrosine 60 of bovine insulin-like growth factor binding protein-2 as a determinant of insulin-like growth factor binding. J Biol Chem 1998; 273:19691–19698.
39. Forbes BE, Turner D, Hodge SJ, McNeil KA, Forsberg G, Wallace JC. Localization of an insulin-like growth factor (IGF) binding site of bovine IGF binding protein-2 using disulfide mapping and deletion mutation analysis of the C-terminal domain. J Biol Chem 1998; 273:4647–4652.
40. Brinkman A, Kortleve DJ, Schuller AG, Zwarthoff EC, Drop SL. Site-directed mutagenesis of the N-terminal region of IGF binding protein 1: analysis of IGF binding capacity. FEBS Lett 1991; 291:264–268.
41. Brinkman A, Kortleve DJ, Zwarthoff EC, Drop SL. Mutations in the C-terminal part of insulin-like growth factor (IGF)-binding protein-1 result in dimer formation and loss of IGF binding capacity. Mol Endocrinol 1991; 5:987–994.
42. Firth SM, Ganeshprasad U, Baxter RC. Structural determinants of ligand and cell surface binding of insulin-like growth factor binding protein-3. J Biol Chem 1998; 273:2631–2638.
43. Twigg SM, Kiefer MC, Zapf J, Baxter RC. Insulin-like growth factor-binding protein 5 complexes with the acid-labile subunit. Role of the carboxy-terminal domain. J Biol Chem 1998; 273:28791–28798.
44. Coverley JA, Baxter RC. Phosphorylation of insulin-like growth factor binding proteins. Mol Cell Endocrinol 1997; 128:1–5.

45. Westwood M, Gibson JM, White A. Purification and characterization of the insulin-like growth factor-binding protein-1 phosphoform found in normal plasma. Endocrinology 1997; 138:1130–1136.
46. Jones JI, D'Erole AJ, Camacho-Hübner C, Clemmons DR. Phosphorylation of insulin-like growth factor (IGF)-binding protein 1 in cell culture and in vivo: effects on affinity for IGF-1. Proc Natl Acad Sci USA 1991; 88:7481–7485.
47. Peterkofsky B, Gosiewska A, Wilson S, Kim YR. Phosphorylation of rat insulin-like growth factor binding protein-1 does not affect its biological properties. Arch Biochem Biophys 1998; 357:101–110.
48. Frost RA, Tseng I. Insulin-like growth factor binding protein-1 is phosphorylated by cultured human endometrial stromal cells and multiple kinases in vitro. J Biol Chem 1991; 266:18082–18088.
49. Jones JI, Busby WH, Wright G, Smith CE, Kimack NM, Clemmons DR. Identification of the sites of phosphorylation in insulin-like growth factor binding protein-1: regulation of its affinity by phosphorylation of serine 101. J Biol Chem 1993; 268:1125–1131.
50. Hoeck WG, Mukku VR. Identification of the major sites of phosphorylation in IGF binding protein-3. J Cell Biochem 1994; 56:262–273.
51. Coverley JA, Baxter RC. Regulation of insulin-like growth factor (IGF) binding protein-3 phosphorylation by IGF-1. Endocrinology 1995; 136:5778–5781.
52. Firth SM, Baxter RC. Characterisation of recombinant glycosylation variants of insulin-like growth factor binding protein-3. J Endocrinol 1999; 160:379–387.
53. Conover CA. Glycosylation of insulin-like growth factor binding protein-3 (IGFBP-3) is not required for potentiation of IGF-I action: evidence for processing of cell-bound IGFBP-3. Endocrinology 1991; 129:3259–3268.
54. Ceda GP, Fielder PJ, Henzel WJ, Louie A, Donovan SM, Hoffman AR, Rosenfeld RG. Differential effects of IGF-I and IGF-II on the expression of IGFBPs in rat neuroblastoma cell line: isolation and characterization of two forms of IGFBP-4. Endocrinology 1991; 128:2815–2824.
55. Bach LA, Thotakura NR, Rechler MM. Human insulin-like growth factor binding protein-6 is *O*-glycosylated. Biochem Biophys Res Commun 1992; 186:301–307.
56. Rees C, Clemmons DR, Horvitz GD, Clarke JB, Busby WH. Protease-resistant form of insulin-like growth factor (IGF) binding protein 4 inhibits IGF-1 actions. Endocrinology 1998; 139:4182–4188.
57. Imai Y, Busby WH, Smith CE, Clarke JB, Garmong AJ, Horwitz GD, Rees C, Clemmons DR. Protease-resistant form of insulin-like growth factor-binding protein 5 in an inhibitor of insulin-like growth factor-I actions on porcine smooth muscle cells in culture. J Clin Invest 1997; 100:2596–2605.
58. Binoux M, Lalou C, Lassarre C, Segovia B. Regulation of IGF bioavailability by IGFBP proteases. In Baxter RC, Gluckman PD, Rosenfeld RG, eds. The Insulin-Like Growth Factors and Their Regulatory Proteins. Elsevier, Amsterdam, 1994, pp. 217–226.
59. Bang P, Nygren J, Carlssonskwirut C, Thorell A, Ljungqvist O. Postoperative induction of insulin-like growth factor binding protein-3 proteolytic activity—relation to insulin and insulin sensitivity. J Clin Endocrinol Metab 1998; 83:2509–2515.
60. Hossenlopp P, Segovia B, Lassarre C, Roghani M, Bredon M, Binoux M. Evidence of enzymatic degradation of insulin-like growth factor-binding proteins in the 150K complex during pregnancy. J Clin Endocrinol Metab 1990; 71:797–805.
61. Suikkari A-M, Baxter RC. Insulin-like growth factor binding protein-3 is functionally normal in pregnancy serum. J Clin Endocrinol Metab 1992; 74:177–183.
62. Baxter RC, Skriver L. Altered ligand specificity of proteolysed insulin-like growth factor binding protein-3. Biochem Biophys Res Commun 1993; 196:1267–1273.
63. Kubler B, Cowell S, Zapf J, Braulke T. Proteolysis of insulin-like growth factor binding proteins by a novel 50-kiloDalton metalloproteinase in human pregnancy serum. Endocrinology 1998; 139:1556–1563.
64. Lalou C, Lassarre C, Binoux M. A proteolytic fragment of insulin-like growth factor (IGF) binding protein-3 that fails to bind IGFs inhibits the mitogenic effects of IGF-I and insulin. Endocrinology 1996; 137:3206–3212.
65. Guler H-P, Zapf J, Froesch ER. Short-term metabolic effects of recombinant human insulin-like growth factor I in healthy adults. N Engl J Med 1987; 317:137–140.
66. Baxter RC. Insulin-like growth factor binding proteins as glucoregulators. Metabolism 1995; 44 (Suppl 4): 12–17.
67. Frystyk J, Skjaerbaek C, Dinesen B, Orskov H. Free insulin-like growth factors (IGF-I and IGF-II) in human serum. FEBS Lett 1994; 348:185–191.

68. Chin E, Zhou J, Dai J, Baxter RC, Bondy CA. Cellular localization and regulation of gene expression for components of the insulin-like growth factor ternary binding protein complex. Endocrinology 1994; 134:2498–2504.
69. Xu S, Cwyfan-Hughes SC, Van der Stappen JWJ, Sanson J, Burton JL, Donnelly M, Holly JMP. Insulin-like growth factors (IGFs) and IGF-binding proteins in human skin interstitial fluid. J Clin Endocrinol Metab 1995; 80:2940–2945.
70. Hodgkinson SC, Spencer GSG, Bass JJ, Davis SR, Gluckman PD. Distribution of circulating insulin-like growth factor-I (IGF-I) into tissues. Endocrinology 1991; 129:2085–2093.
71. Robertson JG, Pickering KJ, Belford DA. Insulin-like growth factor I (IGF-I) and IGF-binding proteins in rat wound fluid. Endocrinology 1996; 137:2774–2781.
72. Lewitt MS, Saunders H, Phuyal JL, Baxter RC. Complex formation by human insulin-like growth factor-binding protein-3 and human acid-labile subunit in growth hormone-deficient rats. Endocrinology 1994; 134:2404–2409.
73. Baxter RC. Glycosaminoglycans inhibit formation of the 140 kilodalton insulin-like growth factor-binding protein complex. Biochem J 1990; 271:773–777.
74. Baxter RC, Suikkari A-M, Martin JL. Characterization of the binding defect in insulin-like growth factor binding protein-3 from pregnancy serum. Biochem J 1993; 294:847–852.
75. Sommer A, Maack CA, Spratt SK, Mascarenhas D, Tressel TJ, Rhodes ET, Lee R, Roumas M, Tatsuno GP, Flynn JA, Gerber N, Taylor J, Cudny H, Nanney L, Hunt TK, Spencer EM. Molecular genetics and actions of recombinant insulin-like growth factor binding protein-3. In: Spencer EM, ed. Modern Concepts of Insulin-Like Growth Factors. Elsevier, New York, 1991, pp. 715–728.
76. Lewitt MS, Saunders H, Baxter RC. Bioavailability of insulin-like growth factors (IGFs) in rats determined by the molecular distribution of IGF-binding protein-3. Endocrinology 1993; 133:1797–1802.
77. Frystyk J, Skjaerbaek C, Zapf J, Orskov H. Increased levels of circulating free insulin-like growth factors in patients with non-islet cell tumour hypoglycaemia. Diabetologia 1998; 41:589–594.
78. Lewitt MS, Denyer GS, Cooney GJ, Baxter RC. Insulin-like growth factor binding protein-1 modulates blood glucose levels. Endocrinology 1991; 129:2254–2256.
79. Rajkumar K, Barron D, Lewitt MS, Murphy LJ. Growth retardation and hyperglycemia in insulin-like growth factor binding protein-1 transgenic mice. Endocrinology 1995; 136:4029–4034.
80. Mortensen DL, Won WB, Siu J, Reifsnyder D, Gironella M, Etcheverry T, Clark RG. Insulin-like growth factor binding protein-1 induces insulin release in the rat. Endocrinology 1997; 138:2073–2080.
81. Lewitt MS, Baxter RC. Insulin-like growth factor-binding protein-1: a role in glucose counter-regulation? Mol Cell Endocrinol 1991; 79:C147–C152.
82. Frystyk J, Hussain M, Skjaerbaek C, Schmitz O, Christiansen JS, Froesch ER, Orskov H. Serum free IGF-I during a hyperinsulinemic clamp following 3 days of administration of IGF-I vs saline. Am J Physiol 1997; 36:E507–E513.
83. Frystyk J, Grofte T, Skjaerbaek C, Orskov H. The effect of oral glucose on serum free insulin-like growth factor-I and -II in healthy adults. J Clin Endocrinol Metab 1997; 82:3124–3127.
84. Baxter RC, Hizuka N, Takano K, Holman SR, Asakawa K. Responses of insulin-like growth factor binding protein-1 (IGFBP-1) and the IGFBP-3 complex to administration of insulin-like growth factor-I. Acta Endocrinologica 1993; 128:101–108.
85. Baxter RC. Circulating binding proteins for the insulin-like growth factors. Trends Endocrinol Metab 1993; 4:91–96.
86. Young SCJ, Miles MV, Clemmons DR. Determination of the pharmacokinetic profiles of insulin-like growth factor binding proteins-1 and -2 in rats. Endocrinology 1992; 131:1867–1873.
87. Mohan S, Libanati C, Dony C, Lang K, Srinivasan N, Baylink DJ. Development, validation, and application of a radioimmunoassay for insulin-like growth factor binding protein-5 in human serum and other biological fluids. J Clin Endocrinol Metab 1995; 80:2638–2645.
88. Ono T, Kanzaki S, Seino Y, Baylink DJ, Mohan S. Growth hormone (GH) treatment of GH-deficient children increases serum levels of insulin-like growth factors (IGFs), IGF-binding protein-3 and -5, and bone alkaline phosphatase isoenzyme. J Clin Endocrinol Metab 1996; 81:2111–2116.
89. Cornell HJ, Herington AC. Evidence for a biologically active, carrier-bound form of non-suppressible insulin-like activity (NSILA) in human serum. Int J Biochem 1983; 15:553–558.
90. Blum WF, Jenne EW, Reppin F, Kietzmann K, Ranke MB, Bierich JR. Insulin-like growth factor I (IGF-I)-binding protein complex is a better mitogen than free IGF-I. Endocrinology 1989; 125:766–772.

91. Tesch GH, Cornell HJ, Herington AC, Oakes S. Effects of insulin-like growth factor binding protein complexes on human fibroblast growth. Growth Regul 1993; 3:151–159.
92. De Mellow JS, Baxter RC. Growth hormone-dependent insulin-like growth factor (IGF) binding protein both inhibits and potentiates IGF-I-stimulated DNA synthesis in human skin fibroblasts. Biochem Biophys Res Commun 1988; 156:199–204.
93. Lin T, Wang D, Nagpal ML, Shimasaki S, Ling N. Expression and regulation of insulin-like growth factor-binding protein-1, -2, -3, and -4 messenger ribonucleic acids in purified rat Leydig cells and their biological effects. Endocrinology 1993; 132:1898–1904.
94. Chochinov RH, Mariz IK, Hajek AS, Daughaday WH. Characterization of a protein in mid-term human amniotic fluid which reacts in the somatomedin-C radioreceptor assay. J Clin Endocrinol Metab 1997; 44:902–908.
95. Busby WJ, Klapper DG, Clemmons DR. Purification of a 31,000-dalton insulin-like growth factor binding protein from human amniotic fluid. Isolation of two forms with different biologic actions. J Biol Chem 263: 1988;14203–14210.
96. Liu L, Brinkman A, Blat C, Harel L. IGFBP-1, an insulin like growth factor binding protein, is a cell growth inhibitor. Biochem Biophys Res Commun 1991; 174:673–679.
97. Frauman AG, Tsuzaki S, Moses AC. The binding characteristics and biological effects in FRTL5 cells of placental protein-12, an insulin-like growth factor-binding protein purified from human amniotic fluid. Endocrinology 1989; 124:2289–2296.
98. Motani A, Rutherford C, Anggard EE, Ferns GA. Insulin-like growth factor binding protein-1 inhibits arterial smooth muscle cell proliferation in vitro but does not reduce the neointimal response to balloon catheter injury. Atherosclerosis 1995; 118:57–66.
99. Burch WM, Correa J, Shively JE, Powell DR. The 25-kilodalton insulin-like growth factor (IGF)-binding protein inhibits both basal and IGF-I mediated growth of chick embryo pelvic cartilage in vitro. J Clin Endocrinol Metab 1990; 70:173–180.
100. Angervo M, Koistinen R, Suikkari AM, Seppala M. Insulin-like growth factor binding protein-1 inhibits the DNA amplification induced by insulin-like growth factor I in human granulosa-luteal cells. Human Reprod 1991; 6:770–773.
101. Lavranos TC, O'Leary PC, Rodgers RJ. Effects of insulin-like growth factors and binding protein 1 on bovine granulosa cell division in anchorage-independent culture. J Reprod Fertil 1996; 107:221–228.
102. Grellier P, Feliers D, Yee D, Woodruff K, Abboud SL. Interaction between insulin-like growth factor-I and insulin-like growth factor-binding proteins in TC-1 stromal cells. J Endocrinol 1996; 149:519–529.
103. Koistinen R, Angervo M, Leinonen P, Seppala M. Phosphorylation of insulin-like growth factor-binding protein-1 from different sources. Growth Regul 1993; 3:34–37.
104. Frost RA, Mazella J, Tseng L. Insulin-like growth factor binding protein-1 inhibits the mitogenic effect of insulin-like growth factors and progestins in human endometrial stromal cells. Biol Reprod 1993; 49:104–111.
105. Blat C, Delbe J, Villaudy J, Chatelain G, Golde A, Harel L. Inhibitory diffusible factor 45 bifunctional activity as a cell growth inhibitor and as an insulin-like growth factor I binding protein. J Biol Chem 1989; 264:12449–12454.
106. Martin JL, Baxter RC. Insulin-like growth factor-binding proteins (IGF-BPs) produced by human skin fibrobasts: immunological relationship to other human IGF-BPs. Endocrinology 1988; 123:1907–1915.
107. Coleman ME, Etherton TD. Porcine insulin-like growth factor (IGF)-binding protein-3 elicits bi-phasic effects on IGF-I stimulated DNA synthesis in neonatal porcine skin fibroblasts. Domest Anim Endocrinol 1994; 11:299–305.
108. Cohen P, Lamson G, Okajima T, Rosenfeld RG. Transfection of the human insulin-like growth factor binding protein-3 gene into Balb/c fibroblasts inhibits cellular growth. Mol Endocrinol 1993; 7:380–386.
109. Goldstein S, Moerman EJ, Baxter RC. Accumulation of insulin-like growth factor binding protein-3 in conditioned medium of human fibroblasts increases with chronologic age of donor and senescence in vitro. J Cell Physiol 1993; 156:294–302.
110. Schmid C, Rutishauser J, Schlapfer I, Froesch ER, Zapf I. Intact but not truncated insulin-like growth factor binding protein-3 (IGFBP-3) blocks IGF-I-induced stimulation of osteoblasts: control of IGF signalling to bone cells by IGFBP-3-specific proteolysis? Biochem Biophys Res Commun 1991; 179:579–585.

111. Andress DL, Birnbaum RS. Human osteoblast-derived insulin-like growth factor (IGF) binding protein-5 stimulates osteoblast mitogenesis and potentiates IGF action. J Biol Chem 1992; 267:22467–22472.
112. Knauer DJ, Smith GL. Inhibition of biological activity of multiplication-stimulating activity by binding to its carrier protein. Proc Natl Acad Sci USA 1980; 77:7252–7256.
113. Bourner MJ, Busby WJ, Siegel NR, Krivi GG, McCusker RH, Clemmons DR. Cloning and sequence determination of bovine insulin-like growth factor binding protein-2 (IGFBP-2): comparison of its structural and functional properties with IGFBP-1. J Cell Biochem 1992; 48:215–226.
114. Cohen P, Noveral JP, Bhala A, Nunn SE, Herrick DJ, Grunstein MM. Leukotriene D4 facilitates airway smooth muscle cell proliferation via modulation of the IGF axis. Am J Physiol 1995; 13:L151–L157.
115. Feyen JH, Evans DB, Binkert C, Heinrich GF, Geisse S, Kocher HP. Recombinant human [Cys281]insulin-like growth factor-binding protein 2 inhibits both basal and insulin-like growth factor I-stimulated proliferation and collagen synthesis in fetal rat calvariae. J Biol Chem 1995; 266:19469–19474.
116. Woodward TL, Turner JD, Hung HT, Zhao X. Inhibition of cellular proliferation and modulation of insulin-like growth factor binding proteins by retinoids in a bovine mammary epithelial cell line. J Cell Physiol 1996; 167:488–499.
117. Mouhieddine OB, Cazals V, Kuto E, Le BY, Clement A. Glucocorticoid-induced growth arrest of lung alveolar epithelial cells is associated with increased production of insulin-like growth factor binding protein-2. Endocrinology 1996; 137:287–295.
118. Horney MJ, Shirley DW, Kurtz DT, Rosenzweig SA. Elevated glucose increases mesangial cell sensitivity to insulin-like growth factor I. Am J Physiol 1998; 43:F1045–F1053.
119. Park JH, McCusker RH, Vanderhoof JA, Mohammadpour H, Harty RF, MacDonald RG. Secretion of insulin-like growth factor II (IGF-II) and IGF-binding protein-2 by intestinal epithelial (IEC-6) cells: implications for autocrine growth regulation. Endocrinology 1992; 131:1359–1368.
120. Corkins MR, Vanderhoof JA, Slentz DH, MacDonald RG, Park JH. Growth stimulation by transfection of intestinal epithelial cells with an antisense insulin-like growth factor binding protein-2 construct. Biochem Biophys Res Commun 1995; 211:707–713.
121. Guo YS, Townsend CJ, Jin GF, Beauchamp RD, Thompson JC. Differential regulation by TGF-beta 1 and insulin of insulin-like growth factor binding protein-2 in IEC-6 cells. Am J Physiol 1995; 31:E1199–E1204.
122. Mohan S, Nakao Y, Honda Y, Landale E, Leser U, Dony C, Lang K, Baylink DJ. Studies on the mechanisms by which insulin-like growth factor (IGF) binding protein-4 (IGFBP-4) and IGFBP-5 modulate IGF actions in bone cells. J Biol Chem 1995; 270:20424–20431.
123. Mohan S, Bautista CM, Wergedal J, Baylink DJ. Isolation of an inhibitory insulin-like growth factor (IGF) binding protein from bone cell-conditioned medium: a potential local regulator of IGF action. Proc Natl Acad Sci USA 1989; 86:8338–8342.
124. Amarnani S, Merriman HL, Linkhart TA, Baylink DJ, Mohan S. Autocrine regulators of MC3T3-E1 cell proliferation. J Bone Mineral Res 1993; 8:157–165.
125. Cheung PT, Smith EP, Shimasaki S, Ling N, Chernausek SD. Characterization of an insulin-like growth factor binding protein (IGFBP-4) produced by the B104 rat neuronal cell line: chemical and biological properties and differential synthesis by sublines. Endocrinology 1991; 129:1006–1015.
126. Bradshaw SL, Naus CC, Zhu D, Kidder GM, Han VK. Insulin-like growth factor binding protein-4 gene expression is induced by transfection of gap junction connexin43 gene in a C6 glioma cell line. Growth Regul 1993; 3:26–29.
127. Kiefer MC, Schmid C, Waldvogel M, Schlapfer I, Futo E, Masiarz FR, Green K, Barr PJ, Zapf J. Characterization of recombinant human insulin-like growth factor binding proteins 4, 5, and 6 produced in yeast. J Biol Chem 1992; 267:12692–12699.
128. Conover CA, Kiefer MC. Regulation and biological effect of endogenous insulin-like growth factor binding protein-5 in human osteoblastic cells. J Clin Endocrinol Metab 1993; 76:1153–1159.
129. Reeve JG, Guadano A, Xiong J, Morgan J, Bleehen NM. Diminished expression of insulin-like growth factor (IGF) binding protein-5 and activation of IGF-I mediated autocrine growth in simian virus 40-transformed human fibroblasts. J Biol Chem 1995; 270:135–142.
130. Bach LA, Salemi R, Leeding KS. Roles of insulin-like growth factor (IGF) receptors and IGF-binding proteins in IGF-II-induced proliferation and differentiation of L6A1 rat myoblasts. Endocrinology 1995; 136:5061–5069.
131. Kato M, Ishizaki A, Hellman U, Wernstedt C, Kyogoku M, Miyazono K, Heldin CH, Funa K. A human keratinocyte cell line produces two autocrine growth inhibitors, transforming growth factor-beta and

insulin-like growth factor binding protein-6, in a calcium- and cell density-dependent manner. J Biol Chem 1995; 270:12373–12379.
132. Grellier P, Degalle B, Babajko S. Expression of insulin-like growth factor-binding protein 6 complementary DNA alters neuroblastoma cell growth. Cancer Res 1998; 58:1670–1676.
133. Buckbinder L, Talbott R, Velasco MS, Takenaka I, Faha B, Seizinger BR, Kley N. Induction of the growth inhibitor IGF-binding protein 3 by p53. Nature 1995; 377:646–649.
134. Friedlander P, Haupt Y, Prives C, Oren M. A mutant p53 that discriminates between p53-responsive genes cannot induce apoptosis. Mol Cell Biol 1996; 16:4961–4971.
135. Ludwig RL, Bates S, Vousden KH. Differential activation of target cellular promoters by p53 mutants with impaired apoptotic function. Mol Cell Biol 1996; 16:4952–4960.
136. Chun SY, Billig H, Tilly JL, Furuta I, Tsafriri A, Hsueh AJ. Gonadotropin suppression of apoptosis in cultured preovulatory follicles: mediatory role of endogenous insulin-like growth factor I. Endocrinology 1994; 135:1845–1853.
137. Flaws JA, DeSanti A, Tilly KI, Javid RO, Kugu K, Johnson AL, Hirshfield AN, Tilly JL. Vasoactive intestinal peptide-mediated suppression of apoptosis in the ovary: potential mechanisms of action and evidence of a conserved antiatretogenic role through evolution. Endocrinology 1995; 136:4351–4359.
138. Guthrie HD, Grimes RW, Hammond JM. Changes in insulin-like growth factor-binding protein-2 and -3 in follicular fluid during atresia of follicles grown after ovulation in pigs. J Reprod Fertil 1995; 104:225–230.
139. Neuenschwander S, Schwartz A, Wood TL, Roberts CJ, Henninghausen L, LeRoith D. Involution of the lactating mammary gland is inhibited by the IGF system in a transgenic mouse model. J Clin Invest 1996; 97:2225–2232.
140. Tonner E, Quarrie L, Travers M, Barber M, Logan A, Wilde C, Flint D. Does an IGF-binding protein (IGFBP) present in involuting rat mammary gland regulate apoptosis? Prog Growth Factor Res 1995; 6:409–414.
141. Guenette RS, Tenniswood M. The role of growth factors in the suppression of active cell death in the prostate: an hypothesis. Biochem Cell Biol 1994; 72:553–559.
142. Gockerman A, Prevette T, Jones JI, Clemmons DR. Insulin-like growth factor (IGF)-binding proteins inhibit the smooth muscle cell migration responses to IGF-I and IGF-II. Endocrinology 1995; 136:4168–4173.
143. Abrass CK, Berfield AK, Andress DL. Heparin binding domain of insulin-like growth factor binding protein-5 stimulates mesangial cell migration. Am J Physiol 1997; 42:F 899–F 906.
144. Bach LA, Hsieh S, Brown AL, Rechler MM. Recombinant human insulin-like growth factor (IGF)-binding protein-6 inhibits IGF-II-induced differentiation of L6A1 myoblasts. Endocrinology 1994; 135:2168–2176.
145. Ewton DZ, Florini JR. IGF binding proteins-4, -5 and -6 may play specialized roles during L6 myoblast proliferation and differentiation. J Endocrinol 1995; 144:539–553.
146. Rotwein P, James PL, Kou K. Rapid activation of insulin-like growth factor binding protein-5 gene transcription during myoblast differentiation. Mol Endocrinol 1995; 9:913–923.
147. James PL, Stewart CE, Rotwein P. Insulin-like growth factor binding protein-5 modulates muscle differentiation through an insulin-like growth factor-dependent mechanism. J Cell Biol 1996; 133:683–693.
148. Stewart CE, James PL, Fant ME, Rotwein P. Overexpression of insulin-like growth factor-II induces accelerated myoblast differentiation. J Cell Physiol 1996; 169:23–32.
149. Ui M, Shimonaka M, Shimasaki S, Ling N. An insulin-like growth factor-binding protein in ovarian follicular fluid blocks follicle-stimulating hormone-stimulated steroid production by ovarian granulosa cells. Endocrinology 1989; 125:912–916.
150. Bicsak TA, Shimonaka M, Malkowski M, Ling N. Insulin-like growth factor-binding protein (IGF-BP) inhibition of granulosa cell function: effect on cyclic adenosine 3′,5′-monophosphate, deoxyribonucleic acid synthesis, and comparison with the effect of an IGF-I antibody. Endocrinology 1990; 126:2184–2189.
151. Mason HD, Willis D, Holly JM, Cwyfan-Hughes S, Seppala M, Franks S. Inhibitory effects of insulin-like growth factor-binding proteins on steroidogenesis by human granulosa cells in culture. Mol Cell Endocrinol 1992; 89:R1–R4.
152. Iwashita M, Adachi T, Katayama E, Kudo Y, Takeda Y. Regulation and physiological role of insulin-like-growth-factor-binding protein-1 in human granulosa cells. Horm Res 1994; 1:22–28.

153. Yoshimura Y, Nagamatsu S, Ando M, Iwashita M, Oda T, Katsumata Y, Shiokawa S, Nakamura Y. Insulin-like growth factor binding protein-3 inhibits gonadotropin-induced ovulation, oocyte maturation, and steroidogenesis in rabbit ovary. Endocrinology 1996; 137:438–446.
154. Menuelle P, Binoux M, Plas C. Regulation by insulin-like growth factor (IGF) binding proteins of IGF-II-stimulated glycogenesis in cultured fetal rat hepatocytes. Endocrinology 1995; 136:5305–5310.
155. Okajima T, Iwashita M, Takeda Y, Sakamoto S, Tanabe T, Yasuda T, Rosenfeld RG. Inhibitory effects of insulin-like growth factor (IGF)-binding proteins-1 and -3 on IGF-activated glucose consumption in mouse BALB/c 3T3 fibroblasts. J Endocrinol 1993; 136:457–470.
156. Walton PE, Gopinath R, Etherton TD. Porcine insulin-like growth factor (IGF) binding protein blocks IGF-I action on porcine adipose tissue. Proc Soc Exp Biol Med 1989; 190:315–319.
157. Peterkofsky B, Gosiewska A, Kipp DE, Shah V, Wilson S. Circulating insulin-like growth factor binding proteins (IGFBPs) 1 and 2 induced in vitamin C-deficient or fasted guinea pigs inhibit IGF-I action in cultured cells. Growth Factors 1994; 10:229–241.
158. Osborne CK, Hobbs K, Trent JM. Biological differences among MCF-7 human breast cancer cell lines from different laboratories. Breast Cancer Res Treat 1987; 9:111–121.
159. Figueroa JA, Sharma J, Jackson JG, McDermott MJ, Hilsenbeck SK, Yee D. Recombinant insulin-like growth factor binding protein-1 inhibits IGF-I, serum, and estrogen-dependent growth of MCF-7 human breast cancer cells. J Cell Physiol 1993; 157:229–236.
160. Pratt SE, Pollak MN. Insulin-like growth factor binding protein 3 (IGF-BP3) inhibits estrogen-stimulated breast cancer cell proliferation. Biochem Biophys Res Commun 1994; 198:292–297.
161. Huynh H, Yang X, Pollak M. Estradiol and antiestrogens regulate a growth inhibitory insulin-like growth factor binding protein 3 autocrine loop in human breast cancer cells. J Biol Chem 1996; 271:1016–1021.
162. Oh Y, Muller HL, Ng L, Rosenfeld RG. Transforming growth factor-beta-induced cell growth inhibition in human breast cancer cells is mediated through insulin-like growth factor-binding protein-3 action. J Biol Chem 1995; 270:13589–13592.
163. Gucev ZS, Oh Y, Kelley KM, Rosenfeld RG. Insulin-like growth factor binding protein 3 mediates retinoic acid- and transforming growth factor beta 2-induced growth inhibition in human breast cancer cells. Cancer Res 1996; 56:1545–1550.
164. Baxter RC, Maitland JE, Raison RL, Reddel RR, Sutherland RL. High molecular weight somatomedin-C/IGF-I from T47D human mammary carcinoma cells: Immunoreactivity and bioactivity. In: Spencer E, ed. Insulin-Like Growth Factors/Somatomedins. Walter de Gruyter, Berlin, 1983, pp. 615–618.
165. Fontana JA, Burrows MA, Clemmons DR, LeRoith D. Retinoid modulation of insulin-like growth factor-binding proteins and inhibition of breast carcinoma proliferation. Endocrinology 1991; 128:1115–1122.
166. Martin JL, Coverley JA, Pattison ST, Baxter RC. Insulin-like growth factor-binding protein-3 production by MCF-7 breast cancer cells: stimulation by retinoic acid and cyclic adenosine monophosphate and differential effects of estradiol. Endocrinology 1995; 136:1219–1226.
167. Yee D, Jackson JG, Kozelsky TW, Figueroa JA. Insulin-like growth factor binding protein 1 expression inhibits insulin-like growth factor I action in MCF-7 breast cancer cells. Cell Growth Differ 1994; 5:73–77.
168. Tennant MK, Thrasher JB, Twomey PA, Birnbaum RS, Plymate SR. Insulin-like growth factor-binding proteins (IGFBP)-4, -5, and -6 in the benign and malignant human prostate: IGFBP-5 messenger ribonucleic acid localization differs from IGFBP-5 protein localization. J Clin Endocrinol Metab 1996; 81:3783–3792.
169. Tennant MK, Thrasher JB, Twomey PA, Birnbaum RS, Plymate SR. Insulin-like growth factor-binding protein-2 and -3 expression in benign human prostate epithelium, prostate intraepithelial neoplasia, and adenocarcinoma of the prostate. J Clin Endocrinol Metab 1996; 81:411–420.
170. Kimura G, Kasuya J, Giannini S, Honda Y, Mohan S, Kawachi M, Akimoto M, Fujita YY. Insulin-like growth factor (IGF) system components in human prostatic cancer cell-lines: LNCaP, DU145, and PC-3 cells. Int J Urol 1996; 3:39–46.
171. Iwamura M, Sluss PM, Casamento JB, Cockett ATK. Insulin-like growth factor-1 - action and receptor characterization in human prostate cancer cell lines. Prostate 1993; 22:243–252.
172. Perkel VS, Mohan S, Baylink DJ, Linkhart TA. An inhibitory insulin-like growth factor binding protein (In-IGFBP) from human prostatic cell conditioned medium reveals N-terminal sequence identity with bone derived In-IGFBP. J Clin Endocrinol Metab 1990; 71:533–535.

173. Srinivasan N, Edwall D, Linkhart TA, Baylink DJ, Mohan S. Insulin-like growth factor-binding protein-6 produced by human PC-3 prostate cancer cells: isolation, characterization and its biological action. J Endocrinol 1996; 149:297–303.
174. Angelloz-Nicoud P, Binoux M. Autocrine regulation of cell proliferation by the insulin-like growth factor (IGF) and IGF binding protein-3 protease system in a human prostate carcinoma cell line (PC-3). Endocrinol 1995; 136:5485–5492.
175. Figueroa JA, Lee AV, Jackson JG, Yee D. Proliferation of cultured human prostate cancer cells is inhibited by insulin-like growth factor (IGF) binding protein-1: evidence for an IGF-II autocrine growth loop. J Clin Endocrinol Metab 1995; 80:3476–3482.
176. Garrouste F, Remacle-Bonnet M, Culouscou JM, Marvaldi J, Pommier G. Type-II insulin-like growth-factor receptor in conditioned medium from HT-29 human colon carcinoma cell line. Int J Cancer 1991; 47:760–764.
177. Culouscou JM, Shoyab M. Purification of a colon cancer cell growth inhibitor and its identification as an insulin-like growth factor binding protein. Cancer Res 1991; 51:2813–2819.
178. Singh P, Dai B, Dhruva B, Widen SG. Episomal expression of sense and antisense insulin-like growth factor (IGF)-binding protein-4 complementary DNA alters the mitogenic response of a human colon cancer cell line (HT-29) by mechanisms that are independent of and dependent upon IGF-I. Cancer Res 1994; 54:6563–6570.
179. Guo YS, Beauchamp RD, Jin GF, Townsend CJ, Thompson JC. Insulinlike growth factor-binding protein modulates the growth response to insulinlike growth factor 1 by human gastric cancer cells. Gastroenterology 1993; 104:1595–1604.
180. Reeve JG, Morgan J, Schwander J, Bleehen NM. Role for membrane and secreted insulin-like growth factor-binding protein-2 in the regulation of insulin-like growth factor action in lung tumors. Cancer Res 1993; 53:4680–4685.
181. Karas M, Danilenko M, Fishman D, LeRoith D, Levy J, Sharoni Y. Membrane-associated insulin-like growth factor-binding protein-3 inhibits insulin-like growth factor-I-induced insulin-like growth factor-I receptor signaling in Ishikawa endometrial cancer cells. J Biol Chem 1997; 272:16514–16520.
182. Durham SK, Kiefer MC, Riggs BL, Conover CA. Regulation of insulin-like growth factor binding protein 4 by a specific insulin-like growth factor binding protein 4 proteinase in normal human osteoblast-like cells: implications in bone cell physiology. J Bone Mineral Res 1994; 9:111–117.
183. Gillespie C, Read LC, Bagley CJ, Ballard FJ. Enhanced potency of truncated insulin-like growth factor-I (des(1–3)IGF-I) relative to IGF-I in *lit/lit* mice. J Endocrinol 1990; 127:401–405.
184. Tomas FM, Knowles SE, Owens PC, Read LC, Chandler CS, Gargosky SE, Ballard FJ. Increased weight gain, nitrogen retention and muscle protein synthesis following treatment of diabetic rats with insulin-like growth factor (IGF)-I and des(1–3)IGF-I. Biochem J 1991; 276:547–554.
185. Sandra A, Boes M, Dake BL, Stokes JB, Bar RS. Infused IGF-I/IGFBP-3 complex causes glomerular localization of IGF-I in the rat kidney. Am J Physiol 1998; 38:E 32–E 37.
186. Tanaka H, Moriwake T, Matsuoka Y, Nakamura T, Seino Y. Potential role of rIGF-I/IGFBP-3 in maintaining skeletal mass in space. Bone 1998; 22:S145–S147.
187. Conover CA, Ronk M, Lombana F, Powell DR. Structural and biological characterization of bovine insulin-like growth factor binding protein-3. Endocrinology 1990; 127:2795–2803.
188. Conover CA. Potentiation of insulin-like growth factor (IGF) action by IGF-binding protein-3: studies of underlying mechanism. Endocrinology 1992; 130:3191–3199.
189. Conover CA, Clarkson JT, Bale LK. Factors regulating insulin-like growth factor-binding protein-3 binding, processing, and potentiation of insulin-like growth factor action. Endocrinology 1996; 137:2286–2292.
190. Conover CA, Powell DR. Insulin-like growth factor (IGF)-binding protein-3 blocks IGF-I-induced receptor down-regulation and cell desensitization in cultured bovine fibroblasts. Endocrinology 1991; 129:710–716.
191. Elgin RG, Busby WJ, Clemmons DR. An insulin-like growth factor (IGF) binding protein enhances the biologic response to IGF-I. Proc Natl Acad Sci USA 1987; 84:3254–3258.
192. Koistinen R, Itkonen O, Selenius P, Seppala M. Insulin-like growth factor-binding protein-1 inhibits binding of IGF-I on fetal skin fibroblasts but stimulates their DNA synthesis. Biochem Biophys Res Commun 1990; 173:408–415.
193. Jones JI, Gockerman A, Busby WHJ, Camacho-Hubner C, Clemmons DR. Extracellular matrix contains insulin-like growth factor binding protein-5: potentiation of the effects of IGF-I. J Cell Biol 1993; 121:679–687.

194. Clemmons DR, Busby WH, Arai T, Nam TJ, Clarke JB, Jones JI, Ankrapp DK. Role of insulin-like growth factor binding proteins in the control of IGF actions. Prog Growth Factor Res 1995; 6:357–366.
195. Parker A, Rees C, Clarke J, Busby WH, Clemmons DR. Binding of insulin-like growth factor (IGF)-binding protein-5 to smooth-muscle cell extracellular matrix is a major determinant of the cellular response to IGF-I. Mol Biol Cell 1998; 9:2383–2392.
196. Ernst M, Rodan GA. Increased activity of insulin-like growth factor (IGF) in osteoblastic cells in the presence of growth hormone (GH): positive correlation with the presence of the GH-induced IGF-binding protein BP-3. Endocrinology 1990; 127:807–814.
197. Andress DL. Comparison studies of IGFBP-5 binding to osteoblasts and osteoblast-derived extracellular matrix. Prog Growth Factor Res 1995; 6:337–344.
198. Andress DL. Heparin modulates the binding of insulin-like growth factor (IGF) binding protein-5 to a membrane protein in osteoblastic cells. J Biol Chem 1995; 270:28289–28296.
199. Chen JC, Shao ZM, Sheikh MS, Hussain A, LeRoith D, Roberts CTJ, Fontana JA. Insulin-like growth factor-binding protein enhancement of insulin-like growth factor-1 (IGF-I)-mediated DNA synthesis and IGF-I binding in a human breast carcinoma cell line. J Cell Physiol 1994; 158:69–78.
200. Angelloz-Nicoud P, Harel L, Binoux M. Recombinant human insulin-like growth factor (IGF) binding protein-3 stimulates prostate carcinoma cell proliferation via an IGF-dependent mechanism. Role of serine proteases. Growth Regul 1996; 6:130–136.
201. Lee YR, Oshita Y, Tsuboi R, Ogawa H. Combination of insulin-like growth factor (IGF)-I and IGF-binding protein-1 promotes fibroblast-embedded collagen gel contraction. Endocrinology 1996; 137:5278–5283.
202. Bar RS, Booth BA, Boes M, Dake BL. Insulin-like growth factor-binding proteins from vascular endothelial cells: purification, characterization, and intrinsic biological activities. Endocrinology 1989; 125:1910–1920.
203. Bourner MJ, Busby WHJ, Siegel NR, Krivi GG, McCusker RH, Clemmons DR. Cloning and sequence determination of bovine insulin-like growth factor binding protein-2 (IGFBP-2): comparison of its structural and functional properties with IGFBP-1. J Cell Biochem 1992; 48:215–226.
204. Baxter RC, Martin JL, Wood MH. Two immunoreactive binding proteins for insulin-like growth factors in human amniotic fluid: relationship to fetal maturity. J Clin Endocrinol Metab 1987; 65:423–431.
205. Wang J-F, Hampton B, Mehlman T, Burgess WH, Rechler MM. Isolation of a biologically active fragment from the carboxy terminus of the fetal rat binding protein for insulin-like growth factors. Biochem Biophys Res Commun 1988; 157:718–726.
206. Jones JI, Gockerman A, Busby WH, Camacho-Hübner C, Clemmons DR. Extracellular matrix contains insulin-like growth factor binding protein-5: potentiation of the effects of IGF-I. J Cell Biol 1993; 121:679–687.
207. Roghani M, Hossenlopp P, Lepage P, Balland A, Binoux M. Isolation from human cerebrospinal fluid of a new insulin-like growth factor-binding protein with a selective affinity for IGF-II. FEBS Lett 1989; 255:253–258.

11 IGF-Independent Actions of the IGF Binding Proteins

Youngman Oh, PhD *and Ron G. Rosenfeld,* MD

Contents

INTRODUCTION

The insulin-like growth factor binding proteins (IGFBPs) are members of the IGF signaling system, along with the ligands (insulin, IGF-I, and IGF-II) and a family of transmembrane receptors (the insulin, type 1 IGF and type 2 IGF receptors) *(1–11)*. The human IGFBP superfamily consists of six high-affinity IGF binders (IGFBPs 1–6) and four low-affinity IGF binders (IGFBP-related proteins [IGFBP-rPs] 1–4) *(12–15)* (Table 1; *see* Chapter 14 for more detail).

In retrospect, the existence of an IGF binding protein was initially suggested more than 30 years ago, by the observation that when plasma was fractionated by gel chromatography under neutral conditions, the nonsuppressible insulin-like activity (NSILA; later termed IGF) migrated in high molecular weight (M_r) fractions (an estimated M_r between 70,000 and 150,000), whereas under acidic conditions, the NSILA "dissociated into one uniform species of molecules," with an estimated M_r between 6000 and 10,000 *(16,17)*. These early studies suggested that IGF peptides were complexed with larger plasma carrier proteins. It was demonstrated subsequently that there existed a large, growth hormone (GH)-dependent binding protein and a smaller, GH-independent binding protein *(18)*. Further studies from various groups led to the cloning and sequencing of cDNAs for six IGF binding proteins, designated IGFBPs 1–6. The relatively high sequence similarity (>50%) and structural similarity (cluster of 10–12 and 6 conserved cysteines in the amino and [N]- carboxy [C]-terminal regions, respectively) among the six IGFBPs indicate that their individual genes may have been derived from successive

From: *Contemporary Endocrinology: The IGF System*
Edited by: R. Rosenfeld and C. Roberts Jr © Humana Press, Inc., Totowa, NJ

Table 1
Structural Characteristics of the Human IGFBP Superfamily

IGFBP	*mRNA size (kb)*	*Molecular weight*	*Number of Amino acids*	*Number of Cysteines*	*N-Linked Glycosylation*	*Chromosomal Localization*
High-affinity IGFBPs						
IGFBP-1	1.6	25,271	234	18	No	7p
IGFBP-2	1.5	31,355	289	18	No	2q
IGFBP-3	2.4	28,717	264	18	Yes	7p
IGFBP-4	1.7	25,957	237	20	Yes	17q
IGFBP-5	1.7, 6.0	28,533	252	18	No	2q
IGFBP-6	1.1	22,847	216	16	No	12
Low-affinity IGFBP related proteins						
IGFBP-rP1	?	251	18	Yes	4q	
IGFBP-rP2	?	349 (pre)	39	Yes	6q	
IGFBP-rP3	?	357 (pre)	41	? (No)	8q	
IGFBP-rP4	?	379 (pre)	35	? (No)	?	

gene duplication of a common ancestor gene during evolution. IGFBPs 1–6 bind IGF-I and IGF-II with high affinity and serve to transport the IGFs, prolong their half-lives, and modulate their proliferative and anabolic effects on target cells. The molecular mechanisms involved in the interaction of the IGFBPs with the IGFs and their receptors remain unclear, but these molecules appear, at least, to regulate the availability of free IGFs for interaction with IGF receptors (IGF-dependent actions of IGFBPs) *(10,11)*.

The complexity of the IGF–IGFBP system was increased by the observation and identification of proteolytic fragments of IGFBPs in vitro and in vivo *(19–23)*. These posttranslationally modified IGFBPs appear to exhibit significantly reduced affinity for IGFs, thereby potentially increasing the availability of IGFs to cell-membrane receptors. This suggests a possible mechanism for the liberation of IGFs from the IGF–IGFBP complex and delivery of IGFs to IGF receptors in targeted tissues *(24)*.

The last 5 yr have seen several new developments in our understanding of IGFBPs: (1) IGFBPs appear to be capable of exerting biological actions in an IGF-independent manner *(25–39)*; (2) a new concept of an IGFBP superfamily has emerged from the identification of low-affinity binders of IGFs *(12–14,40)*, and (3) IGFBPs appear to bind insulin and, in some cases, modulate insulin receptor activation *(41,42)*. This growing body of evidence indicates that each IGFBP might have its own, yet to be elucidated, biological actions, beyond its ability to regulate IGF actions. Undoubtedly, these findings bring with them a wealth of unanswered questions, providing clear and exciting directions for our future research on IGFBPs.

In this chapter, updated information on the IGF-independent actions of IGFBPs and the new low-affinity members of the IGFBP superfamily, IGFBP-rPs 1–4, is presented. In addition, the potential role of IGFBPs as insulin-binding proteins is further discussed, as a subset of IGF-independent actions of IGFBPs.

IGF-INDEPENDENT ACTIONS OF IGFBPS

In addition to the structural and sequence homology among IGFBPs, some IGFBPs appear to possess unique characteristics: (1) IGFBP-1 and IGFBP-3 can be phosphorylated on their serine residues *(43–48)*; (2) IGFBP-1 and IGFBP-2 contain integrin recognition sequences *(8)*; (3) IGFBP-3, -5, and -6 contain heparin binding motifs *(8,10,11)*; and (4) IGFBP-3 has a nuclear localization sequence and can be translocated to the nucleus *(49,50)*. At present, it remains unclear whether these characteristics are important for the ability of IGFBPs to modulate IGF actions or for the IGF-independent actions of IGFBPs. In the following section, those unique characteristics of IGFBPs that appear to be involved in IGF-independent actions of IGFBPs are discussed in detail.

Interaction with $\alpha 5\beta 1$ Integrin: IGFBP-1

The cloning and sequence analysis of cDNAs for IGFBP-1 and IGFBP-2 revealed that the proteins contain an integrin recognition sequence, Arg–Gly–Asp (RGD), near the COOH C-terminus, which is the minimum requirement for interaction with a family of integrins on cell surfaces, suggesting that IGFBP-1 and -2 might bind to cell surfaces through their RGD sequences, without necessarily interacting with the IGF–IGF receptor axis *(8,51)*. Supporting this hypothesis was the observation that addition of a synthetic RGD peptide inhibited the binding of IGFBP-1 to the cell surface *(52)*. Specific binding of IGFBP-1 to the $\alpha 5\beta 1$ integrin (fibronectin receptor) and the functional significance of this in Chinese hamster ovary (CHO) cells has been demonstrated *(53)*. The transfection of human IGFBP-1 or addition of exogenous IGFBP-1 resulted in stimulation of CHO cell migration in a monolayer binding assay. Affinity chromatography of ^{125}I-labeled CHO cell membrane proteins, using IGFBP-1 coupled to agarose, identified the $\alpha 5\beta 1$ integrin as the only cell-surface molecule capable of binding IGFBP-1 in an RGD-dependent manner. In contrast, the mutation of the RGD sequence to Trp–Gly–Asp (WGD) in IGFBP-1 prevented it from binding to the $\alpha 5\beta 1$ integrin or stimulating cell migration. The addition of a synthetic RGD-containing peptide abrogated IGFBP-1-induced cell migration. The effect of IGFBP-1 on cell migration appears to be independent of IGFs, as (1) the WGD mutation showed no change in binding affinity for IGF-I, (2) addition of IGF-I had no effect on cell migration, and (3) no endogenous IGF-I or IGF-II or their mRNAs could be detected. These studies demonstrate an IGF-independent effect of IGFBP-1 and functional significance of the RGD sequence in the IGFBP-1 molecule. In the same context, IGFBP-2, which also contains the RGD sequence, may also bind to integrins and exert IGF-independent actions, although this has not yet been demonstrated.

Cell Surface Association: IGFBP-5

IGFBP-5, IGFBP-3, and IGFBP-6 each contains putative heparin binding motifs: residues ^{133}KKDRRKKL^{140} and ^{205}YKRKQCKP^{212} in IGFBP-5; residues ^{149}KKGHA^{153} and ^{219}YKKKQCRP^{226} in IGFBP-3; and residues ^{172}YRKRQCR^{178} in IGFBP-6. This suggests that these IGFBPs may have the ability to bind to heparin- or heparin sulfate-containing proteoglycans in the extracellular matrix (ECM). In addition, the involvement of heparin- and heparin sulfate-containing proteoglycans in IGF and IGFBP physiology has been previously observed *(54,55)*, and may be relevant to the IGF-potentiating actions of IGFBPs. Jones et al. showed that intact IGFBP-5 binds to fibroblast ECM and potentiates the effect of IGF-I on fibroblast cell growth, whereas a 21-kDa proteolyzed fragment of IGFBP-5 does not potentiate the cell growth response to IGF-I *(56)*. The affinity of IGFBP-5 for IGF-I was lowered approximately seven fold when it was associated with ECM, suggesting that ECM-associated IGFBP-5 would be more capable of releasing IGF to cell-surface IGF receptors. Further studies from Parker et al. demonstrated that two basic regions of IGFBP-5 (amino acids 131–141 and 201–218) contribute to its binding to ECM, but the region containing amino acids 201–218 has the most significance *(57)*. These studies demonstrate that IGFBP-5 is able to bind to proteoglycans in ECM, thereby potentiating IGF actions.

Andress et al. demonstrated that a carboxyl-truncated form of IGFBP-5 (a 23-kDa fragment), derived from U2 human osteosarcoma cells, enhanced the mitogenic action of IGF-I in cultured mouse osteoblasts *(58)*. In addition to its IGF-potentiating action, human recombinant carboxyl-truncated IGFBP-5^{1-169} has been demonstrated to bind to osteoblast monolayers and stimulate mitogenesis, in the absence of exogenous IGF-I, similar to the effects of the native carboxyl-truncated form of IGFBP-5, suggesting the presence of IGF-independent action of IGFBP-5 in osteoblasts *(59)*. Although the truncated IGFBP-5^{1-169} fragment contains a heparin-binding motif (^{133}KKDRRKKL^{140}), osteoblast monolayer binding of ^{125}I-IGFBP-5^{1-169} was not inhibited in the presence of heparin, suggesting that this IGF-independent action of the IGFBP-5 fragment is not mediated through the its binding to cell-surface heparin-containing proteoglycans. Indeed, a high molecular weight membrane protein (about 420 kDa) was crosslinked with both ^{125}I-IGFBP-5 and ^{125}I-IGFBP-5^{1-169}, and partially purified by use of IGFBP-5 affinity chromatography. It is unknown, however, whether this membrane protein mediates IGF-independent actions of IGFBP-5.

IGF-INDEPENDENT ACTION OF IGFBP-3

Historical Perspective of IGF-Independent Action of IGFBP-3

IGFBP-3 was initially identified by Hintz et al. in 1977 as a large, GH-dependent binding protein in human serum *(18)*. The complete primary structure of IGFBP-3 was provided by Wood et al., following the cloning of its cDNA from an adult human liver library in 1988 *(60)*. Subsequently, this GH-dependent protein has been extensively studied and identified as an IGF regulator and carrier protein, which forms part of the 150-kDa ternary complex with IGFs and an acid labile subunit (88-kDa glycoprotein) in circulation.

The first hint of IGF-independent action of IGFBP-3 came from outside the IGF research field in 1991: Villaudy et al. *(26)* showed that a growth inhibitor, IDF45 (inhibitory diffusible factor of 45 kDa), purified from conditioned medium of Swiss 3T3 mouse fibroblasts and subsequently identified as mouse IGFBP-3 by N-terminal amino acid sequence analysis by Blat et al. *(25)* inhibited either serum- or fibroblast growth factor (FGF)-induced DNA synthesis in chick embryo fibroblasts (CEFs). Furthermore, Liu et al. reported that the same protein inhibited the stimulation of CEF DNA synthesis by a serum fraction from which IGFs had been removed by acid gel chromatography *(27)*. These studies suggested that mouse IGFBP-3 could inhibit responsiveness to growth factors and serum components other than IGFs in chicken cells. Cohen et al. reported that overexpression of a transfected human IGFBP-3 cDNA in Balb c/3T3 mouse fibroblasts inhibited cell proliferation, with or without added IGF or insulin *(28)*. They suggested that IGFBP-3 might exert a non-IGF-mediated action, based on the observation that insulin, which has mitogenic activity in these cells but does not bind to IGFBP-3, could not overcome IGFBP-3-induced cell growth inhibition.

Corroborating evidence from Oh et al. *(29)* showed that exogenous recombinant human IGFBP-3 potently inhibits human breast cancer cell growth in an IGF-independent manner. In particular, specific cell-surface binding of IGFBP-3 and an IGF-independent effect of IGFBP-3 on DNA synthesis and cell replication of breast cancer cells were demonstrated. For these experiments, IGF analogs that have differential affinities for IGF receptors or IGFBP-3 were employed with the Hs578T human breast cancer cell line, which lacks biological responsiveness to IGFs, despite the presence of IGF receptors. Further studies supported this new concept of IGF-independent action of IGFBP-3 by showing that: (1) IGFBP-3 binds to specific proteins on cell surfaces and this interaction is strongly correlated with the ability of IGFBP-3 to inhibit cell growth *(30)*; (2) the growth rate is significantly reduced in human IGFBP-3-transfected mouse fibroblast cells (R^- cells) derived from mouse embryos homozygous for a targeted disruption of the type 1 IGF receptor gene *(32)*; (3) IGFBP-3 mediates transforming growth factor-β (TGF-β)-, retinoic acid (RA)-, and antiestrogen-induced growth inhibition in breast cancer cells *(31,35,36)*; (4) regulation of IGFBP-3 gene expression plays a role in signaling by p53, a potent tumor-suppressor protein *(33)*; (5) IGFBP-3 can be translocated to the nucleus *(49,50)*; (6) exogenous IGFBP-3 induces programmed cell death in PC-3 human prostate cancer cells *(37)*; (7) IGFBP-3 fragments derived from limited proteolysis inhibit the stimulation of DNA synthesis induced either by IGF-I or insulin in CEFs or by FGF in R^- cells *(34,38)*; (8) IGFBP-3 binds to the type V TGF-β receptor and competes with TGF-β1 for receptor binding *(39)*; and (9) the kinetics of IGFBP-3 binding to breast cancer cells is typical of ligand–receptor interactions, indicating the presence of an IGFBP-3 receptor in these cells *(61)*.

Although these intensive efforts from various laboratories have supported the concept of IGF-independent action of IGFBP-3, essential information is still lacking, such as definitive identification of the IGFBP-3 receptor and elucidation of the biological mechanisms of these actions.

Cell-Surface Interaction of IGFBP-3

IGFBP-3 has been recognized as "a sticky protein," based on its ability to adhere to test tubes and tissue culture plates, presumably due to its highly basic sequences. It has also been observed that IGFBP-3 associates with different cell types, and that this binding is competitively inhibited by heparin, demonstrating specificity *(62–64)*. Amino acid sequence analysis of IGFBP-3 reveals that it contains two heparin-binding motifs (residue ^{149}KKGHA^{153} representing a short heparin-binding domain, BBXBX, where B is a basic amino acid and X is a nonbasic amino acid, and residue ^{219}YKKKQCRP^{226}, representing a long heparin-binding domain, XBBBXXBX), suggesting that IGFBP-3 can interact with heparin sulfate-containing proteogylcans on cell surfaces *(64–66)*. However, in some cell systems, complete removal of heparin sulfate and other glycosaminoglycans from the cell surface had a minimal effect on IGFBP-3 binding to cell surfaces, strongly suggesting that cell-surface binding of IGFBP-3 cannot be entirely attributed to interaction with heparin sulfate-containing proteoglycans *(67,68)*. Alternatively, the binding of heparin to IGFBP-3 might cause conformational changes in the protein that prevent IGFBP-3 binding to cell surfaces.

The initial demonstration of IGFBP-3 binding directly to cells came from studies on CEFs by Delbe et al. *(69)*. Their studies showed that mouse IGFBP-3 can bind to CEF membranes with low affinity. Binding of IGFBP-3 to the membranes was specific, and binding sites per cell were estimated at 60,000. Using an estrogen receptor (ER)-negative human cell line, Hs578T, Oh et al. showed that IGFBP-3 can be identified in membrane preparations *(70)*, and subsequently demonstrated the existence of IGFBP-3 on monolayer cell surfaces, using immunoperoxidase staining with an IGFBP-3-specific antibody (αIGFBP-3g1). Immunoreactive IGFBP-3 was detectable on the cell surface of Hs578T cells treated with αIGFBP-3g1, but not on cells treated with nonimmune rabbit serum. Cell-surface staining was clearly decreased by treatment with IGF-I, suggesting that IGF binds to IGFBP-3 and prevents IGFBP-3 binding to the cell membranes. Conversely, treatment with [Gln^3,Ala^4,Tyr^{15},Leu^{16}]IGF-I, an IGF-I analog with reduced affinity for IGFBPs, resulted in immunoreactive cell-surface staining equivalent to control, reflecting the analog's decreased affinity for IGFBP-3. These studies indicate that the cell-surface binding of IGFBP-3 is inhibited by the binding of IGF peptides to IGFBP-3.

Further studies showed that binding of IGFBP-3 to the cell surface was increased by $CaCl_2$ and $MnCl_2$ in a dose-dependent manner *(29)*. In addition, IGFBP-3 binding was not mediated through either fibronectin receptors or through mechanisms responsible for IGFBP-1 association with the cell surface *(29)*. Rather, the data support the existence of specific proteins on the Hs578T cell surface that are responsible for the cell-surface association of IGFBP-3.

Demonstration of Putative IGFBP-3 Receptors

Affinity crosslinking of Hs578T monolayers and cell lysates with ^{125}I-IGFBP-3$^{E. coli}$ has been employed to investigate possible mechanisms for the cell-surface interaction of IGFBP-3 *(30)*. One major radiolabeled band of apparent molecular mass 55 kDa was detected on monolayer cell surfaces by affinity crosslinking. Conversely, studies employing affinity crosslinking to Hs578T cell lysates suggested the presence of as many as three specific bands (55, 58, and 85 kDa). The additional two proteins may reflect either intracellular proteins contained within the whole cell lysates, components of an IGFBP-3 receptor complex, or precursor or degradation forms of the IGFBP-3 receptor. Nonetheless, these three bands appeared to be specific for IGFBP-3, as demonstrated by dose-dependent displacement with unlabeled IGFBP-3 and immunoprecipitation with IGFBP-3 specific antibody, but not with nonimmune serum. In addition, IGF-II, but not IGF-II analogs with reduced affinity for IGFBP-3, prevented IGFBP-3 binding to these specific cell-surface association proteins, by forming [IGF-II–IGFBP-3] complexes, consistent with previous data employing monolayer binding assays *(29)*. The specificity of IGFBP-3 binding to cell-surface association proteins was further supported by Western ligand blot (WLB) experiments with ^{125}I-IGFBP-3. WLB of Hs578T cell lysates revealed multiple bands, including 20, 23, and 50 kDa as major species. These bands correspond to those shown by crosslinking, after subtraction of the molecular mass of nonglycosylated IGFBP-3 *(30)*. Rajah et al. also reported that IGFBP-3 appears to bind to several cell-surface association proteins in PC-3 prostate cancer cells *(37)*.

IGFBP-3 binding sites on breast cancer cell membranes were further evaluated through competitive binding studies with unlabeled IGFBPs 1–6 and various forms of IGFBP-3, including synthetic IGFBP-3 fragments *(61)*. Scatchard analysis revealed the existence of high-affinity sites for IGFBP-3 in ER-negative Hs578T human breast cancer cells ($K_d = 8.19 \pm 0.97 \times 10^{-9}$ M with $4.92 \pm 1.51 \times 10^5$ binding sites/cell) and 30-fold fewer receptors in ER-positive MCF-7 cells ($K_d = 8.49 \pm 0.78 \times 10^{-9}$ M with $1.72 \pm 0.31 \times 10^4$ binding sites/cell), using a one-site model. These observations demonstrate binding characteristics of typical receptor–ligand interactions, providing evidence for specific, high-affinity IGFBP-3 receptors on Hs578T cell membranes.

Recently, Leal et al. postulated that the type V TGF-β receptor is the putative IGFBP-3 receptor *(39)*. They demonstrated that IGFBP-3 binds to the type V TGF-β receptor specifically and can compete with TGF-β for receptor binding in mink lung epithelial cells, suggesting that IGFBP-3 is a functional ligand for the type V TGF-β receptor. However, at the present time, it is unclear whether this type V TGF-β receptor represents the primary IGFBP-3 receptor, whether it is identical to the protein observed in breast or prostate cancer cell systems, and whether it plays a role in the growth-inhibitory actions of IGFBP-3 in those cell systems. Nonetheless, this new finding provides further evidence for "IGF-independent anti-proliferative actions of IGFBP-3."

Nuclear Localization of IGFBP-3

Recent structural studies from Radulescu predicted that IGFBP-3 and -5 contain bipartite nuclear localization signals (NLSs) in the basic region of their C-terminal segments, suggesting nuclear localization of those IGFBPs *(71)*. Indeed, Jaques et al. identified IGFBP-3 in the nucleus and cytoplasm of A549 non-small-cell lung cancer cells by use of immunohistochemistry and immuno- or ligand blotting of intact cells or the purified nuclear extracts *(50)*. Li et al. further demonstrated that, in proliferating opossum kidney cells, IGFBP-3 and IGF-I, added separately or together, are transported to the nucleus, whereas in resting cells, IGF-I and IGFBP-3 both internalize and localize to the endosomal compartment *(49)*. These studies suggest that nuclear translocation of IGFBP-3, either by itself or complexed with IGFs, depends on cell cycle stage, whether IGFBP-3 is endogenously produced or exogenously added. These studies raise additional possibilities for mechanisms of both IGF-dependent and -independent actions of IGFBP-3: (1) IGFBP-3 may transport IGFs to the nucleus, where IGF alone or the IGF-IGFBP-3 complex may exert some biological actions (IGF-dependent action), and (2) nuclear IGFBP-3 may regulate the transcription of critical growth-inhibitory genes, or, as Radulescu postulated, may act as a nuclear tumor suppressor, similar to retinoblastoma protein (RB), and interact with RB-binding proteins, such as RBP-1 and/or RBP-2 (IGF-independent action) *(72)*. Although further studies are necessary to elucidate the possible functions of nuclear IGFBP-3, many possibilities exist for IGF-independent actions of IGFBPs in this context.

Cell Growth Regulation by IGFBP-3

Direct Effect of IGFBP-3 on Growth Inhibition in Various Cell Systems

In the Hs578T breast cancer cell system, it has been observed that insulin, IGFs, and epidermal growth factor (EGF) have no mitogenic effects *(73–75)*, presumably because these cells possess the oncogenic ras protein, which was derived from a point mutation at position 12 of the Hs578T ras protein. As shown by Oh et al., both native IGF-II and [Gln^6,Ala^7,Tyr^{18},Leu^{19},Leu^{27}]IGF-II ([QAYL-L]IGF-II), which does not bind to IGFBPs, have no effect on Hs578T cell monolayer growth at peptide concentrations up to 100 n*M* *(29)*. They concluded that the failure of IGF-II to stimulate Hs578T cells cannot be attributed to interference by endogenous IGFBPs, as both IGF-II and the IGF-II analog with reduced affinity for IGFBPs were equivalently impotent. Exogenously added IGFBP-3, however, still showed a significant inhibitory effect on monolayer growth of Hs578T cells. This inhibitory effect was specific for IGFBP-3; no significant inhibitory action was seen with addition of IGFBP-1 *(29)*. In contrast, coincubation of IGFBP-3 with 20 n*M* [QAYL-L]IGF-II did not result in any attenuation of the IGFBP-3 inhibitory effect, even though this analog, with decreased affinity for IGFBPs, is equipotent with IGF-II for binding to the type 2 IGF receptor. Similar findings were observed with IGF-I and IGF-I analogs, indicating that IGFBP-3 action cannot be attributed to the blocking of IGF access to the type 1 IGF receptor. These data indicate that the IGFBP-3 inhibitory effect on cell growth does not result from blocking the mitogenic actions of IGF-II or IGF-I. Rather, the results suggest a cell-surface-specific action of IGFBP-3 itself *(29)*. The ability of IGF-II, but not the analog, to attenuate the inhibitory effect of IGFBP-3 correlates with the affinity of IGF-II for IGFBP-3, thereby preventing cell-surface binding of IGFBP-3. Similar growth inhibitory effects of IGFBP-3 were also observed in MCF-7 *(36)* and MDA-MB-231 human breast cancer cells *(35)*.

Other studies have confirmed the direct inhibitory action of IGFBP-3 in various cell systems: (1) the growth rate is significantly reduced in human IGFBP-3-transfected fibroblast cells derived from mouse embryos homozygous for a targeted disruption of the type 1 IGF receptor gene (R^- cells) *(37)*; (2) purified mouse IGFBP-3 binds to the chick embryo fibroblast cell surface and inhibits cell growth *(27,69)*; (3) recombinant IGFBP-3 inhibits basal and estradiol-stimulated MCF-7 cell proliferation *(36)*; (4) a 16-kDa fragment of IGFBP-3, generated by limited proteolysis of a nonglycosylated form of human recombinant IGFBP-3 by plasmin, inhibits IGF-I- and insulin-induced DNA synthesis in CEFs, despite its lack of binding affinity for IGF-I *(34)*, and also inhibits FGF-induced mitogenic action in R^- cells *(38)*.

IGFBP-3 appears to be also acting as a growth inhibitor in human prostate cancer cells, in a manner similar to that observed in breast cancer cell and other cell systems *(76)*. Moreover, Rajah et al. suggested that the mechanism for IGFBP-3 inhibitory action in prostate cancer cells was programmed cell death (apoptosis) in PC-3 human prostate carcinoma cells *(37)*. The IGFBP-3-induced apoptosis was attenuated by IGF-I, but not by an IGF-I analog that has reduced affinity for IGFBP-3. In addition, in type 1 IGF receptor-negative R- cells, exogenous IGFBP-3, as well as transfection of IGFBP-3 cDNA, induced apoptosis, suggesting that neither IGF-I nor the type 1 IGF receptor is required for IGFBP-3 induced-apoptosis.

The Role of IGFBP-3 in the Inhibition of Breast Cancer Cell Growth by TGF-β, RA, and Antiestrogens

TGF-β and RA have been identified as potent antiproliferative factors in human breast cancer cells in vitro *(77,78)*. Moreover, TGF-β and RA have been found to stimulate IGFBP-3 production in various cell systems in which they inhibit proliferation, suggesting that IGFBP-3 may act as a mediator of their antiproliferative effects *(78–81)*. In ER-negative human breast cancer cells (Hs578T and MDA-MB-231), Oh et al. have demonstrated that TGF-β and RA stimulate IGFBP-3 gene expression (twofold) and production (two- to threefold) prior to their inhibition of cell growth *(31,35)*. To more directly test whether IGFBP-3 mediates TGF-β and RA action, two different strategies were employed to block the IGFBP-3 effect: (1) blocking TGF-β-induced IGFBP-3 expression by IGFBP-3 antisense oligodeoxynucleotide (ODN); and (2) preventing IGFBP-3 binding to the cell surface by treatment with IGFs and IGF-II analogs, thereby blocking IGFBP-3-mediated action. Inhibition of IGFBP-3 gene expression using an IGFBP-3 antisense ODN not only blocks TGF-β and RA stimulation of IGFBP-3 production by up to 90%, but also inhibits their antiproliferative effects by 40–60% *(31,35)*.

Treatment with IGF-II and [Leu^{27}]IGF-II (an IGF-II analog with full affinity for IGFBPs, but no affinity for the type 1 IGF receptor) also diminished TGF-β inhibitory effects, presumably by blocking TGF-β-induced binding of IGFBP-3 to the cell surface, whereas treatment with [QAYL-L]IGF-II (an IGF-II analog with reduced affinity for

IGFBPs) resulted in no attenuation of TGF-β inhibitory effects. These findings suggest that endogenously produced IGFBP-3 is an important direct inhibitor of proliferation of human breast cancer cells and, importantly, that a major mechanism for the inhibitory actions of TGF-β and RA on these cells appears to be their stimulation of cellular production of IGFBP-3, which then could exert direct antiproliferative effects via IGFBP-3 receptors *(31,35)*.

Similar strategies were employed in PC-3 human prostate carcinoma cells, demonstrating that TGF-β-induced apoptosis was prevented by IGFBP-3-neutralizing antibodies or an IGFBP-3 antisense ODN, suggesting that IGFBP-3 mediates TGF-β-induced apoptosis in PC-3 cells *(37)*.

In addition, it has been reported that the mitogenic effect of estrogen is accompanied by inhibition of IGFBP-3 expression, whereas antiestrogens such as tamoxifen and ICI 182,780 stimulate expression of IGFBP-3 in MCF-7, ER-positive breast cancer cells *(81)*. Further studies employing IGFBP-3 antisense ODN in MCF-7 cells demonstrated that IGFBP-3 mediates antiestrogen-induced growth inhibition *(36)*.

A FAMILY OF LOW-AFFINITY IGF BINDERS: IGFBP-RPS 1–4

A new concept of an IGFBP superfamily was introduced after the identification of several low-affinity binders of IGFs (*13,14; see* Chapter 14 for details. *See also* Table 1 for provisional nomenclature.). The concept of an IGFBP superfamily, including low-affinity IGFBP-related proteins (IGFBP-rPs), such as IGFBP-rP1 (MAC25), IGFBP-rP2 (CTGF), and potential IGFBP-rP3 (NOV) and IGFBP-rP4 (CYR61), supports the existence of IGF-independent actions of IGFBPs, because both MAC25 and CTGF (IGFBP-rP1 and -rP2, respectively) have been characterized as low-affinity IGFBPs, using baculovirus-expressed recombinant human MAC25 and CTGF *(13,14,40)*.

The *mac25* gene, residing on chromosome 4q12 and encoding a preprotein of 277 amino acids, was originally cloned from leptomeningial cells and subsequently reisolated through differential display as a sequence preferentially expressed in senescent human mammary epithelial cells, suggesting that IGFBP-rP1 is involved in regulation of cell growth *(12,82)*. Indeed, it has been demonstrated that expression of IGFBP-rP1 mRNA and protein are up-regulated by TGF-β in Hs578T breast cancer cells *(83)*. Moreover, treatment with IGFBP-rP1^{bac} resulted in inhibition of DNA synthesis and cell proliferation in a dose-dependent manner in human breast cancer cells, even in the absence of IGF peptides. Given that IGFBP-rP1 mRNA was down-regulated in human breast cancer cells, and up-regulated in normal, growing mammary epithelial cells by RA, it is clear that, although IGFBP-rP1 constitutes a low-affinity member of the IGFBP family, it primarily functions as a modulator of cell growth in an IGF-independent manner, similar to the action observed with IGFBP-3 in breast cancer cells.

On the other hand, studies from outside of the IGF field revealed that two proteins, "tumor-derived adhesion factor" (TAF) and "prostacyclin-stimulating factor" (PSF), that were identified from EJ-1 human bladder carcinoma cells and human diploid fibroblasts, respectively, are identical to IGFBP-rP1 *(84,85)*. With respect to biological function, the former was found to promote the attachment and spreading of BRL rat liver cells and ECV-304 human umbilical vein endothelial cells, whereas the latter stimulates prostacyclin (PGI_2) production in bovine aortic endothelial cells. These studies provide strong evidence that IGFBP-rP1 clearly has a range of biological activities that may be independent of the ability of this protein to bind IGF.

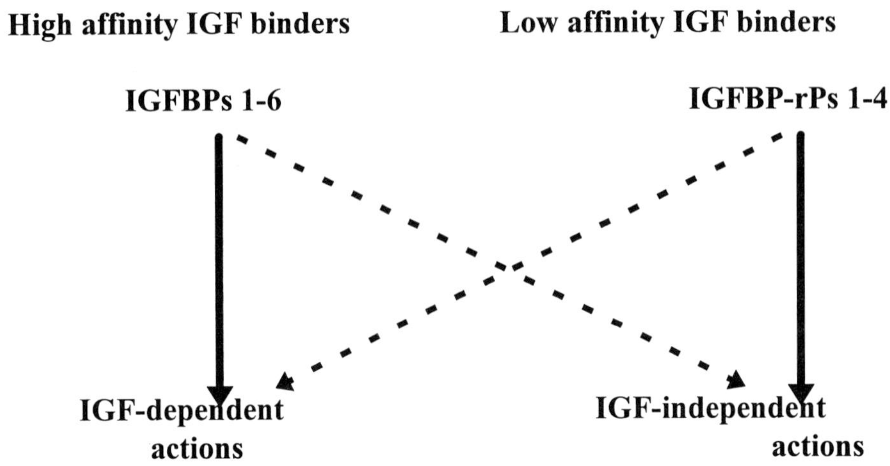

Fig. 1. Schematic diagram of mechanisms for IGF-independent and -dependent antiproliferative actions of IGFBP superfamily.

IGFBP-rP2/CTGF, as well as highly related cDNAs (*nov*-oncogene, *cyr61*), have been identified in human tissues; the predicted proteins are products of "immediate-early genes" expressed after induction by serum, growth factors or certain oncogenes *(86,87)*. *ctgf/nov/cyr61* fall into the group of immediate-early genes that code for secreted, extracellular proteins.

Little is known about the function and regulation of IGFBP-rP2/CTGF, although partially purified CTGF from media conditioned by human umbilical vein endothelial cells demonstrated mitogenic and chemotactic activities *(86)*. To date, information on the roles of IGFBP-rP2/CTGF has been inferred from studies of regulation of *CTGF* mRNA, which is up-regulated by TGF-β, a known growth stimulator in human fibroblasts, thus supporting the notion that CTGF regulates mitogenic activity *(86,88,89)*. On the other hand, mRNA and protein levels of IGFBP-rP2/CTGF are up-regulated by known growth inhibitors, such as TGF-β and RA, in ER-negative breast cancer cells, as well as in prostate cancer cells *(90)*. Furthermore, treatment with IGFBP-rP2^{bac} resulted in inhibition of ER-negative and ER-positive breast cancer cell growth in an IGF-independent manner, similar to the action observed with IGFBP-rP1.

These results led to speculation that the IGFBP superfamily is derived from an ancestor gene/protein that was critically involved in the regulation of cell growth, probably in an IGF-independent manner. Over the course of evolution, the ability to bind IGF peptides with high or low affinity may have been a secondary role acquired by some of the IGFBPs, thereby conferring upon them the ability to influence cell growth (as in the breast) by both IGF-dependent and IGF-independent means (Fig. 1).

IGFBPS AS INSULIN BINDING PROTEINS

Recent studies from Yamanaka et al. demonstrated that IGFBPs can bind insulin, as well as IGFs, subsequently modulating insulin receptor activation by inhibiting insulin binding to its receptor *(41)*. These studies demonstrated that IGFBP-rP1 is a high-affinity insulin binding protein. IGFBP-rP1 not only binds insulin specifically, but also modulates insulin binding to its receptor, subsequently inhibiting the early steps in insulin

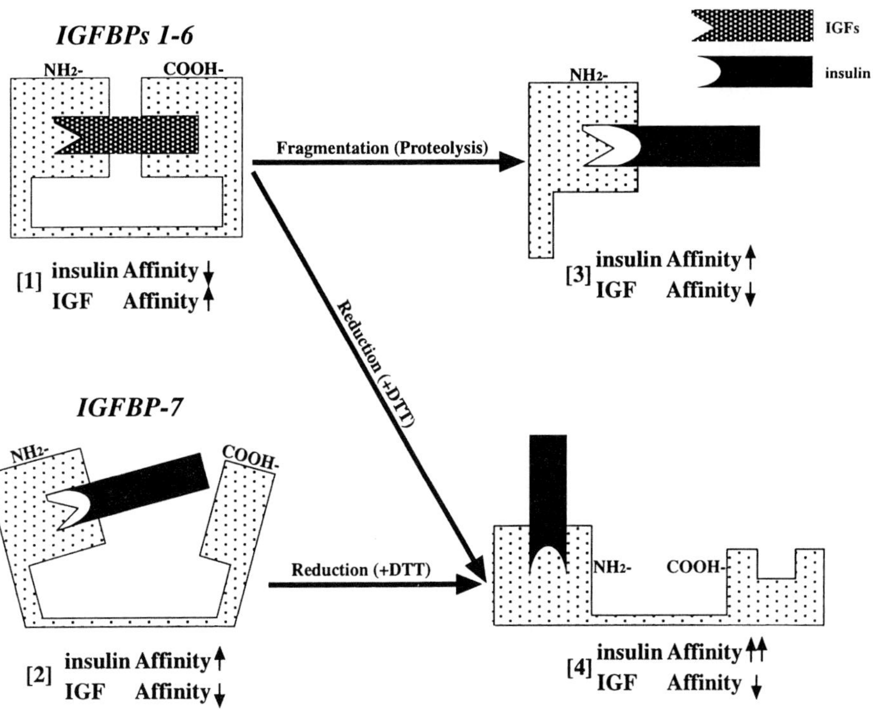

Fig. 2. Schematic representation of IGFBP binding to insulin and IGFs.

action, such as autophosphorylation of the insulin receptor β-subunit and phosphorylation of IRS-1. These data raise the possibility that IGFBPs, and especially IGFBP-rP1, are functional insulin binding proteins. Indeed, the affinity of IGFBPs for insulin can be enhanced by modifications that disrupt disulfide bonds or remove the conserved C-terminus (Fig. 2). Like intact IGFBP-rP1, an N-terminal fragment of IGFBP-3 derived from plasmin digestion, baculovirus-expressed IGFBP-$3^{1\text{-}87}$ and IGFBP-$3^{1\text{-}97}$, and IGFBP-3 fragments in normal human urine, also bind insulin with high affinity and block its action *(41,42)*. These observations suggest that IGFBPs with enhanced affinity for insulin (especially following proteolysis of IGFBPs by specific proteases in vivo) might contribute to the insulin resistance of pregnancy, type 2 (non-insulin-dependent) diabetes mellitus, and other pathological conditions.

These new findings are somewhat against the existing dogma, which proposes that a major difference between insulin and IGFs is that only IGFs have ability to bind to IGFBPs with high affinity. In retrospect, however, there is precedent for IGFBP interaction with insulin: (1) overexpression of a transfected human IGFBP-3 cDNA in Balb c/3T3 mouse fibroblasts inhibits insulin-induced cell proliferation *(28)*; and (2) a 16-kDa proteolytic fragment of IGFBP-3 (IGFBP-$3^{1\text{-}92}$) inhibits insulin-stimulated DNA synthesis in CEFs *(34)*. Whether these effects of IGFBP-3 and its fragments on insulin action can be explained by IGFBP-3 binding to insulin remains to be elucidated. It is unlikely, however, to be the full explanation of IGF-independent actions of IGFBP-3, in light of the ability of IGFBP-3 fragments to inhibit FGF action *(38)*, as well as the actions of insulin and IGF. At present, it is not clear whether N-terminal fragments of other IGFBPs behave similarly to N-terminal fragments of IGFBP-3 for insulin binding. Overall, this new concept for IGFBPs as insulin binding proteins potentially enriches the diversity of biological activities of IGFBPs that are IGF independent.

SUMMARY

The human IGFBP superfamily consists of six high-affinity IGF binders (IGFBPs 1–6) and at least four low-affinity IGF binders (IGFBP-rPs 1–4). The observations that IGFBP-3, as well as IGFBP-1 and -5, bind to various cell surfaces with high specificity and affinity and exhibit specific biological activities within those cell systems, provide strong foundation for IGF-independent actions of IGFBPs. Further findings, such as the various biological functions of IGFBP-rPs and the insulin binding ability of IGFBPs and IGFBP-rPs, provide a growing body of evidence for biological actions of each IGFBP beyond its ability to regulate IGF actions. We are, indeed, only beginning to appreciate the complexity of the IGFBP system.

ACKNOWLEDGMENTS

This work was supported in part by U.S. Army Grant No. DAMD17-96-1-6204 and Oregon Medical Research Foundation Grant No. 976OH298, and Grants CA58110 and DK51513 from the NIH.

REFERENCES

1. Salmon WD, Daughaday WH. A hormonally controlled serum factor which stimulates sulfate incorporation by cartilage in vitro. J Lab Clin Med 1957; 49:825–836.
2. Daughaday WH, Hall K, Raben MS, Salmon WD, Van den Brande JLV, Van Wyk JJ. Somatomedin: proposed designation for sulphation factor. Nature 1972; 235:107–108.
3. Daughaday WH, Rotwein P. Insulin-like growth factors I and II: peptide messenger ribonucleic acid and gene structures serum and tissue concentrations. Endocr Rev 1989; 10:68–91.
4. Baxter RC, Martin JL. Binding proteins for the insulin-like growth factors: structure, regulation and function. Prog Growth Factor Res 1989; 1:49–68.
5. Rosenfeld RG, Lamson GL, Pham H, Oh Y, Conover C, DeLeon DD, Donovan SM, Ocrant I, Giudice LC. Insulin-like growth factor-binding proteins. Recent Prog Horm Res 1991; 46:99–159.
6. Werner H, Woloschak M, Stannard B, Shen-Orr Z, Roberts CT Jr, LeRoith D. The insulin-like growth factor receptor: molecular biology, heterogeneity and regulation. In: LeRoith D ed. Insulin-like Growth Factors: Molecular and Cellular Aspects. CRC Press, Boca Raton, FL, 1991, pp. 17–47.
7. Lowe L. Biological actions of the insulin-like growth factors. In: LeRoith D ed. Insulin-Like Growth Factors: Molecular and Cellular Aspects. CRC Press, Boca Raton, FL, 1991, pp. 49–85.
8. Shimasaki S, Ling N. Identification and molecular characterization of insulin-like growth factor binding proteins (IGFBP-1, -2, -3, -4, -5, and -6). Prog Growth Factor Res 1991; 3:243–266.
9. Oh Y, Muller HL, Neely EK, Lamson G, Rosenfeld RG. New concepts in insulin-like growth factor receptor physiology. Growth Regul 1993; 3:113–123.
10. Kelley KM, Oh Y, Gargosky SE, Gucev Z, Matsumoto T, Hwa V, Ng L, Simpson D, Rosenfeld RG. Insulin-like growth factor-binding proteins (IGFBPs) and their regulatory dynamics. Int J Biochem Cell Biol 1995; 28:619–637.
11. Jones JI, Clemmons DR. Insulin-like growth factors and their binding proteins: biological actions. Endocr Rev 1995; 16:3–34.
12. Swisshelm K, Ryan K, Tsuchiya K, Sager R. Enhanced expression of an insulin-like growth factor-like binding protein (mac25) in senescent human mammary epithelial cells and induced with retinoic acid. Proc Natl Acad Sci USA 1995; 92:4472–4476.
13. Oh Y, Nagalla SN, Yamanaka Y, Kim H, Wilson L, Rosenfeld RG. Identification and characterization of insulin-like growth factor binding protein (IGFBP)-7: recombinant human mac25 protein specifically binds IGF-I and -II. J Biol Chem 1996; 271:30322–30325.

14. Kim H-S, Nagalla SR, Oh Y, Wilson EM, Roberts CT Jr, Rosenfeld RG. Identification of a family of low affinity insulin-like growth factor binding proteins (IGFBPs): characterization of connective tissue growth factor (CTGF) as a member of the IGFBP superfamily. Proc Natl Acad Sci USA 1998; 94:12981–12986.
15. Rosenfeld RG, Oh Y. The blind men and the elephant: a parable for the study of IGFBPs. Endocrinology 1998; 139:5–7.
16. Daughaday WH, Kipnis DM. The growth-promoting and anti-insulin actions of somatotropin. Recent Prog Horm Res 1966; 22:49–93.
17. Burgi H, Muller WA, Humbel RE, Labhart A, Froesch ER. Non-suppressible insulin-like activity of human serum. I. Physicochemical properties, extraction and partial purification. Biochim Biophys Acta 1966; 121:349–359.
18. Hintz RL, Liu F. Demonstration of specific plasma protein binding sites for somatomedin. J Clin Endocrinol Metab 1977; 45:988–995.
19. Hossenlopp P, Segovia B, Lassare C, Roghani M, Bredon M, Binoux M. Evidence of enzymatic degradation of the insulin-like growth factor binding proteins in the 150 K complex during pregnancy. J Clin Endocrinol Metab 1990; 71:797–805.
20. Giudice LC, Farrell EM, Pham H, Lamson G, Rosenfeld RG. Insulin-like growth factor binding proteins in maternal serum throughout gestation and in the puerperium: effects of a pregnancy-associated serum protease activity. J Clin Endocrinol Metab 1990; 71:806–816.
21. Conover CA, De Leon DD. Acid-activated insulin-like growth factor-binding protein-3 proteolysis in normal and transformed cells. Role of cathepsin D. J Biol Chem 1994; 269:7076–7080.
22. Cohen P, Graves HCB, Peehl DM, Kamarei M, Giudice LC, Rosenfeld RG. Prostate specific antigen (PSA) is an IGF binding protein-3 (IGFBP-3) protease found in seminal plasma. J Clin Endocrinol Metab 1993; 75:1046–1053.
23. Lee DY, Park SK, Yorgin PD, Cohen P, Oh Y, Rosenfeld RG. Alteration of insulin-like growth factor-binding proteins (IGFBPs) and IGFBP-3 protease activity in serum and urine from acute and chronic renal failure. J Clin Endocrinol Metab 1994; 79:1376–1382.
24. Cohen P, Peehl DM, Lamson G, Rosenfeld RG. Insulin-like growth factors (IGFs), IGF receptors and IGF binding proteins in primary cultures of prostate epithelial cells. J Clin Endocrinol Metab 1991; 73:401–407.
25. Blat C, Bohlen P, Villaudy J, Chatelain G, Golde A, Harel L. Isolation and amino-terminal sequence of a novel cellular growth inhibitor (inhibitory diffusible factor 45) secreted by 3T3 fibroblasts. J Biol Chem 1989; 264:6021–6024.
26. Villaudy J, Delbe J, Blat C, Desauty G, Golde A, Harel L. An IGF binding protein is an inhibitor of FGF stimulation. J Cell Physiol 1991; 149:492–496.
27. Liu L, Delbe J, Blat C, Zapf J, Harel L. Insulin-like growth factor binding protein (IGFBP-3), an inhibitor of serum growth factors other than IGF-I and -II. J Cell Physiol 1992; 153:15–21.
28. Cohen P, Lamson G, Okajima T, Rosenfeld RG. Transfection of the human insulin-like growth factor binding protein-3 gene into Balb/c fibroblasts inhibits cellular growth. Mol Endocrinol 1993; 7:380–386.
29. Oh Y, Muller HL, Lamson G, Rosenfeld RG. Insulin-like growth factor (IGF)-independent action of IGF-binding protein-3 in Hs578T human breast cancer cells. J Biol Chem 1993; 268:14964–14971.
30. Oh Y, Muller HL, Pham H, Rosenfeld RG. Demonstration of receptors for insulin-like growth factor binding protein-3 on Hs578T human breast cancer cells. J Biol Chem 1993; 268:26045–26048.
31. Oh Y, Muller HL, Ng L, Rosenfeld RG. TGF-β-induced cell growth inhibition in human breast cancer cells is mediated through IGFBP-3 action. J Biol Chem 1995; 270:13589–13592.
32. Valentinis B, Bhala A, DeAngelis T, Baserga R, Cohen P. The human insulin-like growth factor (IGF) binding protein-3 inhibits cell growth of fibroblasts with a targeted disruption of the IGF-I receptor gene. Mol Endocrinol 1995; 9:361–367.
33. Buckbinder L, Talbott R, Velasco-Miguel S, Takenaka I, Faha B, Seizinger BR, Kley N. Induction of the growth inhibitor IGF-binding protein 3 by p53. Nature 1995; 377:646–649.
34. Lalou C, Lassarre C, Binoux M. A proteolytic fragments of insulin-like growth factor (IGF) binding protein-3 that fails to bind IGFs inhibits the mitogenic effects of IGF-I and insulin. Endocrinology 1996; 137:3206–3212.
35. Gucev ZS, Oh Y, Kelley KM, Rosenfeld RG. Insulin-like growth factor binding protein-3 (IGFBP-3) mediates retinoic acid (RA)- and transforming growth factor-β2 (TGF-β2)-induced growth inhibition in human breast cancer cells. Cancer Res 1996; 56:1545–1550.

36. Huynh H, Yang X, Pollak M. Estradiol and antiestrogens regulate a growth inhibitory insulin-like growth factor binding protein 3 autocrine loop in human breast cancer cells. J Biol Chem 1996; 271:1016–1021.
37. Rajah R, Valentinis B, Cohen P. Insulin-like growth factor-binding protein-3 induces apoptosis and mediates the effects of transforming growth factor-β1 on programmed cell death through a p53- and IGF-independent mechanism. J Biol Chem 1997; 272:12181–12188.
38. Zadeh SM and Binoux M. The 16-kDa proteolytic fragment of insulin-like growth factor (IGF) binding protein-3 inhibits the mitogenetic action of fibroblast growth factor on mouse fibroblasts with a targeted disruption of the type 1 IGF receptor gene. Endocrinology 1997; 138:3069–3072.
39. Leal SM, Liu Q, Huang SS, Huang JS. The type V transforming growth factor beta receptor is the putative insulin-like growth factor-binding protein 3 receptor. J Biol Chem 1997; 272:20572–20576.
40. Wilson EM, Oh Y, Rosenfeld RG. Generation and characterization of an IGFBP-7 antibody: identification of 31-kDa IGFBP-7 in human biological fluids and Hs578T human breast cancer conditioned media. J Clin Endocrinol Metab 1997; 82:1301–1303.
41. Yamanaka Y, Wilson EM, Rosenfeld RG, Oh Y. Inhibition of insulin receptor activation by insulin-like growth factor binding proteins. J Biol Chem 1997; 272:30729–30734.
42. Vorwerk P, Yamanaka Y, Spagnoli A, Oh Y, Rosenfeld RG. Insulin and IGF binding by IGFBP-3 fragments derived from proteolysis, baculovirus expression and normal human urine. J Clin Endocrinol Metab 1998; 83:1392–1395.
43. Busby WH, Klapper DG, Clemmons DR. Purification of a 31000 dalton insulin-like growth factor binding protein from human amniotic fluid. J Biol Chem 1988; 263:14203–14210.
44. Jones JI, D'Ercole AJ, Camacho-Hubner C, Clemmons DR. Phosphorylation of insulin-like growth factor binding protein in cell culture and *in vivo*: effects on affinity for IGF-I. Proc Natl Acad Sci USA 1991; 88:7481–7485.
45. Frost RA, Tseng L. Insulin-like growth factor binding protein-1 is phosphorylated by cultured human endometrial stromal cells and multiple protein kinases in vitro. J Biol Chem 1991; 266:18082–18088.
46. Westwood M, Gibson JM, Davies AJ, Young RJ, White A. The phosphorylation pattern of insulin-like growth factor-binding protein-1 in normal plasma is different from that in amniotic fluid and changes during pregnancy. J Clin Endocrinol Metab 1994; 79:1735–1741.
47. Hoeck WG, Mukku VR. Identification of the major sites of phosphorylation in IGF binding protein-3. J Cell Biochem 1994; 56:262–273.
48. Coverley JA, Baxter RC. Regulation of insulin-like growth factor (IGF) binding protein-3 phosphorylation by IGF-I. Endocrinology 1995; 136:5778–5781.
49. Li W, Fawcett J, Widmer HR, Fielder PJ, Rabkin R, Keller GA. Nuclear transport of insulin-like growth factor-I and insulin-like growth factor binding protein-3 in opossum kidney cells. Endocrinology 1997; 138:1763–1766.
50. Jaques G, Noll K, Wegmann B, Witten S, Kogan E, Radulescu RT, Havemann K. Nuclear localization of insulin-like growth factor binding protein 3 in a lung cancer cell line. Endocrinology 1997; 138:1767–1770.
51. Brewer MT, Stetler GL, Squires CH, Thompson RC, Busby WH, Clemmons DR. Cloning, characterization and expression of a human insulin-like growth factor binding protein. Biochem Biophys Res Commun 1988; 152:1289–1297.
52. Clemmons DR. The role of insulin-like growth factor binding proteins in controlling the expression of IGF actions. In: LeRoith D, Raizada MK, eds. Molecular and Cellular Biology of Insulin-Like Growth Factors and Their Receptors. Plenum Publishing, New York, 1989, pp. 381–394.
53. Jones JI, Gockerman A, Busby WH, Wright G, Clemmons DR. Insulin-like growth factor binding protein 1 stimulates cell migration and binds to the α5β1 integrin by means of its Arg-Gly-Asp sequence. Proc Natl Acad Sci USA 1993; 90:10553–10557.
54. Clemmons DR, Underwood LE, Chatelain PG, Van Wyk JJ. Liberation of immunoreactive somatomedin-C from its binding proteins by proteolytic enzymes and heparin. J Clin Endocrinol Metab 1983; 56:384–389.
55. Baxter RC. Glycosaminoglycans inhibit formation of the 140 kDa insulin-like growth factor-binding protein complex. Biochem J 1990; 271:773–777.
56. Jones JI, Gockerman A, Busby WHJ, Camacho-Hubner C, Clemmons DR. Extracellular matrix contains insulin-like growth factor binding protein-5: potentiation of the effects of IGF-I. J Cell Biol 1993; 121:679–687.
57. Parker A, Clarke JB, Busby WHJ, Clemmons DR. Identification of the extracellular matrix binding sites for insulin-like growth factor-binding protein 5. J Biol Chem 1996; 271:13523–13529.

58. Andress DL, Birnbaum RS. Human osteoblast-derived insulin-like growth factor (IGF) binding protein-5 stimulates osteoblast mitogenesis and potentiates IGF action. J Biol Chem 1992; 267:22467–22472.
59. Andress DL. Heparin modulates the binding of insulin-like growth factor (IGF) binding protein-5 to a membrane protein in osteoblast cells. J Biol Chem 1995; 270:28289–28296.
60. Wood WI, Cachianes G, Henzel WJ, Winslow GA, Spencer SA, Hellmiss R, Martin JL, Baxter RC. Cloning and expression of the GH dependent insulin like growth factor binding protein. Mol Endocrinol 1988; 2:1176–1185.
61. Yamanaka Y, Fowlkes JL, Rosenfeld RG, Oh Y. Characterization of insulin-like growth factor binding protein-3 (IGFBP-3) binding to human breast cancer cells. J Clin Endocrinol Metab 1999, in press.
62. Booth BA, Boes M, Dake BL, Linhardt RJ, Caldwell EE, Weiler JM, Bar RS. Structure–function relationships in the heparin-binding C-terminal region of insulin-like growth factor (IGF) binding protein-3. Growth Regul 1996; 6:206–213.
63. Fowlkes JL, Serra DM. Characterization of glycosaminoglycan-binding domains present in insulin-like growth factor-binding protein-3. J Biol Chem 1996; 271:14676–14679.
64. Grimes RW, Manni A, Hammond JM. Postsynthetic regulation of insulin-like growth factor-binding protein-3 by MCF-7 human breast cancer cells. Breast Cancer Res Treat 1996; 39:187–196.
65. Martin JL, Ballesteros M, Baxter RC. Insulin-like growth factor-I (IGF-I) and transforming growth factor-beta 1 release IGF-binding protein-3 from human fibroblasts by different mechanisms. Endocrinology 1992; 131:1703–1710.
66. Smith EP, Lu L, Chernausek SD, Klein DJ. Insulin-like growth factor-binding protein-3 (IGFBP-3) concentration in rat Sertoli cell-conditioned medium is regulated by a pathway involving association of IGFBP-3 with cell surface proteoglycans. Endocrinology 1994; 135:359–364.
67. Oh Y, Gucev Z, Ng L, Muller HL, Rosenfeld RG. Antiproliferative actions of insulin-like growth factor binding protein (IGFBP)-3 in human breast cancer cells. Prog Growth Factor Res 1995; 6:503–512.
68. Yang YM, Yanagishita M, Rechler MM. Heparin inhibition of insulin-like growth factor-binding protein-3 binding to human fibroblasts and rat glioma cells: role of heparan sulfate proteoglycans. Endocrinology 1996; 137:4363–4371.
69. Delbe J, Blat C, Desauty G, Harel L. Presence of IDF45 (mIGFBP-3) binding sites on chick embryo fibroblasts. Biochem Biophys Res Commun 1991; 179:495–501.
70. Oh Y, Muller HL, Pham HM, Lamson G, Rosenfeld RG. Non-receptor mediated, post-transcriptional regulation of insulin-like growth factor binding protein (IGFBP)-3 in Hs578T human breast cancer cells. Endocrinology 1992; 131:3123–3125.
71. Radulescu RT. Nuclear localization signal in insulin-like growth factor-binding protein type 3. Trends Biochem Sci 1994; 19:278.
72. Radulescu RT. From insulin, retinoblastoma protein and the insulin receptor to a new model on growth factor specificity: the nucleocrine pathway. J Endocrinol 1995; 146:365–368.
73. Kraus MH, Yuasa Y, Aaronson SA. A position 12-activated H-*ras* oncogene in all Hs578T mammary carcinosarcoma cells but not normal mammary cells of the same patient. Proc Natl Acad Sci 1984; 81:5384–5388.
74. Kozma S, Boggard M, Buser K, Saurer S, Bos J, Groner B, Hynes N. The human c-*Kirsten ras* gene is activated by a novel mutation in codon 13 in the breast carcinorma cell line MDA-MD-231. Nucleic Acids Res 1987; 15:5963–5970.
75. Davidson NE, Gelmann EP, Lippman ME, Dickson RB. Epidermal growth factor receptor gene expression in estrogen receptor positive and negative human breast cancer cell lines. Mol Endocrinol 1987; 1:216–223.
76. Kaicer E, Blat C, Imbenotte J, Troalen F, Cussenot O, Calvo F, Harel L. IGF binding protein-3 secreted by the prostate adenocarcinoma cells (PC-3): differential effect on PC-3 and normal prostate cell growth. Growth Regul 1993; 3:180–189.
77. Knabbe C, Lippman ME, Wakefield LM, Flanders KC, Kasid A, Derynck R, Dickson RB. Evidence that transforming growth factor-beta is a hormonally regulated negative growth factor in human breast cancer cells. Cell 1987; 48:417–428.
78. Fontana JA, Burrows-Mezu A, Clemmons DR, LeRoith D. Retinoid modulation of insulin-like growth factor-binding proteins and inhibition of breast carcinoma proliferation. Endocrinology 1990; 128:1115–1122.
79. Martin JL, Ballesteros M, Baxter RC. Insulin-like growth factor-I (IGF-I) and transforming growth factor-beta 1 release IGF-binding protein-3 from human fibroblasts by different mechanisms. Endocrinology 1992; 131:1703–1710.

80. San Roman GA, Magoffin DA. Insulin-like growth factor binding proteins in ovarian follicles from women with polycystic ovarian disease: cellular source and levels in follicular fluid. J Clin Endocrinol Metab 1992; 75:1010–1016.
81. Pratt SE, Pollak MN. Estrogen and antiestrogen modulation of MCF-7 human breast cancer cell proliferation is associated with specific alterations in accumulation of insulin-like growth factor-binding proteins in conditioned media. Cancer Res 1993; 53:5193–5198.
82. Murphy M, Pykett MJ, Harnish P, Zang KD, George DL. Identification and characterization of genes differentially expressed in meningiomas. Cell Growth Differ 1993; 4:715–722.
83. Oh Y, Wilson EM, Kim H-S, Yang D-H, Rutten MJ, Graham DL, Deveney CW, Hwa V, Rosenfeld RG. Regulation and biological action of IGFBP-7 in human breast cancer cells. In: Proceedings of the 79th Annual Meeting of the Endocrine Society, 1997, pp. 351.
84. Akaogi K, Okabe Y, Funahashi K, Yoshitake Y, Nishikawa K, Yasumitsu H, Umeda M, Miyazaki K. Cell adhesion activity of a 30 kD major secreted protein from human bladder carcinoma cells. Biochem Biophys Res Commun 1994; 198:1046–1053.
85. Yamauchi T, Umeda F, Masakado M, Mizushima S, Nawata H. Purification and molecular cloning of prostacyclin-stimulating factor from serum-free conditioned medium of human diploid fibroblast cells. Biochem J 1994; 303:591–598.
86. Bradham DM, Igarashi A, Potter RL, Grotendorst GR. Connective tissue growth factor: a cysteine-rich mitogen secreted by human vascular endothelial cells is related to the SRC-induced immediate early gene product cef-10. J Biol Chem 1991; 114:1285–1294.
87. Bork P. The modular architecture of a new family of growth regulators related to connective tissue growth factor. FEBS Lett 1993; 327:125–130.
88. Igarashi A, Okochi H, Bradham M, Grotendorst GR. Regulation of connective tissue growth factor gene expression in human skin fibroblasts and during wound repair. Mol Biol Cell 1993; 4:637–645.
89. Igarashi A, Nashiro K, Kikuchi K, Sata S, Ihn H, Fujimoto M, Grotendorst GR, Takehara K. Connective tissue growth factor gene expression in tissue sections from localized scleroderma, keloid, and other fibrotic skin disorders. J Invest Dermatol 1996; 106:729–733.
90. Yang D-H, Kim H-S, Wilson EM, Rosenfeld RG, Oh Y. Identification of glycosylated 38-kDa IGFBP-8/CTGF/IGFBP-rP2 in human biological fluids and up-regulation of IGFBP-rP2 expression by TGF-β2 in Hs578T human breast cancer cells. J Clin Endocrinol Metab 1998; 83:2593–2596.

12 IGF Binding Proteins and Extracellular Matrix

David R. Clemmons, MD

CONTENTS

INTRODUCTION

The insulin-like growth factors (IGFs) were discovered based on their capacity to stimulate synthesis of proteoglycans in extracellular matrix (ECM) of cartilage *(1)*. During the early years after their discovery there were multiple articles published showing that they stimulated synthesis of proteoglycans, glycosaminoglycans, and other ECM components *(2,3)*. Similarly, when studies with cultured cells began multiple articles were published showing that increases in specific ECM proteins such as various types of collagen and other proteins that were components of ECM could be stimulated by exposure of cells to IGF-I *(4–7)*. The third point of historical note is that almost all the cell types that were shown to be responsive to IGF-I are anchorage-dependent cells; therefore, they are continuously exposed to ECM. In spite of these findings, the connection between stimulation of ECM synthesis and the role of ECM composition in controlling the cellular modification and its role in controlling cellular responsiveness to IGF-I was not analyzed.

FORMS OF IGF BINDING PROTEINS IN MATRIX

In attempting to define the biologic role of IGF binding proteins (IGFBPs) on IGF-I actions, this area of investigation has been reexamined. Initially, it was determined by immunohistochemistry that IGFBPs were present in ECM of tissue sections and in basement membranes of epithelial cells *(8)*. This suggested that they might interact with specific proteins within the ECM. This interaction was further confirmed when in attempting to analyze the role of IGFBPs that were synthesized by human fibroblasts, ECM was prepared and analyzed for IGFBP content. When the fibroblast ECM was analyzed,

From: *Contemporary Endocrinology: The IGF System*
Edited by: R. Rosenfeld and C. Roberts Jr © Humana Press, Inc., Totowa, NJ

the most abundant form of IGFBP that was detected, was IGFBP-5 *(9)*. Although intact IGFBP-5 was not detectable in the culture medium, a large amount of intact IGFBP-5 could be detected in the ECM. The absence of the protein in the interstitial fluid was explained by the fact that these cells also produced an IGFBP-5 protease that cleaved it into a 22-kDa, non-IGF binding fragment *(9)*. In spite of rapid cleavage of IGFBP-5 following its synthesis and secretion into the medium, abundant intact IGFBP-5 was deposited in the ECM. More importantly, the protein that was detected in the ECM had not been cleaved, therefore, ECM binding appeared to protect it from cleavage. This represented a potential important site for binding IGF-I and -II and for accumulation of these factors within the ECM. When other binding proteins were analyzed, however, different results were obtained. These fibroblasts were shown to synthesize IGFBP-3 and IGFBP-4 *(9)*. IGFBP-3 could be detected in ECM *(10)*, but only minimal amounts were detected. This was not due to a lesser amount being synthesized because when synthesis was measured approximately six times as much IGFBP-3 was synthesized as IGFBP-5. In spite of this, the abundance of IGFBP-3 was only about 5% as great as IGFBP-5. This suggests that the region of IGFBP-3 that binds to ECM proteins is not surface exposed or that the affinities of the proteins that are present in fibroblast matrix that bind IBFBP-3 are substantially lower than that those that bind IGFBP-5. IGFBP-4 is also synthesized by fibroblasts, but it is not detectable in ECM.

DETERMINANTS OF MATRIX BINDING

Several investigations have detected IGFBP-5 in the ECM of other cell types. Mohan and co-workers had shown that IGFBP-5 accumulates in bone ECM and that hydroxy-appatite within bone matrix is an important component for binding IGFBP-5 *(11)*. They also showed that bone ECM contains fivefold more IGFBP-5 compared to IGFBP-3 *(12)*. Others have shown that it is more abundant than IGFBP-2 *(13)*. Andress has shown that cultured osteoblast ECM binds intact IBFBP-5 and that a fragment containing amino acids 1-169 is also present *(14,15)*. Enyzmatic treatment of ECM to remove GAG side chains from proteoglycans results in a 30% increase in IGFBP-5 binding *(16)*. This was interpreted as allowing better IGFBP-5 access to non-GAG-containing proteins in the ECM. Other workers have shown that IGFBP-5 is present in chondrocyte matrix that does not contain hydroxyapatite, implying that other substances accounted for binding.

To determine the molecular substituents of ECM that bound IGFBP-5, we initially analyzed the regions of IGFBP-5 that were important for binding. To do this we prepared four synthetic peptides that spanned the IGFBP-5 sequence, two of which had very high charge-to-mass ratios. These peptide sequences are underlined in Fig. 1. The peptides corresponding the region A and region B have the highest charge-to-mass ratios. These peptides were used to compete for binding with IGFBP-5 to the ECM *(17)*. The peptides that were the most active were peptides A and B. Peptides C and D had no effect on ECM binding. The peptide A region of IGFBP-5 is also contained in IGFBP-3; however, it was much less potent in competing with IGFBP-3 for binding. Because peptides A and B were so active we assumed that the ECM proteins that bound IGFBP-5 would also contain highly charged regions that mediated binding. The most likely candidates were proteoglycans that contained glycosaminoglycan side chains. To determine if glycosaminoglycans would bind to IGFBP-5, detailed studies were performed. Heparan sulfate, heparin, and dermatan sulfate all bound to IGFBP-5. IGFBP-5 could be coprecipitated with heparin Sepharose and increasing concentrations of each of these glycosaminogly-

```
                                                  ↓
           M V L L T A V L L L L A A Y A G P A Q S L   1
           G S F V H C E P C D E K A L S M C P P S P   22
           L G C E L V K E P G C G C C M T C A L A E   43
           G Q S C G V Y T E R C A Q G L R C L P R Q   64
Peptide C  D E E K P L H A L L H G R G V C L N E K S   85
           Y R E Q V K I E R D S R E H E E P T T S E   106
Peptide D  M A E E T Y S P K I F R P K H T R I S E L   127
Peptide B  K A E A V K K D R R K K L T Q S K F V G G   148
           A E N T A H P R I I S A P E M R Q E S E Q   169
           G P C R R H M E A S L Q E L K A S P R M V   190
Peptide A  P R A V Y L P N C D R K G F Y K R K Q C K   211
           P S R G R K R G I C W C V D K Y G M K L P   232
           G M E Y V D G D F Q C H T F D S S N V E
```

Fig. 12.1

cans competed for binding *(18,19)*. The molecular determinants of binding were identified by showing that the synthetic sugars that contained O-linked sulfate groups in two or three position of the idurionic acid ring also had activity *(18)*. In contrast, if the four or six position was sulfated, there was no capacity to bind IGFBP-5. Therefore, chondroitin sulfate and keratan sulfate had no effect. That this was physiologically significant was proven by showing that the proteoglycan *tenascin* bound IGFBP-5 *(17)*. Digestion of this protein with heparinase resulted in a significant reduction in binding, proving that heparan sulfate continuing proteoglycans accounted for some ECM binding of IGFBP-5. In addition, nonproteoglycans account for some ECM binding. Plasminogen activator inhibitor-I also binds IGFBP-5 with high affinity *(20)*. This protein is an abundant component of fibroblast ECM and therefore may account for a substanial fraction of the fibroblast ECM binding activity.

The amino acids within IGFBP-5 that account for glycosaminoglycan and ECM binding have been defined *(17,19)*. To determine the amino acids that comprised the ECM binding motif, in vitro mutagenesis was utilized. Because the peptide A region of the IGFBP-5 was the most active in mediating extracellular binding we prepared several mutants in which substitutions were made for single amino acids within this region. Substitution for basic amino acids at positions 202, 206, and 207 resulted in a form of IGFBP-5 that bound poorly to fibroblast ECM *(17)*. Likewise, for substitutions for basic amnio acids at positions 211, 214, 217, and 218 resulted in a form with low binding activity. In contrast mutants with substitutions at positions 201, 202, or 217, 218 or 211 only had either no reduction in binding or minimal changes. This suggests that amino acids 207 and 214 are the most important for binding although this has not been definitely proven. Importantly these substitutions did not result in an alteration of the affinity of the mutant forms of IGFBP-5 for IGF-I. Therefore, these mutants may be useful for dissecting out the role of IGFBP-5 binding to ECM binding in determining IGF-I actions.

Bone ECM contains hydroxyappatite which binds to IGFBP-5 *(11,21)*. Importantly, IGFBP-5 content has been shown to decrease by 28% in elderly subjects and this may be related to a reduction in hydroxyappatite *(22)*. However, the other substituents of bone ECM that account for binding IGFBP-5 have not been determined. As for fibroblast ECM IGFBP-5 binding to osteoblast, ECM protects it from proteolysis *(21)*. Chondrocytes and smooth muscle cells can also bind IGFBP-5, but the components of their ECM that account for binding have not been determined.

FUNCTIONAL ALTERATIONS

The biologic significance of binding of IGFBP-5 to ECM has been ascertained by two methodologies: (1) layering IGFBP-5 onto isolated ECM, then plating cells on top of this matrix and determining its biologic activity, (2) preparing IGFP-5 mutants with reduced ECM binding and transfecting them into cells then overexpression of the mutant protein and ascertaining its effects. The first observation that allowed significant conclusions regarding the biologic significance of ECM association was the observation that IGFBP-5 affinity for IGF-I was lowered by approximately eightfold when it associated with fibroblast ECM *(10)*. This change appeared to be due to binding specific ECM proteins because when IGFBP-5 bound to heparin its affinity for IGF-I was lowered approx 15-fold. This reduction in affinity suggested that the IGF-I that was bound to IGFBP-5 in the ECM might be in better equilibration with receptor. Because the affinity of IGFBP-5 in solution is at least 10-fold greater than the IGF-I receptor, lowering its affinity would allow IGF-I that is bound to the ECM associated IGFBP-5 to be in a more favorable equilibrium with the receptor. To verify this hypothesis the ECM was prepared from fibroblast cultures and then IGFBP-5 layered onto this ECM. Following equilibration fibroblasts were plated on top of ECM and their ability to divide and response to IGF-I were determined. Layering IGFBP-5 onto the ECM resulted in a significant (2.1 fold) increase in the cellular replication response to IGF-I *(10)*. That this was due to IGFBP-5 was further confirmed by embedding IGFBP-5 in an artificial matrix containing type IV collagen. When fibroblasts were plated on this IGFBP-5-enriched matrix they had a 2.5-fold enhancement in their growth response to IGF-I compared to control cultures plated on type IV collagen alone. To further verify the importance of ECM binding the mutants that were described previously that had low ECM binding, were transfected into porcine aortic smooth muscle cells. Analysis of the ECM from the cells expressing the 202, 206, 207 mutant and the 211, 214, 217, 218 mutant showed that that they bound poorly to porcine smooth muscle cells (PSMC)/ECM. In contrast, a control four-point mutation with substitutions for amino acids at positions 201, 202, 206, and 208 had no reduction ECM binding. When increasing concentrations of IGF-I were added the DNA synthesis response was attenuated in the cells that were expressing two forms of IGFBP-5 that bound poorly to ECM. It was not attenuated in the cells expressing the forms of IGFBP-5 that bound well to the ECM. Likewise mock transfected cells showed no attenuation of IGF-I responsiveness. This suggests that ECM-associated IGFBP-5 potentiates the cellular growth responsive to IGF-I.

Several attempts have been made to identify whether other IGF binding proteins bind to ECM. Some immunohistochemistry studies have shown other proteins such as IGFBP-2 can be localized in basement membranes of epithelial cells *(23)*. This suggests that although IGFBP-2 was not detected in fibroblast ECM it may bind to other types of ECM such as basement membrane. To determine if other components were required for IGFBP-2 and assessed binding to ECM we incubated high concentrations of IGF-I with IGFBP-2 and assessed binding to proteoglycans. IGFBP-2 bound avidly to heparin and heparin sulfate if it was saturated with IGF-I but if IGF-I was not added it did not bind *(24)*. This is in contrast to IGFBP-3 and -5, which will bind to ECM without IGF-I being present. Furthermore, IGFBP-2 could be shown to be localized in fibroblast ECM in the presence of high concentrations of IGF-I *(24)*. Werther and co-workers have shown that IGFBP-2 is highly focally localized in the olfactory bulb of rat brain and can be detected in ECM surrounding these tissues *(25)*. Because this area of the brain expresses high concentrations of IGF-I, this suggests that there is in vivo correlation with our in vitro findings and that abundant secretion of IGF-I in the olfactory bulb accounts for IGFBP-2 colocalization in the olfactory bulb ECM. The biologic significance of IGFBP-2 localization in this brain region has not been fully elucidated.

IGFBP-1 has not been easily localizable in ECM; however, unlike other forms of binding proteins, IGFBP-1 has been shown to interact with specific integrin receptor. IGFBP-1 contains an Arg–Gly–Asp sequence near its C-terminus that binds specifically to the $\alpha5\beta1$ integrin *(26)*. When cells expressing a mutated from of IGFBP-1 (WGD) were analyzed, this mutant form of IGFBP-1 was shown not to bind to the $\alpha5\beta1$ integrin. Binding was significant, as wild-type IGFBP-1 stimulated cell migration in cells that had $\alpha5\beta1$ integrin receptors and that the WGD mutant will not stimulate cell migration. To determine if IGFBP-1 was important for modifying IGF-I actions in vivo IGFBP-1 and IGF-I in combination were applied to dermal ulcers in both normal *(27,28)* and diabetic *(29)* rabbits. The IGFBP-1 mutant (WGD) was also tested. The results showed no improvement in the wounds that received WGD IGFBP-I but the wounds that received the wild-type IGFBP-1 had an improved response *(27,28)*. Therefore, in the presence of the same concentration of IGF-I, engagement of the $\alpha5\beta1$ integrin by IGFBP-1 is required for normal wound repair. This interaction is important for IGF-I modulation of wound healing.

In summary, at least three IGF binding proteins have been shown to bind to ECM. IGFBP-5 binds avidly to ECM from cartilage, bone, and connective tissue. In contrast these matrices do not bind IGFBP-2 unless high concentrations of IGF-1 are present. The ECM binding domain of IGFBP-3 and -5 appears to be a highly basic region located in the C-terminus of each protein. This highly basic region contains specific amino acids that mediate binding and when binding is attenuated, cell responsiveness is also decreased. IGFBP-3, in contrast to IGFBP-5, binds less avidly to ECM. The significance of IGFBP-3 association with ECM has not been well defined.

ACKNOWLEDGMENTS

The author wishes to thank Mr. George Mosley for his editorial assistance. These studies were supported by a grant from the National Institutes of Health HL-56580.

REFERENCES

1. Daughaday WH, Mariz IK. Conversion of proline-C^{14} to labeled hydroxyproline by rat cartilage in vitro: effects of hypophysectomy, growth hormone, and cortisol. J Lab Clin Med 1962; 59:741–752.
2. Hall K, Uthne K. Some biological properties of purified sulfation factor (SF) from human plasma. Acta Med Scand 1971; 190:137–143.
3. Salmon Jr WD, Du Vall MR, Thompson EY, Stimulation by insulin in vitro of incorporation of [^{35}S] sulfate and [^{14}C] leucine into protein-polysaccharide complexes, [^{3}H] uridine into RNA, and [^{3}H] thymidine into DNA of costal cartilage from hypophysectomized rats. Endocrinology 1968; 82:493–499.
4. Canalis E, McCarthy T, Centrella M. Growth factors and the regulation of bone remodeling. J Clin Invest 1988; 81:277–281.

5. Tyler JA. Insulin like growth factor-I can decrease degradation and promote synthesis of proteoglycan in cartilage exposed to cytokines. Biochem J 1989; 260:543–548.
6. Ashton TK, Francis MJO. Response of chondrocytes isolated from foetal cartilage to plasma somatomedin activity. J Endocrinol 1978; 76:473–477.
7. Sah RL, Chen CC, Grodzinsky AJ, Trippel SB. Differential effects of bFGF and IGF-I on matrix metabolism in calf and adult bovine cartilage explants. Arch Biochem Biophys 1994; 308:(1):137–147.
8. Hill DJ, Clemmons DR, Wilson S, Hall VKM, Strain AJ, Milner DG. Immunological distribution of one form of insulin-like growth factor (IGF) binding protein and IGF peptides in human fetal tissues. J Mol Endocrinol 1989; 2:31–38.
9. Camacho-Hubner C, Busby WH, McCusker RH, Wright G, Clemmons DR. Identification of the forms of insulin-like growth factor binding proteins produced by human fibroblasts and the mechanisms that regulate their secretion. J Biol Chem 1992; 267:11949–11956.
10. Jones JI, Gockerman A, Busby WH, Camacho-Hubner C, Clemmons DR. Extracellular matrix contains insulin-like growth factor binding protein-5: potentiation of the effects of IGF-I. J Cell Biol 1993; 121:679–687.
11. Bautista CM, Baylink DJ, Mohan S. Isolation of a novel insulin-like growth factor (IGF) binding protein from human bone: a potential candidate for fixing IGF-II in human bone. Biochem Biophys Res Commun 1991; 176:756–763.
12. Mohan S. Insulin-like growth factor binding proteins in bone cell regulation. Growth Regul 1993; 3:67–70.
13. Hakeda Y, Kawaguchi H, Hurley M, Pilbeam CC, Abreu C, Linkhardt TA, Mohan S, Kumegawa M, Raisz LG. Intact insulin-like growth factor binding protein-5 (IGFBP-5) associates with bone matrix and the soluble fragments of IGFBP-5 accumulated in cultured medium of neonatal mouse calvariae by parathyroid hormone and prostaglandin E2 treatment. J Cell Physiol 1996; 166:370–379.
14. Andress DL, Birnbaum RS. Human osteoblast-derived insulin-like growth factor (IGF) binding protein-5 stimulates osteoblast mitogenesis and potentiates IGF action. J Biol Chem 1992; 267:22467–22472.
15. Andress DL, Loop SM, Zapf J, Kiefer MC. Carboxy truncated insulin like-growth factor binding protein-5 stimulates mitogenesis in osteoblast like cells. Biochem Biophys Res Commun 1993; 195:25–30.
16. Andress DL. Comparison studies of IGFBP-5 binding to osteoblasts and osteoblast-derived extracellular matrix. Prog Growth Factor Res 1995; 6:(2–4):337–344.
17. Parker A, Busby WH, Clemmons DR. Identification of the extracellular matrix binding site for insulin like growth factor binding protein-5. J Biol Chem 1996; 271:13523–13529.
18. Arai T, Parker AJ, Busby WH, Clemmons DR. Heparin, heparan sulfate, and dermatan sulfate regulate formation of the insulin-like growth factor-I and insulin-like growth factor-binding protein complexes. J Biol Chem 1994; 269:20388–20393.
19. Arai T, Clarke JB, Parker A, Busby W Jr, Nam TJ, Clemmons DR. Substitution of specific amino acids in insulin-like growth factor-binding protein-5 alters heparin binding and its change in affinity for IGF-I in response to heparin. J Biol Chem 1996; 271:(11):6099–6106.
20. Nam TJ, Busby WH, Clemmons DR. Insulin-like growth factor binding protein-5 binds to plasminogen activator inhibitor-I. Endocrinology 1997; 138:2972–2998.
21. Mohan S, Nakao Y, Honda Y, Landale EC, Leser U, Dony C, Lang K, Baylink DJ. Studies on the mechanisms by which insulin-like growth factor (IGF) binding protein-4 (IGFBP-4) and IGFBP-5 modulate IGF actions in bone cells. J Biol Chem 1995; 270:(35):20424–20431.
22. Nicolas V, Mohan S, Honda Y, Prewett A, Finkelman RD, Baylink DJ, Farley JR. An age-related decrease of insulin-like growth factor binding protein-5 in human cortical bone. Calcif Tissue Int 1995; 57:206–212.
23. Hill DJ, Clemmons DR. Similar distribution of insulin-like growth factor binding proteins 1, 2 and 3 in human fetal tissues. Growth Factors 1992;6:315–326.
24. Arai A, Busby WH, Clemmons DR. Binding of insulin-like growth factor I or II to IGF binding protein enables it to bind to heparin or extracellular matrix. Endocrinology 1996; 137:4571–4575.
25. Russo VC, Bach LA, Fosang AJ, Baker NL, Werther GA. Insulin-like growth factor binding protein-2 binds to cell surface proteoglycans in the rat brain olfactory bulb. Endocrinology 1997; 138:(13);4858–4867.
26. Jones JI, Gockerman A, Busby WH Jr, Wright G, Clemmons DR. Insulin-like growth factor binding protein 1 stimulates cell migration and binds to the $\alpha 5\beta 1$ integrin by means of its Arg-Gly-Asp sequence. Proc Natl Acad Sci USA 1993; 90:10553–10557.

27. Juang R, Mustoe T, Busby W, Clemmons D. Increased wound breaking strength induced by insulin like growth factor I in combination with IGF binding protein-1. Surgery 1994; 115:223–228.
28. Galiano RD, Zhao L, Clemmons DR, Roth SI, Lin X, Mustoe TA. Interaction between the insulin-like growth factor family and the integrin receptor family in tissue repair processes. J Clin Invest 1996; 98:2462–2468.
29. Tsuboi R, Shi CM, Sato C, Cox GN, Ogawa H. Coadministration of insulin like growth factor (IGF-I) and IGF binding protein-1 stimulates wound healing in animal models. J Invest Dermatol 1995; 104:199–203.

13 Biological Actions of Proteolytic Fragments of the IGF Binding Proteins

Michel Binoux, MD, *Claude Lalou*, PhD,
and Sarah Mohseni-Zadeh, PhD

CONTENTS

INTRODUCTION

The existence of insulin-like growth factor binding protein (IGFBP) fragments was first discerned during purification of IGFBP-1 from human placenta *(1)* and IGFBP-3 from human *(2,3)* and rat serum *(4,5)* and human cerebrospinal fluid *(6)*. A truncated form of IGFBP-5 was isolated in advance of the intact protein *(6,7)*. The amino (N)-terminal sequences of these fragments were found to correspond to forms lacking the carboxy (C)-terminal third of the molecule. However, in the case of IGFBP-2, a fragment corresponding to the C-terminus of the native binding protein was isolated from the conditioned medium of a fetal rat liver cell line *(8)*. At the same time, it was discovered that IGFBP-3 undergoes limited proteolysis during pregnancy *(9–11)*. The resulting functional alteration was revealed by the disappearance of the characteristic 42–39-kDa doublet in Western ligand blotting and the concomitant appearance of a 30-kDa fragment in the chromatographic fragments normally containing the 140-kDa complexes that comprise IGFBP-3, IGF-I or -II, and the acid-labile subunit (ALS). From these observations, it seemed that the fragment must play some physiological role. Further studies confirmed its weak affinities for the IGFs, which would explain their facilitated dissociation from the 140-kDa complexes and their enhanced bioavailability *(12,13)*.

It very soon became evident that IGFBP proteolysis occurs in all biological fluids, as well as in cell culture media. All except IGFBP-6 are sensitive to this limited degradation which is physiologically regulated by a variety of proteases with different degrees of selectivity (see reviews in *refs. 14–16* and in Chapters 16 and 26 of this book).

From: *Contemporary Endocrinology: The IGF System*
Edited by: R. Rosenfeld and C. Roberts Jr © Humana Press, Inc., Totowa, NJ

IGFBP fragments have been identified by sodium dodecyl sulfate-polyacrylamide gel electrophoresis (SDS-PAGE) analysis of radiolabeled IGFBPs incubated in medium containing protease activity or following affinity crosslinking with ^{125}I-IGF, but especially in Western immunoblotting. This last methodology has become standard since the development of chemiluminescence techniques that markedly increase sensitivity. Fragments have been purified from biological media and following in vitro proteolysis of recombinant IGFBPs by either pure proteases or biological samples with known protease activity. The biological activities of some of the fragments have been investigated. Through N-terminal sequencing of isolated peptides, the sites of IGFBP cleavage have been identified and some primary structures have been determined. Recombinant peptides have started to be generated and transfection achieved of cDNA plasmids corresponding to IGFBP fragments so as to study their properties and biological activities. The results of these various studies are analyzed in this chapter.

The fundamental question concerning the IGFBP fragments remains their physiological significance. The IGFBPs appear to possess different functional domains, some engaged in IGF binding, some implicated in interaction with the cell surface and extracellular matrix. These contribute toward regulating the bioavailability of the IGFs and modulating their cellular action. Data indicating a regulatory role for the IGFBP fragments as regards IGF bioavailability therefore support the notion of physiological significance. Also, certain of their biological effects would account for some of the more recently reported IGF-independent actions of intact IGFBPs, discussed in Chapter 12. Further unexpected effects of the fragments suggest functions and mechanisms of action that go beyond the realm of the IGF system.

STRUCTURAL AND FUNCTIONAL PROPERTIES OF IGFBP FRAGMENTS

The primary source of data in determining the molecular characteristics of IGFBP fragments has been Western immunoblotting. Separation of the proteolytic fragments by SDS-PAGE is usually carried out in the absence of reducing agent, which means that after transfer to membranes the secondary structure of the proteins is recovered. This is crucial for binding to labeled IGFs (which is possible only with intact or minimally modified IGFBPs). The utility of nonreducing conditions is that the IGFBPs can be revealed on the same membranes first by ligand blotting, then by immunoblotting. The disadvantage is that estimation of molecular mass is perforce imprecise because molecular markers need to migrate under reducing conditions. At present, the major limitation of Western immunoblotting is that it is possible to detect only fragments possessing epitopes recognized by antibodies obtained using intact IGFBPs. These antibodies are specific because they are raised against the mid-regions of the proteins, which are the least conserved. However, it is in these regions that proteolytic cleavage is the most frequent, particularly toward the C-terminal end. After sequencing of the proteolytic fragments of recombinant human (rh) IGFBP-3 generated by plasmin, it emerged that the C-terminal third of IGFBP-3 resulting from primary cleavage by virtually all proteases cannot be detected by anti-IGFBP-3 polyclonal and monoclonal antibodies *(17)*. In contrast, a fragment containing the C-terminal domain of IGFBP-2 has been identified in human milk using a polyclonal antibody also recognizing the N-terminal region *(18)*.

In analysis of fragments derived from radiolabeled, purified, or recombinant IGFBP-3 *(19,20)*, only those with one or more iodinated tyrosines are recognized. In the case of

IGFBP-3, the N-terminal end does not become labeled, whereas the C-terminal end is heavily so *(unpublished data)*. This means that electrophoretic findings cannot be superimposed with data from immunoblotting.

In affinity labeling using radiolabeled IGF, SDS-PAGE analysis of crosslinked complexes can reveal not only fragments that are undetectable by the methods previously described, but also bridges between fragments that are incompletely dissociated.

IGFBP Fragments Identified in Biological Fluids and Cell Culture Media

Table 1 summarizes the major findings published. These illustrate the diverse media in which proteolytic fragments of IGFBPs have been detected: all biological fluids analyzed (particularly serum, but also lymph, cerebrospinal fluid, skin interstitial fluid, synovial fluid, seminal fluid, follicular fluid, amniotic fluid, milk, urine); a wide variety of cell models, whether in primary culture (hepatocytes, endometrial cells) or as cell lines (fibroblasts, osteoblasts, smooth muscle cells, liver cells, granulosa cells); and especially lines of tumoral origin (breast cancer cells, prostate carcinoma cells, neuroblastoma cells).

Table 1 also highlights the differences between the IGFBPs as regards proteolysis. IGFBP-6 is not proteolyzed and IGFBP-1 rarely so. This may reflect structural differences, or a degree of protease specificity, or tissue specificity in the expression of IGFBPs, or proteases and/or their inhibitors. In pregnancy serum where virtually all of the IGFBP-3 is proteolyzed, some IGFBP-4 and little IGFBP-2 are altered and IGFBP-1 is intact *(10,11)*. In the endometrium, where IGFBP-1 expression is closely related to the secretory phase of the cycle and pregnancy, the absence of proteolysis may have physiological significance. Giudice has suggested that decidual IGFBP-1 production at the trophoblast/decidual interface may be a maternal "restraint" on trophoblast invasion *(21)*.

The molecular masses of the fragments estimated from SDS-PAGE are relatively constant for each IGFBP, irrespective of the tissue of origin or the species concerned. As far as the proteases are concerned, some are known: plasmin, prostate-specific antigen (PSA), matrix metalloproteases (MMPs), cathepsin D; but most remain unidentified and are characterized in terms of their sensitivity to inhibitors, their dependence on calcium or zinc, and their molecular mass. Certain observations indicate that they are specific to or more active upon one or another IGFBP and are expressed in several cell systems, particularly those affecting IGFBP-4 and IGFBP-5 *(22–29)*. However, protease activity is frequently attributable to a combination of proteases *(30–33)* and some appear preferentially to attack fragments (secondary proteolysis). This multiplicity naturally complicates the task of characterization and identification and may reflect a cascade of enzymatic triggers. Nevertheless, despite the complexity, there appears to be a relatively small number of proteases that degrade IGFBPs and they seem to be fairly specific. Control of their expression and action would predictably be closely associated with regulation of the IGF system.

Fragments Generated from Native or Recombinant IGFBPs Submitted to Protease Action

Incubation of purified or recombinant IGFBPs either with cultured cells secreting known or unidentified proteases, or in vitro with cell-conditioned media, has led to characterization of the fragments generated by these proteases *(31,34)*. The use of purified proteases has revealed their specificities for the different IGFBPs. With SDS-PAGE identification of the fragments, comparisons have been possible with the forms found in

Table 1
Proteolytic Fragments of IGFBPs in Biological Fluids and Cell-Conditioned Media

IGFBP	*Origin of the IGFBP Fragments*	*Molecular Mass*	*Proteases or Inhibitors*	*Reference*
IGFBP-1	Human decidua	21 and 17–20 kDa	Not determined	*1*
	Human amniotic fluid	20 and 16 kDa	Not determined	*59*
	Fetal rat hepatocytes	16 kDa	Not determined	*123*
IGFBP-2	Human pregnancy serum	Not determined	Ca^{2+}-dependent serine protease	*10,11*
	Human milk	24 and 14–16 kDa	Not determined	*18*
	Human and ovine follicular fluid	23 kDa	EDTA; 1,10-phenanthrolin; Zn^{2+}- and Ca^{2+}-dependent protease	*22,65*
	Human cerebrospinal fluid	25 kDa	Not determined	*59*
	Human neuroblastoma cells	16 and 20 kDa	PA/plasmin	*73*
	Fetal rat liver cell line	21 kDa	Not determined	*8*
	Human airway smooth muscle cells	22 and 17 kDa	MMP-1	*124*
	Porcine aortic smooth muscle cells	16 and 25 kDa	36-kDa serine protease; EDTA	*125*
IGFBP-3	Human normal and pregnancy serum	29–30, 20, 18, and 15 kDa	Ca^{2+}-dependent serine protease; plasmin; >150-kDa protease; 50-kDa metalloproteinase	*10,11,126 30,127 128*
	Human lymph	30 kDa	Not determined	*126*
	Human urine	17.7 kDa	Not determined	*44*
	Human skin interstitial fluid	30 and 21 kDa	Ca^{2+}-dependent serine protease activated by plasminogen	*129*
	Human synovial fluid	30 kDa	EDTA	*130*
	Human and ovine follicular fluid	29 kDa	EDTA; 1,10-phenanthrolin; Zn^{2+}- and Ca^{2+}-dependent protease	*64,65,70*
	Human seminal fluid	23-25 and 19 kDa	Prostate-specific antigen (PSA)	*131,132*
	Human normal prostatic and prostate carcinoma cells, seminal fluid	23-30kDa (rh-IGFBP-3: 20 and 16 kDa)	Cathepsin D and PSA	*25,36*
	Human prostate carcinoma cell line	30, 20, and 16 kDa	Plasmin	*71*
	Porcine granulosa cells	(rhIGFBP-3:14 and 18–22 kDa)	Not determined	*133*

(*continued*)

Table 1 (*Continued*)

IGFBP	*Origin of the IGFBP Fragments*	*Molecular Mass*	*Proteases or Inhibitors*	*Reference*
	Human breast cancer cell line	30 and 20 kDa	25–30-kDa acid protease; EDTA; leupeptin	*31*
	Human dermal fibroblasts	30 kDa	MMP-1 and MMP-3	*134*
	Human osteoblastlike cells	30, 20, and 16 kDa	Plasmin	*34*
IGFBP-4	Human pregnancy serum	10 and 17 kDa	Several, including a 50-kDa MMP	*10,11,128*
	Ovine follicular fluid	15 and 17 kDa	EDTA; 1,10-phenanthrolin	*23,70*
	Human endometrial stromal cells	18 kDa	EDTA; aprotinin; 1,10-phenanthrolin	*89,134*
	Human fibroblasts	17 kDa	EDTA; 1,10-phenanthrolin; serine protease inhibitors	*26,136*
	Porcine aortic smooth muscle cells	19 and 14 kDa	EDTA; 1,10-phenanthrolin; α_1-antichymotrypsin (probably a 48-kDa serine protease)	*27*
	Human osteoblastlike cells	18 and 14 kDa	Not determined	*137,138*
	Rat neuroblastoma cell line	16 kDa	Ca^{2+}-dependent serine protease;	*139*
IGFBP-5	Human pregnancy serum	22 and 15 kDa	50-kDa MMP	*20,128*
	Human amniotic fluid	22 and 15 kDa	Cation-dependent, >100-kDa serine protease	*20*
	Ovarian follicular fluid	18 and 20 kDa	EDTA; 1,10-phenanthrolin	*23,70*
	Rat granulosa cells and human follicles	19.5 and 17.5 kDa	EDTA; 1,10-phenanthrolin; >100-kDa protease	*140*
	Human fibroblasts	22, 20 and 17 kDa	Ca^{2+}-sensitive, 92-kDa serine protease	*33,141,142*
	Human osteocarcinoma cell line	23 kDa	Not determined	*79*
	Murine osteoblast cell line	22 and 14 kDa	52-kDa MMP-1; 60-kDa MMP-2	*32*

Table 2
Fragments Generated Following In Vitro Proteolysis of IGFBPs by Characterized Proteases

	Proteases	*IGFBP Fragments*	*Reference*
rh-IGFBP-1	Cathepsin D	12 kDa	*143*
	Stromelysin-3 (MMP-11)	9, 16 kDa	*144*
	MMP-2, -3, -7	9, 16, 21 kDa	*144*
rh-IGFBP-2	Cathepsin D	23, 18 kDa	*143*
rh-IGFBP-3			
Glycosylated	Plasmin	30, 20 kDa	*43*
Nonglycosylated	Plasmin	22/25, 16 kDa	*43,145*
Nonglycosylated	Thrombin	27, 24 kDa	*145*
Nonglycosylated	Cathepsin D	15, 12 kDa	*143*
Nonglycosylated	MMP-1, -3, -9	20/22, 16 kDa	*134*
Nonglycosylated	PSA	16, 22 kDa	*146*
Nonglycosylated	γ-NGF	22 kDa	*113*
Glycosylated	γ-NGF	35, 30, 25, 17 kDa	*113*
rh-IGFBP-4	γ-NGF	16–17 kDa	*113*
	Cathepsin D	10 kDa	*143*
IGFBP-5			
rh	Cathepsin D	15–23 kDa	*143*
Purified human	Thrombin	22–23, 20 kDa	*29*

rh, recombinant human.

biological fluids or culture media and a first step has been made toward their purification. Table 2 shows the results obtained to date. Two types of observation emerge: data obtained for proteases shown to be present in biological fluids or culture media (plasmin, MMP-1, -3, -9, PSA); and those obtained for proteases whose activity has been demonstrated only in vitro (stromelysin 3, thrombin). In the latter case, physiological significance remains to be proved. As regards cathepsin D, which intervenes in tumoral progression, its effects on IGFBP-3 have been documented after acidification of media conditioned by breast *(35)* and prostate *(25,36)* cancer cell lines. However, other acid pH-dependent proteases, especially cysteine proteinases, may be involved, as demonstrated in mice with a targeted disruption of the cathepsin D gene *(37)*. Acidification may therefore trigger a cascade of enzymatic activities for which cathepsin D is just one of many activators.

Proteolytic Cleavage Sites on the IGFBPs

The sites of proteolytic cleavage have been identified following purification of fragments generated by IGFBP proteolysis and their N-terminal sequencing (see Fig. 1 and Table 3).

For all IGFBPs that are proteolyzed, one or two cleavages always occur toward the end of the variable domain which is rich in basic amino acids. The resulting fragments comprising the N-terminal two thirds of the protein (130–160 amino acids) migrate as 18–23-kDa forms (30 kDa in the case of IGFBP-3 which is glycosylated) in SDS-PAGE under nonreducing conditions. Cleavage sites have also been located at the beginning of the variable domains of IGFBP-3 (residues 95–98) and IGFBP-4 (residue 90), yielding

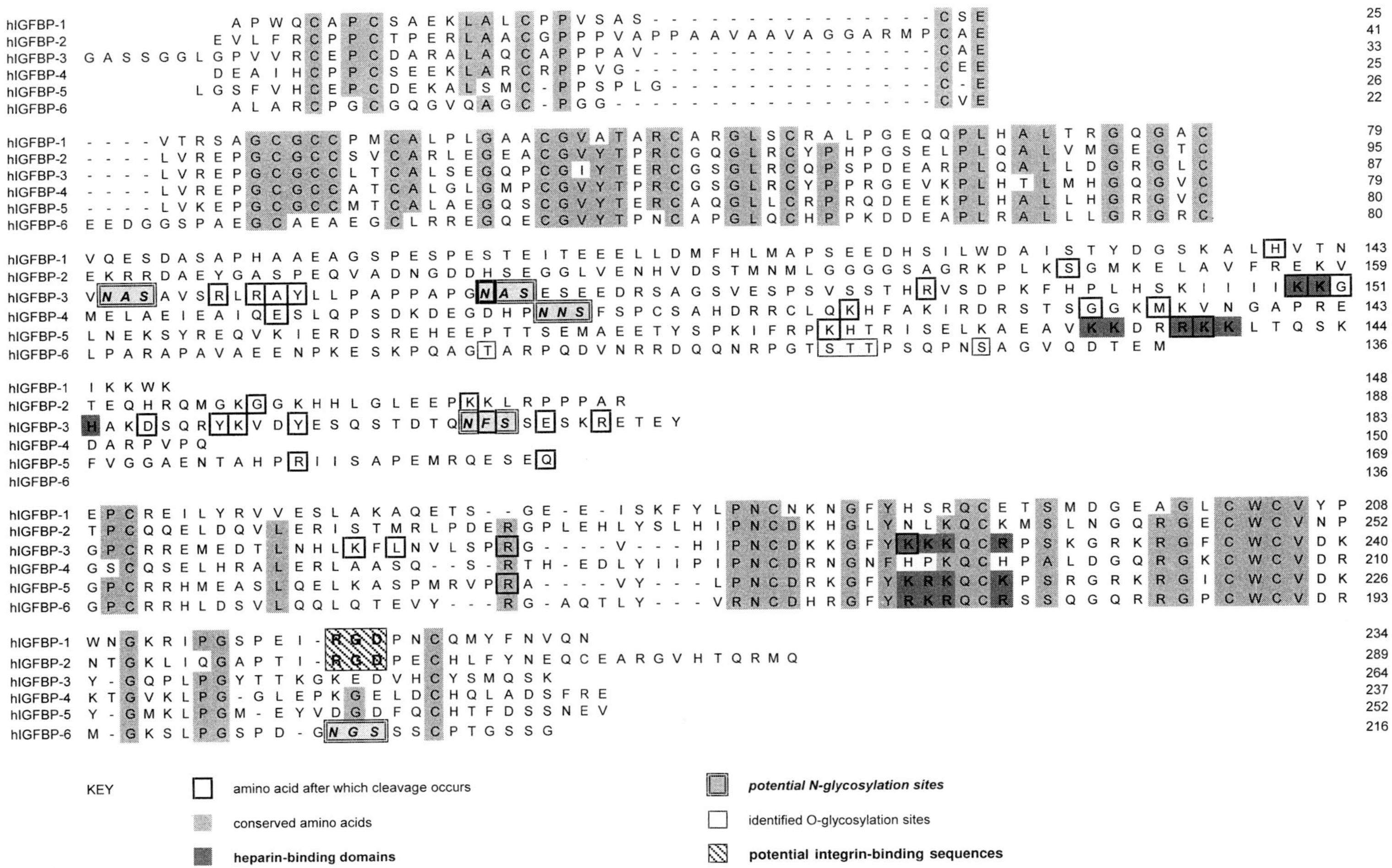

Fig. 1. Location of proteolytic cleavage sites in the primary sequences of IGFBPs. Cleavage occurs after the amino acid (boxed in black). The proteases involved or their origin are indicated in Table 3.

Table 3
Location of Cleavage Sites in the Primary Sequences of IGFBPs According to the Protease Involved

	Proteases (Type or Origin)	*Cleavage Sites*	*Reference*
IGFBP-1	Stromelysin-3	140	*144*
IGFBP-2	Fetal rat liver cell line	147	*8*
	Human milk	169, 181	*18*
IGFBP-3	Plasmin	95, 97, 149, 150, 151, 155, 160, 198, 220	*17,145*
	PSA	97, 132, 159, 173, 179	*146*
	Human serum	97, 148, 149, 150, 154	*145*
	MMP-1	99	*134*
	MMP-2 (human fibroblasts)	99	*134*
	MMP-3	99, 109, 176	*134*
	Cathepsin D	98, 159, 163, 200	*143*
	Thrombin	97, 206	*145*
IGFBP-4	Rat neuronal cells	90, 120, 132	*28*
	Human fibroblasts	135	*147*
	Cathepsin D	135	*143*
IGFBP-5	Thrombin	120, 156, 192	*29*
	Porcine aortic smooth muscle cells	137, 138	*148*
	Human osteoblasts	169	*80*

fragments of approximately 15 kDa or 20 kDa (glycosylated IGFBP-3). These cleavages at the two extremities of the variable region therefore release the three different domains of IGFBP-3 and IGFBP-4. Fragments of 14–17 kDa have also been identified for IGFBP-5 (Table 1), but no sequencing data are as yet available to substantiate the existence of such cleavages.

The conserved N-terminal regions of IGFBPs are insensitive to proteolysis and the C-terminal domains are only slightly affected. Degradation of the C-terminal domain has been observed only for IGFBP-3 and IGFBP-5 in vitro, hence under conditions in which proteolysis is not regulated. Antibodies specific to the C-terminal domains remain to be developed before fragments resulting from these cleavages can be shown to exist in biological fluids. In all likelihood, the relative protease insensitivity of the conserved domains is linked to the numerous sulfide bridges that preserve their secondary structure.

An exception among the IGFBPs is IGFBP-6 which is resistant to proteolytic attack. This may be explained in terms of its small variable domain which comprises 56 amino acids (as opposed to the 70–90 in the other IGFBPs), scarce potential cleavage sites (few basic amino acids and no hydrophobic residues such as tyrosine, phenylalanine, or tryptophan) and, especially, the five sites of O-glycosylation recently identified in this region *[38]*).

Affinities of IGFBP Fragments for IGFs

Proteolysis of IGFBPs was discovered owing to their loss of affinity for IGFs *(9–11)*, as reflected in Western ligand blotting where the bands corresponding to the intact forms decrease or disappear. To this extent, Western ligand blotting constitutes a test of the functional integrity of the protein *(39)*, although sometimes proteolytic fragments may give weak signals if their concentrations in the samples analyzed are sufficiently high. A variety of studies have shown that both the N- and C-terminal domains of IGFBPs are involved in IGF binding, but for the characteristically strong affinity of the IGFBPs for IGFs (10^9–10^{11} M^{-1}) to apply, the protein needs to be intact, therefore the mid-region present. Spencer and Chan have justifiably suggested that this disulfide bond-free region may constitute a slightly mobile connecting peptide between the N- and C-terminal domains (which are heavily crosslinked by disulfide bonds), giving the protein conformational flexibility *(40)*. The necessity of the mid-region for ligand binding was recently proved for IGFBP-3 by Firth et al. using deletion variants *(41)*. This mid-region would be considered as a hinge-joint between the two globular regions, whose purpose is to adapt the three-dimensional structure of the IGFBP such that optimal IGF binding is achieved. The residual ability of IGFBP fragments to bind IGF-I and -II may or may not be physiologically significant. This would depend on the extent to which their concentrations in the cellular environment, despite their weak affinities for IGFs, might allow of competition with IGF receptors or other IGFBPs.

Quantitative data for the affinities of the IGFBP fragments for IGFs have been obtained from competitive binding studies in solution assays or by affinity crosslinking. The latter method is required if radiolabeled IGF binding proves too weak in the solution binding assay, although it is molecular association that is visualized, not dissociation. Short-lived complexes between weak-affinity proteins, or even between proteins with no affinity but at high concentration, may be trapped during incubation with crosslinking reagents. These do not represent conditions of reaction equilibrium such as those obtained in solution binding assays. The findings provided by Western ligand blotting are essentially qualitative. Controls comprising competition between unlabeled and labeled IGFs are frequently not run, which means that the intensities of the signals depend both on the affinities and the quantities of fragment applied to the gel. No interpretation can be ventured without a comparison of the signal obtained with that yielded by an equivalent amount of intact protein.

Table 4 is a summary of the major findings to date. In all competitive binding studies, the affinities of the IGFBP fragments for IGF-I and IGF-II are reduced. The single exception is the 21-kDa carboxy-truncated fragment of IGFBP-1 *(1)*, where its affinity was compared with that of the intact protein determined in a different experiment.

Data for IGFBP-2, IGFBP-4, and IGFBP-5 fragments are limited. In the case of IGFBP-2, only fragments comprising the C-terminal region beginning from Gly 148 *(8)*, Gly 169, and Lys 181 *(18)* have been tested. No information is available concerning the corresponding fragments containing the N-terminal region. In the case of IGFBP-4 and IGFBP-5, by contrast, it is only the N-terminal fragments that have been studied. All the findings published indicate affinities for IGF-I and IGF-II between 5 and 20 times weaker than those of the intact proteins.

In the case of IGFBP-3, the affinities of its fragments for IGFs have been studied in purified serum fractions, or following treatment of recombinant IGFBP-3 with plasmin, or in the culture media of cells expressing deleted IGFBP-3 variants. The major carboxy-truncated fragment of 30 kDa resulting from proteolysis of circulating IGFBP-3 *(10,12,42)* or rh-IGFBP-3 *(17)* has weakened affinity for both IGFs, but particularly for IGF-I. The affinity loss of the fragment isolated from pregnancy plasma *(12)* appears relatively small. This is because the comparison was made with IGFBP-3 isolated from

Table 4
Affinities of IGFBP Fragments for IGFs: Western Ligand Blotting Data and Quantitative Data from Solution Binding Assays and Affinity Crosslinking Studies

	IGFBP Origin	*IGFBP Fragment*	*IGF-Binding Activity*	*Competitive Binding Studies*	*Reference*
IGFBP-1	Human placenta	Purified, 21-kDa carboxyl-truncated	No data	Solution binding assay Affinity for IGF-I similar to that of the intact protein	*1*
	rh-IGFBP-1 variants expressed in COS-1 cells	24- and 18-kDa N-terminal variants	Not detected by ligand blotting; solution binding assay not done.	Not done	*149*
IGFBP-2	Fetal rat liver cell line-conditioned medium	Purified, 21 kDa starting at Gly 148	Detected in HPLC fractions by solution binding assay and affinity labeling	Solution binding assay 5–15 times weaker affinity for IGF-I and IGF-II	*8*
	Human milk	Purified, 14–16-kDa starting at Gly 169 and Lys 181	Detected in HPLC fractions by ligand blotting (not detected in crude milk)	Solution binding assay 10 times weaker affinity for IGF-II	*18*
	Human neuroblastoma cell line-conditioned medium	Nonpurified, 16–20 kDa	Not detected in ligand blotting	Solution binding assay 5 times weaker affinity for IGF-II	*73*
IGFBP-3	Pregnancy plasma	Partially purified, 30-kDa carboxyl-truncated	Detected in chromatographic fractions (but not in plasma) by ligand blotting	Solution binding assay 10 and 2 times weaker affinity for IGF-I and IGF-II, respectively	*10,12*
	Serum from patients with chronic renal failure	Purified, 29-kDa glycosylated from peritoneal dialysate	Not detected in serum by ligand blotting	Solution binding assay 200 and 4–10 times weaker affinity for IGF-I and IGF-II, respectively	*42*
	rh-nonglycosylated IGFBP-3 (*E. coli*) proteolyzed by plasmin	Purified, 22/25-kDa, residues 1–160	Detected in HPLC fractions by ligand blotting (weak signal compared with immunoblotting)	Solution binding assay 50 and 20 times weaker affinity for IGF-I and IGF-II, respectively	*12,17*
	rh-IGFBP-3 variants expressed in yeast	30 kDa glycosylated, residues 1-147 and 1-167	Detected by ligand blotting and in chromatographic fractions by ^{125}I-IGF-I binding	Not done	*40*
	rh-IGFBP-3 variants expressed in CHO cells	Glycosylated 30–35-kDa, residues 1–184	Not detected by ligand blotting (10–20 ng); no binding up to 15 ng/mL in solution assay	Competitive affinity cross-linking 17 and 40 times weaker affinity for IGF-I and IGF-II, respectively	*41*

(continued)

Table 4 (*Continued*)

	IGFBP Origin	*IGFBP Fragment*	*IGF-Binding Activity*	*Competitive Binding Studies*	*Reference*
	rh-nonglycosylated IGFBP-3 (*E. coli*) proteolyzed by plasmin	Purified, residues 161–264	Not detected in HPLC fractions by ligand blotting; solution binding assay not done	Not done	*17*
	rh-IGFBP-3 variants expressed in yeast	Glycosylated, residues 151–263	Detected in chromatographic fractions by ^{125}I-IGF-I binding	Not done	*40*
	rh-nonglycosylated IGFBP-3 (*E. coli*) proteolyzed by plasmin	Purified, 16-kDa, residues 1–95	Not detected in HPLC fractions by ligand blotting or affinity labeling; <2% IGF binding in solution assay at 200 ng/mL	Not done	*12,17*
	Human urine	Partially purified, glycosylated 17.7 kDa	Detected in chromatographic fractions by affinity labeling (300 ng IGF-I or -II) followed by Western immunoblotting (weak signal compared with un-crosslinked fragment)	Not done	*44*
	rh-IGFBP-3 variants expressed in *CHO* cells	15-kDa, nonglycosylated, residues 1–88	Not detected by ligand blotting (10–20 ng); no binding in solution assay up to 15 ng/mL	Competitive affinity crosslinking 60 and 40 times weaker affinity for IGF-I and IGF-II, respectively	*41*
	rh-IGFBP-3 variants expressed in baculovirus	10–14-kDa, residues 1–87 and 1–97	IGFBP-$3^{1\text{-}87}$ detected by ligand blotting at 3 µg (weak signal compared with 30 ng intact IGFBP-3)	Competitive affinity crosslinking Competition by IGF-I	*45,46*
IGFBP-4	Rat neuroblastoma cell line-conditioned medium	Purified, 16-kDa, carboxyl-truncated	Not detected in conditioned medium by ligand blotting	Competitive affinity crosslinking 13 and 20 times weaker affinity for IGF-I and IGF-II, respectively	*139*
IGFBP-5	Human osteosarcoma cell line	Purified, 23-kDa, carboxyl-truncated	Not detected by ligand blotting	Solution binding assay 15 times weaker affinity for IGF-I and IGF-II compared with intact IGFBP-3	*79*
	rh-IGFBP-5 variants expressed in yeast	24-kDa, residues 1–169 and 18-kDa, carboxyl-truncated	In solution assay 4 times weaker IGF-I binding at 100 times the concentration of intact IGFBP-5	Not done	*80*

rh, recombinant human.

normal plasma in which intact IGFBP-3 is associated with proteolyzed IGFBP-3 (20–30% of total IGFBP-3 according to immunoblot analysis). Unlike earlier findings, the variant comprising residues 1–184, the size of which is similar to that of the proteolytic fragment of rh-IGFBP-3 (1–160), exhibits less affinity loss for IGF-I than for IGF-II *(41)*. The complementary fragment (residues 161–264) to the N-fragment generated after the first plasmin-induced cleavage of nonglycosylated rh-IGFBP-3 was identified by sequencing of the high-performance liquid chromatography (HPLC)-purified fragments *(17)*. This is detectable neither by anti-IGFBP-3 antibodies nor by ligand blotting. A preliminary report, concerning a similar variant (residues 151–263) incubated with ^{125}I-IGF-I, mentions elution from gel filtration in the proximity of rh-IGFBP-3, but no competition with unlabeled IGF *(40)*.

A 16-kDa fragment comprising residues 1–95 results from secondary cleavage by plasmin of the 1–160 fragment of nonglycosylated rh-IGFBP-3 *(17)* and has a corresponding glycosylated form of 20 kDa (immunoblotting) in plasma and osteoblast-like cell culture media *(17,34,43)*. This is undetectable in ligand blotting and in solution binding assay at nanomolar concentrations, at which it possesses biological activity (*ref. 43* and the third section of this chapter). Nevertheless, at least in its glycosylated form, this fragment appears to retain some residual affinity for IGFs, as a fragment with similar migration is detected by ligand blotting in pregnancy serum *(10)*. An apparently identical fragment of 17.7 kDa has been detected in human serum and urine *(44)*. Concentrated urine samples were fractionated by size-exclusion chromatography and the eluates containing the fragments incubated with IGF-I or IGF-II. After crosslinking, the fragments were revealed by immunoblotting, but the signal remained extremely weak. Recently, the same laboratory obtained baculovirus-expressed peptides corresponding to this N-terminal fragment and detected a weak signal in ligand blotting as compared with that of IGFBP-3 at 100-fold lower concentration. Affinity crosslinking revealed competition with unlabeled IGF *(45,46)*. Using a similar approach, Firth et al. obtained expression of IGFBP-3^{1-88} in Chinese hamster ovary cells and analyzed its affinity for IGFs, at concentrations likely to exist under physiological conditions *(41)*. No signal was obtained in ligand blotting and no ^{125}I-IGF-I or -II binding was detectable in solution assay up to 15 ng/mL peptide. However, binding of radiolabeled IGF could be detected by affinity cross-linking. There was competition by unlabeled IGF-I and IGF-II, but with 60- and 40-fold weaker potencies than that of intact IGFBP-3. The authors correctly indicate that the failure of binding assays to detect IGF binding to the deletion mutant unless stabilized by crosslinking would suggest that the relatively low-affinity interactions "may have quite rapid off-rates" *(41)*.

BIOLOGICAL ACTIONS OF IGFBP FRAGMENTS

Although there have been numerous reports of the functional repercussions of IGFBP proteolysis, whether demonstrated or deduced, there have to date been few studies describing the activities attributable to the resulting fragments. Only progressively has the hypothesis of a possible functional role emerged from in vivo studies of biological fluids or diverse cultured cell models revealing the appearance or concentration changes of IGFBP fragments. Logically, the next step has been to isolate the fragments and study their biological activities. One of the essential clues, as described previously, has been their loss of affinity for the IGFs resulting from the structural alteration of the intact IGFBP. Another pointer has been the discovery of IGF-independent actions involving different functional domains of the native protein that may be present in the proteolytic fragments. In this section, a review is presented of the effects of IGFBPs fragments, whether related or unrelated to their interactions with IGFs (*see* Table 5), and their mechanisms of action.

Effects of IGFBP Fragments Related to Their Reduced Affinities for IGFs

IGFBP Fragments as Modulators of Endocrine IGF Action

The extensive in vivo observations of IGFBP proteolysis are analyzed in Chapter 26. Data providing grounds to attribute particular roles to particular IGFBP fragments relate essentially to regulation of the bioavailability of circulating IGFs that are produced by the liver.

Effects of IGFBP-3 Proteolysis on the Distribution of IGFs in Serum. The most abundant IGFBP in the bloodstream is IGFBP-3. Western immunoblot analysis of human serum IGFBPs, in both normal and pathological states, has consistently revealed a proteolytic fragment of IGFBP-3 of about 30 kDa. This corresponds to the N-terminal two-thirds of the protein *(see above).* Its proportional quantities relative to the intact 42- and 39-kDa forms vary with physiopathological status. For instance, in normal pregnancy, the intact protein virtually disappears (reviews in *refs. 15,47*). Clear evidence of the functional alteration of this 30-kDa fragment comes from its lack of or limited detectability in Western ligand blotting. In the original work of Hossenlopp et al. on pregnancy serum *(10)*, the fragment was detected in the 140-kDa complex isolated by neutral pH gel filtration and radiocompetition experiments revealed its loss of affinity for IGF-I and IGF-II. It was concluded at the time that this structural alteration of IGFBP-3 facilitated dissociation of the IGFs from the 140-kDa complexes, thus increasing the bioavailability of the IGFs. However, this interpretation met with extensive debate. The counter-argument was that in pregnancy serum most of the IGF-I and IGF-II in the bloodstream is found in the 140-kDa complexes, as is the case in normal serum *(48,49)*. Moreover, Suikkari and Baxter showed that ternary complex formation is possible using acidified pregnancy serum in the presence of ALS, inferring that IGFBP-3 in pregnancy serum is functionally normal *(50)*. However, Davenport et al. demonstrated that the half-life of serum IGF-I is strongly reduced in rat pregnancy, whereas urinary clearance is minimal. They concluded that tissue uptake of IGF-I is increased *(51)*, which would be consistent with the hypothesis of Hossenlopp et al.

Lassarre and Binoux undertook a detailed analysis of the binding characteristics of IGFBP-3 and IGF distribution in pregnancy serum *(12)*. The affinity of IGFBP-3 for IGF-I proved to be ten times weaker than that of normal serum IGFBP-3, and that for IGF-II, twice as weak. In studies at 37°C of the kinetics of dissociation of ^{125}I-IGF-I and -II bound to IGFBP-3, IGF-II was released approximately six times faster and IGF-I ten times faster from pregnancy serum IGFBP-3 than from its normal counterpart. IGFs were measured in the eluates containing 140-kDa complexes, 40-kDa complexes, or free IGFs following gel filtration of plasma. Chromatography was run following preincubation of the plasma samples at 37°C in the presence of EDTA, so as to allow the IGF pools to reequilibrate in closer-to-physiological conditions. The proportion of IGF-I in free form was found to be nearly three times larger in pregnancy than in normal plasma, whereas that of IGF-II was slightly smaller. Relative to total IGF-I, the proportion in the 140-kDa complexes was significantly smaller in pregnancy than in normal plasma, whereas that of IGF-II was larger. In the 40-kDa complexes, the IGF-I proportions were similar in pregnancy and normal plasma, but IGF-II proportions in pregnancy plasma were significantly

Table 5
Biological Actions of Proteolytic Fragments of the IGFBPs

	Effects of IGFBP Fragments Related to Their Reduced Affinities for IGFs	*Ref.*
IGFBP-3	*Circulating form, endocrine actions*	
Human serum N-terminal 30-kDa fragment	IGF release from 140-kDa complexes	*12,53*
	Accelerated IGF dissociation from truncated IGFBP-3	*12*
	Reduction of half-life of IGF-I	*51*
	Increase of free IGF-I	*12*
	Enhancement of IGF bioavailability	*13,54*
	Actions at cellular level	
Rat serum purified N-terminal 30-kDa fragment	Potentiation of IGF-I-stimulated DNA synthesis	*76*
Human purified N-terminal 30-kDa fragment	Weak inhibition of IGF-II-stimulated DNA synthesis	*42*
	Mitogenic potency on epithelial endometrial cells	*78*
rh-nonglycosylated 22–25-kDa fragment (residues 1-160)	Weak inhibition of IGF-stimulated DNA synthesis	*43,77*
	Potentiation of the mitogenic action of secreted IGF-II	*77*
IGFBP-5	*Actions at cellular level*	
Human purified N-terminal 23-kDa fragment	Potentiation of IGF-stimulated DNA synthesis	*79*
rh- 24-kDa 1–169 variant	Potentiation of IGF-stimulated DNA synthesis	*80*

(*continued*)

Table 5 (*Continued*)

	Effects of IGFBP Fragments Unrelated to Their Interaction with IGFs	*Ref.*
IGFBP-3	*Actions at cellular level*	
Human serum 16–18-kDa glycosylated fragment	Inhibition of IGF (but not insulin) bioactivity in rat adipocytes	*84*
	Inhibition of IGF sulphation activity in rat costal cartilage	
	Inhibition of IGF-stimulated DNA synthesis	
rh-nonglycosylated 16-kDa fragment (residues 1–95)	Inhibition of the mitogenic action of exogenous and endogenous IGFs	*43,77*
	Inhibition of the mitogenic action of FGF in fibroblasts with nonfunctional type 1 IGF receptor	*86*
IGFBP	*Regulation of IGFBP proteolysis*	
rh-nonglycosylated IGFBP-3 fragment, residues 161-264	Inhibition of IGFBP-4 proteolysis	*Unpublished*
Synthetic peptides with the heparin-binding sequences of h-IGFBP-3, -5, -6	Inhibition of IGFBP-4 proteolysis	*92*
Synthetic peptides corresponding to h-IGFBP-5 201–208 sequence	Inhibition of IGFBP-5 proteolysis	*96*

smaller. This means that in pregnancy plasma, free IGF-I is enriched at the expense of the fraction in the 140-kDa complexes. Relatively, IGF-II is concentrated in the 140-kDa complexes at the expense of the 40-kDa complexes where IGFBP-4 and, to a lesser extent, IGFBP-2, which has strong affinity for IGF-II, also undergo limited proteolysis. These findings supported the notion that IGFBP-3 is structurally and functionally altered by pregnancy-associated protease(s). The result is reduced affinity for the IGFs, particularly IGF-I, hence accelerated kinetics of dissociation and redistribution of the IGFs among the three circulating pools, with increased IGF-I in free form. From data on the half-lives of IGFs in humans *(52)*, the production rate of free IGF-I was calculated to be almost quadrupled in pregnant women. The results of a subsequent study by Lee and Rechler supported this interpretation *(53)*. Ternary complexes were reconstituted from recombinant h-IGFBP-3, rat ALS, and ^{125}I-IGF-I and these incubated at 37°C with adult rat serum which contains specific IGFBP-3 protease activity. h-IGFBP-3 was proteolyzed and ^{125}I-IGF-I released from the complexes and transferred to endogenous rat IGFBPs.

Effects of the N-terminal 30-kDa Fragment of IGFBP-3 on the Bioavailability of Serum IGFs. To prove that accelerated dissociation of IGF-I bound to the 30-kDa proteolytic fragment results in increased IGF-I bioavailability, the stimulatory activities of pregnancy and normal sera were compared in a chick embryo fibroblast assay. Two pools were made up, one of pregnancy serum and one of normal serum, from samples with known concentrations of IGF-I and IGF-II, mixed in proportions to yield identical IGF concentrations in the two pools. Stimulation of DNA synthesis by pregnancy serum was twice that by normal serum. Addition of excess monoclonal anti-IGF-I and -II antibodies reduced the stimulatory activities of both pools and to the same degree. Addition of rh-IGFBP-3 dose-dependently reduced the stimulation, although more so in the case of normal serum at low IGFBP-3 concentrations *(13)*. These results indicated that IGFs are released more readily from pregnancy serum. They also demonstrated the functional repercussions of serum IGFBP-3 proteolysis at cellular level. Similarly, Holly et al. had previously reported that IGFBP-3 proteolysis in pregnancy serum could amplify the IGF biological activity measured in a cell line-based bioassay *(54)*.

Many studies of catabolic states such as severe illness *(55)* and insulin-dependent (type 1) *(56)* or non-insulin-dependent (type 2) diabetes mellitus *(57)* have revealed increased IGFBP-3 proteolysis. The increase is interpreted as a mechanism of adaptation so as to enhance IGF bioavailability. Similarly, IGFBP-3 proteolysis is relatively diminished in constitutionally short children *(58)* and augmented in constitutionally tall children *(59)*, again indicating that the 30-kDa IGFBP-3 fragment plays some specific role in regulating IGF-I bioavailability and growth rate in these subjects. Also, the increased proportion of 30-kDa IGFBP-3 fragment in the sera of growth hormone (GH)-deficient patients *(60)* and the enhanced IGFBP-3 protease activity in GH-deficient dwarf mice *(61)* may also be viewed as adaptive responses, albeit insufficient, to IGF-I deficiency. These findings all provide indirect evidence of the physiological significance of the C-terminal truncated IGFBP-3 visualized in immunoblotting as the 30-kDa fragment.

Durham et al. *(42)* recently purified to homogeneity the N-terminal fragment of IGFBP-3 that accumulates in the 30-kDa material of serum and peritoneal dialysates from children with chronic renal failure (CRF). The aim was to determine its contribution to the low IGF bioactivity of serum in these patients, despite their normal or high IGF-I and IGF-II levels. Compared with the affinities of native IGFBP-3, the affinity of the purified truncated peptide is reduced 4- to 10-fold for IGF-II and 200-fold for IGF-I. Consistently with binding data, the 30-kDa IGFBP-3 weakly inhibits IGF-II-stimulated thymidine incorporation in chondrosarcoma cells and has no effect on the action of IGF-I. The authors conclude that the truncated IGFBP-3 may modulate the effects of IGF-II, but not of IGF-I, on target tissues, but that its role in the growth failure of children with CRF remains to the determined. As in the case of GH deficiency, the generation of the 30-kDa fragment by limited proteolysis of IGFBP-3, with the attendant accelerated dissociation of IGFs and particularly IGF-I, may be interpreted as an adaptive, although inefficacious, response to the various deficiencies contributing toward growth retardation.

An immunofunctional assay for IGFBP-3 has recently been developed *(62)*, in which the intact and truncated forms in serum are adsorbed to test tubes coated with anti-IGFBP-3 monoclonal antibodies, then incubated with radiolabeled IGF-I that binds only to intact IGFBP-3. Although total IGFBP-3 measured by radioimmunoassay (RIA) remains stable through the nycthemeron, the proportion of intact protein rises during the night prior to the peak in IGFBP-1. The causes of this change remain unknown, but they may involve a drop in the ratio of IGFBP-3 protease(s)/protease inhibitor(s), as indicated by Cotterill et al. in patients with Laron-type dwarfism *(63)*. This would suggest that during nocturnal fasting, some cooperation exists between IGFBP-3 and IGFBP-1 to limit IGF bioavailability. Such an interpretation would fit well with the notion of interaction between insulin and the IGF system in controlling substrate utilization.

The overall picture arising from these findings confirms the essential part played by the 140-kDa complexes in regulating the bioavailability of IGFs in the bloodstream. Their role as steady-state IGF reservoir now appears more as IGF reserves of modulable levels, from which the release of IGFs is enzymatically regulated by the balance between the native and truncated forms of IGFBP-3. The former by its strong affinity sequesters IGFs in the high molecular weight complexes, the latter by its weaker affinity, particularly for IGF-I, releases them.

As regards the 18- to 20- and 16-kDa fragments of IGFBP-3 generated by secondary proteolysis of the 30-kDa fragment *(17)*, which are inconsistently detected in serum, no data are as yet available to indicate any endocrine role. Similarly, the possible physiological significance of the IGFBP-2 and IGFBP-4 fragments detected in serum remains unknown and it seems possible that they are generated outside of the vascular compartment.

Roles of IGFBP Fragments in Determining the Bioavailability of Serum IGFs to the Ovarian Follicle. A unique model for investigating the IGF system is the ovarian follicle that can be sampled *in situ* at different stages of maturation and atresia, then analyzed in vitro. The results of the various studies in human *(64,65)* and other mammalian species *(22,66–70)* are convergent in attributing a determining role to IGFBPs and their proteolytic fragments. Both control the bioavailability of IGFs provided by the blood owing to the amplified vascularization that is characteristic of follicular growth. Besnard *(22,70)* has shown that during this phase in ovines the bioavailability of serum IGFs released from proteolyzed IGFBP-3 is promoted by restricted local synthesis of IGFBP-2 and -5 and degradation of IGFBP-4 by a specific protease detected in follicular fluid. IGFBP-4 proteolysis is also observed in humans and can be inhibited by unsaturated IGFBP-3, although IGFBP-3 is not cleaved by this activity and therefore does not act as a competitive substrate *(65)*. During follicular atresia in the ewe, IGFBP-3 proteolysis increases, but the bioavailability of released IGF is diminished. This is due to both the increased synthesis of IGFBP-2 and IGFBP-5 which inhibit or sequester it, and the action of IGFBP-4 which is restored to normal functionality by arrest of its proteolysis *(22,70)*. Monget et al. have obtained results indicating that the C-terminal fragment of IGFBP-3 (generated by cleavage of the central portion of the protein *[17]* and increased in follicular fluid during atresia) inhibits IGFBP-4 proteolysis in the same way as intact IGFBP-3 *(manuscript submitted).*

IGFBP Fragments as Modulators of Autocrine/Paracrine IGF Action

In investigating the effects of IGFBP fragments on IGF action at cellular level, virtually all findings have come from cell systems used as sources of or targets for IGFBP fragments. Three types of local action have been described: increase of IGF bioavailability and potentiation or inhibition of IGF activities.

Increase of IGF Bioavailability. As stated previously, the primary functional consequence of limited IGFBP proteolysis is reduced affinity of the fragments for IGFs. With characterization of the major fragment that retains weak affinity for IGF-I and/or IGF-II and observation of its appearance or increase in culture media concomitant with stimulation of IGF-dependent cellular activity, it can safely be concluded that it is responsible for accelerated dissociation of bound IGF and hence increased IGF bioavailability. This was shown in a human prostate carcinoma cell line that produces IGF-II, IGFBP-3, and urokinase-type plasminogen activator (uPA). Addition of plasminogen to the culture medium provokes an increase in the concentration of a 30-kDa proteolytic fragment of IGFBP-3 and stimulation of cell proliferation. The stimulation is totally blocked in the presence of anti-type 1 IGF receptor antibody. Reciprocally, in the presence of uPA inhibitor, the concentration of the 30-kDa fragment generated by IGFBP-3 proteolysis decreases at the same time as proliferation *(71)*. Likewise, in prostate epithelial cells, PSA reverses the inhibitory effect of rh-IGFBP-3 on IGF-II-induced mitogenesis *(72)*.

Comparable effects have been observed for a 20-kDa proteolytic fragment of endogenously secreted IGFBP-2 generated by plasmin in a neuroblastoma cell line. The fragment binds IGF-II with an affinity that is one fifth that of native IGFBP-2. In the presence of plasminogen, the concentration of the 20-kDa fragment in the culture medium increases concomitantly with stimulation of proliferation. Here again, the effect is inhibited by anti-IGF-II or anti-type 1 IGF receptor antibodies *(73)*.

In the case of IGFBP-4, experiments using human fibroblasts indicate cleavage into 18- and 14-kDa fragments by a specific protease. Intact rh-IGFBP-4 potently inhibits IGF-I-stimulated aminoisobutyric acid (AIB) uptake by bovine fibroblasts, whereas proteolyzed rh-IGFBP-4 has no inhibitory effect *(74)*. Chernausek et al. *(28)* identified a 16-kDa fragment of rat IGFBP-4, corresponding to residues 1–120. This binds IGFs with reduced affinity and no longer inhibits IGF-I-stimulated thymidine uptake in the cell test used.

These findings are consistent with those from in vivo studies, suggesting that proteolysis of IGFBP-2, -3, and -4 generates fragments that remain capable of binding IGFs, but with weak affinity. In all the cell systems studied, they enhance IGF bioavailability and bioactivity at the cell surface.

Potentiation of IGF Action. A number of studies using cultured cells have suggested or demonstrated that certain IGFBP fragments potentiate the mitogenic effects of the IGFs.

Conover investigated the effects of IGFs on $[^3H]$AIB uptake in bovine fibroblast monolayers. IGF action is inhibited if IGF is coincubated with rh-IGFBP-3, but potentiated if the cells are preincubated with IGFBP-3 for 24–72 h. The different effects are related to processing of cell-associated IGFBP-3 into lower molecular size molecules *(75)*.

Schmid et al. found that the 30-kDa fragment of IGFBP-3 (lacking the C-terminal third of the molecule) purified from rat serum enhances IGF-I-stimulated DNA synthesis in rat osteoblastic cells, whereas intact IGFBP-3 blocks the stimulation *(76)*.

Angelloz-Nicoud and Binoux, in their work on prostate carcinoma (PC-3) cells cited previously, observed enhanced mitogenicity of rh-IGF-II added to cell-free media preconditioned by 48 h of culture in the presence of plasminogen (owing to the enlarged proportion of 30-kDa proteolytic fragment of secreted IGFBP-3). By contrast, this mitogenic activity is diminished when cells are cultured in the presence of serine protease inhibitor *(71)*. The fragment corresponding to the 30-kDa form was isolated by HPLC following limited proteolysis of nonglycosylated rh-IGFBP-3 by plasmin in vitro *(43)*. This 22–25-kDa fragment stimulates PC-3 cell proliferation in the absence of added IGF, but the effect is suppressed in the presence of anti-type 1 IGF receptor antibody *(77)*. Because IGF-II is known to be responsible for autocrine proliferation in PC-3 cells, these results suggest that this N-terminal IGFBP-3 fragment, which is capable of binding IGF-II although with 20 times less affinity than the intact protein, potentiates the action of secreted IGF-II via uptake followed by release in the vicinity of the IGF receptor.

Koustilieris et al. purified two approximately 30-kDa N-terminal fragments of IGFBP-3 from human peritoneal fluid collected at laparoscopy. These have mitogenic effects on primary cultures of epithelial cells isolated from the endometrium *(78)*. Although the implication of the IGFs in this mitogenic activity was not determined, it would seem highly likely in view of the results cited previously.

Andress and Birnbaum purified a 23-kDa carboxyl-truncated fragment of IGFBP-5 from cultures of human osteoblast-like cells. This has significantly reduced affinity for IGFs, but stimulates IGF-induced osteoblast mitogenesis *(79)*. Similar results are obtained using a recombinant truncated IGFBP-5 comprising residues 1–169 *(80)*. Apparently divergent results were obtained by Jones et al. using a line of fetal human fibroblasts and a purified preparation of human IGFBP-5 *(81)*. They found that IGFBP-5 is incorporated in intact form into the extracellular matrix (ECM), where it is protected from proteolytic attack and can potentiate the mitogenic action of IGF-I. By contrast, when IGFBP-5 is present only in the medium, it is degraded into a 22-kDa fragment that has no effect on IGF-I-stimulated growth. In fact, these two sets of results are not necessarily contradictory, but rather reflect the different cell models used. Andress has since noted that both intact and carboxyl-truncated IGFBP-5 preferentially bind to the cell surface of osteoblast-like cells *(82)*. It would therefore seem that the prerequisite for potentiation of IGF action is adherence to the ECM or cell surface.

Inhibition of IGF Action. Compared with that of intact IGFBP-3, the affinity of the 22–25-kDa fragment isolated after plasmin-induced degradation of nonglycosylated rh-IGFBP-3 is reduced 20-fold for IGF-II and 50-fold for IGF-I. Coherent with its weak affinity is its weak inhibition of IGF-I-induced DNA synthesis in a chick embryo fibroblast assay *(43)*. Similarly, it weakly inhibits PC-3 cell proliferation induced by exogenous IGF-II. Nevertheless, as mentioned previously, when added alone to the culture medium, it stimulates the endogenous IGF-II-dependent autocrine proliferation of these cells *(77)*. This would indicate, as Conover showed previously *(75)*, that opposite effects may be observed depending on experimental conditions and the cell type involved, where different mechanisms of action may come into play.

Effects of IGFBP Fragments Unrelated to Their Interaction with IGFs

Regulation of Cell Growth

Both stimulatory and inhibitory effects have been reported.

Mitogenic Effects of the 23-kDa Fragment of IGFBP-5. Apart from the enhanced osteoblast mitogenesis observed when the carboxyl-truncated fragment is coincubated with IGF-I or -II described previously, Andress and Birnbaum reported in the same article that the fragment is also capable of stimulating mitogenesis in the absence of added IGFs. The effect persists in the presence of excess IGFBP-3 sufficient to sequester the IGFs secreted by the cells. Here it seems that a second pathway may be involved that is independent of IGF receptor activation *(79)*. Nonetheless, considering the capacity of IGFBP-5 produced by osteoblasts to store IGFs in the ECM *(83)* and the affinity of the 23-kDa fragment for the osteoblast surface *(82)*, the possibility cannot be excluded that the fragment interacts with IGFs released from the ECM.

Inhibitory Effects of the Small N-Terminal Fragment of IGFBP-3. Controlled in vitro proteolysis of nonglycosylated rh-IGFBP-3 by plasmin yields fragments of 22–25 and 16 kDa, as estimated by Western immunoblotting. These were isolated by reverse-phase chromatography and subsequently sequenced *(17,43)*. With analysis of these fragments' properties, it should be possible by analogy to deduce the biological functions of the corresponding glycosylated forms of about 30 kDa and 20 kDa that are found in serum and other biological fluids, as well as in culture media. As mentioned in the second section, the 16-kDa fragment is obtained by secondary cleavage of the 22–25-kDa form and the two fragments correspond to residues 1–95 and 1–160 of IGFBP-3.

Inhibition of the Biological Action of IGFs. It was stated previously that the 22–25-kDa IGFBP-3 fragment retains weak affinity for IGFs and weakly inhibits IGF-I stimulation of DNA synthesis in a chick embryo fibroblast assay. Unexpectedly, the 16-kDa fragment, which in Western ligand blotting, affinity crosslinking, and solution binding assays lacks affinity for IGFs, proved to be as potent an inhibitor of IGF-I stimulation as intact rh-IGFBP-3. It also inhibits stimulation of DNA synthesis by insulin used at concentrations 100 times those of IGF-I to ensure interaction with the type 1 IGF receptor. In contrast, intact IGFBP-3 and the 22–25-kDa fragment have no effect on insulin stimulation *(43)*. These results were confirmed using the prostate adenocarcinoma cell line PC-3, in which the 16-kDa fragment blocks both IGF-II-dependent autocrine proliferation and growth induced by exogenous IGF-II *(77)*. It also inhibits IGF-I- and estradiol-induced proliferation in MCF-7 cells *(unpublished observations)*. Despite its lack of affinity for IGFs, this N-terminal 1–95 fragment of IGFBP-3 therefore behaves as an antagonist of IGF mitogenic action.

In retrospect, it seems likely that the glycosylated 18–20-kDa form of the IGFBP-3^{1-95} fragment *(17,43)* corresponds to the "specific inhibitor" identified in 1981 by Herington and Kuffer in a partially purified human serum fraction, which they described as an acid-stable, heat-labile glycoprotein of 16–18 kDa *(84)*. In vitro, this protein dose-dependently inhibited the insulin-like bioactivity of partially purified IGF (but not of insulin) in rat adipocytes and its cartilage sulfation activity in rat costal cartilage and also depressed IGF-stimulated DNA synthesis in cultured human fibroblasts. Subsequently, Ooi and Herington showed that the purified serum fraction containing IGF-inhibitory activity could be crosslinked to ^{125}I-IGF and revealed in SDS-PAGE as a doublet of 21.5–25 kDa *(85)*. The discrepancies between the different reports of the ability of the inhibitory N-terminal fragments of IGFBP-3 to bind IGFs were discussed previously.

In view of the similarities between the possible signaling pathways following activation of the type 1 IGF receptor and the insulin receptor, it seemed that IGFBP-3^{1-95} may inhibit insulin action. However, in fetal rat hepatocytes in primary culture, it inhibits the mitogenic action of IGF-I by 80–100%, whereas its maximal inhibition of insulin-stimulated glycogenesis is only 50% *(62)*. This would indicate a relative specificity for the IGF signaling pathway. In another set of experiments, similar peptides (rh-IGFBP-3$^{1\text{-}87}$ and $^{1\text{-}97}$) expressed in baculovirus have been observed to inhibit insulin-stimulated autophosphorylation of the insulin receptor β-subunit and phosphorylation of IRS-1 *(45,46)*. The authors attribute the inhibition to the ability of the peptides to bind to insulin, but this would seem unlikely because the interpretation is based on the results of Western ligand blotting and affinity crosslinking using very high peptide concentrations (*see* p. 292). Further studies will no doubt reveal the points of impact of IGFBP-3$^{1\text{-}87}$, $^{1\text{-}95}$, and $^{1\text{-}97}$ in the signaling pathways of the two receptors as related to their biological effects.

Inhibition of the mitogenic action of FGFs. The biological activity of the 16-kDa (non-glycosylated) 1–95 fragment of rh-IGFBP-3 was recently tested in mouse fibroblasts with a targeted disruption of the type 1 IGF receptor gene *(86)*. In the cell clone used, basic fibroblast growth factor (bFGF) (but not IGF, epidermal growth factor [EGF], or platelet-derived growth factor [PDGF]) is mitogenic in serum-free medium. The 16-kDa fragment dose-dependently (20–100 ng/mL) inhibits (up to 100%) maximal stimulation induced by 25 ng/mL bFGF, whereas intact IGFBP-3 has virtually no effect. Similar results are obtained with control wild-type cells where the mitogenic activity of 1% fetal calf serum (equal to that of 25 ng/mL bFGF) is only 25–30% inhibited by 100 ng/mL of 16-kDa fragment or 200 ng/mL of rh-IGFBP-3. This agrees with an antagonistic action, affecting the mitogenic activity of serum that is attributable to IGFs. IGFBP-3$^{1\text{-}95}$ therefore appears to be a potent inhibitor of the mitogenic signals resulting from activation of both the type 1 IGF and FGF receptors. The absence of effect by intact IGFBP-3 in these experiments is noteworthy for two reasons. First, it shows that a proteolytic fragment may exhibit a type of activity that is lacking in its parent protein. Second, it indicates a discrepancy with the results of Villaudy et al., who found that IGFBP-3 inhibits FGF-induced DNA synthesis in chick embryo fibroblasts *(87)*. However, because FGF is capable of stimulating IGF-I expression *(88)*, it is possible that this inhibition reflects sequestration by IGFBP-3 of the IGF secreted by the fibroblasts.

Preliminary results have been obtained in a hamster pulmonary fibroblast cell line (CCL 39) stably transfected with a cDNA plasmid encoding the 1–95 amino acid sequence of human IGFBP-3, placed under the control of the constitutive cytomegalovirus promoter. Here, the corresponding peptide secreted by the cells inhibits FGF-induced stimulation of cell proliferation *(manuscript in preparation)*.

Inhibition of IGFBP Proteolysis

Several studies have indicated that IGFs promote IGFBP-4 degradation by specific proteases: they either induce conformational changes by binding to IGFBP-4, or directly activate the protease(s) involved *(26,74,89)*. IGFBP-3, by contrast, reverses this effect *(65,90,91)*. Its C-terminal region containing a highly basic heparin-binding domain is involved, as the 161–264 fragment of rh-IGFBP-3 generated in vitro by plasmin inhibits IGFBP-4 proteolysis (*see* p. 297). Also, IGFBP-5 and IGFBP-6, which have homologous domains, and synthetic peptides with the heparin-binding consensus sequences (XBBBXXBX) corresponding to those of IGFBP-3, -5, and -6 (*see* Fig. 1) are all capable of inhibiting IGFBP-4 proteolysis *(91,92)*.

In the case of IGFBP-5, unlike IGFBP-4, proteolysis is partially prevented in the presence of IGFs *(93)*. In addition, its heparin-binding domain of the C-terminal region is involved in protection against IGFBP-5 proteolysis. Addition of heparin, heparan sulfate, or dermatan sulfate to the medium of fibroblast monolayer cultures inhibits

IGFBP-5 degradation. This suggests a potential role for proteoglycans of the ECM or cell surface in regulating IGFBP-5 proteolysis *(94)*. A synthetic peptide corresponding to the 201–218 sequence of IGFBP-5 that contains the heparin-binding motif binds to the plasminogen activator inhibitor, PAI-1, which is present in the ECM. It seems that such a mechanism may be involved in protecting IGFBP-5 from the proteolytic action of plasmin *(95)*. Campbell and Andress noted that this synthetic peptide blocks plasmin binding to IGFBP-5 and transformation of plasminogen into plasmin by acting as a competitive inhibitor to plasminogen activator *(96)*. From this, it would be plausible to hypothesize that the C-terminal fragment of IGFBP-5 exerts negative feedback on plasmin-induced IGFBP-5 proteolysis and, indirectly, on the mitogenic action of the N-terminal fragment.

Mechanisms of Action of the IGFBP Fragments

The different biological actions of the IGFBP fragments reflect, on the one hand, their conservation or relative loss of a known function of the native protein and, on the other hand, new properties that are apparently unrelated to those of their parent protein.

Mechanisms of Action of the IGFBP Fragments Related to Their Reduced Affinities for IGFs

In the second section it was seen that the primary sites of proteolytic cleavage are situated in the mid-region of the protein. The attendant structural alteration results in greater or lesser loss of affinity for IGF-I and IGF-II (evidenced primarily in Western ligand blotting). This reduced affinity at least in part accounts for the biological activities attributed to the N-terminal fragments of IGFBP-3 and IGFBP-5 generated by this cleavage. In the bloodstream and extracellular media, the effect of this loss, as demonstrated for IGFBP-3 *(12)*, is to facilitate dissociation of IGFs bound to the fragments concerned and increase IGF bioavailability.

Potentiation of IGF action by IGFBP-3 appears to necessitate association with the cell surface *(75,97)*. One of the mechanisms described is proteolysis of IGFBP-3 at the cell surface. This was demonstrated in fibroblasts by Conover, who interpreted it as a means of providing optimal presentation of IGF to its receptor *(75,98)*. As regards potentiation of IGF action by the 30-kDa N-terminal IGFBP-3 fragment purified from rat serum *(76)* or by the corresponding nonglycosylated 22–25-kDa form (residues 1–160) generated by proteolysis of rh-IGFBP-3 *(77)*, association with the cell surface has yet to be proved. There may be some interaction with the 149–153 KKGHA sequence which Fowlkes and Serra have shown to bind to heparin *(99)*, although Conover et al. found that heparin fails to inhibit potentiation of IGF action *(100)*.

IGFBP-5 and its 1–169 fragment exert their mitogenic effects via interaction with a binding site at the cell surface. Andress identified this site as a 420-kDa band in reduced SDS-PAGE in crosslinking studies using the two radioiodinated proteins and membrane extracts from mouse osteoblasts *(101)*. Intact and 1–169 IGFBP-5 also bind to the ECM, although with lower affinity than that for the osteoblast surface. Binding to the ECM is enhanced (particularly of the 1–169 form) after enzymatic treatment to remove glycosaminoglycan (GAG) side chains of proteoglycans, but binding to the cell surface remains unchanged. Therefore, both IGFBP-5 and its carboxyl-truncated fragment preferentially bind to the cell surface, their binding to the ECM is not mediated by GAGs, and loss of GAGs within osteoblast-derived ECM can result in extracellular sequestration of IGFBP-5, especially of its truncated form *(82,101)*.

Mechanisms of Action of the IGFBP Fragments Unrelated to Their Interaction with IGFs

The C-terminal regions of IGFBP-3 and IGFBP-5 are essential for cell association. As stated previously, they are relatively rich in basic amino acids which may interact with negatively charged molecules at the cell surface. Booth et al. demonstrated that 18-amino-acid synthetic peptides corresponding to portions of IGFBP-3 and IGFBP-5 that include the heparin-binding domain (see Fig. 1) decrease IGFBP-3 and -5 binding to endothelial cell monolayers *(102)*. Firth et al. showed that a mutant IGFBP-3 deprived of the basic sequence 228–232 is incapable of associating with the cell surface *(41)*. The 161–264 fragment generated by proteolysis of rh-IGFBP-3 and labeled with ^{125}I was found to bind MCF-7 cells with strong affinity *(62)*.

The intervention of the C-terminal regions of IGFBP-3 and IGFBP-5 or their corresponding fragments in protecting IGFBP-4 and IGFBP-5 against proteolysis was discussed in the third section. Here, the molecular mechanisms remain a subject of debate.

As regards the N-terminal 16-kDa (nonglycosylated) or 18–20-kDa (glycosylated) fragment of IGFBP-3, which as seen previously inhibits the mitogenic and/or metabolic actions of IGFs and FGFs in a variety of cell types, important questions arise as to its mode of action, sites of interaction with its target cells, and the signaling pathways employed.

It is established that its mode of action is unrelated to its molecular interaction with IGFs. Even if in crosslinking experiments some authors have observed binding to IGFs *(45,85)*, whereas others have not *(43)*, its affinity for IGFs is extremely weak, as its binding to ^{125}I-IGF in solution binding assay is virtually nil. Interaction with IGFs could therefore not account for the inhibition of IGF action observed at nanomolar concentrations. Furthermore, it suppresses the mitogenic effects of FGF in cells deprived of a functional type 1 IGF receptor *(86)*.

The site(s) of interaction with the cell surface, necessarily the first step in transduction of the inhibitory signals, remain(s) unknown. Unlike the middle and C-terminal regions, the N-terminal region of IGFBP-3 contains no GAG-binding sequences. Attempts to iodinate the one tyrosine residue (at position 57) of the nonglycosylated 1–95 fragment have failed, most probably because it is inaccessible in the tightly compact structure of the fragment that contains the 12 cysteine residues of the N-terminal region. No data are available in support of any interaction with the IGFBP-3 binding sites identified by Oh et al. *(103)* or the type V transforming growth factor-β (TGF-β) receptor recently described as the putative IGFBP-3 receptor *(104)*. There appears to be no interaction with the type 1 IGF receptor, at least as regards competition for ^{125}I-IGF binding *(105)*.

Recent results from our laboratory using MCF-7 cells indicate that IGFBP-3$^{1-95}$ inhibits IGF-I stimulation of PI-3 kinase activity and IRS-1 phosphorylation *(unpublished observation)*. This would be consistent with data suggesting that PI-3 kinase activity is required for transduction of the mitogenic IGF-I signal in MCF-7 cells *(106)*.

PHYSIOLOGICAL SIGNIFICANCE OF IGFBP FRAGMENTS

Fine Regulation of IGF Action. Data describing the actions of the IGFBP fragments and the relatively few findings concerning their mechanisms of action have shed some light on their physiological significance. It was seen in the previous section that adherence of the IGFBPs to the cell surface or ECM is one of the determining events in making them accessible to proteases. The relative balance between IGFBP proteases and their inhibitors present in the cellular environment will regulate the relative proportions of intact and proteolyzed IGFBP. These proportions in turn modulate the bioavailability of the IGFs, both in the bloodstream and at cellular level. In the case of IGFBP-3, secondary generation of the 1–95 fragment from the 1–160 fragment as a result of prolonged or intensified proteolytic attack constitutes a secondary line of control in the transduction of signals arising from activation of the type 1 IGF receptor. Diverse mechanisms further contribute toward these regulatory loops: IGFs stimulate PAI synthesis; to a certain extent, IGFs are capable of protecting IGFBP-3 and -5 against proteolytic degradation by either stabilizing the IGFBPs or interacting with their proteases and/or protease inhibitors; heparin and the GAGs protect IGFBP-5 against proteolysis. Conversely, IGFBP-2 and -4 proteolysis is enhanced in the presence of IGFs; IGFBP-4 proteolysis is inhibited by IGFBP-3 or its C-terminal fragment; and IGFBP-5 proteolysis is suppressed by its C-terminal region. Therefore, the system comprising the six IGFBPs—with their varying expression from one tissue to another and differing affinities for IGF-I and -II, their posttranslational modifications, their relative sensitivities to different types of protease, the generation of fragments with reduced affinities for IGFs or with intrinsic activities divorced from their interaction with IGFs—forms a composite permitting flexible, versatile and subtle regulation of the signals presented by the IGFs to their target cells.

IGFBP-3 in the Bloodstream. Although it now appears established that the major 30-kDa fragment (residues 1–160) of IGFBP-3 has a biological role in modulating the bioavailability of circulating IGFs, the physiological significance of the 1–95 fragment in the bloodstream seems less certain. It is inconsistently revealed by Western immunoblotting as the glycosylated 20-kDa form, even in pregnancy serum where most of IGFBP-3 is proteolyzed. The (probably identical) IGFBP-3 fragment estimated at 17.7 kDa by Spagnoli et al. is found at low concentrations in the serum of GH-deficient children and increases during GH therapy *(44)*. In our laboratory, its signals are more intense in the sera of constitutionally tall children, where IGFBP-3 proteolysis is particularly marked *(unpublished observation)*. In clinical terms, it seems unlikely that the circulating form of IGFBP-$3^{1\text{-}95}$ functions as a growth inhibitor, such as it does in cell culture. Spagnoli et al. also noted that it is the predominant form in urine, suggesting that it is more rapidly cleared from the bloodstream than the other fragments. Whichever the case, its generation in the circulation is established. Sommer et al. report having purified from Cohn fraction IV an IGFBP-3 fragment belonging to its *N*-terminal domain, migrating at approx 18 kDa in SDS-PAGE *(107)*. In a later publication, Spencer and Chan mentioned a personal communication from Sommer indicating that this fragment contains the 96 N-terminal residues of IGFBP-3 and that it is glycosylated *(40)*.

Multifunctionality. The IGF-independent actions of the IGFBPs at the cellular level necessitate interaction with the cell surface, but association with the cell surface or ECM exposes the IGFBPs to proteolytic attack. It is therefore conceivable that at least some of the intrinsic activities of the IGFBPs are attributable to their fragments. This may be true of the inhibitory effects of rh-glycosylated IGFBP-3 described by Oh et al. in human breast cancer cells *(108)* that secrete proteases capable of degrading IGFBP-3. It is also conceivable that the intact protein and its fragments have different actions, either if the domain of the fragment required for interaction is masked in the intact molecule, or if the interaction involves several domains, in which case the protein needs to be intact. This appears to be true of the interaction between IGFBP-3 and the type 1 IGF receptor recently reported by Mohseni-Zadeh and Binoux, where the receptor loses affinity for its ligand. None of the fragments (1–95, 1–160, or 161–264) have this effect *(105)*. In contrast, IGFBP-3 $^{1\text{-}95}$ inhibits the mitogenic action of FGF on mouse fibroblasts with a targeted disruption of the type 1 IGF receptor gene, whereas the intact protein does not *(86)*. However, in another laboratory, the same cell line transfected with IGFBP-3 cDNA exhibits a reduced growth rate. This would indicate that the IGFBP-3 produced by the cells acts via a mechanism unrelated to the IGF receptor *(109)*. Nevertheless, in the protective effects of intact IGFBP-3 and its C-terminal fragment against IGFBP-4 proteolysis (*see* p. 301), it is evidently the same functional domain that is involved. Different but complementary mechanisms may therefore come into play in the cellular and extracellular actions of IGFBP-3 and its proteolytic fragments. These diverse mechanisms of action are at the root of the multifunctional nature of IGFBP-3.

Generation in Processes Involving Proteases. The physiological significance of IGFBP proteolysis and generation of fragments with fewer, enhanced, or new functions, compared with the native protein, needs to be viewed in the wider perspective of the roles played by proteases and their inhibitors in cell migration, angiogenesis, hemostasis, tissue resorption and repair, and tumoral invasion. The existence of truly specific IGFBP proteases, although probable, remains to be demonstrated. However, the PA/PAI system, thrombin, PSA, the MMPs, the cathepsins, and other secondarily activated proteases have wider or narrower spectra of actions bringing into play recognition of specific sites that may be present on the IGFBPs. These then would be among the protease substrates in the above processes. Plasmin and other serine proteases, envisaged in this chapter as IGFBP proteases, have long been known to affect the mobilization and/or activation of growth factors such as FGF and TGF-β *(110,111)*. Likewise, the γ-subunits of nerve growth factor (NGF), which are proteases taking part in the processing of the NGF-β precursor and modulation of its biological activity *(112)*, are capable of cleaving IGFBPs and may contribute toward a synergistic action between NGF and the IGFs *(113)*.

Potential Mediation in Other Growth Regulatory Systems. Another potentially important aspect of the IGF-independent actions of IGFBP-3 is its participation in mediating at least some of the growth inhibitory effects of TGF-b, retinoic acid, and antiestrogens in human breast cancer cells *(114–116)* and TGF-b induction of apoptosis in PC-3 cells *(117)*. With analysis of the mechanisms of action of IGFBP-$3^{1\text{-}95}$, it may be possible to establish some link between its antimitogenic effects in MCF-7 and PC-3 cells and this mediatory action of IGFBP-3.

PROSPECTS

Several lines of research remain open in view of the numerous questions that remain unanswered concerning IGFBP fragments.

1. The major fragments of IGFBP-3 and IGFBP-5 have been identified and some of their actions described. Further research would entail: (a) identification and characterization of specific binding sites on the cell surface; (b) as regards the IGF-independent actions of the N-terminal fragments: determination of their relationships with cell proliferation,

differentiation and apoptosis and study of the signaling pathways involved, and in vivo investigation of their physiological effects and their roles in tumorigenesis; (c) as regards the C-terminal fragments: preparation of specific antibodies; identification of possible cellular action, including internalization and transport to the nucleus related to the presence of a nuclear localization signal. This provocative question is evoked by the discovery of IGFBP-3 in the cytoplasm and nucleus of lung carcinoma and kidney cells *(118,119)*.

2. IGFBP-2 and IGFBP-4 proteolysis has been documented in diverse biological fluids and cell culture media, but identification of their fragments remains preliminary and their possible functions remain unknown.
3. For structural reasons, IGFBP-6 is protected from protease action and IGFBP-1 proteolysis is rare. Nevertheless, it would be of interest to identify a C-terminal fragment of IGFBP-1, considering the protein's ability to potentiate IGF action and stimulate cell migration following association with an $\alpha 5\beta 1$ integrin receptor, which involves the RGD sequence at its C-terminal end *(120,121)*.
4. The structural similarities between the N-terminal regions of the IGFBPs and IGFBP-related proteins (which have weak or no affinity for the IGFs) provide justification for the question raised by Oh et al. *(122)*. It is tempting to consider that some link may exist between the known and putative biological effects of the IGFBP-related proteins and the intrinsic actions of the IGFBPs or their N-terminal proteolytic fragments.

ACKNOWLEDGMENTS

This work was supported by the Institut National de la Santé et de la Recherche Médicale, the Association pour la Recherche sur le Cancer, and an Institut Lilly-ALFEDIAM grant. We thank Genée Allanson for help in preparing this chapter.

REFERENCES

1. Huhtala ML, Koistinen R, Palomaki P, Partanen P, Bohn H, Seppala M. Biologically active domain in somatomedin-binding protein. Biochem Biophys Res Commun 1986; 141:263–270.
2. Martin JL, Baxter RC. Insulin-like growth factor binding protein from human plasma. Purification and characterization. J Biol Chem 1986; 261:8754–8760.
3. Zapf J, Kiefer M, Merryweather J, Masiarz F, Bauer D, Born W, Fischer JA, Froesch ER. Isolation from adult human serum of four insulin-like growth factor (IGF) binding proteins and molecular cloning of one of them that is increased by IGF I administration and in extrapancreatic tumor hypoglycemia. J Biol Chem 1990; 265:14892–14898.
4. Zapf J, Born W, Chang JY, James P, Froesch ER, Fischer JA. Isolation and NH_2-terminal amino acid sequences of rat serum carrier proteins for insulin-like growth factors. Biochem Biophys Res Commun 1988; 156:1187–1194.
5. Shimonaka M, Schroeder R, Shimasaki S, Ling N. Identification of a novel binding protein for insulin-like growth factors in adult rat serum. Biochem Biophys Res Commun 1989; 165:189–195.
6. Roghani M, Segovia B, Whitechurch O, Binoux M. Purification from human cerebrospinal fluid of insulin-like growth factor binding proteins (IGFBPs). Isolation of IGFBP-2, an altered form of IGFBP-3 and a new IGFBP species. Growth Regul 1991; 1:125–130.
7. Andress DL, Birnbaum RS. A novel human insulin-like growth factor binding protein secreted by osteoblast-like cells. Biochem Biophys Res Commun 1991; 176:213–218.
8. Wang JF, Hampton B, Mehlman T, Burgess WH, Rechler MM. Isolation of a biologically active fragment from the carboxy terminus of the fetal rat binding protein for insulin-like growth factors. Biochem Biophys Res Commun 1988; 157:718–726.
9. Hossenlopp P, Bredon M, Frankenne F, Binoux M. IGF circulants et grossesse: déstabilisation des complexes 150K par inactivation enzymatique des protéines de liaison. Ann Endocrinol (Paris) 1988; 49:18C.
10. Hossenlopp P, Segovia B, Lassarre C, Roghani M, Bredon M, Binoux M. Evidence of enzymatic

degradation of insulin-like growth factor-binding proteins in the 150K complex during pregnancy. J Clin Endocrinol Metab 1990; 71:797–805.
11. Giudice LC, Farrel EM, Pham H, Lamson G, Rosenfeld RG. Insulin-like growth factor binding proteins in maternal serum throughout gestation and in the puerperium: effects of pregnancy-associated serum protease activity. J Clin Endocrinol Metab 1990; 71:806–816.
12. Lassarre C, Binoux M. Insulin-like growth factor binding protein-3 is functionally altered in pregnancy plasma. Endocrinology 1994; 134:1254–1262.
13. Blat C, Villaudy J, Binoux M. *In vivo* proteolysis of serum insulin-like growth factor (IGF) binding protein-3 results in increased availability of IGF to target cells. J Clin Invest 1994; 93:2286–2290.
14. Jones JI, Clemmons DR. Insulin-like growth factors and their binding proteins: biological actions. Endocr Rev 1995; 16:3–34.
15. Kelley KM, Oh Y, Gargosky SE, Gucev Z, Matsumoto T, Hwa V, Ng L, Simpson DM, Rosenfeld RG. Insulin-like growth factor-binding proteins (IGFBPs) and their regulatory dynamics. Int J Biochem Cell Biol 1996; 28:619–637.
16. Fowlkes JL. Insulin-like growth factor-binding protein proteolysis—an emerging paradigm in insulin-like growth factor physiology. Trends Endocrinol Metab 1997; 8:299–306.
17. Lalou C, Sawamura S, Segovia B, Ogawa Y, Binoux M. Proteolytic fragments of insulin-like growth factor binding protein-3: N-terminal sequences and relationships between structure and biological activity. CR Acad Sci Paris/Life Sci 1997; 32:621–628.
18. Ho PJ, Baxter RC. Characterization of truncated insulin-like growth factor-binding protein-2 in human milk. Endocrinology 1997; 138:3811–3818.
19. Lamson G, Giudice LC, Rosenfeld RG. A simple assay for proteolysis of IGFBP-3. J Clin Endocrinol Metab 1991; 72:1391–1393.
20. Claussen M, Zapf J, Braulke T. Proteolysis of insulin-like growth factor binding-5 by pregnancy serum and amniotic fluid. Endocrinology 1994; 134:1964–1966.
21. Giudice, L.C. The roles of IGFs and IGFBP-1 in non-pregnant human endometrium and at the maternal:placental interface during human pregnancy. In: Takano K, Hizuka N, Takahashi SI, eds. Molecular Mechanisms to Regulate the Activities of Insulin-Like Growth Factors. Elsevier, Amsterdam, 1998, pp. 195–204.
22. Besnard N, Pisselet C, Monniaux D, Locatelli A, Benne F, Gasser F, Hatey F, Monget P. Expression of messenger ribonucleic acids of insulin-like growth factor-binding protein-2, -4, and -5 in the ovine ovary: localization and changes during growth and atresia of antral follicles. Biol Reprod 1996; 55:1356–1367.
23. Besnard N, Pisselet C, Monniaux D, Monget P. Proteolytic activity degrading insulin-like growth factor-binding protein-2, -3, -4, and -5 in healthy growing and atretic follicles in the pig ovary. Biol Reprod 1997; 56:1050–1058.
24. Conover CA. Insulin-like growth factor binding protein proteolysis in bone cell model. Prog Growth Factor Res 1995; 6:301–309.
25. Conover CA, Perry JE, Tindall DJ. Endogenous cathepsin D-mediated hydrolysis of insulin-like growth factor binding proteins in cultured human prostatic carcinoma cells. J Clin Endocrinol Metab 1995; 80:987–993.
26. Fowlkes JL, Freemark M. Evidence for a novel insulin-like growth factor (IGF)-dependent protease regulating IGF-binding protein-4 in dermal fibroblasts. Endocrinology 1992; 131:2071–2076.
27. Parker A, Gockerman A, Busby Jr WH, Clemmons DR. Properties of an insulin-like growth factor-binding protein-4 protease that is secreted by smooth muscle cells. Endocrinology 1995; 136:2470–2476.
28. Chernausek SD, Smith CE, Duffin KL, Busby WH Jr., Wright G, Clemmons DR. Proteolytic cleavage of insulin-like growth factor binding protein-4 (IGFBP-4). Localization of cleavage site to non-homologous region of native IGFBP-4. J Biol Chem 1995; 270:11377–11382.
29. Zheng B, Clarke JB, Busby WH Jr, Duan CM, Clemmons DR. Insulin-like growth factor-binding protein-5 is cleaved by physiological concentrations of thrombin. Endocrinology 1998; 139:1708–1714.
30. Bang P, Fielder PJ. Human pregnancy serum contains at least two distinct proteolytic activities with the ability to degrade insulin-like growth factor binding protein-3. Endocrinology 1997; 138:3912–3917.
31. Salahifar H, Baxter RC, Martin JL. Insulin-like growth factor binding protein (IGFBP)-3 protease activity secreted by MCF-7 breast cancer cells: Inhibition by IGFs does not require IGF-IGFBP interaction. Endocrinology 1997; 138:1683–1690.

32. Thrailkill KM, Quarles LD, Nagase H, Suzuki K, Serra DM, Fowlkes JL. Characterization of insulin-like growth factor-binding protein 5-degrading proteases produced throughout murine osteoblast differentiation. Endocrinology 1995; 136:3527–3533.
33. Nam TJ, Busby WH Jr, Clemmons DR. Characterization and determination of the relative abundance of two types of insulin-like growth factor binding protein-5 proteases that are secreted by human fibroblasts. Endocrinology 1996; 137:5530–5536.
34. Lalou C, Silve C, Rosato R, Segovia B, Binoux M. Interactions between insulin-like growth factor-I (IGF-I) and the system of plasminogen activators and their inhibitors in the control of IGF-binding protein-3 production and proteolysis in human osteosarcoma cells. Endocrinology 1994; 135:2318–2326.
35. Conover CA, De Leon DD. Acid-activated insulin-like growth factor binding protein-3 proteolysis in normal and transformed cells. J Biol Chem 1994; 269:7076–7080.
36. Nunn SE, Peehl DM, Cohen P. Acid-activated insulin-like growth factor binding protein protease activity of cathepsin D in normal and malignant prostatic epithelial cells and seminal plasma. J Cell Physiol 1997; 171:196–204.
37. Braulke T, Claussen M, Saftig P, Wendland M, Neifer K, Schmidt B, Zapf J, von Figura K, Peters C. Proteolysis of IGFBPs by cathepsin D in vitro and in cathepsin-D-deficient mice. Prog Growth Factor Res 1995; 6:265–271.
38. Neumann GM, Marinaro JA, and Bach LA. Identification of *O*-glycosylation sites and partial characterization of carbohydrate structure and disulfide linkages for human insulin-like growth factor binding protein-6. Biochemistry 1998; 37:6572–6585.
39. Hossenlopp P, Seurin D, Segovia-Quinson B, Hardouin S, Binoux M. Analysis of serum insulin-like growth factor binding proteins using Western blotting: use of the method for titration of the binding proteins and competitive binding studies. Anal Biochem 1986; 154:138–143.
40. Spencer EM and Chan K. A 3-dimensional model for the insulin-like growth factor binding proteins (IGFBPs): supporting evidence using the structural determinants of the IGF binding site on IGFBP-3. Prog Growth Factor Res 1995; 6:209–214.
41. Firth SM, Ganeshprasad U, Baxter RC. Structural determinants of ligand and cell surface binding of insulin-like growth factor binding protein-3. J Biol Chem 1998; 273:2631–2638.
42. Durham SK, Mohan S, Liu F, Baker BK, Lee PDK, Hintz RL, Conover CA, Powell DR. Bioactivity of a 29-kilodalton insulin-like growth factor binding protein-3 fragment present in excess in chronic renal failure serum. Pediatr Res 1997; 42:335–341.
43. Lalou C, Lassarre C, Binoux M. A proteolytic fragment of insulin-like growth factor (IGF) binding protein-3 that fails to bind IGFs inhibits the mitogenic effects of IGF-I and insulin. Endocrinology 1996; 137:3206–3212.
44. Spagnoli A, Gargosky SE, Spadoni GL, Macgillivray M, Oh Y, Boscherini B, Rosenfeld RG. Characterization of a low molecular mass form of insulin-like growth factor binding protein-3 (17.7 kilodaltons) in urine and serum from healthy children and growth hormone (GH)-deficient patients: relationship with GH therapy. J Clin Endocrinol Metab 1995; 80:3668–3676.
45. Vorwerk P, Yamanaka Y, Spagnoli A, Oh Y, Rosenfeld RG. Insulin and IGF binding by IGFBP-3 fragments derived from proteolysis, baculovirus expression and normal human urine. J Clin Endocrinol Metab 1998; 83:1392–1395.
46. Yamanaka Y, Wilson EM, Rosenfeld RG, Oh Y. Inhibition of insulin receptor activation by insulin-like growth factor binding proteins. J Biol Chem 1997; 272:30729–30734.
47. Rajaram S, Baylink DJ, Mohan S. Insulin-like growth factor binding proteins in serum and other biological fluids: regulation and functions. Endocr Rev 1997; 18:801–831.
48. Davies SC, Holly JM, Coulson VJ, Cotterill AM, Abdulla AF, Whittaker PG, Chard T, Wass JA. The presence of cation-dependent proteases for insulin-like growth factor binding proteins does not alter the size distribution of insulin-like growth factors in pregnancy. Clin Endocrinol (Oxf) 1991; 34:501–506.
49. Gargosky SE, Owens PC, Walton PE, Owens JA, Robinson JS, Wallace JC, Ballard FJ. Most of the circulating insulin-like growth factor-I and factor-II are present in the 150 kDa complex during human pregnancy. J Endocrinol 1991; 131:491–497.
50. Suikkari AM, Baxter RC. Insulin-like growth factor-binding protein-3 is functionally normal in pregnancy serum. J Clin Endocrinol Metab 1992; 74:177–183.
51. Davenport ML, Clemmons DR, Miles MV, Camacho-Hübner C, D'Ercole AJ, Underwood LE. Regulation of serum insulin-like growth factor-I (IGF-I) and IGF binding proteins during rat pregnancy. Endocrinology 1990; 127:1278–1286.

52. Guler HP, Zapf J, Schmid C, Froesch ER. Insulin-like growth factors I and II in healthy man. Estimations of half-lives and production rates. Acta Endocrinol (Copenh) 1989; 121:753–758.
53. Lee CY, Rechler MM. Proteolysis of insulin-like growth factor (IGF)-binding protein-3 (IGFBP-3) in 150-kilodalton IGFBP complexes by a cation-dependent protease activity in adult rat serum promotes the release of bound IGF-I. Endocrinology 1996; 137:2051–2058.
54. Holly JMP, Claffey DCP, Cwyfan-Hugues SC, Frost VJ, Yateman ME. Proteases acting on IGFBPs; their occurrence and physiological significance. Growth Regul 1993; 3:88–91.
55. Davies CS, Wass JAH, Ross RJM, Cotterill AM, Buchanan CR, Coulson VJ, Holly JMP. The induction of a specific protease for insulin-like growth factor binding protein-3 in the circulation during severe illness. J Endocrinol 1991; 130:469–473.
56. Bereket A, Lang CH, Blethen SL, Fan J, Frost RA, Wilson TA. Insulin-like growth factor binding protein-3 proteolysis in children with insulin-dependent diabetes mellitus: possible role for insulin in the regulation of IGFBP-3 protease activity. J Clin Endocrinol Metab 1995; 80:2282–2288.
57. Bang P, Brismar K, Rosenfeld RG. Increased proteolysis of insulin-like growth factor binding protein-3 (IGFBP-3) in noninsulin-dependent diabetes mellitus serum, with elevation of a 29-kilodalton (kDa) glycosylated IGFBP-3 fragment contained in the approximately 130- to 150-kDa ternary complex. J Clin Endocrinol Metab 1994; 78:1119–1127.
58. Lindgren BF, Segovia B, Lassarre C, Binoux M, Gourmelen M. Growth retardation in constitutionally short children is related both to low serum levels of insulin-like growth factor-I and to its reduced bioavailability. Growth Regul 1996; 6:158–164.
59. Binoux M, Lalou C, Lassarre C, Segovia B. Regulation of IGF bioavailability by IGFBP proteases. In: Baxter RC, Gluckman PD, Rosenfeld RG, eds. The Insulin-Like Growth Factors and Their Regulatory Proteins. Excerpta Medica, Elsevier, Amsterdam, 1994, pp. 217–226.
60. Lassarre C, Perin L, Binoux M. Protease-induced alteration of insulin-like growth factor binding protein-3 as detected by radioimmunoassay. Agreement with ligand blotting data. Growth Regul 1994; 4:48–55.
61. Koedam JA, Hoogerbrugge CM, Van Buul-Offers SC. Insulin-like growth factor binding protein-3 protease activity in Snell normal and Pit-1 deficient dwarf mice. J Endocrinol 1998; 157:295–303.
62. Binoux M, Lalou C, Mohseni-Zadeh S, Angelloz-Nicoud P, Daubas C, Babajko S. Proteolytic fragments of IGF binding protein-3: physiological significance. In: Takano K, Hizuka N, Takahashi SI, eds. Molecular Mechanisms to Regulate the Activities of Insulin-Like Growth Factors. Elsevier, Amsterdam, 1998, pp. 99–106.
63. Cotterill AM, Holly JMP, Taylor AM, Davies SC, Coulson VJ, Preece MA, Wass JAH, Savage MO. The insulin-like growth factor binding proteins and insulin-like growth factor-bioactivity in Laron-type dwarfism. J Clin Endocrinol Metab 1992; 74:56–63.
64. Cataldo NA, Giudice LC. Insulin-like growth factor binding protein profiles in human ovarian follicular fluid correlate with follicular functional status. J Clin Endocrinol Metab 1992; 74:821–829.
65. Cwyfan Hughes SC, Mason HD, Franks S, Holly JMP. Modulation of the insulin-like growth factor binding proteins by follicule size in the human ovary. J Endocrinol 1997; 154:35–43.
66. Howard HJ, Ford JJ. Relationship among concentrations of steroids, inhibin, insulin-like growth factor-I (IGF-I) and IGF-binding proteins during follicular development. Biol Reprod 1992; 47:193–201.
67. Monget P, Monniaux D, Pisselet C, Durand P. Changes in insulin-like growth factor-I (IGF-I), IGF-II, and their binding proteins during growth and atresia of ovine ovarian follicles. Endocrinology 1993; 132:1438–1446.
68. Grimes RW, Guthrie HD, Hammond JM. Insulin-like growth factor-binding protein-2 and -3 are correlated with atresia and preovulatory maturation in the porcine ovary. Endocrinology 1994; 135:1996–2000.
69. Echternkamp SE, Howard HJ, Roberts AJ, Grizzle J, Wise T. Relationships among concentrations of steroids, insulin-like growth factor -I and insulin-like growth factor binding proteins in ovarian follicular fluid of beef cattle. Biol Reprod 1994; 51:971–981.
70. Besnard N, Pisselet C, Zapf J, Hornebeck W, Monniaux D, Monget P. Proteolytic activity is involved in changes in intrafollicular insulin-like growth factor-binding protein levels during growth and atresia of ovine ovarian follicles. Endocrinology 1996; 137:1599–1607.
71. Angelloz-Nicoud P, Binoux M. Autocrine regulation of cell proliferation by the insulin-like growth factor (IGF) and IGF binding protein-3 protease system in a human prostate carcinoma cell line (PC-3). Endocrinology 1995; 136:5485–5492.
72. Cohen P, Peehl DM, Graves HCB, Rosenfeld RG. Biological effects of prostate specific antigen as an insulin like growth factor-binding protein-3 protease. J Endocrinol 1994; 142:407–415.

73. Menouny M, Binoux M, Babajko S. Role of insulin-like growth factor binding protein-2 and its limited proteolysis in neuroblastoma cell proliferation: modulation by transforming growth factor-β and retinoic acid. Endocrinology 1997; 138:683–690.
74. Conover CA, Kiefer MC, Zapf J. Posttranslational regulation of insulin-like growth factor binding protein-4 in normal and transformed human fibroblasts. J Clin Invest 1993; 91:1129–1137.
75. Conover CA. Glycosylation of insulin-like growth factor binding protein-3 (IGFBP-3) is not required for potentiation of IGF-I action: evidence for processing of cell-bound IGFBP-3. Endocrinology 1991; 129:3259–3268.
76. Schmid C, Rutishauser J, Schläpfer I, Froesch ER, Zapf J. Intact but not truncated insulin-like growth factor binding protein-3 (IGFBP-3) blocks IGF I-induced stimulation of osteoblasts: control of IGF signalling to bone cells by IGFBP-3-specific proteolysis? Biochem Biophys Res Commun 1991; 179:579–585.
77. Angelloz-Nicoud P, Lalou C, Binoux M. Prostate carcinoma (PC-3) cell proliferation is stimulated by the 22–25 kDa proteolytic fragment (1–160) and inhibited by the 16 kDa fragment (1–95) of recombinant human insulin-like growth factor binding protein-3. Growth Horm IGF Res 1998; 8:71–75.
78. Koustilieris M, Akoum A, Lazure C, Frenette G, Lemay A. N-terminal truncated forms of insulin-like growth factor binding protein-3 in the peritoneal fluid of women without laparoscopic evidence of endometriosis. Fertil Steril 1996; 68:314–321.
79. Andress DL, Birnbaum RS. Human osteoblast-derived insulin-like growth factor (IGF) binding protein-5 stimulates osteoblast mitogenesis and potentiates IGF action. J Biol Chem 1992; 267:22467–22472.
80. Andress DL, Loop SM, Zapf J, Kiefer MC. Carboxy-truncated insulin-like growth factor binding protein-5 stimulates mitogenesis in osteoblast-like cells. Biochem Biophys Res Commun 1993; 195:25–30.
81. Jones JI, Gockerman A, Busby WH Jr, Camacho-Hübner C, Clemmons DR. Extracellular matrix contains insulin-like growth factor binding protein-5: potentiation of the effects of IGF-I. J Cell Biol 1993; 121:679–687.
82. Andress DL. Comparison studies of IGFBP-5 binding to osteoblasts and osteoblast-derived extracellular matrix. Prog Growth Factor Res 1995; 6:337–344.
83. Bautista CM, Baylink DJ, Mohan S. Isolation of a novel insulin-like growth factor (IGF) binding protein from human bone: a potential candidate for fixing IGF-II in human bone. Biochem Biophys Res Commun 1991; 176:756–763.
84. Herington AC, Kuffer AD. Identification of a specific inhibitor of nonsuppressible insulin-like activity in a partially purified human serum fraction. Endocrinology 1981; 109:1634–1640.
85. Ooi GT, Herington AC. Covalent cross-linking of insulin-like growth factor-I to a specific inhibitor from human serum. Biochem Biophys Res Commun 1986; 137:411–417.
86. Mohseni Zadeh S, Binoux M. The 16-kDa proteolytic fragment of insulin-like growth factor (IGF) binding protein-3 inhibits the mitogenic action of fibroblast growth factor on mouse fibroblasts with a targeted disruption of the type 1 IGF receptor gene. Endocrinology 1997; 138:3069–3072.
87. Villaudy J, Delbé J, Blat C, Desauty G, Goldé A, Harel L. An IGF binding protein is an inhibitor of FGF stimulation. J Cell Physiol 1991; 149:492–496.
88. Clemmons DR. Multiple hormones stimulate the production of somatomedin by cultured human fibroblasts. J Clin Endocrinol Metab 1984; 58:850–856.
89. Myers SE, Cheung PT, Handwerger S, Chernausek SD. Insulin-like growth factor-I (IGF-I) enhanced proteolysis of IGF-binding protein-4 in conditioned medium from primary cultures of human decidua: independence from IGF receptor binding. Endocrinology 1993; 133:1525–1531.
90. Donnelly MJ, Holly JMP. The role of IGFBP-3 in the regulation of IGFBP-4 proteolysis. J Endocrinol 1996; 149:R1–R7.
91. Fowlkes JL, Serra DM, Rosenberg CK, Thrailkill KM. Insulin-like growth factor (IGF)-binding protein-3 (IGFBP-3) functions as an IGF-reversible inhibitor of IGFBP-4 proteolysis. J Biol Chem 1995; 270:27481–27488.
92. Fowkles JL, Thrailkill KM, Carlos-George N, Rosenberg CK, Serra DM. Heparin-binding highly basic regions within the thymglobulin type 1 repeat of insulin-like growth factor binding proteins (IGFBPs)-3, -5 and -6 inhibit IGFBP-4 degradation. Endocrinology 1997; 138:2280–2285.
93. Conover CA, Kiefer MC. Regulation and biological effect of endogenous insulin-like growth factor binding protein-5 in human osteoblastic cells. J Clin Endocrinol Metab 1993; 76:1153–1159.
94. Arai T, Arai A, Busby WH Jr, Clemmons DR. Glycosaminoglycans inhibit degradation of insulin-like growth factor-binding protein-5. Endocrinology 1994; 135:2358–2363.

95. Nam TJ, Busby Jr WH, Clemmons DR. Insulin-like growth factor binding protein-5 binds to plasminogen activator inhibitor-I. Endocrinology 1997; 138:2972–2978.
96. Campbell PG, Andress DL. Plasmin degradation of insulin-like growth factor-binding protein-5 (IGFBP-5): regulation by IGFBP-5-(201–218). Am J Physiol 1997; 273 (Endocrinol Metab 36), E996-E1004.
97. McCusker RH, Camacho-Hübner C, Bayne ML, Cascieri MA, Clemmons DR. Insulin-like growth factor (IGF) binding to human fibroblast and glioblastoma cells, the modulating effect of cell released IGF binding proteins (IGFBPs). J Cell Physiol 1990; 144:244–253.
98. Conover CA. Potentiation of insulin-like growth factor (IGF) action by IGF-binding protein-3: studies of underlying mechanism. Endocrinology 1992; 130:3191–3199.
99. Fowlkes JL, Serra DM. Characterization of glycosaminoglycan-binding domains present in insulin-like growth factor-binding protein-3. J Biol Chem 1996; 271:14676–14679.
100. Conover CA, Clarkson JT, Bale LK. Factors regulating insulin-like growth factor binding protein-3 binding, processing, and potentiation of insulin-like growth factor action. Endocrinology 1996; 137:2286–2292.
101. Andress DL. Heparin modulates the binding of insulin-like growth factor (IGF) binding protein-5 to a membrane protein in osteoblastic cells. J Biol Chem 1995; 270:28289–28296.
102. Booth BA, Boes M, Andress DL, Dake BL, Kiefer MC, Maaks C, Linhardt RJ, Bar K, Caldwell EEO, Weiler J, Bar RS. IGFBP-3 and IGFBP-5 association with endothelial cells: role of C-terminal heparin binding domain. Growth Regul 1995; 5:1–17.
103. Oh Y, Müller HL, Pham H, Rosenfeld RG. Demonstration of receptors for insulin-like growth factor binding protein-3 on Hs578T human breast cancer cells. J Biol Chem 1993; 268:26045–26048.
104. Leal SM, Liu QJ, Huang SS, Huang JS. The type V transforming growth factor beta receptor is the putative insulin-like growth factor-binding protein 3 receptor. J Biol Chem 1997; 272:20572–20576.
105. Mohseni Zadeh S, Binoux M. Insulin-like growth factor (IGF) binding protein-3 interacts with the type 1 IGF receptor, reducing the affinity of the receptor for its ligand: an alternative mechanism in the regulation of IGF action. Endocrinology 1997; 138:5645–5648.
106. Dufourny B, Alblast J, Van Teeffelen HAAM, Van Shaik FMA, Van der Burg B, Steenbergh PH, Sussenbach JS. Mitogenic signaling of insulin-like growth factor I in MCF-7 human breast cancer cells requires phosphatidylinositol 3- kinase and is independent of mitogen-activated protein kinase. J Biol Chem 1997; 272:971–981.
107. Sommer A, Maack CA, Spratt SK, Mascarenhas D, Tressel TI, Rhodes ET, Lee R, Roumas M, Tatsuno GP, Flynn JA, Gerber N, Taylor J, Cudny H, Nanney L, Hunt TK, Spencer EM. Molecular genetics and actions of recombinant insulin-like growth factor binding protein-3. In: Spencer EM, ed. Modern Concepts of Insulin-Like Growth Factors. Elsevier, Amsterdam, 1991, pp. 715–728.
108. Oh Y, Müller HL, Lamson G, Rosenfeld RG. Insulin-like growth factor (IGF)-independent action of IGF-binding protein-3 in Hs578T human breast cancer cells. Cell surface binding and growth inhibition. J Biol Chem 1993; 268:14964–14971.
109. Valentinis B, Bhala A, Deangelis T, Baserga R, Cohen P. The human insulin-like growth factor (IGF) binding protein-3 inhibits the growth of fibroblasts with a targeted disruption of the IGF-I receptor gene. Mol Endocrinol 1995; 9:361–367.
110. Bashkin P, Doctrow S, Klagsbrun M, Svahn CM, Folkman J, Vlodavsky I. Basic fibroblast growth factor binds to subendothelial extracellular matrix and is released by heparitinase and heparin-like molecules. Biochemistry 1989; 28:1737–1743.
111. Sato Y, Rifkin DB. Inhibition of endothelial cell movement by pericytes and smooth muscle cells: activation of a latent transforming growth factor-betal-like molecule by plasmin during co-culture. J Cell Biol 1989; 109:309–315.
112. Fahnestock M, Woo JE, Lopez GA, Snow J, Walz DA, Arici MJ, Mobley WC. Beta-NGF-endopeptidase: structure and activity of kallikrein encoded by the gene mGK-22. Biochemistry 1991; 30:3443–3450.
113. Rajah R, Bhala A, Nunn SE, Peehl DM, Cohen P. 7S Nerve growth factor is an insulin-like growth factor-binding protein protease. Endocrinology 1996; 137:2676–2682.
114. Oh Y, Müller HL, Ng L, Rosenfeld RG. Transforming growth factor-beta-induced cell growth inhibition in human breast cancer cells is mediated through insulin-like growth factor binding protein-3 action. J Biol Chem 1995; 270:13589–13592.
115. Gucev ZS, Oh Y, Kelley KM, Rosenfeld RG. Insulin-like growth factor binding protein-3 mediates retinoic acid- and transforming growth factor beta 2-induced growth inhibition in human breast cancer cells. Cancer Res 1996; 56:1545–1550.

116. Huynh H, Yang XF, Pollak M. Estradiol and antiestrogens regulate a growth inhibitory insulin-like growth factor binding protein 3 autocrine loop in human breast cancer cells. J Biol Chem 1996; 271:1016–1021.
117. Rajah R, Valentinis B, Cohen P. Insulin-like growth factor (IGF)-binding protein-3 induces apoptosis and mediates the effects of transforming growth factor-beta 1 on programmed cell death through a p53- and IGF-independent mechanism. J Biol Chem 1997; 272:12181–12188.
118. Jaques G, Noll K, Wegmann B, Witten S, Kogan E, Radulescu RT, Havemann K. Nuclear localization of insulin-like growth factor binding protein 3 in a lung cancer cell line. Endocrinology 1997; 138:1767–1770.
119. Li WL, Fawcett J, Widmer HR, Fielder PJ, Rabkin R, Keller GA. Nuclear transport of insulin-like growth factor-I and insulin-like growth factor binding protein-3 in opossum kidney cells. Endocrinology 1997; 138:1763–1766.
120. Elgin RG, Busby WH Jr, Clemmons DR. An insulin-like growth factor binding protein enhances the biologic response to IGF-I. Proc Natl Acad Sci USA 1987; 84:3254–3258.
121. Jones JI, Gockerman A, BusbyWH Jr, Wright G, Clemmons DR. Insulin-like growth factor binding protein 1 stimulates cell migration and binds to the $\alpha 5\beta 1$ integrin by means of its Arg-Gly-Asp sequence. Proc Natl Acad Sci USA 1993; 90:10553–10557.
122. Oh Y, Yamanaka Y, Kim HS, Vorwerk P, Wilson E, Hwa V, Yang DH, Spagnoli A, Wanek D, Rosenfeld RG. IGF-independent actions of IGFBPs. In: Takano K, Hizuka N, Takahashi SI, eds. Molecular Mechanisms to Regulate the Activities of Insulin-Like Growth Factors. Elsevier, Amsterdam, 1998, pp. 125–133.
123. Menuelle P, Binoux M, Plas C. Regulation by insulin-like growth factor (IGF) binding proteins of IGF-II-stimulated glycogenesis in cultured fetal rat hepatocytes. Endocrinology 1995; 136:5305–5310.
124. Rajah R, Nunn SE, Herrick DJ, Grunstein MM, Cohen P. Leukotriene D4 induces MMP-1 which functions as an IGFBP protease in human airway smooth muscle cells. Am J Physiol 1996; 271 (Lung Cell Mol Physiol) L1014-L1022.
125. Gockerman A, Clemmons DR. Porcine aortic smooth muscle cells secrete a serine protease for insulin-like growth factor binding protein-2. Circ Res 1995; 76:514–521.
126. Lalou C, Binoux M. Evidence that limited proteolysis of insulin-like growth factor binding protein-3 (IGFBP-3) occurs in the normal state outside the bloodstream. Regul Pept 1993; 48:179–188.
127. Frost VJ, Macaulay VM, Wass JA, Holly JM. Proteolytic modification of insulin-like growth factor-binding proteins: comparison of conditioned media from human cell lines, circulating proteases and characterized enzymes. J Endocrinol 1993; 138:545–554.
128. Kubler B, Cowell S, Zapf J, Braulke T. Proteolysis of insulin-like growth factor binding proteins by a novel 50-kilodalton metalloproteinase in human pregnancy serum. Endocrinology 1998; 139:1556–1563.
129. Xu S, Savage P, Burton JL, Sansom J, Holly JMP. Proteolysis of insulin-like growth factor-binding protein-3 by human skin keratinocytes in culture in comparison to that in skin interstitial fluid: the role and regulation of components of the plasmin system. J Clin Endocrinol Metab 1997; 82:1863–1868.
130. Matsumoto Y, Gargosky SE, Iwasaki K, Rosenfeld R. Identification and characterization of insulin-like factors (IGFs), IGF-binding proteins (IGFBPs) and IGFBP proteases in human synovial fluid. J Clin Endocrinol Metab 1996; 81:150–155.
131. Cohen P, Graves HCB, Peehl DM, Kamarei M, Giudice LC, Rosenfeld RG. Prostate-specific antigen (PSA) is an insulin-like growth factor binding protein-3 protease found in seminal plasma. J Clin Endocrinol Metab 1992; 75:1046–1053.
132. Lee KO, Oh YM, Giudice LC, Cohen P, Peehl DM, Rosenfeld RG. Identification of insulin-like growth factor-binding protein-3 (IGFBP-3) fragments and IGFBP-5 proteolytic activity in human seminal plasma: a comparison of normal and vasectomized patients. J Clin Endocrinol Metab 1994; 79:1367–1372.
133. Grimes RW, Hammond JM. Proteolytic degradation of insulin-like growth factor (IGF) binding protein-3 by porcine granulosa cells in culture: regulation by IGF-I. Endocrinology 1994; 134:337–343.
134. Fowlkes JL, Enghild JJ, Suzuki K, Nagase H. Matrix metalloproteinases degrade insulin-like growth factor binding protein-3 in dermal fibroblast cultures. J Biol Chem 1994; 269:25742–25746.
135. Irwin JC, Dsupin BA, Giudice LC. Regulation of insulin-like growth factor-binding protein-4 in human endometrial stromal cell cultures: evidence for ligand-induced proteolysis. J Clin Endocrinol Metab 1995; 80:619–626.
136. Neely EK, Rosenfeld RG. Insulin-like growth factors (IGFs) reduce IGF-binding protein-4 (IGFBP-4)

and IGFBP-3 independently of IGF receptors in human fibroblasts and epidermal cells. Endocrinology 1992; 130:985–993.
137. Durham SK, Riggs BL, Conover CA. The insulin-like growth factor-binding protein-4 (IGFBP-4)-protease system in normal human osteoblast-like cells: regulation by transforming growth factor-beta. J Clin Endocrinol Metab 1994; 79:1752–1758.
138. Durham SK, Riggs BL, Harris SA, Conover CA. Alterations in insulin-like growth factor (IGF)-dependent IGF-binding protein-4 proteolysis in transformed osteoblastic cells. Endocrinology 1995; 136:1374–1380.
139. Cheung P, Wu J, Banach W, Chernausek S. Glucocorticoid regulation of an insulin-like growth factor-binding protein-4 protease produced by a rat neuronal cell line. Endocrinology 1994; 135:1328–1335.
140. Resnick CE, Fielder PJ, Rosenfeld RG, Adashi EY. Characterization and hormonal regulation of a rat ovarian insulin-like growth factor binding protein-5 endopeptidase: an FSH-inducible granulosa cell-derived metalloprotease. Endocrinology 1998; 139:1249–1257.
141. Camacho-Hübner C, Busby WH Jr, McCusker RH, Wright G, Clemmons DR. Identification of the forms of insulin-like growth factor-binding proteins produced by human fibroblasts and the mechanisms that regulate their secretion. J Biol Chem 1992; 267:11949–11956.
142. Nam TJ, Busby WH Jr, Clemmons DR. Human fibroblasts secrete a serine protease that cleaves insulin-like growth factor binding protein-5. Endocrinology 1994; 135:1385–1391.
143. Claussen M, Kubler B, Wendland M, Neifer K, Schmidt B, Zapf J, Braulke T. Proteolysis of insulin-like growth factors (IGF) and IGF binding proteins by cathepsin D. Endocrinology 1997; 138:3797–3803.
144. Manes S, Mira E, Barbacid M, Ciprès A, Fernandez-Resa P, Buesa JM, Mérida I, Aracil M, Marquez G, Martinez AC. Identification of insulin-like growth factor-binding protein-1 as a potential physiological substrate for human stromelysin-3. J Biol Chem 1997; 272:25706–25712.
145. Booth BA, Boes M, Bar RS. IGFBP-3 proteolysis by plasmin, thrombin, serum: heparin binding, IGF binding, and structure of fragments. Am J Physiol 1996; 271 (Endocrinol Met 34) :E465–E470.
146. Fielder PJ, Rosenfeld RG, Graves HCB, Grandbois K, Maack CA, Sawamura S, Ogawa Y, Sommer A, Cohen P. Biochemical analysis of prostate specific antigen-proteolyzed insulin-like growth factor binding protein-3. Growth Regul 1994; 4:164–172.
147. Conover CA, Durham SK, Zapf J, Masiarz FR, Kiefer MC. Cleavage analysis of insulin-like growth factor (IGF)-dependent IGF-binding protein-4 proteolysis and expression of protease-resistant IGF-binding protein-4 mutants. J Biol Chem 1995; 270:4395–4400.
148. Imai Y, Busby WH Jr, Smith CE, Clarke JB, Garmong AJ, Horwitz GD, Rees C, Clemmons DR. Protease-resistant form of insulin-like growth factor-binding protein 5 is an inhibitor of insulin-like growth factor-I actions on porcine smooth muscle cells in culture. J Clin Invest 1997; 100:2596–2605.
149. Brinkman A, Kortleve DJ, Groffen CAH, Drop SLS. The C-terminal region of the low molecular weight IGF binding protein, IGFBP-1, is essential for ligand binding. In: Drop SLS, Hintz RL, eds. Insulin-Like Growth Factor Binding Proteins. Elsevier, Amsterdam, 1989, pp. 11–18.

14 The IGF Binding Protein Superfamily

Vivian Hwa, PHD, *Youngman Oh,* PHD,
Christine P. Burren, MD, *Won Kyu Choi,* MD,
Donna L. Graham, Angela Ingermann,
Ho-Seong Kim, MD, *Abel Lopez-Bermejo,* MD,
Giuseppe Minniti, MD, *Srinivasa R. Nagalla,* MD,
Kisoo Pai, MD, *Anna Spagnoli,* MD,
Peter Vorwerk, MD, *Don L. V. Wanek,*
Elizabeth M. Wilson, Yoshitaka Yamanaka, MD,
Doo Hyun Yang, MD, *and Ron G. Rosenfeld,* MD

CONTENTS

INTRODUCTION

The conventional definition of an insulin-like growth factor binding protein (IGFBP) is a protein whose major function is to act as a carrier for the insulin-like growth factors (IGFs). Based on this simple criterion, six IGFBPs that bind IGFs with high affinity, designated IGFBP-1 to -6 *(33)* have been identified to date from numerous biological systems. Although distinct structural differences exist among the six IGFBPs, a key conserved feature is the high number of cysteines (16–20 cysteines) found in IGFBPs. The clustering of the invariant cysteines at the amino (N)-terminal and carboxy (C)-terminal thirds (Fig. 1) of the proteins has led to models proposing that the N-terminus and C-terminus are two domains that, together, form the tertiary site necessary for high-affinity IGF binding. This striking structural feature, as well as the capacity of the proteins

From: *Contemporary Endocrinology: The IGF System*
Edited by: R. Rosenfeld and C. Roberts Jr © Humana Press, Inc., Totowa, NJ

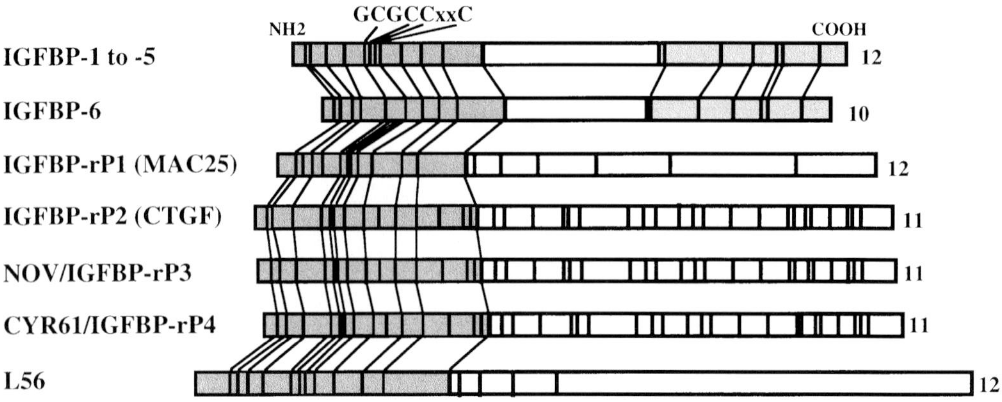

Fig. 1. Schematic primary structures of IGFBP-1 to -6, and new, IGFBP-related proteins. Numbers of conserved cysteines are indicated on the *right* of each structure, and locations of the cysteines are indicated as *vertical lines*. The N-terminal domains are shown as *dark, stippled bars;* the conserved C-terminal domains of IGFBP-1 to -6 are shown as *light, stippled bars*. Nonconserved segments between the IGFBPs and new IGFBP-related proteins are represented by *white bars*. The consensus motif GCGCCxxC in the N-terminal domains is as indicated.

to bind IGFs with high affinity, have become the signature criteria for determining whether a protein is a member of the IGFBP family. However, this conventional dogma for IGFBPs has been challenged recently; a group of cysteine-rich proteins *(8,28,45)* has been identified whose N-terminus is homologous to the IGFBP N-terminal domain, but that deviate from the common IGFBP structure in the mid-region and C-terminus (Fig. 1). The relatedness of these proteins to the IGFBPs is substantiated further by the demonstration that they can also bind IGFs, albeit at a lower affinity than observed with IGFBP-1 to -6 *(18,30)*. These findings suggest that the IGFBP family should be expanded to include those proteins that share significant domain conservation as well as demonstrable abilities to bind IGFs. Furthermore, it is proposed that these new IGFBP-related proteins, together with the conventional IGFBPs, constitute an IGFBP superfamily *(18)*. The concept of a superfamily denotes that members are united by some common structural and functional features that are conserved through evolution. We suggest that the proteins of the IGFBP superfamily be subgrouped into those that bind IGF with high affinity (IGFBP-1 to -6), and those that bind IGF with low affinity (new members).

STRUCTURE AND EVOLUTION

IGFBP-1 to -6 are well characterized (for reviews, *see refs. 14,16,33)*. They share a high degree of similarity in their primary protein structure, particularly in their N- and C-terminal regions. Overall, they are characterized by approx 36% similarity (defined as the direct comparisons of amino acid sequence without accounting for phylogenetic relationships, using the Clustral method *[38]*, DNA Star program). A prominent feature of these proteins is the number of conserved cysteines (16–20 cysteines in the prepeptides), clustered at the N-terminus and the C-terminus (Fig. 2). The N-terminal third of the protein (80–93 amino acid residues after the signal peptide), which shows a high similarity of 58% among the IGFBPs, contains 12 of the 16–20 cysteines. In IGFBP-1

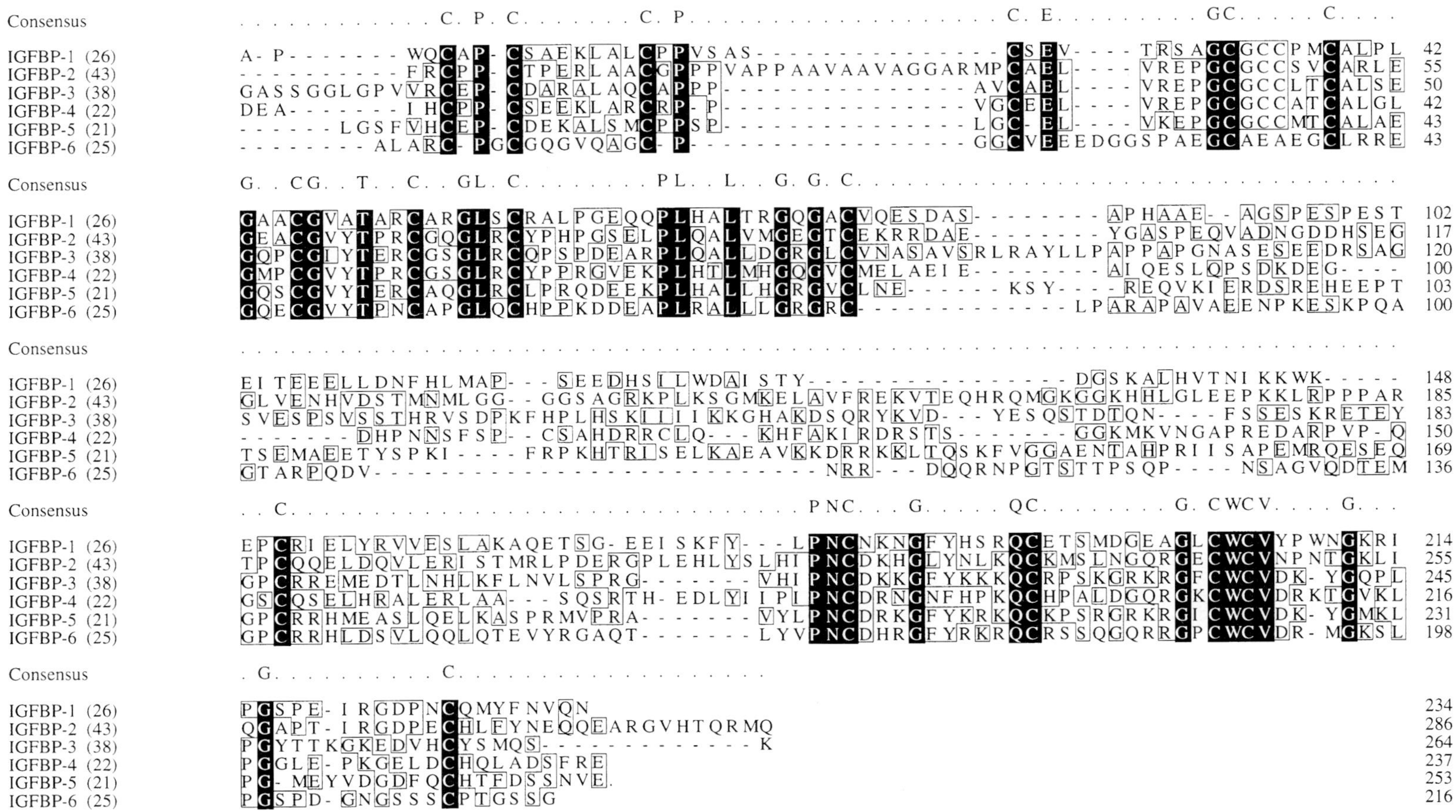

Fig. 2. Amino acid sequence alignment of human IGFBP-1 to -6. The signal peptides (number of amino acid residues are indicated in *brackets*) were not included in the analysis. The numbering system *(right of the figure)* is that of the mature peptide. Alignment was performed using the Clustral method in the DNA STAR program. Small gaps were introduced to optimize alignment. Consensus amino acid residues are as indicated*(shaded solid black); boxed* amino acid residues indicate that the residue is found in the majority of peptides.

to -5, these 12 cysteines are fully conserved, whereas in IGFBP-6, 10 out of the 12 cysteines are invariant. Within this N-terminal region a local motif (GCGCCxxC) is well conserved among the IGFBPs, with the exception of IGFBP-6, which substitutes a GCAEAEGC motif. The significance of this motif is as yet unknown, but it is hypothesized that, because the context of the GCGCCxxC motif displays a high degree of conservation among the IGFBPs, it is likely to be important for interactions with IGFs.

The C-terminus of the IGFBPs is also highly conserved, and is separated from the N-terminus by a mid-protein segment of 55–95 amino acid residues, depending on the IGFBP. The similarity among these mid-protein segments is poor (about 15%), whereas the C-termini have a significant similarity of approx 34% (Fig. 2). The remaining 6 cysteines (of the 16–20 cysteines) in IGFBPs are found in the C-terminus and are strictly conserved. Interestingly, the C-termini of IGFBP-1 to -6 contain a domain that shares 37% similarity with the thyroglobulin-type I domain *(17)*. The thyroglobulin-type I domain consists of about 65 amino acid residues that are repeated 10 times in the N-terminal part of thyroglobulin *(20)*. Its function(s) is unknown, but the domain is found in a number of proteins with varying physiological functions in different organisms. These include the major histocompatibility complex class II-associated p41 invariant chain fragment *(5)*, nidogen *(21)*, saxiphilin *(25,26)*, a tumor-associated cell surface antigen known also as GA733 *(37)*, a cysteine protease inhibitor from the egg of Chum salmon *(42)*, and equistatin, a new inhibitor of cysteine proteinases from sea anemones *(19)*. In IGFBPs, the role of this domain has yet to be determined, but is likely to affect binding to IGFs, and perhaps participate in the binding of IGFBPs to cell surfaces via heparin binding sites (e.g., IGFBP-3 *[6,10,35]*).

The recent identification of proteins that contain domains with significant homology to the N-terminus of the IGFBPs suggests the existence of other potential IGF binding proteins. The first such protein to demonstrate significant similarity with the N-terminus of IGFBPs and affinity for IGFs was the product of a cDNA designated *mac25*, which was originally cloned from leptomeninges, and predicted to encode a 31-kDa protein *(28)*. The same protein was subsequently purified from a human bladder carcinoma cell line (TAF, tumor-derived adhesion factor *[1]*) and from human diploid fibroblast cells (PSF, prostacyclin-stimulating factor *[43]*. The N-terminus of Mac25/TAF/PSF contains 11 or all 12 conserved cysteines noted in IGFBP-1 to -5 (depending on whether small gaps were introduced to optimize the protein sequence alignment, as was done for the alignment shown in Fig. 2), and shares approx 45% similarity with the IGFBP N-terminal domains. Oh et al. recently cloned the cDNA for Mac25/TAF/PSF (henceforth referred to as PSF) from Hs578T breast cancer cells and expressed the protein in a baculovirus expression system *(30)*. This protein was demonstrated to be capable of binding IGFs, albeit with at least 20- to 25-fold lower affinity than IGFBP-3. On the basis of these observations, they renamed this IGFBP-related protein IGFBP-7. The fact that PSF/IGFBP-7 has a lower affinity for IGFs, despite its high level of similarity to the IGFBP family at the N-terminus, suggests that more than the N-terminus is required for high-affinity binding to the IGFs.

Another group of novel, highly related proteins that contain a domain with significant similarity to the N-terminus of the IGFBPs is a group of cysteine-rich proteins collectively termed the "CCN family" by Bork *(7)*. The CCN family includes human connective tissue growth factor, CTGF *(8)*; an oncogene associated with MAV (*m*yeloblastosis-*a*ssociated *v*irus type-I)-induced nephroblastomas, *novH* (*n*ephroblastomas *ov*er-expression gene *[13,23]*), and a growth factor-inducible immediate-early gene, *cyr61* *(12,29)*. Protein structures for NovH and Cyr61 were deduced from their respective cDNA sequences. Overall, the CCN proteins share a high similarity of 46%. They are the products of immediate-early genes that are transcribed within minutes of exposure of cells to serum or growth factors, and are involved in regulation of cellular proliferation. The N-termini of these proteins demonstrate a significant similarity of 44% with the N-terminal domains of IGFBP-1 to -6, and 39% similarity with that of PSF/IGFBP-7 (Fig. 3). Importantly, of the 12 cysteines found in the N-terminus of the IGFBPs, 11 are conserved in CTGF, NovH, and Cyr61; valine is substituted for the 12th cysteine. Furthermore, the local motif GCGCCxxC is also conserved. A search of the protein data bank (BLAST search, SwissProt sequences) revealed several other proteins containing the GCGCCxxC motif, including a cysteine-rich protein found in the eggshell of the silk moth (chorion protein; *[31]*). However, an alignment of the chorion protein with the IGFBPs reveals a similarity of less than 15%. The GCGCCxxC motif thus appears to be highly conserved in, but not unique to, IGFBPs. A comparison of amino acid sequences in the immediate vicinity of the motif (26–35 amino acids, Fig. 3) indicates that IGFBP-1 to -5, PSF/IGFBP-7, and the CCN proteins have high degrees of similarity (40–61%). Interestingly, in this region, IGFBP-6 shows only about 27% similarity with these proteins. The biological significance of this region is not known, but could be an important part of the tertiary structure required for interaction with IGFs.

The IGFBP N-terminal domain can also be found in the predicted protein structure of a recently described cDNA encoding a 51-kDa putative serine protease *(45)*. The N-terminus of this protein, designated L56, shares 39% similarity with the IGFBP N-terminal domain and contains all of the 12 conserved cysteines, as well as the (A)CGCCxxC motif (Fig. 3). The ability of L56 to bind IGFs via its N-terminal domain has yet to be demonstrated.

The similarity among IGFBPs and all of the IGFBP-related proteins (PSF/IGFBP-7, CTGF, NovH, Cyr61, and L56) is significantly decreased (to less than 15%) beyond the N-terminal domain (Fig. 3). PSF/IGFBP-7, CTGF, NovH, Cyr61, and L56 do not contain the thyroglobulin-type I domain at the C-terminus. Analysis of the PSF/IGFBP-7 amino acid sequence using the Gapped BLAST database search program reveals that immediately adjacent to the N-terminal domain is a stretch of 30–45 amino acid residues that has 30% similarity to the Kazal family of serine proteinase inhibitors, including the (human) pancreatic secretory trypsin inhibitor *(3)*. Interestingly, this same region is also found in L56 at the analogous position (*see also* Fig. 3). The remainder of the PSF/IGFBP-7 sequence shares 41% similarity with the immunoglobulin (Ig)-like domains found in heparin sulfate proteoglycan (HSPG; *[15,27]*) and in the fibroblast growth factor receptor *(9)*. The functions of these regions with respect to the biological activities of PSF/IGFBP-7 are not known. The C-terminus of L56 is highly similar (58%) to the family of HtrA/Do serine proteases found in bacteria *(45)*, leading to the hypothesis that L56 might function as a serine protease.

The C-terminal half of the CCN proteins contain three domains identified as the Von Willebrand factor type C repeat (VWC), the thrombospondin type I repeat, and the C-terminal (CT) domain *(7)*. The functions of these three domains are still unclear, although they are believed to be involved in dimerization, oligermerization, and interactions with the extracellular matrix. Hence, the common feature among the IGFBPs and the IGFBP-related proteins is the N-terminus with its conserved cysteines and the GCGCCxxC motif; the remaining protein domains have clearly diverged.

Analysis of known gene structures for the IGFBPs (IGFBP-1 to -5) and IGFBP-related proteins (PSF/IGFBP-7, CTGF, NovH) supports the concept of the conservation of the

Fig. 3. Amino acid sequence alignment of human IGFBP-1 to -6, and human IGFBP-related proteins. Signal peptides (number of amino acid residues are indicated in *brackets*) were not included in the analysis, and the first amino acid residue shown is that for the mature peptide. Alignment was performed using the Clustral method (DNA STAR program), as for Fig. 2. Small gaps were introduced to optimize alignment. Consensus amino acid residues are shaded *solid black; boxed* amino acid residues indicate that the residue is found in the majority of peptides. IGFBP-rP1, Mac25/PSF/TAF/IGFBP-7; IGFBP-rP2, CTGF; IGFBP-rP3, Nov H; IGFBP-rP4, Cyr61; IGFBP-rP5, L56 (*see also* Table 1). Note that the last 98 amino acid residues (C-terminus) of the predicted IGFBP-rP5 (L56) protein are not shown in the alignment.

N-terminal domain. Each gene consists of four or five exons. Most striking is that the DNA sequence in exon 2 of the human *novH* gene *(22)* and gene for CTGF *(11)* encodes the N-terminal domain only (amino acid sequence $V^2AAT \rightarrow TGICT^{76}$, *see* Fig. 3); the DNA sequence for the NovH signal peptide is found within exon 1. For the IGFBPs and for PSF/IGFBP-7, the 5′ untranslated region, the signal peptide and the N-terminal domain are encoded by exon 1 of the respective genes.

The question arises whether there is an evolutionary relationship between the IGFBPs and these related proteins. The conventional IGFBPs appear to be conserved during evolution, as IGFBP activity has been detected in nonmammalian sera, including lamprey serum *(39)*. The structures of these primitive IGFBPs are largely unknown, but, presumably, contain conserved N-termini and, possibly, C-terminal regions. Of the IGFBP-related proteins, CTGF-like cDNAs have been isolated from the frog *(44)* and from the fruit fly *(24)*; *nov (13)* and *cyr61* (*cef-10*, *(34)*) cDNAs were originally isolated from chicken. Analysis of the N-termini of the predicted protein structures based on these cDNAs indicates that a high degree of similarity exists among the IGFBPs and these human and nonmammalian IGFBP-related proteins. Even the CTGF-like protein (also called twisted gastrulation protein, Tsg *[24]*) from the fruit fly, which shows an overall similarity of only 13% to the human and frog CTGFs, demonstrates a significant similarity of about 34% and 39% to the N-terminus of IGFBPs and IGFBP-related proteins, respectively. Because arthropods (such as the fruit fly) evolved well before vertebrates, it is clear that the N-terminal domain is ancient. As the amino acid residues of the N-terminus share a significant similarity, a phylogenetic tree of the N-terminal domains can be generated (Cladogram, DNA Star program) that indicates the number of amino acid substitution events leading to divergence (Fig. 4). Interestingly, PSF/IGFBP-7 diverged from the rest of the proteins early on the evolutionary scale, whereas the point of divergence for the CCN proteins, Tsg (from the fruit fly), L56, and the IGFBPs probably occurred later, and at relatively close points in evolutionary time, based on an estimated 52 substitution events for each branch in total.

The ancientness of the N-terminal domain indicate there were constraints in the evolution of the N-terminal domain. In contrast, the lack of similarity in the C-terminal among members of the IGFBP superfamily suggests that the evolution of the C-termini is complex and may be independent from that of the N-terminal domain. The evolutionary sequence of events that could account for the array of IGFBPs and IGFBP-related proteins includes: (1) the IGFBPs and related proteins share a common ancestral gene that over time selectively retained the sequences encoding the N-terminal domain while the remaining sequences evolved to encode for the various mid-segments and C-termini of the IGFBPs and IGFBP-related proteins; (2) the N-terminal domain was originally part of the gene encoding the IGFBP family and, through gene fusion, was subsequently acquired by the genes encoding the IGFBP-related proteins; (3) a gene encoding for an IGFBP-related protein originally carried the N-terminal domain DNA sequences that were later acquired by the genes encoding for the IGFBP family and other IGFBP-related proteins; or (4) the DNA sequences for the N-terminal domain is a module that, in the course of evolution, were disseminated to genes by various DNA recombinational events, i.e., the IGFBP-related proteins and the IGFBP family independently gained the N-terminal domain module. It is nevertheless apparent that the IGFBPs and IGFBP-related proteins do share a structural, functional (*see below*), and evolutionary relationship through the N-terminal domain.

NOMENCLATURE

Based on the conservation of the N-terminal domains and the ability to bind IGFs, we propose that PSF, L56, and the CCN group of proteins, together with IGFBP-1 to -6,

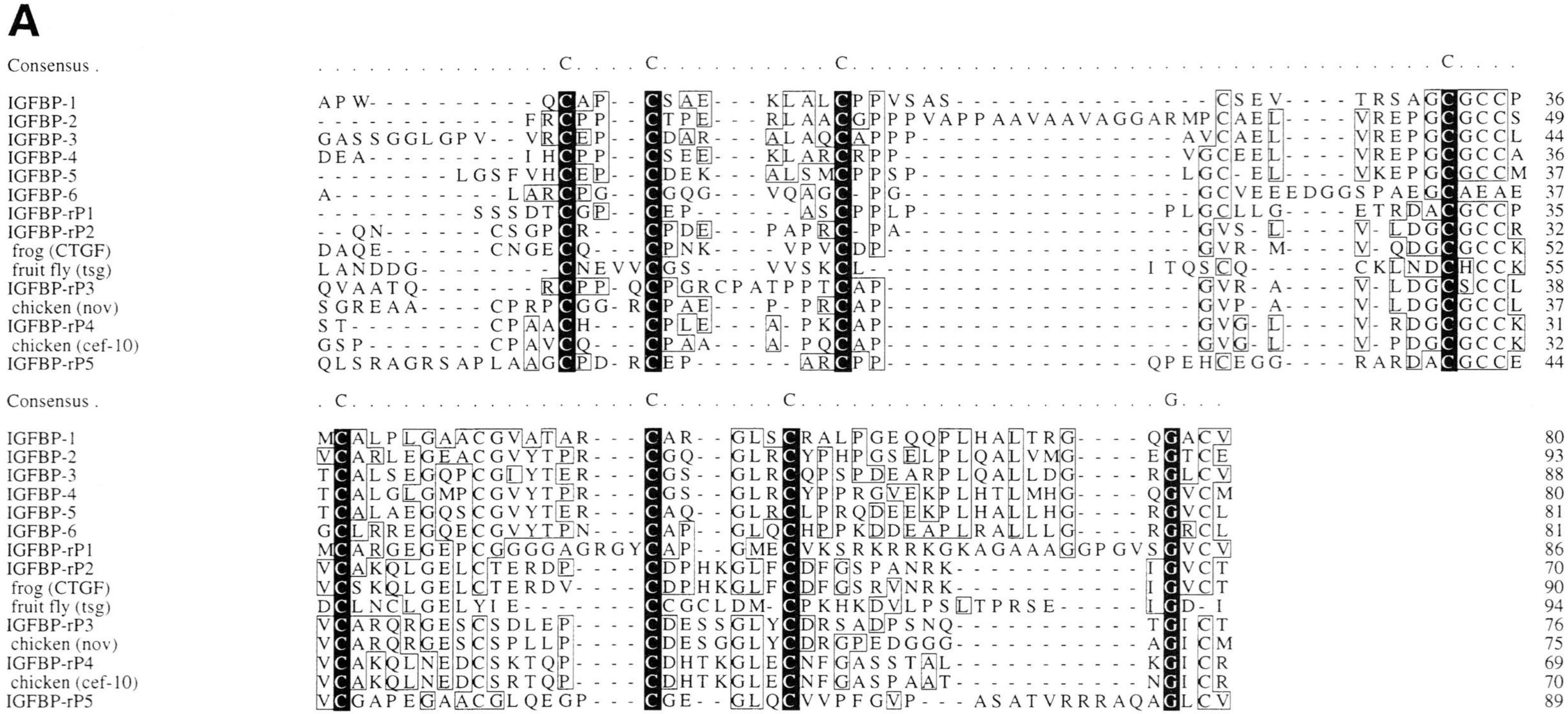

Fig. 4. Alignment and phylogenetic tree (cladogram) of the N-terminal domains of the IGFBP superfamily. **(A)** The N-terminal domains were aligned using the Clustral method. Included are the N-terminal domains from nonhuman IGFBP-rPs. Small gaps were introduced to optimize alignment. **(B)** A cladogram was generated, based on the alignment (DNA STAR program). The x-axis indicates the number of amino acid substitution events.

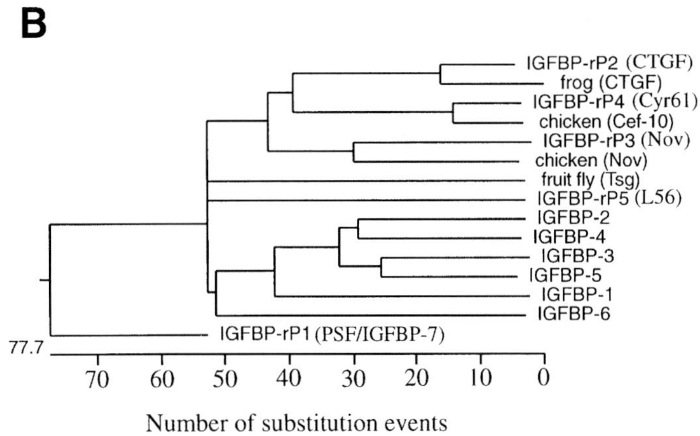

Fig. 4. (*continued*)

constitute an IGFBP superfamily. The issue of nomenclature, however, remains unsettled at this time (Table 1). Mac25 was originally identified on the basis of a cDNA derived from leptomeninges; the *mac25* cDNA was subsequently expressed in a baculovirus system, shown to bind IGFs, and renamed IGFBP-7 *(30)*. It was subsequently determined that this derived protein was identical to PSF and TAF *(32)*. The CCN group includes the CTGF protein and the *novH* and *cyr61* cDNAs. At the time of this writing, only CTGF has been expressed in baculovirus and shown to bind IGFs; similar studies *novH* and *cyr61* are pending.

The physiological role(s) of PSF, CTGF, NovH, Cyr61, and L56 in the IGF system remains undefined at this time. Nevertheless, their structural relationship with IGFBP-1 to -6 is unequivocal, and the ability of PSF and CTFG to bind IGFs has been established. The nomenclature options that are currently being considered are summarized in Table 1, and include: (1) retention of the multiple published names; (2) naming the proteins IGFBP-7 through -11; and (3) naming the proteins IGFBP-related proteins (IGFBP-rP)-1 to -5. The latter nomenclature is recommended *(4)* and is employed for the remainder of this chapter.

HIGH-AFFINITY VS LOW-AFFINITY IGF BINDERS

The ability of IGFBPs to bind IGFs with high affinity is well established, although the precise amino acid residues involved are still unknown. Based on the degree of similarity within the IGFBP family, it is postulated that both the N-terminus and the C-terminus participate in the high-affinity binding of IGFs. Consistent with this hypothesis, IGFBP-rP1 (PSF/IGFBP-7) and IGFBP-rP2 (CTGF/IGFBP-8), which are homologous with the IGFBPs only in the N-terminus, were shown by Oh et al. *(30)* and Kim et al. *(18)*, respectively, to have at least 20–25-fold lower affinity for IGFs. Nevertheless, IGFBP-rP1 and IGFBP-rP2 bind IGFs with some specificity, as both prolactin and growth hormone have been shown to be unable to bind the IGFBP-rPs *(41)*. The decrease in affinity for IGFs may be attributable to the lack of the appropriate C-terminus, and can be tested by generating heterologous proteins, fusing, for example, the C-terminus of IGFBP-3 to the N-terminus of any of the IGFBP-rPs, and determining if affinity for IGFs has now increased. Consistent with this hypothesis is the observation that the N-terminus of

Table 1
Proposed Nomenclature for New Members of the IGFBP Family

Published Names	*IGFBP Name*	*IGFBP-Related Protein*
Mac25 *(28)*	IGFBP-7 (provisional) *(30)*	IGFBP-rP1
TAF *(1)*		
PSF *(43)*		
CTGF *(8)*	IGFBP-8 (provisional) *(18)*	IGFBP-rP2
NovH *(23)*	IGFBP-9 (provisional)	IGFBP-rP3 (provisional)
Cyr61 *(12)*	IGFBP-10 (provisional)	IGFBP-rP4 (provisional)
L56 *(45)*		IGFBP-rP5 (provisional)

IGFBP-3 *(40,41)* is capable of binding IGF, but with reduced affinity. The fact that IGFBP-rPs appear to have at least a 20–25-fold lower affinity for IGF, compared to the IGFBP, suggests that there are two distinct classes of IGF binders in the IGFBP superfamily: the low-affinity IGF binders (IGFBP-rP1-5), and the high-affinity IGF binders (IGFBP-1 to -6). The role(s) of the high-affinity IGF binders in modulating IGF activity is well established. In contrast, the role(s) of the low-affinity IGF binders in modulating IGF activity is less clear. It is likely that regulation of IGF activity is not the major function of the IGFBP-rPs, and that the IGFBP-rP may, in fact, be more involved in functions independent of their effects on IGFs. In this regard, it is worth pointing out that IGF-independent actions have also been demonstrated for some of the conventional IGFBPs (*see* Chapter 11).

An unexpected action of IGFBP-rP1 is its ability to bind insulin at an affinity at least equal to its ability to bind IGFs *(2,41)*. To date, no insulin-binding proteins have been described, and none of the IGFBPs were believed to bind insulin with any degree of significance. In retrospect, it was only the lack of an appropriate detection technique that led to this incorrect conclusion. Yamanaka et al. *(41)* have recently demonstrated that, in a Western ligand blot using reduced, denaturing conditions, IGFBP-1 to -6 are capable of binding ^{125}I-insulin, although with lower affinity than IGFBP-rP1 has for insulin. This has led to a model for the secondary/tertiary structure of the IGFBPs that would accommodate IGF-binding, as well as insulin-binding, active sites. The model suggests that the folding of IGFBPs is such that the N- and C-termini together form a high-affinity IGF-binding site, which simultaneously partially hides an insulin-binding site deep within the protein. Disruption of this tertiary structure through proteolysis of the IGFBPs, or by reducing the protein through the use of agents such as dithiothreitol (DTT), not only disrupts the high-affinity IGF-binding site, but also exposes the hidden insulin-binding site. The result is an increase in affinity for insulin, with a concomitant decrease in affinity for IGFs. Consistent with this hypothesis, the in vitro-generated N-terminus of IGFBP-3 *(40,41)* as well as in vivo-proteolyzed N-terminal fragments of IGFBP-3 from human urine, demonstrate reduced affinity for IGFs, and display enhanced insulin binding *(36,40)*. For IGFBP-rP1, and perhaps also for the other IGFBP-rPs, the presence of a different C-terminus (as in IGFBP-rP1) results in essentially the same effect, and explains both the decrease in affinity of IGFBP-rP1 for IGFs and its ability to bind insulin *(41)*. In addition, it has been shown that the synthetic N-terminus (amino acid residues 1–97) *(40)* as well as IGFBP-rP1 *(41)* could effectively inhibit the autophosphorylation of the β-subunit of the insulin receptor (IR) in IR-overexpressing NIH3T3 cells, indicating a biological effect of the N-terminus in the inhibition of insulin action. It is thus suggested that the N-terminus of IGFBP-3 (and of the IGFBP-rPs) may have important biological roles in the modulation of both IGF and insulin actions.

SUMMARY

The evidence to date—that is, the similarity in primary structures, evolutionary conservation of the N-terminus, and functional studies—supports the existence of an IGFBP superfamily. Not discussed in this chapter is the potential significance and biological role(s) of the conserved IGFBP C-terminal domain (C-terminal third of the IGFBPs) with respect to the binding, and modulation, of IGFs. It is expected that the C-terminus participates in the high-affinity binding of IGFs. Because the IGFBP C-terminal third is composed of predominantly the thyroglobulin type-1 domain, it is valid to question whether proteins containing this domain should also be considered part of the IGFBP superfamily. We await data that would definitively demonstrate the involvement of the thyroglobulin type-1 domain in IGF binding.

Based on the evidence that is available, we speculate that the IGFBP superfamily is derived from an ancestor gene/protein that was critically involved in the regulation of cell growth and was capable of binding IGF peptides. Over the course of evolution, some members evolved into high-affinity IGF binders and others into low-affinity IGF binders, thereby conferring on the IGFBP superfamily the ability to influence cell growth by both IGF-dependent and IGF-independent means.

ACKNOWLEDGMENTS

This work was supported by NIH Grants CA58110 and DK51513-01A2, and by US Army Grant DAMD 17-96-/-2604.

REFERENCES

1. Akaogi K, Okabe Y, Funahashi K, Yoshitake Y, Nishikawa K, Yasunitsu H, Umeda M, Miyazaki K. Cell adhesion activity of a 30-kDa major secreted protein from human bladder carcinoma cells. Biochem Biophys Res Commun 1994; 198:1046–1053.
2. Akaogi K, Sato J, Okabe Y, Sakamoto Y, Yasumitsu H, Miyazaki K. Synergistic growth stimulation of mouse fibroblasts by tumor-derived adhesion factor with insulin-like growth factors and insulin. Cell Growth Differ 1996; 7:1671–1677.
3. Bartelt D, Shapanka R, Greene L. The primary structure of the human pancreatic secretory trypson inhibitor. Amino acid sequence of the reduced S-aminoethylated protein. Arch Biochem Biophys 1997; 179:189–199.
4. Baxter RC, Binoux M, Clemmons DR, Conover C, Drop SLS, Holly JMP, Mohan S, Oh Y, Rosenfeld RG. Recommendations for nomenclature for the insulin-like growth factor binding protein (IGFBP) superfamily. J Clin Endocrinol Metab 1998; 83:8213.
5. Bevec T, Stoka V, Pungercic G, Dolenc I, Turk V. Major histocompatibility compex class II-associated p41 invariant chain fragment is a strong inhibitor of lysosomal cathepsin L. J Exp Med 1996; 183:1331–1338.
6. Booth B, Boes M, Dake B, Linhardt R, Caldwell E, Weiler J, Bar R. Structure–function relationships in the heparin-binding C-terminal region of insulin-like growth factor binding protein-3. Growth Regul 1996; 6:206–213.
7. Bork P. The modular architecture of a new family of growth regulators related to connective tissue growth factor. FEBS Lett 1993; 327:125–130.
8. Bradham DM, Igarashi A, Potter RL, Grotendorst GR. Connective tissue growth factor: a cysteine-rich mitogen secreted by human vascular endothelial cells is related to the SRC-induced immediate early gene product CEF-10. J Cell Biol 1991; 114:1285–1294.
9. Dionne CA, Crumley G, Bellot F, Kaplow JM, Searfoss G, Ruta M, Burgess WH, Jaye M, Schlessinger J. Cloning and expression of two distinct high-affinity receptors cross-reacting with acidic and basic fibroblast growth factors. EMBO J 1990; 9:2685–2692.

10. Fowlkes J, Serra D. Characterization of glycosaminoglycan-binding domains present in insulin-like growth factor-binding protein-3. J Biol Chem 1996; 271:14676–14679.
11. Grotendorst GR. Connective tissue growth factor: a mediator of TGF-β action on fibroblasts. Cytokine Growth Factor Rev 1997; 8:171–179.
12. Jay P, Berge-Lefranc J, Marsollier C, Mejean C, Taviaux S, Berta P. The human growth factor-inducible immediate early gene, *CYR61*, maps to chromosome 1p. Oncogene 1997; 14:1753–1757.
13. Joliot V, Martinerie C, Dambrine G, Plassiart G, Brisac M. Proviral rearrangements and overexpression of a new cellular gene *(nov)* in myeloblastosis-associated virus type I-induced nephroblastomas. Mol Cell Biol 1992; 12:10–21.
14. Jones JI, Clemmons DR. Insulin-like growth factors and their binding proteins: biological actions. Endocr Rev 1995; 16:3–34.
15. Kallunki P, Tryggvason K. Human basement membrane heparan sulfate proteoglycan core protein: a 467-kD protein containing multiple domains resembling elements of the low density lipoprotein receptor, laminin, neural cell adhesion molecules, and epidermal growth factor. J Cell Biol 1992; 116:559–571.
16. Kelley K, Oh Y, Gargosky S, Gucev Z, Matsumoto T, Hwa V, Ng L, Simpson D, Rosenfeld R. Insulin-like growth factor-binding proteins (IGFBPs) and their regulatory dynamics. Int J Biochem Cell Biol 1996; 28:619–637.
17. Kiefer MC, Marsiarz FR, Bauer DM, Zapf J. Identification and molecular cloning of two new 30-kDa insulin-like growth factor binding proteins isolated from adult human serum. J Biol Chem 1991; 266:9043–9049.
18. Kim H-S, Nagalla SR, Oh Y, Wilson E, Charles J, Roberts T, Rosenfeld RG. Identification of a family of low-affinity insulin-like growth factor binding proteins (IGFBPs): characterization of connective tissue growth factor as a member of the IGFBP superfamily. Proc Natl Acad Sci USA 1997; 94:12981–12986.
19. Lenarcic B, Ritonja A, Strukelj B, Turk B, Turk V. Equistatin, a new inhibitor of cysteine proteinases from *Actinia equina*, is structurally related to thyroglobulin type-I domain. J Biol Chem 1997; 272:13899–13903.
20. Malthiery Y, Lissitzky S. Primary structure of human thyroglobulin deduced from the sequence of its 8448-base complementary DNA. Eur J Biochem 1987; 165:491–498.
21. Mann K, Deutzmann R, Aumailley M, Timpl R, Raimondi L, Yamada Y, Pan T, Conway D, Chu M. Amino acid sequence of mouse nidogen, a multidomain basement membrane protein with binding activity for laminin, collagen IV and cells. EMBO J 1989; 8:65–72.
22. Martinerie C, Huff V, Joubert I, Badzioch M, Saunders G, Strong L, Perbal B. Structural analysis of the human nov proto-oncogene and expression in Wilms tumors. Oncogene 1994; 9:2729–2732.
23. Martinerie C, Viegas-Pequignot E, Guenard I, Dutrillaux B, Nguyen V, Bernheim A, Perbal B. Physical mapping of human loci homologous to the chicken *nov* proto-oncogene. Oncogene 1992; 7:2529–2534.
24. Mason E, Konrad K, Webb C, Marsh J. Dorsal midline fate in *Drosophila* embryos requires twisted gastrulation, a gene encoding a secreted protein related to human connective tissue growth factor. Genes Dev 1994; 8:1489–1501.
25. Morabito MA, Moczydlowski E. Molecular cloning of bullfrog saxipholin: a unique relative of the transferrin family that binds saxitoxin. Proc Natl Acad Sci USA 1994; 91:2478–2482.
26. Morabito MA, Moczydlowski E. Molecular cloning of bullfrog saxiphilin: a unique relative of the transferrin family that binds saxitoxin. *(Correction).* Proc Natl Acad Sci USA 1995; 92:6651.
27. Murdoch AD, Dodge GR, Cohen I, Tuan RS, Iozzo RV. Primary structure of the human heparan sulfate proteoglycan from basement membrane (HSPG2/Perlecan). J Biol Chem 1992; 267:8544–8557.
28. Murphy M, Pykett MJ, Harnish P, Zang KD, George DL. Identification and characterization of genes differentially expressed in Meningiomas. Cell Growth Differ 1993; 4:715–722.
29. O'Brien TP, Yang GP, Sanders L, Lau LF. Expression of cyr61, a growth factor-inducible immediate-early gene. Mol Cell Biol 1990; 10:3569–3577.
30. Oh Y, Nagalla SR, Yamanaka Y, Kim H-S, Wilson E, Rosenfeld RG. Synthesis and characterization of insulin-like growth factor binding protein (IGFBP-7). J Biol Chem 1996; 271:30322–30325.
31. Rodakis G, Moschonas N, Regier J, Kafatos F. The B multigene family of chorion proteins in saturniid silkmoths. J Mol Evol 1983; 19:322–332.
32. Rosenfeld RG. Editorial: The blind men and the elephant—a parable for the study of insulin-like growth factor binding proteins. Endocrinology 1998; 139:5–7.
33. Shimasaki S, Ling N. Identification and molecular characterization of insulin-like growth factor binding proteins (IGFBP-1, -2, -3, -4, -5 and -6). Prog Growth Factor Res 1991; 3:243–266.

34. Simmons D, Levy D, Yannoni Y, Erikson R. Identification of a phorbol ester-repressible v-*src*-inducible gene. Proc Natl Acad Sci USA 1989; 86:1178–1182.
35. Smith E, Lu L, Chernausek S, Klein D. Insulin-like growh factor-binding protein-3 (IGFBP-3) concentration in rat Sertoli cell-conditioned medium is regulated by a pathway involving association of IGFBP-3 with cell surface proteoglycans. Endocrinology 1994; 135:359–364.
36. Spagnoli A, Rosenfeld RG. The mechanisms by which growth hormone brings about growth. Endocrinol Metab Clin North Am 1996; 25:615–631.
37. Szala S, Froehlich M, Scollon M, Kasai Y, Steplewski S, Koprowski H, Linnenbach A. Molecular cloning of cDNA for the carcinoma-associated antigen GA733-2. Proc Natl Acad Sci USA 1990; 87:3542–3546.
38. Thompson JD, Higgins DG, Gibson TJ. CLUSTAL W: improving the sensitivity of progressive multiple sequence alignment through sequence weighting, position-specific gap penalties and weight matrix choice. Nucleic Acids Res 1994; 22:4673–4680.
39. Upton A, Chan S, Steiner D, Wallace J, Ballard F. Evolution of insulin-like growth factor binding proteins. Growth Regul 1993; 3:29–32.
40. Vorwerk P, Yamanaka Y, Spagnoli A, Oh Y, Rosenfeld RG. Insulin and IGF binding by IGFBP-3 fragments derived from proteolysis, baculovirus expression and normal human urine. J Clin Endocrinol Metab 1998; 83:1392–1395.
41. Yamanaka Y, Wilson EM, Rosenfeld RG, Oh Y. Inhibition of insulin receptor activation by insulin-like growth factor binding proteins. J Biol Chem 1997; 272:30729–30734.
42. Yamashita M, Konagaya S. A novel cysteine protease inhibitor of the egg of Chum salmon, containing a cysteine-rich thyroglobulin-like motif. J Biol Chem 1996; 271:1282–1284.
43. Yamauchi T, Umeda F, Masakado M, Isaji M, Mizushima S, Nawata H. Purification and molecular cloning of prostacyclin-stimulating factor from serum-free conditioned medium of human diploid fibroblast cells. Biochem J 1994; 303:591–598.
44. Ying Z, King ML. Isolation and characterization of *xnov*, a *Xenopus laevis* ortholog of the chicken nov gene. Gene 1996; 171:243–248.
45. Zumbrunn J, Trueb B. Primary structure of a putative serine protease specific for IGF-binding proteins. FEBS Lett 1996; 398:187–192.

15 IGFs and Cell Growth

Renato Baserga, MD, *Marco Prisco,* PHD,
and Atsushi Hongo, PHD

CONTENTS

INTRODUCTION

A tissue or an organ grows when the number of cells produced per unit time exceeds the number of cells that die in the same period. This rather obvious statement often has been obscured by the fact that most investigators use cells in culture for basic studies. Such a use has innumerable advantages, but, on the other hand, under optimal conditions, cells in culture grow exponentially, with very little cell death, which is therefore largely neglected. However, in animals, physiological cell death occurs constantly, so that the size of any population of cells, whether normal or abnormal, depends on both the rate of cell division and the rate of cell death *(1)*. More precisely, an increase in cell number of any given cell population depends on three parameters: The first is the length of the cell cycle. A shortening of the cell cycle will cause the cells to divide more frequently, resulting in an increase in cell number. The second is the growth fraction *(2),* i.e., the fraction of cells in a population that are actively dividing in the cell cycle. In many cell populations, a sizable fraction of cells is in G0, a distinct phase of the cell cycle, in which the cells are quiescent, and do not traverse the cell cycle, although they are alive and capable, under certain conditions, of responding to proliferative stimuli. An increase in

From: *Contemporary Endocrinology: The IGF System*
Edited by: R. Rosenfeld and C. Roberts Jr © Humana Press, Inc., Totowa, NJ

the growth fraction, i.e., a decrease in the G0 fraction, will also result in an increased production of cells. The third is a decrease in the rate of cell death. The third group was all but ignored until recently, when apoptosis suddenly became the fashionable way for cells to die. But it had been known for several years that the rate of cell proliferation (the first and second parameters) was not sufficient to explain why tumors grow and normal tissues in the adult animal do not. Baserga and Kisieleski *(3)* were the first to show that the cells of one of the fastest growing tumors of mice did not proliferate faster than the normal cells of the lining epithelium of the small intestine. The obvious explanation was that the epithelial cells of the intestinal mucosa constantly undergo renewal, with a cell dying at the tip of the villi for every cell generated in the crypts. In other words, the size of the population of the very actively dividing epithelial cells of the intestinal mucosa does not increase, because the number of cells produced per unit time equals the number of cells that die in the same time period. Later, Steel et al. *(4)* convincingly showed that, in tumor growth, some extent of cell death had to be postulated. Finally, Bresciani et al. *(5)* showed that, in human squamous cell carcinomas of the skin, tumor cells also die in large amounts, but the tumors grew because the rate of cell division always exceeded the rate of cell death.

Given that the growth of a cell population depends on a balance between cell division and cell death, the next question is what regulates this balance. Growth in any population of cells, whether normal or abnormal, depends on the environmental signals, among which growth factors (either stimulatory or inhibitory) are perhaps the most important. Until recently, growth factors were looked upon merely as mitogenic agents, or, in the case of inhibitory growth factors, as anti-mitogenic agents. But in recent years a new aspect of growth factors' action has been emerging, i.e., the ability of growth factors and/or their receptors to protect cells from cell death in general and apoptosis in particular. An ideal paradigm for studying how growth factors regulate the delicate balance between cell division and cell death is offered by the insulin-like growth factor I receptor (IGF-IR), activated by its ligands. The IGF-IR is known to be mitogenic in vivo and in vitro, it promotes growth in size of the cell, it is one of the most powerful anti-apoptotic receptors, and is quasi-obligatory for the transformation of cells. Thus, the IGF-IR acts in at least four different ways to control the size of a cell population.

There is another reason why the IGF-IR is an attractive topic for the student of cell proliferation. In general, most cells require more than one growth factor for cell division, and each cell type has its preferences. Epithelial cells prefer epidermal growth factor (EGF), fibroblasts like platelet-derived growth factor (PDGF), neuronal cells respond best to nerve growth factor, and each subtype of hemopoietic cells has its own constellation of primary growth factors (interleukins, CSF-1 etc.). But, in the great majority of cell types, the second growth factor is IGF-I, or IGF-II, and these cell types include fibroblasts, epithelial cells, smooth muscle cells, endothelial cells, T lymphocytes, myeloid cells, chondrocytes, osteoblasts, and the stem cells of the bone marrow *(6)*. It was previously thought that lymphopoietic cells may not need the IGF-IR for growth, but a recent review by Clark *(7)* has made a compelling case for the involvement of the IGF system in lymphopoiesis and immune functions. There are a few exceptions to the ubiquitous distribution of the IGF-IR, e.g., hepatocytes, but by and large, most cell types have IGF-IRs and respond to stimulation by IGFs. This means that the IGF-IR is involved in regulating the growth of a large number of cell populations. This chapter discusses the mechanism(s) by which the IGF-IR and its ligands regulate the size of these populations.

HISTORICAL BACKGROUND

The mitogenicity of IGF-I has been known for a long time, indeed even before IGF-I and IGF-II acquired their present name. Considerable evidence on the mitogenic activity of the IGFs, especially IGF-I, had already been marshalled when IGF-I was still known by its original name of somatomedin-C *(8)*. Its growth-promoting effects had already been described, both in vivo and in vitro. Thus, pediatricians were already aware that, in patients with acromegaly or hypopituitarism, clinical assessment of disease activity correlated with blood levels of somatomedin-C better than they did with growth hormone levels. In vitro, it was shown that as little as 5 ng/mL of somatomedin-C could stimulate the growth of fibroblasts in culture. The role of IGF-I (to use its present name) in the growth of cells in culture was then clearly established by the seminal work of the Boston group *(9,10)*, who demonstrated that IGF-I was required, with PDGF, for the growth of 3T3 cells. These authors developed a theory, based on the assumption that PDGF was a "competence factor" required for initiating the mitogenic process in quiescent cells, while IGF-I was a "progression factor," required for entry into S phase and mitosis, after the cells had been primed by PDGF. The competence-progression theory is no longer fashionable, but the importance of their findings is undiminished, because their data showed that, for optimal growth, normal cells require more than one growth factor. Neither PDGF nor IGF-I, singly, can stimulate the growth of 3T3 cells, but in combination they do. Also, as mentioned previously, IGF-I turns out to be the second growth factor for a variety of cell types.

To complete this very short historical survey, we should mention the results of Leof et al. *(11)* and Campisi et al. *(12)*, both groups showing that IGF-I was required for the transition of cells from a restriction point in G1 into S phase. Campisi et al. *(12)* first used insulin, but at supraphysiologic concentrations, which are known to activate the IGF-IR *(13)*. In a later article, Campisi and Pardee *(14)* conclusively demonstrated that it was IGF-I that regulated the entry into S, intriguingly, by a posttranscriptional mechanism. These experiments focused on IGF-I as the "progression factor." This was probably a limited view of the role of IGF-I in cell cycle progression, because, as we shall see later, we now know that the IGF-IR is required in all phases of the cell cycle for optimal growth in vitro.

DEFINITION OF MITOGENICITY

It seems useful at this point to clarify some issues that relate to the mitogenicity of the IGFs and insulin. The first problem is the definition of mitogenicity. Most investigators in this field use, as a measure of mitogenicity, the incorporation of [^{3}H]thymidine into an acid-insoluble fraction of cells. There are two problems here, one conceptual and one technical. The conceptual problem is that there are many instances in which cells, stimulated by growth factors, synthesize DNA but do not divide. The occurrence, in vivo as well as in vitro, of cells with a tetraploid amount of DNA is a clear demonstration that DNA synthesis can be dissociated from mitosis (*see ref. 1* for examples). Its occurrence in vivo has been best documented in cardiac myocytes *(15)*, in keratinocytes *(16)*, cells of the nervous system (especially Purkinje cells), and in hepatocytes *(17)*. It occurs frequently also in culture: as an example (but others can be furnished), cells devoid of IGF-IRs and overexpressing the PDGF receptor cannot grow in PDGF only, but they can enter DNA synthesis [18]). The dissociation between DNA synthesis and mitosis is discussed further in the following section.

The technical problem is that the incorporation of [^{3}H]thymidine into an acid-insoluble fraction of the cell is only distantly related to mitogenicity, and is not even an accurate measurement of how many cells are actually entering S phase. Growth factors, by changing the permeability of cell membranes *(19),* increase the uptake of nucleotides, and in addition, they stimulate kinases, such as thymidine kinase. The net result is a change in the deoxyribonucleoside triphosphate pool *(20),* so that a measurement of thymidine incorporation into an acid-insoluble fraction can tell us as much about membrane function as about DNA synthesis. In addition, it is known that IGF-I can increase the uptake of small molecules such as amino acids *(21).* Measurement of total protein is not satisfactory either. There is substantial evidence in the literature that, under certain conditions, IGF-I can increase cell size (and therefore protein amounts), without even inducing DNA synthesis, let alone cell division (*see* the fourth section). To classify a growth factor, or its receptor, as mitogenic, one has to show that it induces cell division, and the best way to show this in tissue cultures is by the very old-fashioned method of counting cell numbers. In theory, the number of mitoses could also be an indicator of mitogenicity. However, mitoses are short-lived, and an increase in the duration of mitosis in certain cells could give the illusion of an increase in cell proliferation, where no increase is actually present. Let us not forget that colchicine, an alkaloid that blocks cells in mitosis, was originally considered to be a mitogen because, when administered to animals, it markedly increased the frequency of mitoses in all tissues. It would be helpful, therefore, to avoid misunderstandings, if we all agreed that mitogenicity means an increase in the number of cells, and that [^{3}H]thymidine incorporation is only an inaccurate index of DNA synthesis.

GROWTH IN SIZE AND GROWTH IN NUMBER

As mentioned previously, there is often a dissociation between DNA synthesis and mitosis. Cells with a tetraploid amount of DNA are essentially cells that have completed a round of DNA synthesis, but have stopped in G2. This happens frequently during development and postnatal growth, and sometimes the cell may undergo several rounds of DNA synthesis without dividing. Liver cells (hepatocytes), neurons in the central nervous system and cardiac myocytes often have $4n$, $8n$, and even $16n$ amounts of DNA (see the third section for some references). A classic example of growth in size and growth in number is given by the postnatal growth of liver, as described in the rat by Fukuda and Sibatani *(17),* whose data are summarized in Fig. 1. In essence, postnatal growth of the rat liver depends both on an increase in cell number and an increase in cell size, which is usually accompanied by an increase in ploidy. The respective contributions of these two processes, however, are disparate; cell number increases more than 10-fold, while cell size increases somewhat less than fourfold. Interestingly, if the animals are subjected to caloric restrictions, cell number and DNA/cell remain at control levels, while cell mass decreases by about 50%, with equal decreases in RNA and protein amounts *(17).* Incidentally, this is a good example of what are the priorities of a cell. In general, during growth, increase in cell number predominates (Fig. 1) as dramatically illustrated by the growth of the human, who starts from a single fertilized egg and develops into an adult whose body contains roughly 5×10^{13} cells. Even in the growth of tumors, an increase in cell number is the most important determinant of tumor mass, but growth in size cannot be simply ignored, because neglecting it may often lead to the wrong conclusions.

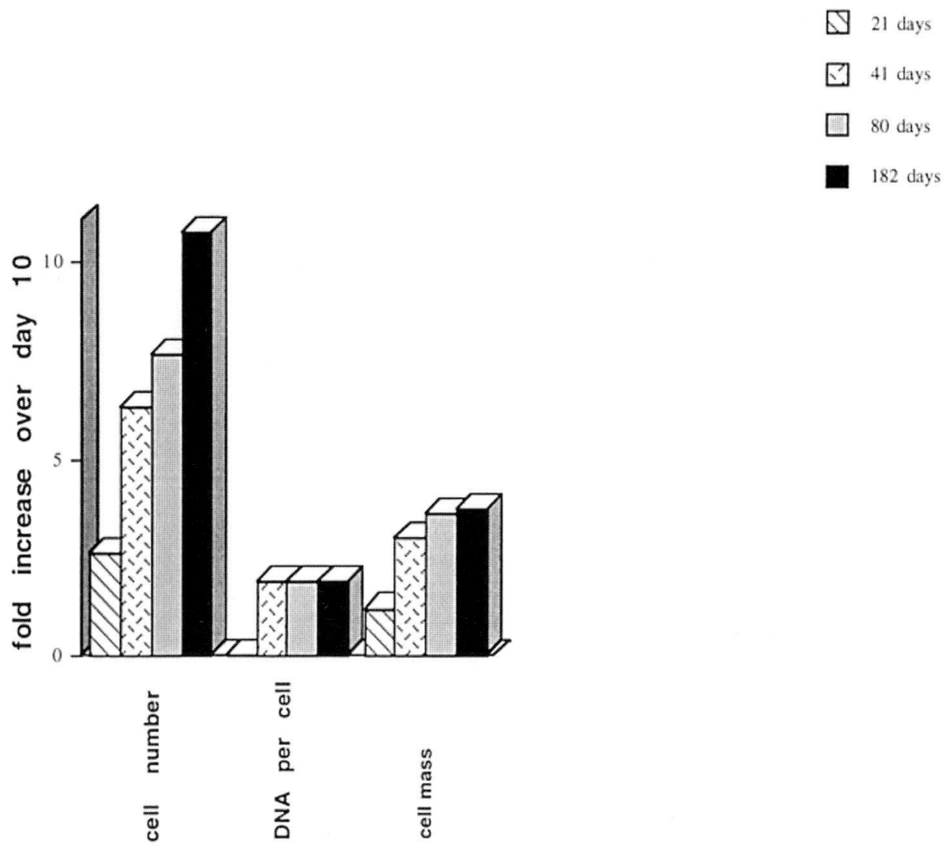

Fig. 1. Growth in size and growth in number. Data from Fukuda and Sibatani *(17)*. Between d 10 and d 182 after birth, the number of cells in the rat liver increases by almost 11-fold, while the cell mass (cell size) increases by almost 4-fold. On this basis, one would expect an increase in liver weight of about 40-fold; the actual measurements gave a 39-fold increase. The amount of DNA/cell doubles (increase in ploidy) between d 21 and d 41, then stabilizes.

This is especially true in the case of IGF-I, which can certainly induce an increase in cell size, sometimes separate from cell division. But it is also true of other situations, in which growth in size (often called by the term used by pathologists, hypertrophy) is not accompanied by cell division *(1)*. An example of the relationship between cell size and IGF-I stimulation is given by the finding of Surmacz et al. *(22)*, who showed that IGF-I activated the rDNA promoter, and ribosomal RNA amounts are one of the best measures of cell size. Another example is given by cardiac myocytes that can be stimulated to enter S phase by IGF-I, and thus increase in size, but do not divide *(23)*. Similarly, in favor of an effect of IGF-I on the size of cells is the finding that the activation of the IGF-IR is required not just in G1, but in all phases of the cell cycle, as demonstrated by the finding that R^- cells have a prolonged cell cycle, in which all phases of the cycle are elongated *(24)*. Because G2 progression and arrest before mitosis also imply growth in size (*see above*), the requirement for IGF-I for the G2 to M transition indirectly confirms the ability of IGF-I to stimulate growth in size *(25)*. Other examples are given below.

WHICH RECEPTOR? WHICH LIGAND?

There are three ligands and five receptors, and a recurring problem in the literature is the assignment of a precise responsibility in the mitogenic process to each ligand and to each receptor. For the moment, the ligands are three: insulin, IGF-I, and IGF-II. The last one comes in different isoforms, all of which have been shown to be mitogenic *(26)*, but, for simplicity, we will consider IGF-II as a single ligand. These ligands interact, with varying affinity, with the three major receptors: the IR, the type 1 IGF-IR, and the IGF-IIR *(27)*. The IGF-IIR, otherwise known as the cation-dependent mannose-6-phosphate receptor *(28)*, is not mitogenic: mice with a targeted disruption of the active IGF-II receptor gene actually grow to a larger size than wild-type littermates *(29)*. The conclusion is that the IGF-IIR down-regulates IGF-II levels, and thus serves to maintain optimal amounts of the ligand in the environment. 3T3 cells with a targeted disruption of the IGF-IR genes, called R^- cells *(24,30)*, are totally unresponsive to IGF-II (as well as to IGF-I), although they do have IGF-II receptors *(31)*.

The other two receptors can be mitogenic. There is no question that the IGF-IR is mitogenic: it may require the cooperation of another growth factor, but it certainly participates in the regulation of cell growth (*see above*). It is activated either by IGF-I or IGF-II, at concentrations of the same order of magnitude. It is also activated by insulin to transmit a mitogenic signal, but only at supraphysiological concentrations of insulin *(13)*, which creates another problem. One often reads in the literature that a certain cell line requires insulin for growth, only to find out that the amount of insulin used is in the order of grams, and, at these concentrations, insulin activates the IGF-IR (*see above*). It would be useful to know a concentration of insulin, which is still mitogenic through the IR, but not through the IGF-IR. Unfortunately, this is probably difficult to answer with precision, because it may vary with the levels of IGF-IRs, the levels of IRs, and, finally, the cell context which I define (in a somewhat light vein) as something that the cell knows, and we don't.

Is the IR mitogenic? At physiological levels, probably very little *(32)*; when overexpressed, it can certainly transmit a mitogenic signal. A convincing demonstration comes from the experiments of Morrione et al. *(31)*, who used the above-mentioned R^- cells, 3T3-like cells generated from mouse embryos with a targeted disruption of the IGF-IR genes *(33,34)*. These cells have no IGF-IRs, and do not respond to any of the three ligands, but if they are stably transfected with a plasmid overexpressing the IR, they can be stimulated to grow by insulin alone. Interestingly, they are stimulated even better by IGF-II, but not by IGF-I *(31)*, although autophosphorylation of the IR is higher with insulin than with IGF-II. This raises some questions about the role of the receptors autophosphorylation in the transmission of a mitogenic signal; the literature has examples of receptors, including the IGF-IR and the IR, that have a markedly decreased autophosphorylation and can still be functional *(35–37)*.

In Fig. 2, I have tried to summarize the situation as I see it. The other two receptors are the insulin receptor-related receptor (IRRR) and c-ros. The IRRR has substantial homology to the IGF-I and insulin receptors *(38)*. In chimeric constructs with the extracellular subunits of either the insulin or the IGF-I receptor, its tyrosine kinase domain transmits a mitogenic signal, but the ligand for the IRRR has not yet been identified. Neither insulin nor the mature IGFs, and not even serum, can activate the IRRR *(39)*. The last receptor, c-ros *(40)*, also has homologies to the insulin and IGF-I receptors, but its function in mitogenesis has not been firmly established.

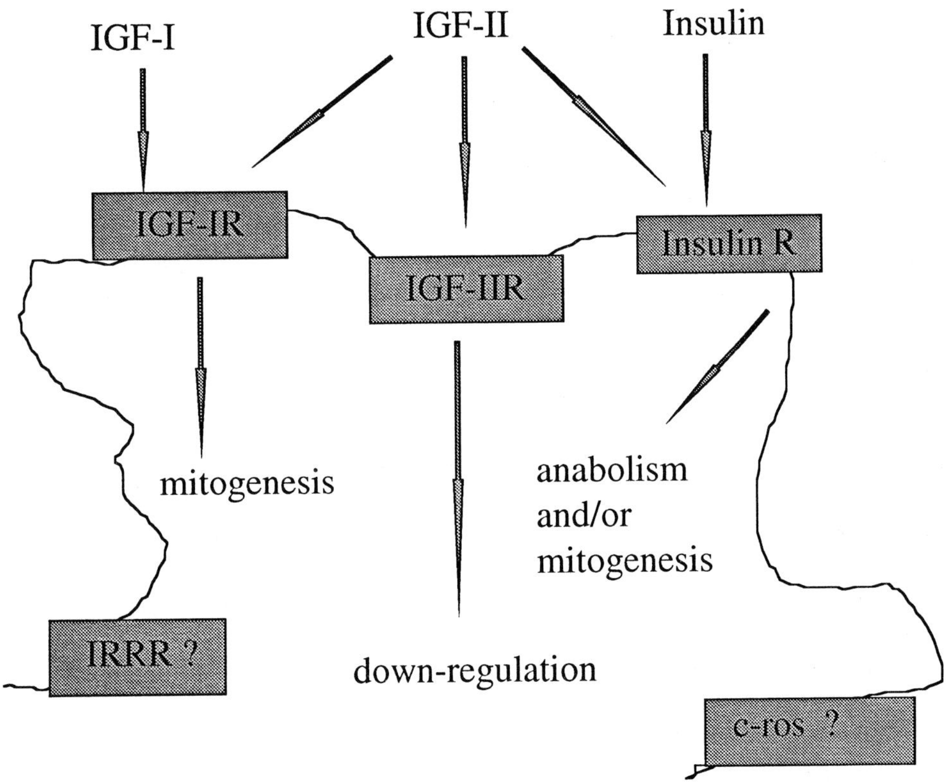

Fig. 2. Ligands, receptors, and mitogenicity. IGF-II occupies a central role, being capable to elicit a mitogenic stimulus from both the IGF-I receptor and the insulin receptor. IGF-I is strongly mitogenic with the type 1 receptor, and insulin weakly mitogenic with its own receptor. The ligands for the other two receptors (the insulin receptor-related receptor and c-ros) still have not been identified, but the β-subunit of the IRRR is mitogenic.

IGFs AND GROWTH

In Vivo

The role of the IGF system in the development of the murine embryo has been elucidated by the seminal experiments of Efstratiadis and co-workers *(29,33,34)*. Their results are summarized in Table 1. To understand these data, one has to remember that IGF-II is the predominant ligand of the IGF-IR in mouse embryos, which express negligible amounts of IGF-I. When both the IGF-II and the IGF-IR genes are disrupted by homologous recombination, the homozygous mutant embryos weigh at birth only 30% the weight of wild-type littermates. A targeted disruption of the imprinted *Igf2r* gene causes an overgrowth of 135% in respect to wild-type littermates, together with elevated serum and tissue levels of IGF-II. These mutant mice die perinatally of cardiac abnormalities, possibly because, in the absence of IGF-IIR down-regulation of IGF-II, the IGF-IR is overstimulated. Consistent with this hypothesis is the finding that *Igf2r* mutants are of normal size and viable, when they carry a second mutation, eliminating either the IGF-II gene or the IGF-IR gene. Triple mutants lacking IGF-IR, IGF-IIR, and IGF-II are nonviable dwarfs *(29)*, 30% in size, like the double mutants lacking IGF-II and IGF-IR. The results

Table 1
The IGF System in the Development of the Mouse Embryo

Genotype			
IGF-IR	*IGF-2R*	*IGF-II*	*Phenotype*
++	++	++	Normal
– –	++	++	46%
++	– –	++	140%
++	++	+*p*-	61.5%
– –	– –	++	105%
– –	++	+*p*-	34%
+ +	– –	+*p*-	74%
– –	– –	+*p*-	34%

Adapted from Ludwig et al. *(29)*. Each + or – indicates the presence or absence of the specific alleles, *p*– means that the paternal allele has been deleted. The percentages indicate the animal weight in percent of normal controls.

of knockout experiments are summarized in Table 1. They indicate that IGF-II also signals through another receptor, which has been recently identified as the IR, both in vitro *(31)*, and in vivo (A. Efstratiadis, *personal communication*).

The role of the IGF system in normal growth and development has been established also in other animals, including humans. This topic has been reviewed recently by Baserga et al. *(41)* and is not detailed here. I wish, however, to mention some interesting observations. First, there is a good correlation between IGF-I levels and body size in different breeds of dogs *(42)* and in mice. Plasma IGF-I concentrations may vary from 40 ng/mL in Dachsund to 400 ng/mL in Newfoundland, a giant breed. Deeb and Wolf *(43)* have shown that life expectancy has an inverse correlation to body size, with the giant breeds dying mostly in the first 7 years of life, while Toy poodles and Chihuahua often reach 15 years of age, and sometimes older. The reduced life expectancy of giant breeds is accompanied by a marked increase in tumor incidence in early life.

The second point is that transgenic mice confirm the importance of the IGF system in body growth. Mathews and co-workers *(44)* were the first to report that transgenic mice overexpressing IGF-I have an increased body weight. The overexpression of IGF-I can also be targeted, by the use of appropriate promoters, to a specific tissue. For instance, in a transgenic mouse line, in which the IGF-I construct was engineered to be exclusively expressed in the heart, the heart was doubled in size in respect to wild-type littermates *(45)*. Overexpression of IGF-I in skin results in hyperplasia of the epidermis, dermal abnormalities, and tumor formation in transgenic mice *(46)*.

In Vitro

The importance of IGF-I (or IGF-II) for the growth of cells in culture in serum-free medium has been known for a long time (*see* Historical Background). Under these conditions, a requirement for a functional IGF-IR has been formally demonstrated by the use of R^- cells *(24,30)*, generated by a 3T3-like protocol from mouse embryos with a targeted disruption of the IGF-IR genes *(33,34)*. R^- cells do not grow in SFM supplemented by the growth factors that sustain the growth of mouse embryo cells derived from wild-type littermates or of other 3T3 cells. An important finding with R^- cells was that, contrary to common belief, in serum-free medium, activation of the IGF-IR was necessary throughout each phase of the cell cycle *(24,25)*, and not just for the entry into S phase. An indirect confirmation of a role of the IGF-IR in the S and G2 phases of the cell cycle has been recently provided by Morrione et al. *(47)*. However, R^- cells grow in 10% serum, albeit at a lower rate than cells with a physiologic number of IGF-IRs, clearly demonstrating that the IGF-IR is not an absolute requirement for growth, although it may be important for optimal growth. These results parallel those obtained in vivo with the growth rate of mouse embryos null for the IGF-IR, but are seemingly at variance with the statement made above that IGF-I may be the second growth factor for many types of cells. Clearly, serum can bypass, at least partially, the requirement for a functional IGF-IR. It also indicates the existence in serum of growth factor(s) that have not yet been identified; on the other hand, it is possible that the IGFs may still be essential for optimal growth, even in vivo, perhaps through the IR (*see above*). Alternatively, it may indicate that, in vivo, especially in the adult animal, the main function of the IGF-IR may be with transformation and survival of cells, rather than with normal growth.

Although signaling from the IGF-IR is discussed in another chapter, it should be mentioned here that the overwhelming evidence is that the IRS proteins (especially IRS-1) and Shc proteins are necessary for mitogenesis *(48–51)*. However, growing evidence indicates that the ras pathway that originates from these substrates may be required, but is not sufficient for IGF-I-mediated mitogenesis, as an overexpressed, activated ras fails to induce growth in R^- cells *(24)*.

THE IGF-I RECEPTOR AND TRANSFORMATION

Transforming Activity

Overexpression and/or constitutive activation of IGF-IR in a variety of cell types leads to ligand-dependent growth in serum-free medium and to the establishment of a transformed phenotype, i.e., ability to form colonies in soft agar and/or to produce tumors in mice *(24,52–56)*. However, a large number of overexpressed gene products can transform cells, including proto-oncogenes, activated cellular oncogenes, signal transducing molecules, and glycolytic enzymes; in fact, transformation can be considered as a common outcome of the overexpression of gene products. Growth factor receptors are no exception. What makes IGF-IR different from other growth factor receptors is that R^- cells (i.e., mouse embryo cells with a targeted disruption of the IGF-IR genes) are refractory to transformation by certain viral and cellular oncogenes that readily transform mouse embryo cells with a physiological number of IGF-IRs, such as 3T3 cells of various derivations. The list of oncogenes that fail to transform R^- cells include the SV40 large T antigen *(30)*, an activated ras or a combination of T antigen and ras *(24)*, the bovine papilloma virus E5 protein *(57)*, an activated c-*src* *(58)*, and overexpressed growth factor receptors, such as the EGF receptor *(56)*, the PDGF receptor *(18)*, and the insulin receptor *(59)*. R^- cells do transform spontaneously, but at a lower rate than other 3T3 cells. The fact that R^- cells occasionally can transform spontaneously should not be surprising, because rodent cells have a propensity to chromosomal rearrangements and mutations. Any gain-of-function mutation in a signal transducing molecule downstream of the IGF-IR would bypass the IGF-IR requirement for transformation. Indeed, the oncogene v-*src* seems to do that, because it can transform R^- cells *(58)*, while an activated c-*src* fails to do so: both the activated c-*src* and v-*src* cause tyrosyl phosphorylation of several substrates in R^- cells, but only v-*src* seems to be capable of inducing tyrosyl phosphorylation of the protein p130cas.

Table 2
Inhibition of Tumor Growth and Metastases by Down-Regulation of the IGF System

Strategy	*Cell Type*	*Reference*
Antisense	Human melanoma	*63*
	Human rhabdomyosarcoma	*64*
	Human lung cancer cells	*66*
	Rat glioblastoma	*61,76*
	Rat rhabdomyosarcoma	*76*
	Murine melanoma	*77*
	Hamster mesothelioma	*67*
	Murine lung carcinoma	*75*
Antibodies	Human breast carcinoma	*68*
	Human rhabdomyosarcoma	*70*
Dominant negatives	Rat cells	*71*
	Rat glioblastoma	*74*

Reversal of the transformed phenotype refers to colony formation in soft agar, and/or inhibition of tumorigenicity in syngeneic animals or nude mice, and inhibition of metastases.

Another property of the IGF-IR (shared, however, with other growth factor receptors) is that the transformed phenotype can be reversed to a nontransformed phenotype in a variety of tumor cell lines by decreasing the number of IGF-IRs, or by interfering with its function. Different approaches have been used, including antisense expression plasmids or antisense oligodeoxynucleotides against either IGF-I *(60,61),* or the IGF-IR *(62–67),* antibodies to the IGF-IR *(68–70),* and dominant negative mutants of the IGF-IR *(71–74).* All these procedures can reverse the transformed phenotype, and/or inhibit tumorigenesis, and induce loss of the metastatic phenotype *(75).* These findings are summarized in Table 2. The list is incomplete, but it clearly shows that targeting of the IGF-IR can reverse the transformed phenotype in several types of tumor cells, from humans and rodents. As usual, one should be careful not to generalize. Although the IGF-IR may have transforming activity, it does not necessarily mean that IGF-I is always transforming. Thus, in the case of granulosa cells of the ovary, IGF-I may actually have the opposite effect, because it stimulates the production of inhibin *(78).* Targeted disruption of the inhibin genes causes ovarian tumors in animals *(79),* suggesting that, in this case, IGF-I acts as an indirect suppressor of tumor formation.

Anchorage-Independence vs Monolayers

There is another intriguing aspect of IGF-IR action: it seems that the receptor, partially dispensable for cells growing in monolayer cultures, becomes quasi-obligatory when the cells are growing in anchorage-independent conditions. Although it has been repeatedly pointed out that the IGF-IR is more important for transformation than for growth in monolayer *(41,80),* we would like to illustrate this with another example, that clearly shows the increased susceptibility to apoptosis of cells in anchorage-independence. Figure 3 shows a number of clones of a human glioblastoma cell line, T98G, stably transfected with a plasmid, MyCF, expressing the C-terminal sequences of the IGF-IR, from proline residue 1223 to the end; other clones were obtained by transfection of T98G with the MyCF plasmid carrying amino acid substitutions, as indicated

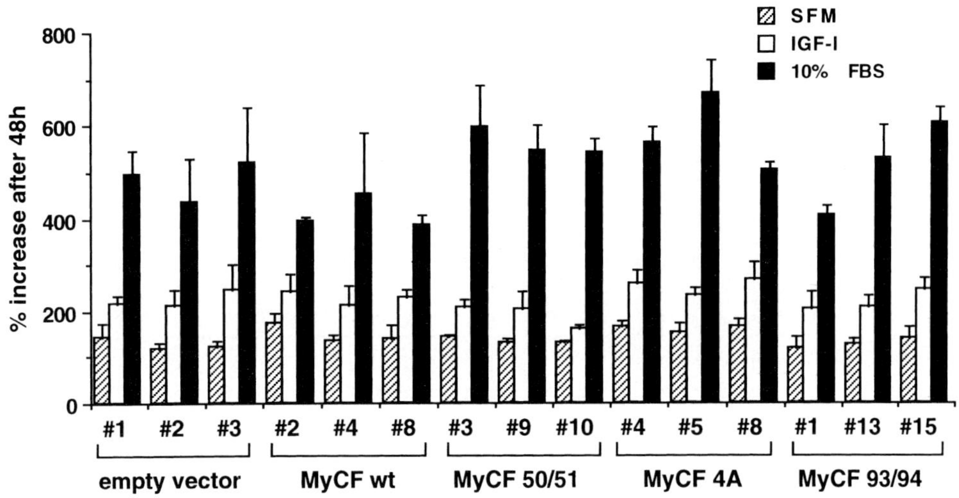

Fig. 3. Growth in monolayer of T98G cells expressing the MyCF plasmids. T98G cells were stably transfected with a plasmid expressing the C-terminus sequence of the IGF-I receptor *(81)*. The sequence, designated as MyCF, was either the wild-type C-terminus sequence of the IGF-I receptor or a mutant sequence. The various mutations included amino acid substitutions of the residues 1293–1294 (MyCF 93/94), or the serine quartet at 1280–1284 (MyCF 4A), or tyrosines 1250/1251 (MyCF 50/51). All sequences carried a myristylation signal for localization to the membrane. The wild-type and mutant sequences were equally expressed in all cell lines (*not shown*). Growth, expressed as percentage increase in cell number in 48 h, was determined 48 h after plating.

in Fig. 3. All the sequences (wild-type or carrying amino acid substitutions) had been fused to a myristylation signal for membrane localization *(81)*. Clones generated from these transfections were tested for their ability to grow in monolayer cultures under three different conditions: serum-free medium (SFM) and SFM supplemented either with IGF-I (50 ng/mL) or 10% fetal calf serum. T98G are known to grow, albeit modestly, in SFM; IGF-I slightly increases their growth rate, and 10% serum gives optimal growth, confirmed in the clones transfected with the empty vector in Fig. 3. Of the various clones expressing either the wild-type MyCF or its mutants, some grow more and some grow less than the parental cell line or the empty vector cells, but this seems to be due essentially to clonal variations rather than the effect of an expressed C-terminus. The important point is that one would find it difficult to separate these clones on the basis of their growth in monolayer cultures. A totally different picture emerges when the same cellular clones are tested in soft agar (Table 3). Clones expressing the wild-type MyCF were markedly inhibited, while the cells expressing the 93/94 MyCF mutant formed a number of colonies that were only slightly fewer than the T98G cells transfected with the empty vector. In these experiments, we concentrated on the clonogenicity of the cells expressing the C-terminus sequences, although these sequences can also induce large-scale apoptosis *(81)*. The point we want to make is that a mutation at residues 1293/1294 of the IGF-IR C-terminus inactivates the inhibitory effect of the expressed myristylated C-terminus sequence on colony formation in soft agar (anchorage-independence), but that this effect could not have been predicted from the results obtained in monolayer cultures.

Table 3
Colony Formation in Soft Agar of T98G Cells and Derivatives

Cell line	*Number of Colonies in Soft Agar After 1 wk*
Empty vector clones	104 (74–109)
wt clones	6.1 (2–9)
50/51 Clones	11.5 (2–19)
4A Clones	15.0 (8–20)
93/94 Clones	86.8 (69–98)

All clones were derived from T98G cells. The cells expressing the wild-type and mutant plasmids are the same as in Fig. 3. The soft agar plates were also checked after 2 wk; the number of colonies was increased in all of the plates, but the proportions were the same as at 1 wk.

Tumor Suppressors

If the IGF-IR is crucial for transformation by a variety of oncogenes, an obvious corollary would be that oncogenes will up-regulate and anti-oncogenes down-regulate the IGF system. This is indeed borne out by the literature. The IGF autocrine or paracrine loop is up-regulated by other growth factors, such as basic PDGF, basic fibroblast growth factor (bFGF), and EGF *(82–84),* hormones such as growth hormone and estrogens, and oncogenes such as SV40 T antigen *(85),* c-*myb (86),* the Ewing's sarcoma-Wilms' tumor 1 fusion protein responsible for desmoplastic small round cell tumor *(87),* and the hepatitis B virus *(88).*

Conversely, tumor suppressor genes such as *WT1 (89)* and *p53 (90,91),* as well as interferon *(92),* cause a decrease in IGF-IR levels through a repression of transcription from the IGF-IR promoter. *p53* has other connections to the IGF system: it represses transcription from the IGF-II promoter *(93)* and from the insulin receptor promoter *(94).* It also induces the expression of the IGF-binding protein 3 *(95)* that antagonizes the effects of IGF-I and IGF-II. The fact that the IGF-IR promotes cell division and transformation and inhibits apoptosis seems to invite its down-regulation by genes that suppress growth and favor apoptosis.

The fact that oncogenes may up-regulate the IGF system and tumor suppressor genes down-regulate it is only part of a body of rapidly increasing evidence that cellular and viral oncogenes and anti-oncogenes may act through the regulation of growth factors and their receptors. This concept was first introduced with the findings that cellular oncogenes such as c-*myb (86)* and viral oncogenes such as SV40 T antigen *(85)* induced the expression and the secretion of IGF-I. These findings indicated that the decrease in growth factor requirements of transformed cells can be simply explained by an increased production of endogenous growth factors. This has been discussed in a review by Baserga *(80),* and more extensively and more recently in a review by Drummond-Barbosa and DiMaio *(96),* who coined the expression "virocrine transformation," to indicate the importance of growth factor regulation in the transformation by viral and cellular oncogenes.

THE IGF-I RECEPTOR IN APOPTOSIS

The IGF-IR activated by its ligands is emerging as a powerful inhibitor of apoptosis induced by a variety of agents. Thus, addition of IGF-I inhibits apoptosis induced by IL-3 withdrawal *(53,97),* c-*myc* overexpression *(98),* ICE proteases expression *(99),* serum withdrawal *(100),* anticancer drugs *(101),* and transforming growth factor-β (TGF-β) *(102).* An overexpressed and activated IGF-IR protects cells from apoptosis induced by etoposide *(103),* IL-3 withdrawal *(53,91,104),* osmotic shock *(105),* tumor necrosis factor *(106), p53 (91),* ionizing and nonionizing radiations *(107–109),* and okadaic acid *(110).* Conversely, down-regulation of the IGF-IR function, either by antisense strategies *(75–77,111),* by dominant negatives *(71,74),* or by triple-helix formation *(112)* causes apoptosis of tumor cells in vivo and/or abrogation of tumorigenesis and metastases. The variety of the procedures used to induce apoptosis suggests that the overexpressed wild-type IGF-IR may have a widespread anti-apoptotic effect.

While targeting of the IGF-IR may cause apoptosis of cells even in monolayer cultures, again its effect is more dramatic when the cells are in anchorage-independence conditions. Untransformed mammalian cells are dependent on adhesion to a substratum, usually referred to as the extracellular matrix (ECM), for survival. When the cells are detached from the ECM, they undergo programmed cell death *(113–115).* In contrast, transformed cells are generally capable of growing in the absence of contact with the ECM, a property that is usually referred to as anchorage-independence *(116).* The ability of tumor cells to grow in the absence of contact with the ECM should not be considered as just an artifact of cells in culture: anchorage-independence correlates quite well with tumorigenicity *(117),* and is probably the property that allows tumor cells in the animal (including humans) to infiltrate surrounding tissues (local recurrences) and to establish distant metastases.

While tumor cells can grow in the absence of contact with the ECM, they also undergo apoptosis if their anchorage independence is perturbed by agents that have little effect on their growth in monolayer *(118,119).* Similarly, the anti-apoptotic effect of the activated IGF-IR is much more dramatic under anchorage-independence conditions than on cells in monolayers *(74,76,77,120).* This brings out an important consideration with respect to the specificity of IGF-IR targeting: if the functional down-regulation of the IGF-IR is more effective in inducing apoptosis in anchorage-independent cells than when cells are growing in monolayer, then it is possible that the targeting of the IGF-IR may discriminate between normal (anchorage-dependent) and tumor (anchorage-independent) cells. Experiments in animals have confirmed this prediction *(41).* Extrapolation to human tumors would be, at the moment of writing, premature.

DOMAINS OF THE IGF-I RECEPTOR β-SUBUNIT

Extensive mutational analyses have been carried out to determine the domains of the IGF-IR that are required for IGF-I-mediated mitogenesis, for transformation, and for protection from apoptosis. These data are summarized in Table 4. Although some mutations have the same effect as mutations on the corresponding residues of the IR, there are also important differences. Before proceeding to an analysis of the various IGF-IR mutants, it is important to establish some cardinal rules.

In the first place, mitogenicity means ability to confer to cells the capacity to respond with mitogenesis to IGF-I (IGF-I-mediated mitogenesis). We shall never tire to repeat that in 10% serum, in monolayer cultures, the IGF-IR is not required for growth, although it may be required for optimal growth *(24,30).* There is a factor (or factors) in serum that stimulates cell proliferation in the absence of IGF-IRs, and, almost certainly, of the receptors for other known growth factors. Second, transformation is assayed, usually, in

Table 4
Summary of the Mitogenic, Transforming, and Anti-apoptotic Activities of the IGF-I Receptor and its Mutants

Receptor	*Mitogenicity*	*Transformation*	*Anti-apoptosis*
wt	+++	+++	+++
d1229	+++	–	++++
d1245	+++	–	++++
d1270	ND	+++	ND
d1293	+++	+	+
d1310	+++	+++	+++
K 1003–	–	–	–
Y950F	–	–	+
Y3F	–	–	+
Y1250F	+++	+++	+++
Y1251F	+++	–	–
S1280-84S	+++	–	++++
1293/94	+++	+	++
Y1316F	+++	+++	++++

Mitogenicity is the ability to make cells grow in SFM supplemented solely with IGF-I; transforming activity is the ability to make cells form colonies in soft agar; and protection from apoptosis as protection from a variety of apoptotic injuries (see text for references). ND, not done. d means deletion of the aminoacid sequence C-terminal to the indicated residue. The – sign means a complete absence of that activity, the + signs are a rough quantitative estimate of activities.

soft agar, where the cells need 10% serum to form colonies. True, there have been reports of some cell lines making colonies in soft agar in SFM; although we have had an extensive experience in our laboratory with colony formation in soft agar, we have never been able to obtain colonies with any cell line in SFM. Perhaps some cell lines produce large amounts of growth factors that may eliminate the need for 10% serum, but we believe that, in the great majority of cases, serum addition is required, which complicates things. As for apoptosis, one should always keep in mind that the IGF-IR may protect cells from a variety of apoptotic injuries, but not all.

With this in mind, Table 4 (in which the functions of the wild-type IGF-IR are given as – or +) offers some very clear conclusions, and some less clear results. The first general conclusion is that the three functions can be mapped to different domains of the β-subunit. A second general conclusion is that some domains have partial functions, i.e., a mutation in that domain causes a decrease but not an abrogation of a function, be it mitogenesis, transformation, or protection from apoptosis. The C-terminus stands out: it is dispensable for mitogenesis, but is required for transformation *(121)*, and its elimination actually may increase the anti-apoptotic effect of the IGF-IR. This finding, incidentally, is the best example that transforming activity does not necessarily correspond to protection from apoptosis. They may be going hand in hand, sometimes; but they can also be dissociated. Within the C-terminus, there are several residues that play an important part in its function, and that probably act through pathways that have not yet been identified. Although signaling (as we have previously mentioned) is left to another contributor, we would like to point out that the IGF-IR protects from apoptosis even cells that do not have IRS-1 or IRS-2, such as 32D cells *(91)*, although these substrates may transmit an anti-apoptotic signal *(107,122)* clearly, it is not the only one originating from the IGF-IR *(91,100)*.

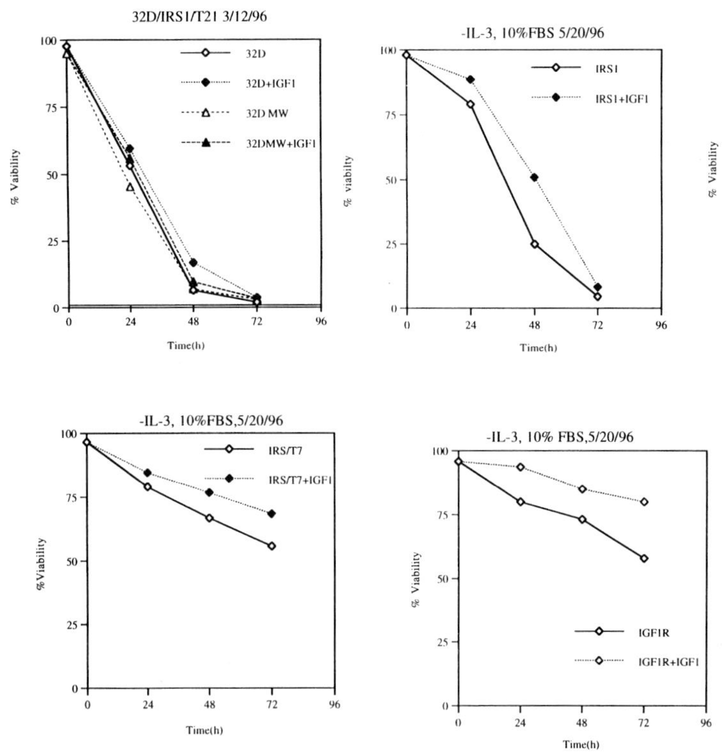

Fig. 4. Effect of IRS-1 and T antigen on survival on 32D cells. 32D cells and transfectants were grown in IL-3 containing medium, then they were carefully washed and cultured in 10% serum without IL-3 for the times indicated in the abscissa. Percentage of viable cells (as percent of plated cells) was determined by standard methods. IRS-1 means 32D cells overexpressing IRS-1, T indicates cells expressing the Simian virus large T antigen, and IRS/T cells expressing both (the numbers refer to clone numbers). IGF-IR indicates cells overexpressing the IGF-I receptor (see *ref. 123* for details).

Curiously, IRS-1, *per se,* has only a limited protective effect against apoptosis, and the SV40 large T antigen not at all, but, in combination, they are as effective as the IGF-IR itself *(123)*. An illustration of this finding is given in Fig. 4, where the cells used were 32D cells and transfectants, and apoptosis was induced by IL-3 withdrawal. Like the parent 32D cells, transfectants expressing only the T antigen or only IRS-1 died when IL-3 was withdrawn, although the latter cells were partially protected. Cells expressing both survived under these conditions, almost as well as 32D cells overexpressing the IGF-IR. These observations raise the question whether the SV40 T antigen mimics the IRS-1-independent pathway.

ROLE OF THE IGF-I RECEPTORS NUMBER

Background

There are distinct differences between what an overexpressed growth factor receptor can do and what the same receptor at physiological levels can do. This is true of almost any growth factor receptor: obviously, growth factor receptors, at wild-type levels, are nontransforming, because they are present in every cell, and, if they were transforming,

Table 5
IGF-I Receptor Number and Cell Behavior

Cell Line	*Receptor Number*	*DNA Synthesis*	*Mitogenicity*	*Transformation*	*Apoptosis*
R–	0	No	No	No	No
R12	3	No	No	No	No
R508	15	Yes	No	No	No
R503	22	Yes	Yes	+	Yes
R600	30	Yes	Yes	+	Yes
p6	500	Yes	Yes	++++	Yes

Receptor number *(124)* is $\times 10^3$. DNA synthesis and mitogenicity refer to the ability of the individual cell lines to enter S phase and/or divide in serum-free medium supplemented solely with IGF-I. Transformation is ability to form colonies in soft agar: a single + means that the cell line requires supplementation with IGF-I. Apoptosis is resistance to okadaic acid-induced apoptosis *(110)* or to forced anchorage-independence (*unpublished data*).

all the cells would be transformed. Yet, these same receptors, when overexpressed, transform cells (*see 56* for references). This is an intriguing phenomenon, because it implies that transformation is simply increased mitogenic stimulus, or, alternatively, increased protection from apoptosis in anchorage-independent conditions. On the other hand, some investigators have simply discarded this observation, because, in most instances, the growth factor receptors were grossly overexpressed, inviting the criticism that, under these conditions, the physiology of the cell is completely perturbed. However, careful observation shows that even small increases in growth factor receptor number can cause profound alterations in biological responses *(124).* To define the role of receptor numbers in IGF-I-mediated mitogenesis, Rubini et al. *(124)* used R^- cells (which have no endogenous IGF-IRs) and transfected them with a human IGF-IR cDNA under the control of a rat IGF-IR promoter, obtaining a large number of clones with different numbers of IGF-IRs. All these clones were derived from a single parental cell line, and had therefore a similar genetic background. From the many clones that were tested, some conclusions could be drawn: (1) all clones with fewer than 15×10^3 receptors per cell did not grow in SFM supplemented solely with IGF-I, or with IGF-I and PDGF; (2) all clones with 22×10^3 or more receptors per cell, grew in IGF-I only; (3) a clone, R508, with 15×10^3 receptors per cell did not grow in IGF-I only, but grew in IGF-I plus PDGF, when its receptor number had increased to 23×10^3; (4) 22×10^3 receptors per cell are also sufficient for growth in soft agar, although only after supplementation of 10% serum with IGF-I; and (5) the same number of receptors, 22×10^3, is also sufficient to protect cells from okadaic acid-induced apoptosis *(110).* In Table 5, we have taken clones representative of these four classes and have summarized their behavior in terms of mitogenic response to IGF-I, colony formation in soft agar, and resistance to apoptosis. These results are critically important in more than one way: (1) they show that small increments in IGF-IR number (a 50% increase) can change the mitogenic response to IGF-I from negative to positive; (2) there is a good correlation between receptor number on one side and mitogenicity (to IGF-I), transforming activity, and protection from apoptosis on the other; and (3) they explain the cooperation between IGF-I and other growth factors such as PDGF, EGF, estrogens, and bFGF, all of which induce an increase in the number of IGF-I receptors *(41).* These increases are almost invariably small, in the order of two- to

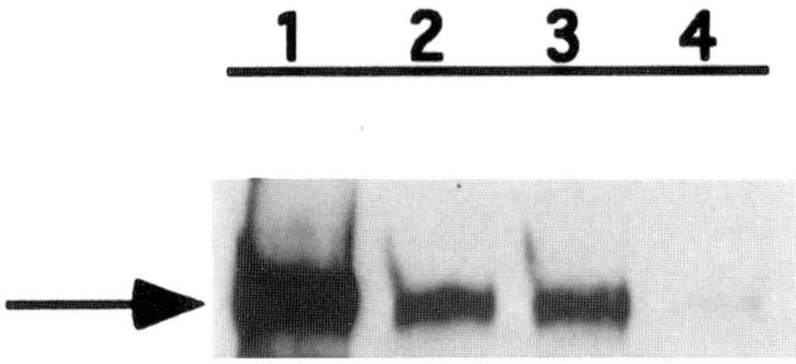

Fig. 5. IGF-I receptor levels in selected cell lines. Whole cell lysates were immunoprecipitated, and the gels were blotted with an antibody to the β-subunit of the IGF-I receptor. *Lanes:* 1, p6 cells (5×10^5 receptors per cell); 2, R600 cells (30×10^3 receptors per cell); 3, R508 cells (15×10^3 receptors); 4, CaOV3 cells (3000 receptors per cell). *Arrow* indicates β-subunit; to visualize the receptors of other cell lines, the p6 lane is grossly overexposed.

threefold, but we now know that they are sufficient to induce a change in the biological response to IGF-I.

Illustration

As an illustration, I have chosen four cell lines that have different numbers of IGF-I receptors (Fig. 5). By Scatchard analysis, the number of IGF-I receptors per cell ($\times 10^3$) are: 500 for p6 cells, 30 for R600 cells, and 22 for R503 cells. With CaOV3 cells, a Scatchard analysis is not practical, because these cells secrete very large amounts of IGF binding proteins. But the Western blot of Fig. 5 clearly indicates that CaOV3 cells have a very low level of IGF-IRs, approx. $4–5 \times 10^3$ receptors per cell. These cells all grow in 10% serum in monolayers, where the role of the IGF-I receptor is optional *(30);* but, when these cells are exposed to 10 n*M* okadaic acid, there is a difference in sensitivity. CaOV3 cells (with the lowest levels of IGF-I receptors) actually decrease in number, while the other three cell lines, although partially inhibited, are still growing (Table 6). The influence of the receptor number becomes even more accentuated if the cells are tested for colony formation in soft agar (a reasonable measure of transformation), as clearly shown in Table 6. For instance, at 25 n*M* okadaic acid, R503 no longer form colonies, while R600 cells still form a few colonies. The difference in receptor number between these two cell lines is modest: R600 cells have about 30% more IGF-I receptors than R503 cells. Thus, differences can be seen even at levels of receptor overexpression that are within physiological ranges. Similar results were obtained when these and other cell lines were tested in monolayers for their resistance to apoptosis induced by okadaic acid: R503 cells, for example, were resistant to apoptosis induced by 20–25 n*M* okadaic acid, but R508 cells were not *(110).* Therefore, the functions of the IGF-IR depend not only on precise domains of the receptor, but also on the number of receptors, and the change in number does not have to be of a large order of magnitude to cause significant differences. The results of Rubini et al. *(124)* also explain another puzzle. G/0 mouse embryo cells (unresponsive to IGF-I only) do have functional IGF-I receptors, which can be autophosphorylated and can transmit a signal that is not mitogenic, but induces growth in size *(125)* and the expression of several genes *(22,126,127).* To transmit a mitogenic signal, a modest but reproducible increase in IGF-IR number is necessary and sufficient. The question at this point is a question that has vexed cell and molecular biologists for a long time: does overexpression simply increase the length or the intensity of the same stimulus, or does

Table 6
Growth of Selected Cell Lines in Monolayers and in Soft Agar

	Cell Number × 10^3	
Cell Line	*10% Serum*	*10% Serum Plus* 10 *nM* okadaic acid
p6	71.4	50.2
R600	54.9	42.5
R503	52.8	40.7
CaOV3	7.34	0.8

All cells were plated at a density of 10 × 10^3 cells per dish, and the number of cells were counted after 3 d. For the levels of IGF-I receptors, *see* Fig. 5.

	Number of Colonies in Soft Agar			
Treatment	*p6*	*R600*	*R503*	*CaOV3*
10% Serum	192	83	43	93
plus 10 n*M* okadaic acid	112	17	6	0
plus 25 n*M* okadaic acid	8	2	0	0
plus 30 n*M* okadaic acid	1	0	0	0

The number of colonies in soft agar was determined after 3 wk.

it open new pathways? Obviously, at some point, a new pathway is open, as some cells will not grow and others will, but the subtler question is whether the activation of the DNA synthesizing machinery and the progression to mitosis are due to a prolonged stimulation of the same signaling pathway of the receptors, or to a separate pathway.

CONCLUSIONS

Our main conclusions can be summarized as follows.

1. The IGF-IR regulates growth, either in vivo or in vitro, in four different ways. It stimulates cell growth (size), and cell division. It is quasi-obligatory for transformation, and it protects cells from apoptosis. These four functions all lead to an increase in the size of a cell population.

2. The IR has probably an extremely limited mitogenic activity, where mitogenesis is meant to be as defined in this chapter, i.e. actual cell division and not thymidine incorporation. The IR, however, may become important when IGF-II is present and abundant, especially if the receptor is overexpressed.

3. The proliferative effects of the IGF-IR are determined by certain specific domains, but are also exquisitely dependent on the number of receptors. For mitogenesis, the Y950 and the tyrosine kinase domain seem to be, if not essential, certainly very important, while the C-terminus is dispensable. The question of whether an IGF-IR is mitogenic in the absence of autophosphorylation remains unresolved.

The number of receptors is crucial. We are not referring to absolute values, as different cells may have different abilities to amplify a signal originating from the IGF-IR. But, given a certain type of cell, the receptor number determines whether the cell will not respond at all to IGF-I, will respond only in cooperation with other growth factors, or will respond with full mitogenesis to IGF-I only. The differences in receptor number do not have to be extravagant: mouse embryo fibroblasts with 15×10^3 IGF-IRs per cell do not grow in IGF-I, but 22×10^3 receptors per cell are sufficient for IGF-I-mediated mitogenesis.

4. The same conclusions apply to the transforming activity of the IGF-IR. However, at variance with mitogenicity, the presence of a C-terminus is a requisite for transformation, and several residues in the C-terminus contribute to its transforming ability. Again, the number of receptors is important; judged by the ability of forming colonies in soft agar, 30×10^3 receptors per cell are sufficient for transformation.

5. As to the anti-apoptotic activity of the IGF-IR, surprisingly, it seems to correlate more with its mitogenicity than with its transforming ability. In fact, the C-terminus (required for transformation) is not only dispensable for protection from apoptosis, but it may actually send a pro-apoptotic signal that is usually neutralized by the rest of the receptor.

6. These separate domains and the different responses all indicate that there are other pathways, in addition to the well established pathway that from IRS-1 and Shc leads to ras and eventually to the nucleus. It is a ras-independent pathway *(24),* it could be a totally unknown one, or it could branch out at a certain point from the better known pathway. The identification of this or these pathways is one of the most interesting topics in the biology of IGF-I-mediated growth.

7. The IGF-IR seems to be a good target for anti-cancer therapy, because its targeting induces apoptosis, a host response, and has only modest effects on normal cells *(41).* It should be a good target theoretically, and it actually is so in experimental animals bearing transplantable tumors. Whether the promising results obtained in animals are valid for humans is a question that does not yet have an answer. Targeting the IGF-IR is more efficient than targeting its ligands. The adult rodent has negligible plasma levels of IGF-II, but other adult animals, including humans, have substantial circulating concentrations of both IGFs *(128).* Targeting one ligand leaves the other one free to activate the IGF-IR.

All these considerations make the IGF-IR an attractive subject of investigation, both at the basic and applied levels. Only a few years ago, the IGF-IR was the "poor relative" of the IR, a kind of redundant receptor that did not work as well as its more famous cousin. It is now generally accepted that the IGF-IR has an important role of its own, a role that has more to do with cell proliferation and cell survival than with metabolism. The elucidation of the mechanism by which the IGF-IR regulates the size of cell populations ought to be a very fertile area of investigations in the near future.

REFERENCES

1. Baserga R. The Biology of Cell Reproduction. 1985. Harvard University Press, Cambridge, MA.
2. Mendelsohn ML. Autoradiographic analysis of cell proliferation in spontaneous breast cancer in C3H mouse. III. The growth fraction. J Natl Cancer Inst 1962; 28:1015–1029.
3. Baserga R, Kisieleski W. Comparative study of the kinetics of cellular proliferation of normal and tumorous tissues with the use of tritiated thymidine. J Natl Cancer Inst 1962; 28:331–339.
4. Steel GG, Adams K, Barrett JC. Analysis of the cell population kinetics of transplated tumours of widely differing growth rates. Br J Cancer 1966; 20:784–800.
5. Bresciani F, Paoluzzi R, BenAssi M, Nervi C, Casale C, Ziparo E. Cell kinetics and growth of squamous cell carcinomas in man. Cancer Res 1974; 34:2405–2415.

6. Rubin R, Baserga R. Biology of disease. Insulin-like growth factor I receptor. Lab Invest 1995; 73:311–331.
7. Clark R. The somatogenic hormones and insulin-like growth factor I: stimulators of lymphopoiesis and immune function. Endocr Rev 1997; 18:157–179.
8. Clemmons DR, Van Wyk JJ. Somatomedin: physiological control and effects on cell proliferation. In Baserga, R, ed. Tissue Growth Factors. Springer-Verlag, Berlin, 1981, pp. 161–208.
9. Scher CD, Shephard RC, Antoniades HN, Stiles CD. Platelet derived growth factor and the regulation of the mammalian fibroblasts cell cycles. Biochim Biophys Acta 1979; 560:217–241.
10. Stiles CD, Capone GT, Scher CD, Antoniades NH, Van Wyk JJ, Pledger WJ. Dual control of cell growth by somatomedins and platelet derived growth factor. Proc Natl Acad Sci USA 1979; 76:1279–1283.
11. Leof EB, Wharton W, Van Wyk JJ, Pledger WJ. Epidermal growth factor EGF and somatomedin C regulated G1 progression in competent BALB/c3T3 cells. Exp Cell Res 1982; 141:107–115.
12. Campisi J, Morreo G, Pardee AB. Kinetics of G1 transit following brief starvation for serum factors. Exp Cell Res 1984; 152:459–466.
13. Flier JS, Usher P, Moses AC. Monoclonal antibody to the type 1 insulin-like growth factor (IGF-I) receptor blocks IGF-I receptor-mediated DNA synthesis: clarification of the mitogenic mechanisms of IGF-I and insulin in human skin fibroblasts. Proc Natl Acad Sci USA 1986; 83:664–668.
14. Campisi J, Pardee AB. Posttranscriptional control of the onset of DNA synthesis by an insulin-like growth factor. Mol Cell Biol 1984; 4:1807–1814.
15. Fischer B, Schluter G, Adler CP, Sandritter W. Zytophometrische DNS, histon-, und Nicht-histonprotein bestimmungen an zell kernen von menschlischen herzen. Beitr Path 1970; 141:238–26061.
16. Gelfant S. A new concept of tissue and tumor cell proliferation. Cancer Res 1977; 37:3845–3862.
17. Fukuda M, Sibatani A. Biochemical studies on the number and the composition of liver cells in postnatal growth of the rat. J Biochem 1953; 40:95–110.
18. DeAngelis T, Ferber A, Baserga R. The insulin-like growth factor I receptor is required for the mitogenic and transforming activities of the platelet-derived growth factor receptor. J Cell Physiol 1995; 164:214–221.
19. Pardee A, Dubrow R, Hamlin JL, Kletzien RF. Animal cell cycle. Annu Rev Biochem 1978; 47:715–750.
20. Walters RA, Tobey RA, Ratcliff RL. Cell cycle-dependent variations of deoxyribonucleoside triphosphate pools in Chinese hamster cells. Biochim Biophys Acta 1973; 319:336–347.
21. Karl PJ. Insulin-like growth factor 1 stimulates amino acid uptake by the cultured human placental trophoblast. J Cell Physiol 1995; 165:83–88.
22. Surmacz E, Kaczmarek L, Ronning O, Baserga R. Activation of the ribosomal DNA promoter in cells exposed to insulin-like growth factor 1. Mol Cell Biol 1987; 7:657–663.
23. Reiss K, Cheng W, Pierzchalski P, Kodali S, Li B, Wang S, Liu Y, Anversa P. Insulin-like growth factor 1 receptor and its ligand regulate the reentry of adult ventricular myocytes into the cell cycle. Exp Cell Res 1997; 235:198–209.
24. Sell C, Dumenil G, Deveaud C, Miura M, Coppola D, DeAngelis T, Rubin R, Efstratiadis A, Baserga R. Effect of a null mutation of the type 1 IGF receptor gene on growth and transformation of mouse embryo fibroblasts. Mol Cell Biol 1994; 14:3604–3612.
25. Valentinis B, Porcu P, Quinn K, Baserga R. The role of the insulin-like growth factor I receptor in the transformation by simian virus-40 T antigen. Oncogene 1994; 9:825–831.
26. Perdue JF, LeBon TR, Kato J, Hampton B, Fujita-Yamaguchi Y. Binding specificities and transducing function of the different molecular weight forms of insulin-like growth factor II (IGF-II) on IGF-I receptors. Endocrinology 1991; 129:3101–3108.
27. Lowe WL Jr. Biological actions of the insulin-like growth factors. In: LeRoith D, ed. Insulin-Like Growth Factors: Molecular and Cellular Aspects. CRC Press, Boca Raton, FL, 1991, pp. 49–85.
28. Nissley P, Kiess W, Sklar MM. The insulin-like growth factor/mannose-6-phosphate receptor. In: LeRoith D, ed. Insulin-Like Growth Factors: Molecular and Cellular Aspects. CRC Press, Boca Raton, FL, 1991, pp. 111–150.
29. Ludwig T, Eggenschwiler J, Fisher P, D'Ercole JP, Davenport ML, Efstratiadis A. Mouse mutants lacking the type 2 IGF receptor (IGF2R) are rescued from perinatal lethality in Igf2 and Igflr null backgrounds. Dev Biol 1996; 177:517–535.
30. Sell C, Rubini M, Rubin R, Liu J-P, Efstratiadis A, Baserga R. Simian virus 40 large tumor antigen is unable to transform mouse embryonic fibroblasts lacking type-1 IGF receptor. Proc Natl Acad Sci USA 1993; 90:11217–11221.
31. Morrione A, Valentinis B, Xu S-Q, Yumet G, Louvi A, Efstratiadis A, Baserga R. Insulin-like growth

factor II stimulates cell proliferation through the insulin receptor. Proc Natl Acad Sci USA 1977; 94:3777–3782.
32. Moses AC, Tsuzaki S. Is insulin a growth factor? In: LeRoith D, ed. Insulin-Like Growth Factors: Molecular and Cellular Aspects. CRC Press, Boca Raton, FL, 1991, pp. 245–270.
33. Liu J-P, Baker J, Perkins AS, Robertson EJ, Efstratiadis A. Mice carrying null mutations of the genes encoding insulin-like growth factor I (igf-1) and type 1 IGF receptor (Igflr). Cell 1993; 75:59–72.
34. Baker J, Liu J-P, Robertson EJ, Efstratiadis A. Role of insulin-like growth factors in embryonic and postnatal growth. Cell 1993; 75:73–82.
35. Sung CK, Maddux BA, Hawley DM, Goldfine ID. Monoclonal antibodies mimic insulin activation of ribosomal protein S6 kinase without activation of insulin receptor tyrosine kinase. J Biol Chem 1989; 264:18951–18959.
36. Lichtner RB, Wiedemuth M, Kittmann A, Ullrich A, Schirrmacher V, Khazaie K. Ligand-induced activation of epidermal growth factor receptor in intact rat mammary adenocarcinoma cells without detectable receptor phosphorylation. J Biol Chem 1992; 267:11872–11880.
37. Blakesley VA, Kato H, Roberts CT Jr, LeRoith D. Mutation of a conserved amino acid residue (tryptophan 1173) in the tyrosine kinase domain of the IGF-I receptor abolishes autophosphorylation but does not eliminate biological function. J Biol Chem 1995; 270:2764–2769.
38. Shier P, Watt VM. Primary structure of a putative receptor for a ligand of the insulin family. J Biol Chem 1989; 264:14605–14608.
39. Zhang B, Roth RA. The insulin receptor-related receptor. J Biol Chem 1992; 267:18320–18328.
40. Chen J, Hanafusa T, Wang LH. Ala Gly mutation in the putative catalytic loop confers temperature sensitivity on ros, insulin receptor and insulin-like growth factor I receptor protein-tyrosine kinases. Proc Natl Acad Sci USA 1994; 91:321–325.
41. Baserga R, Hongo A, Rubini M, Prisco M, Valentinis B. The IGF-I receptor in cell growth, transformation and apoptosis. Biochim Biophys Acta 1997; 1332:105–126.
42. Eigenman JE, Amador A, Patterson DF. Insulin-like growth factor I levels in proportionate dogs, chondrodystrophic dogs and in giant dogs. Acta Endocrinol 1988; 118:105–108.
43. Deeb B, Wolf N. Studying longevity and mortality in giant and small breeds of dogs. Vet Med 1994; 89:702–713.
44. Mathews LS, Hammer RE, Behringer RR, D'Ercole AJ, Bell GI, Brinster RL, Palmiter RD. Growth enhancement of transgenic mice expressing human insulin-like growth factor I. Endocrinology 1988; 123:2827–2833.
45. Reiss K, Cheng W, Ferber A, Kajstura J, Li P, Li B, Olivetti G, Homcy CJ, Baserga R, Anversa P. Overexpression of insulin-like growth factor I in the heart is coupled with myocyte proliferation in transgenic mice. Proc Natl Acad Sci USA 1996; 93:8630–8635.
46. Bol DK, Kiguchi K, Gimenez-Conti I, Rupp T, DiGiovanni J. Overexpression of insulin-like growth factor I induces hyperplasia, dermal abnormalities and spontaneous tumor formation in transgenic mice. Oncogene 1997; 14:1725–1734.
47. Morrione A, Valentinis B, Resnicoff M, Xu S-Q, Baserga R. The role of mGrb10 in insulin-like growth factor I-mediated growth. J Biol Chem 1997; 272:26382–26387.
48. Myers MJ, Sun XJ, Cheatham B, Jachna BR, Glasheen EM, Backer JM, White MF. IRS-1 is a common element in insulin and insulin-like growth factor-I signaling to the phosphatidylinositol 3′-kinase. Endocrinol 1993; 132:1421–1430.
49. Myers MG Jr, Grammer TC, Wang LM, Sun XJ, Pierce JH, Blenis J, White MF. Insulin receptor substrate-1 mediates phosphatidykinositol 3′-kinase and p70S6K signaling during insulin, insulin-like growth factor-I and interleukin-4 stimulation. J Biol Chem 1994; 269:28783–28789.
50. Waters SB, Yamauchi K, Pessin JE. Functional expression of insulin receptor substrate-1 is required for insulin-stimulated mitogenic signal. J Biol Chem 1993; 268:22231–22234.
51. Rose DW, Saltiel AR, Majumdar M, Decker SJ, Olefsky JM. Insulin receptor substrate 1 is required for insulin-mediated mitogenic signal transduction. Proc Natl Acad Sci USA 1994; 91:797–801.
52. Kaleko M, Rutter WG, Miller AD. Overexpression of the human insulin-like growth factor 1 receptor promotes ligand-dependent neoplastic transformation. Mol Cell Biol 1990; 10:464–473.
53. McCubrey JA, Stillman LS, Mayhew MW, Algate PA, Dellow RA, Kaleko M. Growth promoting effects of insulin-like growth factor 1 (IGF-1) on hematopoietic cells. Overexpression of introduced IGF-1 receptor abrogates interleukin-3 dependency of murine factor dependent cells by ligand dependent mechanism. Blood 1991; 78:921–929.
54. Pietrzkowski Z, Sell C, Lammers R, Ullrich A, Baserga R. Role of insulin-like growth factor I (IGF-I)

and the IGF-I receptor in epidermal growth factor-stimulated growth of 3T3 cells. Mol Cell Biol 1992; 12:3883–3889.
55. Liu D, Zong C, Wang LH. Distinctive effects of the carboxyl-terminal sequence of the insulin-like growth factor I receptor on its signaling functions. J Virol 1993; 67:6835–6840.
56. Coppola D, Ferber A, Miura M, Sell C, D'Ambrosio C, Rubin R, Baserga R. A functional insulin-like growth factor I receptor is required for the mitogenic and transforming activities of the epidermal growth factor receptor. Mol Cell Biol 1994; 14:4588–4595.
57. Morrione A, DeAngelis T, Baserga R. Failure of the bovine papilloma virus to transform mouse embryo fibroblasts with a targeted disruption of the insulin-like growth factor I receptor gene. J Virol 1995; 69:5300–5303.
58. Valentinis B, Morrione A, Taylor SJ, Baserga R. Insulin-like growth factor 1 receptor signaling in transformation by src oncogenes. Mol Cell Biol 1997; 17:3744–3754.
59. Miura M, Surmacz E, Burgaud JL, Baserga R. Different effects on mitogenesis and transformation of a mutation at tyrosine 1251 of the insulin-like growth factor I receptor. J Biol Chem 1995; 270:22639–22644.
60. Trojan J, Blossey BK, Johnson TR, Rudin SD, Tykocinski M, Ilan J, Ilan I. Loss of tumorigenicity of rat glioblastoma directed by episome-based antisense cDNA transcription of insulin-like growth factor 1. Proc Natl Acad Sci USA 1992; 89:4874–4878.
61. Trojan J, Johnson TR, Rudin SD, Ilan J, Tykocinski ML, Ilan I. Treatment and prevention of rat glioblastoma by immunogenic C6 cells expressing antisense insulin-like growth factor I RNA. Science 1993; 256:94–97.
62. Resnicoff M, Coppola D, Sell C, Rubin R, Ferrone S, Baserga R. Growth inhibition of human melanoma cells in nude mice by antisense strategies to the type 1 insulin-like growth factor receptor. Cancer Res 1994; 54:4848–4850.
63. Resnicoff M, Sell C, Rubini M, Coppola D, Ambrose D, Baserga R, Rubin R. Rat glioblastoma cells expressing an antisense RNA to the insulin-like growth factor I (IGF-I) receptor are non-tumorigenic and induce regression of wild-type tumors. Cancer Res 1994; 54:2218–2222.
64. Shapiro DN, Jones BG, Shapiro LH, Dias P, Houghton PJ. Antisense-mediated reduction in insulin-like growth factor 1 receptor expression suppresses the malignant phenotype of a human alveolar rhabdomyosarcoma. J Clin Invest 1994; 94:1235–1242.
65. Neuenschwander S, Roberts CT Jr, LeRoith D. Growth inhibition of MCF-7 breast cancer cells by stable expression of an insulin-like growth factor 1 receptor antisense ribonucleic acid. Endocrinology 1995; 136:4298–4303.
66. Lee CT, Wu S, Gabrivolich D, Chen H, Nadaf-Rahrov S, Ciernick IF, Carbone DP. Antitumor effect of an adenovirus expressing antisense insulin-like growth factor I receptor on human lung cancer cell lines. Cancer Res 1996; 56:3038–3041.
67. Pass HL, Mew DJY, Carbone M, Matthews WA, Domington JS, Baserga R, Walker CL, Resnicoff M, Steinberg SM. Inhibition of hamster mesothelioma tumorigenesis by an antisense expression plasmid to the insulin-like growth factor I receptor. Cancer Res 1996; 56:4044–4048.
68. Arteaga CL. Interference of the IGF system as a strategy to inhibit breast cancer growth. Breast Cancer Res Treat 1992; 22:101–106.
69. Arteaga CL, Kitten LJ, Coronado EB, Jacobs S, Kull FC, Allred DC, Osborne CK. Blockade of the type 1 somatomedin receptor inhibits growth of human breast cancer cells in athymic mice. J Clin Invest 1989; 84:1418–1423.
70. Kalebic T, Tsokos M, Helman LJ. In vivo treatment with antibody against the IGF-I receptor suppresses growth of human rhabdomyosarcoma and down regulates p34/cdc2. Cancer Res 1994; 54:5531–5534.
71. Prager D, Li HL, Asa S, Melmed S. Dominant negative inhibition of tumorigenesis in vivo by human insulin-like growth factor receptor mutant. Proc Natl Acad Sci USA 1994; 91:2181–2185.
72. Li S, Resnicoff M, Baserga R. Effects of mutations at serines 1280–1283 on the mitogenic and transforming activities of the insulin-like growth factor I receptor. J Biol Chem 1996; 271:12254–12260.
73. Burgaud J-L, Resnicoff M, Baserga R. Mutant IGF-I receptors as dominant negatives for growth and transformation. Biophys Biochem Res Commun 1995; 214:475–481.
74. D'Ambrosio C, Ferber A, Resnicoff M, Baserga R. A soluble insulin-like growth factor I receptor that induces apoptosis of tumor cells in vivo and inhibits tumorigenesis. Cancer Res 1996; 56:4013–4020.
75. Long L, Rubin R, Baserga R, Brodt P. Loss of the metastatic phenotype in murine carcinoma cells expressing an antisense RNA to the insulin-like growth factor I receptor. Cancer Res 1995; 55:1006–1009.

76. Resnicoff M, Abraham D, Yutanawiboonchai W, Rotman H, Kajstura J, Rubin R, Zoltick P, Baserga R. The insulin-like growth factor I receptor protects tumor cells from apoptosis in vivo. Cancer Res 1995; 55:2463–2469.
77. Resnicoff M, Burgaud J-L, Rotman HL, Abraham D, Baserga R. Correlation between apoptosis, tumorigenesis and levels of insulin-like growth factor I receptors. Cancer Res 1995; 55:3739–3741.
78. Zhiwen Z, Carson RS, Herington AC, Lee VWK, Burger HG. Follicle-stimulating hormone and somatomedin-C stimulate inhibin production by rat granulosa cells in vitro. Endocrinology 1987; 120:1633–1638.
79. Matzuk MM, Finegold MJ, Su J-GJ, Hsueh AJW, Bradley A. Inhibin is a tumor suppressor gene with gonadal specificity in mice. Nature 1992; 360:313–319.
80. Baserga R. Oncogenes and the strategy of growth factors. Cell 1994; 79:927–930.
81. Liu Y, Lehar S, Corvi C, Payne G, Chittenden T, O'Connor R. Expression of the IGF-I receptor C-terminus as a myristylated protein leads to induction of apoptosis in tumor cells. Cancer Res 1998; 58:570–576.
82. Clemmons DR. Multiple hormones stimulate the production of somatomedin by cultured human fibroblasts. J. Clin Endocrinol Metab 1984; 58:850–856.
83. Rubini M, Werner H, Gandini E, Roberts CT Jr, LeRoith D, Baserga R. Platelet-derived growth factor increases the activity of the promoter of the insulin-like growth factor-1 (IGF-1) receptor gene. Exp Cell Res 1994; 211:374–379.
84. Rosenthal SM, Brown EJ, Brunetti A, Goldfine ID. Fibroblast growth factor inhibits insulin-like growth factor II (IGF-II) gene expression and increases IGF-I receptor abundance in BC3H-1 muscle cells. Mol Endocrinol 1991; 5:678–684.
85. Porcu P, Ferber A, Pietrzkowski Z, Roberts CT, Adamo M, LeRoith D, Baserga R. The growth stimulatory effect of simian virus 40 T antigen requires the interaction of insulin-like growth factor I with its receptor. Mol Cell Biol 1992; 12:3883–3889.
86. Travali S, Reiss K, Ferber A, Petralia S, Mercer WE, Calabretta B, Baserga R. Constitutively expressed c-myb abrogates the requirement for insulin-like growth factor 1 in 3T3 fibroblasts. Mol Cell Biol 1991; 11:731–736.
87. Karnieli E, Werner H, Rauscher FJ III, Benjamin LE, LeRoith D. The IGF-I receptor gene promoter is a molecular target for the Ewing's sarcoma-Wilms' tumor I fusion protein. J Biol Chem 1996; 271:19304–19309.
88. Kim SO, Park JG, Lee YI. Increased expression of the insulin-like growth factor I (IGF-I) receptor gene in hepatocellular carcinoma cell lines: implications of IGF-I receptor gene activation by hepatitis B virus X gene product. Cancer Res 1996; 56:3831–3836.
89. Werner H, Shen-Orr Z, Rauscher FJ, Morris JF, Roberts CT Jr, LeRoith D. Inhibition of cellular proliferation by the Wilms' tumor suppressor WT1 is associated with suppression of insulin-like growth factor I receptor gene expression. Mol Cell Biol 1995; 15:3516–3522.
90. Werner H, Karnieli E, Rauscher FJ III, LeRoith D. Wild type and mutant p53 differentially regulate transcription of insulin-like growth factor I receptor gene. Proc Natl Acad Sci USA 1996; 93:8318–8323.
91. Prisco M, Hongo A, Rizzo MG, Sacchi A, Baserga R. The IGF-I receptor as a physiological relevant target of p53 in apoptosis caused by interleukin-3 withdrawal. Mol Cell Biol 1997; 17:1084–1092.
92. Thulasi R, Dias P, Houghton PJ, Houghton JA. a2a interferon-induced differentiation of human alveolar rhabdomyosarcoma cells: correlation with down-regulation of the insulin-like growth factor type 1 receptor. Cell Growth Diff 1996; 7:531–541.
93. Zhang L, Kashanchi F, Zhan Q, Brady JN, Fornace AJ, Seth P, Helman LJ. Regulation of insulin-like growth factor II P3 promoter by p53: a potential mechanism for tumorigenesis. Cancer Res 1996; 56:1367–1373.
94. Webster NJG, Resnik JL, Reichart DB, Strauss B, Haus M, Seely BL. Repression of the insulin receptor promoter by the tumor suppressor gene product p53: a possible mechanism for receptor overexpression in breast cancer. Cancer Res 1996; 56:2781–2788.
95. Buckbinder L, Talbott R, Velasco-Miguel S, Takenaka I, Faha B, Seizinger BR, Kley N. Induction of the growth inhibitor IGF-binding protein 3 by p53. Nature 1995; 377:646–649.
96. Drummond-Barbosa D, DiMaio D. Virocrine transformation. Biochim Biophys Acta 1997; 1332: M1–M17.
97. Rodriguez-Tarduchy G, Collins MKL, Garcia I, Lopez-Rivas A. Insulin-like growth factor I inhibits apoptosis in IL-3 dependent hemopoietic cells. J Immunol 1992; 149:535–540.
98. Harrington EA, Bennett MR, Fanidi A, Evan GI. c-*myc*-induced apoptosis in fibroblasts is inhibited by specific cytokines. EMBO J 1994; 13:3286–3295.

99. Jung YK, Miura M, Yuan J. Suppression of interleukin-1 beta converting enzyme-mediated cell death by insulin-like growth factor. J Biol Chem 1996; 271:5112–5117.
100. Parrizas M, Saltiel AR, LeRoith D. Insulin-like growth factor I inhibits apoptosis using the phosphatidylinositol 3-kinase and mitogen-activated protein kinase pathways. J Biol Chem 1997; 272:154–161.
101. Dunn SE, Hardman RA, Kari FW, Barrett CJ. Insulin-like growth factor 1 (IGF-I) alters drug sensitivity of HBL100 human breast cancer cells by inhibition of apoptosis induced by diverse anticancer drugs. Cancer Res 1997; 57:2687–2693.
102. Hsing AY, Kadomatsu K, Bonham NJ, Danielpour D. Regulation of apoptosis induced by transforming growth factor-β in nontumorigenic and tumorigenic rat prostatic epithelial cell lines. Cancer Res 1996; 56:5146–5149.
103. Sell C, Baserga R, Rubin R. Insulin-like growth factor I (IGF-I) and the IGF-I receptor prevent etoposide-induced apoptosis. Cancer Res 1995; 55:303–306.
104. O'Connor R, Kauffmann-Zeh A, Liu Y, Lehar S, Evan GI, Baserga R, Blattler WA. The IGF-I receptor domains for protection from apoptosis are distinct from those required for proliferation and transformation. Mol Cell Biol 1997; 17:427–435.
105. Singleton JR, Randolph AE, Feldman EL. Insulin-like growth factor I receptor prevents apoptosis and enhances neuroblastoma tumorigenesis. Cancer Res 1996; 56:4522–4529.
106. Wu Y, Tewari M, Cui S, Rubin R. Activation of the insulin-like growth factor I receptor inhibits tumor necrosis factor-induced cell death. J Cell Physiol 1996; 168:499–509.
107. Kulik G, Klippel A, Weber MJ. Antiapoptotic signaling by the insulin-like growth factor I receptor, phosphatidylinositol 3-kinase, and Akt. Mol Cell Biol 1997; 17:1595–1606.
108. Turner BC, Haffty BG, Narayanan L, Yuan J, Havre PA, Grumbs AA, Kaplan L, Burgaud JL, Carter D, Baserga R, Glazer PM. Insulin-like growth factor 1 receptor overexpression mediates cellular resistance and local breast cancer recurrence after lumpectomy and radiation. Cancer Res 1997; 57:3079–3083.
109. Nakamura S, Watanabe H, Miura M, Sasaki T. Effect of the insulin-like growth factor 1 receptor on ionizing radiation-induced cell death in mouse embryo fibroblasts. Exp Cell Res 1997; 235:287–294.
110. D'Ambrosio C, Valentinis B, Prisco M, Reiss K, Rubini M, Baserga R. Protective effect of the IGF-I receptor on apoptosis induced by okadaic acid. Cancer Res 1997; 57:3264–3271.
111. Valentinis B, Baserga R. The IGF-I receptor protects tumor cells from apoptosis induced by high concentrations of serum. Biochem Biophys Res Commun 1996; 224:362–368.
112. Rininsland F, Johnson TR, Chernicky CL, Schulze E, Burfeind B, Ilan J, Ilan J. Suppression of insulin-like growth factor I receptor by a triple-helix strategy inhibits IGF-I transcription and tumorigenic potential of rat C6 glioblastoma cells. Proc Natl Acad Sci USA 1997; 94:5854–5859.
113. Frisch SM, Francis H. Disruption of epithelial cell-matrix interactions induces apoptosis. J Cell Biol 1994; 124:619–626.
114. Day ML, Foster RG, Day KC, Zhao X, Humphrey P, Swanson P, Postigo AA, Zhang SH, Dean DC. Cell anchorage regulates apoptosis through the retinoblastoma tumor suppressor/E2F pathway. J Biol Chem 1997; 272:8125–8128.
115. Khwaja A, Rodriguez-Viciana P, Wennstrom S, Warne PH, Downward J. Matrix adhesion and ras transformation both activate a phosphoinositide 3-OH kinase and protein kinase B/AAkt cellular survival pathway. EMBO J 1997; 16:2783–2793.
116. Macpherson I, Montagnier L. Agar suspension culture for the selective assay of cells transformed by polyoma virus. Virology 1964; 23:291–294.
117. Aaronson SA, Todaro GJ. Basis for the acquisition of malignant potential by mouse cells cultivated in vitro. Science 1968; 162:1024–1026.
118. Rak J, Mitsuhashi Y, Erdos V, Huang S, Filmus J, Kerbel R. Massive programmed cell death in intestinal epithelial cells induced by three-dimensional growth conditions: suppression of mutant c-H-ras oncogene expression. J Cell Biol 1995; 131:1587–1598.
119. Lebowitz PF, Sakamuro D, Prendergast G. Farnesyl transferase inhibitors induce apoptosis of ras-transformed cells denied substratum attachment. Cancer Res 1997; 57:708–713.
120. Baserga R. The price of independence. Exp Cell Res 1997; 236:1–3.
121. Surmacz E, Sell C, Swantek J, Kato H, Roberts CT Jr, LeRoith D, Baserga R. Dissociation of mitogenesis and transforming activity by C-terminal truncation of the insulin-like growth factor I receptor. Exp Cell Res 1995; 218:370–380.
122. Zamorano J, Wang HY, Wang L-M, Pierce JH, Keegan AD. IL-4 protects cells from apoptosis via the insulin receptor substrate pathway and a second independent signaling pathway. J Immunol 1996; 157:4926–4934.

123. Zhou-Li F, Xu S-Q, Dews M, Baserga R. Co-operation of simian virus 40 T antigen and insulin receptor substrate-1 in protection from apoptosis induced by interleukin-3 withdrawal. Oncogene 1997; 15:961–970.
124. Rubini M, Hongo A, D'Ambrosio C, Baserga R. The IGF-I receptor in mitogenesis and transformation of mouse embryo cells: role of receptor number. Exp Cell Res 1997; 230:284–292.
125. Zetterberg A, Engstrom W, Dafgard E. The relative effects of different types of growth factors on DNA replication, mitosis and cellular enlargement. Cytometry 1984; 5:368–375.
126. Damante G, Cox F, Rapoport B. IGF-1 increases c-fos expression in FRTL5 rat thyroid cells by activating the c-fos promoter. Biochem Biophys Res Commun 1988; 151:1194–1199.
127. Santisteban P, Acebron A, Polycarpou-Schwarz M, DiLauro R. Insulin and insulin-like growth factor 1 regulate a thyroid-specific nuclear protein that binds to the thyroglobulin promoter. Mol Endocrinol 1992; 6:1310–1317.
128. Conlon MA, Tomas FM, Owens PC, Wallace JC, Howarth GS, Ballard FJ. Long R/3 insulin-like growth factor I (IGF-I) infusion stimulates organ growth but reduces plasma IGF-I, IGF-II and IGF binding protein concentrations in the guinea pig. J Endocrinol 1995; 146:247–253.

16 Posttranslational Modification of the IGF Binding Proteins

Cheryl A. Conover, PhD

CONTENTS

INTRODUCTION

Primary sequences for the six currently characterized insulin-like growth factor binding proteins (IGFBPs) range in length from 201 amino acids for rat IGFBP-6 to 289 amino acids for human IGFBP-2 *(1)*. Sequence analyses indicate extensive homology among the IGFBPs in the amino(N)- and carboxy(C)-terminal domains with their conserved cysteines, domains that confer the ability to bind IGF ligand with high affinity. The homology within the IGF-binding end portions of the molecule defines the family of IGFBPs. However, it is the unique central domain that makes each IGFBP structurally and ultimately functionally distinctive.

The IGFBPs are expressed in multiple tissues under developmental and hormonal influences. However, IGFBP availability and bioactivity is determined not only by specific IGFBP gene expression, but also through structure/function modifications to the translated protein. These modifications vary among the six characterized IGFBP family members, and include glycosylation, phosphorylation, proteolytic processing, and interaction with components of the cell membrane and matrix (Table 1). Perhaps not surprisingly, these posttranslational modifications almost exclusively involve the central nonconserved region of the IGFBPs. This chapter reviews each of these modifications in turn with respect to what we know about characterization, regulation, and biological consequences of the altered IGFBPs. Some aspects are only briefly noted as they are covered in more detail in accompanying chapters. The reader is also encouraged to read other reviews of the IGFBPs and their regulation *(1–4)*.

From: *Contemporary Endocrinology: The IGF System*
Edited by: R. Rosenfeld and C. Roberts Jr © Humana Press, Inc., Totowa, NJ

Table 1
Post-Translational Modifications of the IGFBPs

	Glycosylation	*Phosphorylation*	*Proteolysis*	*Cell/ECM*[a]
IGFBP-1	–	+	+	+
IGFBP-2	–	–	+	+
IGFBP-3	+	+	+	+
IGFBP-4	+	–	+	–
IGFBP-5	+	+	+	+
IGFBP-6	+	–	+	–

[a] Cell or extracellular matrix association.

Table 2
Predicated Glycosylation Sites in Human IGFBP-1 to -6

	N-linked[a]	*O-linked*[b]
IGFBP-1	–	Thr^{27}, Ser^{95}
IGFBP-2	–	–
IGFBP-3	Asn^{89}, Asn^{109}, Asn^{172}	–
IGFBP-4	Asn^{204}	–
IGFBP-5	–	Thr^{103}, Thr^{104}, Thr^{111}
IGFBP-6	–	Thr^{119}, Thr^{121}, Thr^{122}, Thr^{212}, Ser^{120}, Ser^{124}, Ser^{208}

[a] ScanProsite (http://expasy.hcuge.ch/sprot/prosite.html)
[b] NetOGlyc 2.0 (http://www.cbs.dtu.dk/services/NetOGlyc/)
Courtesy of Dr. S. Firth.

GLYCOSYLATION

Glycosylation was the first posttranslational modification identified for IGFBPs. This came to light with the discrepancy in IGFBP molecular size between that predicted for the protein core from cDNA sequence and that indicated by migration of the native protein on sodium dodecyl sulfate-polyacrylamide gel electrophoresis (SDS-PAGE). Potential glycosylation sites predicted from amino acid sequence of the human IGFBPs are shown in Table 2. The existence of in vivo glycosylated isoforms of IGFBP-3, -4, -5, and -6 has been confirmed.

IGFBP-3 Glycosylation

One third of the molecular weight of IGFBP-3 is contributed by N-linked sugars. Consequently, IGFBP-3 from mammalian sources (predicted molecular mass of 29 kDa) migrates as two or more glycoprotein forms on SDS-PAGE with apparent molecular mass of 37–48 kDa *(5)*. IGFBP-3 has several N-linked glycosylation sites located in the nonconserved midregion of the molecule, which for human IGFBP-3 are at Asn^{89}, Asn^{109}, and Asn^{172} *(6)*. Naturally occurring glycoforms of human IGFBP-3 use either all three sites or Asn^{89} and Asn^{109} only, accounting for the doublet of 38/45 kDa on SDS-PAGE. Rat IGFBP-3 has four potential *N*-glycosylation sites and usually appears as three bands on SDS-PAGE. However, glycosylation *per se* is not essential for secretion, for binding to IGF-I and -II, or for the ability of IGFBP-3 to form the 150-kDa ternary complex with IGF and an acid-labile subunit (ALS) *(6,7)*. Furthermore, fully glycosylated and nonglycosylated forms of IGFBP-3 have similar biological effects in vitro, i.e., inhibition of IGF action, potentiation of IGF action, and IGF-independent inhibition of tumor cell growth *(6–9)*. On the other hand, glycosylation may modulate the susceptibility of IGFBP-3 to proteolytic degradation. We have shown that nonglycosylated *E. coli*-derived IGFBP-3 is rapidly degraded in bovine fibroblast cultures, whereas glycosylated Chinese hamster ovary (CHO) cell-derived IGFBP-3 remains intact *(7)*. Claussen et al. *(10)* reported that cathepsin D hydrolyzes nonglycosylated IGFBP-3 faster than glycosylated IGFBP-3. Because many of the identified proteolytic cleavage sites are located in the nonconserved central region of IGFBP-3 that contains the *N*-glycosylation sites, it may be speculated that the oligosaccharide chains interfere with enzyme recognition sites. Also, as discussed later on in this chapter, specific proteolysis is an important determinant of the circulating half-life of IGFBP-3 as well as its bioactivity at the target cell.

IGFBP-4 Glycosylation

IGFBP-4 has one potential N-linked glycosylation site, but the glycosylated product is usually present as a minor form. Both glycosylated (28 kDa) and nonglycosylated (24 kDa) IGFBP-4 bind IGF-I and IGF-II, as can be inferred by Western ligand blotting of biological fluids containing the two isoforms *(11)*. The significance of IGFBP-4 glycosylation is unknown, as all of the biological studies to date have employed predominantly, if not exclusively, 24-kDa nonglycosylated IGFBP-4. This form of IGFBP-4 has been shown to inhibit IGF action in a variety of systems *(12–16)*. In follicular fluid, glycosylated IGFBP-4 appears to be more resistant to proteolysis than nonglycosylated IGFBP-4 *(17)*, which may have implications for IGFBP-4 function in vivo.

IGFBP-5 Glycosylation

IGFBP-5 is variably *O*-glycosylated *(18)*. As determined by Western ligand blotting of culture medium conditioned by U-2 human osteosarcoma cells, IGFBP-5 appears as an IGF-binding triplet of 29/32/34 kDa. It can be reduced to its 29-kDa protein core size by sequential treatment with neuraminidase and *O*-glycanase, whereas treatment with *N*-glycanase or endoglycosidase F has no effect. Similar 29- to 34-kDa forms of IGFBP-5 have been identified in normal human fibroblast and osteoblastic cell-conditioned media *(12,19,20)*.

IGFBP-6 Glycosylation

The predicted molecular mass of IGFBP-6 based on cDNA structure is 23 kDa, but IGFBP-6 migrates with an apparent molecular mass of 28–34 kDa on SDS-PAGE. Bach et al. *(21)* demonstrated that IGFBP-6 purified from human cerebrospinal fluid contains O-linked but not N-linked oligosaccharide chains. Although the physiological significance of differential *O*-glycosylation of IGFBP-6 remains unknown, it may modulate proteolysis. *O*-Glycosylation has been reported to protect IGFBP-6 from proteolysis by trypsin and proteinase K, but to enhance proteolysis by chymotrypsin *(22)*. In addition, *O*-glycosylation does not seem to contribute to the preferential binding affinity of IGFBP-6 for IGF-II over IGF-I; however, it may alter absolute IGF-I affinity. In a preliminary report *(22)*, *O*-glycosylated IGFBP-6 had fivefold higher affinity for IGF-I than nonglycosylated IGFBP-6.

PHOSPHORYLATION

The IGFBP family members can be included among those recently recognized secreted proteins that can be phosphorylated. All of the IGFBPs contain potential sites for phosphorylation by serine/threonine kinases *(23),* but only IGFBP-1, -3, and -5 phosphoisoforms have been corroborated so far. Phosphorylation, with its regulatable introduction of negative charge into the molecule, can alter the biochemical and physiological properties of the IGFBPs in ways that have only begun to be explored.

IGFBP-1 Phosphorylation

IGFBP-1 was the first IGFBP shown to be phosphorylated. In 1988, Busby et al. *(24)* demonstrated that two isoforms of IGFBP-1 with similar molecular weights but with differing charges could be isolated and separated by anion-exchange chromatography from amniotic fluid. Subsequently, several IGFBP-1 isoforms have been separated by nondenaturing PAGE and identified as serine-phosphovariants of IGFBP-1. The three main sites of phosphorylation are on serine residues at 101, 119, and 169 in the central nonconserved portion of the molecule *(25).*

Several investigators *(26,27)* were unable to demonstrate phosphorylation of IGFBP-1 by a cAMP-dependent ectokinase, and suggested that IGFBP-1 was secreted as a phosphoprotein and subsequently dephosphorylated. Studies to identify the responsible intracellular kinases have implicated members of the casein kinase family *(25,27,28).* Phosphorylation of IGFBP-1 appears to be constitutive (e.g., not regulated by IGFs, insulin, protein kinase A, or protein kinase C in vitro *[26]*), and it may be that the dephosphorylation reaction is more tightly regulated. Phosphatases that may be involved in this regulation have not as yet been identified.

The state of phosphorylation modulates the effect of IGFBP-1 on IGF action. Nonphosphorylated forms of IGFBP-1 can potentiate IGF-I stimulated DNA synthesis in vitro and enhance the effectiveness of IGFs in wound healing in vivo *(24,29,30).* Phosphorylation of IGFBP-1 increases the affinity of IGFBP-1 for IGF-I approximately sixfold *(25,26),* converting it into an inhibitor of IGF action. Although the exact mechanism is unknown, phosphate groups may alter the molecular charge of the protein and/or promote a conformational change in IGFBP-1 altering its affinity for IGF-I. An inhibitory function for the phosphorylated form of IGFBP-1 was also suggested by the work of Frost and Tseng *(27).* These studies showed that heightened IGFBP-1 phosphorylation, in the absence of changes in IGFBP-1 protein concentration, coincided with growth arrest and differentiation of endometrial stromal cells.

In normal adult plasma, IGFBP-1 exists in a single high phosphorylation state *(31).* This is because liver is the major source of circulating IGFBP-1, and adult liver exclusively secretes the phosphorylated form of IGFBP-1. In general, regulators of hepatic IGFBP-1 (insulin, glucagon, sex steroids, somatostatin, IGF) alter levels of protein expression but not the phosphorylation status of the IGFBP-1 *(32,33).* Highly phosphorylated IGFBP-1 is more resistant to proteolysis *(34),* which may help to explain the relative preservation of serum levels of IGFBP-1 in the face of circulating IGFBP proteases (see section following).

In the maternal circulation during pregnancy, nonphosphorylated and lesser phosphorylated forms of IGFBP-1 predominate *(31).* In addition, human amniotic fluid contains a large proportion of nonphosphorylated IGFBP-1. Decidua is the primary source of nonphosphorylated IGFBP-1 in the maternal circulation and in amniotic fluid *(35).*

The biological significance of nonphosphorylated serum IGFBP-1 during pregnancy is unclear. Westwood et al. *(36)* showed that affinity for IGF-I of the highly phosphorylated IGFBP-1 purified from normal adult serum was 10-fold greater than that of the nonphosphorylated IGFBP-1, but the affinity of the two forms for IGF-II did not differ. Thus, changes in IGFBP-1 phosphorylation status during pregnancy may favor increased IGF-I bioavailability.

IGFBP-3 Phosphorylation

IGFBP-3 is secreted as a phosphoprotein from fibroblasts and transfected CHO cells *(37)*. IGFBP-3 is reported to be phosphorylated at two major serine residues at positions 111 and 113 *(38)*. Hoeck and Mukku *(38)* suggested that phosphorylation of IGFBP-3 takes place intracellularly. Consistent with an intracellular location, we could not demonstrate ectokinase activity toward endogenous or exogenous IGFBP-3 in human fibroblasts. The protein kinases and phosphatases responsible for regulating IGFBP-3 phosphorylation status have not been identified, and little is known about the effect of phosphorylation on IGFBP-3 physiology. Studies by Mukku et al. *(37)* suggest that phosphorylation does not affect the affinity of IGFBP-3 for IGFs, but that it may influence the interactions of IGFBP-3 with ALS or the cell surface via charge alterations. Recently, Coverley and Baxter *(39)* reported that IGF-I, acting through the type I IGF receptor, stimulates IGFBP-3 phosphorylation in human fibroblasts. Their data also suggested that cell-associated IGFBP-3 in human fibroblasts either is not phosphorylated or is dephosphorylated upon release from the cell surface. It is intriguing to speculate that nonphosphorylated IGFBP-3 represents the potentiating form of IGFBP-3 *(7,8,40)*, analogous to the situation found with nonphosphorylated IGFBP-1 *(24,29)*. Phosphorylation of IGFBP-3, like glycosylation, may also affect its susceptibility to proteolytic modifications, which has important implications for IGFBP-3 physiology.

IGFBP-5 Phosphorylation

IGFBP-5 secreted by human fibroblasts is highly phosphorylated *(41)*. IGFBP-5 is preferentially found in the extracellular matrix produced by cultured human fibroblasts, and in this form is associated with lowered IGF affinity and potentiation of IGF-stimulated DNA synthesis *(42)*. Furthermore, it is likely that phosphorylation promotes the affiliation of IGFBP-5 with hydroxyapatite in bone matrix. This feature of IGFBP-5 is important because it is this form of IGFBP-5 that anchors IGF-II in the bone matrix, providing a pericellular reservoir of a critical bone growth factor *(43)*.

PROTEOLYSIS

Studies in 1990 reporting discrepancies between Western ligand blotting and radioimmunoassay measurements for IGFBP-3 in serum of pregnant women introduced the notion of IGFBP proteases *(44,45)*. Since then, proteolytic activities have been described in many different clinical situations and biological fluids that cleave IGFBP-1 through -6 with varying specificities *(46)*.

Many of the responsible proteases are turning out to be previously characterized enzymes that will cleave several IGFBPs and include plasmin, prostate-specific antigen (PGA), nerve growth factor (NGF), matrix metalloproteases (MMP), and cathepsin D. Others are apparently novel proteases, not yet chemically defined, that are relatively specific for a single form of IGFBP. Control of IGFBP proteolysis is complex and occurs at the level of enzyme expression as well as by activation, inhibition, and compartmentalization of the expressed protease. Although it is well accepted that IGFBPs regulate the half-life of IGFs in the circulation and their availability to cell receptors, the converse seems to be true as well, i.e., that IGFs can alter the stability and local bioavailability through the regulation of specific proteases. Furthermore, there is usually more than one protease in the different biological systems, and they are likely to be highly interactive. The importance of IGFBP regulation by proteases is only briefly explored in this chapter. It is indicated by resultant modification in IGFBP affinity for IGFs or complete destruction of IGF binding potential. Moreover, IGFBP proteolysis may produce cleavage products with unique function, e.g., intracellular signaling independent of IGF binding. As with the other posttranslational modifications discussed, most of the proteolytic cleavage sites that have been identified to date are within the nonconserved region of the IGFBP.

IGFBP-1 Proteolysis

IGFBP-1 is susceptible to proteolysis by cathepsin D, stromolysin-3 (MMP-11), and plasmin in vitro *(10,47–49),* but IGFBP-1 proteolysis or IGFBP-1 fragments have not been observed in the circulation or in physiological fluids in vivo. A preliminary report by Gibson et al. *(34)* indicates that the highly phosphorylated form of IGFBP-1 is more resistant to proteolysis than the nonphosphorylated form, which may explain preservation of serum IGFBP-1 found primarily in the highly phosphorylated form *(31).* Furthermore, radiolabeled IGFBP-1 was proteolyzed into two fragments of 14 and 17 kDa during incubation in medium conditioned by $HepG_2$ cells and decidualized endometrial cells and in amniotic fluid, although it was not proteolyzed by normal plasma or plasma obtained during pregnancy *(34).* These data suggest that there are naturally occurring circulating inhibitors preventing IGFBP-1 proteolysis in serum, but that IGFBP-1 proteolysis may occur extravascularly at the tissue level.

IGFBP-2 Proteolysis

IGFBP-2 Proteolysis In Vivo

IGFBP-2 generally circulates as the intact 34-kDa protein. However, immunoreactive IGFBP-2 fragments of ~22 and 14 kDa were identified in the serum of newborn fasted pigs, a 23-kDa fragment was found in the follicular fluid of dominant follicles, and 14- and 16-kDa immunoreactive IGFBP-2 fragments were recently identified in human milk *(17,50,51).* In none of these biological fluids was an accompanying IGFBP-2 protease detected, suggesting that the fragments were released as the result of IGFBP-2 proteolysis occurring extravascularly or at the vascular endothelial surface.

IGFBP-2 Proteolysis In Vitro

In studies of Cohick et al. *(52)* and Gockerman and Clemmons *(53),* porcine vascular smooth muscle cells were shown to secrete a protease that cleaves IGFBP-2 into fragments of ~23 and 14-kDa. It was specific for IGFBP-2 in that it did not degrade other forms of IGFBPs. This IGFBP-2 proteolytic activity was inhibited by aprotinin, α_1-antichymotrypsin, and EDTA, suggesting a cation-dependent serine protease. It binds weakly to heparin-Sepharose, allowing it to be separated from a high-affinity heparin-binding IGFBP-5 protease also secreted by these cells *(53).* The IGFBP-2 protease in this system appears to be secreted constitutively, but activity is enhanced by serum deprivation. It is also stimulated by addition of IGFs to the culture, with IGF-II being more effective than IGF-I. The mechanism of this IGF effect has not been elucidated, but it could be replicated in cell-free assay, indicating that it does not require IGF binding to receptors on cells. It may be of relevance that the fragment sizes produced by IGFBP-2 proteolysis in porcine vascular smooth muscle cell conditioned medium are similar to those identified in the serum of newborn pigs fasted for 48 h *(50)*. Because the fragments were not generated by a circulating IGFBP-2 protease, it may be speculated that proteolysis occurs on the endothelial surface by an enzyme such as that produced by the vascular smooth muscle cells in vitro. The significance of this enzyme activity or the consequent IGFBP-2 cleavage products and possible involvement in physiological response mechanisms remain to be determined.

A neuroblastoma cell line secretes tissue plasminogen activator (tPA) and exhibits plasmin-induced IGFBP-2 proteolysis when plasminogen is added to the cells *(54)*. Because tPA binds to the cell surface, IGFBP-2 proteolysis is effectively restricted to the pericellular environment. Plasma-induced IGFBP-2 proteolysis results in enhanced autocrine IGF-II-mediated cell proliferation.

IGFBP-3 Proteolysis

IGFBP-3 Proteolysis In Vivo

Circulating IGFBP-3 proteases were initially identified during pregnancy *(44,45)*, and then subsequently in association with a variety of catabolic states or states of physiologic stress such as postoperative recovery, severe illness, GH resistant and deficient states, malnutrition, malignancy, and diabetes *(55–61)*. Serum IGFBP-3 protease activity is also induced with brief, high-intensity exercise and by acute metabolic acidosis *(62,63)*. Probably more than one protease is responsible for these observed activities, but inhibitor profiles suggest predominantly serine proteases active at neutral pH. Bang and Fielder *(64)* recently reported that IGFBP-3 protease activity in human pregnancy serum could be separated into two fractions. Activity in the 70- to 90-kDa fraction was purified and shown to represent plasminogen. This activity was also identified in normal adult serum, although there it was masked until activated by the chromatographic procedure. IGFBP-3 protease activity specific for pregnancy serum chromatographed as >150 kDa. Serum obtained during human pregnancy also showed an increase in 50- to 60-kDa MMP-1 and -3 and a concomitant decrease in a tissue inhibitor of matrix metalloproteases (TIMP-1), but it was concluded as unlikely that these enzymes/inhibitor account for the increased IGFBP-3 proteolytic activity. On the other hand, addition of TIMP-1 clearly inhibited IGFBP-3 proteolytic activity in serum obtained from pregnant rats *(65)*. Apparent discrepancies may reflect species specificity. Pigs, baboons, and rhesus monkeys have little or no pregnancy-associated IGFBP-3 protease activity *(66)*. Interestingly, rhesus monkey serum contains a protease that degrades human IGFBP-3, but it does not affect rhesus monkey IGFBP-3. Rhesus monkey IGFBP-3 is 2 kDa larger in size than human IGFBP-3 and may have a unique carbohydrate structure that renders it resistant to proteolysis.

The identification and regulation of serum IGFBP-3 protease activity is an area of active investigation. Both enzymes and enzyme inhibitors need to be considered. The source of pregnancy-associated serum IGFBP-3 protease appears to be decidua, and its appearance and disappearance from the maternal circulation parallels that of placentally derived hormones *(44,66)*. Fetal distress seems to regulate maternal IGFBP-3 protease activity, in that there is increased IGFBP-3 protease activity in maternal serum during pregnancies complicated by uteroplacental insufficiency or multiple fetuses *(67)*.

Obviously, decidua cannot account for serum IGFBP-3 protease activity observed in men and women in the various clinical conditions associated with catabolism and stress.

IGFBP-3 protease activity in these conditions seems to be under metabolic regulation because it is increased with fasting or undernutrition and decreased with nutritional support *(55,57)*. Bang *(68)* hypothesized that insulin resistance or its associated metabolic and hormonal alterations induce IGFBP-3 proteolysis to increase IGF availability to compensate for lack of insulin action. The IGFBP-3 protease does not appear to be of hepatic origin because it increases even after total hepatectomy. Tonner et al. *(69)* demonstrated that cultured sheep thymus cells secrete IGFBP-3 proteolytic activity. A component of the immune system is an attractive candidate as bioresponse sensor to metabolic stress. Also, plasminogen in serum is increased in states of stress and may be at least partially responsible for circulating IGFBP-3 protease activity in these conditions *(64)*.

IGFBP-3 proteolysis may reflect an adaptive process that allows IGFs bound to serum IGFBP-3 within the 150-kDa ternary complex to be more available for extravascular fluids/tissues—in the case of pregnancy to the fetus, in the case of catabolic states to the tissues in need *(70–72)*. Protease-induced alteration of IGFBP-3 may be a fundamental mechanism for mobilizing the store of IGFs, especially IGF-I, in the bloodstream as a compensatory response. But it could also be counterproductive. Systemic treatment of breast cancer patients with the antitumor drug suramin induced near complete proteolysis of serum IGFBP-3 with a concomitant 2- to 13-fold increase in serum levels of "free IGF-1 *(73)*. A concern was registered that this decrease in intact IGFBP-3 and increase in free IGF-I could reflect an alteration in IGF availability to target tissues, including the tumor tissue.

IGFBP-3 proteolytic activity is also found in other biological fluids outside the vascular compartment: urine, vitreous humor, lymph, blister fluid, wound fluid, amniotic fluid, cerebrospinal fluid, peritoneal fluid, seminal plasma, and synovial fluid *(74–79)*. The proteolytic pattern of IGFBP-3 fragments differs among the fluids, suggesting multiple IGFBP-3 proteases that may be tissue- or fluid-specific. In these cases, IGFBP-3 proteases may be produced and function locally.

The presence of IGFBP-3 fragments but not IGFBP-3 protease activity in serum in certain situations indicates IGFBP-3 proteolysis taking place in interstitial fluids or extravascular tissues with resultant cleavage products being released into the circulation. Lalou and Binoux *(80)* compared IGFBP-3 protease activity in lymph versus serum, and found higher activity in lymph. Fetal serum carries abundant IGFBP-3 fragments, but has no IGFBP-3 protease activity *(81)*. Presumably the fragments are generated by the maternal protease. The physiologic importance of IGFBP-3 fragments is unclear. A 29-kDa IGFBP-3 fragment isolated from pregnancy serum and peritoneal dialysates binds IGF-I with ~100-fold and IGF-II with ~2-fold reduced affinity compared to intact IGFBP-3 *(82,83)*. This reduced affinity would result in increased IGF, especially IGF-I, availability. Also, intrinsic mitogenic activity of some purified IGFBP-3 fragments in vivo has been reported *(84)*.

IGFBP-3 Proteolysis In Vitro

Cell-derived IGFBP-3 proteases have also been described for a number of normal and malignant cell types, and generally have been ascribed to known proteases as discussed below.

Plasmin. Plasmin is a broad-spectrum cation-independent serine protease, which appears to play an important role in the proteolysis of IGFBP-3 in the cellular microenvironment. Plasmin is involved in the proteolysis of IGFBP-3 produced by MG63 human osteosarcoma and PC3 prostate carcinoma cells *(85,86)*. Addition of plasminogen to these cells, which produce urokinase plasminogen activator (uPA), induces IGFBP-3 proteolysis and increases proliferation. IGF-I can inhibit IGFBP-3 proteolysis when it is added to the cells but it does not inhibit plasmin-induced proteolysis under cell-free conditions. IGF-I, acting through the type I IGF receptor on cells, appears to either decrease uPA or increase plasminogen activator inhibitor (PAI) *(85)*. The authors of these studies conclude that autoocrine control of MG63 and PC3 cell proliferation is dependent upon IGFBP-3 proteolysis. Furthermore, plasmin action requires cell surface interaction. Protected from soluble protease inhibitors, cell-bound plasmin can hydrolyze IGFBP-3, thereby increasing IGF availability. This localizes the reaction and the generated IGFBP-3 fragments. Lalou et al. *(87)* found that plasmin activity produced IGFBP-3 fragments of 22–25 and 16 kDa. The 22- to 25-kDa IGFBP-3 fragment had decreased affinity for IGFs and was weakly inhibitory toward IGF action; this fragment had no effect on insulin action. On the other hand, the 16-kDa IGFBP-3 fragment, which did not bind IGFs, inhibited IGF and insulin action as well as the actions of basic fibroblast growth factor in a type I IGF receptor-null cell line *(88)*, suggesting that the effects of the 16-kDa IGFBP-3 fragment are independent of IGF binding and of type I IGF receptor signaling.

Prostate-Specific Antigen. Cohen et al. *(89)* first demonstrated that PSA is the IGFBP-3 protease in seminal plasma. The cleavage pattern and inhibitor profile for PSA-induced IGFBP-3 proteolysis were different from pregnancy-associated IGFBP-3 protease(s). All of the PSA cleavage sites are in the mid-region of the IGFBP-3 molecule between amino acids 88 and 186 *(90)*. IGFs do not protect IGFBP-3 from PSA hydrolysis whether added in culture or in cell-free assay. Although IGFBP-3 was the first IGFBP substrate identified, PSA will also proteolyze IGFBP-5, but has weak if any activity against IGFBP-2 and -4 *(46,89)*. 7S NGF (NGF subunit) with sequence homology to PSA was also shown to degrade IGFBP-3, -4, -5, and -6 *(91)*.

Matrix Metalloproteases. MMP-1, -2, and -3 degrade IGFBP-3 (and IGFBP-5) in pregnancy-associated serum and in medium conditioned by human fibroblasts and airway smooth muscle cells *(64,65,92,93)*. IGFBP-3 fragments generated by MMP-3 indicate cleavages sites within the midregion after Tyr^{99}, Asn^{109}, and Glu^{176} *(92)*.

Cathepsin D. We identified acid-activatable IGFBP-3 protease activity in medium conditioned by human fibroblasts, osteosarcoma cells, MCF-7 human breast carcinoma cells, and prostate cancer cells *(94,95)*. pH profile, inhibition by pepstatin, and immunodepletion experiments identified the activity as belonging to cathepsin D. Cathepsin D secretion/acid-activated IGFBP-3 protease activity was enhanced by estrogen treatment in breast cancer cells but was not regulated by androgens in prostate cancer cells. Cathepsin D cleavage sites are in the nonconserved central region of IGFBP-3 *(10)*. Cathepsin D secreted by prostate carcinoma cells degrades all endogenous IGFBPs *(95)*, and pure cathepsin D will degrade exogenous IGFBP-1 through -6 *(10)*.

Novel IGFBP-3 Proteases. Salahifar et al. *(96)* recently characterized proteolysis of IGFBP-3 by an enzyme secreted by MCF-7 human breast cancer cells that appears to be different from the above-mentioned proteases in three major ways: pH optimum, inhibitor profile, and regulation by IGFs. MCF-7 serum-free conditioned medium was capable of proteolyzing exogenous IGFBP-3 and generating a 21-kDa N-terminal fragment of IGFBP-3 during incubation under cell-free conditions, but only at pH 4.5–5.5. Proteolytic activity was inhibited with EDTA and leupeptin and not with aprotinin or benzamidine, suggesting a cation-dependent cysteine protease. Gel chromatography indicated an approximate molecular mass for the proteolytic activity at 25–35 kDa. Although the pH optimum, inhibitor profile, and molecular size are consistent with cathepsin B, there was no inhibition of IGFBP-3 proteolysis with cathepsin B inhibitors, and pure cathepsin B was not an effective IGFBP-3 protease in vitro. Moreover, MCF-7 cell-conditioned medium contained little or no immunoreactive cathepsin B. Thus, the exact biochemical nature of this enzyme activity remains to be determined.

The possible mechanism by which this IGFBP-3 protease was regulated by IGFs was of particular interest. Addition of IGF-I to the cell-free assay prevented IGFBP-3 proteolysis *(96)*. Similar receptor-independent actions of IGF have been reported, and in most cases found to involve IGF binding to IGFBP-3 *(97–99)*. What was different here was that an IGF-I analog, [long Arg^3] ([LR^3]IGF-I), which does not bind IGFBP-3, was capable of inhibiting IGFBP-3 proteolysis in this cell-free system. These data suggest the possibility of a direct interaction between IGF and the protease. Grimes and Hammond *(100)* also noted that IGF-I and [LR^3]IGF-I attenuate IGFBP-3 degradation during incubation in medium from cultured porcine ovarian granulosa cells. A putative IGF regulatory domain in a proteolytic enzyme has recently been described in the literature *(101)*. A preliminary report further suggests that estrogen treatment increases and transforming growth factor (TGF) treatment decreases this IGFBP-3 cysteine protease activity in MCF-7 cells *(102)*.

IGFBP-4 Proteolysis

IGF-Dependent IGFBP-4 Proteolysis

We and others identified proteolytic activity secreted by normal human fibroblasts and osteoblasts in culture that cleaves the IGFBP-4 molecule into ~18- and 14-kDa fragments *(103–106)*. The divalent cation chelator EDTA and the zinc metalloprotease inhibitor 1,10-phenanthroline were the most effective chemical inhibitors of IGFBP-4 proteolysis in these systems and their effects could be reversed with Ca^{2+} and Zn^{2+}, respectively. Serine protease inhibitors such as benzamidine and aprotinin produced variable effects, and there was no inhibition by TIMPs. Based on these and other data, this IGFBP-4 protease appears to be a Ca^{2+}-requiring metalloprotease (or metallo-serine protease), but evidently is not an MMP. IGFBP-4 is the only IGFBP substrate (of the six known IGFBPs) for this protease, which is active in a broad pH range of 5.5–9. The signature feature of IGFBP-4 proteolysis in these systems is its strict dependence on IGFs for functional activity. Only very low concentrations of IGFs are needed, and, in general, IGF-II is more effective than IGF-I in activating IGFBP-4 proteolysis. A large number of other peptides (including structurally homologous insulin) and steroids have been shown to be incapable of inducing IGFBP-4 proteolysis in these systems during cell-free incubation. Even some of the apparently constitutive IGFBP-4 proteolysis as originally reported was subsequently determined to be due to endogenous IGFs *(105,107–109)*. These properties (metalloprotease, neutral pH, IGFBP-4 specificity, IGF dependence) are also shared by IGFBP-4 proteases described in cultures of smooth muscle cells *(110)*, endometrial stromal cells *(111)*, decidual cells *(112)*, granulosa cells *(113)*, and certain malignant cell lines *(114,115)*.

The exact mechanism underlying this IGF dependency of IGFBP-4 proteolysis is unknown. It has been suggested that IGF binding to IGFBP-4 changes its conformation in such a way as to make the IGFBP-4 molecule more susceptible to proteolysis. Data of Irwin et al. *(111)* would support this mechanism, as they showed that IGFBP-4 crosslinked with IGF-II was proteolyzed by endometrial stromal cell cultures in the absence of added IGF-II. Furthermore, a number of laboratories, including our own, have shown that insulin and IGF analogs that do not bind IGFBP-4 do not induce IGFBP-4 proteolysis suggesting a need for IGF–IGFBP interaction. However, relative affinities of different IGFs for IGFBP-4 do not strictly correlate with ability to induce proteolysis, suggesting that IGF peptides may directly activate a latent IGFBP-4 protease. The plausibility of such an interaction is suggested by the recent isolation of a cDNA clone encoding a novel serine protease with an IGF binding motif *(101)*. Data also support a third possible mechanism—repression of protease inhibition. Fowlkes et al. *(116)* showed that addition of IGFBP-3 can inhibit IGFBP-4 proteolysis in MC3T3-E1 cell cultures, and that IGF binding to IGFBP-3 reverses the inhibition. Thus, IGFBP-3 may be an inhibitor of the protease and confer IGF dependence on the system. Indeed, the ability of IGFs to induce IGFBP-4 proteolytic activity correlates with their affinity for IGFBP-3 (IGF-II > IGF-I > [des1-3]IGF-I >>> [LR^3]IGF-I, insulin *[103,117]*), but it is not clear whether this mechanism would operate in all systems, i.e., those that do not have IGFBP-3. Perhaps other IGFBPs serve this purpose in systems that do not express IGFBP-3 *(118)*. Furthermore, these three mechanisms are not necessarily mutually exclusive; for example, IGFBP-3 may not be a direct inhibitor of the protease but may modulate IGF availability for interaction with the protease or IGFBP-4 substrate. Purification of the IGFBP-4 protease will be required to resolve the issue of IGF-dependence.

A number of studies are now emerging that further suggest regulation of IGF-dependent IGFBP-4 proteolysis at the level of production/secretion of the enzyme. TGF-β treatment of normal human osteoblasts *(119)*, dexamethasone treatment of B104 neuroblastoma cells *(108)*, and follicle-stimulating hormone (FSH) treatment of granulosa cells *(113)* have been shown to increase IGF-dependent IGFBP-4 proteolysis. Unlike IGFs, none of these reagents act directly to activate IGFBP-4 proteolysis in cell-free assay. Rather, they enhance the expression and/or secretion of the IGF-dependent IGFBP-4 protease. In addition, there is a preliminary report of an IGFBP-4 protease induced by vitamin C treatment in differentiating mouse MC3T3-E1 osteoblast cell cultures *(120)*. This proteolysis produced ~22-kDa IGFBP-4 fragments and apparently is not regulated by IGFs. There are also inhibitors of IGFBP-4 proteolysis. Phorbol ester tumor promoters induce a cycloheximide-sensitive inhibitor of IGFBP-4 proteolysis in human fibroblasts and osteoblasts *(121,122)*. Estrogen treatment decreases IGFBP-4 proteolysis in estrogen-responsive bone cells *(123)*, but it was not determined whether this inhibition was due to a decrease in protease and/or an increase in inhibitor. FSH-induced IGFBP-4 proteolysis can be blocked by gonadotropin-releasing hormone *(124)* and TGF-α *(125)*.

Sequencing of the 18- and 14-kDa reaction products of IGF-dependent IGFBP-4 proteolysis was undertaken to determine the cleavage site. N-terminal sequence analysis of the 14-kDa fragment indicated that this fragment represented the C-terminal half of the human IGFBP-4 molecule and suggested cleavage after Met at position 135 in the mid-region of the mature protein *(126)*. The proteolytic cleavage site in rat IGFBP-4 was reported to be between Lys^{120} and His^{121}, with a secondary site between Met^{131} and Lys^{132} *(109)*. Synthetic peptides containing the region of the putative cleavage site for rat IGFBP-4 inhibited proteolytic activity toward IGFBP-4. Cleavage of IGFBP-4 abolishes its ability to inhibit IGF action in a variety of systems, but whether the 18- or 14-kDa IGFBP-4 cleavage products have functions independent of the IGFs is unknown. The 18-kDa N-terminal half of IGFBP-4 has decreased affinity for IGFs, although it can be detected by Western ligand blot using ^{125}I-IGF-II. In comparison to the intact protein, the isolated N-terminal IGFBP-4 fragment was not able to inhibit IGF-stimulated DNA synthesis in B104 cells *(109)*. Single amino acid substitutions around the putative cleavage sites yield mutant IGFBP-4s relatively resistant to proteolysis that can alter IGF action in vitro *(109,126)*.

IGFBP-4 protease activity is also associated with developing follicles *(127,128)*. Iwashita et al. *(127)* reported that treatment of human luteinized granulosa cells with FSH or estrogen increases IGFBP-4 protease activity. It is not known in these cells whether estrogen and FSH directly stimulate production of the protease or whether the effect is mediated through increased IGF production. IGFBP-4 proteolysis is absent in androgen-dominant follicular fluid, in follicles developing under the influence of pharmacologically induced androgen dominance, and in atretic follicles *(128,129)*. In vivo, IGFBP-4 proteolysis may reduce the inhibitory effect of IGFBP-4, thus providing the dominant follicle with more available IGFs to synergize with gonadotropins.

IGFBP-4 AS SUBSTRATE FOR KNOWN PROTEASES

IGFBP-4 can also serve as substrate for plasmin, MMPs, and cathepsin D *(10,130,131)*. Purified cathepsin D cleaves IGFBP-4 at Met^{156} *(10)*. There also appears to be IGFBP-4 proteolytic activity in pregnancy serum, which is different from pregnancy-associated IGFBP-3 proteolytic activity *(44,45,130,132)*. It is unclear whether this IGFBP-4 proteolytic activity can be accounted for by plasmin or MMP in the serum. Remacle-Bonnet et al. *(131)* present an interesting model of interplay between the IGF and plasmin systems in HT29-DY human colon carcinoma cells in culture. In this model, activation of cell surface plasmin leads to the selective hydrolysis of IGFBP-4 promoting autocrine IGF-II bioavailability. It was of note that the other IGFBPs produced by these cells, although sensitive to proteolysis by soluble plasmin, were not altered by cell-bound plasmin. In vivo, it is the cell-bound form of plasmin that is physiologically relevant.

IGFBP-5 Proteolysis

The existence of an IGFBP-5 protease was indicated by the identification and purification by Andress and Birnbaum in 1991 *(133)* of a 23-kDa truncated IGFBP-5 abundant in medium conditioned by U-2 human osteosarcoma cells. We subsequently characterized the proteolytic activity in U-2 conditioned medium *(18)*. Activity was specific for IGFBP-5; no proteolysis of IGFBP-1, -2, -3, -4, or -6 occurred during cell-free incubation in U-2 conditioned medium, whereas 29-kDa recombinant human IGFBP-5 was rapidly and completely degraded into immunoreactive fragments of ~16 kDa. Endogenously generated IGFBP-5 fragments appeared as ~22-, 17-, and 16-kDa immunoreactive forms reflecting the *O*-glycosylated nature of the native protein. IGFBP-5 proteolysis in U-2 conditioned medium was inhibited most effectively by serine protease inhibitors (aprotinin, PMSF, DFP, TLCK), EDTA, and 1,10-phenanthroline, indicating a member of the cation-dependent class of serine proteases as has been described for IGFBP-5-degrading proteases in conditioned media from human fibroblasts *(134)*, osteoblasts *(106)*, rat articular chondrocytes *(135)*, and porcine smooth muscle cells *(53)*. IGFBP-5 proteolytic activity is acid labile (no activity < pH 5.5) and optimal at pH 6–8. Protease inhibitor and pH profiles indicate a protease different than plasmin, MMP, PSA, and cathepsin D, all of which have been shown to degrade IGFBP-5. Most of the biochemical characterization has come from the work of Clemmons and colleagues in human fibroblasts *(134,136,137)*. Taking advantage of the tight adherence of IGFBP-5 proteolytic activity to heparin–Sepharose, they have partially purified a 92-kDa serine protease from human fibroblast conditioned medium. It is unclear at this time whether this protease is the same as that responsible for the IGFBP-5-specific degradation in U-2 conditioned medium and other biological fluids.

Cellular exposure to IGFs results in markedly increased media levels of intact IGFBP-5. In U-2 osteosarcoma cells and human fibroblasts, this IGF-induced increase is not reflected in increased IGFBP-5 mRNA expression, and only those IGFs that bind IGFBP-5 stimulate IGFBP-5 accumulation *(18,19)*. Results in cell-free assays have been variable, calling into question whether the mechanism is solely by IGF binding and protecting the IGFBP-5 substrate from degradation. Kanzaki et al. *(106)* could find no effect of exogenous IGF-II on IGFBP-5 proteolysis during cell-free incubation in human osteoblast medium, and Nam et al. *(134)* found only a minimal effect in human fibroblast conditioned medium. These negative findings may be due in part to the nature of the protease and to assay conditions. We found that IGF-I altered the kinetics of proteolysis increasing the half-life of exogenous IGFBP-5 from 1.5 h to 3.2 h *(47)*; at >8 h the protective effect of IGFs was less apparent. Matsumoto et al. *(135)* found that preincubation of ^{125}I-IGFBP-5 with IGF-I or IGF-II inhibited proteolysis during cell-free incubation, whereas simultaneous incubation of IGFs, conditioned medium and the ^{125}I-IGFBP-5 did not.

IGFBP-5 proteolysis is also regulated by glycosaminoglycans in fibroblasts *(137)*, FSH treatment of rat granulosa cells *(138)*, and interleukin-6 treatment of rat fetal calvaria-derived osteoblasts *(139)*. IGFBP-5 proteolysis increases in a time-dependent manner in MC3T3-E1 cells during culture, peaking at the transition from replication to differentiation, which may in part be attributed to MMP activity *(140)*.

In some systems, IGFBP-5 may undergo a gain of function through proteolysis. The IGFBP-5 fragment purified from U-2 conditioned medium, as well as a carboxy-truncated recombinant IGFBP-5 has been shown to associate with osteoblast cell surfaces and stimulate mitogenesis *(141,142)*.

IGFBP-6 Proteolysis

IGFBP-6 is generally not subjected to proteolysis under physiological conditions. Thus, IGFBP-6 is not degraded by pregnancy serum, amniotic fluid, cathepsin D, or PSA, and it is relatively insensitive to plasmin. However, IGFBP-6 can serve as substrate for 7S NGF in vitro *(91)*.

CELL SURFACE AND MATRIX ASSOCIATION

Discussion in this section is limited to those IGFBPs that undergo apparent modification of structure/function upon association with the cell surface or the extracellular matrix. So far, such modifications to IGFBP-2, IGFBP-3, and IGFBP-5 have been described. For information regarding aspects of IGFBP binding and IGF-independent signaling through integrins and putative IGFBP receptors on cells, the reader is invited to turn to other chapters in this book.

IGFBP-2

IGFBP-2 *per se* has little attraction for extracellular matrix components. However, Arai et al. *(143)* recently reported that, when complexed with IGF, especially IGF-II, IGFBP-2 binding to heparin and extracellular matrix produced by human fibroblasts was marked. Although the biological consequences of these interactions were not determined, a means for localizing IGFBP-2 and IGF-II is implied. Precedent set by studies of cell-associated IGFBP-3 and IGFBP-5 (*see below*) would suggest conformational and affinity changes in extracellular matrix-associated IGFBP-2 promoting delivery of IGF to receptors on nearby cells.

IGFBP-3

Addition of IGFBP-3 inhibits IGF action in a variety of cell types by virtue of its ability to bind IGF IGF with high affinity and effectively prevent receptor interaction. However, pretreatment of fibroblasts with IGFBP-3 results in potentiation of subsequent IGF-stimulated cell growth *(7,8,40)*. Investigating the underlying mechanism of this effect, we found IGFBP-3 cell-association to be requisite, but subsequent modifications to the cell-bound IGFBP-3 molecule or its environment also appeared necessary for an IGF-I-enhancing effect *(7,8)*. These subsequent modifications are inhibited with chloroquine, a weak base, or bafilomycin, a specific inhibitor of ATP-dependent hydrogen ion pumps *(144)*. Furthermore, heparin did not inhibit IGFBP-3 potentiation of IGF-I action, indicating that glycosaminoglycans were not the relevant cell binding sites for this effect. Cell-bound IGFBP-3 has reduced affinity for IGF-I compared to soluble IGFBP-3 *(7,8,145)*, and alters type I IGF receptor signaling through effects at or downstream from phosphatidylinositol 3-kinase (*8*, and *unpublished data*). We propose a model by which IGFBP-3 is cleaved by a pH-dependent membrane protease, resulting in a form with decreased affinity for IGF-I but enhanced affinity for an IGFBP-3 "receptor." IGFBP-3 receptor activation produces an intracellular signal that impinges on the type I IGF receptor signaling pathway. Thus, cell-associated IGFBP-3 may provide a mechanism for optimal presentation of IGF-I to its receptor as well as a means to heighten receptor reactivity to IGF-I and related peptides. Recent studies in human colon cancer cells also suggested that IGFBP-3 associated with the cell layer enhances responses to IGF-I *(146)*. The mechanism for IGFBP-3 potentiation of IGF action appears to be fundamentally different from the IGF-independent inhibitory effects of IGFBP-3 reported in tumor cells *(9,147)*, which may be mediated by the type V TGF-β receptor *(148)*.

IGFBP-5

IGFBP-5 binds extracellular matrix in preference to other IGFBPs via its natural attraction for glycosaminoglycans and hydroxyapatite, as well as by binding PAI-1 and vitronectin *(149,150)*. Although extracellular matrix-associated IGFBP-5 is structurally intact and, indeed, is relatively resistant to proteolysis, it has 7- to 10-fold reduced affinity for IGF-I compared to soluble IGFBP-5, suggesting a conformational change upon matrix binding. Intact soluble IGFBP-5 is inhibitory *(15,18)*, whereas intact extracellular matrix-associated IGFBP-5 potentiates IGF action *(42)*.

CONCLUDING REMARKS

All of the characterized IGFBPs exhibit posttranslational modification. IGFBP-3, -4, -5, -6 are variably glycosylated; IGFBP-1, -3, and -5 are secreted as phosphorylated proteins; all of the characterized IGFBPs serve as proteolytic substrates; and IGFBP-2, -3, and -5 undergo changes in structure/function upon association with the cells and/or extracellular matrix. By and large, the responsible kinases, phosphatases, proteases, protease inhibitors, and membrane matrix interactions have yet to be determined. Regulation is complex, but in many cases the IGFs themselves are involved as positive or negative modulators, suggesting feedback management in the system. Moreover, the physiological significance of posttranslational modification to the IGFBPs is unclear, although the emerging data reveal new functions of the IGFBPs beyond their IGF binding role including targeting of the IGFs and active participation in cell-specific and condition-specific cell growth.

Obviously much more work is needed before we fully understand the various posttranslational modifications to the IGFBPs, how they serve the IGF system, how they might coordinate the IGF system with other systems through common enzymes, inhibitors and cellular components, and, indeed, how they may allow the IGFBPs to come into their own as important growth factors.

ACKNOWLEDGMENTS

I extend a special thanks to my fellows and technicians (L. Bale, J. Clarkson, S. Durham, K. Kalli, J. Lawrence) and those of others cited in this chapter, who have contributed so much to this exciting field of research.

REFERENCES

1. Shimasaki S, Ling N. Identification and molecular characterization of insulin-like growth factor binding proteins (IGFBP-1, -2, -3, -4, -5 and -6). Prog Growth Factor Res 1991; 3:243–266.
2. Baxter RC. Insulin-like growth factor binding proteins in the human circulation: a review. Horm Res 1994; 42:140–144.
3. Kelly KM, Oh Y, Gargosky SE, Gucev Z, Matsumoto T, Hwa V, Ng L, Simpson DM, Rosenfeld RG. Insulin-like growth factor-binding proteins (IGFBPs) and their regulatory dynamics. Int J Biochem Cell Biol 1996; 28:619–637.
4. Clemmons DR. Insulin-like growth factor binding proteins and their role in controlling IGF actions. Cytokine Growth Factor Rev 1997; 8:45–62.
5. Martin JL, Baxter RC. Insulin-like growth factor-binding protein from human plasma: purification and characterization. J Biol Chem 1986; 261:8754–8760.
6. Firth SM, Baxter RC. The role of glycosylation in the action of IGFBP-3. Prog Growth Factor Res 1995; 6:223–229.
7. Conover CA. Glycosylation of insulin-like growth factor binding protein-3 (IGFBP-3) is not required for potentiation of IGF-I action: evidence for processing of cell-bound IGFBP-3. Endocrinology 1991; 129:3259–3268.
8. Conover CA. Potentiation of insulin-like growth factor (IGF) action by IGF-binding protein-3; studies of underlying mechanism. Endocrinology 1992; 130:3191–3199.
9. Oh Y, Muller HL, Lamson G, Rosenfeld RG. Insulin-like growth factor (IGF)-independent action of IGF-binding protein-3 in Hs578T human breast cancer cells. J. Biol Chem 1993; 268:14964–14971.
10. Claussen M, Kubler B, Wendland M, Niefer K, Schmidt B, Zapf J, Braulke T. Proteolysis of insulin-like growth factors (IGF) and IGF binding proteins by cathepsin D. Endocrinology 1997; 138:3797–3803.
11. Conover CA, Liu F, Powell D, Rosenfeld RG, Hintz RL. Insulin-like growth factor binding proteins from cultured human fibroblasts: characterization and hormonal regulation. J Clin Invest 1989; 83:851–859.
12. Conover CA, Kiefer MC, Zapf J. Posttranslational regulation of insulin-like growth factor binding protein-4 in normal and transformed human fibroblasts: insulin-like growth factor dependence and biological studies. J Clin Invest 1993; 91:1129–1137.
13. Cheung PT, Smith EP, Shimasaki S, Ling N, Chernausek SD. Characterization of an insulin-like growth factor binding protein (IGFBP-4) produced by the B104 rat neuronal cell line: chemical and biological properties and differential synthesis by sublines. Endocrinology 1991; 129:1006–1015.
14. Mohan S, Bautista CM, Wergedal J, Baylink DJ. Isolation of an inhibitory insulin-like growth factor (IGF) binding protein from bone cell-conditioned medium: a potential local regulatory of IGF action. Proc Natl Acad Sci USA 1989; 86:8338–8342.
15. Kiefer MC, Schmid C, Waldvogel M, Schlapfer I, Futo E, Masiarz FR, Green K, Barr PJ, Zapf J. Characterization of recombinant human insulin-like growth factor binding proteins 4, 5, and 6 produced in yeast. J Biol Chem 1992; 267:12692–12699.
16. Culouscou J-M, Shoyab M. Purification of a colon cancer cell growth inhibitor and its identification as an insulin-like growth factor binding protein. Cancer Res 1991; 51:2813–2819.

17. Hughes SCC, Mason HD, Franks S, Holly JMP. Modulation of the insulin-like growth factor-binding proteins by follicle size in the human ovary. J Endocrinol 1997; 154:35–43.
18. Conover CA, Kiefer MC. Regulation and biological effect of endogenous insulin-like growth factor binding protein-5 in human osteoblastic cells. J Clin Endocrinol Metab 1993; 76:1153–1159.
19. Camacho-Hubner C, Busby WH Jr, McCusker RH, Wright G, Clemmons DR. Identification of the forms of insulin-like growth factor-binding proteins produced by human fibroblasts and the mechanisms that regulate their secretion. J Biol Chem 1992; 267:11949–11956.
20. Hassager C, Fitzpatrick LA, Spencer EM, Riggs BL, Conover CA. Basal and regulated secretion of insulin-like growth factor binding proteins in osteoblast-like cells is cell line specific. J Clin Endocrinol Metab 1992; 75:228–233.
21. Bach LA, Thotakura NR, Rechner MM. Human insulin-like growth factor binding protein-6 is *O*-glycosylated. Biochem Biophys Res Commun 1992; 186:301–307.
22. Marinaro JA, Bach LA. *O*-Glycosylation of human IGFBP-6 modulates proteolysis and IGF-I binding. In: Proceedings of the Fourth International Symposium on Insulin-Like Growth Factors, Tokyo, Japan, October 21–24, 1997, p. 107.
23. Coverly JA, Baxter RC. Phosphorylation of insulin-like growth factor binding proteins. Mol Cell Endocrinol 1997; 28:1–5.
24. Busby WH Jr, Klapper DG, Clemmons DR. Purification of a 31,000-dalton insulin-like growth factor binding protein from human amniotic fluid: isolation of two forms with different biologic actions. J Biol Chem 1988; 263:14203–14210.
25. Jones JI, Busby WH Jr, Wright G, Smith CE, Kimack NM, Clemmons DR. Identification of the sites of phosphorylation in insulin-like growth factor binding protein-1: regulation of its affinity by phosphorylation of serine 101. J Biol Chem 1993; 268:1125–1131.
26. Jones JI, D'Ercole AJ, Camacho-Hubner C, Clemmons DR. Phosphorylation of insulin-like growth factor (IGF)-binding protein 1 in cell culture and in vivo: Effects on affinity for IGF-I. Proc Natl Acad Sci USA 1991; 88:7481–7485, 1991.
27. Frost RA, Tseng L. Insulin-like growth factor-binding protein-1 is phosphorylated by cultured human endometrial stromal cells and multiple protein kinases in vitro. J Biol Chem 1991; 266:18082–18088.
28. Ankrapp DP, Jones JI, Clemmons DR. Characterization of insulin-like growth factor binding protein-1 kinases from human hepatoma cells. J Cell Biochem 1996; 60:387–399.
29. Elgin RG, Busby WH Jr, Clemmons DR. An insulin-like growth factor (IGF) binding protein enhances the biologic response to IGF-I. Proc Natl Acad Sci USA 1987; 84:3254–3258.
30. Jyung RW, Mustoe TA, Busby WH, Clemmons DR. Increased wound-breaking strength induced by insulin-like growth factor I in combination with insulin-like growth factor binding protein-1. Surgery 1994; 115:233–239.
31. Westwood M, Gibson JM, Davies AJ, Young RJ, White A. The phosphorylation pattern of insulin-like growth factor-binding protein-1 in normal plasma is different from that in amniotic fluid and changes during pregnancy. J Clin Endocrinol Metab 1994; 79:1735–1741.
32. Westwood M, Gibson JM, Williams AC, Clayton PE, Hamberg O, Flyvbjerg A, White A. Hormonal regulation of circulating insulin-like growth factor-binding protein-1 phosphorylation status. J Clin Endocrinol Metab 1995; 80:3520–3527.
33. Frost RA, Bereket A, Wilson TA, Wojnar MM, Lang CH, Gelato MC. Phosphorylation of insulin-like growth factor binding protein-1 in patients with insulin-dependent diabetes mellitus and severe trauma. J Clin Endocrinol Metab 1994; 78:1533–1535.
34. Gibson JM, Westwood M, Siddals K, Shaikh T, White A. IGFBP-1 is proteolysed by decidual cell medium, HEP G2 medium and amniotic fluid. In: Proceedings of the Fourth International Symposium on Insulin-Like Growth Factors, Tokyo, Japan, October 21–24, 1997, p. 135.
35. Martina, NA, Kim E, Chitkara Y, Wathen NC, Chard T, Giudice LC. Gestational age-dependent expression of insulin-like growth factor-binding protein-1 (IGFBP-1) phosphoisoforms in human extraembryonic cavities, maternal serum, and decidua suggests decidua as the primary source of IGFBP-1 in these fluids during early pregnancy. J Clin Endocrinol Metab 1997; 82:1894–1898.
36. Westwood M, Gibson JM, White A. Purification and characterization of the insulin-like growth factor-binding protein-1 phosphoform found in normal plasma. Endocrinology 1997; 138:1130–1136.
37. Mukku VR, Chu H, Baxter R. Insulin-like growth factor binding protein, IGFBP-3 is secreted as a phospho-protein. In: Proceedings of the Second International Symposium on Insulin-Like Growth Factors, San Francisco, CA, January 12–16, 1991. Abstract, p. 237.
38. Hoeck WG, Mukku VR. Identification of the major sites of phosphorylation in IGF binding protein-3. J Cell Biochem 1994; 56:262–273.

39. Coverley JA, Baxter RC. Regulation of insulin-like growth factor (IGF) binding protein-3 phosphorylation by IGF-I. Endocrinology 1995; 136:5778–5781.
40. DeMellow JSM, Baxter RC. Growth hormone-dependent insulin-like growth factor (IGF) binding protein both inhibits and potentiates IGF-I-stimulated DNA synthesis in human skin fibroblasts. Biochem Biophys Res Commun 1988; 156:199–204.
41. Jones JI, Gockerman A, Clemmons DR. Insulin-like growth factor binding protein-5 (IGFBP-5) binds to extracellular matrix and is phosphorylated. In: Proceedings of the 74th Meeting of the Endocrine Society, San Antonio, TX, June 24–27, 1992, p. 372.
42. Jones JI, Gockerman A, Busby WH Jr, Camacho-Hubner C, Clemmons DR. Extracellular matrix contains insulin-like growth factor binding protein-5: potentiation of the effects of IGF-I. J Cell Biol 1993; 121:689–687.
43. Bautista CM, Baylink DJ, Mohan S. Isolation of a novel insulin-like growth factor (IGF) binding protein from human bone: a potential candidate for fixing IGF-II in human bone. Biochem Biophys Res Commun 1991; 176:756–763.
44. Giudice LC, Farrell EM, Pham H, Lamson G, Rosenfeld RG. Insulin-like growth factor binding proteins in maternal serum throughout gestation and in the puerperium: effects of a pregnancy-associated serum protease activity. J Clin Endocrinol Metab 1990; 71:806–816.
45. Hossenlopp P, Segovia B, Lassarre C, Roghani M, Bredon M, Binoux M. Evidence of enzymatic degradation of insulin-like growth factor-binding proteins in the 150K complex during pregnancy. J Clin Endocrinol Metab 1990; 71:797–805.
46. Collett-Solberg PF, Cohen P. The role of the insulin-like growth factor binding proteins and the IGFBP proteases in modulating IGF action. Endocrinol Metab Clin North Am 1996; 25:591–614.
47. Conover CA. Insulin-like growth factor binding protein proteolysis in bone cell models. Prog Growth Horm Res 1995; 6:301–309.
48. Manes S, Mira E, del Mar Barbacid M, Cipres A, Fernandez-Resa P, Buesa JM, Merida I, Aracil M, Marquez G, Martinez AC. Identification of insulin-like growth factor-binding protein-1 as a potential physiological substrate for human stromelysin-3. J Biol Chem 1997; 272:25706–25712.
49. Campbell PG, Novak JF, Yanosick TB, McMaster JH. Involvement of the plasmin system in dissociation of the insulin-like growth factor-binding protein complex. Endocrinology 1992; 130:1401–1412.
50. McCusker RH, Cohick WS, Busby WH, Clemmons DR. Evaluation of the developmental and nutritional changes in porcine insulin-like growth factor-binding protein-1 and -2 serum levels by immunoassay. Endocrinology 1991; 129:2631–2638.
51. Ho PJ, Baxter RC. Characterization of truncated insulin-like growth factor-binding protein-2 in human milk. Endocrinology 1997; 138:3811–3818.
52. Cohick WS, Gockerman A, Clemmons DR. Regulation of insulin-like growth factor (IGF) binding protein-2 synthesis and degradation by platelet-derived growth factor and the IGFs is enhanced by serum deprivation in vascular smooth muscle cells. J Cell Physiol 1995; 164:187–196.
53. Gockerman A, Clemmons DR. Porcine aortic smooth muscle cells secrete a serine protease for insulin-like growth factor binding protein-2. Circ Res 1995; 76:514–521.
54. Menouny M, Binoux M, Babajko S. Role of insulin-like growth factor binding protein-2 and its limited proteolysis in neuroblastoma cell proliferation: modulation by transforming growth factor-β and retinoic acid. Endocrinology 1997; 138:683–690.
55. Davies SC, Wass JAH, Ross RJM, Cotterill AM, Buchanan CR, Coulson VJ, Holly JMP. The induction of a specific protease for insulin-like growth factor binding protein-3 in the circulation during severe illness. J Endocrinol 1991; 130:469–473.
56. Davenport ML, Isley WL, Pucilowska JB, Pemberton LB, Lyman B, Underwood LE, Clemmons DR. Insulin-like growth factor-binding protein-3 proteolysis is induced after elective surgery. J Clin Endocrinol Metab 1992; 75:590–595.
57. Pucilowska JB, Davenport ML, Kabir I, Clemmons DR, Thissen J-P, Butler T, Underwood LE. The effect of dietary protein supplementation on insulin-like growth factors (IGFs) and IGF-binding proteins in children with shigellosis. J Clin Endocrinol Metab 1993; 77:1516–1521.
58. Bang P, Brismar K, Rosenfeld RG, Hall K. Fasting affects serum insulin-like growth factors (IGFs) and IGF-binding proteins differently in patients with noninsulin-dependent diabetes mellitus versus healthy nonobese and obese subjects. J Clin Endocrinol Metab 1994; 78:960–967.
59. Muller HL, Oh Y, Gargosky SE, Wilson KF, Lehrnbecher T, Rosenfeld RG. Insulin-like growth factor binding protein-3 concentrations and insulin-like growth factor binding protein-3 protease activity in sera of patients with malignant solid tumors or leukemia. Pediatr Res 1994; 35:720–724.
60. Fielder PJ, Guevara-Aguirre J, Rosenbloom AL, Carlsson L, Hintz RL, Rosenfeld RG. Expression of

serum insulin-like growth factors, insulin-like growth factor-binding proteins, and the growth hormone-binding protein in heterozygote relatives of Ecuadorian growth hormone receptor deficient patients. J Clin Endocrinol Metab 1992; 74:743–750.
61. Bereket A, Lang CH, Blethen SL, Fan J, Frost RA, Wilson TA. Insulin-like growth factor binding protein-3 proteolysis in children with insulin-dependent diabetes mellitus: a possible role for insulin in the regulation of IGFBP-3 protease activity. J Clin Endocrinol Metab 1995; 80:2279–2281.
62. Schwart AJ, Brasel JA, Hintz RL, Mohan S, Cooper DM. Acute effect of brief low- and high-intensity exercise on circulating insulin-like growth factor (IGF) I, II and IGF-binding protein-3 and its proteolysis in young healthy men. J Clin Endocrinol Metab 1996; 81:3492–3497.
63. Bereket A, Wilson TA, Kolasa AJ, Fan J, Lang CH. Regulation of the insulin-like growth factor system by acute acidosis. Endocrinology 1996; 137:2238–2245.
64. Bang P, Fielder PJ. Human pregnancy serum contains at least two distinct proteolytic activities with the ability to degrade insulin-like growth factor binding protein-3. Endocrinology 1997; 138:3912–3917.
65. Fowlkes JL, Suzuki K, Nagase H, Thrailkill KM. Proteolysis of insulin-like growth factor binding protein-3 during rat pregnancy: a role for matrix metalloproteinases. Endocrinology 1994; 135:2810–2813.
66. Giudice LC, Dsupin BA, De Las Fuentes L, Gargosky SE, Rosenfeld RG, Zelinski-Wooten MB, Stouffer RL, Fazleabas AT. Insulin-like growth factor binding proteins in sera of pregnant nonhuman primates. Endocrinology 1993; 132:1514–1526.
67. Langford KS, Nicolaides KH, Jones J, Abbas A, McGregor AM, Miell JP. Serum insulin-like growth factor-binding protein-3 (IGFBP-3) levels and IGFBP-3 protease activity in normal, abnormal, and multiple human pregnancy. J Clin Endocrinol Metab 1995; 80:21–27.
68. Bang P. Serum proteolysis of IGFBP-3. Prog Growth Factor Res 1995; 6:285–292.
69. Tonner E, Beattie J, Flint DJ. Production of insullin-like growth factor binding protein (IGFBP), IGFBP-3 protease, and expression of IGF-I receptors by cells of the sheep immune system. Eur J Endocrinol 1995; 132:118–122.
70. Blat C, Villaudy J, Binoux M. In vivo proteolysis of serum insulin-like growth factor (IGF) binding protein-3 results in increased availability of IGF to target cells. J Clin Invest 1994; 93:2286–2290.
71. Lassarre C, Binoux M. Insulin-like growth factor binding protein-3 is functionally altered in pregnancy plasma. Endocrinology 1994; 134:1254–1262.
72. Davenport ML, Clemmons DR, Miles MV, Camacho-Hubner C, D'Ercole AJ, Underwood LE. Regulation of serum insulin-like growth factor-I (IGF-I) and IGF binding proteins during rat pregnancy. Endocrinology 1990; 127:1278–1286.
73. Lawrence JB, Conover CA, Haddad TC, Ingle JN, Reid JM, Ames MM, Suman VJ, Marks RS, Erlichman C, Hartmann LC. Evaluation of continuous infusion suramin in metastatic breast cancer: impact on plasma levels of insulin-like growth factors (IGFs) and IGF-binding proteins. Clin Cancer Res 1997; 3:1713–1720.
74. Gargosky SE, Pham HM, Wilson KF, Liu F, Giudice LC, Rosenfeld RG. Measurement and characterization of insulin-like growth factor binding protein-3 in human biological fluids: discrepancies between radioimmunoassay and ligand blotting. Endocrinology 1992; 131:3051–3060.
75. Schoen TJ, Waldbillig RJ, Searcy G, Gaudet SJ, Jones BE, Chader GJ, Moshyedi P. Identification and partial characterization of a proteinase specific for insulin-like growth factor binding protein-3 in aqueous and vitreous humors. Curr Eye Res 1994; 14:127–135.
76. Xu S, Cwyfan-Hughes SC, Van Der Stappen JWJ, Sansom J, Burton JL, Donnelly M, Holly JMP. Insulin-like growth factors (IGFs) and IGF-binding proteins in human skin interstitial fluid. J Clin Endocrinol Metab 1995; 80:2940–2945.
77. Robertson JG, Pickering KJ, Belford DA. Insulin-like growth factor I (IGF-I) and IGF-binding proteins in rat wound fluid. Endocrinology 1996; 137:2774–2781.
78. Lee D-Y, Park S-K, Yorgin PD, Cohen P, Oh Y, Rosenfeld RG. Alteration in insulin-like growth factor-binding proteins (IGFBPs) and IGFBP-3 protease activity in serum and urine from acute and chronic renal failure. J Clin Endocrinol Metab 1994; 79:1376–1382.
79. Matsumoto T, Gargosky SE, Iwasaki K, Rosenfeld RG. Identification and characterization of insulin-like growth factors (IGFs), IGF-binding proteins (IGFBPs), and IGFBP proteases in human synovial fluid. J Clin Endocrinol Metab 1996; 81:150–155.
80. Lalou C, Binoux M. Evidence that limited proteolysis of insulin-like growth factor binding protein-3 (IGFBP-3) occurs in the normal state outside of the bloodstream. Regul Pept 1993; 48:179–188.

81. Bang P, Stangenberg M, Westgren M, Rosenfeld RG. Decreased ternary complex formation and predominance of a 29-kDa IGFBP-3 fragment in human fetal serum. Growth Regul 1994; 4:68–76.
82. Liu F, Baxter RC, Hintz RL. Characterization of the high molecular weight insulin-like growth factor complex in term pregnancy serum. J Clin Endocrinol Metab 1992; 75:1261–1267.
83. Durham SK, Mohan S, Liu F, Baker BK, Lee PDK, Hintz RL, Conover CA, Powell DR. Bioactivity of a 29-kilodalton insulin-like growth factor binding protein-3 fragment present in excess in chronic renal failure serum. Pediatr Res 1997; 42:335–341.
84. Koutsilieris M, Frenette G, Akoum A, Lemany A, Lazure C. N-terminal truncated forms of insulin-like growth factor binding protein-3 in the peritoneal fluid of women without laparoscopic evidence of endometriosis. Fertil Steril 1995; 63:314–321.
85. Lalou C, Silve C, Rosato R, Segovia B, Binous M. Interactions between insulin-like growth factor-I (IGF-I) and the system of plasminogen activators and their inhibitors in the control of IGF-binding protein-3 production and proteolysis in human osteosarcoma cells. Endocrinology 1994; 135:2318–2326.
86. Angelloz-Nicoud P, Binoux M. Autocrine regulation of cell proliferation by the insulin-like growth factor (IGF) and IGF binding protein-3 protease system in a human prostate carcinoma cell line (PC-3). Endocrinology 1995; 136:5485–5492.
87. Lalou C, Lassarre C, Binoux M. A proteolytic fragment of insulin-like growth factor (IGF) binding protein-3 that fails to bind IGFs inhibits the mitogenic effects of IGF-I and insulin. Endocrinology 1996; 137:3206–3212.
88. Zadeh SM, Binoux M. The 16-kDa proteolytic fragment of insulin-like growth factor (IGF) binding protein-3 inhibits the mitogenic action of fibroblast growth factor on mouse fibroblasts with a targeted disruption of the type 1 IGF receptor gene. Endocrinology 1997; 138:3069–3072.
89. Cohen P, Graves HCB, Peehl DM, Kamarei M, Giudice LC, Rosenfeld RG. Prostate-specific antigen (PSA) is an insulin-like growth factor binding protein-3 protease found in seminal plasma. J Clin Endocrinol Metab 1992; 75:1046–1053.
90. Fielder PJ, Rosenfeld RG, Graves HCB, Grandbois K, Maack CA, Sawamura S, Ogawa Y, Sommer A, Cohen P. Biochemical analysis of prostate specific antigen-proteolyzed insulin-like growth factor binding protein-3. Growth Regul 1994; 1:164–172.
91. Rajah R, Bhala A, Nunn SE, Peehl DM, Cohen P. 7S nerve growth factor is an insulin-like growth factor-binding protein protease. Endocrinology 1996; 137:2676–2682.
92. Fowlkes JL, Enghild JJ, Suzuki K, Nagase H. Matrix metalloproteinases degrade insulin-like growth factor-binding protein-3 in dermal fibroblast cultures. J Biol Chem 1994; 269:25742–25746.
93. Cohen P, Noveral JP, Bhala A, Nunn SE, Herrick DJ, Grunstein MM. Leukotriene D_4 facilitates airway smooth muscle cell proliferation via modulation of the IGF axis. Am J Physiol 1995; 269:L151–L157.
94. Conover CA, DeLeon DD. Acid-activated insulin-like growth factor-binding protein-3 proteolysis in normal and transformed cells: role of cathepsin D. J Biol Chem 1994; 269:7076–7080.
95. Conover CA, Perry JE, Tindall DJ. Endogenous cathepsin D-mediated hydrolysis of insulin-like growth factor-binding proteins in cultured human prostatic carcinoma cells. J Clin Endocrinol Metab 1995; 80:987–993.
96. Salahifar H, Baxter RC, Martin JL. Insulin-like growth factor binding protein (IGFBP)-3 protease activity secreted by MCF-7 breast cancer cells: inhibition by IGFs does not require IGF-IGFBP interaction. Endocrinology 1997; 138:1683–1690.
97. Conover CA. A unique receptor-independent mechanism by which insulinlike growth factor I regulates the availability of insulinlike growth factor binding proteins in normal and transformed human fibroblasts. J Clin Invest 1991; 88:1354–1361.
98. Neely EK, Rosenfeld RG. Insulin-like growth factors (IGFs) reduce IGF-binding protein-4 (IGFBP-4) concentration and stimulate IGFBP-3 independently of IGF receptors in human fibroblasts and epidermal cells. Endocrinology 1992; 130:985–993.
99. Martin JL, Ballesteros M, Baxter RC. Insulin-like growth factor-I (IGF-I) and transforming growth factor-β_1 release IGF-binding protein-3 from human fibroblasts by different mechanisms. Endocrinology 1992; 131:1703–1710.
100. Grimes RW, Hammond JM. Proteolytic degradation of insulin-like growth factor (IGF)-binding protein-3 by porcine ovarian granulosa cells in culture: regulation by IGF-I. Endocrinology 1994; 134:337–343.
101. Zumbrunn J, Trueb B. Primary structure of a putative serine protease specific for IGF-binding proteins. FEBS Lett 1996; 398:187–192.

102. Salahifar H, Martin JL, Baxter RC. Estrogen stimulates, and transforming growth factor β_1 (TGFβ_1) inhibits, IGFBP-3 protease activity secreted by MCF-7 breast cancer cells. In: Proceedings of the Fourth International Symposium on Insulin-Like Growth Factors, Tokyo, Japan, October 21–24, 1997, p. 136.
103. Conover CA, Kiefer MC, Zapf J. Posttranslational regulation of insulin-like growth factor binding protein-4 in normal and transformed human fibroblasts. Insulin-like growth factor dependence and biological studies. J Clin Invest 1993; 91:1129–1137.
104. Fowlkes J, Freemark M. Evidence for a novel insulin-like growth factor (IGF)-dependent protease regulating IGF-binding protein-4 in dermal fibroblasts. Endocrinology 1992; 131:2071–2076.
105. Durham SK, Kiefer MC, Riggs BL, Conover CA. Regulation of insulin-like growth factor binding protein 4 by a specific insulin-like growth factor binding protein 4 proteinase in normal human osteoblast-like cells: implications in bone cell physiology. J Bone Miner Res 1994; 9:111–117.
106. Kanzaki S, Hilliker S, Baylink DJ, Mohan S. Evidence that human bone cells in culture produce insulin-like growth factor-binding protein-4 and -5 proteases. Endocrinology 1994; 134:383–392.
107. Durham SK, DeLeon DD, Okazaki R, Riggs BL, Conover CA. Regulation of insulin-like growth factor (IGF)-binding protein-4 availability in normal human osteoblast-like cells: role of endogenous IGFs. J Clin Endocrinol Metab 1995; 80:104–110.
108. Cheung P-T, Wu J, Banach W, Chernausek SD. Glucocorticoid regulation of an insulin-like growth factor-binding protein-4 protease produced by a rat neuronal cell line. Endocrinology 1994; 135:1328–1335.
109. Chernausek SD, Smith CE, Duffin KL, Busby WH, Wright G, Clemmons DR. Proteolytic cleavage of insulin-like growth factor binding protein 4 (IGFBP-4). Localization of cleavage site to non-homologous region of native IGFBP-4. J Biol Chem 1995; 270:11377–11382.
110. Parker A, Gockerman A, Busby WH, Clemmons DR. Properties of an insulin-like growth factor-binding protein-4 protease that is secreted by smooth muscle cells. Endocrinology 1995; 136:2470–2476.
111. Irwin JC, Bsupin BA, Giudice LC. Regulation of insulin-like growth factor-binding protein-4 in human endometrial stromal cell cultures: evidence for ligand-induced proteolysis. J Clin Endocrinol Metab 1995; 80:619–626.
112. Myers SE, Cheung PT, Handwerger S, Chernausek SD. Insulin-like growth factor-I (IGF-I) enhanced proteolysis of IGF-binding protein-4 in conditioned medium from primary cultures of human decidua: independence from IGF receptor binding. Endocrinology 1993; 133:1525–1531.
113. Liu X-J, Malkowski M, Gui Y, Erickson GF, Shimasaki S, Ling N. Development of specific antibodies to rat insulin-like growth factor-binding proteins (IGFBP-2 to -6): analysis of IGFBP productioon by rat granulosa cells. Endocrinology 1993; 132:1176–1183.
114. Price WA, Boats-Staats BM, Stiles AD. Insulin-like growth factor-I (IGF-I) regulates IGFBP-3 and IGFBP-4 by multiple mechanisms in A549 human adenocarcinoma cells. Cell Mol Biol 1995; 13:466–476.
115. Noll K, Wegmann BR, Havemann K, Jaques G. Insulin-like growth factors stimulate the release of insulin-like growth factor-binding protein-3 (IGFBP-3) and degradation of IGFBP-4 in nonsmall cell lung cancer cell lines. J Clin Endocrinol Metab 1996; 81:2653–2662.
116. Fowlkes JL, Serra DM, Rosenberg CK, Thrailkill KM. Insulin-like growth factor (IGF)-binding protein-3 (IGFBP-3) functions as an IGF-reversible inhibitor of IGFBP-4 proteolysis. J Biol Chem 1995; 270:27481–27488.
117. Donnelly MJ, Holly JMP. The role of IGFBP-3 in the regulation of IGFBP-4 proteolysis. J Endocrinol 1996; 149:R1–R7.
118. Fowlkes JL, Thrailkill KM, George-Nascimento C, Rosenberg CK, Serra DM. Heparin-binding, highly basic regions within the thyroglobulin type-1 repeat of insulin-like growth factor (IGF)-binding proteins (IGFBPs) -3, -5, and -6 inhibit IGFBP-4 degradation. Endocrinology 1997; 138:2280–2285.
119. Durham SK, Riggs BL, Conover CA. The insulin-like growth factor-binding protein-4 (IGFBP-4)-IGFBP-4 protease system in normal human osteoblast-like cells: regulation by transforming growth factor-β. J Clin Endocrinol Metab 1994; 79:1752–1758.
120. Bachrach B, Wenstrup RJ, Smith EP, Chernausek SD. Vitamin C dependent induction of IGFBP-4 (BP-4) proteolysis by osteoblasts: evidence for a new BP-4 protease. In: 79th Annual Meeting of the Endocrine Society, Minneapolis, MN, June 11–14, 1997, Abstract.
121. Conover CA, Clarkson JT, Bale LK. Phorbol ester tumor promoters regulate insulin-like growth factor-binding protein-4 proteolysis. Endocrinology 1993; 133:1347–1351.
122. Durham SK, Riggs BL, Harris SA, Conover CA. Alterations in insulin-like growth factor (IGF)-

dependent IGF-binding protein-4 proteolysis in transformed osteoblastic cells. Endocrinology 1995; 136:1374–1380.
123. Kassem M, Okazaki R, DeLeon D, Harris SA, Robinson JA, Spelsberg TC, Conover CA, Riggs BL. Potential mechanism of estrogen-mediated decrease in bone formation: estrogen increases production of inhibitory insulin-like growth factor-binding protein-4. Proc Assoc Am Physicians 1996; 108:155–164.
124. Erickson GF, Li D, Sadrkhanloo R, Liu X-J, Shimasaki S, Ling N. Extrapituitary actions of gonadotropin-releasing hormone: stimulation of insulin-like growth factor-binding protein-4 and atresia. Endocrinology 1994; 134:1365–1372.
125. Piferrer F, Shimasaki DLS, Erickson GF. Transforming growth factor-α stimulates insulin-like growth factor binding protein-4 (IGFBP-4) expression and blocks follicle-stimulating hormone regulation of IGFBP-4 production in rat granulosa cells. Mol Cell Endocrinol 1997; 133:9–17.
126. Conover CA, Durham SK, Zapf J, Masiarz FR, Kiefer MC. Cleavage analysis of insulin-like growth factor (IGF)-dependent IGF-binding protein-4 proteolysis and expression of protease-resistant IGF-binding protein-4 mutants. J Biol Chem 1994; 270:4394–4400.
127. Iwashita M, Kudo Y, Hoshimura Y, Adachi T, Katayama E, Takeda Y. Physiological role of insulin-like-growth-factor-binding protein-4 in human folliculogenesis. Horm Res 1996; 46 (Suppl 1), 31–36.
128. Chandrasekher YA, Van Dessel HJHM, Fauser BCJM, Giudice IC. Estrogen- but not androgen-dominant human ovarian follicular fluid contains an insulin-like growth factor binding protein-4 protease. J Clin Endocrinol Metab 1995; 80:2734–2739.
129. Yap OW, Thierry van Dessel HJHM, Chandrasekher YA, Fauser BCJM, Giudice LC. Insulin-like growth factors and insulin-like growth factor binding proteins in androgen-dominant ovarian follicles from testosterone-treated female-to-male trans-sexuals. Fertil Steril 1997; 68:252–258.
130. Kubler B, Zward O, Saftig P, Roth W, Schmidt B, Scharf J, Hartmann H, Murphy G, Zapf J, von Figura K, Peters C, Braulke T. Insulin-like growth factor binding protein proteolysis in serum and cathepsin D- and cathepsin L-deficient mice. In: Proceedings of the Fourth International Symposium on Insulin-Like Growth Factors, Tokyo, Japan, October 21–24, 1997, p. 137.
131. Remacle-Bonnet MM, Garrouste FL, Pommier GJ. Surface-bound plasmin induces selective proteolysis of insulin-like-growth-factor (IGF)-binding protein-4 (IGFBP-4) and promotes autocrine IGF-II bio-availability in human colon-carcinoma cells. Int J Cancer 1997; 72:835–843.
132. Davies SC, Holly JMP, Coulson VJ, Cotterill AM, Abdulla AF, Whittaker PG, Chard T, Wass JAH. The presence of cation-dependent proteases for insulin-like growth factor binding proteins does not alter the size distribution of insulin-like growth factors in pregnancy. Clin Endocrinol 1991; 34:501–506.
133. Andress DI, Birnbaum RS. A novel human insulin-like growth factor binding protein secreted by osteoblast-like cells [published erratum appears in Biochem Biophys Res Commun 1991 Jun 14; 177 (2):895]. Biochem Biophys Res Commun 1991; 176:213–218.
134. Nam TJ, Busby WH Jr, Clemmons DR. Human fibroblasts secrete a serine protease that cleaves insulin-like growth factor-binding protein-5. Endocrinology 1994; 135:1385–1391.
135. Matsumoto T, Gargosky SE, Kelley K, Rosenfeld RG. Characterization of an insulin-like growth factor binding protein-5 protease produced by rat articular chondrocytes and a neuroblastoma cell line. Growth Regul 1996; 6:185–190.
136. Nam TJ, Busby WH Jr, Clemmons DR. Characterization and determination of the relative abundance of two types of insulin-like growth factor binding protein-5 proteases that are secreted by human fibroblasts. Endocrinology 1996; 137:5530–5536.
137. Arai T, Arai A, Busby WH Jr, Clemmons DR. Glycosaminoglycans inhibit degradation of insulin-like growth factor-binding protein-5. Endocrinology 1994; 135:2358–2363.
138. Fielder PJ, Pham H, Adashi EY, Rosenfeld RG. Insulin-like growth factors (IGFs) block FSH-induced proteolysis of IGF-binding protein-5 (BP-5) in cultured rat granulosa cells. Endocrinology 1993; 133:415–418.
139. Franchimont N, Durant D, Canalis E. Interleukin-6 and its soluble receptor regulate the expression of insulin-like growth factor binding protein-5 in osteoblast cultures. Endocrinology 1997; 138:3380–3386.
140. Thrailkill KM, Quarles LD, Nagase H, Suzuki K, Serra DM, Fowlkes JL. Characterization of insulin-like growth factor-binding protein 5-degrading proteases produced throughout murine osteoblast differentiation. Endocrinology 1995; 136:3527–3533.
141. Andress DL, Birnbaum RS. Human osteoblast-derived insulin-like growth factor (IGF) binding protein-5 stimulates osteoblast mitogenesis and potentiates IGF action. J Biol Chem 1992; 267:22467–22472.

142. Mohan S, Nakao Y, Honda Y, Landale E, Leser U, Dony C, Lang K, Baylink DJ. Studies on the mechanisms by which insulin-like growth factor (IGF) binding protein-4 (IGFBP-4) and IGFBP-5 modulate IGF actions in bone cells. J Biol Chem 1995; 270:20424–20431.
143. Arai T, Busby W Jr, Clemmons DR. Binding of insulin-like growth factor (IGF) I or II to IGF-binding protein-2 enables it to bind to heparin and extracellular matrix. Endocrinology 1996; 137:4571–4575.
144. Conover CA, Clarkson JT, Bale LK. Factors regulating insulin-like growth factor-binding protein-3 binding, processing, and potentiation of insulin-like growth factor action. Endocrinology 1996; 137:2286–2292.
145. McCusker RH, Camacho-Hubner C, Baye ML, Cascieri MA, Clemmons DR. Insulin-like growth factor (IGF) binding to human fibroblast and glioblastoma cells: the modulating effect of cell released IGF binding proteins. J Cell Physiol 1990; 144:244–253.
146. Michell NP, Dent S, Langman MJS, Eggo MC. Insulin-like growth factor binding proteins as mediators of IGF-I effects on colon cancer cell proliferation. Growth Factors 1997; 14:269–277.
147. Valentinis B, Bhala A, DeAngelis T, Baserga R, Cohen P. The human insulin-like growth factor (IGF) binding protein-3 inhibits the growth of fibroblasts with a targeted disruption of the IGF-I receptor gene. Mol Endocrinol 1995; 9:361–367.
148. Leal SM, Liu Q, Huang SS, Huang JS. The type V transforming growth factor-β receptor is the putative insulin-like growth factor-binding protein 3 receptor. J Biol Chem 1997; 272:20572–20576.
149. Parker A, Clarke JB, Busby WH Jr, Clemmons DR. Identification of the extracellular matrix binding sites for insulin-like growth factor-binding proteins 5. J Biol Chem 1996; 271:13523–13529.
150. Nam TJ, Busby W Jr, Clemmons DR. Insulin-like growth factor binding protein-5 binds to plasminogen activator inhibitor-I. Endocrinology 1997; 138:2972–2978.

III IGF Physiology

17 IGFs in the Female Reproductive System

Linda C. Giudice, MD, PhD

Contents

OVERVIEW

The female reproductive system is characterized by remarkably predictable changes that occur during the menstrual cycle that involve autocrine, paracrine, and endocrine actions of IGFs. These changes set the stage for ovulation of a single oocyte from the ovary, its fertilization, and transport of an embryo in the Fallopian tube, and implantation in the endometrium. In the absence of implantation, the corpus luteum undergoes regression, and the endometrium is shed as the menses. The primary goal of these events, which occur cycle to cycle, is successful establishment of a pregnancy. Dysfunctions in ovarian follicle development, oocyte maturation, the process of ovulation, ovum pickup and transport of gametes and an embryo through the Fallopian tubes, and nonreceptivity of the endometrium to implantation all have a common endpoint of undermining efforts to establish a normal pregnancy. To couples desiring fertility, abnormalities in one or more of these processes (as well as in the male) can pose formidable barriers to achieving their goals, whereas seekers of contraception have exploited several of these possibilities. In addition to an impact on fertility, dysfunction in the female reproductive system can lead to clinically significant sequelae for women. Anovulation predisposes women to dysfunctional uterine bleeding, endometrial hyperplasia, and/or endometrial cancer. Blocked Fallopian tubes can result in severe lower abdominal pain. Poorly developed endometrium can result in abnormal uterine bleeding and repeated miscarriage. Endometriosis, a benign gynecologic condition in which endometrial tissue is found outside of the uterus, responds

From: *Contemporary Endocrinology: The IGF System*
Edited by: R. Rosenfeld and C. Roberts Jr

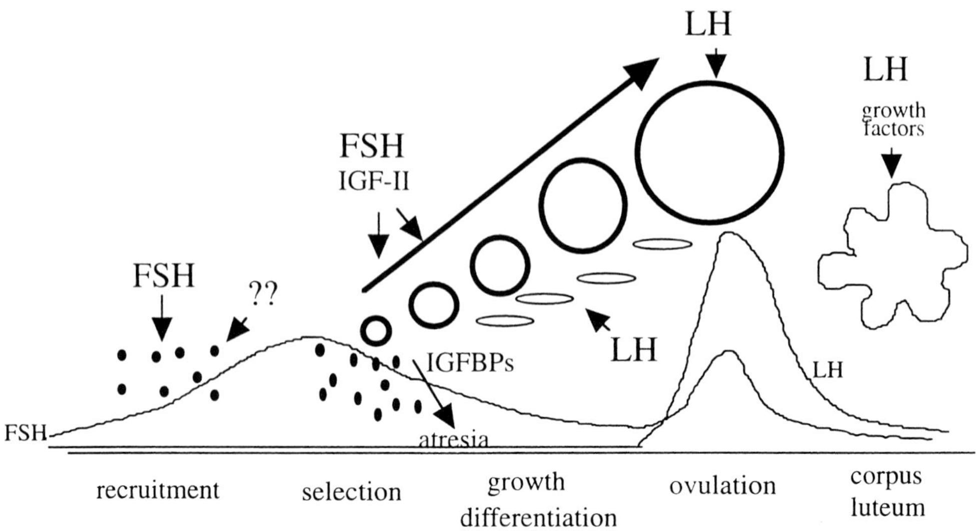

Fig. 1. Schematic representation of growth factor involvement in ovarian follicular development (*see text*).

to changes in circulating ovarian steroids and is associated with severe dysmenorrhea (pain with menses) and infertility. Uterine myometrium is most commonly thought of as a active participant in the process of labor and as a structural bystander in the nonpregnant state, other than its potential contributions to dysmenorrhea. However, leiomyomas or "fibroids," benign smooth muscle tumors of this compartment of the female reproductive tract, often impair fertility and have other side effects that are detrimental to women's health. These include abdominal/pelvic discomfort, excessive uterine bleeding, and repeated miscarriage. Leiomyomas comprise a leading cause of hysterectomy in women of reproductive age in the United States.

Mechanisms underlying normal reproductive function involve autocrine, paracrine, endocrine, and/or juxtacrine interactions of growth modulators. In this chapter, the roles of the insulin-like growth factor (IGF) family in normal and abnormal human ovarian follicle development, oviductal function, endometrial development, implantation, endometriosis, and in the growth of uterine leiomyomas are reviewed. Our information is at best incomplete and borrows heavily from animal models, although some reproductive disorders do not have natural animal homologues.

OVARY

Normal Ovarian Physiology

Ovarian estradiol is essential for development of secondary sexual characteristics at puberty and for growth of steroid-responsive tissues such as the breast and the uterus. Within 1–3 yr of menarche, regular, monthly ovulatory cycles are established that are highly dependent on a normally functioning hypothalamic–pituitary axis. Ovarian follicular development (Fig. 1) is independent of gonadotropin action up until the early antral stage. Thereafter, growth and acquisition of steroidogenic capability are dependent upon the presence of pituitary-derived follicle-stimulating hormone (FSH) and luteinizing hormone (LH) (*[1,2]*, reviews). In the absence of pregnancy, in the late luteal phase, FSH plays a key role in recruiting follicles for the next cycle (Fig. 1). FSH promotes follicle growth until the 4- to 6-mm stage in the mid-follicular phase. Thereafter, FSH levels fall, although the dominant follicle continues to grow and produce estradiol. The remainder of the cohort then undergoes atresia (*[2,3]*, reviews). The production of sufficient amounts of estradiol is essential for further follicle development and prevention of atresia *(5)*, and only in the follicle that has gained dominance is a substantial increase in granulosa aromatase gene expression and estradiol production observed (*[6]*, review). Enhancement of FSH action by local growth modulators, in the face of declining circulating FSH levels, is believed to be crucial for the major increase in aromatase activity in the follicle gaining dominance and in the growth of the follicles *per se.* Mechanisms underlying follicular selection are not well understood, although atresia of the remaining cohort occurs by apoptotic mechanisms that may also be under the control of growth factors and related peptides *(3)*. Ovulation in response to LH at mid-cycle (Fig. 1) also involves intraovarian growth modulators, as do corpus luteum function and subsequent regression. One of the best-characterized sets of growth modulators in human ovary (and in other species) is the IGF family. This chapter focuses on this family in human ovary. A recent review focuses on the IGF family and insulin in the ovary of humans and other species *(6)*.

IGFs in the Ovary

The IGF family, comprised of the IGF peptides, IGF binding proteins (IGFBPs), IGFBP proteases, and IGF receptors, is cycle dependent and highly compartmentalized in human ovary. A summary of the expression of the IGFs, IGF receptors, IGFBPs, and IGFBP proteases in human ovary is presented in Table 1. IGF-II is the *major* IGF peptide in human ovary. Human preovulatory granulosa abundantly express IGF-II mRNA, as do granulosa-luteal cells and the corpus luteum *(7–11)*. In contrast, IGF-1 mRNA is barely detectable in the adult ovary and not in the granulosa layer at any stage of follicle development, and it is controversial whether IGF-I is expressed at all by the thecal compartment *(9–11)*. It is unlikely that GH actions on human ovary are mediated via IGF-I, as human granulosa, which contain GH receptors *(12)*, do not express IGF-I *(7–11)*, and human theca do not have GH receptors *(13)*. The absence of IGF-II mRNA expression in prepubertal ovary and the in vitro observation that gonadotropins regulate the expression of IGF-II mRNA as well as secretion of IGF-II peptide by cultured human granulosa-luteal cells *(14,15)* suggest that ovarian IGF-II gene expression is regulated by gonadotropins.

IGFs in Follicular Fluid

Ovarian follicular fluid (FF) from women receiving gonadotropin therapy contains IGF-I and IGF-II peptides *(16–21)*. In normally cycling women, FF IGF-I levels are not statistically different in FF from estrogen- vs androgen-dominant follicles and do not correlate with follicular size *(20)*. These observations suggest that the origin of IGF-I in FF may be by transudation from serum and local production in the theca. FF IGF-II levels, however, are higher in estrogen-dominant, compared to androgen-dominant, follicles and correlate positively with follicle size and cycle day *(20)*. These observations suggest that FF IGF-II levels derive primarily from local production in the granulosa (and perhaps the theca), with some additional contribution from the circulation. There is no cycle dependence of circulating IGF-I and IGF-II levels in normally cycling women, underscoring the importance of the local intraovarian IGF system *(20)*. That circulating IGF-I

Table 1
Summary of IGFs, IGFBPs, IGFBP Proteases, and IGF Receptors (Ir,IIr) in Human Ovary[a]

		IGF-I	*IGF-II*	*IGFBP-1*	*IGFBP-2*	*IGFBP-3*	*IGFBP-4*	*IGFBP-5*	*Protease IGFBP-2*	*Protease IGFBP-3*	*Protease IGFBP-4*	*Protease IGFBP-5*	*Type Ir*	*Type IIr*
Atretic follicle	G	–/–	–/1+	–/–	3+/4+	–/–	2+/	1+/					2+/3+	2+
	T	2+/–	1+/1+	–/–	3+/4+	2+/	4+/	2+/					–/2+	2+/
	S	–/–	–/	–/–		–/–	4+/4+	4+/4+					–/	–/
	V			–/–			/4+							
Early follicle (3–5 mm)	G	–/–	2+/	–/–	/4+								3+/3+	3+/
	T	2+/–	2+	–/–	/4+								–/2+	3+/
	S	–/–		–/–									–/	–/
	V			–/–										
Late follicle (7–20 mm)	G	–/–	4+/4+	4+/	–/4+	3+/–	–/4+	–/–4+cu	+	+	+++	?	4+/4+	4+/
	T	–/–	–	–/–	4+/4+	3+/	2+/2+	2+/	+	+	+++		–/2+	3+/
	S	–/–		–/–	–/	–/–	4+/3+	4+/4+	+	+	+++		–/	–/
	V				/4+	/4+	/4+							
Corpus luteum (and granulosa luteal cells)	G	–	4+/4+	4+/					?	+	+++		–/*	4+/*
	T	–		–/–					?	+	+++		/	3+/
	S	–		–/–					?	+	+++		–/	–/
	V					/4+								

[a] Data are from *refs. 8–11,79,80,86*. Because there are discrepancies between the groups using *in situ* hybridization, the results are reported from refs. *9* and *10*. Data are presented as strong hybridization (4+) to weak hybridization (1+). If no number appears, the data were not reported. G, granulosa; T, theca; S, stroma; V, vascular, endothelium; cu, cumulus, ?, not known.

[b] Type I receptor mRNA expression present in granulosa-luteal cells, from *ref. 8*. (*see text*).

Table 2
IGF Actions in Human Ovary

Granulosa (and Granulosa/Luteal) Cells	*Theca Cells/Explants*	*Follicles*
+ Aromatase activity and mRNA	+ Androstenedione production	+ Oocyte maturation
+ Basal E_2 and P secretion	+ Testosterone production	
+ FSH-stimulated E_2 and P secretion	+ DNA synthesis	
+ DNA synthesis		
+ Granulosa proliferation		
– IGFBP-1, IGFBP-2 production		
+ IGFBP-4 proteolysis		
+ IGFBP-5 production		
+ IGFBP-2 proteolysis?		

+, stimulation; –, inhibition.

is not requisite for normal ovarian follicular development in humans is supported by the findings of patients with Laron-type dwarfism (or growth hormone receptor deficiency [GHRD]). In these patients, the GH receptor is absent or defective *(22,23),* although ovulation and fertility are not compromised and spontaneous conceptions have been reported *(24–27).* Furthermore, normal follicular development occurs in response to injectable gonadotropins, and ovulation, and conception occurred in a woman whose serum GH levels were markedly elevated, who was GH receptor deficient, and where serum and FF IGF-I levels were barely detectable *(25,27).* Serum IGF-II levels were about 25% of normal, and FF IGF-II, however, was not measured. These clinical observations suggest that circulating IGF-I is not essential for normal follicular development and that IGF-II is the major bioavailable IGF in human ovary. At the other extreme is the setting of elevated serum IGF-I levels as in acromegaly. It is currently not known whether FF levels of IGF-I are elevated in acromegaly, in parallel with elevated serum levels commonly observed in this disorder. In normal women treated with pharmacologic doses of IGF-I, enhancement of the preovulatory LH surge has been reported *(28),* suggesting a role for this peptide in the events surrounding ovulation.

IGF Receptors in the Ovary

IGF receptor mRNAs are expressed in granulosa and theca in human ovary *(8,10,11),* and specific binding sites have been demonstrated in granulosa cells *(29,30).* Functional type I IGF receptors have been described in theca *(31).*

IGF Actions in the Ovary

Most in vitro studies on the effects of IGFs on ovarian cellular function have used IGF-I, the primary IGF in the rat *(5),* although the endogenous ligand in humans in vivo is IGF-II. Most of the effects of the IGFs in human ovary are likely effected through the type I receptor, although some actions of IGF-II may also be mediated via the type II IGF receptor *(32).* IGF actions on ovarian cellular constituents include mitogenic effects and augmentation of steroidogenesis (Table 2). IGF-I stimulates DNA synthesis in human granulosa and granulosa-luteal cells *(33–35).* It also stimulates basal estradiol secretion by granulosa and granulosa-luteal cells *(36).* Furthermore, in human granulosa, IGF-I also synergies with FSH and human chorionic gonadotropin (hCG) in augmenting estradiol *(37,38)* and progesterone *(39)* production. In granulosa cells from unstimulated preovulatory follicles and gonadotropin-stimulated follicles, IGF-I alone and synergistically with gonadotropins stimulates cytochrome P_{450} aromatase activity and mRNA *(37,41–43)*. IGF-I inhibits IGFBP-1 production by luteinizing granulosa *(36,44–46)*. IGF-II stimulates basal progesterone and estradiol secretion by cultured human granulosa *(47–49)*. The effect of IGF-II on estradiol is most pronounced if the cells are preincubated with insulin, possibly due to insulin-induced upregulation of type I IGF receptors, formation of hybrid insulin/IGF-I receptors, or inhibition of IGFBP production *(49)*. IGF-II also stimulates aromatization of androgen precursors *(31)* and inhibits IGFBP-1 *(46)* and IGFBP-2 *(32)*. In addition, IGF-II also stimulates human granulosa proliferation in vitro *(50)*.

IGFs also have effects on thecal cells and oocytes. IGF-I enhances DNA and androgen synthesis in human thecal monolayer cultures *(51)*, and synergizes with LH in androstenedione production *(52)*. IGF-II also increases androgen secretion by human theca in culture *(31)*. IGF-I also augments in vitro maturation of immature human oocytes *(53)*. The IGF system has also been implicated in follicular atresia (*[3]*, review). Follicular atresia is associated with apoptosis and regular cleavage of nuclear DNA by endonucleases *(54)*. IGF-I and gonadotropins suppress this process in rat and chicken ovarian follicles in vitro, and it is enhanced by IGFBPs *(55)*. The cellular mechanisms underlying this action by IGF-I are beginning to be understood. Recently, apoptosis was shown to occur in human ovary, in androgen-dominant, but not estrogen-dominant, follicles *(56)*. It is likely that apoptotic mechanisms involving the IGF system also operate in human ovary.

IGFBPs and IGFBP Proteases in the Ovary

IGFBPs in Ovary

Human ovary expresses mRNAs encoding IGFBP-1, -2, -3, -4, and -5 *(10,11)*. *In situ* hybridization and immunohistochemical analyses have revealed distinctive patterns of mRNA expression and corresponding immunoreactive protein in antral follicles *(10,11)*. IGFBP-1 is expressed in granulosa cells of the dominant follicle following the LH surge and is highly expressed in the corpus luteum *(11,57)*. IGFBP-2 is expressed by granulosa of small, nondominant antral follicles, and IGFBP-3 is expressed in granulosa of only dominant follicles. IGFBP-4 is expressed in slightly greater amounts in granulosa of dominant, compared to small antral follicles. IGFBP-5 is also expressed by granulosa. IGFBP-2, -3, -4, and -5 are expressed in human theca from small antral follicles and dominant follicles (Table 1, *[10,11]*).

IGFBPs in Follicular Fluid

IGFBP-1, -2, and -3 are detectable in FF from gonadotropin-stimulated, luteinizing follicles *(19,57–59)*. In FF from normally cycling women, IGFBP-1 (*59a*) and IGFBP-2, IGFBP-3, and IGFBP-4 *(60)* are present. Different IGFBP profiles are characteristic of estrogen-dominant, healthy, growing follicles and androgen-dominant, atretic or growth-arrested follicles *(60–62)*. Androgen-dominant follicles (with low estrogen to androgen ratios) have high levels of IGFBP-2 and IGFBP-4, compared to healthy estrogen-dominant follicles that have barely detectable levels of these IGFBPs by ligand binding techniques. The unique expression of IGFBPs in follicles destined for growth, atresia, or arrested development suggests a role for the IGFBPs in these processes (*vide infra*).

Production, Regulation, and Action of Ovarian IGFBPs

Luteinizing granulosa cells synthesize IGFBP-1 and IGFBP-2 *de novo (32,63)* and secrete IGFBP-1, IGFBP-2, and IGFBP-3 in vitro *(64–67)*. Cultures of human thecal tissue produce IGFBP-2, -3, and -4 *(68–70)*. Thecal IGFBP-3 accumulation was markedly stimulated by LH/hCG in one study *(70)*, but not in others *(69,71)*. Thecal expression of IGFBP-5 mRNA, but not -1, -2, -3, or -4, is stimulated by LH *(71)*. Each IGFBP produced by luteinizing granulosa is uniquely regulated. IGFBP-1 is inhibited by insulin, IGF-I, and IGF-II, FSH, and octreotide, a somatostatin analog, and increased by LH, epidermal growth factor, prostaglandins, and phorbol esters *(45,46,64,72–75)*. IGF-I and IGF-II acting via the type I receptor are more potent in inhibiting IGFBP-1 than insulin, which acts via its homologous receptor *(45,46)*. IGFBP-2 production by luteinizing granulosa is inhibited by LH/hCG via cAMP *(32)*. This is reversed by activin-A or interferon-γ *(76)*. IGFBP-2 is also inhibited by IGF-II, but not IGF-I, suggesting possible action via the type II IGF receptor *(32)*. Granulosa IGFBP-3 is stimulated by cAMP, although it also appears to be degraded by an IGFBP-3 protease (*vide infra*). IGFs appear to increase IGFBP-3 in medium conditioned by luteinizing granulosa, either by releasing IGFBP-3 from the cell surface or by inhibiting an IGFBP-3 protease *(32,66)*. IGFBP-4 secretion is inhibited by LH, and IGFBP-5 mRNA in granulosa is stimulated by activin A *(32,38,76,77)*.

An important role for IGFBPs as modulators of IGF action in the ovary in vivo is likely, based on their action in vitro, which is primarily inhibitory. In human granulosa, IGFBP-1 inhibits IGF-I-stimulated [^{3}H]thymidine incorporation *(72)*. IGFBP-1 and IGFBP-3 inhibit IGF-I-stimulated granulosa estradiol and progesterone production *(78)* (and also *see below*). It has been postulated that inhibitory IGFBPs contribute to arrested follicular development. In normally cycling women, the IGFBP profile in FF is dependent on the functional status of the follicle (*vide supra*). It is likely that elevated levels of inhibitory IGFBPs in androgen-dominant follicles in normally cycling women and in those with PCOS (*vide infra*) decrease intrafollicular levels of bioavailable IGF peptides. This is presumed to contribute to arrested follicular development in the ovary.

IGFBP Proteases

In addition to the regulation of gonadotropin and IGF action in the ovary by control of IGFBPs, modulation of IGF action is further influenced by IGFBP proteases which affect the affinity of the IGFBPs for the IGFs. In human ovary, an IGFBP-4 protease is present in estrogen-dominant, but not androgen-dominant, FF *(79)*. It is a metal-dependent enzyme with a pH optimum of 6–8 and is made by nonluteinizing granulosa as well as by luteinizing granulosa. The degree of proteolysis of IGFBP-4 is inversely proportional to the androgen:estrogen ratio within the follicle *(80)*. Its activity is stimulated by gonadotropins and by IGF-II, and FSH and IGF-II synergistically stimulate this activity in nonluteinizing granulosa *(79–81)*. This protease activity may be one of several mechanisms in estrogen-dominant follicles which increases the levels of bioavailable IGF peptides for synergism with gonadotropins in steroidogenesis and follicle development. Recent evidence from three groups demonstrates that IGFBP-4 inhibits IGF-I-stimulated estradiol production by human granulosa cells, with inhibition exceeding 80% compared to similar in vitro experiments in which IGFBP-2 or IGFBP-3 inhibited granulosa steroidogenesis by about 20% *(81,82)*. In contrast to the inhibitory effects of recombinant human IGFBP-4 on granulosa estradiol production, addition of proteolyzed IGFBP-4 was without effect on estradiol production *(80,83,84)*, supporting an important role for the IGFBP-4/IGFBP-4 protease system in regulation of steroidogenesis in human ovary.

A recent study has shown that IGFBP-3 unsaturated with IGFs inhibits IGFBP-4 proteolysis, whereas, when saturated, it permits IGFBP-4 proteolysis *(85,86)*. The net result is that in estrogen-dominant follicles, where IGF levels are high and IGFBP-3 is presumably saturated, IGFBP-4 proteolysis increases IGF bioavailability from the pool of IGFs bound to this binding protein. In contrast, in androgen-dominant follicles, where IGFBP-3 is assumed to be unsaturated due to low levels of IGFs, any IGFBP-4 proteolytic activity is inhibited by unsaturated IGFBP-3.

IGFBP-3 protease was first observed in estrogen-dominant follicular fluid (FF_e) derived from patients undergoing in vitro fertilization, by Gargosky et al. *(87)*. Iwashita et al. *(84)* further demonstrated an IGFBP-3 protease in FF_e that cleaved radiolabeled IGFBP-3 into smaller fragments and was stimulated in media conditioned by luteinizing granulosa, in response to increasing doses of FSH. A low molecular weight fragment of IGFBP-3 (29 kDa) has been identified in FF from dominant, compared to small antral, follicles *(86,87)*, consistent with the presence of an IGFBP-3 protease. With regard to IGFBP-2, Western immunoblotting has revealed almost exclusively a 23-kDa IGFBP-2 fragment and minimally detectable intact IGFBP-2 in dominant follicle FF, compared with nearly exclusively intact IGFBP-2 and minimal fragments in FF from small cohort follicles *(60,86)*. These observations are consistent with an IGFBP-2 protease in FF_e, although specific IGFBP-2 proteolysis has not yet been demonstrated in these follicles.

Limited information is available regarding IGFBP proteases in the theca/stroma compartment of the human ovary or ovaries of other species. Mason and colleagues *(70)* have shown that human thecal explants secrete IGFBP-3, IGFBP-2, and glycosylated and native IGFBP-4. Low molecular weight forms of IGFBP-3 are present in medium conditioned by thecal explants *(86)*. The putative IGFBP-3 protease, present in conditioned medium of thecal explants is partially inhibited by metal chelators *(70,85)*.

MODEL

Because IGFs are potent stimulators of steroidogenesis and follicle growth, regulation of these peptides by IGFBPs and IGFBP proteases is temporally and spatially related within most tissues, with the ovary being no exception. This is likely to provide timed availability of IGF action and inhibition of action during follicular development or limited follicular growth or steroidogenesis, respectively. The consistent expression of IGFBP-4 protease, e.g., among different species, is remarkable (*[6]*, review), and attests to the importance of this protease in ovarian function. A model of gonadotropin, IGF, IGFBP, and IGFBP protease actions in human granulosa and theca is presented in Fig. 2. In healthy, growing, estrogenic follicles, androgen from theca is used as substrate for gonadotropin- and IGF-stimulated aromatase activity for the production of estradiol. High levels of bioavailable IGFs are effected within the follicle by a combination of decreased production of IGFBPs (mainly of IGFBP-2) and increased degradation by IGFBP-specific proteases (mainly IGFBP-4). Granulosa-derived IGFBPs are in turn regulated by IGFs, insulin, cAMP, and other modulators (*vide supra*). In androgen-dominant follicles, however, the amount of bioavailable IGFs is markedly decreased by high levels of inhibitory IGFBPs.

Polycystic Ovarian Syndrome (PCOS)

PATHOPHYSIOLOGY

Polycystic ovary syndrome (PCOS), the most common cause of anovulation in women, is a heterogeneous syndrome characterized by hyperandrogenism, persistent anovula-

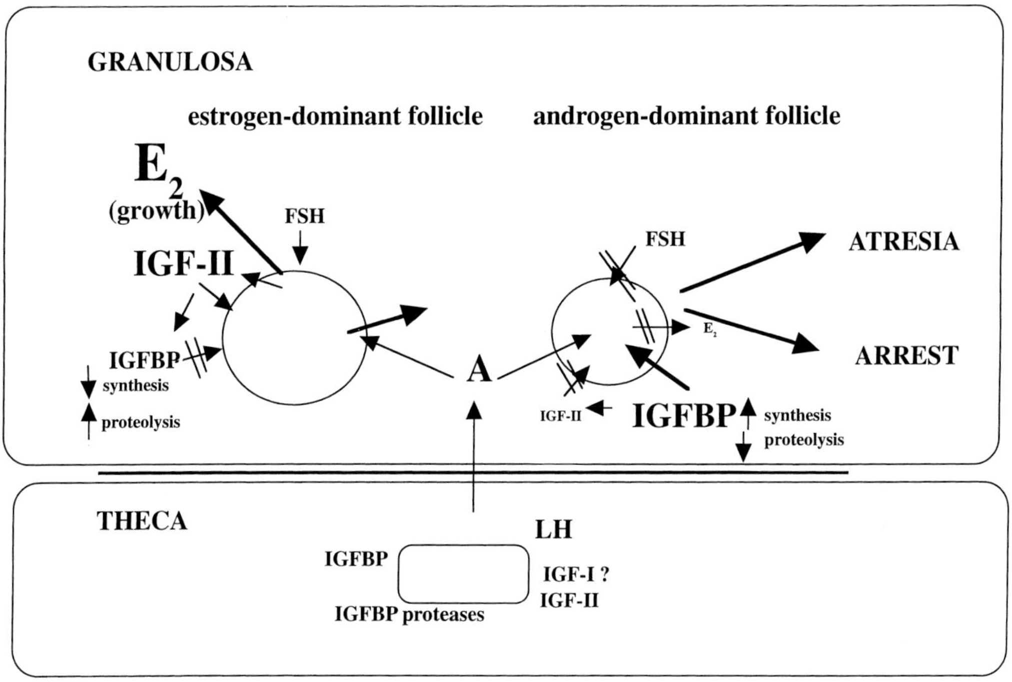

Fig. 2. A model depicting the role of members of the IGF family in ovarian follicle development and atresia (*see text*).

tion, oligomenorrhea or amenorrhea, hirsutism, elevated LH levels, and the accumulation of small cysts in the ovaries *(88–91)*. Many women with PCOS are obese and insulin-resistant, and hyperinsulinism has also been found in lean PCOS patients (*[91]*, review). There is also a subset of rare syndromes of extreme insulin-resistance which all have in common hyperandrogenism, insulin-resistance, and polycystic ovaries (*[91–93]*, reviews). In PCOS, the initial stages of follicle development, i.e., recruitment and growth to the small antral stage, are not impaired, although selection of a dominant preovulatory follicle does not occur. This leads to the accumulation of many small antral follicles in the ovary *(90)*. Ovulation can be induced in most patients with an anti-estrogen, e.g., clomiphene citrate, although about 30% of PCOS women are clomiphene-resistant and require gonadotropins for ovulation induction. Resistance to exogenously administered gonadotropins is also often observed clinically. Although the subject of extensive research, the basic mechanisms underlying PCOS are still largely unknown. Recently, the ovary has become the primary focus of investigation, as more information has been gleaned regarding intraovarian growth factors that coregulate or mediate gonadotropin actions. Serum and follicular fluid levels of FSH in anovulatory women with PCOS are within the normal, preovulatory range *(90,94,95)*. Also, granulosa cells of PCOS patients show normal *(41,96,97)* or even elevated *(98)* FSH-induced aromatase activity in vitro. Because FSH stimulation and granulosa aromatase activity are normal when PCOS granulosa are removed from the PCOS ovarian environment, it has been postulated that locally produced growth factors or inhibitors are involved in arrested follicular growth in women with PCOS. Such factors likely lead to a higher follicular FSH threshold *(99)*, with the consequence of relative gonadotropin resistance and anovulation. Both elevated LH (acting on theca and increasing ovarian androgen production) and hyperinsulinism are believed to contribute to PCOS, and the potential contributions of the IGF system in the pathogenesis of PCOS are reviewed herein.

IGF Family

IGFs stimulate granulosa aromatase and cellullar proliferation (*vide supra*), processes that are characteristically absent in PCOS follicle in vivo. The lack of stimulation of aromatase may be due to intrinsic defects in PCOS granulosa aromatase, inadequate levels of intrafollicular IGFs or FSH, or inhibition of IGF and/or FSH action within the PCOS follicle. Follicular fluid in PCOS follicles contains physiological levels of both FSH *(90,95)* and IGF-I *(100,100)*. Although PCOS granulosa cells produce little aromatase in vivo, they respond to FSH and/or IGF-I in vitro in a normal *(96,97)* or even elevated *(98)* fashion. Thus, the defect in PCOS is not a lack of stimulating hormones or a decreased responsiveness of the granulosa to IGFs (or FSH) with regard to aromatase production *per se*. Therefore, it has been postulated that locally active inhibitors of IGFs and/or FSH in follicullar fluid of small antral PCOS follicles block stimulation of aromatase. IGFBPs are likely candidates. Indeed, IGFBP profiles in PCOS follicles are very similar to androgen-dominant, small antral (atretic) follicles obtained from regularly cycling women. There are high levels of IGFBP-2 and IGFBP-4 in the androgen-dominant follicles of the PCOS ovary *(62,67,102)*, similar to androgen-dominant follicles in normally cycling women and in marked contrast to the near absence of these IGFBPs in estrogen-dominant follicles, detected by ligand binding techniques. *In situ* hybridization studies support these observations *(11)*. In addition, the IGFBP-4 protease is not detectable in the PCOS follicle *(21)*, similar to small, androgen-dominant follicles in normally cycling women and in marked contrast to healthy, estrogen-dominant follicles *(79)*. In view of the mostly inhibitory actions of IGFBPs on IGF action, these observations are consistent with this class of molecules being one of several groups of inhibitors of ovarian follicular development in the pathogenesis of PCOS. However, the lack of distinction between androgen-dominant, arrested follicles in PCOS and androgen-dominant, atretic follicle in normally cycling women would argue against a unique role for IGFBPs in the PCOS ovary.

Insulin Resistance, IGFs, and Hyperandrogenemia

An association between hyperinsulinism and hyperandrogenism is well established *(91,93)*. In vitro theca-interstitial cells obtained from women with hyperandrogenism release significantly more androgen per milligram of tissue than do those from non-hyperandrogenic women *(103)*. While it has been postulated that insulin may cross-react at the type I IGF receptor to augment LH-stimulated thecal androgen production, levels would have to be several-fold higher than are observed clinically in women with PCOS and hyperinsulinism. Recent studies show that the insulin receptor mediates insulin action in granulosa derived from women with PCOS *(104)*. Why there is not excessive stimulation of the granulosa, which are richly endowed with the type I IGF receptor, have abundant substrate for aromatase, and bathe in physiologic FSH and IGF levels, is not certain at this time, although restriction of IGF bioavailability by high levels of intrafollicular inhibitory IGFBPs may contribute in part to this resistance to aromatase stimulation (*vide supra*). Thus, both elevated LH and insulin and perhaps IGFs lead to hyperandrogenism observed in this disorder. In addition to direct action on the ovary, altered levels of SHBG can also contribute to increased bioavailable androgens in the circulation.

With regard to the IGF system and its potential role in the pathogenesis of hyperandrogenism observed in women with PCOS, elevated circulating levels of IGF-I may have effects similar to those proposed for insulin. Total circulating levels of IGFs have been reported to be higher in PCOS than in normo-ovulatory women in one study *(105),* but not in others *(106–108).* Free but not total circulating IGF-I was found to be elevated in women with PCOS in another study *(109).* An alternative approach to elevate IGF biooactivity is by changing relative levels of IGFBPs. Because IGFs in serum are bound to a large extent to IGFBPs, and IGFBPs mostly have inhibitory effects on IGF actions, decreased IGFBP serum concentrations may lead to increased bioavailability of IGF within serum and thus in the ovary. Lower IGFBP-1 levels are seen in PCOS women compared to controls, particularly if they are obese *(20,105,106,109–112),* likely a consequence of hyperinsulinism. Fasting serum IGFBP-1 concentrations negatively correlate with insulin in all subjects, including those with PCOS *(108,113–115).* In women with PCOS, IGFBP-1 levels decline during both oral and intravenous glucose tolerance tests, mirroring the insulin levels *(115,116).* Hyperandrogenism is not correlated with serum IGFBP-1 in obese women. In lean women, PCOS is associated with lower IGFBP-1 levels. Furthermore, normalization of serum IGFBP-1 in PCOS correlates with normalization of insulin levels, but not androgen levels. Ovarian electrocautery, which improves ovulatory function and decreases hyperandrogenemia, and also GnRH agonist suppression of ovarian steroid production have no effect on IGFBP-1 serum levels or insulin sensitivity *(117–119).* These observations suggest that it is hyperinsulinemia, not hyperandrogenemia, that leads to lower IGFBP-1 levels and that the liver in PCOS responds normally to elevated insulin by decreasing IGFBP-1 synthesis *(105,115).* Serum IGFBP-1 levels reflect short-term fluctuations in insulin levels *(120)* and the extent of peripheral insulin resistance. In PCOS women, IGFBP-1 may be useful clinically as a marker for insulin resistance *(117).*

Levels of the main circulating carrier of IGFs, IGFBP-3, are the same in normo-ovulatory women and women with PCOS *(102,109,121–123).* Higher levels of bioavailable IGFs (in serum or locally in the theca and stroma) may synergize with LH in stimulating thecal androgen production, resulting in the hyperandrogenism characteristic of this syndrome. High levels of free IGFs, however, may not synergize with FSH in stimulating granulosa aromatase activity, due to high intrafollicular levels of IGFBPs in the androgen-dominant PCOS follicles *(102).* Whether higher levels of free IGFs are present in the circulation or in the ovarian microenvironment (theca and stroma) of PCOS patients, compared to normo-ovulatory women, remains controversial and is worthy of further investigation.

OVIDUCT

Cyclic changes in the oviduct occur to assist gamete and embryo trafficking at the appropriate time, and perhaps also for embryo nourishment. The critical role of the oviduct, beyond that of a conduit, is underscored by higher pregnancy rates in advanced reproductive technologies that include embryo or gamete transfer into the oviduct, compared to direct transfer into the uterine cavity. This section reviews the limited data available on the IGF family in the oviduct, which may fulfill paracrine roles in gamete trafficking and preimplantation embryonic development. IGF-I and IGF-II are present in oviductal fluid and in media conditioned by oviductal cells *(124–126).* Oviductal cells express IGF-I and IGF-II mRNAs, both IGF receptors, and IGFBPs *(127–132).*

Immunohistochemical studies have confirmed the presence of IGF-I and the type I IGF receptor in human and rat oviductal epithelium *(132,133),* and IGF-I mRNA and protein are differentially expressed in the late proliferative phase of the cycle *(131,133).* The data suggest that IGF-I may mediate estradiol action in this tissue, as it appears to do in uterine endometrium *(vide infra).*

ENDOMETRIUM

Normal Endometrial Physiology and Function

Uterine endometrium is a dynamic tissue whose cellular components proliferate during the aptly named proliferative phase and undergo differentiation and some proliferation in the secretory phase of the menstrual cycle. It is comprised of a variety of cell types, including glandular and surface epithelium, stroma, vascular endothelium and smooth muscle cells, and a variety of immune cells (*[134]* review). The endometrium's mission is to provide a receptive environment to the conceptus. This includes appropriate apposition of the embryo to the surface epithelium, transition through the epithelium, and invasion into the endometrial stroma and maternal vasculature. It must protect the conceptus from rejection by the maternal immune system and provide a nourishing environment. In the absence of implantation, a series of events is initiated, prompted by waning corpus luteum function, that permits orderly tissue shedding, efficient hemostasis, and tissue regeneration. These changes in endometrium are regulated by steroid hormones and growth factors and related peptides that coregulate or mediate the effects of the gonadal steroids on this tissue. Although growth factors are found in many tissues, and endometrium is no exception, what is striking about endometrium is variation in levels of several growth factors and their receptors in this tissue throughout the menstrual cycle. Recently, growth factors in human endometrium have been extensively reviewed *(134).* The IGF system is an important participant in endometrial proliferation, development, and implantation and is reviewed herein.

The IGF Family in Cycling Endometrium

Interest in this growth factor system in the uterus arose from two landmark studies. The first revealed high levels of IGF-I mRNA expressed in rat uterus that was estradiol (E_2)-dependent, increasing nearly 20-fold after E_2 administration *(135).* The second study revealed that a major secretory protein of decidualized human endometrium was an IGF binding protein, IGFBP-1 *(136–138).*

Members of the IGF family undergo unique changes in human endometrium throughout the menstrual cycle *(139–145).* IGF-I mRNA is expressed in stromal cells in mid-late proliferative and early secretory endometrium, whereas IGF-II mRNA is preferentially expressed in secretory, compared to proliferative, endometrium. Because of its temporal expression and E_2 dependence, IGF-I has been postulated to mediate the mitotic actions of E_2 in this tissue. IGF-I is stimulated by E_2 in endometrium of other species and has been implicated in E_2-dependent neoplasms of human endometrium *(146).* Thus, IGF-I is likely one of several estromedins during the proliferative phase, acting as a mitogenic stimulus to effect rapid endometrial growth. In contrast, IGF-II, expressed abundantly in mid-late secretory endometrium, may be a mediator of progesterone (P) action *(147).* The type I IGF receptor is expressed primarily in glandular epithelium and lower amounts are expressed in stroma. Because the IGF peptides are stromally derived, they are believed to participate in stromal proliferation, by autocrine mechanisms, and in epithelial proliferation, by paracrine mechanisms. IGF-I and IGF-II are mitogenic to in vitro cultured human endometrial cells and regulate secretory functions of stromal cells *(147,148),* supporting roles for these peptides in endometrium in vivo.

Messenger RNAs encoding IGFBP-1-6 are expressed primarily in stroma in human endometrium. IGFBP-5 is the only IGFBP that is preferentially expressed in the proliferative phase *(145)*. It may facilitate transport of the IGFs from their sites of synthesis (stroma) to their sites of action (glandular epithelium and stroma) in proliferative phase. This conjectural function of IGFBP-5 is based on the observation that IGFBP-5 is found in many tissues in the extracellular matrix and proteolytic mechanisms exist for "off-loading" IGFs from this IGFBP (*[149],* review). IGFBP-1, -2, -3, -4, and -6 are differentially expressed in secretory endometrium. IGFBP-1, however, is by far the most abundantly expressed. While it is likely that IGFBP-2, -3, and -4 regulate IGF actions in both phases of the menstrual cycle, their compartmentalization may reflect regulation in select regions of the endometrium. In contrast, IGFBP-1 is exclusively expressed in the secretory, but not proliferative, phase and likely plays a major role in regulating IGF availability to receptors on both glandular epithelium and stroma during this phase of the cycle.

An IGFBP-1 transgenic mouse model is an ideal model to explore the hypothesis that IGF-I is a mediator of E_2 action in the mouse uterus *(150)*. In IGFBP-1 transgenic mice, IGFBP-1 mRNA and protein were expressed in the glandular epithelium. Impaired E_2 action (decreased DNA synthesis) in the uterus was observed, compared to wild type controls, suggesting that IGF-I is a mediator of E_2 action, and an IGFBP can regulate its actions *(150)*. The temporal and spatial relationships of IGF-I, IGF-II, and IGFBP-1 expression suggest that these members of the IGF family play major roles in endometrium during the menstrual cycle and in early pregnancy (*vide infra*).

The IGF Family at the Decidual–Trophoblast Interface

In addition to their roles in cycling endometrial development, the IGFs and IGFBP-1 are believed to play a major role in the process of implantation. As in cycling endometrium, IGF-II mRNA is expressed in the endometrium of pregnancy (decidua), but only when the pregnancy is extrauterine (e.g., tubal ectopic pregnancy) *(143)*. With an intrauterine pregnancy, IGF-II is exclusively expressed in the placenta *(151)*. The most abundant expression of IGF-II is in the columns of the intermediate (invading) trophoblasts in the anchoring villi, and there is a gradient of IGF-II mRNA expression in the cytotrophoblast columns, with the greatest levels expressed at the invading front, suggesting a role for IGF-II in trophoblast invasion *(151)* (Fig. 3). IGF receptor mRNAs are expressed in placental trophoblasts *(151),* suggesting these cells are targets for IGF actions. With regard to IGFBPs, trophoblasts do not express IGFBP mRNAs (except for weak expression of IGFBP-3 mRNA). This is in contrast to secretory endometrium and decidua, in which IGFBP-1 is the major IGFBP expressed and is a major protein product of these tissues *(143,151–155)*. When cocultured with human embryos, secretion of IGFBPs by endometrial cells is stimulated by embryonic signals and is hormonally dependent *(156),* underscoring the maternal–embryonic crosstalk during implantation.

IGFBP-1 Production and Regulation in Endometrium and Decidua

IGFBP-1, formerly known as placental protein-12 and α_1-progesterone-dependent endometrial globulin (α_1-PEG) *(137,157),* is made in the liver, granulosa-luteal cells,

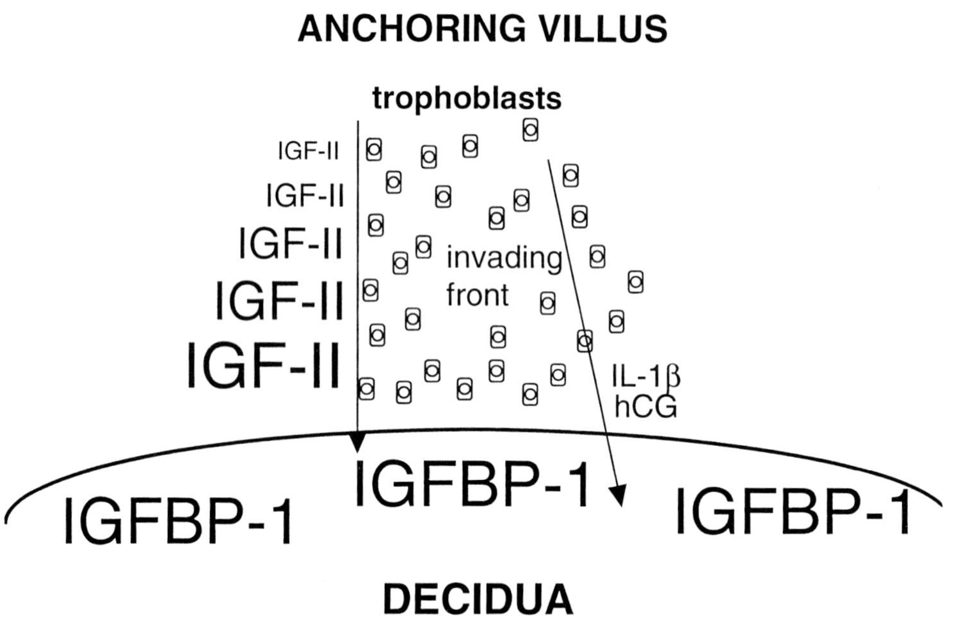

Fig. 3. Schematic of the interactions occurring at the decidual–trophoblast interface (*see text*).

secretory endometrium, and decidua in the adult human. IGFBP-1 has been immunolocalized to the extracellular matrix and stromal cells of decidualizing endometrium, in the periarteriolar regions, and on the villous trophoblast, but not on placental fibroblasts *(137,158–162).* In vivo human endometrial stromal cells undergo proliferation and differentiation in response to progesterone, after E_2 priming. Endometrial stromal cells can be decidualized in vitro with progesterone or progestins, or stimulators of cyclic AMP (cAMP), and E_2 or epidermal growth factor (EGF) *(163–165).* This in vitro model has provided an opportunity to investigate production and regulation of IGFBP-1, whose mRNA is expressed primarily in this cell type in endometrium in vivo. IGFBP production in endometrium is dependent on stromal differentiation *(142,148),* and although decidualized stromal cells increase their production of all IGFBPs upon decidualization in vitro, most striking are the markedly high levels of IGFBP-1 produced by these cells upon decidualization (25.3 + 3.2 μg/day per 10^6 cells). This is in contrast, e.g., to another decidual protein, prolactin (40 μg/10^6 cells/day) *(142,148,166).* Furthermore, progesterone regulates IGFBP-1 protein and mRNA expression in decidualized endometrial stromal cells, and the progesterone receptor antagonist RU486 is inhibitory *(142,166).* The IGFBP-1 gene has a glucocorticoid response element, and it is likely that progesterone, via this promoter element, exerts its effects on IGFBP-1 gene expression (*[113]* review). Besides progesterone and cAMP, other stimulators of IGFBP-1 gene expression include glucocorticoids *(112),* hCG (likely via cAMP) *(167,168),* HCG α-subunit *(169),* and hypoxia *(170).* In contrast, insulin, IGF-I, IGF-II, and IL-1β are inhibitory to IGFBP-1 gene expression *(148,171–174).* It is of interest that IL-1β and IGF-II are in the maternal–fetal interface, and thus their regulation of IGFBP-1 may serve to control trophoblast invasiveness in vivo (*see below).* The effects of interleukin-1β (IL-1β), transforming growth factor-β (TGF-β), stem cell factor (SCF), colony stimulating factor-1 (CSF-1), leukemia inhibitory factor (LIF), and IGF-II on IGFBP-1 production by decidualized endometrial stromal IGFBP-1 production in vitro reveal that only IGF-II and IL-1β have an effect on IGFBP-1 secretion by decidualized endometrial stromal cells in culture *(172)*. They are both inhibitory, with ED_{50} values consistent with actions through their respective receptors. IL-1β inhibits the process of decidualization of endometrial stromal cells *(173,174)*. Also, because IL-1β stimulates production of trophoblast matrix metalloproteinase-9 (MMP-9), which promotes the invasive trophoblast phenotype *(175)*, it is likely that inhibition of IGFBP-1 in the decidua by this cytokine further promotes trophoblast invasion, by inhibiting this decidual "restraint protein" (*vide infra*).

IGFBP-1 and Implantation

The spatial pattern and relative abundance of IGFBP-1 in decidua suggest that it interacts with the IGF-II-expressing, invading cytotrophoblast. In human implantation, invasion of the trophoblast is aggressive, and there is intermingling of the trophoblasts within the maternal decidua. There are likely multiple, redundant mechanisms in place to limit and facilitate the invasive process *(176)*. IGFBP-1 and IGF-II seem well poised to participate in the extensive maternal–placental dialogue at the molecular level. In vitro studies support modulatory roles for IGFBP-1 in trophoblast cellular function. IGFBP-1 inhibits the binding and biological activity of IGF-I on choriocarcinoma cells, suggesting that abundant decidual IGFBP-1 may regulate trophoblast-derived IGF autocrine and paracrine actions *(177)*. Besides its IGF binding properties, IGFBP-1 also has IGF-independent actions that include binding to cell membranes and altering cellular motility *(178)*. IGFBP-1 contains an Arg–Gly–Asp (RGD) motif, which is a recognition site for several cell-adhesion molecules, including the $\alpha_5\beta_1$ integrin. In Chinese hamster ovary (CHO) cells, IGFBP-1 binds to the $\alpha_5\beta_1$ integrin, presumably via its RGD sequence, and stimulates motility of these cells in vitro *(178)*. The invading trophoblast at the maternal–fetal interface in humans uniquely expresses the $\alpha_5\beta_1$ integrin, among all trophoblast phenotypes *(179)*. IGFBP-1 specifically binds to human trophoblasts and specifically to the $\alpha_5\beta_1$ integrin in trophoblasts. Furthermore, it inhibits trophoblast attachment to fibronectin, another RGD ligand in the placental bed *(180)*. Cytotrophoblast interaction with fibronectin, through the $\alpha_5\beta_1$ integrin, restrains invasion *(181)*. When human cytotrophoblasts are cocultured with human endometrial stromal cells decidualized in vitro and producing high amounts of IGFBP-1, trophoblast invasion into stromal cell multilayers is inhibited *(180)*. Furthermore, insulin inhibition of IGFBP-1 production results in trophoblast invasion into the stromal multilayers. The invasion is inhibited by the addition of exogenous IGFBP-1 into the coculture system, in a dose-dependent fashion *(180)*. These studies suggest that IGFBP-1 is likely one of several maternal "restraints" on trophoblast invasion. Whether it does this via direct interactions with the trophoblast (as an $\alpha_5\beta_1$ ligand or a ligand for other trophoblast cell-surface proteins) or by inhibiting IGF-II actions at the maternal–fetal interface is currently not known. IGFBP-1 has also been shown to stimulate migration of passaged human trophoblasts on a plastic surface *(182)*. These seemingly opposing observations of IGFBP-1's effects on trophoblast behavior may reflect differences in cell populations, in vitro culture conditions, and/or intermediate mediators present in cocultures vs singly cultured cells. Cumulatively, the data suggest a role for IGFBP-1, either as a direct modulator of trophoblast invasion or as a binding protein of IGF-II, at the maternal–fetal interface in humans (Fig. 3).

ENDOMETRIOSIS

Clinical Significance

Endometriosis is a benign gynecologic condition in which uterine endometrial tissue is found in extrauterine sites and responds to cycling steroid hormones. It is associated with incapacitating pelvic pain and refractory infertility (*[183]*, review). This common gynecologic disorder affects up to 10% of women in their reproductive years and up to 50% of women with infertility. Although it has been intensively studied, the pathogenesis of this disorder is not fully understood. The retrograde menstruation theory is the most commonly accepted, with endometrial cells attaching to and proliferating and differentiating on the pelvic peritoneum and other pelvic organs *(184)*. Growth factors and cytokines likely play a role in the pathogenesis and progression of growth of ectopic endometrium, and the IGF family is included in this group of growth mediators *(185)*.

The IGF Family and Endometriosis

Growth factors in peritoneal fluid (PF), which bathes the pelvic and abdominal cavity, are likely to be stimuli to ectopic endometriosis implant growth (*[186]*, review) and are derived from several sources, including macrophages, sloughed endometrium (by retrograde menstruation), ovarian follicular fluid, peritoneal cells, and endometriosis tissue itself. Among the plethora of growth factors in peritoneal fluid (*[185]*, review) are members of the IGF family. Human PF contains IGF-I; IGF-II; IGFBP-1, -2, -3, and -4; and an IGFBP-3 protease *(187)*. IGFBP-2 and -3 levels are ~ 50% of those in paired in paired serum samples *(187)*, likely reflecting the hydrostatic pressure gradient between plasma and the peritoneal compartment. The IGFs are present in concentrations > 50% of serum levels, suggesting that they may derive from one or more of the sources listed above. PF is mitogenic to endometrial cells, and about one third of this mitogenic activity is attributable to IGFs, as this activity is inhibited by a blocking antibody to the type I IGF receptor *(187)*. IGFBP-3 in PF exists primarily in fragments derived from proteolysis of this binding protein by an IGFBP-3 protease *(187)*. Although the identity of this enzyme has not yet been determined, it is a serine protease *(187)* other than plasmin *(188)*. This protease is not present in paired serum samples *(187)* and appears to be distinct from the pregnancy-serum IGFBP-3 protease *(189,190)*. Isolation of several mitogens from human PF *(191,192)* resulted in identification of one of the mitogens as an IGFBP-3 fragment *(193)*. Therefore, in addition to IGFs acting as mitogens on endometrial cells, IGFBP-3 fragments are also mitogens, and both may mediate E_2-induced growth of endometriotic foci. The expression and regulation of IGF components in endometriotic foci has not yet been reported, although it is likely that they will parallel those observed in normal endometrium. IGFBPs are present in higher levels in PF during the luteal, compared to the follicular phase *(194)*, but whether IGFs and IGFBP-3 fragments and protease levels differ in PF from women with and without endometriosis remains to be determined.

MYOMETRIUM AND UTERINE LEIOMYOMATA

Clinical Significance

Uterine leiomyomata ("fibroids") are benign smooth muscle cell tumors of myometrial origin and are the most common female reproductive tract neoplasms. Leiomyomas occur in approx 25% of Caucasian women and 50% of black women during their reproductive years, especially during the perimenopause, and many grow during pregnancy

(*[195]*, review). These observations, epidemiologic data, and the decrease in size that accompanies hypoestrogenic states (menopause or treatment with GnRH analogues) all suggest that these tumors are steroid-hormone (primarily estradiol)-dependent. Clinically, these benign tumors can cause abnormal uterine bleeding, lower abdominal and pelvic discomfort, and spontaneous abortion, depending on their size, location, and number. After surgical removal (myomectomy), additional leiomyomata often grow, and hysterectomy is a common treatment for these tumors. Of the 650,000 hysterectomies performed in the United States in 1985, 27% were conducted because of symptomatic leiomyomata. Despite their common occurrence, risks, and cost, precise mechanisms for growth of these benign tumors and explanations for racial differences have not been well defined. Although steroid hormones have been long implicated in normal and tumorous myometrial growth, roles for growth factors and related peptides in the pathogenesis of the growth of these tumors is beginning to be appreciated.

Insulin-Like Growth Factor Family

IGF-I and IGF-II mRNAs are expressed in myometrium, leiomyoma, and leiomyosarcoma *(140,196–198)*. The relative levels of IGF-I gene expression in these tissues is not resolved. Two studies report increased IGF-I expression in leiomyomas, compared to myometrium from the same uterus *(140,196)*, whereas two others suggest equivalent expression *(197,198)*. One report identifies reduced IGF-I mRNA in leiomyosarcoma *(197)*. IGF-I mRNA expression in leiomyomas was found to be dependent on the in vivo estrogen status in one *(199)*, but not another *(198)* study. However, in support of the estrogen dependence of IGF-I biosynthesis in leiomyomas is the observation that explants of leiomyomas (and myometrium) obtained at surgery from subjects pretreated with a GnRH agonist (resulting in hypoestrogenism) secrete markedly lower amounts of IGF-I, compared to explants of tissues obtained from placebo-treated controls *(200)*. In contrast to IGF-I, there is general agreement that IGF-II mRNA expression is significantly increased in leiomyoma, compared to myometrium, and is even more abundantly expressed in leiomyosarcoma *(140,197,198)*. The latter has led to elevated circulating IGF-II or "big IGF-II" and subsequent severe hypoglycemia *(201)*, due to cross-reactivity of IGF-II at the insulin receptor. Aberrant expression of IGF-II in sarcomas of the uterus has been linked to hypomethylation of the IGF-II gene *(202)*.

Myometrium, leiomyoma, and leiomyosarcoma contain binding sites for IGF-I and -II *(203–205)*. IGF-I binding is higher in leiomyoma than in myometrium *(203,204)*, due to increased receptor numbers *(203)*. Scatchard analysis suggests higher IGF-II binding in uterine leiomyosarcoma compared to either myometrium or benign leiomyomata *(205,206)*. Messenger RNA levels of both the type I and type II IGF receptor in leiomyomas are not menstrual cycle- or in vivo estrogen status-dependent *(199)*. With regard to IGFBPs, IGFBP-2 through -6 have been identified in myometrium and leiomyoma *(198,199)*. One study reports lower expression of IGFBP-3 in leiomyoma compared to myometrium *(198)*, although the effect that this may have on bioavailable IGF levels in leiomyoma tissue has yet to be determined. Likewise, differential expression of IGFBP-4, -5, and -6 in these tissues, and the activity of an IGFBP-4 protease found in uterine and leiomyoma smooth muscle cells *(207)* have not to date been explored with regard to a potential role in the pathogenesis of leiomyoma growth. While IGFs are potent mitogens for smooth muscle cells in vitro (Irwin and Giudice, *unpublished observations*, *208*) in vitro doubling times are much more rapid than the slow growth generally observed for leiomyomas in vivo. It is likely that growth factors and their inhibitors and other mechanisms of tumor growth (e.g., increased extracellular matrix production *[209]*) are operational in the slow growth of these benign tumors.

CHALLENGES FOR THE FUTURE

During the reproductive years, the diverse processes that occur in the female reproductive tract culminate in coordinated and repetitive events that are necessary for the continuation of the species. The absence of appropriate temporal and spatial coordination of underlying mechanisms can contribute to abnormal reproductive processes and morbidity to women of reproductive and postreproductive age. Although there is still much to be learned about members of the IGF family in general and in normal and abnormal ovarian follicular development, oviductal function, endometrial development, implantation, and endometriosis, as well as leiomyomata, additional challenges remain for the future. These include using our knowledge about basic mechanisms underlying IGF and IGFBP action and regulation to alter abnormal processes and enhance normal processes in the female reproductive system to the advantage of the organism. In addition, therapies targeted to members of the IGF family may be useful in achieving fertility, fertility regulation, and nonsurgical therapies for uterine fibroids and endometriosis.

ACKNOWLEDGMENTS

This work was supported in part by NIH HD31579 and HD25220.

REFERENCES

1. Gougeon A. Dynamics of follicular growth in the human: a model from preliminary results. Hum Reprod 1986; 1:81–85.
2. Richards J. Hormonal control of gene expression in the ovary. Endocr Rev 1994; 15:725–751.
3. Hsueh AJW, Billig H, Tsafriri A. Ovarian follicle atresia: a hormonally controlled apoptotic process. Endocr Rev 1994; 15:707–724.
4. McNatty KP, Moore-Smith D, Makris A, Osathanondh R, Ryan KJ. The micro-environment of the human antral follicle: interrelationship among the steroid levels in antral fluid, the population of granulosa cells, and the status of the oocyte in vitro and in vivo. J Clin Endocrinol Metab 1979; 49:851–860.
5. Adashi EY. In: Yen SSC, Jaffe RB, eds. Reproductive Endocrinology. Saunders, Philadelphia, 1991, pp. 181–237.
6. Poretsky L, Cataldo NA, Rosenwaks Z, Giudice LC. The insulin-related regulatory system in the ovary in health and disease. Endocr Rev 1999, in press.
7. Geisthovel F, Moretti-Rojas IM, Asch RH, Rojas FJ. Expression of insulin-like growth factor-II (IGF-II) messenger ribonucleic acid (mRNA), but not IGF-I mRNA in human preovulatory granulosa cells. Hum Reprod 1989; 4:899–903.
8. Hernandez ER, Hurwitz A, Vera A, Pellicer A, Adashi EY, LeRoith D, Roberts CT Jr. Expression of the genes encoding the insulin-like growth factors and their receptors in the human ovary. J Clin Endocrinol Metab 1992; 74:419–425.
9. El-Roeiy A, Chen X, Roberts VJ, LeRoith D, Roberts CT, Yen SSC. Expression of insulin-like growth factor-I (IGF-I) and IGF-II and the IGF-I, IGF-II, and insulin receptor genes and localization of the gene products in the human ovary. J Clin Endocrinol Metab 1993; 77:1411–1418.
10. Zhou J, Bondy C. Anatomy of the human ovarian insulin-like growth factor system. Biol Reprod 1993; 48:467–470.
11. El-Roeiy A, Chen X, Roberts VJ, Shimasaki S, Ling N, LeRoith D, Roberts CT, Yen SSC. Expression of the genes encoding the insulin-like growth factors, the IGF and insulin receptors, and IGF-binding proteins-1-6 and the localization of their gene products in normal and polycystic ovary syndrome ovaries. J Clin Endocrinol Metab 1994; 78:1488–1496.
12. Carlsson B, Bergh C, Bentham J, Olsson J-H, Norman MR, Billig H, Roos P, Hillensjo T. Expression of functional growth hormone receptors in human granulosa cells. Hum Reprod 1992; 7:1205–1209.

13. Katz M, Riciarelli E, Adashi EY. The potential relevance of growth hormone to female reproductive physiology and pathophysiology. Fertil Steril 1993; 59:8–34.
14. Ramasharma K, Li CH. Human pituitary and placental hormones control human insulin-like growth factor II secretion in human granulosa cells. Proc Natl Acad Sci USA 1987; 84:2643–2648.
15. Voutilainen R, Miller WL. Coordinate trophic hormone regulation of mRNAs for insulin-like growth factor II and the cholesterol side-chain cleavage enzyme, P450ssc, in human steroidogenic tissues. Proc Natl Acad Sci USA 1987; 84:1590–1595.
16. Ramasharma K, Cahrera CM, Li CH. Identification of insulin-like growth factor II in human seminal and follicular fluids. Biochem Biophys Res Commun 1986; 140:536–540.
17. Geisthovel F, Moretti-Rojas IM, Rojas FJ, Asch RH. Immunoreactive insulin-like growth factor I in human follicular fluid. Hum Reprod 1989; 4:35–40.
18. Rabinovici J, Dandekar P, Angle M, Rosenthal S, Martin M. Insulin-like growth factor I levels in follicular fluid from human preovulatory follicles: correlation with serum IGF-I levels. Fertil Steril 1990; 54:428–433.
19. Giudice LC, Farrell EM, Pham H, Rosenfeld RG. Identification of insulin-like growth factor binding protein-3 (IGFBP-3) and IGFBP-2 in human follicular fluid. J Clin Endocrinol Metab 1990; 71:1330–1338.
20. van Dessel TJHM, Chandrasekher YA, Fauser BCJM, Giudice LC. Serum and follicular fluid levels of insulin-like growth factor (IGF)-I, IGF-II, and IGF binding proteins-1 and -3 during the normal menstrual cycle. J Clin Endocrinol Metab 1996; 81:1224–1231.
21. Yap OWS, van Dessel TJHM, Chandrasekher YA, Fauser BCJM, Giudice LC. Similar insulin-like growth factor binding protein profiles in androgen-dominant follicular fluid from normal, polycystic ovary syndrome, and androgen-treated female-to male transsexual human ovaries. Fertil Steril 1997; 68:252–258.
22. Laron Z, Pertzelan A, Mannheimer S. Genetic pituitary dwarfism with high serum concentrations of growth hormone—a new inborn error of metabolism? Isr J Med Sci 1996; 2:152–155.
23. Godowski PJ, Leung DW, Meacham LR. Characterization of the human growth hormone receptor gene and demonstration of a partial gene deletion in two patients with Laron-type dwarfism. Proc Natl Acad Sci USA 1989; 86:8083.
24. Laron Z. In: von P Frick H, von Harnack GA, Kochsiek K, Martini GA, Prader A, eds. Advances in Internal Medicine and Pediatrics. Springer-Verlag, Berlin, 1984, 132–143.
25. Lunenfeld B, Menashe Y, Dor Y, Pariente C, Insler V. The effect of growth hormone, growth factors, and their binding globulins on ovarian responsiveness. Contracept Fertil Sex 1991; 19:133–136.
26. Menashe Y, Sack J, Mashiach S. Spontaneous pregnancies in two women with Laron-type dwarfism: are growth hormone and circulating insulin-like growth factor mandatory for induction of ovulation? Hum Reprod 1991; 6:670–671.
27. Dor J, Ben-Shlomo I, Lunenfeld B, Pariente C, Levran D, Karasik A, Seppala M, Mashiach S. Insulin-like growth factor-I (IGF-I) may not be essential for ovarian follicular development: evidence from IGF-I deficiency. J Clin Endocrinol Metab 1992; 74:539–542.
28. Duffy DM, Lindheim SR, Ary BA, Saad M, Vijod MA, Chang L, Lobo RA. The effects of growth hormone releasing factor (GRF) and insulin-like growth factor-I IGF-I) on folliculogenesis in normal women. Program of the American Fertility Society, 1993, Abs. 0-155, p. S-75.
29. Gates GS, Bayer S, Seibel M, Poretsky L, Flier JS, Moses AC. Characterization of insulin-like growth factor binding to human granulosa cells obtained during *in vitro* fertilization. J Recept Res 1987; 7:885–889.
30. Balboni GC, Vannelli GB, Barni T, Orlando C, Serio M. Transferrin and somatomedin C receptors in human ovarian follicles. Fertil Steril 1987; 48:706–710.
31. Nahum R, Thong KJ, Hillier SG. Metabolic regulation of androgen production by human thecal cells in vitro. Hum Reprod 1995; 10:75–81.
32. Cataldo NA, Woodruff TK, Giudice LC. Regulation of insulin-like growth factor binding protein production by human luteinizing granulosa cells cultured in defined medium. J Clin Endocrinol Metab 1993; 76:207–215.
33. Olsson J-H, Carlsson B, Hillensjo T. Effect of insulin-like growth factor-I on deoxyribonucleic acid synthesis in cultured human granulosa cells. Fertil Steril 1990; 54:1052–1057.
34. Angervo M, Koistinen R, Suikkari A-M, Seppala M. Insulin-like growth factor binding protein-1 inhibits the DNA amplification induced by insulin-like growth factor I in human granulosa-luteal cells. Hum Reprod 1991; 6:770–773.
35. Yong EL, Baird DT, Yates R, Reichert LE, Hillier SG. Hormonal regulation of the growth and steroidogenic function of human granulosa cells. J Clin Endocrinol Metab 1992; 74:842–850.

36. Wood AM, Lambert A, Hooper MAK, Mitchell GG, Robertson WR. Exogenous steroids and the control of estradiol secretion by human granulosa-lutein cells by follicle stimulating hormone and insulin-like growth factor-I. Hum Reprod 1994; 9:19–25.
37. Erickson GF, Garzo VG, Magoffin DA. Insulin-like growth factor-I regulates aromatase activity in human granulosa and granulosa-luteal cells. J Clin Endocrinol 1989; 69:716–720.
38. Mason H, Margara R, Winston RML, Seppälä M, Koistinen R, Franks S. Insulin-like growth factor-1 (IGF-1) inhibits production of IGF-binding protein-1 while stimulating estradiol secretion in granulosa cells from normal and polycystic human ovaries. J Clin Endocrinol Metab 1993; 76:1275–1279.
39. Erickson GF, Garzo VG, Magoffin DA. Progesterone production by human granulosa cells cultured in serum free medium: effects of gonadotrophins and insulin-like growth factor-I (IGF-I). Hum Reprod 1991; 6:1074–1081.
40. Steinkampf MP, Mendelson CR, Simpson ER. Effects of epidermal growth factor and insulin-like growth factor I on the levels of mRNA encoding aromatase cytochrome P-450 of human ovarian granulosa cells. Mol Cell Endocrinol 1988; 59:93–98.
41. Erickson GF, Magoffin DA, Cragun RJ, Chang RJ. The effects of insulin and insulin-like growth factors-I and -II on estradiol production by granulosa cells of polycystic ovaries. J Clin Endocrinol Metab 1990; 70:894–902.
42. Christman GM, Randolph JG, Peegel H, Menon KM. Differential responsiveness of luteinized human granulosa cells to gonadotropins and insulin-like growth factor-I for induction of aromatase activity. Fertil Steril 1991; 55:1099–1105.
43. Bergh C, Olsson J-H, Hillensjo T. Effect of insulin-like growth factor-I on steroidogenesis in cultured human granulosa cells. Acta Endocrinol (Copenh) 1991; 125:177–181.
44. Martin MC. Collection and report of assisted reproductive technology outcome data: problems and solutions. Fertil Steril 1993; 57:762–763.
45. Poretsky L, Chandrasekher YA, Bai C, Liu HC, Rosenwaks Z, Giudice L. Insulin receptor mediates inhibitory effect of insulin, but not of insulin-like growth factor (IGF)-I, on IGF binding protein-1 (IGFBP-1) production in human granulosa cells. J Clin Endocrinol Metab 1996; 81:493–496.
46. Poretsky L, Chun L, Liu B, Rosenwaks Z. Insulin-like growth factor II (IGF-II) inhibits insulin-like growth factor binding protein-1 (IGFBP-1) production in luteinized human granulosa cells with a potency similar to insulin-like growth factor I (IGF-I). J Clin Endocrinol Metab 1996; 81:3412–3414.
47. Kamada S, Kubota T, Taguchi M, Wen-Rong H, Sakamoto S, Aso T. Effects of insulin-like growth factor-II on proliferation and differentiation of ovarian granulosa cells. Horm Res 1992; 37:141–145.
48. Kubota T, Kamada S, Ohara M, Taguchi M, Sakamoto S, Shimizu Y, Aso Y. Insulin-like growth factor-II in follicular fluid of the patients with in vitro fertilization and embryo transfer. Fertil Steril 1993; 59:844–849.
49. Mason HD, Willis DS, Holly JMP, Franks S. Insulin preincubation enhances insulin-like growth factor-II (IGF-II) action on steroidogenesis in human granulosa cells. J Clin Endocrinol 1994; 78:1265–1231.
50. Di Blasio AM, Vigano P, Ferrari A. Insulin-like growth factor-II stimulates human granulosa-luteal cell proliferation in vitro. Fertil Steril 1994; 61:483–487.
51. Hillier SG, Miró F. Inhibin, activin, and follistatin: potential roles in ovarian physiology. Ann NY Acad Sci 1993; 687:29–38.
52. Bergh C, Carlsson B, Olsson JH, Selleskog U, Hillensjo T. Reglation of androgen production in cultured human thecal cells by insulin-like grwoth factor I and insulin. Fertil Steril 1993; 59:323–331.
53. Gomez E, Tarin JJ, Pellicer MD. Oocyte maturation in humans: the role of gonadotropins and growth factors. Fertil Steril 1993; 60:40–46.
54. Tilly JL, Kowalski KI, Schomberg DW, Hsueh AJW. Apoptosis in atretic ovarian follicles is associated with selective decreases in messenger ribonucleic acid transcripts for gonadotropin receptors and cytochrome P450 aromatase. Endocrinology 1992; 131:1670–1675.
55. Chun S-Y, Billig H, Tilly JL, Furuta I, Tsafriri A, Hsueh AJW. Gonadotropin suppression of apoptosis in cultured preovulatory follicles: mediatory role of endogenous insulin-like growth factor-I. Endocrinology 1994; 135:1845–1853.
56. Yuan W, Giudice LC. Programmed cell death in human ovary is a function of follicle and corpus luteum status. J Clin Endocrinol Metab 1997; 82:3148–3155.
57. Seppala M, Wahlstrom T, Koskimes AI, et al. Human preovulatory follicular fluid, luteinized cells of hyperstimulated preovulatory follicles, and corpus luteum contain placental protein 12. J Clin Endocrinol Metab 1984; 58:505–510.
58. Hartsthorne GM, Bell SC, Waites GT. Binding proteins for insulin-like growth factors in the human ovary: identification, follicular fluid levels and immunohistological localization of the 29–32 kd type 1 binding protein, IGF-bp1. Hum Reprod 1990; 5:649–660.

59. Hamori M, Blum WF, Torok A, Stehle R, Waibel E, Cledon P, Ranke MB. Insulin-like growth factors and their binding proteins in human follicular fluid. Hum Reprod 1991; 6:313–318.
59a. Holly JMP, Eden JA, Alaghband-Zadeh J, Carter GD, Jemmott RC, Cianfarani S, Chard T, Wass JA. Insulin-like growth factor binding protein in follicular fluid from normal dominant and cohort follicles, polycystic, and multicystic ovaries. Clin Endocrinol 1990; 33:53–64.
60. Cataldo NA, Giudice LC. Insulin-like growth factor binding protein profiles in human ovarian follicular fluid correlate with follicular functional status. J Clin Endocrinol Metab 1992; 74:821–829.
61. San Roman GA, Magoffin DA. Insulin-like growth factor binding proteins in healthy and atretic follicles during natural menstrual cycles. J Clin Endocrinol Metab 1993; 76:625–632.
62. Schuller AGP, Lindenbergh-Kortleve DJ, Pache TD, Zwarthoff EC, Fauser BCJM, Drop SLS. Insulin-like growth factor binding protein-2, 28 kDa and 24 kDa IGF-BP levels are decreased in fluid of dominant follicles, obtained from normal and polycystic ovaries. Regul Pept 1993; 48:157–163.
63. Suikkari AM, Jalkanen J, Koistinen R, Butzow R, Ritvos O, Ranta T, Seppala M. Human granulosa cells synthesize low molecular weight insulin-like growth factor-binding protein. Endocrinology 1989; 124:1088–1090.
64. Jalkanen J, Suikkari A-M, Koistinen R, et al. Regulation of insulin-like growth factor binding protein-1 production in human granulosa-luteal cells. J. Clin. Endocrinol Metab 1989; 69:1174–1180.
65. Giudice LC, Milki AM, Milkowski DA, El-Danasouri I. Human granulosa contain messenger ribonucleic acids (mRNAs) encoding insulin-like growth factor binding proteins (IGFBPs) and in culture secrete IGFBPs. Fertil Steril 1991; 56:475–480.
66. Hamori M, Blum WF, Torok A, Stehle R, Waibel E, Cledon P, Ranke MB. Immunoreactive insulin-like growth factor binding protein-3 in the culture of human luteinized granulosa cells. Acta Endocrinol (Copenh) 1991; 124:685–690.
67. San Roman GA, Magoffin DA. Insulin-like growth factor binding proteins in ovarian follicles from women with polycystic ovarian disease: cellular source and levels in follicular fluid. J Clin Endocrinol Metab 1992; 75:1010–1016.
68. Hamori M, Blum WF, Stehle R, Waibel E, Cledon P, Ranke MB. Immunoreactive insulin-like growth factor binding protein-3 in the culture of human luteinized granulosa cells. Acta Endocrinol (Copenh) 1991; 124:685–691.
69. Chandrasekher YA, Clark CR, Faessen GH, Giudice LC. Insulin-like growth factor binding protein profile in theca explant cultures from normal human ovaries. Proc Endocr Soc 1995; 166(Abstract).
70. Mason HD, Cwyfan-Hughes SC, Heinrich G, Franks S, Holly JMP. Insulin-like growth factor (IGF) I and II, IGF-binding proteins, and IGF-binding protein proteases are produced by theca and stroma of normal and polycystic human ovaries. J Clin Endocrinol Metab 1996; 81:276–284.
71. Voutilainen R, Franks S, Mason HD, Martikainen H. Expression of insulin-like growth factor (IGF), IGF-binding protein, and IGF receptor messenger ribonucleic acids in normal and polycystic ovaries. J Clin Endocrinol Metab 1996; 81:1003–1008.
72. Angervo M, Koistinen R, Seppala M. Epidermal growth factor stimulates production of insulin-like growth factor-binding protein-1 in human granulosa-luteal cells. J Endocrinol 1992; 134:12–131.
73. Adachi T, Iwashita M, Kuroshima A, Takeda Y. Regulation of IGF binding proteins by FSH in human luteinizing granulosa cells. J Assist Reprod Genet 1995; 12:639–643.
74. Holst N, Kierulf KH, Seppala M, Koistinen R, Jacobsen MB. Regulation of insulin-like growth factor-binding protein-1 and progesterone secretion from human granulosa-luteal cells: effects of octreotide and insulin. Fertil Steril 1997; 68:478–482.
75. Yap OWS, Chandrasekher YA, Giudice LC. Growth factor regulation of IGFBP regulation by cultured human granulosa-luteal cells. Fertil Steril 1998; 70:535–540.
76. Cataldo NA, Fujimoto VY, Jaffe RB. Interferon-gamma and activin-A promote insulin-like growth factor-binding protein-2 and -4 accumulation by human luteinizing granulosa cells, and interferon-gamma promotes their apoptosis. J Clin Endocrinol Metab 1998; 83:179–186.
77. Mason H, Willis D, Seppala M, Franks S. Insulin inhibits insulin-like growth factor binding protein-1 (IGFBP-1) production by granulosa, theca, and stroma from human ovaries. J Endocrinol 1993; 137:179–185.
78. Mason HD, Willis D, Holly JMP, Cwyfan-Hughes SC, Seppala M, Franks S. Inhibitory effects of insulin-like growth factor binding proteins on steroidogenesis by human granulosa cells in culture. Mol Cell Endocrinol 1992; 89:R1.
79. Chandrasekher YA, Van Dessel HJ, Fauser BC, Giudice LC. Estrogen- but not androgen-dominant human ovarian follicular fluid contains an insulin-like growth factor binding protein-4 protease. J Clin Endocrinol Metab 1995; 80:2734–2739.
80. Iwashita M, Kudo Y, Yoshimura Y, Adachi T, Katayama E, Takeda Y. Physiological role of insulin-

like-growth-factor-binding protein-4 in human folliculogenesis. Horm Res 1996; 46 (Suppl): 31–36.
81. Chandrasekher YA, Clark CR, Faessen GH, Giudice LC. Role of insulin-like growth factor binding protein-4 (IGFBP-4) and IGFBP-4 protease in human granulosa cell function. Proc Soc Study Reprod 1995; 82 (Abstract).
82. Mason HD, Cwyfan-Hughes S, Holly MP, Franks S. Potent inhibition of human ovarian steroidogenesis by insulin-like growth factor binding protein-4 (IGFBP-4). J Clin Endocrinol Metab 1998; 83:284–287.
83. Iwashita M, Adachi T, Katayama E, Kudo Y, Takeda Y. Regulation and physiological role of insulin-like-growth-factor-binding protein-I in human granulosa cells. Horm Res 1994; 41:22–28.
84. Iwashita M, Adachi T, Kudo T, Takeda Y. Gonadotropins regulate insulin-like growth factor binding proteins in human luteinizing granulosa cells. In: Fujimoto S, Hsueh AJ, Strauss JFIII, Tanaka T, eds. New Achievements in Research of Ovarian Function. Rome: Ares-Serono Symposia, 1995, pp. 141–146.
85. Mason HD, Cwyfan-Hughes S, Voutilainen R, Martikainen H, Franks S, Holly J. Insulin-like growth factors (IGFs) and IGF binding proteins (IGFBPs) in normal and polycystic ovaries. Front Endocrinol 1996; 19:207–218.
86. Cwyfan-Hughes S, Mason HD, Franks S, Holly JM. Modulation of the insulin-like growth factor-binding proteins by follicle size in the human ovary. J Endocrinol 1997; 154:35–43.
87. Gargosky SE, Pham HM, Wilson KF, Liu F, Giudice LC, Rosenfeld RG. Measurement and characterization of insulin-like growth factor binding protein-3 in human biological fluids: discrepancies between radioimmunoassay and ligand blotting. Endocrinology 1992; 131:3051–3060.
88. Yen SSC. The polycystic ovary syndrome. Clin Endocrinol 1980; 12:177–182.
89. Franks S, Mason HD, Polson DW, Winston RM, Margara R, Reed MJ. Mechanism and management of ovulatory failure in women with polycystic ovary syndrome. Hum Reprod 1988; 3:531–534.
90. Fauser BCJM. Observations in favor of normal early follicle development and disturbed dominant follicle selection in polycystic ovary syndrome. Gynecol Endocrinol 1994; 8:1–8.
91. Dunaif A. Insulin resistance and the polycystic ovary syndrome mechanism and implications for pathogenesis. Endocr Rev 1997; 18:774–800.
92. Poretsky L. On the paradox of insulin-induced hyperandrogenism in insulin-resistant states. Endocr Rev 1991; 12:3–10.
93. Poretsky L, Piper B. Insulin resistance, hypersecretion of LH, and a dual-defect hypothesis for the pathogenesis of polycystic ovary syndrome. Obstet Gynecol 1994; 84:613–621.
94. Fauser BCJM, Pache TD, Lamberts SW, Hop WC, De Jong FC, Dahl KD. Serum bioactive and immunoractive luteinizing hormone and follicle-stimulating hormone levels in women with cycle abnormalities, with or without polycystic ovarian disease. J Clin Endocrinol Metab 1991; 73:811–817.
95. Erickson GF, Magoffin DA, Garzo VG, Cheung AP, Chang RJ. Granulosa cells of polycystic ovaries: are they normal or abnormal? Hum Reprod 1992; 7:293–299.
96. Erickson GF, Hsueh AJW, Quigley ME, Rebar RW, Yen SSC. Functional studies of aromatase activity in human granulosa cells from normal and polycystic ovaries. J Clin Endocrinol Metab 1979; 49:514–519.
97. Mason HD, Margara R, Winston RML, Beard RW, Reed MJ, Franks S. Inhibition of oestradiol production by epidermal growth in human granulosa cells of normal and polycystic ovaries. Clin Endocrinol 1990; 33:511–517.
98. Mason HD, Willis DS, Beard RW, Winston RML, Margara R, Franks S. Estradiol production by granulosa cells of normal and polycystic ovaries: relationship to menstrual cycle history and concentrations of gonadotropins and sex steroids in follicular flluid. J Clin Endocrinol Metab 1994; 79:1355–1360.
99. Brown JB. Pituitary control of ovarian function—concepts derived from gonadotrophin therapy. Aust N Z J Obstet Gynaecol 1978; 18:47–54.
100. Eden JA, Jones J, Carter GD, Alaghband-Zadeh J. A comparison of follicular fluid levels of insulin-like growth factor-I in normal dominant and cohort follicles, polycystic and multicystic ovaries. Clin Endocrinol 1988; 29:327–336.
101. Eden JA, Jones J, Carter GD, Alaghband-Zadeh J. Follicular fluid concentrations of insulin-like growth factor 1, epidermal growth factor, transforming growth factor-alpha and sex-steroids in volume matched normal and polycystic human follicles. Clin Endocrinol 1990; 32:395–405.
102. Cataldo NA, Giudice LC. Follicular fluid insulin-like growth factor binding protein profiles in polycystic ovary syndrome. J Clin Endocrinol Metab 1992; 74:695–697.

103. Barbieri RL, Makris S, Randall RW, Daniels G, Kistner RW, Ryan KJ. Insulin stimulates androgen accumulation in incubations of ovarian stroma obtained from women with hyperandrogenism. J Clin Endocrinol Metab 1986; 62:904–910.
104. Willis D, Franks S. Insulin action in human granulosa cells from normal and polycystic ovaries is mediated by the insulin receptor and not the type-I insulin-like growth factor receptor. J Clin Endocrinol Metab 1995; 80:3788–3790.
105. Iwashita M, Mimuro T, Watanabe M, Setoyama T, Matsuo A, Adachi T, Takeda Y, Sakamoto S. Plasma levels of insulin-like growth factor-1 and its binding protein in polycystic ovary syndrome. Horm Res 1990; 33(Suppl 2):21–26.
106. Conway GS, Jacobs HS, Holly JM, Wass JA. Effects of luteinizing hormone, insulin, insulin-like growth factor-I and insulin-like growth factor small binding protein 1 in the polycystic ovary syndrome. Clin Endocrinol 1990; 33:593–603.
107. Suikkari A-M, Tiitinen A, Stenman U-H, Seppala M, Laatikainen T. Oral contraceptives increase insulin-like growth factor binding protein-1 concentration in women with polycystic ovarian disease. Fertil Steril 1991; 55:895–899.
108. Homburg R, Pariente C, Lunenfeld B, Jacobs HS. The role of insulin-like growth factor-1 (IGF-1) and IGF binding protein-1 (IGFBP-1) in the pathogenesis of polycystic ovary syndrome. Hum Reprod 1992; 7:1379–1383.
109. van Dessel HJ, Chandrasekher YA, Faessen G, Lee PD, Fauser BC, Giudice LC. Elevated serum levels of free insulin-like growth factor (IGF-I) in polycystic ovary syndrome. American Society for Reproductive Medicine, Boston, November 2–6, 1996, Abstract P-0217.
110. Pekonen F, Laatikainen T, Buyalos R, Rutanen EM. Decreased 34k insulin-like growth factor binding protein in polycystic ovarian disease. Fertil Steril 1989; 51:972–975.
111. Suikkari AM, Ruutiainen K, Erkkola R, Seppala M. Low levels of low molecular weight insulin-like growth factor-binding protein in patients with polycystic ovarian disease. Hum Reprod 1989; 4:136–139.
112. Lanzone A, Fulghesu AM, Pappalardo S, Proto C, Le Donne M, Andreani CL, Muscatello R, Caruso A, Mancuso S. Growth hormone and somatomedin-C secretion in patients with polycystic ovarian disease. Their relationships with hyperinsulinism and hyperandrogenism. Gynecol Obstet Invest 1990; 2:149–153.
113. Lee PDK, Jensen MD, Divertie GD, Heiling VJ, Katz HH, Conover CA. Insulin-like growth factor-binding protein-1 response to insulin during suppression of endogenous insulin secretion. Metabolism 1993; 42:409–414.
114. Lee PDK, Guidice LC, Conover CA, Powell DR. Insulin-like growth factor binding protein-1. Recent findings and new directions. Proc Soc Exp Biol Med 1997; 216:319–357.
115. Buyalos RP, Pekonen F, Halme JK, Judd HL, Rutanen EM. The relationship between circulating androgens, obesity, and hyperinsulinemia on serum insulin-like growth factor binding protein-1 in the polycystic ovarian syndrome. Am J Obstet Gynecol 1995; 172:932–939.
116. Hamilton-Fairley D, Kiddy D, Anyaoku V, Koistinen R, Seppala M, Franks S. Response of sex hormone binding globulin and insulin-like growth factor binding protein-I to an oral glucose tolerance test in obese women with polycystic ovary syndrome before and after calorie restriction. Clin Endocrinol 1993; 39:363–367.
117. Tiitinen AE, Tenhunen A, Seppala M. Ovarian electrocauterization causes LH-regulated but not insulin-regulated endocrine changes. Clin Endocrinol 1993; 39:181–184.
118. Greenblatt E, Casper RF. Endocrine changes after laparoscopic ovarian cautery in polycystic ovarian syndrome. Am J Obstet Gynecol 1987; 156:279–285.
119. Homburg R, Orvieto R, Bar-Hava I, Ben-Rafael Z. Serum levels of insulin-like growth factor-1, IGF binding protein-1 and insulin and the response to human menopausal gonadotrophins in women with polycystic ovary syndrome. Hum Reprod 1996; 11:716–719.
120. Hamilton-Fairley D, White D, Griffiths M, Anyaoku V, Koistinen R, Seppala M, Franks S. Diurnal variation of sex hormone binding globulin and insulin-like growth factor binding protein-I in women with polycystic ovary syndrome. Clin Endocrinol 1995; 43:159–165.
121. El-Roeiy A, Baxter RC, Chen X, Laughlin G, Yen SSC. The role of Growth hormone, insulin growth factors (IGF-I and IGF-II) and their binding proteins (IGFBP-3 and IGFBP-1) in PCO syndrome. Society for Gynecologic Investigation, San Antonio, TX, March, 1991, Abstract No. 517.
122. Morris RS, Carmina E, Vijod MA, Stanczyk FZ, Lobo RA. Alterations in the sensitivity of serum insulin-like growth factor I and insulin-like growth factor binding protein-3 to octreotide in polycystic ovary syndrome. Fertil Steril 1995; 63:742–746.

123. Morales AJ, Laughlin GA, Butzow T, Maheshwari H, Baumann G, Yen SS. Insulin, somatotropic and luteinizing hormone axes in lean and obese women with polycystic ovary syndrome: common and distinct features. J Clin Endocrinol Metab 1996; 81:2854–2864.
124. Wiseman DL, Henricks DM, Eberhardt DM, & Bridges WC. Identification and content of insulin-like growth factors in porcine oviductal fluid. Biol Reprod 1992; 47:126–132.
125. Lee CT, Mitchell KR, Diehl JR, Henricks DM. Secretion and expression of IGF-I in cultured porcine oviductal cells. Biol Reprod 1992; 46(Suppl 1):251(Abstract).
126. Diehl JR, Henricks DM, Gray SL. EGF and EGF-I in the uterine and oviductal fluids of pregnant and nonpregnant pigs from day 10 to day 18. [Abstract] Biol Reprod 1994; 50(Suppl 1):272.
127. Williams JE, Butler JE. Expression of insulin-like growth factor-I (IGF-I) gene by ovine oviductal cells. J Anim Sci 1991; 69(Suppl 1):406(Abstract).
128. Kincy VL, Williams JE, Butler JE. Expression of the insulin-like growth factor-II (IGF-II) gene by ovine oviductal cells. Biol Reprod 1992; 46(Suppl 1):68(Abstract).
129. Andrews GK, Dalton T, Kover K. Analysis of the expression of growth factor, interleukin-1, and lactoferrin gene and the distribution of inflammatory leukocytes in the preimplantation mouse oviduct. Biol Reprod 1994; 51:597–606.
130. Giudice LC, Dsupin BA, Irwin JC, Eckert RL. Identification of insulin-like growth factor binding proteins in human oviduct. Fertil Steril 1992; 57:294–301.
131. Giudice LC, Dsupin BA, Jin JH, Vu TH, Hoffman AR. Insulin-like growth factor (IGF), IGF binding protein, and IGF receptor gene expression in human oviductal fimbria and ampulla during the menstrual cycle. [Abstract 318] 40th Annual Meeting of the Society for Gynecologic Investigation, Toronto, March, 1993.
132. Carlsson B, Hillensjo T, Nilsson A, Tornell J, Billig H. Expression of insulin-like growth factor (IGF-I) in the rat fallopian tube: possible autocrine and paracrine action of fallopian tube-derived IGF-I on the fallopian tube and on the preimplantation embryo. Endocrinology 1993; 133:2031–2039.
133. Pfeifer T, Chegini N. Immunohistochemical localization of insulin-like growth factor (IGF-I), IGF-I receptor, and IGF binding proteins 1–4 in human fallopian tube at various reproductive stages. Biol Reprod 1994; 50:281–289.
134. Giudice LC. The Endometrium. In: Knobil E, Neill JD, eds. Enyclopedia of Reproduction. Academic Press, San Diego, 1998; 1067–1077.
135. Murphy LJ, Murphy LC, Friesen HG. A role for the insulin-like growth factors as estromedins in the rat uterus. Trans Assoc Am Physicians 1987; 99:204–214.
136. Koistinen R, Kalkinnen N, Huhtala M-L, Seppala M, Bohn H, Rutanen E-M. Placental protein 12 is a decidual protein that binds somatomedin and has an identical N-terminal amino acid sequence with somatomedin-binding protein from human amniotic fluid. Endocrinology 1986; 118:1375–1378.
137. Bell SC, Patel SR, Jackson JA, Waites GT. Major secretory protein of human decidualized endometrium in pregnancy is an insulin-like growth factor binding protein. J Endocrinol 1988; 118:317–328.
138. Julkunen M, Koistinen R, Aalto-Setala K, Seppalla M, Janne OA, Kontula K. Primary structure of human IGFBP/placental protein 12 and tissue expressing of its mRNA. FEBS Lett 1988; 236:295–301.
139. Julkunen M, Koistinen R, Suikkari A-M, Seppala M, Janne OA. Identification by hybridization histochemistry of human endometrial cells expressing mRNAs encoding a uterine β-lactoglobulin homologue and insulin-like growth factor-binding protein-1. Mol Endocrinol 1990; 4:700–707.
140. Boehm KD, Daimon M, Gorodeski IG, Sheean LA, Utian WH, Ilan J. Expression of the insulin-like and platelet-derived growth factor genes in human uterine tissues. Mol Reprod Dev 1990; 27: 93–101.
141. Giudice LC, Milkowski DA, Lamson G, Rosenfeld RG, Irwin JC. Insulin-like growth factor binding proteins (IGFBP-2 and IGFBP-3) in human endometrium: steroid-dependent messenger RNA expression and protein synthesis. J Clin Endocrinol Metab 1991; 72:779–787.
142. Giudice LC, Dsupin BA, Irwin JC. Steroid and peptide regulation of insulin-like growth factor binding proteins secreted by human endometrial stromal cells is dependent on stromal differentiation. J Clin Endocrinol Metab 1992; 75:1235–1241.
143. Giudice LC, Dsupin BA, Jin IH, Vu TH, Hoffman AR. Differential expression of mRNAs encoding insulin-like growth factors and their receptors in human uterine endometrium and decidua. J Clin Endocrinol Metab 1993; 76:1115–1122.
144. Giudice LC, Irwin JC, Dsupin BA, de las Fuentes L, Jin IH, Vu TH, Hoffman AR. Insulin-like growth factors (IGFs), IGF binding proteins (IGFBPs), and IGFBP protease in human uterine endometrium: their potential relevance to endometrial cyclic function and maternal-embryonic interactions. In: Baxter

RC, Gluckmann PD, Rosenfeld RG, eds. The Insulin-Like Growth Factors and Their Regulatory Proteins. Excerpta Medica International Congress Series 1056, Amsterdam, 1994, pp. 351–361.
145. Zhou J, Dsupin BA, Giudice LC, Bondy CA. Insulin-like growth factor system gene expression in human endometrium during the menstrual cycle. J Clin Endocrinol Metab 1994; 79:1723–1734.
146. Rutanen E-M, Nyman T, Lehtovirta P, Ammala M, Pekonen F. Suppressed expression of insulin-like growth factor binding protein-1 mRNA in the endometrium: a molecular mechanisms associating endometrial cancer with its risk factors. Int J Cancer 1994; 59:307–312.
147. Frost RA, Mazella J, Tseng L. Insulin-like growth factor binding protein-1 inhibits the mitogenic effect of insulin-like growth factors and progestins in human endometrial stromal cells. Biol Reprod 1993; 49:104–111.
148. Irwin JC, De las Fuentes L, Dsupin BA, Giudice LC. Insulin-like growth factor (IGF) regulation of human endometrial stromal cell function: coordinate effects on IGFBP-1, cellular proliferation, and prolactin secretion. Regul Pept 1993; 48:165–177.
149. Jones JI, Clemmons DR. Insulin-like growth factors and their binding proteins. Endocr Rev 1994; 18:1–31.
150. Rajkumar K, Dheen T, Krsek M, Murphy LJ. Impaired estrogen action in the uterus of insulin-like growth factor binding protein-1 transgenic mice. Endocrinology 1996; 137:1258–1264.
151. Han VKM, Bassett N, Walton J, Challis JRG. The expression of insulin-like growth factor (IGF) and IGF-binding protein (IGFBP) genes in the human placenta and membranes: evidence for IGF-IGFBP interactions at the feto-maternal interface. J Clin Endocrinol Metab 1996; 81:2680–2693.
152. Rutanen E-M, Koistinen R, Wahlstrom T, Sjoberg J, Stenman U-H, Seppala M. Placental protein 12 (PP12) in the human endometrium: tissue concentration in relation to histology and serum levels of PP12, progesterone, and oestradiol. Br J Obstet Gynecol 1984; 91:377–381.
153. Rutanen E-M, Koistinen R, Wahlstrom T, Bohn H, Ranta T, Seppala M. Synthesis of placental protein 12 by human decidua. Endocrinology 1985, 116:1304–1309.
154. Rutanen E-M, Koistinen R, Sjoberg J, Julkunen M, Wahlstrom T, Bohn H, Seppala M. Synthesis of placental protein 12 by human endometrium. Endocrinology 1986; 118:1067–1071.
155. Rutanen E-M, Menabawey M, Isaka K, Bohn H, Chard T, Grudzinskas JG. Synthesis of placental protein 12 by decidua from early pregnancy. J Clin Endocrinol Metab 1986; 63:675–679.
156. Liu HC, Mele C, Catz D, Noyes N, Rosenwaks Z. Production of insulin-like growth factor binding proteins (IGFBPs) by human endometrial stromal cells is stimulated by the presence of embryos. J Assist Reprod 1995; 12:78–87.
157. Seppala M, Julkunen M, Riittinen L, et al. Endometrial proteins: a reappraisal. Hum Reprod 1991; 7:31–40.
158. Waites GT, James RFL, Bell SC. Immunohistological localization of the human pregnancy-associated endometrial α_1-globulin, an insulin-like growth factor-binding protein, during the menstrual cycle. J Endocrinol Metab 1988; 67:1100–1104.
159. Waites GT, James RFL, Bell SC. Human pregnancy-associated endometrial α-globulin, an insulin-like growth factor-binding protein: immunohistological localization in the decidua and placenta during pregnancy employing monoclonal antibodies. J Endocrinol 1989; 120:351–357.
160. Rutanen E-M, Gonzalez E, Said J, Braunstein GD. Immunohistochemical localization of the insulin-like growth factor binding protein-1 in female reproductive tissues by monoclonal antibodies. Endocr Pathol 1991; 2:132–138.
161. Bryant-Greenwood GD, Rutanen E-M, Partanen S, Coelho TK, Yamamoto SY. Sequential appearance of relaxin, prolactin, and IGFBP-1 growth and differentiation of the human endometrium. Mol Cell Endocrinol 1993; 95:23–29.
162. Giudice LC, Martina NA, De Las Fuentes L, Crystal RA, Druzin ML. Insulin-like growth factor binding protein-1 (IGFBP-1) at the maternal-fetal interface and IGF-I, IGF-II, and IGFBP-1 in the circulation of women with severe pre-eclampsia. Am J Obstet Gynecol 1997; 176:751–757.
163. Bell SC, Jackson JA, Ashore J, Zhu HH, Tseng L. Regulation of insulin-like growth factor-binding protein-1 synthesis and secretion by progestin and relaxin in long term cultures of human endometrial stromal cells. J Clin Endocrinol Metab 1991; 72:1014–1019.
164. Irwin JC, Kirk D, King RJB, Quigley MM, Gwatkin RBL. Hormonal regulation of human endometrial stromal cells in culture: as in vitro model for decidualization. Fertil Steril 52:761–768.
165. Irwin JC, Utian WH, Eckert RL. Sex steroids and growth factors differentially regulate the growth and differentiation of cultured human endometrial stromal cells. Endocrinology 1991; 129:2385–2392.
166. Rosenberg M, Mazella J, Tseng L. Relative potency of relaxin, insulin-like growth factors, and insulin

on the prolactin production in progestin-primed human endometrial stromal cells in long-term culture. Ann NY Acad Sci 1991; 622:138–44.
167. Tang B, Gurpide E. Direct effects of gonadotropins on decidualization of human endometrial stromal cells. J Steroid Biochem Mol Biol 1993; 47:115–121.
168. Han SW, Lei ZM, Sanfilippo JS, Rao ChV. Human chorionic gonadotropin as a new regulator of human endometrial cell differentiation into decidua. In: 77th Annual Meeting of the US Endocrine Society, Washington, DC, Abstract P2–84.
169. Moy E, Kimzey LM, Nelson LM, Blithe DL. Glycoprotein hormone a-subunit functions synergistically with progesterone to stimulate differentiation of cultured human endometrial stromal cells to decidualized cells: a novel role for free a-subunit in reproduction. Endocrinology 1996; 137:1332–1339.
170. Tazuke SI, Mazure N, Crystal RA, Carland G, Faessen GH, Suen LF, Sugawara J, Powell D, Giaccia A, Giudice LC. Hypoxia regulation of IGFBP-1 gene expression and its relevance to intrauterine fetal hypoxia in human pregnancy. Proc Natl Acad Sci (USA) 1998; 95:10188–10193.
171. Thrailkill KM, Clemmons DR, Busby WH Jr, Handwerger S. Differential regulation of IGFBP secretion from human decidual cells by IGF-I, insulin, and relaxin. J Clin Invest 1990; 86:878–883.
172. Mark SP, Martina NA, Irwin JC, Giudice LC. Cytokine regulation of IGFBP-1 production in decidualized endometrial stromal cells. American Society for Reproductive Medicine Annual Meeting, Boston, November, 1996, p. 84, Abstract No. 0-007.
173. Kariya M, Kanzaki H, Takakura K, Imai K, Okamoto N, Emi N, Kariya Y, Mori T. Interleukin-1 inhibits *in vitro* decidualization of human endometrial stromal cells. J Clin Endocrinol Metab 1991; 73:1170–1174.
174. Frank GR, Brar AK, Jikihara H, Cedars MI, Handwerger S. Interleukin-1β and the endometrium: an inhibitor of stromal cell differentiation and possible autoregulator of decidualization in humans. Biol Reprod 1995; 52:184–191.
175. Librach J, Feigenbaum SL, Bass KE, Cui K-Y, Versastas N, Sadovsky Y, Quigley JP, Franch DL, Fisher SJ. Interleukin-1β regulates human cytotrophoblast metalloproteinase activity and invasion in vitro. J Biol Chem 1994; 269:17125–17131.
176. Giudice LC. Intimate conservations between the implanting conceptus and its maternal host. ORGYN 1997; VIII:48–51.
177. Ritvos O, Ranta T, Jalkanen J, Suikkari A-M, Voutilainen R, Bohn H, Seppala M. IGFBP from human decidua inhibits the binding and biological action of IGF-I in cultured choriocarcinoma cells. Endocrinology 1989; 122:2150–2157.
178. Jones JI, Gockerman A, Busby WH, Clemmons DR. IGFBP-1 stimulates cell migration and binds to the $\alpha_5\beta_1$ integrin by means of its Arg-Gly-Asp sequence. Proc Natl Acad Sci USA 1993; 90:10553.
179. Fisher SJ, Damsky CH. Human cytotrophoblast invasion. Cell Biol 1993; 4:183–188.
180. Irwin JC, Giudice LC. IGFBP-1 binds to the $\alpha_5\beta_1$ integrin in human cytotrophoblasts and inhibits their invasion into decidualized endometrial stromal cells *in vitro*. Growth Hormone and IGF Research 1998; 8:21–31.
181. Damsky CH, Librach C, Lim K-H, Fitzgerald ML, McMaster MT, Janatpour M, Zhou Y, Logan SK, Fisher SJ. Integrin switching regulates trophoblast invasion. Development 1994; 19:3657–3666.
182. Irving JA, Lala PK. Functional role of cell surface integrins on human trophoblast cell migration: regulation by TGF-b, IGF-II, and IGFBP-1. Exp Cell Res 1985; 217:419–429.
183. Schenken RS. Modern concepts of endometriosis: classification and its consequences for therapy. J Reprod Med 1998; 43:269–275.
184. Sampson JA. Peritoneal endometriosis due to menstrual dissemination of endometrial tissue into the peritoneal cavity. Am J Obstet Gynecol 1927; 14:422–469.
185. Giudice LC, Tazuke SI, Swiersz L. Status of current research on endometriosis. J Reprod Med 1998; 43:252–262.
186. Ramey JW, Archer DF. Peritoneal fluid: its relevance to the development of endometriosis. Fertil Steril 1993; 60:1–14.
187. Giudice LC, Dsupin BA, Gargosky SE, Rosenfeld RG, Irwin JC. The insulin-like growth factor system in human peritoneal fluid: its effects on endometrial stromal cells and its potential relevance to endometriosis. J Clin Endocrinol Metab 1994; 79:1284–1293.
188. Irwin JC, Cheng B-H, Giudice LC. Human peritoneal fluid contains a specific insulin-like growth factor binding protein-3 (IGFBP-3) protease which is distinct from plasmin. International Fertility Federation Societies (IFFS)/American Society for Reproductive Medicine (ASRM), San Francisco, Oct 9–12, 1998, Abstract.

189. Hossenlopp P, Segovia B, Lassarre C, Roghani M, Bedon M, Binoux M. Evidence of enzymatic degradation of insulin-like growth factor binding proteins in the 150K complex during pregnancy. J Clin Endocrinol Metab 1990; 71:759–805.
190. Giudice LC, Farrell EM, Pham H, Lamson GL, Rosenfeld RG. Insulin-like growth factor binding proteins in maternal serum throughout gestation and in the puerperium: effects of a pregnancy-associated serum protease activity. J Clin Endocrinol Metab 1990; 71:806–816.
191. Koutsilieris M, Allaire-Michaud L, Fortier M, Lemay A. Mitogen(s) for endometrial-like cells can be detected in human peritoneal fluid. Fertil Steril 1991; 56:888–893.
192. Koutsilieris M, Niklinski W, Frenette G, Lemay A. Heparin-Sepharose binding growth factors in peritoneal fluid of women with endometriosis. Fertil Steril 1993; 59:93–97.
193. Koutsilieris M, Akoum A, Lazure C, Frenette G, Lemay A. N-terminal truncated forms of insulin-like growth factor binding protein-3 (IGFBP-3) purified from the peritoneal fluid of normal women. Endocrine Soc 76th Annual Meeting, Anaheim, June 1994, Abstract No. 1049.
194. Taskin O, Dsupin BA, Wiehle RD, Giudice LC. Insulin-like growth factor binding protein-3 (IGFBP-3) and IGFBP-3 protease in peritoneal fluid of women with minimal and mild endometriosis. Hum Reprod 1996; 11:312–320.
195. Tazuke S, Giudice LC. Myometrium and Leiomyoma. In: Knobil E, Neill JD, eds. Encyclopedia of Reproduction. Academic Press, San Diego, 1998, in press.
196. Hoppener JWM, Mosselman S, Roholl PJM, Lambrechts C, Slebos RJ, de Pagter-Holthuizen P, Lips CJ, Jansz HS, Sussenbach JS. Expression of insulin-like growth factor-I and -II genes in human smooth muscle tumors. EMBO J 1988; 7:1379–1385.
197. Gloudemans T, Prinsen I, Van Unnik JA, Lips CJ, Den Otter W, Sussenbach JS. Insulin-like growth factor gene expression in human smooth muscle tumors. Cancer Res 1990; 50:6689–6695.
198. Vollenhoven BJ, Herington AC, Healy DL. Messenger ribonucleic acid expression of the insulin-like growth factors and their binding proteins in uterine fibroids and myometrium. J Clin Endocrinol Metab 1993; 76:1106–1110.
199. Giudice LC, Irwin JC, Dsupin BA, Pannier EM, Jin IH, Vu TH, Hoffman AR. Insulin-like growth factor (IGF), IGF binding protein (IGFBP), and IGF receptor gene expression and IGFBP synthesis in human uterine leiomyomata. Hum Reprod 1993; 8:1796–1806.
200. Rein MS, Friedman AJ, Pandian MR, Heffner LJ. The secretion of insulin-like growth factors I and II from explant cultures of fibroids and myometrium from women treated with a gonadotropin-releasing hormone agonist. Obstet Gynecol 1990; 76:388–394.
201. Daughaday WH, Emanuele MA, Brooks MH, Barbato AL, Kapadia M, Rotwein P. Synthesis and secretion of insulin-like growth factor II by a leiomyosarcoma with associated hypoglycemia. N Engl J Med 1988; 319:1434–1440.
202. Gloudemans T, Pospiech I, Van Der Ven LT, Lips CJ, Schneid H, Den Otter W, Sussenbach JS. Expression and CpG methylation of the insulin-like growth factor II gene in human smooth muscle tumors. Cancer Res 1992; 52:6516–6521.
203. Tommola P, Pekonen F, Rutanen E-M. Binding of epidermal growth factor and insulin-like growth factor I in human myometrium and leiomyomata. Obstet Gynecol 1989; 74:658–662.
204. Chandrasekher YA, Heiner J, Osuamkpe MS, Nagamani M. Insulin-like growth factor I and II binding in human myometrium and leiomyomas. Am J Obstet Gynecol 1992; 166:64–69.
205. Nagamani M, Stuart CA, Huang HQ. Specific binding and growth effect of insulin and insulin-like growth factors in a human leiomyosarcoma cell line. Society for Gynecologic Investigation 41st Annual Meeting, March, 1994, Chicago, Abstract P250, p. 319.
206. Nagamani M, Stuart CA. Specific binding sites for insulin in the human myometrium and leiomyomas of the uterus. Fertil Steril 1992; 58:120–123.
207. Pannier EM, Irwin JC, Ulaner GA, Kim E, Giudice LC. Benign, but not malignant, human uterine smooth muscle cells secrete an IGFBP-4 protease that is regulated by IGFs, under review.
208. Rossi MJ, Chegini N, Masterson BJ. Presence of epidermal growth factor, platelet-derived growth factor, and their receptors in human myometrial tissue and smooth muscle cells: their action in smooth muscle cells in vitro. Endocrinology 1992; 130:1716–1727.
209. Rein MS, Nowak RA. Biology of uterine myomas and myometrium *in vitro*. Semin Reprod Endocrinol 1992; 10:310–319.

18 The IGF Axis in the Male Reproductive System

Stuart A. Weinzimer, MD *and Pinchas Cohen,* MD

Contents

INTRODUCTION

The insulin-like growth factor (IGF) system is a multilayered network of molecules involved in the regulation of cell growth and death in many organs and tissues, normal and pathologic. The IGF system also participates in the modulation of important cell- and tissue-specific physiological and metabolic functions, via endocrine, paracrine, and autocrine interactions. The IGFs, IGF-receptors (IGF-R), IGF binding proteins (IGFBPs), and the IGFBP-receptors and proteases all feature prominently in the healthy prostate and in the common conditions of abnormal prostate growth, benign prostatic hypertrophy (BPH), and prostatic carcinoma (CaP). The IGF system similarly participates in the proliferation and hypertrophy of the bladder, both in congenital obstructive conditions and in acquired diseases of the urinary system. In the testis, the IGF system interacts with gonodatropins, androgens, and multiple other peptide hormones in the coordination of growth, spermatogenesis, and steroidogenesis.

The human prostate consists of both glandular and nonglandular elements within a single capsule. The glandular tissue, which is located anteromedially, may be subdivided into four zones, based on relative distance from the prostatic urethra. The peripheral zone, which comprises about 70% of the prostate, and the central zone, which makes up another 25%, are related to the distal urethra. The remaining 5% constitutes the transitional and periurethral zones, which approximate the proximal urethra. The diseases of prostate, BPH and CaP, localize to specific anatomic and histologic compartments. BPH originates from prostate stromal cells (PC-S) located almost exclusively in the transitional and periurethral zones. CaP develops from prostate epithelial cells (Pc-E) from the peripheral zone *(1)*.

From: *Contemporary Endocrinology: The IGF System*
Edited by: R. Rosenfeld and C. Roberts Jr © Humana Press, Inc., Totowa, NJ

The functional units of the testis are the seminiferous tubules, in which spermatogonia located in the basal portion of the tubule give rise to primary and secondary spermatocytes, spermatids, and spermatozoa located toward the luminal surface. The Sertoli cells rest on the basement membrane of the tubule and provide structural and metabolic support for the developing germ cells. They also produce müllerian-inhibiting substance. Sertoli cells proliferate only during the fetal and neonatal periods. The basement membrane surrounding the tubules is lined by peritubular myoid cells. Between the tubules lie the interstitial cells, which contain fibroblasts, macrophages, blood vessels, and the Leydig cells, the source of sex steroid production in the testis. Leydig cells proliferate prenatally but decrease in number markedly after birth. Proper development of the testis requires the presence of an intact sex-determining region on the Y chromosome, transcription of downstream sex-specific genes, and influences by gonadotropins and testosterone. Inhibins and activins are glycoprotein hormones secreted by the testis that, in turn, modulate the release of gonadotropins. A comprehensive review of the regulation of the testis by peptide hormones has recently been published *(2)*.

The IGF Axis in the Male Ejaculate

Several of the anatomic components of the male genital tract contribute to the contents of the ejaculate. Many components of the IGF axis have been detected in seminal plasma including IGF-I *(3)*, IGF-II *(4)*, IGFBPs *(5)*, the circulating form of the type II IGF-R *(6)*, and IGFBP proteases *(5,7)*. The main protease in seminal plasma is prostate-specific antigen (PSA), and it is active primarily against IGFBP-3 and IGFBP-5 *(7)*. Additional proteases may also be present in seminal plasma including cathepsins, metalloproteinases, and serine proteinases of unknown sources. It is difficult to assess the exact source of all of these molecules. It is clear, however, that the prostate gland is the main contributor of both IGFBPs and their proteases, while the testis may be the main source of IGF-I. It is known that IGF-I levels in seminal plasma fall dramatically after vasectomy, while the levels of IGFBP-2 and -4 remain the same; and IGFBP-3 levels fall slightly *(8,9)*. It also appears that seminal plasma IGF-I levels are growth-hormone (GH) dependent *(10)*.

THE IGF AXIS IN THE PROSTATE

IGFs, IGF Receptors, and IGFBPs in the Prostate

Although human seminal fluid contains both IGF-I and IGF-II, PC-S cells produce readily detectable levels of IGF-II but not IGF-I *(11)*, and PC-E cells do not synthesize or secrete significant amounts of either IGF-I or IGF-II *(12)*. Nonetheless, both PC-S and PC-E express type I-IGF receptors and respond to the mitogenic effect of the IGFs *(12,13)*. In fact, the relative mitogenic potency of growth factors added to primary PC-E cultures reflects their relative affinities for the type I IGF-R: IGF-I > IGF-II > insulin *(12)*. The type II IGF-R, on the other hand, is not found in substantial amounts in either PC-E or PC-S *(12)*.

The prostatic IGF-IGF-R interaction is modulated by the IGFBPs, of which at least six have been characterized in the human prostate. The IGFBPs not only alter the availability of free IGFs for interaction with the IGF-R, but also exhibit some IGF-independent growth-mediating effects *(14)*. The IGFBPs may be growth enhancing, especially IGFBP-5 *(15)*, or growth-inhibitory, such as IGFBP-3, which also participates in apoptosis *(14,16)*. In the prostate, IGFBP-1 peptide has not been detected in the cultured medium of either normal PC-E or PC-S, and its mRNA was absent on Northern blots of

Table 1
Summary of the Known Prostatic IGF System Components

IGF Molecule	*PC-E*	*PC-S*	*Prostate Cell Lines*	*Circulation*
IGFs	None	IGF-II	IGF-II (in PC-3)	IGF-I, IGF-II
IGF action	Proliferation	Unknown	Proliferation (all) survival (PC-3)	Unknown
IGF receptors	Type I	Type I	Type I (all lines)	None
IGFBPs	IGFBP-2,4,5	IGFBP-2,3,4,5	IGFBP-2-6	IGFBP-3,-2
IGFBP action	IGF-inhibitory (IGFBP-2, -3)	Unknown	Apoptosis:IGFBP-3 in PC-3	IGF-inhibitory
IGF axis regulators	None found	TGF-β, WT-1, androgens	TGF-β (PC-3), DHT (LNCaP)	Prostate cancer age
Putative IGFBP receptors	Unknown	35, 50, and 60 kDa	20, 60, and 300 kDa	Currently being examined
IGFBP proteases	PSA, hK-2, cathepsin D	MMP-1	Cathepsins, MMPs, urokinase	MMPs, others
Protease action	Block IGFBP inhibition	Degrade IGFBP-5	Promote growth	Cleave IGFBPs

PC-E, prostate epithelial cell; PC-S, prostate stromal cell; LNCaP and PC-3, prostate cancer cell lines; DHT, dihydrotestosterone; MMP-1, matrix metalloproteinase-1; hK-2, human kallikrein 2.

both cell types *(11,12)*. IGFBP-2 and IGFBP-4, the major IGFBPs in seminal fluid *(17)*, are both synthesized and secreted predominantly by normal PC-E *(12)* and also by PC-S *(11)*. IGFBP-3, however, is definitively made by PC-S *(11,18)* and possibly by PC-E *(12,18)* as well as by several CaP cell lines. IGFBP-5 and -6 are made in cultured prostatic cells and are detected in histologic section of prostate tissue *(19)*. mac-25 (IGFBP-7) is expressed in PC-S [Cohen et al., *personal communication*). IGFBP production by PC-E appears to be constitutive, as treatment of PC-E with various growth regulators and media additives did not change IGFBP levels; PC-S IGFBP production, on the other hand, is influenced by a variety of cytokines (Table 1).

IGFBP Proteases: Prostatic Growth Modulators

Although IGFBP-3 is produced by prostate cells, its detection in seminal plasma can be achieved only by radioimmunoassay and not by Western ligand blotting; IGFBP-3 proteolysis by seminal plasma, even in azoospermic samples, has been confirmed using labeled IGFBP-3 in protease *(20)* and immunoblot assays *(21)*. PSA, produced by the luminal PC-E, was further identified as a seminal IGFBP-3 protease *(20)*. A significant inverse correlation was found between PSA and nondigested IGFBP-3 in prostatic massage fluid *(22)*.

The role of PSA as a prostatic growth modulator is directly linked to its IGFBP-3 protease function. Cleavage of IGFBP-3 by PSA caused a 10 to 100-fold decrease in the affinity of IGFBP-3 fragments for the IGFs *(16)*. Exposure of PC-E in serum-free culture conditions to IGFs stimulated a doubling of cell number; this stimulation was blocked by addition of IGFBP-3 and subsequently reinstated by further addition of PSA *(16)*. Thus, PSA functions as a co-mitogen with the IGFs, by freeing them from IGFBP binding and

Table 2
Prostatic IGFBP Proteases and Their Characteristics

Protease	*Class*	*IGFBPs Cleaved*	*Other Substrates*	*Known Inhibitor*
PSA (HK-3)	Kallikrein	3, 5	Fibronectin, Laminin	α_1-ACT
HK-2	Kallikrein	1–5	Fibronectin	PCI, kallistatin
Urokinase	Serine	3	Fibronectin	PAI
Plasmin	Serine	1–6	Fibrin	Anti-plasmin
Cathepsin D	Aspartyl	1–5	T-kininogen, fibronectin	Pepstatin
MMP-2	Metallo	3, 5	Collagen	TIMPs

hK-2, human kallikrein 2; TIMP, tissue inhibitor of metalloproteinase; PCI, protease C inhibitor; PAI, plasminogen activator inhibitor.

thereby increasing their potential interaction with the IGF receptors. This effect may be compounded by PSA reduction of IGFBP-3, which in and of itself participates in apoptosis of prostate cancer cells.

PSA is not the only IGFBP protease active in the prostate, nor is IGFBP-3 the only IGFBP cleaved *(21)*. The prostatic IGFBP proteases can be divided into three categories: kallikrein enzymes (such as PSA), matrix metalloproteinases (MMPs), and the cathepsins *(23–25)*. Whereas PSA is functional at neutral pH, the cathepsins, lysosomal enzymes implicated in malignant processes, are acid-activated *(26)*. All of these proteins appear to be secreted into seminal plasma, but their role there is not yet known. The prostatic IGFBP proteases are summarized in Table 2.

Regulation of the Prostatic IGF Axis

The prostatic IGF system is regulated by androgens, cytokines, and several transcription factors and oncogenes. Androgens are the key circulating factors that promote prostate growth, and androgen ablation therapy is the hallmark of nonsurgical approaches to prostatic disease *(27–29)*. Paradoxically, the incidence of prostate disease increases with age, as androgen levels are waning, and neoplastic prostatic tissues often become androgen independent. Two-month treatment with a gonadotropin-releasing hormone analog, producing subcastration serum levels of testosterone in men with BPH, significantly increased their prostatic IGF-R binding capacity, both low and high affinity,when compared to that of non-androgen-ablated BPH prostates. Furthermore, expression of type I-IGF-Rs was found not only in the basal epithelial cells, as in the untreated controls, but also extending to the glandular epithelium *(30)*. When a constitutively active androgen receptor was stably transfected into PC-3 cells, an androgen receptor-negative prostate carcinoma line, IGFBP-3 transcription and protein levels decreased *(31)*. Further studies are needed to elucidate the effects of androgens on the different IGF system components in the normal prostate.

Cytokines, particularly transforming growth factor-β (TGF-β) and tumor necrosis factor-α (TNF-α), have been found by our lab to regulate the prostatic IGF system. Both TGF-β and TNF-α modulate the production of IGFBP-3 by normal PC-S cells. TGF-β also inhibits normal PC-S growth to 30% of baseline, and it seems that IGFBP-3 induced apoptosis is involved in the growth inhibition *(32)*.

Most recently, transcription factors, such as the tumor suppressors Wilms tumor gene product (WT-1) and p53, have been linked to the prostatic IGF system. WT-1 represses transcription of both the type I IGF-R *(33)* and IGF-II *(34)*. p53, on the other hand, induces IGFBP-3 gene expression *(35)*, and mutant p53s from a variety of nonprostate cancers that fail to activate IGFBP-3 demonstrate impaired apoptosis *(36,37)*. Peehl and colleagues found that exposing normal, BPH, or CaP tissues to ionizing radiation, a form of genotoxic stress, does not increase p53 in PC-E but does produce a three- to nine-fold augmentation of p53 in PC-S. In parallel, the PC-E cells did not undergo cell cycle arrest as did the PC-S cells *(38)*. The role of IGFBP-3 in the mediation of p53 effects in the prostate cell is currently under investigation.

Effects of IGFs and IGFBPs on the Prostatic Cell Cycle

The IGFs in the prostate are primarily mitogenic and act through the IGF-R. The type I IGF-R is a tyrosine kinase that activates the mitogen-activated protein (MAP) kinase pathway *(39)*. The IGFBPs not only influence the amount of free IGF available for receptor interaction, but also produce growth effects via IGF-independent pathways. For example, IGFBP-3 promotes apoptosis. Apoptosis-inducing agents are reported to mediate increases in the bax family or reductions in the bcl-2 family of proteins. These opposing molecules form dimers, and the balance between the two (mainly through the levels of free bax or bcl-2) determines the degree of apoptosis incurred during specific conditions *(40–42)*. bax, in turn, activates the interleukin (IL)-1 β-converting enzyme (ICE) family of cysteine proteases, which cleave numerous proteins and lead to apoptosis *(43)*. Our laboratory recently demonstrated that IGFBP-3 mediates apoptosis by inducing bcl-2 serine phosphorylation, and hence, ICE activation.

The IGF System in BPH

BPH, the most commonly occurring benign proliferative abnormality found in any internal organ, has been described as an abnormal proliferation of prostatic epithelial ducts and acini, probably preceded by hyperproliferation of the stroma in the prostatic transition zone *(44)*. This histologic observation has borne out in molecular studies. No abnormalities in the IGF axis were detected in PC-E from BPH prostates *(11)*; BPH PC-S, however, demonstrated a 10-fold increase in IGF-II transcription as well as a tripling in type I-IGF-R transcription. These findings are most likely the result of a significant diminution of WT-1 expression to 0–20% of normal *(45)*, as it is well described that WT-1 is a tumor suppressor that inhibits transcription of multiple growth-promoting genes including IGF-II and type I IGF-R. In BPH PC-S, IGFBP-2 expression was also markedly reduced and was replaced by IGFBP-5 expression, as revealed by both mRNA and peptide analyses *(11)*.

The genes for IGFBP-2 and -5 are tightly linked on chromosome 2q33-34 and may follow coordinated regulation *(46)*. Thus, it is tempting to postulate that a genetic event leads to diminished expression of IGFBP-2, an IGF-inhibitory molecule, and augmented expression of IGFBP-5, an IGF-enhancing peptide. In addition, IGFBP-3, another growth-inhibitory IGFBP, was found by our laboratory to not only be fourfold lower at baseline in BPH vs normal PC-S, but to manifest a blunted induction by TGF-β in BPH PC-S; this phenomenon was associated with failed TGF-β-induced growth inhibition of BPH cells *(47)*. This IGFBP imbalance, coupled with the IGF-II overproduction, constitutes an autocrine stimulus for stromal proliferation and a paracrine signal for epithelial growth. The IGF system disturbances in BPH are illustrated in Fig. 1.

The IGF System in CaP

Prostate cancer is the most commonly diagnosed cancer in American men and the second most lethal. Each of the IGF system levels delineated earlier has been reported

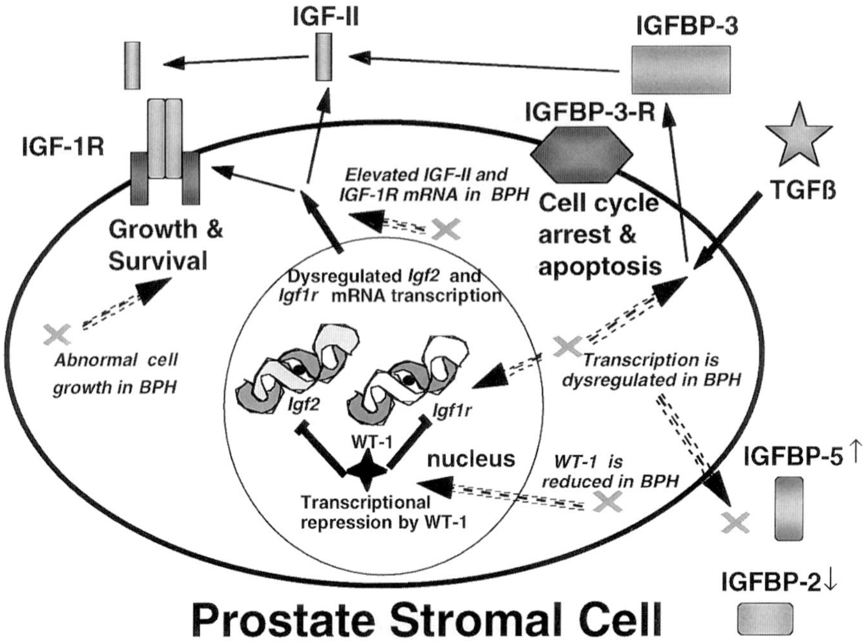

Fig. 1. IGF axis disturbances in BPH. Decreased WT-1 expression appears to lead to increased IGF-R and IGF-II expression, which may result in accelerated proliferation of BPH PC-S. In addition, loss of IGFBP-2 but gain of IGFBP-5 secreted from PC-S influence the mitogenic stimulus. The blunted TGF-β induction of IGFBP-3 associated with a decreased TGF-β induced growth inhibition in PC-S from BPH may also contribute to the hyperplastic phenotype and may involve binding of IGFBP-3 to its receptors (IGFBP-3-R).

affected in CaP. CaP cells are responsive to the mitogenic effects of the IGF/IGF-R signal. For example, adding exogenous IGF-I or IGF-II to DU 145, an androgen-independent CaP cell line that expresses IGF-Rs, stimulates monolayer and anchorage-independent cell growth *(48)*. Whereas normal PC-E do not synthesize or secrete significant amounts of either IGF-I or IGF-II, IGF-I secretion has been found in the established metastatic cancer cell lines PC-3, LNCaP, and DU 145 *(49)*. This may reflect a late genetic change, as IGF-I production has not been confirmed in any PC-E from CaP tissues. IGF-II production has been detected in DU 145 *(48)*, PC-3 *(50)*, SV-40 immortalized PC-E *(51)*, and CaP tissues *(52)*. Furthermore, monoclonal anti-IGF-antibodies and anti-IGF-R-antibodies inhibit PC-3 cell growth dose-dependently up to 80% *(50)*. Reduction in IGF-R has been reported in SV-40 immortalized cell lines *(51)* and CaP tissues *(52)*, but this down-regulation may not be a primary event in neoplastic transformation; it may result from a downstream mutation, making the cells IGF-R-independent.

Changes in the IGFBP profile have also been associated with CaP. In one study comparing benign PC-E with prostate intraepithelial neoplasia (PIN) and prostatic adenocarcinoma, IGFBP-2 mRNA and protein increased with progressive malignancy; IGFBP-3 mRNA levels did not differ, but protein immunoreactivity changed PIN: > benign > adenocarcinoma *(53)*. IGFBP-4 and IGFBP-5 also increased with progressive malignancy, and there was no change in IGFBP-6 levels *(19)*. A direct relationship of these findings to the progression of CaP has not been demonstrated, and they may represent secondary phenomena. Our laboratory has shown that IGFBP-3 dose-dependently induces apoptosis through an IGF-IGF-R-independent pathway in PC-3 cells *(32)*, suggesting that decreased IGFBP-3-induced apoptosis might be involved in prostate cancer progression.

IGFBP alterations have been linked to protease perturbations. The volume of prostate cancer is correlated to the elevation in serum PSA level, which has been correlated with increased serum IGFBP-2 *(54–56)* and decreased intact IGFBP-3 *(55,56)*. Because the pattern of IGFBP-3 cleavage fragments in serum differs from that created by seminal PSA *(20,54)* and serum PSA is inactivated by inhibitors *(57)*, PSA does not likely cause proteolysis of serum IGFBP-3. PSA is active, however, in seminal fluid and presumably in the prostate as well. Whereas PSA is normally sequestered in the prostatic lumens, malignancy may disrupt the acinar architecture and thereby lead to PSA leakage into the surrounding stroma, and eventually, the bloodstream. Thus, IGFBP-3 proteolysis by PSA is more likely a local effect within the prostate or metastatic foci whose consequences contribute to the local propagation of neoplasia or metastasis. Cathepsin D may also contribute to the local and metastatic progression of malignancy by similar mechanisms. Although cathepsin D secretion did not differ between cancerous and normal PC-E in cell culture, early malignant changes, including increased sialic acid content in surface glycoproteins and exaggerated proton release by membrane-bound proteins, may create an acidic microenvironment that would activate cathepsin D protease activity *(26)*. Urokinase and plasminogen-activator proteolysis of IGFBP-3 were found to impact growth of cultured PC-3 cells *(50)*.

Cytokines, particularly TGF-β1 and TNF-α, which are known to induce apoptosis, also induce IGFBP-3 expression in the p53-negative prostate cancer cell line PC-3 *(30)*. This effect of TGF-β1 is prevented by cotreatment with IGFBP-3 neutralizing antibodies or IGFBP-3-specific antisense thiolated oligonucleotides, suggesting that IGFBP-3 mediates TGF-β1-induced apoptosis in PC-3 cells *(32)*. Our current understanding of the IGF system in CaP is illustrated in Fig. 2; an overview of the prostatic IGF axis is summarized in Table 1.

THE IGF AXIS IN THE TESTIS

Sertoli Cells

IGF-I production in the testis was first detected in Sertoli and peritubular cells of the immature rat *(58,59)*, and has also been immunolocalized to porcine *(60)* and human *(61)* Sertoli cells. Type I IGF-Rs, with binding affinity for IGF-1 > IGF-II > insulin, have been demonstrated in rodent *(62,63)*, pig *(64)*, and human *(65)* Sertoli cells. Type II IGF-Rs, with binding affinity IGF-II > IGF-I >> insulin, have been found in Sertoli cells of mice *(62)*. Of the IGFBPs, prepubertal rat Sertoli cells produce predominantly IGFBP-3 *(66,67)*, while human Sertoli cells produce IGFBP-2 *(68)*.

Leydig Cells

IGF-I has been immunolocalized to Leydig cells in the immature rat *(69,70)*, and in the human *(61)*. Immunostaining for IGF-II in the rat testis is positive only in Leydig and some interstitial cells *(71)*. IGF-Rs have been demonstrated in rat Leydig cells *(63)*. Both rat and human Leydig cells secrete IGFBP-2 and -4 *(68,72)*. In addition, human Leydig cells also produce low levels of IGFBP-5. In the murine Leydig cell line TM-3, Hasegawa et al. have demonstrated that IGFBP-4 is the main IGFBP expressed, with minimal expression of IGFBP-3, -5, and -6 *(73–75)*. They have shown that IGFBP-4 is both secreted and membrane bound *(74)*.

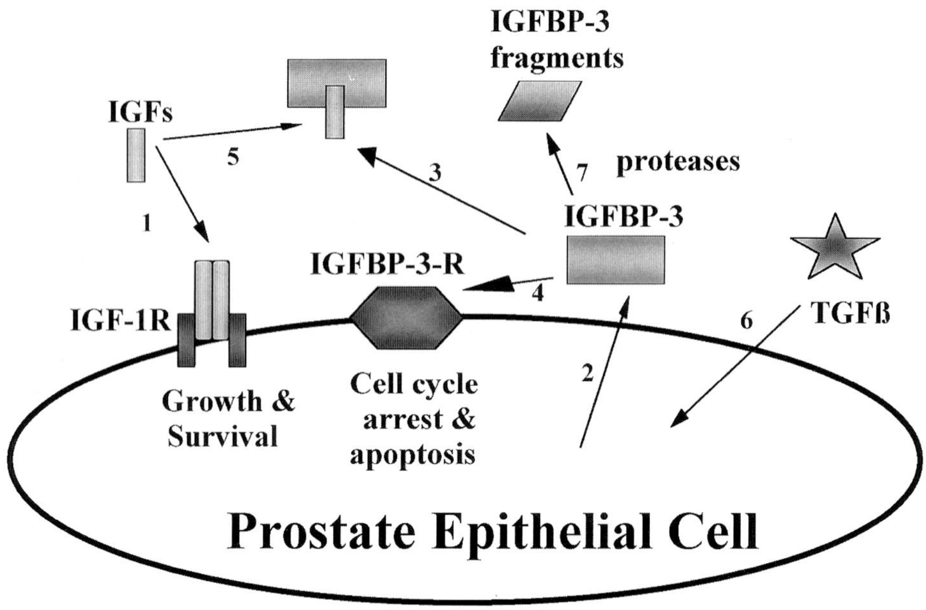

Fig. 2. The IGF axis in CaP. (1) IGFs may lead to increased growth and survival of CaP cells by binding to IGF-R. (2) IGFBPs are secreted from prostate cells and (3) block IGF action. (4) IGFBP-3 can also inhibit cell growth and induce apoptosis by binding its own receptor; it may be reduced in p53 mutant CaP. (5) IGFs can block IGFBP-3 by preventing it from binding to its receptor. (6) TGF-β1 induction of IGFBP-3 activates cell cycle arrest and apoptosis; this, too, may be blunted in CaP. (7) Augmented protease activity in CaP increases IGFBP-3 fragmentation and diminishes its functioning.

Other Cells

In the human, IGF-I has been immunolocalized to the spermatocyte series *(76)*, while IGF-II has been localized to the vascular and peritubular cells *(68)*. In the rat, IGF-II is seen in multiple cell types *(77)*. Types I and II IGF-Rs in the human testis are found in the spermatogonia and spermatocytes *(61,76)*. In the rat, IGFBP-2 is produced in peritubular cells and seminiferous tubules *(72)*. IGFBP-2 is the predominant IGFBP in testicular interstitial fluid *(78)*. In the human, IGFBP-3 has been detected only in vascular endothelial cells. Interstitial connective tissue contains IGFBP-4, abundant IGFBP-5, and IGFBP-6. Peritubular cells also produce IGFBP-6 *(68)*.

Regulation of Testicular IGF Axis

Several developmental patterns in IGF-I expression have been observed. In the rat, IGF-I mRNA is maximally expressed at birth and then declines with age *(79)*, while in pigs there is a postnatal increase in IGF-I mRNA *(80)*. IGF-I is expressed in all cell types at birth but decreases rapidly within the first postnatal month in Sertoli and Leydig cells; spermatogenic cells have a rapid increase in IGF-I content at puberty, and in the adult rat IGF-I immunoreactivity is confined to spermatocytes *(81)*. Fetal testicular IGF-II decreases with advancing gestational age *(82)*. IGF-IR expression does not appear to be developmentally regulated *(83)*, but increased IGF-IR has been found in cryptorchid testes, localized especially to the Sertoli cells *(84)*. Developmental regulation of IGFBP expression has not been noted.

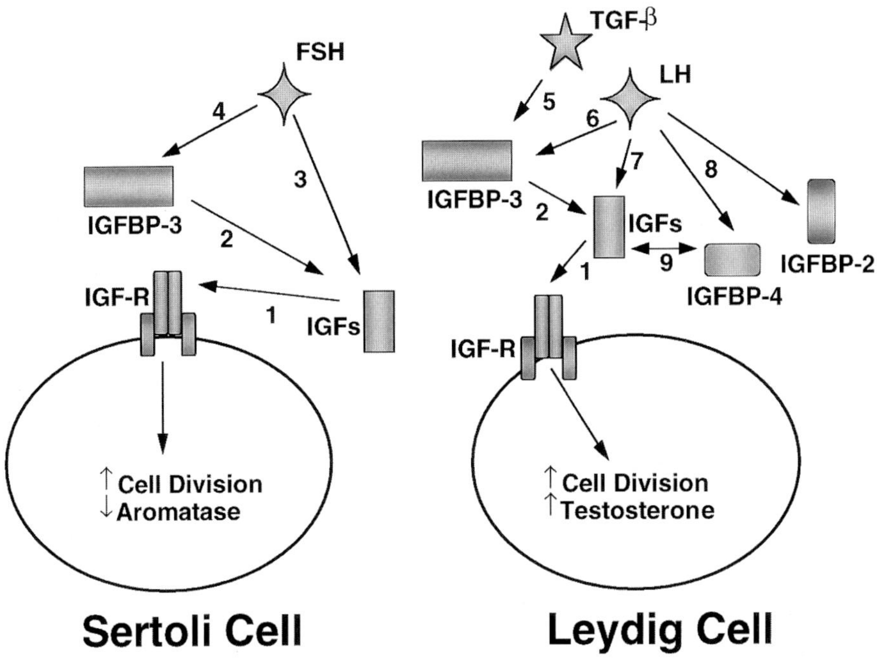

Fig. 3. The IGF axis in the testis. IGF binding to the IGF-R (1) increases cell division and modulates androgen expression in Sertoli and Leydig cells. IGFBP-3 inhibits these functions by binding IGF-I and preventing its interaction with the IGF-R (2). In the Sertoli cell, FSH stimulates IGF-I (3) and inhibits IGFBP-3 (4). In the Leydig cell, TGF-β stimulates IGFBP-3 (5), as does LH (6), which is also stimulatory to IGF-I (7) and inhibitory to IGFBP-2 and -4 (8). These binding proteins have the capacity to block IGF action and mediate contact inhibition of Leydig cells (9).

In the immature pig, follicle-stimulating hormone (FSH) and fibroblast growth factor (FGF) stimulate IGF-I synthesis in Sertoli cells *(85)*, while pubertal rat Sertoli cells increase IGF-I production after treatment with FSH or GH *(59,69)*. FGF and luteinizing hormone (LH) increase IGF-I synthesis in rat Leydig cells *(69)* and immature pig Leydig cells *(85)*. Closset et al. *(86)* found an increase in testicular (but not plasma) IGF-I mRNA in hypophysectomized rats after treatment with FSH, LH, or GH; IGF-II levels were unchanged. Gonadotropins and GH also increase IGF-IR number in Leydig cells *(87,88)*.

Regulation of the IGF axis at the level of the IGFBPs is complex. Cailleau *(69)* reported a stimulatory effect of LH on IGFBP-3 expression in Leydig cells but an inhibitory effect of FSH on IGFBP-3 expression in Sertoli cells. Lin et al. *(72)* have observed that levels of IGFBP-2 through -4 are decreased 22%, 80%, and 90%, respectively, in rat Leydig cells after hypophysectomy, and hCG also inhibits IGFBP-2 in a dose-dependent fashion *(89)*. Smith *(66)* and Rappaport *(67)* found that FSH treatment of rat Sertoli cells dose-dependently lowered mRNA levels of IGFBP-3. However, they observed an increase in IGFBP-3 levels following hypophysectomy or with IGF-I treatment. In TM-3 cells IGFBP-4 is down-regulated by gonadotropins and IGF-II, but up-regulated by IGF-I *(73,74)*. Hasegawa et al. have also shown that IGFBP-4 secretion is regulated by the cell density and is much lower in proliferating cells than in confluent cells, leading them to believe that IGFBP-4 may be involved in the mediation of contact inhibition in that cell type *(75)*. The expression and regulation of the IGF axis components in the testis are summarized in Fig. 3 and Table 3.

Table 3
Summary of Testicular IGF System Components

IGF Molecule	*Sertoli Cell*	*Leydig Cell*	*Other Cells*
IGFs	IGF-I	IGF-I, -II	IGF-I (spermatogonia) IGF-II (endothelium)
IGF action	↓ Aromatase ↑ FSH binding ↑ Hexose transport	↑ T production ↑ LH, hCG binding	↑ Sperm differentiation ↑ Spermatogonial DNA synthesis
IGF receptors	Type I	Type I	Type I, II (spermatogonia)
IGFBPs	IGFBP-2, -3, -4	IGFBP-2–6	IGFBP-2 (peritubular) IGFBP-4—6 (CT)
IGFBP action	IGF-inhibitory	IGF-inhibitory	IGF-inhibitory
IGF regulators	FGF, FSH, GH	FGF, GH, LH	
IGFBP regulators	FSH, TNF-α	LH, TGF-β	

CT, connective tissue; T, testosterone.

IGFs and Testicular Growth

IGF-I is mitogenic in testicular cell culture, increasing cell number in rat *(90)* and porcine *(91)* Leydig cells. Both IGF-I and IGF-II induce proliferation of rat and pig prepubertal Sertoli cells *(62,92)*. IGF-I also potentiates the growth-stimulatory effect of FGF on Sertoli cells in culture *(93)*. Both insulin and IGF-I increased thymidine uptake in intact Sertoli/germ cell cocultures, and Sertoli cell monolayers showed a dose-dependent increase in DNA synthesis after the addition of IGF-I *(94)*.

IGFs and Testicular Steroidogenesis

IGF-I increases both basal and gonadotropin-stimulated testosterone production in rat *(95–97)*, pig *(87,91)*, and human *(98)* Leydig cells, and increases both LH and human chorionic gonadotropin (hCG) binding to Leydig cells *(91)*. This IGF potentiation of LH and hCG is more prominent in immature than in adult cells *(97,98)*. IGF-I participates in the maturation of Leydig cells *(99)*. IGF-II also stimulates testosterone production, although at 10% potency of IGF-I; this probably occurs through IGF-II binding to the IGF-IR *(96)*. Treatment of Leydig cells with IGF-I results in a dose-dependent increase in hCG receptors *(87,100)*. IGF-I increases LH receptor mRNA and receptor density in pig Leydig cells and also increases mRNA expression of the steroidogenic enzymes cytochrome P-450scc, 17 α-hydroxylase, and 3β-hydroxysteroid dehydrogenase *(101–102)*. IGF-I inhibits the FSH-stimulated aromatization of androstenedione and decreases aromatase mRNA in Sertoli cells *(103)*. Treatment of immature dwarf Snell mice with recombinant GH or IGF-I similarly increased testicular LH receptors and the acute steroidogenic response to hCG, suggesting that the GH effect on Leydig cell maturation acts through IGF-I *(93)*. However, neither GH nor IGF-I treatment produced a similar response in primates *(104)*.

TGF-β was found by Lin et al. *(105)* to inhibit Leydig cell steroidogenesis. This effect is mediated through the induction of IGFBP-3, which inhibits in a dose-dependent manner the IGF-I stimulation of testosterone production in Leydig cells *(72,106)*, most likely through prevention of IGF-I binding to the IGF-IR. IGFBP-4 had a similar but smaller inhibitory effect on testosterone production, while IGFBP-2 had no effect whatsoever *(72)*. IL-1β is a potent inhibitor of Leydig cell steroidogenesis, whose effect is presumably mediated through the inhibition of IGF-I expression *(107)*.

IGFS and Spermatogenesis

IGF-I and IGF-II stimulate spermatogonial DNA synthesis *(108)*, and IGF-I increases the transport of hexose, an important substrate in germ cell metabolism, into Sertoli cells and the germinal epithelium *(109)*. IGF-I and TGF-β stimulate differentiation of type A spermatogonia *(110)*. IGF-I stimulates FSH binding to porcine Sertoli cells, a process that can be antagonized by TNF-α through induction of IGFBP-3 *(111)*. Transgenic mice homozygous for IGF-I deletions are infertile dwarfs with poor postnatal growth, small testes, immature Leydig cells, low basal and LH-stimulated testosterone production, and less than 20% spermatogenesis compared to controls *(83)*. IGF-II knockout mice, however, while dwarfs at birth, have normal postnatal growth and fertility *(112)*. Interestingly, while transgenic mice homozygous for deletions in the GH releasing hormone (GHRH) receptor *(113)*, post-GHRH receptor signal transduction *(114)*, or GH receptor *(115)* all have low serum IGF-I levels, none have reduced fertility. Similarly, GH treatment of hypophysectomized animals did not increase testicular IGF-I levels *(79)*. This suggests that intratesticular rather than circulating IGF-I is critical for testicular development and function.

THE IGF AXIS IN THE MALE URINARY BLADDER

The bladder responds to distension or obstruction with smooth muscle hypertrophy and cellular hyperplasia. The IGF axis has been shown to be important in the regulation of both urothelial and smooth muscle proliferation in conditions of normal and aberrant growth.

In animal models of bladder hypertrophy induced by urethral ligation or diuretic- or diabetes-induced polyuria *(116–118)*, IGF-I mRNA was increased threefold after 3 d and IGFBP-2 and -4 mRNA was increased fivefold after 10 d. Regression of the obstruction in the ligated rats resulted in normalization of IGF-I and IGFBP mRNA levels. IGF-IR levels remained unchanged. Conversely, targeted overexpression of IGF-I in transgenic mice resulted in increased bladder weight and concentric hyperplasia of the muscular layer *(119)*. Increased mucosal RNA expression of IGF-II and type I IGF-R coincides with reepithelialization and urothelial proliferation following acute injury *(120)*. Increased levels of IGF-II and type I IGF-R are seen in bladder carcinomas with loss of imprinting of the IGF-II transcriptional regulator H19 *(121,122)*.

Our laboratory has demonstrated expression and dynamic regulation of IGFBP-2, -3, -4, and -5 using Western ligand and immunoblotting techniques in human and sheep bladder smooth muscle (BLSM) cell culture strains *(123)*. IGFBP-3 stimulates BLSM cell proliferation and potentiates IGF-I-induced cell proliferation. TGF-β1 induces endogenous IGFBP-3 production in BLSM, and both TGF-β and TNF-α inhibit IGFBP-2, -4, and -5 production. Thus, cytokines modulate growth of BLSM through induction of growth-stimulatory IGFBPs and suppression of growth-inhibitory IGFBPs.

CONCLUSIONS

As our understanding of the prostatic and testicular IGF systems increases, so too does our awareness of its complexity. In the prostate, this carefully balanced network of stimulatory and inhibitory molecules keeps cellular growth in check. When the IGF system is disturbed, aberrant growth such as BPH and cancer may ensue. Cell cycle control points lie not only with the IGFs themselves, their receptors, and binding proteins;

IGFBP proteases, upstream regulators, and downstream response molecules all provide additional sites for IGF-related growth modulation. In the testis, IGF axis components have important paracrine and autocrine functions; they modulate the effects of pituitary-derived gonadotropins and locally derived cytokines. We are just beginning to elucidate the mechanisms regulating the modulation of cell proliferation in the bladder. Further research will likely reveal yet unknown components and more complex interactions. The male reproductive system is an ideal model for IGF investigation, because it not only contains the many part of the IGF axis, but this axis may hold the key for the successful treatment of BPH, CaP, and bladder hypertrophy, clinical conditions with tremendous societal impact.

REFERENCES

1. McNeal JE, Prostate. In: Sternberg SS, ed., Histology for Pathologists. Raven Press, New York, 1992.
2. Gnessi L, Fabbri A, Spera G. Gonadal peptides as mediators of developmental and functional control of the testis: an integrated system with hormones and local environment. Endocr Rev 1997; 18:541–609.
3. Baxter, RC, Martin JL, Handelsman DJ. Identification of human semen insulin-like growth factor-I/somatomedin-C immunoreactivity and binding protein. Acta Endocrinol (Copenh) 1984; 106: 420–427.
4. Rosenfeld RG, Pham H, Oh Y, Lamson G, Giudice LC. Identification of insulin-like growth factor-binding protein-2 (IGFBP-2) and a low molecular weight IGFBP in human seminal plasma. J Clin Endocrinol Metab 1990; 70:551–553.
5. Cohen P, Graves HCB, Peehl DM, Kamarei M, Giudice LC, Rosenfeld RG. Prostate specific antigen (PSA) is an IGF binding protein (IGFBP-3) protease found in seminal plasma. J Clin Endocrinol Metab 1992; 75:1046–53.
6. Desnoyers L, Manjunath P. Major proteins of bovine seminal fluid bind to insulin-like growth factor-II. J Biol Chem 1994; 269:5776–5780.
7. Nunn SE, Beers-Gibson T, Cohen P. Regulation of prostate cell growth by the IGFBPs and their proteases. Endocrine 1997; 7:115–118.
8. Glander HJ, Kratzch J, Weisbrich C, Birkenmeier. Insulin-like growth factor-I and alpha 2-macroglobulin in seminal plasma correlate with semen quality. Hum Reprod 1996; 11:2454–2460.
9. Ovesen P, Flyvbjerg A, Orskov H. Insulin-like growth factor-I (IGF-I) and IGF binding proteins in seminal plasma before and after vasectomy in normal men. Fertil Steril 1995; 63:913–918.
10. Breier BH, Vickers MH, Gravance CG, Casey PJ. Growth hormone (GH) therapy markedly increases the motility of spermatozoa and the concentration of insulin-like growth factor-I in seminal vesicle fluid in the male GH-deficient dwarf rat. Endocrinology 1996; 137:4061–4064.
11. Cohen P, Peehl DM, Hintz RL, Rosenfeld RG. IGF axis abnormalities in cultures of prostate stromal cells from patients with benign prostatic hypertrophy. J Clin Endocrinol Metab 1994; 79:1410–1415.
12. Cohen P, Peehl DM, Lamson G, Rosenfeld RG. Insulin-like growth factors (IGFs), IGF receptors and IGF binding proteins in primary cultures of prostate epithelial cells. J Clin Endocrinol Metab 1991; 73:491–407.
13. Iwamura M, Sluss PM, Casamento JB, Cockett AT. Insulin-like growth factor I: action and receptor characterization in human prostate cancer cell lines. Prostate 1996; 22:243–52.
14. Valentinis B, Bhala A, Baserga R, Cohen P. Transfection of the human insulin-like growth factor (IGF) binding protein-3 gene into fibroblasts from IGF-receptor gene-targeted mice inhibits cellular growth. Mol Endocrinol 1995; 9:361–367.
15. Conover CA. Potentiation of insulin-like growth factor (IGF) action by IGF-binding protein-3: studies of underlying mechanism. Endocrinology 1992:130:3191–9.
16. Cohen P, Peehl DM, Graves HCB, Rosenfeld RG. Biological effects of prostate specific antigen (PSA) as an IGF binding protein-3 (IGFBP-3) protease. J Clin Endocrinol 1994; 142:407–415.
17. Rosenfeld RG, Pham H, Oh Y, Lamson G, Giudice LC. Identification of insulin-like growth factor-binding protein-2 (IGF-BP-2) and a low molecular weight IGF-BP in human seminal plasma. J Clin Endocrinol Metab 1990; 70:551–3.
18. Birnbaum RS, Ware JL, Plymate SR. Insulin-like growth factor-binding protein-3 expression and secretion by cultures of human prostate epithelial cells and stromal fibroblasts. J Endocrinol 1994; 141:535–40.

19. Tennant MK, Thrasher JB, Twomey PA, Birnbaum RS, Plymate SR. Insulin-like growth factor-binding proteins (IGFBP)-4, -5, and -6 in the benign and malignant human prostate: IGFBP-5 messenger ribonucleic acid localization differs from IGFBP-5 protein localization. J Clin Endocrinol Metab 1996; 81:3783–3792.
20. Cohen P, Graves HC, Peehl DM, Kamarei M, Giudice LC, Rosenfeld RG. Prostate-specific antigen (PSA) is an insulin-like growth factor binding protein-3 protease found in seminal plasma. J Clin Endocrinol Metab 1992; 75:1046–53.
21. Lee KO, Oh Y, Cohen P, Peehl DM, Rosenfeld RG. Identification of IGFBP-5 protease activity in human seminal plasma. J Clin Endocrinol Metab 1994; 79:1367–1372.
22. Plymate SR, Rosen CJ, Paulsen CA, Ware JL, Chen J, Vessella RE, Birnbaum. Proteolysis of insulin-like growth factor-binding protein-3 in the male reproductive tract. J Clin Endocrinol Metab 1996; 81:618–624.
23. Frenette G, Tremblay RR, Lazure C, Dube JY. Prostatic kallikrein hK2, but not prostate-specific antigen (hK3), activates single-chain urokinase-type plasminogen activator. Int J Cancer 1997; 71: 897–899.
24. Deperthes D, Frenette G, Brillard-Bourdet M, Bourgeois L, Gauthier F, Tremblay RR, Dube JY. Potential involvement of kallikrein hK2 in the hydrolysis of the human seminal vesicle proteins after ejaculation. J Androl 1996; 17:659–665.
25. Lalou C, Silve C, Rosato R, Segovia B, Binoux M. Interactions between insulin-like growth factor-I (IGF-I) and the system of plasminogen activators and their inhibitors in the control of IGF-binding protein-3 production and proteolysis in human osteosarcoma cells. Endocrinology 1994; 135:2318–2326.
26. Nunn S, Peehl DM, Cohen P. Cathepsin D is an acid activated IGFBP protease produced in normal and malignant prostate epithelial cells. J Cell Physiol 1997; 171:196–204.
27. Kim J, Kim H, Zelner DJ, Ahn HJ, Sensibar JA, Lee C. Transforming growth factor-beta 1 is a mediator of androgen-regulated growth arrest in an androgen-responsive prostatic cancer cell line, LNCaP. Endocrinology 1996; 137:991–999.
28. Sells D, Wood P, Joshi-Barve SS, Muthukumar S, Jacob RJ, Crist SA, Humphreys S, Rangnekar VM. Commonality of the gene programs induced by effectors of apoptosis in androgen-dependent and -independent prostate cells. Cell Growth Differ 1994; 5:457–466.
29. Levine AC. Pathogenesis and medical management of benign prostatic hyperplasia. Trends Endocrinol Metab 1995; 6:128–132.
30. Fiorelli G, DeBellis A, Longo A, Giannini S, Natali A, Costantini A, Vannelli GB, Serio M. Insulin-like growth factor-I receptors in human hyperplastic prostate tissue: characterization, tissue localization, and their modulation by chronic treatment with a gonadotropin-releasing hormone analog. J Clin Endocrinol Metab 1991; 72:740–746.
31. Marcelli M, Haidacher SJ, Plymate SR, Birnbaum RS. Altered growth and insulin-like growth factor-binding protein-3 production in PC3 prostate carcinoma cells stably transfected with a constitutively active androgen receptor complementary deoxyribonucleic acid. Endocrinology 1995; 136:1040–1048.
32. Rajah R, Valentinis B, Cohen P. IGFBP-3 induces apoptosis and mediates the effects of TGF-β on cell death. J Biol Chem 1997; 272:2231–8.
33. Werner H, Rauscher FJ, Sukhatme VP, Drummond IA, Roberts CT, LeRoith D. Transcriptional repression of the insulin-like growth factor I receptor (IGF-R) gene by the tumor suppressor WT-1 involves binding to sequences both upstream and downstream of the IGF-I-R gene transcription start site. J Biol Chem 1994; 269:12577–82.
34. Drummond IA, Madden SL, Rohwer-Nutter P, Bell GI, Sukhatme VP, Rauscher FJ. Repression of the insulin-like growth factor II gene by the Wilms tumor suppressor WT-1. Science 1992; 257:674–81.
35. Buckbinder L, Talbott R, Velasco-Miguel S, Takenaka I, Faha B, Seizinger BR, Kley N. Induction of the growth inhibitor IGF-binding protein 3 by p53. Nature 1995; 377:646–649.
36. Ludwig RL, Bates S, Vousden KH. Differential activation of target cellular promoters by p53 mutants with impaired apoptotic function. Mol Cell Biol 1996; 16(9):4952–4960.
37. Friedlander P, Haupt Y, Prives C, Oren M. A mutant p53 that discriminates between p53-responsive genes cannot induce apoptosis. Mol. Cell Biol 1996; 16(9):4961–4971.
38. Girinsky C, Koumenis T, Graeber G, Peehl DM, Giaccia AJ. Attenuated response of p53 and p21 in primary cultures of human prostatic epithelial cells exposed to DNA-damaging agents. Cancer Res 1995; 55:3726–3731.
39. D'Ercole AN. Insulin-like growth factors and their receptors in growth. Endocrinol Metab Clin North Am 1996; 25(3):573–90.

40. Tu K, McConnell M, Martin C, Campbell ML, Fernandez A, Von Eschenbach AC, McDonnell TJ. Combination adriamycin and suramin induces apoptosis in bcl-2 expressing prostate carcinoma cells. Cancer Lett 1995; 93:147–155.
41. Okazawa H, Shimizu J, Kamei M, Imafuku I, Hamada H, Kanazawa I. Bcl-2 inhibits retinoic acid-induced apoptosis during the neural differentiation of embryonal stem cells. J Cell Biol 1996; 132: 955–968.
42. Tsukada K, Eguchi K, Migita K, Kawabe Y, Kawakami A, Matsuoka N, Takashima H, Mizokami A, Nagataki S. Transforming growth factor beta 1 induces apoptotic cell death in cultural human umbilical vein endothelial cells with down-regulated expression of bcl-2. Biochem Biophys Res Commun 1995; 2105:1076–1082.
43. Jung Y, Miura M, Yuan J. Suppression of interleukin-1 beta-converting enzyme-mediated cell death by insulin-like growth factor. J Biol Chem 1996; 271(9):5112–7.
44. McNeal JE. Pathology of benign prostatic hyperplasia: insight into etiology. Urol Clin North Am 1990; 17:477–486.
45. Dong G, Rajah R, Vu T, Hoffman AR, Rosenfeld RG, Roberts CT, Peehl DM, Cohen P. Decreased expression of Wilms' tumor gene WT-1 and elevated expression of insulin growth factor-II (IGF-II) and type I IGF receptor genes in prostatic stromal cells from patients with benigh prostatic hyperplasia. J Clin Endocrinol Metab 1997; 82:2198–2203.
46. Allander SV, Larsson C, Ehrenborg E, Suwanichkul A, Weber G, Morris SL, Bajalica S, Kiefer MC, Luthman H, Powell DR. Characterization of the chromosomal gene and promoter for human insulin-like growth factor binding protein-5. J Biol Chem 1994; 269:10891–10898.
47. Nunn S, Peehl DM, Cohen P. Abnormal regulation of IGFBP-3 by TGF-β1 in prostatic stromal cell cultures from BPH patients. In: 10th International Congress on Endocrinology Progress and Abstracts, 1996:776.
48. Figueroa JA, Lee AV, Jackson JG, Yee D. Proliferation of cultured human prostate cancer cells is inhibited by insulin-like growth factor (IGF) binding protein-1: evidence for an IGF-II autocrine growth loop. J Clin Endocrinol Metab 1995; 80:3476–3482.
49. Pietrzkowski Z, Mulholland G, Gomella L, Jameson BA, Wernicke D, Baserga R. Inhibition of growth of prostatic cancer cell lines by peptide analogues of insulin-like growth factor-1. Cancer Res 1993; 53:1102–6.
50. Angelloz Nicoud P, Binoux M. Autocrine regulation of cell proliferation by the insulin-like growth factor (IGF) and IGF binding protein-3 protease system in a human prostate carcinoma cell line (PC-3). Endocrinology 1995; 136:5485–5492.
51. Plymate SR, Tennant M, Birnbaum RS, Thrasher JB, Chatta G, Ware JL. The effect on the insulin-like growth factor system in human prostate epithelial cells of immortalization and transformation by simian virus-40 antigen. J Clin Endocrinol Metab 1996; 81:3709–3716.
52. Tennant MK, Thrasher JB, Twomey PA, Drivdahl RH, Birnbaum RS, Plymate SR. Protein and messenger ribonucleic acid (mRNA) for the type I insulin-like growth factor (IGF) receptor is decreased and IGF-II mRNA is increased in human prostate carcinoma compared to benign prostate epithelium. J Clin Endocrinol Metab 1996; 81:3774–3782.
53. Tennant MK, Thrasher JB, Twomey PA, Birnbaum RS, Plymate SR. Insulin-like growth factor binding protein-2 and -3 expression in benign human prostate epithelium, prostate intraepithelial neoplasia, and adenocarcinoma of the prostate. J Clin Endocrinol Metab 1996; 81:411–420.
54. Stamey TA, Kabalin JN, McNeal JE, Johnstone IM, Freiha F, Redwine EA, Yang N. Prostate specific antigen in the diagnosis and treatment of adenocarcinoma of the prostate. II. Radical prostatectomy treated patients. J Urol 1989; 141:1076–1083.
55. Cohen P, Peehl DM, Stamey TA, Wilson K, Clemmons DR, Rosenfeld RG. Elevated levels of IGF binding protein-2 in the serum of prostate cancer patients. J Clin Endocrinol Metab 1993; 76:1031–1038.
56. Kanety H, Madjar Y, Dagan Y, Levi J, Papa MZ, Pariente C, Goldwasser B, Karasik A. Serum insulin-like growth factor-binding protein-2 (IGFBP-2) is increased and IGFBP-3 is decreased in patients with prostate cancer: correlation with serum prostate-specific antigen. J Clin Endocrinol Metab 1993; 77:229–233.
57. Christensson A, Laurell CB, Lilja H. Enzymatic activity of prostate-specific antigen and its reactions with extracellular serine proteinase inhibitors. Eur J Biochem 1990; 194:755–63.
58. Handelsman DJ, Spaliviero JA, Scott CD, Baxter RC. Identification of insulin-like growth factor-I and its receptors in the rat testis. Acta Endocrinol (Copenh) 1985; 109:543–549.
59. Tres LL, Smith EP, Van Wyk JJ, Kierszenbaum AL. Immunoreactive sites and accumulation of somatomedin-C in rat Sertoli-spermatogenic cell co-cultures. Exp Cell Res 1986; 162:33–50.

60. Chatelain PG, Naville D, Saez JM. Somatomedin-C/insulin-like growth factor 1-like material secreted by porcine Sertoli cells in vitro: characterization and regulation. Biochem Biophys Res Commun 1987; 146:1009–1017.
61. Vannelli BG, Barni T, Orlando C, Natali A, Serio M, Balboni GC. Insulin-like growth factor-I (IGF-I) and IGF-I receptor in human testis: an immunohistochemical study. Fertil Steril 1988; 49:666–669.
62. Borland K, Mita M, Oppenheimer CL, Blinderman LA, Massague J, Hall PF, Czech MP. The actions of insulin-like growth factors I and II on cultured Sertoli cells. Endocrinology 1984; 114:240–246.
63. Oonk RB, Grootegoed JA. Insulin-like growth factor I (IGF-I) receptors on Sertoli cells from immature rats and age-dependent testicular binding of IGF-I and insulin. Mol Cell Endocrinol 1988; 55:33–43.
64. Perrard-Sapori MH, Chatelain PG, Rogemond N, Saez JM. Modulation of Leydig cell functions by culture with Sertoli cells or with Sertoli cell-conditioned medium: effect of insulin, somatomedin-C and FSH. Mol Cell Endocrinol 1987; 50:193–201.
65. Forti G, Vannelli GB, Barni T, Balboni GC, Orlando C, Serio M. Sertoli-germ cell interactions in the human testis. J Steroid Biochem Mol Biol 1992; 43:419–422.
66. Smith EP, Dickson BA, Chernausek SD. Insulin-like growth factor binding protein-3 secretion from cultured rat Sertoli cells: dual regulation by follicle stimulating hormone and insulin-like growth factor-I. Endocrinology 1990; 127:2744–2751.
67. Rappaport MS, Smith EP. Insulin-like growth factor (IGF) binding protein 3 in the rat testis: follicle-stimulating hormone dependence of mRNA expression and inhibition of IGF-I action on cultured Sertoli cells. Biol Reprod 1995; 52:419–425.
68. Zhou J, Bondy C. Anatomy of the insulin-like growth factor system in the human testis. Fertil Steril 1993; 60:897–904.
69. Cailleau J, Vermeire S, Verhoeven G. Independent control of the production of insulin-like growth factor I and its binding protein by cultured testicular cells. Mol Cell Endocrinol 1990; 69:79–89.
70. Lin T, Wang DL, Calkins JH, Guo H, Chi R, Housley PR. Regulation of insulin-like growth factor-I messenger ribonucleic acid expression in Leydig cells. Mol Cell Endocrinol 1990; 73:147–152.
71. Koike S, Noumura T. Immunohistochemical localization of insulin-like growth factor-II in the perinatal rat gonad. Growth Regul 1995; 5:185–189.
72. Lin T, Wang D, Nagpal ML, Shimasaki S, Ling N. Expression and regulation of insulin-like growth factor-binding protein-1, -2, -3, and -4 messenger ribonucleic acids in purified rat Leydig cells and their biological effects. Endocrinology 1993; 132:1898–1904.
73. Hasegawa T, Hasegawa Y, Aso T, Kotoh S, Tsuchia Y, Cohen P, Rosenfeld RG. Multiple hormonal regulation of IGFBP-4 in Leydig cells. Clin Pediatr Endocrinol 1994; 3:179–190.
74. Hasegawa T, Cohen P, Rosenfeld RG. Characterization of the IGF axis in TM-3 Leydig cells.Growth Regul 1995; 5:151–159.
75. Hasegawa T, Hasegawa Y, Rosenfeld RG, Cohen P. IGFBP-4 accumulation is negatively correlated with growth rate in TM-3 cells. Growth Hormone and IGF Research, 1998; 8:277–282.
76. Forti G, Barni T, Orlando C, Natali A, Serio M, Balboni GC. Sertoli cell proteins in the human seminiferous tubule. J Steroid Biochem 1989; 32:135–144.
77. Dombrowicz D, Hooghe-Peters EL, Gothot A, Sente B, Vanhaelst L, Closset J, Hennen G. Cellular localization of IGF-I and IGF-II mRNAs in immature hypophysectomized rat testis and epididymis after in vivo hormonal treatment. Arch Int Physiol Biochim Biophys 1992; 100:303–308.
78. Moore A, Morris ID. The involvement of insulin-like growth factor-I in local control of steroidogenesis and DNA synthesis of Leydig and non-Leydig cells in the rat testicular interstitium. J Endocrinol 1993; 138:107–114.
79. Berry SA, Pescovitz OH. Ontogeny and pituitary regulation of testicular growth hormone-releasing hormone-like messenger ribonucleic acid. Endocrinology 1990; 127:1404–1411.
80. Clark AM, Samaras SE, Hammond JM, Hagen DR. Changes in the messenger ribonucleic acid for insulin-like growth factor-I and -II in the porcine testis during and between two waves of testicular development. Biol Reprod 1994; 50:993–999.
81. Hansson HA, Billig H, Isgaard J. Insulin-like growth factor I in the developing and mature rat testis: immunohistochemical aspects. Biol Reprod 1989; 40:1321–1328.
82. Voutilainen R, Miller WL. Developmental and hormonal regulation of mRNAs for insulin-like growth factor II and steroidogenic enzymes in human fetal adrenals and gonads. DNA 1988; 7:9–15.
83. Baker J, Hardy MP, Zhou J, Bondy C, Lupu F, Bellve AR, Efstratiadis. Effects of an Igf1 gene null mutation on mouse reproduction. Mol Endocrinol 1996; 10:903–918.
84. Antich M, Fabian E, Sarquella J, Bassas L. Effect of testicular damage induced by cryptorchidism on insulin-like growth factor I receptors in rat Sertoli cells. J Reprod Fertil 1995; 104(2):267–275.

85. Naville D, Chatelain PG, Avallet O, Saez JM. Control of production of insulin-like growth factor I by pig Leydig and Sertoli cells cultured alone or together. Cell-cell interactions. Mol Cell Endocrinol 1990; 70:217–224.
86. Closset J, Gothot A, Sente B, Scioppo ML, Igout A, Vandenbroeck M, Dombrowicz D, Hennen G. Pituitary hormones dependent expression of insulin-like growth factors I and II in the immature hypophysectomized rat testis. Mol Endocrinol 1989; 3:1125–1131.
87. Perrard-Sapori MH, Chatelain PG, Jaillard C, Saez JM. Characterization and regulation of somatomedin-C/insulin-like growth factor I (Sm-C/IGF-I) receptors on cultured pig Leydig cells. Effects of Sm-C/IGF-I on luteotropin receptors and steroidogenesis. Eur J Biochem 1987; 165:209–214.
88. Lin T, Blaisdell J, Haskell JF. Hormonal regulation of type I insulin-like growth factor receptors of Leydig cells in hypophysectomized rats. Endocrinology 1988; 123:134–139.
89. Wang D, Nagpal ML, Lin T, Shimasaki S, Ling N. Insulin-like growth factor-binding protein-2: the effect of human chorionic gonadotropin on its gene regulation and protein secretion and its biological effects in rat Leydig cells. Mol Endocrinol 1994; 8:69–76.
90. Khan S, Teerds K, Dorrington J. Growth factor requirements for DNA synthesis by Leydig cells from the immature rat. Biol Reprod 1992; 46:335–341.
91. Benahmed M, Morera AM, Chauvin MC, de Peretti E. Somatomedin C/insulin-like growth factor 1 as a possible intratesticular regulator of Leydig cell activity. Mol Cell Endocrinol 1987; 50:69–77.
92. Jaillard C, Chatelain PG, Saez JM. In vitro regulation of pig Sertoli cell growth and function: effects of fibroblast growth factor and somatomedin-C. Biol Reprod 1987; 37:665–674.
93. Chatelain PG, Sanchez P, Saez JM. Growth hormone and insulin-like growth factor I treatment increase testicular leutinizing hormone receptors and steroidogenic responsiveness of growth hormone deficient dwarf mice. Endocrinology 1991; 128:1857–1862.
94. Dubois W, Callard GV. Culture of intact Sertoli/germ cell units and isolated Sertoli cells from Squalus testis. II. Stimulatory effects of insulin and IGF-I on DNA synthesis in premeiotic stages. J Exp Zool 1993; 267:233–244.
95. Lin T, Haskell J, Vinson N, Terracio L. Direct stimulatory effects of insulin-like growth factor-I on Leydig cell steroidogenesis in primary culture. Biochem Biophys Res Commun 1986; 137:950–956.
96. Kasson BG, Hsueh AJ. Insulin-like growth factor-I augments gonadotropin-stimulated androgen biosynthesis by cultured rat testicular cells. Mol Cell Endocrinol 1987; 52:27–34.
97. Gelber SJ, Hardy MP, Mendis-Handagama SM, Casella SJ. Effects of insulin-like growth factor-I on androgen production by highly purified pubertal and adult rat Leydig cells. J Androl 1992; 13:125–130.
98. Syed V, Khan SA, Nieschlag E. Epidermal growth factor stimulates testosterone production of human Leydig cells in vitro. J Endocrinol Invest 1991; 14:93–97.
99. Benton L, Shan L-X, Hardy MP. Differentiation of adult Leydig cells. J Steroid Biochem Mol Biol 1995; 53:61–68.
100. Nagpal ML, Wang D, Calkins JH, Chang WW, Lin T. Human chorionic gonadotropin up-regulates insulin-like growth factor-I receptor gene expression of Leydig cells. Endocrinology 1991; 129(6): 2820–2826.
101. Moore A, Chen CL, Davis JR, Morris ID. Insulin-like growth factor-I mRNA expression in the interstitial cells of the rat testis. J Mol Endocrinol 1993; 11:319–324.
102. Chuzel F, Clark AM, Avallet O, Saez JM. Transcriptional regulation of the lutropin/human choriogonadotropin receptor and three enzymes of steroidogenesis by growth factors in cultured pig Leydig cells. Eur J Biochem 1996; 239:8–16.
103. Rappaport MS, Smith EP. Insulin-like growth factor I inhibits aromatization induced by follicle-stimulating hormone in rat Sertoli cell culture. Biol Reprod 1996; 54:446–452.
104. Crawford BA, Handelsman DJ. Recombinant growth hormone and insulin-like growth factor-I do not alter gonadotrophin stimulation of the baboon testis in vivo. Eur J Endocrinol 1994; 131:405–412.
105. Lin T, Blaisdell J, Haskell JF. TGF-β inhibits Leydig cell steroidogenesis in primary culture. Biochem Biophys Res Commun 1987; 146:387–394.
106. Besset V, Collette J, Chauvin MA, Franchimont P, Benahmed M. Effect of transforming growth factor-beta 1 on the insulin-like growth factor system in cultured porcine Leydig cells. Mol Cell Endocrinol 1994; 99:251–257.
107. Lin T, Wang D, Nagpal ML, Chang W, Calkins JH. Down-regulation of Leydig cell insulin-like growth factor-I gene expression by interleukin-1. Endocrinology 1992; 130:1217–1224.
108. Soder O, Bang P, Wahab A, Parvinen M. Insulin-like growth factors selectively stimulate spermatogonial, but not meiotic, deoxyribonucleic acid synthesis during rat spermatogenesis. Endocrinology 1992; 131:2344–2350.

109. Mita M, Borland K, Price JM, Hall PF. The influence of insulin and insulin-like growth factor-I on hexose transport by Sertoli cells. Endocrinology 1985; 116:987–992.
110. Tajima Y, Watanabe D, Koshimizu U, Matsuzawa T, Nishimune Y. Insulin-like growth factor-I and transforming growth factor-alpha stimulate differentiation of type A spermatogonia in organ culture of adult mouse cryptorchid testes. Int J Androl 1995; 18:8–12.
111. Besset V, Le Magueresse-Battistoni B, Collette J, Benahmed M. Tumor necrosis factor alpha stimulates insulin-like growth factor binding protein 3 expression in cultured porcine Sertoli cells. Endocrinology 1996; 137:296–303.
112. DeChiara TM, Efstratiadis, Robertson EJ. A growth-deficiency phenotype in heterozygous mice carrying an insulin-like growth factor II gene disrupted by targeting. Nature 1990; 345:78–80.
113. Chubb C. Sexual behavior and fertility of little mice. Biol Reprod 1987; 37:564–569.
114. Downs TR, Frohman LA. Evidence for a defect in growth hormone-releasing factor signal transduction in the dwarf (dw/dw) rat pituitary. Endocrinology 1991; 129:58–67.
115. Nogami H, Takeuchi T, Suzuki K, Okuma S, Ishikawa H. Studies on prolactin an growth hormone gene expression in the pituitary gland of spontaneous dwarf rats. Endocrinology 1989; 125:964–970.
116. Chen Y, Bornfeldt KE, Arner A, Jennische E, Malmvist U, Uvelius B, Arnqvist HJ. Increases in insulin-like growth factor I in hypertrophying smooth muscle. Am J Physiol 1994; 266(2 Pt 1):E224–E229.
117. Chen Y, Arner A, Bornfeldt KE, Uvelius B, Arnqvist HJ. Development of smooth muscle hypertrophy is closely associated with increased gene expression of insulin-like growth factor binding protein-2 and -4. Growth Regul 1995; 5:45–52.
118. Chen Y, Gustafsson B, Arnqvist HJ. IGF-binding protein-2 is induced during development of urinary bladder hypertrophy in the diabetic rat. Am J Physiol 1997; 272(2 Pt 1):E297–E303.
119. Wang J, Niu W, Nikiforov Y, Naito S, Chernausek S, Witte D, LeRoith D, Strauch A, Fagin JA. Targeted overexpression of IGF-I evokes distinct patterns of organ remodeling in smooth muscle cell tissue beds of transgenic mice. J Clin Invest 1997; 100:1425–1439.
120. deBoer WI, Schuller AG, Vermey M, van der Kwast TH. Expression of growth factors and receptors during specific phases in regenerating urothelium after acute injury in vivo. Am J Pathol 1994; 145:1199–1207.
121. Elkin M, Shevelev A, Schultze E, Tykocinski M, Cooper M, Ariel I, Pode D, Kopf E, de Groot N, Hochberg A. The expression of the imprinted H19 and IGF-2 genes in human bladder carcinoma. FEBS Lett 1995; 374:57–61.
122. Cooper MJ, Fischer M, Komitowski D, Shevelev A, Schultze E, Ariel I, Tykocinski ML, Miron S, Ilan J, de Groot N, Hochberg A. Developmentally imprinted genes as markers for bladder tumor progression. J Urol 1996; 155:2120–2127.
123. Weinzimer SA, Macarak E, and Cohen P. The IGF-IGFBP axis in the regulation of bladder smooth muscle growth. In: 1997 International Research Symposium on Interstitial Cystitis Progress Abstracts, 1997; 90.

19 IGFs and the Nervous System

Phillip S. Leventhal, PhD, *James W. Russell,* MD, *and Eva L. Feldman,* MD, PhD

Contents

EXPRESSION OF THE INSULIN-LIKE GROWTH FACTORS, IGF RECEPTORS, AND IGF BINDING PROTEINS IN THE NERVOUS SYSTEM

Insulin-like growth factors-I and -II (IGF-I and IGF-II) are peptide growth factors closely related in sequence to insulin *(1,2)*. In the nervous system, the IGFs, IGF receptors, and IGF binding proteins (IGFBPS) are widely expressed and promote proliferation, survival, and differentiation of neuronal and nonneuronal cells. In addition, high levels of IGFs and IGFBPs are found in a variety of nervous system tissues during development. Because of their potent survival and differentiation activities, the IGFs are currently being tested as therapeutics for a variety of neuronal pathologies, including various neuropathies, motor neuron disease, and physical or hypoxic–ischemic insults. The physiological effects of the IGFs in the nervous system are thought to be mediated by the IGF-I receptor, while the IGF-II receptor, in conjunction with the IGFBPs, regulates IGF bioavailability.

Developmental Expression in the Central Nervous System

Developmental Expression of the IGFs in the Central Nervous System

IGF-I is widely expressed in the central nervous system (CNS) (*see* Table 1 for a summary). During embryonic development of the rat, IGF-I mRNA is present in the

From: *Contemporary Endocrinology: The IGF System*
Edited by: R. Rosenfeld and C. Roberts Jr © Humana Press, Inc., Totowa, NJ

Table 1
Expression of IGFs in the Nervous System

Tissue/System	*Subregion*	*Growth Factor*	*References*
Cranial nerves	Trigeminal and cranial ganglia	IGF-I	*5,7,8*
Brain	Hippocampus	IGF-I	*7*
	Corpus striatum	IGF-I	*7*
	Diencephalon	IGF-I	*7*
	Hypothalamus	IGF-II	*5*
	Cerebellum	IGF-I	*3,4,6,7,9,11*
		IGF-II	*6*
	Brainstem	IGF-I	*6,7*
		IGF-II	*6*
	Olfactory bulb	IGF-I	*5,7*
	Choroid plexus	IGF-I	*7*
		IGF-II	*4,5,8,12–14*
	Leptomeninges	IGF-II	*4,5,12–14*
	Vasculature	IGF-II	*4,5,8,12*
	Ventricles	IGF-II	*5*
Spinal cord		IGF-I	*6*

cerebellar neurons, including Purkinje cells, as well as in retinal, sensory, and trigeminal ganglia *(3–5)*. Generally, during fetal development, a high level of IGF-I expression is found in the olfactory bulb, midbrain, cerebral cortex, and hippocampus *(6,7)*. IGF-I mRNA is also found in the spinal cord *(6)*. Several studies suggest that the production of IGF-I in the CNS of the rat is found in areas corresponding to axonal targets or is associated with differentiating and not proliferating neurons *(8,9)*.

Expression of IGF-I in the rat cerebellum, spinal cord, olfactory bulb, midbrain, and cerebellum is high during embryonic development, peaking around d 14 or 15, with levels decreasing prior to birth and just after birth *(8–10)*. Most studies show that in the cerebellum and other brain regions, the level of IGF-I protein is very low in the adult *(6,9)*. However, one study indicated that IGF-I protein in the rat cerebellum increases after the initial postnatal decline, with a second peak of expression in early adulthood *(11)*. A second study in the adult rat also found a high level of IGF-I expression in sets of sensory and cerebellar projection neurons *(4)*.

Like IGF-I, IGF-II is widely expressed in the CNS (*see* Table 1). However, in contrast to IGF-I, IGF-II is most highly expressed in nonneuronal tissues of the nervous system. Indeed, the highest levels of expression of IGF-II in the embryonic or adult rat brain is found in the choroid plexus, leptomeninges, microvasculature, and myelin sheaths *(4,5,8,12–14)*. A similar pattern of expression is observed during fetal and early postnatal development of the human brain *(15)*. Like IGF-I, levels of IGF-II mRNA in the rat brain peak around d 14, with an approximately three- to fourfold decrease to levels found in the adult *(6)*. However, high levels of expression persist in the choroid plexus and leptomeninges. Likewise, in the developing human brain, expression of IGF-II mRNA is highest during fetal development, with levels in most of the brain dropping off rapidly after 2 mo, with high levels remaining only in the choroid plexus and leptomeninges *(15)*.

Table 2
Expression of IGF Receptors in the Nervous System

Tissue/System	*Subregion*	*Receptor Type*	*References*
Brain	Hippocampus	IGF-IR	*206*
		IGF-IIR	*13*
	Hypothalamus	IGF-IR	*206*
		IGF-IIR	*25*
	Cerebellum	IGF-IR	*4,206,207*
	Cerebral cortex	IGF-IR	*4,25,207*
	Brainstem	IGF-IIR	*208*
	Olfactory bulb	IGF-IR	*206,207*
	Choroid plexus	IGF-IR	*4,207*
		IGF-IIR	*13,25*
	Meninges	IGF-IR	*4*
		IGF-IIR	*13*
	Vasculature	IGF-IR	*4*
		IGF-IIR	*25*
Visual	Retina	IGF-IIR	*25*

Developmental Expression of the IGF Receptors in the CNS

The biological actions of the IGFs are mediated by the type I IGF receptor (IGF-IR), a receptor tyrosine kinase. Specific aspects of IGF-IR structure and signaling as it relates to neuronal function of the IGFs is discussed in the second section of this chapter. More detailed reviews of the IGF-IR receptor can be found elsewhere *(16,17)*.

IGF-IR mRNA is widely expressed in the rat CNS (*see* Table 2), specifically in the ganglion cells of the retina, hippocampus, Purkinje cells of the cerebellum, projection neurons of the cerebellar nuclei, interneurons of the hippocampus and neocortex, and brain stem nuclei *(7,18)*. In the brain, expression of IGF-IR mRNA is highest in the developing cerebellum, midbrain, olfactory bulb, and in the ventral floorplate of the hindbrain *(6,8,19)*. The level of IGF-IR is highest during late embryogenesis (embryonic days 15–20) and declines to adult levels soon after birth *(20,21)*. After birth, and into adulthood, the expression of the IGF-IR is most abundant in the choroid plexus, meninges, and vascular sheaths *(4)*.

The IGF-II receptor (IGF-IIR) is structurally unrelated to the IGF-IR and does not possess a significant intracellular domain. It has a low affinity for IGF-I, but a high affinity for IGF-II, and also binds mannose-6-phosphate. There is little evidence that the IGF-IIR plays a role in IGF signal transduction. The IGF-IIR appears to have a major role in protein sorting by the Golgi *(22)*. Its biological function is not well understood, although when expressed at the extracellular surface, it may act like the IGFBPs (*see* pp. 428–430) to modulate IGF-II bioavailability *(23,24)*.

In the rat CNS, overall expression of the IGF-IIR is highest in the pyramidal cell layers of the hippocampus, the granule layer of the dentate gyrus, the choroid plexus, and in the cerebral vasculature, ependymal cells, retina, and pituitary *(13)*. There is a transient high level of the IGF-IIR in rat brain during development, particularly in the cortex and hypothalamus that decreases toward birth, and drops to its lowest level by adulthood. However, expression in some brain areas, including the choroid plexus, cerebral vasculature, ependymal cells, and retina remain high throughout development *(25)*. In the human, less is known about the IGF-IIR, but there appears to be a relatively low level of IGF-IIR protein during fetal and early postnatal CNS development *(26,27)*.

Developmental Expression of the IGF Binding Proteins in the CNS

The IGFs are most often found complexed with a family of binding proteins designated the IGFBPs. There are currently seven IGFBPs (IGFBP-1 through -7). The IGFBPs are thought to regulate IGF bioavailability by preventing their degradation, modulating levels of free IGFs, and possibly by delivering them to target tissues. The structure, function, and biological role of the IGFBPs are reviewed elsewhere *(28,29)*. The expression of many of the IGFBPs in the CNS has been described (*See* Table 3 for a summary). Of particular interest in the CNS are IGFBP-2 and -5. In many parts of the CNS, expression or localization of these binding proteins during development coincides with that of the IGFs. For example, in the developing rat, IGFBP-2 and IGF-I expression appears to be coordinated during retinal and cerebellar maturation *(3,30)*. IGFBP-2 is also coexpressed with IGF-II in the leptomeninges, choroid plexus, and parenchymal microvasculature, and is colocalized in the myelin sheaths of individual axons and in myelinated nerve tracts *(12,31)*. Similarly, IGFBP-5 and IGF-I are coexpressed during development of olfactory bulb; medial and dorsal lateral geniculate bodies; and ventral tier, cochlear, leminiscal, and vestibular nuclei. There is also temporal coordination of IGFBP-6 and IGF-I in the developing rat cerebellar cortex and hippocampus *(32,33)* and in an IGF-I transgenic mouse model, IGF-I promotes IGFBP-5 expression *(34)*. Whether these studies on IGFBP expression and localization in the CNS can be related to human development is not clear. One study, for example, reports that little or no IGFBP-5 is expressed in the human brain during fetal development *(35)*. In general, however, it is thought that IGFBPs are key regulators of IGF bioavailability in the developing nervous system.

Developmental Expression of the IGFs, IGF Receptors, and IGF Binding Proteins in the Peripheral Nervous System

Little is known about the expression of IGFs, IGF receptors, and IGF binding proteins in the peripheral nervous system (PNS). The summaries in Tables 1–3 include the current state of knowledge in this regard. Preliminary experiments suggest that many elements of the IGF system are expressed in peripheral neurons. For example, *in situ* hybridization experiments demonstrate that IGF-IR mRNA is localized in the axons of embryonic rat dorsal root ganglion neurons, and, in human dorsal root ganglions, IGF-IR is abundant in the perikaryon *(1,60)*. In the rat, IGFBP-5 is expressed in fetal peripheral nerves as well as in adult peripheral Schwann cells and axons *(33)*. Cultured Schwann cells also express IGFBP-5 and IGF-IR *(36)*. Finally, expression of IGF-II is at a high level in human fetal spinal ganglia *(37)* and in muscle during innervation by motor neurons *(38)*, and IGF-II protein is localized in adult rat motor neurons *(39)*. In general, much of the evidence supporting a role for the IGFs, IGFBPs, and IGF receptors in the PNS comes from studies showing that these proteins are up-regulated during sciatic nerve regeneration in the rat (*see* pp. 428–430).

Expression of the IGFs, IGF Receptors, and IGF Binding Proteins During Neuronal Regeneration

Up-regulation of IGFs and IGFBPs is seen in the CNS in response to a variety of insults. For example, cerebral hypoxic–ischemic injury in the juvenile rat results in a loss

Table 3
Expression of IGF Binding Proteins in the Nervous System Development

Tissue/System	*Subregion*	*IGFBP*	*References*
Brain	Cerebral cortex	-4,-5	*32,209*
	Hippocampus	-2,-4,-5	*30,32,34,209*
	Diencephalon	-4,-5	*34,209*
	Intermediate lobe of pituitary	-2	*210*
	Cerebellum	-2,-5	*3,30,33,34,211*
	Brainstem	-5	*32–34*
	Olfactory bulb	-2,-4,-5	*32,211,212*
	Choroid plexus	-2,-4	*12,30,31,209,213*
	Leptomeninges	-2	*12,30*
	Vasculature	-2	*12*
	Myelin	-2	*12*
Spinal cord		-5	*33*
Olfactory	Olfactory placodes	-2	*31*
	Olfactory nerve	-5	*32*
	Olfactory peduncle	-4	*209*
Auditory	Otic placodes	-2	*31*
Visual	Retina	-2	*3,30*
Peripheral nerve		-5	*33*

of IGFBP-2, -4 and -5; IGF-I; and IGF-IR expression in the first 24 h, followed 48 h later by an increase in IGFBP-5 and IGF-I expression *(40,41)*. Likewise, cryogenic spinal cord injury in the rat causes an increase in IGF-I expression within 3 d, followed by a much later enhancement of IGFBP-2 and IGF-IR expression. In this case, IGF-I expression is localized to astrocyte-like cells in the spinal cord whereas, IGF-IR expression is found in oligodendrocytes *(42)*. IGFBP-2 and IGF-I expression is also enhanced after cytotoxic lesion of the rat dentate gyrus after 4 d. In this case, the expression of IGF-I is localized to microglia *(43)*. Similarly, a marked increase in IGFBP-2 and IGF-I expression is observed after cerebral contusion *(44)*. After a penetrating brain injury in rat, an acute (first 7 d) increase in the expression of IGF-I as well as IGFBP-1, -2, -3, and -6 and IGF-IR is found in astrocytes, neurons, and monocytic cells. Increased production of IGFBP-3 is also found in the microvascular endothelium *(45)*. Together, these findings suggest that increased production of IGFBP-2 and IGF-I, and perhaps IGF-IR and other IGFBPs, participates in neuronal regeneration in the CNS after a variety of insults.

IGFs, IGF-IR, and IGFBPs are also up-regulated during PNS regeneration. The most common system for examining the role of the IGF system in PNS regeneration is sciatic nerve crush or transection in the rat. For example, one study showed that after sciatic nerve crush, both IGF-I and IGF-II mRNAs were increased in the sciatic nerve distal to the crush site. At the crush site, IGF-I expression is increased after a 4-d delay, peaking after 10 d. IGF-II gene expression is unchanged at the site of injury, but is increased in more distal, intramuscular reaches of the nerves *(46)*. Another study by the same group demonstrated that the increased expression of IGF-II mRNA after sciatic nerve crush is lost after reestablishment of functional neuromuscular synapses *(47)*. In a similar experimental system, increased IGF-I and IGF-IR expression is observed after sciatic nerve transection. At earlier times (7 or fewer days), the increase in IGF-I expression is localized mostly in Schwann cells. After this point, the majority of IGF-I is found in macrophages invading the distal nerve stump *(36)*. Transection of the facial nerve also results in enhanced expression of IGF-I after a similar time period (4–7 d). In this instance, IGF-I appears in astrocytic processes *(48)*. Collectively, these results indicate that, as in the CNS, enhancement of IGFs and perhaps IGF-IR expression may help mediate neuronal regeneration in the PNS. However, unlike the observations in the CNS, there is currently little information regarding the role of IGFBPs in the response to PNS injuries.

That IGFs and other components of the IGF system are up-regulated in response to injury suggests that IGFs may be useful for treating PNS or CNS injuries or disorders. The results of some studies using IGFs as therapeutics are promising (*see* the fifth section). However, care must be taken in interpreting these in vivo results because of coexpression of multiple growth factors, and the complexity of cellular responses and interactions following injury.

THE IGFS ARE NEUROTROPHIC FACTORS

In conjunction with studies on the use of IGFs as therapeutics, many investigations have examined the cellular and molecular mechanisms of IGF signaling in nervous system disease. In this regard, the IGFs have been shown to be neutrophic factors, i.e., they promote the survival and differentiation of sensory *(49)*, sympathetic *(49,50)*, and motor neurons *(49)*. In fact, the IGFs are the only known growth factors that support both sensory and motor nerve regeneration in adult animals *(51–56)*. In addition, the IGFs have survival, chemotactic, and proliferative effects on nonneuronal cells in the nervous system (*see* the third and fourth sections).

IGFs Promote the Survival of Neuronal Cells

Several studies demonstrate that IGFs promote survival of neurons in vitro. For example, IGF-I has been shown to enhance the survival of d 7 rat embryo cerebellar granule cells. That IGF-I is a neurotrophic factor is also shown by its ability to stimulate terminal differentiation of the cells into cerebellar granule neurons *(57)*. In addition, IGF-I promotes the survival of rat cerebellar, hypothalamic, and hippocampal neurons in vitro *(11,58,59)*. Finally, IGF-I and -II rescue chick lumbar neurons from injury-induced and developmentally regulated death *(60)*.

IGF-I Prevents Neuronal Apoptosis

The ability of IGF-I (and IGF-II) to promote neuronal survival is associated with the ability of these factors to prevent apoptosis. Apoptosis, a form of programmed cell death, is characterized by a specific set of intracellular changes, including condensation of chromatin, DNA fragmentation, and membrane blebbing or budding, and it occurs without cell lysis and does not induce an immune response *(61)*. In addition, specific biochemical pathways appear to be required for apoptosis, such as altered expression of the *bcl* and *bax* oncogenes *(62)* and activation of the interleukin-1β converting enzyme (ICE)/caspase family of cysteine proteases *(63,64)*. Apoptosis can be induced in cells in vitro by providing noxious stimuli or withdrawal of a required factor *(65–69)*. Apoptosis contrasts with necrosis, which occurs by a process of cellular swelling and lysis. In addition, unlike apoptosis, necrotic cell death elicits an immune response *(61)*.

IGF-I appears to be a potent agent for rescuing neurons from apoptosis. For example, IGF-I prevents *N*-methyl-D-aspartate (NMDA) and nitric oxide-induced apoptosis in

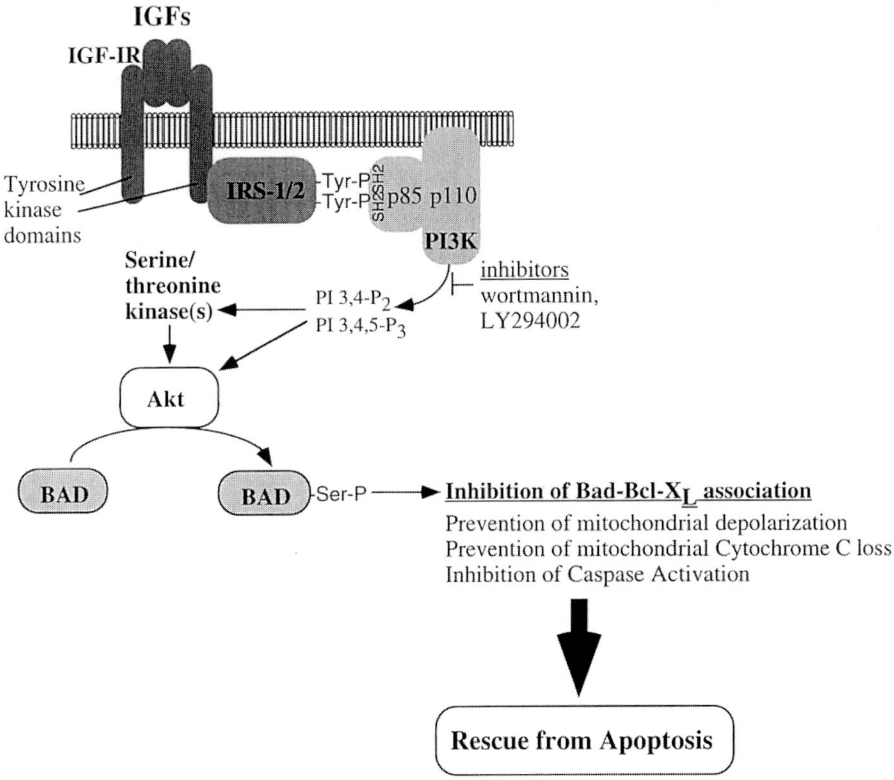

Fig. 1. Summary of intracellular pathways whereby IGFs prevent neuronal apoptosis. IGF binding by the IGF-IR leads to activation of the receptor's intrinsic tyrosine kinase domain, resulting in receptor autophosphorylation and IRS protein phosphorylation. IRS tyrosine phosphorylation promotes the activity of PI3K via an SH2-dependent interaction. This results in the production of phosphatidylinositols phosphorylated at the 3-position of the inositol ring. These newly synthesized phosphatidylinositol phosphates activate the serine/threonine kinase Akt by a dual mechanism, including direct binding to Akt and activation of kinases that phosphorylate specific serine and threonine residues on Akt. Activated Akt, in turn, phosphorylates BAD, and prevents it from interacting with Bcl–X_L at the mitochondrial membrane. This is thought to prevent several changes required for apoptosis, including mitochondrial depolarization, mitochondrial loss of cytochrome *C*, and activation of the caspase protease cascade.

hippocampal and cortical neurons from spontaneously hypertensive rats *(70,71)*. Rat hippocampal neurons are also rescued by IGF-I from the induction of apoptosis by amyloid-derived peptides *(72)*. A similar anti-apoptotic effect of IGF-I is observed in rat cerebellar neurons subjected to serum or potassium withdrawal *(73)*, okadaic acid treatment *(74)*, or high KCl levels *(68,75)*, and in rat motor neurons during normal development or following axotomy or spinal transection *(76)*. Finally, SH-SY5Y human neuroblastoma cells, a well-characterized neuronal cell line, are rescued from hyperosmotic shock-induced apoptosis by IGF-I *(77,78)*.

Biochemical Mechanisms by which IGFs Inhibit Neuronal Apoptosis

Recent studies by a number of laboratories have focused on the biochemical mechanisms whereby IGF-I rescues neurons from apoptosis. A summary of the pathways currently implicated in IGF-I prevention of apoptosis in shown in Fig. 1. In this section we have focused only on the IGF-IR events clearly implicated in the inhibition of apoptosis in neuronal cells. As in many of the IGF-IR signaling pathways, a primary response to activation of the IGF-IR is the tyrosine phosphorylation of the insulin receptor substrate (IRS) proteins. The tyrosine phosphorylation of IRS proteins generates binding sites for certain proteins containing src homology 2 (SH2) domains such as the p85 regulatory subunit of PI3K *(79,80)*. This association between the regulatory subunit of PI3K and tyrosine phosphates on IRS proteins stimulates the p110 catalytic subunit of PI3K, enhancing the production of phosphoinositides phosphorylated at the 3-position of the inositol sugar ring *(81)*. That PI3K plays a primary role in the ability of IGFs to prevent apoptosis is supported by the ability of the PI3K inhibitor wortmannin to prevent IGF-I blockage of apoptosis in serum-starved cerebellar granule cells and in SH-SY5Y cells exposed to hyperosmotic stress *(82,83)*. Likewise, at concentrations known to stimulate the IGF-IR *(17)*, insulin prevents apoptosis in PC12 cells by a PI3K-dependent mechanism *(84)*.

In terms of the downstream targets of PI3K, there is recent evidence for participation of the serine/threonine kinase Akt. Lipid products of PI3K appear to activate Akt by both directly binding to the enzyme *(85,86)* and by directly stimulating PDK1 (phosphoinositide-dependent kinase-1) and other kinases to phosphorylate Akt at a crucial threonine residue *(87)*. Consistent with a primary role for Akt and PI3K in the effects of IGFs, Akt activation by IGF-I in cerebellar granule cells is blocked by PI3K inhibitors, and a dominant-negative mutant of Akt prevents IGF-I rescue from apoptosis *(88)*. In these cerebellar granule cells, IGF-I stimulation of Akt results in phosphorylation of BAD, which prevents BAD from binding the Bcl-2 protein, and inhibits the ability of Bcl-2 to initiate mitochondrial changes required for apoptosis *(89)*. In addition, studies in SH-SY5Y human neuroblastoma cells show that IGF-I prevents the activation of two proteases in the apoptosis pathway, caspases-1 and -3, by a PI3K-dependent mechanism *(83)*. The mechanism by which PI3K regulates caspase activation is not currently known, but may involve phosphorylation of BAD by Akt *(89)*. Together, these results show that PI3K plays a key role in IGF-I prevention of neuronal apoptosis by regulating signaling pathways, protease cascades, and mitochondrial integrity.

IGFs Promote Axonal Growth and Regeneration

In addition to supporting neuronal survival, IGF neurotrophic actions include the ability to promote axonal growth and regeneration in both the CNS and PNS. For example, IGFs appear to promote axonal growth and innervation at neuromuscular junctions. Glazner and Ishii reported that IGF-II expression is up-regulated during motor neuron reinnervation of muscle after sciatic nerve crush, but that IGF-II is down-regulated after reestablishment of functional synapses *(47,90)*. Similarly, IGF-I expression is enhanced in muscle after snake venom toxin-induced paralysis *(91)*. The idea that IGFs promote motor neuron regeneration is supported by the ability of IGFs to enhance neurite sprouting by motor neurons in culture *(91,92)*. Also, IGFBP-4, which neutralizes IGFs, reduces nerve sprouting in botulinum toxin-paralyzed muscle *(91)*. Thus, both IGF-I and IGF-II are implicated in peripheral nerve regeneration and muscle reinnervation. However, there may be some differences in the roles of these two growth factors; IGF-I expression is enhanced in neurons while increased IGF-II expression is found mostly in the denervated muscle *(93)*.

In addition to promoting muscle reinnervation, IGFs promote axonal regrowth. In transected dorsal root ganglion neurons in culture, IGF-I promotes axonal regeneration

(94). Also, locally infused IGF-II enhances the rate of regeneration in sciatic nerves after nerve crush. Furthermore, anti-IGF-II antiserum reduces the rate of spontaneous regeneration, supporting a role for IGF-II in sciatic nerve regeneration in vivo *(53)*. Likewise, IGF-I in combination with platelet-derived growth factor promotes the regeneration of transected and anastomosed rat sciatic neurons *(95)*. Finally, IGF-I expression is enhanced after facial nerve transection. In this case, IGF-I expression is localized in the reactive astrocytes, suggesting a supportive role for astrocytes in nerve regeneration *(48)* (*see* the fourth section for a more detailed discussion of IGFs and astrocyte function).

Basic Cytoskeletal Mechanisms of Axon Regeneration and Growth

During development and regeneration the outgrowth of axonal processes, or neurites, begins with the extension of actin-rich filopodia and lamellipodia *(96–98)*. Filopodia are spike-like protrusions that contain a core of parallel actin filaments as well as concentrations of signaling molecules that allow them to respond to extracellular matrix and soluble factors *(99)*. Lamellipodia are highly dynamic veil-like membranes that mediate cellular protrusion *(100)*.

At the tip of a neurite is a specialized structure known as the growth cone. The growth cone is a highly motile region rich in filopodia and lamellipodia that mediates neurite advance and guidance *(101)*. Interactions of growth cone filopodia with the extracellular matrix and soluble guidance cues determines the direction and extent of neurite outgrowth *(99,102,103)*. The actin cytoskeleton underlying the growth cone filopodia is particularly important for this guidance function, as disruption with inhibitors of actin polymerization cause a loss of appropriate neurite pathfinding *(104,105)*. Lamellipodia on the growth cone are also essential for neurite motility, acting as protrusive structures that allow the growth cone advance *(106)*. Like the filopodia, the lamellipodia are dependent on actin filament substructures. In this regard, polymerization of actin at the leading edge of lamellipodia generates a protrusive force that drives membrane advance *(107)*. Similarly, avoidance of some extracellular matrix and soluble factors is mediated by depolymerization of the growth cone actin cytoskeleton *(108,109)*.

Another key aspect of growth cone guidance and advance is the formation of adhesions between the growth cone and the extracellular matrix *(101)*. This is thought to allow generation of tension within the growth cone, a factor that may be critical for growth cone migration. Furthermore, the differential ability to form adhesions on various extracellular matrix proteins and the different signaling pathways activated by specific adhesion receptors help direct changes in actin cytoskeleton that determine whether a region of a growth cone advances or collapses. Thus, the regulation of the actin cytoskeleton by environmental cues is a central aspect of growth cone guidance and advance.

Growth cone advance and guidance is dependent not only on the actin cytoskeleton, but also on microtubules. Microtubules help make up the body of the neurite and the interior core of the growth cone whereas the actin cytoskeleton is concentrated at the periphery in lamellipodia and filopodia *(97,101)*. The initial extension of a neurite occurs when extracellular matrix or soluble cues instruct invasion of filopodia with cytoplasm containing microtubules and other cytoskeletal elements *(97)*. Given this central localization of microtubules and the role of microtubules in initial neurite extension, it is not surprising that microtubule polymerization inhibitors prevent neurite outgrowth *(97,110)*. Finally, microtubules are required not only for the elaboration of neurites, but also for growth cone turning *(111)*.

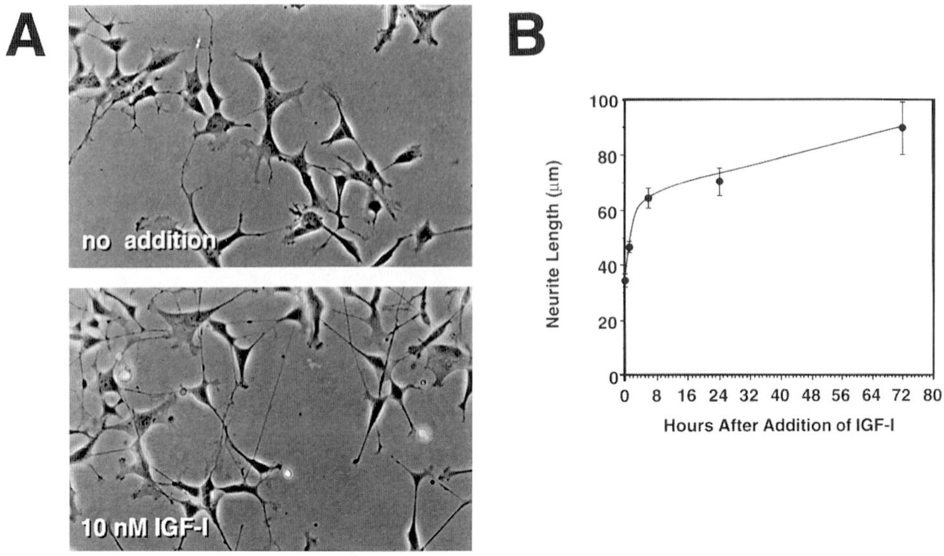

Fig. 2. IGF-I stimulates neurite outgrowth by SH-SY5Y human neuroblastoma cells. Serum-starved SH-SY5Y human neuroblastoma cells were treated for the indicated amounts of time with 10 n*M* IGF-I. **(A)** Phase-contrast photomicrographs of SH-SY5Y cells grown for 72 h in serum-free medium containing no addition or IGF-I. **(B)** The average length of neurites ± SEM as function of time after IGF-I addition. These results show that untreated SH-SY5Y cells possess some neurites, but IGF-I causes an increase in neurite length that can be detected within a few hours.

IGFs as Promoters of Neurite Outgrowth

The IGFs promote neurite outgrowth in a variety of in vitro systems such as chick sympathetic neurons *(50)*, rat brain cortical neurons *(112)*, and guinea pig myenteric plexus neurons *(113)*, and SH-SY5Y human neuroblastoma cells (Fig. 2). In addition, exposure of innervated skeletal muscle to IGFs enhances the sprouting of neurites by chick motor neurons *(114)*.

That IGFs play a role in neurite outgrowth in vivo is supported by several lines of evidence. First, IGF binding proteins suppress nerve sprouting by chick motor neurons in botulinum toxin-paralyzed skeletal muscle *(92)*. A function for IGFs in enhanced neurite outgrowth at the neuromuscular junction is further suggested by the fact that IGF-I expression is increased in botulinum toxin-paralyzed skeletal muscle *(115)*. Enhanced expression of IGF-I is also observed in reactive microglia within deafferented hippocampal laminae just prior to and during neurite outgrowth *(116)*. Likewise, expression of IGF-I and IGF-II is increased in E7 sympathetic ganglia during the period of neurogenesis, suggesting a role in the promotion of neurite outgrowth *(50)*. This ability of IGFs to stimulate neurite outgrowth is thought to be critical for axonal regeneration. For example, IGF-II expression is increased after hypoxic brain injury *(117)* and locally infused IGF-II increases the speed of sensory axon regeneration in crushed rat sciatic nerve. Also, IGF-I enhances neurite regeneration after axotomy in adult rat dorsal root ganglion explants *(94)*. That the IGF-I receptor participates in neurite outgrowth is not surprising given its expression in the central and PNS (*see* the first section) as well as its localization in neuronal growth cones *(118–120)*.

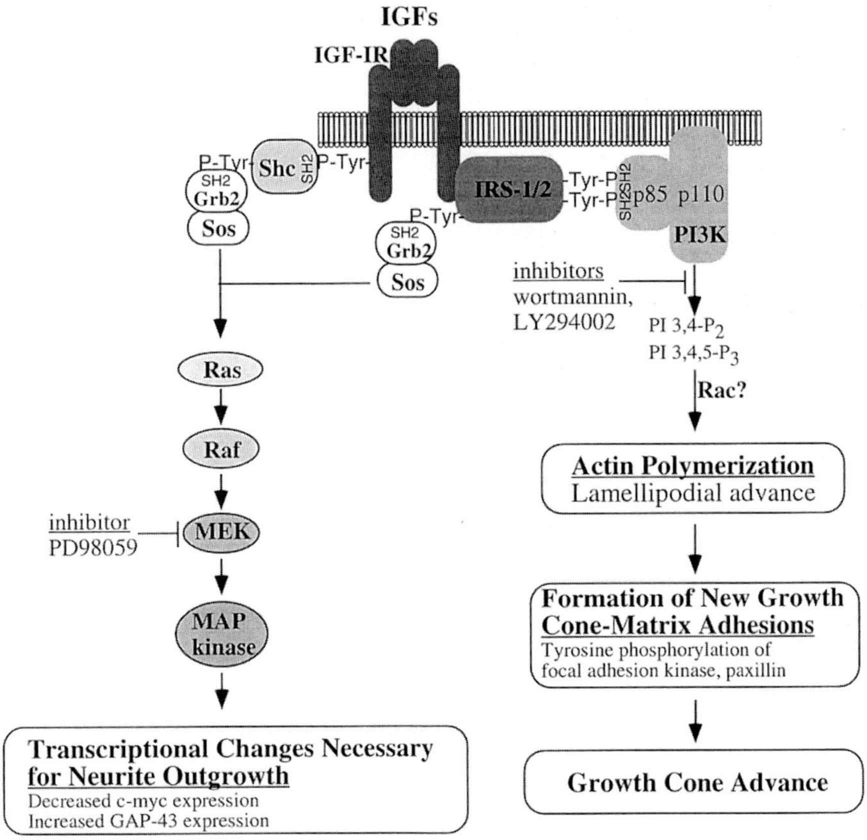

Fig. 3. Summary of intracellular pathways whereby IGFs promote neurite outgrowth. IGF binding by the IGF-IR leads to activation of the receptor's intrinsic tyrosine kinase domain, resulting in receptor autophosphorylation and IRS protein phosphorylation. These phosphorylations result in MAP kinase activation, downstream of an SH2-dependent interaction with Shc and/or Grb2. MAP kinase activation, in turn, drives changes in transcription required for neurite outgrowth. IRS tyrosine phosphorylation also promotes the activity of PI3K via an SH2-dependent interaction. This results in the production of phosphatidylinositols phosphorylated at the 3-position of the inositol ring. These newly synthesized phosphatidylinositol phosphates drive actin polymerization, possibly via the small GTPase Rac. Because the pathway leading to PKC activation is not well understood it is not shown here, but it is also thought that PKC may participate in the induction of actin polymerization.

Signaling and Cytoskeletal Mechanisms in IGF-Stimulated Neurite Outgrowth

The use of cultured neuronal cell lines has allowed the elucidation of the signaling mechanisms underlying IGF-stimulated neurite outgrowth (*See* Fig. 3 for a summary of the known mechanisms). One well-characterized model system is the SH-SY5Y human neuroblastoma cell line *(121)*. In these cells, IGF-I and IGF-II promote neurite outgrowth through the IGF-I receptor *(122)*. Studies in SH-SY5Y cells demonstrate that the initial effect of IGFs on neuronal cells is the stimulation of actin polymerization in growth cones and at the cell's leading edge, which drives the advance of lamellipodia *(123)* (Figs. 4

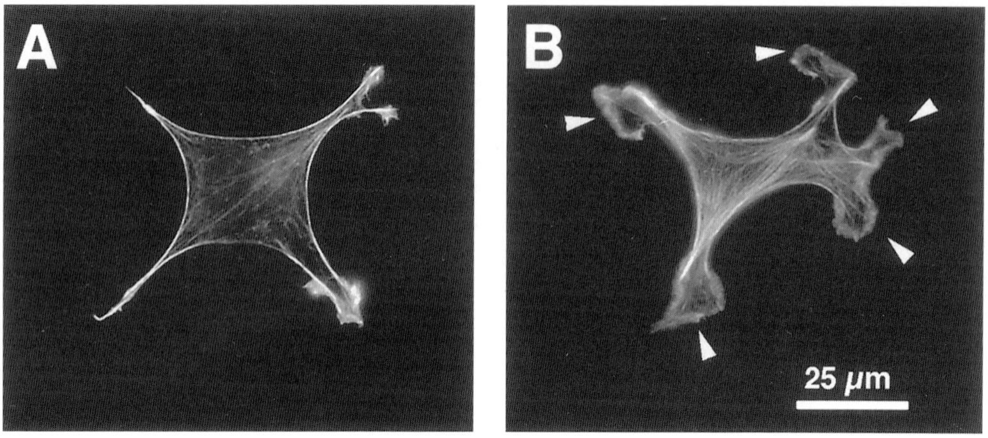

Fig. 4. IGF-I stimulates leading edge actin polymerization and lamellipodial advance in SH-SY5Y human neuroblastoma cells. Rhodamine–phalloidin staining of SH-SY5Y cells (**A**) untreated and (**B**) treated for 30 min with 10 n*M* IGF-I shows that IGF-I causes a rapid induction of actin polymerization at the cell's leading edge (*arrowheads*). This leading edge actin polymerization is thought to drive lamellipodial protrusion leading to cell motility and growth cone advance. (Reprinted by permission from Leventhal et al. J Biol Chem 1997; 272:5214–5218.)

and 5). A rapid stimulation of actin polymerization and lamellipodial advance in response to IGFs has similarly been observed in nonneuronal cells such as rat glomerular mesangial cells *(124)* and human epidermoid carcinoma cells *(125,126)*.

The signaling pathway whereby IGFs promote leading-edge actin polymerization begins with the binding of growth factor by the IGF-IR *(123,125)* (*see* Fig. 3), which stimulates the activity of the IGF-I receptor's intracellular tyrosine kinase domain, and results in the downstream tyrosine phosphorylation of the insulin receptor substrate (IRS) proteins *(123)* (B. Kim, *unpublished observations*). The tyrosine phosphorylation of IRS proteins generates binding sites for certain proteins containing SH2 domains such as the p85 regulatory subunit of PI3K *(79,80)*. Indeed, promotion of actin polymerization and lamellipodial advance by IGF-I is well known to require IRS tyrosine phosphorylation and subsequent PI3K activation *(127)*. The ability of the PI 3-kinase inhibitors wortmannin and LY294002 to prevent IGF-I-stimulated neurite outgrowth demonstrates the essential role for this enzyme in growth cone advance and guidance (B. Kim, *unpublished observations*) *(128)*. Similarly, these inhibitors or a dominant-negative mutant of PI3K prevent neurite outgrowth in response to nerve growth factor *(129)*. Thus, it is likely that PI3K plays a central role in regulating neurite formation, guidance, and advance by controlling actin polymerization at the cell and growth cone leading edge. As described on pp. 431–432, this requirement for PI3K activation is shared with the anti-apoptotic signaling pathway. Although Akt appears to play a key role in mediating the effects of PI3K in the anti-apoptosis pathway, there is currently no evidence whether or not this enzyme participates in the regulation of the actin cytoskeleton. A protein that has been suggested as a mediator for PI3K-dependent actin polymerization is the small GTPase Rac, a which is known to be required for actin polymerization and lamellipodial advance *(130)*. In fact, recent studies have shown that PI3K products may promote the activity of Rac-specific GTP/GDP exchange factors *(131,132)*. However, at this time, it remains to be determined whether Rac is a direct mediator of the PI3K-actin cytoskeleton regulatory pathway.

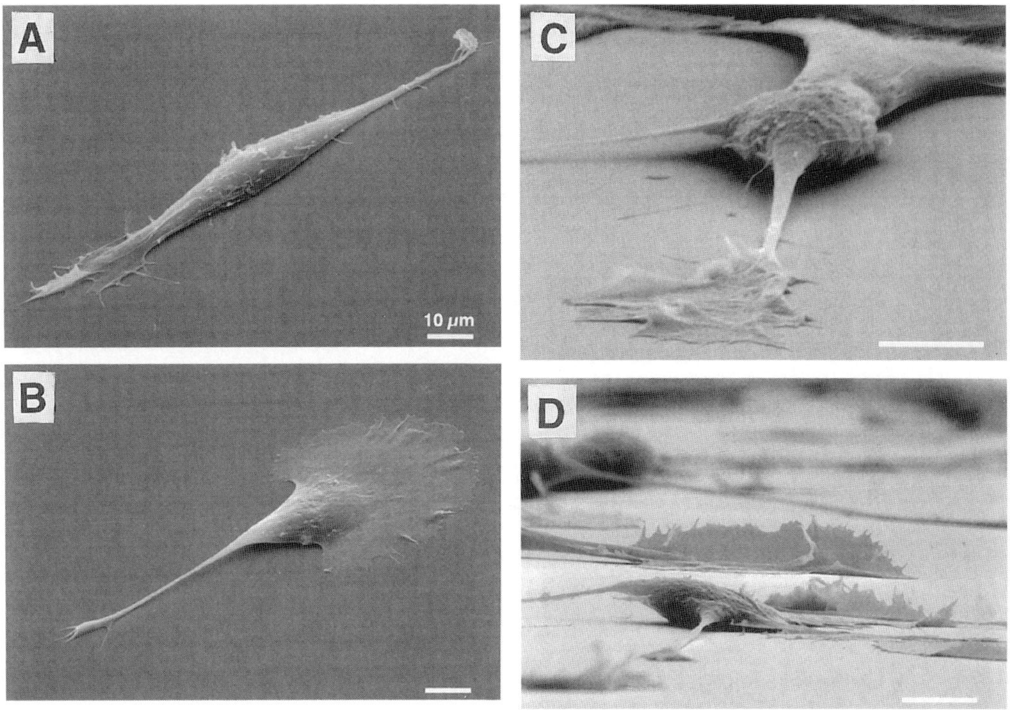

Fig. 5. IGF-I induces lamellipodial advance and membrane ruffling in SH-SY5Y human neuroblastoma cells. Comparison of scanning electron micrographs of SH-SY5Y cells that were **(A)** untreated or **(B–D)** treated for 30 min with 10 n*M* IGF-I shows that IGF-I causes the extension of lamellipodia (flat, veil-like membrane protrusions) as well as membrane ruffles (perpendicular membrane protrusions). This lamellipodial protrusion is essential for cell motility and growth cone advance. (Reprinted by permission from Leventhal and Feldman, Trends Endocrinol Metab 1997; 8:1–6.)

In addition to leading-edge actin polymerization, formation of new adhesions between the growth cone and the extracellular matrix is required for growth cone motility and guidance. Formation of cell–extracellular matrix adhesions requires binding of extracellular matrix proteins to transmembrane adhesion receptors, such as the integrins, and the activation of intracellular protein tyrosine phosphorylation *(100)*. In SH-SY5Y cells, for example, IGF-I stimulates the formation of phosphotyrosine-rich adhesion complexes at the leading edge of extending lamellipodia *(123)* (Fig. 6). Two of the proteins in these adhesion complexes that are substrates for IGF-I-stimulated tyrosine phosphorylation are focal adhesion kinase and paxillin. Paxillin and phosphotyrosine staining has also been identified in adhesion complexes in the growth cones of chick sympathetic neurons *(133)*. In general, tyrosine phosphorylation of adhesion-related proteins is thought to contribute to the hierarchy of intracellular protein complexes that form at adhesion sites *(134)*. These complexes, in turn, may anchor the actin cytoskeleton to cell–extracellular matrix adhesion sites. Consistent with this idea, tyrosine kinase inhibitors prevent neurite outgrowth in SH-SY5Y cells by disrupting the actin cytoskeleton *(135)*. Similarly, tyrosine kinase inhibitors prevent the formation of filopodia and lamellipodia after axotomy in *Aplysia* neurons by preventing actin polymerization *(96)*. Studies in *Aplysia* neurons have further identified phosphotyrosine at the tips of growth cone filopodia, but

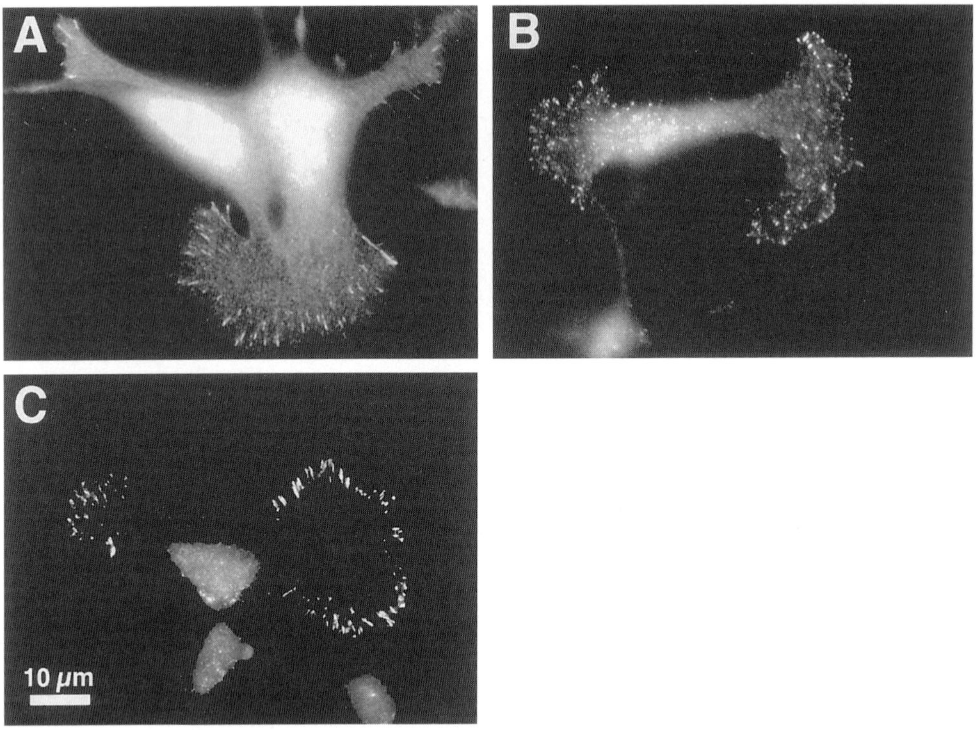

Fig. 6. Focal adhesion proteins and phosphotyrosine are concentrated in the lamellipodia of IGF-I-treated SH-SY5Y human neuroblastoma cells. Immunoreactive focal adhesion kinase (**A**), paxillin (**B**), and phosphotyrosine (**C**) are found in focal adhesion-like distribution within the lamellipodia of SH-SY5Y cell treated for 30 min with 10 n*M* IGF-I. Note that staining for phosphotyrosine and thus tyrosine-phosphorylated forms of FAK and paxillin, is found just at the leading edge of the lamellipodia, consistent with a role in lamellipodial protrusion. (Reprinted by permission from Leventhal et al. J Biol Chem 1997; 272:5214–5218.)

this may be distinct from adhesion-related tyrosine phosphorylations *(96,136)*. In general, the stimulation of adhesion-related tyrosine phosphorylations by IGFs appears to be associated with lamellipodial advance and reorganization of the actin cytoskeleton during neurite formation and growth cone advance.

Activation of protein kinase C (PKC) family members (serine/threonine kinases) may also participate in IGF-stimulated neurite outgrowth. In SH-SY5Y cells, a selective PKC inhibitor, GF 109203X, prevents the induction of neurite outgrowth by a combination of IGF-I and basic fibroblast growth factor (bFGF) *(137)*. These studies further implicated the ε isoform of PKC in the IGF-I/bFGF-stimulation of neurite outgrowth. Indeed, PKC-ε was identified in the growth cone of SH-SY5Y cells *(138)*. Because these studies were performed in cells treated with a combination of bFGF and IGF-I, it is difficult to determine the contribution of PKC to IGF-I signaling leading to neurite outgrowth. However, the PKC inhibitor was shown to prevent the extension of filopodia under these conditions *(137)*, suggesting a possible role in regulating the actin cytoskeleton in the growth cone.

Transcription Regulator Mechanisms in IGF-Stimulated Neurite Outgrowth

Clearly, a major role for IGF-I in neuronal cells is in inducing signaling pathways leading to a rapid reorganization of the actin cytoskeleton. In addition, IGFs produce long-term effects in neuronal cells associated with increased neurite outgrowth. For example, IGF treatment of SH-SY5Y cells leads to an induction of GAP-43 mRNA expression and a decrease in c-*myc* mRNA expression *(139)*. The decrease in c-*myc* expression is consistent with a transition from a proliferating to a differentiating phenotype. Expression of GAP-43 is also expected, as this protein is a marker of neuronal differentiation. In this regard, GAP-43 is thought to help in the bundling of actin filaments in filopodia *(140,141)*. IGF-I regulation of GAP-43 and c-*myc* expression in the cells requires activation of the Ras/Raf/mitogen-activated protein (MAP) kinase pathway downstream of the IGF-I receptor. Furthermore, MAP kinase inhibition with PD98059 prevents IGF-I-stimulated neurite outgrowth *(142)* (Fig. 7). It is likely that activation of MAP kinase, like phosphatidylinositol (PI) 3-kinase, is a common feature of the induction of neurite outgrowth, as PD98059 also inhibits nerve growth factor-stimulated neurite outgrowth in PC12 cells *(143)*.

Like GAP-43, expression of tubulin is regulated by IGFs. IGF-I treatment of SH-SY5Y cells enhances the expression of tubulin by stabilizing tubulin mRNA *(144)*. This may enhance the ability of the cells to extend neurites because neurite outgrowth and growth cone turning required microtubule polymerization *(97,111)*.

THE IGFS AS PROMOTERS OF MYELINATION IN THE CNS AND PNS

IGF-I has been clearly implicated as a critical factor in CNS myelination in vivo. In IGF-I-overexpressing mice myelin content in the brain is 30% greater than in controls and their brains are 55% larger *(145)*. This suggests that IGF-I is a potent inducer of CNS myelination and growth. Furthermore, in this transgenic animal the number of ensheathed axons is increased compared to normal littermates. Both small and large axons were ensheathed in the presence of IGF-I, suggesting that IGF-I-induced myelination is independent of axon diameter *(146)*. This concept is contrary to the traditional view that myelination and axonal size are linked *(147)*, possibly through phosphorylation of neurofilaments or other structures in the axon *(148)*. This model assumes that larger axons are more likely to be myelinated. Thus, IGF-I may be able to overcome the limitations on myelination imposed by axon size and type. Although IGF-I overexpression has been found to enhance sensory cortex size, animals that overexpress IGFBP-1, and thus have decreased free IGF-I, have a significantly decreased number of myelinated fibers and thickness of myelin sheaths *(149)*. The murine IGF-I knockout is a perinatal lethal mutation and little is known about myelination; thus, little is known about myelination in this animal.

These findings in transgenic animals have led to investigation of the role of IGF-I in animal models of human CNS disease. In a model of spinal cord transection, IGF-I and platelet-derived growth factor in combination have been shown to enhance myelination of spinal cord axons post-transection, although there was no increase in axonal regeneration into the Schwann cell grafts *(150)*. Early in spinal cord injury there is up-regulation of astrocyte-induced IGF-I and IGFBP-2 associated with remyelination *(42)*. In addition IGF-I has clearly been shown to decrease the extent of inflammation, demyelination, and the degree of clinical deficit in animals with experimental allergic encephalomyelitis (EAE) *(151–153)*. Furthermore, IGF-I has been shown to actually enhance remyelination in this model of the human disease, multiple sclerosis *(151)*. In yet another model of CNS demyelination, that following cuprizone treatment, IGF-IR mRNA is found to be increased early in recovery *(154)*. Astrocyte IGF-I mRNA was most increased in areas

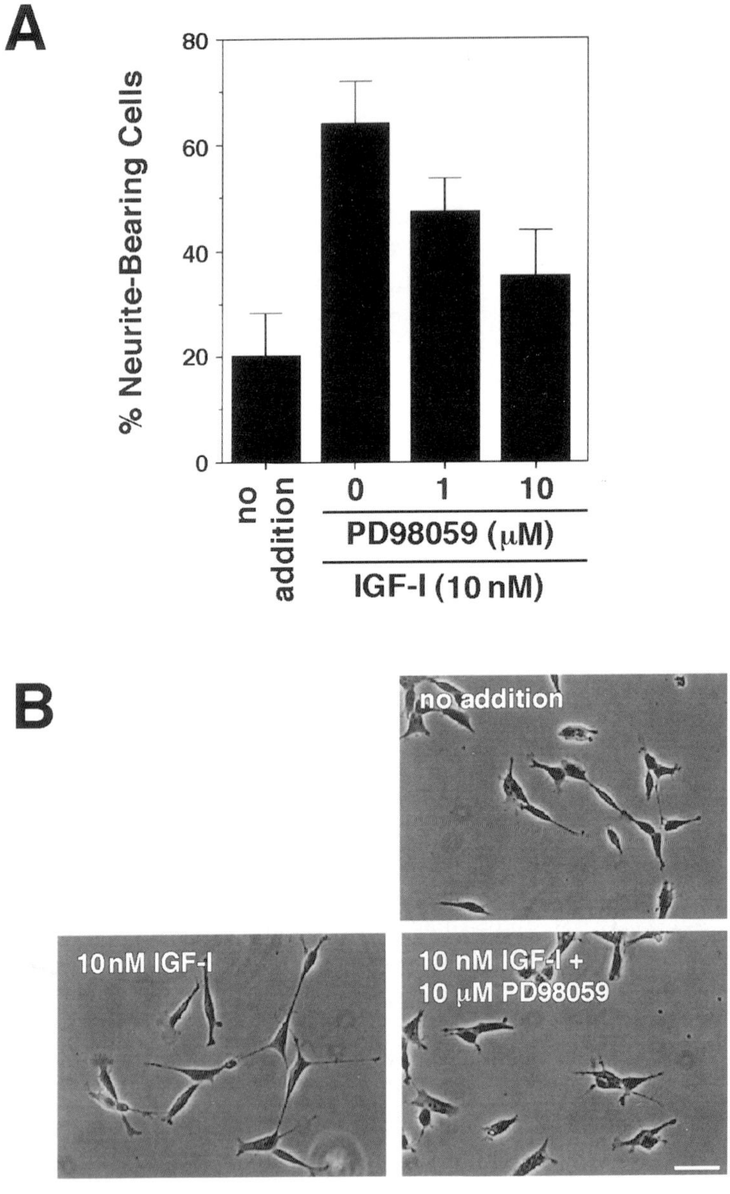

Fig. 7. The MEK inhibitor PD98059 blocks the stimulation of neurite outgrowth by IGF-I in SH-SY5Y human neuroblastoma cells. Incubation of SH-SY5Y cells with 10 n*M* IGF-I stimulates neurite outgrowth and is reduced by coincubation with the MEK inhibitor PD98059 as shown by **(A)** measurement of the percent of neurite-bearing cells and **(B)** by phase-contrast microscopy of the cells. This is consistent with a role of the MAP kinase pathway in IGF-I-stimulated neurite outgrowth. (Reprinted by permission Kim et al. J Biol Chem 1987; 272:21268–21273.)

of severe myelin breakdown and returned to normal with remyelination *(154)*. Thus IGF-I and IGF-IR appear to be intricately involved in recovery from acquired demyelinating disorders. Based on this promising animal work, treatment trials with IGF-I in humans with multiple sclerosis are currently being undertaken.

In organotypic CNS culture, IGF-I has similarly been shown to induce myelination of axons by oligodendrocytes. In monolayer cultures of rat brain, IGF-I has been shown to primarily increase the number of oligodendrocytes, thus increasing the total amount of myelin produced, rather than the amount of myelin synthesized per oligodendrocyte *(155)*. This finding is entirely consistent with our knowledge of the mitotic and anti-apoptotic properties of IGF-I. Furthermore, in addition to astrocyte-derived IGF-I, there is an autocrine expression of IGF-I in developing oligodendrocytes *(156)*. As with the in vivo models, addition of anti-white matter serum and complement is able to induce demyelination in mouse spinal cord explant–oligodendrocyte cocultures. In comparison, addition of IGF-I increases remyelination *(157)*.

Little is known about the downstream signaling pathways following IGF-IR activation in oligodendrocytes. It is known that IGF-I-mediated oligodendrocyte survival is dependent on PI3K signaling and can be directly inhibited by the potent PI3K inhibitors, LY294002 or wortmannin *(158)*. In addition to IGF-I signaling through PI3K, *N*-acetyl cysteine has been found to enhance oligodendrocyte survival, even with suboptimal concentrations of IGF-I *(159)*. It is not known if myelination *per se*, as opposed to oligodendrocyte survival, is affected by inhibition of PI3K signaling, nor has the role of the MAP kinase pathway been elucidated in this paradigm.

In the PNS even less is known about the role of IGF-I in myelination. Schwann cells and axons are known to express both IGF-IR, and IGFBP-5 *(36,160)*, and there is colocalization of the two in both late embryonic and adult rats. It has long been known that in the presence of axonal contact Schwann cells express the major myelin protein P_0, a preliminary event in myelination *(161–163)*. Myelin basic protein, another important component of myelin, is regulated in a similar way *(164)*. Interestingly, in the absence of axonal contact, IGF-I in the presence of cAMP potently stimulates P_0 expression in Schwann cells, although cAMP induction alone has little effect on P_0 levels *(165)*. Interestingly, IGF-I in the presence of acidic or basic FGF can also act as a survival factor for embryonic rat Schwann cells *(166)*. Finally, our own initial studies with IGF-I in the PNS indicate that IGF-I increases Schwann cell P_0 expression, induces attachment of Schwann cells to bare sensory axons, and is able to promote myelination in peripheral axons using a chemically defined serum and insulin-free culture medium *(167)* (Figs. 8 and 9). Collectively, these findings indicate that IGF-I may play a role in Schwann cell myelin production and cell survival.

IGF-I is thus able to promote axonal attachment of Schwann cells or oligodendrocytes, and induce myelination both the CNS and PNS. The critical role of IGF-I in developmental myelination is clearly demonstrated by changes in CNS myelin in transgenic animals. The potent effects of this growth factor in animal models of human CNS disease make it likely that IGF-I will be important in ameliorating some human demyelinating disorders.

EFFECTS OF IGFS ON ASTROCYTES IN THE CNS

Astrocytes are a group of related CNS cell types that play a supportive role in neuronal survival, differentiation, and regeneration by providing growth factors and extracellular matrix proteins and by phagocytosing cellular debris and foreign materials *(168,169)*. Physical injury or demyelination in the brain induces the expression of IGF-I by astrocytes *(154,170)*. IGFs also may be critical in mediating the effects of other growth factors in the CNS. For example, astrocyte production of IGF-I is enhanced by epidermal growth factor and bFGF *(171)*. Also, antibodies against IGF-I block epidermal growth factor-stimulated astrocyte proliferation *(171,172)*.

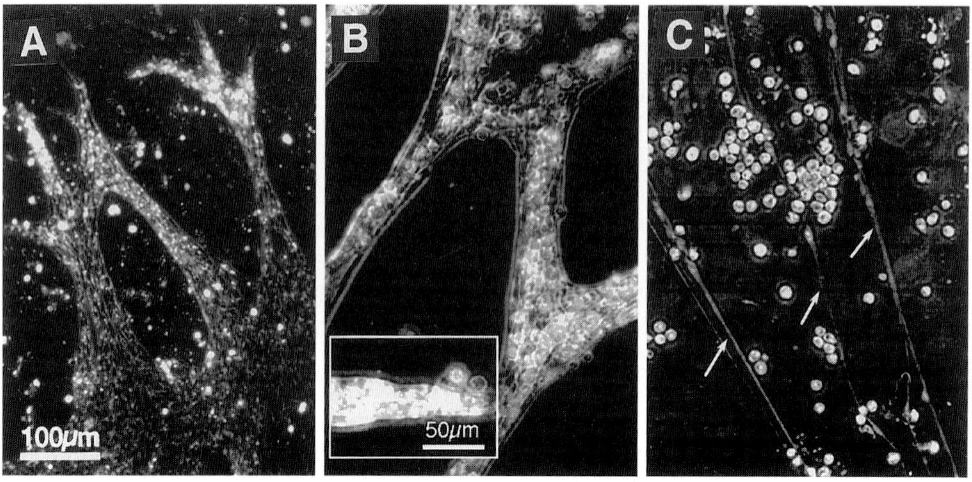

Fig. 8. IGF-I promotes the myelination of dissociated sensory neurons. Dissociated embryonic day 15 rat dorsal root ganglion neurons were cultured in FUDR for 7 d to remove endogenous Schwann cells. Rat MT4H1 Schwann cells (Gift G. Tennekoon, University of Pennsylvania) were labeled with DiI and added to the dissociated axons. The cocultures were allowed to myelinate in serum/insulin-free myelinating medium (SIFMM) with 10 ng/mL of NGF for 72 h containing **(A)** 5% serum (which contains IGF-I), **(B)** 10 n*M* IGF-I, or **(C)** no addition. Cells were imaged by fluorescence microscopy. In the presence of serum or IGF-I, there is differentiation, attachment, and ensheathment of sensory axons by the MT4H1 Schwann cells. IGF-I is more effective in this regard than serum. In the absence of IGF-I or serum, the MT4H1 Schwann cells are undifferentiated and failed to ensheath the axons.

Because both IGF-I and IGF-II promote astrocyte proliferation and DNA synthesis *(172,173)*, it is possible that IGF-I is an autocrine factor for astrocyte growth. IGF-I produced by astrocytes also is likely to act as a paracrine factor to stimulate neuron survival and differentiation and to promote axonal myelination by oligodendrocytes. Furthermore, IGFs induce astrocytes to produce tissue plasminogen activator *(174)* and erythropoietin *(175)*, which may promote neuronal cell survival and differentiation. Signaling pathways that may participate in these effects of IGFs on astrocytes include increased expression and activity of the Na^+/K^+-ATPase *(176)*, stimulation of protein kinase C-α *(173)*, activation of ribosomal S6 kinase *(177)*, and intracellular tyrosine phosphorylation leading to PI3K activation *(178)*. Finally, IGF-I has anabolic effects in astrocytes, promoting glucose uptake and enhancing production of the Glut-1 glucose transporter *(179)*. Thus, IGFs appear to function as autocrine and paracrine growth factors in the CNS to regulate glucose levels, to promote neuronal survival and differentiation, to enhance axonal myelination, and to stimulate CNS regeneration.

USE OF THE IGFS AS THERAPEUTICS IN CNS AND PNS DISEASE (TABLE 4)

As described in this chapter, IGFs are potent neurotrophic factors, promoting neuronal survival and differentiation. Also, the IGFs promote axonal myelination in the CNS and PNS. For these reasons, the IGFs have been proposed, and are currently being tested, in the treatment of a variety of central and peripheral nervous system pathologies, including various neuropathies, motor neuron disease, cerebral ischemia/stroke, and neuronal trauma.

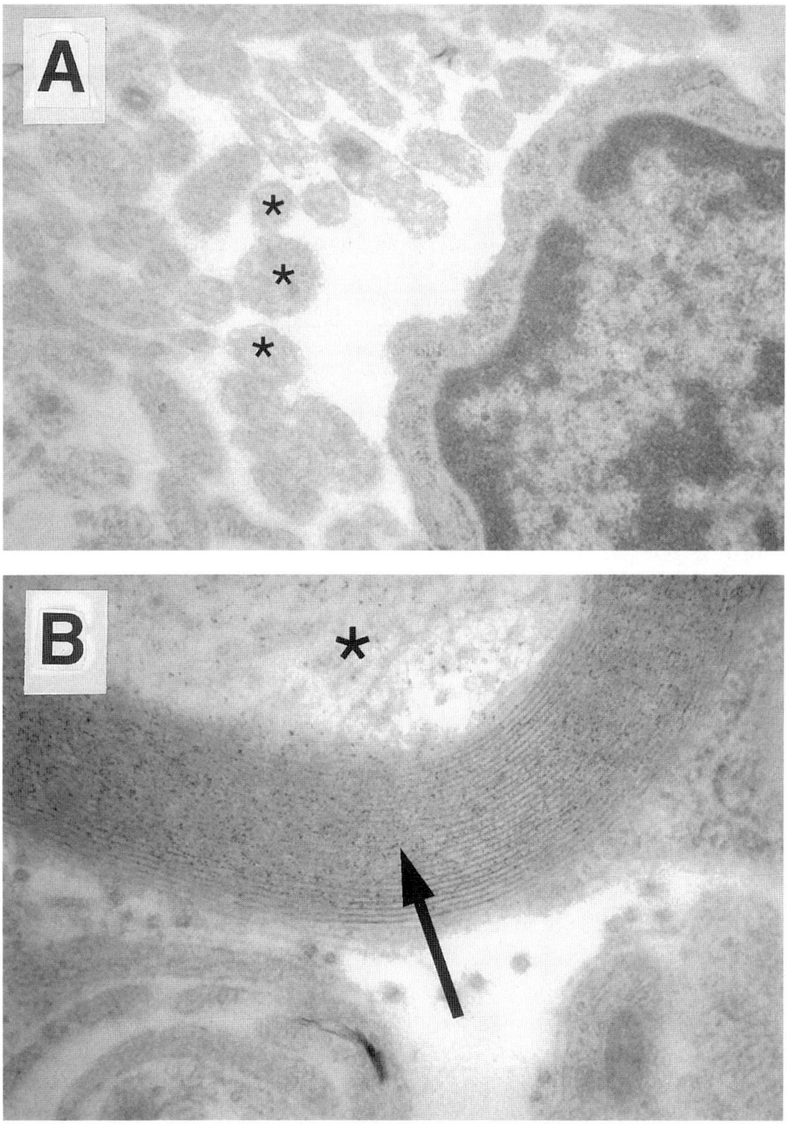

Fig. 9. IGF-I promotes Schwann cell myelination of dorsal root ganglion axons. Dissociated E15 rat dorsal root ganglion neurons were treated to remove endogenous Schwann cells and then were allowed to myelinate for 21 d in SIFMM with 10 ng/mL of NGF in the presence of added secondary rat Schwann cells. Shown are transmission electron micrographs of cocultures containing **(A)** no addition or **(B)** 10 n*M* IGF-I. Axons (*) in the absence of IGF-I remain unmyelinated whereas addition of IGF-I causes abundant myelination of axons with distinct myelin lamellae and clear major dense lines (*arrow*). Magnifications 57,000× in **A** and 100,000× in **B**.

Use of IGFs in the Treatment of Brain Hypoxia–Ischemia

Perinatal hypoxia–ischemia (HI) in the brain is a major cause of death and morbidity. Selective and delayed neuronal death, including both apoptosis and necrosis, contributes to the neuronal loss. A similar selective and delayed neuronal loss also occurs after focal brain ischemia (stroke). Similar to HI, in stroke much of the delayed neuronal loss appears to be due to selective apoptosis *(180,181)*. Because IGF-I is a neuronal survival

Table 4
Use of IGFs as Therapeutics in CNS and PNS Disease

Disease	*References*
Hypoxia/ischemia	*70,182–186*
Physical Trauma	*188–190*
Diabetic neuropathy	*193,196*
Peripheral neuropathy induced by chemotherapeutic agents	*197,198*
Amyotrophic lateral sclerosis (ALS)	*203*
Experimental autoimmune encephalomyelitis (EAE)	*151,153,204,205*

factor and can rescue neurons from apoptosis in response to a variety of insults (*see* pp. 430–432), several studies have examined whether it might prevent neuronal loss after HI insult or stroke. In one popular model, a transient and unilateral HI is induced in adult rats by ligation of one of the carotid arteries. Several investigators have shown that central administration of recombinant human (rh)-IGF-I 1–2 h after the HI insult reduces neuronal loss throughout the brain and also decreases the incidence of cerebral infarction *(70,182–185)*. In one of these studies, the effect of IGF-I was shown to be due, at least in part, to a direct reduction in delayed neuronal apoptosis *(70)*. In another study, using a sheep model of cerebral ischemia, rhIGF-I administered through the cerebral ventricle decreased neuronal loss, slowed the onset and reduced the incidence of seizures, and delayed the cytotoxic edema that typically occurs after cerebral ischemia *(186)*. Only one study contradicted these protective effects of IGF-I by showing that intracerebroventricular injection of IGF-I prior to and after transient cerebral ischemia did not reduce neuronal damage in the rat *(187)*. However, this study likely used too low of a dose of IGF-I; only 2 μg of IGF-I was administered per rat *(187)*, whereas the other studies utilized at least 20 μg IGF-I per rat *(182–185)*. Together, these studies indicate that even a 1–2-h delayed administration of IGF-I can reduce neuronal loss after stroke or HI insult, and suggest that this growth factor can be a clinically useful therapeutic agent.

IGFs as Promoters of Neuronal Regeneration After Physical Trauma

Another area where IGFs have been explored as potential therapies is in the treatment of physical insults to neuronal tissues. For example, IGF-I administration to the sciatic nerve promotes regeneration after nerve crush in the rat *(188)*. In the frog, sciatic nerve regeneration is enhanced by IGF-II, but not IGF-I *(189)*. Also, IGF-I improves motor neuron and cognitive function in rats following an experimental brain injury *(190)*. Moreover, as described on pp. 428–430, the IGFs are expressed in the CNS and PNS after physical injury as part of the healing process. Although these findings suggest that IGFs may be effective, only one study has examined the use of IGFs for treating nervous system traumas in man. Specifically, initial (phase II) clinical trials indicate that intravenous IGF-I administration may improve the clinical outcome of patients with moderate to severe head injuries. Overall, much more work is necessary to determine if IGFs may be useful for treating physical trauma to neuronal tissues.

IGFs in the Treatment of Peripheral Neuropathies and Motor Neuron Disease

Peripheral neuropathies are disorders that affect various types of sensory, autonomic, or motor neurons and are characterized by a loss of tissue innervation, decreased neuronal survival, and reduced ability to regenerate. Peripheral neuropathies are often found in advanced cases of diabetes and can also be produced by a variety of antitumor agents. Because of their ability to promote neuronal survival and sprouting in animal models and in vitro, there has been great interest in utilizing the IGFs as therapeutic agents for peripheral neuropathies *(191,192)*. Moreover, reduction in IGF expression during diabetes is thought to contribute to the etiology of diabetic neuropathy *(193–195)*. That IGFs may be useful for treating diabetic neuropathy has been shown in several animal studies. For example, Ishii and co-workers have shown that in rats with non-insulin-dependent diabetes mellitus, subcutaneous infusion of IGF-II restores pain/pressure thresholds toward normal, suggesting that it reduces diabetic neuropathy *(193)*. It was also shown in a separate study that injection of IGF-I or IGF-II near an injury site promotes sciatic nerve regeneration after nerve crush in the diabetic rat *(196)*. Recent investigations further support the use of IGFs in prevention of peripheral neuropathies caused by chemotherapeutic agents. For example, intraperitoneal injection of mice with IGF-I produces a dose-dependent amelioration of vincristine-induced neuropathy *(197,198)*. IGF-I also reduces peripheral neuropathy caused by cisplatinum and taxol *(198)*. Because peripheral neuropathy is often the dose-limiting factor, IGF treatment may allow for both the prevention of neuropathy as well as the delivery of higher levels of antitumor agents.

Motor neuron disease is a heterogeneous group of disorders that selectively affects upper and/or lower motor neurons *(199)*. Patients with classic amyotrophic lateral sclerosis (ALS), also known as Lou Gehrig's disease, present with spasticity, hyperflexia, and extensor plantar responses (upper motor neuron signs) in the setting of progressive muscle wasting and weakness (lower motor neuron signs) *(199,200)*. ALS accounts for approx 80% of all cases of motor neuron diseases and has an estimated incidence of 2 per 100,000 *(199,200)*. An increase in IGF-IR is observed in the spinal cord of ALS patients, suggesting that IGF-I may help the body to respond to ALS *(201)*. That IGF-I may help treat ALS or other motor neuron diseases is supported by the ability of rhIGF-1 to reduce motor neuron deterioration in the wobbler mouse *(202)*. Currently, there is no effective treatment for the motor neuron loss in ALS, but recent studies have explored the use of IGF-I. An initial double-blind, placebo-controlled, randomized study of 266 patients in North America showed that subcutaneous administration of rhIGF-I can slow the progression of motor neuron functional impairment and increase longevity of patients with ALS *(203)*. For this reason, IGF-I is currently being considered by the Food and Drug Administration as a potential treatment for ALS.

Treatment of Demyelinating Diseases with IGFs

The ability of IGFs to promote myelination in the CNS and PNS (*see* the third section) indicates that IGFs may be useful in treating demyelinating diseases such as multiple sclerosis (MS). To this end, several laboratories have investigated the efficacy of IGFs in preventing EAE, a rat model of MS. Two studies demonstrated that intravenous or subcutaneous injection of IGF-I reduces the clinical deficits observed in EAE *(153,204)*. In addition, IGF-I treatment enhances the production of myelin as shown by increased levels of myelin basic protein, proteolipid protein, and 2′,3′-cyclic nucleotide 3′-phosphodiesterase *(151,153)*. Furthermore, IGF-I treatment reduces the number and size of demyelinated spinal cord lesions, increases the number of remyelinating segments, and enhances the proliferation of oligodendrocyte-like cells in the lesions *(151,204)*. In addition to these effects on myelination, IGF-I also appears to reduce immune cell responses that occur during EAE and may occur in MS *(205)*. Given these findings, it is possible that IGFs may be useful for treating MS and other demyelinating diseases.

SUMMARY AND CONCLUSIONS

A large body of evidence indicates that the IGFs are important growth factors in the nervous system. The IGFs are expressed at high levels in the CNS and PNS during development and in response to a variety of insults. Moreover, the IGFs are potent neurotrophic factors, with an apparently specific ability to rescue neurons from apoptosis, to promote neuronal sprouting, and to enhance myelination. Finally, the IGFs appear to have an excellent potential as therapeutic agents to assist neuronal regeneration after physical trauma; to reduce diabetic, chemical-induced, and other neuropathies; and for treating ALS and demyelinating diseases such as MS. Some initial studies have begun to examine the cellular and biochemical mechanisms whereby the IGFs elicit these effects, but this work is still in its infancy. Continuation of these studies is essential to determine the efficacy of IGFs for the treatment of neuronal injury and pathologies as well as for the design of more potent and specific therapeutic agents.

ACKNOWLEDGMENTS

The authors are grateful to Catherine Delaney and Kelli Sullivan for assistance in writing of this manuscript. This work was supported by a grant from the Millie Schembechler Adrenal Research Fund of the University of Michigan Comprehensive Cancer Center (to PSL and ELF), National Institutes of Health Grants R29 NS32843 (to ELF) and K08 NS019388 (to JWR), and grants from the American Diabetes Association, Juvenile Diabetes Foundation, and Muscular Dystrophy Association (to ELF).

REFERENCES

1. Daughaday WH, Rotwein P. Insulin-like growth factor I and II. Peptide, messenger ribonucleic acid and gene structures, serum, and tissue concentrations. Endocr Rev 1989; 10:68–91.
2. Sara VR, Hall K. Insulin-like growth factors and their binding proteins. Physiol Rev 1990; 70:591–614.
3. Lee W-H, Javedan S, Bondy CA. Coordinate expression of insulin-like growth factor system components by neurons and neuroglia during retinal and cerebellar development. J Neurosci 1992; 12:4737–4744.
4. Bondy C, Werner H, Roberts CT Jr, LeRoith D. Cellular pattern of type-I insulin-like growth factor receptor gene expression during maturation of the rat brain: comparison with insulin-like growth factors I and II. Neuroscience 1992; 46:909–923.
5. Ayer-Le Lievre C, Ståhlbom P-A, Sara VR. Expression of IGF-I and -II mRNA in the brain and craniofacial region of the rat fetus. Development 1991; 111:105–115.
6. Rotwein P, Burgess SK, Milbrandt JD, Krause JE. Differential expression of insulin-like growth factor genes in rat central nervous system. Proc Natl Acad Sci USA 1988; 85:265–269.
7. García-Segura LM, Pérez J, Pons S, Rejas MT, Torres-Alemán I. Localization of insulin-like growth factor I (IGF-I)-like immunoreactivity in the developing and adult rat brain. Brain Res 1991; 560: 167–174.
8. Bondy C, Werner H, Roberts CT Jr, LeRoith D. Cellular pattern of insulin-like growth factor-I (IGF-I) and type I IGF receptor gene expression in early organogenesis: comparison with IGF-II gene expression. Mol Endocrinol 1990; 4:1386–1398.
9. Anderson IK, Edwall D, Norstedt G, Rozell B, Skottner A, Hansson H-A. Differing expression of insulin-like growth factor I in the developing and in the adult rat cerebellum. Acta Physiol Scand 1988; 132:167–173.
10. Devaskar SU, Sadiq HF, Holtzclaw L, George M. The development pattern of rabbit brain insulin and insulin-like growth factor receptor expression. Brain Res 1993; 605:101–109.
11. Torres-Aleman I, Pons S, Arevalo MA. The insulin-like growth factor I system in the rat cerebellum: developmental regulation and role in neuronal survival and differentiation. J Neurosci Res 1994; 39:117–126.
12. Logan A, Gonzalez A-M, Hill DJ, Berry M, Gregson NA, Baird A. Coordinated pattern of expression

and localization of insulin-like growth factor-II (IGF-II) and IGF-binding protein-2 in the adult rat brain. Endocrinology 1994; 135:2255–2264.
13. Couce ME, Weatherington AJ, McGinty JF. Expression of insulin-like growth factor-II (IFG-II) and IGF-II/mannose-6-phosphate receptor in the rat hippocampus: an *in situ* hybridization and immunocytochemical study. Endocrinology 1992; 131:1636–1642.
14. Stylianopoulou F, Herbert J, Soares MB, Efstratiadis A. Expression of the insulin-like growth factor II gene in the choroid plexus and the leptomeninges of the adult rat central nervous system. Proc Natl Acad Sci USA 1988; 85:141–145.
15. McKelvie PA, Rosen KM, Kinney HC, Villa-Komaroff L. Insulin-like growth factor II expression in the developing brain. J Neuropathol Exp Neurol 1992; 51:464–471.
16. LeRoith D, Werner H, Beitner-Johnson D, Roberts CT Jr. Molecular and cellular aspects of the insulin-like growth factor I receptor. Endocr Rev 1995; 16:143–163.
17. De Meyts P, Wallach B, Christoffersen CT, Urso B, Gronskov K, Latus L-J, Yakushiji F, Ilondo MM, Shymko RM. The insulin-like growth factor-I receptor. Structure, ligand-binding mechanism and signal transduction. Horm Res 1994; 42:152–169.
18. Bondy C, Lee W-H. Patterns of insulin-like growth factor and IGF receptor gene expression in the brain: functional implications. Ann NY Acad Sci 1993; 692:33–43.
19. Simmen FA, Simmen RCM. Peptide growth factors and proto-oncogenes in mammalian conceptus development. Biol Reprod 1991; 44:1–5.
20. Van Evercooren AB, Olichon-Berthe C. Expression of IGF-I and insulin receptor genes in the rat central nervous system: a developmental, regional, and cellular analysis. J Neurosci Res 1991; 28: 244–253.
21. Kar S, Chabot J-G, Quirion R. Quantitative autoradiographic localization of [^{125}I]insulin-like growth factor I, [^{125}I]insulin-like growth factor II, and [^{125}I]insulin receptor binding sites in developing and adult rat brain. J Comp Neurol 1993; 333:375–397.
22. Dahms NM. Insulin-like growth factor II cation-independent mannose 6-phosphate receptor and lysosomal enzyme recognition. Biochem Soc Trans 1996; 24:136–141.
23. Kiess W, Yang Y, Kessler U, Hoeflich A. Insulin-like growth factor II (IGF-II) and the IGF-II/mannose-6-phosphate receptor: the myth continues. Horm Res 1994; 41:66–73.
24. Van Obberghen E. Signaling through the insulin receptor and the insulin-like growth factor-I receptor. Diabetologia 1994; 37:S125–S134.
25. Valentino KL, Ocrant I, Rosenfeld RG. Developmental expression of insulin-like growth factor-II receptor immunoreactivity in the rat central nervous system. Endocrinology 1990; 126:914–920.
26. Funk B, Kessler U, Eisenmenger W, Hansmann A, Kolb HJ, Kiess W. Expression of the insulin-like growth factor-II/mannose-6-phosphate receptor in multiple human tissues during fetal life and early infancy. J Clin Endocrinol Metab 1992; 75:424–431.
27. Sklar MM, Thomas CL, Municchi G, Roberts CT Jr, LeRoith D, Kiess W, Nissley P. Developmental expression of rat insulin-like growth factor-II/mannose 6-phosphate receptor messenger ribonucleic acid. Endocrinology 1992; 130:3484–3491.
28. Clemmons DR. Insulin-like growth factor binding proteins and their role in controlling IGF actions. Cytokine Growth Factor Rev 1997; 8:45–62.
29. Collett-Solberg PF, Cohen P. The role of the insulin-like growth factor binding proteins and the IGFBP proteases in modulating IGF action. Endocrinol Metab Clin North Am 1996; 25:591–614.
30. Lee W-H, Michels KM, Bondy CA. Localization of insulin-like growth factor binding protein-2 messenger RNA during postnatal brain development: correlation with insulin-like growth factors I and II. Neuroscience 1993; 53:251–265.
31. Sullivan KA, Feldman EL. Immunohistochemical localization of insulin-like growth factor II and insulin-like growth factor binding protein-2 during development in the rat. Endocrinology 1994; 135:540–547.
32. Bondy C, Lee W-H. Correlation between insulin-like growth factor (IGF)-binding protein 5 and IGF-I gene expression during brain development. J Neurosci 1993; 13:5092–5104.
33. Cheng H-L, Sullivan KA, Feldman EL. Immunohistochemical localization of insulin-like growth factor binding protein-5 in the developing rat nervous system. Dev Brain Res 1996; 92:211–218.
34. Mewar R, McMorris FA. Expression of insulin-like growth factor-binding protein messenger RNAs in developing rat oligodendrocytes and astrocytes. J Neurosci Res 1997; 59:721–728.
35. Han VKM, Matsell DG, Delhanty PJD, Hill DJ, Shimasaki S, Nygard K. IGF-binding protein mRNAs in the human fetus: tissue and cellular distribution of developmental expression. Horm Res 1996; 45:160–166.

36. Cheng H-L, Randolph A, Yee D, Delafontaine P, Tennekoon G, Feldman EL. Characterization of insulin-like growth factor-I (IGF-I), IGF-I receptor and binding proteins in transected nerves and cultured Schwann cells. J Neurochem 1996; 66:525–536.
37. Braulke T, Gotz W, Claussen M. Immunohistochemical localization of insulin-like growth factor binding protein-1, -3, and -4 in human fetal tissues and their analysis in media from fetal tissue explants. Growth Regul 1996; 6:55–65.
38. Ishii DN. Relationship of insulin-like growth factor II gene expression in muscle to synaptogenesis. Proc Natl Acad Sci USA 1989; 86:2898–2902.
39. Hansson H-A, Nilsson A, Isgaard J, Billig H, Isaksson O, Skottner A, Andersson IK, Rozell B. Immunohistochemical localization of insulin-like growth factor I in the adult rat. Histochemistry 1988; 89:403–410.
40. Morano S, Sensi M, Di Gregorio S, Pozzessere G, Petrucci AFG, Valle E, Pugliese G, Caltabiano V, Vetri M, Di Mario U, Purrello F. Peripheral, but not central, nervous system abnormalities are reversed by pancreatic islet transplantation in diabetic Lewis rats. Eur J Neurosci 1996; 8:1117–1123.
41. Klempt M, Klempt ND, Gluckman PD. Hypoxia and hypoxia/ischemia affect the expression of insulin-like growth factor binding protein 2 in the developing rat brain. Brain Res Mol Brain Res 1993; 17:347–350.
42. Yao D-L, West NR, Bondy CA, Brenner M, Hudson LD, Zhou J, Collins GH, Webster HD. Cryogenic spinal cord injury induces astrocytic gene expression of insulin-like growth factor I and insulin-like growth factor binding protein 2 during myelin regeneration. J Neurosci Res 1995; 40:647–659.
43. Breese CR, D'Costa A, Rollins YD, Adams C, Booze RM, Sonntag WE, Leonard S. Expression of insulin-like growth factor-1 (IGF-1) and IGF-binding protein 2 (IGF-BP2) in the hippocampus following cytotoxic lesion of the dentate gyrus. J Comp Neurol 1996; 369:388–404.
44. Nordqvist ACS, von Holst H, Holmin S, Sara VR, Bellander BM, Schalling M. Increase of insulin-like growth factor (IGF)-1, IGF binding protein-2 and -4 mRNAs following cerebral contusion. Brain Res Mol Brain Res 1996; 38:285–293.
45. Walter HJ, Berry M, Hill DJ, Logan A. Spatial and temporal changes in the insulin-like growth factor (IGF) axis indicate autocrine/paracrine actions of IGF-I within wounds of the rat brain. Endocrinology 1997; 138:3024–3034.
46. Pu SF, Zhuang HX, Ishii DN. Differential spatio-temporal expression of the insulin-like growth factor genes in regenerating sciatic nerve. Brain Res Mol Brain Res 1995; 34:18–28.
47. Glazner GW, Ishii DN. Insulinlike growth factor gene expression in rat muscle during reinnervation. Muscle Nerve 1995; 18:1433–1442.
48. Gehrmann J, Yao D-L, Bonetti B, Bondy CA, Brenner M, Zhou J, Kreutzberg GW, Webster HD. Expression of insulin-like growth factor-I and related peptides during otoneuron regeneration. Exp Neurol 1994; 128:202–210.
49. Ishii DN, Glazner GW, Pu S-F. Role of insulin-like growth factors in peripheral nerve regeneration. Pharmacol Ther 1994; 62:125–144.
50. Zackenfels K, Oppenheim RW, Rohrer H. Evidence for an important role of IGF-I and IGF-II for the early development of chick sympathetic neurons. Neuron 1995; 14:731–741.
51. Houenou LJ, Li L, Lo AC, Yan Q, Oppenheim RW. Naturally occurring and axotomy-induced motoneuron death and its prevention by neurotrophic agents: a comparison between chick and mouse. In: van Pelt J, Corner MA, Uylings HBM, Lopes da Silva, FH, eds. Progress in Brain Research. Elsevier, Amsterdam, 1994, pp. 217–226.
52. Contreras PC, Steffler C, Yu EY, Callison K, Stong D, Vaught JL. Systemic administration of rhIGF-I enhanced regeneration after sciatic nerve crush in mice. J Pharmacol Exp Ther 1995; 274:1443–1449.
53. Glazner GW, Lupien S, Miller JA, Ishii DN. Insulin-like growth factor II increases the rate of sciatic nerve regeneration in rats. Neuroscience 1993; 54:791–797.
54. Ishii D. Implication of insulin-like growth factors in the pathogenesis of diabetic neuropathy. Brain Res Brain Res Rev 1995; 20:47–67.
55. Ishii D, Marsh D. On the therapeutic potential for insulin-like growth factor use in motor neuron disease. Neurology 1993; 124:96–99.
56. Lewis ME, Vaught JL, Neff NT, Grebow PE, Callison KV, Yu E, Contreras PC, Baldino F Jr. The potential of insulin-like growth factor-I as a therapeutic for the treatment of neuromuscular disorders. Ann NY Acad Sci 1993; 692:201–208.
57. Lin X, Bulleit RF. Insulin-like growth factor I (IGF-I) is a critical trophic factor for developing cerebellar granule cells. Brain Res Dev Brain Res 1997; 99:234–242.
58. Torres-Aleman I, Naftolin F, Robbins RJ. Trophic effects of basic fibroblast growth factor on fetal rat

hypothalamic cells: interactions with insulin-like growth factor-I. Brain Res Dev Brain Res 1990; 52:253–257.
59. Lindholm D, Carroll P, Tzimagiogis G, Thoenen H. Autocrine-paracrine regulation of hippocampal neuron survival by IGF-1 and neurotrophins BDNF, NT-3, and NT-4. Eur J Neurosci 1996; 8:1452–1460.
60. Neff NT, Prevette D, Houenou LJ, Lewis ME, Glicksman MA, Yin Q-W, Oppenheim RW. Insulin-like growth factors: putative muscle-derived trophic agents that promote motoneuron survival. J Neurobiol 1993; 24:1578–1588.
61. Majno G, Joris I. Apoptosis, oncosis, and necrosis. An overview of cell death. Am J Pathol 1995; 146:3–15.
62. Reed JC. Bcl-2 and the regulation of programmed cell death. J Cell Biol 1994; 124:1–6.
63. Jacobson MD, Evan GI. Breaking the ice. Structural and functional similarities have been discovered between two mammalian proteins, Bcl-2 and interleukin 1b-converting enzyme, and proteins encoded by nematode cell-death genes. Curr Biol 1994; 4:337–340.
64. Baur AM, Gamberger TI, Weerda HG, Gjuric M, Tamm ER. Laminin promotes differentiation, adhesion and proliferation of cell cultures derived from human acoustic nerve schwannoma. Acta Oto-Laryngol 1995; 115:517–521.
65. Pittman RN, Wang S, DiBenedetto AJ, Mills JC. A system for characterizing cellular and molecular events in programmed neuronal cell death. J Neurosci 1993; 13:3669–3680.
66. Eastman A, Grant S, Lock R, Tritton T, Van Houten N, Yuan J. Cell death in cancer and development: AACR special conference in cancer research. Cancer Res 1994; 54:2812–2818.
67. Enokido Y, Akaneya Y, Niinobe M, Mikoshiba K, Hatanaka H. Basic fibroblast growth factor rescues CNS neurons from cell death caused by high oxygen atmosphere in culture. Brain Res 1992; 599: 261–271.
68. D'Mello SR, Galli C, Ciotti T, Calissano P. Induction of apoptosis in cerebellar granule neurons by low potassium: inhibition of death by insulin-like growth factor I and cAMP. Proc Natl Acad Sci USA 1993; 90:10989–10993.
69. Rukenstein A, Rydel RE, Greene LA. Multiple agents rescue PC12 cells from serum-free cell death by translation- and transcription-independent mechanisms. J Neurosci 1991; 11:2552–2563.
70. Tagami M, Ikeda K, Fujino H, Kubota A, Numano F, Yamori Y. Insulin-like growth factor-1 attenuates apoptosis in hippocampal neurons caused by cerebral ischemia and reperfusion in stroke-prone spontaneously hypertensive rats. Lab Invest 1997; 76:613–617.
71. Tagami M, Yamagata K, Nara Y, Fujino H, Kubota A, Numano F, Yamori Y. Insulin-like growth factors prevent apoptosis in cortical neurons isolated from stroke-prone spontaneously hypertensive rats. Lab Invest 1997; 76:603–612.
72. Dore S, Kar S, Quirion R. Insulin-like growth factor I protects and rescues hippocampal neurons against beta-amyloid- and human amylin-induced toxicity. Proc Natl Acad Sci USA 1997; 94:4772–4777.
73. Villalba M, Bockaert J, Journot L. Concomitant induction of apoptosis and necrosis in cerebellar granule cells following serum and potassium withdrawal. NeuroReport 1997; 8:981–985.
74. Fernandez-Sanchez MT, Garcia-Rodriguez A, Diaz-Trelles R, Novelli A. Inhibition of protein phosphatases induces IGF-1-blocked neurotrophin-insensitive neuronal apoptosis. FEBS Lett 1996; 398:106–112.
75. Galli C, Meucci O, Scorziello A, Werge TM, Calissano P, Schettini G. Apoptosis in cerebellar granule cells is blocked by high KCl, forskolin, and IGF-I through distinct mechanisms of action: the involvement of intracellular calcium and RNA synthesis. J Neurosci 1995; 15:1172–1179.
76. Lewis ME, Neff NT, Contreras PC, Stong DB, Oppenheim RW, Grebow PE, Vaught JL. Insulin-like growth factor-I: potential for treatment of motor neuronal disorders. Exp Neurol 1993; 124:73–88.
77. Matthews CC, Odeh H, Feldman EL. Insulin-like growth factor-I is an osmoprotectant in human neuroblastoma cells. Neuroscience 1997; 79:525–534.
78. Matthews CC, Feldman EL. Insulin-like growth factor I rescues SH-SY5Y human neuroblastoma cells from hyperosmotic induced programmed cell death. J Cell Physiol 1996; 166:323–331.
79. Giorgetti S, Ballotti R, Kowalski-Chauvel A, Tartare S, Van Obberghen E. The insulin and insulin-like growth factor-I receptor substrate IRS-1 associates with and activates phosphatidylinositol 3-kinase in vitro. J Biol Chem 1993; 268:7358–7364.
80. Waters SB, Pessin JE. Insulin receptor substrate 1 and 2 (IRS1 and IRS2): what a tangled web we weave. Trends Cell Biol 1996; 6:1–4.
81. Carpenter CL, Cantley LC. Phosphoinositide 3-kinase and the regulation of cell growth. Biochim Biophys Acta 1996; 1288:M11–M16.

82. Miller TM, Tansey MG, Johnson EM Jr., Creedon DJ. Inhibition of phosphatidylinositol 3-kinase activity blocks depolarization- and insulin-like growth factor I-mediated survival of cerebellar granule cells. J Biol Chem 1997; 272:9847–9853.
83. Singleton JR, Dixit VM, Feldman EL. Type I insulin-like growth factor receptor activation regulates apoptotic proteins. J Biol Chem 1996; 271:31791–31794.
84. Yao R, Cooper GM. Requirement for phosphatidylinositol-3 kinase in the prevention of apoptosis by nerve growth factor. Science 1995; 267:2003–2006.
85. Klippel A, Kavanaugh WM, Pot D, Williams LT. A specific product of phosphatidylinositol 3-kinase directly activates the protein kinase Akt through its plekstrin homology domain. Mol Cell Biol 1997; 17:338–344.
86. Franke TF, Kaplan DR, Cantley LC, Toker A. Direct regulation of the *Akt* proto-oncogene product by phosphatidylinositol-3, 4-bisphosphate. Science 1997; 275:665–668.
87. Alessi DR, James SR, Downes CP, Holmes AB, Gaffney PR, Reese CB, Cohen P. Characterization of a 3-phosphoinositide-dependent protein kinase which phosphorylates and activates protein kinase B_{α}. Curr Biol 1997; 7:261–269.
88. Dudek H, Datta SR, Franke TF, Birnbaum MJ, Yao R, Cooper GM, Segal RA, Kaplan DR, Greenberg ME. Regulation of neuronal survival by the serine-threonine protein kinase Akt. Science 1997; 275: 661–664.
89. Datta SR, Dudek H, Tao X, Masters S, Fu H, Gotoh Y, Greenberg ME. Akt phosphorylation of BAD couples survival signals to the cell-intrinsic death machinery. Cell 1997; 91:231–241.
90. Ishii DN. Relationship of insulin-like growth factor II gene expression in muscle to synaptogenesis. Proc Natl Acad Sci USA 1989; 86:2898–2901.
91. Jennische E, Hansson HA. Regenerating skeletal muscle cells express insulin-like growth factor I. Acta Physiol Scand 1987; 130:327–332.
92. Caroni P, Schneider C, Kiefer MC, Zapf J. Role of muscle insulin-like growth factors in nerve sprouting: suppression of terminal sprouting in paralyzed muscle by IGF-binding protein 4. J Cell Biol 1994; 125:893–902.
93. Glazner GW, Morrison AE, Ishii DN. Elevated insulin-like growth factor (IGF) gene expression in sciatic nerves during IGF-supported nerve regeneration. Brain Res Mol Brain Res 1994; 25:265–272.
94. Akahori Y, Horie H. IGF-I enhances neurite regeneration but is not required for its survival in adult DRG explant. NeuroReport 1997; 8:2265–2269.
95. Welch JA, Kraus KH, Wells MR, Blunt DG, Weremowitz J. Effect of combined administration of insulin-like growth factor and platelet-derived growth factor on the regeneration of transected and anastomosed sciatic nerve in rats. Am J Vet Res 1997; 58:1033–1037.
96. Goldberg DJ, Wu D-Y. Inhibition of formation of filopodia after axotomy by inhibitors of protein tyrosine kinases. J Neurobiol 1995; 27:553–560.
97. Smith CL. Cytoskeletal movements and substrate interactions during initiation of neurite outgrowth by sympathetic neurons *in vitro*. J Neurosci 1994; 14:384–398.
98. Ono K, Shokbuni T, Nagata I, Togunaga A, Yasui Y, Nagatsuji N. Filopodia and growth cones in the vertically migrating granule cells of the postnatal mouse. Exp Brain Res 1997; 117:17–29.
99. Davenport DW, Dou P, Rehder V, Kater SB. A sensory role for neuronal growth cone filopodia. Nature 1993; 361:721–724.
100. Lauffenberger DA, Horwitz AF. Cell migration: a physically integrated molecular process. Cell 1996; 84:359–369.
101. Letourneau PC. The cytoskeleton in nerve growth cone motility and axonal pathfinding. Perspect Dev Neurobiol 1996; 4:111–123.
102. Gomez TM, Letourneau PC. Filopodia initiate choices made by sensory neuron growth cones at laminin/fibronectin borders *in vitro*. J Neurosci 1994; 14:5959–5972.
103. Zheng JQ, Wan JJ, Pooo MM. Essential role of filopodia in chemotropic turning of nerve growth cone induced by a glutamate gradient. J Neurosci 1996; 16:1140–1149.
104. Chien CB, Rosenthal DE, Harris WA, Hoth CE. Navigational errors made by growth cones without filopodia in the embryonic *Xenopus* brain. Neuron 1993; 11:237–251.
105. Bentley D, Toroian-Raymond A. Disoriented pathfinding by pioneer neurone growth cones deprived of filopodia by cytochalasin treatment. Nature 1986; 323:712–715.
106. Kleitman N, Johnson MI. Rapid growth cone translocation on laminin is supported by lamellipodial not filopodial structures. Cell Motil Cytoskel 1996; 13:288–300.
107. Okabe S, Hirokawa N. Actin dynamics in growth cones. J Neurosci 1991; 11:1918–1929.

108. Fan J, Mansfield SG, Redmond T, Gordon-Weeks PR, Raper J. The organization of F-actin and microtubules in growth cones exposed to a brain-derived collapsing factor. J Cell Biol 1993; 121:867–878.
109. Lankford KL, Letourneau PC. Evidence that calcium may control neurite outgrowth by regulating the stability of actin filaments. J Cell Biol 1989; 109:1229–1443.
110. Leventhal PS, Feldman EL. Tyrosine phosphorylation and enhanced expression of paxillin during neuronal differentiation *in vitro*. J Biol Chem 1996; 271:5957–5960.
111. Challacombe JF, Show DM, Letourneau PC. Actin filament bundles are required for microtubule reorientation during growth cone turning to avoid an inhibitory guidance cure. J Cell Sci 1996; 109:2031–2040.
112. Aizenman Y, de Vellis J. Brain neurons develop in a serum and glial free environment: effects of transferrin, insulin, insulin-like growth factor-I and thyroid hormone on neuronal survival, growth and differentiation. Brain Res 1987; 406:32–42.
113. Mulholland MW, Romanchuk G, Simeone DM, Flowe K. Stimulation of myenteric plexus neurite outgrowth by insulin and IGF I and II. Life Sci 1992; 51:1789–1796.
114. Caroni P, Grandes P. Nerve sprouting in innervated adult skeletal muscle induced by exposure to elevated levels of insulin-like growth factors. J Cell Biol 1990; 110:1307–1317.
115. Caroni P, Schneider C. Signaling by insulin-like growth factors in paralyzed skeletal muscle: Rapid induction of IGF1 expression in muscle fibers and prevention of interstitial cell proliferation by IGF-BP5 and IGF-BP4. J Neurosci 1994; 14:3378–3388.
116. Guthrie KM, Nguyen T, Gall CM. Insulin-like growth factor-1 mRNA is increased in deafferented hippocampus: spatiotemporal correspondence of a trophic event with axon sprouting. J Comp Neurol 1995; 352:147–160.
117. Beilharz EJ, Bassett NS, Sirimanne ES, Williams CE, Gluckman PD. Insulin-like growth factor II is induced during wound repair following hypoxic-ischemic injury in the developing rat brain. Brain Res Mol Brain Res 1995; 29:81–91.
118. Quiroga S, Garofalo RS, Pfenninger KH. Insulin-like growth factor I receptors of fetal brain are enriched in nerve growth cones and contain a β-subunit variant. Proc Natl Acad Sci USA 1995; 92:4309–4312.
119. Mascotti F, Cáceres A, Pfenninger KH, Quiroga S. Expression and distribution of IGF-I receptors containing a β-subunit variant (β_{gc}) in developing neurons. J Neurosci 1997; 17:1447–1459.
120. Cheng N, Sahyoun N. Neuronal tyrosine phosphorylation in growth cone glycoproteins. J Biol Chem 1990; 265:2417–2420.
121. Påhlman S, Mamaeva S, Meyerson G, Mattsson MEK, Bjelfman C, Örtoft E, Hammerling U. Human neuroblastoma cells in culture: a model for neuronal cell differentiation and function. Acta Physiol Scand Suppl 1990; 592:25–37.
122. Recio-Pinto E, Ishii DN. Insulin and insulinlike growth factor receptors regulating neurite formation in cultured human neuroblastoma cells. J Neurosci Res 1988; 19:312–320.
123. Leventhal PS, Shelden EA, Kim B, Feldman EL. Tyrosine phosphorylation of paxillin and focal adhesion kinase during insulin-like growth factor-I-stimulated lamellipodial advance. J Biol Chem 1997; 272:5214–5218.
124. Berfield AK, Spicer D, Abrass CK. Insulin-like growth factor I (IGF-I) induces unique effects in the cytoskeleton of cultured rat glomerular mesangial cells. J Histochem Cytochem 1997; 45:583–593.
125. Kadowaki T, Koyasu S, Nishida E, Sakai H, Takaku F, Yahara I, Kasuga M. Insulin-like growth factors, insulin, and epidermal growth factor cause rapid cytoskeletal reorganization in KB cells. J Biol Chem 1986; 261:16141–16147.
126. Izumi T, Saeki Y, Akanuma Y, Takaku F, Kasuga M. Requirement for receptor-intrinsic tyrosine kinase activities during ligand-induced membrane ruffling of KB cells. J Biol Chem 1988; 263:10386–10393.
127. Kotani K, Yonezawa K, Hara K, Ueda H, Kitamura Y, Sakaue H, Ando A, Chavanieu A, Calas B, Grigorescu F, Nishiyama M, Waterfield MD, Kasuga M. Involvement of phosphoinositide 3-kinase in insulin- or IGF-1-induced membrane ruffling. EMBO J 1994; 13:2313–2321.
128. Lavie Y, Agranoff BW. Inhibition of neuroblastoma cell phosphatidylinositol 3-kinase by CDP-diacylglycerol and phosphatidate. J Neurochem 1996; 66:811–816.
129. Jackson TR, Blader IJ, Hammonds-Odie LP, Burga CR, Cooke F, Hawkins PT, Wolf AG, Heldman KA, Theibert AB. Initiation and maintenance of NGF-stimulated neurite outgrowth requires activation of a phosphoinositide 3-kinase. J Cell Sci 1996; 109:289–300.
130. Nobes CD, Hawkins P, Stephens L, Hall A. Activation of the small GTP-binding proteins rho and rac by growth factor receptors. J Cell Sci 1995; 108:225–233.

131. Han J, Luby-Phelps K, Das B, Shu X, Xia Y, Mosteller RD, Murali Krishna U, Falck JR. Role of substrates and products of PI 3-kinase in regulating activation of rac-related guanosine triphosphates by Vav. Science 1998; 279:558–560.
132. Nimnual AS, Yatsula BA, Bar-Sagi D. Coupling of Ras and Rac guanosine triphosphatases through the Ras exchanger Sos. Science 1998; 279:560–563.
133. Yagihashi S. Nerve structural defects in diabetic neuropathy: Do animals exhibit similar changes? Neurosci Res Commun 1997; 21:25–32.
134. Miyamoto S, Teramoto H, Coso OA, Gutkind JS, Burbelo PD, Akiyama SK, Yamada KM. Integrin function: molecular hierarchies of cytoskeletal signaling molecules. J Cell Biol 1995; 131:791–805.
135. Leventhal PS, Feldman EL. The tyrosine kinase inhibitor methyl 2,5-dihydroxycinnimate disrupts changes in the actin cytoskeleton required for neurite formation. Brain Res Mol Brain Res 1996; 43:338–340.
136. Goldberg DJ, Wu D-Y. Inhibition of formation of filopodia after axotomy by inhibitors of protein tyrosine kinases. J Neurobiol 1995; 27:553–560.
137. Fagerstrom S, Pahlman S, Gestblom C, Nanberg E. Protein kinase C-epsilon is implicated in neurite outgrowth in differentiating human neuroblastoma cells. Cell Growth Differ 1996; 7:775–785.
138. Parrow V, Fagerstrom S, Meyerson G, Nanberg E, Pahlman S. Protein kinase C-alpha and -epsilon are enriched in growth cones of differentiating SH-SY5Y human neuroblatoma cells. J Neurosci Res 1995; 41:782–791.
139. Sumantran VN, Feldman EL. Insulin-like growth factor I regulates c-*myc* and GAP-43 messenger ribonucleic acid expression in SH-SY5Y human neuroblastoma cells. Endocrinology 1993; 132:2017–2023.
140. Aigner L, Caroni P. Absence of persistent spreading, branching, and adhesion in GAP-43-depleted growth cones. J Cell Biol 1995; 128:647–660.
141. Zuber MX, Goodman DW, Karns LR, Fishman MC. The neuronal growth-associated protein GAP-43 induces filopodia in non-neuronal cells. Science 1989; 244:1193–1195.
142. Kim B, Leventhal PS, Saltiel AR, Feldman EL. Insulin-like growth factor-I-mediated neurite outgrowth in vitro requires MAP kinase activation. J Biol Chem 1997; 272:21268–21273.
143. Pang L, Sawada T, Decker SJ, Saltiel AR. Inhibition of MAP kinase kinase blocks the differentiation of PC-12 cells induced by nerve growth factor. J Biol Chem 1995; 270:13585–13588.
144. Fernyhough P, Mill JF, Roberts JL, Ishii DN. Stabilization of tubulin mRNAs by insulin and insulin-like growth factor I during neurite formation. Brain Res Mol Brain Res 1989; 6:109–120.
145. Carson MJ, Behringer RR, Brinster RL, McMorris FA. Insulin-like growth factor I increases brain growth and central nervous system myelination in transgenic mice. Neuron 1993; 10:729–740.
146. Ye P, Carson J, D'Ercole AJ. Insulin-like growth factor-I influences the initiation of myelination: studies of the anterior commissure of transgenic mice. Neurosci Lett 1995; 201:235–238.
147. Windebank AJ, Wood P, Bunge RP, Dyck PJ. Myelination determines the caliber of dorsal root ganglion neurons in culture. J Neurosci 1985; 5:1563–1569.
148. Starr R, Attema B, DeVries GH, Monteiro MJ. Neurofilament phosphorylation is modulated by myelination. J Neurosci Res 1996; 44:328–337.
149. Ye P, Carson J, D'Ercole AJ. *In vivo* actions of insulin-like growth factor-I (IGF-I) on brain myelination: studies of IGF-I and IGF binding protein-1 (IGFBP-1) transgenic mice. J Neurosci 1995; 15:7344–7356.
150. Oudega M, Xu XM, Guénard V, Kleitman N, Bunge MB. A combination of insulin-like growth factor-I and platelet-derived growth factor enhances myelination but diminishes axonal regeneration into Schwann cell grafts in the adult rat spinal cord. Glia 1997; 19:247–258.
151. Yao D-L, Liu X, Hudson LD, De F. Webster H. Insulin-like growth factor I treatment reduces demyelination and up-regulates gene expression of myelin-related proteins in experimental autoimmune encephalomyelitis. Proc Natl Acad Sci USA 1995; 92:6190–6194.
152. Webster HD. Growth factors and myelin regeneration in multiple sclerosis. Mult Scler 1997; 3:113–120.
153. Yao DL, Liu X, Hudson LD, Webster HD. Insulin-like growth factor-I given subcutaneously reduces clinical deficits, decreases lesion severity and upregulates synthesis of myelin proteins in experimental autoimmune encephalomyelitis. Life Sci 1996; 58:1301–1306.
154. Komoly S, Hudson LD, Webster HD, Bondy CA. Insulin-like growth factor I gene expression is induced in astrocytes during experimental demyelination. Proc Natl Acad Sci USA 1992; 89:1894–1898.
155. Mozell RL, McMorris FA. Insulin-like growth factor I stimulates oligodendrocyte development and myelination in rat brain aggregate cultures. J Neurosci Res 1991; 30:382–390.

156. Shinar Y, McMorris FA. Developing oligodendroglia express mRNA for insulin-like growth factor-I, a regulator of oligodendrocyte development. J Neurosci Res 1995; 42:516–527.
157. Roth GA, Spada V, Hamill K, Bornstein MB. Insulin-like growth factor I increases myelination and inhibits demyelination in cultured organotypic nerve tissue. Brain Res Mol Brain Res 1995; 88: 102–108.
158. Vemuri GS, McMorris FA. Oligodendrocytes and their precursors require phosphatidylinositol 3-kinase signaling for survival. Development 1996; 122:2529–2537.
159. Noble M, Mayer-Proschel M. On the track of cell survival pharmaceuticals in the oligodendrocyte type-2 astrocyte lineage. Perspect Dev Neurobiol 1996; 3:121–131.
160. Russell JW, Windebank AJ, Schenone A, Feldman EL. Insulin-like growth factor-I prevents apoptosis in neurons after nerve growth factor withdrawal. J Neurobiol 1998; 36:455–467.
161. Brockes JP, Raff MC, Nishiguchi DJ, Winter J. Studies on cultured rat Schwann cells, III. Assays for peripheral myelin proteins. J Neurocytol 1980; 9:67–77.
162. Trapp BD, Itoyama Y, Sternberger NH, Quarles RH, Webster HD. Immunocytochemical localization of P_0 protein in Golgi complex membranes and myelin of developing rat Schwann cells. J Cell Biol 1981; 90:1–6.
163. Baron P, Shy M, Kamholz J, Scarlato G, Pleasure D. Expression of P0 protein mRNA along rat sciatic nerve during development. Brain Res Dev Brain Res 1994; 83:285–288.
164. Cheng LL, Mudge AW. Cultured schwann cells constitutively express the myelin protein P_0. Neuron 1996; 16:309–319.
165. Stewart HJ, Bradke F, Tabernero A, Morrell D, Jessen KR, Mirsky R. Regulation of rat Schwann cell Po expression and DNA synthesis by insulin-like growth factors in vitro. Eur J Neurosci 1996; 8: 553–564.
166. Gavrilovic J, Brennan A, Mirsky R, Jessen KR. Fibroblast growth factors and insulin-like growth factors combine to promote survival of rat Schwann cell precursors without induction of DNA synthesis. Eur J Neurosci 1995; 7:77–85.
167. Russell JW, Cheng HL, Feldman EL. Insulin-like growth factor-I (IGF-I) induces Schwann cell differentiation and neuronal myelination. Mol Biol Cell 1997; 8:330a.
168. Powell EM, Meiners S, DiProspero NA, Geller HM. Mechanisms of astrocyte-directed neurite guidance. Cell Tissue Res 1997; 290:385–393.
169. Norenberg MD. Astrocyte responses to CNS injury. J Neuropathol Exp Neurol 1994; 53:213–220.
170. Garcia-Estrada J, Garcia-Segura LM, Torres-Aleman I. Expression of insulin-like growth factor I by astrocytes in response to injury. Brain Res 1992; 592:343–347.
171. Chernausek SD. Insulin-like growth factor-I (IGF-I) production by astroglial cells: regulation and importance for epidermal growth factor-induced cell replication. J Neurosci Res 1993; 34:189–197.
172. Han VKM, Smith A, Myint W, Nygard K, Bradshaw S. Mitogenic activity of epidermal growth factor on newborn rat astroglia: interaction with insulin-like growth factors. Endocrinology 1992; 131:1134–1142.
173. Tranque PA, Calle R, Naftolin F, Robbins R. Involvement of protein kinase-C in the mitogenic effect of insulin-like growth factor-I on rat astrocytes. Endocrinology 1992; 131:1948–1954.
174. Tranque P, Naftolin F, Robbins R. Differential regulation of astrocyte plasminogen activators by insulin-like growth factor-I and epidermal growth factor. Endocrinology 1994; 134:2606–2613.
175. Masuda S, Chikuma M, Sasaki R. Insulin-like growth factors and insulin stimulate erythropoietin production in primary cultured astrocytes. Brain Res 1997; 746:63–70.
176. Matsuda T, Murata Y, Kawamura N, Hayashi M, Tamada K, Takuma K, Maeda S, Baba A. Selective induction of a1 isoform of (Na^+ + K^+)-ATPase by insulin/insulin-like growth factor-I in cultured rat astrocytes. Arch Biochem Biophys 1993; 307:175–182.
177. Pierre M, Toru-Delbauffe D, Gavaret JM, Pomerance M, Jacquemin C. Activation of S6 kinase activity in astrocytes by insulin, somatomedin C and TPA. FEBS Lett 1986; 206:162–166.
178. Pomerance M, Gavaret JM, Breton M, Pierre M. Effects of growth factors on phosphatidylinositol-3-kinase in astroglial cells. J Neurosci Res 1995; 40:737–746.
179. Masters BA, Werner H, Roberts CT, Jr, LeRoith D, Raizada MK. Developmental regulation of insulin-like growth factor-I-stimulated glucose transporter in rat brain astrocytes. Endocrinology 1991; 128:2548–2557.
180. Scott RJ, Hegyi L. Cell death in perinatal hypoxic ischemic brain injury. Neuropathol Appl Neurobiol 1997; 23:307–314.
181. Johnson EM, Greenlund LJ, Akins PT, Hsu CY. Neuronal apoptosis: current understanding of molecular mechanisms and potential role in ischemic brain injury. J Neurotrauma 1995; 12:843–852.

182. Guan J, Williams CE, Skinner SJM, Mallard EC, Gluckman PD. The effects of insulin-like growth factor (IGF)-1, IGF-2, and des-IGF-1 on neuronal loss after hypoxic-ischemic brain injury in adult rats: Evidence for a role for IGF binding proteins. Endocrinology 1996; 137:893–898.
183. Guan J, Williams C, Gunning M, Mallard C, Gluckman P. The effects of IGF-1 treatment after hypoxic-ischemic brain injury in adult rats. J Cereb Blood Flow Metab 1993; 13:609–616.
184. Gluckman P, Klempt N, Guan J, Mallard C, Sirimanne E, Dragunow M, Klempt M, Singh K, Williams C, Nikolics K. A role for IGF-1 in the rescue of CNS neurons following hypoxic-ischemic injury. Biochem Biophys Res Commun 1992; 182:593–599.
185. Zhu CZ, Auer RN. Intraventricular administration of insulin and IGF-1 in transient forebrain ischemia. J Cereb Blood Flow Metab 1994; 14:237–242.
186. Johnston BM, Mallard EC, Williams CE, Gluckman PD. Insulin-like growth factor-1 is a potent neuronal rescue agent after hypoxic-ischemic injury in fetal lambs. J Clin Invest 1996; 97:300–308.
187. Bergstedt K, Wieloch T. Changes in insulin-like growth factor I receptor density after transient cerebral ischemia in the rat. Lack of protection against ischemic brain damage following injection of insulin-like growth factor 1. J Cereb Blood Flow Metab 1993; 13:895–898.
188. Kanje M, Skottner A, Sjoberg J, Lundborg G. Insulin-like growth factor I (IGF-I) stimulates regeneration of the rat sciatic nerve. Brain Res 1989; 486:396–398.
189. Edbladh M, Fex-Svenningsen Å, Ekström PA, Edström A. Insulin and IGF-II, but not IGF-I, stimulate the in vitro regeneration of adult frog sciatic sensory axons. Brain Res 1994; 641:76–82.
190. Saatman KE, Contreras PC, Smith DH, Raghupathi R, McDermott KL, Fernandez SC, Sanderson KL, Voddi M, McIntosh TK. Insulin-like growth factor-I (IGF-I) improves both neurological motor and cognitive outcome following experimental brain injury. Exp Neurol 1997; 147:418–427.
191. Apfel SC, Kessler JA. Neurotrophic factors in the treatment of peripheral neuropathy. Ciba Found Symp 1996; 196:98–108.
192. Ishii DN. Implication of insulin-like growth factors in the pathogenesis of diabetic neuropathy. Brain Res Brain Res Rev 1995; 20:47–67.
193. Zhuang HX, Wuarin L, Fei ZJ, Ishii DN. Insulin-like growth factor (IGF) gene expression is reduced in neural tissues and liver from rats with non-insulin-dependent diabetes mellitus, and IGF treatment ameliorates diabetic neuropathy. J Pharmacol Exp Ther 1997; 283:366–374.
194. Wuarin L, Namdev R, Burns JG, Fei ZJ, Ishii DN. Brain insulin-like growth factor-II mRNA content is reduced in insulin-dependent and non-insulin-dependent diabetes mellitus. J Neurochem 1996; 67:742–751.
195. Migdalis IN, Kalogeropoulou K, Kalantzis L, Nounopoulos C, Bouloukos A, Samartzis M. Insulin-like growth factor-I and IGF-I receptors in diabetic patients with neuropathy. Diab Med 1995; 12:823–827.
196. Ishii DN, Lupien SB. Insulin-like growth factors protect against diabetic neuropathy: effects on sensory nerve regeneration in rats. J Neurosci Res 1995; 40:138–144.
197. Contreras PC, Vaught JL, Gruner JA, Brosnan C, Steffler C, Arezzo JC, Lewis ME, Kessler JA, Apfel SC. Insulin-like growth factor-I prevents development of a vincristine neuropathy in mice. Brain Res 1997; 774:20–26.
198. Songyang Z, Baltimore D, Cantley LC, Kaplan DR, Franke TF. Interleukin 3-dependent survival by the Akt protein kinase. Proc Natl Acad Sci USA 1997; 94:11345–11350.
199. Williams DB, Windebank AJ. Motor neuron disease. In: Dyck PJ, Thomas PK, Griffin JW, Low PA, Poduslo JF, eds. Peripheral Neuropathy 3rd edit. WB Saunders, Philadelphia, 1993, pp. 1028–1050.
200. Williams DB, Windebank AJ. Motor neuron disease (amyotrophic lateral sclerosis). Mayo Clin Proc 1991; 66:54–82.
201. Adem A, Ekblom J, Gillberg P-G, Jossan SS, Höög A, Winblad B, Aquilonius S-M, Wang L-H, Sara V. Insulin-like growth factor-I receptors in human spinal cord: changes in amyotrophic lateral sclerosis. J Neural Transm Gen Sect 1994; 97:73–84.
202. Vaught JL, Contreras PC, Glicksman MA, Neff NT. Potential utility of rhIGF-1 in neuromuscular and/or degenerative disease. Ciba Found Symp 1996; 196:18–27.
203. Lai EC, Felice KJ, Festoff BW, Gawel MJ, Gelinas DF, Kratz R, Murphy MF, Natter HM, Norris FH, Rudnicki SA. Effect of recombinant human insulin-like growth factor-I on progression of ALS. A placebo-controlled study. The North American ALS/IGF-I Study Group. Neurology 1997; 49:1621–1630.
204. Liu X, Yao DL, Webster J. Insulin-like growth factor I treatment reduces clinical deficits and lesion severity in acute demyelinating autoimmune encephalomyelitis. Mult Scler 1995; 1:2–9.
205. Liu X, Linnington C, Webster HD, Lassmann S, Yao DL, Hudson LD, Wekerle LD, Kreutzberg GW.

Insulin-like growth factor-I treatment reduces immune cell responses in acute non-demyelinative experimental autoimmune encephalomyelitis. J Neurosci Res 1997; 47:531–538.

206. Aguado F, Rodrigo J, Cacicedo L, Mellstrom B. Distribution of insulin-like growth factor-I receptor mRNA in rat brain. Regulation in the hypothalamo-neurohypophysial system. J Mol Endocrinol 1993; 11:231–239.
207. Marks JL, Porte D, Jr, Baskin DG. Localization of type I insulin-like growth factor receptor messenger RNA in the adult rat brain by *in situ* hybridization. Mol Endocrinol 1991; 5:1158–1168.
208. Nagano T, Sato M, Mori Y, Du Y, Takagi H, Tohyama M. Regional distribution of messenger RNA encoding the insulin-like growth factor type 2 receptor in the rat lower brainstem. Brain Res Mol Brain Res 1995; 32:14–24.
209. Brar AK, Chernausek SD. Localization of insulin-like growth factor binding protein-4 expression in the developing and adult rat brain: analysis by in situ hybridization. J Neurosci Res 1993; 35:103–114.
210. Zhou X, Herman JP, Paden CM. Expression of insulin-like growth factor binding protein-2 during postnatal development of the rat neurointermediate pituitary. Neuroendocrinology 1997; 66:17–27.
211. Green BN, Jones SB, Streck RD, Wood TL, Rotwein P, Pintar JE. Distinct expression patterns of insulin-like growth factor binding proteins 2 and 5 during fetal and postnatal development. Endocrinology 1994; 134:954–962.
212. Russo VC, Edmondson SR, Mercuri FA, Buchanan CR, Werther GA. Identification, localization, and regulation of insulin-like growth factor binding proteins and their messenger ribonucleic acids in the newborn rat olfactory bulb. Endocrinology 1994; 135:1437–1446.
213. Delhanty PJD, Han VKM. The expression of insulin-like growth factor (IGF)-binding protein-2 and IGF-II genes in the tissues of the developing ovine fetus. Endocrinology 1993; 132:41–52.

20 IGF System Components and Their Role in Bone Metabolism

Subburaman Mohan, PhD
and David J. Baylink, MD

Contents

INTRODUCTION

Bone is a dynamic tissue that is capable of repair after injury and remodeling in response to changes in physical strains that occur throughout life. This regenerative property of the bone is essential for the maintenance of the structural mass, strength, and shape of bone and for fracture healing. The volume of bone and ultimately its strength are determined by the balance between two opposing processes, osteoclastic bone resorption and osteoblastic bone formation, that together constitute bone remodeling. The balance may be shifted to favor more formation than resorption, leading to net gain in bone as seen during puberty or during increased mechanical loading. The balance can also be shifted to favor less formation than resorption as seen during disuse or during chronic disease states such as osteoporosis. Studies in a number of laboratories have focused on understanding the regulation of the bone resorption and bone formation processes based on the premise that insights on molecular signaling messenger molecules that are responsible for controlling these two processes could lead to new treatments for osteoporosis. The following section briefly describes the bone cell parameters involved in bone formation and bone resorption processes and the corresponding regulatory signaling molecules that are responsible for regulating these bone cell parameters.

From: *Contemporary Endocrinology: The IGF System*
Edited by: R. Rosenfeld and C. Roberts Jr © Humana Press, Inc., Totowa, NJ

Basic studies have shown that bone resorption at local sites depends on both the number and activity of osteoclasts, which are derived from hemopoietic stem cells belonging to the colony-forming unit (CFU)-derived monocyte-macrophage family *(1)*. The activity and differentiation of osteoclasts from their precursor cells are tightly regulated by osteotropic hormones and cytokines *(1)*. Recent studies have shown that some stimuli (e.g., interleukins) may act on osteoclast line cells to initiate resorption, while other stimuli (e.g., parathyroid hormone [PTH]) may direct osteoblasts to produce a chemical signal that may further cue osteoclast line cells to increase bone resorption. It has also been shown that some stimuli (e.g., PTH) may act by primarily increasing the activity of existing osteoclasts while others (e.g., macrophage colony stimulating factor [MCSF]) may act by primarily stimulating the differentiation of new osteoclasts from their precursors. Recent studies also demonstrate that systemic and local factors (e.g., transforming growth factor [TGF], insulin-like growth factor [IGF]) regulate bone resorption in part by regulating osteoclastic apoptosis *(1)*. Thus, the amount of bone resorbed at the remodeling sites may be regulated by three individual control mechanisms that modulate the rate of osteoclast formation, osteoclast activity, or osteoclast apoptosis.

As in the case of bone resorption, the rate of bone formation at any given site depends on the number and activity of bone-forming osteoblasts, which are derived from undifferentiated pluripotent mesenchymal cells *(2)*. Studies on the factors affecting growth and differentiation of osteoblast line cells revealed evidence of involvement of both systemic hormones and local growth factors. Studies in our laboratory and other laboratories have shown that a number of systemic hormones, including PTH, 1,25-dihydroxycholecalciferol, testosterone, progesterone, and glucocorticoids, may act on osteoblast line cells in part via regulating local production of growth factors. Of the various known systemic and local regulators of osteoblast cells, there are several features of the IGF system that suggest that it may play a major role in the regulation of proliferation and differentiation of osteoblasts. In addition, the IGF system may also play a role in the regulation of osteoclast recruitment, activity, and apoptosis. Therefore, we have focused our efforts toward describing the regulation of production and actions of the components of the IGF system in bone and elucidating the potential role of IGF system components in the pathogenesis of primary and secondary types of osteoporosis. In this chapter, the primary focus is on the role of IGF system components in human bone metabolism.

COMPONENTS OF THE IGF SYSTEM IN BONE

Recent findings demonstrate that the local actions of IGFs in the bone microenvironment may be regulated by multiple components of the IGF regulatory system, including IGF-I, IGF-II, type I and type II IGF receptors, IGF binding proteins (IGFBPs), and extracellular IGFBP proteases (Fig. 1). The different components of the IGF system expressed in bone are described below.

Expression of IGF Peptides

Animals treated systemically with agents that stimulate bone resorption show evidence of increased bone formation in addition to the expected increase in bone resorption. This paradoxical increase in bone formation has been postulated to represent a counterregulatory mechanism for the maintenance of bone volume *(3)*. Subsequent studies revealed that this increase in the rate of bone formation was (1) proportional to the

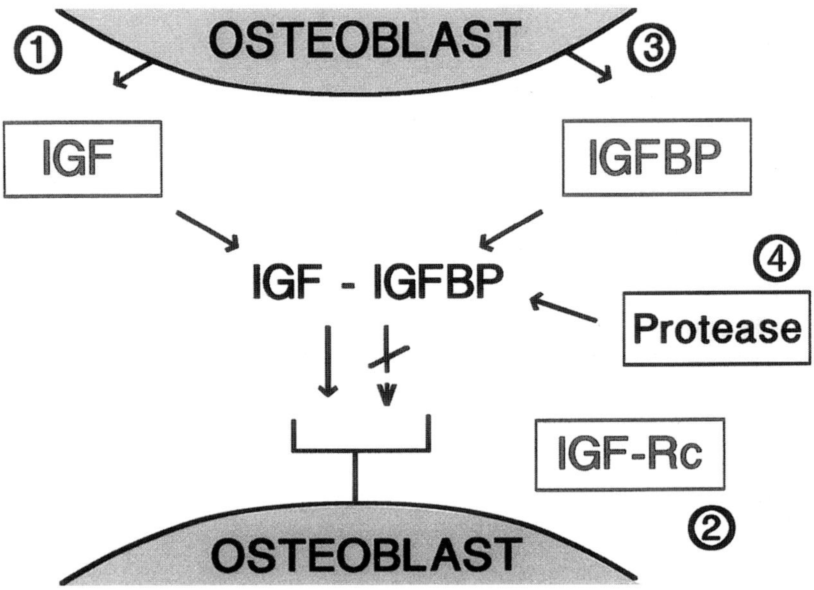

Fig. 1. Components of the IGF system in bone. The activities of IGFs in bone cell microenvironment are determined not only by (1) synthesis of IGFs but also by (2) the presence of IGF receptors, (3) inhibitory and stimulatory IGFBPs, and (4) IGFBP proteases.

increase in the rate of bone resorption, (2) site specific, and (3) may have been mediated in part by increased production of local bone growth factors *(3)*. Subsequent studies to identify the active principle, termed skeletal growth factor (SGF), from extracts of human bone revealed that SGF was present as high molecular weight forms in nondissociative extracts and serum-free human bone cell conditioned medium, apparently in complexes with binding proteins. SGF was subsequently purified to homogeneity, and structural studies of the amino acid sequences from the amino (N) terminal region and several tryptic fragments of human bone-derived SGF were shown to be identical to the corresponding sequences of IGF-II purified from human serum. Based on structural, biological, and immunological studies, it was concluded that SGF was very similar if not identical to IGF-II *(4)*. Human bones also contained IGF-I, but the concentration of IGF-I was several-fold less compared with IGF-II *(4,5)*. It was also found that IGF-I and IGF-II are evolutionarily conserved and are present in the skeletal tissues of mammals through cartilaginous fish *(6)*.

During the time that we were purifying IGF-II from human bone extract, Canalis and his co-workers were isolating growth factors from medium conditioned by 21-d-old fetal rat calvariae. Based on their findings that conditioned medium derived from cultures of fetal rat calvaria contained significant levels of IGF-I, it was concluded that IGF-I, and not IGF-II, was the key regulator of osteoblast cell proliferation in rats *(7)*. Consistent with these data, studies on the relative distribution of IGF-I and IGF-II in the serum and skeletal extracts of 10 animals belonging to 5 vertebrate classes revealed that the levels of IGF-II were generally higher than IGF-I in all classes, with the exception of rodents, in which the reverse was true *(6)*. Similarly, adult human osteoblasts produce much more IGF-II than IGF-I, whereas the reverse is true for adult rat osteoblasts *(5,8–11)*. Thus, these results suggest that humans and rodents differ in the relative rates of production of

Table 1
IGF and IGFBP Expression in Various Osteoblast Cell Models

Cell model	*IGF-I*	*IGF-II*	*IGFBP-1*	*IGFBP-2*	*IGFBP-3*	*IGFBP-4*	*IGFBP-5*	*IGFBP-6*
Normal hOBs	+	+++	+	+/–	++++	++++	++++	++
HOBIT	–	+++	+		+++	++++	+	
SaOS-2	–	+++	–		+++	++++	+	+
TE85-1	–	+++	–	+++	–	++	–	
TE85-2	–	+++	–	–	+ +	++	+	
MG63	–	–	–	+++	+	+	–	
U2	–	++	–	–	–	–	++++	+
rOBs	+++	+	–	+++	+	+	+	+
UMR	++		–	–	–	+++	+++	–
ROS	++		–	–	–	+++	–	–
MC3T3	++	+	–	+	–	++	+	+

Abbreviations: hOBs, untransformed normal human osteoblasts; HOBIT, SV40-immortalized hOBs; TE85-1, human osteosarcoma cells obtained from ATCC; SaOS-2, TE85, MG63, and U2, human osteosarcoma cell lines TE85-1 and TE85-2 represent cell lines obtained from ATCC and Sloan Kettering Institute respectively *(9,44);* rOB, primary cultures of rat osteoblasts derived from fetal or neonatal calvaria; UMR106 and ROS, rat osteosarcoma cell lines; MC3T3-E, clonal mouse osteoblastic cell line; +, present; – not detectable.

IGF-I and IGF-II by osteoblasts, as well as in the relative distribution of these two growth factors in the serum and skeletal extracts. Based on these differences in the relative distribution of IGF-I and IGF-II between rodents and humans, one should be cautious in extrapolating results from a rodent model system to humans.

Osteoblasts

Although human osteoblasts in culture produce a number of growth factors, past studies have shown that IGF-II is the most abundant growth factor produced by human osteoblasts. Our recent studies on the relative expression of IGF-I and IGF-II in human osteoblasts derived from different skeletal sites revealed that human osteoblasts in serum-free culture express 8- to 100-fold greater levels of IGF-II than IGF-I *(12,13).* These studies also revealed that the production of IGF-I and IGF-II by human osteoblasts is significantly different between the skeletal sites of their origin. In contrast to normal human osteoblasts, which express both IGF-I and IGF-II, studies in our laboratory as well as other laboratories have shown that a number of established human osteoblast-like osteosarcoma cell lines do not express measurable levels of IGF-I (Table 1). Many of the human osteosarcoma cell lines tested, on the other hand, expressed significant levels of IGF-II *(9,13,14).* Based on these findings, it can be concluded that the relative rates of IGF-I and IGF-II expression vary not only between untransformed normal osteoblasts and transformed osteoblasts, but also between osteoblasts derived from different human skeletal sites.

In contrast to human osteoblasts, adult rodent osteoblasts have been shown to express primarily IGF-I *(8–10,15,16).* Recent studies have also shown that the IGF expression pattern may change depending on the stage of osteoblast differentiation. Birnbaum et al. *(17)* have investigated the changes in IGF-I and IGF-II expression and secretion during the proliferation and differentiation of normal rat osteoblasts in vitro. They have reported that secretion of IGF-I occurred principally when preosteoblasts are proliferating and when the cultures are mineralizing but not when the cells are differentiating. In contrast to IGF-I, the levels of IGF-II were undetectable in the conditioned medium, although the IGF-II message was expressed by proliferating osteoblasts. Because proliferating human osteoblasts as well as rat stromal cells express IGF-II *(5,9,18)*, it remains to be established if the lack of detectable IGF-II in the conditioned medium of proliferating rat osteoblasts represents an assay artifact or a difference in IGF-II expression between proliferating human and rat osteoblasts.

Based on what is known about IGF expression in various osteoblast cell types, the following conclusions can be drawn: (1) Human osteoblasts produce primarily IGF-II while rat osteoblasts produce primarily IGF-I. (2) Relative rates of IGF-I and IGF-II expression in osteoblasts may vary depending on whether cells used are normal or transformed, stage of differentiation, site from which osteoblasts are derived, and culture conditions. Thus, one needs to be aware of the differences in IGF expression among the various osteoblast cell models when selecting cell models for a chosen study or when interpreting data on IGF expression using selected cell models.

Stromal Cells

The formation of bone is dependent on an ordered proliferation of stromal cells and subsequent differentiation into their descendant preosteoblasts. Although much is known about the expression of various IGF system components in osteoblasts, studies on the expression pattern of IGFs in stromal cells are limited. Stromal preosteoblasts derived from both rodents and humans have been shown to produce IGF-I and IGF-II *(18,19)*. The expression of IGFs have been shown to change during differentiation of stromal cells into osteoblasts in vitro. It has been reported that dexamethasone, which induces differentiation of stromal cells, causes an increase in the expression of IGF-II and a decrease in the expression of IGF-I *(20)*. The significance of changes in IGF expression pattern during stromal cell differentiation, however, remains to be established.

Osteoclasts

Studies on growth factor expression in osteoclasts have been limited for several reasons, including: (1) Osteoclasts are much fewer in number compared to osteoblasts in bone. (2) Osteoclasts are fragile and it is difficult to isolate adequate numbers of purified osteoclasts for in vitro studies. (3) Osteoclasts, unlike osteoblasts, do not proliferate. (4) Osteoclasts undergo apoptosis both in vitro and in vivo. Two recent reports provide evidence that osteoclasts also serve as an important source for IGFs in bone. First, Lazowski et al. *(21)* have reported that osteoclasts contain immunoreactive IGF-I in rat bone, thus suggesting that these cells also may produce IGFs. Second, Middleton et al. *(22)* investigated the expression of IGF-I mRNA and IGF-II mRNA by osteoclasts in adult human osteophyte tissue by *in situ* hybridization. They have found that osteoclasts express mRNA for both IGF-I and IGF-II. These data are consistent with the possibility that the IGFs produced by osteoclasts may act on osteoclast line cells in an autocrine manner to control bone resorption and/or that the IGFs produced by osteoclasts may act on osteoblast line cells in a paracrine manner to regulate the coupled increase in bone formation.

Expression of IGF Receptors

Osteoblasts

IGF actions in osteoblasts are mediated through binding of IGF peptides to specific plasma membrane receptors. Two types of IGF receptors, namely type I and type II IGF receptors, have been identified to date *(23,24)*. The type I IGF receptor is a tyrosine kinase signaling receptor structurally related to the insulin receptor, and binds IGF-I and IGF-II with high affinity and insulin with low affinity. The type II IGF receptor has two binding sites, one for IGF-II and another for proteins containing mannose-6-phosphate. The involvement of the type I IGF receptor in mediating the biological actions of IGF-I and IGF-II in osteoblasts has been well documented *(25–29)*. However, the involvement of type II IGF receptor in mediating the biological actions of IGF-II is controversial. Analogs of IGF-II with selective binding affinity for either type I or type II receptors have been used in various cell models to identify the potential role of type I and type II IGF receptors in mediating the actions of IGFs. In this regard, we have found that [Leu^{27}] and [Leu^{43}] IGF-II analogs, which bind type II IGF receptor but not type I IGF receptor, had very little mitogenic activity in mouse and human osteoblasts *(30)*. These data demonstrate evidence for lack of involvement of type II IGF receptors in mediating the biological effects of IGF-II. On the other hand, we also found that the [Arg^{54},Arg^{55}] analog, which binds to type I but not to type II IGF receptors, exhibited significantly less mitogenic activity compared to wild-type IGF-II in stimulating osteoblast cell proliferation. Thus, further studies are needed to evaluate if the type I IGF receptor mediates all of the biological effects of IGF-II on osteoblasts.

Osteoclasts

The current knowledge of receptors on osteoclasts is very limited owing to difficulties in preparing highly purified preparations of mature mammalian osteoclasts for biochemical studies. Middleton et al. showed by *in situ* hybridization that osteoclasts in adult human osteophyte tissue express mRNA for the type I IGF receptor *(22)*. Consistent with this finding, Hou et al. *(31)* have recently reported that mature osteoclasts from normal rabbit long bone express mRNA for IGF-I receptor as detected by *in situ* hybridization. The successful translation of IGF-I receptor message into active binding sites on osteoclasts was confirmed by autoradiography using a heterogeneous bone cell population, and by isolation of the receptor–ligand complex from purified osteoclasts *(31)*. Based on these findings and the findings that IGF-I has significant biological effects on osteoclast recruitment and/or activity *(32,33)*, it can be speculated that IGFs produced by osteoclasts may have a functional role in regulating osteoclastic bone resorption.

Expression of IGF Binding Proteins

During the course of our studies on the regulation of IGF production in osteoblasts, we came across a discrepancy that purified IGF-II but not human osteoblast cell conditioned medium that contained IGF-II was mitogenic for human osteoblasts. To explain this discrepancy, we proposed the hypothesis that human osteoblasts in culture produce, in addition to IGF-II, an inhibitor of IGF-II activity. Our subsequent studies on the isolation of this inhibitor led to the identification of a new protein that was originally termed as inhibitory IGF binding protein *(34)*. We subsequently isolated a cDNA clone that encoded this protein from a human bone cell cDNA library and determined the full-length sequence of this protein *(35)*. Because this protein was the fourth IGF binding protein to be completely sequenced, inhibitory IGF binding protein is now known as IGFBP-4.

During the course of our studies on IGFBP-4, we were simultaneously investigating why IGF-II did not bind to hydroxyapatite, although it was released upon demineralization of human bone (which meant IGF-II was bound to hydroxyapatite). We therefore proposed that IGF-II is fixed in bone by means of another protein *(4,36)*. Our subsequent studies on the mechanisms by which IGF-II is fixed in bone in such large amounts led to the discovery of IGFBP-5. We found that purified IGFBP-5, but not IGF-II, binds to hydroxyapatite. In addition, we found that the complex of IGF-II + IGFBP-5 bound to hydroxyapatite and that human osteoblasts in culture produce IGFBP-5 *(37)*. These data, together with our finding that IGFBP-5 has several-fold higher affinity for IGF-II than IGF-I, could partly explain why IGF-II is 10–15 times more abundant than IGF-I in human bone *(37)*.

Osteoblasts

Based on studies in our laboratory and other laboratories, it is now known that human osteoblasts in culture produce all six high-affinity IGFBPs (Table 1). However, the relative rates of IGFBP expression appear to vary depending on cell type, differentiation status, and culture conditions *(12,38–41)*. Recent studies in our laboratory on the relative expression of IGFBP-3, IGFBP-4, and IGFBP-5 in human osteoblasts derived from various skeletal sites revealed that IGFBP-3 expression was highest in the vertebral cells and lowest in the mandibular cells whereas IGFBP-4 expression was highest in the mandibular cells and lowest in the stromal cells. IGFBP-5 expression was found to be highest in the calvarial cells and lowest in the rib cells *(12)*. It remains to be established whether other mechanisms (e.g., differences in mechanical loading, extracellular matrix production) besides stage of differentiation contribute to the site-dependent differences in the production of IGF system components.

Comparison of the relative rates of expression of the six known IGFBPs in various human and rat osteosarcoma cell lines reveal that the IGFBP expression varies depending on the osteosarcoma cell type *(38,39,42–46)*. Of the various osteosarcoma cell types tested, only MG63 human osteosarcoma cells appear to express IGFBP-1 (Table 1). IGFBP-2 appears to be expressed in TE85 (ATCC) and MG63 human osteosarcoma cell types. IGFBP-3 is expressed in TE85 (Sloan-Kettering) and MG63 human osteosarcoma cell types. All of the osteosarcoma cell types tested appear to express IGFBP-4, except U2 human osteosarcoma cells, which express large amounts of IGFBP-5. U2 osteosarcoma cells express IGFBP-6 but at much lower levels compared to IGFBP-5. Of the two rat osteosarcoma cell types, ROS cells in culture express only IGFBP-4, whereas UMR cells express both IGFBP-4 and IGFBP-5. Mouse osteoblasts, MC3T3-E1, have been shown to express IGFBP-2, IGFBP-4, IGFBP-5, and IGFBP-6. Based on these findings, it can be concluded that different osteosarcoma cell lines exhibit unique patterns of IGFBP expression and thus provide models for studies on the mechanism of IGFBP expression for given hormonal stimuli, provided the results are confirmed using untransformed normal osteoblasts.

Two recent studies demonstrate evidence for differential expression of IGFBPs during osteoblast development. Birnbaum and Wiren *(47)* showed that maximal expression of IGFBP-2 and IGFBP-5 mRNA and protein occurred in proliferating rat preosteoblasts, whereas maximal expression of IGFBP-3, IGFBP-4, and IGFBP-6 occurred in mature rat osteoblasts. Both proliferating preosteoblasts and mature osteoblasts showed no detectable expression of IGFBP-1. Consistent with these data, Thrailkill et al. *(48)* have reported evidence for temporal differences in the expression of IGFBP-2, IGFBP-4, and IGFBP-5 in MC3T3-E1 mouse osteoblasts. The expression of IGFBP-2 and IGFBP-4 increased with time, with higher levels of expression occurring in mature osteoblasts compared to replicating preosteoblasts whereas the expression of IGFBP-5 was higher in replicating preosteoblasts compared to mature osteoblasts. Although the mechanisms that contribute to the differential expression of IGFBPs during osteoblast development or the potential significance of the changes in IGFBP expression in modulating osteoblast development are currently unknown, these data suggest that the developmental stage of the osteoblast is an important determinant of IGFBP expression.

Stromal Cells

Little is known about the expression of IGFBPs in stromal cells. Recent studies by Cheng et al. have shown that human bone marrow stromal cells in culture under basal conditions produce IGFBPs 2–6, and that dexamethasone treatment increased steady-state levels of IGF-II mRNA and decreased steady-state levels of IGFBP-3 and IGFBP-5 mRNA *(20)*. The potential role of stromal cell produced IGFBPs in mediating differentiation of stromal cells remain to be established.

Expression of IGF Binding Protein Proteases

IGFBP proteases represent the newest component of the IGF regulatory system in bone. IGFBP proteases are capable of cleaving IGFBPs into forms that have either significantly reduced or no affinity for IGFs *(12,38,46,49–52)*. Recent studies also demonstrate that degradation of IGFBP by proteases may be as important as synthesis in determining IGFBP abundance in the local extracellular fluid. Although the proteases produced by osteoblasts have not been completely characterized, studies to date reveal evidence that osteoblasts in culture produce, in addition to IGFBP-specific proteases, other proteases capable of degrading not only IGFBPs but also other proteins *(12,38, 46,49–52)*. IGFBP-specific and nonspecific proteases produced by osteoblasts that degrade IGFBPs are briefly described in the sections that follow.

IGFBP-Specific Proteases

Studies in our laboratory and Conover's laboratory have provided evidence that human osteoblasts in culture produce an IGFBP-4-specific protease capable of degrading IGFBP-4 but not the other known IGFBPs *(51,53–56)*. This protease appears to cleave IGFBP-4 at the middle variable region at a site that is in close proximity to the IGF-binding domain *(49,57)*. One of the unique features of the IGFBP-4 protease produced by osteoblasts is the requirement of IGF-II for its functional activity, with IGF-II being more active than IGF-I. Based on the characteristics of IGFBP-4 protease (high specificity, IGF-II and Ca^{2+} dependence, fragmentation pattern), it is likely that the IGF-II-dependent IGFBP-4 protease produced by human osteoblasts is novel.

In addition to the IGFBP-4 protease, recent studies also demonstrate that osteoblasts in culture produce a specific protease capable of degrading IGFBP-5 but not the other IGFBPs *(51,58)*. Studies on the characterization of this IGFBP-5 protease from the conditioned medium of bone cells reveal that the IGFBP-5-specific protease produced by human osteoblasts is a metal-dependent serine protease. Based on the findings that the activity of IGFBP-5-specific protease changes in response to treatment with growth factors and osteoregulatory agents in osteoblasts *(43,48,51,58–60)* and that IGFBP-5 proteolytic fragments are much less active compared to intact IGFBP-5 in stimulating osteoblast cell proliferation *(61)*, it can be proposed that the IGFBP-5-specific protease is an important regulatory component of the IGF system in human osteoblasts.

Nonspecific IGFBP Proteases

Since the discovery of a pregnancy-associated protease that was active from the second trimester of pregnancy in humans, subsequent studies have shown that a number of purified proteins, including matrix metalloprotease, cathepsin, and plasmin, are capable of degrading multiple IGFBPs *(9,38,50,60,62,63)*. These proteases have broad protein specificity, as they can degrade other proteins besides IGFBPs. Campbell et al. *(46,64)* used chromatographic analysis to show that plasmin dissociates ^{125}I-IGF-I and -II from complexes formed with IGFBPs released into MG63 human osteosarcoma cell conditioned medium. Plasmin is also known to degrade multiple IGFBPs including the major IGFBPs produced by human osteoblasts, i.e., IGFBP-3, IGFBP-4, and IGFBP-5 *(46,63,64)*. Osteoblasts have also been known to produce urokinase plasminogen activator and tissue-type plasminogen activators and their inhibitors *(46,63,64)*. Based on these findings, it has been speculated that the plasmin system plays an important role in regulating limited proteolysis of IGFBPs in the vicinity of their sites of action, and hence the cellular availability of IGFs.

Thrailkill et al. *(60)* have demonstrated by using ^{125}I-IGFBP-5 substrate zymography that MC3T3-E1 mouse osteoblasts secrete proteases with molecular weights of 52–72 and 92 kDa. They have also reported that antisera to matrix metalloproteases-1 and -2 immunoprecipitated IGFBP-5-degrading proteases with molecular masses of 52 and 69/72 kDa, respectively, suggesting that mouse osteoblasts in culture secrete matrix metalloproteases capable of degrading IGFBP-5. Based on these data, it has been speculated that matrix metalloproteases produced by osteoblasts degrade IGFBP-5 and may control local IGFBP-5 actions. Conover and DeLeon *(65)* have shown that MG-63 human osteosarcoma cells secrete an acid-activated ^{125}I-IGFBP-3 protease identified as aspartic protease, cathepsin D, based on acidic pH optimum, inhibition by peptstatin, distinctive proteolytic pattern, and immunoreactivity with cathepsin D antisera. Acid-activated proteases have been shown to degrade multiple IGFBPs *(9,38)*. Based on the identification of both IGFBP-specific and nonspecific IGFBP proteases in osteoblast cell conditioned medium, it can be speculated that the proteolysis of IGFBPs by these proteases could serve as an important control mechanism to regulate IGF actions. The relative contributions of various proteases in regulating IGFBP actions in osteoblasts remain to be studied.

ACTIONS OF IGF SYSTEM COMPONENTS

Studies in a number of laboratories have shown that IGFs are important regulators of bone metabolism in vitro and in vivo (Table 2). In vitro actions of IGFs on responding cells using cell culture model systems and in vivo actions of IGFs on bone turnover using in vivo model systems are described in the sections that follow.

In Vitro Effects on Bone Cells

IGFs

IGF-I and IGF-II have been shown to stimulate proliferation in serum-free cultures of human osteoblasts derived from iliac crest, calvarial osteoblasts derived from embryonic chick, fetal mouse, and fetal rat *(4,25,66–68)*. In addition to its effects on osteoblast cell proliferation, IGF-II has been shown to stimulate synthesis of type I collagen in untransformed normal human osteoblasts (Table 2). IGF-I and IGF-II have been shown to increase bone collagen synthesis and decrease collagen degradation in intact fetal rat calvaria in vitro *(69,70)*. It is not known whether the mitogenic and differentiative

Table 2
Effects of IGF System Components on Bone Cells In Vitro and Bone Metabolism In Vivo

IGF system component	*In vitro effects*	*In vivo effects*
IGF-I	↑ OB proliferation, ↑ OB activity ↑ OC generation, ↓ OC apoptosis	↑ BF, ↑ BR
IGF-II	↑ OB proliferation, ↑ OB activity ↓ OC apoptosis	↑ BF
IGFBP-1	↓ OB proliferation	N.D.
IGFBP-2	↓ OB proliferation	N.D.
IGFBP-3	↑↓ OB proliferation	↑ BF
IGFBP-4	↓ OB proliferation	N.D.
IGFBP-5	↑ OB proliferation	↑ BF (local)
IGFBP-6	↓ OB proliferation	N.D.

N.D., Not determined; OB, osteoblast; OC, osteoclast; BF, bone formation; BR, bone resorption.

functions of IGFs depend on the target cell type (e.g., preosteoblasts and osteoblasts) on which they act, or whether the IGFs induce both of these effects on a single target cell type (e.g., preosteoblasts). It is also not known whether the molecular signaling involved in the differentiation functions of IGFs is similar or different from the molecular signaling involved in the mitogenic action of IGFs.

In addition to the well-established effects of IGFs on osteoblasts, recent studies also demonstrate that IGFs exert significant biological effects on osteoclasts. Mochizuki et al. *(32)* reported that IGF-I induces the formation of osteoclast-like cells from murine hematopoietic precursor cells. In contrast to these results, Hill et al. *(33)* have reported that both IGF-I and IGF-II, in a dose-dependent manner, increase the number of tartrate-resistant acid phosphatase (TRAP)-positive multinucleated cells capable of forming resorption lacunae when cultured on ivory slices. In addition, we have recently found that both IGF-I and IGF-II increase the survival rate of osteoclasts in rabbit marrow cultures *(71)*. Consistent with these data, Hou et al. *(31)* have recently reported that induction of nuclear fragmentation in serum-depleted cultures of purified mature osteoclasts was dose-dependently inhibited by IGF-I in the picomolar range, but not by 1 n*M* insulin, suggesting that the IGF-I actions are mediated via IGF-I receptors. These data provide evidence of a direct role for IGFs in regulating bone resorption by enhancing osteoclast formation and/or decreasing the apoptosis of osteoclasts.

IGFBPs

In addition to functioning as classic binding proteins (i.e., act as transport proteins and prolong the half-life of IGFs), recent studies demonstrate that IGFBPs play an integral role in modulating IGF responses in either an inhibitory or stimulatory manner in a variety of cell types. Studies on the biological effects of purified IGFBPs using various in vitro model systems have provided evidence that different binding proteins may modulate IGF action differently and the same binding protein can have both an IGF inhibiting and IGF potentiating role under different conditions. The factors that determine these differences appear to be many, including IGFBP phosphorylation, IGFBP cell surface association, IGFBP proteolysis, and the ratio of IGF to IGFBP *(23,40,72)*. The following section briefly describes our current understanding of the biological effects of various IGFBPs in osteoblasts.

In regard to the biological actions of various IGFBPs (Table 2), Campbell and Novak *(64)* have reported that purified IGFBP-1 inhibited IGF-I-induced cell proliferation in human MG63 human osteosarcoma cells. In contrast to these results, we found that IGFBP-1 up to 30 ng/mL had no effect on chick osteoblast cell proliferation *(34)*. Purified recombinant human IGFBP-2 inhibited IGF-I-induced bone cell proliferation at an apparent dose ratio of 10:1 *(73)*. In contrast to IGFBP-1 and IGFBP-2, which have been shown to inhibit IGF actions in osteoblasts, the effects of IGFBP-3 seem to vary depending on the dose of IGFBP-3 culture conditions used. Ernst and Rodan *(74)* have shown that IGFBP-3 augmented the effects of IGF-I on rat osteoblast cell proliferation. We found that IGFBP-3 potentiated the IGF-I effect at low concentrations, whereas it inhibited the IGF-I effect at high concentrations in osteoblasts *(40)*.

With regard to IGFBP-4, our previous studies have shown that IGFBP-4 inhibits IGF-I and IGF-II-induced cell proliferation in embryonic chick calvaria cells and MC3T3-E1 mouse osteoblasts under all culture conditions tested *(34,61,75)*. Exogenous addition of IGFBP-4 also inhibited IGF-II-induced cell proliferation in serum-free cultures of normal human osteoblasts *(61)*. In addition, IGFBP-4 has been shown to inhibit the growth of embryonic chick pelvic cartilage in vitro *(76)*. The inhibitory effect of IGFBP-4 on IGF-induced cell proliferation does not appear to be specific to osteoblasts because IGFBP-4 inhibited IGF-induced proliferation in a number of other cell types *(23)*. In contrast to IGFBP-4, studies on the biological actions of IGFBP-5 revealed that IGFBP-5 is a stimulator of IGF-induced cell proliferation in embryonic chick calvarial cells, MC3T3-E1 mouse osteoblasts, and normal human osteoblasts *(37,61)*. Similar potentiating effects of IGFBP-5 from U2 human osteosarcoma cell conditioned medium have been reported by Andress and Birnbaum *(42,77,78)*. Although Kiefer et al. *(79)* have demonstrated that recombinant human IGFBP-5 produced in yeast inhibited DNA synthesis in SaOS-2/B10 human osteosarcoma cells, subsequent studies from the same laboratory have shown that recombinant human IGFBP-5 stimulated DNA synthesis in rat osteoblasts *(80)*. In contrast to IGFBP-5, IGFBP-6 has been shown to be more potent in inhibiting IGF-II-induced osteoblast cell proliferation than IGF-I-induced cell proliferation *(81)*. Based on the findings that IGFBPs modulate IGF actions either positively or negatively, and that the production of IGFBPs is regulated by various physiological agents (*see* the fourth section), it can be speculated that the balance between stimulatory and inhibitory classes of IGFBPs determines the degree and extent of IGF-induced cellular responses in target tissues such as bone.

Although it has been known for several years that IGFBPs modulate IGF actions in osteoblasts and other cell types, not enough is known on the molecular mechanisms by which IGFBPs mediate their effects on osteoblasts. Studies on the molecular mechanisms of IGFBP-4 actions revealed that coincubation of osteoblasts with ^{125}I-IGF-I or ^{125}I-IGF-II decreased the binding of both of these ligands in a dose-dependent manner *(61)*. In addition, IGFBP-4 decreased the binding of IGF-I tracer to purified type I IGF receptor. These data, in conjunction with data showing that IGFBP-4 had no effect on cell proliferation induced by analogs of IGF-I or IGF-II that exhibited >100-fold reduced affinity for binding to IGFBP-4, suggest that IGFBP-4 may inhibit IGF action by preventing the binding of IGF ligand to its membrane receptor (Fig. 2). In contrast to IGFBP-4, IGFBP-5 increased the binding of ^{125}I-IGF-I or ^{125}I-IGF-II tracer to bone cells. Studies on the mechanism by which IGFBP-5 increased the binding of ^{125}I-IGF tracer to bone cells suggest that IGFBP-5 could facilitate IGF binding by a mechanism in which IGFBP-5 has cell surface binding independent of IGF receptors *(61,77,82)*. These data, in conjunction

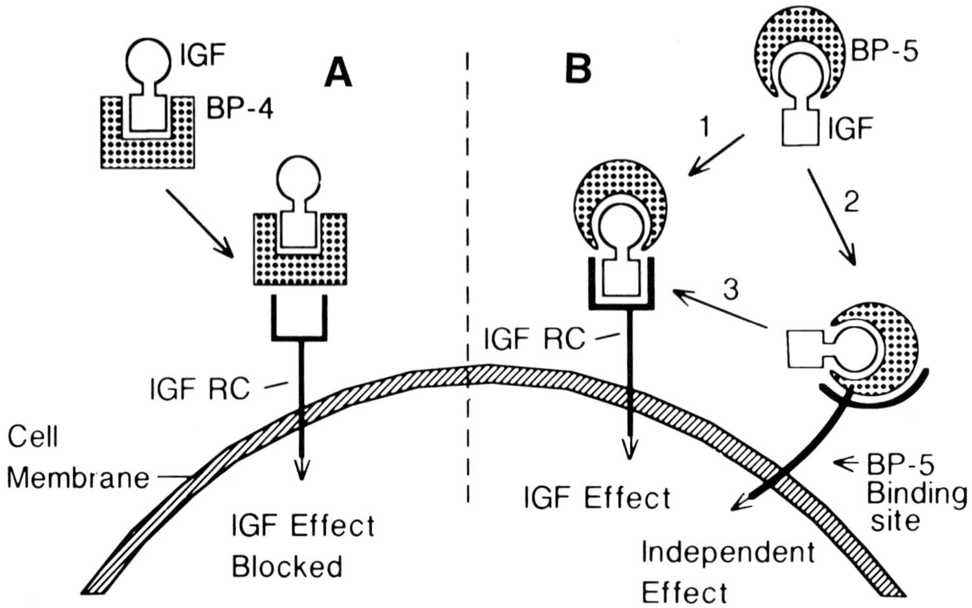

Fig. 2. Proposed models of IGFBP-4 and IGFBP-5 actions. IGFBP-4 inhibits IGF binding to IGF receptors. With regard to IGFBP-5, three alternate models are proposed. In model 1, the complex of IGFBP-5+IGF binds to IGF receptors. In model 2, IGFBP-5 binds to putative IGFBP-5 receptors and stimulates cell proliferation by an IGF-independent pathway. In model 3, IGFBP-5 binding sites may capture and shuttle IGFs at the surface of IGF-responsive cells, enabling the ligand to easily bind to IGF receptors. (Reproduced with permission.)

with the findings that IGFBP-5 potentiated cell proliferation even in the presence of those same IGF analogs that exhibited >200-fold reduced affinity for binding to IGFBP-5, suggest that IGFBP-5 may in part stimulate bone cell proliferation by an IGF-independent mechanism involving IGFBP-5-specific cell-surface binding sites (Fig. 2). In addition, Jones et al. *(23,83)* have proposed that association of IGFBP-5 with proteins on the cell surface or in the extracellular matrix results in an increase in the local concentration of IGFs in the vicinity of the IGF receptor. Association of IGFBP-5 with the cell surface or with extracellular matrix lowers the affinity of IGFBP-5 for IGF, thereby allowing IGF to bind to receptors.

Similar IGF-independent effects have also been reported for IGFBP-3 in a number of cell types *(84,85)*. Consistent with this idea, two recent reports provide evidence for nuclear localization of IGFBP-3 *(86,87)*. In addition, IGFBP-5 contains a nuclear localization signal. The significance of the presence of nuclear localization signal in IGFBP-3 and IGFBP-5 is not known. However, this exciting finding may clarify the direct intrinsic actions of some of the IGFBPs on cells. Thus, the explosion of IGFBP research during the past several years has provided evidence that IGFBPs may have both IGF-dependent and IGF-independent actions. It is anticipated that future studies will provide experimental data needed to verify the IGF-independent actions of IGFBPs.

IGFBP Proteases

IGFBP proteases produced by local tissues appear to play an important role in regulating the actions of IGFBPs in local body fluids. Based on the findings that the IGFBPs modulate IGF actions both positively and negatively, it can be speculated that IGFBP proteases for inhibitory and stimulatory classes of IGFBPs should have opposite effects in regulating IGF actions. For example, activation of IGFBP-4 protease results in the degradation of IGFBP-4 into fragments, which have markedly reduced the affinity for IGFs for IGFBP-4 fragments *(49)*. This accelerates the dissociation of IGFs from IGFBP-4, thereby increasing the proportion of IGFs available to bind to the type I receptors. In contrast, the activation of IGFBP-5 protease may not only reduce the actions of IGFs but may also inhibit the amount of IGFs deposited in bone matrix for future actions *(23,61)*. Consistent with the idea that IGFBP proteases may play a role in regulating local IGF actions, studies in our laboratory and other laboratories have shown that the relative amounts of IGFBP protease produced by osteoblasts vary depending on the skeletal site *(12,53)*. Based on the findings that the production and activity of IGFBP proteases change in response to treatment with osteoregulatory agents, it can be concluded that the breakdown of IGFBPs by IGFBP proteases may be equally as important as synthesis in controlling the actions of IGFBPs in the local body fluids.

In Vivo Effects on Bone Metabolism

IGFs

The importance of IGFs in mediating the stimulatory effect of GH on longitudinal bone growth is well known *(88,89)*. Consistent with the idea that IGFs are important stimulators of bone formation, several animal studies demonstrate that IGFs may be administered systemically to increase bone formation and bone repair *(90–95)*. Froesch and his collaborators *(96)* were the first to demonstrate that both IGF-I and IGF-II increased bone formation in hypophysectomized rats. Subsequent studies have mostly utilized IGF-I for in vivo studies because IGF-I, and not IGF-II, is the primary mitogen in adult bone metabolism in rats and mice. Spencer et al. *(95)* have shown that continuous infusion of IGF-I for 14 d increased both cortical and trabecular bone formation in adult female rats. Consistent with the bone-forming effects of IGF-I in normal rats, studies have shown that IGF-I administration increased bone formation in rats with osteopenia induced by ovariectomy. Kalu et al. *(90)* have shown that daily subcutaneous injections of IGF-I for 5 wk caused partial restoration of femoral calcium and trabecular bone volume with no effect on osteoblasts or osteoclasts. Ammann et al. *(91)* have shown that delivery of IGF-I via an osmotic minipump implanted subcutaneously for 6 wk in ovariectomized rats caused a dose-dependent increase in bone mineral density in the lumbar spine and proximal and mid-shaft tibia. In addition, Machwate et al. *(92)* have shown that continuous infusion of IGF-I enhanced the recruitment of osteoblastic cells, increased trabecular formation, and prevented trabecular bone loss in unloaded rats. More recently, Verhaeghe et al. *(97)* have shown that administration of human growth hormone (GH) and IGF-I to aged ovariectomized rats prevented further loss of bone mass and strength at sites containing trabecular bone. Based on these and other studies, it is clear that systemic administration of IGF-I increases bone formation in both ovary-intact and ovariectomized rats. However, the magnitude of the bone-forming effect of IGF-I appeared to vary in different studies. The apparent reasons for the observed differences in bone-forming effect of IGF-I probably include variations in dose, time, route of administration, intermittent vs sustained administration, and site of bone studied.

If the IGFs are important regulators of bone formation, we should see a deficiency in bone formation in IGF knockout mice, and indeed, this observation has been made in mice lacking a functional IGF-I gene *(98,99)*. Thus, studies in various animal model systems are consistent with an anabolic role for IGF-I in stimulating bone formation.

Table 3
Summary of Clinical Studies on the Effects of IGF-I Systemic Administration on Bone Metabolism in Humans

Study	*Patient population*	*No. of subjects*	*Dose (μg/kg/d)*	*Duration*	*BF*	*BR*
Johansson et al., 1992	Idiopathic osteoporosis	one male	80 × 2	7 d	↑	↑
Ebeling et al., 1993	Healthy postmenopausal	18 Women	30, 60, 120, 180	6 d	↑	↑
Rubin et al., 1994	Postmenopausal and Werner syndrome	1 Woman	30–75	6 mo	↑	↑
Grinspoon et al., 1995	Young and healthy	14 Women	100 × 2	6 d	↑	↔
Johansson et al., 1995	Idiopathic osteoporosis	24 Men	20, 40, 80	6 wk	↑	↑
Ghiron et al., 1995	Healthy elderly	16 Women	15, 60	28 d	↑	↑
Johansson et al., 1996	Idiopathic osteoporosis	12 Men	80	7 d	↑	↑

The promising results on the anabolic effects of IGF-I in several experimental animal models and the recent availability of recombinant human IGF-I have permitted the use of IGF-I in human clinical trials (Table 3). Johansson et al. *(100)* first reported that biochemical markers of bone formation and bone resorption in serum and urine were substantially increased during the treatment of osteoporotic men with rhIGF-I at a dose of 160 μg/kg/d divided into two subcutaneous injections for 7 d. The increase in markers of bone formation occurred rapidly (2–3 wk) and remained elevated for a period of several weeks after the last IGF-I administration. Subsequently, Ebeling et al. *(101)* have shown that daily administration of IGF-I by single subcutaneous injection for 6 d caused a dose-dependent increase in the concentrations of markers of collagen synthesis (procollagen peptide [PCP]) and breakdown (urinary excretion of deoxypyridinoline) in healthy postmenopausal women. These authors demonstrated that the low dose of IGF-I (30 μg/d) increased the level of bone formation and not bone resorption markers. These data confirmed that low doses of IGF-I could selectively stimulate bone formation while avoiding significant side effects associated with high doses of IGF-I (e.g., hypoglycemia, tachycardia).

Subsequent to these initial studies, several trials have been performed investigating the possibility of using rhIGF-I for treatment of osteoporosis (Table 3). The findings from these clinical trials are briefly described below: (1) Grinspoon et al. *(102)* studied young healthy women who received daily subcutaneous injections of 200 μg/kg/d rhIGF-I for 6 d after an initial fasting period of 4 d. In the baseline fasting period bone turnover indices declined. IGF-I treatment significantly increased serum osteocalcin and PCP levels but had no effect on urinary excretion of deoxypyridinoline. The reasons why bone resorption was suppressed in these women who underwent fasting remain to be studied. (2) Johansson et al. *(103,104)* studied the effect of three doses of rhIGF-I (20, 40, or 80 μg/kg/d) on bone turnover markers in 24 men with idiopathic osteoporosis. In this study, it was found that IGF-I administration increased markers of bone turnover dose-dependently in serum and urine. At the 20 μg/kg dose, increase in serum PCP levels remain elevated during the entire 6 wk of injections. In contrast, urine deoxypyridinoline levels were not altered by the low dose of IGF-I. These data demonstrate that a low dose of IGF-I can stimulate bone formation for a sustained period of time in osteoporotic individuals. (3) Ghiron et al. *(105)* evaluated the effects of rhIGF-I on bone turnover markers in 16 healthy elderly women. IGF-I at 60 μg/kg/d caused a significant increase in both bone formation and bone resorption markers. On the other hand, with 15 μg/kg/d IGF-I, no change was observed in resorption markers whereas bone formation markers (osteocalcin, PCP) increased progressively. These data also demonstrate that IGF-I at low doses increased bone formation activity with minimal evidence for increased resorption.

Based on the findings from IGF-I clinical trials in humans (Table 3), a number of conclusions can be drawn: (1) rhIGF-I administration caused a dose-dependent increase in bone turnover in both normal and osteoporotic individuals. (2) The effects of rhIGF-I on bone turnover are sustained over a period of several weeks. (3) Generally higher doses of IGF-I are associated with a high frequency of side effects. (4) Low doses of IGF-I appeared to cause an increase in bone formation but not bone resorption. Based on these preliminary data, further long-term studies utilizing low doses of IGF-I are warranted to evaluate the potential anabolic effects of IGF-I for treatment of osteoporosis. Although IGF-II is the predominant form of IGF produced by human osteoblasts and stored in bone, there have been no studies on the effects of IGF-II administration on bone metabolism in humans. Future studies are needed to address the question of whether IGF-II-induced effects on bone metabolism are qualitatively and quantitatively similar to or different from those of IGF-I.

IGFBPs

Although IGFBPs have been shown to exert important effects on osteoblast cell proliferation in both an IGF-dependent and an IGF-independent manner, very little is known concerning the in vivo effects of IGFBPs on bone formation parameters. The systemic effects of IGFBPs on bone formation has been thus far tested only with IGFBP-3. Bagi et al. *(94,106)* treated ovariectomized rats (8 wk after surgery) with daily subcutaneous injections of IGFBP-3 + IGF-I complex or IGF-I alone for a period of 8 wk. Three doses of IGFBP-3 + IGF-I complexes were tested in this study. The authors found that bone formation parameters were significantly greater at femoral metaphysis and epiphysis in the group treated with the highest dose of IGFBP-3/IGF-I complex (7.5 and 2.6 mg/kg/d) compared to vehicle-treated controls *(94,106)*. In subsequent studies, the same group of investigators showed that the highest dose of IGFBP-3/IGF-1 complex increased body weight, lean body mass, femoral bone mineral content, and bone mineral density as measured by dual energy X-ray absorptiometry (DXA). In addition, the authors showed that the newly formed bone at cortical envelopes exhibited normal lamellar structure and improved mechanical activity. It is not known at this time whether IGFBP-3 mediates its effects on bone formation by increasing the endocrine actions of IGFs and/or by IGF-independent effects *(93,107)*.

In vivo studies on the effects of other IGFBPs on bone formation are lacking owing to unavailability of adequate recombinant proteins for these studies. In regard to the other IGFBPs, there has been only one published report thus far that describes the anabolic effects of IGFBP-5 in mice. Kling et al. *(108)* have recently shown that local application of IGFBP-5 on calvarial bone increased bone mass in mice. In addition, our own recent findings demonstrate that systematic administration of IGFBP-5 increases bone formation parameters in mice (unpublished). Based on the in vitro findings that IGFBPs regulate IGF actions both in a positive and a negative manner and the findings that the

Table 4
In Vitro Effects of Systemic and Local Agents on Cell Proliferation and Expression of Various IGF System Components in Normal Human Osteoblasts

Effector	*Effect on IGF system components*	*Effect on proliferation*
Systemic		
PTH	↑ IGFBP-4	No effect
Glucocorticoid	↓ IGF-I, IGF-II, IGFBP-3, IGFBP-5	Decrease
1-25-Vitamin D_3	↑ IGFBP-4	Decrease
Progesterone	↑ IGF-II, IGFBP-5	Increase
Estradiol	↑ IGFBP-4	Not known
Testosterone	↑ IGF-II receptor	Increase
Local		
Mechanical loading, EMF	↑ IGF-II	Increase
PGE_2 (high doses)	↓ IGF-II, ↑ IGFBP-4	Decrease
Bone growth factors		
IGFs	↑ IGFBP-3 and IGFBP-5 ↓ IGFBP-4	Increase
TGF-β_1	↑ IGFBP-3 and IGFBP-5 ↓ IGFBP-4	Increase
BMP-7	↑ IGF-II, IGFBP-3, and IGFBP-5 ↓ IGFBP-4	Increase

production of IGFBPs is regulated by osteoregulatory agents (*see below*), it is important to examine the in vivo effects of various IGFBPs in the absence and presence of IGFs on bone formation to optimize IGF therapy for osteoporosis.

REGULATION OF IGF SYSTEM COMPONENTS

If IGF system components function to regulate local bone formation, then one might expect that, in general, agents that stimulate bone formation would increase the effective concentration of stimulatory IGF system components and/or decrease the effective concentration of inhibitory IGF system components while agents which inhibit bone formation would have the opposite effects. Studies on bone cell regulation of IGF system components in the conditioned medium in vitro and serum regulation of IGF system components in vivo provide some indirect support for this concept.

Bone Cell Regulation of IGF System Components

Studies on regulation of IGF system components using osteoblasts (derived from humans, rats, and mice) demonstrate that a number of local and systemic factors influence the level of IGFs and IGFBPs in the conditioned medium. These studies also demonstrate that the regulation of expression of various IGF system components is complex, involving both transcriptional and posttranscriptional mechanisms. The findings that the expression of IGF system components in osteoblasts is regulated by a variety of local and systemic agents that control bone metabolism (Table 4) are consistent with the idea that IGF system is one of the most important growth factor systems regulating bone formation.

IGFs

As discussed earlier, IGF-I is the predominant IGF expressed by mouse and rat osteoblasts whereas IGF-II is the predominant IGF expressed by human osteoblasts. Therefore, the majority of studies on the expression of IGF-I have focused on the rat and mouse osteoblast model systems. Many of the major hormones that regulate the skeleton exert significant effects on IGF-I expression. For example, PTH and estradiol have been shown to up-regulate the expression of IGF-I in rat osteoblasts *(109–114)*. PTH has been shown to exert its effects on IGF-I synthesis via increased cAMP production. In addition, recent studies also demonstrate that both PTH and prostaglandin E_2 (PGE_2) (also increases cAMP) increase IGF-I synthesis via increases in IGF-I gene transcription *(115)*. Based on the findings that the stimulation of the alkaline phosphatase (ALP) activity and the expression of osteocalcin mRNA by 6-h intermittent exposure to PTH were almost completely inhibited by the addition of neutralizing antibody to IGF-I, it has been speculated that the anabolic effects of PTH on bone formation in rats may involve increased IGF-I production *(116)*. In contrast to these studies, our studies in human osteoblasts have shown that PTH had no significant effect on the expression of IGF-I *(117)*. Because the actions of IGFs are regulated by a complicated mechanism including interactions with IGFs, type I IGF receptors, and IGFBPs and IGFBP proteases, further studies are needed to determine conclusively the role of IGFs in mediating the anabolic effect of PTH on the skeleton.

Treatment of rat osteoblasts with estradiol increased IGF-I level in the conditioned medium of rat osteosarcoma cells *(113)*. Consistent with this finding, Ernst and Rodan *(112)* reported that 17β-estradiol enhanced IGF-I synthesis at the transcriptional level in rat bone cells transfected with estrogen receptors. In contrast to estrogen, treatment of rat bone cells with glucocorticoids decreased IGF-I synthesis in primary cultures of rat osteoblast cultures *(118,119)*. In addition, we have found that 1,25-dihydroxy vitamin D_3 ($1,25(OH)_2D_3$) decreased IGF-I expression in mouse osteoblasts *(120)*. Thus, although a number of steroid hormones influence IGF-I expression, very little is known about the molecular mechanism by which these steroid hormones differentially modulate IGF-I expression in rat osteoblasts.

In addition to systemic regulators, IGF-I expression in rat and mouse osteoblasts has also been shown to be regulated by locally produced growth factors. For example, interleukin-I treatment has been shown to increase IGF-I expression in mouse organ culture, in part by a prostaglandin-dependent mechanism *(121)*. In addition, Canalis et al. *(8,16,122)* have shown that treatment with basic fibroblast growth factor (bFGF), transforming growth factor-β (TGF-β), and platelet-derived growth factor (PDGF) decrease IGF-I synthesis by reducing IGF-I transcript levels in rat osteoblasts. Based on the findings that IGF-I expression in rat and mouse osteoblasts in vitro is regulated by both systemic and local effectors of bone metabolism *(8,123,124)*, it has been speculated that IGF-I expression in the local bone microenvironment is regulated by complex interplay between systemic and local factors that are present at any given time.

In contrast to the extensive work on IGF-I expression in rodent osteoblasts, very little is known about modulation of IGF-I expression in human osteoblasts, as human osteoblasts in culture produce much less IGF-I compared to IGF-II *(12,13,28,125,126)*. Recent studies have shown that treatment of human osteoblasts with forskolin (a stimulator of cAMP) and TGF-β_1 increased IGF-I mRNA levels as determined by reverse transcription-polymerase chain reaction (RT-PCR) *(125)*. However, it remains to be determined whether the increase in IGF-I mRNA by forskolin and TGFβ_1 treatment is associated with corresponding changes in IGF-I protein levels.

Studies on IGF-II expression in human osteoblasts have shown evidence for regulation by both systemic and local effectors of bone metabolism. Treatment of human osteoblasts with progesterone has been shown to increase IGF-II expression *(127)* while dexamethasone

Table 5
Regulation of IGF and IGFBP Expression in Human and Rat Osteoblasts

Effector	*Human osteoblasts*	*Rat osteoblasts*
Systemic		
PTH	↑ BP-4	↑ IGF-1, ↑ BP-4, ↑ BP-5
Glucocorticoid	↓ IGF-1, ↓ IGF-II, ↓ BP-3, ↓ BP-5	↓ IGF-I, ↓ BP-5
Retinoic acid	↓ BP-5, ↑ BP-6	↑ BP-5
Estradiol	↑ BP-4	↑ IGF-1, ↑ BP-3
Local		
IGFs	↑ BP-3, ↓ BP-4 ↑ BP-5	↑ BP-3, ↑ BP-4, ↑ BP-5
TGF-β_1	↑ BP-3, ↓ BP-4 ↑ BP-5	↓ BP-5
FGF	↓ BP-5	↓ BP-5
BMP-7	↑ IGF-II, ↑ BP-3, ↓ BP-4, ↑ BP-5	↑ BP-3, ↓ BP-4, ↓ BP-5
PGE_2	↑ BP-4	↑ BP-4, ↑ BP-4

treatment has been shown to decrease expression of both IGF-I and IGF-II *(128)*. PGE_2, at doses that elevate intracellular cAMP levels, has been shown to decrease IGF-II expression in human osteoblasts. Consistent with these data, both dibutyryl cAMP and forskolin have been shown to reduce IGF-II expression in human osteoblasts *(44,45)*. Treatment of human osteoblasts with bone morphogenic protein (BMP)-7 has been shown to increase expression of IGF-II *(59)*. Although a number of osteoregulatory agents regulate expression of IGF-II, the molecular mechanisms by which the various systemic and local regulators influence IGF-II gene expression remain to be studied.

Studies on regulation of IGF expression in human and rat osteoblasts by osteoregulatory agents have shown evidence for both similarities and differences between the two species. For example, dexamethasone treatment has been shown to decrease IGF-I expression in both human and rat osteoblasts *(128,129)*. Similarly, BMP-7 treatment produced an increase in IGF-II expression in both human and rat osteoblasts *(59,130)*. On the other hand, TGF-β_1 caused an increase in IGF-I expression in human osteoblasts while it decreased IGF-I expression in rat osteoblasts *(125,131)*. PTH treatment caused an increase in the expression of IGF-I in rat osteoblasts but not in human osteoblasts *(109,117)*. Based on the known differences between regulation of IGF expression by osteoregulatory agents between human and rat osteoblasts (Table 5), one must be careful in interpreting data on IGF expression using selected models.

The only true negative feedback system regulating bone formation is that of mechanical strain, and there is evidence that the effect of mechanical strain to increase bone formation is mediated, at least in part, by the locally produced IGFs *(132,133)*. We have found that treatment of human osteoblasts with a low level electric field (EMF) at a specific frequency (potential mediators of mechanical loading) increased cell proliferation and IGF-II expression, and that the EMF-induced increase in cell proliferation could be blocked by IGF-II neutralizing antibodies but not by osteocalcin antibodies *(134,135)*. Consistent with the EMF effects, Zaman et al. *(132)* have shown that strain caused an immediate release of prostacyclin and increased IGF-II transcript levels. This finding, together with the previous finding that prostacyclin increased IGF-II levels in the conditioned medium of bone cells *(133)*, suggests a possible role for prostacyclin and IGF-II in the early strain-related response of load-related bone modeling/remodeling.

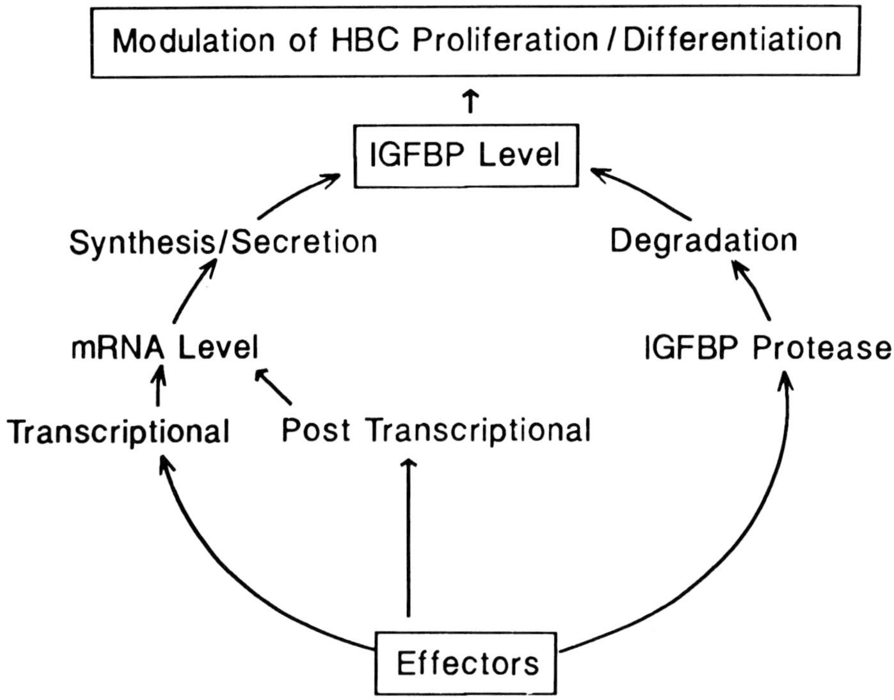

Fig. 3. Model of potential control points for modulation of IGFBP concentration in extracellular fluid by systemic and local effectors. The effect of a given IGFBP on local IGF action will depend on its concentration or level which is determined by (1) the synthesis/secretion rate of IGFBP and (2) the clearance which includes the extracellular degradation of the IGFBP. A given effector may modulate the synthesis/secretion at multiple regulatory sites, including transcriptional as well as posttranscriptional mechanisms. In addition, the given effector may modulate degradation by influencing the production and/or activity of the corresponding IGFBP protease.

IGFBPs

Studies during the past few years on regulation of IGFBP levels in the conditioned medium of osteoblasts derived from humans, rats, and mice demonstrate that a number of systemic and local factors influence the concentration of various IGFBPs in the conditioned medium. Because IGFBPs exert both stimulatory and inhibitory effects on osteoblasts, it is essential that the production of stimulatory and inhibitory classes of IGFBPs is differentially regulated by osteoregulatory agents to produce an optimal effect. Consistent with this hypothesis, studies on the regulation of IGFBPs in osteoblasts provide evidence for not only differential regulation of various IGFBPs in osteoblasts but also for complex regulation of IGFBP levels in the conditioned medium by mechanisms involving both synthesis and degradation (Fig. 3). Studies on regulation of IGFBP levels in osteoblast conditioned medium by systemic and local effectors of bone metabolism are summarized below.

Systemic Effectors. PTH, a systemic calcium-regulating hormone, increased production of IGFBP-4 both at the protein and mRNA levels in normal human osteoblasts and in human and rat osteosarcoma cells *(35,43,136)*. PTH has been shown to exert its effects on osteoblasts by intracellular mechanisms involving activation of protein kinase-A (PKA) and/or protein kinase-C (PKC) pathways. Of these two pathways, a number of findings suggest that the PTH effect on IGFBP-4 production is mediated by a cAMP-dependent PKA pathway: (1) Agents that stimulate cAMP production (e.g., forskolin, PGE_2, isobytylmethyl xanthine), as well as dibutyryl cAMP, increased IGFBP-4 production, whereas agents that stimulate PKC activity (e.g., PMA) had no significant effect on IGFBP-4 production *(44,137),* and (2) dibutyryl cAMP increased IGFBP-4 gene transcription and IGFBP-4 gene promoter activity *(138).* The PTH-induced increase in IGFBP-4 production has been speculated to play a role in the pathogenesis of senile type II osteoporosis (*see* the sixth section).

The effect of PTH on IGFBP-5 production has not been established in human osteoblasts. However, studies in primary cultures of rat osteoblasts and in UMR106 rat osteosarcoma cells have revealed that PTH also induces IGFBP-5 mRNA and protein expression by a cAMP-dependent mechanism *(43,139).* In these bone cell cultures, the PTH effect on IGFBP-5 protein accumulation in the conditioned medium appears to be mediated by two mechanisms: (1) PTH increased IGFBP-5 mRNA expression and (2) the PTH-induced increase in IGF-I decreased IGFBP-5 proteolysis *(9,43).* Further studies are needed to evaluate if PTH-induced changes in the types (inhibitory and stimulatory) and amounts of IGFBPs contribute to the anabolic (intermittent administration) vs catabolic (prolonged increase in PTH during secondary hyperparathyroidism) effects of PTH.

1,25-Dihydroxy vitamin D_3, another major calcium-regulating hormone, has been shown to increase IGFBP-4 production both at the mRNA and the protein level and decrease cell proliferation in MC3T3-E1 mouse osteoblasts and MG63 human osteosarcoma cells *(140,141).* Consistent with these in vitro findings, treatment of human subjects with oral 1,25-dihydroxyvitamin D_3 for psoriasis resulted in a significant increase in serum IGFBP-4 concentration compared to pretreatment levels *(141).* These observations suggest that 1,25-dihydroxy vitamin D_3 plays an important role in the regulation of IGFBP-4 secretion in vitro and in vivo.

In addition to calcium-regulating hormones, a number of steroid hormones have also been shown to influence production of IGFBPs. Estradiol treatment of human osteoblasts overexpressing, the estradiol receptor caused an increase in IGFBP-4 expression which has been speculated to mediate the inhibitory effects of estradiol on bone formation *(142).* In contrast to estradiol, treatment of human and rat osteoblasts with glucocorticoids decreases expression of a number of IGFBPs, including IGFBP-3 and IGFBP-5 *(143–145).* In addition, retinoic acid treatment increased expression of inhibitory IGFBPs (e.g., IGFBP-4 and IGFBP-6) and decreased expression of stimulatory IGFBPs (e.g., IGFBP-5) in human osteoblasts *(146,147).* Both glucocorticoid and retinoic acid treatment decreased osteoblast cell proliferation. In contrast to these agents, progesterone treatment stimulated cell proliferation and increased IGFBP-5 mRNA and protein levels in both normal human osteoblasts as well as human osteosarcoma cells *(127,148).* The potential role of various IGFBPs in mediating the effects of various steroid hormones on osteoblast cell proliferation and differentiation require further investigations.

Local Effectors. Studies from a number of laboratories have provided evidence regarding the regulation of IGFBPs by local skeletal factors such as IGFs, TGF-β, BMPs, FGF, PDGF, and interleukins *(51,59,131,140,149–151).* Treatment of human osteoblasts with IGF-II decreased intact IGFBP-4 levels, primarily by a mechanism involving increased IGFBP-4 proteolysis *(38,51).* In contrast, IGFs increased IGFBP-5 levels via both increased expression and decreased proteolysis *(12,38,51).* Treatment of human osteoblasts with both TGF-β and BMP-7 increased expression of IGFBP-3 and IGFBP-5 and decreased expression of IGFBP-4 (Fig. 4). The expression of IGFBPs by osteo-

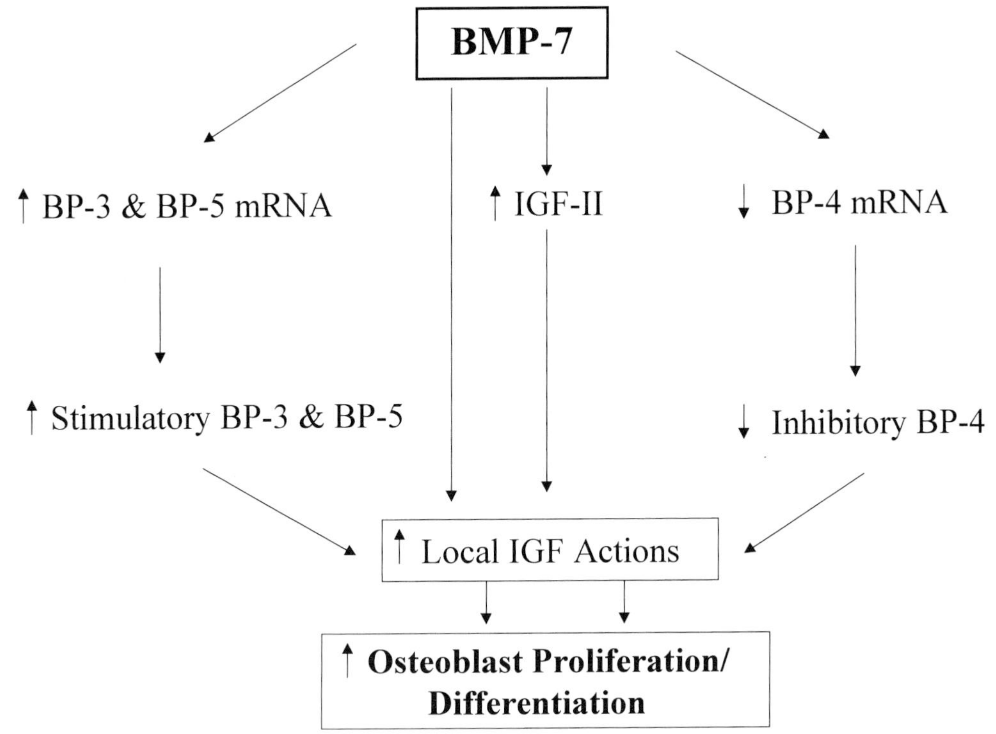

Fig. 4. Regulation of IGF system components in human osteoblasts by BMP-7. Multifactorial model to explain the mechanism by which a local growth factor (e.g., BMP-7) may in part mediate its effects on the proliferation and differentiation of osteoblast line cells.

blasts has also been shown to be under the control of cytokines. Both TNF-α and IL-Iα increased IGFBP-4 expression in osteoblasts *(140,150)*. Studies on in vitro regulation of IGFBP expression have provided evidence that, in general, agents that increase osteoblast cell proliferation increase expression of stimulatory IGFBPs and decrease expression of inhibitory IGFBPs, whereas agents that inhibit cell proliferation exhibit opposite effects. Although these data emphasize the possibility that the IGFBPs could play a role in mediating the effects of local growth factors, further studies are needed to establish the cause and effect relationship between changes in growth factor-induced IGFBP expression vs osteoblast cell proliferation.

Studies on the regulation of IGFBP expression in human and animal model systems also reveal differences between osteoblasts derived from various species. For example, retinoic acid treatment increased IGFBP-5 expression in rat osteoblasts while it decreased IGFBP-5 expression in human osteoblasts *(147,152)*. BMP-7, on the other hand, decreased IGFBP-5 expression in rat osteoblasts while it increased IGFBP-5 expression in human osteoblasts *(59,130)*. The potential significance of these differences in IGFBP expression between human and rat osteoblasts and the molecular mechanisms that contribute to these differences remain unknown.

IGFBP Proteases

The findings that human osteoblasts in culture produce proteases capable of degrading one or more IGFBPs and that the production and/or activity of IGFBP proteases can be regulated by a number of osteoregulatory agents raise interesting possibilities for the involvement of IGFBP proteases in regulating the local actions of IGFs. Among the important modulators of IGFBP protease activity are the IGFs themselves. Treatment of human osteoblasts with IGFs increased IGFBP-4 proteolysis *(38,52)*. This coordinated regulation of IGFBP-4 proteolysis by IGFs would enhance the biological activity of locally produced IGFs. Studies on the mechanisms by which IGFs increase IGFBP-4 proteolysis revealed that addition of IGFs to cell-free conditioned medium increased IGFBP-4 proteolysis *(51,53)*. Because IGFBP-4 proteolysis is not induced by the addition of insulin, des(1–3)IGF-I, or des(1–6)IGF-II, which bind IGFBP-4 at extremely low affinity, it has been proposed that the binding of IGF to IGFBP-4 may enhance the susceptibility of IGFBP-4 to proteolytic degradation. In contrast to the effects of IGFs, TGF-β treatment of human osteoblasts enhanced IGFBP-4 protease activity by a mechanism involving increased IGFBP-4 protease expression and/or secretion in human osteoblasts *(54)*. IGFBP-4 protease may also be regulated by inhibitors produced by osteoblasts *(153)*. These data suggest that IGFBP-4 proteolysis in the osteoblast cell microenvironment may be controlled by complex mechanisms involving synthesis of proteases as well as activators and inhibitors of IGFBP-4 protease. The addition of both IGF-I and IGF-II to osteoblasts greatly attenuated the degradation of IGFBP-5 *(38,51)*. Based on the findings that exogenous addition of IGFs to osteoblast cell conditioned media had no effect on IGFBP-5 proteolysis, it has been speculated that the mechanisms by which IGFs regulate IGFBP-4 vs IGFBP-5 proteolysis are different. In addition to IGFs, the proteolysis of IGFBP-5 is also influenced by BMP-7 treatment and during osteoblast differentiation *(59,60)*. The potential mechanism by which IGFBP proteases may act to regulate the actions of IGFs in bone microenvironment is discussed in the fifth section.

Two general conclusions can be drawn based on our current understanding of the regulation of IGF system components in osteoblasts by various osteoregulatory agents: (1) the regulation of IGF system by osteoregulatory agents appears to be well coordinated and (2) the redundancy of IGF system component regulation may provide the necessary means to serve the multiple effector inputs that regulate bone formation and bone resorption processes.

Serum Regulation

IGFs

A number of recent studies demonstrate that the GH–IGF axis plays an important role in maintenance of bone mass in adult as well as longitudinal growth of bone in childhood. Consistent with a role for systemic circulating IGFs in the regulation of bone formation, several cross-sectional studies have demonstrated a strong linear relationship between serum IGF-I and bone mineral density (BMD) in postmenopausal women *(154–158)*. The association between serum levels of IGF-I and BMD remained significant even when age and body size were held constant. Reed et al. *(159)* have shown that men and women with idiopathic osteoporosis have significantly reduced serum IGF-I concentrations compared to age-matched controls, and that serum IGF-I levels showed significant positive correlations with the extent of osteoblastic surface determined from bone biopsy analyses. Johansson et al. *(160)* have recently reported that men with idiopathic osteoporosis exhibited significant negative bone balance caused by pronounced decrease in wall thickness (representing decreased bone formation) compared to age-matched controls. The men with idiopathic osteoporosis also had low serum levels of IGFBP-3 compared to age-matched controls. In addition, studies of GH-deficient patients have revealed that serum

IGF-I concentrations are closely correlated with reduced bone density, and that GH replacement increases both serum IGF-I and bone mineral density. These data suggest an important role for circulating IGF-I in the maintenance of adult bone mass.

Although much is known regarding serum levels of IGF-I and BMD, studies on the relationship between serum levels of IGF-II and BMD are scarce owing to difficulty in developing valid assays for measurement of IGF-II without the binding protein interference. By using a Bio-Spin protocol for complete separation of IGFs from IGFBPs, we have found that the serum IGF-II levels were significantly lower in the osteoporotic women compared to age-matched normal women, and that the correlation between serum levels of IGF-II and BMD remained significant, even when age was held constant *(155)*. Although the role of IGF-II in the pathogenesis of bone loss is not well established, most of the published work on IGF-I provides indirect evidence for a role of circulating IGFs in bone metabolism.

IGFBPs

In serum, most of the IGFs (about 75–80%) circulate as a 150–200-kDa complex while the remaining IGFs circulate as a 50-kDa complex (about 20–25%). Only a small amount of IGFs (<1%) circulate as a free form *(161)*. The 150–200-kDa IGF complex consists of 7.5-kDa IGF-I or IGF-II plus 38–43-kDa IGFBP-3 and the 80–90-kDa non-IGF binding acid-labile component termed acid-labile subunit (ALS). The 50-kDa complex consists of IGF-I or IGF-II bound to one of the remaining five IGFBPs. Comparison of the relative amounts of IGFBP-4 and IGFBP-5 in adult human serum with the relative abundance of other known IGFBPs revealed that the levels of these two IGFBPs are higher than the levels of IGFBP-1, IGFBP-2, and IGFBP-6, but less than 15% of the IGFBP-3 level *(161)*. There is a 50% molar excess of IGFBPs over IGFs in serum, which implies that only a small percentage of IGFs remain in the free form.

If the inhibitory and stimulatory IGFBPs act in a reciprocal manner to regulate IGF actions, it would be important for these two classes of IGFBPs to be regulated by different mechanisms to produce a complementary effect in response to changes in physiological and pathological conditions. Otherwise, parallel changes in stimulatory and inhibitory classes of IGFBPs would cancel each other out. Consistent with the idea that different mechanisms may regulate serum levels of inhibitory and stimulatory classes of IGFBPs, we and others have found that serum levels of stimulatory IGFBP-3 and IGFBP-5 show a significant positive correlation with both IGF-I and IGF-II concentrations, while serum levels of inhibitory IGFBP-2 and IGFBP-4 showed a weak negative correlation or no correlation *(72,136,161–163)*.

Recent studies have examined the relationship between serum levels of various IGFBPs and BMD. In a recent report, Sugimoto et al. *(157)* have found a significant negative correlation between serum levels of inhibitory IGFBP-2 and BMD of lumbar spine, femoral neck, and mid-radius. In contrast to IGFBP-2, a number of studies have shown significant positive relationship between serum levels of IGFBP-3 and BMD at various skeletal sites. The relationship between serum levels of IGFBP-3 and BMD at mid radius remained significant even when age and body size were held constant. The clinical utility of changes in serum levels of these IGFBPs to predict osteoporosis remains to be established.

In vitro studies on regulation and actions of IGFBP-4 and IGFBP-5 in osteoblasts are consistent with an important role for these two IGFBPs in the regulation of bone formation.

Table 6
Changes in IGF System Component Levels During Various Physiological and Pathological Situations and Associated Changes in Bone Formation

Metabolic condition	*Changes in serum IGF system component level*	*BF change*
Puberty	↑ IGF-1, ↑ IGF-II, ↓ BP-1, ↑ BP-3, ↑ BP-5	↑ BF
Aging	↓ IGF-1, ↓ IGF-II, ↑ BP-1, ↑ BP-2, ↓ BP-3, ↑ BP-4, ↓ BP-5	↓ BF
Growth hormone deficiency	↓ IGF-1, ↓ IGF-II, ↓ BP-3, ↓ BP-4, ↓ BP-5	↓ BF
Glucocorticoid excess	↓ IGF-I, ↓ IGF-II, ↓ BP-3, ↓ BP-5	↓ BF
Chronic renal failure	↑ BP-1, ↑ BP-2, ↑ BP-4, ↑ BP-6	↓ BF

If IGFBP-4 and IGFBP-5 function to regulate local bone formation by modulating IGF actions, then one might expect that agents that stimulate bone formation would increase the effective concentration of IGFBP-5 and/or decrease the effective concentration of IGFBP-4, while agents that inhibit bone formation have the opposite effects. In addition, we anticipate that serum levels of IGFBP-4 and IGFBP-5 would show considerable changes in response to clinical disease states affecting bone metabolism. Our recent findings on the perturbations in serum levels of IGFBP-4 and IGFBP-5 in three different contexts (i.e., changes during aging, osteoporosis, and GH treatment), as discussed later, are consistent with the idea that IGFBP-4 and IGFBP-5 are important components of the IGF system in bone (Table 6).

Changes During Aging. We found that serum IGFBP-4 levels showed an age-related increase in pooled data from 102 men and women between ages 23 and 87 yr ($r = 0.54$, $p < 0.001$). The mean serum IGFBP-4 level was 35% higher in the 61–87-yr age group compared with the 23–40-yr age group *(162)*. In contrast to IGFBP-4, the serum levels of IGFBP-5 showed an age-related decrease ($r = -0.48, p < 0.001$). The mean serum level of IGFBP-5 was 35% lower in the 61–87 yr age group compared with the 21–40-yr age group. Because serum level of the stimulatory IGFBP decreased with age while levels of the inhibitory IGFBP increased with age, we determined the molar ratio of IGFBP-5 to IGFBP-4 in young and old subjects. Changes in this ratio are a measure of the combined growth-promoting actions of these two IGFBPs. The molar ratio of IGFBP-5 to IGFBP-4 declined significantly from 1 to 0.54 with advance in age, which suggests that the endocrine IGF actions may decrease with age.

Because multiple tissues contribute to circulating serum levels of IGFBP-4 and IGFBP-5, we evaluated if similar age-related changes in IGFBP levels occur in bone. As a means of testing this hypothesis, we measured skeletal content of IGFBP-5 in human femoral cortical bone of 44 men and 20 men between ages 20 and 64 yr *(164)*. We found that the skeletal levels of IGFBP-5 decreased by 28% between the ages of 20 and 29 yr and 54 and 64 yr *(164)*. Moreover, we found that IGFBP-5 levels in both serum and bone substance showed significant positive correlations with IGF-I and IGF-II *(164,165)*. Based on these data, we speculate that the decrease in the amount of stimulatory IGFBP-5 in serum and skeleton (i.e., as skeletal IGFBP-5 is released during bone resorption, it could promote bone formation at the remodeling site) could contribute, in part, to the age-related impairment in bone formation.

Changes in Osteoporotic Subjects. To determine if changes in serum level of IGFBP-4 and IGFBP-5 could be involved in the pathogenesis of age-related osteoporosis, we measured circulating concentrations of IGFBP-4 and IGFBP-5 in 50 elderly women with femoral neck fractures and 50 healthy age-matched controls *(166)*. Although the mean serum IGFBP-4 level did not differ significantly between the control and fracture groups, an age-associated increase in serum IGFBP-4 was found in the fracture but not in the control group. In contrast to serum IGFBP-4 levels, the mean serum IGFBP-5 level was reduced by 45% ($p < 0.001$) in the fracture group compared to the control group. In addition, serum levels of IGFBP-5 were found to be predictive of serum osteocalcin and femoral neck bone density when pooled data from normal and osteoporotic populations were analyzed by age-adjusted multiple regression models. These data are consistent with the hypothesis that changes in the concentration of IGFBP-5 may cause an impairment in bone formation in osteoporotic subjects.

We also evaluated serum levels of IGF system components in 19 patients with trochanteric fracture of the proximal femur and 19 age-matched controls. We found that in trochanteric fracture patients compared to age-matched controls, serum levels of IGF-I, IGF-II, IGFBP-3, and IGFBP-5 were decreased, while serum levels of PTH and IGFBP-4 were increased. Partial correlation analysis showed that serum levels of stimulatory IGF system components correlated strongly with trochanteric BMD after adjusting for age, bone mineral index (BMI), PTH, and 25(OH)D *(166–168)*. Although these findings are consistent with the hypothesis that age-related bone loss is in part due to a deficit of IGF-stimulatory components, intervention studies with stimulatory IGF system components are required to demonstrate the cause-and-effect relationship between changes in serum levels of IGF system components and bone loss.

Changes During GH Treatment. A number of in vitro and in vivo studies have demonstrated an important effect of GH on the regulation of bone formation *(88,89,169)*. The finding that GH deficiency in childhood and adolescence results in retarded growth and that treatment of children with GH deficiency with GH increases growth supports an important role for GH in bone metabolism. To test the hypothesis that GH mediates its effect on bone not only by up-regulating the production of IGF-I but also by increasing the production of IGFBP-5, we measured serum levels of IGF system components before and after GH treatment in GH deficient children *(170)*. We found that serum levels of IGFBP-5 were significantly increased compared to pretreatment values during the entire 12-mo treatment period. Moreover, the increases in serum IGFBP-5 levels showed significant positive correlation with the increase in serum levels of bone alkaline phosphatase, thus suggesting that the GH-induced increase in serum IGFBP-5 level may, in part, mediate the anabolic effects of GH.

MODEL OF IGF ACTIONS IN THE REGULATION OF BONE METABOLISM

The IGFs are unique regulators of bone metabolism in that they can act both systemically and locally to regulate bone turnover. Regarding systemic regulation, several studies have shown that systemic administration of IGF-I, which increases serum levels of IGF-I, causes a marked increase in both bone formation and bone resorption in both animals and in humans *(10,171)*. In addition, Bagi et al. *(94,172)* have shown that administration of IGF-I + IGFBP-3 binary complex is more effective than IGF-I alone in stimulating bone formation parameters in ovariectomized rats. The findings that systemic administration of IGFs exhibit significant effects on the skeleton, and that a large readily available source of IGFs is available in blood, are consistent with the possibility that IGFs act systemically in an endocrine manner. In addition, IGFs produced by bone cells have been shown to act locally. These actions of IGFs are summarized below.

Endocrine IGF Actions

Circulating IGFs may serve to promote general growth in various tissues, including bone. The following examples provide support to the endocrine effects of IGFs on bone. First, Rosen et al. *(173)* have recently demonstrated that a mouse strain, C3H/HeJ, with high BMD has 30% greater serum IGF-I level compared to another mouse strain, C57Bl/6J, with low BMD. These studies demonstrate a significant positive correlation between BMD and serum IGF-I level. These data, together with findings from several cross-sectional studies that show a strong positive linear relationship between serum IGF-I and BMD in postmenopausal women. Second, we and others have found that increases in serum levels of stimulatory IGF system components during GH therapy and during puberty correlate with increases in bone formation markers, emphasizing that circulating IGFs could serve as potential regulators of bone formation *(170)*. Third, acute treatment of chronic obstructive pulmonary disease patients with prednisone resulted in acute decreases in bone formation, as reflected by progressive decreases in serum osteocalcin and procollagen peptide, which returned to normal 7 d after withdrawal of prednisone therapy *(174)*. We also found that changes in prednisone-induced bone formation were associated with significant reductions in serum levels of IGF-I, IGF-II, IGFBP-3, and IGFBP-5. Multiple regression analysis between serum osteocalcin and stimulatory IGF system components gave an r value of 0.5, suggesting that at least some of the effects of glucocorticoids to depress bone formation may be mediated, in part, by the effect of glucocorticoids to depress serum levels of the stimulatory IGF system components.

Because the majority of IGFs circulate in serum as IGF-IGFBP complexes, there must be mechanisms by which IGFs in the circulation could be made bioavailable to local target tissues such as bone *(72,161)*. One of the potential control mechanisms for regulation of endocrine IGF actions is by IGFBP proteases. In this regard, it has been speculated that IGFBP proteases produced by vascular endothelial cells or IGFBP proteases present in the circulation may disrupt the ALS–IGFBP–3-IGF ternary complex and release IGFs (Fig. 5). IGFs thus released may be transported into the tissue space or may bind to other IGFBPs such as IGFBP-4 and cross the vascular endothelium. The binding of IGFs to small molecular weight IGFBPs may protect IGFs from degradation and may also increase the half-life of IGFs in the circulation. Upon transport to the target tissues such as bone, IGFBP-4 protease produced by target cells may release IGFs to bind to IGF receptors to produce a growth-promoting response. Based on the findings that a number of local growth factors (e.g., TGF-β) can regulate IGFBP-4 proteolysis *(54)*, it can be speculated that tissue-specific regulation of IGFBP proteolysis may provide a mechanism to increase site-specific bioavailability of serum IGFs depending on the local needs. Although there are no experimental data available to support the hypothesis that IGFBP proteases act to regulate IGF bioavailability, the findings that the activity of IGFBP proteases change during various physiological and pathological situations are consistent with a role for IGFBP proteases in modulating IGF action.

Local IGF Actions

IGFs are also available to skeletal tissues through *de novo* synthesis by osteoblasts and osteoclasts and by release of stored growth factors from bone matrix during osteoclastic bone resorption *(5,161,175–177)*. These IGFs can act in an autocrine (IGF system components produced by one cell type act on the same line of cells which produces them) and/or paracrine manner (IGF system components produced by one cell type act on a different cell type). These actions of IGFs are described below.

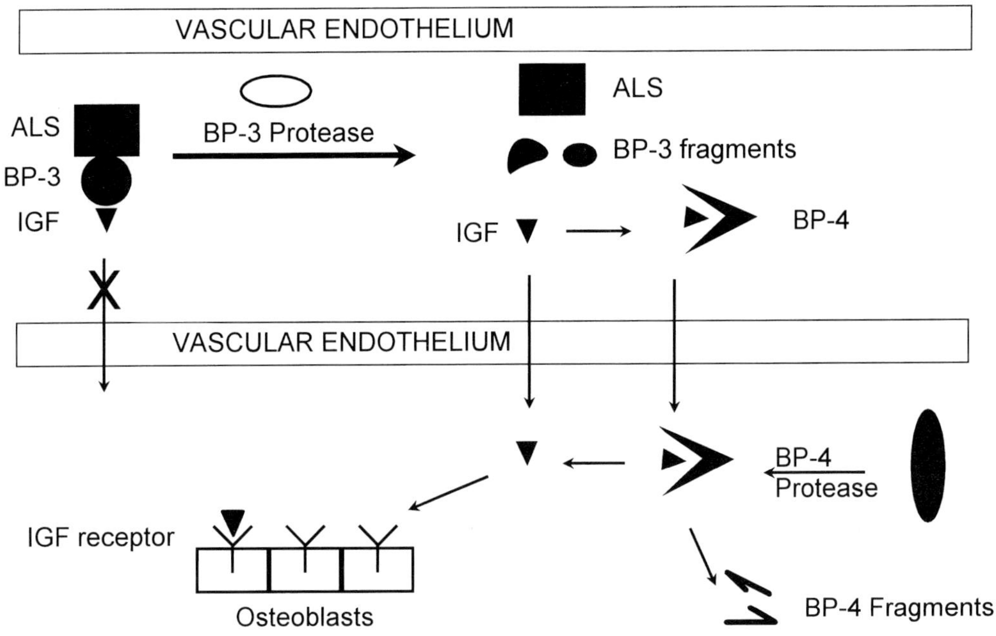

Fig. 5. Potential role of IGFBP proteases in modulating IGF bioavailability to local tissues. Proteolysis of IGFBP-3 by protease may result in disruption of this complex and release of IGFs which may be transported into the tissue space in a free form or may bind to other IGFBPs such as IGFBP-4 and cross the endothelium into the tissue space as an IGFBP-4/IGF complex. Upon transport into the tissue space, the IGFBP-4 protease produced by target cells may release the IGFs to bind to the IGF receptor and elicit growth-promoting responses. (With permission.)

The quantity and quality of growth factors released from bone and produced by contemporary osteoblasts probably represent one of the key determinants of the rate of bone formation at the remodeling site. In this regard, our past studies have shown that IGFs, the most abundant growth factors in bone, are fixed in bone by means of IGFBP-5 which binds to hydroxyapatite and to certain extracellular matrix proteins *(161)*. In regard to the functional significance of the IGFs and IGFBP-5 fixed in bone, we have advanced the concept that sequestration of IGFs and IGFBP-5 could provide a mechanism whereby osteoclastic bone resorption during the remodeling phase would automatically be followed by a formation phase (i.e., coupling of bone formation to resorption) by virtue of the release of IGFs and stimulatory IGFBP-5 from bone substance by osteoclastic bone desolution (Fig. 6). Thus the actions of bone matrix-released IGFs and IGFBP-5 would allow for formation of new bone by a delayed paracrine mechanism. Consistent with the idea that growth factors in bone may have a functional role, recent studies have shown that the concentration of growth factors and their binding proteins in bone vary during disease states *(164,165,178,179)*.

Another source of growth factors are those that are released from osteoclast and osteoblast line cells (Fig. 7). In this regard, the growth factors produced by osteoclasts may act on osteoblast line of cells in a paracrine manner. In addition, growth factors released by contemporary osteoblasts are thought to expand the osteoblastic cell pool, thereby ensuring that adequate quantities of bone will be formed. IGFs produced by osteoblasts may act on themselves in an autocrine manner or on nearby stromal cells in a paracrine manner. The finding that osteoblast cell production of IGF system components

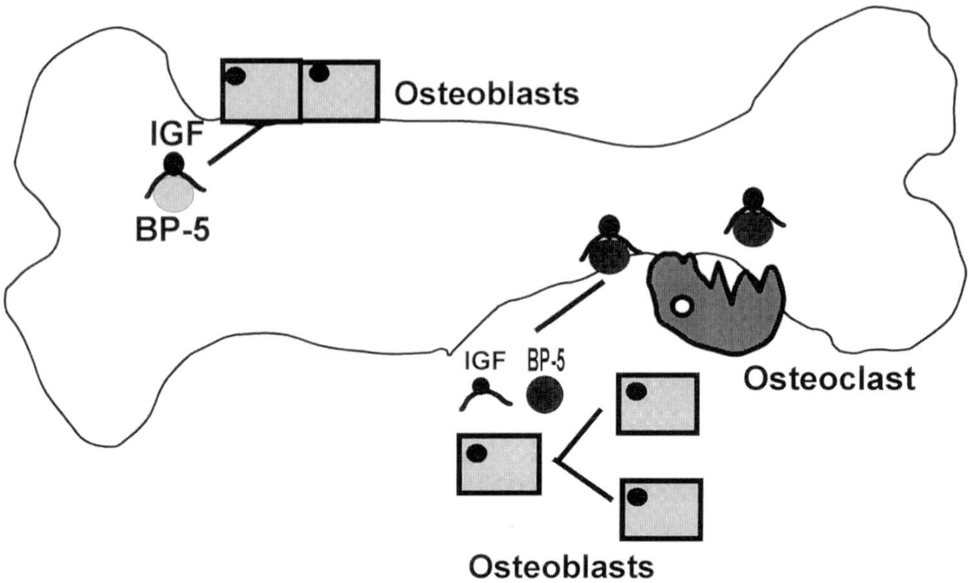

Fig. 6. Model illustrating the delayed paracrine actions of IGFs fixed in bone by IGFBP-5. Osteoclastic bone resorption of an area of bone releases IGFs and the corresponding IGFBP-5, both of which can act on osteoblast line cells to ensure site-specific bone replacement.

is regulated at both systemic and local levels suggest that IGFs produced by osteoblasts are important local regulators of bone metabolism. The following example provides experimental evidence as to how a systemic hormone can modulate bone formation via modulating local production of growth factors. Elevated glucocorticoids have been shown to reduce the mean wall thickness of trabecular bone *(119)*. Mean wall thickness is a bone formation parameter that reflects the total amount of bone formed during a remodeling cycle on trabecular bone. In conditions associated with glucocorticoid excess, the decrease in mean wall thickness leads to an underfilling of the resorption bay and thus a net negative balance of bone at each remodeling site. In vitro evidence shows that glucocorticoids decrease the production of IGF-I, IGF-II, and stimulatory IGFBPs in isolated osteoblasts *(143,145)*. In addition, as discussed earlier, treatment of chronic obstructive pulmonary disease patients with glucocorticoid causes generalized down-regulation of serum stimulatory IGF system components *(174)*. Based on these studies, it can be speculated that glucocorticoids may decrease bone formation by not only decreasing the local osteoblast cell production of IGFs and stimulatory IGFBP-5, but also by decreasing the endocrine actions of IGFs.

The only true negative feedback system regulating bone formation is that of mechanical strain, and there is evidence that the effect of mechanical strain to increase bone formation is mediated, at least in part, by the IGF system. In this regard, studies in our laboratory and in other laboratories provide evidence for locally produced IGF-II in mediating the effects of mechanical loading as discussed earlier *(135)*. In addition, we and others have found that 10 min of high-intensity exercise increased circulating serum levels of IGF-I, IGF-II, and IGFBP-3 within 10 min after exercise *(180,181)*. The mechanism by which mechanical loading causes an acute increase in the serum levels of IGF system components remains unknown at this time. In any case, these

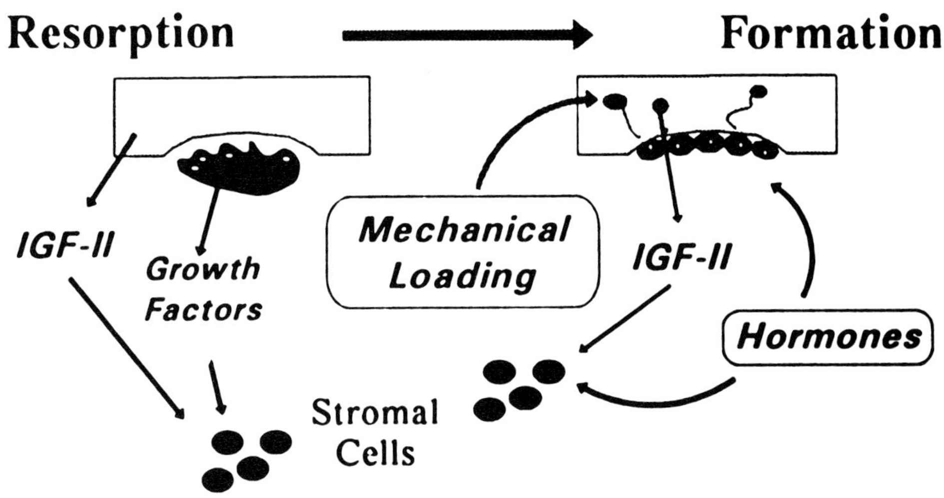

Fig. 7. Models of IGF actions in bone. The stimulation of bone formation during the coupling of bone formation to resorption by IGFs may be regulated by three mechanisms: (1) IGFs released from bone during osteoclastic bone resorption may function as a delayed paracrine agent to act on osteoblast-line cells to ensure site-specific bone replacement. (2) IGFs produced by osteoclasts may act as paracrine agents to create a population of bone-forming cells in proportion to the volume of bone resorbed. (3) IGFs produced by osteoblasts may act as autocrine agents and regulate bone resorption cavity refill. Systemic (e.g., PTH, 1,25-dihydroxyvitamin D_3) and local (mechanical loading, growth factors) agents may mediate their actions in part by modulating osteoblast cell production of the various IGF system components.

findings are consistent with the idea that the action of exercise to increase bone formation may involve increase in both the endocrine and local actions of growth factors, especially IGF-II (Fig. 7).

If the IGFs (systemic and local) are important regulators of bone formation, we should see a deficiency in bone formation in clinical situations where IGF production is altered. Indeed, it has been found that patients with GH deficiency exhibit low BMD at various skeletal sites and that GH replacement therapy led to an increase in net bone turnover and a net increase in bone BMD *(88,182–185)*. In addition, bone formation is significantly impaired in mice lacking functional IGF-I gene *(98,99)*. Although there is a large body of evidence that suggests that the IGF system is an important growth factor system in bone, the extent to which the various IGF system components contribute to regulate bone formation remains to be studied.

POTENTIAL ROLE OF IGF SYSTEM COMPONENTS IN THE PATHOGENESIS OF SENILE OSTEOPOROSIS

Based on certain in vitro and in vivo findings, we have proposed a model that implicates the age-related changes in the IGF system components and vitamin D-PTH axis in the pathogenesis of type II senile osteoporosis. According to this model (Fig. 8), the decrease in circulating levels of IGF-I, IGFBP-3, IGFBP-5, and, to a lesser extent IGF-II, that occurs with age might result in decrease in osteoblast cell proliferation, not only by reducing skeletal exposure to endocrine IGF actions, but also by inhibiting skeletal synthesis of stimulatory IGFBP-5. In vitro, both IGF-I and IGF-II have been found to be

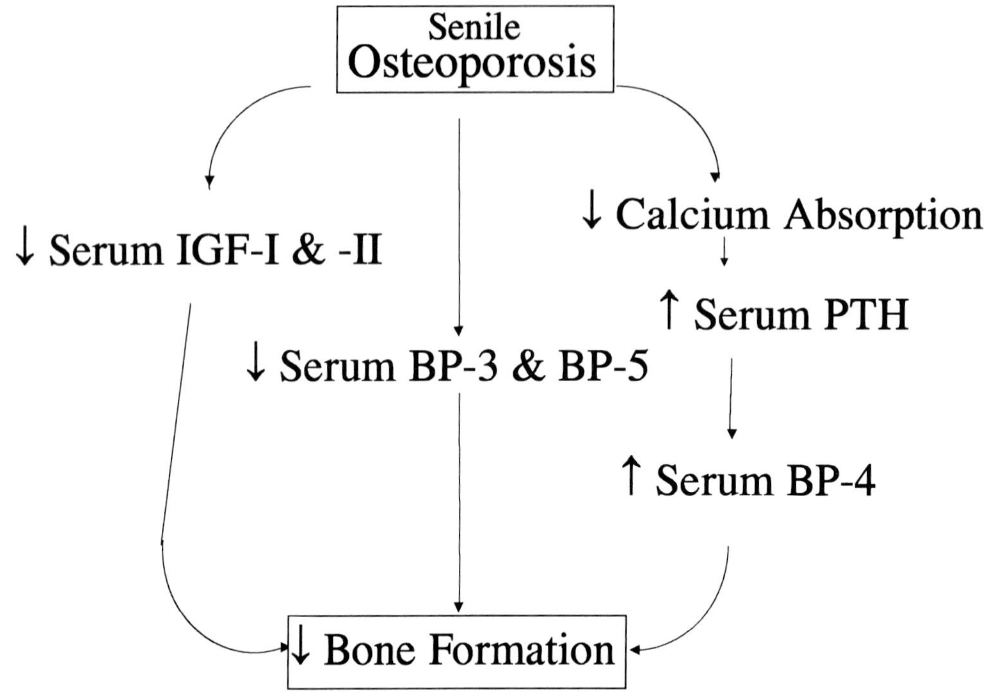

Fig. 8. Model illustrating how age-related changes in IGF system components could lead to impairment of bone formation and contribute to the pathogenesis of senile osteoporosis. Senile osteoporosis is a disease characterized by low bone mass and microarchitectural deterioration of bone tissue leading to enhanced bone fragility and a consequent increase in fracture risk. The impaired bone formation which is a major contributor to this disease may in part be caused by the age-related deficit in the local and endocrine actions of IGFs (*see text for details*).

major regulators of IGFBP-5 expression in osteoblasts. In aggregate, these effects could contribute to a decrease in IGF-mediated osteoblast proliferation and bone formation.

The skeletal capacity for osteoblast cell proliferation could be inhibited by additional mechanisms, in particular the increasing prevalence of vitamin D and calcium deficiency with aging. In this regard, recent in vitro evidence indicates that osteoblast cell proliferation and IGF-II production are dependent on the level of extracellular calcium, suggesting that calcium deficiency in the elderly could lead to a decrease in bone formation by a mechanism involving underproduction of IGF-II *(186)*. Consistent with this possibility, we have recently found that serum levels of stimulatory IGF system components correlate positively with serum levels of vitamin D in fracture patients and age-matched control subjects, suggesting that calcium deficiency in the elderly is associated with a deficit of serum stimulatory IGF system components.

Moreover, it has been established that the decrease in serum calcium levels in elderly women who have inadequate calcium intakes leads to an increase in the production of PTH to reestablish serum calcium homeostasis. We found that the serum levels of PTH correlated with serum levels of IGFBP-4 *(136,187)*, and we also found in vitro that PTH treatment increases production of IGFBP-4 in human osteoblasts *(136)*. These findings could explain, at least in part, our clinical finding of increased serum IGFBP-4 levels with advancing age. Thus, the underproduction of the stimulatory components and the overproduction of an inhibitory component of the IGF system occurs as a consequence of aging, and these changes lead to age-associated changes in the local as well as hormonal actions of IGFs, all of which could contribute to the pathogenesis of type II osteoporosis.

In addition to the age-related changes in the production of IGF system components, two studies have shown age-dependent decrease in human osteoblastic responsiveness to IGF-I. Pfeilschilfter et al. *(188)* have shown that osteoblasts from older donors require approx 10-fold higher concentrations of IGF-I to yield comparable increases in DNA synthesis than cells from younger donors. D'Avis et al. *(189)* have recently shown that osteoblast-like cells derived from younger patients are more responsive to rhIGF-I than cells derived from older patients. In addition, these authors have shown an age-related decline in the synthesis of extracellullar matrix components in response to IGF-I. Thus, the age-related decline in bone formation may be due to both a deficit in the production of, and responsiveness to, stimulatory IGF system components.

CONCLUDING REMARKS

IGFs are the most abundant mitogens produced by osteoblasts and stored in bone. The actions of IGFs in the bone microenvironment are regulated not only by the rate of IGF synthesis but also by other components of the IGF system, including IGF receptors, inhibitory IGFBPs, stimulatory IGFBPs, and IGFBP proteases. Based on the findings that the production of various IGF system components in osteoblasts are regulated by a variety of osteoregulatory agents, including calcium-regulating hormones, steroid hormones, and locally produced cytokines, it is likely that IGF actions in the bone microenvironment are regulated by a complex interplay between systemic skeletal factors and locally acting growth factors.

IGFs are unique in that they can act both as endocrine hormones and local growth factors for a variety of tissues, including bone. The findings that there is a marked impairment in both the endocrine and local actions of IGF system components in clinical disease states including both primary (e.g., senile type II) and secondary (e.g., glucocorticoid-induced) types of osteoporosis are consistent with the hypothesis that the IGF system is an important growth factor system for bone, and that a deficit of stimulatory IGF system components results in bone loss. Future intervention studies are needed to test the hypothesis that age-related and glucocorticoid-induced bone loss is associated with a deficit in endocrine and local IGF actions.

Regarding the potential utility of IGF system components for osteoporosis therapy, GH and IGFs have been considered promising candidates because of the established in vitro and in vivo effects of these agents on bone-forming osteoblasts and bone-resorbing osteoclasts. Because IGF-I causes an increase in both formation and bone resorption, future studies should be aimed in identifying a dose of IGF either alone or in combination with a stimulatory IGFBP (e.g. IGFBP-5) such that the bone-forming effects of IGFs are maximized while the bone-resorbing effects of IGFs are minimized. The development of such anabolic therapies which increase bone density are essential to treat patients who have already lost large amounts of bone.

ACKNOWLEDGMENTS

This work was supported by funds from National Institutes of Health (AR31062) and the Veterans Administration.

REFERENCES

1. Roodman GD. Advances in bone biology: the osteoclast. Endocr Rev 1996; 17:308–332.
2. Aubin JE, Liu F. The osteoblast lineage. In: Bilezikian JP, Raisz LG, Rodan GA, eds. Principles in Bone Biology. Academic Press, San Diego, 1996, pp. 51–68.
3. Mohan S, Linkhart T, Farley J, Baylink D. Bone-derived factors active on bone cells. Calcif Tissue Int 1984; 36 (Suppl 1): S139–45.
4. Mohan S, Jennings JC, Linkhart TA, Baylink DJ. Primary structure of human skeletal growth factor: homology with human insulin-like growth factor-II. Biochim Biophys Acta 1988; 966:44–55.
5. Mohan S, Baylink DJ. Bone growth factors. [Review]. Clin Orthop Rel Res 1991; 30–48.
6. Bautista CM, Mohan S, Baylink DJ. Insulin-like growth factors I and II are present in the skeletal tissues of ten vertebrates. Metab Clin Exp 1990; 39:96–100.
7. Canalis E, McCarthy T, Centrella M. Isolation and characterization of insulin-like growth factor I (somatomedin-C) from cultures of fetal rat calvariae. Endocrinology 1988; 122:22–27.
8. Canalis E, Pash JM, Verghese S. Skeletal growth factors. Crit Rev Euk Gene Expr 1993; 3:155–166.
9. Conover CA. The role of insulin-like growth factors and binding proteins in bone cell biology. In: Bilezikian JP, Rodan GA, eds. Principles of Bone Biology. Academic Press, San Diego, 1996; pp. 607–618.
10. Baylink DJ, Finkelman RD, Mohan S. Growth factors to stimulate bone formation. [Review]. J Bone Miner Res 1993; 8 (Suppl 2): S565–72.
11. Shinar DM, Endo N, Halperin D, Rodan GA, Weinreb M. Differential expression of insulin-like growth factor-I (IGF-I) and IGF-II messenger ribonucleic acid in growing rat bone. Endocrinology 1993; 132:1158–1167.
12. Malpe R, Baylink DJ, Linkhart TA, Wergedal JE, Mohan S. Insulin-like growth factor (IGF)-I, -II, IGF binding proteins (IGFBP)-3, -4, and -5 levels in the conditioned media of normal human bone cells are skeletal site-dependent. J Bone Miner Res 1997; 12:423–430.
13. Mohan S, Bautista CM, Herring SJ, Linkhart TA, Baylink DJ. Development of valid methods to measure insulin-like growth factors-I and -II in bone cell-conditioned medium. Endocrinology 1990; 126:2534–2542.
14. Okazaki R, Conover CA, Harris SA, Spelsberg TC, Riggs BL. Normal human osteoblast-cells constitutively express genes for insulin-like growth factors-I and II but transformed human osteoblast cell lines do not. J Bone Miner Res 1995; 10:788–795.
15. McCarthy TL, Centrella M, Canalis E. Constitutive synthesis of insulin-like growth factor-II by primary osteoblast-enriched cultures from fetal rat calvariae. Endocrinology 1992; 130:1303–1308.
16. Canalis E. Insulin like growth factors and the local regulation of bone formation. [Review]. Bone 1993; 14:273–276.
17. Birnbaum RS, Bowsher RR, Wiren KM. Changes in IGF-I and -II expression and secretion during the proliferation and differentiation of rat osteoblasts. J Endocrinol 1995; 144:251–259.
18. Zhang RW, Simmons DJ, Crowther RS, Mohan S, Baylink DJ. Contribution of marrow stromal cells to the regulation of osteoblast proliferation in rats: evidence for the involvement of insulin-like growth factors. Bone Miner 1991; 13:201–215.
19. Hayden JM, Hart KA, Baylink DJ, Johnstone D, Thompson G, Mohan S. Basal and regulated secretion of IGF system components differ between human bone marrow stromal cells and normal human osteoblasts. In: 10th International Congress of Endocrinology, San Francisco, CA, June 12–15, 1996 (Abstract).
20. Cheng SL, Zhang SF, Mohan S, Lecanda F, Fausto A, Hunt AH, Canalis E, Avioli LV. Regulation of IGF-I, IGF-II and their binding proteins in human bone marrow stromal cells by dexamethasone. J Cell Biochem 1998; 71:449–458.
21. Lazowski DA, Fraher LJ, Hodsman A, Steer B, Mdodrowski D, Han VKM. Regional variation of insulin-like growth factor-I gene expression in mature rat bone and cartilage. Bone 1994; 15:563–576.
22. Middleton J, Arnott N, Beresford J. Osteoblasts and osteoclasts in adult human osteophyte tissue express the mRNAs for insulin-like growth factors I and II and the type I IGF receptor. Bone 1995; 16:287–293.
23. Jones JI, Clemmons DR. Insulin-like growth factors and their binding proteins: biological actions. [Review]. Endocr Rev 1995; 16:3–34.
24. LeRoith D, Werner H, Beitner-Johnson D, Roberts CT. Molecular and cellular aspects of the insulin-like growth factor I receptor. Endocr Rev 1995; 16:143–163.
25. Slootweg MC, Hoogerbrugge CM, de Poorter TL, Duursma SA, Van Buul-Offers SC. The presence of

classical insulin-like growth factor (IGF) type-I and type-II receptors on mouse osteoblasts: autocrine/paracrine growth effects of IGFs. J Endocrinol 1990; 125:271–277.
26. Mohan S, Baylink DJ. Characterization of the IGF regulatory system in bone. [Review]. Adv Exp Med Biol 1993; 343:397–406.
27. Centrella M, McCarthy TL, Canalis E. Receptors for insulin-like growth factors-I and -II in osteoblast-enriched cultures from fetal rat bone. Endocrinology 1990; 126, 39–44.
28. Mohan S, Baylink DJ. The role of insulin-like growth factor-II in the coupling of bone formation to resorption. In: Spencer EM, ed. Modern Concepts of Insulin-Like Growth Factors. Elsevier, New York, 1991, pp. 169–184.
29. Raile K, Kessler HU, Pfuender M, Blum WF, Kolb H, Schwarz HP, Kiess W. Human osteosarcoma (U-2 OS) cells express both insulin-like growth factor-I (IGF-I) receptors and insulin-like growth factor-II/mannose-6-phosphate (IGF-II/M6P) receptors and synthesize IGF-II: autocrine growth stimulation by IGF-II via IGF-I receptor. J Cell Physiol 1994; 159:531–541.
30. Nakao Y, Takeda Y, Yamaguchi Y, Baylink DJ, Mohan S. Mitogenic and IGF binding protein binding activity of IGF-II analogs. J Bone Miner Res 1993; 8:S199 (Abstract).
31. Hou P, Sato T, Hofstetter W, Foged NT. Identification and characterization of the insulin-like growth factor I receptor in mature rabbit osteoclasts. J Bone Miner Res 1997; 12:534–540.
32. Mochizuki H, Hakeda Y, Wakatsuki N, Usui N, Akashi S, Sato T, Tanaka K, Kumegawa M. Insulin-like growth factor-I supports formation and activation of osteoclasts. Endocrinology 1992; 131:1075–1080.
33. Hill PA, Reynolds JJ, Meikle MC. Osteoblasts mediate insulin-like growth factor-I and -II stimulation of osteoclast formation and function. Endocrinology 1995; 136:124–131.
34. Mohan S, Bautista CM, Wergedal J, Baylink DJ. Isolation of an inhibitory insulin-like growth factor (IGF) binding protein from bone cell-conditioned medium: a potential local regulator of IGF action. Proc Natl Acad Sci USA 1989; 86:8338–8342.
35. LaTour D, Mohan S, Linkhart TA, Baylink DJ, Strong DD. Inhibitory insulin-like growth factor-binding protein: cloning, complete sequence, and physiological regulation. Mol Endocrinol 1990; 4:1806–1814.
36. Mohan S, Jennings JC, Linkhart TA, Baylink DJ. Isolation and purification of a low-molecular-weight skeletal growth factor from human bones. Biochim Biophys Acta 1986; 884:234–242.
37. Bautista CM, Baylink DJ, Mohan S. Isolation of a novel insulin-like growth factor (IGF) binding protein from human bone: a potential candidate for fixing IGF-II in human bone. Biochem Biophys Res Commun 1991; 176:756–763.
38. Conover CA. Insulin-like growth factor binding protein proteolysis in bone cell models. Prog Growth Factor Res 1995; 6:301–309.
39. Hassager C, Fitzpatrick LA, Spencer EM, Riggs BL, Conover CA. Basal and regulated secretion of insulin-like growth factor binding proteins in rat osteoblast-like cell is cell specific. J Clin Endocrinol Metab 1992; 75:228–233.
40. Mohan S. Insulin-like growth factor binding proteins in bone cell regulation. [Review]. Growth Regul 1993; 3:67–70.
41. McCarthy TL, Casinghino S, Centrella M, Canalis E. Complex pattern of insulin-like growth factor binding protein expression in primary rat osteoblast enriched cultures: regulation by prostaglandin E2, GH and the insulin-like growth factors. J Cell Physiol 1994; 160:163–175.
42. Andress DL. A novel human insulin-like growth factor binding protein secreted by human osteoblast-like cells. Biochem Biophys Res Commun 1991; 176:213–218.
43. Conover CA, Bale LK, Clarkson JT, Torring O. Regulation of insulin-like growth factor binding protein-5 messenger ribonucleic acid expression and protein availability in rat osteoblast-like cells. Endocrinology 1993; 132:2525–2530.
44. Mohan S, Baylink DJ. Evidence that the inhibition of TE85 human bone cell proliferation by agents which stimulate cAMP production may in part be mediated by changes in the IGF-II regulatory system. Growth Regul 1991; 1:110–118.
45. Mohan S, Strong DD, Hilliker S, Malpe R, Lee K, Farley J, Baylink DJ. Dibutyryl cyclic adenosine monophosphate differentially regulates cell proliferation in low and high alkaline phosphatase SaOS-2 human osteosarcoma cells: evidence for mediation by the insulin-like growth factor-II system. J Cell Physiol 1993; 156:462–468.
46. Campbell PG, Wines K, Yanosick TB, Novak JF. Binding and activation of plasminogen on the surface of osteosarcoma cells. J Cell Physiol 1994; 159:1–10.
47. Birnbaum RS, Wiren KM. Changes in insulin-like growth factor binding protein expression and

secretion during the proliferation, differentiation and mineralization of primary cultures of rat osteoblasts. Endocrinology 1994; 135:223–230.
48. Thrailkill KM, Siddhanti SR, Fowlkes JL, Quarles LD. Differentiation of MC3T3-E1 osteoblasts is associated with temporal changes in the expression of IGF-I and IGFBPs. Bone 1995; 17:307–313.
49. Conover CA, Durham SK, Zapf J, Masiarz FR, Kiefer MC. Cleavage analysis of insulin-like growth factor (IGF)-dependent IGF-binding protein-4 proteolysis and expression of protease-resistant IGF-binding protein-4 mutants. J Biol Chem 1995; 270:4395–4400.
50. Delany AM, Rydziel S, Canalis E. Autocrine down-regulation of collagenase-3 in rat bone cell cultures by insulin-like growth factors. Endocrinology 1996; 137:4665–4670.
51. Kanzaki S, Hilliker S, Baylink DJ, Mohan S. Evidence that human bone cells in culture produce insulin-like growth factor-binding protein-4 and -5 proteases. Endocrinology 1994; 134:383–392.
52. Mohan S, Strong DD, Linkhart TA, Baylink DJ. Regulation and actions of insulin-like growth factor binding protein (IGFBP)-4 and IGFBP-5 in bone: physiological and clinical implications. In: Baxter RC, Gluckman PD, Rosenfeld RG, eds. The Insulin-Like Growth Factors and Their Regulatory Proteins. Excerpta Medica, Amsterdam, 1994, pp. 205–215.
53. Durham SK, Kiefer MC, Riggs BL, Conover CA. Regulation of insulin-like growth factor binding protein 4 by a specific insulin-like growth factor binding protein 4 proteinase in normal human osteoblast-like cells: implications in bone cell physiology. J Bone Miner Res 1994; 9:111–117.
54. Durham SK, Riggs BL, Conover CA. The insulin-like growth factor-binding protein-4 (IGFBP-4)-IGFBP-4 protease system in normal human osteoblast-like cells: regulation by transforming growth factor-beta. J Clin Endocrinol Metab 1994; 79:1752–1758.
55. Durham SK, De Leon DD, Okazaki R, Riggs BL, Conover CA. Regulation of insulin-like growth factor (IGF)-binding protein-4 availability in normal human osteoblast-like cells: role of endogenous IGFs. J Clin Endocrinol Metab 1995; 80:104–110.
56. Durham SK, Riggs BL, Harris SA, Conover CA. Alterations in insulin-like growth factor (IGF)-dependent IGF-binding protein-4 proteolysis in transformed osteoblastic cells. Endocrinology 1995; 136:1374–1380.
57. Qin X, Strong DD, Pham T, Baylink DJ, Mohan S. Studies on the mechanism by which insulin-like growth factor (IGF)-II enhances IGF binding protein (IGFBP)-4 proteolysis in human osteoblast cell conditioned medium. J Bone Miner Res 1997; 12:S160 (Abstract).
58. Conover CA, Kiefer MC. Regulation and biological effect of endogenous insulin-like growth factor binding protein-5 in human osteoblastic cells. J Clin Endocrinol Metab 1993; 76:1153–1159.
59. Knutsen R, Honda Y, Strong DD, Sampath TK, Baylink DJ, Mohan S. Regulation of insulin-like growth factor system components by osteogenic protein-1 in human bone cells. Endocrinology 1995; 136:857–865.
60. Thrailkill KM, Quarles LD, Nagase H, Suzuki K, Serra DM, Fowlkes JL. Characterization of insulin-like growth factor-binding protein 5-degrading proteases produced throughout murine osteoblast differentiation. Endocrinology 1995; 136:3527–3533.
61. Mohan S, Nakao Y, Honda Y, Landale E, Leser U, Dony C, Lang K, Baylink DJ. Studies on the mechanisms by which insulin-like growth factor (IGF) binding protein-4 (IGFBP-4) and IGFBP-5 modulate IGF actions in bone cells. J Biol Chem 1995; 270:20424–20431.
62. Campbell PG, Novak JF, Wines K, Walton PE. Localization of plasmin activity on osteosarcoma cells: cell surface proteolysis of insulin-like growth factor binding proteins. Growth Regul 1993; 3:95–98.
63. Lalou C, Silve C, Rosato R, Segovia B, Binoux M. Interactions between insulin-like growth factor-I (IGF-I) and the system of plasminogen activators and their inhibitors in the control of IGF-binding protein-3 production and proteolysis in human osteosarcoma cells. Endocrinology 1994; 135:2318–2326.
64. Campbell PG, Novak JF. Insulin-like growth factor binding protein inhibits IGF action in human osteosarcoma cells. J Cell Physiol 1991; 149:293–300.
65. Conover CA, DeLeon DD. Acid-activated insulin-like growth factor binding protein-3 proteolysis in normal and transformed cells. Role of cathepsin D. J Biol Chem 1994; 269:7076–7080.
66. Wergedal JE, Mohan S, Lundy M, Baylink DJ. Skeletal growth factor and other growth factors known to be present in bone matrix stimulate proliferation and protein synthesis in human bone cells. J Bone Miner Res 5:179–186.
67. Linkhart TA, Jennings JC, Mohan S, Wakley GK, Baylink DJ. Characterization of mitogenic activities extracted from bovine bone matrix. Bone 1986; 7:479–487.
68. Canalis E, Centrella M, Burch W, McCarthy TL. Insulin-like growth factor I mediates selective anabolic effects of parathyroid hormone in bone cultures. J Clin Invest 1989; 83:60–65.

69. Canalis E, Rydziel S, Delany AM, Varghese S, Jeffrey JJ. Insulin-like growth factors inhibit interstitial collagenase synthesis in bone cell cultures. Endocrinology 1995; 136:1348–1354.
70. Strong DD, Beachler AL, Wergedal JE, Linkhart TA. Insulin-like growth factor-II and transforming growth factor beta regulate colllagen expression in osteoblast-like cells in vitro. J Bone Miner Res 1991; 6:15–23.
71. Takeda Y, Baylink DJ, Linkhart TA, Edwall D, Kumegawa M, Mohan S. Calcitrophic hormones and growth factors regulate osteoclast survival in vitro. J Bone Miner Res 1994; 9:S311 (Abstract).
72. Rajaram S, Baylink DJ, Mohan S. Insulin-like growth factor binding proteins in serum and other body fluids: regulation and functions. Endocr Rev 1997; 18:801–803.
73. Feyen JHM, Evans DB, Binkert C, Heinrich G, Geiss S, Kocher HP. Recombinant human (cys281) insulin-like growth factor binding protein 2 inhibits both basal and insulin-like growth factor-1 stimulated proliferation and collagen synthesis in fetal rat calvariae. J Biol Chem 1991; 266:19469–19474.
74. Ernst M, Rodan GA. Increased activity of insulin-like growth factor (IGF) in osteoblastic cells in the presence of GH (GH): positive correlation with the presence of GH-induced IGF binding protein BP-3. Endocrinology 1990; 127:807–814.
75. Amarnani S, Merriman HL, Linkhart TA, Baylink DJ, Mohan S. Autocrine regulators of MC3T3-E1 cell proliferation. J Bone Miner Res 1993; 8:157–165.
76. Schiltz PM, Mohan S, Baylink DJ. Insulin-like growth factor binding protein-4 inhibits both basal and IGF-mediated chick pelvic cartilage growth in vitro. J Bone Miner Res 1993; 8:391–396.
77. Andress DL, Birnbaum RS. Human osteoblast-derived insulin-like growth factor (IGF) binding protein-5 stimulates osteoblast mitogenesis and potentiates IGF action. J Biol Chem 1992; 267:22467–22472.
78. Andress DL, Loop SM, Zapf J, Kiefer MC. Carboxy-truncated insulin-like growth factor binding protein-5 stimulates mitogenesis in osteoblast-like cells. Biochem Biophys Res Commun 1993; 195:25–30.
79. Kiefer MC, Schmid C, Waldvogel M, Schlapfer I, Futo E, Masiarz FR, Green K, Barr PJ, Zapf J. Recombinant human insulin-like growth factor binding proteins 4, 5, and 6: biological and physiochemical characterization. Growth Regul 1993; 3:56–59.
80. Schmid C, Schalpfer I, Gostelipeter MA, Froesch ER, Zapf J. Effects and fate of human insulin-like growth factor binding protein-5 in rat osteoblast cultures. Am J Physiol 1996; 271:E1029–1035.
81. Srinivasan N, Edwall D, Linkhart TA, Baylink DJ, Mohan S. Insulin-like growth factor binding protein-6 produced by human PC-3 prostate cancer cells: isolation, characterization and its biological action. J Endocrinol 1996; 149:297–303.
82. Andress DL. Heparin modulates the binding of insulin-like growth factor (IGF) binding protein-5 to a membrane protein in osteoblastic cells. J Biol Chem 1995; 270:28289–28296.
83. Jones JI, Gockerman A, Busby WH Jr, Camacho-Hubner C, Clemmons DR. Extracellular matrix contains insulin-like growth factor binding protein-5: potentiation of the effects of IGF-I. J Cell Biol 1993; 121:679–687.
84. Oh Y, Muller HL, Lamson G, Rosenfeld RG. Insulin-like growth factor (IGF)-independent action of IGF-binding protein-3 in Hs578T human breast cancer cells. Cell surface binding and growth inhibition. J Biol Chem 1993; 268:14964–14971.
85. Rajah R, Valentinis B, Cohen P. Insulin like growth factor (IGF)-binding protein-3 induces apoptosis and mediates the effects of transforming growth factor-b1 on programmed cell death through a p53- and IGF-independent mechanism. J Biol Chem 1997; 272:12181–12188.
86. Li WL, Fawcett J, Widmer HR, Fielder PJ, Rabkin R, Keller GA. Nuclear transport of insulin-like growth factor-I and insulin-like growth factor binding protein-3 in opossum kidney cells. Endocrinology 1997; 138:1763–1766.
87. Jaques G, Noll K, Wegmann B, Witten S, Kogan E, Radulescu RT, Havemann K. Nuclear localization of insulin-like growth factor binding protein 3 in a lung cancer cell line. Endocrinology 1997; 138:1767–1770.
88. Eriksen EF, Kassem M, Langdahl B. GH, insulin-like growth factors and bone remodelling. Eur J Clin Invest 1996; 26:525–534.
89. Rosen CJ. GH, insulin-like growth factors, and the senescent skeleton: Ponce de Leon's Fountain revisited? J Cell Biochem 1994; 56:348–356.
90. Kalu DN, Liu CC, Slaerno E, Salih M, Ray M, Hollis BW. IGF-I partially prevents ovariectomy-induced bone loss: a comparative study with PTH. J Bone Miner Res 1991; 6:548.
91. Ammann P, Rizzoli R, Meyer JM, Slosman D, Bonjour JP. IGF-I and palmidronate increase bone mineral density in ovariectomized adult rats. Am J Physiol 1993; 265:E770–E776.
92. Machwate M, Zerath E, Holy X, Pastoureau P, Marie PJ. Insulin-like growth factor-I increases trabe-

cular bone formation and osteoblastic cell proliferation in unloaded rats. Endocrinology 1994; 134:1031–1038.

93. Bagi CM, DeLeon E, Brommage R, Adams S, Rosen D, Sommer A. Systemic administration of rhIGF-I or rhIGF-I/IGFBP-3 increases cortical bone and lean body mass in ovariectomized rats. Bone 1995; 16:263S–269S.
94. Bagi CM, Brommage R, Deleon L, Adams S, Rosen D, Sommer A. Benefit of systemically administered rhIGF-I and rhIGF-I/IGFBP-3 on cancellous bone in ovariectomized rats. J Bone Miner Res 1994; 9:1301–1312.
95. Spencer EM, Liu CC, Si ECC, Howard GA. In vivo actions of insulin-like growth factor I on bone formation and resorption in rats. Bone 1991; 12:21–26.
96. Zapf J, Froesch ER. Insulin-like growth factors/somatomedins: structure, secretion and biological actions and physiological role. Horm Res 1986; 24:121–130.
97. Verhaeghe J, van Bree R, van Herck E, Thomas H, Skottner A, Dequeker J, Mosekilde LI, Einhorn TA, Bouillon R. Effects of recombinant human GH and insulin-like growth factor-I, with or without 17β-estradiol, on bone and mineral homeostasis of aged ovariectomized rats. J Bone Miner Res 1996; 11:1723–1735.
98. Baker J, Liu J, Robertson EJ, Efstratiadis A. Role of insulin-like growth factors in embryonic and postnatal development. Cell 1993; 75:73–82.
99. Liu J, Baker J, Perkins AS, Robertson EJ, Efstratiadis A. Mice carrying null mutations of the genes encoding insulin-like growth factor I (IGF-I) and type 1 IGF receptor (IGF1r). Cell 1993; 75:59–72.
100. Johansson AG, Lindh E, Ljunghall S. IGF-I stimulates bone turnover in osteoporosis. Lancet 1992; 339:1619.
101. Ebeling PR, Jones JD, O'Fallon WM, Janes CH, Riggs BL. Short-term effects of recombinant human IGF-I on bone turnover in normal women. J Clin Endocrinol Metab 1993; 77:1384–1387.
102. Grinspoon S, Baum H, Peterson S, Klibanski A. Effects of rhIGF-I administration on bone turnover during short-term fasting. J Clin Invest 1995; 96:900–906.
103. Johansson G, Lindh E, Kollerup G, Ljunghall S. Dose-dependent effects of IGF-I on metabolism of collagen type I in male osteoporosis. J Bone Miner Res 1995; 10:S393.
104. Johansson AG, Lindh E, Blum WF, Kollerup G, Sorenson OH, Ljunghall S. Effects of GH and IGF-I in men with idiopathic osteoporosis. J Clin Endocrinol Metab 1996; 81:44–48.
105. Ghiron LJ, Thompson JL, Holloway L, Hintz RL, Butterfield GE, Hoffman AR, Marcus R. Effects of recombinant insulin-like growth factor-I and GH on bone turnover in elderly women. J Bone Miner Res 1995; 10:1844–1852.
106. Bagi CM, Brommage R, Deleon L, Adams S, Rosen D, Sommer A. Benefit of systemically administered rhIGF-I and rhIGF-I/IGFBP-3 on cancellous bone in ovariectomized rats. J Bone Miner Res 1994; 9:1301–1312.
107. Bagi CM, DeLeon E, Brommage R, Rosen D, Sommer A. Treatment of ovariectomized rats with the complex of rhIGF-I/IGFBP-3 increases cortical and cancellous bone mass and improves structure in the femoral neck. Calcif Tissue Int 1995; 57:40–46.
108. Kling L, Dony C, Leser U, Popp F, Bauss F, Lang K. Insulin-like growth factor binding protein-5 increases the osteoanabolic effects of insulin-like growth factor-I on calvarial bone in mice after local administration. J Bone Miner Res 1996; 11:S153 (Abstract).
109. McCarthy TL, Centrella M, Canalis E. Parathyroid hormone enhances the transcript and polypeptide levels of insulin-like growth factor I in osteoblast-enriched cultures from fetal rat bone. Endocrinology 1989; 124:1247–1253.
110. Linkhart TA, Mohan S. Parathyroid hormone stimulates release of insulin-like growth factor-I (IGF-I) and IGF-II from neonatal mouse calvaria in organ culture. Endocrinology 1989; 125:1484–1491.
111. McCarthy TL, Ji CH, Shu H, Casinghino S, Crothers K, Rotwein P, Centrella M. 17β-Estradiol potently suppresses cAMP-induced insulin-like growth factor-I gene activation in primary rat osteoblast cultures. J Biol Chem 1997; 272:18132–18139.
112. Ernst M, Rodan GA. Estradiol regulation of insulin-like growth factor-I expression in osteoblastic cells: evidence for transcriptional control. Mol Endocrinol 1991; 5:1081–1089.
113. Gray TK, Mohan S, Linkhart TA, Baylink DJ. Estradiol stimulates in vitro the secretion of insulin-like growth factors by the clonal osteoblastic cell line, UMR106. Biochem Biophys Res Commun 1989; 158:407–412.
114. McCarthy TL, Centrella M, Canalis E. Cyclic AMP induces insulin-like growth factor I synthesis in osteoblast-enriched cultures. J Biol Chem 1990; 265:15353–15356.
115. Thomas MJ, Umayahara Y, Shu H, Centrella M, Rotwein P, McCarthy TL. Identification of the cAMP

response element that controls transcriptional activation of the insulin-like growth factor-I gene by prostaglandin E2 in osteoblasts. J Biol Chem 1996; 271:21835–21841.
116. Ishizuka T, Yokose S, Hori M, Noda T, Suda T, Yoshiki S, Yamaguchi A. Parathyroid hormone exerts disparate effects on osteoblast differentiation depending on exposure time in rat osteoblastic cells. J Clin Invest 1997; 99:2961–2970.
117. Finkelman RD, Mohan S, Linkhart TA, Abraham SM, Boussy JP, Baylink DJ. PTH stimulates the proliferation of TE-85 human osteosarcoma cells by a mechanism not involving either increased cAMP or increased secretion of IGF-I, IGF-II or TGF beta. Bone Miner 1992; 16:89–100.
118. Chen TL, Chang LY, Bates RL, Perlman AJ. Dexamethasone and 1,25-dihydroxyvitamin D3 modulation of insulin-like growth factor binding proteins in rat osteoblast-like cell cultures. Endocrinology 1991; 128:73–80.
119. Canalis E. Mechanisms of glucocorticoid action in bone: implications to glucocorticoid-induced osteoporosis. J Clin Endocrinol Metab 1996; 81:3441–3447.
120. Scharla SH, Strong DD, Mohan S, Baylink DJ, Linkhart TA. 1,25-Dihydroxyvitamin D3 differentially regulates the production of insulin-like growth factor I (IGF-I) and IGF-binding protein-4 in mouse osteoblasts. Endocrinology 1991; 129:3139–3146.
121. Linkhart TA, MacCharles DC. Interleukin-1 stimulates release of insulin-like growth factor-I from neonatal mouse calavaria by a prostaglandin synthesis-dependent mechanism. Endocrinology 1992; 131:2297–2305.
122. Gabbitis B, Pash J, Canalis E. Regulation of insulin-like growth factor-II synthesis in bone cell cultures by skeletal growth factors. Endocrinology 1994; 135:284–289.
123. Tremollieres FA, Strong DD, Baylink DJ, Mohan S. Insulin-like growth factor II and transforming growth factor beta 1 regulate insulin-like growth factor I secretion in mouse bone cells. Acta Endocrinol 1991; 125:538–546.
124. Diez JJ, Iglesias P, Sastre J, Mendez J, Selgas R, Gómez-Pan A. GH responses to GH-releasing hormone and clonidine before and after erythropoietin therapy in CAPD patients. Nephron 1996; 74:548–554.
125. Okazaki R, Durham SK, Riggs BL, Conover CA. Transforming growth factor-beta and forskolin increase all classes of insulin-like growth factor-I transcripts in normal human osteoblast-like cells. Biochem Biophys Res Commun 1995; 207:963–970.
126. Strom P, Ljunghall S, Melhus H. Adult human osteoblast-like cells do not express insulin-like growth factor-I. Biochem Biophys Res Commun 1994; 199:78–82.
127. Tremollieres FA, Strong DD, Baylink DJ, Mohan S. Progesterone and promegestone stimulate human bone cell proliferation and insulin-like growth factor-2 production. Acta Endocrinol 1992; 126:329–337.
128. Mohan S, Chevalley T, Linkhart TA, Douglas D, Libanati C, Baylink DJ. Reduction in IGF-I and IGF-II production may mediate glucocorticoid inhibition of human bone cell proliferation. J Bone Miner Res 1995; 10:S492(Abstract).
129. McCarthy TL, Centrella M, Canalis E. Cortisol inhibits the synthesis of insulin-like growth factor-I in skeletal cells. Endocrinology 1990; 126:1569–1575.
130. Yeh LC, Adamo ML, Kitten AM, Olson MS, Lee JC. Osteogenic protein-1-mediated insulin-like growth factor gene expression in primary cultures of rat osteoblastic cells. Endocrinology 1996; 137:1921–1931.
131. Canalis E, Pash J, Gabbitas B, Rydziel S, Varghese S. Growth factors regulate the synthesis of insulin-like growth factor-I in bone cell cultures. Endocrinology 1993; 133:33–38.
132. Zaman G, Suswillo RFL, Cheng ZM, Tavares LA, Lanyon LE. Early responses to dynamic strain change and prostaglandins in bone derived cells in culture. J Bone Miner Res 1997; 12:769–777.
133. Rawlinson SC, Mohan S, Baylink DJ, Lanyon LE. Exogenous prostacyclin, but not prostaglandin E2, produces similar responses in both G6PD activity and RNA production as mechanical loading, and increases IGF-II release, in adult cancellous bone in culture. Calcif Tissue Int 1993; 53:324–329.
134. Fitzsimmons RJ, Strong DD, Mohan S, Baylink DJ. Low-amplitude, low-frequency electric field-stimulated bone cell proliferation may in part be mediated by increased IGF-II release. J Cell Physiol 1992; 150:84–89.
135. Fitzsimmons RJ, Ryaby JT, Mohan S, Magee FP, Baylink DJ. Combined magnetic fields increase insulin-like growth factor-II in TE-85 human osteosarcoma bone cell cultures. Endocrinology 1995; 136:3100–3106.
136. Honda Y, Landale EC, Strong DD, Baylink DJ, Mohan S. Recombinant synthesis of insulin-like growth factor-binding protein-4 (IGFBP-4): development, validation, and application of a radioimmunoassay

for IGFBP-4 in human serum and other biological fluids. J Clin Endocrinol Metab 1996; 81:1389–1396.
137. Mohan S, Srinivasan N, Pham T, Linkhart TA, Baylink DJ. Opposing effects of TGF beta1 on IGF binding protein (IGFBP)-4 and IGFBP-5 level in the conditioned medium of normal human osteoblasts: evidence for mediation via different post receptor signaling mechanism. In: 79th Annual Meeting of the Endocrine Society, Minneapolis, MN, June 11–14, 1997 (Abstract).
138. Qin X, Morales S, Mohan S, Linkhart TA, Baylink DJ, Strong DD. Structural and functional analysis of IGF binding protein-4 promoter: demonstration of functional cAMP response elements. J Bone Miner Res 1996; 11:S309(Abstract).
139. Pash JM, Canalis E. Transcriptional regulation of insulin-like growth factor-binding protein-5 by prostaglandin E2 in osteoblast cells. Endocrinology 1996; 137:2375–2382.
140. Scharla SH, Strong DD, Mohan S, Chevalley T, Linkhart TA. Effect of tumor necrosis factor-alpha on the expression of insulin-like growth factor I and insulin-like growth factor binding protein 4 in mouse osteoblasts. Eur J Endocrinol 1994; 131:293–301.
141. Scharla SH, Strong DD, Rosen C, Mohan S, Holick M, Baylink DJ, Linkhart TA. 1,25-Dihydroxyvitamin D3 increases secretion of insulin-like growth factor binding protein-4 (IGFBP-4) by human osteoblast-like cells in vitro and elevates IGFBP-4 serum levels in vivo. J Clin Endocrinol Metab 1993; 77:1190–1197.
142. Kassem M, Okazaki R, DeLeon DD, Harris SA, Robinson JA, Spelsberg TC, Conover CA, Riggs BL. Potential mechanism of estrogen mediated decrease in bone formation: estrogen increases production of inhibitory insulin-like growth factor binding protein-4 in human osteoblastic cell line with high levels of estrogen receptors. Proc Assoc Am Physicians 1996; 108:154–164.
143. Okazaki R, Riggs BL, Conover CA. Glucocorticoid regulation of insulin-like growth factor-binding protein expression in normal human osteoblast-like cells. Endocrinology 1994; 134:126–132.
144. Gabbitas B, Canalis E. Cortisol enhances the transcription of insulin-like growth factor-binding protein-6 in cultured osteoblasts. Endocrinology 1996; 137:1687–1692.
145. Chevalley T, Strong DD, Mohan S, Baylink DJ, Linkhart TA. Evidence for a role for insulin-like growth factor binding proteins in glucocorticoid inhibition of normal human osteoblast-like cell proliferation. Eur J Endocrinol 1996; 134:591–601.
146. Gabbitas B, Canalis E. Retinoic acid stimulates the transcription of insulin-like growth factor binding protein-6 in skeletal cells. J Cell Physiol 1996; 169:15–22.
147. Zhou YH, Mohan S, Linkhart TA, Baylink DJ, Strong DD. Retinoic acid regulates insulin-like growth factor-binding protein expression in human osteoblast cells. Endocrinology 1996; 137:975–983.
148. Boonyaratankornkit V, Strong DD, Mohan S, Baylink DJ, Beck C, Linkhart TA. A CACCC box motif in the IGF binding protein-5 promoter mediates progesterone transactivation of IGFBP-5 transcription in human osteoblasts. J Bone Miner Res 1996; 11:S131(Abstract).
149. Hayden JM, Strong DD, Baylink DJ, Powell DR, Sampath TK, Mohan S. Osteogenic protein-1 stimulates production of insulin-like growth factor binding protein-3 nuclear transcripts in human osteosarcoma cells. Endocrinology 1997; 138:4240–4247.
150. Bailey D, Mohan S, Linkhart T. IL-1 alpha increases IGF binding protein-4 production in normal human bone cells via a prostaglandin dependent mechanism. J Bone Miner Res 1996; 11:S309 (Abstract).
151. Mohan S, Strong DD, Lempert UG, Tremollieres F, Wergedal JE, Baylink DJ. Studies on regulation of insulin-like growth factor binding protein (IGFBP)-3 and IGFBP-4 production in human bone cells. Acta Endocrinologica 1992; 127:555–564.
152. Dong Y, Canalis E. Insulin-like growth factor (IGF) I and retinoic acid induce the synthesis of IGF-binding protein 5 in rat osteoblastic cells. Endocrinology 1995; 136:2000–2006.
153. Conover CA, Clarkson JT, Bale LK. Phorbol ester tumor promoters regulate insulin-like growth factor-binding protein-4 proteolysis. Endocrinology 1993; 133:1347–1351.
154. Rosen CJ, Donahue LR. Insulin-like growth factors: potential therapeutic options for osteoporosis. Trends Endocrinol Metab 1995; 6:235–240.
155. Mohan S, Baylink DJ. Development of a simple valid method for the complete removal of insulin-like growth factor (IGF)-binding proteins from IGFs in human serum and other biological fluids: comparison with acid-ethanol treatment and C18 Sep-Pak separation. J Clin Endocrinol Metab 1995; 80:637–647.
156. Romagnoli E, Minisola S, Carnevale V, Rosso R, Pacitti MT, Scarda A, Scarnecchia L, Mazzuoli G. Circulating levels of insulin-like growth factor binding protein 3 (IGFBP-3) and insulin-like growth factor I (IGF-I) in perimenopausal women. Osteopor Int 1994; 4:305–308.

157. Sugimoto T, Nishiyama K, Kuribayashi F, Chihara K. Serum levels of insulin-like growth factor (IGF) I, IGF-binding protein (IGFBP)-2, and IGFBP-3 in osteoporotic patients with and without spinal fractures. J Bone Miner Res 1997; 12:1272–1279.
158. Nasu M, Sugimoto T, Chihara M, Hiraumi M, Kurimoto F, Chihara K. Effect of natural menopause on serum levels of IGF-I and IGF-binding proteins: relationship with bone mineral density and lipid metabolism in perimenopausal women. Eur J Endocrinol 1997; 136:608–616.
159. Reed B, Zerwekh JE, Sakhaee K, Breslau NA, Gottschalk F, Pak CYC. Serum IGF-I is low and correlated with osteoblastic surface in idiopathic osteoporosis. J Bone Miner Res 1995; 10:1218–1224.
160. Johannsson G, Eriksen EF, Lindh E, Langdahl B, Blum WF, Lindahl A, Ljunggren Ö, Ljunghall S. Reduced serum levels of the GH-dependent insulin-like growth factor binding protein and a negative bone balance at the level of individual remodeling units in idiopathic osteoporosis in men. J Clin Endocrinol Metab 1997; 82:2795–2798.
161. Mohan S, Baylink DJ. Editorial: Insulin-like growth (IGF)-binding proteins in serum: do they have additional roles besides modulating the endocrine IGF actions? J Clin Endocrinol Metab 1996; 81:3817–3820.
162. Mohan S, Farley JR, Baylink DJ. Age-related changes in IGFBP-4 and IGFBP-5 levels in human serum and bone: implications for bone loss with aging. Prog Growth Factor Res 1995; 6:465–473.
163. Mohan S, Libanati C, Dony C, Lang K, Srinivasan N, Baylink DJ. Development, validation, and application of a radioimmunoassay for insulin-like growth factor binding protein-5 in human serum and other biological fluids. J Clin Endocrinol Metab 1995; 80:2638–2645.
164. Nicolas V, Mohan S, Honda Y, Prewett A, Finkelman RD, Baylink DJ, Farley JR. An age-related decrease in the concentration of insulin-like growth factor binding protein-5 in human cortical bone. Calcif Tissue Int 1995; 57:206–212.
165. Nicolas V, Prewett A, Bettica P, Mohan S, Finkelman RD, Baylink DJ, Farley JR. Age-related decreases in insulin-like growth factor-I and transforming growth factor-beta in femoral cortical bone from both men and women: implications for bone loss with aging. J Clin Endocrinol Metab 1994; 78:1011–1016.
166. Boonen S, Mohan S, Dequeker J, Aerssens J, Bouillon R, Baylink DJ. Deficit of the serum stimulatory components of insulin-like growth factor system in patients with osteoporosis and femoral neck fracture. J Bone Miner Res 1996; 11:S328(Abstract).
167. Boonen S, Mohan S, Dequeker J, Aerssens J, Bouillon R, Baylink DJ. Prospective study of the serum insulin-like growth factor system components (IGFs) in patients with senile osteoporosis and trochanteric fracture compared with age-matched controls. J Bone Miner Res 1997; 12:S122 (Abstract).
168. Boonen S, Lesaffre E, Aerssens J, Pelemans W, Dequeker J, Bouillon R. Deficiency of the GH-insulin-like growth factor-I axis potentially involved in age-related alterations in body composition. Gerontology 1996; 42:330–338.
169. Pfeilschifter J. GH and bone metabolism: the role of the vitamin D system. Eur J Endocrinol 136:30–32.
170. Ono T, Kanzaki S, Seino Y, Baylink DJ, Mohan S. Growth hormone (GH) treatment of GH-deficient children increases serum levels of insulin-like growth factors (IGFs), IGF binding protein-3 and -5, and bone alkaline phosphatase isoenzyme. J Clin Endocrinol Metab 1996; 81:2111–2116.
171. Rosen CJ, Donahue LR, Hunter SJ. Insulin-like growth factors and bone: the osteoporosis connection. Proc Soc Exp Biol Med 1994; 206:83–102.
172. Bagi C, van der Meulen M, Brommage R, Rosen D, Sommer A. The effect of systemically administered rhIGF-I/IGFBP-3 complex on cortical bone strength and structure in ovariectomized rats. Bone 1995; 16:559–565.
173. Rosen CJ, Dimai HP, Vereault D, Donahue LR, Beamer WG, Farley J, Linkhart S, Linkhart T, Mohan S, Baylink DJ. Circulating and skeletal insulin-like growth factor-I (IGF-I) concentrations in two inbred strains of mice with different bone mineral densities. Bone 1997; 21:217–223.
174. Libanati C, Rajaram S, Baylink DJ, Mohan S. In a longitudinal prospective study, glucocorticoid (GC) therapy causes a marked reduction in bone formation markers and stimulatory insulin-like growth factor (IGF) system components. J Bone Miner Res 1996; 11:S150(Abstract).
175. Mohan S, Baylink DJ. Insulin-like growth factor system components and the coupling of bone formation to resorption. Horm Res 1996; 45:59–62.
176. Farley JR, Tarbaux N, Murphy LA, Masuda T, Baylink DJ. In vitro evidence that bone formation may be coupled to resorption by release of mitogen(s) from resorbing bone. Metab Clin Exp 1987; 36:314–321.
177. Hayden JM, Mohan S, Baylink DJ. The insulin-like growth factor system and the coupling of formation to resorption. [Review]. Bone 1995; 17:93S–98S.
178. Finkelman RD, Linkhart TA, Mohan S, Lau KH, Baylink DJ, Bell NH. Vitamin D deficiency causes

a selective reduction in deposition of transforming growth factor beta in rat bone: possible mechanism for impaired osteoinduction. Proc Natl Acad Sci USA 1991; 88:3657–3660.
179. Dequeker J, Mohan S, Finkelman RD, Aerssens J, Baylink DJ. Generalized osteoarthritis associated with increased insulin-like growth factor types I and II and transforming growth factor beta in cortical bone from the iliac crest. Possible mechanism of increased bone density and protection against osteoporosis. Arthritis Rheum 1993; 36:1702–1708.
180. Cappon J, Brasel JA, Mohan S, Cooper DM. Effect of brief exercise on circulating insulin-like growth factor I. J Appl Physiol 1994; 76:2490–2496.
181. Schwartz AJ, Brasel JA, Hintz RL, Mohan S, Cooper DM. Acute effects of brief low- and high-intensity exercise on circulating IGF-I, IGF-II, and IGF binding protein-3 and the proteolysis in young healthy men. J Clin Endocrinol Metab 1996; 81:3492–3497.
182. Ghiron LJ, Thompson JL, Holloway L, Hintz RL, Butterfield GE, Hoffman AR, Marcus R. Effects of recombinant insulin-like growth factor-I and GH on bone turnover in elderly women. J Bone Miner Res 1995; 10:1844–1852.
183. Mauras N, Doi SQ, Shapiro JR. Recombinant human insulin-like growth factor I, recombinant human GH, and sex steroids: effects on markers of bone turnover in humans. J Clin Endocrinol Metab 1996; 81:2222–2226.
184. Johansson AG, Forslund A, Hambraeus L, Blum WF, Ljunghall S. GH-dependent insulin-like growth factor binding protein is a major determinant of bone mineral density in healthy men. J Bone Miner Res 1994; 9:915–921.
185. Hintz RL. Current and potential therapeutic uses of GH and insulin-like growth factor I. Endocrinol Metab Clin North Am 1996; 25:759–773.
186. Honda Y, Fitzsimmons RJ, Baylink DJ, Mohan S. Effects of extracellular calcium on insulin-like growth factor II in human bone cells. J Bone Miner Res 1995; 10:1660–1665.
187. Rosen C, Donahue LR, Hunter S, Holick M, Kavookjian H, Kirschenbaum A, Mohan S, Baylink DJ. The 24/25-kDa serum insulin-like growth factor-binding protein is increased in elderly women with hip and spine fractures. J Clin Endocrinol Metab 1992; 74:24–27.
188. Pfeilschifter J, Diel I, Pilz U, Brunotte K, Naumann A, Ziegler R. Mitogenic responsiveness of human bone cells in vitro to hormones and growth factors decrease with age. J Bone Miner Res 1993; 8:707–717.
189. D'Avis PY, Frazier CR, Shapiro JR, Fedarko NS. Age-related changes in effects of insulin-like growth factor I on human osteoblast-like cells. Biochem J 1997; 324:753–760.

21 IGFs and Skeletal Muscle

Stephen M. Rosenthal, MD

CONTENTS

INTRODUCTION

The insulin-like growth factors (IGFs) are unique in that they are the only known mitogens that, when free of other serum components, stimulate both the proliferation and differentiation of skeletal muscle cells. This dual effect of IGFs is of particular interest because these two processes are believed to be mutually exclusive in this tissue and in a wide variety of cell types *(1)*. Thus, exploring this model of IGF action in skeletal muscle provides an opportunity to address a fundamental issue in cell biology: understanding the mechanisms that influence the decision of a particular cell type to proliferate or differentiate. The focus of this chapter is on the interface of IGF biology with that of skeletal muscle embryogenesis/differentiation and cell cycle regulation, with particular emphasis on exploring the mechanisms by which IGFs influence the decision of skeletal muscle cells to proliferate or differentiate.

SKELETAL MUSCLE EMBRYOGENESIS AND DIFFERENTIATION

Early Embryology: Somite Formation and Cell Migration

During the last 10 yr, remarkable advances in skeletal muscle biology have permitted insights into fundamental questions of embryogenesis and cell biology, in particular: How does a multipotential cell become committed to a particular lineage, and, once committed, what are the mechanisms that influence the decision of a cell to either proliferate or differentiate?

With the exception of some muscles of the head, derived from cephalic mesoderm, all skeletal muscle in vertebrates are derived from somites *(2)*. Somites are transiently

From: *Contemporary Endocrinology: The IGF System*
Edited by: R. Rosenfeld and C. Roberts Jr © Humana Press, Inc., Totowa, NJ

occurring, paired epithelial spheres that form in a rostrocaudal progression by segmentation of paraxial mesoderm adjacent to the neural tube *(2)*. In response to signals from surrounding tissues, somites become compartmentalized into distinct regions: the ventral region, or sclerotome, gives rise to cells that ultimately become the ribs, vertebrae, and intervertebral discs; the dorsal region, or dermamyotome, gives rise to muscle and dermis *(2)*. The dermamyotome subsequently further compartmentalizes: the dorsomedial region becoms the myotome, giving rise to cells that become the axial muscles of the back; the ventrolateral region gives rise to cells that migrate to the developing limb buds, ultimately becoming the body wall and limb musculature; the dorsal-most region of the dermamyotome becomes the dermatome, giving rise to dermis *(2)*.

Myogenic Commitment and Differentiation

Although it has been known for some time that inductive signals from surrounding tissues lead to myogenesis in somites, recent studies have identified specific factors that play an essential role in this process *(3–5)*. Members of the *Wnt* family produced in the dorsal neural tube in combination with sonic hedgehog, produced in the notochord and ventral floor plate of the neural tube, as well as with other ectodermal factors, are thought to play principal roles in the induction of myogenic precursor cells in the somites *(3–5)*. The mechanisms by which these inductive signals activate the myogenic pathway are unknown.

A significant advance in our understanding of the mechanisms by which mesodermal stem cells are initially committed to the myogenic lineage and subsequently differentiate into fully mature myocytes and multinucleated myotubes has stemmed from the discovery of a family of skeletal muscle-specific transcription factors homologous to the protooncogene c-*myc* *(5,6)*. When transfected into a variety of nonmuscle cells, these myogenic basic helix–loop–helix factors, which include MyoD, myogenin, myf-5, and MRF-4, all have the ability to convert these cells to skeletal muscle cells *(5,6)*. These peptides dimerize with ubiquitously expressed *E2A* gene products, bind to a CANNTG consensus sequence (known as an E-box) in the regulatory regions of skeletal muscle-specific genes, and activate their transcription *(5,6)*. The ability of MyoD and related factors to bind DNA is influenced by the presence of the c-*myc*-related Id peptide *(7)*. Proliferating myoblasts express high levels of Id protein that inhibit DNA binding of the myogenic bHLH factors by competing for their dimerization partners, the *E2A* gene products *(7)*. Down-regulation of Id expression is required for differentiation, and overexpression of Id protein inhibits myogenesis *(7,8)*.

Although each of the myogenic bHLH factors can activate the entire myogenic program when transfected into a nonmuscle cell, insights into the roles of these individual factors has emanated from studies of temporal expression patterns in the developing vertebrate embryo and from targeted mutation studies of each of these genes. In the mouse, *myf-5* is the first of the myogenic bHLH factors to be expressed during embryogenesis, with transcripts appearing at approx 8 d of development in the early somite, prior to its compartmentalization into dermamyotome and sclerotome *(3,5,6)*. *Myogenin* is first expressed in the developing mouse at approx 8.5 d as myogenic cells in the myotome begin the differentiation process, while *MRF-4* and *MyoD* are expressed in the somite at 9.5–10.5 d of development *(3,5,6)*.

Targeted mutation studies of the myogenic bHLH factors in mice have provided insights into the roles of these individual factors during embryogenesis. Gene knockout experiments of either *myf-5* or *MyoD* demonstrate essentially normal skeletal muscle, whereas *myf-5* and *MyoD* double knockout mice die at birth, completely lacking skeletal muscle as well as any detectable precursor myoblasts *(9)*. These studies suggest a redundancy in the actions of *myf-5* and *MyoD* and indicate that these factors play an essential role in the early stages of myogenesis and myoblast specification *(9)*. In mice lacking the *myogenin* gene, myoblasts are present and are found in the appropriate locations, yet there is a severe reduction in the number of myoblasts that have undergone differentiation, also resulting in death within minutes of birth *(10,11)*. These studies indicate that *myogenin* is "downstream" of *myf-5* and *MyoD*, and support a primary and unique role for *myogenin* in the process of differentiation of myoblasts into myocytes. Targeted mutation of the *MRF-4* gene results in essentially normal skeletal muscle development *(12)*. Although *MRF-4* does not appear to play an essential role in muscle embryogenesis, *MRF-4* is the principal myogenic bHLH factor expressed in adult skeletal muscle *(13)*.

Although *myf-5* and *MyoD* are the first muscle-specific markers expressed in undifferentiated myoblasts, it is unclear whether the expression of these factors is a consequence of commitment to the myogenic lineage or whether these factors play a role in the commitment process itself. A principal focus of current investigation, therefore, is the identification of regulatory factors "upstream" of *myf-5* and *MyoD*. Recent studies have focused on the potential role of *Pax-3* in the induction of the earliest expression of the myogenic bHLH factors in the myogenic precursor cells of the somite *(14–16)*. *Pax-3* is a mammalian homolog of the *Drosophila paired* gene, encoding a paired box-containing transcription factor *(14–16)*. *Pax-3* is initially expressed in unsegmented paraxial mesoderm and throughout the early somite; subsequently, *Pax-3* expression is localized to the dermamyotome, with the highest levels in the lateral region of the dermamyotome *(14–16)*. Mice lacking a functional *Pax-3* gene do not develop limb muscles, and *Pax-3/myf-5* double-knockout mice have absent skeletal muscle formation in the trunk and limbs *(14)*. Although *MyoD* is expressed in *myf-5* knockouts, *MyoD* is not expressed in *Pax-3/myf-5* double knockouts, suggesting that *Pax-3* is an "upstream" regulator of *MyoD* *(14)*. In support of this concept, recent studies have demonstrated that ectopic expression of *Pax-3* in some nonmuscle cell types was able to induce *MyoD* expression and subsequently activate the process of myogenic differentiation *(15)*.

Skeletal muscle-specific gene expression requires not only activation by the myogenic bHLH factors, but also cooperative interaction by members of the myocyte enhancer factor-2 *(MEF2)* family *(17)*. *MEF2* gene products bind to A/T-rich regions in the promoters and enhancers of muscle-specific genes and are thought to play a role in amplification and maintenance of skeletal muscle-specific gene expression *(17)*. *MEF2* gene products, however, are not skeletal muscle specific and, unlike the myogenic bHLH factors, are unable to induce myogenesis when transfected alone into fibroblasts *(17,18)*. *MEF2* gene expression does occur in the somite myotome at approx 9 d of postcoitum, making it downstream of *myf-5* and *myogenin* in the developmental cascade *(17)*.

Proliferation and Differentiation Are Mutually Exclusive Events in Skeletal Muscle: Coupling of Cell Cycle Withdrawal with Terminal Differentiation

The decision of a cell to grow or differentiate is linked to mechanisms that regulate the cell cycle. A significant advance in our understanding of skeletal muscle cell differentiation is the recent discovery that muscle-specific transcription factors interact directly with the retinoblastoma gene product (Rb), a key element of the cell cycle machinery *(19)*. Rb is an ubiquitous nuclear factor that cycles between phosphorylated and un- (or hypo-)phosphorylated forms *(19–21)*. Rb functions as a growth suppressor when unphosphorylated in the early G1 phase of the cell cycle *(19–21)*. It is in this form that Rb interacts directly with myogenin or MyoD and promotes differentiation *(19)*. However, in mid-to-late G1, just just prior to the initiation of DNA synthesis, Rb becomes phosphorylated, is inactive as a growth suppressor, can no longer bind to myogenin or MyoD, and differentiation is inhibited *(22)*.

Although the mechanisms that regulate Rb phosphorylation are not fully understood, the G1 cyclin-dependent kinases *(cdk)*s *4* and *6* and their partners, the *group D cyclins (D1, D2, D3)*, are thought to play a principal role in this process *(23–25)*. The activation of cdk4 and 6 occurs not only by binding to cyclin D, but also by phosphorylation of a conserved threonine by cdk-activating kinase (CAK), an enzyme complex composed of cyclin H, cdk7, and a newly described CAK assembly factor, p36 *(26–28)*. CAK activity, however, does not appear to change during the cell cycle and also appears to be invariant in proliferating vs quiescent cells *(26,27)*. CAK activity is therefore not considered to be rate limiting in cdk4 or 6 activation *(26,27)*. In contrast, cdk4 and 6 activity are known to be negatively regulated by two classes of cdk inhibitors: (1) the family of p15, p16, p18, and p19, which are specific inhibitors of cdk4 and 6 *(29,30)*; and (2) the $p21^{CIP1}$, $p27^{KIP1}$, $p57^{KIP2}$ family which inhibits not only cdk4 and 6 activity, but in the case of p21, p27, and p57, also inhibits cdk2 activity *(31–34)*.

p18, p19, p21, and *p27* are all expressed in skeletal muscle cells *(35–37)*. Of these for cdk inhibitors, however, *p18, p21,* and *p27* are minimally expressed in proliferating myoblasts, and their expression increases markedly once cells have left the cell cycle, essentially in parallel with the induction of *myogenin* gene expression *(35–37)*. In contrast, *p19* is the only cdk4/cdk6 inhibitor that is expressed at relatively high levels in muscle cells during the time that the differentiation process is initiated *(35)*. This suggests that although p18, p21, and p27 may play a role in maintaining terminal differentiation once myoblasts have left the cell cycle, p19 is unique in that it may play a role in initiating cell cycle arrest, leading to terminal differentiation.

EXPRESSION AND REGULATION OF IGF PEPTIDES, RECEPTORS, AND BINDING PROTEINS IN SKELETAL MUSCLE

As in a wide variety of tissues, skeletal muscle cells express IGF peptides, receptors, and binding proteins, enabling a significant component of IGF action in muscle through autocrine/paracrine mechanisms. Before focusing on two principal but divergent actions of IGFs in skeletal myoblasts, proliferation and differentiation, the expression and regulation of various components of the IGF signaling system in skeletal muscle will be reviewed. The studies summarized in the following sections have been carried out in cloned murine cell lines, primary skeletal myoblasts, or whole tissue from a variety of species.

Peptides

Although both IGF peptides are expressed in skeletal muscle, IGF-II, as detailed below, is the principal IGF peptide expressed in this tissue. The expression of IGF peptides in muscle is developmentally regulated, is tightly linked to the differentiation process, and is influenced by a variety of additional factors. *In situ* hybridization studies in fetal rats demonstrated IGF-II mRNA in myotomes, whereas IGF-I mRNA was perinatally expressed in myocytes *(38,39)*. In primary cultures, fetal rat myoblasts have been shown to secrete both IGF-I and IGF-II *(40)*. Subsequent studies in rats demonstrated high levels of IGF-II mRNA in fetal and early neonatal skeletal muscle samples, with a marked reduction in IGF-II gene expression after the second postnatal day *(41)*. Skeletal muscle from adult rats contained low but detectable amounts of both IGF-I and IGF-II mRNA *(42)*. In similar developmental studies in mice, IGF-II mRNA was present in greatest abundance during embryonic life and the early postnatal period, with virtually undetectable levels in adult skeletal muscle *(43)*.

As noted previously, IGF peptide expression in skeletal muscle is closely linked to the differentiation state of the cells. Studies assessing the association of IGF peptide expression and differentiation have been carried out principally in cloned murine skeletal muscle cells, including parent and subclones of the mouse C2 *(44)*, rat L6 *(45)*, and mouse BC3H1 *(46)* and Sol 8 *(47)* cell lines. These cells express the biochemical and morphologic features of primary muscle cells and have been used extensively as model systems for studying the events that occur during myoblast proliferation and subsequent differentiation *(44–47)*. In high serum-supplemental medium, these cells proliferate as myoblasts; however, when serum is withdrawn, these myoblasts leave the cell cycle and (with the exception of BC3H-1 cells) undergo terminal differentiation in a manner similar to that seen in vivo *(44–47)*.

IGF-II mRNA is detectable at low levels in skeletal myoblasts and increases markedly upon differentiation induced by serum withdrawal *(43,48–50)*. Differentiation-associated increases in IGF-II mRNA are associated with increases in IGF-II peptide secretion *(48–50)*. Notably, a positive correlation was demonstrated between the degree of IGF-II peptide secretion by myoblasts and the rate of myoblast differentiation following serum withdrawal *(50)*. The differentiation-associated increase in IGF-II mRNA has been shown in C2 cells to be a consequence of transcriptional activation of the IGF-II gene as demonstrated by direct analysis of nuclear IGF-II precursor mRNA and by nuclear run-on assays *(51)*. Transcriptional activation of the IGF-II gene during C2 myoblast differentiation has been reported to occur principally through promoter 3 *(51)*.

In addition to activation of IGF-II gene expression in a differentiation-dependent manner, IGF-II expression is regulated by a variety of cytokines in skeletal myoblasts. In L6A1 muscle cells, IGF-II gene expression was inhibited by exogenous IGF-I or IGF-II and by insulin, with IGF-I being the most potent inhibitor *(52)*. In addition, exogenous IGF peptides and insulin also inhibited IGF-II mRNA accumulation in primary cultures of human skeletal myoblasts, although in these cells, IGF-I and IGF-II were equipotent inhibitors of IGF-II gene expression *(52)*. Furthermore, basic fibroblast growth factor (bFGF), a known inhibitor of myogenic differentiation, markedly inhibited IGF-II gene expression and IGF-II secretion in BC3H-1 muscle cells *(53)*.

In contrast to IGF-II, IGF-I peptide expression is minimal in skeletal myoblasts *(49,54)*. Following initiation of differentiation induced by serum withdrawal, IGF-I secretion by C2 cells increased two- to threefold over concentrations measured in unconditioned media, while IGF-II secretion in the same cultures increased approx 20-fold *(54)*. IGF-I gene expression was also much less than that of IGF-II in differentiating BC3H-1 myocytes *(49)*.

In addition to developmental and differentiation-regulated expression of IGF peptides in skeletal muscle, IGF peptide expression in this tissue is also influenced by muscle injury *(55–58)* and growth hormone treatment *(59)*. Ischemic injury to adult rat skeletal muscle resulted in increased IGF-I mRNA in this tissue during the regeneration process *(56)*. *In situ* hybridization studies in regenerating skeletal muscle demonstrated increased IGF-I mRNA in proliferating myoblasts and in muscle satellite cells *(56)*. An increase in

IGF-I mRNA has also been reported in injured skeletal muscle from hypophysectomized rats *(56)*. In a more recent study, skeletal muscle regeneration in young and adult rats following bupivacaine-induced muscle fiber degeneration was associated with enhanced expression of both IGF-I and IGF-II *(57)*. In studies of fetal rat hindlimbs, transection of the sciatic nerve resulted in up-regulation of IGF-II mRNA in hindlimb skeletal muscle and correlated with the ability of the denervated skeletal muscle to become reinnervated *(58)*. In addition, growth hormone-stimulated skeletal muscle hypertrophy in female rats was associated with up-regulation of both IGF-I and IGF-II mRNA in this tissue *(59)*.

Receptors

Skeletal muscle cells express "classic" IGF-I and IGF-II receptors, a variety of "atypical" IGF-I and IGF-II receptors with principally β-subunit and, less commonly, α-subunit heterogeneity, and also express hybrid IGF-I–insulin receptors. As with IGF peptides, the expression of IGF receptors in skeletal muscle is developmentally regulated, tightly linked with the differentiation state of the cells, and is also regulated by a variety of cytokines and other factors.

Developmental studies have demonstrated the presence of distinct IGF-I and IGF-II receptors in rat skeletal muscle from the prenatal period through adulthood *(60)*. The IGF-II receptor was reported to be the predominant IGF-receptor in fetal skeletal muscle, being approx 50-fold more abundant than IGF-I receptors *(60)*. Shortly after birth, muscle IGF-II receptor number decreased markedly, whereas IGF-I receptor number remained constant in the early weeks after birth and then gradually declined *(60)*. Cultured human skeletal muscle cells also express specific IGF-I and IGF-II receptors *(61)*.

Expression of the IGF-I receptor, thought to be the principal mediator of IGF-I and IGF-II action in skeletal muscle cells, as reviewed in the next section of this chapter, is closely associated with the differentiation process. IGF-I receptors are abundant in myoblasts, and, by Scatchard analysis, decrease significantly during differentiation without change in receptor affinity, as reported in L6 and BC3H-1 cells *(49,62)*. In C2 cells, IGF-I receptor content was reported to increase twofold during the early differentiation process with a subsequent decrease to values seen in undifferentiated myoblasts *(54)*. The mechanisms for the differentiation-associated decrease in IGF-I receptor expression were studied in BC3H-1 cells *(49)*. Differentiation from myoblasts to myocytes resulted in a 60% decrease in IGF-I receptor content (Scatchard analysis) with parallel decreases in IGF-I receptor mRNA and IGF-I receptor biosynthesis as determined in metabolic labeling experiments *(49)*.

Conflicting observations have been made with respect to IGF-II receptor expression and muscle differentiation and may reflect differences in study design time points. Crosslinking studies and Scatchard analyses demonstrated a three- to fourfold decrease in IGF-II receptor content after 18 d of differentiation in L6 cells *(62)*. In contrast, IGF-II receptor content was reported to increase almost sixfold in C2 cells after 24 h of differentiation and remained elevated through 96 h of differentiation *(48)*.

Both β-subunit and α-subunit IGF-I receptor heterogeneity have been described in skeletal muscle cells. In addition to a β-subunit of M_r 95,000 characteristic of receptors in adult muscle cells, fetal rat muscle contains a β-subunit with M_r 105,000 *(63)*. Expression of this larger β-subunit is highest in fetal muscle and decreases in early postnatal life *(63)*. Similarly, in BC3H-1 muscle cells, which express IGF-I receptor β-subunits of M_r 97,000 and 95,000, the larger subunit is predominant in myoblasts and decreases markedly with differentiation *(49)*. Both β- and α-IGF-I receptor subunit heterogeneity have been described in specific subclones of C2 myoblasts *(64,65)*. "Permissive" C2 cells, which differentiate spontaneously upon serum withdrawal, contain β-subunits of M_r 95,000 and 101,000 in IGF-I receptors in both myoblasts and differentiated myocytes *(64)*. "Inducible" C2 cells, which require exogenous IGFs (or high doses of insulin) for differentiation, contain IGF-I receptors with β-subunits that change during differentiation *(64)*. "Inducible" C2 myoblasts contain IGF-I receptors with a single β-subunit of M_r 106,000, which is lost upon differentiation concurrent with the appearance of M_r 95,000 and 101,000 β-subunits *(64)*. Interestingly, this atypical IGF receptor in "inducible" C2 cells has a tetrameric structure similar to that of the IGF-I and insulin receptors, yet it preferentially binds IGF-II with high affinity, and its β-subunit autophosphorylation is more responsive to IGF-II than to IGF-I or insulin *(64,65)*. In addition, α-subunit heterogeneity has also been reported in these C2 subclones: "permissive" myoblasts contained an apparent IGF-I receptor α-subunit of M_r 130,000, whereas "inducible" myoblasts contained an apparent IGF-I receptor α-subunit of M_r 140,000 *(65)*.

Although the molecular mechanisms that account for the above-noted heterogeneity in IGF-I receptor β-subunit size are not well understood, the mechanism for another form of IGF-I receptor β-subunit heterogeneity has been elucidated: two forms of IGF-I receptor β-subunit mRNA are expressed in muscle and a variety of other tissues that differ in the presence or absence of three nucleotides (CAG) beginning at position 2829 of the coding region *(66,67)*. Deletion of the 3-basepair results in an amino acid change from Thr–Gly to Arg in the extracellular portion of the β-subunit near the transmembrane domain *(66,67)*.

In addition to "classic" and atypical IGF receptors, skeletal muscle also contains IGF-I receptor–insulin receptor hybrids *(68)*. These receptors bind IGF-I with high affinity, bind insulin with lower affinity, and display IGF-I-induced receptor autophosphorylation *(68)*. Using selective immunoadsorption and immunoblotting, recent studies of microsomal membranes from rabbit and human skeletal muscle indicated that the majority (70–80%) of receptors binding IGF-I were actually IGF-I receptor–insulin receptor hybrids *(68)*. It is interesting to note, however, that several lines of undifferentiated myoblasts (C2, BC3H-1) have minimal expression of insulin receptors, which increase markedly upon differentiation *(69)*. Thus, while IGF-I signaling may be mediated in part by IGF-I receptor–insulin receptor hybrids in differentiated skeletal muscle cells, such hybrids would not be expected to play a significant role in IGF-I signaling in these undifferentiated myoblasts.

IGF-I receptor expression in skeletal muscle cells is also regulated by cytokines and by other factors. As in a variety of cells, IGF-I receptors in myoblasts are down-regulated by IGF peptides *(49,70,70a)*. Exogenous IGF peptides in BC3H-1 myoblasts resulted in decreased IGF-I receptor mRNA, decreased IGF-I receptor biosynthesis, and increased IGF-I receptor degradation as demonstrated by pulse–chase experiments of metabolically labeled receptors *(49,70)*. As noted previously, skeletal muscle differentiation is associated with marked up-regulation of IGF-II peptide gene expression, which precedes the decrease in IGF-I receptor expression *(49)*. These studies in BC3H-1 cells suggest, therefore, that IGF-I receptor expression during muscle cell differentiation may be regulated, at least in part, through autocrine production of IGF-II *(49,70)*.

In contrast to the down-regulating effect of IGF peptides on IGF-I receptor gene expression, bFGF not only inhibits IGF-II expression but also up-regulates IGF-I receptor mRNA in skeletal myoblasts *(53)*. Subsequent studies in skeletal muscle cells have demonstrated that bFGF increases IGF-I receptor gene expression by increasing IGF-I receptor promoter activity mediated by an apparent bFGF response element located between nucleotides –476 and –188 in the 5′-flanking region of the IGF-I receptor promoter *(71)*. In addition, up-regulation of IGF-I receptor mRNA has been described in regenerating skeletal muscle following injury *(57)*.

Binding Proteins

Skeletal muscle cells have been reported to secrete IGFBP-2, BP-4, BP-5, and BP-6 *(54,72–82)*. The expression and secretion of skeletal muscle-derived IGF binding proteins (IGFBPs) is differentiation dependent and is regulated by cytokines.

The expression and regulation of IGFBPs in skeletal muscle cells vary among species and cell line subclones. Mouse C2 cells have been reported to express a single BP, BP-5, demonstrated by ligand blot, immunoblot, and Northern analysis *(54,75)*. BP-5 mRNA is present in C2 myoblasts, and both BP-5 mRNA and protein accumulate during differentiation *(75)*. The "inducible" C2 subclone, a clonal derivative of C2 cells that is dependent on IGF-I or insulin to initiate terminal differentiation, also expresses only BP-5 through differentiation-associated activation of BP-5 gene transcription *(78)*. In a separate report, C2 cells have been reported to express BP-2, BP-4, and BP-5 *(82)*. The C2C12 subclone of C2 cells has been reported to express BP-2 mRNA and protein which are present in myoblasts and decrease with differentiation *(74)*. Mouse BC3H-1 cells express only BP-5 *(72,73,77)*.

Rat L6 skeletal muscle cells secrete BP-4 and BP-5 *(72,73,77)*, whereas both L6E9 and L6A1c subclones secrete BP-4, BP-5, and BP-6 *(79,80)*. In both L6E9 and L6A1c cells, BP-4 and BP-6 accumulated in conditioned media during differentiation induced by serum withdrawal *(79,80)*.

IGFBPs have also been described in human fetal skeletal muscle fibers *(76)* and in turkey clonally derived skeletal muscle satellite cells *(81)*. In 14–18-wk gestation human fetal skeletal muscle fibers, BP-4, BP-5, and BP-6 were detected by *in situ* hybridization *(76)*. In turkey skeletal muscle satellite cells, IGFBP-2 mRNA and protein were detected in proliferating cells and decreased with differentiation *(81)*.

As with IGF peptides and receptors, IGFBPs are also regulated by cytokines in skeletal muscle cells, as determined from studies carried out in BC3H-1 cells and parent and subclones of L6 cells *(77)*. In BC3H-1 cells, the secretion of the single IGFBP, BP-5, is increased by IGF-I and insulin, and decreased by transforming growth factor-β_1 (TGF-β_1), bFGF, epidermal growth factor (EGF), and thrombin (the latter of which did not cause apparent IGFBP-5 proteolysis despite its being a serine protease) *(77)*. In the parent L6 cell line, both IGFBP-4 and BP-5 secretion are increased by IGF-I and insulin, bFGF and suramin, and decreased by TGF-β_1, thrombin and dexamethasone *(77)*. EGF had no apparent effect on L6 BP-4 and BP-5 secretion *(77)*. In both L6A1c and L6E9 cells, IGFs led to an increase of BP-4, BP-5, and BP-6 in conditioned media *(79,80)*. In parent L6 cells as well as L6E9 and L6A1 subclones, treatment with IGF-I analogs with markedly reduced affinities for IGFBPs but with affinities for the IGF-I receptor comparable to native IGF-I also increased IGFBP secretion *(73,79,80)*.

IGFS SEQUENTIALLY STIMULATE PROLIFERATION AND DIFFERENTIATION IN SKELETAL MUSCLE CELLS

The decision of skeletal myoblasts to proliferate or differentiate is strongly linked to the cellular environment, influenced by cell–extracellular matrix interactions *(83)*, direct cell–cell interactions *(84)*, and by soluble growth factors *(85)*. Although growth factors, in general, inhibit myogenic differentiation *(85)*, IGFs are unique in that they are the only known mitogens that, when free of other serum components, sequentially stimulate proliferation and differentiation of skeletal muscle cells *(86–89)*. The mechanisms by which IGFs influence the decision of skeletal myoblasts to proliferate or differentiate are incompletely understood. In particular, the mechanisms responsible for the "switch" in IGF action from a mitogenic to a differentiation response are unknown and are a principal focus of investigation. This section reviews target genes of IGF action, the role of IGF receptor subtypes, and the role of IGF binding proteins in influencing the decision of skeletal myoblasts to proliferate or differentiate.

Sequential Stimulation of Proliferation and Differentiation

Although it has been established for approx 20 yr that exogenous IGFs stimulate myoblast proliferation and differentiation *(40,90–98)*, it has been subsequently established that these responses are sequential *(86–89)*. Recent studies indicate that the mitogenic response to IGF-I in skeletal myoblasts is limited to passage through a single cell cycle; subsequently, IGFs induce cell cycle exit and initiation of the differentiation process *(89,99,100)*. This suggests that an event occurs during the initial cell cycle that limits entry into additional cell cycles, as discussed in a following subsection.

Although exogenous IGFs have a time-dependent biphasic effect on myogenic proliferation and differentiation, endogenous IGF production by skeletal muscle cells is required for survival and differentiation in tissue culture *(50,89,101)*. When proliferating myoblasts are removed from serum, muscle differentiation occurs spontaneously in association with induction of IGF peptide production by these cells *(48–50)*. In fact, skeletal myoblasts that differentiate most rapidly following serum withdrawal are those that spontaneously secrete relatively greatest amounts of IGF-II, the principal IGF peptide secreted by muscle cells *(50)*. In addition, overexpression of IGF-II *(102)* or IGF-I *(103)* in murine skeletal myoblasts resulted in accelerated differentiation, and in in vivo studies, use of an avian skeletal α-actin promoter to direct expression of hIGF-I in skeletal muscle resulted in myofiber hypertrophy in transgenic mice *(103)*. In contrast, IGF antisense oligodeoxynucleotides or cRNAs inhibit the muscle differentiation process *(50,89,101)* and (in IGF antisense cRNA-expressing cells) result in apoptosis *(89)*.

Taken together, these data suggest that IGFs function as "maintenance factors" in skeletal muscle cells: when subconfluent, proliferating myoblasts are exposed to exogenous IGFs, these IGFs initially maintain the cycling state of these cells. However, once myoblasts leave the cell cycle, autocrine production of IGFs is induced, which promotes and maintains survival and the differentiated state of these cells. These studies suggest that, in developing skeletal muscle in vivo, endocrine and/or paracrine sources of IGFs contribute initially to a proliferative response in myoblasts. Subsequently, when environmental cues favor differentiation (e.g., reaching a critical threshold of cell density), an autocrine source of IGFs is induced that potentiates the differentiated state. A proliferative role for IGFs during early muscle development is supported by the marked skeletal muscle hypoplasia observed in transgenic mice homozygous for a targeted mutation of either the IGF-I peptide *(104)* or the IGF-I receptor *(105)*.

Target Genes of IGF Action that Influence the Decision of Skeletal Myoblasts to Proliferate or Differentiate

Several laboratories have focused on *myogenin* as a target gene of IGF action, as it is the only member of the *MyoD* family of skeletal muscle-specific transcription factors that is induced at the time when myoblasts first leave the cell cycle and begin to differentiate *(1)*. The importance of *myogenin* in muscle differentiation is supported by studies in which mice with targeted disruption of the *myogenin* gene had ample myoblasts that largely failed to undergo differentiation *(10,11)*. In addition, when cultured skeletal myoblasts were treated with myogenin antisense oligodeoxynucleotides, IGF-I was unable to stimulate differentiation in these cells *(106)*.

While skeletal myoblasts differentiate merely upon serum withdrawal, IGFs increase this process in association with increased levels of myogenin mRNA *(79,97)*. With the exception of BC3H-I cells *(98)*, studies demonstrating IGF-induced increases in myogenin mRNA, however, involved treatment periods of greater than 24 h *(79,97)*. In contrast, when cells were treated for less than 24 h with equivalent doses of IGF peptides, levels of myogenin mRNA markedly decreased *(87,88)*. Preliminary studies indicate that the inhibitory effect of IGF-I on myogenin steady-state mRNA is a consequence of decreased myogenin transcription mediated by sequence(s) contained from –94 to –9 of the *myogenin* gene *(107)*. It is noteworthy that both the early IGF-induced inhibition *(87)* and subsequent stimulation *(93)* of muscle differentiation occur through processes that are independent of the mitogenic effect of IGFs.

In addition to the biphasic effects of IGFs on myogenin expression, IGFs regulate components of the cell cycle machinery that influence the decision of skeletal myoblasts to proliferate or differentiate. As previously noted, the phosphorylation state of the ubiquitous nuclear Rb protein plays an important role in determining whether myoblasts proliferate or differentiate: phosphorylated Rb promotes mitogenesis, whereas un-(or hypo-)phosphorylated Rb promotes cell cycle exit and differentiation *(19)*. In parallel with the early IGF-induced inhibition of myogenin expression and concurrent mitogenic response, IGF treatment results in persistence of Rb phosphorylation, typical of proliferating cells, and up-regulation of the gene expression of cyclin-dependent kinase 4 and cyclin D1, components of a holoenzyme that plays a principal role in mediating Rb phosphorylation *(87)*. After 24 h, when IGF-I switches from stimulating proliferation to stimulating differentiation, Rb becomes dephosphorylated, and cyclin D1 and cdk4 gene expression are no longer maintained *(87,88)*. Figure 1 demonstrates the inverse relationship between the effects of IGF-I on cyclin D1/cdk4 mRNA and on myogenin mRNA at each time point studied between 6 and 24 h. In addition, this figure illustrates the observation that IGF-I markedly inhibits myogenin gene expression for up to 24 h, and subsequently switches to stimulating myogenin gene expression.

It has been previously established that the IGF switch from stimulating proliferation to stimulating differentiation of skeletal myoblasts is not a consequence of cell confluence, as it occurs over a wide range of cell densities *(92)*, but rather is a time-dependent phenomenon *(89,99,100)*. Recent observations from several laboratories in nonsynchronized as well as synchronized cells indicate that the mitogenic response of myoblasts to IGF-I is limited to progression through a single cell cycle, with subsequent onset of differentiation *(89,99,100)*. These findings suggest that an event or events occur during the initial cell cycle that preclude entry into an additional cell cycle. It is known that the un-(or hypo-)phosphorylated form of Rb promotes cell cycle exit and differentiation *(19)*. Thus, early IGF treatment could lead to expression of a cdk4/cdk6 inhibitor during the initial mitogenic response which, by promoting the hypophosphorylated form of Rb, could limit further cell cycle entry. *p19* is the only cdk4/cdk6 inhibitor known to be expressed at relatively high levels in muscle cells during the time that the differentiation

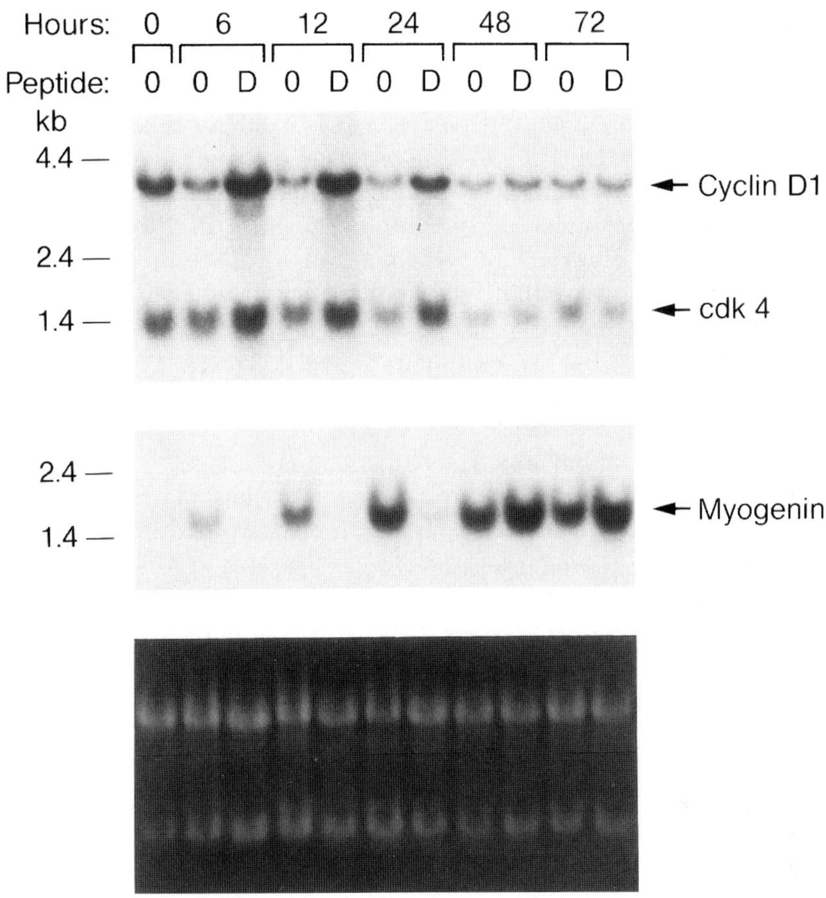

Fig. 1. Northern blots of cyclin D1, cdk4, and myogenin mRNA in control (*lanes 0*) and des(1–3)IGF-I (*lanes D*)-treated (20 ng/mL) L6E9 myoblasts over time. Ethidium bromide staining of the gel is shown below. Note that the IGF-I analog initially inhibits myogenin gene expression for up to 24 h, then switches to stimulating myogenin mRNA. (Reprinted from Rosenthal SM, Cheng Z-Q. Proc Natl Acad Sci USA 1995;92:10307–10311. Copyright [1995] National Academy of Sciences, USA.)

process is initiated *(35)*. Furthermore, *p19* is periodically expressed in macrophages with maximal synthesis during S phase *(30)*.

In a recent preliminary study in synchronized skeletal myoblasts, *p19* gene expression was also found to be S phase-dependent *(100)*. In addition, IGF-I stimulated p19 mRNA during S phase in these synchronized skeletal myoblasts *(100)*. Thus, p19 is a potential candidate for mediating the IGF switch from proliferation to differentiation in skeletal myoblasts. Further studies to evaluate more definitively the role of p19 in this switch are a focus of current investigation. Gene expression of the cdk inhibitor p21 is also increased in skeletal myoblasts following treatment with IGFs *(108)*. Because p21 is not expressed significantly in proliferating myoblasts but rather is induced once cells have left the cell cycle and begun to differentiate *(36,37)*, p21 may play a role in maintaining the differentiated state of myoblasts, whereas p19 may play a role in initiating the switch from proliferation to differentiation.

The Role of IGF Receptor Subtypes in Skeletal Muscle Cell Proliferation and Differentiation

The IGF-I receptor is thought to play a principal role in myoblast proliferation and differentiation induced by IGF-I or IGF-II. This conclusion emanates from studies examining relative potencies of native IGF peptides, studies using receptor-selective IGF analogs, and from studies in transgenic mice lacking functional IGF-I or IGF-II receptors. The initial studies implicating the IGF-I receptor in myoblast proliferation and differentiation demonstrated that native IGF-I was more potent than IGF-II in mediating these IGF actions *(109)*. In addition, [Thr^{59}]IGF-I and [N-Met]IGF-I, both of which have high affinity for IGF-I receptors but low affinity for IGF-II receptors, were virtually equipotent with IGF-I in stimulating proliferation and subsequent differentiation *(109)*. Furthermore, [Leu^{24}, 1–62]IGF-I, with markedly reduced affinity for IGF-I receptors, was much less potent than IGF-I in stimulating differentiation *(86)*.

Receptor-selective IGF-II analogs have also demonstrated that IGF-II acts principally through the IGF-I receptor to stimulate both proliferation and differentiation in myoblasts. [Arg^{54}, Arg^{55}]IGF-II, which does not bind significantly to IGF-II receptors but has high affinity for IGF-I receptors, was equipotent with IGF-II in stimulating myoblast proliferation and differentiation *(110)*. In contrast, [Leu^{27}]IGF-II, which binds IGF-II receptors with high affinity but has low affinity for IGF-I receptors, was much less potent than IGF-II in stimulating proliferation and differentiation *(86,110,111)*. Although IGF-II was more potent than [Leu^{27}]IGF-II in stimulating differentiation, the [Leu^{27}]IGF-II analog, however, did stimulate differentiation at doses where there was no detectable interaction with IGF-I receptors by competitive binding and crosslinking studies *(110,111)*. An interpretation of these data is that IGF-II signals differentiation principally through the IGF-I receptor, with the IGF-II receptor playing a smaller role, or that the [Leu^{27}]IGF-II analog interacts with IGF-I receptors, albeit at very low affinity. Although the IGF-II receptor does not likely play a significant role in signaling IGF-induced proliferation or differentiation in myoblasts, this receptor does play a principal role in IGF-II degradation in skeletal myoblasts *(112)*.

Additional insights into the role of IGF receptors in myoblast proliferation and differentiation emanate from in vivo and in vitro studies. Mice lacking functional IGF-I receptors demonstrate severe skeletal muscle hypoplasia, although qualitative assays for muscle differentiation-specific markers showed no differences from control mice *(105)*. When endogenous IGF-I receptors were functionally inactivated with a dominant negative, kinase-inactive IGF-I receptor, muscle differentiation was markedly delayed in association with prolonged phosphorylation of the cell cycle regulatory Rb protein and a greater percentage of cells in S phase *(113)*. In contrast, when wild-type human IGF-I receptors were overexpressed in bovine or rodent skeletal cells, an IGF-I-induced proliferative response was augmented in comparison to control cells, and differentiation occurred more rapidly than in control cells following serum withdrawal *(114,115)*. Of note, consistent with an earlier report demonstrating a dose-dependent biphasic effect of IGF-I on myoblast differentiation *(116)*, myoblasts overexpressing IGF-I receptors also demonstrated a dose-dependent biphasic effect of IGF-I on differentiation *(114,115)*. Taken together, these studies indicate that although the IGF-I receptor may not be essential for muscle differentiation, it does play a critical role in muscle proliferation and in the timing of myoblast differentiation. While the skeletal muscle phenotype in mice lacking functional IGF-II receptors was not specifically described, such mice demonstrated generalized fetal overgrowth with an increase in cell number associated with increased circulating and tissue levels of IGF-II *(117,118)*.

In view of the role of IGF-I receptors in mediating IGF-induced proliferation as well as in mediating inhibition and subsequent stimulation of myoblast differentiation, studies by a number of laboratories have focused on elucidating IGF-I receptor signaling pathways that mediate these actions. These studies have principally employed chemical inhibitors reported to be selective for signaling pathways.

With respect to the mitogenic effect of IGFs in myoblasts, DNA synthesis and cell proliferation occur in association with increased c-Fos and c-Jun mRNA, increased cyclin D1 mRNA and protein, as well as persistent Rb phosphorylation *(87,88,108,119,120)*. These mitogenic effects of IGFs in myoblasts have been reported to occur concurrently with increased mitogen-activated protein (MAP) kinase phosphorylation *(108,121)*, increased IGF-I receptor-associated phosphatidylinositol (PI) 3 kinase activity *(120)*, and increased phosphorylation of Shc proteins *(120)*. The MAP kinase kinase inhibitor PD098059 markedly inhibited IGF-I-induced myoblast proliferation as well as IGF-I-induced increases in c-Fos mRNA and cyclin D1 protein in L6A1 cells *(108)*. In C2C12 cells, however, inhibition of MAP kinase activation with PD098059 resulted in a more modest inhibition of IGF-I-induced DNA synthesis, suggesting a MAP-kinase-independent pathway that may play a role in the mitogenic effects of IGF-I in skeletal myoblasts *(121)*. Although inhibition of PI 3-kinase activity by wortmannin (WM) had no apparent effect on IGF-I-induced DNA synthesis in C2C12 myoblasts, inhibition of PI 3-kinase activation with LY294002 did partially inhibit the mitogenic effect of IGF-I in L6A1 myoblasts *(108,121)*. Taken together, these studies suggest a significant, if not principal, role of the MAP kinase pathway in IGF-I-induced myoblast proliferation, although MAP kinase-independent pathway(s) appear to play some role as well.

With respect to myoblast differentiation, several studies not involving IGF treatment have evaluated the role of the MAP kinase or PI 3-kinase pathways. Increased activation of the MAP kinase pathway by transfection with *ras* oncogenes inhibits myoblast differentiation *(122,123)*, whereas blocking this pathway with PD098059 augments differentiation that occurs spontaneously following serum withdrawal *(124)*. In one report, however, myoblast differentiation induced by serum withdrawal was associated with sustained MAP kinase activation *(125)*. Stimulation of the PI 3-kinase pathway appears to be essential for muscle differentiation induced by serum withdrawal, as PI 3-kinase inhibitors LY294002 or WM blocked this process *(126)*.

With respect to signaling pathways that mediate IGF-induced inhibition and subsequent IGF-induced stimulation of differentiation, a preliminary report in L6E9 cells indicates that the early IGF-I-induced inhibition of myogenin gene expression is not mediated by either the MAP kinase or PI 3-kinase pathways *(124)*. In contrast, as reported in L6A1 cells, IGF stimulation of differentiation is augmented by blocking the MAP kinase pathway and inhibited by blocking the PI 3-kinase pathway *(108)*.

The Role of IGF Binding Proteins in Skeletal Myoblast Proliferation and Differentiation

Studies examining the role of IGF binding proteins in skeletal myoblast proliferation and differentiation have utilized two principal approaches. The role of endogenously derived muscle IGFBPs has been evaluated by comparing the potencies of native IGF peptides with IGF analogs having reduced affinity for IGF binding proteins *(78–80)* and by inhibiting endogenous BP expression with antisense strategies *(127)*. The role of exogenous or ectopically derived IGFBPs in muscle proliferation and differentiation has been evaluated by treatment of cell cultures with purified BPs *(110,128)* or by overexpression using plasmid transfection and viral infection strategies *(82,127)*. These studies, carried out by several laboratories, have uniformly demonstrated that IGFBPs inhibit IGF-induced proliferation and differentiation in skeletal myoblasts.

With respect to the role of endogenous IGFBPs in myoblast proliferation and differentiation, studies carried out in different subclones of L6 cells and a subclone of C2 cells demonstrated that des(1–3)IGF-I, which has markedly reduced affinity for IGF-BPs but has an affinity for the IGF-I receptor that is comparable to native IGF-I,was more potent than native IGF-I in stimulating mitogenesis and differentiation *(78–80)*. In addition, LongR^{3}IGF-I, another IGF-I analog with reduced affinity for IGF-BPs, was more potent than native IGF-I in stimulating proliferation as well as differentiation *(80)*. [Thr^{48}, Ser^{49}, Ile^{50}]IGF-II, which binds to the IGF-I receptor with slightly lower affinity than native IGF-II but that has decreased affinity for IGFBPs, was more potent than native IGF-II in stimulating both proliferation and differentiation *(110)*. Inhibition of endogenous BP-5 expression in C2 cells using an antisense strategy accelerated the differentiation process that occurs in association with serum withdrawal and induction of IGF peptide expression *(127)*. Taken together, these studies demonstrate that IGFBPs expressed by skeletal muscle cells inhibit proliferation and differentiation induced by IGF peptides.

The inhibitory role of IGFBPs on myoblast proliferation and differentiation has also been demonstrated in studies with exogenous or ectopically derived IGFBPs. Recombinant human (rh) BP-6 inhibited IGF-II-induced proliferation and differentiation *(110,128)*. No effect of rhBP-6 was seen on IGF-I-induced differentiation, presumably because of the relatively low affinity of BP-6 for IGF-I *(128)*. However, purified rat IGFBP-2 inhibited both IGF-I and IGF-II-induced myoblast differentiation *(128)*. Overexpression of IGFBP-5 impaired differentiation normally induced by serum withdrawal, although IGFBP-5 overexpressing myoblasts demonstrated prolonged survival in low serum conditions in comparison to control cells *(127)*. In addition, ectopic expression of IGFBP-4 in myoblasts inhibited both proliferation and differentiation in the presence or absence of exogenous IGF peptides *(82)*.

SUMMARY

IGFs play a unique role in skeletal muscle in that they are the only known growth factors that, when free of other serum components, sequentially stimulate proliferation and differentiation. Although IGFs influence the expression of skeletal muscle-specific genes as well as various components of the cell cycle machinery, the target genes of IGF action that influence the decision of myoblasts to proliferate or differentiate and that cause the switch in IGF response are still unknown. In addition, although the IGF-I receptor plays a principal role in IGF signaling in skeletal muscle, post-receptor signaling pathways involved in proliferation, survival, and in the biphasic effects of IGFs on differentiation are incompletely understood. The role of IGFBPs, whether muscle-derived or exogenous, appears to be inhibitory with respect to proliferation as well as differentiation induced by IGFs.

REFERENCES

1. Olson EN. Interplay between proliferation and differentiation within the myogenic lineage. Dev Biol 1992; 154:261–272.
2. Buckingham M. Molecular biology of muscle development. Cell 1994; 78:15–21.
3. Lassar A, Münsterberg AE. Wiring diagrams: regulatory circuits and the control of skeletal myogenesis. Curr Opin Cell Biol 1994; 6:432–442.
4. Lassar A, Münsterberg AE. The role of positive and negative signals in somite patterning. Curr Opin Neurobiol 1996; 6:57–63.
5. Molkentin JD, Olson EN. Defining the regulatory networks for muscle development. Curr Opin Genet Dev 1996; 6:445–453.
6. Olson EN, Klein WH. bHLH factors in muscle development: dead lines and commitments, what to leave in and what to leave out. Genes Dev 1994; 8:1–8.
7. Benezra R, Davis RL, Lockshon D, Turner DL, Weintraub H. The protein Id: a negative regulator of helix–loop–helix DNA binding proteins. Cell 1990; 61:49–59.
8. Jen Y, Weintraub H, Benezra R. Overexpression of Id protein inhibits the muscle differentiation program: in vivo association of Id with E2A proteins. Genes Dev 1992; 6:1466–1479.
9. Rudnicki MA, Schneglesberg PNJ, Stead RH, Braun T, Arnold H-H, Jaenisch R. MyoD or Myf-5 is required for the formation of skeletal muscle. Cell 1993; 75:1351–1359.
10. Hasty P, Bradley A, Morris JH, Edmondson DG, Venuti JM, Olson EN, Klein WH. Muscle deficiency and neonatal death in mice with a targeted mutation in myogenin gene. Nature 1993; 364:501–506.
11. Nabeshima Y, Hanaoka K, Hayasaka M, Esumi E, Li S, Nonaka I, Nabeshima Y. Moygenin gene disruption results in perinatal lethality because of severe muscle defect. Nature 1993; 364:532–535.
12. Olson EN, Arnold H-H, Rigby PWJ, Wold BJ. Know your neighbors: three phenotypes in null mutants of the myogenic bHLH gene *MRF4*. Cell 1996; 85:1–4.
13. Rhodes SJ, Konieczny SF. Identification of MRF4: A new member of the muscle regulatory factor gene family. Genes Dev 1989; 3:2050–2061.
14. Tajbakhsh S, Rocancourt D, Cossu G, Buckingham M. Redefining the genetic hierarchies controlling skeletal myogenesis: *Pax-3* and *Myf-5* act upstream of *MyoD*. Cell 1997; 89:127–138.
15. Maroto M, Reshef R, Münsterberg AE, Koester S, Goulding M, Lassar AB. Ectopic *Pax-3* activates *MyoD* and *Myf-5* expression in embryonic mesoderm and neural tissue. Cell 1997; 89:139–148.
16. Rawls A, Olson EN. MyoD meets its maker. Cell 1997; 89:5–8.
17. Molkentin JD, Black BL, Martin JF, Olson EN. Cooperative activation of muscle gene expression by MEF2 and myogenic bHLH proteins. Cell 1995; 83:1125–1136.
18. Molkentin JD, Olson EN. Combinatorial control of muscle development by basic helix–loop–helix and MADS-box transcription factors. Proc Natl Acad Sci USA 1996; 93:9366–9373.
19. Gu W, Schneider JW, Condorelli G, Kaushal S, Mahdavi V, Nadal-Ginard B. Interaction of myogenic factors and the retinoblastoma protein mediates muscle cell commitment and differentiation. Cell 1993; 72:309–324.
20. Chen PL, Scully P, Shew JY, Wang JY, Lee WH. Phosphorylation of the retinoblastoma gene product is modulated during the cell cycle and cellular differentiation. Cell 1989; 58:1193–1198.
21. Laiho M, DeCaprio JA, Ludlow JW, Livingston DM, Massague J. Growth inhibition by TGF-beta linked to suppression of retinoblastoma protein phosphorylation. Cell 1990; 62:175–185.
22. Sherr CJ. C1 phase progression: cycling on cue. Cell 1994; 79:551–555.
23. Matsushime H, Ewen ME, Strom DK, Kato J-Y, Hanks SK, Roussel MF, Sherr CJ. Identification and properties of an atypical catalytic subunit ($p34^{PSK-J3}$/cdk4) for mammalian D type G1 cyclins. Cell 1992; 71:323–334.
24. Kato J-Y, Matsushime H, Hiebert SW, Ewen ME, Sherr CJ. Direct binding of cyclin D to the retinoblastoma gene product (pRb) and pRb phosphorylation by the cyclin D-dependent kinase CDK4. Genes Dev 1993; 7:331–342.
25. Matsushime H, Quelle DE, Shurtleff SA, Shibuya M, Sherr CJ, Kato J-Y. D-type cyclin-dependent kinase activity in mammalian cells. Mol Cell Biol 1994; 14:2066–2076.
26. Fisher RP, Morgan DO. A novel cyclin associates with MO15/CDK7 to form the CDK-activating kinase. Cell 1994; 78:713–724.
27. Matsuoka M, Kato J-Y, Fisher RP, Morgan DO, Sherr CJ. Activation of cyclin-dependent kinase 4 (cdk4) by mouse MO15-associated kinase. Mol Cell Biol 1994; 14:7265–7275.
28. Fisher RP, Jin P, Chamberlin HM, Morgan DO. Alternative mechanisms of CAK assembly require an assembly factor or an activating kinase. Cell 1995; 83:47–57.

29. Quelle DE, Ashmun RA, Hannon GJ, Rehberger PA, Trono D, Richter KH, Walker C, Beach D, Sherr CJ, Serrano M. Cloning and characterization of murine p16^{INK4A} and p15^{INK4B} genes. Oncogene 1995; 11:635–645.
30. Hirai H, Roussel MF, Kato J-Y, Ashmun RA, Sherr CJ. Novel INK4 proteins, p19 and p18, are specific inhibitors of the cyclin D-dependent kinases CDK4 and CDK6. Mol Cell Biol 1995; 15:2672–2681.
31. Xiong Y, Hannon GJ, Zhang H, Casso D, Kobayashi R, Beach D. p21 is a universal inhibitor of cyclin kinases. Nature 1993; 366:701–704.
32. Polyak K, Lee M-H, Erdjument-Bromage H, Koff A, Roberts JM, Tempst P, Massaque J. Cloning of P27^{Kip1}, a cyclin-dependent kinase inhibitor and a potential mediator of extracellular antimitogenic signals. Cell 1994; 78:59–66.
33. Lee M-H, Reynisdottir I, Massague J. Cloning of p57^{KIP2}, a cyclin-dependent kinase inhibitor with unique domain structure and tissue distribution. Genes Dev 1995; 9:639–649.
34. Sherr CJ, Roberts JM. Inhibitors of mammalian G1 cyclin-dependent kinases. Genes Dev 1995; 9:1149–1163.
35. Franklin DS, Xiong Y. Induction of p18INK4c and its predominant association with CDK4 and CDK6 during myogenic differentiation. Mol Biol Cell 1996; 7:1587–1599.
36. Halevy O, Novitch BG, Spicer DB, Skapek SX, Rhee J, Hannon GJ, Beach D, Lassar AB. Correlation of terminal cell cycle arrest of skeletal muscle with induction of p21 by MyoD. Science 1995; 267:1018–1021.
37. Guo K, Wang J, Andres V, Smith RC, Walsh K. MyoD-induced expression of p21 inhibits cyclin-dependent kinase activity upon myocyte terminal differentiation. Mol Cell Biol 1995; 15:3823–3829.
38. Stylianopoulou F, Efstratiadis A, Herbert J, Pintar J. Pattern of the insulin-like growth factor II gene expression during rat embryogenesis. Dev Biol 1988; 103:497–506.
39. Bondy CA, Werner H, Roberts CT Jr, LeRoith D. Cellular pattern of insulin-like growth factor-I (IGF-I) and type I IGF receptor gene expression in early organogenesis: comparison with IGF-II gene expression. Mol Endocrinol 1990; 4:1386–1398.
40. Hill DJ, Crace CJ, Nissley SP, Morrell D, Holder AT, Milner RDG. Fetal rat myoblasts release both rat somatomedin-C (SM-C)/insulin-like growth factor I (IGF I) and multiplication-stimulating activity *in vitro*: partial characterization and biological activity of myoblast-derived SM-C/IGF I. Endocrinology 1985; 117:2061–2072.
41. Brown AL, Graham DE,. Nissley SP, Hill DJ, Strain AJ, Rechler MM. Developmental regulation of insulin-like growth factor II mRNA in different rat tissues. J Biol Chem 1986; 261:13144–13150.
42. Murphy LJ, Bell GI, Friesen HG. Tissue distribution of insulin-like growth factor I and II messenger ribonucleic acid in the adult rat. Endocrinology 1987; 120:1279–1282.
43. Rosen KM, Wentworth BM, Rosenthal N, Villa-Komaroff L. Specific, temporally regulated expression of the insulin-like growth factor II gene during muscle cell differentiation. Endocrinology 1993; 133:474–481.
44. Yaffe D, Saxel O. Serial passaging and differentiation of myogenic cells isolated from dystrophic mouse muscle. Nature 1977; 270:725–727.
45. Yaffe D. Retention of differentiation potentialities during prolonged cultivation of myogenic cells. Proc Natl Acad Sci USA 1968; 61:477–483.
46. Schubert D, Harris AJ, Devine CE, Heinemann S. Characterization of a unique muscle cell line. J Cell Biol 1974; 61:398–413.
47. Mulle C, Benoit P, Pinset C, Roa M, Changeux JP. Calcitonin gene-related peptide enhances the rate of desensitization of the nicotinic acetylcholine receptor in cultured mouse muscle cells. Proc Natl Acad Sci USA 1988; 85:5728–5732.
48. Tollefsen SE, Sadow JL, Rotwein P. Coordinate expression of insulin-like growth factor II and its receptor during muscle differentiation. Proc Natl Acad Sci USA 1989; 86:1543–1547.
49. Rosenthal SM, Brunetti A, Brown EJ, Mamula PW, Goldfine ID. Regulation of insulin-like growth factor (IGF) I receptor expression during muscle cell differentiation. J Clin Invest 1991; 87:1212–1219.
50. Florini JR, Magri KA, Ewton DZ, James PL, Grindstaff K, Rotwein PS. "Spontaneous" differentiation of skeletal myoblasts is dependent upon autocrine secretion of insulin-like growth factor-II. J Biol Chem 1991; 266:15917–15923.
51. Kou K, Rotwein P. Transcriptional activation of the insulin-like growth factor-II gene during myoblast differentiation. Mol Endocrinol 1993; 7:291–302.
52. Magri KA, Benedict MR, Ewton DZ, Florini JR. Negative feedback regulation of insulin-like growth factor-II gene expression in differentiating myoblasts *in vitro*. Endocrinology 1994; 135:53–62.

53. Rosenthal SM, Brown EJ, Brunetti A, Goldfine ID. Fibroblast growth factor inhibits insulin-like growth factor-II (IGF-II) gene expression and increases IGF-I receptor abundance in BC3H-1 muscle cells. Mol Endocrinol 1991; 5:678–684.
54. Tollefsen SW, Lajara R, McCusker RH, Clemmons DR, Rotwein P. Insulin-like growth factors (IGF) in muscle development. Expression of IGF-I, the IGF-I receptor, and an IGF binding protein during myoblast differentiation. J Biol Chem 1989; 264:13810–13817.
55. Jennische E, Skottner A, Hansson H-A. Satellite cells express the trophic factor IGF-I in regenerating skeletal muscle. Acta Physiol Scand 1987; 129:9–15.
56. Edwall D, Schalling M, Jennische E, Norstedt G. Induction of insulin-like growth factor I messenger ribonucleic acid during regeneration of rat skeletal muscle. Endocrinology 1989; 124:820–825.
57. Marsh DR, Criswell DS, Hamilton MT, Booth FW. Association of insulin-like growth factor mRNA expressions with muscle regeneration in young, adult, and old rats. Am J Physiol 1997; 273 (Regul Integrat Comp Physiol 42):R353-R358.
58. Ishii DN. Relationship of insulin-like growth factor II gene expression in muscle to synaptogenesis. Proc Natl Acad Sci USA 1989; 86:2898–2902.
59. Turner JD, Rotwein P, Novakofski J, Bechtel PJ. Induction of mRNA for IGF-I and -II during growth hormone-stimulated muscle hypertrophy. Am J Physiol 1988; 255 (Endocrinol Metab 18): E513–E517.
60. Alexandrides T, Moses AC, Smith RJ. Developmental expression of receptors for insulin, insulin-like growth factor I (IGF-I), and IGF-II in rat skeletal muscle. Endocrinology 1989; 124:1064–1076.
61. Shimizu M, Webster C, Morgan DO, Blau HM, Roth RA. Insulin and insulinlike growth factor receptors and responses in cultured human muscle cells. Am J Physiol 1986; 251 (Endocrinol Metab 14): E611–E615.
62. Beguinot F, Kahn CR, Moses AC, Smith RJ. Distinct biologically active receptors for insulin, insulin-like growth factor I, and insulin-like growth factor II in cultured skeletal muscle cells. J Biol Chem 1985; 260:15892–15898.
63. Alexandrides TK, Smith RJ. A novel fetal insulin-growth factor (IGF) I receptor. J Biol Chem 1989; 264:12922–12930.
64. Barenton B, Domeyne A, Garandel V, Garofalo RS. A developmentally regulated form of insulin-like growth factor receptor β-subunit in C2 myoblasts exhibiting altered requirements for differentiation. Endocrinology 1993; 133:651–660.
65. Domeyne A, Pinset C, Montarras D, Garandel V, Rosenfeld RG, Barenton B. Preferential binding of insulin-like growth factor-II (IGF-II) to a putative α2β2 IGF-II receptor type in C2 myoblasts. Eur J Biochem 1992; 208:273–279.
66. Yee D, Lebovic GS, Marcus RR, Rosen N. Identification of an alternate type-I insulin-like growth factor receptor β subunit mRNA transcript. J Biol Chem 1989; 264:21439–21441.
67. Abbott AM, Bueno R, Pedrini MT, Murray JM, Smith RJ. Insulin-like growth factor I receptor gene structure. J Biol Chem 1992; 267:10759–10763.
68. Bailyes EM, Navé BT, Soos MA, Orr SR, Hayward AC, Siddle K. Insulin receptor/IGF-I receptor hybrids are widely distributed in mammalian tissues: quantification of individual receptor species by selective immunoprecipitation and immunoblotting. Biochem J 1997; 327:209–215.
69. Brunetti A, Maddux BA, Wong KY, Goldfine ID. Muscle cell differentiation is associated with increased insulin receptor biosynthesis and messenger RNA levels. J Clin Invest 1989; 83:192–198.
70. Rosenthal SM, Brown EJ. Mechanisms of insulin-like growth factor (IGF)-II-induced IGF-I receptor down-regulation in BC3H-1 muscle cells. J Endocrinol 1994; 141:69–74.
70a. De Vroede MA, Romanus JA, Standaert ML, Pollet RJ, Nissley SP, Rechler MM. Interaction of insulin-like growth factors with a nonfusing mouse muscle cell line: binding, action, and receptor down-regulation. Endocrinology 1984; 114:1917–1929.
71. Hernández-Sánchez C, Werner H, Roberts CT Jr, Woo EJ, Hum DW, Rosenthal SM, LeRoith D. Differential regulation of insulin-like growth factor-I (IGF-I) receptor gene expression by IGF-I and basic fibroblastic growth factor. J Biol Chem 1997; 272:4663–4670.
72. McCusker RH, Clemmons DR. Insulin-like growth factor binding protein secretion by muscle cells: effect of cellular differentiation and proliferation. J Cell Physiol 1988; 137:505–512.
73. McCusker RH, Camacho-Huber C, Clemmons DR. Identification of the types of insulin-like growth factor-binding proteins that are secreted by muscle cell in vitro. J Biol Chem 1989; 264:7795–7800.
74. Ernst CW, McCusker RH, White ME. Gene expression and secretion of insulin-like growth factor-binding proteins during myoblast differentiation. Endocrinology 1992; 130:607–615.
75. James PL, Jones SB, Busby WH Jr, Clemmons DR, Rotwein P. A highly conserved insulin-like growth

factor-binding protein (IGFBP-5) is expressed during myoblast differentiation. J Biol Chem 1993; 268:22305–22312.
76. Delhanty PJD, Hill DJ, Shimasaki S, Han VKM. IGFBP-4, -5 and -6 mRNAs in the human fetus: localization to sits of growth and differentiation? Growth Regul 1993; 3:8–11.
77. McCusker RH, Clemmons DR. Effects of cytokines on insulin-like growth factor-binding protein secretion by muscle cells *in vitro*. Endocrinology 1994; 134:2095–2102.
78. Rotwein P, James PL, Kou K. Rapid activation of insulin-like growth factor-binding protein-5 gene transcription during myoblast differentiation. Mol Endocrinol 1995; 9:913–923.
79. Silverman LA, Cheng Z-Q, Hsiao D, Rosenthal SM. Skeletal muscle cell-derived insulin-like growth factor (IGF) binding proteins inhibit IGF-I induced myogenesis in rat L6E9 cells. Endocrinology 1995; 136:720–726.
80. Ewton DZ, Florini JR. IGF binding proteins-4, -5 and -6 may play specialized roles during L6 myoblast proliferation and differentiation. J Endocrinol 1995; 144:539–553.
81. Ernst CW, McFarland DC, White ME. Expression of insulin-like growth factor II (IGF-II), IGF binding protein-2 and myogenin during differentiation of myogenic satellite cells derived from the turkey. Differentiation 1996; 61:25–33.
82. Damon SE, Haugk KL, Birnbaum RS, Quinn LS. Retrovirally mediated overexpression of insulin-like growth factor binding protein 4: evidence that insulin-like growth factor is required for skeletal muscle differentiation. J Cell Physiol. 1998; 175:109–120.
83. Heino J, Massague J. Cell adhesion to collagen and decreased myogenic gene expression implicated in the control of myogenesis by transforming growth factor beta. J Biol Chem 1990; 265:10181–10184.
84. Rosen GD, Sanes JR, LaChance R, Cunningham JM, Roman J, Dean DC. Roles for the integrin VLA-4 and its counter receptor VCAM-1 in myogenesis. Cell 1990; 69:1107–1119.
85. Florini JR, Ewton DZ, Magri KA. Hormones, growth factors, and myogenic differentiation. Annu Rev Physiol 1991; 53:201–216.
86. Ewton DZ, Roof SL, Magri KA, McWade FJ, Florini JR. IGF-II is more active than IGF-I in stimulating L6A1 myogenesis: greater mitogenic actions of IGF-I delay differentiation. J Cell Physiol 1994; 16: 277–284.
87. Rosenthal SM, Cheng Z-Q. Opposing early and late effects of insulin-like growth factor I on differentiation and the cell cycle regulatory retinoblastoma protein in skeletal myoblasts. Proc Natl Acad Sci USA 1995; 92:10307–10311.
88. Engert JC, Berglund EB, Rosenthal N. Proliferation precedes differentiation in IGF-I-stimulated myogenesis. J Cell Biol 1996; 135:431–440.
89. Stewart CEH, Rotwein P. Insulin-like growth factor-II is autocrine survival factor for differentiating myoblasts. J Biol Chem 1996; 271:11330–11338.
90. Florini JR, Nicholson ML, Dulak NC. Effects of peptide anabolic hormones on growth of myoblasts in culture. Endocrinology 1977; 101:32–41.
91. Ewton DZ, Florini JR. Relative effects of the somatomedins, multiplication-stimulating activity, and growth hormone on myoblasts and myotubes in culture. Endocrinology 1980; 106:577–583.
92. Ewton DZ, Florini JR. Effects of the somatomedins and insulin on myoblast differentiation in vitro. Dev Biol 1981; 86:31–39.
93. Turo KA, Florini JR. Hormonal stimulation of myoblast differentiation in the absence of DNA synthesis. Am J Physiol 1982; 243 (Cell Physiol 12):C278–C284.
94. Schmid C, Steiner T, Froesch ER. Preferential enhancement of myoblast differentiation by insulin-like growth factors (IGF I and IGF II) in primary cultures of chicken embryonic cells. FEBS Lett 1983; 161:117–121.
95. Dodson MV, Allen RE, Hossner KL. Ovine somatomedin, multiplication-stimulating activity, and insulin promote skeletal muscle satellite cell proliferation *in vitro*. Endocrinology 1985; 117:2357–2363.
96. Allen RE, Boxhorn LK. Regulation of skeletal muscle satellite cell proliferation and differentiation by transforming growth factor-beta, insulin-like growth factor I, and fibroblast growth factor. J Cell Physiol 1989; 138:311–315.
97. Florini JR, Ewton DZ, Roof SL. Insulin-like growth factor-I stimulates terminal myogenic differentiation by induction of myogenin gene expression. Mol Endocrinol 1991; 5:718–724.
98. Brown EJ, Hsiao D, Rosenthal SM. Induction and peak gene expression of insulin-like growth factor II follow that of myogenin during differentiation of BC3H-1 muscle cells. Biochem Biophys Res Commun 1992; 183:1084–1089.
99. Milasincic DJ, Dhawan J, Farmer SR. Anchorage-dependent control of muscle-specific gene expression in C_2C_{12} mouse myoblasts. In Vitro Cell Dev Biol Anim 1996; 32:90–99.

100. Adi S, Wu NY, Rosenthal SM. Insulin-like growth factor (IGF)-I stimulates cyclin-dependent kinase (CDK)4 inhibitor p19 gene expression: a potential mechanism which limits the early mitogenic effect of IGFs and promotes subsequent differentiation in skeletal myoblasts. In: Fourth International Symposium on Insulin-Like Growth Factors, Tokyo, Oct, 1997.
101. Montarras D, Pinset C, Perez MC, Ilan J, Gros F. Muscle differentiation: insulin-like growth factors as positive modulators of myogenic regulatory genes? CR Acad Sci Paris Ser III Sci Vie 1993; 316:1029–1031.
102. Stewart CEH, James PL, Fant ME, Rotwein P. Overexpression of insulin-like growth factor-II induces accelerated myoblast differentiation. J Cell Physiol 1996; 169:23–32.
103. Coleman ME, DeMayo F, Yin KC, Lee HM, Geske R, Montgomery C, Schwartz RJ. Myogenic vector expression of insulin-like growth factor I stimulates muscle cell differentiation and myofiber hypertrophy in transgenic mice. J Biol Chem 1995; 270:12109–12116.
104. Powell-Braxton L, Hollingshead P, Warburton C, Dowd M, Pitts-Meek S, Dalton D, Gillett N, Stewart TA. IGF-I is required for normal embryonic growth in mice. Genes Dev 1993; 7:2609–2617.
105. Liu J-P, Baker J, Perkins AS, Robertson EJ, Efstratiadis A. Mice carrying null mutations of the genes encoding insulin-like growth factor I (IGF-1) and type 1 IGF receptor (IGF1r). Cell 1993; 75:59–72.
106. Florini JR, Ewton DZ. Highly specific inhibition of IGF-I-stimulated differentiation by an antisense oligodeoxyribonucleotide to myogenin mRNA. J Biol Chem 1990; 265:13435–13437.
107. Adi S, Zhang P, Wu NY, Mellon SH, Rosenthal SM. Insulin-like growth factor (IGF)-I inhibits myogenin promoter activity through element(s) contained within the sequence –94 to –9 of the myogenin gene. In: Fifth Joint Meeting, European Society for Pediatric Endocrinology and Lawson Wilkins Pediatric Endocrinology Society, Stockholm, June, 1997.
108. Coolican SA, Samuel DS, Ewton DZ, McWade FJ, Florini JR. The mitogenic and myogenic actions of insulin-like growth factors utilize distinct signaling pathways. J Biol Chem 1997; 272:6653–6662.
109. Ewton DZ, Falen SL, Florini JR. The type II insulin-like growth factor (IGF) receptor has low affinity for IGF-I analogs: pleiotypic actions of IGFs on myoblasts are apparently mediated by the type I receptor. Endocrinology 1987; 120:115–123.
110. Bach LA, Salemi R, Leeding KS. Roles of insulin-like growth factor (IGF) receptors and IGF-binding proteins in IGF-II-induced proliferation and differentiation of L6A1 rat myoblasts. Endocrinology 1995; 136:5061–5069.
111. Rosenthal SM, Hsiao D, Silverman LA. An insulin-like growth factor-II (IGF-II) analog with highly selective affinity for IGF-II receptors stimulates differentiation, but not IGF-I receptor down-regulation in muscle cells. Endocrinology 1994; 135:38–44.
112. Kiess W, Haskell JF, Lee L, Greenstein LA, Miller BE, Aarons AL, Rechler MM, Nissley SP. An antibody that blocks insulin-like growth factor (IGF) binding to the type II IGF receptor is neither an agonist nor an inhibitor of IGF-stimulated biologic responses in L6 myoblasts. J Biol Chem 1987; 262:12745–12751.
113. Rosenthal SM, Cheng Z-Q, Hsiao D, Woo EJ, Filvaroff EH, Gustafson TA. Functional inactivation of the insulin-like growth factor-I receptor delays differentiation of skeletal muscle cells. Manuscript submitted, 1999.
114. Quinn LS, Steinmetz B, Maas A, Ong L, Kaleko M. Type-1 insulin-like growth factor receptor overexpression produces dual effects on myoblast proliferation and differentiation. J Cell Physiol 1994; 159:387–398.
115. Quinn LS, Haugk KL. Overexpression of the type-1 insulin-like growth factor receptor increases ligand-dependent proliferation and differentiation in bovine skeletal myogenic cultures. J Cell Physiol 1996; 168:34–41.
116. Florini JR, Ewton DZ, Falen SL, Van Wyk JJ. Biphasic concentration dependency of stimulation of myoblast differentiation by somatomedins. Am J Physiol 1986; 250 (Cell Physiol 19):C771–C778.
117. Lau MMH, Stewart CEH, Liu Z, Bhatt H, Rotwein P, Stewart CL. Loss of the imprinted IGF2/cation-independent mannose 6-phosphate receptor results in fetal overgrowth and perinatal lethality. Genes Dev 1994; 8:2953–2963.
118. Ludwig T, Eggenschwiler J, Fisher P, D'Ercole AJ, Davenport ML, Efstratiadis A. Mouse mutants lacking the type 2 IGF receptor (IGF2R) are rescued from perinatal lethality in *Igf2* and *Igf1r* null backgrounds. Dev Biol 1996; 177:517–535.
119. Ong J, Yamashita S, Melmed S. Insulin-like growth factor I induced c-*fos* messenger ribonucleic acid in L6 rat skeletal muscle cells. Endocrinology 1987; 120:353–357.
120. Giorgino F, Smith RJ. Dexamethasone enhances insulin-like growth factor-I effects on skeletal muscle cell proliferation. J Clin Invest 1995; 96:1473–1483.

121. Milasincic DJ, Calera MR, Farmer SR, Pilch PF. Stimulation of C2C12 myoblast growth by basic fibroblast growth factor and insulin-like growth factor 1 can occur via mitogen-activated protein kinase-dependent and -independent pathways. Mol Cell Biol 1996; 16:5964–5973.
122. Olson EN, Spizz G, Tainsky MA. The oncogenic forms of N-*ras* or H-*ras* prevent skeletal myoblast differentiation. Mol Cell Biol 1987; 7:2104–2111.
123. Gossett LA, Zhang W, Olson EN. Dexamethasone-dependent inhibition of differentiation of C2 myoblasts bearing steroid-inducible N-*ras* oncogenes. J Cell Biol 1988; 106:2127–2137.
124. Wu NY, Adi S, Rosenthal SM. Evidence that insulin-like growth factor(IGF)-I-induced inhibition of myogenin gene expression is not mediated by the ras,mitogen-activated protein (MAP) kinase signaling pathway. In: 79th Annual Meeting, Endocrine Society, June 1997.
125. Sarbassov DD, Jones LG, Peterson CA. Extracellular signal-regulated kinase-1 and -2 respond differently to mitogenic and differentiative signaling pathways in myoblasts. Mol Endocrinol 1997; 11:2038–2047.
126. Kaliman P, Vinals F, Testar X, Palacin M, Zorzano A. Phosphatidylinositol 3-kinase inhibitors block differentiation of skeletal muscle cells. J Biol Chem 1996; 271:19146–19151.
127. James PL, Stewart CEH, Rotwein P. Insulin-like growth factor binding protein-5 modulates muscle differentiation through an insulin-like growth factor-dependent mechanism J Cell Biol 1996; 133: 683–693.
128. Bach LA, Hsieh SP, Brown AL, Rechler MM. Recombinant human insulin-like growth factor (IGF)-binding protein-6 inhibits IGF-II-induced differentiation of L6A1 myoblasts. Endocrinology 1994; 135:2168–2176.

22 IGFs and the Digestive Tract

P. Kay Lund, PhD

CONTENTS

INTRODUCTION

The gastrointestinal tract is a major target organ for growth hormone (GH) and the insulin-like growth factors (IGFs) *(1,2)*. GH and/or IGF-I may have therapeutic benefits in patients with bowel disease *(3)*. The IGFs may, however, have a dual identity in that they may promote normal tissue growth and repair, but in excess, may exacerbate disease processes such as fibrosis and tumorigenesis. This chapter emphasizes known and emerging actions of GH/IGF-I in intestinal adaptation, inflammatory bowel disease, and colon cancer.

GH and the IGFs act on the gastrointestinal (GI) tract in an endocrine manner. IGFs that are expressed locally within the bowel in a GH-dependent or independent manner may have autocrine or paracrine effects on specific cell types within the bowel. Defining the relative contributions of circulating IGFs and locally expressed IGFs remains a central question in understanding IGF action in the bowel as well as other organs. Available information about the known endocrine and local actions of the IGFs in the bowel and possible strategies aimed at further dissecting these different modes of IGF action are described. A hallmark of IGF action is that alone they often have weak effects on cell proliferation or differentiated function but the IGFs strongly potentiate the actions of

From: *Contemporary Endocrinology: The IGF System*
Edited by: R. Rosenfeld and C. Roberts Jr © Humana Press, Inc., Totowa, NJ

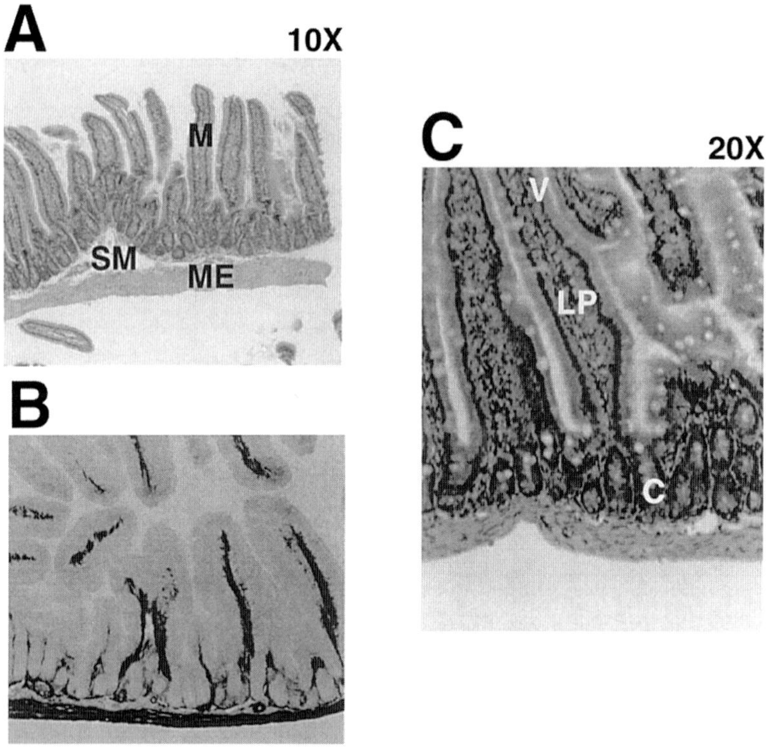

Fig. 1. Photomicrographs to illustrate the microanatomy of the small intestine. **(A)** Low-power photomicrograph of a section through the wall of the small intestine. M, mucosa; SM, submucosa; ME, muscularis externa. **(B)** Localization of mesenchymal cells that show positive immunostaining for α smooth muscle actin illustrates positively stained pericryptal myofibroblasts, a core of positive stained cells within the villi, a thin layer of positively stained smooth muscle cells separating the mucosa and submucosa (muscularis mucosa), and the intensely stained muscularis externa. **(C)** High-power photomicrograph to illustrate crypts (C), villi (V), and lamina propria (LP).

other hormones and growth factors. Information about the interactions of the IGFs with other hormones and growth factors in the GI tract are therefore reviewed.

GROWTH OF THE DIGESTIVE TRACT

Structure of the Bowel Wall

The intestinal mucosa or mucosal lining of the bowel consists of an epithelial cell layer that faces the lumen of the bowel, the underlying lamina propria and a thin layer of smooth muscle termed the muscularis mucosa (Figs. 1 and 2). The mucosal lining of the bowel serves several important functions. The epithelial layer produces electrolytes and enzymes that are essential for digestion of ingested food. Epithelial cells also express transport proteins involved in nutrient absorption. The bowel mucosa serves as a barrier to prevent entry of microorganisms and ingested toxins into the body. In the lamina propria a layer of mesenchymal cells (fibroblasts/myofibroblasts) lies directly adjacent to the epithelial cells and the two cell types are separated by basement membrane (Figs. 1 and 2). Interactions between epithelial cells and adjacent mesenchymal cells are integral to normal

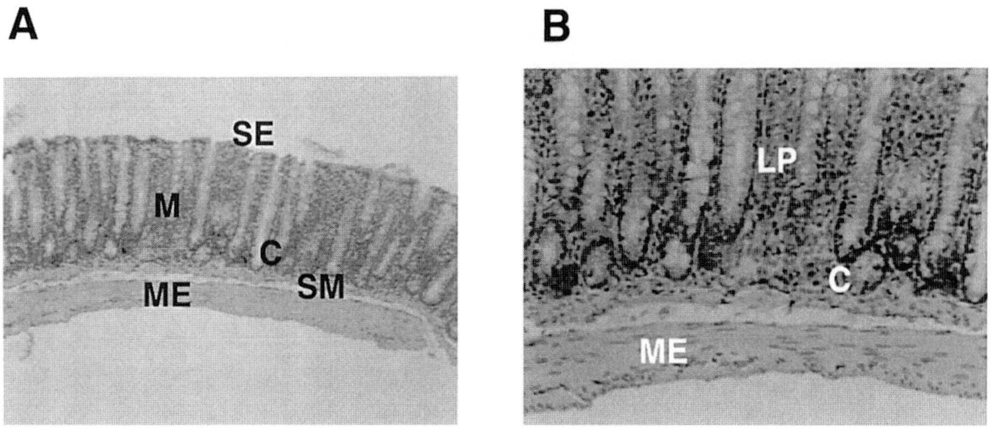

Fig. 2. Low- (**A**) and high (**B**)-power photomicrographs to illustrate the microanatomy of the colon; SE, surface epithelial cells; L, lamina propria; SM, submucosa; ME, muscularis externa.

development, differentiation, and morphogenesis of the bowel epithelium *(4,5)*. In addition to mesenchymal cells, the lamina propria contains immune cells, smooth muscle cells (Fig. 1), and capillaries. The submucosa is a connective tissue layer lying below the mucosa and is composed of primarily fibroblasts and collagen. The submucosa separates the mucosa from underlying enteric smooth muscle, the muscularis externa. In most regions of bowel, the enteric smooth muscle is comprised of an inner circular smooth muscle layer and an outer longitudinal smooth muscle layer (Fig. 1). Coordinated contractions of the enteric smooth muscle determine appropriate mixing of ingested food and aboral transit of food in a time frame that permits efficient digestion and absorption.

Continuous Renewal of the Epithelial Lining

The epithelial lining of the bowel is continuously renewed in all regions from the esophagus to the colon. The processes underlying epithelial renewal have been most extensively studied in the small intestine and colon *(6–8)*. In the mammalian small intestine and colon there is complete turnover of the majority of the epithelial cell population in a period of 24–96 h *(7)*. The small intestinal epithelium is organized into a crypt–villus axis (Figs. 1 and 3). The crypts contain a compartment of continuously proliferating cells. Cells migrate from the crypts onto the villi. As they exit the crypts, proliferation ceases and the cells differentiate into multiple epithelial subtypes. These include: (1) columnar epithelial cells that express proteins involved in electrolyte secretion, nutrient digestion, and nutrient absorption; (2) goblet cells that secrete mucus; and (3) enteroendocrine cells that secrete a wide range of bioactive hormones and peptides that control gut function and metabolism. The columnar epithelial cells and goblet cells migrate up the villus until they are sloughed off at the villus tips (Fig. 3). The rate of migration and turnover of enteroendocrine cells is slower than that of other epithelial subtypes *(7)*.

The colon is organized into crypts and surface epithelium and, as in small bowel, contains a compartment of proliferating cells within the crypts that undergo terminal differentiation into columnar epithelial cells, goblet cells, and enteroendocrine cells as they migrate to the surface epithelium (Fig. 2). Surface colonic enterocytes are sloughed off into the lumen (Fig. 3). Along the length of the small intestine and colon there are distinct populations and proportions of particular epithelial cell types.

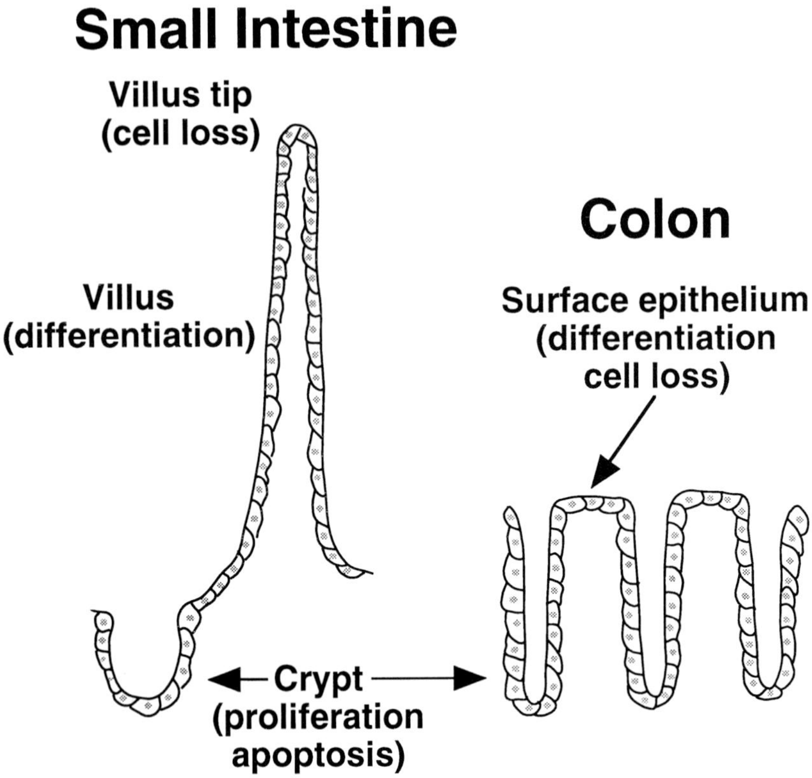

Fig. 3. Schematic of the small intestine and colon to show the location of proliferating crypt cells, differentiated cells, and cells that undergo spontaneous apoptosis in the crypts.

The high rate of turnover of epithelial cells within the small intestine and colon probably reflects exposure of surface epithelial cells to digestive enzymes, bacteria, and ingested toxins. Loss of cells at the villus tips in the small intestine and surface epithelium in the colon is thought to occur by programmed senescence that involves loss of cell adhesion and/or programed cell death/apoptosis although the role of apoptosis is still a matter of some debate *(8)*. The rate of cell proliferation and production of new cells within the crypts must be finely regulated to match the rates of cell loss from the luminal surface to maintain a steady-state functional mass of the intestinal epithelium. Cell differentiation also must be tightly regulated to maintain the normal spectrum of different cell types that dictate the specific functional characteristics of the mucosal epithelium within a particular bowel segment. Although the contribution of apoptosis to cell loss into the intestinal lumen is not yet well defined it is well established that a small subpopulation of cells within the intestinal crypts undergo spontaneous apoptosis *(8,9)*. Spontaneous crypt cell apoptosis may reflect excessive cell production during the normal process of renewal of the mucosal epithelium. The high rates of cell proliferation within the crypts and/or exposure of the crypt cells to the toxic environment of the intestinal lumen also may result in unavoidable production of genetically aberrant cells. Spontaneous apoptosis may provide a mechanism to dispose of these genetically damaged crypt cells. It is hypothesized that the high rates of crypt cell apoptosis protect against tumorigenesis *(8)*. Indirect support for this possibility stems from observations that the small bowel shows much higher rates of spontaneous crypt cell apoptosis than the colon and tumors of the small bowel are very rare compared with a relatively high incidence of tumors in the colon *(8)*. Rates of crypt cell apoptosis are increased by DNA damaging cytotoxins and gamma irradiation *(8)*. Apoptosis in this setting occurs by p53-dependent mechanisms. Recent evidence indicates a role for the IGF system in regulating apoptosis in the intestine and a link between the IGF system and P53 (*see below*).

INTESTINAL ADAPTATION

The small intestine, and to a lesser extent the large intestine, have an exquisite capacity to regulate functional mass and functional capabilities to match physiological demand *(10)*. This phenomenon is termed intestinal adaptation and has been studied extensively in animals *(10)*. Intestinal adaptation does occur in humans although has been less extensively studied. Defining the mechanisms that regulate intestinal adaptation is relevant to a number of bowel diseases *(10)*.

Nutrient-Mediated Intestinal Adaptation

If animals are fasted or fed entirely by intravenous nutrition (total parenteral nutrition [TPN]), there is the rapid development of mucosal hypoplasia and atrophy characterized by reduced crypt depth, reduced villus height, reduced DNA and protein content per unit length of bowel, and reduced digestive and absorptive capabilities (Fig. 4; *[10,11]*). There is an aboral gradient of this effect with small bowel showing greater reductions in mass than the large bowel *(10)*. Conversely, refeeding after a fast or the reintroduction of enteral nutrient after parenteral nutrition results in rapid reversal of the hypoplasia *(10)*. If animals are made hyperphagic, the bowel mucosa shows adaptive hyperplasia so that the increased mass matches the increased physiological demand *(10)*.

Adaptive Growth in Response to Loss of Functional Bowel Mass

After surgical removal of portions of the bowel, the remaining bowel has the ability to undergo adaptive increases in growth to restore functional mass. This is the most extensively studied and clinically relevant example of intestinal adaptation *(10)*. Surgical removal or resection of portions of the bowel is required in a number of clinical situations. These include congenital disorders; inflammatory disorders such as necrotizing enterocolitis, Crohn's disease, and ulcerative colitis; and other problems including intestinal neoplasia, ischemia, or trauma. Short bowel syndrome is a malabsorption syndrome that occurs following resection of major portions of the small intestine *(12,13)*. Short bowel syndrome is a significant problem in an increasing number of premature infants who develop necrotizing enterocolitis due to ischemic injury of the small intestine and colon *(13)*. These infants cannot tolerate oral nutrition and must be fed by TPN, often for several years and sometimes for a lifetime. Short bowel syndrome occurs in a significant number of adult patients who undergo bowel resection for Crohn's disease, neoplasia, or ischemic disorders. Many of these adult patients must undergo long-term parenteral nutrition. As well as the practical, psychological, and social problems associated with parenteral nutrition, the atrophy of the mucosa induced by TPN probably limits the capacity of the bowel for adaptive growth. A commonly used animal model of intestinal adaptation is massive proximal small bowel resection. After resection of major portions of proximal small bowel, the remnant bowel exhibits dramatic adaptive growth *(10)*. This involves mucosal hyperplasia such that there can be as much as a fourfold increase in the mucosal surface area and compensatory increases in digestive and absorptive capabilities per unit

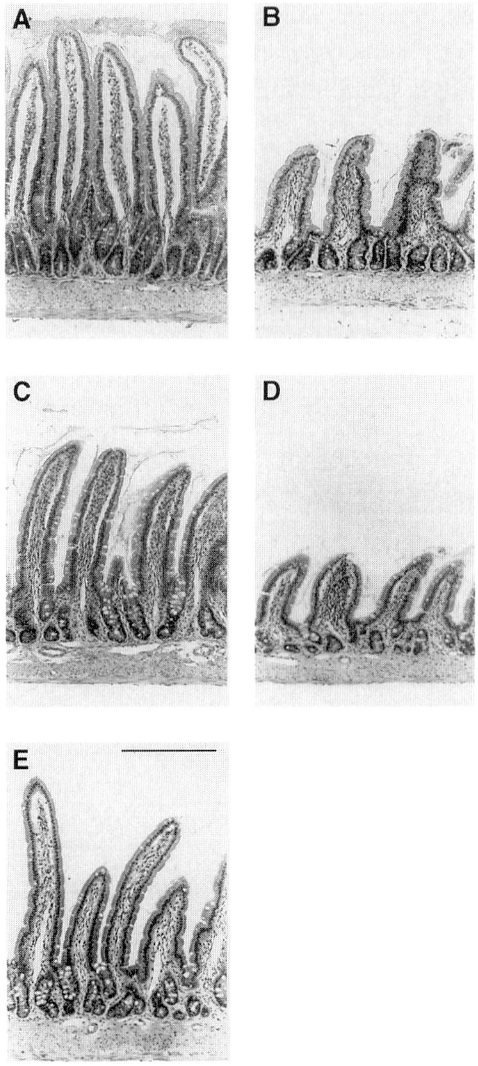

Fig. 4. Photomicrographs of jejunum of *ad libitum* fed rat **(A)** TPN-fed animal given TPN alone **(B)** or TPN supplemented with GH **(C)**, IGF-I **(D)**, or GH + IGF-I **(E)**. Note the mucosal atrophy and hypoplasia induced by TPN and reversal of this effect by IGF-I alone or GH + IGF-I but not GH alone. (These data were kindly provided by Dr. Denise Ney, University of Wisconsin and are part of a study reported by Peterson et al. Am J Physiol 1997; 272:G1100–G1108).

length of intestine. There also are adaptive increases in bowel length *(12)*. Significant adaptive growth after bowel resection requires ongoing enteral nutrition *(10,12)*. In humans after bowel resection adaptive growth of intestine is evident but is less dramatic than observed in animal models *(10,12)*. In patients with short bowel syndrome, strategies to promote adaptive growth are desirable. Recent evidence suggests that GH and/or IGF-I promote intestinal adaptation in an endocrine and paracrine manner (*see below*).

Cellular Mechanisms of Intestinal Adaptation

It is well established that altered rates of crypt cell proliferation contribute to intestinal adaptation. Altered rates of crypt/villus cell migration or altered differentiated function of villus/surface epithelial cells may contribute to adaptive changes in functional capabilities of the bowel *(6,10)*. Altered rates of apoptosis of crypt and/or villus cells may contribute to intestinal adaptation although this has not yet been studied extensively *(6,9)*. Factors that inhibit apoptosis of intestinal crypt cells could presumably promote adaptive hyperplasia within the bowel but may also predispose to tumorigenesis. Thus, a complete understanding of the relative risks and benefits of therapy with GH and IGF-I in patients with short bowel syndrome requires an understanding of the cellular mechanisms by which they promote bowel growth.

Most studies of bowel growth, adaptation, and disease focus on the mucosal epithelium. An important component of adaptive growth and functional capability is bowel length. After small bowel resection, altered mass or motility of smooth muscle layers could contribute to adaptive increases in functional capability of remnant bowel. Recent evidence indicates that enteric smooth muscle is an important target organ for IGF-I.

Increased bowel length is a desirable outcome in patients with short bowel syndrome and GH and IGF-I appear to promote increases in bowel length (*see below*).

ENDOCRINE EFFECTS OF GH/IGF-I ON INTESTINAL ADAPTATION

Fasting and Refeeding

Serum IGF-I declines during fasting and returns to normal levels with refeeding *(14,15)*. These changes in serum IGF-I show a high degree of positive correlation with the time course and magnitude of fasting induced adaptive hypoplasia of bowel mucosa and refeeding induced adaptive hyperplasia *(15)*.

TPN

In rats, reduced serum concentrations of IGF-I accompany the atrophy/hypoplasia of the small bowel induced by TPN, indicating that reduced trophic effects of circulating IGF-I may contribute to TPN induced reductions in bowel mass *(15)*. Given these observations, it seems possible that administration of IGF-I and/or GH may prevent TPN-induced mucosal atrophy and loss of function. A recent series of studies directly analyzed the effects of GH and IGF-I on growth and function of the jejunum in rats undergoing TPN after surgical stress. GH and IGF-I each attenuate the body weight loss during TPN and increase nitrogen retention and serum IGF-I concentrations in an additive manner *(16)*. IGF-I, but not GH, has dramatic effects to reverse the TPN-induced atrophy of bowel mucosa (Fig. 4; *[11]*). The lack of effect of GH on the mucosa appears to be specific to growth of the mucosal layer of the bowel because both GH and IGF-I increase protein synthesis in the external smooth muscle layers of the jejunum of TPN-fed rats *(16)*. Some mucosal responses to GH occur in TPN-fed rats because GH alters ion transport, sucrase activity, and glucose transport in these animals *(11)*. Circulating levels of IGF-I are similar in TPN-fed rats given GH or IGF-I and GH receptor (GHR) expression is normal in small bowel of this TPN model *(17)*. It seems possible, therefore, that the catabolic state induced by surgical stress/TPN causes a decrease in the responsiveness of the jejunal mucosa to the growth-promoting actions of GH at the level of the GHR signaling mechanisms that induce growth. Alternatively, there may be differences in bioavailability of circulating IGF-I for actions on the bowel mucosa in GH compared with IGF-I infused animals because of different effects of GH and IGF-I on circulating acid labile subunit (ALS), circulating IGF binding proteins (IGFBPs), or IGFBPs expressed locally within the bowel mucosa *(17)*. Whatever the mechanism for the reduced effects of GH on mucosal growth relative to IGF-I, these studies add to a growing body of evidence that IGF-I mediates many of the actions of GH on growth of bowel mucosa and may be more potent than GH, especially in a situation of TPN. These observations in rats are relevant to humans undergoing long-term TPN. Recent clinical trials indicate that administration of GH and glutamine in patients undergoing long-term TPN for short bowel syndrome promotes normal nutrient absorption and reduces the requirements for TPN *(3)*. It will be of interest to establish if this effect of GH is due to reversal of TPN-induced mucosal atrophy and to compare the effects of GH and IGF-I on bowel growth and function in patients undergoing TPN.

Bowel Resection

It has been known for many years that in rat models, GH enhances adaptive growth of the small bowel after proximal small bowel resection *(18)*. In rat models, systemic IGF-I administration improves weight gain and enhances growth of small bowel mucosa after proximal small bowel resection *(19,20)*. Systemic IGF-I enhances colonic mucosal growth and function in rats after resection of 60% small bowel and cecum, indicating potential benefits of IGF-I in reversing colonic dysfunction associated with short bowel syndrome *(21)*. IGF-I administration improves growth and absorption of glucose and water in transplanted rat small intestine *(22)*. In patients with catabolic disease, combined GH/IGF-I therapy promotes greater anabolic effects than GH and reduces the insulin resistance and hyperglycemia that are associated with GH therapy alone *(23)*. Recent studies indicate nitrogen-sparing effects of short-term IGF-I therapy in patients after radical large bowel resection *(24)*. Together, these findings indicate that GH and/or IGF-I will be used increasingly as therapy for patients with short bowel syndrome.

LOCAL EXPRESSION OF THE IGF SYSTEM IN INTESTINAL ADAPTATION

Local IGF-I Expression

IGF-I and/or IGF-II are expressed throughout human and rodent small bowel *(25,26)*. In rats, IGF-II predominates in bowel during embryonic development whereas IGF-I predominates postnatally *(25)*. In humans, IGF-I and IGF-II are expressed at high levels in fetal bowel *(26,27)*. Levels of IGF-I and IGF-II expression have not been systematically compared in human bowel postnatally but in contrast to rats, significant levels of IGF-II mRNA are detected in adult human bowel (Lund and Pucilowska, *unpublished observations*).

A number of lines of evidence indicate that locally expressed IGF-I contributes to adaptive growth of small bowel. In rats, local increases in IGF-I expression in different bowel segments after refeeding correlate with the time course and magnitude of adaptive growth of these segments in response to refeeding *(15)*. Myenteric denervation of a segment of small bowel in rats elicits dramatic local increases in growth of mucosa and enteric smooth muscle *(28,29)*. This localized adaptive growth is accompanied by dramatic increases in local expression of IGF-I within the mucosa and smooth muscle without effects on circulating IGF-I *(30)*. Increased IGF-I expression occurs in transplanted small bowel and may contribute to growth and function of the transplant *(22)*. Some *(21,31)*, but not all *(31)*, studies indicate increased local expression of IGF-I in small bowel or colon during adaptive growth in response to bowel resection. Available evidence indicates that mesenchymal cells (fibroblasts, myofibroblasts, or smooth muscle cells) within the lamina propria or submucosa are the major sites of local IGF-I or IGF-II synthesis in normal small bowel in vivo *(15,25,26,30)*. The major role of intestinal mesenchyme in intestinal growth during fetal development is well established *(5)* but little is known about the role of mesenchyme in regulating bowel growth postnatally or about the precise mesenchymal cell derived factors that regulate growth. The increased local IGF-I expression during adaptive growth of bowel epithelium in adult rats indicates that IGF-I may be an important mesenchymal cell derived growth factor that exerts paracrine actions on mucosal growth or autocrine effects on growth of intestinal mesenchymal cells themselves. Recent in vitro studies demonstrate directly that IGFs are a component of intestinal fibroblast derived mitogenic activity for intestinal epithelial cells *(32,33)*.

Locally Expressed IGF Binding Proteins (IGFBPs) and Intestinal Adaptation

IGFBP-3, IGFBP-4, and IGFBP-5 are probably most relevant to intestinal growth and adaptation after birth as they are the major IGFBPs found in the bowel of normal adult rats and mice *(15,30,34,35)* and humans (Lund, Pucilowska, and Mohapatra, *unpublished data*). In the small bowel or colon of adult rats or mice in vivo, *in situ* hybridization does not reveal detectable levels of IGFBP-3, IGFBP-4, and IGFBP-5 mRNAs in epithelial cells. Each of these IGFBP mRNAs is, however, detected in the lamina propria of small bowel mucosa and/or enteric smooth muscle layers *(15,30,35)*. These findings and observations that cultured intestinal fibroblasts, myofibroblasts, or smooth muscle cells in vitro express IGFBP-3, IGFBP-4, and IGFBP-5 *(32,36,37)* indicate that mesenchymal cells are a primary source of IGFBPs in the small intestine. Mesenchymal cell derived IGFBPs may act in a paracrine manner to modulate IGF action on neighboring intestinal epithelial cells or have autocrine effects on mesenchymal cells. IGFBP-3 is predominantly if not exclusively expressed in the mucosal lamina propria in normal rat or mouse small bowel *(15,34)*. Expression of IGFBP-3 is reduced in lamina propria of small bowel in three different animals models of adaptive hyperplasia of the mucosa: proximal small bowel resection *(2,31)*, fasting following by refeeding *(15)*, and myenteric denervation *(30)*. Decreased expression of lamina propria derived IGFBP-3 may increase the bioavailability of circulating or locally expressed IGF-I for trophic actions on neighboring intestinal epithelial cells. Observations that IGF analogs with reduced affinity for IGFBPs have more potent trophic effects on the bowel than IGF-I support this possibility *(38,39)*.

IGFBP-4 is expressed in lamina propria and muscularis externa of adult rat small bowel but available evidence does not indicate altered expression in situations of adaptive growth although this has been studied only at the mRNA level.

In rat small bowel or colon, IGFBP-5 is expressed at high levels in the muscularis externa *(30,35,37)*. IGFBP-5 expression is increased in muscularis externa during adaptive overgrowth in response to myenteric denervation, concomitant with elevated expression of IGF-I *(30)*. As IGFBP-5 potentiates IGF action in some systems *(40)* this increase in IGFBP-5 expression may enhance the trophic effects of IGF-I on adaptive growth of intestinal smooth muscle. In TPN-fed rats, IGF-I and not GH induces IGFBP-5 expression in the mucosa, in cells within the lamina propria *(41)*. This IGF-I specific induction of mucosal IGFBP-5 and mucosal growth in TPN-fed rats provides indirect support for a hypothesis that IGFBP-5 enhances the trophic actions of IGF-I on mucosa during adaptive growth *(17,41)*. Defining the relevance of altered IGFBP expression in bowel in vivo during intestinal adaptation requires a more complete understanding of the effects of different IGFBPs on IGF-I action on specific cell types in bowel. Additional studies in intestinal cell lines in culture and in transgenic and gene knockout models with altered IGFBP expression *(42)* will be required to elucidate the role of IGFBPs in intestinal adaptation.

THE IGF SYSTEM AND INFLAMMATORY BOWEL DISEASE

Inflammatory Bowel Disease (IBD)

Ulcerative colitis and Crohn's disease are common, immunologicaly mediated, inflammatory diseases of the gastrointestinal tract that are characterized by chronic inflammation with periods of clinical remission and reactivation *(43,44)*. Ulcerative colitis is limited to the colon and its characterized by acute and chronic inflammation of the lamina propria with or without epithelial cell destruction, epithelial ulceration, and crypt abscesses *(43,44)*. Crohn's disease similarly involves mucosal damage, ulceration, and inflammation but differs from ulcerative colitis in that it may affect any region of the gastrointestinal tract and is characterized by transmural inflammation and fibrosis *(43,44)*. Fibrosis involves disorganized hyperplasia and collagen deposition within the lamina propria, muscularis mucosa, submucosa, and the external muscle layers *(44,45)*. Fibrosis, fistulae, and stricture formation, with subsequent bowel obstruction, are frequent complications of Crohn's disease that lead to surgical intervention and development of short-bowel syndrome in a significant number of patients *(44,45)*.

Growth factors may have multiple and potentially opposing roles in IBD. Endogenous or exogenously administered growth factors may have beneficial effects in IBD owing to trophic actions on intestinal epithelial or mesenchymal cells. Remission of symptoms in active IBD involves repair or regeneration of damaged mucosa *(44)*. Growth factors could potentially limit or protect against mucosal damage during active inflammation or could promote better or more rapid mucosal restitution/regeneration during disease remission. Small bowel resection is commonly used to treat complications of Crohn's disease *(10,44)*. By stimulating adaptive growth of remnant bowel after resection, growth factors could promote enhanced or optimal recovery. After mucosal or transmural damage during active inflammation, restoration of normal bowel architecture requires mesenchymal cell proliferation and collagen deposition to promote tissue repair. Growth factors could enhance these processes.

Potentially beneficial actions of growth factors in IBD must be balanced against their possible detrimental effects. Growth factors could mediate or exacerbate inflammation by chemotactic or mitogenic effects on macrophages or immune cells or by stimulating the production of proinflammatory cytokines. In excess, growth factors could promote aberrant proliferation of mucosal epithelial cells or could lead to enhanced susceptibility to colon cancer. Growth factors also could promote excessive mesenchymal cell responses and worsen the complications of fibrosis associated with Crohn's disease *(2,45)*. Patients with IBD show increased incidence of colon cancer and the underlying cause is not defined *(46)*.

IGFS AND INTESTINAL INFLAMMATION AND INJURY

Macrophages, Immune Cells, and Cytokines

GH and IGF-I regulate immune function *(47)*. Both GH and IGF-I prime neutrophils for superoxide anion secretion, which is increased in active Crohn's disease *(48)*. IGF-I is expressed in lymphocytes and stimulates proliferation of immature T lymphocytes in an autocrine manner *(49)*. Proinflammatory cytokines induce IGF-I expression in some circumstances. Macrophages that migrate to sites of tissue injury express IGF-I *(50–52)*. Receptor-mediated uptake of advanced glycosylation end products (AGEs) found in diabetes and aging induces macrophages to synthesize and secrete pro-IGF-I *(53)*. IL-1 is the mediator of AGE-stimulated expression and secretion of IGF-I precursor *(53)*. Hyaluronate activation of CD44 induces IGF-I expression in murine macrophages by a tumor necrosis factor-α (TNF-α)-dependent mechanism *(54)*. In contrast, interferon-β or -γ suppresses TNF-α induction of IGF-I in macrophages *(55)*. Recent studies indicate that IGF-I expression is up-regulated in colon of mice with acute colitis induced by oral ingestion of dextran sulfate (DSS), concomitant with increased expression of interleukin-1 and TNF-α, markers of acute inflammation *(56)*. Nutrient restriction inhibits DSS-induced inflammation and prevents the increase in IL-1 and IGF-I expression *(56)*. These findings suggest that induction of IGF-I is an early event in acute colitis in vivo and provide indirect evidence for an association between IL-1 and IGF-I induction during acute colitis. Additional studies will be required to establish whether IL-1 has a direct effect on IGF-I expression during acute inflammation of the bowel and to establish if IGF-I has protective effects or promotes wound healing after acute inflammation.

Wound Healing and Mesenchymal Cell Responses to Inflammation

There are precedents for a role of IGF-I in wound healing and collagen deposition. This is relevant to potential roles of the IGFs in restoration of normal bowel architecture and function after acute bowel inflammation or in the excessive fibrogenic complications that occur due to chronic inflammation in Crohn's disease. IGF-I stimulates proliferation and collagen synthesis in cultured skin fibroblasts and aortic smooth muscle cells *(57–59)*. Insulin at concentrations that activate the type 1 IGF receptor stimulates collagen synthesis in intestinal fibroblasts from patients with Crohn's Disease *(60)*. IGF-I acts in an autocrine manner to stimulate wound repair in cultured endothelial cells *(61)*. Recent studies provide direct evidence that IGF-I stimulates proliferation of intestinal fibroblasts and intestinal smooth muscle cells in culture *(36,37)*.

In a model of freeze–thaw injury, IGF-I is produced in macrophages during initial stages of injury, but later is increased in endothelial cells of newly formed blood vessels and in chondroblasts and chondrocytes *(62)*. By analogy with this model, during acute phase bowel inflammation, cytokines may regulate IGF-I or IGF-II production by macrophages or lymphocytes and the IGFs, in turn, could regulate proliferation of these or other cell types. During the progression from acute to chronic inflammation, cytokines from immune cells may induce IGF production by mesenchymal cells and local production of IGF-I may trigger proliferation, phenotypic modification, and collagen deposition in an autocrine or paracrine manner. At present there are little nor no data about cellular sites of IGF expression during acute inflammation of bowel although studies in models of chronic inflammation associated with fibrosis indicate that IGF-I expression is induced in mesenchymal cells and areas of fibrosis (*see below*).

Altered IGF Expression Associated with Chronic Bowel Inflammation and Fibrosis

In bowel of rats with chronic granulomatous inflammation induced by intramural injection of peptidoglycan polysaccharides (PG-PS) there are dramatic local increases in

IGF-I expression in bowel and no significant change in circulating IGF-I levels *(63)*. Sites of increased IGF-I mRNA expression in the inflamed bowel are submucosal mesenchymal cells, which lie adjacent to IL-1 expressing cells at the core of granulomas *(63,64)*. These findings and data in other sytems demonstrating that IL-1 or TNF-α induce IGF-I expression in macrophages *(53,54)* support a hypothesis that proinflammatory cytokines induce IGF-I expression in intestinal mesenchymal cells during chronic inflammation or during the progression from acute to chronic inflammation.

During chronic PG-PS induced intestinal inflammation, the cells that show increased IGF-I expression have fibroblast/smooth muscle-like morphology and represent sites of increased collagen deposition and fibrosis *(63,64)*. These IGF-I expressing cells also show positive immunostaining for α-smooth muscle actin *(63,64)*, indicating that IGF-I expression is increased in modified fibroblasts (myofibroblasts) or smooth muscle cells in areas of fibrosis during granulomatous inflammation. Other more recent data indicate that increased IGF-I expression is a characteristic feature of chronic bowel inflammation associated with fibrosis. In a different animal model of chronic inflammation induced by intracolonic instillation of ethanol/trinitrobenzene sulfonic acid (TNBS), IGF-I expression is increased in lamina propria and in smooth muscle layers, particularly in areas of transmural inflammation and disorganization of smooth muscle *(35)*.

These observations in TNBS- and PG–PS-treated animals indicate that IGF-I may mediate mesenchymal cell repair responses in at least two different models of chronic bowel inflammation. Recent findings that IGF-I stimulates collagen synthesis in cultured rat intestinal smooth muscle cells support the concept that IGF-I plays a role in tissue repair and collagen deposition *(37)*. Additional studies suggest that findings in the animal models are relevant to Crohn's disease. In ulcerated or strictured bowel of patients with Crohn's disease obtained at surgical resection, levels of IGF-I and IGF-II mRNAs are increased relative to levels in uninvolved bowel *(65)*. In contrast, in patients with ulcerative colitis who do not show fibrosis of the bowel, levels of IGF-I and IGF-II mRNAs are similar in grossly involved and uninvolved bowel *(65)*. Recent observations indicate that IGF-I immunoreactivity is elevated in gavage fluid obtained from patients with Crohn's disease but not ulcerative colitis *(66)*. Together these data provide evidence that increased local IGF expression may be a distinguishing feature of Crohn's disease and ulcerative colitis. If, as in animal models of chronic inflammation, the sites of increased IGF expression in bowel of patients with Crohn's disease correspond to regions of fibrosis, this will add strong support to our hypothesis that IGFs mediate or contribute to fibrosis in Crohn's disease. To definitively test this hypothesis, it will be essential to define the precise cell types that are responsible for fibrosis in Crohn's disease and to further analyze IGF action on proliferation and collagen synthesis in different intestinal mesenchymal cell subtypes.

The precise cell type(s) mediating the fibrotic response in intestinal inflammation are not conclusively defined *(2)*. Graham and colleagues have obtained strong evidence to suggest that phenotypically modified intestinal smooth muscle cells underlie stricture and fibrosis of Crohn's disease *(45)*. Other groups implicate subepithelial and/or submucosal fibroblasts as sites of increased collagen deposition in Crohn's disease *(60,67)*. A third possibility is that intestinal myofibroblasts underlie the fibrosis of Crohn's disease. Alpha smooth muscle actin expression is characteristic of cells in strictures associated with Crohn's disease *(45,68)* and could reflect smooth muscle or myofibroblast phenotype. Myofibroblasts play a role in many fibrotic conditions that involve reorganization of connective tissues including would healing, hepatic fibrosis, vascular injury, pulmonary fibrosis, and stromal dell responses to neoplasia *(69)*. During a number of situations of wound repair on connective tissue reorganization, cytokines, growth factors, and/or specific extracellular matrix components appear to induce preferential "selection" or "clonal expansion" of cells with myofibroblast phenotype that are characterized by expression of vimentin and α smooth muscle actin but lack of expression of desmin and smooth muscle myosin *(69–71)*. Other studies indicate that smooth muscle cells or fibroblasts can assume myofibroblast-like phenotype during injury or proliferation *(69)*. Meaningful evaluation of the role of IGF-I in inflammation induced fibrosis will require definitive identification of the precise mesenchymal cells types that overexpress IGF-I in bowel of animal models of chronic bowel inflammation and fibrosis such as the PG-PS model and in bowel of patients with Crohn's disease. In this regard, development of additional markers of fibroblast, myofibroblast, and smooth muscle phenotype would be useful.

Transforming growth factor-β (TGF-β) is implicated as a potential mediator of fibrosis in Crohn's disease and other diseases *(72)*. TGF-β is up-regulated in animals models of inflammation-induced fibrosis and in bowel of patients with Crohn's disease *(73–75)*. TGF-β increases collagen synthesis in cultured intestinal smooth muscle cells and intestinal fibroblasts *(45,73–75)*. A number of observations indicate that TGF-β and IGF-I may interact in regulating mesenchymal cell phenotype, proliferation, and/or fibrosis during chronic inflammation of the bowel. In a human intestinal fibroblast cell line TGF-β but not IGF-I induces a transition to myofibroblast phenotype assessed by positive fibrillar staining for α smooth muscle actin *(36)*. Preliminary data indicate also that IGF-I can stimulate proliferation of TGF-β-treated, α actin-positive intestinal fibroblasts although this is associated with decreased immunostaining for fibrillar α smooth muscle actin (Pucilowska, Simmons, and Lund, *unpublished observations*). These latter observations indicate that IGF-I may reverse TGF-β-mediated transition to myofibroblast phenotype, at least in vitro. Several models of TGF-β and IGF-I interaction during inflammation-induced fibrosis could be proposed based on the available data. In one model, increased expression of TGF-β may first induce a phenotypic transformation of intestinal fibroblasts and/or smooth muscle cells to a myofibroblast phenotype followed by clonal expansion or hyperproliferation of these cells in response to locally induced IGFs. Alternatively, locally induced IGFs could first stimulate hyperproliferation of intestinal fibroblasts or smooth muscle cells and then TGF-β may induce a transition to a myofibroblast and/or collagen producing phenotype. Another possibility that is not mutually exclusive and has not yet been fully explored is that TGF-β and IGF-I may interact to induce more potent collagen deposition or fibrosis than either peptide alone. Recent findings suggest that this is the case in some other systems *(76)*. To test these possible models of the interaction between TGF-β and IGF-I during the course of inflammation-induced fibrosis in the bowel it will be necessary to compare the time course and cellular localization of TGF-β and IGF-I during progression from acute phase to chronic phase inflammation and during development of associated fibrosis in animal models such as the PG-PS model. Recent characterization of cultured fibroblasts *(36,77)*, myofibroblasts *(78)*, and smooth muscle cells *(37,45)* derived from normal or inflamed bowel also will provide useful models to assess directly the interactions between TGF-β and IGF-I. Insight from such studies may have relevance to tissue repair and fibrosis in other systems as well as the bowel and to potential therapies to limit fibrosis.

The IGF System and Therapy in IBD

Growth failure in children with IBD that results from malabsorption, malnutrition, or steroid therapy has been linked to decreases in circulating IGF-I *(79)*. Nutritional therapy that normalizes circulating IGF-I concentrations appears to reverse growth failure in children with IBD *(79)*. With availability of recombinant GH and IGF-I, replacement therapy in children with IBD to reverse growth failure is a possible future direction. Tests of recombinant GH therapy in adults after bowel surgery for IBD or during parenteral nutrition are already in progress *(3,80,81)*. In patients receiving TPN due to gastrointestinal disease, 1–4 wk of GH therapy significantly improved nitrogen balance and nutrient utilization and effects were associated with increases in circulating IGF-I *(3,80,81)*. Systemically administered GH/IGF-I may promote mucosal growth, restitution, or enhanced digestive and absorptive capabilities in patients with IBD as well as having general effects on nutrient status. Recent studies in a rat model of colonic damage and acute experimental colitis induced by oral dextran sulfate indicate that systemic IGF-I may reduce epithelial damage and/or submucosal inflammation *(82)*. If GH and/or IGF-I therapy become more widely used in patients with IBD, it will be important, especially in patients with Crohn's disease, to monitor fibrogenic complications. Although speculative, it seems possible that only locally produced IGF-I exerts fibrogenic effects. In this regard it is important that relative or specific effects of circulating IGF and locally produced IGF-I be examined in models of mucosal damage, inflammation, and fibrosis. A recently developed transgenic mouse model where IGF-I expression is driven by an α smooth muscle actin promoter *(83)* should help provide insights into the role of mesenchymal cell derived IGF-I in intestinal inflammation and fibrosis (*[84]; see below*). A number of lines of evidence indicate that IGF-I is secreted from mesenchymal cells as a precursor *(57,85)*. IGF precursor immunoreactivity is increased at sites of IGF mRNA induction in animal models of inflammation-induced fibrosis *(63,64)*. It is tempting to speculate that effects of circulating IGF-I may differ from those of locally expressed IGF precursor during inflammation and fibrosis. The biological actions of precursor forms of IGF-I remains, as yet, an underexplored topic. Information about the biological actions of IGF precursors vs IGF-I could be very important in defining the specific role(s) of locally expressed IGFs in bowel disease.

Recombinant IGFBPs are increasingly available and are being used in conjunction with IGFs in other systems such as burn injury *(57)* to promote an optimal balance between tissue repair and scarring. Although speculative at present, inhibitory IGFBPs could represent a potential therapeutic approach to control or limit fibrogenic complications of Crohn's disease. Expression of IGFBP-4 and IGFBP-5 mRNAs is increased in animal models of inflammation-induced fibrosis *(35)*. In vitro and in vivo studies are required to establish the role of IGFBPs in this process.

THE IGF SYSTEM AND CANCER OF THE GASTROINTESTINAL TRACT

There is now a wealth of information to suggest that the IGFs may play a role in carcinogenesis in many tissues including the bowel and accessory organs such as pancreas and liver. A complete review of this literature is outside the scope of this chapter and readers are referred to several excellent reviews *(86–90)*.

Colon cancer is a major cause of cancer related death in the United States *(91)*. Relative to the colon, tumors of the small bowel are very rare *(8)*. Expression of IGF-II and to a lesser extent IGF-I is up-regulated in colon cancers *in situ (92)*. Many colon cancer cell lines express IGF-II *(88,90)*. In colon cancer cells in culture, IGF-II exerts autocrine actions to stimulate proliferation and inhibit differentiation *(93,94)*. The type 1 IGF receptor is expressed in primary colon cancers and colon cancer cell lines *(95)* and overexpression of the type 1 IGF receptor is associated with cell transformation *(86)*. To date, there is little information to establish definitively a role of elevated IGF or type 1 IGF expression in the development or progression of colon cancer or other bowel derived cancers in vivo. Acromegaly is associated with increased incidence of colon cancer *(96)*, implicating excess GH and/or IGF-I as factors that may promote increased susceptibility to colon cancer. On the other hand transgenic mice engineered to show chronic GH overexpression (and consequently overexpression of endogenous IGF-I) do not show increased incidence of spontaneous bowel cancers (*[97]*; Lund, *unpublished observations*). These GH transgenics do show abnormal histopathology in liver and spleen *(98)*. Transgenic mice that have chronic IGF-I overexpression associated with long-term elevation of circulating IGF-I and ectopic overexpression of IGF-I in small bowel and colonic epithelium *(34)* also do not show increased incidence of spontaneous bowel cancer. Thus, there is, as yet, little in vivo evidence to implicate GH or IGF-I as direct causative agents in bowel cancer. It seems likely, however, that GH and/or IGF-I/IGF-II may serve as cofactors with other molecules, cellular abnormalities, or insults to promote initiation, growth, or progression of tumors. Evidence in support of this possibility stems from observations in a mouse model of tumorigenesis of pancreatic β-cell induced by insulin promoter mediated overexpression of SV40 T antigen. In this model IGF-II appears to cooperate with SV40 T antigen to induce pancreatic islet cell tumors in vivo *(99)*. Future strategies aimed at targeting of IGF-I, IGF-II, or type I IGF receptor overexpression specifically to the colonic epithelium in vivo or at targeting of dominant negative mutants of the type 1 IGF receptor to bowel in vivo will likely provide some major insights into the role of locally expressed IGFs and the type 1 IGF receptor in bowel cancer. Recent studies indicate that high levels of plasma IGF-I show a strong positive association with increased risk of prostate cancer *(100)*. This underscores the importance of further investigation of the association between circulating IGF levels and cancer of other organs including the colon.

IGFs, Apoptosis, and Bowel Cancer

Mutations in genes that induce DNA repair or apoptosis are associated with increased risk of colon cancer *(8)*. IGF-I inhibits cytotoxin-induced apoptosis of colon cancer cell lines *(101–103)* and this effect is associated with increased expression of mRNAs encoding the multidrug resistance gene *mdr*-I, c-H-*ras*, and manganese superoxide dismutase *(103)*. Transgenic mice with overexpression of IGF-I in bowel epithelium (MT-hIGF-I transgenics; *see below*) show reduced levels of spontaneous and irradiation induced apoptosis in small bowel crypts in vivo *(104)*. IGF-II promotes pancreatic β-cell tumors in vivo via effects to reduced apoptosis *(99)*.

The IGF System, p53, and Colon Cancer

Deletion or mutation in the *p53* gene which promotes DNA repair and/or apoptosis of genetically damaged cells is associated with increased risk of colon cancer *(8,105)* and cancer of other organs. Mice that are homozygous for *p53* gene deletion do not show increased incidence of colon cancer but this probably reflects death due to earlier onset of cancer in other organs *(8)*. Mice that are heterozygous for *p53* deletion show a small

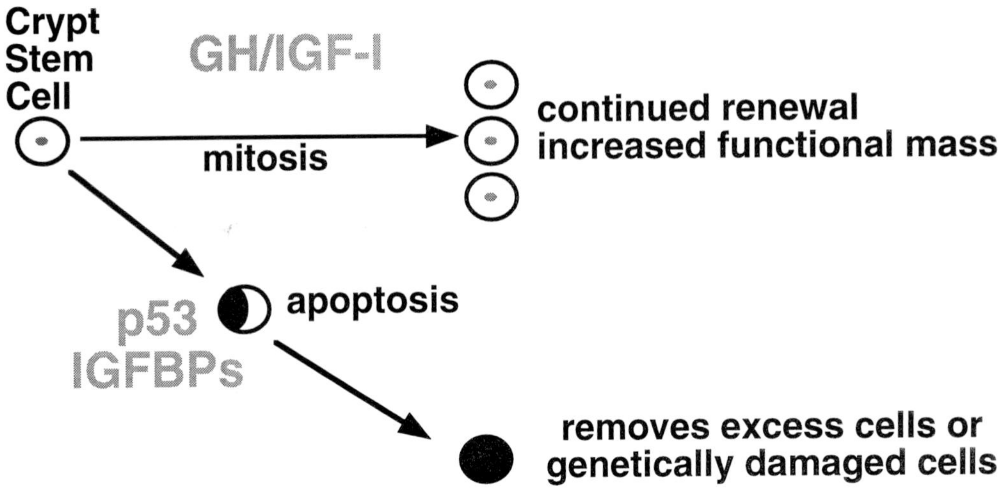

Fig. 5. Schematic to show the putative "selection point" where crypt cells may undergo proliferation or apoptosis and the potential role of IGFs, *p53*, and IGFBP-3 in regulating this selection point.

but significant increase in the incidence of colon cancer *(8)*. Loss of both copies of the *p53* gene is associated with dramatic reductions in spontaneous and irradiation-induced apoptosis in the small bowel and colon *(106,107)*. Recent studies link the *p53* gene and the IGF system. In colon cancer cells in vitro, overexpression of *p53* induces IGFBP-3 and this is associated with cell cycle arrest *(108)*. To assess whether this link is operative in the bowel in vivo, we analyzed IGFBP-3 mRNA in the bowel of mice subject to gamma irradiation at times when induction of *p53* and irradiation-induced apoptosis are maximal. We found no evidence that irradiation/*p53* induction increases IGFBP-3 expression in bowel in vivo. In fact, IGFBP-3 expression is lower in bowel of irradiated than nonirradiated animals *(104)*. One caveat with these data is that the approaches used or time points studied may not have revealed IGFBP-3 induction in a small population of crypt cells, although these approaches did reveal irradiation induced induction of the "cell death" gene, *Bax*, in crypt cells *(104)*. In other in vitro studies, *p53* has been shown to down-regulate the expression of the type 1 IGF receptor and IGF-II *(109,110)*. These links between the IGF system and *p53* support a hypothesis that mutation or loss of *p53* may promote tumor progression or expansion via reduced expression of IGFBP-3 which may normally inhibit IGF-dependent tumor growth and by favoring autocrine actions of IGFs to promote tumor cell proliferation and inhibit tumor cell apoptosis (Fig. 5). Additional studies, especially in vivo, will be required to definitively test this hypothesis. One recent in vivo study in mice with only one copy of the *p53* gene links the IGF system and cytotoxin-induced bladder cancer *(14)*. In these mice, dietary restriction reduced the levels of cytotoxin-induced progression of bladder cancers, increased tumor cell apoptosis, and reduced plasma levels of IGF-I; IGF-I infusion into the diet restricted mice restored tumor progression and reduced the levels of tumor cell apoptosis *(14)*. As well as supporting a link between the IGF system and tumorigenesis, this study also raises the issue that the IGF system may represent an important link between diet and tumor progression.

In summary, an increasing amount of evidence suggests that the anti-apoptotic actions of the IGFs may contribute to or underlie their role in tumorigenesis in bowel and other systems. IGFBPs may also represent agents that may inhibit tumor progression by their inhibitory actions on IGF action or via IGF-independent actions *(40)*. This is currently an avenue of active investigation by many groups. It seems likely that the outcome of further studies in this area will be highly relevant to defining the role of the endogenous IGF system in colon cancer growth and progression and to understanding the relative risks and benefits associated with GH and/or IGF-I therapy, especially long-term therapy.

TRANSGENIC MOUSE MODELS

Transgenic mice with widespread or targeted overexpression of specific genes represent useful in vivo systems to study hormone and growth factor action in bowel or molecular mechanisms of action. In these models, the consequences of a single genetic perturbation for normal or adaptive growth of bowel may be analyzed. One advantage of transgenic models is that they can be used to study the long-term effects of hormone/growth factor perturbation on bowel growth or function at different times in development in an in vivo situation where tissue architecture and cellular interactions are preserved. Some transgenic models offer the additional advantage that they may be used to analyze the actions of locally expressed growth factors that are not readily assessed by other in vivo approaches. To date, bowel growth and function have been examined in detail in only a few mouse models of perturbed IGF status and these are reviewed below.

MT-bGH and MT-hIGF-I Transgenics

These transgenic mice have excess GH or IGF-I as a result of germ line transmission of a bovine GH transgene or a human IGF-I cDNA driven by the mouse metallothionein-1 promoter (MT-1). The MT-1 promoter directs widespread, high-level expression of transgenes. Relative to wild-type (WT) mice, MT-bGH transgenic mice show high levels of circulating GH and IGF-I. The levels of IGF-I mRNA expressed in small bowel also are moderately increased in MT-bGH transgenics *(97)*. Thus, MT-bGH transgenics are a model of excess GH and excess endogenous IGF-I in the circulation and bowel. MT-hIGF-I transgenics have elevated serum levels of IGF-I that are similar to those in MT-bGH transgenics but have undetectable levels of GH in the circulation and dramatic reductions in GH levels in the pituitary gland, presumably due to negative feedback effects of the high levels of serum IGF-I *(34)*. The MT-hIGF-I transgenics are therefore a useful in vivo model of IGF-I excess and relative GH deficiency that have provided some insights into the relative actions of GH and IGF-I on bowel growth.

Controls for differences in enteral nutrient intake are essential in any study of bowel growth because of the inherent dependence of the mass and rates of continuous renewal of the mucosal epithelium on enteral nutrition *(10)*. One obvious way to control for nutrient intake is to pair-feed transgenics and WT. This is not a totally straightforward issue in animal models such as MT-bGH and MT-hIGF-I transgenics that show differences in body size. Our studies in these models indicate that food intake per gram body weight is the same in MT-bGH transgenics or MT-hIGF-I transgenics as in their WT littermates, so pair feeding of the transgenics with WT actually places them in a situation of nutrient restriction. With this in mind we examined bowel growth in transgenics and WT littermates allowed to feed *ad libitum* and also in transgenics that were pair-fed the same amount of daily food intake as WT littermates for 4 d before study. Four days of pair-feeding was chosen because prior studies demonstrate that maximum atrophy of bowel mucosa occurs in response to nutrient restriction for this period of time *(97)*. The

Table 1
Summary of the Findings on Bowel Growth
in the MT-bGH and MT-hIGF-I Transgenics

	MT-bGH transgenics	*MT-IGF-I transgenics*
Body weight	↑ 60%	↑ 23%
Bowel length	↑ 20%	↑ 44%
Bowel mass	↑	↑
Crypt cell proliferation	No change	↑
Sucrase activity	↑	No change
Crypt cell apoptosis	No change	↓

rational was that any differences in bowel growth between pair-fed transgenics and WT would reflect the actions of excess GH/IGF-I in a situation that would not usually favor increased bowel growth.

In bowel of *ad libitum* or pair-fed MT-bGH transgenics there are increases in bowel length and bowel mass (DNA content and protein content) and increases in crypt depth and villus height relative to bowel of WT littermates *(97)*. *Ad libitum* or pair-fed MT-hIGF-I transgenics also show increases in all these measures of bowel growth relative to their WT littermates and effects are similar to or quantitatively greater than observed in MT-bGH transgenics (Table 1; *[34]*). The effect of IGF-I on bowel length is particularly striking as the MT-hIGF-I transgenics show a 44% increase in small bowel length whereas the MT-bGH animals show only a 20% increase in small bowel length (*[34,97]*; Table 1). The enterotrophic effects in MT-hIGF-I transgenics are even more compelling when one considers that these mice show only modest, 23% increases in body weight relative to WT whereas MT-bGH mice are 60% heavier than WT littermates (Table 1; *[34,97]*). Thus the bowel is particularly responsive to excess IGF-I and these data in the transgenic models support the notion that IGF-I can mediate most if not all of the enterotrophic effects of GH even in the absence of detectable GH in the circulation.

Some specific effects on bowel are unique to either MT-bGH transgenics or MT-hIGF-I transgenics and raise some intriguing questions about GH and IGF-I interaction in bowel. MT-bGH transgenics show increased activity of sucrase, a marker of villus cell differentiation and digestive function in bowel mucosa, whereas MT-IGF-I transgenics do not *(34,97)*. This is consistent with a specific effect of GH to promote differentiated function of enterocytes and supports the dual effector theory of GH/IGF action in bowel as has been proposed in other systems *(34,97)*. This theory predicts primarily differentiative actions of GH and primarily mitogenic actions of IGF-I. Despite increases in bowel mass, adult MT-bGH transgenics do not show increases in crypt cell mitoses (a measure of cells at G2/M phase of the cell cycle) or DNA synthesis (a measure of numbers of crypt cells entering S phase of the cell cycle). This indicates that adult MT-bGH mice do not show a sustained increase in crypt cell proliferation even though they show a sustained increase in bowel mass relative to WT littermates. MT-bGH mice do show increases in crypt cell proliferation immediately after weaning, a developmental time point when mice first become responsive to GH and are first exposed to excess GH and excess endogenous IGF-I *(111)*. This suggests that excess GH/IGF-I initially induces an increase in crypt cell proliferation that probably underlies an initial increase in bowel mass, but that with continued GH/IGF-I excess a new steady state is reached and the increase in small bowel mass is somehow maintained without continued increases in crypt cell proliferation. One mechanism that could account for this is increased lifespan of villus epithelial cells in MT-bGH transgenics but this has not been directly tested. However, the absence of increases in crypt cell proliferation in MT-bGH transgenics despite continued high levels of circulating IGF-I also suggests that some compensatory mechanism somehow prevents sustained increases in crypt cell proliferation. In contrast to MT-bGH transgenics, MT-hIGF-I transgenics show maintained increases in crypt cell proliferation at adulthood indicating that compensatory mechanisms are unable to counteract the proliferative effects of prolonged IGF-I excess *(34)*.

Recent preliminary studies indicate that MT-hIGF-I transgenics show reduced levels of spontaneous and irradiation-induced crypt cell apoptosis and reduced effects of irradiation to induce arrest of crypt cell proliferation relative to their WT littermates (*[104]*; Table 1). As discussed previously, these observations add to the growing body of evidence obtained primarily in vitro that IGF-I has anti-apoptotic activities, and extend the observations to the in vivo situation. Interestingly, however, adult MT-bGH transgenics, show no difference in spontaneous or irradiation induces crypt cell apoptosis relative to WT *(97,104)*. These differences in crypt cell apoptosis in MT-bGH and MT-IGF-I transgenics do not reflect differences in levels of circulating IGF-I as levels of circulating IGF-I are similar in the two models *(34,97)*. IGFBP-3 expression is higher in bowel of MT-IGF-I transgenics than in MT-bGH transgenics *(34)*, so that differences in IGFBP-3 do not account for the differences in apoptosis between the models. Several other possibilities may account for the differences in crypt cell apoptosis in MT-bGH and MT-IGF-I transgenics. Circulating levels of acid-labile subunit (ALS) have not been evaluated in these models. Because GH induces ALS expression in liver it is possible that there are higher levels of circulating ALS in MT-bGH than Mt-hIGF-I transgenics which could limit the bioavailability of circulating IGF-I for mitogenic or anti-apoptotic actions on bowel. Another difference between the two transgenic models is that MT-IGF-I transgenics show very high level expression of the IGF-I transgene within the small bowel, much higher than the increase in endogenous IGF-I mRNA that occurs in the bowel of MT-bGH transgenics *(34)*. Furthermore, in MT-hIGF-I mice, the IGF-I transgene is expressed primarily within the mucosal epithelial cells whereas endogenous IGF-I is expressed in intestinal mesenchymal cells *(34)*. These differences in the transgenic models indicate that either the greater magnitude of local IGF-I overexpression or the fact that IGF-I is overexpressed in epithelial cells may underlie the reduced crypt cell apoptosis in MT-hIGF-I transgenics but not MT-bGH transgenics. Additional studies in different models will be required to distinguish among these possibilities. Such studies are important because they may point to major differences in the actions of circulating compared with locally expressed IGFs or differences in the actions of IGFs that are overexpressed at ectopic sties (as occurs in colon cancer, for example) compared with IGF-I that is overexpressed at normal, mesenchymal sites of expression in the bowel. The observation that MT-bGH transgenics do not show reduced crypt cell apoptosis despite elevated circulating levels of IGF-I suggests that further investigation of the role of GH in apoptosis in bowel or other organs is warranted. There is little or no information in the literature about GH effects on apoptosis, probably because most cells in culture lose GH responsiveness. The crypt cell apoptosis data in MT-bGH and MT-hIGF-I transgenics raise the intriguing possibility that excess GH may in some way counteract the anti-apoptotic actions of elevated circulating or locally expressed IGF-I. This should be amenable to testing in other in vivo models such as GH and/or IGF-I infusion models or in cells in vitro that are responsive to both peptides and is important in terms of defining the relative risks and benefits of GH compared with IGF therapy.

SMP-IGF-I Transgenics with Mesenchyme-Specific Overexpression of IGF-I

Novel transgenic mice (SMP-IGF-I mice) that overexpress IGF-I in mesenchymal cells via an α smooth muscle actin (α-SM) promoter were recently developed by Dr. James Fagin and colleagues. Even though serum IGF-I is not elevated in these mice, they show overgrowth of a number of smooth muscle beds, including the muscularis externa of the small bowel *(83)*. The SMP-IGF-I transgenic mice also show dramatic increases in length of small bowel *(83)*. The phenotypic changes in intestine in the SMP-IGF-I mice provide compelling evidence that IGF-I can exert local, autocrine, or paracrine effects on growth of enteric smooth muscle in vivo. The potential paracrine effects of transgene derived IGF-I on mucosal epithelium have not yet been examined in detail. As the α-SM actin promoter should direct expression of IGF-I to pericryptal myofibroblasts that are implicated as sources of trophic support for crypt epithelial cells, the SMP-IGF transgenic mice represent a potentially useful model to assess whether IGF-I derived from mucosal mesenchymal cells in vivo has paracrine effects on mucosal epithelial cells or autocrine effects on mucosal mesenchymal cells *(84)*. It is noteworthy that the α-SM actin promoter could be used in future studies to target overexpression of other growth factors or growth factor inhibitors to intestinal mesenchymal cells. This may provide a generally useful in vivo approach to elucidate the actions of mesenchymal cell derived growth factors in regulating bowel growth and function in vivo.

MECHANISMS OF IGF-I ACTION IN BOWEL AND INTERACTIONS WITH OTHER GROWTH FACTORS

Although the bowel is recognized as an important target organ for GH and the IGFs a complete mechanistic understanding of the role of the IGF system in proliferation, differentiation, and development of the bowel is lacking. This requires suitable in vitro systems to study proliferation and differentiated function in bowel derived cells, particularly epithelial cells.

In Vitro Models of Bowel Epithelial Cells

Model in vitro systems offer the advantage that they can be used to assess the direct actions of hormones and growth factors and their mechanisms of action at the cellular and molecular level. A large number of colon cancer cell lines have been characterized for proliferative responses to IGF-I and production of endogenous IGFBPs *(112–114)*. Although highly relevant to understanding the role of the IGFs/IGFBPs in colon cancer, the colon cancer cells are not as useful for analyses of IGF action in normal bowel. Colon cancer cell lines often express high levels of growth factors including the IGFs *(88,90)* and this may create difficulties in interpretation of the outcome of treatment with exogenous IGF. The Caco-2 colon cancer cell line has been studied extensively for IGF action. This cell line expresses IGF-II and spontaneously differentiates. Available evidence in Caco cells indicates that the IGFs promote proliferation and inhibit differentiation and that IGFBP-3 may be a marker of cell differentiation *(113,114)*. Compared to colon cancer derived cells there is much less information about IGF expression and action in normal colonic epithelium or in small bowel epithelial cells. This is because these cells survive poorly in culture and lose phenotypic characteristics. Explants of fetal small bowel have been used extensively by several groups to provide important insights into growth and differentiation of bowel during development *(4,5)*. More recently primary cultures of human fetal enterocytes have been used as model systems to study factors that promote differentiated function *(115)*. Limited amounts of tissue, very limited viability, and limited growth capabilities make it difficult to use fetal explants or primary cultures of fetal enterocytes as model systems to analyze the cellular/molecular mechanisms of growth factor dependent proliferation or growth factor effects on differentiation. To date, there is little or no information about the actions of the IGF system in fetal explants or primary enterocyte cultures. Relatively few cell lines derived from small bowel are available as model systems. The most commonly used are IEC-6 cells which are nontransformed, diploid cells that share some of the phenotypic characteristics of small bowel crypt cells *(116)*. IEC-6 cells have the major advantage that they express little or no IGF-I or IGF-II *(117)* which reflects the situation in normal small bowel epithelium. IEC-6 cells or clonal derivatives of these cells have been widely used to study the proliferative actions of growth factors including IGF-I *(117–120)*. IEC-6 cells have limited differentiative capabilities and so are not very useful for studies of the role of the IGF system in differentiated function of the bowel. A recent report demonstrates that IEC-6 cells transfected with an expression construct for the *Cdx-2* gene express brush border enzymes and show differentiated phenotype typical of small bowel enterocytes *(121)*. These transfected IEC-6 cells could therefore provide an alternate system to study the actions of IGFs/IGFBPs on enterocyte differentiation. Alternatives to IEC-6 cells are conditionally immortalized cell lines transfected with inducible oncogenes or temperature-sensitive oncogenes *(112,122)*. The latter have been found to express brush border enzymes when propagated at permissive temperature, then brought down to nonpermissive temperature *(122)* and so many represent a useful system to analyze growth factor effects on differentiation.

Mechanisms of IGF Interaction with Other Growth Factors

Regulation of small bowel growth likely involves multiple hormones and growth factors. Increasing evidence points to interactions between epidermal growth factor (EGF)/TGF-α and IGF-I in regulating proliferation of intestinal epithelial cells *(6,119,123)*. The EGF/TGF-α family of peptides stimulate proliferation of intestinal crypt cells and enhance adaptive growth of the intestinal mucosa *(6,119,123)*. In several bowel-derived cell types, including IEC-6 cells, IGF-I or IGF-II is required in combination with EGF or another growth factor to stimulate cell proliferation or progression through the cell cycle *(117,120,124)*. In IEC-6 cells EGF, basic fibroblast growth factor (bFGF), or IGF-I alone induce weak mitogenic effects but when in combination with IGF-I either EGF or bFGF induce potent synergistic mitogenic effects. Synergistic mitogenic effects are observed when cells are pretreated with EGF or bFGF for several hours followed by IGF-I but not when cells are pretreated with IGF-I followed by EGF or bFGF *(117)*. EGF may promote IGF responsiveness of IEC-6 cells or other cell types by enhancing or maintaining expression of type 1 IGFR or by down-regulating IGFBPs *(117,125)*. Synergistic mitogenic interactions between EGF and IGF-I in bowel may result from amplification of common signaling pathways or activation of divergent signaling pathways. AP-1 is comprised of heterodimers of the Fos/Jun families or Jun homodimers and is a key intracellular mediator of growth factor induced cell proliferation

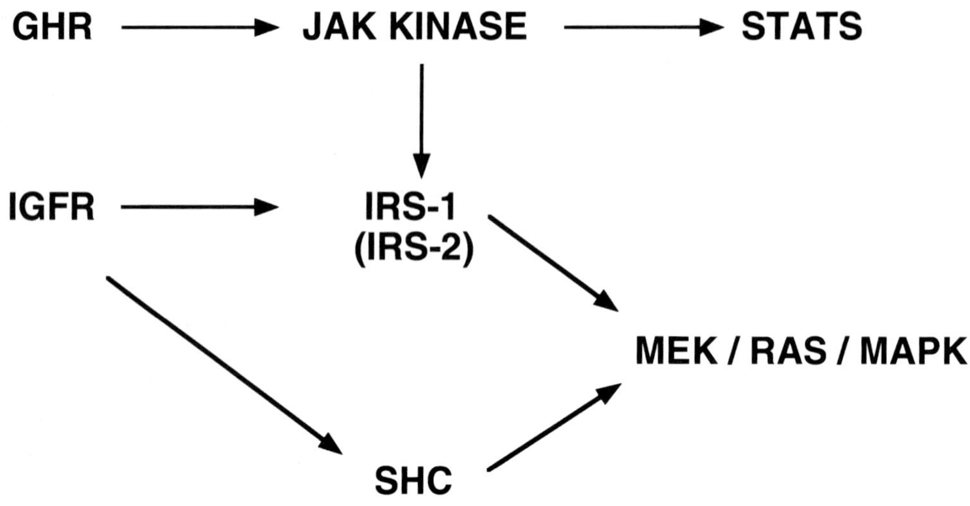

Fig. 6. Simplified schematic of the signal transduction pathways for GH and IGF-I.

(126,127). Our prior studies indicate that EGF and IGF-I have different effects on pathways leading to activation of AP-1 in IEC-6 cells *(117)*. EGF and IGF-I both induce c-Fos where EGF, and not IGF, induces synthesis of c-Jun and transcriptional activation of AP-1-driven reporter genes in IEC-6 cells. This transcriptional activation is greater with EGF and IGF in combination than EGF alone and this cannot be explained by effects on AP-1 synthesis *(117)*. Growth factor dependent AP-1 synthesis and transcriptional activation occurs via the Ras/MEK/MAPK pathway *(126,127)*. A parallel pathway for transcriptional activation of c-Jun involves c-Jun nuclear kinases (JNK1 and JNK2; alternatively called stress-activated protein kinases SAPK) *(128)*. JNK are activated by different upstream kinases (MKK4 or SEK 1) than MAPK *(128)*. Although EGF can activate JNK in some systems, the JNK pathway is most commonly associated with stress and cytokine receptors *(128)*. A transfection based assay indicates that IGF-I, and not EGF, activates JNK in IEC-6 cells *(117)*. The elucidation of signaling pathways that mediate growth factor interactions is a rapidly expanding and challenging field *(128)* that has major relevance to understanding the compelx pathways that regulate intestinal growth and differentiation. As well as in vitro systems, the increasing availability of mice with targeted deletion of key intracellular signaling molecules that mediate type 1 IGF receptor action *(129)* should facilitate rapid progress in this area.

Mechanisms of Interaction of IGF-I with GH

As mentioned in the preceding sections, IGF-I appears to mediate most but not all of the actions of GH on bowel growth. Data in the MT-bGH and MT-hIGF transgenics support the concept that GH may promote differentiated function of the enterocyte whereas IGF-I may exert predominantly mitogenic and anti-apoptotic actions. The known pathways for signaling by the type 1 IGF receptor are addressed in detail in other chapters. Figure 6 summarizes a simplified version of the pathways for GH/IGF signaling, indicating unique and common signaling intermediates. Known actions of GH on transcriptional activation appear to be mediated by the JAK-STAT pathway. GH-dependent JAK-kinase activation is, however, associated with tyrosine phosphorylation of insulin-receptor substrate-I (IRS-1), a major mediator of the actions of IGF-I *(130)*; yet the role of GH dependent IRS-I activation in GH action or interactions with IGF-I is poorly defined. Recent studies indicate that GH may induce insulin/IGF resistance by effects on IRS-1 *(131)*. A recent report indicates that in liver, the EGF receptor is a major intracellular target of GH-activated JAK kinase, and mediates many of the downstream effects of GH receptor activation, at least in vitro *(132)*. Whether or not such interactions occur in bowel and their impact on growth and differentiation in intestine remains an interesting future avenue of investigation.

ACKNOWLEDGMENTS

The author's work is supported by NIH Grants DK40247, DK 47769, and CA44684 and a grant from the Crohn's and Colitis Foundation of America. The author wishes to thank Dr. Denise Ney and Dr. James Fagin for sharing unpublished data during the preparation of this chapter. The author also extends thanks to those past and current members of her laboratory who have conducted much of the work reviewed here including Randy Fuller, Eileen Hoyt, Niru Mohapatra, Kinuko Ohneda, Jolanta Pucilowska, Jim Simmons, Doug Winesett, Martin Ulshen, Kristen Williams, and Ellen Zimmermann.

REFERENCES

1. Lund PK. Insulin-like growth factors. In: Dockray G, Walsh JH, eds. Gut Peptides: Biochemistry and Physiology. Raven Press, New York, 1994, pp. 587–613.
2. Lund PK, Zimmermann EM. Insulin-like growth factors and inflammatory disease. In: Goodlad R, Wright N, eds. Cytokines and Growth Factors in Gastroenterology. Bailliére's Clinical Gastroenterology, London, 1996, pp. 83–96.
3. Byrne TA, Persinger RL, Young LS, Ziegler TR, Wilmore, DW. A new treatment for patients with short-bowel syndrome. Growth hormone, glutamine, and a modified diet. Ann Surg 1995; 222:243–254; discuss.
4. Louvard D, Kedinger M, Hauri HP. The differentiating intestinal epithelial cell: establishment and maintenance of functions through interactions between cellular structures. Annu Rev Cell Biol 1992; 8:157–195.
5. Simon-Assmann P, Kedinger M. Heterotypic cellular cooperation in gut morphogenesis and differentiation. Semin Cell Biol 1993; 4:221–230.
6. Podolsky DK. Regulation of intestinal epithelial proliferation: a few answers, many questions. Am J Physiol 1993; 264:G179–G186.
7. Potten CS. Structure, function and proliferative organization of the mammalian gut. In Potten CS, Hendry JH, eds. Radiation and Gut. Elsevier, Amsterdam,1995, pp. 1–31.
8. Potten CS, Wilson JW, Booth C. Regulation and significance of apoptosis in the stem cells of the gastrointestinal epithelium. Stem Cells 1997; 15:82–93.
9. Que FG, Gores GJ. Cell death by apoptosis: basic concepts and disease relevance for the gastroenterologist. Gastroenterology 1996; 110:1238–1243.
10. Jenkins AP, Thompson RP. Mechanisms of small intestinal adaptation. Digest Dis 1994; 12:15–27.
11. Peterson CA, Carey HV, Hinton PL, Lo HC, Ney DM. GH elevates serum IGF-I levels but does not alter mucosal atrophy in parenterally fed rats. Am J Physiol 1997; 272:G1100–G1108.
12. Nightingale JM, Lennard-Jones JE. The short bowel syndrome: what's new and old? Digest Dis 1993; 11:12–31.
13. Vanderhoof JA. Short bowel syndrome in children and small intestinal transplantation. Pediatr Clin North Am 1996; 43:533–550.
14. Dunn SE, Kari FW, French J, Leininger JR, Travlos G, Wilson R. Dietary restriction reduces insulin-like growth factor I levels, which modulates apoptosis, cell proliferation, and tumor progression in p53-deficient mice. Cancer Res 1997; 57:4667–4672.
15. Winesett DE, Ulshen MH, Hoyt EC, Mohapatra NK, Fuller CR, Lund PK. Regulation and localization of the insulin-like growth factor system in small bowel during altered nutrient status. Am J Physiol 1995; 268:G631–G640.
16. Lo HC, Ney DM. GH and IGF-I differentially increase protein synthesis in skeletal muscle and jejunum of parenterally fed rats. Am J Physiol 1996; 271:E872–E878.

17. Yang H, Ney DM, Peterson CA, Lo HC, Carey HV, Adamo ML. Stimulation of intestinal growth is associated with increased insulin-like growth factor-binding protein 5 mRNA in the jejunal mucosa of insulin-like growth factor-I-treated parenterally fed rats. Proc Soc Exp Biol Med 1997; 216:438–445.
18. Taylor B, Murphy GM, Dowling RH. Pituitary hormones and the small bowel: effect of hypophysectomy on intestinal adaptation to small bowel resection in the rat. Eur J Clin Invest 1979; 9:115–127.
19. Lemmey AB, Martin AA, Read LC, Tomas FM, Owens PC, Ballard FJ. IGF-I and the truncated analogue des-(1–3)IGF-I enhance growth in rats after gut resection. Am J Physiol 1991; 260:E213–E219.
20. Vanderhoof JA, Langnas AN, Pinch LW, Thompson JS, Kaufman SS. Short bowel syndrome. J Pediatr Gastroenterol Nutr 1992; 14:359–370.
21. Mantell MP, Ziegler TR, Adamson WT, Roth JA, Zhang W, Frankel W, Bain A, Chow JC, Smith RJ, Rombeau JL. Resection-induced colonic adaptation is augmented by IGF-I and associated with upregulation of colonic IGF-I mRNA. Am J Physiol 1995; 269:G974–G980.
22. Zhang W, Frankel WL, Adamson WT, Roth JA, Mantell MB, Bain A, Zeigler TR, Smith RJ, Rombeau JL. Insulin-like growth factor-I improves mucosal structure and function in transplanted rat small intestine. Transplantation 1995; 59:755–761.
23. Clemmons DR, Underwood LE. Clinical review 59: uses of human insulin-like growth factor-I in clinical conditions. J Clin Endocrinol Metab 1994; 79:4–6.
24. Leinskold T, Permert J, Olaison G, Larsson J. Effect of postoperative insulin-like growth factor I supplementation on protein metabolism in humans. Br J Surg 1995; 82:921–925.
25. Lund PK, Moats-Staats BM, Hynes MA, Simmons JG, Jansen M, D'Ercole AJ, Van Wyk JJ. Somatomedin-C/insulin-like growth factor-I and insulin-like growth factor-II mRNAs in rat fetal and adult tissues. J Biol Chem 1986; 261:14539–14544.
26. Han VK, D'Ercole AJ, Lund PK. Cellular localization of somatomedin (insulin-like growth factor) messenger RNA in the human fetus. Science 1987; 236:193–197.
27. Han VK, Hill DJ, Strain AJ, Towle AC, Lauder JM, Underwood LE. Identification of somatomedin/insulin-like growth factor immunoreactive cells in the human fetus. Pediatr Res 1987; 22:245–249.
28. See NA, Epstein ML, Schultz E, Pienkowski TP, Bass P. Hyperplasia of jejunal smooth muscle in the myenterically denervated rat. Cell Tissue Res 1988; 253:609–617.
29. See NA, Epstein ML, Dahl JL, Bass P. The myenteric plexus regulates cell growth in rat jejunum. J Auton Nerv Syst 1990; 31:219–229.
30. Mohapatra NK, Ulshen MH, Fuller CR, Zimmermann EM, Haldeman K, Lund PK. Cell specific expression and regulation of insulin-like growth factor I (IGF-I) and IGF binding proteins (IGFBPs) during adaptive growth of small intestine. Endocr Soc Abstr OR, 1995; 37–5.
31. Albiston AL, Taylor RG, Herington AC, Beveridge DJ, Fuller PJ. Divergent ileal IGF-I and IGFBP-3 gene expression after small bowel resection: a novel mechanism to amplify IGF action? Mol Cell Endocrinol 1992; 83:R17–R20.
32. Simmons JG, Pucilowska J, Lund PK. Insulin-like growth factor-II (IGF-II) expressed by intestinal fibroblasts is a potent mitogen for intestinal epithelial cells. Gastroenterology 1997; 112:A-906.
33. Simmons JG, Pucilowska JB, Lund PK. Insulin-like growth factor II interacts with other growth factors secreted by intestinal fibroblasts in regulating proliferation of intestinal epithelial cells or fibroblasts. Am J Physiol, in press.
34. Ohneda K, Ulshen MH, Fuller CR, D'Ercole AJ, Lund PK. Enhanced growth of small bowel in transgenic mice expressing human insulin-like growth factor I. Gastroenterology 1997; 112:444–454.
35. Zhee JM, Mohapatra N, Lund PK, Eysselein VE, McRoberts JA. Differential expression and localization of IGF-I and IGF binding proteins in inflamed rat colon. J Recept Signal Transd Res 1998; 18: 265–280.
36. Pucilowska J, Simmons JG, Sartor RB, Lund PK. IGF-I and TGF-β have different effects on phenotype and proliferation of intestinal fibroblasts: a potential role of IGF binding proteins (IGFBPs). Gastroenterology 1997; 112:A-1064.
37. Zimmermann EM, Li L, Hou TY, Cannon M, Christman GM, Bitar KN. Insulin-like growth factor-I induces collagen and insulin-like growth factor binding protein-5 mRNAs in rat intestinal smooth muscle. Am J Physiol 1997; 273:G875–882.
38. Steeb CB, Shoubridge CA, Tivey DR, Read LC. Systemic infusion of IGF-I or LR(3)IGF-I stimulates visceral organ growth and proliferation of gut tissues in suckling rats. Am J Physiol 1997; 272:G522–G533.
39. Steeb CB, Trahair JF, Read LC. Administration of insulin-like growth factor-I (IGF-I) peptides for three days stimulates proliferation of the small intestinal epithelium in rats. Gut 1995; 37:630–638.

40. Jones JI, Clemmons DR. Insulin-like growth factors and their binding proteins: biological actions. Endocr Rev 1995; 16:3–34.
41. Ney, DM, Peterson C, Gillingham M, Carey HV, Mohapatra NK, Lund PK. IGF-I stimulates intestinal growth and localized expression of IGF binding protein 3 and 5 mRNAs in parenterally-fed rats. Endocrine Society Abstracts 79th Annual Meeting, 1997, P1–166.
42. Pintar JE, Cerro JA, Wood TL. Genetic approaches to the function of insulin-like growth factor-binding proteins during rodent development. Horm Res 1996; 45:172–177.
43. Sartor RB. Current concepts of the etiology and pathogenesis of ulcerative colitis and Crohn's disease. Gastroenterol Clin North Am 1995; 24:475–507.
44. Owen DA. Pathology of inflammatory bowel disease. In: MacDermott RP, Stenson WF, eds. Inflammatory Bowel Disease. Elsevier, New York, 1992, pp. 493–524.
45. Graham MF. Stricture formation: pathophysiologic and therapeutic concepts. In: MacDermott RP, Seerson WF, eds. Inflammatory Bowel Disease. Elsevier, New York, 1992, pp. 323–336.
46. Jain SK, Peppercorn MA. Inflammatory bowel disease and colon cancer: a review. Digest Dis 1997; 15:243–252.
47. Kelley KW. Growth hormone, lymphocytes and macrophages. Biochem Pharmacol 1989; 38:705–713.
48. Fu YK, Arkins S, Wang BS, Kelley KW. A novel role of growth hormone and insulin-like growth factor-I. Priming neutrophils for superoxide anion secretion. J Immunol 1991; 146:1602–1608.
49. Gjerset RA, Yeargin J, Volkman SK, Vila V, Arya J, Haas M. Insulin-like growth factor-I supports proliferation of autocrine thymic lymphoma cells with a pre-T cell phenotype. J Immunol 1990; 145:3497–3501.
50. Rappolee DA, Mark D, Banda MJ, Werb Z. Wound macrophages express TGF-alpha and other growth factors in vivo: analysis by mRNA phenotyping. Science 1988; 241:708–712.
51. Nagaoka I, Someya A, Iwabuchi K, Yamashita T. Expression of insulin-like growth factor-IA and factor-IB mRNA in human liver, hepatoma cells, macrophage-like cells and fibroblasts. FEBS Lett 1991; 280:79–83.
52. Rom WN, Basset P, Fells GA, Nukiwa T, Trapnell BC, Crysal RG. Alveolar macrophages release an insulin-like growth factor I-type molecule. J Clin Invest 1988; 82:1685–1693.
53. Kirstein M, Aston C, Hintz R, Vlassara H. Receptor-specific induction of insulin-like growth factor I in human monocytes by advanced glycosylation end product-modified proteins. J Clin Invest 1992; 90:439–446.
54. Noble PW, Lake FR, Henson PM, Riches DW. Hyaluronate activation of CD44 induces insulin-like growth factor-I expression by a tumor necrosis factor-alpha-dependent mechanism in murine macrophages. J Clin Invest 1993; 91:2368–2377.
55. Lake FR, Noble PW, Henson PM, Riches DW. Functional switching of macrophage responses to tumor necrosis factor-alpha (TNF alpha) by interferons. Implications for the pleiotropic activities of TNF alpha. J Clin Invest 1994; 93:1661–1669.
56. Savendahl L, Underwood LE, Haldeman KM, Ulshen MH, Lund PK. Fasting prevents experimental murine colitis produced by dextran sulfate sodium and decreases interleukin-1-beta and insulin-like growth factor I messenger ribonucleic acid. Endocrinology 1997; 138:734–740.
57. Mueller RV, Spencer EM, Sommer A. The role of IGF-I and IGFBP-3 in wound healing. In: Spencer E, ed. Modern Concepts of Insulin-Like Growth Factors. Elsevier, New York, 1991, pp. 185–192.
58. Clemmons DR, Shaw DS. Purification and biologic properties of fibroblast somatomedin. J Biol Chem 1986; 261:10293–10298.
59. Clemmons DR, Van Wyk JJ. Evidence for a functional role of endogenously produced somatomedinlike peptides in the regulation of DNA synthesis in cultured human fibroblasts and porcine smooth muscle cells. J Clin Invest 1985; 75:1914–1918.
60. Stallmach A, Schuppan D, Riese HH, Matthes H, Riecken EO. Increased collagen type III synthesis by fibroblasts isolated from strictures of patients with Crohn's disease. Gastroenterology 1992; 102:1920–1929.
61. Taylor WR, Alexander RW. Autocrine control of wound repair by insulin-like growth factor I in cultured endothelial cells. Am J Physiol 1993; 265:C801–C805.
62. Jennische E, Skottner A, Hansson HA. Dynamic changes in insulin-like growth factor I immunoreactivity correlate to repair events in rat ear after freeze-thaw injury. Exp Mol Pathol 1987; 47:193–201.
63. Zimmermann EM, McNaughton K, Sartor RB, Lund PK. IGF-1 is overexpressed in cells with a smooth muscle phenotype in peptidoglycan-polysaccharide induced chronic enterocolitis in the rat. Gastroenterology 1993; 104:A-808.
64. Zimmermann EM, Sartor RB, McCall RD, Pardo M, Bender D, Lund PK. Insulinlike growth factor I

and interleukin 1 beta messenger RNA in a rat model of granulomatous enterocolitis and hepatitis. Gastroenterology 1993; 105:399–409.
65. Cohen JA, Zimmermann EM, Sartor RB, Lund PK. IGF-I and IGF-II are overexpressed in inflamed and strictured intestine in Crohn's disease. Gastroenterology 1993; 104:A-683.
66. Ghosh S, Humphreys K, Papachrysostomou M, Ferguson A. Detection of insulin-like growth factor-I and transforming growth factor-beta in whole gut lavage fluid: a novel method of studying intestinal fibrosis. Eur J Gastroenterol Hepatol 1997; 9:505–508.
67. Matthes H, Herbst H, Schuppan D, Stallmach A, Milani S, Stein H. Cellular localization of procollagen gene transcripts in inflammatory bowel diseases. Gastroenterology 1992; 102:431–442.
68. Graham MF, Gluck UM, Shah BV. b-Tropomyosin and a-actin are phenotypic markers for human intestinal smooth muscle cells in vitro. Mol Cell Differ 1994; 2:45–60.
69. Sappino AP, Schurch W, Gabbiani G. Differentiation repertoire of fibroblastic cells: expression of cytoskeletal proteins as marker of phenotypic modulations. Lab Invest 1990; 63:144–161.
70. Desmouliere A, Rubbia-Brandt L, Abdiu A, Walz T, Macieira-Coelho A. Alpha-smooth muscle actin is expressed in a subpopulation of cultured and cloned fibroblasts and is modulated by gamma-interferon. Exp Cell Res 1992; 201:64–73.
71. Desmouliere A, Rubbia-Brandt L, Grau G, Gabbiani G. Heparin induces alpha-smooth muscle actin expression in cultured fibroblasts and in granulation tissue myofibroblasts. Lab Invest 1992; 67:716–726.
72. Rogler G, Andus T. Cytokines in inflammatory bowel disease. World J Surg 1998; 22:382–389.
73. Babyatsky MW, Rossiter G, Podolsky DK. Expression of transforming growth factors alpha and beta in colonic mucosa in inflammatory bowel disease. Gastroenterology 1996; 110:975–984.
74. van Tol EAF, Kong FM, Rippe R, Yamauchi M, Lund PK, Sartor RB. Bacterial cell wall polymers promote intestinal fibrosis by direct stimulation of myofibroblasts. Am J Physiol 1999; in press.
75. Perr H, Oh P, Johnson D. Developmental regulation of transforming growth factor beta-mediated collagen synthesis in human intestinal muscle cells. Gastroenterology 1996; 110:92–101.
76. Yaeger PC, Masi TL, de Ortiz JL, Binette F, Tubo R, McPherson JM. Synergistic action of transforming growth factor-beta and insulin-like growth factor-I induces expression of type II collagen and aggrecan genes in adult human articular chondrocytes. Exp Cell Res 1997; 237:318–325.
77. Fritsch C, Simon-Assmann P, Kedinger M, Evans GS. Cytokines modulate fibroblast phenotype and epithelial-stroma interactions in rat intestine. Gastroenterology 1997; 112:826–838.
78. Valentich JD, Popov V, Saada JI, Powell DW. Phenotypic characterization of an intestinal subepithelial myofibroblast cell line. Am J Physiol 1997; 272:C1513–C1524.
79. Kirschner BS, Sutton MM. Somatomedin-C levels in growth-impaired children and adolescents with chronic inflammatory bowel disease. Gastroenterology 1986; 91:830–836.
80. Ward HC, Halliday D, Sim AJ. Protein and energy metabolism with biosynthetic human growth hormone after gastrointestinal surgery. Ann Surg 1987; 206:56–61.
81. Ziegler TR, Young LS, Manson JM, Wilmore DW. Metabolic effects of recombinant human growth hormone in patients receiving parenteral nutrition. Ann Surg 1988; 208:6–16.
82. Howarth GS, Xian CJ, Read LC. Insulin-like growth factor-1 partially attenuates colonic damage in rats with experimental colitis induced by oral dextran sulphate sodium. Scand J Gastroenterol 1998; 33:180–190.
83. Wang J, Nir W, Nikiforov Y, Naito S, Chernausek S, Witte D, LeRoith D, Strauch A, Fagin JA. Targeted overexpression of IGF-I evokes distinct patterns of organ remodeling in smooth muscle cell tissue beds of transgenic mice. J Clin Invest 1997; 100:1425–1439.
84. Lund PK. The alpha-smooth muscle actin promoter—a useful tool to analyse autocrine and paracrine roles of mesenchymal cells in normal and diseased bowel. Gut 1998; 42:320–322.
85. Hoyt EC. A study of multiple signal peptides of insulin-like growth factor-1. Master's Thesis, UNC—Chapel Hill, Dept. of Physiology.
86. Baserga R, Hongo A, Rubini M, Prisco M, Valentinis B. The IGF-I receptor in cell growth, transformation and apoptosis. Biochim Biophys Acta 1997; 1332:F105–F126.
87. Blakesley VA, Stannard BS, Kalebic T, Helman LJ, LeRoith D. Role of the IGF-I receptor in mutagenesis and tumor promotion. J Endocrinol 1997; 152:339–344.
88. Singh P, Rubin N. Insulinlike growth factors and binding proteins in colon cancer. Gastroenterology 1993; 105:1218–1237.
89. Rubin R, Baserga R. Insulin-like growth factor-I receptor. Its role in cell proliferation, apoptosis, and tumorigenicity. Lab Invest 1995; 73:311–331.
90. Singh P, Dai B, Dhruva B, Widen SG. Episomal expression of sense and antisense insulin-like growth factor (IGF)- binding protein-4 complementary DNA alters the mitogenic response of a human colon

cancer cell line (HT-29) by mechanisms that are independent of and dependent upon IGF-I. Cancer Res 1994; 54:6563–6570.
91. Hoops TC, Traber PG. Molecular pathogenesis of colorectal cancer. Hematol Oncol Clin North Am 1997; 11:609–633.
92. Tricoli JV, Rall LB, Karakousis CP, Herrera L, Petrelli NJ, Bell GI. Enhanced levels of insulin-like growth factor messenger RNA in human colon carcinomas and liposarcomas. Cancer Res 1986; 46:6169–6173.
93. Zarrilli R, Pignata S, Romano M, Gravina A, Casola S, Bruni CB. Expression of insulin-like growth factor (IGF)-II and IGF-I receptor during proliferation and differentiation of CaCo-2 human colon carcinoma cells. Cell Growth Differ 1994; 5:1085–1091.
94. Singh P, Dai B, Yallampalli U, Lu X, Schroy PC. Proliferation and differentiation of a human colon cancer cell line (CaCo2) is associated with significant changes in the expression and secretion of insulin-like growth factor (IGF) IGF-II and IGF binding protein-4: role of IGF-II. Endocrinology 1996; 137:1764–1774.
95. Guo YS, Narayan S, Yallampalli C, Singh P. Characterization of insulinlike growth factor I receptors in human colon cancer. Gastroenterology 1992; 102:1101–1108.
96. Jenkins PJ, Fairclough PD, Richards T, Lowe DG, Monson J, Grossman A, Wass JA, Besser M. Acromegaly, colonic polyps and carcinoma. Clin Endocrinol 1997; 47:17–22.
97. Ulshen MH, Dowling RH, Fuller CR, Zimmermann EM, Lund PK. Enhanced growth of small bowel in transgenic mice overexpressing bovine growth hormone. Gastroenterology 1993; 104:973–980.
98. Quaife CJ, Mathews LS, Pinkert CA, Hammer RE, Brinster RL, Palmiter RD. Histopathology associated with elevated levels of growth hormone and insulin-like growth factor I in transgenic mice. Endocrinology 1989; 124:40–48.
99. Christofori G, Naik P, Hanahan D. A second signal supplied by insulin-like growth factor II in oncogene-induced tumorigenesis. Nature 1994; 369:414–418.
100. Chan JM, Stampfer MJ, Giovannucci E, Gann PH, Ma J, Wilkinson P, Hennekens CH, Pollak M. Plasma insulin-like growth factor-I and prostate cancer risk: a prospective study. Science 1998; 279:563–566.
101. Wong BC, Mitchell NM, Langman MJ, Eggo MC. Induction of apoptosis by protein kinase C-B1 activator in colon-205 human colonic carcinoma cells and its prevention by insulin-like growth factor I. Gastroenterology 1996; 110:A615–A610.
102. Wong BC, Mitchell NM, Langman MJ, Eggo MC. Insulin-like growth factor I protects human colonic cancer cells from indomethacin induced apoptosis. Gastroenterology 1996; 110:A615–A610.
103. Guo YS, Jin GF, Houston CW, Thompson JC, Townsend CM Jr. Insulin-like growth factor-I promotes multidrug resistance in MCLM colon cancer cells. J Cell Physiol 1998; 175:141–148.
104. Williams KL, Ohneda K, Fuller CR, Lund PK. Differential effects of chronic IGF-I and GH excess on apoptosis in the small intestine. Endocrine Society Abstracts 80th Annual Meeting, OR26-4.
105. Lane DP. Cancer. p53, guardian of the genome. Nature 1992; 358:15–16.
106. Merritt AJ, Potten CS, Kemp CJ, Hickman JA, Balmain A, Lane DP, Hall PA. The role of p53 in spontaneous and radiation-induced apoptosis in the gastrointestinal tract of normal and p53-deficient mice. Cancer Res 1994; 54:614–617.
107. Clarke AR, Gledhill S, Hooper ML, Bird CC, Wyllie AH. p53 dependence of early apoptotic and proliferative responses within the mouse intestinal epithelium following gamma-irradiation. Oncogene 1994; 9:1767–1773.
108. Buckbinder L, Talbott R, Velasco-Miguel S, Takenaka I, Faha B, Kley N. Induction of the growth inhibitor IGF-binding protein 3 by p53. Nature 1995; 377:646–649.
109. Werner H, Karnieli E, Rauscher FJ, LeRoith D. Wild-type and mutant p53 differentially regulate transcription of the insulin-like growth factor I receptor gene. Proc Nat Acad Sci USA 1996; 93:8318–8323.
110. Zhang L, Kashanchi F, Zhan Q, Zhan S, Brady JN, Fornace AJ, Seth P, Helman LJ. Regulation of insulin-like growth factor II P3 promotor by p53: a potential mechanism for tumorigenesis. Cancer Res 1996; 56:1367–1373.
111. Ulshen MH, Fuller CR, Lund PK. Effects of excess growth hormone (GH) on small bowel growth in weanling transgenic mice. Gastroenterology 1995; 108:A759–A750.
112. Singh P, Dai B, Yallampalli C, Xu Z. Expression of IGF-II and IGF-binding proteins by colon cancer cells in relation to growth response to IGFs. Am J Physiol 1994; 267:G608–G617.
113. Hoeflich A, Yang Y, Huber S, Rascher W, Koepf G, Blum WF, Heinz-Erian P, Kolb HJ, Kiess W. Expression of IGFBP-2, -3, and -4 mRNA during differentiation of Caco-2 colon epithelial cells. Am J Physiol 1996; 271:E922–E931.

114. Oguchi S, Walker WA, Sanderson IR. Profile of IGF-binding proteins secreted by intestinal epithelial cells changes with differentiation. Am J Physiol 1994; 267:G843–G850.
115. Sanderson IR, Ezzell RM, Kedinger M, Erlanger M, Xu ZX, Pringault E, Leon-Robine S, Louvard D, Walker WA. Human fetal enterocytes in vitro: modulation of the phenotype by extracellular matrix. Proc Nat Acad Sci USA 1996; 93:7717–7722.
116. Quaroni A, Wands J, Trelstad RL, Isselbacher KJ. Epithelioid cell cultures from rat small intestine. Characterization by morphologic and immunologic criteria. J Cell Biol 1979; 80:248–265.
117. Simmons JG, Hoyt EC, Westwick JK, Brenner DA, Pucilowska JB, Lund PK. Insulin-like growth factor-I and epidermal growth factor interact to regulate growth and gene expression in IEC-6 intestinal epithelial cells. Mol Endocrinol 1995; 9:1157–1165.
118. Kurokowa M, Lynch K, Podolsky DK. Effects of growth factors on an intestinal epithelial cell line: transforming growth factor beta inhibits proliferation and stimulates differentiation. Biochem Biophys Res Commun 1987; 142:775–782.
119. Suemori S, Ciacci C, Podolsky DK. Regulation of transforming growth factor expression in rat intestinal epithelial cell lines. J Clin Inves 1991; 87:2216–2221.
120. Wolpert S, Wong ML, Bass BL. Matrix alters the proliferative response of enterocytes to growth factors. Am J Surg 1996; 171:109–112.
121. Suh E, Traber PG. An intestine-specific homeobox gene regulates proliferation and differentiation. Mol Cell Biol 1996; 16:619–625.
122. Paul EC, Hochman J, Quaroni A. Conditionally immortalized intestinal epithelial cells: novel approach for study of differentiated enterocytes. Am J Physiol 1993; 265:C266–C278.
123. Chaet MS, Arya G, Ziegler MM, Warner BW. Epidermal growth factor enhances intestinal adaptation after massive small bowel resection. J Pediatr Surg 1994; 29:1035–1038.
124. Qureshi FG, Tchorzewski MT, Duncan MD, Harmon JW. EGF and IGF-I synergistically stimulate proliferation of human esophageal epithelial cells. J Surg Res 1997; 69:354–358.
125. Pietrzkowski Z, Sell C, Lammers R, Ullrich A, Baserga R. Roles of insulinlike growth factor 1 (IGF-1) and the IGF-1 receptor in epidermal growth factor-stimulated growth of 3T3 cells. Mol Cell Biol 1992; 12:3883–3889.
126. De Meyts P, Wallach B, Christoffersen CT, Urso B, Gronskov K, Latus LJ, Yakushiji F, Ilondo MM, Shymko RM. The insulin-like growth factor-I receptor. Structure, ligand-binding mechanism and signal transduction. Horm Res 1994; 42:152–169.
127. van der Geer P, Hunter T, Lindberg RA. Receptor protein-tyrosine kinases and their signal transduction pathways. Annu Rev Cell Biol 1994; 10:251–337.
128. Kyriakis JM, Banerjee P, Nikolakaki E, Dai T, Rubie EA, Ahmad MF, Avruch J, Woodgett JR. The stress-activated protein kinase subfamily of c-Jun kinases. Nature 1994; 369:156–160.
129. Araki E, Lipes MA, Patti ME, Bruning JC, Haag B, Johnson RS. Alternative pathway of insulin signalling in mice with targeted disruption of the IRS-1 gene. Nature 1994; 372:186–190.
130. Postel-Vinay MC, Finidori J. Growth hormone receptor: structure and signal transduction. Eur J Endocrinol 1995; 133:654–659.
131. Smith TR, Elmendorf JS, David TS, Turinsky J. Growth hormone-induced insulin resistance—role of the insulin receptor, IRS-1, GLUT-1, and GLUT-4. Am J Physiol Endocrinol Metab 1997; 35:E1071–E1079.
132. Yamauchi T, Ueki K, Tobe K, Tamemoto H, Sekine N, Wada M, Honjo M, Takahashi M, Takahashi T, Hirai H, Tushima T, Akanuma Y, Fujita T, Komuro I, Yazaki Y, Kadowaki T. Tyrosine phosphorylation of the EGF receptor by the kinase Jak2 is induced by growth hormone. Nature 1997; 390:91–96.

23 Actions of IGF System Proteins from Studies of Transgenic and Gene Knockout Models

A. Joseph D'Ercole, MD

CONTENTS

INTRODUCTION

Studies of mice that either overexpress proteins or have disrupted expression of a specific gene(s) have proven invaluable to our understanding of the actions of the insulin-like growth factors (IGF-I and IGF-II) and their primary cell-surface receptor, the type 1 IGF receptor (IGF1R). They also have provided major insights into the actions of the type 2 IGF receptor (identical to the cation-independent mannose-6-phosphate receptor; IGF2R/ci-M6PR) and the growth-promoting potential of the insulin receptor (InR). Although investigation of mice with altered IGF binding protein (IGFBP) expression has yielded important information, it has not as yet revealed comparable information regarding the fundamental actions of these proteins.

This chapter summarizes the knowledge that has accrued from the study of mice with altered expression of IGF system proteins (i.e., IGFs, IGF binding proteins [IGFBPs], and IGFRs). Table 1 lists each of the reported transgenic (Tg) animal models involving IGF system proteins, the characteristics of the transgenes used in their generation, and the abbreviations used in this chapter *(1–45)*. Table 2 lists similar data for the mouse lines generated with gene disruptions of IGF system proteins and the InR *(46–64)*.

From: *Contemporary Endocrinology: The IGF System*
Edited by: R. Rosenfeld and C. Roberts Jr © Humana Press, Inc., Totowa, NJ

Table 1
Transgenic Mouse Lines that Overexpress IGF System Proteins

Protein	*Transgene Promoter*	*Transgene Coding DNA*	*Abbreviation in Text*	*References*
IGF-I	m Metallothionein-I(mMT-I)	hIGF-IAcDNA	mMT-I/hIGF-I	*(1–13)*
	m IGF-II 5′ flanking region	hIGF-IAcDNA	mIGF-5′/hIGF-I	*(13,14)*
	b α_{S1} Casein (b α_{S1} Cas)[a]	synthetichIGF-I and [Gln^{58}}hIGF-I	αCas/hIGF-I	*(15,16)*
	m Whey acidic protein (WAP)	rIGF-IAcDNA	mWAP/rIGF-I	*(17)*
	Avian Skeletal α Actin (aSKαACT)	hIGF-IAcDNA	aSkαAct/hIGF-I	*(18)*
	rWAP	hIGF-IAcDNA and des(1–3)IGF-IA	rWAP/hIGF-I rWAP/deshIGF-I	*(19)*
	m Smooth muscle α actin flanking fragments (mSMA)	rIGF-IcDNA	mSMA/rIGF-I	*(20)*
	r α Myosin heavy chain (rαMHC)	hIGF-IBcDNA	rαMHC/hIGF-I	*(21,22)*
	r IGF-15′ Flanking region	Firefly luciferase	rIGF-I5′/Luc	*(23)*
IGF-II	m H-2K^b	hMinigenes	mH-2K^b/hIGF-II	*(24–26)*
	m Major urinary protein (MUP)	hIGF-IIcDNA	mMUP/hIGF-II	*(27)*
	o β-Lactoglobulin (oβ-Lact)	mIGF-IIgene	oβ-Lact/mIGF-II	*(28)*
	b Keratin 10 (bKer)	mIGF-IIgene	bKer/mIGF-II	*(29,30)*
	r Phosphoenolpyruvate carboxykinase (PEPCK)	hIGF-IIcDNA	rPEPCK/hIGF-II	*(31–33)*
IGFBP-1	mMT-I	hIGFBP-1cDNA	mMT-I/hIGFBP-1	*(7,11,34,35)*
	m Phosphoglycerate kinase-1 (PGK)	rIGFBP-1gene	mPGK/rIGFBP-1	*(36–42)*
	h α_1-Antitrypsin (hα1-AT)	hIGFBP-1cDNA	hα1-AT/hIGFBP-1	*(43)*
IGFBP-3	mMT-I	hIGFBP-3cDNA	mMT-I/hIGFBP-3	*(41,44)*
	mWAP	hIGFBP-3cDNA	mWAP/hIGFBP-3	*(17)*
IGFBP-4	mSMA	rIGFBP-4cDNA	mSMA/rIGFBP-4	*(45)*

[a]Tg rabbits.

THE ROLE OF IGF SYSTEM PROTEINS IN SOMATIC GROWTH

Detailed studies of growth in mice with ablated gene expression have established the role of the IGFs and IGF1R in normal growth. Tables 3 and 4 summarize the data on growth and viability of mice from these experiments. This work has defined many of the actions and the relative importance of IGF-I and IGF-II in embryonic and postnatal growth, as well as that of the IGF1R, the IGF2R/ci-MPR, and the InR. These experiments were made possible by exploiting the phenomenon of homologous recombination in pluripotent, embryonic stem cells to generate mice with disrupted IGF and IGF receptor genes that cannot express their protein product. The mice resulting from these procedures, therefore, are commonly referred to as "knockout" (KO) mice.

Table 2
Transgenic Mouse Lines with Disputed IGF System Protein Genes

Gene	*Abbreviation Used in Text*	*References*
IGF-I	IGF-IKO	*(46–50)*
Mutant Midi-IGF-I[a]	midi-IGF-I	*(49,51)*
IGF-II	IGF-IIKO	*(52–54)*
IGF1R	IGF1RKO	*(46,47,55)*
IGF2R/ci-MPR	IGF2R/ci-MPRKO	*(56–59)*
InR	InRKO	*(60,61)*
IGFBP-2	IGFBP-2KO	*(62–64)*

[a]Midi-IGF-I mice were generated using a homologous recombination strategy to ablate IGF-I gene function, but the disrupted IGF-I gene retained some function, such that IGF-I expression is markedly reduced, but not completely ablated *(49,51)*.

Actions in Fetal Somatic and Placental Growth

IGF-I

Mice with complete ablation of IGF-I expression, that is, mice that are homozygous for the disrupted IGF-I gene, exhibit marked *in utero* and postnatal growth retardation *(46,47,49)*. Growth rates in mice without IGF-I expression first slow between e12.5 and e13.5 (e = embryonic day, i.e., days gestation; in the mouse, gestation is ~19 d) and continue at much reduced rates through the remainder of gestation, such that birth weights are reduced ~60% of normal (*[47]*; Tables 3 and 4). Mice heterozygous for the disrupted gene, that is, mice with one normal and one disrupted allele, either have no or minimal growth retardation, indicating that a single IGF-I gene can express sufficient IGF-I for normal or near-normal growth *(46)*. IGF-I is not essential for placental growth, because placentas from IGF-I KO mice exhibit normal growth kinetics. This finding also indicates that placental insufficiency does not contribute to the growth failure of IGF-I KO fetuses. Depending upon the strain of mice carrying the disrupted IGF-I gene, 32–95% of IGF-I KO mice die immediately or shortly after birth, probably due to respiratory insufficiency *(46,49)*. The reasons for the differences in survival among strains of mice is not known. Among the possibilities are strain differences in the capacity to increase the expression of alternative ligands, such as IGF-II, or to activate the IGF1R, surrogate receptors, or crucial intracellular mediators of IGF-I action.

IGF-II

Knockout of the IGF-II gene also results in fetal growth retardation. The character of this altered growth, however, is distinct from that of IGF-I KO mice, and the reduction in growth rate is confined to the *in utero* period *(52)*. IGF-II, therefore, exerts its most important control of growth in mice at developmental times different from IGF-I. Mice lacking IGF-II expression exhibit a dramatic decrease in growth velocity during a brief period during gestation (between ~e10.5 and e11.5) and grow at only a minimally reduced velocity for the remainder of fetal life. At birth, IGF-II KO mice have weights that are reduced by ~40% (i.e., 60% of normal). They also exhibit placental growth retardation beginning at ~e13. The reduced placental growth velocity continues through e16 to e17. This failure of normal placental growth may cause, or at least contribute to, the fetal growth retardation occurring late in gestation (e13 to e19).

Table 3
In Utero Somatic and Placental Growth Characteristics of IGF System Protein KO Mice

	Fetal Growth Alterations			Placental Growth Alterations		
Mutation	*Nature*	*Onset*	*Duration*[a]	*Nature*	*Onset*	*Duration*
IGF-I	↓	~e13	e13–19		None	
IFG-II	↓ Slight↓	~e10.5	e10.5–11 e11–19	↓	~e13	e13–16
IGF1R	↓	~e10.5	e10.5–11 e12.5–19		None	
IGF-I and IGF-II or IGF-II and IGF1R	↓	~e10.5	e10.5–11 e12.5–19	↓	~e13	e13–16
IGF2R/ci-MPR	↑	~e13	e13–19	↓	~e13	e13–16
IGF1R and IGF2R/ci-MPR		Near normal		↑	~e13	e13–16
InR	↓	~e18.5	e18.5–19		None	
InR and IGF1R	↓	~e10.5	e10.5–11 e12.5–19 ~e13.5 and e16.5–19[b]		None	

Data are from refs *(46,47,58,61)*.
[a]Duration: The time periods listed are those when growth rates change and are different from normal.
[b]Indicates times when the growth of InR/IGF1R KO appears to be slower than that of IGF1RKO mice.

IGF1R

Disruption of the IGF1R gene results in more profound *in utero* growth retardation than does disruption of either the IGF-I or IGF-II genes *(47)*. Birth weights of mice homozygous for the disrupted IGF1R gene are 45% of normal, whereas those of mice heterozygous for the disruption are normal, indicating that a single functional gene is sufficient for normal growth. Mice without the IGF1R have retarded growth that begins at ~e10.5 and initially is identical in character to that of IGF-II gene disrupted mice. Growth velocities, however, continue to decline throughout gestation, in a manner parallel to that of IGF-I KO mice. As a result, birth weights to IGF1R KO mice are reduced by ~55%, indicating a major role for the IGF1R in mediating the growth-promoting actions of both IGFs. Even this degree of growth retardation, however, does not account for the loss of function of both IGF-I and IGF-II, because: (1) it is not as great as the sum of the growth retardation that results from IGF-I and IGF-II gene knockouts—a 40% reduction in birth weights in each IGF KO, i.e., an 80% reduction, nor (2) is it as great as the 70% reduction in birth weight occurring in mice carrying disruptions of both IGF genes (as a result as of cross-mating each line of KO mice).

The growth and phenotypic characteristics of mice carrying disruptions of both the IGF-I and IGF1R gene, however, are not obviously different from those of IGF1R KO mice (Tables 3 and 4). In contrast, mice with disruptions of both the IGF-II and the IGF1R genes have greater reductions in birth weight than IGF1R KO mice. It appears, therefore, that the IGF1R mediates virtually all of the growth-promoting actions of IGF-I, but not

Table 4
Growth and Viability of IGF System Protein KO Mice

Mutation	*Perinatal Viability*	*Birth Weight (% Normal)*	*Postnatal Growth*
Single mutations			
IGF-I	<5–~70%	~60	Very poor: neglible P0-3, ↓ velocity during suckling and ↓↓ velocity thereafter
IGF-II	Normal	~60	Normal velocity, but no catchup growth
IGF1R	0	~45	None
IGF2R/ci-MPR	0 (with exceptions)	135–140	Near normal in survivors
InR	0	~90	None
Multiple mutations			
IGF-I and IGF-II	0	25–35	None
IGF1R and IGF2R/ci-MPR	Normal	~100	Subnormal velocity P15–30, but catchup by P70
IGF-I and IGF1R	0	~45	None
IGF-II and IGF1R	0	~30	None
InR and IGF1R	0	~30	None
IGF-II and IGF2R/ci-MPR	Normal	~75	Near normal, but no catchup
IGF-II, IGF1R, and IGF2R/ci-MPR	0	~35	None

Data from refs. *(46,47,58,60,61)*.

all those of IGF-II *(47,58)*. These findings suggest that another receptor(s) can mediate the effects of the IGFs, especially those of IGF-II (*see* pp. 550–551). Furthermore, IGF1R KO mice exhibit no placental growth retardation, indicating that another receptor(s) mediates IGF-II's actions on the growth of the placenta. Importantly, no IGF1R KO mice survive significantly past birth, a finding that points to the crucial role of the IGF1R and of IGF-I in postnatal development.

Fibroblasts lacking the IGF1R, derived from IGF1R KO mice, have proven invaluable in the study of IGF1R signal transduction and function (*see* review: *[65]*). Use of these fibroblasts has demonstrated that the IGF1R is crucial to a normal rate of proliferation, mediates IGF-I's capacity to inhibit apoptosis *(66)*, and appears necessary for fibroblast transformation with oncogenes *(67)*.

IGF2R/ci-MPR

Disruption of the IGF2R/ci-MPR *(56–58)* does not result in growth retardation, indicating that this receptor does not transduce growth-promoting signals. This finding is consistent with many other experiments that have failed to link the IGF2R/ci-MPR to IGF signaling *(68)*. The IGF2R/ci-MPR, nonetheless, has a role in growth and development.

Ablation of its expression results in moderate fetal and placental overgrowth (~25–40% increases: *[56–58]*) and cardiac hypertrophy, which is presumably the reason that these mice die during the perinatal period (*see* the fourth section). The IGF2R/ci-MPR has two known functions: (1) it traffics or internalizes proteins containing mannose-6-phosphate moieties, such as lysosomal enzymes, to the lysosome, and (2) it is responsible for the turnover of IGF-II, and possibly IGF-I, by endocytosis (*see* review; *[69]*). The absence of the IGF2R/ci-MPR, thus, would be expected to result in faulty degradation of mannose-6-phosphate-containing proteins and increased levels of IGF-II. The latter prediction has been confirmed by the findings of 2.5- to 4.4-fold increases in serum IGF-II concentrations *(56,58)*, and an apparent increase in the amount of IGF-II in the tissues of IGF2R/ci-MPR KO mice *(58)*.

These findings suggest that excess IGF-II is responsible for both the overgrowth and death of these KO mice. Cross-mating experiments confirm this speculation. In the absence of IGF-II expression, IGF2R/ci-MPR KO mice are completely rescued—that is, they do not exhibit perinatal lethality. As expected, they also are growth retarded, because they lack IGF-II *(58)*. The "toxic" effects of IGF-II appear to be mediated by the IGF1R, at least in part, because in the absence of the IGF1R, IGF2R/ci-MPR KO mice also are completely rescued. Remarkably, the latter double-KO mice exhibit normal fetal growth. This finding definitively demonstrates that IGFs can stimulate fetal growth through a receptor different from their IGF1R (*see* pp. 550–551). Consistently, when the IGF-II null mutation is bred into the naturally occurring mutant mouse line, T_{me}, with a chromosomal deletion (Thp) that includes the gene for the IGF2R/ci-MPR *(54)*, the T^{hp} phenotype is partially rescued and lethality is markedly reduced. It is important to understand that the T^{hp} phenotype is more severe than that of the IGF2R/ci-MPR KO mice, presumably because more than a single gene is deleted (*see* pp. 564–565). Finally, as expected from in vitro studies, the cation-dependent mannose-6-phosphate receptor (cd-MPR) does not appear to influence IGF physiology or fetal growth *(58)*.

Insulin and/or X Receptor(s)

The elegant studies of Baker et al. *(47)* and Ludwig et al. *(58)* provide clear genetic evidence that an undefined receptor(s) mediates some IGF-II growth -promoting signals during fetal life and can subserve the IGF1R's growth-promoting functions, at least under some circumstances. As reviewed previously, mice lacking the IGF2R/ci-MPR have supernormal IGF-II levels and exhibit *in utero* somatic and placental overgrowth *(58)*. The growth-stimulating effects of IGF-II cannot be entirely mediated by the IGF1R, however, because KO mice lacking both the IGF1R and the IGF2R/ci-MPR have normal *in utero* somatic and placental growth and are viable *(58)*. Furthermore, IGF-II must be responsible for most, if not all, the signaling, because triple-KO mice lacking IGF-II expression in addition to that of both receptors (i.e., the IGF1R and IGF2R/ci-MPR) are markedly growth retarded (30% of normal at birth) and not viable.

To understand these experiments it is useful to remember that an absence of either the IGF1R or the IGF2R/ci-MPR is lethal. This indicates that, if one of these receptors is not expressed, the lack of the other somehow promotes survival. Increased IGF-II, and possibly IGF-I, results from the absence of the IGF2R/ci-MPR. This physiologic alteration, caused by an absence of IGF2R/ci-MPR, must be responsible for survival in the absence of the IGF1R by providing sufficient ligand to transmit growth-promoting signals through an alternative receptor(s). In this context, it important to remember that the alternative receptor is effective in stimulating normal growth only when an excess of ligand is present.

This indicates that the alternative receptor(s) does not bind IGFs avidly and is unlikely to be a primary IGF receptor. Furthermore, evidence accruing from the isolated ablation of IGF-I, IGF-II, or the IGF1R makes it clear that the IGF1R is the major mediator of IGF signals under normal physiologic circumstances.

Two lines of evidence indicate that the InR is an alternative receptor responsible for transmitting the growth-promoting signal of IGF-II. The first evidence comes from experiments utilizing fibroblasts lacking the IGF1R, which were derived from IGF1R KO mice *(55)*. Unlike normal mouse fibroblasts, these cells do not proliferate in serum-free media supplemented with purified IGFs. When the InR is overexpressed in these cells (by transfection with an InR expression plasmid), they grow normally in serum-free media supplemented with either IGF-II or insulin, but not with IGF-I.

The second line of evidence for the InR mediating IGF-stimulated growth-promoting actions comes from the study of mice that lack both the IGF1R and the InR. These mice were generated by serial cross-breeding of mice heterozygous for the null mutation of each gene *(61)*. InR KO mice exhibit near normal *in utero* growth (~90% of normal), but die in the first several days of postnatal life from the consequences of diabetic ketoacidosis *(60)*. Mice lacking both the IGF1R and the InR have birth weights similar to those of mutants lacking expression of both IGFs and to those lacking both IGF-II and IGF1R (~30% of normal; *see* Table 4). The InR, therefore, appears to mediate IGF-II's fetal somatic growth-promoting signals, because its absence accounts for the discrepancy between the growth retardation in IGF1R KO mice (45% of normal) and the double mutant, IGF-II/IGF1R KO mice (30% of normal). The growth retardation of the InR/IGF1R double-mutants is greater than that of IGF1R KO mice at e13.5 and becomes progressively greater from e16.5 onward. Edema in the InR/IGF1R KO mice probably accounts for the finding that their weights are similar between e14 and e16. As would be expected, InR/IGF1R KO mice die at about the time of birth. Placental growth in InR/IGF1R KO mice is normal, indicating that the InR does not mediate the stimulation of placental growth by IGF-II.

Although these experiments provide strong evidence that the InR is responsible for mediating IGF-II stimulation of fetal somatic growth in the absence of the IGF1R and the IGF2R/ci-MPR, they do not exclude a function for other receptors. In fact, the finding of normal placental growth in the absence of both the InR and the IGF1R *(61)* provides strong evidence for the existence of an as yet unidentified receptor, such as one similar to the insulin-receptor-related receptor (IRR: *[70]*) or a mammalian homologue of daf-2 *(71)*. A growth-stimulating role for IGF-I also cannot be excluded when both the IGF1R and the IGF2R/ci-MPR are absent, because mice lacking both IGF-II and the IGF2R/ci-MPR exhibit greater *in utero* growth than do IGF-II KO mice. IGF-I, thus, also appears to be capable of stimulating growth through an alternative receptor. Given that IGF-I cannot stimulate proliferation through the InR in fibroblasts lacking the IGF1R *(55)*, the InR seems to be excluded and the responsible receptor has yet to be identified.

IGFBPs

Because IGFBPs modulate IGF actions and because some IGFBPs are exuberantly expressed during *in utero* life (*see* review: *[72]*), the IGFBPs seem certain to have a role in the IGF regulation of *in utero* growth. To date, however, studies of mice with altered gene function have not produced a great deal of new information on the *in utero* actions of IGFBPs. IGFBP-1, a known inhibitor of IGF action when present in molar excess, is highly expressed *in utero* and its serum levels are increased in intrauterine growth retardation (*see* reviews: *[72,73]*). Data from IGFBP-1-overexpressing Tg mice are consistent with the view that IGFBP-1 inhibits IGF action *in utero*, because birth weights are reduced (8–17%) in the only line of Tg mice that overexpress IGFBP-1 prior to birth (mPGK/rIGFBP-1 Tg mice) *(37)*.

IGFBP-2 KO mice have no apparent phenotype *(63,64)*, despite the high *in utero* expression of this IGFBP, suggesting that other IGFBPs can compensate for a loss of IGFBP-2 function. Equally surprising is the preliminary finding that IGFBP-4 KO mice have somewhat reduced *in utero* growth rates, resulting in birth weights that are reduced by 5–10% of normal (J. Pintar, *personal communication*). This finding calls into question the assumption that IGFBP-4 acts exclusively by inhibiting IGF actions.

Actions in Postnatal Somatic Growth

Actions of IGF-I and IGF-II

IGF-I and IGF-II KO Mice. IGF-I KO mice that survive the perinatal period grow very poorly *(47)*. They do not exhibit perceptible growth during the first several postnatal days, but achieve a subnormal growth velocity during the remainder of the suckling period (first 3 wk of life). Thereafter they grow minimally, such that as adults they are ~30% of normal adult size. Their growth pattern is similar to that of growth hormone (GH)-deficient mice, but their growth velocity is more impaired. These findings point to IGF-I as the major mediator of the growth-promoting activity of GH in postnatal life. Postnatally, growth velocities of IGF-II KO mice are normal, but no catchup growth occurs, and therefore IGF-II KO mice remain about 60% of normal size as adults. Unlike IGF-I KO mice, they exhibit no apparent increase in morbidity.

IGF Overexpressing Tg Mice. Tg models of IGF overexpression have greatly contributed to our understanding of IGF actions in postnatal somatic growth. These studies extend the knowledge gained from in vivo administration of IGFs, because they provide long-term IGF exposure. They also are capable of increasing IGF expression in specific organs, rather than delivering IGFs from the circulation, and thus they better replicate the physiologic, local actions of the IGFs. Overexpression experiments often offer advantages over gene KO experiments. For example, overexpression of a gene is likely to amplify the effect of its actions, whereas the same action may be masked in a KO experiment because an alternative gene is capable of compensating for the KO gene's loss. Also because mice lacking IGF-I or the IGF1R have poor survival, their study in adult life is restricted or not possible.

IGF-I Overexpressing Tg Mice. Mice overexpressing IGF-I, using a transgene driven by the mMT-I promoter, express IGF-I in multiple tissues postnatally and have elevated serum IGF-I levels *(1,2)*. These mice become significantly larger than their non-Tg siblings at 6 wk of age, but a trend toward overgrowth is observed by 3–4 wk of postnatal life. This pattern of overgrowth is similar to that occurring in GH overexpressing Tg mice *(1,2)*. As adults these Tg mice weigh about 30% more than normals. Disproportional overgrowth of some organs (brain, carcass, kidney, pancreas and spleen; *see* the third section) accounts for part of the somatic overgrowth, but the modest 21% increase in carcass size accounted for 75% of the increase in body weight. This indicates that IGF-I overexpression has significant effects on skeletal, connective, and adipose tissue and/or muscle growth. Despite this finding, these Tg mice do not have a convincing increase in linear or skeletal growth, as assessed by the measurement of tail lengths, cumulative tibial width growth, or long bone length. The DNA content of all overgrown tissues is increased to the same extent as body weight, indicating that hyperplasia accounts for the increases in organ size.

IGF-I Tg mice do not overgrow to the same extent as GH-overexpressing Tg mice, who typically are twice normal size as adults. One possible reason for the less marked overgrowth of IGF-I Tg mice is that endocrine IGF-I overexpression is modest in these mice, with blood IGF-I concentrations being increased only ~50%. A more likely explanation for the modest overgrowth is that the total amount of IGF-I expressed is not significantly increased in most tissues. For example, endogenous liver IGF-I mRNA abundance is decreased. This is likely due to a decrease in GH secretion. Pituitary GH mRNA expression is reduced by 60% (probably secondary to negative feedback inhibition by the transgene IGF-I), and thus, GH stimulation of the endogenous IGF-I gene is suppressed in all GH-responsive tissues. It may be, therefore, that in many tissues of these Tg mice total IGF-I expression is not markedly increased, and is derived in large part from the transgene. The latter speculation also suggests that locally synthesized IGF-I is more important to growth than is blood-borne IGF-I.

Evidence from rαMHC/IGF-I Tg mice argues against a crucial role for circulating IGF-I in the stimulation of somatic growth *(21)*. In these Tg mice the expression of the transgene is restricted to the heart, and results in cardiac hypertrophy (*see* p. 559). The heart IGF-I overexpression also results in an 84% increase in adult serum IGF-I levels, but despite these increases, no changes in bone length and only small increases in the weights of some organs were observed. The relatively low expression of this transgene during the first several weeks of life may in part explain the lack of somatic overgrowth. Nonetheless, this study, taken together with many others, suggests that locally expressed IGF-I better stimulates tissue growth than does IGF-I delivered in an endocrine fashion.

IGF-I's Mediation of GH Action. To determine whether IGF-I stimulates linear growth in the absence of GH, IGF-I Tg mice were cross-mated with mice having ablated pituitary GH expression (mice carrying a transgene that fused the GH promoter to the gene for diphtheria toxin, such that somatotrophs were killed by the cell-specific expression of the toxin) *(2)*. Mice carrying both transgenes express little or no GH, but have normal circulating IGF-I and grow normally. Both their lengths and weights were indistinguishable from their normal littermates, i.e., mice carrying no transgene. IGF-I, therefore, can replicate the somatic growth-promoting actions of GH.

IGF-II-Overexpressing Tg Mice. Postnatal overexpression of IGF-II does not result in increased somatic growth. Using a variety of promoters (*see* Table 1) to drive IGF-II overexpression, elevations of serum IGF-II have been achieved in several Tg mouse lines, usually those expressing the transgene in liver (mH-2K^b/hIGF-II, *[24]*); mMUP/hIGF-II *[27]*; and rPEPCK/hIGF-II *[31]* Tg mice), but also in one line without liver transgene expression (bKer/mIGF-II; *[30]*). Despite these elevations of circulating IGF-II (2 to 30-fold depending on the line of mice), no enhancement of somatic growth has been noted. Overgrowth of some organs that express the transgene, however, has been impressive (*see* pp. 557–562). Furthermore, when mH-2Kb/hIGF-II Tg mice are bred to Snell dwarf mice, i.e., mice that are GH deficient, no increase in somatic growth results. In two IGF-II Tg mouse lines *(27,30)*, elevations in serum IGF-II levels are associated with modest reductions in body weight that appear to increase in magnitude with advancing postnatal age (*see* p. 568). The decreased weight in these Tg mice is due to a marked decrease in fat mass, and possibly a decrease in lean body mass *(27)*. This effect of IGF-II must be mediated by endocrine mechanisms, because neither IGF-II transgene is expressed in adipose tissue.

While these results argue against a somatic growth-promoting role for IGF-II during postnatal life, it is worth pointing out that rodents, unlike humans, express very little IGF-II postnatally, and thus the role of IGF-II in rodents may be unique. Nonetheless, the endocrine actions of IGF-II on adipose tissue and the local growth-promoting actions on certain organs/tissues (*see* pp. 555–562), as well as its potential role in tumorigenesis (*see* pp. 568–569), cannot be ignored.

IGF1R, IGF2R/ci-MPR, and Other Receptors

Direct study of IGF1R-mediated actions on postnatal growth in vivo are not possible, because IGF1R KO mice do not survive after birth *(46,47)*. This fact points out that the IGF1R is essential for postnatal growth under normal physiologic circumstances. The finding that some IGF-I KO mice survive postnatally, albeit with poor growth and function, suggests that the IGF1R interacts with IGF-II or possibly other unidentified ligands, in addition to its interaction with IGF-I. The near-normal postnatal growth in the minority of IGF2R/ci-MPR KO mice that survive indicates that this receptor does not mediate IGF signaling *(58)*. As reviewed previously, evidence from KO mice lacking both IGF1R and IGF2R/ci-MPR indicates that the InR, and possibly another unidentified receptor(s), can subserve the function of the IGF1R *(58)*. These double-receptor mutants survive postnatally and exhibit growth kinetics that are somewhat unique. They grow normally in the first 2 wk of postnatal life, have a subnormal growth velocity from about d 15–30, and then have catchup growth, such that they are normal in size by about d 70 of life. Any interpretation of these data is totally speculative at this time.

IGFBPs

mPGK/rIGFBP-1 *(36)* and hα2-AT/hIGFBP-1 *(43)* Tg have subnormal weight gain postnatally, such that their adult weights are reduced by 10–25%. In mPGK/rIGFBP-1 Tg mice, weight gain is poorest around the time of weaning, whereas in hα2-AT/hIGFBP-1 Tg mice the reduced weight gain appears to occur after weaning. Weight gain in mMT-I/hIGFBP-1 Tg mice is similar to that of mPGK/rIGFBP-1 Tg mice, but these alterations did not meet tests of significance *(34)*. Linear growth is apparently normal in mMT/hIGFBP-1 Tg mice *(34)*; measures of linear growth have not been reported in the other lines of IGFBP-1-overexpressing Tg mice. Each IGFBP-1 Tg mouse line exhibits significant elevations in serum IGFBP-1 levels, suggesting that endocrine effects of IGFBP-1 overexpression are possible. In addition, each line exhibits reduced size of specific organs, especially spleen and brain (*see* pp. 555–562 and Table 5). Taken together, it is clear that IGFBP-1 can inhibit IGF action during postnatal growth, but these actions appear to be restricted to weight gain (which are likely related to lipid and/or possibly glucose metabolism; *see* pp. 568 and 567–568 respectively) and to local actions in specific organs.

IGFBP-3 Tg mice, made with a mMT-I-driven transgene, do not exhibit alterations of somatic growth *(44)*. The weight of some organs, however, is increased *(44)*; thus, IGFBP-3 may have a small influence on postnatal growth. mSMA/rIGFBP-4-overexpressing Tg mice show marked hypoplasia of smooth muscle where this transgene is expressed, but there also is no discernible effect on overall somatic growth *(45)*. The postnatal growth of IGFBP-2 KO mice is normal *(63,64)*, and unexpectedly, that of IGFBP-4 KO mice is somewhat reduced, such that they have 15–20% reductions in body weight as adults (J. Pintar, *personal communication*). Given the assumed IGF inhibitory actions of these IGFBPs, an enhancement of growth was anticipated. These KO experiments emphasize that we do not as yet have a satisfactory understanding of IGFBP actions.

Table 5
Organ/Tissue Growth Influenced by Overexpression of IGFs or IGFBPs

Organ/Tissue	*Alteration in Expression*	*Growth Effect*	*References*
Adipose	↑IGF-I	↑ Dermal and subdermal fat	*(3)*
	↑IGF-II	↓ Fat mass	*(27,29,30)*
Adrenals	↑IGF-II	↑ Weight	*(31,33)*
Brain	↑IGF-I	↑ Weight of most regions	*(1,2,7)*
		↑ Myelin	*(6–8)*
		↑ Oligodendrocyte no.	*(7)*
		↑ Neuron no. in cerebral cortex	*(10)*
		↑ Cerebellum weight with ↑ granule and Purkinje cell no.	*(14)*
		↑ Spinal cord weight	*(11)*
	↑IGFBP-1	↓ Weight	*(7,34,35,37,43)*
		↓ Myelin	*(7)*
Cartilage/bone	↑IGF-I	↑ Size in GH-deficient mice	*(2)*
Gut	↑IGF-I	↑ Small intestine weight and DNA	*(12)*
	↑IGF-II	↑ Colon/appendix weight and DNA	*(29)*
Heart	↑IGF-I	↑ Myocyte no. and heart weight	*(21)*
	↑IGFBP-1	↓ Weight	*(43)*
	↑IGFBP-3	↑ Weight	*(44)*
Kidney	↑IGF-I	↑ Glomerular size	*(3,5)*
	↑IGF-II	↑ Weight	*(31,33)*
	↑IGFBP-1	↓ Weight	*(43)*
Liver	↑IGFBP-3	↑ Weight	*(44)*
Mammary gland	↑IGF-I, des (1–3) IGF-I & IGFBP-3	Incomplete involution	*(17,19)*
Muscle: skeletal	↑IGF-I	Myofibril hypertrophy	*(18)*
: smooth	↑IGF-I	↑ In aorta, bladder, intestine, stomach and uterus	*(20)*
	↑IGFBP-4	↓ In aorta, bladder, intestine, stomach and uterus	*(45)*
Pancreas	↑IGF-I	↓ Weight	*(1,3)*
Skin	↑IGF-II	Hyperkeratosis and ↑ epidermal size	*(29)*
Spleen	↑IGF-I, IGF-II, & IGFBP-3	↑ Weight	*(1,3,26,44)*
Thymus	↑IGF-II	↑ Weight, ↑ thymocyte and T cell no.	*(24–26)*
Uterus	↑IGF-II	↑ Weight	*(29)*

ACTIONS OF IGF SYSTEM PROTEINS ON SPECIFIC ORGANS/TISSUES

Alterations in the growth of many organs/tissues have been observed in IGF and IGFBP-overexpressing Tg mice (Table 5) and in IGF and IGF1R KO mice (Table 6). The review that follows is divided, somewhat arbitrarily, according to whether the weight of evidence favors a vital role for the IGFs and IGF1R in normal growth and development. An organ/tissue was categorized as "dependent" on the IGFs and IGF1R, if it met two criteria: (1) it exhibits hypoplasia, immaturity, and/or growth retardation in IGF and/or

Table 6
Organ/Tissue Growth Influenced by Ablation of IGF or IGF1R Expression

Organ/Tissue	*Alteration in Expression*	*Growth Effect*	*References*
Brain and spinal cord	↓IGF-I and IGF1R	↑ Cell density (↓ neuropil)	*(46)*
		↓ Oligodendrocyte no. and myelin	*(50)*
		↓ Dentate gyrus granule cell no.	
		↓ Cortical parvalbumin cells	
Cartilage/bone	↓IGF-I and IGF1R	Markedly delayed ossification	*(46)*
Heart	↓IGF2R/ci-MPR (i.e., ↑IGF-II)	Marked ↑ weight and hypertrophy	*(58)*
	midi-IGF-I	↓ Weight and left ventricular mass	*(51)*
Muscle (skeletal)	↓IGF-I and IGF1R	Hypoplasia	*(46)*
Placenta	↓IGF-II	↓ Weight	*(46,52)*
	↓IGF2R/ci-MPR (i.e., ↑IGF-II)	↑ Weight	*(58)*
Reproductive system	↓IGF-I		*(48)*
Male: testes		41% Normal weight	
epididymis		23% Normal weight	
vas deferens		10% Normal weight	
prostate		13% Normal weight	
Female: ovary		21–28% Normal weight and absent Graaffian follicles	
uterus		13% Normal weight	
Skin	↓IGF-I and IGF1R	Thin epidermal stratum and spinosum, and ↓ hair follicle no.	*(46)*

IGF1R KO mice; and (2) it is hyperplastic or overgrown in an IGF-overexpressing Tg mouse line. The reproductive system clearly meets these criteria, and is discussed in a separate section (*see* pp. 562–564).

Organs/tissues not included in the latter group also may be dependent on the IGFs and IGF1R for normal development, but the existing evidence from KO and Tg mice is not as clear. There are a variety of reasons for this: (1) Data from IGF1R KO mice provide the most direct evidence of IGF dependence. The perinatal death of these mice, however, makes detailed studies of some organs/tissues difficult or impossible, oftentimes because the organs/tissues of interest undergo significant maturation and growth postnatally (e.g., gut, kidney, pancreas, etc.). (2) In IGF KO mice, the normal growth-promoting actions of the knocked-out IGF may not be obvious, because alternative mechanisms have compensated for its absence—that is, the expressed IGF (or other factors) can subserve the role of the IGF that isn't expressed. (3) Overgrowth of an organ/tissue in IGF-overexpressing Tg mice does not necessarily indicate IGF-dependence, because some of alterations in organ/tissue growth, resulting from high transgene expression, may represent actions occurring only with protein overexpression, rather than being an amplification of an IGF system protein function during normal growth and development. (4) In contrast, the absence of a specific growth alteration in IGF and/or IGFBP-overexpressing Tg mice may result because the transgene either is not expressed in an organ/tissue or is not expressed at a sufficiently early developmental time to affect growth. The latter also is important in interpreting the specific growth effects of IGF system proteins. In some organs/tissues the influences of IGF system proteins are clearly autocrine or paracrine, either because the transgene is directed by a tissue-specific promoter and/or there is no evidence of an alteration of the transgene protein in the blood. In other cases endocrine effects may be operative, because the level of the transgene protein is altered in the circulation. In this situation it often is difficult to distinguish between local and endocrine actions.

Organs/Tissues in Which Normal Growth Appears Dependent on IGF and IGF1Rs

Brain

Our understanding of the role of IGF in brain growth has advanced significantly through study of mice with altered IGF system protein expression. The brains of IGF-I and IGF1R KO mice are small relative to controls, although their weight is not as decreased as their body weight *(50)*. At birth the striking histologic feature of these mice is their increase in neuronal density *(46)*, which is likely due to a paucity of neuropil or neuritic outgrowth, i.e., dendrites and axons. A paucity of myelin and an apparent decrease in oligodendrocyte development and myelination occurs predominately during the first month of postnatal life in the mouse, myelination changes are not likely to be obvious in KO mice with lethal mutations.

Consistent with these findings is the decrease in myelination in mice ectopically expressing IGFBP-1 *(7,35,36)*. IGFBP-1, an inhibitor of IGF action when present in molar excess, does not appear to be normally expressed in the brain. When expression is promoted ectopically by transgene expression, IGFBP-1 blocks endogenous IGF action *(35)*, thus mimicking the findings in IGF-I and IGF1R KO mice. The brains of these mice are small *(7,35,36,43)*, and the time when the brain growth retardation occurs appears to depend upon the promoter used in the transgene. For example, brain growth retardation is apparent earlier and is greater when the transgene is expressed during fetal life, as in mPGK/rIGFBP-1 Tg mice *(36)* compared to mMT-I/hIGFBP-1 Tg mice *(35)*. Oligodendrocyte number appears to be decreased, and together with a marked decrease in myelin protein gene expression (and possibly alterations in other metabolic factors), leads to a decrease in myelination, which is characterized by a decrease in the number of myelinated axons and the thickness of their myelin sheaths *(7,8)*. As would be expected, overexpression of IGF-I leads to an increase in brain myelin content *(6)*. Morphologically, IGF-I Tg mice have an increased number of myelinated axons with thicker myelin sheaths *(7)*. Furthermore, smaller axons are myelinated *(8)*. A modest increase in the number of oligodendrocytes and a marked increase in myelin protein gene expression accounts at least in part for these changes *(7)*.

In addition to its marked effects on myelination, studies of IGF-I Tg mice indicate that IGF-I promotes the growth of many brain regions, with the cerebral cortex *(7)* and cerebellum *(14)* possibly being the most sensitive to IGF-I stimulation. In MT-I/IGF-I Tg mice *(7)* all regions of the brain are significantly increased in size: cerebral cortex > hippocampus > diencephalon > brain stem > cerebellum (in a separate study spinal cord also was found to be significantly increased in weight; *[11]*), and the degree of transgene expression correlates well with the magnitude of overgrowth *(7)*. Although increases in myelination undoubtedly account for much of this increase, DNA also is increased *(2)*. Our detailed studies of the posterior medial barrel subfield, a portion of the cerebral cortex that integrates sensory input from the whiskers, indicate that neuron number is modestly increased (~20%), that neuron cell body size is increased, and perhaps most importantly, that neurite outgrowth is markedly increased *(10)*. These findings are consistent with the apparent decrease in neuropil in IGF1R KO mice *(46)* and with in vitro data indicating that IGF-I stimulates neuron outgrowth (see review: *[74]*).

In other IGF-I overexpressing Tg mice, in which the mIGF-II 5′ flanking region drives IGF-I expression, we found maximal transgene expression in the cerebellum that was accompanied by marked cerebellar overgrowth (1.9- and 2.4-fold increases in weight and DNA content, respectively) *(14)*. The granule cells were increased 82% in number, and Purkinje cell number was increased by 20%. Although we found evidence of increased granule cell proliferation postnatally when the transgene is expressed, IGF-I inhibition of apoptosis also could account for some of the increase in granule cell number. We believe that the increase in Purkinje cell number is due to inhibition of apoptosis, because Purkinje cells develop prenatally prior to the expression of the transgene.

Taken together, studies of mice with altered IGF system gene expression provide consistent and complementary data indicating that IGF-I stimulates oligodendrocyte proliferation and/or survival, myelination, neuritic outgrowth, and proliferation and/or survival of some neuronal populations, e.g., cortical sensory neurons *(10)*, dentate gyrus granule cells and striatal parvalbumin containing neurons *(50)*, and cerebellar granule cells and Purkinje cells *(14)*. These actions are undoubtedly paracrine/autocrine, because the local changes in expression correspond with the specific changes in growth. In addition, circulating IGFs cross the blood–brain barrier poorly, making endocrine actions unlikely. In the case of the IGF-I Tg mice, studies were performed in lines in which circulating IGF-I levels were either normal or low (presumably due to inhibition of pituitary GH secretion secondary to pituitary transgene IGF-I expression and direct inhibition of somatotroph function).

Cartilage/Bone

In IGF-I and IGF1R KO mice, the formation of ossification centers is markedly delayed—by 1–4 d of gestation depending on the portions of the skeleton studied *(46)*. Although not quantified, the skeletal structures in these mice appear much smaller than those in normals. IGF-II KO mice, on the other hand, exhibit no consistent delay in ossification, and skeletal size does not differ markedly from normal. Both the gross and histologic structure of cartilage and bone appears normal in all three types of mutant mice. These data suggest that IGF-I, and possibly IGF-II (but to a lesser extent), acts through the IGF1R to stimulate the rate of skeletal formation, rather than to promote its differentiation and/or pattern formation. Consistent with this interpretation are the findings that the bones of adult IGF-I-overexpressing Tg mice are not convincingly larger than those of normal mice *(1)*, despite the observation that they may have somewhat increased linear growth as measured by tail length *(2)*. In addition, the skeleton of IGF-I Tg mice exhibit no structural abnormalities *(3)*. IGF-I expression in the absence of GH, however, normalizes skeletal growth (*see* pp. 552–554), indicating that IGF-I mediates promotion of skeletal growth by GH *(2)*. In contrast, when IGF-II is overexpressed in the absence of GH, no enhanced growth can be documented *(24)*. Furthermore, increases in skeletal size have not been observed in any line of IGF-II-overexpressing Tg mice, despite detailed examination in one such line *(32)*. On the other hand, an apparent advancement in ossification was observed in one line of IGF2R/ci-MPR KO mice, suggesting that the increased IGF-II resulting from this mutation hastens skeletal maturation *(57)*. This same line of KO mice also appeared to have some skeletal deformities *(57)*, which were not observed in other IGF2R/ci-MPR KO mice *(56,58)*.

Muscle

Strong data from IGF system protein KO or Tg mice indicate that IGF-I is essential for normal muscle cell proliferation during development, as well as being capable of stimulating hypertrophy postnatally. IGF-II also may be important in cardiomyocyte proliferation, or at least can function as an IGF-I surrogate in heart. IGFs, however, do not appear essential for muscle differentiation.

Skeletal Muscle. Hypoplasia of skeletal muscle is one of the more striking histologic features of IGF-I and IGF1R KO mice *(46)*. This hypoplasia is characterized by a striking decrease in myocyte number, possibly as much as fourfold in some muscle groups of IGF1R KO mice, the mice with the more severe phenotype. Fetal myocytes in KO mice, however, appear normal. Because these KO mice appear to die from respiratory failure, it seems likely that muscle hypoplasia, specifically in the intercostal muscles and diaphragm (which are severely affected), is responsible for the perinatal death. Consistent with this speculation is the finding that the lungs of dead KO mice, especially IGF1R KO mice, do not appear to inflate. Other than atelectasis, which is the expected result of respiratory failure, the lungs appear histologically normal and contain a normal amount of surfactant apoproteins. There also is evidence that IGF-I overexpression stimulates muscle growth. In aSkαAct/hIGF-I Tg mice, marked myofibril hypertrophy occurs with local IGF-I overexpression *(18)*. These mice, however, do not exhibit muscle hyperplasia, most likely because the promoter used does not drive transgene expression until after the time myocyte differentiation and proliferation occur.

Heart. Strong evidence for stimulation of heart growth by IGF-II comes from IGF2R/ci-MPR KO mice *(56–58)*. As reviewed on pp. 549–550, these KO mice have elevations in IGF-II levels that presumably stimulate the dramatic cardiomegaly observed. In each line of KO mice, cardiomegaly is marked, and death is likely secondary to outflow obstruction due to the hypertrophy. For example in one line of these mice, heart weights at e18.5 are increased 4.17-fold *(58)*. This increase in size is characterized by 65–70% increases in ventricular wall thickness and is due to hyperplasia, because cardiomyocyte size is normal. It is not clear whether the cardiac enlargement represents a normal IGF-II effect or one that only occurs with overexpression.

Evidence from midi IGF-I mutant mice *(51)* and rαMHC/IGF-I Tg mice *(21,22)* supports a role for IGF-I in heart growth. As adults, midi IGF-I mice, who have reduced IGF-I gene expression, exhibit heart weights and left ventricular masses that are 30% of normal. Interestingly, these mice also have elevated blood pressure and increased left ventricular contractility. Increased adenylyl cyclase activity, without an increase in β adrenergic receptors, appears to mediate the increased contractility. rαMHC/IGF-I Tg mice express IGF-I in cardiac myocytes, and this results in a marked increase in myocyte number and heart weight that occurs as the expression of the transgene increases in young adulthood *(21)*. Furthermore, when myocardial infarctions are experimentally induced in these Tg mice, they do not exhibit the myocyte apoptosis or necrosis that occurs in the myocardium of normal mice subjected to the same procedure *(22)*. The latter finding suggests that IGF-I may stimulate heart growth by inhibiting the apoptosis of myocytes, rather than by stimulating their proliferation.

Smooth Muscle. When a mouse smooth muscle actin 5′-flanking region (mSMA) is used to drive expression of either rIGF-I or rIGFBP-4, striking changes in smooth muscle of many organs occur without changes in body weight *(20,45)*. In mSMA/rIGF-I Tg mice weights of bladder, stomach, intestine, uterus, and aorta are increased, owing to smooth muscle hyperplasia and tissue remodeling *(20)*. The bladder and stomach show concentrically thickened smooth muscle, while the intestine and uterus grow longer. mSMA/rIGFBP-4 Tg mice exhibit a phenotype reciprocal to the mSMA/rIGF-I Tg mice *(45)*. Despite normal body size, they have lighter bladders, intestines, aortas, uteri, and stomachs, each of which is characterized by smooth muscle hypoplasia.

Skin

The skin of IGF-I and IGF1R KO mice is remarkable for its translucence during fetal life, and is characterized by thinning of the epidermal stratum and spinosum and by a paucity of hair follicles *(46)*. Because surviving IGF-I Tg mice appear to have relatively normal fur and skin, these observations appear to represent a delay in skin maturation. Consistently, IGF-II overexpression in skin, driven by a keratin gene promoter, results in "wrinkled skin" due to hyperkeratosis and a thickened epidermis *(29)*. In the latter mice, skin DNA content is increased, indicating IGF-II stimulation of epidermal cellular proliferation. Finally, IGF-I Tg mice have been reported to have degenerative dermal changes *(3)*, a finding that is difficult to reconcile with the above data.

Organs/Tissues Influenced by Alterations in IGF Systems Protein Expression

Gastrointestinal System

IGF-I and IGF-II overexpression stimulates gastrointestinal growth *(12,29)*. Small bowel length and mass are increased in IGF-I Tg mice, as they are in GH-overexpressing Tg mice *(12)*. Villas height, crypt depth, and the number of mitotic crypt cells are each increased in jejunum of these mice, where the transgene is especially well expressed in the villus epithelia. In contrast, increased crypt cell proliferation is not observed in GH Tg mice, and in these mice GH appears to affect differentiated villus function more so than does IGF-I. These data suggest that each IGF-I and GH have specific functions in bowel development, and that local endogenous IGF-I expression may mediate the capacity of GH to stimulate bowel growth. Similarly, transgene IGF-II overexpression driven by a keratin promoter, which is well expressed throughout the gastrointestinal tract, results in gut hyperplasia *(29)*. It is not clear, however, whether IGF-II has a physiologic role in gut growth, or whether IGF-II is subserving the role of IGF-I under these circumstances.

Kidney

As with the administration of IGF-I *(75)*, IGF-I is associated with modest increases in kidney size that seem to result from hyperplasia *(1)*. Histologically, glomerular size is increased *(3,5)*. Interestingly, GH-overexpressing Tg mice exhibit twofold increases in kidney size and glomerular hypertrophy that progresses to glomerular sclerosis *(5)*, strongly indicating that IGF-I and GH exert different actions in the kidney. Increases in kidney weight also have been observed in IGF-II overexpressing Tg mice *(31,33)*. In each of these IGF Tg mouse lines the transgene is expressed in kidney, and in neither are the circulating IGF levels markedly elevated, arguing that the IGF-stimulated kidney growth is affected in a paracrine/autocrine fashion. Small reductions in kidney size, proportional to those of the body, have been observed in IGFBP-1 Tg mice *(37,43)*, again consistent with IGFs having a stimulatory influence on kidney growth.

Mammary Gland

A number of Tg animal lines have been generated using specific-promoter-driven fusion genes to target IGF system protein expression to the mammary gland. In each case there is no alteration in the circulating levels of the transgene proteins, indicating that all the effects observed can be attributed to local transgene expression. The first of these utilized the bovine αS1 casein promoter to direct IGF-I expression to the rabbit mammary gland *(16,76)*. The milk from these rabbits proved to be a suitable source for the purification of biologically active IGF-I. No changes in the quantity or quality of the milk produced or in the histology of the breast were detected, possibly because the actions of IGF-I were blunted by the increased expression of IGFBP-2 (which was presumably induced by IGF-I overexpression) *(16)*.

Others have used the promoter for whey acidic protein (WAP) to direct mammary expression of IGF-I or des(1–3)IGF-I in mice *(17,19)*. Like the milk from Tg rabbits, milk from these mice had an increase in an IGFBP that appeared to be IGFBP-2 *(19)*. Breasts of mice that express des(1–3)IGF-I, a truncated form of IGF-I that does not bind to IGFBPs, exhibited incomplete involution following lactation that was characterized by ductile hypertrophy and loss of secretory lobules *(19)*. Other lines of IGF-I Tg mice made with a similar WAP-driven fusion gene also exhibited incomplete involution that appeared to be secondary to a decrease in apoptosis *(17)*. Tg mice made to overexpress hIGFBP-3 in the breast, again utilizing the mWAP promoter, show similar changes in involution and apoptosis *(17)*, suggesting that both IGF-I and IGFBP-3 inhibit programmed cell death, although the mechanism(s) is not clear.

Using the ovine β-lactoglobulin promoter fused to the mIGF-II gene, a number of lines of Tg mice were found to have an increased incidence of mammary tumors, occurring most frequently after 6 mo of age *(28)*. This is in obvious contrast to the modest structural changes in other animals overexpressing IGF-I in the breast. Perhaps a more extensive study at an older age would reveal an increased incidence in the IGF-I Tg mice, or alternately, IGF-II may have more tumorigenic effects than IGF-I (*see* the seventh section). The nature of the promoter or the genetic background of these mice also could contribute to the tumorigenesis in the oβLac/IGF-II Tg mice.

Spleen

The spleen is increased in weight in mMT-I/hIGF-I Tg mice (~2.2 fold; *[11]*), mMT-I/hIGFBP-3 Tg mice (two- to fourfold; *[41,44]*), and when the mH-2Kb/hIGF-II transgene is expressed in Snell Dwarf mice (25–75%; *[24]*). Histologically, IGF-I Tg mouse spleens have an apparent increase in megakaryocytes and extramedullary hematopoiesis in the red pulp *(3)*, and IGF-II Tg spleens exhibit increased trabecular size *(26)*. Presumably, IGFBP-3 effects on spleen are the result of an augmentation of IGF-I actions, but alternative explanations are possible. The transgenes likely are expressed in the spleens of each mouse line (but this was not reported for mMT-I/hIGFBP-3 Tg mice), making local actions possible. Because spleen is invariably increased following IGF-I administration, however, endocrine effects are equally possible. In IGFBP-1 Tg mouse lines, the spleen is reduced in size, but in proportion to the modest reduction in body size *(37,43)*.

Thymus

Studies of mH-2Kb/hIGF-II Tg mice demonstrate that IGF-II can have a major influence on thymic growth during postnatal development *(24)*. Thymus weights are increased as much as 50% at 1 mo of age, and remain somewhat increased despite thymus involution at 3–4 mo of age. When this transgene is expressed in GH-deficient, Snell Dwarf mice, the effects of IGF-II are even more dramatic (2.5- and 2-fold increases in thymus weight at 4 and 8 wk, respectively). The increase in thymus size is due to increased cellularity, characterized by increases in thymocyte number and the progressive maturation of normal, predominately CD4+, T cells *(25)*. Although this transgene is more intensely expressed in the thymic medulla, the thymic cortex exhibits the greatest increase in size *(26)*. We also have noted increases in the thymic weight of young mMT-I/hIGF-I Tg mice, and our preliminary studies indicate that a decrease in apoptosis accounts for the increased cellularity (*unpublished observations*).

Others

No disproportionate changes in liver size have been reported in IGF system KO mice. Given the paucity of IGF1R expression in liver *(77)*, this is not surprising. However, rαMHC/IGF-I Tg mice develop modest increases in liver weight with increasing age *(21)*, and IGFBP-3 Tg mice have been reported to have ~50% increases in liver weight. These increased weights must be due to endocrine effects because neither transgene is expressed in the liver *(21,44)*. It is tempting to speculate that the increased liver size is due to the stimulation of extramedullary hematopoiesis. The increased spleen weights in both Tg mouse lines make effects on hemopoietic tissues especially attractive. In the case of IGFBP-3 overexpression, an augmentation of IGF-I-stimulated extramedullary hematopoiesis seems likely, but direct effects of IGFBP-3, independent of the IGFs, cannot be excluded.

Pancreases of mMT-I/hIGF-I Tg mice have been reported to be increased nearly twofold in weight, but are histologically normal *(1,3)*. In our studies of these mice, we have noted no change in islet number or size (*unpublished observations*). Modest overgrowth of the adrenals also has been observed *(31–33)*.

ACTIONS OF IGF SYSTEM PROTEINS ON FERTILITY AND REPRODUCTION

Growth of the Reproductive System Organs

IGF-I KO Mice

Detailed studies of surviving IGF-I KO mice have documented the essential role that IGF-I plays in the growth and function of both the male and female reproductive systems *(48)*. In reproductive organs of both sexes, growth is impaired in a fashion consistent with lack of normal maturation and with deficiencies in steroidogenesis. Because significant species differences exist in IGF system protein expression, however, precise extrapolation of the findings in IGF-I KO mice to humans may not be valid.

Male. The testes of IGF-I KO mice weigh ~40% of normal, but this reduction is not as marked as that of the body as a whole (~28% of normal in adult KO mice studied). Histologically, Sertoli and germ cells are morphologically normal and exhibit a normal cellular arrangement. Furthermore, germ cells at each of stages of spermatogenesis are present. Leydig cells, however, are small (half normal size), and contain lipid inclusions that differ from those in normal Leydig cells in that they are not surrounded by smooth endoplasmic reticulum and are arranged in small clusters. Because the endoplasmic reticulum is involved in Leydig cell testosterone production, steroidogenesis was examined in vitro and found to be impaired. Basal and luteinizing hormone (LH)-stimulated testosterone production was 62% and 28% of normal, respectively. A relative decrease in 17β-hydroxysteroid dehydrogenase activity is likely, because androstenedione production is six- to eightfold increased under the same conditions. The latter findings are reminiscent of steroid production in immature Leydig cells. This aberrant Leydig cell function results in serum testosterone levels that are only 18% of normal.

With the exception of the caput epididymus (which is 27% of normal weight, and thus, reduced in size proportionately to the somatic growth retardation), all other accessory organs are disproportionately small, being 10–16% of normal. Given the known androgen dependence of testicular accessory organs, the reduced testosterone is most likely responsible for their small size. Sperm number is also decreased, but motility appears normal.

Female. Ovaries from IGF-I KO mice are proportionately reduced in size, being 21–30% of normal weight, but histologically are distinctly abnormal *(48)*. Graafian follicles are nearly absent, as are corpora lutea and corpora albcantia, indicating that ovulation does not occur. Furthermore, after gonadotrophin stimulation there is little indication of ovulation. Although a number of Graafian follicles appear, corpora lutia do not and eggs cannot be recovered from the oviduct. The total number of follicles is likely normal, however, because follicle density in these small ovaries is 2.4-fold increased. Taken together, these findings are consistent with IGF-I stimulating normal follicle maturation and facilitating a normal response to gonadotrophins. As would be expected, serum estrogen levels are depressed (~50% normal). The uterus is markedly reduced in weight (~13% of normal). Although this may be in part due to the low estrogen production, it is more likely due to the absence of uterine IGF-I expression (*see below*).

Other IGF System Protein Tg Mice

No alterations in reproductive organ size have been reported in IGF-II KO mice. In contrast, bKer/mIGF-II TG mice *(29)* exhibit high uterine transgene expression and have uteruses that are 7.7-fold increased in weight, mostly because of edema, as evidenced by their modest increase (43%) in DNA content *(29)*. These findings do not make a strong argument for IGF-II as a major stimulator of uterine growth. Study of mPGK/rIGFBP-1 Tg mice, however, further support a role of IGF-I in uterine growth *(38)*. Uterine DNA synthetic capacity and weight gain in response to estrogen treatment is reduced in these Tg mice. The increase in IGF-I mRNA, however, is not affected, indicating that estrogens increase uterine IGF-I, but that the actions of IGF-I are blocked by the rIGFBP-1 expressed in the uterus. Uterine EGF mRNA also increases in response to estrogen, but this increase is reduced compared to that in normal mice, suggesting that IGF-I in part mediates the estrogen induction of EGF expression.

Fertility

All data from study of IGF system protein KO and Tg mice point to IGF-I being vital both to fertility and to reproductive productivity. In surviving IGF-I KO males, a lack of appropriate mating behavior appears to be primarily responsible for the infertility *(48)*. Sperm from IGF-I KO males can fertilize eggs in vitro, and one in vivo impregnation has been documented. The lack of mounting behavior, likely secondary to the low testosterone levels, seems certain to account for the infertility of these males. A lack of ovulation is the primary reason for the infertility of female IGF-I KO mice, but were they to ovulate occasionally, their infantile uteruses would certainly prevent a sustained pregnancy. The IGF1R presumably mediates IGF-I actions, but because IGF1R KO mice do not survive, direct study of its role is not possible. Because IGF-II KO mice exhibit normal fertility and reproduction, an important role for IGF-II in rodent reproduction is unlikely. A role for IGF-II in human fertility, however, cannot be excluded.

Litter size is reduced in mPGK/rIGFBP-1 Tg mice *(38)*, hα1-AT/hIGFBP-1 Tg mice *(43)*, and in one line of mMT-I/hIGFBP-1 Tg mice *(34)*. Uterine expression of the transgene could account for this in mPGK/rIGFBP-1 and mMT-I/hIGFBP-1 Tg mice, but this does not appear to be the case in hα1-AT/hIGFBP-I Tg mice (this transgene is not likely to be expressed in the uterus). Homozygote hα1-AT/hIGFBP-I females appear to have decreased fertility, but because there is increased fetal and perinatal mortality, even in matings between homozygous males and normal females, the effects of the transgene expression on the progeny may be the primary reason for the reduced reproductivity *(43)*. We have made similar observations in one line of mMT-I/hIGFBP-1 Tg mice (*unpublished data*). Female bKer/mIGF-II Tg mice also exhibit decreased reproductivity, undoubtedly secondary to their uterine abnormalities *(29)*. Finally, we have noted decreased fertility in both males and females from some lines of mMT-I/hIGFBP-I Tg mice. The reasons for this are not clear, but could be related to mating behavior or neuroendocrine function, because each Tg line with decreased fertility exhibits marked brain overgrowth (*see* pp. 557–558).

IMPRINTING OF THE IGF-II AND IGF2R/CI-MPR GENES

Mice carrying a disruption of the IGF-II gene, created by homologous recombination, provided the first direct evidence of parental gene imprinting in mammals *(52,53)*. Unexpectedly, DeChiara, Robertson, and Efstratiadis found that the heterozygous progeny of males carrying the disrupted mutant gene exhibit fetal growth retardation and are phenotypically indistinguishable from mice homozygous for the mutation, while progeny of mothers with the mutant gene mated to normal males are phenotypically normal. These findings indicated that only the paternally inherited gene is normally expressed. Using more detailed genetic analyses, they went on to verify this conclusion by showing that only the paternal IGF-II gene is expressed in most tissues, the choroid plexus and leptomeninges being exceptions *(53)*.

Conceptually, imprinting means that at least one allele of the two IGF-II alleles is marked in a way that determines its capacity for expression. This imprinting could be either positive or negative, i.e., marked for either expression or repression. Because the precise mechanisms of imprinting are not known with certainty, the term "imprinting" has been applied both to genes that are expressed and those that are repressed, presumably depending upon the mechanistic bias of the author. Some, therefore, refer to the IGF-II gene as being maternally imprinted, implying that a repressor of gene transcription is operative (this is likely the case; *see below*), while others refer to it as being paternally imprinted, because the father's gene is expressed, albeit that the mechanism(s) may be passive. Recent work indicates that a genomic element, termed the imprint maintenance element (IME), upstream of the IGF-II gene governs imprinting of the IGF-II gene and at least one other nearby gene *(78)*. Specifically when the IME is deleted, expression of the IGF-II gene occurs.

Further research has shown that IGF-II is one of a cluster of imprinted genes that lie in a 1-megabase region (p15.5) of human chromosome 11 (chromosome 7 of the mouse) *(79)*. This region has been mapped as the site of mutations resulting in the Beckwith–Wiedemann syndrome (BWS), a variable constellation of clinical findings that includes increased birth sizes, organomegaly, omphalocele, and a predisposition to Wilms' tumor during childhood *(80)*. The precise etiology of the various forms of BWS has not been defined, but it appears that erasure of IGF-II imprinting (in this case repression of the maternal IGF-II gene) may be responsible for the overgrowth and may contribute to tumor growth. Other genes in the 11p15.5 cluster that are likely involved in BWS include: (1) *H19*, a maternally expressed gene downstream of the IGF-II gene that encodes an untranslated RNA that regulates the transcription of the IGF II gene *(81)* and (2) *p57*, also a maternally expressed gene, which encodes a cyclin-dependent kinase *(82,83)*. When *H19* is deleted by homologous recombination and maternally inherited, the maternal IGF-II gene is expressed and overgrowth results *(81)*. When the *p57* gene is knocked out, progeny have omphaloceles without overgrowth *(82,83)*. When mutant *H19* mice are mated with mice carrying disruption mutations of the IGF2R/ci-M6PR gene (which also is imprinted; *see below*), mice inheriting the appropriate parental mutations have omphaloceles, like *p57* knockout mice, and overgrow, undoubtedly because IGF-II levels are increased *(59)*. Mutations in the IGF-II gene have not been identified in BWS patients and only a few patients have mutations of the *p57* gene. Nonetheless, it seems certain that interactions of the products of these genes, as well as *H19*, account for BWS, albeit in an undefined fashion.

Generation of KO mice also has established that the IGF2R/ci-M6PR gene is parentally imprinted *(56–58)*. Three laboratories working independently showed that this gene is expressed from the maternal allele. They also showed that ablation of IGF2R/ci-M6PR gene function is in large part responsible for the phenotype of the naturally occurring Tme mutant mouse, a fetal lethal resulting from a genomic deletion that includes the IGF2R/ci-MPR gene *(84)*. While somewhat variable, these KO mice have phenotypes that include fetal overgrowth, cardiac overgrowth (which is thought to result in perinatal death), a hairpin tail, and postaxial polydactyly, i.e., findings very similar to those in the Tme mouse. Ludwig et al. also found imperforate vaginas in surviving females, leading to massive uterine distension and infertility, findings such as the human Kaufman–McKusick syndrome *(58)* and those in bKer/mIGF-II Tg mice *(29,30)*. The reciprocal imprinting of the IGF-II and the IGF2R genes is intriguing and likely has implications regarding the contribution of each parental genome to somatic growth.

IGF SYSTEM PROTEIN ACTIONS IN THE REGULATION OF IGFs, IGFBPs, AND METABOLISM

Regulation of IGFs and IGFBPs

IGF and Insulin Regulation

Considerable information about the regulation of IGFs and IGFBPs has accrued from the study of mice with altered IGF system protein expression. Multiple organ overexpression of IGF-I in mMT-I/hIGF-I Tg mice (with an ~50% increase in serum IGF-I) leads to a 60% reduction in pituitary GH mRNA. In turn, this results in a decrease in serum GH *(1)* and a threefold decrease in endogenous liver IGF-I mRNA *(2)*. Conversely, in midi IGF-I mice with 70% reductions in serum IGF-I levels, serum GH is increased about fivefold *(51)*. These data confirm other evidence that IGF-I exerts negative feedback control on GH expression and secretion. In IGF-I Tg mice, however, it is not clear whether IGF-I in the circulation or alterations in brain or pituitary IGF-I expression exert these effects on pituitary function. Overexpression of IGF-I also influences insulin secretion because serum insulin is decreased in IGF-I-overexpressing Tg mice *(3)*. No information is available on the effects of IGF-I overexpression on IGF-II. Because mouse

IGF-II expression is very low postnatally when the mMT-I/hIGF-I transgene is expressed, discerning changes in IGF-II expression is difficult.

Tg mice also have been used to study the molecular control of IGF-I gene expression *(23)*. By fusing 11.3 kb of the 5′ flanking region of the rat IGF-I gene, including introns 1 and 2, to firefly luciferase, we have investigated IGF-I transcriptional control. While the expression of this transgene replicates near-normal endogenous IGF-I gene expression in brain, it does not do so in liver, perhaps because it lacks crucial positive regulatory elements, or alternatively contains dominant repressor elements important for liver IGF-I expression.

IGF-II overexpression appears to have a modest influence on IGF-I expression, depending upon whether the transgene IGF-II reaches the blood and the levels that are achieved. With serum IGF-II elevations of 50–100 ng/mL in mH-2K^b/hIGF-II *(24)* and bKer/mIGF-II *(30)* Tg mouse lines, serum IGF-I levels are lower than in control mice, although they do not meet tests of statistical significance. When serum IGF-II reaches higher levels, ranging from 150 ng/mL to 450 ng/mL in rPEPCK/hIGF-II *(31–33)* and mMUP/hIGF-II *(27)* Tg mice, respectively, serum IGF-I is reduced by 20–25%. It has been proposed that IGF-II reduces serum IGF-I by displacing it from binding sites on IGFBPs *(31–33)*. One line of mMUP/hIGF-II Tg mice also showed modest decreases in serum insulin levels (~20%), which presumably results from chronic reductions in blood glucose levels.

Overexpression of IGFBP-1 in β_1-antitrypsin-driven Tg mice has been reported to lower serum IGF-I, and this appears to be dependent on the degree of IGFBP-1 transgene expression, because homozygous mice had lower IGF-I levels than did heterozygotes *(43)*. In contrast, we observed a nonsignificant increase in serum IGF-I levels in mMT-I/hIGFBP-1 Tg mice *(34)*. The reason(s) for this discrepancy is not obvious, but may relate to IGFBP-1 actions in liver, where the α_1-antitrypsin-directed transgene is expressed and where most circulating IGF-I is synthesized. Serum IGFBP-1 levels in mMT-I/hIGFBP-1 Tg mice are many fold higher (3–15 ×) than in the α_1-AT/hIGFBP-1 Tg mice, making an endocrine action of IGFBP-1 unlikely to account for the differences in serum IGF-I. Although unlikely, it seems possible that the high serum IGFBP-1 levels in mMT-I/hIGFBP-1 Tg mice could mask an IGFBP-1 effect on IGF-I expression by accumulating IGF-I in the serum.

Regulation of IGFBPs

Significant insights into the regulation of IGFBPs have come from studies comparing IGFBP expression in Tg mice with altered IGF system protein and GH expression. The first such study compared IGFBPs in sera derived from the cross-mating of mMT-I/hIGF-I Tg mice with those made GH-deficient by the expression of a diphtheria toxin transgene in somatotrophs *(4)*. In these studies we demonstrated that IGF-I increases serum IGFBP-2 and IGFBP-3, whereas GH increases the acid-labile subunit in the 150-kDa IGF-IGFBP complex and reduces serum IGFBP-2. More specifically, comparisons of Tg mouse sera showed that IGF-I overexpression results in a 2.1-fold increase in IGFBP-3 and a restoration of IGFBP-3 to normal in GH-deficient mice (IGFBP-3 is 5.5-fold lower in GH-deficient mice). Furthermore, it was shown that the 150-kDa IGF–IGFBP complex was not restored in the absence of GH despite a marked elevation in IGFBP-3, indicating that GH induces the acid-labile subunit of this complex. IGF-I overexpression also increased IGFBP-2 2.9-fold, whereas GH apparently reduces it, because IGFBP-2 levels are higher in GH-deficient mice. In a study of oMT-I/oGH Tg mice liver IGFBP-3 expression increases when the transgene is induced by zinc administration *(85)*. Because hepatic IGF-I expression is much more dramatically increased by transgene induction *(85)*, however, it is possible that the increase in IGF-I expression caused the increased IGFBP-3 expression.

Local mammary gland IGF-I overexpression can stimulate local IGFBP expression. Milk from rabbits expressing a casein promoter-driven IGF-I transgene has increased IGFBP-3 levels *(16)*, while milk from mice with breast IGF-I overexpression appears to have increases in both IGFBP-2 and IGFBP-3 *(19)*. Gut expression of IGF-I in mMT-I/hIGF-I Tg mice increases the local expression of IGFBP-3 mRNA *(12)*, and in brain, IGF-I overexpression markedly stimulates IGFBP-5 expression *(13)*. Overexpression of IGF-II also appears to increase IGFBPs. Serum IGFBP-2, and possibly IGFBP-3, are moderately increased in rPEPCK/hIGF-II Tg mice; IGFBP-2 is likely increased in mMUP/hIGF-II Tg mice *(27)*; and, when the mH-2K^b/hIGF-II transgene is expressed in Snell Dwarf (hypopituitary) mice by cross-mating, serum IGFBP-3 is dramatically increased *(24)*.

Finally, studies of IGFBP-1 transgene expression in liver have provided insights into the molecular regulation of the endogenous IGFBP-1 gene. Huang et al., studying a transgene constructed with the rat IGFBP-1 gene driven by the mPGK promoter, found evidence that the transgene and endogenous gene are regulated similarly in liver *(42)*. They found that developmental age and nutrition regulate both genes similarly. Dai et al., studying mice carrying an mMT-I/hIGFBP-1 transgene constructed with a truncated human cDNA, also found similarities in the regulation of each gene—in that both transgene and endogenous gene expression decrease with age, and fasting increases the expression of both *(34)*. Molecular studies suggested that changes in mRNA stability mediate these changes. An AU-rich region near the end of the coding region may account for this regulation, because such regions are known to be involved in mRNA destabilization. The similarity of the findings in these two distinct lines of Tg mice utilizing different promoters and coding sequences from different species, but sharing the same AU-rich region, supports this conclusion.

Glucose Regulation

Sufficiently increased circulating IGF-I levels are known to cause hypoglycemia, but no IGF-I Tg mouse lines have been shown to have hypoglycemia. Furthermore, in unpublished studies of mMT-I/hIGF-I Tg mice, we have not been able to demonstrate abnormalities of the serum glucose response to intraperitoneal glucose administration. This is likely because most IGF-I Tg mice exhibit only modest elevations in serum levels. Furthermore, even the lines of mMT-I/hIGF-I Tg mice with 50–80% increases in serum IGF-I likely have compensated for the increased levels by suppressing insulin secretion *(3)*. A mild degree of hypoglycemia has been observed in some IGF-II Tg mouse lines *(27,31)*. Compensatory mechanisms, however, are likely to be operative, because glucose levels are only ~20% reduced, despite IGF-II levels that are as much as 30-fold increased *(27)*.

The role of IGFBP-1 in glucose homeostasis has been extensively evaluated in PGK/rIGFBP-1 Tg mice *(37,39,40)*. mPGK/rIGFBP-I Tg mice have elevated fasting glucose (by as much as 67% in some lines) and even more dramatic glucose intolerance in response to intraperitoneal glucose boluses *(39)*. The hypoglycemic response of these mice to IGF-I, but not to des(1–3)IGF-I or to insulin, is reduced, indicating that IGFBP-1 attenuates the hypoglycemic actions of IGF-I *(40)*. The glucose uptake response to IGF-I of adipocytes isolated from these mice also is reduced. In addition, these mice have insulin resistance, because fasting insulin levels are elevated *(39)* and muscles isolated from these mice have a reduced response to insulin *(40)*. The insulin resistance, however, does not appear to be sufficiently severe to account for the hyperglycemia, because: (1) insulin-to-glucose ratios are only modestly elevated during fasting and fall in response to a glucose challenge, and (2) decreases in glucose following insulin administration are near normal *(39)*. It has been suggested, therefore, that the insulin resistance is secondary and develops with time to compensate for the hyperglycemia induced by the decrease in IGF-I action *(40)*. Supporting this argument are the findings that islet cell number and size are increased in the pancreases of these mice, presumably an early response to the insulin demand. On the other hand, total pancreatic insulin content is decreased, possibly the result of prolonged and excessive insulin secretion *(39)*.

mMT-I/hIGFB-1-overexpressing Tg mice do not exhibit glucose intolerance *(34,43)*. This is unlikely to be due to the degree of transgene expression, because mMT-I/hIGFBP-1 Tg have higher serum IGFBP-1 levels than do mPGK/rIGFBP-1 Tg mice *(34,37)*. Rather, it is likely that the difference in the amino acid sequences in the human and rat species account for this discrepancy. Tyrosine residues are in different positions in the two species of IGFBP-1, and our preliminary studies indicate that rodent liver cannot phosphorylate human IGFBP-1. Previous work has shown that phosphorylation dramatically increases the affinity of IGFBP-1 for IGF, and thus, the phospho-IGFBP-1 binds IGFs more avidly making it unavailable for interaction with the IGF1R *(86)*. Thus, the nonphosphorylated transgene hIGFBP-1 does not appear to bind IGF-I avidly enough to block its glucose regulatory actions.

Regulation of Lipid Metabolism

Study of IGF-overexpressing Tg mice suggests a role for IGF in lipid metabolism. Both mMUP/hIGF-II *(27)* and bKer/mIGF-II *(30)* Tg mice exhibit significant decreases in fat mass. These effects are endocrine because neither transgene is expressed in adipose tissue. In adult mMUP/hIGF-II Tg mice, fat mass, as measured by tritiated water distribution, is decreased to 23% and 53% of normal in the two lines studied. This decrease in fat increases with age and is undoubtedly the reason for the reduced body weights of these mice. Decreased food intake does not account for the reduced fat mass. In adult bKer/mIGF-II Tg mice, total carcass lipid is ~50% of control mice *(30)*, and, again, alterations in food intake do not account for this change. All fat deposits measured were decreased in these mice, including the epididymal fat pad (58% of normal), subcutaneous fat (74% of normal), and brown fat (80% of normal). The oxidation of orally administered labeled lipids is increased in these IGF-II-overexpressing mice, and this appears to be due to increased oxidation by brown fat. The latter finding makes it tempting to speculate that the increased activity of brown fat accounts for the reduction in total fat mass, although a reduction in white fat (epipdymal fat pad) lipoprotein lipase activity could contribute to the phenotype by making lipids more available for oxidation in other tissues. Despite these alterations in fat metabolism, serum lipid levels are normal and no alterations in insulin or GH secretion could be detected. mMT-I/hIGF-I Tg mice have no documented alteration in fat mass, but have been reported to exhibit an increase in subcutaneous fat histologically and to have elevated serum tryglycerides, but normal cholesterol *(3)*.

THE ROLE OF IGF SYSTEM PROTEINS IN CANCER

Study of a variety of Tg mouse lines expressing proteins that are not IGF-related have implicated IGF-II in tumor formation, specifically in the development of hepatocarcinomas *(87–89)* and pancreatic islet cell tumors *(90)*. Hepatocarcinomas form in a number of Tg mouse lines created with fusion genes encoding the SV40 early region *(87,88)*, the pre-S/S region of hepatitis B virus *(87)*, and the c-*myc* oncogene driven by the woodchuck hepatitis virus regulatory region *(89)*, and in each cell line the tumors express IGF-II mRNA. This expression is focal, usually most marked in areas that are highly proliferative, and frequently occurring during the late progressive stage of carcinogenesis *(87,89)*, but not usually in liver areas thought to be premalignant *(87)*. These data suggest that IGF-II expression is not causal in hepatocarcinoma, but rather a second signal that promotes tumor proliferation *(89)*. Studies of Tg mice created with a fusion gene linking the insulin promoter to the SV40 large T antigen support this conclusion *(90)*. The T antigen is expressed in all islets of these mice, but only 10% of islets undergo neovascularization and even fewer (1–2%) develop solid tumors. As judged by *in situ* hybridization histochemistry, IGF-II expression correlates with increased islet cell proliferation, and IGF-II is highly expressed, although often focally, in all tumors. When these tumors are cultured, antisense oligonucleotides to IGF-II mRNA decrease DNA synthesis. Furthermore, when these T antigen Tg mice were crossed with IGF-II KO mice, so that IGF-II cannot be expressed, tumors continue to form but in fewer number and with greatly reduced size. This finding, therefore, provides direct evidence that IGF-II does not initiate tumor formation, but appears to stimulate tumor progression. In hepatocarcinoma cells, IGF-II has been immunochemically localized to the perinuclear cytoplasm *(87,89)*, rather than the cell membrane, suggesting that IGF-II actions are not mediated by the cell-surface IGF1R. Because the IGF1R is at best poorly expressed in hepatocytes, this finding is not surprising *(77)*.

Studies of IGF-II-overexpressing Tg mice also are consistent with the view that IGF-II has a secondary role in tumorigenesis *(27,28)*. An increased incidence of tumors was reported in Tg mice expressing a hIGF-II cDNA driven by the mMUP *(27)*, and as mentioned previously, Tg mice expressing the mIGF-II gene in mammary gland *(28)*. Hepatocarcinoma and lymphomas were the most common of a variety of tumors that developed in MUP/IGF-II Tg mice *(27)*. Most tumors developed after 18 mo of age and were more common in homozygous mice that express more IGF-II. In contrast, there are no reports of increased tumor formation in IGF-I-overexpressing mice. We have made a significant effort to examine the brains (a site of high transgene expression) of our IGF-I Tg mice and have found no evidence of tumors; neither have we observed tumors in other tissues. This difference between IGF-I and IGF-II Tg mice raises a number of intriguing conceptual problems. Most notably, it argues against the widely held belief that the IGF1R mediates the signals of both IGFs. It also makes a signaling role for the IGF2R/ci-M6PR seem possible, or alternatively suggests a novel signaling pathway for IGF-II action that is not shared by IGF-I.

FUTURE DIRECTIONS

Studies of Tg and KO mice have provided invaluable information on the role of IGF system proteins in the regulation of growth and development. Despite the great acquisition of new information that has accrued from these studies, further investigation of such models undoubtedly will continue to advance our understanding, both through more study of the existing lines of mice and through the creation of new lines. In the existing mouse lines, more detailed studies of specific organs/tissues and studies directed at regulatory and mechanistic issues (e.g., alterations in receptor and signaling protein expression and function) are virtually certain to yield new information. Our current ability to manipulate the mouse genome makes it possible to generate mouse lines that offer the chance to address nearly any question. Using yeast-derived recombinase mechanisms (Cre-Lox system; *[91]*), the technology is now available to ablate in vivo gene function conditionally, that is, at a site or time of our choice. Such systems are especially appealing for study of IGF1R function during postanatal life and/or in a specific organ/ tissue. Because IGF1R KO mice die at birth *(48)*, the role of the IGF1R during the postnatal development of specific organs/tissues cannot be studied. For example, ablating IGF1R expression only in a specific organ/tissue and/or only at defined developmental times is likely to provide insights into its postnatal growth-promoting functions and/ or into its nondevelopmental functions in that organ/tissue (depending on timing of the gene ablation). Similarly, using a variety of systems in Tg mice, such as tetracycline response elements *(92)*, it is possible to overexpress a protein at a specific site and time. We are limited only by the quality of the questions we ask and our creativity.

ACKNOWLEDGMENTS

The author is grateful to a number of colleagues: to Argiris Efstradiades, John Pintar, and James Fagin for providing access to their data which was not published at the time the manuscript was written, and to Ping Ye, Ali Calikoglu, and Billie Moats-Staats for reviewing the manuscript. The work from the authors laboratory reported in this review was supported by NIH Grants RO1 HD08299 and T32 DK07129.

REFERENCES

1. Mathews LS, Hammer RE, Behringer RR, D'Ercole AJ, Bell GI, Brinster RL, Palmiter RD. Growth enhancement of transgenic mice expressing human insulin-like growth factor I. Endocrinology 1988; 123:2827–2833.
2. Behringer RR, Lewin TM, Quaife CJ, Palmiter RD, Brinster RL, D'Ercole AJ. Expression of insulin-like growth factor I stimulates normal somatic growth in growth hormone-deficient transgenic mice. Endocrinology 1990; 127:1033–1040.
3. Quaife CJ, Mathews LS, Pinkert CA, Hammer RE, Brinster RL, Palmiter RD. Histopathology associated with elevated levels of growth hormone and insulin-like growth factor I in transgenic mice. Endocrinology 1989; 124:40–48.
4. Camacho-Hubner C, Clemmons DR, D'Ercole AJ. Regulation of insulin-like growth factor (IGF) binding proteins in transgenic mice with altered expression of growth hormone and IGF-I. Endocrinology 1991; 129:1201–1206.
5. Doi T, Striker LJ, Gibson CC, Agodoa LYC, Brinster RL, Striker GE. Glomerular lesions in mice transgenic for growth hormone and insulinlike growth factor-I. I. Relationship between increased glomerular size and mesangial sclerosis. Am J Pathol 1990; 137:541–552.
6. Carson MJ, Behringer RR, Brinster RL, McMorris FA. Insulin-like growth factor I increases brain growth and central nervous system myelination in transgenic mice. Neuron 1993; 10:729–740.
7. Ye P, Carson J, D'Ercole AJ. In vivo actions of insulin-like growth factor-I (IGF-I) on brain myelination: studies of IGF-I and IGF binding protein-1 (IGFBP-1) transgenic mice. J Neurosci 1995; 15:7344–7356.
8. Ye P, Carson J, D'Ercole AJ. Insulin-like growth factor-I influences the initiation of myelination: studies of anterior commissure of transgenic mice. Neurosci Lett 1995; 201:235–238.
9. Pucilowski O, Ayensu WK, D'Ercole AJ. Insulin-like growth factor I expression alters acute sensitivity and tolerance to ethanol in transgenic mice. Eur J Pharmacol 1996; 305:57–62.
10. Gutiérrez-Ospina G, Calikoglu AS, Ye P, D'Ercole AJ. In vivo effects of insulin-like growth factor-I on the development of sensory pathways: analysis of the primary somatic sensory cortex (S1) of transgenic mice. Endocrinology 1996; 137:5484–5492.
11. Chen L, Lund PK, Burgess SB, Rudisch BE, McIlwain DL. Growth hormone, insulin-like growth factor I, and motoneuron size. J Neurobiol 1997; 32:202–212.

12. Ohneda K, Ulshen MH, Fuller CR, D'Ercole AJ, Lund PK. Enhanced growth of small bowel in transgenic mice expressing human insulin-like growth factor I. Gastroenterology 1997; 112:444–454.
13. Ye P, D'Ercole AJ. Insulin-like growth factor-I regulates IGF binding protein-5 (IGFBP-5) gene expression in brain. Endocrinology 1998; 139:65–71.
14. Ye P, Xing YZ, Dai ZH, D'Ercole AJ. In vivo actions of insulin-like growth factor-I (IGF-I) on cerebellum development in transgenic mice: evidence that IGF-I increases proliferation of granule cell progenitors. Dev Brain Res 1996; 95:44–54.
15. Raizada MK. Editorial: Insulin receptor-related receptor: an orphan with neurotrophic/neuromodulatory potential. Endocrinology 1993; 133:1–2.
16. Wolf E, Jehle PM, Weber MM, Sauerwein H, Daxenberger A, Breier BH, Besenfelder U, Frenyo L, Brem G. Human insulin-like growth factor I (IGF-I) produced in the mammary glands of transgenic rabbits: yield, receptor binding, mitogenic activity, and effects on IGF-binding proteins. Endocrinology 1997; 138:307–313.
17. Neuenschwander S, Schwartz A, Wood TL, Roberts CT, Jr, Henninghausen L, LeRoith D. Involution of the lactating mammary gland is inhibited by the IGF system in a transgenic mouse model. J Clin Invest 1996; 97:2225–2232.
18. Coleman ME, DeMayo F, Yin KC, Lee HM, Geske R, Montgomery C, Schwartz RJ. Myogenic vector expression of insulin-like growth factor I stimulates muscle cell differentiation and myofiber hypertrophy in transgenic mice. J Biol Chem 1995; 270:12109–12116.
19. Hadsell DL, Greenberg NM, Fligger JM, Baumrucker CR, Rosen JM. Targeted expression of des(1–3) human insulin-like growth factor I in transgenic mice influences mammary gland development and IGF-binding protein expression. Endocrinology 1996; 137:321–330.
20. Wang JW, Niu W, Nikiiforov Y, Naito S, Chernausek S, Witte D, LeRoith D, Strauch A, Fagin JA. Targeted overexpression of IGF-I evokes distinct patterns of organ remodeling in smooth muscle cell tissue beds in transgenic mice. J Clin Invest 1997; 100:1425–1439.
21. Reiss K, Cheng W, Ferber A, Kajstura J, Li P, Li B, Olivetti G, Homcy CJ, Baserga R, Anversa P. Overexpression of insulin-like growth factor-I in the heart is coupled with myocyte proliferation in transgenic mice. Proc Natl Acad Sci USA 1996; 93:8630–8635.
22. Li Q, Li B, Wang X, Leri A, Jana KP, Liu Y, Kajstura J, Baserga R, Anversa P. Overexpression of insulin-like growth factor-1 in mice protects from myocyte death after infarction, attenuating ventricular dilation, wall stress, and cardiac hypertrophy. J Clin Invest 1997; 100:1991–1999.
23. Ye P, Umayahara Y, Ritter D, Bunting T, Autman H, Rotwein P, D'Ercole AJ. Regulation of insulin-like growth factor-I (IGF-I) gene expression in the brain of transgenic mice expressing an IGF-I-luciferase fusion gene. Endocrinology 1997; 138:5466–5475.
24. Van Buul-Offers SC, De Haan K, Reijnen-Gresnigt MG, Meinsma D, Jansen M, Oei SL, Bonte EJ, Sussenbach JS, Van den Brande JL. Overexpression of human insulin-like growth factor-II in transgenic mice causes increased growth of the thymus. J Endocrinol 1995; 144:491–502.
25. Kooijman R, Van Buul-Offers SC, Scholtens LE, Schuurman HJ, Van den Brande JL, Zegers BJM. T cell development in insulin-like growth factor-II transgenic mice. J Immunol 1995; 154:5736–5745.
26. Van Der Ven LTM, Roholl PJM, Reijnen-Gresnigt MG, Bloemen RJ, Van Buul-Offers SC. Expression of insulin-like growth factor II (IGF-II) and histological changes in the thymus and spleen of transgenic mice overexpressing IGF-II. Histochem Cell Biol 1997; 107:193–203.
27. Rogler CE, Yang D, Rosetti L, Donohoe J, Alt E, Chang CJ, Rosenfeld R, Neely K, Hintz R. Altered body composition and increased frequency of diverse malignancies in insulin-like growth factor transgenic mice. J Biol Chem 1994; 269:13779–13784.
28. Bates P, Fisher R, Ward A, Richardson L, Hill DJ, Graham CF. Mammary cancer in transgenic mice expressing insulin-like growth factor II (IGF-II). Br J Cancer 1995; 72:1189–1193.
29. Ward A, Bates P, Fisher R, Richardson L, Graham CF. Disproportionate growth in mice with Igf-2 transgenes. Proc Natl Acad Sci USA 1994; 91:10365–10369.
30. Da Costa THM, Williamson DH, Ward A, Bates P, Fisher R, Richardson L, Hill DJ, Robinson ICAF, Graham CF. High plasma insulin-like growth factor-II and low lipid content in transgenic mice: measurements of lipid metabolism. J Endocrinol 1994; 143:433–439.
31. Wolf E, Kramer R, Blum WF, Foll J, Brem G. Consequences of postnatally elevated insulin-like growth factor-II in transgenic mice: endocrine changes and effects on body and organ growth. Endocrinology 1994; 135:1877–1994.
32. Wolf E, Rapp K, Blum WF, Kolb H, Brem G. Skeletal growth of transgenic mice with elevated levels of circulating insulin-like growth factor-II. Growth Reg 1995; 5:177–183.

33. Blackburn A, Schmitt A, Schmidt D, Wanke R, Hermanns W, Brem G, Wolf E. Actions and interactions of growth hormone and insulin-like growth factor-II: body and organ growth of transgenic mice. Transgen Res 1997; 6:213–222.
34. Dai Z, Xing Y, Boney C, Clemmons DR, D'Ercole AJ. Human insulin-like growth factor binding protein-1 (IGFBP-1) transgenic mice: characterization and insights into the regulation of IGFBP-1 expression. Endocrinology 1994; 135:1316–1327.
35. D'Ercole AJ, Dai Z, Xing Y, Boney C, Wilkie MB, Lauder JM, Han VK, Clemmons DR. Brain growth retardation due to the expression of human insulin-like growth factor binding protein-1 (IGFBP-1) in transgenic mice: an in vivo model for analysis of IGF function in the brain. Dev Brain Res 1994; 82:213–222.
36. Murphy LJ, Barron D, Seneviratne C. Hormonal regulation of insulin-like growth factor binding protein-1 expression and the development of transgenic mouse models to study IGFBP-1 function. Adv Exp Med Biol 1993; 343:279–291.
37. Rajkumar K, Barron D, Lewitt MS, Murphy LJ. Growth retardation and hyperglycemia in insulin-like growth factor binding protein-1 transgenic mice. Endocrinology 1995; 136:4029–4034.
38. Rajkumar K, Dheen T, Krsek M, Murphy LJ. Impaired estrogen action in the uterus of insulin-like growth factor binding protein-1 transgenic mice. Endocrinology 1996; 137:1258–1264.
39. Rajkumar K, Dheen ST, Murphy LJ. Hyperglycemia and impaired glucose tolerance in IGF binding protein-1 transgenic mice. Am J Physiol Endocrinol Metab 1996; 270:E565–E571.
40. Rajkumar K, Krsek M, Dheen ST, Murphy LJ. Impaired glucose homeostasis in insulin-like growth factor binding protein-1 transgenic mice. J Clin Invest 1996; 98:1818–1825.
41. Murphy LJ, Rajkumar K, Molnar P. Phenotypic manifestations of insulin-like growth factor binding protein-1 (IGFPB-1) and IGFBP-3 overexpression in transgenic mice. Prog Growth Factor Res 1996; 6:425– 432.
42. Huang H, Rajkumar K, Murphy LJ. Expression of insulin-like growth factor binding proteins (IGFBPs) in IGFBP-1 transgenic mice. J Endocrinol 1997; 152:99–108.
43. Gay E, Seurin D, Babajko S, Doublier S, Cazillis M, Binoux M. Liver-specific expression of human insulin-like growth factor binding protein-1 in transgenic mice: repercussions on reproduction, ante- and perinatal mortality and postnatal growth. Endocrinology 1997; 138:2937–2947.
44. Murphy LJ, Molnar P, Lu X, Huang H. Expression of human insulin-like growth factor binding protein-3 in transgenic mice. J Mol Endocrinol 1995; 15:293–303.
45. Wang J, Niu W, Witte DP, Chernausek SD, Nikiforov YE, Clemens TL, Shabifi B, Strauch A, Fagin JA. Overexpression of IGFBP-4 in smooth muscle cells of transgenic mice through a smooth muscle Actin -IGFBP-4 fusion gene induces smooth muscle hypoplasia. Endocrinology 1998; 139:2605–2614.
46. Liu J, Baker J, Perkins AS, Robertson EJ, Efstratiadis A. Mice carrying null mutations of the genes encoding insulin-like growth factor I (Igf-1) and type 1 IGF receptor. Cell 1993; 75:59–72.
47. Baker J, Liu J, Robertson EJ, Efstratiadis A. Role of insulin-like growth factors in embryonic and postnatal growth. Cell 1993; 75:73–82.
48. Baker J, Hardy MP, Zhou J, Bondy C, Lupu F, Bellvé AR, Efstratiadis A. Effects of an Igf1 gene null mutation on mouse reproduction. Mol Endocrinol 1996; 10:903–918.
49. Powell-Braxton L, Hollingshead P, Warburton C, Dowd M, Pitts-Meek S, Dalton D, Gillett N, Stewart TA. IGF-I is required for normal embryonic growth in mice. Genes Dev 1993; 7:2609–2617.
50. Beck KD, Powell-Braxton L, Widmer H-R, Valverde J, Hefti F. Igf1 gene disruption results in reduced brain size, CNS hypomyelination and loss of hippocampal and striatal parvalbumin-containing neurons. Neuron 1995; 14:717–730.
51. Lembo G, Rockman HA, Hunter JJ, Steinmetz H, Koch WJ, Ma L, Prinz MP, Ross J, Jr, Chien KR, Powell-Braxton L. Elevated blood pressure and enhanced myocardial contractility in mice with severe IGF-1 deficiency. J Clin Invest 1996; 98:2648–2655.
52. DeChiara TM, Efstratiadis A, Robertson EJ. A growth-deficiency phenotype in heterozygous mice carrying an insulin-like growth factor II gene disrupted by targeting. Nature 1990; 345:78–80.
53. DeChiara TM, Robertson EJ, Efstratiadis A. Parental imprinting of the mouse insulin-like growth factor II gene. Cell 1991; 64:849–859.
54. Filson AJ, Louvi A, Efstratiadis A, Robertson EJ. Rescue of the T-associated maternal effect in mice carrying null mutations in Igf-2 and Igf2r, two reciprocally imprinted genes. Development 1993; 118:731–736.
55. Morrione A, Valentinis B, Xu SQ, Yumet G, Louvi A, Efstratiadis A, Baserga R. Insulin-like growth factor II stimulates cell proliferation through the insulin receptor. Proc Natl Acad Sci USA 1997; 94:3777–3782.

56. Lau MMH, Stewart CEH, Liu Z, Bhatt H, Rotwein P, Stewart CL. Loss of the imprinted IGF2 cation-independent mannose 6-phosphate receptor results in fetal overgrowth and perinatal lethality. Genes Dev 1994; 8:2953–2963.
57. Wang Z-Q, Fung MR, Barlow DP, Wagner EF. Regulation of embryonic growth and lysosomal targeting by the imprinted Igf2/Mpr gene. Nature 1994; 372:464–467.
58. Ludwig T, Eggenschwiler J, Fisher P, D'Ercole AJ, Davenport ML, Efstratiadis A. Mouse mutants lacking the type 2 IGF receptor (IGF2R) are rescued from perinatal lethality in Igf2 and Igf1r null backgrounds. Dev Biol 1996; 177:517–535.
59. Eggenschwiler J, Ludwig T, Fisher P, Leighton PA, Tilghman SM, Efstratiadis A. Mouse mutant embryos overexpressing IGF-II exhibit phenotypic features of the Beckwith–Wiedemann and Simpson-Golabi-Behmel syndromes. Genes Dev 1997; 11:3128–3142.
60. Accili D, Drago J, Lee EJ, Johnson MD, Cool MH, Salvatore P, Asico LD, Jose PA, Taylor SI, Westphal H. Early neonatal death in mice homozygous for a null allele of the insulin receptor gene. Nature Genet 1996; 12:106–109.
61. Louvi A, Accili D, Efstratiadis A. Growth-promoting interaction of IGF-II with the insulin receptor during mouse embryonic development. Dev Biol 1997; 189:33–48.
62. Wood TL, Rogler L, Streck RD, Cerro J, Green B, Grewal A, Pintar JE. Targeted disruption of IGFBP-2 gene. Growth Regulation 1993; 3:5–8.
63. Pintar JE, Cerro JA, Wood TL. Genetic approaches to the function of insulin-like growth factor-binding proteins during rodent development. Horm Res 1996; 45:172–177.
64. Pintar JE, Schuller A, Cerro JA, Czick M, Grewal A, Green B. Genetic ablation of IGFBP-2 suggests functional redundancy in the IGFBP family. Prog Growth Factor Res 1995; 6:437–445.
65. Baserga R, Sell C, Porcu P, Rubini M. The role of the IGF-I receptor in the growth and transformation of mammalian cells. Cell Prolif 1994; 27:63–71.
66. Resnicoff M, Abraham D, Yutanawiboonchai W, Rotman HL, Kajstura J, Rubin R, Zoltick P, Baserga R. The insulin-like growth factor I receptor protects tumor cells from apoptosis in vivo. Cancer Res 1995; 55:2463–2469.
67. Sell C, Rubini M, Rubin R, Liu J-P, Efstratiadis A, Baserga R. Simian virus 40 large tumor antigen is unable to transform mouse embryonic fibroblasts lacking type 1 insulin-like growth factor receptor. Proc Natl Acad Sci USA 1993; 90:11217–11221.
68. Roth RA. Structure of the receptor for insulin-like growth factor II: the puzzle amplified. Science 1988; 239:1269–1271.
69. Seta KA, Kovacina K, Roth RA. The insulin receptor family. In: LeRoith D, Raizada MK, eds. Current Directions in Insulin-Like Growth Factor Research. Plenum Press, New York, 1994, pp. 113–124.
70. Watt VM, Sheir P, Chan J. IRR: A novel member of the insulin receptor family. In: LeRoith D, Raizada MK, eds. Current Directions in Insulin-Like Growth Factor Research. Plenum Press, New York, 1994, pp. 125–132.
71. Kimura KD, Tissenbaum HA, Liu YX, Ruvkun G. daf-2, an insulin receptor-like gene that regulates longevity and diapause in *Caenorhabditis elegans*. Science 1997; 277:942–946.
72. Jones JI, Clemmons DR. Insulin-like growth factors and their binding proteins: biological actions. Endocr Rev 1995; 16:3–34.
73. Lee PDK, Conover CA, Powell DR. Regulation and function of insulin-like growth factor-binding protein-1. Proc Soc Exp Biol Med 1993; 204:4–29.
74. D'Ercole AJ, Ye P, Calikoglu AS, Guiterrez-Ospina G. The role of the insulin-like growth factors in the central nervous system. Mol Neurobiol 1996; 13:227–255.
75. Hammerman MR, Miller SB. Effects of growth hormone and insulin-like growth factor I on renal growth and function. J Pediatr 1997; 131 (Suppl.):S17–S19.
76. Brem G, Hartl P, Besenfelder U, Wolf E, Zinovieva N, Pfaller R. Expression of synthetic cDNA sequences encoding human insulin-like growth factor-I in the mammary gland of transgenic rabbits. Gene 1994; 149:351–355.
77. Lowe WL, Jr, Adamo M, Werner H, Roberts CT, Jr, LeRoith D. Regulation by fasting of rat insulin-like growth factor I and its receptor. Effects on gene expression and binding. J Clin Invest 1989; 84:619–626.
78. Hu JF, Vu TH, Hoffman AR. Genomic deletion of an imprint maintenance element abolishes imprinting of both insulin-like growth factor II and H19. J Biol Chem 1997; 272:20715–20720.
79. Feinberg AP. Multiple genetic abnormalities of 11p15 in Wilms' tumor. Med Pediatr Oncol 1996; 27:484–489.
80. Elliott M, Bayly R, Cole T, Temple IK, Maher ER. Clinical features and natural history of Beckwith–Wiedemann syndrome: presentation of 74 new cases. Clin Genet 1996; 46:168–174.

81. Leighton PA, Ingram RS, Eggenschwiler J, Efstratiadis A, Tilghman SM. Disruption of imprinting caused by deletion of the H19 gene region in mice. Nature 1995; 375:16–17.
82. Zhang P, Liegeois NJ, Wong C, Finegold M, Hou H, Thompson JC, Silverman A, Harper JW, DePinho RA, Elledge SJ. Altered cell differentiation and proliferation in mice lacking p57KIP2 indicates a role in Beckwith–Wiedemann syndrome. Nature 1997; 387:151–158.
83. Yan Y, Frisen J, Lee MH, Massague J, Barbacid M. Ablation of the CDK inhibitor p57Kip2 results in increased apoptosis and delayed differentiation during mouse development. Genes Dev 1997; 11: 973–983.
84. Forejt J, Gregorová S. Genetic analysis of genomic imprinting: an Imprintor-1 gene controls inactivation of the paternal copy of the mouse Tme locus. Cell 1992; 70:443–450.
85. Chow JC, Murray JD, Pomp D, Baldwin RL, Calvert CC, Oberbauer AM. Regulation of insulin-like growth factor-I and binding protein-3 expression in oMT1a-oGH transgenic mice. Transgen Res 1994; 3:127–133.
86. Jones JI, D'Ercole AJ, Camacho-Hubner C, Clemmons DR. Phosphorylation of insulin-like growth factor (IGF)-binding protein 1 in cell culture and in vivo: effects on affinity for IGF-I. Proc Natl Acad Sci USA 1991; 88:7481–7485.
87. Schirmacher P, Held WA, Yang D, Chisari FV, Rustum Y, Rogler CE. Reactivation of insulin-like growth factor II during hepatocarcinogenesis in transgenic mice suggests a role in malignant growth. Cancer Res 1992; 52:2549–2556.
88. Cariani E, Dubois N, Lasserre C, Briand P, Brechot C. Insulin-like growth factor II (IGF-II) mRNA expression during hepatocarcinogenesis in transgenic mice. J Hepatol 1991; 13:220–226.
89. Liu P, Terradillos O, Renard CA, Feldmann G, Buendia MA, Bernuau D. Hepatocarcinogenesis in woodchuck hepatitis virus e-myc mice: sustained cell proliferation and biphasic activation of insulin-like growth factor II. Hepatology 1997; 25:874–883.
90. Christofori G, Naik P, Hanahan D. A second signal supplied by insulin-like growth factor II in oncogene-induced tumorigenesis. Nature 1994; 369:414–418.
91. Lewandoski M, Wassarman KM, Martin GR. Zp3-cre, a transgenic mouse line for the activation or inactivation of loxP-flanked target genes specifically in the female germ line. Curr Biol 1997; 7:148–151.
92. Kistner A, Gossen M, Zimmermann F, Jerecic J, Ullmer C, Lubbert H, Bujard H. Doxycycline-mediated quantitative and tissue specific control of gene expression in transgenic mice. Proc Natl Acad Sci USA 1997; 93:10933–10938.

IV

Clinical Aspects of the IGFs

24 Metabolic Effects of IGFs

Jürgen Zapf, MD, *E. Rudolph Froesch,* MD, *and Christoph Schmid,* MD

CONTENTS

THE IGF–INSULIN RELATIONSHIP

Structural and Evolutionary Aspects; Major Similarities and Differences

The relationship between insulin-like growth factors (IGFs) and insulin has been known since the discovery of nonsuppressible insulin-like activity (NSILA) in serum at the beginning of the 1960s *(1)*, and was derived from the similarities of their biological actions. In the mid-1970s it was found that NSILA consisted of two polypeptides termed IGF-I and IGF-II. They turned out to be not only biologically but also structurally closely related to insulin *(2,3)*. This finding was at first surprising, because the discovery of NSILA had been based on a most prominent feature that essentially distinguished it from insulin: its complete lack of cross-reactivity with insulin antibodies. However, after the elucidation of the primary and tertiary structures of IGF-I and -II *(2–4)*, the reason for this lack of cross-reactivity became apparent (Fig. 1). Insulin contains two main antibody binding sites. One of them comprises the amino (N)-terminal region of the B chain (B2, B3, and B4) plus an adjacent region of the A chain (A8, A9, and A10). This region is completely different in the IGFs. The second antibody binding site of insulin comprises residues A1 and A19–A21 and the carboxyl (C)-terminal region of the B chain, but this region is covered in the IGFs by the C domain and by a C-terminal extension called the D domain.

Despite this distinct immunological difference, IGFs and insulin are close relatives: they not only share important similarities in their structure, their receptors, and their signaling pathways which determine their biological actions, but they also have a common

From: *Contemporary Endocrinology: The IGF System*
Edited by: R. Rosenfeld and C. Roberts Jr © Humana Press, Inc., Totowa, NJ

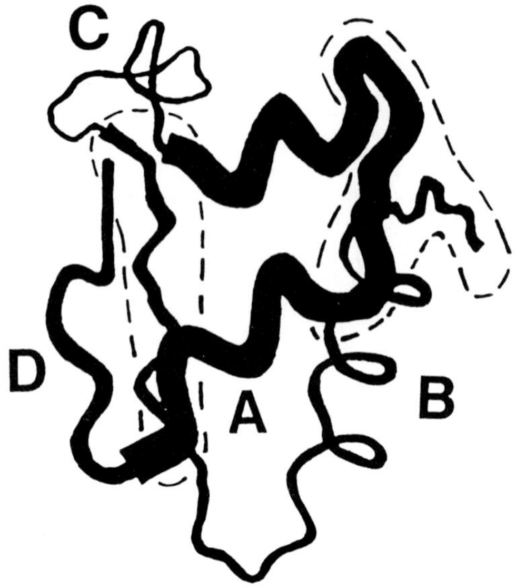

Fig. 1. Insulin antibody binding regions (*broken lines*) of insulin projected on the three-dimensional model of IGF-I. The four domains of the IGF-I molecule are labeled with A, B, C, and D.

ancestor, possibly an old serine protease *(2)*. The ancestor molecule may have stimulated cell and tissue growth after food intake, and this function probably included some "insulin-like activity." The latter seems to have been refined by the emergence of proinsulin, whereas growth promoting activity has been preserved mostly in the IGFs. Based on structural comparisons between IGFs and insulins of different species, and between insulins of the most primitive and the highest vertebrates, one can assume that a gene duplication of the ancestor molecule occurred before the appearance of the first vertebrates on earth, i.e., around 600 million years ago *(2)*. Despite the divergence of their biological functions and the refinement and adaptation to specific purposes, both insulin and the IGFs seem to "remember" their ancestral roots; thus, IGFs are able to exert acute effects on metabolism, and insulin is able to stimulate growth *(5,6)*. Physiologically, however, these related molecules act in a concerted and complementary fashion, insulin being the acute regulator of anabolic metabolism and IGF the long-term regulator of growth. Whether and to what extent IGFs exert acute metabolic effects under physiological conditions is still under debate. All the in vivo metabolic actions of IGFs described in this chapter are pharmacological effects obtained either by acute bolus injection or during constant intravenous or subcutaneous administration of doses that raise total IGF to supraphysiological levels.

Similarities Between IGFs and Insulin (Table 1)

Both the A and B domains of IGF-I and -II contain sequences that can be aligned to identical regions in human insulin: 38% of the sequence of residues 42–70 in IGF-I and 44% of the sequence of residues 41–67 in IGF-II is identical to the sequence of the insulin A-chain. Residues 1–29 in IGF-I and 1–32 in IGF-II display a sequence identity of 48% and 38% respectively, to the insulin B chain. Most importantly, 17 of the 19 invariant amino acid residues occurring in insulins of various animal species are also present in IGF-I and -II. These invariant residues form the hydrophobic core of the molecules and

Table 1
Similarities Between IGF-I and Insulin

Amino acid sequence:	~45% Homology of A and B domain with insulin A and B chain
Three-dimensional structure:	folding identical to proinsulin
Receptor:	~50–60% Overall homology to insulin receptor
Signaling pathway:	Tyrosine kinase → phosphorylation of IRS-1 → activation of phosphatidylinositol-3-kinase and SH_2-containing proteins
Biological actions:	IGF-I mimics **all** biological actions of insulin.

are mainly responsible for the three-dimensional folding. Thus, a three-dimensional model for IGF-I has been derived in which the folding of the polypeptide chain is essentially identical to that in human proinsulin *(4)* (Fig. 1).

The similarity between IGF-I and insulin also pertains to their receptors: the type 1 IGF receptor displays a 50–60% overall homology to the insulin receptor *(7)*. The highest homology (84%) is found in the β-subunit cytoplasmic region which corresponds to the enzymatic domain for tyrosine-specific kinase activity. A 64–67% sequence homology is found in the extracellular α-subunit regions flanking the cysteine-rich sections (*see* Chap. 4). In contrast to the type 1 IGF receptor which mediates also most if not all biological effects of IGF-II, the type 2 IGF receptor shows a completely different structure (*see* Chap. 5). In view of the high sequence homology between IGF and insulin, and the high degree of homology of the α-subunit binding and the intracellular tyrosine kinase regions of the type 1 IGF receptor, it is not surprising that IGF and insulin cross-react at each other's receptor. Both activate receptor tyrosine kinase and trigger the phosphorylation of insulin receptor substrate-1 (IRS-1) resulting in the activation of phosphatidylinositol-3-kinase and phosphorylation of SH_2-containing proteins *(8)*. These common features account for the well-established fact that IGF-I and -II mimic all the biological actions of insulin investigated so far, although their potencies with respect to specific functions differ (*see below*).

Differences Between IGFs and Insulin

The major structural differences between the three-dimensional model of IGF-I and proinsulin occur (1) in the C domain, which is shorter than in proinsulin, but partially complemented by the C-terminal extension (D domain, Fig. 1) and (2) on the hydrophilic surface of the molecule, part of which interacts with the hormone receptor and with antibodies. These dissimilarities explain the differences in the competitive potencies between insulin and IGFs at each other's receptors, the differences in their biological potencies, and the lack of cross-reactivity between insulin and anti-IGF antibodies, and between IGF and anti-insulin antibodies.

Major differences between IGF-I and insulin are listed in Table 2. In contrast to insulin, IGF-I and -II are one-chain molecules in which the C-domains are conserved. The C-domains of human IGF-I and -II are shorter (12 and 8 amino acid residues, respectively) than the C peptide of human proinsulin and do not display any sequence homology with the latter. In addition, IGF-I and -II contain a C-terminal extension (D domain) of 8 and

Table 2
Major Differences Between IGF-I and Insulin

	IGF-I	*Insulin*
Structure	• 1 Chain • C Domain conserved • Additional D domain	• Two chains • C Peptide cleaved off • No D domain
Insulin antibody binding regions	• One completely different from insulin • The other covered by C and D domain	Both regions accessible to insulin antibodies
Receptor affinity	Type 1 IGF receptor: K_d for IGF-I 0.2–1.5 n*M* K_d for insulin 10–100 n*M*	Insulin receptor: K_d for IGF-I 10–100 n*M* K_d for insulin 0.1–0.2 n*M*
Site of production	Most tissues, liver as the main site of production	Pancreatic β-cell
Secretion	Constant, slow release	Emeiocytosis, pulsatile secretion
Main physiological stimulus	Growth hormone	Glucose
Mode of action	Endocrine and auto-/paracrine	Endocrine
Specific BPs	Six (IGFBP-1 to -6)	None No binding to IGFBPs
Half-life in the circulation	~ 15 h (in 150-kDa complex) ~ 20 min (in 40-kDa complex)	~ 10 min
Plasma level	20–40 n*M*, relatively constant, age-dependent >99% Bound, free ~ 0.13 n*M*	0.1–0.5 n*M*, "free," fluctuating
Physiological role	Growth and differentiation (via type 1 IGF receptor)	Fuel homeostasis (via insulin receptor)
Mediation of acute metabolic effects	Mainly type 1 IGF receptor in skeletal muscle, insulin receptor in adipose tissue and liver; K_d (glucose transport, glycolysis, glycogen synthesis) in skeletal muscle: 5–14 n*M*	Insulin receptor; K_d (glucose transport, glycolysis, glycogen synthesis) in skeletal muscle: ~ 0.5 n*M*

6 amino acid residues, respectively, which are lacking in insulin. Despite the high degree of homology between the insulin and the type 1 IGF receptor, and despite the similarities between the receptor binding regions of insulin and IGFs, their respective cross-reactivities at the insulin and the type 1 IGF receptor differs by one to two orders of magnitude. This explains, at least in part, why insulin is a weak growth factor and why IGFs are weak "insulins." An extended and more sophisticated explanation for this difference in mitogenic and metabolic signaling between IGFs and insulin was given by De Meyts *(9)*. It is based on a model in which the ligands bind bivalently and alternatively at opposite sites of the two receptor α-subunits.

There are a number of important biological differences between IGFs and insulin: insulin is a classic hormone with a specific site of production in the pancreatic β-cell. Unlike insulin, IGFs are produced by many tissues of the organism, the liver being a main site of production, and they act in both an endocrine and an auto-/paracrine manner. In contrast to insulin, only little IGF-I and -II circulate in the free form (0.4% and 0.2% of the total IGF-I and -II, respectively, in human serum) *(10)*. The rest is bound to specific IGF binding proteins (IGFBP-1 to -6) (*see* Chap. 10). One of these IGFBPs, IGFBP-3, forms a 150-kDa complex with IGF-I and -II and a third component, the acid-labile subunit (ALS). This complex is responsible for the long half-life of IGFs in the circulation *(11,12)* and for the relatively constant diurnal IGF serum level. IGF-I levels are age dependent. They increase during childhood, peak at puberty, and decrease again during adulthood *(13,14)*. IGF-I is released constantly under the influence of growth hormone whereas insulin secretion occurs by emeiocytosis in a pulsatile manner according to the fluctuations of the serum glucose concentration. The total serum level of IGF-I and -II is ~300 and ~900 times higher, respectively, than basal insulin levels, whereas free IGF levels *(10)* are in a range similar to those of basal serum insulin levels.

The main physiological function of insulin is the regulation of fuel homeostasis, whereas IGF-I and -II are growth and differentiation factors that act on growth through type 1 IGF receptors. However, IGF-I and -II also exert acute metabolic effects in vitro and in vivo (*see below*). The in vitro situation is less complex than the situation in vivo because IGF effects are measured in the absence of IGFBPs. Direct acute metabolic effects in vivo, however, are observed only when the binding capacity of the IGFBPs is overridden and free IGF rises into a range where interaction with the type 1 IGF or the insulin receptor becomes possible. This situation is encountered after intravenous bolus injections or during intravenous infusions for a few hours. A number of metabolic effects in vivo during infusion or injection for several days are secondary to hormonal changes.

The following sections dealing with the metabolic effects of IGFs are not intended to present a comprehensive and complete review of the literature on this topic, but focus on representative aspects as described in a selected number of articles.

In Vitro Effects on Isolated Tissues and Cells: Basis for Understanding In Vivo Actions

Glucose and Lipid Metabolism in Adipose Tissue and Isolated Adipocytes

Rat adipose tissue was the first tissue where the effects of insulin on glucose and lipid metabolism were studied because the rat adipocyte is exquisitely sensitive to insulin. Like insulin, IGF-I and -II enhance glucose uptake and net gas exchange in rat adipose tissue *(1,15)*, and they stimulate glucose uptake, glucose oxidation, and lipid synthesis from glucose in isolated rat fat cells *(15)*. However, on a molar basis the two factors are 50 to 100 times less potent than insulin (Table 3). A small potency difference between IGF-I and -II (two- to threefold) is observed in isolated fat cells but not in whole adipose tissue. It is obviously due to collagenase digestion. This potency difference corresponds to their different potencies in competing for insulin binding to the adipocyte, suggesting that IGF-I and -II stimulate glucose metabolism via the insulin receptor *(15)*. Indeed, short trypsinization of the adipocytes, which destroys the insulin receptor, does not only abolish insulin binding, its stimulatory effects on glucose metabolism, and its antilipolytic effect, but also the corresponding insulin-like effects of IGF-I and -II, whereas specific binding of IGF-I and -II to the trypsinized adipocytes remains nearly unaffected *(15a)*.

Table 3
IGF and Insulin Concentrations Eliciting Half-Maximal Stimulation of Metabolic Effects and Causing Half-Maximal Displacement of ^{125}I-IGF or ^{125}I-Insulin Binding in Rat Adipocytes, in Rat Heart, and Mouse Soleus Muscle and in Adult Rat Hepatocytes

	IGF-I (mol/L)	*IGF-II (mol/L)*	*Insulin (mol/L)*
Rat adipocytes[a]			
Glucose oxidation	6.7×10^{-9}	3.0×10^{-9}	5.3×10^{-11}
Lipogenesis	6.7×10^{-9}	2.1×10^{-9}	6.0×10^{-11}
Displacement of ^{125}I-insulin	1.2×10^{-6}	3.1×10^{-7}	4.1×10^{-9}
Displacement of ^{125}I-IGF I	5.7×10^{-9}	2.7×10^{-10}	None
Displacement of ^{125}I-IGF II	$>1.3 \times 10^{-8}$	8.0×10^{-10}	None
Rat heart[b]			
Glucose uptake	1.1×10^{-9}		5.5×10^{-10}
Displacement of ^{125}I-insulin	2.0×10^{-7}		5.5×10^{-9}
Displacement of ^{125}I-IGF	2.0×10^{-8}		$>10^{-3}$
Mouse soleus muscle[c]			
2-Deoxyglucose uptake	10^{-8}	$\sim 10^{-8}$	5.3×10^{-10}
Glycolysis	5.0×10^{-9}	n.d.	4.6×10^{-10}
Glycogen synthesis	10^{-8}	n.d.	9.0×10^{-10}
Protein synthesis	6.0×10^{-9}	n.d.	5.0×10^{-10}
Displacement of ^{125}I-insulin	$\sim 10^{-6}$	n.d.	9.0×10^{-9}
Displacement of ^{125}I-IGF I	9.0×10^{-9}	n.d.	$>10^{-4}$
Rat hepatocytes[d]			
Glycolysis	3.0×10^{-8}	$\sim 10^{-8}$	$3\text{–}4 \times 10^{-10}$
Inhibition of glycogenolysis	$6\text{–}9 \times 10^{-9}$	$\sim 10^{-8}$	$3\text{–}6 \times 10^{-11}$
Glycogen synthesis	n.d.	$\sim 10^{-8}$	$\sim 2.0 \times 10^{-10}$
AIB transport	n.d.	$\sim 7.0 \times 10^{-9}$	$\sim 5.0 \times 10^{-10}$
Displacement of ^{125}I-insulin	3.0×10^{-8}	8.0×10^{-9}	$2\text{–}5 \times 10^{-10}$
Displacement of ^{125}I-IGF I: no specific binding	—	—	—
Displacement of ^{125}I-IGF II	n.d.	1.5×10^{-9}	none

[a]From ref. *(15)*.
[b]In the heart perfusion experiments, the IGF tracer used was 60% pure and contained ca. 98% IGF I: unlabeled IGF was ca. 1.2% pure and contained a 1:1 mixture of IGF I and II (refs. *22,23*).
[c]Extrapolated from refs. *(24,25)*.
[d]Extrapolated from refs. *(31,32)*.
n.d., not determined.

This finding fits the results obtained by King et al. *(16)*, who showed that blockade of the fat cell insulin receptor by insulin receptor antibodies inhibited the metabolic actions of IGF. Findings in human adipocytes are consistent with the notion that IGF-I and -II stimulate glucose metabolism via the insulin receptor *(17–19)*. In the human adipocyte, IGF-I was 600 to 1000 times less potent than insulin in stimulating glucose transport *(17,18)* and glucose oxidation *(17)*. IGF-II stimulated glucose transport with a 10-fold higher potency than IGF-I but was still 35 to 100 times less potent than insulin *(19)*, and insulin receptor antibodies inhibited both insulin- and IGF-II-stimulated glucose transport *(19)*.

Another typical insulin effect on adipose tissue, inhibition of lipolysis, is also mimicked by IGF-I and -II *(15,20,21)*. Both factors inhibit enhanced lipolysis in adipose tissue of fasted–refed rats as well as epinephrine-stimulated lipolysis in rat adipose tissue and isolated fat cells. Again, the molar potency compared with insulin is around 100 times lower, suggesting that the antilipolytic effect of IGF-I and -II is also mediated through the insulin receptor. A similarly low potency of IGF-I and -II is observed with respect to the reduction of the epinephrine-stimulated cAMP level and protein kinase A activity which accompanies antilipolysis *(21)*. The antilipolytic effect of IGF-I was also demonstrated in human fat cells where IGF-I was 800 to 1000 times less potent than insulin *(17)*.

Glucose and Protein Metabolism in the Isolated Perfused Rat Heart and in Isolated Skeletal Muscle

In the perfused rat heart, IGF stimulates glucose transport and lactate production *(22)*. In contrast to adipose tissue, the biological potency ratio between IGF and insulin in this model was higher than in adipose tissue. Based on this finding and on binding studies with ^{125}I-insulin and ^{125}I-IGF, it has been concluded that insulin and IGF act separately via the insulin and the type 1 IGF receptor, respectively *(23)*.

A similar conclusion was reached from studies in isolated mouse soleus muscle. In this in vitro model, IGF-I stimulated 2-deoxyglucose uptake, glycolysis, and glycogen synthesis; it activated glycogen synthase and it enhanced protein synthesis *(24,25)*. The potency of IGF-I in these experiments was 5–10% that of insulin and thus 5 to 10 times higher than in adipose tissue (Table 3). Comparison of these results with the competitive binding characteristics of ^{125}I-insulin and ^{125}I-IGF-I led to the conclusion that the effects of IGF-I on glucose and protein metabolism in mouse skeletal muscle were mediated via the type 1 rather than via the insulin receptor *(24)*.

The same conclusion was drawn from other studies: in a murine nonfusing mouse muscle cell line where the insulin receptor had been down-regulated by high insulin concentrations, IGF-I-stimulated 2-deoxyglucose uptake was unimpaired, whereas a 100-fold loss in insulin responsiveness was observed *(26)*. In epitrochlearis muscle of young rats IGF-I was reported to be equipotent to insulin in stimulating 2-deoxyglucose uptake and aminoisobutyric (AIB) acid transport, and even more potent than insulin in enhancing protein synthesis, observations that appear incompatible with an insulin receptor-mediated action of IGF-I *(27)*. Comparison between the effects of IGF-I and insulin on glucose disposition in isolated rat soleus muscle lent further support that IGF-I acts via its own receptor *(28)*. IGF-I increased glucose transport, lactate formation, glycogen synthesis, and the flux of glucose to hexose monophosphate, but had no effect on the rate of glucose oxidation (in contrast to insulin) or glycogenolysis. Like insulin, it did not cause changes in the tissue content of ATP, ADP, AMP, phosphocreatine, or citrate. Consistent with the observations of Poggi et al. *(24)*, the potency of IGF-I in stimulating lactate formation and glycogen synthesis was 8–10 times lower than that of insulin.

Dohm et al. investigated the effect of IGF-I in human muscle fibers in vitro obtained from normal and from obese nondiabetic and type 2 (non-insulin-dependent) diabetic patients *(29)*. They showed that normal human muscle contained fourfold less IGF-I than insulin binding sites and that IGF-I binding was unchanged in muscle from obese nondiabetic and diabetic patients. In normal muscle, both insulin and IGF-I stimulated 2-deoxyglucose transport half-maximally at similar concentrations (~3 n*M*), but muscle strips from obese nondiabetic and diabetic subjects were resistant to the action of both IGF-I and insulin. Because IGF-I stimulated glucose transport with the same potency as insulin despite the lower abundance of IGF-I receptors, the authors concluded that IGF-I acted via its own receptor. The resistance of muscle from obese subjects with or without type 2 diabetes was interpreted as a primary defect possibly due to depletion of glucose transporters, to a defect in the ability to translocate glucose transporters to the plasma membrane, or to a loss of intrinsic activity of glucose transporters. The in vitro finding of IGF-I resistance in the patient groups is at variance with several human in vivo trials where IGF-I can overcome insulin resistance (*see below*).

In addition to glucose transport, stimulation of the sodium–potassium pump, another metabolic effect of insulin, is mimicked by IGF-I in isolated rat soleus and extensor digitorum longus muscle *(30)*. Because this effect of IGF-I was observed at concentrations (10^{-8} *M*) that are unlikely to stimulate the insulin receptor, the authors concluded that IGF-I-stimulated sodium/potassium transport is mediated by the type 1 IGF receptor.

Glucose Metabolism and Amino Acid Transport in Cultured Hepatocytes (Table 3)

In cultured adult rat hepatocytes, stimulation of glycolysis was observed at half-maximal concentrations of ~50 n*M* for IGF-I and at ~0.8 n*M* for insulin *(31)*. Like insulin, IGF-I also inhibited basal and glucagon-activated glycogenolysis, again with a 50 to 100-fold lower potency. Acute stimulatory effects on glycogen synthesis, glycolysis, and AIB transport were also reported for IGF-II *(32)*. As observed with IGF-I, half-maximal stimulation of these metabolic indices required IGF-II concentrations that were 50 to 100 times higher than those of insulin. The same potency ratio was observed for inhibition of basal and glucagon-activated glycogenolysis by IGF-II and insulin. The potency differences between IGF-II and insulin corresponded to the difference in the insulin-displacing potencies of the two hormones obtained in competitive binding studies with ^{125}I-insulin. However, insulin did not compete for ^{125}I-IGF II binding, and antibodies against the type 2/mannose-6-phosphate receptor did not affect the metabolic responses to IGF-II. The authors concluded that IGF-II, like IGF-I, exerts acute metabolic effects on the liver by interaction with the insulin receptor. This is consistent with the finding that hepatocytes from adult rats barely express type 1 IGF receptors *(33)*.

In summary, both IGF-I and -II exert acute insulin-like effects in insulin-sensitive tissues or cells in vitro with a potency between 1% and 10% that of insulin. In adipose tissue and liver, these actions are mediated via the insulin receptor, whereas in heart and skeletal muscle IGF-I and -II most likely elicit these actions mainly via the type 1 IGF receptor.

ACUTE EFFECTS OF IGFS IN VIVO: BOLUS INJECTIONS AND CLAMP STUDIES IN ANIMALS AND IN NORMAL HUMAN SUBJECTS

IGF-I and -II are found in the circulation under normal physiological conditions at relatively constant levels that are 30 to 300 times higher than those of the fluctuating insulin levels. Nevertheless, endogenous IGFs, above all IGF-I, which is under more stringent control of growth hormone than IGF-II, are primarily growth factors. As such, they exert their actions on growing tissues via the type 1 IGF receptor. Nevertheless, as judged from their concentrations in the circulation and from their biological potencies in vitro (Table 3), acute insulin-like effects on intermediary metabolism would appear likely. If this were the case, one would expect permanent stimulation of glucose metabolism in skeletal muscle and thus permanent fasting hypoglycemia, permanent glycogen synthesis in muscle and liver, and permanent amino acid uptake and protein synthesis in

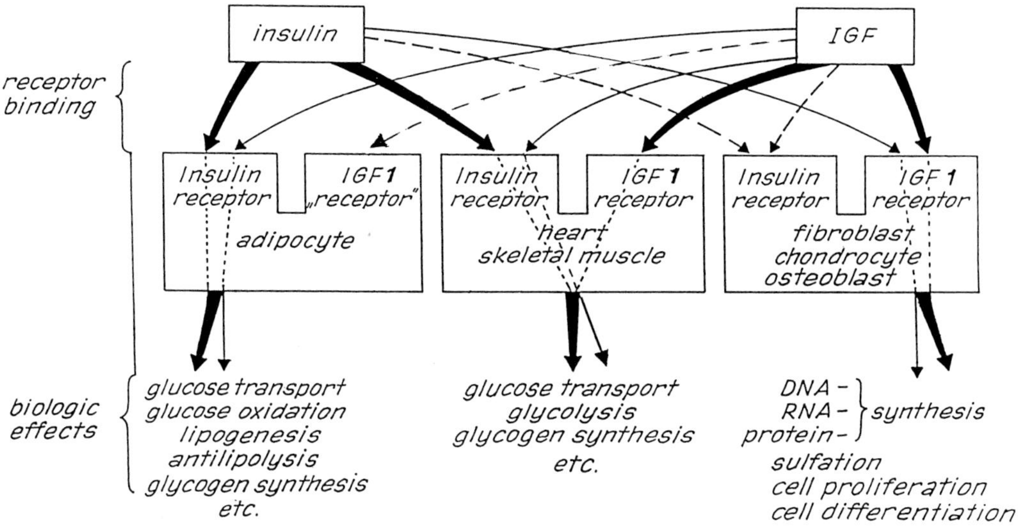

Fig. 2. Complementary anabolic actions of insulin and IGF.

insulin-sensitive tissues. Evolution has provided means to protect the organisms from the huge constant insulin-like potential in the circulation by "introducing" specific IGF binding proteins (IGFBPs) that retain most of the IGF in the circulation and restrict its bioavailability (*see* Chap. 10). The level of free circulating IGF is in a similar range as the fasting insulin level and several times lower than peak insulin levels attained during feeding. Before the discovery of the IGFBPs it had always been puzzling that the high total IGF concentrations in serum did not cause permanent hypoglycemia or prevent diabetes after the destruction of the pancreatic β-cells. However, considering the low free serum IGF levels in comparison with the IGF concentrations required to produce acute metabolic effects (Table 3), it is not surprising that such effects are not observed in vivo under normal physiological conditions. Evolution has reserved acute regulation of anabolic processes for insulin and long-term regulation of growth for IGFs. Because growth and differentiation require the supply of metabolic fuels to the cell, IGFs and insulin represent a complementary system in the regulation of whole body anabolism (Fig. 2). Acute insulin-like effects of IGFs, therefore, represent an artificial situation that can result only when pharmacological doses of IGFs are administered into the circulation and the level of free or bioavailable IGF increases substantially.

Animal Studies

Bolus Injection

More than 30 yr ago, Froesch et al. *(34)* reported that partially purified serum extracts containing IGF-I and -II (then termed NSILA-s), when injected intravenously together with anti-insulin serum into normal rats, lowered blood glucose and enhanced the incorporation of [6-^{14}C]glucose into diaphragm glycogen and into adipose tissue total lipids. When pure native IGF-I and -II isolated from human serum *(2,3)* and recombinant human (rh) IGF-I and -II became available, similar experiments in normal and hypophysectomized (hypox) rats definitely showed that intravenous bolus injections of the pure factors

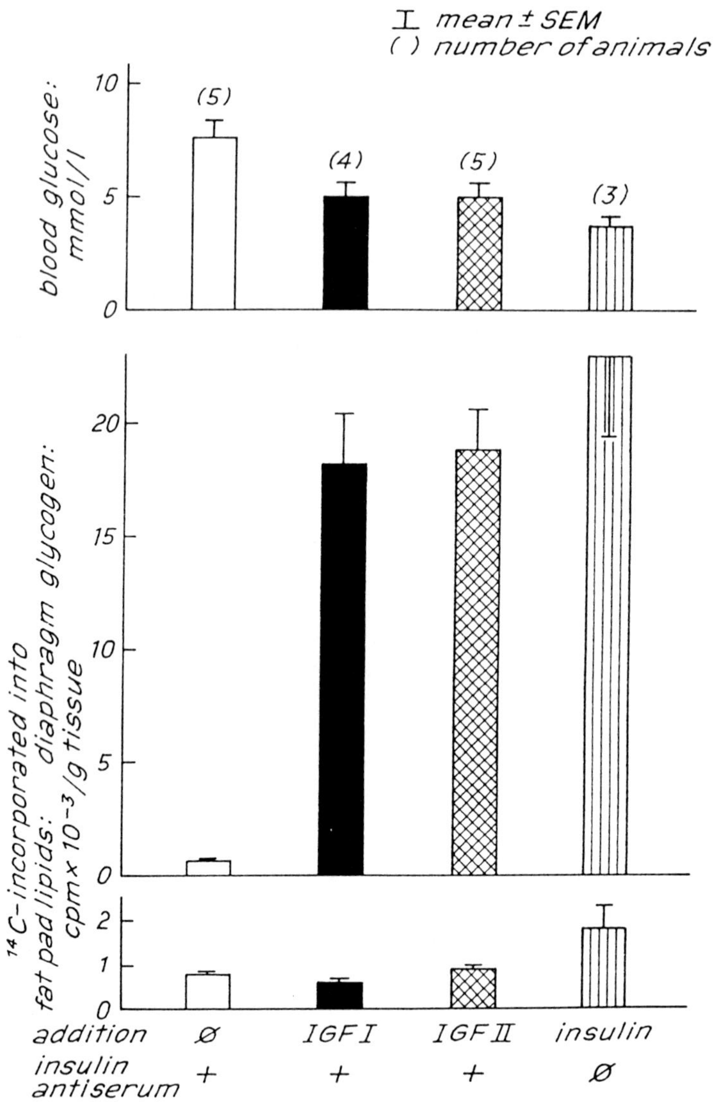

Fig. 3. Acute effects of intravenously injected IGF-I, IGF-II (20 μg together with [U-^{14}C]glucose and anti-insulin serum) or insulin (6 mU without antiserum) on blood glucose, glycogen synthesis in diaphragm, and lipid synthesis in epididymal adipose tissue of hypophysectomized rats 30 min after injection. (Modified from ref. *[12]*.) Details of the experiment are described in ref. *(12)*.

lowered the blood sugar, enhanced the disappearance of radiolabeled glucose from serum, and increased glucose incorporation into diaphragm glycogen in the presence of anti-insulin serum *(12)* (Fig. 3). These effects were accompanied by a marked rise of the free IGF-I and -II levels which decreased rapidly with serum half-lives between 5 (IGF-II) and 15–30 min (IGF-I). As in the in vitro studies, IGF-I and -II were more than 50 times less hypoglycemic than insulin. At the IGF doses used (20 μg/rat), no significant stimulation of lipid synthesis from [U-^{14}C]glucose was observed in epididymal fat pads, whereas insulin (6 mU/rat) markedly enhanced lipid synthesis. In contrast to insulin, which suppressed serum free fatty acid (FFA) levels in both normal and hypox rats,

IGF-I and -II suppressed FFAs only in normal rats, and the effect was of shorter duration than that of insulin *(12)*.

Clamp Studies

The bolus studies were subsequently extended by investigations using the glucose clamp technique and a continuous intravenous infusion of IGF-I for 1 or 2 h.

Jacob et al. investigated the acute effects of IGF-I on glucose, amino acid, and FFA metabolism in awake fasted rats *(35)*. IGF-I infusion (5 μg/kg/min for 90 min after a priming dose of 250 μg/kg) caused a twofold rise in IGF-I levels and a 75–85% decrease in plasma insulin. In the absence of glucose coinfusion, IGF-I lowered the blood sugar level by 30–40 mg/dL. This blood sugar lowering effect was due to a transient increase in glucose uptake, whereas hepatic glucose production and plasma FFA levels remained unchanged. When euglycemic glucose levels were maintained by concomitant glucose infusion, IGF-I produced a sustained rise in glucose uptake and, in contrast to insulin, a marked stimulation of glycogen synthesis in skeletal muscle, heart, and liver. However, it still failed to reduce hepatic glucose production and plasma FFA levels. Insulin infused at a dose of 1.4 U/kg/min stimulated whole body glucose uptake to the same extent as the IGF-I dose, suppressed hepatic glucose production, and lowered plasma FFA levels. IGF-I reduced plasma amino acid levels both during hypoglycemia and euglycemia. This reduction was accompanied by a decrease in leucine flux and by diminished [1-^{14}C]leucine incorporation into muscle and liver protein, suggesting reduced protein breakdown. Identical effects were observed with insulin. As calculated from the IGF-I and insulin doses infused during the clamp and resulting in similar increases of whole body glucose uptake and in similar inhibition of protein breakdown, insulin was 85 times more potent than IGF-I.

In a similar study using the glucose clamp technique, Moxley et al. compared the actions of IGF-I (0.57 nmol/kg/min) and insulin (0.01 nmol/kg/min) in normal fed rats *(36)*. The two hormone doses stimulated whole body glucose uptake and 2-deoxyglucose uptake in quadriceps femoris, soleus, and diaphragm muscles to a similar extent, but IGF-I suppressed hepatic glucose production less than insulin. The preferential action of IGF-I on skeletal muscle as compared to that on the liver was interpreted to agree with the distribution of functional type 1 IGF receptors in the two tissues.

The effect of IGF-I on protein synthesis in skeletal muscle has recently been studied in more detail by Jacob et al. in awake postabsorptive rats *(37)*. The authors confirmed the antiproteolytic effect of IGF-I in skeletal muscle and heart, but in addition showed that IGF-I (1 μg/kg/min, infused intravenously for 4 h) stimulated protein synthesis if the reduction of serum amino acid levels or of the serum insulin level during IGF-I infusion was prevented by the coinfusion of amino acids or insulin. Thus, although IGF-I itself inhibits protein breakdown, amino acids and insulin exert a permissive effect on the stimulation by IGF-I of protein synthesis in heart and skeletal muscle.

The in vivo metabolic actions of IGF-I and -II were compared with those of insulin in two elaborate studies in anesthetized fasted adult rats during a 2-h euglycemic clamp *(38,39)*. These studies provided dose–response relationships between IGF-I/-II (25–200 nmol/kg/h) or insulin (0.5–13 nmol/kg/h) and information on a number of acute actions of the two hormones, such as hypoglycemia and antilipolysis, whole body glucose disposal, glucose uptake and glycogenesis in skeletal muscle, and on lipogenesis in epididymal adipose tissue. In addition to total IGF levels, the authors also measured free serum IGF-I concentrations during the clamp, which rose up to 90% of total levels. With respect to all of the metabolic indices determined in these studies, the authors found that ~50 times higher IGF-I and -II concentrations were required to achieve the same half-maximal responses as with insulin, and no preferential stimulation of glucose metabolism in skeletal muscle was observed. It was, therefore, concluded that in skeletal muscle as well as in adipose tissue, IGF-I and -II stimulated carbohydrate metabolism via the insulin receptor. Although this conclusion appears to be at variance with the results of the above in vitro studies and with those of most other in vivo studies (*see also below*), they seriously raise the possibility that IGF-I and -II exert acute stimulatory effects on glucose metabolism of rat skeletal muscle not only via the type 1 IGF, but also via the insulin receptor.

Giacca et al. investigated the effects of IGF-I and insulin on glucoregulation and fat metabolism in fasted, totally depancreatized dogs *(40)*. The animals were given a primed intravenous infusion of IGF-I (2.6 nmol/kg bolus followed by infusion of 57.4 pmol/kg/min) or insulin (0.06 nmol/kg bolus followed by infusion of 6.4 pmol/kg/min) over 90 min. Moderate hyperglycemia (9.9 m*M*) was maintained by a subbasal intraportal infusion of insulin (1.29 pmol/kg/min). The insulin dose required to induce the same fall of plasma glucose as with the IGF-I infusion was around 10 times lower on a molar basis. IGF-I and insulin differed with respect to the mechanism responsible for the blood sugar lowering effect. Whereas glucose production was much less suppressed, glucose utilization was stimulated more by IGF-I than by insulin, and serum lactate and pyruvate levels increased only with IGF-I but not with insulin. In contrast, IGF-I suppressed glycerol and FFA levels much less than insulin. A comparable decrease of β-hydroxybutyrate, alanine, and glucagon was observed with both hormones. The study showed clear differential effects between IGF-I and insulin on glucose utilization and clearance. This difference is unlikely to be due to effects on the liver and on adipose tissue because IGF-I was less effective than insulin on hepatic glucose production and antilipolysis. Therefore, the authors concluded that muscle is the most likely site for the preferential stimulation of glucose utilization with IGF-I, in keeping with the observation that IGF-I but not insulin caused a significant rise in serum lactate and pyruvate levels. As judged from the blood sugar lowering potency ratio between IGF-I and insulin, which is similar to the in vitro potency ratio in mouse skeletal muscle *(24)*, and from the apparent preferential effect of IGF-I on skeletal muscle as discussed previously, it is likely that the IGF-I effects on muscle are, at least in part, mediated through the type 1 IGF receptor.

Jacob et al. examined the metabolic responses to IGF-I (5 μg/kg/min after a priming bolus of 250 μg/kg) and insulin (2 mU/kg/min after a priming bolus of 60 mU/kg) in spontaneously diabetic and nondiabetic BB/w rats during a 2-h euglycemic clamp *(41)*. The authors found substantial differences between normal and diabetic animals with respect to several metabolic responses following IGF-I or insulin administration. Whereas insulin-stimulated whole body glucose uptake was impaired in the diabetic animals, IGF-I stimulated glucose uptake identically in diabetic and nondiabetic animals. Interestingly, IGF-I significantly suppressed hepatic glucose production in the diabetic but not in the nondiabetic rats, a finding that remained largely unexplained.

A similar study was performed in partially pancreatectomized mildly diabetic (plasma glucose 18.4 ± 0.8 m*M*) rats *(42)*. During a 2-h euglycemic hyperinsulinemic clamp, glucose utilization was reduced in diabetic animals due to defective muscle glycogen synthesis and glycolysis. In contrast, and in agreement with the results of Jacob et al. *(41)*, IGF-I–stimulated glucose uptake and muscle glycogen synthesis and glycolysis were similar in diabetic and in normal animals. This finding further argues for a type 1 IGF receptor-mediated action of IGF-I on skeletal muscle. Another finding in this study consistent with the in vivo studies mentioned previously *(40,41)* was the decreased hepatic sensitivity to IGF-I compared with insulin at doses that were equipotent on glucose uptake. Nevertheless, high IGF-I doses (1.8 mg/kg administered during the 2-h clamp) completely suppressed hepatic glucose production in both the normal and the pancreatectomized rats.

Cascieri et al. reported that intraperitoneal injection of IGF-I was equipotent with insulin in stimulating glycogen synthesis from [U-^{14}C]glucose into diaphragm muscle of lean mice *(43)*. The equipotency of IGF-I and insulin was interpreted to mean that the glycogenic effect of IGF-I on the diaphragm is mediated by the type 1 IGF and not via the insulin receptor. Despite the fact that the number of type 1 IGF receptors and their affinities for IGF I were not different in muscle membranes from lean and *ob/ob* mice, whereas the number of insulin receptors was decreased in the latter, obese mice were resistant to the action of both hormones, consistent with the observations of Poggi et al. in goldthioglucose-obese mice *(24)*. The authors suggested that "the insensitivity of muscle from *ob/ob* mice to insulin and IGF-I is not likely due to a receptor defect, but is due to a post-receptor abnormality in a transduction step common to the action of both peptides" *(43)*.

In fasted lambs, the effects of IGF-I, IGF-II, and insulin on glucose and protein metabolism were investigated during a 5-h intravenous infusion *(44)*. IGF-I and insulin were administered at doses that were equipotent in reducing blood glucose (0.17 nmol/kg/h for insulin and 6.7 nmol/kg/h for IGF-I). IGF-II administered at the same dose as IGF-I increased glucose clearance, but had no significant effect on the blood sugar level, because glucose utilization appeared to be offset by a concomitant increase in the glucose appearance rate. Both IGF-I and insulin reduced net protein loss. The IGF-I but not the insulin dose increased the rate of protein synthesis in skeletal and cardiac muscle and in hepatic tissue. In contrast to IGF-I, the same dose of IGF-II had no significant effect on net protein loss nor on protein synthesis. At a threefold lower dose, IGF-I had no effect on glucose kinetics, whereas it reduced protein loss. The authors concluded from this finding that protein metabolism was more sensitive to circulating IGF-I than was carbohydrate metabolism. In agreement with the studies mentioned previously, IGF-I, in contrast to insulin, did not reduce the rate of glucose production. From their finding that IGF-I and insulin doses at equal hypoglycemic activities exert differential effects on glucose and protein metabolism, the authors speculated that "some of the metabolic effects of circulating IGF-I are mediated by receptors other than those for insulin." In a subsequent study the same authors showed that parenteral nutrition in fasted lambs acted in a synergistic manner to the protein-sparing effects of IGF-I and that the combination of parenteral nutrition and IGF-I resulted in a protein anabolic state as reflected by an increase of the fractional synthetic rates of protein in cardiac muscle, diaphragm, adductor and psoas muscle, and hepatic tissue *(45)*.

Studies in Human Subjects

Bolus Injections

In 1987, Guler et al. demonstrated for the first time that IGF-I exerted acute metabolic effects in humans *(46)*. They injected 100 μg/kg of recombinant human (rh) IGF-I intravenously as a bolus into eight healthy volunteers and observed a pronounced fall of the blood sugar level that reached a nadir 30 min after the injection and then started to increase again (Fig. 4). A 0.15 IU/kg (1 nmol/kg) dose of insulin, i.e., a 13-fold lower dose

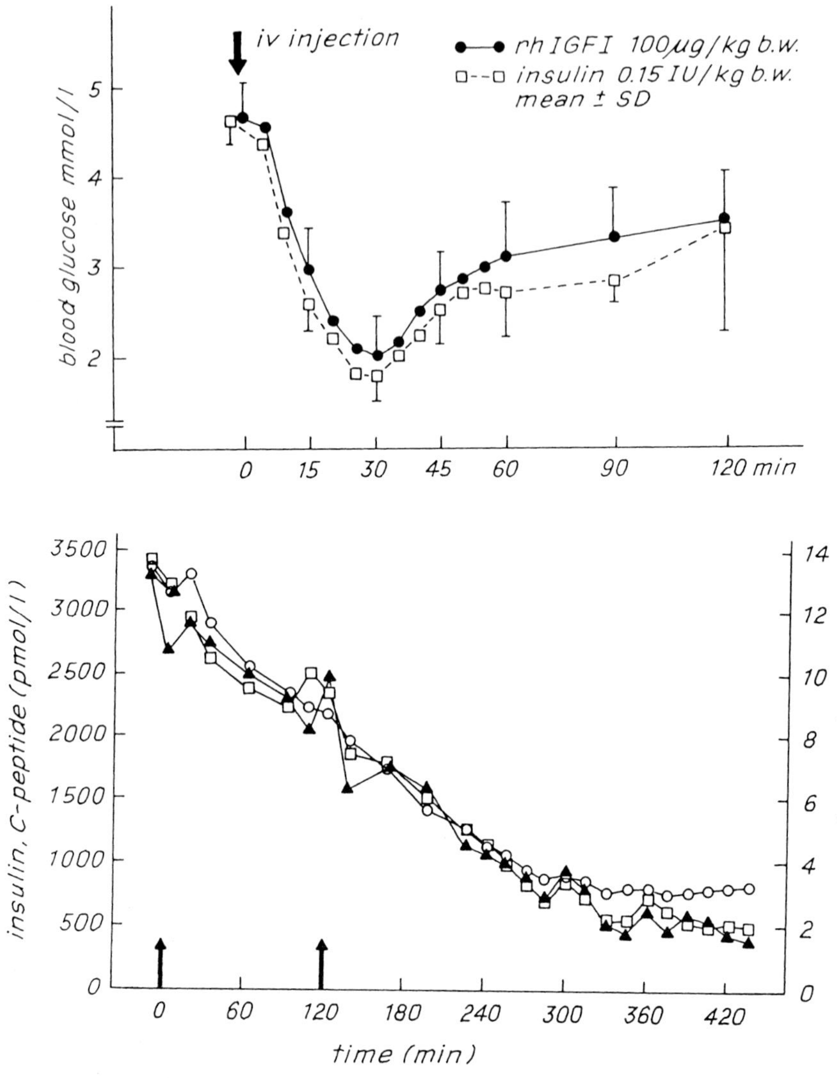

Fig. 4. *Upper panel*: Blood glucose levels in eight healthy human subjects after an intravenous bolus injection of rhIGF-I (100 µg/kg) or insulin (0.15 U/kg). (From ref. *[46]*. With permission. Copyright 1987 Massachusets Medical Society. All rights reserved.) *Lower panel*: Serum levels of glucose (○), insulin (▲), and C peptide (❐) in a patient with type A insulin resistance after intravenous administration of rhIGF-I (100 µg/kg) at time 0 (↑) and 120 min (↑). (From ref. *[48]*. With permission. Copyright 1991 Springer-Verlag.)

than that of IGF-I, elicited an identical hypoglycemic response. The kinetics of this response were the same after IGF-I and insulin injection. The nadir of the blood sugar level was preceded by a dramatic increase of the free IGF-I level and accompanied by a suppression of the serum FFAs *(46)*. The suppressive effect on serum FFAs lasted longer after insulin than after IGF-I administration. Both IGF-I and insulin caused a rise of serum GH that appeared to be of shorter duration after IGF-I than after insulin injection. In retrospect, this difference can be explained by subsequent findings showing that continuous IGF-I infusions in the absence of hypoglycemia causes a pronounced suppression of GH secretion (*see below*). The IGF-I bolus substantially suppressed serum insulin levels

(46), a finding that was initially interpreted to be due the acute hypoglycemic effect (however, *see below*). Besides GH, the response of other counterregulatory hormones (glucagon, epinephrine, norepinephrine, and cortisol) was the same during IGF-I- and insulin-induced hypoglycemia.

Laager and Keller *(47)* later compared the effects of IGF-I and insulin bolus injections (65 μg/kg and 0.1 IU/kg, respectively) on counterregulation during the acute fall of plasma glucose in normal and type 2 diabetic subjects. Similar to the study of Guler et al. *(46)*, plasma glucose fell immediately and reached a nadir 30 min after the injection of both hormones in normal subjects. In the diabetic patients, the decrement of the blood sugar was greater than in normal subjects, but the blood sugar did not reach hypoglycemic levels. In addition, the blood sugar lowering effect lasted longer than in normal subjects and was more pronounced with IGF-I than with insulin. In both normal and diabetic subjects, the fall of the blood sugar was accompanied by a transient decrease of the serum FFA levels and a sustained fall of the serum leucine concentration. In the normal subjects, counterregulatory increases in plasma glucagon, adrenaline, cortisol, and GH concentrations after hypoglycemia were blunted after the IGF-I as compared to the insulin bolus, whereas no difference in the noradrenaline responses was observed. The blunted responses were also apparent in the diabetic group except that plasma cortisol increased only after IGF-I but not after insulin. As observed by Guler et al. *(46)*, serum insulin and C peptide levels were suppressed after the IGF-I and insulin bolus in both groups.

The effect of IGF-I bolus injections on glucose homeostasis was investigated by Schoenle et al. *(48)* in three patients with type A insulin resistance. This syndrome is characterized by glucose intolerance and overt diabetes despite extreme hyperinsulinemia, and is accompanied by acanthosis nigricans, and by virilization in females *(49)*. Patients have one or several mutations of the insulin receptor gene, which may lead to impaired insulin binding and/or post-receptor signaling through the insulin receptor tyrosine kinase. Insulin treatment is therefore unsuccessful. In this study, two subsequent intravenous bolus injections of 100 μg/kg IGF-I caused a gradual fall of the blood sugar into the normal range within 4 h. It was accompanied by a significant decrease of the elevated serum insulin and C peptide levels (Fig. 4). These results suggest that intravenously administered IGF-I and the subsequent rise of free IGF-I obviously circumvent the defective insulin receptor, most likely by interaction with the intact type 1 IGF receptor in skeletal muscle. The sluggish decline of the blood sugar level in these patients contrasts with the immediate rapid fall of the blood sugar in normal subjects *(46)* with intact insulin receptors. The different kinetics in the glucose response to IGF-I in type A insulin resistance was originally interpreted to result from the fact that intravenously injected IGF-I enhances glucose uptake in muscle of normal subjects mainly by interacting with insulin receptors. However, many in vitro and in vivo studies argue for a type 1 IGF receptor-mediated IGF-I action on muscle. The explanation for the different glucose kinetics may, therefore, be more complex. The possibility might be envisaged that type 1 IGF receptor function and/or signaling in type A insulin-resistant patients is not completely normal. This reasoning is also supported by the finding that not all patients with type A insulin resistance respond to IGF-I treatment with a fall of the blood sugar level *(50)*.

The blood sugar lowering effect of an intravenous IGF-I bolus was also shown in another type of insulin resistance, the Mendenhall's syndrome, a congenital disorder characterized by short stature, somatic abnormalities, and severe insulin-resistant diabetes mellitus *(51)*.

In this study, intravenous injection of 100 μg/kg of IGF-I caused a pronounced decrease of the blood sugar level with similar kinetics as observed by Schoenle et al. *(48)*.

Euglycemic Clamp Studies and Short Infusions

Substrate Metabolism. IGF-I infusion studies in humans with or without euglycemic hyperinsulinemic clamp largely confirmed results obtained in animals.

Boulware et al. investigated the effects of IGF-I on glucose, lipid, and amino acid metabolism in healthy nonobese subjects during a 3-h primed continuous intravenous infusion of IGF-1 (20 μg/kg bolus followed by a continuous infusion of 0.4 μg/kg/min) *(52)*. Euglycemia was maintained by an exogenous glucose infusion. IGF-I produced a two- to threefold increase in glucose uptake and a marked suppression of hepatic glucose production. Free IGF levels rose to >70 ng/mL, and insulin, C peptide, and glucagon levels decreased. The authors observed a 40–60% fall of the serum FFAs and of total and branched-chain amino acids. Overall, there was an increase in carbohydrate oxidation and a decrease in fat oxidation.

Similar results were reported by Laager et al. *(53)*, who compared the effect of IGF-I (5 μg/kg/h and 30 μg/kg/h) with that of insulin (0.04 nmol/kg/h) during an 8-h euglycemic clamp. The higher IGF I dose preferentially enhanced peripheral glucose utilization as compared to insulin, but was less effective in suppressing hepatic glucose production. Serum FFAs, acetoacetate, and β-hydroxybutyrate were reduced after 8 h of IGF-I or insulin infusion. In a similar study, the authors reported a dose-dependent fall of the plasma triglyceride levels *(54)*. In both studies, IGF-I suppressed whole body leucine flux and leucine oxidation indicating decreased whole body protein breakdown. All these effects were accompanied by a suppression of insulin secretion. Because the high IGF-I dose exerted preferential effects on glucose consumption and protein metabolism as compared to the administered insulin dose, whereas insulin preferentially suppressed hepatic glucose production, the authors speculated that insulin-like metabolic effects of IGF I in humans are mediated in part via type 1 IGF and in part via insulin receptors. Inhibition of hepatic glucose production and suppression of serum FFA levels reported in short-term IGF-I clamp studies and after intravenous bolus injection of IGF-I in humans were not observed during IGF-I infusions over several days under non-clamping conditions (*see below*).

The effect of IGF-I- and insulin-induced hypoglycemia on counterregulatory hormones was studied by Kerr et al. *(55)*. They infused IGF-I (0.7 μg/kg/min) and insulin (0.8 μU/kg/min) for 2 h in healthy volunteers. Blood glucose was allowed to fall freely during the infusion. With both hormones the kinetics of the decrease of blood glucose was identical and accompanied by an initial increase in whole body glucose uptake, which was slightly greater during insulin than during IGF-I, and by a similar initial suppression of hepatic glucose production. The latter, however, rebounded to higher production rates during the insulin infusion. Both hormone infusions caused a pronounced suppression of the serum C peptide level which remained suppressed 2 h after stopping the IGF-I infusion, whereas it increased after the insulin infusion was discontinued. In contrast to insulin, no glucagon response was observed during IGF-I-induced hypoglycemia, and the GH response during IGF-I was significantly blunted as compared with the insulin infusion. The absence of the glucagon response was interpreted to be consistent with the blunted rebound of hepatic glucose production. Epinephrine and norepinephrine levels showed a similar rise during both infusions but remained higher after cessation of IGF-I than after cessation of insulin. This was associated with a more rapid return of plasma glucose to preinfusion values after insulin than after IGF-I. No differences in the cortisol responses were observed with both hormones. With a different experimental protocol, in which IGF-I was infused continuously subcutaneously over several days, basal glucagon levels remained unchanged and its response to arginine stimulation was not different as compared with saline-infused control subjects *(56)*. Thus, IGF-I appears to suppress the glucagon response to hypoglycemia, but not to arginine at normal blood glucose levels.

The effect of IGF-I on human forearm muscle protein metabolism was investigated by Fryburg *(57)* during a 6-h intraarterial infusion of three different IGF-I doses (1.8, 6.0, and 10.0 µg/kg/h) under nonclamping conditions. Forearm glucose uptake was increased after 3 h with the two higher IGF-I doses, whereas no changes were observed in the contralateral arm. Significant positive shifts in net forearm phenylalanine balance occurred at all three IGF doses, peaking at 6 µg/kg/h, and no changes occurred in the control arm. As calculated from the phenylalanine kinetics, all three IGF doses prompted a similar increase in the disappearance rate of phenylalanine (Phe R_d), but only the two highest IGF-I doses decreased the phenylalanine appearance rate (Phe R_a). Despite elevated systemic total IGF-I levels no significant changes in either Phe R_d or Phe R_a were observed in the control arm. The author concluded that intraarterial IGF-I exerts growth hormone-like effects, i.e., stimulation of protein synthesis at all doses tested, whereas insulin-like effects, i.e., inhibition of proteolysis, are observed only at higher doses. The differences between the responses of the infused and the contralateral arm were suggested to be due to differences in the free IGF-I levels at the two sites.

In a subsequent study, Fryburg et al. investigated the effect of IGF-I on human skeletal muscle protein anabolism in the presence of hyperaminoacidemia during a 6-h intraarterial IGF-I (0.1 µg/kg/min) or insulin (0.035 mU/kg/min) infusion in postabsorptive normal adults *(58)*. Phenylalanine infusion, which raised the systemic phenylalanine concentration twofold, stimulated muscle protein synthesis. Both insulin and IGF-I further improved muscle protein balance, but whereas this anabolic effect of insulin was due solely to inhibition of proteolysis, the IGF-I effect was due to stimulation of protein synthesis only. Although most of these results are comparable to those obtained in fasted lambs by Koea et al. *(45)* the latter group reported that IGF-I had also a pronounced anticatabolic (antiproteolytic) effect during parenteral nutrition. The different experimental approaches of the two studies may explain this discrepancy. Fryburg et al. argue that "the more potent action of IGF-I to increase protein synthesis under hyperaminoacidemic conditions enhances intracellular recycling of amino acids derived from proteolysis back into newly synthesized protein, which makes detection of a decrease in proteolysis more difficult."

Hemeodynamic Effects. Another IGF-I effect observed during intraarterial forearm perfusion was a pronounced (up to threefold) increase in forearm blood flow *(57,58)*. Similar results were reported by Copeland and Nair *(59)*. IGF-I (10 µg/kg/h) infused over 90 min into a brachial artery in healthy young adults caused a 68% increase in forearm blood flow. These hemodynamic effects of IGF-I appear to be mediated via the release of nitric oxide (NO) by the vascular endothelium *(60,61)*.

IGF-I also elicits hemodynamic effects when administered systemically. Donath et al. injected IGF-I subcutaneously (60 µg/kg) in healthy male volunteers and investigated cardiac function and performance by Doppler echocardiography after 2–4 h *(62)*. IGF-I significantly increased stroke volume, cardiac output, and the ejection fraction. During a 4-h intravenous infusion of 60 µg/kg of IGF-I in patients with chronic heart failure *(63)*, cardiac output and stroke volume increased, and systemic vascular resistance, right atrial pressure, and pulmonary artery wedge pressure decreased. Thus, short-term IGF-I administration improved cardiac performance in these patients, most likely by decreasing peripheral vascular resistance, as also observed with GH *(64)*. Furthermore, IGF-I exerts pronounced effects on renal hemodynamics (for review *see* ref. *[65]*). The first direct evidence that IGF-I administration increased renal plasma flow (RPF) in normal human subjects was provided by Guler et al. *(66)*. During a 3-d subcutaneous infusion of 20 μg/kg/h of IGF-I, RPF increased by 22–26%. At the same time, glomerular filtration rate (GFR) rose by 31–32%. Proximal and distal tubular reabsorption of fluid and sodium was elevated to a similar extent. Hirschberg et al. *(67)* confirmed these results. During a 3-d treatment with 60 μg/kg of IGF-I administered subcutaneously three times per day to normal human subjects, RPF and GFR were increased 6 h after starting IGF-I treatment and remained elevated throughout the 3 d of treatment. Both indices returned to baseline 24 h after cessation of the IGF-I injections. Similar results were reported by Giordano and De Fronzo *(68)* during a 3-h intravenous IGF-I infusion that raised serum IGF-I levels threefold.

More detailed information on IGF-I-induced changes in renal hemodynamics came from studies in rats. Hirschberg and Kopple *(69)* demonstrated that after an intravenous IGF-I bolus (25 μg/kg) and a subsequent 20-min intravenous infusion of IGF-I (25 μg/kg) in fasted anesthetized rats, the increase in GFR and RPF was accompanied by a decrease in renal vascular resistance. These findings were confirmed and extended by Baumann et al. *(70)*. They infused between 0.08 and 1.66 nmol/kg/min of IGF I for 20 min during a 200-min euglycemic clamp in unanesthetized rats and observed an immediate rise of the GFR and RPF with half-maximal stimulation at IGF-I serum concentrations of 24 pmol/mL. In contrast to IGF-I, insulin infused at a dose of 60 pmol/kg/min did not influence GFR or RPF. Renal vascular resistance was reduced up to 50% following the IGF-I infusion, and urine flow increased up to threefold. Because stimulation of renal hemodynamics was observed at IGF-I serum levels that did not stimulate total body glucose disposal during the euglycemic clamp, the authors concluded that the stimulatory effect of IGF-I on renal hemodynamics is not mediated by insulin but by type 1 IGF receptors. The latter are present in cultured glomerular mesangial, proximal tubular epithelial, renal vascular endothelial, and smooth muscle cells. To investigate the mechanisms of the IGF-I-induced rise in hemodynamics Hirschberg et al. *(71)* used the nephron micropuncture technique in fasted and nonfasted rats. IGF-I (25 or 75 μg/kg intravenous bolus followed by an intravenous infusion of 75 or 250 μg/kg/h) induced a fall in the efferent arteriolar resistance, a rise in single nephron plasma flow, and an increase of the glomerular ultrafiltration coefficient without affecting the glomerular transcapillary hydraulic pressure difference. The authors concluded that "most likely, IGF-I relaxed the glomerular mesangium, which resulted in expansion of the capillary surface area."

The biochemical mechanism mediating IGF-induced effects on renal hemodynamics seems to involve prostaglandins and NO: In rats, indomethacin, a cyclooxigenase inhibitor, but not somatostatin blocked the IGF-I-induced rise in GFR and RPF in rats *(69,72)*, and the NO-synthase inhibitor N^G-nitro-L-arginine methylester completely prevented vasodilation during acute IGF-I infusion (25 ng/kg/min) which increased RPF in the absence of the inhibitor *(72)*.

METABOLIC EFFECTS OF IGFS IN VIVO DURING SUBCUTANEOUS ADMINISTRATION

So far, we have discussed acute metabolic actions of IGF-I. These insulin-like actions are based on the structural homology between the two peptides. However, the ability of

IGF to mimic insulin actions is limited by circulating IGF binding proteins (IGFBPs, *see* Chap. 10). Therefore, insulin-like effects are observed when the binding capacity of these BPs is exceeded, resulting in increased free IGF levels *(12,46)*, or when the circulating 40-kDa IGFBP complex becomes saturated with IGF, resulting in increased IGF bioavailability to insulin-sensitive tissues. The latter situation is encountered after intravenous injection of the 40-kDa IGFBP complex into hypophysectomized rats *(73)* or during extrapancreatic tumor hypoglycemia where most of the circulating IGF (particularly IGF-II) is shifted from the 150- to the 40-kDa IGFBP complex *(74,75)*. However, as a mediator of the effects of GH on growth, IGFs, particularly IGF-I, exert GH-like actions. With respect to substrate metabolism as well as to the effects on kidney and bone the question arises whether IGF-I also mediates these GH effects. Studies in experimental animals and in humans in which IGF-I was administered during several days show that some of its actions are GH-like whereas others are opposite to those of GH. These studies also indicate that the effects of exogenously administered IGF-I are not necessarily comparable to those of endogenous IGF-I induced in the liver (and other tissues) by GH, as GH may have direct effects that may mask or synergize with the effects of the induced endogenous IGF-I.

In Vivo Metabolic Changes During Constant Subcutaneous Infusion of IGFs Over Several Days

Studies in Rats

Because most of the experimental studies in animals were designed to measure and compare the effects of IGF-I and GH on skeletal and organ growth, metabolic changes during treatment have not always received sufficient attention.

Relatively little information is available concerning metabolic effects of IGF-I administration in normal rats. Spontaneous release of GH has been reported to be suppressed by IGF-I in normal rats *(76)*, but recently Harel and Tannenbaum showed that pure IGF-I or long R^3-IGF-I administered intracerebroventricularly to normal adult male rats did not alter the pulsatile pattern of GH release *(77)*. In contrast to humans, GH administration does not further raise IGF-I serum levels in normal rats *(78,78a)*. Serum insulin levels decrease during IGF-I administration, but remain unchanged with GH (Zapf, *unpublished observations*). We have compared the effects of a 6-d subcutaneous infusion of IGF-I, IGF-II (7 mg/kg/24 h) or hGH (1.4 U/kg/24 h) on different metabolic indices of normal male rats. Both IGF-I and -II reducted the serum urea, triglyceride, and FFA levels and IGF-I increased serum phosphate, but neither IGF-I nor IGF-II changed serum glucose, cholesterol, creatinine, uric acid, or calcium levels (Zapf, *unpublished*). IGF-II suppressed endogenous IGF-I *(78a)*. The reduction of serum triglycerides and FFAs was accompanied by a decease of the epididymal fat pad weight and of the fat pad *ob* mRNA, and by a fall of the serum leptin levels *(78a)*. GH administration reduced serum FFAs, but did not cause changes in fat pad weight or leptin, nor in any of the other indices mentioned previously. Only the IGF-I infused animals showed a significant increase in body weight, but the skeletal growth rate was not different from that in the other three groups.

Tomas et al. used a lower subcutaneous infusion rate of IGF-I (2.3 mg/kg/24 h) during a 14-d treatment of normal female rats *(78)*. They observed a significant increase in body weight and in nitrogen retention without an increase in food intake, suggesting an increased food conversion efficiency. No significant increases in muscle protein synthesis

Table 4
Effects of rhIGF-I (300 μg/d) or rhGH (200 mU/d) on Serum Hormones, IGFBPs, and ALS and on Serum Metabolites During a 6-d Subcutaneous Infusion in Hypox Rats

	rhIGF-I	*GH*
Hormones, ALS, and IGFBPs		
IGF-I	↑	↑
GH	∅	↑
Acid-labile subunit (ALS)	∅	↑
IGFBP-3	↑	↑
150-kDa-complexed IGF-I	∅	↑
40-kDa-complexed IGF-I	↑	↑
Free IGF-I	∅	↑
Insulin	(↓)	↑
Leptin	↓	→
Erythropoietin	↑	↑
Serum Metabolic Indices		
Glucose	→	→
FFAs	→	→
Triglycerides	→(↓)	→
Cholesterol	→	→
Urea	↓	↓
Creatinine	↓	↓
Inorg. phosphorus	↑	↑
Alk. phosphatase	→(↓)	→(↓)
Ca^{2+}	→	→

The results presented in this table are a summary from three different infusion experiments performed in our laboratory.

or myofibrillar protein breakdown were observed during IGF-I treatment. The infusion of GH (1.8 mg/kg/day) did not cause any changes in IGF-I or in body weight, muscle protein synthesis, or breakdown. The authors argue that methodological limitations may have accounted for the discrepancy between the improved nitrogen retention and the lack of the expected increase in muscle protein accretion. In agreement with our observations, carcass fat weight decreased during IGF-I but not during GH treatment. In comparison with IGF-I, the same dose of LR^3-IGF-I, an IGF-I variant that binds poorly to IGFBPs, stimulated body weight gain and nitrogen retention more than IGF-I and significantly inhibited muscle protein breakdown. Thus, IGFBPs may interfere with the action of administered IGF-I on protein metabolism.

In three different experiments we have infused hypox rats subcutaneously with 300 μg/d of IGF-I over 6 d. We did not observe any changes in serum insulin, glucose, FFA, triglyceride, or cholesterol levels with IGF-I or GH. However, transient hypoglycemia occurred a few hours after the IGF-I infusion was started *(79)*. Hypoglycemia disappeared after 16–24 h, coinciding with the reappearance of IGFBP-3, which is lacking in hypox rats. Serum phosphate levels increased consistently, as with GH treatment (200 mU/d). Most conspicuous changes during IGF-I infusion were pronounced decreases in serum urea and serum creatinine (Table 4). Whereas the decrease of the serum creatinine levels mainly reflects increased glomerular filtration rate, decreased blood urea levels essentially reflect positive nitrogen balance consistent with the increased weight gain and growth of the animals. Hypox rats have decreased serum leptin levels (Zapf, *unpublished observations*), and fat pads contain less *ob* mRNA than controls *(80)*. IGF-I suppressed *ob* mRNA to nearly zero and reduced serum leptin, whereas GH had no effect on leptin although it restored serum IGF-I to normal *(80)*. It thus appears that infused IGF-I and GH-induced endogenous IGF-I are not equivalent with respect to leptin regulation.

Increased serum phosphate levels during a 6-d subcutaneous IGF-I infusion (240 µg/d) in hypox rats were also reported by Caverzasio et al. *(81)*. Increased serum phosphate was accompanied by an increase of the plasma level of 1,25-$(OH)_2D_3$, whereas plasma calcium was unaffected. The increase in serum phosphate was due to an increased maximal tubular reabsorption of phosphate (T_mP)/unit volume of glomerular filtrate. GFR as assessed by the clearance of inulin was enhanced. Furthermore, sodium excretion was lower in the IGF-I-treated animals vs vehicle-treated controls. Increased phosphate reabsorption was independent of parathyroid hormone (PTH), as the IGF-I effect was also observed in thyroparathyroidectomized hypox rats. Using isolated renal brush border membrane vesicles from hypox controls and IGF-I-treated hypox rats, the authors showed that the stimulatory mechanism of IGF-I on phosphate reabsorption was located at the level of the luminal membrane of the proximal tubules. Because GH exerts the same effects on renal phosphate handling in hypox rats, the authors considered it likely that IGF-I mediates these GH effects.

Whereas concordant changes of serum metabolites are observed during IGF-I and GH infusion in hypox rats, some of the hormonal changes elicited by IGF-I and GH are opposite (Table 4). Thus, the distribution of infused IGF-I and endogenous GH-induced IGF I between the 150- and 40-kDa IGFBP serum complexes differs: The acid-labile subunit (*see* Chap. 10) which is decreased in hypox rats, increases during GH but not during IGF-I infusion, resulting in the appearance of IGF-I in the 150-kDa IGFBP complex in GH-treated animals. In contrast, the infused IGF-I is bound to the smaller molecular weight 40-kDa IGFBP complex. Free IGF-I is barely detectable *(82)*. This suggests that the metabolic changes reflecting growth during IGF-I infusion in hypox rats are likely to be directly caused by IGF-I bound to the 40-kDa complex.

In contrast to hypox rats, in which the main interest was focused on the growth effects produced by IGF treatment, the IGF studies in diabetic rats were designed to answer the question whether or not IGF treatment can reverse metabolic alterations caused by diabetes. The results of the first study using continuous subcutaneous infusion of IGF-I (2.3 mg/kg/24 h) over 6 d in streptozotocin-diabetic rats showed that hyperglycemia, glucosuria, hyperphagia, and polydypsia persisted. However, body weight, tibial epiphyseal width and thymidine incorporation into cartilage increased to a similar, although not to the same, extent as during treatment with 20 U/kg/24 h of insulin which nearly normalized the blood sugar level *(83)*. Similarly, Tomas et al. *(84,85)* did not observe amelioration of glucosuria in streptozotocin-diabetic rats infused for 7 d with 2.6–4.3 mg/kg/24 h of IGF-I, but a significant increase in body weight and nitrogen retention. IGF-I enhanced muscle protein synthesis but did not inhibit protein breakdown.

We have infused streptozotocin-diabetic rats subcutaneously with high IGF-I and IGF-II doses (13 mg/kg/24 h) for 6 d and compared the metabolic and growth effects with those of insulin (13 U/kg × 24 h) *(85a)*. All three hormones caused a similar rise of the total serum IGF level. Surprisingly, however, IGF-I lowered the elevated blood sugar level by only 20%, and IGF-II did not change the blood sugar level at all. Glucosuria persisted in both IGF-treated groups. The infused insulin dose nearly normalized the blood sugar and nearly abolished glucosuria. In contrast to the weak effect of IGF-I and -II on glucose homeostasis, serum FFAs and triglycerides were reduced to a similar extent as with the insulin dose. Thus, although IGF-stimulated glucose uptake by skeletal muscle appears to be offset by a lacking suppression of hepatic glucose production (*see above*), the infused IGFs obviously inhibited lipolysis and enhanced triglyceride synthesis in adipose tissue whose mass increased during the infusion. The decrease in serum triglyceride levels may have been due to increased lipoprotein lipase resulting in enhanced clearing of triglycerides from the circulation and resynthesis of the released FFAs in adipose tissue. Because insulin enhances lipoprotein lipase synthesis in rat adipocytes *(86)* it is likely that, due to their insulin-like effects at high infusion rates, IGF-I and -II also stimulate synthesis of this enzyme.

Serum urea was elevated nearly twofold in diabetic as compared to normal rats, but in contrast to hypox rats, it was only slightly (although significantly) reduced by IGF-I despite a dramatic effect on weight gain and skeletal growth. This discrepancy may be explained by decreased urea excretion due to impaired renal function in the diabetic animals, which was not reversed by IGF or by insulin treatment. Serum creatinine, uric acid, cholesterol, phosphate, and calcium were not different in normal and diabetic animals and unaffected by IGF-I, IGF-II, or insulin treatment. The observed metabolic effects during high-dose IGF infusion occur in the absence of insulin and are obviously due to insulin-like actions of the infused IGFs. Insulin-like actions of IGFs in vivo usually appear when the binding capacity of IGFPBs in the circulation is overridden, resulting in increased free IGF. Free IGF was increased during the IGF-I but not during the IGF-II infusion, although IGF-II caused a similar reduction in serum FFAs and triglycerides as IGF-I and insulin. During the IGF-II infusion, all of the infused IGF-II was bound to the 150- and 40-kDa IGFBP complexes. Because the capillary passage of the 150-kDa but not of the 40-kDa IGFBP complex is restricted, IGF-II in the 40-kDa complex must have been biologically active.

Hormonal, IGFBP, and Metabolic Changes During Subcutaneous Infusion or Subcutaneous Injection of IGF-I Over Several Days in Humans

Intermediary Metabolism

In 1989 Guler et al. *(87)* demonstrated that continuous subcutaneous infusion of 20 µg/kg/h IGF-I for 6 d in two healthy adults had pronounced effects on several serum hormone levels and metabolites. Some of these effects were similar, but some opposite to those of GH. Importantly, the infused IGF-I suppressed GH, insulin, and C peptide levels without significantly changing blood glucose. The suppression of insulin secretion in the absence of elevated serum glucose levels suggested that IGF-I infusion enhanced insulin sensitivity and was interpreted as "insulin sparing effect" of IGF-I. Total IGF-I reached near-maximal levels within 13–14 h after the infusion was started. Concomitantly, free IGF-I rose and total IGF-II fell *(87)*. Furthermore, IGFBP-2 increased three- to fivefold during the IGF-I infusion with a half-maximal increase after 30 h. Only a minor and transient increase of IGFBP-3 was observed *(88)*. All of these changes were reversible after the infusion was stopped: Serum IGF-I returned to normal after 2–3 d with a half-life of approx 24 h, IGF-II rose to preinfusion values with a half-life of 2–3 d, and elevated IGFBP-2 fell to normal levels with a half-life of 30 h.

Further evidence that IGF-I enhanced insulin sensitivity was provided by Zenobi et al. *(89)*. They infused IGF-I (7 and 14 µg/kg/h) intravenously for 2 d into six healthy adults and measured the responses of glucose, insulin, and C peptide during an oral glucose tolerance test and a meal tolerance test. Glucose tolerance remained unchanged during the IGF-I infusions despite suppressed insulin and C peptide levels, resulting in a decreased insulin/glucose ratio. These results suggested that tissue sensitivity to insulin had increased under the IGF-I infusion.

The inhibitory effect of IGF-I on insulin secretion has recently been analyzed in healthy volunteers *(90)*. To stimulate insulin secretion, subjects received a continuous intravenous glucose infusion (2.5 mg/kg/min) for 75 min before an intravenous infusion of IGF-I was given over 50 min. Insulin secretion rates were decreased via specific inhibition of the insulin secretory burst mass, whereas the burst frequency was not affected. The effect of IGF-I on GH pulsatility was investigated in fasted normal men during a 6-h euglycemic clamp *(91)*. IGF-I (10 μg/kg/h) caused a pronounced decrease of the serum GH concentration within 60 min with a half-life of the GH secretion rate of 16.6 min. The maximal suppressive effect was observed at 2 h, after which GH remained suppressed. Serum insulin and glucagon levels decreased to nearly undetectable levels after 60 min, whereas serum FFAs, β-hydroxybutyrate, and acetoacetate were unaffected during the first 3 h, but decreased thereafter. Deconvolution analysis of the GH secretion rates revealed that the predominant effect of IGF-I on GH secretion was due to a decrease of the mass of GH secreted per pulse and to the number of detectable secretory pulses. It was concluded that the IGF-I effect on both GH and insulin secretion is likely to be mediated through type 1 IGF-I receptors in the pituitary/hypothalamus and in the pancreas, and that these effects were direct and independent of insulin-like metabolic effects *(91)*.

In addition to the hormonal and IGFBP changes, IGF-I infusion in the study of Guler et al. *(87)* caused a substantial fall of the serum creatinine and urea levels and a rise of the creatinine clearance to 30% over the control value, suggesting a profound effect of IGF-I on glomerular filtration and thus on kidney function *(87)*. The same effect on creatinine clearance was observed during 5 d of GH treatment with daily subcutaneous injections of 3 mg *(87)*.

The fact that exogenously administered IGF-I exerted effects that were opposite to those of GH, and others which mimicked GH action (*see* Table 5), suggested that exogenous IGF-I and endogenous IGF-I induced by GH administration were not equivalent in all respects and that GH may have "direct," IGF-independent effects on substrate metabolism. On the other hand, it appeared that during treatment over several days, IGF-I may not directly cause some of the metabolic changes but rather act indirectly via changes of other hormones, such as insulin or GH. These aspects were investigated in more detail by Hussain et al. in normal subjects and in GH-deficient patients *(56,92)*.

In the first study *(56)*, IGF-I (10 μg/kg/h) was administered by subcutaneous infusion during 5 d to healthy adults in a crossover, randomized fashion under an isocaloric diet. On the third and fourth treatment day, an L-arginine stimulation test and an intravenous glucose tolerance test were performed. A euglycemic, hyperinsulinemic clamp combined with indirect calorimetry and glucose tracer infusion followed on the fifth treatment day. The study confirmed several of the observations of Guler et al. *(87)* and of Zenobi et al. *(89)*, such as a significant decrease in serum urea and creatinine levels and a sustained decrease of serum insulin and C peptide (Fig. 5), but in addition led to a number of other important observations. It showed that the insulin, C peptide, and GH response to L-arginine stimulation was suppressed during the IGF-I treatment period (Fig. 6), that intravenous glucose tolerance was unaltered despite reduced insulin secretion, and that energy expenditure and lipid oxidation increased and protein oxidation

Table 5
Comparison Between the Effects of Subcutaneously Administered rhIGF-I and GH in Adult Humans on Serum Levels of Hormones, ALS, IGFBPs; Carbohydrate, Lipid and Protein Metabolism; Bone Turnover and Calcium-Regulating Hormones; and Renal Function

Treatment	*rhIGF-I*	*GH*
I. Serum level of hormones, ALS, IGFBPs		
GH	↓	↑↑
IGF-I	↑↑	↑↑
IGF-II	↓↓	→
Acid-labile subunit (ALS)	↓	↑
IGFBP-2	↑	↓
IGFBP-3	→	↑
Insulin, C peptide	↓	↑
Leptin	↓	↑
II. Carbohydrate, lipid and protein metabolism		
Serum glucose	→(↓)	↑
FFAs, ketone bodies	↑	↑
Triglycerides (VLDL)	↓	↑
LDL cholesterol, Apo B-100	→(↓)	→(↑)
Lp (a)	↓	↑
Serum urea	↓	↓
Insulin sensitivity	↑	↓
Lipid oxidation	↑	↑
Energy expenditure	↑	↑
Protein oxidation	↓	↓
III. Bone Turnover and Calcium-Regulating Hormones		
Serum osteocalcin	↑	↑
Serum procollagen I carboxyterminal peptide	↑	↑
Urinary deoxypyridinoline/creatinine ratio	↑	↑
Serum calcium	→	(↑)
Serum PTH	→	→
Serum $1,25(OH)_2D_3$	↑	↑
IV. Renal function		
Serum urea	↓	↓
Uric acid	↓	→
Serum creatinine	↓	↓
Glomerular filtration rate	↑	↑
Renal plasma flow	↑	↑
Serum P_i	→	↑
T_mP/GFR	→	↑

decreased, whereas carbohydrate oxidation remained unchanged. It also showed that IGF-I treatment raised serum FFA levels, reflecting enhanced lipolysis. Most importantly, it documented that oxidative and nonoxidative glucose disposal in response to insulin was enhanced during the IGF-I infusion, providing direct evidence that exogenously administered IGF-I increases insulin sensitivity.

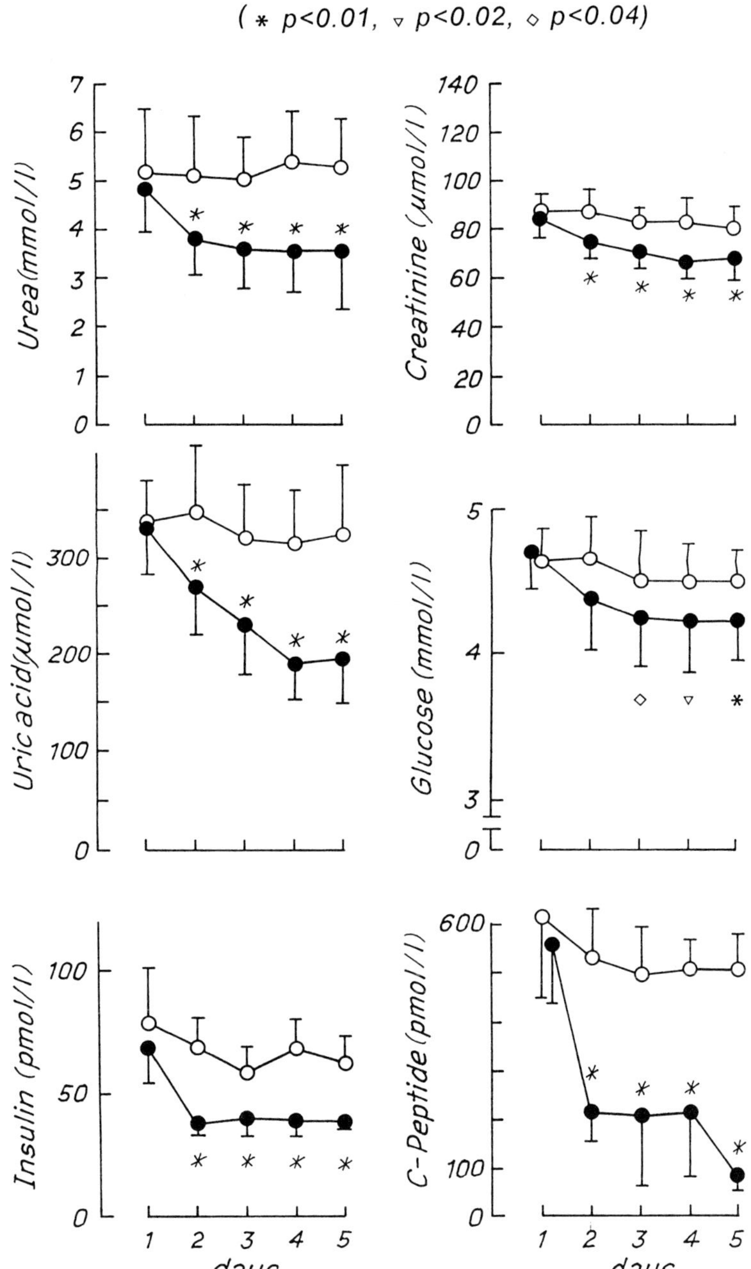

Fig. 5. Fasting serum parameters during a subcutaneous saline (○) or rhIGF-I (10 μg/kg/h) infusion (●) in eight healthy subjects. All points are means ± SD. *p* values vs saline: * <0.01, ▽ <0.02, □ <0.04. (From ref. *[56]*. With permission. Copyright 1993 The American Society for Clinical Investigation.)

The question whether suppressed GH levels during IGF-I treatment are responsible for increased insulin sensitivity was addressed in the second study in which the effects of IGF-I on insulin sensitivity and substrate oxidation were studied in GH-deficient patients and compared with those of GH and combined GH/IGF-I treatment *(92)*. All the effects of IGF-I observed in normal subjects were reproduced in the GH-deficient patients; thus,

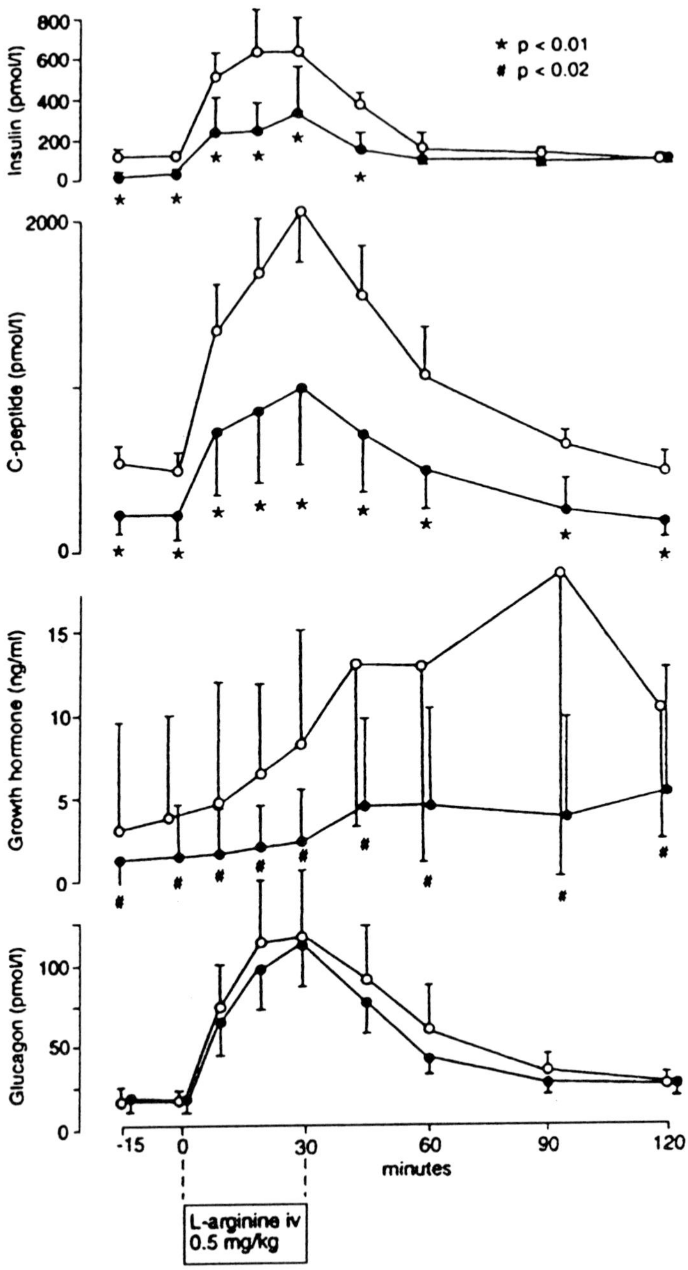

Fig. 6. Insulin, C peptide, GH, and glucagon responses to an intravenous arginine stimulation test on d 3 of a subcutaneous infusion with saline (○) or rhIGF-I (10 μg/kg/h, sc) (●) in eight normal subjects. Points are means ± SD. (From ref. *[56]*. With permission. Copyright 1993, The American Society for Clinical Investigation.)

IGF-I suppressed insulin secretion, enhanced energy expenditure and lipid oxidation (Fig. 7), and raised serum FFA and β-hydroxybutyrate levels (Fig. 8), reduced protein oxidation, and increased oxidative and nonoxidative glucose disposal in response to insulin (Fig. 8). Like IGF-I, GH (2 IU/m^2 injected subcutaneously daily) stimulated energy expenditure and lipid oxidation, raised serum FFA and β-hydroxybutyrate levels,

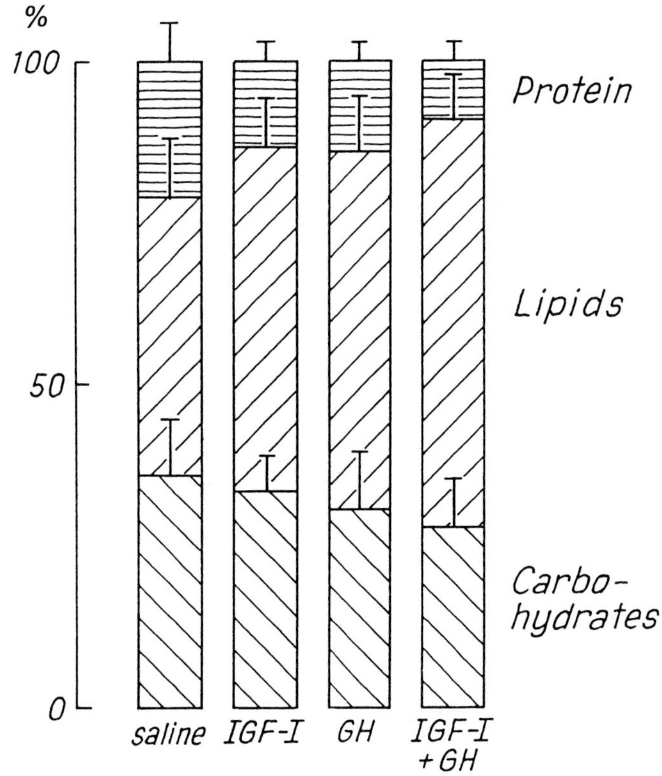

Fig. 7. Contribution of protein, lipid, and carbohydrate oxidation to resting energy expenditure (EE) in percent of EE in eight GH-deficient patients before and on d 7 of treatment with saline, IGF-I (10µg/kg/h, infused subcutaneously), GH (2 IU/m^2, daily subcutaneous injection), or the combination of the two hormones. *Bars* give SD values. (Modified from ref. *[92]*.)

and decreased protein oxidation. During combined IGF/GH treatment, the two hormones exerted synergistic effects on those indices. In contrast to IGF-I, however, GH enhanced insulin secretion, raised serum glucose levels, and decreased insulin sensitivity. Conjoint administration of IGF-I with GH reversed these latter GH effects.

Despite the fact that infused IGF-I mimics the effect of GH on lipid mobilization, the mechanisms by which this occurs are different: GH activates hormone-sensitive triglyceride lipase and enhances the responsiveness of this enzyme to lipolytic hormones, and it decreases the sensitivity of adipose tissue to the antilipolytic effect of insulin *(93)*. In contrast, the lipolytic effect of IGF-I appears to be secondary to the suppression of insulin secretion which releases the brakes on lipolysis and thus leads to increased serum FFA levels. The opposite effects of exogenous IGF-I and GH on insulin sensitivity (which mainly reflects insulin sensitivity of skeletal muscle) may be explained by their opposite actions on insulin secretion and thus by decreased or increased serum insulin levels, respectively. In acromegalic patients, GH oversecretion causes hyperinsulinemia and alters the number and affinity of insulin receptors leading to insulin resistance *(94)*. Similarly, elevated serum insulin levels during GH administration may down-regulate insulin receptors, alter their affinity, and impair postreceptor signaling. By contrast, decreased insulin levels during IGF-I infusion may cause the opposite and thus explain increased insulin sensitivity. However, it cannot be excluded that IGF-I acts directly on skeletal muscle to enhance insulin sensitivity.

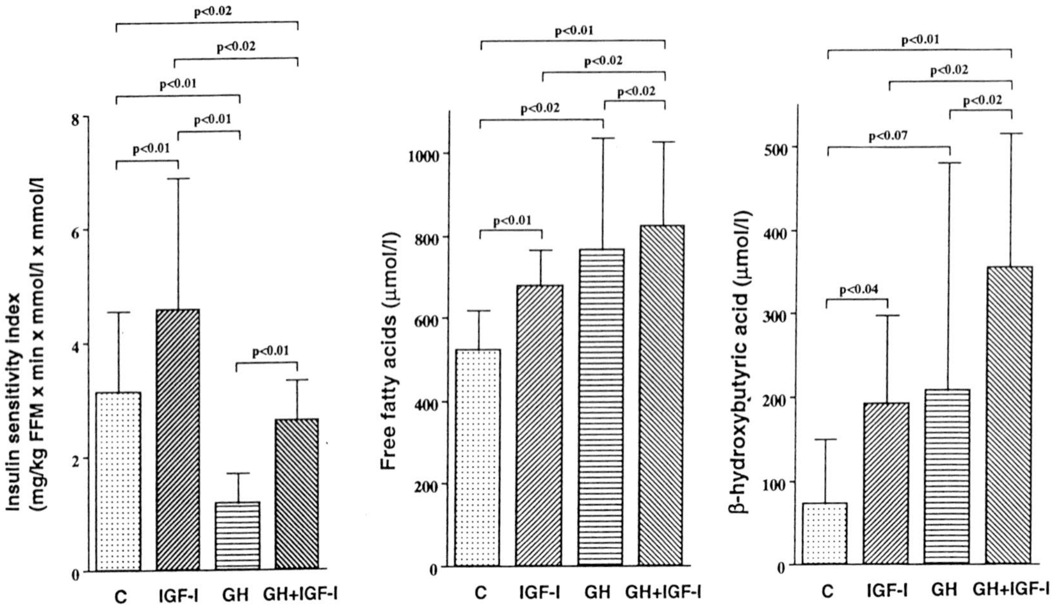

Fig. 8. *Left panel*: Insulin sensitivity index (△R_d/glucose × △insulin) (mg/kg FFM × mmol/L × mmol/L) in eight GH-deficient subjects on d 7 of treatment as indicated in Fig 7. (Derived from ref. *[92]*.) *Middle panel and right panel*: Fasting FFA and β-hydroxybutyrate levels in eight GH-deficient subjects on d 7 of treatment as indicated in Fig. 7. *Columns* represent mean values, *bars* give SD. (From ref. *[92]*. With permission. Copyright 1994, The American Society for Clinical Investgation.)

Whether exogenous IGF-I and GH reduce protein oxidation by the same or by different mechanisms is unclear. The synergistic effect of both hormones on protein oxidation does not necessarily suggest that they act directly and independently of each other on skeletal muscle. Locally infused GH has been reported to stimulate skeletal muscle protein synthesis in the human forearm without suppressing proteolysis *(95)*. On the other hand, both inhibition of proteolysis and stimulation of protein synthesis have been observed during IGF-I infusion in humans *(57)* and in experimental animals *(44,45)*. However, the suppression of insulin secretion and the fall of serum amino acids during IGF-I infusion may mask a stimulatory effect on protein synthesis so that the inhibition of protein breakdown becomes more prominent *(37)*. It would not appear unlikely that an antiproteolytic action of locally produced IGF-I during GH treatment is masked by a direct stimulatory effect of GH on amino acid uptake and protein synthesis.

Hussain et al. measured forearm muscle substrate metabolism in eight healthy subjects who received 10 µg/kg/h of IGF-I during a 3-d subcutaneous infusion *(96)*. FFA and ketone body uptake by forearm muscle was increased during IGF-I administration in the face of elevated plasma levels of these metabolites, whereas basal glucose levels and forearm glucose balance remained unchanged. However, insulin-stimulated whole body and net forearm glucose uptake were more markedly stimulated during IGF-I than during saline infusion. The same study also investigated the effect of stepwise hypoglycemic clamp on counterregulatory hormones. Hypoglycemia-induced GH and glucagon responses were blunted in the IGF-I-infused subjects, cortisol and epinephrine responses were enhanced, and the norepinephrine response remained unchanged. As observed by Kerr et al. *(55)*, recovery from hypoglycemia was delayed in IGF-I-infused subjects. Lipolysis, ketogenesis, and utilization of ketone bodies by forearm muscle were grossly enhanced in response to an intravenous GH pulse (0.5 U) during subcutaneous IGF-I infusion (8 μg/kg/h) in normal subjects *(97)*. The GH bolus did not increase circulating IGF-I levels and did not further influence circulating insulin levels. The exaggerated lipolytic response after the GH pulse is best explained by the severe inhibition of insulin secretion during IGF-I treatment, which is in keeping with the observation that the GH peaks during the night lead to exaggerated FFA and ketone body peaks in insulin-deficient diabetic patients.

Effects on Bone Turnover and Kidney Function

In addition to its actions on intermediary metabolism (Table 5, II), exogenously administered IGF-I exerts effects also on indices of bone turnover (Table 5, III) and, as previously mentioned, on renal function (Table 5, IV).

IGFs and their binding proteins are found at high concentrations in bone. They may derive from the circulation or can be produced locally, and they may thus function in an endocrine, paracrine, or autocrine manner. In human femoral cortical bone, IGF-I concentrations decrease throughout adulthood with increasing age *(98,99)*. Apart from decreased production in bone, such changes may reflect a corresponding decline in serum IGF-I, or they may relate to a decrease in bone IGFBP-5 *(100)*.

The following short-term studies provide evidence that circulating IGF-I affects bone turnover, and suggest that endocrine IGF-I is relevant for the control of bone metabolism. In this context, it should be mentioned that the target cells during IGF infusion studies in adults are not the same as the target cells for the classic effects of IGF-I on longitudinal bone growth, i.e., epiphyseal cartilage cells.

When infused subcutaneously at a rate of 8 μg/kg/h into normal young adults, IGF-I raised serum osteocalcin and procollagen type I C-terminal propeptide (PICP) levels (markers of bone formation) after 36 h, and enhanced the urinary deoxypyridinoline/creatinine and calcium/creatinine ratios (markers of bone resorption) *(101)*. Similar results were obtained in GH-deficient adults during a 5-d subcutaneous infusion of IGF-I (5 μg/kg/h) or during treatment with daily subcutaneous GH injections of 0.03 IU/kg) *(102)*. During IGF-I treatment, the increase in markers of bone resorption was preceded by an increase in markers of bone formation, suggesting that IGF-I exerted a direct anabolic effect on bone forming cells in vivo. It is likely that IGF-I mediates the effects of GH on bone turnover.

In postmenopausal and elderly women receiving IGF-I, both bone formation and bone resorption markers increased *(103,104)*. Ebeling et al. *(103)* treated groups of four or five postmenopausal women for 7 d with 30, 60, 120, and 180 μg rhIGF I/kg/d, and observed a dose-dependent rise of serum PICP and of the urinary excretion of deoxypyridinoline, but not of serum osteocalcin and bone-specific alkaline phosphatase (bALP). Side effects (orthostatic hypotension, sinus tachycardia, edema, and parotid discomfort) were more frequent at the two highest dosages of IGF-I. In elderly women treated with GH (25 μg/kg/d, sc) or IGF-I (15 and 60 μg/kg b.i.d., sc) for 28 d, PICP and osteocalcin, but not bALP, and urinary hydroxyproline, pyridinolines, and serum N-procollagen 1 telopeptides increased in response to GH and to high-dose IGF-I treatment *(104)*, indicating that both hormones activated remodeling osteons. Lower doses of IGF-I appeared to increase osteoblastic function with only marginal changes in resorption markers *(103)*.

Besides enhancing GFR and RPF *(66)*, IGF-I increased 1,25-$(OH)_2D_3$ levels in normal subjects *(101)* and in GH-deficient patients *(102)* but not in osteoporotic men *(105)*. Serum PTH, calcium, and phosphate levels remained unchanged *(101,102)*, suggesting that IGF-I had directly stimulated renal 1α-hydroxylase activity as shown in cultured mammalian kidney cells *(106)*. IGF-I also stimulated phosphate uptake by kidney cells and elevated serum phosphate in hypox rats *(81)*. During a 3-d IGF-I treatment of normal men (three daily subcutaneous injections of 60 μg IGF-I/kg), Hirschberg et al. observed a marked decrease in fractional renal excretion of phosphate. Although the serum phosphate concentrations did not change, the authors speculated that IGF-I mimicked the effect of GH on tubular phosphate reabsorption *(67)*. In contrast, Bianda et al. found that only GH but not IGF-I increased the serum phosphate level and the maximal tubular phosphate reabsorption divided by the glomerular filtration rate (T_mP/GFR) in healthy men *(101)*. Similarly, other authors *(104,105)* reported elevated serum phosphate only in response to GH treatment, and this effect was not mimicked by IGF-I. Thus, in contrast to findings in growing rats *(81)*, circulating IGF-I does not appear to mediate the effect of GH on renal phosphate reabsorption in adult humans.

METABOLIC EFFECTS OF IGF-I ADMINISTRATION IN CATABOLIC STATES AND IN VARIOUS DISEASES

Catabolic States

Based on a number of reports that GH partially reverses catabolic conditions resulting from dietary energy restriction *(107)*, gastrointestinal surgery *(108)*, glucocorticoid excess *(109)*, or aging *(110)*, and based on the somatomedin hypothesis that IGF-I mediates many GH effects, some investigators set out to study the effects of IGF-I treatment in catabolic patients.

Clemmons et al. *(111)* used caloric restriction to produce catabolism in normal adult volunteers who then received either daily 16-h intravenous infusions of IGF-I (12 μg/kg/h) or daily subcutaneous injections of GH (0.05 mg/kg) over 6 d. Both IGF-I and GH improved nitrogen balance to a similar extent, but serum urea nitrogen decreased more during IGF-I than during GH treatment. The intravenous IGF-I infusion lowered fasting blood glucose and suppressed serum insulin and C peptide levels, whereas GH treatment raised blood glucose and serum insulin and C peptide levels. The authors suggested that hyperglycemic catabolic patients might benefit from the blood sugar lowering effect of IGF-I.

In a subsequent study, the same group used a similar experimental design to study the combined effects of IGF-I and GH *(112)* as in the first study. The combination of IGF-I with GH caused a significantly greater nitrogen retention than IGF-I or GH alone and also resulted in a substantial conservation of urinary potassium, suggesting that most of the protein accretion had occurred in muscle and connective tissue. Combined IGF-I/GH administration also caused a higher increase of the serum IGF-I concentrations than IGF-I alone, raised both IGFBP-3 and acid-labile subunit, and attenuated the hypoglycemic effect of IGF-I as well as the suppression of serum insulin. Combination therapy of IGF-I plus GH may therefore have advantages over single hormone therapy in catabolic patients.

Liebermann et al. observed that a single injection of GH in AIDS patients produced a smaller increase in circulating IGF-I than in normal age-matched controls *(113)*. The authors therefore reasoned that IGF-I administration to these patients might overcome this partial GH resistance and performed a daily 12-h intravenous infusion of IGF-I for 10 d in 21 patients with AIDS. A significant but transient increase in daily nitrogen retention was found only with low dose IGF-I treatment. Because IGFBP-3 levels decreased during the IGF-I infusions the authors speculated that this may have contributed to the lack of a sustained anabolic effect. Combined GH/IGF-I therapy has also been assessed in patients with HIV wasting *(114,115)*, but no significant beneficial effects were seen.

Mauras and Beaufrère *(116)* produced a catabolic state by administration of prednisone to healthy volunteers and investigated the effects of IGF-I (100 μg/kg sc twice daily) on protein catabolism induced by the prednisone treatment. The latter treatment group was compared with a group that received only prednisone (oral dose 0.8 mg/kg, three times daily). During the combined treatment, whole body proteolysis and leucine oxidation were increased, but significantly less than in subjects treated with prednisone alone. The combined IGF-I/prednisone treatment did not cause significant changes in postabsorptive glucose concentrations or a marked reduction in circulating serum insulin, whereas prednisone alone raised serum glucose and serum insulin. Thus, IGF-I diminished the protein catabolic effect of glucocorticoids and, at the same time, prevented the diabetogenic effects of prednisone.

Diabetes and Type A Insulin Resistance

Because these aspects are covered in Chap. 30, only brief remarks are made here.

The finding that IGF-I had an insulin-sparing effect *(87,89)* and that it was able to lower the blood sugar in extremely insulin-resistant patients *(48)* raised some hope that IGF-I treatment might become an adjuvant therapy for the treatment of type 1 (insulin-dependent) and type 2 (non-insulin-dependent) diabetes. In a study with eight type 2 diabetics receiving two daily subcutaneous injections of 120 μg/kg of IGF-I, Zenobi et al. *(117)* showed that IGF-I reduced fasting glucose and total triglyceride levels (besides reducing serum insulin and C peptide) (Fig. 9), and that postprandial serum glucose values following a meal tolerance test improved. In a subsequent study, the same authors demonstrated that the reduction of total serum triglycerides was due to the reduction of very low density lipoproteins (VLDL), and that this effect persisted during 3 d after discontinuation of the IGF-I treatment *(118)*. Besides, a significant reduction of fasting total and LDL cholesterol levels and of lipoprotein (a) was observed during IGF-I treatment *(118)*.

That IGF-I, indeed, has therapeutic potential in the treatment of type 1 diabetes was first reported by Dunger et al. *(119)* and subsequently by Carroll et al. *(120)*. Dunger et al. showed that adolescents with type 1 diabetes profit from adjuvant IGF-I therapy. When administered at a dose of 40 μg/kg in the evening, IGF-I reduced exaggerated GH secretion and insulin requirements to sustain euglycemia over night. The reduction in insulin requirement was attributed directly to the effect of IGF-I. More recently, these results were confirmed in adults with type 1 diabetes injecting 50 μg/kg sc of IGF-I twice daily for 19 d together with two daily insulin injections. Insulin requirements were decreased by 45% and the mean overnight GH concentrations decreased more than threefold. At the same time, total cholesterol and triglycerides were reduced during the IGF-I therapy as observed in type 2 diabetics *(113)*.

Altogether, the results of these studies support the idea that IGF-I may lend itself as an adjuvant therapeutic agent for the treatment of type 1 and type 2 diabetes.

Osteoporosis

Since not only IGF-I concentrations in serum and bone *(98,99)* but also osteoblast function appear to decrease with age, it has been speculated that lack of IGF-I contributes to age-related (type II) osteoporosis. IGF-I was also reported to correlate positively with physical fitness; fading physical fitness and/or muscle strength with age may therefore

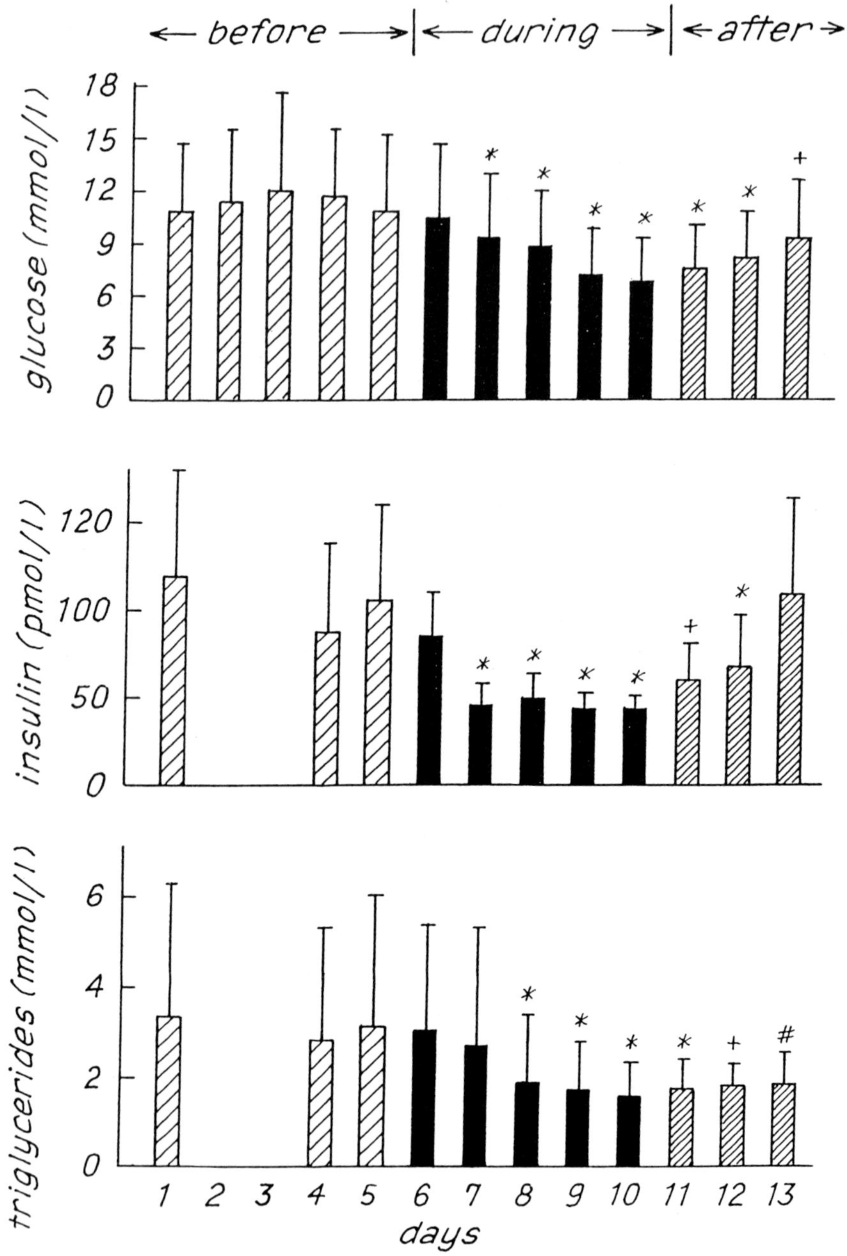

Fig. 9. Effect of rhIGF-I (2 × 120 μg sc/kg/day) on fasting plasma glucose, insulin, and triglyceride levels of eight type 2 diabetic patients. *Columns* represent mean values ± SD. *p* values vs control: * <0.01, + <0.05. (From ref. *[117]*. With permission. Copyright 1992, The American Society for Clinical Investigation.)

relate to decreased bone density *(121)*. In subgroups of individuals, low serum IGF-I has indeed been found to be associated with low bone density. Overall, however, epidemiological data regarding association between serum IGF-I and bone mineral density are not entirely convincing, as osteopenia will more frequently occur in patients with poor health and nutritional status or with liver disease (all resulting in low serum IGF-I levels). Most of the controlled clinical studies found no consistent differences in serum IGF-I or -II levels between women with spinal (type 1, postmenopausal) osteoporosis and age-matched controls.

GH and IGF-I are pivotal for growth and the acquisition of bone mass in childhood and early adulthood. Untreated GH (and IGF-I) deficiency during childhood, particularly during puberty, results in decreased bone size and in lower peak bone mass, with lower bone density persisting throughout life. Clinical observations suggest that GH continues to play a role throughout life in the remodeling of the mature skeleton. Thus, for example, bone turnover is decreased in GH deficiency. Adult-onset GH deficiency is often accompanied by sex hormone deficiency (often uncorrected for several years) and occasionally associated with decreased bone mass. It therefore remains unclear whether adult-onset GH deficiency is a cause of increased fracture frequency. Elderly patients with adult-onset GH deficiency are usually not osteopenic.

Case reports indicate that IGF-I stimulates bone turnover in osteoporosis *(122,123)*. Johansson et al. studied men with idiopathic osteoporosis and found low plasma IGF-I in some of them. In a 59-year-old patient treated with subcutaneous injections of 160 μg rhIGF-I/kg/d over 7 d, they observed an increase in bone formation markers (PICP, osteocalcin, and bALP) as well as an increase in markers for bone resorption (C-terminal telopeptide of collagen type I [ICTP], urinary hydroxyproline and calcium excretion). The anabolic activity was sustained for a few weeks, and the authors suggested that patients with osteoporosis and low IGF-I may profit from the endocrine effects of circulating IGF-I on bone and from intermittent treatment with IGF-I *(122)*. Rubin et al. treated a patient with Werner syndrome, a low serum IGF-I level and established osteoporosis, with rhIGF-I during 6 mo. They also found an increase in bone turnover markers, and an increase in lumbar bone mass without evidence of bone loss in the appendicular skeleton, suggesting a net increase in bone formation *(123)*.

Grinspoon et al. *(124)* tested the effects of rhIGF-I administration on bone turnover during short-term fasting. In line with the finding of Clemmons et al. that diet-induced catabolism can be reversed by rhIGF-I infusion *(111)*, the authors showed that in this setting (low IGF-I and low bone turnover at baseline) IGF-I had a marked and preferential effect on biochemical parameters reflecting bone formation. Similar findings were obtained by the same group in 23 women with anorexia nervosa *(125)*.

Some studies have addressed the question whether GH and IGF-I have similar or distinct effects on bone turnover in human adults *(102,104,105)*. Johansson et al. *(105)* treated 12 men with idiopathic osteoporosis with GH (2 IU/m^2/d) or IGF-I (80 μg/kg) sc for 7 d and monitored a 12-wk washout period. A similar crossover treatment protocol was used by Bianda et al. *(102)*, who treated eight GH-deficient patients with GH (0.03 U/kg/d) and IGF-I (continuous subcutaneous infusion of 5 μg/kg/d) for 5 d and followed the washout period over 8 wk. Both studies confirmed that IGF-I rapidly stimulated bone formation markers (without a previous increase in resorption markers), but that the effects waned within 1–2 wk after the treatment was stopped. The effects of IGF-I were comparable to those of GH. These studies indicate that IGF-I at doses that do not affect blood sugar has stimulatory effects on bone via an endocrine mechanism of action.

There have been no controlled long-term trials regarding the effects of IGF-I on osteoporosis. Therefore, the long-term effect of IGF-I on bone mass, bone structure, and fracture frequency in human adults remains unknown.

Outlook

The IGF-I trials in normal human subjects and in patients with various metabolic diseases have resulted in a number of new and interesting observations. Several of the metabolic effects of exogenously administered IGF-I appear to be of therapeutic relevance.

The suppression of serum insulin levels and the concomitant increase of insulin sensitivity are shared with a class of newly developed drugs for the treatment of type 2 diabetes, the thiazolidinediones. Although medication is easier with the latter, because they can be taken orally, IGF-I, by its different mechanism of action, has other advantages: It improves the serum lipid profile by lowering serum LDL-cholesterol, Apo B-100, and lipoprotein (a) besides reducing VLDL, it reduces vascular resistance, it has no toxic effect on the liver and does not increase fat mass. IGF-I might therefore well be considered for the treatment and prevention of the metabolic syndrome X, which is characterized by insulin resistance, dyslipidemia, and hypertension and an increased risk of death from cardiovascular complications. Similarly, IGF-I may be a reasonable option for transient adjuvant treatment of type 2 diabetes to attain better and more rapid metabolic control. Suppression of exaggerated GH secretion in combination with the previously mentioned metabolic effects may render IGF-I an effective adjuvant therapeutic agent for the treatment of type 1 diabetes to reduce insulin requirements and to avoid large variations of the blood sugar level during the night, above all in brittle diabetes. The hemodynamic effects of IGF-I might well prove beneficial in the long-term treatment of heart failure, not only because of its vasodilatory effects, but also because of its stimulatory effects on the formation of new myofibrils in cardiomyocytes. Its effects on renal hemodynamics make it a potential candidate for the prevention and/or treatment of acute and chronic renal failure. It appears likely that combined IGF-I and GH treatment may be able to reverse severely catabolic states and to arrest or improve osteoporosis.

The basis of envisaging such therapeutic interventions has been laid by several of the trials mentioned in this chapter. Nevertheless, the effectiveness of such therapeutic interventions can be proven only if basic and clinical trials with IGF-I continue. If we stop these trials, we are left with a number of enticing IGF-I effects without any final proof or disproof of therapeutic efficacy. Long-term studies are required to increase our experience to definitely answer the question whether or not IGF-I therapy is a reasonable treatment for one or several of the previously mentioned metabolic disorders.

REFERENCES

1. Froesch ER, Bürgi H, Ramseier EB, Bally P, Labhart A. Antibody-suppressible and nonsuppressible insulin-like activities in human serum and their physiologic significance. An insulin assay with adipose tissue of increased precision and specificity. J Clin Invest 1963; 42:1816–1834.
2. Rinderknecht E, Humbel RE. The amino acid sequence of human insulin-like growth factor I and its structural homology with proinsulin. J Biol Chem 1978; 253:2769–2776.
3. Rinderknecht E, Humbel RE. Primary structure of human insulin-like growth factor II. FEBS Lett 1978; 89:283–286.
4. Blundell TL, Bedarkar S, Rinderknecht E, Humbel RE. Insulin-like growth factor: a model for tertiary structure accounting for immunoreactivity and receptor binding. Proc Natl Acad Sci USA 1978; 75:180–184.
5. Salter J, Best CH. Insulin as a growth hormone. Br Med J 1953; (Aug 15):353–358.
6. Zapf J. The IGF–insulin relationship. J Pediatr Endocrinol Metab 1997; 10:87–95.
7. Ullrich A, Gray A, Tam AW, Yang-Feng T, Tsubokawa M, Collins C, Henzel W, Le Bon T, Kathuria S, Chen E, Jacobs S, Francke U, Ramachandran J, Fujita-Yamaguchi Y. Insulin-like growth factor I

receptor primary structure: comparison with insulin receptor suggests structural determinants that define functional specificity. EMBO J 1986; 5:2503–2512.
8. LeRoith D, Werner H, Beitner-Johnson D, Roberts CT Jr. Molecular and cellular aspects of the insulin-like growth factor I receptor. Endocr Rev 1995; 16:143–163.
9. De Meyts P. The structural basis of insulin-like growth factor I receptor binding and negative co-operativity, and its relevance to mitogenic versus metabolic signalling. Diabetologia 1994; 37:S135–48.
10. Frystyk J, Skjærbæk C, Dinesen B, Ørskov H. Free insulin-like growth factors (IGF I and IGF II) in human serum. FEBS Lett 1994; 348:185–191.
11. Guler H-P, Zapf J, Schmid Chr, Froesch ER. Insulin-like growth factors I and II in healthy man. Estimations of half-lives and production rates. Acta Endocrinol 1989; 121:753–758.
12. Zapf J, Hauri Ch, Waldvogel M, Froesch ER. Acute metabolic effects and half-lives of intravenously administered insulin-like growth factors I and II in normal and hypophysectomized rats. J Clin Invest 1986; 77:1768–1775.
13. Zapf J, Walter H, Froesch ER. Radioimmunological determination of insulinlike growth factors I and II in normal subjects and in patients with growth disorders and extrapancreatic tumor hypoglycemia. J Clin Invest 1981; 68:1321–1330.
14. Hall K, Sara VR. Somatomedin levels in childhood, adolescence and adult life. Clin Endocrinol Metab 1984; 13:91–112.
15. Zapf J, Schoenle E, Froesch ER. Insulin-like growth factors (IGF) I and II: some biological actions and receptor binding characteristics of two purified constituents of nonsuppressible insulin-like activity (NSILA-S) of human serum. Eur J Biochem 1978; 87:285–296.
15a. Zapf J, Schoenle E, Waldvogel M, Sand I, Froesch ER. Effect of trypsin treatment of rat adipocytes on biological effects and binding of insulin and insulin-like growth factors. Further evidence for the action of insulin-like growth factors through the insulin receptor. Eur J Biochem 1981; 113:605–609.
16. King GL, Kahn CR, Rechler MM, Nissley SP. Direct demonstration of separate receptors for growth and metabolic activities of insulin and multiplication-stimulating activity (an insulinlike growth factor) using antibodies to the insulin receptor. J Clin Invest 1980; 66:130–140.
17. Bolinder J, Lindbald A, Engfeldt P, Arner P. Studies of acute effects of insulin-like growth factor I and II in human fat cells. J Clin Endocrinol Metab 1987; 65:732–737.
18. Sinha MK, Buchanan C, Leggett N, Martin L, Khazanie PG, Dimarchi R, Pories WJ, Caro JF. Mechanism of IGF-I-stimulated glucose transport in human adipocytes. Demonstration of specific IGF-I receptors not involved in stimulation of glucose transport. Diabetes 1989; 38:1217–1225.
19. Sinha MK, Buchanan C, Raineri-Maldonado C, Khazanie P, Atkinson S, DiMarchi R, Caro JF. IGF-II receptors and IGF-II stimulated glucose transport in human fat cells. Am J Physiol 1990; 258:E534–E542.
20. Zapf J, Schmid Ch, Froesch ER. Biological and immunological properties of insulin-like growth factors I and II. Clin Endocr Metabol 1984; 13:3–30.
21. Zapf J, Froesch ER, Humbel RE. The insulin-like growth factors (IGF) of human serum: chemical and biological characterization and aspects of their possible physiological role. Curr Top Cell Regul 1981; 19:257–309.
22. Meuli C, Froesch ER. Effects of insulin and of NSILA-S on the perfused rat heart: glucose uptake, lactate production and efflux of 3-*O*-methyl glucose. Eur J Clin Invest 1975; 5:93–99.
23. Meuli C, Froesch ER. Binding of insulin and nonsuppressible insulin-like activity to isolated perfused rat heart muscle. Arch Biochem Biophys 1976; 177:31–38.
24. Poggi C, Le Marchand-Brustel Y, Zapf J, Froesch ER, Freychet P. Effects and binding of insulin-like growth factor I in the isolated soleus muscle of lean and obese mice: comparison with insulin. Endocrinology 1979; 105:723–730.
25. Monier S, Le Cam A, Le Marchand-Brustel Y. Insulin and insulin-like growth factor I. Effects on protein synthesis in isolated muscles from lean and goldthioglucose-obese mice. Diabetes 1983; 32:392–397.
26. Cascieri MA, Chicchi GG, Hayes NS, Strader CD. (Thr-59)-insulin-like growth factor I stimulates 2-deoxyglucose transport in BC3H1 myocytes through the insulin-like growth factor receptor, not the insulin receptor. Biochim Biophys Acta 1986; 886:491–499.
27. Dardevet D, Sornet C, Attaix D, Baracos VE, Grizard J. Insulin-like growth factor-1 and insulin resistance in skeletal muscles of adult and old rats. Endocrinology 1994; 134:1475–1484.
28. Dimitriadis G, Parry-Billings M, Bevan S, Dunger D, Piva T, Krause U, Wegener G. Effects of insulin-like growth factor I on the rates of glucose transport and utilization in rat skeletal muscle in vitro. Biochem J 1992; 285:269–274.

29. Dohm GL, Elton CW, Raju MS, Mooney ND, DiMarchi R, Pories WJ, Flickinger EG, Atkinson SM Jr, Caro JF. IGF-I stimulated glucose transport in human skeletal muscle and IGF-I resistance in obesity and NIDDM. Diabetes 1990; 39:1028–1032.
30. Dorup I, Clausen T. Insulin-like growth factor I stimulates active Na(+) -K+ transport in rat soleus muscle. Am J Physiol 1995; 268:E849–857.
31. Hartmann H, Schmitz F, Christ B, Jungermann K, Creutzfeldt W. Metabolic actions of insulin-like growth factor-I in cultured hepatocytes from adult rats. Hepatology 1990; 12:1139–1143.
32. Hartmann H, Meyer-Alber A, Braulke T. Metabolic actions of insulin-like growth factor II in cultured adult rat hepatocytes are not mediated through the insulin-like growth factor II receptor. Diabetologia 1992; 35:216–223.
33. Caro JF, Poulos J, Ittoop O, Pories WJ, Flickinger EG, Sinha MK. Insulin-like growth factor I binding in hepatocytes from human liver, human hepatoma, and normal, regenerating, and fetal rat liver. J Clin Invest 1988; 81:976–981.
34. Froesch ER, Müller WA, Bürgi H, Waldvogel M, Labhart A. Nonsuppressible insulin-like activity of human serum. II. Biological properties of plasma extracts with non-suppressible insulin-like activity. Biochim Biophys Acta 1966; 121:360–374.
35. Jacob R, Barrett E, Plewe G, Fagin KD, Sherwin RS. Acute effects of insulin-like growth factor I on glucose and amino acid metabolism in the awake fasted rat. J Clin Invest 1989; 83:1717–1723.
36. Moxley RT, Arner P, Moss A, Skottner A, Fox M, James D, Livingston JN. Acute effects of insulin-like growth factor I and insulin on glucose metabolism in vivo. Am J Physiol 1990; 259:E561–E567.
37. Jacob R, Hu X, Niederstock D, Hasan S, McNulty PH, Sherwin RS, Young LH. IGF I stimulation of muscle protein synthesis in the awake rat: permissive role of insulin and amino acids. Am J Physiol 1996; 270:E60–E66.
38. Schmitz F, Hartmann H, Stümpel F, Creutzfeldt W. In vivo metabolic action of insulin-like growth factor I in adult rats. Diabetologia 1991; 34:144–149.
39. Stümpel F, Hartmann H. Acute actions of insulin-like growth factor II on glucose metabolism in adult rats. Diabetologia 1992; 35:932–938.
40. Giacca A, Gupta R, Efendic S, Hall K, Skottner A, Lickley L, Vranic M. Differential effects of IGF-I and insulin on glucoregulation and fat metabolism in depancreatized dogs. Diabetes 1990; 39: 340–347.
41. Jacob RJ, Sherwin RS, Bowen L, Fryburg D, Fagin KD, Tamborlane WV, Shulman GI. Metabolic effects of IGF-I and insulin in spontaneously diabetic BB/w rats. Am J Physiol (Endocrinol Metab) 1991; 260:E262–E268.
42. Rossetti L, Frontoni S, Dimarchi R, DeFronzo RA, Giaccari A. Metabolic effects of IGF-I in diabetic rats. Diabetes 1991; 40:444–448.
43. Cascieri MA, Slater EE, Vicario PP, Green BG, Bayne ML, Saperstein R. Impaired IGF I-mediated stimulation of glucose incorporation into glycogen in vivo in the ob/ob mouse. Diabetologia 1989; 32:342–347.
44. Douglas RG, Gluckman PD, Ball K, Breier B, Shaw JHF. The effects of infusion of insulinlike growth factor (IGF) I, IGF-II, and insulin on glucose and protein metabolism in fasted lambs. J Clin Invest 1991; 88:614–622.
45. Koea JB, Douglas RG, Breier BH, Shaw JHF, Gluckman PD. Synergistic effect of insulin-like growth factor-I administration on the protein-sparing effects of total parenteral nutrition in fasted lambs. Endocrinology 1992; 131:643–648.
46. Guler H-P, Zapf J, Froesch ER. Short-term metabolic effects of recombinant human insulin-like growth factor I in healthy adults. N Engl J Med 1987; 317:137–140.
47. Laager R, Keller U. Effects of recombinant human insulin-like growth factor I and insulin on counterregulation during acute plasma glucose decrements in normal and type 2 (non-insulin-dependent) diabetic subjects. Diabetologia 1993; 36:966–971.
48. Schoenle EJ, Zenobi PD, Torresani T, Werder EA, Zachmann M, Froesch ER. Recombinant human insulin-like growth factor I (rhIGF I) reduces hyperglycemia in patients with extreme insulin resistance. Diabetologia 1991; 34:675–679.
49. Kahn CR, Flier JS, Bar RS, Archer JA, Gorden P, Martin MM et al. The syndromes of insulin resistance and acanthosis nigricans: insulin receptor disorders in man. N Engl J Med 1976; 294:739–745.
50. Bondy CA, Underwood LE, Clemmons DR, Guler HP, Bach MA, Skarulis M. Clinical uses of insulin-like growth factor I. Ann Intern Med 1994; 120:593–601.
51. Quin JD, Fisher BM, Paterson KR, Inoue A, Beastall GH, MacCuish AC. Acute response to recombi-

nant insulin-like growth factor I in a patient with Mendenhall's syndrome. N Engl J Med 1990; 323:1425–1426.
52. Boulware SD, Tamborlane WV, Matthews LS, Sherwin RS. Diverse effects of insulin-like growth factor I on glucose, lipid, and amino acid metabolism. Am J Physiol 1992; 262:E130–E133.
53. Laager R, Ninnis R, Keller U. Comparison of the effects of recombinant human insulin-like growth factor-I and insulin on glucose and leucine kinetics in humans. J Clin Invest 1993; 92:1903–1909.
54. Turkalj I, Keller U, Ninnis R, Vosmeer S, Stauffacher W. Effect of increasing doses of recombinant human insulin-like growth factor-I on glucose, lipid, and leucine metabolism in man. J Clin Endocrinol Metab 1992; 75:1186–1191.
55. Kerr D, Tamborlane WV, Rife F, Sherwin RS. Effect of insulin-like growth factor-1 on the responses to and recognition of hypoglycemia in humans. J Clin Invest 1993; 91:141–147.
56. Hussain M, Schmitz O, Mengel A, Keller A, Christiansen JS, Zapf J, Froesch ER. Insulin-like growth factor I stimulates lipid oxidation, reduces protein oxidation, and enhances insulin sensitivity in humans. J Clin Invest 1993; 92:2249–2256.
57. Fryburg DA. Insulin-like growth factor I exerts growth hormone- and insulin-like actions on human muscle protein metabolism. Am J Physiol 1994; 267:E331–E336.
58. Fryburg DA, Jahn LA, Hill SA, Oliveras DM, Barrett EJ. Insulin and insulin-like growth factor-I enhance human skeletal muscle protein anabolism during hyperaminoacidemia by different mechanisms. J Clin Invest 1995; 96:1722–1729.
59. Copeland KC, Nair KS. Recombinant human insulin-like growth factor-I increases forearm blood flow. J Clin Endocrinol Metab 1994; 79:230–232.
60. Tsukahara H, Gordienko DV, Tonshoff B, Gelato MC, Goligorsky MS. Direct demonstration of insulin-like growth factor-I-induced nitric oxide production by endothelial cells. Kidney Int 1994; 45:598–604.
61. Kiowski W, Stoschitsky K, Kim JH, Hussain MA, Froesch ER. Insulin-like growth factor I causes endothelium-dependent arterial vasodilation. Circulation 1994; 90:1–242, Abstract 1298.
62. Donath MY, Jenni R, Brunner H-P, Anrif M, Kohli S, Glatz Y, Froesch ER. Cardiovascular and metabolic effects of insulin-like growth factor I at rest and during exercise in humans. J Clin Endocrinol Metab 1996; 81:4089–4094.
63. Donath MY, Sütsch LG, Yan XW, Piva B, Brunner HP, Glatz Y, Zapf J, Follath F, Froesch ER, Kiowski W. Acute cardiovascular effects of insulin-like growth factor I in patients with chronic heart failure. Clin Endocrinol Metab 1998; 83:3177–3183.
64. Fazio S, Cittadini A, Biondi B, Riccio G, Sabatini D, Saccà L. Heart failure caused by dilated cardiomyopathy: rationale for treatment with growth hormone. Endocrinol Metab 1997; 4:33–36.
65. Feld S, Hirschberg R. Growth hormone, the insulin-like growth factor system, and the kidney. Endocr Rev 1996; 17:423–480.
66. Guler H-P, Eckardt K-U, Zapf J, Bauer C, Froesch ER. Insulin-like growth factor I increases glomerular filtration rate and renal plasma flow in man. Acta Endocrinol 1989; 121:101–106.
67. Hirschberg R, Brunori G, Kopple JD, Guler H-P. Effects of insulin-like growth factor I on renal function in normal men. Kidney Int 1993; 43:387–397.
68. Giordano M, DeFronzo RA. Acute effect of human recombinant insulin-like growth factor I on renal function in humans. Nephron 1995; 71:10–15.
69. Hirschberg R, Kopple JD. Evidence that insulin-like growth factor I increases renal plasma flow and glomerular filtration rate in fasted rats. J Clin Invest 1989; 83:326–330.
70. Baumann U, Eisenhauer T, Hartmann H. Increase of glomerular filtration rate and renal plasma flow by insulin-like growth factor-I during euglycaemic clamping in anaesthetized rats. Eur J Clin Invest 1992; 22:204–209.
71. Hirschberg R, Kopple JD, Blantz RC, Tucker BJ. Effects of recombinant human insulin-like growth factor I on glomerular dynamics in the rat. J Clin Invest 1991; 87:1200–1206.
72. Haylor J, Singh I, el Nahas AM. Nitric oxide synthesis inhibitor prevents vasodilation by insulin-like growth factor I. Kidney Int 1991; 39:333–335.
73. Zapf J, Hauri Ch, Futo E, Hussain M, Rutishauser J, Maack Chr A, Froesch ER. Intravenously injected insulin-like growth factor (IGF) I/IGF binding protein-3 complex exerts insulin-like effects in hypophysectomized, but not in normal rats. J Clin Invest 1995; 95:179–186.
74. Daughaday WH, Kapadia M. Significance of abnormal serum binding of insulin-like growth factor II in the development of hypoglycemia in patients with non-islet cell tumors. Proc Natl Acad Sci USA 1989; 86:6778–6782.

75. Zapf J, Futo E, Peter M, Froesch ER. Can "big" IGF II in serum of tumor patients account for the development of extrapancreatic tumor hypoglycemia (EPTH)? J Clin Invest 1992; 90:2574–2584.
76. Abe H, Molitch ME, van Wyk JJ, Underwood LE. Human growth hormone and somatomedin C suppress the spontaneous release of growth hormone in unanesthetized rats. Endocrinology 1983; 113:1319–1327.
77. Harel Z, Tannenbaum GS. Synergistic interaction between insulin-like growth factors-I and -II in central regulation of pulsatile growth hormone secretion. Endocrinology 1992; 131:758–764.
78. Tomas FM, Knowles SE, Chandler CS, Francis GL, Owens PC, Ballard FJ. Anabolic effects of insulin-like growth factor-I (IGF-I) and an IGF-I variant in normal female rats. J Endocrinol 1993; 137: 413–421.
78a. Böni-Schnetzler M, Hauri C, Zapf J. Leptin is suppressed during infusion of recombinant human insulin-like growth factor I (rhIGF I) in normal rats. Diabetologia 1999; 42:160–166.
79. Zapf J, Hauri C, Waldvogel M, Futo E, Häsler H, Binz K, Guler HP, Schmid C, Froesch ER. Recombinant human insulin-like growth factor I induces its own specific carrier protein in hypophysectomized and diabetic rats. Proc Natl Acad Sci USA 1989; 86:3813–3817.
80. Böni-Schnetzler M, Gosteli-Peter MA, Moritz W, Froesch ER, Zapf J. Reduced ob mRNA in hypophysectomised rats is not restored by growth hormone (GH), but further suppressed by exogenously administered insulin-like growth factor (IGF) I. Biochem Biophys Res Commun 1996; 225:296–301.
81. Caverzasio J, Montessuit C, Bonjour JP. Stimulatory effect of insulin-like growth factor -1 on renal Pi transport and plasma 1,25-dihydroxyvitamin D_3. Endocrinology 1990; 127:453—459.
82. Zapf J, Schmid C, Froesch ER. Metabolic actions of insulin-like growth factors. In: Molecular mechanisms to regulate the activities of insulin-like growth factors. Proceedings of the 4th International Symposium on insulin-like growth factors, Tokyo, Oct. 21–24, 1997 (K. Takano, N. Hizuka, S.-I. Takahashi, eds.), Elsevier Science, Amsterdam 1998, 341–349.
83. Scheiwiller E, Guler H-P, Merryweather J, Scandella C, Maerki W, Zapf J, Froesch ER. Growth restoration of insulin-deficient diabetic rats by recombinant human insulin-like growth factor I. Nature 1986; 323:169–171.
84. Tomas FM, Knowles SE, Owens PC, Read LC, Chandler CS, Gargosky SE, Ballard FJ. Increased weight gain, nitrogen retention and muscle protein synthesis following treatment of diabetic rats with insulin-like growth factor (IGF)-I and des(1–3) IGF-I. Biochem J 1991; 276:547–554.
85. Tomas FM, Knowles SE, Owens PC, Chandler CS, Francis GL, SE, Ballard FJ. Insulin-like growth factor-I and more potent variants restore growth of diabetic rats without inducing all characteristic insulin effects. Biochem J 1993; 291:781–786.
85a. Zapf J. Growth promotion by insulin-like growth factor I in hypophysectomized and diabetic rats. Mol Cell Endocrinol 1998; 140:143–149.
86. Ong JM, Kirchgessner TG, Schotz MC, Kern PA. Insulin increases the synthetic rate and messenger RNA level of lipoprotein lipase in isolated rat adipocytes. J Biol Chem 1988; 263:12933–12938.
87. Guler H-P, Schmid C, Zapf J, Froesch ER. Effects of recombinant insulin-like growth factor I on insulin secretion and renal function in normal human subjects. Proc Natl Acad Sci USA 1989; 86:2868–2872.
88. Zapf J, Schmid Ch, Guler HP, Waldvogel M, Hauri Ch, Futo E, Hossenlopp P, Binoux M, Froesch ER. Regulation of binding proteins for insulin-like growth factors (IGF) in humans. Increased expression of IGF binding protein 2 during IGF I treatment of healthy adults and in patients with extrapancreatic tumor hypoglycemia. J Clin Invest 1990; 86:952–961.
89. Zenobi PD, Graf S, Ursprung H, Froesch ER. Effects of insulin-like growth factor-I on glucose tolerance, insulin levels, and insulin secretion. J Clin Invest 1992; 89:1908–1913.
90. Pørksen N, Hussain MA, Bianda TL, Nyholm B, Christiansen JS, Butler PC, Veldhuis JD, Froesch ER, Schmitz O. IGF-I inhibits burst mass of pulsatile insulin secretion at supraphysiological and low IGF-I infusion rates. Am J Physiol (Endocrinol Metab 35) 1997; 272:E352–E358.
91. Hartman ML, Clayton PE, Johnson ML, Celniker A, Perlman AJ, Alberti KGMM, Thorner MO. A low dose euglycemic infusion of recombinant human insulin-like growth factor I rapidly suppresses fasting-enhanced pulsatile growth hormone secretion in humans. J Clin Invest 1993; 91:2453–2462.
92. Hussain MA, Schmitz O, Mengel A, Glatz Y, Christiansen JS, Zapf J, Froesch ER. Comparison of the effects of growth hormone and insulin-like growth factor I on substrate oxidation and on insulin sensitivity in growth hormone-deficient humans. J Clin Invest 1994; 94:1126–1133.
93. Dietz J, Schwartz J. Growth hormone alters lipolysis and hormone-sensitive lipase activity in 3T3-F442A adipocytes. Metab Clin Exp 1991; 40:800–806.
94. Muggeo M, Bar RS, Roth J, Kahn CR, Gorden P. The insulin resistance of acromegaly: evidence

for two alterations in the insulin receptor on circulating monocytes. J Clin Endocrinol Metab 1979; 48:17–25.

95. Fryburg DA, Gelfand RA, Barrett EJ. Growth hormone acutely stimulates forearm muscle protein synthesis in normal humans. Am J Physiol (Endocrinol Metab 23) 1991; 260:E499–E504.
96. Hussain MA, Schmitz O, Christiansen JS, Christensen NJ, Alberti KGMM, Froesch ER. IGF-I alters skeletal muscle substrate metabolism and blunts recovery from insulin-induced hypoglycemia. Am J Physiol (Endocrinol Metab 33) 1996; 270:E545–E551.
97. Bianda TL, Hussain MA, Keller A, Glatz Y, Schmitz O, Christiansen JS, Alberti KGMM, Froesch ER. Insulin-like growth factor-I in man enhances lipid mobilization and oxidation induced by a growth hormone pulse. Diabetologia 1996; 39:961–969.
98. Nicolas V, Prewett A, Bettica P, Mohan S, Finkelman RD, Baylink DJ, Farley JR. Age-related decreases in insulin-like growth factor-I and transforming growth factor-β in femoral cortical bone from both men and women: implications for bone loss with aging. J Clin Endocrinol Metab 1994; 78:1011–1016.
99. Boonen S, Aerssens J, Dequeker J, Nicholson P, Cheng X-G, Lowet G, Verbeke G, Bouillon R. Age-associated decline in human femoral neck cortical and trabecular content of insulin-like growth factor I: potential implications for age-related (type II) osteoporotic fracture occurrence. Calcif Tissue Int 1997; 61:173–178.
100. Nicolas V, Mohan S, Honda Y, Prewett A, Finkelman RD, Baylink DJ, Farley JR. An age-related decrease in the concentration of insulin-like growth factor binding protein-5 in human cortical bone. Calcif Tissue Int 1995; 57:206–212.
101. Bianda T, Hussain MA, Glatz Y, Bouillon R, Froesch ER, Schmid C. Effects of short-term insulin-like growth factor-I or growth hormone treatment on bone turnover, renal phosphate reabsorption and 1,25 dihydroxyvitamin D_3 production in healthy man. J Intern Med 1992; 241:143–150.
102. Bianda T, Glatz Y, Bouillon R, Froesch ER, Schmid C. Effects of short-term insulin-like growth factor-I (IGF-I) or growth hormone (GH) treatment on bone metabolism and on production of 1,25-dihydroxycholecalciferol in GH-deficient adults. J Clin Endocrinol Metab 1998; 98:81–87.
103. Ebeling PR, Jones JD, O'Fallon WM, Janes CH, Riggs BL. Short-term effects of recombinant human insulin-like growth factor I on bone turnover in normal women. J Clin Endocrinol Metab 1993; 77:1384–1387.
104. Ghiron LJ, Thompson JL, Holloway L, Hintz RL, Butterfield GE, Hoffman AR, Marcus R. Effects of recombinant insulin-like growth factor-I and growth hormone on bone turnover in elderly women. J Bone Miner Res 1995; 10:1844–1852.
105. Johansson AG, Lindh E, Blum WF, Kollerup G, Sørensen OH, Ljunghall S. Effects of growth hormone and insulin-like growth factor I in men with idiopathic osteoporosis. J Clin Endocrinol Metab 1996; 81:44–48.
106. Menaa C, Vrtovsnik F, Friedlander G, Corvol M, Garabédian M. Insulin-like growth factor I, a unique calcium-dependent stimulator of 1,25 dihydroxyvitamin D_3 production. J Biol Chem 1995; 270:25461–25467.
107. Snyder DK, Clemmons DR, Underwood LE. Treatment of obese, diet-restricted subjects with growth hormone for 11 weeks: effects of anabolism, lipolysis and body composition. J Clin Endocrinol Metab 1988; 67:54–61.
108. Ward HC, Halliday D, Sim AJW. Protein and energy metabolism with biosynthetic human growth hormone after gastrointestinal surgery. Ann Surg 1987; 206:56–61.
109. Horber FF, Haymond MW. Human growth hormone prevents protein catabolic side effects of prednisone in humans. J Clin Invest 1990; 86:265–272.
110. Rudman D, Feller AG, Nagraj HS, Gergans GA, Lalitha PY, Goldberg AF, Schlenker RA, Cohn L, Rudman IW, Mattson DE. Effects of human growth hormone in men over 60 years old. N Engl J Med 1990; 323:52–54.
111. Clemmons DR, Smith-Banks A, Underwood LE. Reversal of diet-induced catabolism by infusion of recombinant insulin-like growth factor-I in humans. J Clin Endocrinol Metab 1992; 75:234–238.
112. Kupfer SR, Underwood LE, Baxter RC, Clemmons DR. Enhancement of the anabolic effects of growth hormone and insulin-like growth factor I by use or both agents simultaneously. J Clin Invest 1993; 91:391–396.
113. Lieberman SA, Butterfield GE, Harrison D, Hoffman AR. Anabolic effects of recombinant insulin-like growth factor-I in cachectic patients with the acquired immunodeficiency syndrome. J Clin Endocrinol Metab 1994; 78:404–410.
114. Lee PDK, Pivarnik JM, Bukar JG, Muurahainen N, Berry PS, Skolnik PR, Nerad JL, Kudsk KA, Jackson L, Ellis KJ, Gesundheit N. A randomized, placebo-controlled trial of combined insulin-like

growth factor I and low dose growth hormone therapy for wasting associated with human immunodeficiency virus infection. J Clin Endocrinol Metab 1996; 81:2968–2975.
115. Waters D, Danska J, Hardy K, Koster F, Qualls C, Nickell D, Nightingale S, Gesundheit N, Watson D, Schade D. Recombinant human growth hormone, insulin-like growth factor 1, and combination therapy in AIDS-associated wasting. Ann Intern Med 1996; 125:865–872.
116. Mauras N, Beaufrère B. Recombinant human insulin-like growth factor-I enhances whole body protein anabolism and significantly diminishes the protein catabolic effects of prednisone in humans without a diabetogenic effect. J Clin Endocrinol Metab 1995; 80:868–874.
117. Zenobi PD, Jaeggi-Groisman SE, Riesen WF, Røder ME, Froesch ER. Insulin-like growth factor-I improves glucose and lipid metabolism in type 2 diabetes mellitus. J Clin Invest 1992; 90:2234–2241.
118. Zenobi PD, Holzmann P, Glatz Y, Riesen WF, Froesch ER. Improvement of lipid profile in type 2 (non-insulin-dependent) diabetes mellitus by insulin-like growth factor I. Diabetologia 1993; 36:465–469.
119. Dunger DB, Cheetham TD, Crowne EC. Insulin-like growth factors (IGFs) and IGF-I treatment in the adolescent with insulin-dependent diabetes mellitus. Metabolism 1995; 44:119–123.
120. Carroll PV, Umpleby M, Ward GS, Imuere S, Alexander E, Dunger D, Sönksen PH, Russell-Jones DL. rhIGF-I administration reduces insulin requirements, decreases growth hormone secretion, and improves the lipid profile in adults with IDDM. Diabetes 1997; 46:1453–1458.
121. Kelly PJ, Elsman JA, Stuart MC, Pocock NA, Sambrook PN, Gwinn TH. Somatomedin-C, physical fitness, and bone density. J Clin Endocrinol Metab 1990; 70:718–723.
122. Johansson AG, Lindh E, Ljunghall S. Insulin-like growth factor I stimulates bone turnover in osteoporosis. Lancet 1992; 339:1619.
123. Rubin CD, Reed B, Sakhaee K, Pak CYC. Treating a patient with the Werner syndrome and osteoporosis using recombinant human insulin-like growth factor. Ann Intern Med 1994; 121:665–668.
124. Grinspoon SK, Baum HBA, Peterson S, Klibanski A. Effects of rhIGF I administration on bone turnover during short-term fasting. J Clin Invest 1995; 96:900–906.
125. Grinspoon S, Baum H, Lee K, Anderson E, Herzog D, Klibanski A. Effects of short-term recombinant human insulin-like growth factor I administration on bone turnover in osteopenic women with anorexia nervosa. J Clin Endocrinol Metab 1996; 81:3864–3870.

25 Clinical Uses of IGF-I and IGF Binding Protein Assays

Louis E. Underwood, MD

CONTENTS

Measurements of the insulin-like growth factors (IGFs) and the IGF binding proteins (IGFBPs) have considerable value in diagnosing disease and in assessing the clinical status of patients with known diagnoses. This chapter reviews applications of these measurements and provides an overview of their clinical utility.

IGFS AND IGFBPS IN THE DIAGNOSIS OF DISORDERS OF GROWTH HORMONE SECRETION OR ACTION

Diagnosis of Growth Hormone Deficiency

A benefit of measurement of serum concentrations of IGF-I, IGFBP-3, or the acid-labile subunit (ALS) of the circulating IGF-I complex is that their concentrations often reflect the integrated concentration of secreted growth hormone (GH). Each of these peptides is responsive to GH; each is decreased in GH deficiency; each is elevated in GH excess; and if GH secretion is low, each is promptly increased when GH is replaced. Unlike the pulsatile secretion and short half-life of GH in serum (± 20 min), IGF-system peptides exhibit serum half-lives of 12–16 h.

Diagnosis of GH Deficiency in Children

The diagnosis of GH deficiency in children is sometimes problematic because the spectrum of severity of the deficiency among affected patients may be wide. Pediatric endocrinologists have grappled for years with the problem of diagnosis, which results from the lack of tools to make (or exclude) the diagnosis reliably. GH responses to provocative secretagogues often fail to discriminate between sufficient and insufficient GH secretion. Likewise, repeated, frequent, or continuous sampling of blood for GH may

From: *Contemporary Endocrinology: The IGF System*
Edited by: R. Rosenfeld and C. Roberts Jr © Humana Press, Inc., Totowa, NJ

not provide a diagnosis. Serum concentrations of immunoreactive IGF-I are determined to a significant degree by GH secretion and correlate in short children with GH production *(1–3)*. Blum et al. *(4)* observed in a group of prepubertal children that GH secretion correlated with serum IGF-I concentrations (r = 0.78), and that IGF-I correlated with height (r = 0.66). In healthy children *(5)*, IGF-I correlates with body mass index and with stage of puberty, reaching maximum values when pubertal stages reach Tanner III or IV *(6)*.

Several studies report on the relationship between GH secretion and serum IGF-I concentrations in short children, examining the value of the latter in diagnosis. Hasegawa et al. reported on results obtained from 309 normal, 73 GH-deficient, and 99 short, presumably not GH-deficient children *(7)*. They observed that 80% of the IGF-I values in the GH-deficient children were below age-related normal values, whereas 81% of the normal short children had IGF-I values within the normal range. Nunez et al. *(8)* observed that serum IGF-I correlated with both serum GH responses to provocative agents and to levels of GH secreted overnight. They determined that IGF-I results combined with IGFBP-3 measurements are helpful in selecting children for whom GH testing can be avoided. Using –1 SD as a cutoff for the two peptides, they observed that IGF-I/IGFBP-3 testing would identify 88% of the GH-deficient patients. Furthermore, 68% of the idiopathic short stature patients would be above the –1 SD cutoff and would not need GH testing.

Despite relatively good correlations within groups of patients between GH secretion and IGF-I concentrations the problems in assessing GH status by measurement of IGF-I in individual patients are substantial. The reasons for this are as follows:

- IGF-I values are quite low early in life and one's ability to separate the values obtained from GH-deficient children vs those obtained from normal children is limited.
- Poor nutrition, which accompanies many illnesses causing growth failure, lowers IGF-I despite normal or even elevated levels of GH.
- Measurements of serum GH values are not reliable indicators of GH secretory status. Therefore, there is no "gold standard" by which to assess the reliability of IGF-I values with confidence.
- Specifically, the values used to describe normal vs abnormal serum GH responses to provocative stimuli are arbitrary and have been changed over the years. There are no rules for resolving conflicting results obtained from two or more GH tests. Little attention has been directed at describing the effects of age, sex, and pubertal status on serum GH responses. The radioimmunoassays (RIAs) used to measure GH are not well standardized with regard to the details of methods and the potency of the standards used *(9)*.

These problems, combined with the variability in severity in GH among patients, denies IGF-I measurements a "stand-alone" position as a diagnostic test for GH deficiency.

Efforts have been made to determine whether measurements of the free fraction of IGF-I in serum might provide a better indicator of GH status than more traditional measurements of the total IGF-I. No improvement in diagnostic reliability is observed when free IGF-I is measured *(10)*.

Many of the investigators who have explored the possible use of IGF-I measurements for the diagnosis of GH deficiency in children have examined concurrently the value of other serum GH response peptides. A study by Rosenfeld et al. *(11)* assessed the value of measurements of IGF-II along with IGF-I. The logic of this is sound in that IGF-II, although not elevated in GH excess, is usually reduced in patients with GH deficiency.

These investigators observed that there was moderate improvement in diagnostic accuracy when both peptides were measured. Testing with a combination of IGF-I and IGF-II, however, has not been actively pursued or validated by others.

As with IGF-I, serum concentrations of IGFBP-3 are reduced in GH deficiency, increased with GH treatment, and elevated in states of GH excess. The ALS in the circulating IGF/IGFBP-3 molecular complex is also regulated by GH. IGFBP-3 has the theoretical advantage over IGF-I that it provides better separation in young children between those who have sufficient GH secretion from those who are deficient. In general, studies examining the diagnostic value of IGFBP-3 show that when used in conjunction with IGF-I measurements, diagnostic precision is improved somewhat.

Hasegawa et al. *(7)*, in the study cited previously, showed that in a group of 73 GH-deficient children, a slightly larger number (93%) had low IGFBP-3 values than had low IGF-I values (80%). The number of normal short children with normal IGFBP-3 values (83%) was virtually the same as the number with normal IGF-I values (81%). They observed that while IGFBP-3 measurements were more sensitive in children who were less than 10 yr of age, when it correlated completed with the assessment by direct measures of GH secretion, the correlation declined to 79% among children over 10 yr of age *(7)*.

In the report by Nunez et al. *(8)*, the addition of IGFBP-3 measurements to IGF-I also improve accuracy of diagnosis in those patients in whom both values were diagnostically consistent. We reported on the use of serum IGF-I and IGFBP-3 values in 57 GH-deficient children, 23 children with idiopathic short stature (ISS) and 415 normal children (Figs. 1 and 2) *(12)*. The children with presumed GH deficiency had failed to respond (peak serum GH values <1 ng/mL) to two provocative tests. Of these, 49 had low IGF-I; 4 had borderline values; and 4 had normal values. Of the 53 whose IGF-I values were below normal, 51 had low IGFBP-3. Of the 23 children with ISS (peak serum GH response in provocative tests, >10 ng/mL), 13 had normal IGF-I and IGFBP-3; 7 had low IGF-I and 3 had borderline low IGF-I and low IGFBP-3. We conclude that these tests are sensitive, but lack the desired specificity. Others have reported similar results *(13)*.

We measured the serum IGFBP-2 in these groups of children, taking advantage of the observation that IGFBP-2 values are elevated in GHD (Fig. 3) *(12)*. Virtually all the GH-deficient children who had low IGF-I had IGFBP-2 values greater than 1 SD above the mean for age-related normals. When the relationship between normals and GH-deficient children was assessed as IGF-I/IGFBP-2 ratio, the separation was improved further (Fig. 4). Nevertheless, none of the variables analyzed (IGF-I, IGFBP-3, IGFBP-2, or IGF-I/IGFBP-2 ratio) allowed an absolute separation of the presumed GH-deficient children from those presumed to have ISS.

Studies of the value of measurements of serum concentrations of ALS in diagnosing GH deficiency in children have not been reported. There is reason to believe, however, that ALS may have value as it is GH dependent, and ALS levels are reduced in GH-deficient children *(14)*. In adults, however, ALS concentrations seem not to discriminate sufficiently to be of great value in the diagnosis in a given patient of GH deficiency or of acromegaly *(15)*.

One of the principal obstacles confounding the interpretation of IGF system peptides in short children is that a variety of conditions associated with short stature and growth failure cause these peptides to be reduced. Mohnike et al. *(16)* reported that serum concentrations of IGF-I, IGF-II, and IGFBP-3 were regularly reduced in patients with acute lymphoblastic leukemia. In addition, many of these patients exhibited elevated IGFBP-2

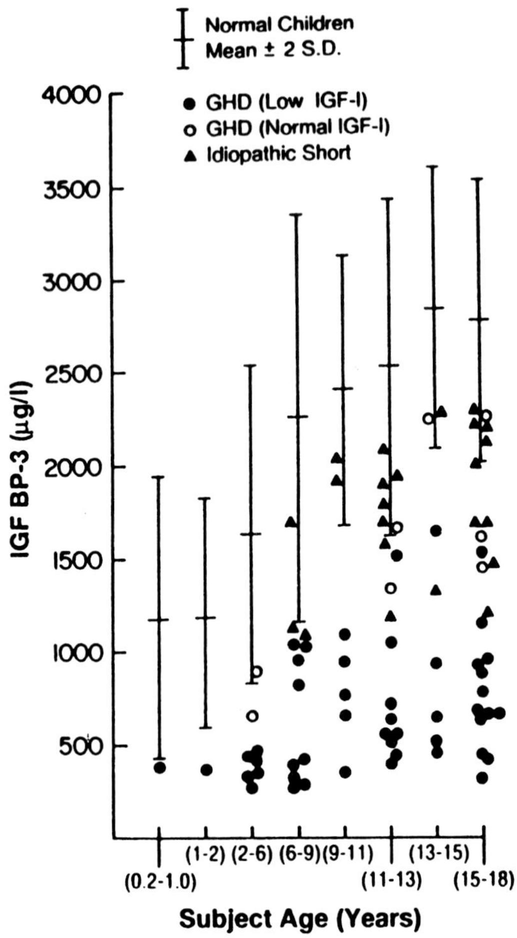

Fig. 1. Concentrations of IGFBP-3 by RIA in the serum of 415 normal children, 49 children with GHD (●), 8 children with GHD and normal IGF-I (○), and 23 with ISS (▲). Results of normal children are expressed as means ± 2 SD. (From Smith WJ, et al. J Clin Endocrinol Metab 1993; 77:1294–1299, with permission.)

values, as seen in GH deficiency. Similar changes in these peptides are observed in other chronic diseases, particularly when supplies of nutrients are inadequate.

In summary, measurements of single components of the IGF system cannot be used in all short children for confirming or excluding the diagnosis of GH deficiency. On the other hand, these tests are useful if interpreted in light of each other, and when other diagnostic tools are used concurrently. In the final analysis, diagnosis and decisions about GH therapy require consideration of multiple factors, such as auxologic data, medical history, physical findings, and serum GH values, as well as serum values of IGF system peptides.

Diagnosis of GH Deficiency in Adults

Scrutiny of IGF system peptides to assess their value in the diagnosis of GH deficiency in adults is limited compared to diagnosis in children. Baum et al. *(17)* compared 23 men with GH deficiency with 23 age-matched control men. Among the GH-deficient men, 20/23 had low IGF-I and 16/23 had low IGFBP-3 values. Among the controls, 3/23 also

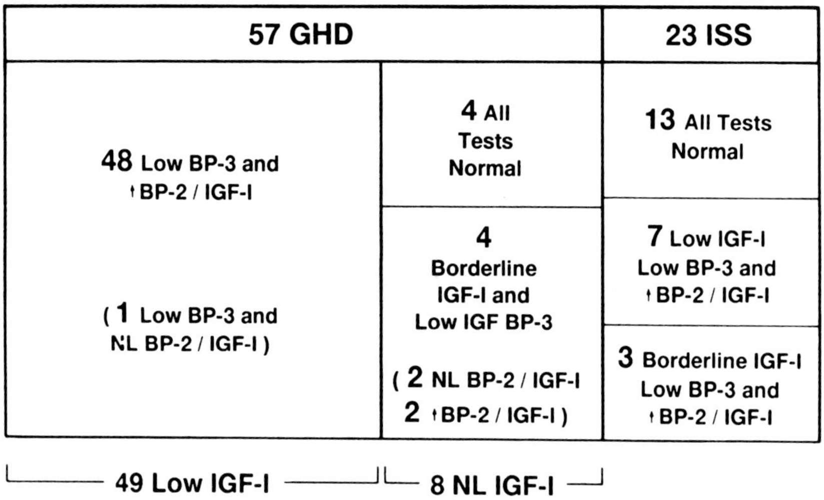

Fig. 2. Diagram of the concordance/discordance among the IGFBP-2/IGF-I ratio, IGFBP-3, IGF-I, and GH provocative testing results in the 80 children with GHD or ISS. The largest group is the 49 children with GHD, low IGF-I, increased IGFBP-2/IGF-I ratios, and decreased IGFBP-3. The 8 children with GHD and normal IGF-I are subdivided into groups with borderline IGF-I values (5th–10th percentile; $n = 4$) and normal (10th–95th percentile) IGF-I values ($n = 4$). The children with ISS are also subdivided into a group with normal IGF-I values ($n = 13$) and those with low ($n = 7$) or borderline ($n = 3$) IGF-I values. (From Smith WJ, et al. J Clin Endocrinol Metab 1993; 77:1294–1299, with permission.)

had low IGF-I. Ghigo et al. *(18)*, comparing 54 adults with GH deficiency with 325 controls, observed that only 57% of the GH-deficient adults had serum IGF-I values below the third percentile of the normals. Also, many of the control adults over the age of 40 yr had relatively low values. Thissen et al. *(19)* measured IGF-I and IGFBP-3 in 25 GH-deficient adults (20–76 yr of age) and compared their values with those made on 81 age- and sex-stratified controls. The mean IGF-I values of the GH-deficient patients were decreased by 37%, and the IGFBP-3 by 70%. Using the mean –2 SD of the control subjects as a cutoff point, the sensitivity of the IGF-I measurement was low (52%), and for the IGFBP-3 measurements the sensitivity was only 36%. Similar findings had been reported by Hoffman et al. *(20)*, Bates et al. *(21)*, and Svensson et al. *(22)*.

Investigating whether measurements of free IGF-I in serum might provide more useful diagnostic information than measurement of total IGF-I, Skjaerbaek et al. *(23)* compared 27 GH-deficient adults with 42 healthy volunteers, free IGF-I being measured after ultrafiltration of the serum samples. The mean free IGF-I (700 ± 100 ng/L; 1 SEM) was approx 70% of control values (1010 ± 70 ng/L), with considerable overlap of individual subjects. In both GH-deficient and control subjects, the free IGF-I correlated significantly ($p < 0.05$) with total values (GH-deficient, $r = 0.78$; controls, $r = 0.42$). The authors concluded that measurements of free IGF-I provided no advantage in diagnosis of GH deficiency over measurement of total IGF-I.

Although IGF-I measurements are a more sensitive index of GH deficiency in adults than IGFBP-3, the finding of a normal IGF-I value in an adult in whom GH deficiency is being considered does not exclude this possibility. The tendency to underdiagnose GH deficiency using IGF-I is, therefore, of considerable concern. It is generally agreed that

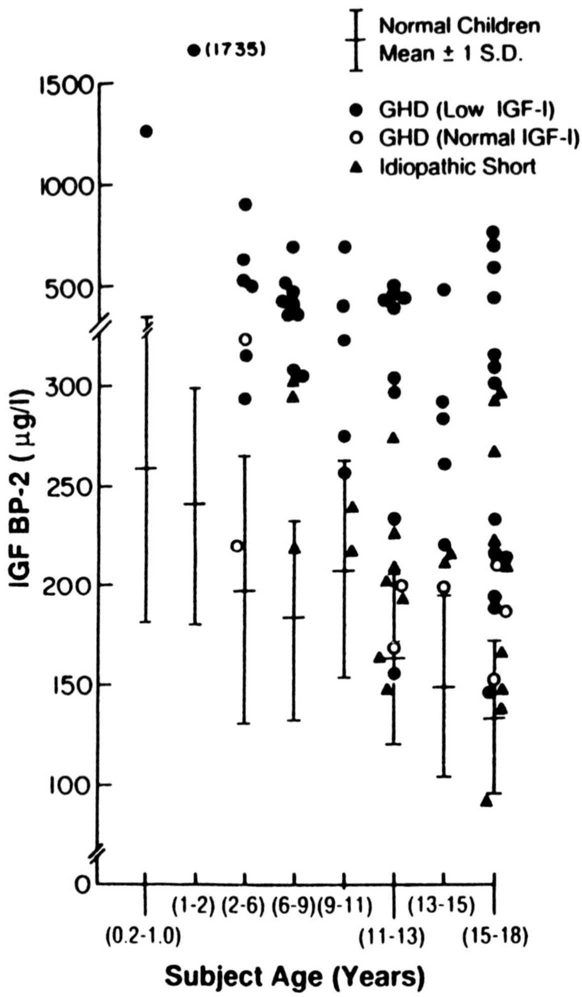

Fig. 3. Serum concentrations of immunoreactive of IGFBP-2 in 415 normal children (mean ± 1 SD), 57 children with GHD and low IGF-I (●; $n = 49$) or GHD and normal IGF-I (○; $n = 8$), and 23 boys with ISS who had normal GH responses to provocative stimuli and normal serum IGF-I values (▲). Among the normal subjects, the divisions by age are on the birthday of the years indicated. (From Smith WJ, et al. J Clin Endocrinol Metab 1993; 77:1294–1299, with permission.)

measurement of IGFBP-3 in adults for diagnosis of GH deficiency is of little value. It seems highly probable that the reason for the poor performance of the IGFBP-3 assay is related to the fact that factors in addition to GH regulate IGFBP-3 secretion. For example, sex steroids increase IGFBP-3 in serum of children with GH deficiency. In such children, IGFBP-3 rises during puberty, suggesting that sex steroids do not require GH to raise IGFBP-3 *(24)*.

IGF-I Values During GH Therapy

Increased serum IGF-I concentrations in GH-deficient individuals during GH therapy is a well-established event. When GH is administered therapeutically at intervals of 24 hr, the level of IGF-I in serum is relatively constant. In children, little emphasis has been placed on the serum IGF-I response to GH as a measure of GH efficacy, because efficacy

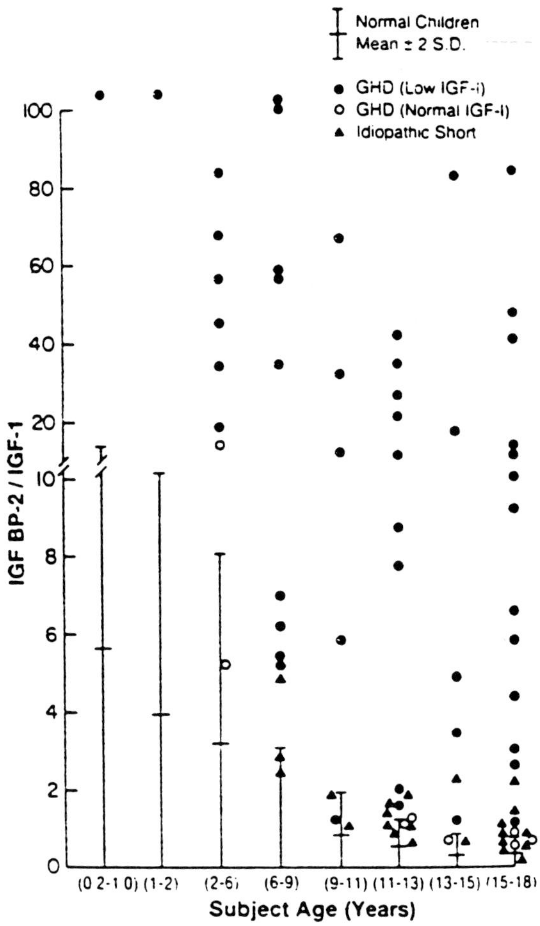

Fig. 4. Ratio of IGFBP-2 to IGF-I for 415 normal children (mean ± 2 SD), 49 children with GHD and low IGF-I (●), 8 children with GHD and normal IGF-I (○), and 23 children with ISS (▲). The results for normal children are expressed as the means ± 2 SD. (From Smith WJ, et al. J Clin Endocrinol Metab 1993; 77:1294–1299, with permission.)

has been monitored by assessing statural growth. In adults, in whom growth cannot be monitored, and in whom undesirable side effects of GH therapy are more common than in children, IGF-I measurements are believed to be useful in customizing the GH dose for the patient. Most prescribers adjust the GH dose so as to keep the serum IGF-I within the normal range.

Diagnosis of GH Excess

Measurement of IGF-I in patients with suspected acromegaly has significant diagnostic value, as the peptide is consistently elevated and there is little overlap with the values of normal individuals *(25–28)*. Indeed, in some patients IGF-I measurements appear to be a more reliable indicator of GH excess than measurement of GH itself *(26)*. For example, GH values in some patients are not strikingly elevated, although they may fluctuate little and have higher than normal between-peak, nadir values. These patients will usually have markedly elevated IGF-I values *(26)*.

Serum IGF-I concentrations seem to correlate better than GH values with the clinical improvement that occurs with treatment of acromegaly *(29)*. A recent report by Barkan et al. *(30)* points out that within 10 yr of radiation therapy, 80% of patients with acromegaly have GH levels below 5 ng/mL. Of 38 patients who had received the radiation treatment 6.8 ± 0.8 yr previously (range 1–19 yr), only two patients had normal IGF-I values. The noncured patients had a mean GH value of 4.6 ± 1.1 ng/mL, but mean IGF-I values that were 219 ± 26% of the upper normal limit. A random GH concentration below 1.5 ng/mL was associated with pathologically high IGF-I in 43% of these patients. In addition to showing that GH measurements after treatment are of limited value, the investigators raised questions about the efficacy of radiation treatment in facilitating cure of patients with GH excess *(30)*.

IGFBP-3 values are also sensitive indicators of GH excess in acromegaly. Greenspoon et al. *(28)* showed that IGFBP-3 is uniformly elevated in patients with untreated acromegaly, and is predictably elevated in patients who have had incomplete removal of their GH-producing pituitary adenoma. Thissen et al. *(19)* observed that IGFBP-3 measurements were nearly as good for the diagnosis of acromegaly as IGF-I. Among 25 untreated patients with acromegaly, the mean IGFBP-3 value was two times the mean for control subjects (IGF-I was four times control). The sensitivity of the IGFBP-3 measurements for the diagnosis of acromegaly (using + 2 SD of the controls as a cutoff) was 76% (100% using extracted IGF-I values). The finding that IGFBP-3, although useful, is not as sensitive as IGF-I in detecting GH excess is in agreement with the observations of Wass *(31)* and Paramo et al. *(32)*.

IGFS AND IGFBPS IN ASSESSMENT OF NUTRITIONAL STATUS

Twenty-five years ago, Grant et al. *(33)* reported that somatomedin bioactivity was decreased in children with kwashiorkor. Hintz et al. *(34)* observed that activity is also decreased in marasmus. These findings have been confirmed using RIAs for IGF-I *(35)*. Serum IGF-II values are also decreased in chronic malnutrition *(35)* while serum GH is increased *(33,35,36)*. Teenagers with anorexia nervosa, inflammatory bowel disease, celiac disease, and HIV infection also have reduced IGF-I *(37–41)*. Fasting produces a prompt decline in IGF-I *(42)*, with values reaching hypopituitary levels in adult humans within 7–10 d. Several studies done by our group indicate that the maintenance of IGF-I levels in serum is dependent on sufficient supplies of both energy and protein *(43–45)*, and that at least 12 kcal of energy/kg/d is needed to maintain IGF-I in normal adults. On the other hand, IGF-I responses to the supply of dietary protein is graded—with incremental decreases occurring in response to reductions of the protein ingested. We have observed also that the proportion of essential amino acids making up the dietary protein is important in the regulation of IGF-I *(46)*. After subjecting normal volunteers to fasting, we observed that refeeding a low-protein, energy-adequate diet (0.48 g/protein/kg/d) in which the protein was low in essential amino acids (80% nonessential amino acids) produces attenuation of the rise of IGF-I toward normal. Alternatively, a diet of similar composition, except that 80% of the amino acids were essential amino acids, resulted in a significantly better return of IGF-I toward normal *(46)*.

The measurement of IGF-I in serum has value in assessing nutritional status. It provides advantages over serum albumin, transferrin, and retinol-binding protein and prealbumin because of the insensitivity of the latter proteins to short-term changes and their relatively low specificity. Unterman et al. *(47)* observed that serum immunoactive

IGF-I among malnourished, hospitalized patients was 40% below control values, a greater reduction than observed for albumin (66% of controls, and transferrin 59% of controls). Donahue and Phillips *(48)* reported that hospitalized patients with protein- or protein-calorie malnutrition had reduced serum IGF-I, whereas patients who had caloric deficits but were protein sufficient had normal IGF-I values. Interpretation of IGF-I values in patients in whom malnutrition is suspected is problematic, because the underlying disease process itself may affect IGF-I by mechanisms other than nutrient insufficiency. For example, IGF-I may be reduced in patients with liver disease *(49,50)*. Alternatively, IGF-I may be raised *(51)* or lowered *(52)* in patients with chronic renal failure.

Measurement of IGF-I appears to have greater value as a means of monitoring response to nutritional support than in determining nutritional status. We have observed that nutritional rehabilitation (for 10–16 d) of chronically malnourished patients (10% below ideal body weight) produces a prompt and marked increase in serum IGF-I concentrations, such that the mean value at d 10 of rehabilitation is 2.7 times higher than the baseline value. In the same study, we saw no significant change in transferrin, prealbumin, or retinol-binding protein. These observations have been confirmed by Donahue and Phillips *(48)*, who showed a 264% increase in IGF-I compared to the 50% increase in transferrin and a 9% increase in albumin. As in our study, the changes in IGF-I correlated with changes in nitrogen balance ($r = 0.45$; $p < 0.005$). In patients with chronic renal failure, nutritional support caused serum IGF-I to increase significantly within 4 d *(53)*.

As with IGF-I, nutritional status and dietary manipulation also affect the IGFBPs. Changes in IGFBP-3 in various nutritional states/manipulations parallel the changes in IGF-I closely. IGFBP-3 is reduced in malnutrition and with dietary restriction *(54,55)*. Concurrently IGFBP-2 rises under these conditions *(39,56)*. IGFBP-I, however, fluctuates acutely in response to dietary intake. It rises several-fold with overnight fasting, declines in minutes after ingestion of food, and correlates inversely with insulin secretion *(57–60)*.

We have examined the effects of dietary manipulations on eight healthy adults and six or eight prepubertal children volunteers by subjecting them to diet manipulations over periods of 6 d *(54)*. In one 6-d study period we reduced caloric intake by 50%. This caused a significant decrease in nitrogen balance and in IGF-I in serum. Serum IGFBP-2 values did not change in response to this calorie restriction. IGFBP-3 declined significantly in children (from 3189 ± 90 ng/mL to 2843 ± 96 ng/mL; $p < 0.05$), but not in adults. In the other test period, we restricted dietary protein (reduced from 1 g/kg down to 0.66 g/kg/d) in both the adults and the children. This also produced negative nitrogen balance in both patient groups. Here, however, IGFBP-2 increased significantly (adults: basal 131 ± 15 to 164 ± 15 ng/mL; $p < 0.005$; and in children: 126 ± 13 at baseline to 158 ± 15 ng/mL after 6 d; $p < 0.05$). IGFBP-3 was reduced slightly in response to protein restriction in adults, but not in children. The findings confirm the parallel nature of regulation of IGF-I and IGFBP-3 by nutrients and point up the dominance of protein intake in the regulation of IGFBP-2.

We also have examined the influence of calorie and protein intake on IGF-I, IGFBP-3, and IGFBP-2 in premature newborns *(61)*. We observe a consistent, predictable relationship between caloric intake and protein intake on IGF-I and IGFBP-3. We observe that in newborns the serum concentration of IGF-I is determined by the sufficiency of both calories and protein. The IGFBP-3 concentration is, however, less dependent on dietary protein than is IGF-I. The IGFBP-2 serum concentration is effected little by calories but is quite sensitive to the protein ingested, rising almost twofold over 6 d.

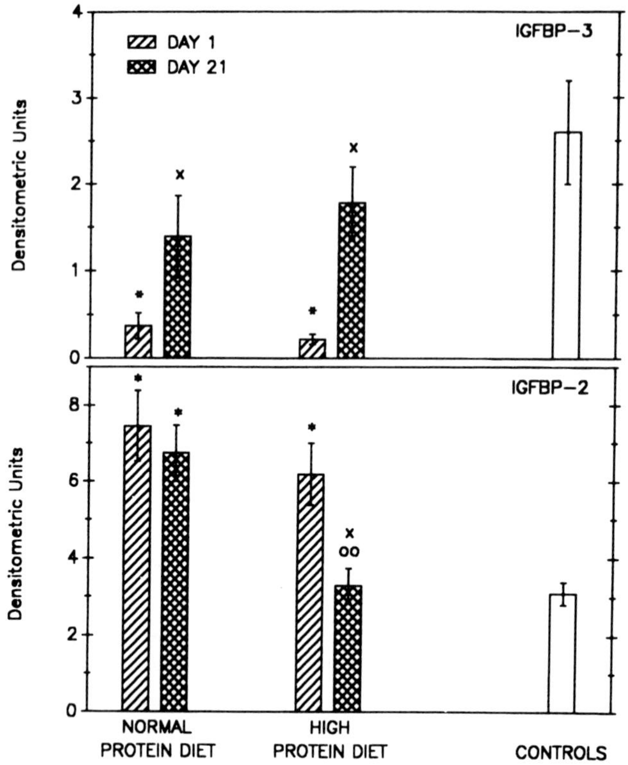

Fig. 5. Densitometric analysis of ligand blots for serum IGFBP-3 (*upper panel*) and IGFBP-2 (*lower panel*) in controls ($n = 25$) and Bangladeshi children before and after refeeding the normal protein ($n = 10$) and the high protein ($n = 12$) diets. Values are shown as the means ± SEM. Upper panel, IGFBP-3 levels on day 1 were lower than control values (*$p < 0.007$) and increased on both dietary regimens by d 21, at which time they did not differ from control values. However, only the increase in the high protein diet group was statistically significant (X, $p < 0.007$). Lower panel, on d 1, IGFBP-2 levels were higher than control values (*, $p < 0.007$), were lower than IGFBP-2 in the other group (○, $p < 0.007$), and did not differ from control values. (From Pucilowska JP, et al. J Clin Endocrinol Metab 1993; 77:1516–1521, with permission.)

We observed marked reductions in IGF-I and IGFBP-3 and elevated IGFBP-2 values in malnourished children *(62)*. In a refeeding experiment carried out on 22 chronically undernourished children in Bangladesh, we observe that refeeding a protein-enriched (15% protein), calorie-sufficient diet for 21 d was more effective in normalizing IGF-I and IGFBPs than refeeding a normal protein (6% protein) diet. IGF-I levels in serum increased significantly better with the protein-rich diet (after 21 d: IGF-I in normal protein = 128 ± 70 vs 298 ± 124 ng/mL in the subjects fed the protein-rich diet; $p < 0.01$). Likewise IGFBP-3 values were significantly improved and IGFBP-2 values in the patients fed the protein-enriched diet were significantly lower than those fed the regular protein diet (Fig. 5).

In summary, evidence is not sufficient to show that measurement of IGFBP-2 or IGFBP-3 on a single serum sample from a given patient will result in the precise definition of nutritional status. It is possible, however, that knowing the concentrations of these IGFBPs will complement the information gained from other observations on the status of the patient. As with IGF-I, serial measurements of IGFBP-3 and IGFBP-2 during efforts at nutritional repletion may be of value, in that they may provide an indication of the adequacy of the regimen.

THE IGF SYSTEM IN DISORDERS OF CARBOHYDRATE METABOLISM

Diabetes Mellitus

Although numerous reports indicates that serum concentrations of IGF-I and IGFBP-3 may be reduced in patients with type 1 (insulin-dependent) diabetes mellitus *(63–66),* it is not clear whether these reductions are mediated directly or represent the result of subtle, suboptimal intracellular nutritional deficit. The focus of study of IGF system components in patients with diabetes has been on IGFBP-I. Here we address the question of the possible clinical utility of IGFBP-I measurements in assessing the status and treatment of patients with diabetes mellitus. Serum concentrations of IGFBP-I are inversely related to insulin secretion and to the severity of a patient's hyperinsulinism *(67).* In hyperinsulinemic adults who have non-insulin-dependent diabetes, IGFBP-I in serum does not increase during fasting *(68).* The concept has been advanced that the reduced serum IGFBP-I concentration in hyperinsulinemic obese individuals might be a stimulus for obesity itself, as it would increase the levels of free IGF-I. More free IGF-I would, in turn, suppress GH secretion and thereby increase the balance toward insulin-like activity in the serum *(69–72).*

Reduced serum IGFBP-I, which has been shown to be a reliable indicator of hyperinsulinism in obese women *(73),* has also been shown to correlate with risk factors for cardiovascular disease in patients with non-insulin-dependent diabetes *(74).* Therefore, if one adheres to the concept that hyperinsulinism is a significant factor in the development of cardiovascular disease, measurement of IGFBP-I will be useful in estimating the likelihood that a given individual will be affected. Along these lines, Gibson et al. *(74)* have shown that among 74 patients with non-insulin-dependent diabetes, IGFBP-I values in serum were a predictor of risk for coronary artery disease ($R^2 = 0.47$).

Hypoglycemia Secondary to Hyperinsulinism

Hyperinsulinism is a common cause of sustained, intractable hypoglycemia among infants, and confirming that hyperinsulinism is indeed present is often difficult. The diagnosis of hyperinsulinism is usually reached by excluding other possible causes for the hyperglycemia, and by showing that insulin levels are inappropriate for the concurrent low blood glucose level. It is difficult in many of the affected infants to obtain unequivocally elevated, diagnostic insulin values, in part because concentrations fluctuate rapidly. Levitt-Katz et al. *(75)* have reported on the use of IGFBP-I measurements in the diagnostic evaluation of infants with hypoglycemia. They measured IGFBP-I levels in 13 infants with hyperinsulinism, 16 with ketotic hypoglycemia or no diagnosed disorder, 3 with hypopituitarism or isolated GH deficiency, 2 with glycogen storage disease, and 2 patients with disorders of fatty acid oxidation. In the children with ketotic hypoglycemia or no disorder (not hyperinsulinemic), the IGFBP-I value increased during fasting to 343.8 ± 71.3 ng/mL in the sample obtained at the time of hypoglycemia. On the other hand, the mean IGFBP-I in the patient's with hyperinsulinism was only 52.4 ± 11.5 ng/mL ($p < 0.0001$, vs ketotic group). The children with hyperinsulinism and poorly controlled blood sugar values had a mean IGFBP-I of 30.1 ± 10.4 ng/mL, while those who were able to fast 4 h had a mean value of 71.5 ± 16.9 ng/mL. The patients with other nonhyperinsulinemic causes of hypoglycemia had IGFBP-I values similar to those of the ketotic patients. The authors conclude that a reduction in IGFBP-I in hypoglycemic children is a reliable indicator of excess insulin action and that measurement of IGFBP-I will aid in the diagnosis of hypoglycemic conditions.

Hypoglycemia Associated with Malignant Tumors

Rare patients have been described who have malignant extrahepatic, non-insulin-secreting tumors that are accompanied by profound hypoglycemia *(76)*. These patients exhibit signs of excess insulin (inhibition of hepatic glucose production, increased peripheral consumption of glucose and decreased lipolysis), yet insulin levels are low. Their tumor secretes an excess of IGF-II in the form of a bioactive "big IGF-II" *(77–79)*, which is the product of an incompletely processed pro-IGF-II peptide. This molecule has insulin-like actions. It shifts the IGF-II in the circulation from its usual binding to the 150–200-kDa complex to the smaller 50-kDa complex. Because this 50-kDa complex can cross the vascular endothelium to tissues, it is believed that the bioavailability of IGF-II is increased and the target tissue is subjected to excessive insulin-like action. In any case, IGF-II measurements in patients with hypoglycemia in whom malignancy is suspected should be helpful in diagnosing and managing this condition.

OTHER CONDITIONS

Alterations in serum components of the IGF system have been reported in a variety of diseases. For the most part, however, the value of measuring these components in individual patients for the purpose of diagnosis and for management is not clear. An abbreviated list follows:

1. Serum IGF-I levels have been reported to be higher in a group of men at risk for prostatic cancer, compared to controls *(80)*. Men whose IGF-I values were in the highest quartile of the sample (total $n = 152$) had a relative risk of 4.3 (95% confidence interval 1.8–10.6), compared with men in the lowest quartile.
2. Cohen et al. *(81)* have reported that IGFBP-2 values are increased in men with prostatic adenocarcinoma. This observation may prove useful diagnostically, if used with the understanding that other cancers may cause IGFBP-2 levels to increase. The finding has been confirmed *(82,83)*.
3. Genetic forms of GH insensitivity: Laron syndrome, an inherited defect of the GH receptor, is characterized by markedly reduced IGF-I values, low IGFBP-3, and increased IGFBP-I and IGFBP-2. Treatment with IGF-I causes the IGFBP-I to decline and IGFBP-3 to increase *(84–86)*. Measurement of IGF-I has value in the diagnosis of Laron syndrome and similar disorders, when evaluated in the context of the patient presentation and the GH concentrations in serum.

REFERENCES

1. Ranke MB, Blum WF, Bierich JR. Clinical relevance of serum measurements of insulin-like growth factors and somatomedin binding proteins. Acta Paediatr Scand (Suppl) 1988; 347:114–126.
2. Rosenfeld RG, Wilson DM, Lee PDK, Hintz RL. Insulin-like growth factors I and II in evaluation of fetal growth retardation. J Pediatr 1986; 109:428–433.
3. Underwood LE and Van Wyk JJ. Normal and Aberrant Growth, in Williams Textbook of Endocrinology (Foster DW, Wilson JP, eds.), 8th Ed. Philadelphia, WB Saunders, 1991; pp.1079–1138.

4. Blum WF, Albertsson-Wikland K, Rosberg S, Ranke MB. Serum levels of insulin-like growth factor I (IGF-I) and IGF binding protein 3 reflect spontaneous growth hormone secretion. J Clin Endocrinol Metab 1993; 76:1610–1616.
5. Juul A, Dalgaard P, Blum WF, Bang P, Hall K, Michaelsen KF, Muller J, Shakkebaek NE. Serum levels of insulin-like growth factor (IGF) binding protein-3 (IGFBP-3) in healthy infants, children and adolescents: the relation to IGF-I, IGF-II, IGFBP-1, IGFBP-2, age, sex, body mass index, and pubertal maturation. J Clin Endocrinol Metab 1995; 80:2534–2542.
6. Underwood LE, Smith EP, Van Wyk JJ, Clemmons DR, D'Ercole AJ, Pandian MR, Preece MA, Moore WV. Somatomedin-C/Insulin-like growth factor I: regulation and clinical applications. In: Raiti S and Tolman RA (eds), Human Growth Hormone, Plenum Publishing, New York, pp. 609–619.
7. Hasegawa Y, Hasegawa T, Aso T, et al. Comparison between insulin-like growth factor-I (IGF-I) and IGF binding protein-3 (IGFBP-3) measurement in the diagnosis of growth hormone deficiency. Endocr J 1996; 40:185–190.
8. Nunez SB, Municchi G, Barnes KM, Rose SR. Insulin-like growth factor I (IGF-I) and IGF binding protein-3 concentrations compared to stimulated and night growth hormone in the evaluation of short children: a clinical research center study. J Clin Endocrinol Metab 1996; 81:1927–1932.
9. Rosenfeld RG, Albertsson-Wikland K, Cassorla F, Frasier SD, Hasegawa Y, Hintz RL, Lafranchi S, Lippe B, Loriaux L, Melmed S, Preece MA, Ranke MB, Reiter EO, Rogol AD, Underwood LE, Werther GA. Diagnosis of growth hormone deficiency revisited. J Clin Endocrinol Metab 1995; 80:1532–1540.
10. Juul A, Holm K, Kastrup KW, Pedersen SA, Michaelsen KF, Scheike T, Rasmussen S, Muller J, Shakkebaek NE. Free insulin-like growth factor I serum levels in 1430 healthy children and adults, and its diagnostic value in patients suspected of growth hormone deficiency. J Clin Endocrinol Metab 1997; 82:2497–2502.
11. Rosenfeld RG, Wilson DM, Lee PDK, Hintz RL. Insulin-like growth factors I and II in the evaluation of growth retardation. J Pediatr 1986; 109:428–433.
12. Smith WJ, Nam TJ, Underwood LE, Busby WH, Celnicker A, Clemmons DR. Use of insulin-like growth factor binding protein-2 (IGFBP-2), IGFBP-3 and IGF-I for assessing growth hormone status in short children. J Clin Endocrinol Metab 1993; 77:1294–1299.
13. Rasat R, Livesey JL, Espiner EA, Abbott GD, Donald RA. IGF-I and IGFBP-3 screening for disorders of growth hormone secretion. New Zealand Med J 1996; 109:156–159.
14. Baxter RC: Editorial. The binding proteins' binding protein–clinical applications of acid-labile subunit (ALS) measurement. J Clin Endocrinol Metab 1997; 82:3941–3943.
15. Khosravi MJ, Diamondi A, Mistry J, Krishna RG, Khare A. Acid-labile subunit of human insulin-like growth factor-binding protein complex: measurement, molecular and clinical evaluation. J Clin Endocrinol Metab 1997; 82:3944–3951.
16. Mohnike K, Kluba U, Blum WF, Aumann V, Vorwek P, Mittler U. Serum concentrations of insulin-like growth factors (IGF-I) and IGF-II and IGF binding proteins (IGFBP)-2 and IGFBP-3 in 49 children with ALL, NHL or solid tumors. Klin Padiatr 1995; 207:225–229.
17. Baum HBA, Biller BM, Katznelson L, Oppenheim DS, Clemmons DR, Cannistraro KB, Schoenfeld DA, Best SA, Klibanski A. Assessment of growth hormone (GH) secretion in men with adult-onset GH deficiency compared with that in normal men: a clinical research center study. J Clin Endocrinol Metab 1996; 81:84–92.
18. Ghigo E, Aimaretti G, Gianotti L, Bellone J, Arvat E, Camanni F. New approach to the diagnosis of growth hormone deficiency in adults. Eur J Endocrinol 1996; 134:352–356.
19. Thissen J-P, Ketelslegers J-M, Maiter D. Use of insulin-like growth factor I (IGF-I) and IGF binding protein-3 in the diagnosis of acromegaly and growth hormone deficiency in adults. Growth Regul 1996; 6:222–229.
20. Hoffman DM, O'Sullivan AJ, Baxter RC, Ho KKY. Diagnosis of growth hormone deficiency in adults. Lancet 1994; 343:1064–1068.
21. Bates AS, Evans AJ, Jones P, Clayton RN. Assessment of GH status in adults with GH deficiency using serum growth hormone, serum insulin-like growth factor-I and urinary growth hormone secretion. Clin Endocrinol 1995; 42:425–430.
22. Svensson J, Johannsson G, and Bengtsson BA. Insulin-like growth factor-I in growth hormone-deficient adults: relationship to population-based normal values, body composition and insulin tolerance test. Clin Endocrinol 1997; 46:579–586.
23. Skjaerbaek C, Vahl N, Frystyk J, Hansen TB, Jorgensen JO, Hagen C, Chriastiansen JS, Orskov H. Serum free insulin-like growth factor-I in growth hormone-deficient adults before and after growth hormone replacement. Eur J Endocrinol 1997; 137:132–137.

24. Adan L, Souberbielle JC, Brauner R. Diagnostic markers of permanent idiopathic growth hormone deficiency. J Clin Endocrinol Metab 1994; 78:353–358.
25. Melmed S. Acromegaly. N Engl J Med 1990; 322:966–977.
26. Clemmons DR, Van Wyk JJ, Underwood LE, Ridgway EC, Kliman B, Kjellberg RN. Evaluation of acromegaly by radioimmunoassay of somatomedin-C. N Engl J Med 1979; 301:1138–1142.
27. Barkan AL, Beitins IZ, Kelch RP. Plasma insulin-like growth factor-I/somatomedin-C in acromegaly; correlation with the degree of growth hormone hypersecretion. J Clin Endocrinol Metab 1988; 67:69–73.
28. Grinspoon S, Clemmons D, Swearingen B, Kllibanski A. Serum insulin-like growth factor-binding protein-3 levels in the diagnosis of acromegaly. J Clin Endocrinol Metab 1995; 80:927–932.
29. Wass JAR, Clemmons DR, Underwood LE. Changes in circulating somatomedin-C levels in bromocriptine treated acromegaly. Clin Endocrinol 1982; 17:369–377.
30. Barkan AL, Halasz I, Dornfeld KJ, Jaffe CA, Friberg RD, Chandler WE, Sandler HM. Pituitary irradiation is ineffective in normalizing plasma insulin-like growth factor I in patients with acromegaly. J Clin Endocrinol Metab 1997; 82:3187–3191.
31. Wass JAR. Growth hormone, insulin-like growth factor-I and its binding proteins in the follow-up of acromegaly. J Endocrin 155 (Supp 1), 1997; S17–19.
32. Paramo C, Andrade OMA, Fluiters E, Luna R, de la Fuente J, Garcia-Mayor RV. Comparative study of insulin-like growth factor-I (IGF-) and IGF-binding protein-3 (IGFBP-3) level and IGF-I/IGFBP-3 ratio measurements and their relationship with an index of clinical activity in the management of patients with acromegaly. Metabolism 1997; 46:494–498.
33. Grant DB, Hambley J, Becker D, Pimstone BL. Reduced sulphation factor in undernourished children. Arch Dis Child 1973; 48:596–600.
34. Hintz RL, Suskind R, Amatayakul K, Thanangkul O, Olson R. Plasma somatomedin and growth hormone values in children with protein-calorie malnutrition. J Pediatr 1978; 92:153–156.
35. Soliman AT, Hassan AE, Aref MK, Hintz RL, Rosenfeld RG, Rogol AD. Serum insulin-like growth factor I and II concentrations and growth hormone and insulin responses to arginine infusion in children with protein-energy malnutrition before and after nutritional rehabilitation. Pediatr Res 1986; 20:1122–1130.
36. Smith SR, Egar PJ, Pozefsky T, Chhetri MK, Prout TE. Growth hormone in adults with protein calorie malnutrition. Endocrinology 1974; 39:53–62.
37. Rappaport R, Prevot C, Czernichow P. Somatomedin activity and growth hormone secretion. I. Changes related to body weight in anorexia nervosa. Acta Paediatr Scand 1980; 69:37–41.
38. Tanaka T, Maesaka H, and Suwa S. Changes in somatomedin activity in anorexia nervosa. Endocrinol Jpn 1985; 32:891–897.
39. Counts DR, Gwirtsman H, Carlsson LM, Lesem M, Cutler GB Jr. The effect of anorexia nervosa and refeeding on growth hormone-binding protein, the insulin-like growth factors (IGFs), and the IGF-binding proteins. J Clin Endocrinol Metab 1992; 75:762–767.
40. Kirschner BS, Sutton MM. Somatomedin-C levels in growth-impaired children and adolescents with chronic inflammatory bowel disease. Gastroenterology 1986; 91:830–836.
41. Lecornu M, David L, and Francois R. Low serum somatomedin activity in celiac disease. Helv Paediat Acta 1978; 33:509–516.
42. Clemmons DR, Klibanski A, Underwood LE, McArthur JW, Ridgway EC, Beitins IZ, Van Wyk JJ. Reduction of plasma immunoreactive somatomedin-C during fasting in humans. J Clin Endocrinol Metab 1981; 53:1247–1250.
43. Isley WL, Underwood LE and Clemmons DR. Dietary components that regulate serum somatomedin-C concentrations in humans. J Clin Invest 1983; 71:175–182.
44. Isley WL, Underwood LE, Clemmons DR. Changes in plasma somatomedin-C in response to ingestion of diets with variable protein and energy content. JPEN 1984; 8:407–411.
45. Snyder DK, Clemmons DR, Underwood LE. Dietary carbohydrate content determines responsiveness to growth hormone in energy-restricted humans. J Clin Endocrinol Metab 1989; 69:745–752.
46. Clemmons DR, Seek MM, Underwood LE. Supplemental essential aminoacids augment the somatomedin-C/insulin-like growth factor-I response to refeeding after fasting. Metabolism 1985; 34:391–395.
47. Unterman TG, Vazquez RM, Sla AJ, Martyn PA, Phillips LS. Nutrition and somatomedin. XII. Usefulness of somatomedin-C in nutritional assessment. Am J Med 1985; 78:228–234.
48. Donahue SP, Phillips LS. Response of IGF-I to nutritional support in malnourished hospital patients: a possible indicator of short-term changes in nutritional status. Am J Clin Nutr 1989; 50:962–969.

49. Caufriez A, Reding P, Urbain D, Golstein J, Copinschi G. Insulin-like growth factor-I: a good indicator of a functional hepatocellular capacity in alcoholic liver cirrhosis. J Endocrinol Invest 1991;14: 317, 321.
50. Caregaro L, Alberino F, Amodio P, Merkel C, Angeli P, Plebani M, Bolognesi M, Gatta A. Nutritional and prognostic significance of insulin-like growth factor I in patients with liver cirrhosis. Nutrition 1997; 13:185–190.
51. Takano K, Hall K, Kastrup KW, Hizuka N, Shizume K, Kawai K, Akimoto M, Takuma T, Sugino N. Serum somatomedin A in chronic renal failure. J Clin Endocrinol Metab 1979; 48:371–376.
52. Goldberg AC, Trivedi B, Delmez JA, Harter HR, Daughaday WH. Uremia reduces serum insulin-like growth factor-I, increases insulin-like growth factor-II, and modified their serum protein binding. J Clin Endocrinol Metab 1982; 55:1040–1045.
53. Vehe KL, Brown RO, Moore LW, Acchiardo SR, Luther RW. The efficacy of nutrition support in infected patients with chronic renal failure. Pharmacotherapy 1991; 11:303–307.
54. Smith WJ, Underwood LE, Clemmons DR. Effects of caloric or protein restriction on insulin-like growth factor-I (IGF-I) on IGF binding protein in children and adults. J Clin Endocrinol Metab 1995; 80:443–449.
55. Clemmons DR, Thissen J-P, Maes M, Ketelslegers JM, Underwood LE. Insulin-like growth factor-I (IGF-I) infusion into hypophysectomized or protein-deprived rats induces specific IGF binding proteins in serum. Endocrinology 1989; 125:2967–2972.
56. Clemmons DR, Busby WH, Snyder DK. Variables controlling the secretion of insulin-like growth factor binding protein-2 in normal human subjects. J Clin Endocrinol Metab 1991; 73:727–733.
57. Busby WH, Snyder DK, Clemmons DR. Radioimmunoassay of a 26,000 dalton plasma insulin-like growth factor binding protein: control by nutritional variables. J Clin Endocrinol Metab 1988; 67:1225–1230.
58. Baxter RC, Cowell CT. Diurnal rhythm of growth hormone-independent binding protein for insulin-like growth factors in human plasma. J Clin Endocrinol Metab 1987; 65:432–440.
59. Snyder DK, Clemmons DR. Insulin dependent regulation of IGFBP-I. J Clin Endocrinol Metab 1991; 71:1532–1636.
60. Cohick WS, McGuire MA, Clemmons DR, Bauman DE. Regulation of IGF binding proteins in serum, and lymph of lactating cows by somatotropin. Endocrinology 1992; 130:1508–1514.
61. Smith WJ, Underwood LE, Keyes L, Clemmons DR. Use of insulin-like growth factor I (IGF-I) and IGF-binding protein measurements to monitor feeding of premature infants. J Clin Endocrinol Metab 1997; 82:3982–3988.
62. Pucilowska JB, Davenport ML, Kabir I, Clemmons DR, Thissa JP, Butler T, Underwood LE. The effect of dietary protein supplementation on IGFs and IGFBPs in children with Shigellosis. J Clin Endocrinol Metab 1993; 77:1516–1521.
63. Ismail IS, Miell JP, Scanlon MF, Peters JR. Effects of cholinergic modulation on serum insulin-like growth factor-I and its binding proteins in normal and diabetic subjects. Clin Endocrinol 1995; 42:147–152.
64. Tapanainen P, Kaar ML, Leppaluoto J, Huttunen NP, Knip M. Normal stimulated growth hormone secretion but low peripheral levels of insulin-like growth factor I in prepubertal children with insulin-dependent diabetes mellitus. Acta Paediatr 1995; 84:6546–6650.
65. Hilding A, Brismar K, Degerglad M, Thoren M, Hall K. Altered relation between circulating levels of insulin-like growth factor binding protein-1 and insulin in growth hormone deficient patients and insulin-dependent diabetic patients compared to that in healthy subjects. J Clin Endocrinol Metab 1995; 80:2646–2652.
66. Bereket A, Lang CH, Blethen SL, Wilson TA. Insulin-like growth factor binding protein-2 and insulin: studies in children with type I diabetes mellitus and maturity-onset diabetes of the young. J Clin Endocrinol Metab 1995; 80:3647–3652.
67. Conover CA, Lee PD, Kanaley JA, Clarkson JT, Jensen MD. Insulin regulation of insulin-like growth factor binding protein-1 in obese and non-obese humans. J Clin Endocrinol Metab 1992; 74:1355–1360.
68. Bang P, Brismar K, Rosenfeld RG, Hall K. Fasting affects serum insulin-like growth factors (IGFs) and IGF-binding protein differently in patients with noninsulin-dependent diabetes mellitus versus healthy and nonobese and obese subjects. J Clin Endocrinol Metab 1994; 78:960–967.
69. Jorgensen JO, Pedersen SB, Borglum J, Frystyk J, Ho KK, Christiansen JS, Orskov H, Blum WF, Richelsen B. Serum concentrations of insulin-like growth factors (IGFs), IGF binding proteins 1 and 3 and growth hormone binding protein in obese women and the effects of growth hormone administration: a double-blind, placebo-controlled study. Eur J Endocrinol 1995; 133:65–70.

70. Lee PDK, Conver CA, Powell DR. Regulation and function of insulin-like growth factor-binding protein-1. Proc Soc Exp Biol Med 1993; 204:4–29.
71. Lee PD, Giudice LC, Conover CA, Powell DR. Insulin-like growth factor binding protein-1: Recent findings and new directions. Proc Soc Exptl Biol Med 1997; 216:319–357.
72. Baxter RC. Insulin-like growth factor binding proteins as glucoregulators. Metabolism 1995; 44:12–17.
73. Mogul HR, Marshall M, Frey M, Burke HB, Wynn PS, Wilker S, Southern AL, Gambert SR. Insulin-like growth factor binding protein-1 as a market for hyperinsulinemia in obese menopausal women. J Clin Endocrinol Metab 1996; 81:4492–4495.
74. Gibson JM, Westwood M, Young RJ, White A. Reduced insulin-like growth factor binding protein-1 (IGFBP-1) levels correlate with increased cardiovascular risk in non-insulin dependent diabetes mellitus (NIDDM). J Clin Endocrinol Metab 1996; 81:860–863.
75. Levitt-Katz LE, Satin-Smith MS, Collett-Solberg P, Thornton PS, Baker L, Stanley CA, Cohen P. Insulin-like growth factor binding protein-I in the diagnosis of hypoglycemia caused by hyperinsulinism. J Pediatr 1997; 131:193–199.
76. Froesch ER and Zapf J. Tumor hypoglycemia and various forms of nonsuppressible insulin-like activity. In: Andreani D, Lefebvre PJ, and Marks V, eds., Current Views on Hypoglycemia and Glycogen, Academic Press, New York, 1980; 163–171.
77. Zapf J, Schoenle EJ, Waldvogel M, Sand I, Froesch ER. Effect of trypsin-treatment of rat adipocytes on biological effects and binding of insulin and insulin-like growth factors. Further evidence for the action of insulin-like growth factors through the insulin receptor. Eur J Biochem 1981; 113:605–609.
78. Daughaday WH, Kapadia M. Significance of abnormal serum binding of insulin-like growth factor II in the development of hypoglycemia in patients with non-islet-cell tumors. Proc Natl Acad Sci USA 1989; 86:6778–6782.
79. Daughaday WH, Trivedi B, and Baxter RC. Serum "big insulin-like growth factor II" from patients with tumor hypoglycemia lacks normal E-domain O-linked glycosylation, a possible determinant of normal propeptide processing. Proc Natl Acad Sci USA 1993; 90:5823–5827.
80. Chan JM, Stampfer MJ, Giovannucci E, Gann PH, Ma J, Wilkinson P, Hennekens CH, Pollak M. Plasma insulin-like growth factor-I and prostate cancer risk: a prospective study. Science 1998; 279:563–566.
81. Cohen P, Peehl DM, Stamey TA, Wilson KF, Clemmons DR, Rosenfeld RG. Elevated levels of insulin like growth factor binding protein-2 in prostate cancer patients. J Clin Endocrinol Metab 1993; 76:1031–1035.
82. Kanety H, Madjar Y, Dagan Y, Levi J, Papa MZ, Pariente C, Goldwasser B, Karasik A. Serum insulin-like growth factor binding protein-2 (IGFBP-2) is increased and IGFBP-3 is decreased in patients with prostatic cancer: correlation with serum prostate specific antigen. J Clin Endocrinol Metab 1993; 77:229–233.
83. Ho PJ and Baxter RC. Insulin-like growth factor binding Protein-2 in patients with prostate carcinoma and benign prostatic hyperplasia. Clin Endocrinol 1997; 46:333–342.
84. Cotterill AM, Holly JM, Taylor AM, Davies SC, Coulson VJ, Preece MA, Wass JA, Savage MO. The insulin-like growth factor (IGF)-binding proteins and IGF bioactivity in Laron-type dwarfism. J Clin Endocrinol Metab 1992; 74:56–63.
85. Kanety H, Karasik A, Klinger B, Silbergeld A, Laron Z. Long term treatment of Laron type dwarfs with insulin-like growth factor-I increases serum insulin-like growth factor binding protein-3 in the absence of growth hormone activity. Acta Endocrinol (Copenh) 1993; 128:144–149.
86. Savage MO, Blum WF, Ranke MB, Postel-Vinay MC, Cotterill AM, Hall K, Chatelain PG, Preece MA, Rosenfeld RG. Clinical features and endocrine status in patients with growth hormone insensitivity (Laron syndrome). J Clin Endocrinol Metab 1993; 77:1465–1471.

26 IGF Binding Protein Proteolysis in Various Clinical States

Laura A. Maile, PhD, *Anna L. Crown,* MD,
and Jeff M.P. Holly, PhD

Contents

INTRODUCTION

The insulin-like growth factors (IGFs) are generally present throughout the body at concentrations far higher than are required for maximal stimulation of their cell receptors. These high IGF concentrations are maintained by the six IGF binding proteins (IGFBPs), which are ubiquitously present in various combinations, generally at concentrations in excess of the IGFs. The affinities with which the IGFBPs bind the IGFs are much higher than those for the type 1 IGF receptor, which mediates most of their cellular actions. Factors that could lower the affinities of the IGFBPs may therefore play a critical role in regulating IGF availability. A general mechanism whereby this can be achieved is the limited proteolysis of the IGFBPs. If each IGFBP has its own specific, tightly regulated protease, then this could represent a very sophisticated mechanism to control IGF bioavailability.

IGFBP-3 is the most abundant IGFBP in the circulation, where it forms an 150-kDa ternary complex with IGFs and an acid-labile subunit (ALS). This ternary complex is restricted to the circulation by the capillary endothelium. There are no intracellular stores of IGF, and the ternary complex appears to serve as a circulating reservoir *(1)*. The precise mechanism by which the affinity of the ternary complex for IGFs is reduced, to mobilize this store, is not clearly understood. However, early on it was shown that serum proteolytic enzymes could reduce the affinity with which IGFs were bound in these complexes *(2,3)*. The first description that proteolysis may play an important physiological

From: *Contemporary Endocrinology: The IGF System*
Edited by: R. Rosenfeld and C. Roberts Jr © Humana Press, Inc., Totowa, NJ

role in determining IGF bioavailability came from two articles, published simultaneously, describing IGFBP-3 proteolysis in the circulation during pregnancy *(4,5)*. Following these two initial reports, it has become clear that proteolysis of IGFBPs is a general mechanism for controlling their biological actions, and that this regulation is perturbed in many clinical states. Some of these changes appear to be adaptations, but some may contribute to the pathology.

CIRCULATING PROTEASES

Physiological

Pregnancy

During pregnancy, serum IGF-II levels remain constant, whereas those of IGF-I, ALS, and IGFBP-3 all increase as measured by radioimmunoassay *(6–8)*. By Western ligand blot (WLB), however, no IGFBP-3 is detectable in the serum of pregnant women. This discrepancy suggested that a modification of IGFBP-3 occurred that altered its ability to be detected by IGF binding. The enzymatic nature of this modification was suggested by mixing experiments of serum from pregnant and nonpregnant subjects, which resulted in loss of intact IGFBP-3 from the latter as detected by WLB *(4,5)*. This IGFBP-3 proteolytic activity is detected in the circulation from 6 wk gestation, is heat labile and pH dependent. The activity can be inhibited by EDTA, aprotinin, and 4-(2-aminoethyl) benzenesulfonyl fluoride (AEBSF), indicating that it is a cation-dependent serine protease. Fragments of IGFBP-3 can be detected in the sera from pregnant women *(9)*, and protease activity can be demonstrated in an in vitro assay. Mixing serum from pregnant women with radiolabeled IGFBP-3 results in loss of intact IGFBP-3 and the appearance of smaller molecular weight fragments *(10)*.

Initial observations suggested that the modification of serum IGFBP-3 during pregnancy reduced its affinity for the IGFs. Pregnancy serum IGFBP-3, stripped of endogenous IGFs, had a 80–90% reduction in affinity for radiolabeled IGFs compared with control serum *(4)*. Dissociation of IGF-II from IGFBP-3 in pregnancy serum was reported to be six times faster, and the dissociation of IGF-I to be 10 times faster than their dissociation from IGFBP-3 in normal serum *(11)*. However, other studies suggested that IGFBP-3 from pregnancy serum was functionally normal. The maintenance of high levels of IGFs in pregnancy supports the idea that IGFBP-3 is still functional as a carrier, as the ligand otherwise would be rapidly cleared from the circulation. Comparison of the IGF distribution between the ternary and binary complexes suggested that there was no major change in pregnancy *(12)*. Suikkari and Baxter *(13)* observed that radiolabeled ALS was able to form a ternary complex with IGFBP-3 from pregnancy serum and native IGF, but was unable to form such a complex with IGFBP-3 and radiolabeled IGF. This suggested that the proteolysis of IGFBP-3 in pregnancy resulted in a loss of affinity of the binding protein for radiolabeled IGF, but did not affect its ability to bind native IGF and form the 150-kDa ternary complex, albeit with reduced affinity.

Although IGFBP-3 is undoubtedly modified, how this affects IGF distribution and availability in vivo is not yet entirely clear. The potential implications of IGFBP-3 protease activity, for IGF bioavailability were, however, demonstrated by the increased biological effect of serum IGF in a cell line bioassay when pregnancy serum was included in the assay *(14)*. Similar results were reported in a different in vitro bioassay by Blat et al. *(15)*. However, these artificial in vitro assays do not necessarily reflect the in vivo situation in pregnancy. The only in vivo data derive from animal studies using radiolabeled

IGF-I, which also may not be a true reflection in view of the differences demonstrated between this and native IGF-I. Radiolabeled IGF-I injected into pregnant rats was cleared from the serum five times faster than that injected into nonpregnant rats *(16)*. Urinary clearance was not significantly increased and radiolabeled IGF-I did not cross the placenta. This provided evidence to support the idea that proteolysis of IGFBP-3 and the subsequent reduction in affinity may increase the availability of IGF-I. This increase in bioavailability may assist the mother in meeting the increased demands of pregnancy, and, possibly, by interacting directly with receptors on the placenta, increase transport of glucose and amino acids across the placenta, hence affecting fetal growth.

Proteolysis of IGFBP-3 has been demonstrated in maternal serum and in amniotic fluid; however, no increase in IGFBP-3 proteolysis can be detected in human fetal serum from 19–35 wk gestation *(17)*. This suggests that the pregnancy serum protease is restricted to the maternal compartment and is therefore produced by nonfetal tissues.

In addition to the much studied IGFBP-3 protease in pregnancy serum, there are also proteases active toward IGFBP-2 and IGFBP-4, although IGFBP-1 appears resistant *(12)*. These proteases differ in their sensitivity to inhibitors and are most probably different enzymes.

Exercise

Under normal physiological conditions, exercise is the other most obvious situation when, to meet increased metabolic demands, an induction of protease activity to increase IGF availability might be anticipated. In a number of studies of both acute and chronic exercise, we have failed to find convincing evidence for major changes in circulating IGFBP-3 protease (Cwyfan-Hughes and Holly, *unpublished observations*). There has, however, been one report of such an exercise-induced change. In a study of patients taking part in 10-min periods of high- and low-intensity exercise, a small increase in serum IGFBP-3 proteolytic activity was detected at the end of the exercise period. This increased level of activity was maintained during the first 10 min of recovery but had returned to normal by 20 min. However, WLB of serum samples taken at the same time points revealed no variation in IGFBP-3 levels, and the significance of the transient appearance of activity is not clear *(18)*.

Pathophysiology

Altered Growth Hormone Status

An increase in IGFBP-3 protease activity has been seen in the serum of patients with growth hormone (GH) receptor insensitivity *(19,20)*. Proteolysis of circulating IGFBP-3 has been reported to be inversely related to GH status *(21)*. This suggested a possible feedback mechanism whereby IGFBP-3 proteolysis is altered according to IGF-I availability. However, a recent study *(22)* reported that GH therapy administered to GH-deficient patients and somatostatin administration to acromegalic patients had no effect on IGFBP-3 fragmentation in vivo or any effect on proteolytic activity when measured in an in vitro protease assay. This suggests that GH is probably not a major direct regulator of the protease.

Catabolic States

Specific IGFBP-3 protease activity has been described in the sera of patients with various acute catabolic conditions such as sepsis, trauma, and postoperatively *(23–25)*. In patients presenting with acute traumas such as head injuries, increased specific

IGFBP-3 protease activity and decreased IGF-I concentrations were evident immediately on admission to intensive care units. In patients recovering from such traumas, the protease activity decreased over a number of days and the serum IGF-I concentration returned to normal. In patients who did not recover, there was no such reduction in protease or increase in IGF-I *(25)*.

Although the pattern of IGFBP-3 fragmentation and the inhibitor profile of the proteolytic activity in the serum from these catabolic states were the same as that of the pregnancy-associated protease, there were a number of significant differences. The most noticeable distinction was the specificity of the protease for IGFBP-3 in the catabolic conditions (except for bypass surgery, in which at one time point there was evidence of a protease specific for IGFBP-4 that was not seen after other elective surgical procedures), this is distinct from pregnancy where IGFBP-2/3/4 proteases were active. In addition, whereas the presence of serum protease in pregnancy was accompanied by a rise in IGF-I levels, the appearance of IGFBP-3 protease following trauma or surgery was accompanied by a rapid decrease in IGF-I concentration. Coincident with the induction in protease activity following surgery, there was a shift in the relative amount of IGF-I in the ternary complex with a reduction of 48–63%; this is in contrast to pregnancy, in which there was no alteration in IGF distribution. However, this shift was not due to protease activity *per se,* as the addition of IGFBP-3 and protease inhibitors did not result in ternary complex formation *(23)*. It would seem, therefore, that there is concomitant destabilization of the ternary complex with the induction of a protease which may be important for altering the availability and delivery of IGFs to counter the catabolic state.

Metabolic Regulation of Circulating Proteases

In patients with severe illness, Davies et al. *(23)* showed that the circulatory IGFBP-3 protease was regulated according to nutritional status; activity was more apparent with fasting and decreased following commencement of parenteral nutrition. Healthy age/sex-matched volunteers on the same feeding program did not have any demonstrable activity, indicating that simple fasting alone does not induce protease activity. The influence of nutritional status was supported by the observation that, following refeeding in severely malnourished children, there was a decrease in IGFBP-3 protease activity *(26)*. A recent study *(27)* reported that IGFBP-3 proteolytic activity was associated with malnourishment and diarrhea; however, in contrast to previous studies, in children with the greatest weight gain after 1 wk of nutritional support there was a significant rise in proteolytic activity.

In adult patients with non-insulin-dependent (type 2) diabetes (NIDDM) a marked increase in serum IGFBP-3 proteolysis has been reported *(28,29)*, although in those with insulin-dependent (type 1) diabetes (IDDM), the presence of such proteolysis appeared to be more variable *(30)*. In a more recent study of subjects with NIDDM, insulin administration induced cation-dependent serine-like IGFBP-3 protease activity; however, insulin resistance *per se* did not appear to result in induction of such activity *(31)*.

In newly diagnosed children with IDDM, initiation of insulin therapy was associated with a decrease in IGFBP-3 proteolysis and a direct causal relationship was inferred *(32)*. However, these children were ketotic and probably to some extent catabolic and, as both of these conditions would also have been normalized with insulin therapy, a direct effect of insulin was not proven. Alterations in the protease were not apparent until a week after commencing insulin therapy.

The exact nature of the relationship between insulin and regulation of the IGFBP-3 protease is unclear; however, a link is apparent. In insulin-resistant states, increased

IGFBP-3 proteolytic activity and subsequent increase in IGF availability may represent an attempt to compensate for reduced insulin activity.

BURNS

In a study of patients with severe burns, under basal conditions with adequate nutrition, there was a marked reduction in circulating intact IGFBP-3 and an increase in its major fragment, which suggested an increase in IGFBP-3 proteolytic activity. These alterations were not affected by administration of insulin. There was also a decrease in circulatory IGF-I concentration and net uptake of IGF-I by leg muscle from the circulation. Despite these changes, impaired muscle metabolism was observed with a negative balance between protein synthesis and degradation despite sufficient nutrient uptake. The proteolysis of IGFBP-3 in these circumstances may reflect a mechanism to release IGF in an attempt to maintain muscle homeostasis *(33)*.

CANCER

Proteolytic modification of IGFBP-3 has been described in the serum from a variety of cancer patients. We have shown, in the majority of patients with cancer of the head/neck, breast, or lung that there was demonstrable serum IGFBP-3 proteolytic activity that was absent from healthy controls. As with the activity observed in acute catabolic states and pregnancy, this activity appeared to be due to a metal ion-dependent serine protease, which yielded a similar pattern of fragmentation and demonstrated similar specificity for IGFBP-3. In contrast to the circulating proteases, we have shown that the activity in media conditioned by a variety of transformed cells was inhibited by normal serum and would therefore seem not to be active in the presence of endogenous circulating inhibitors. The IGFBP-3 fragmentation pattern was variable but activity was generally inhibited by EDTA. From these results, we concluded that the activity seen in the circulation was different from that produced by these cancer cell lines although they were both cation-dependent proteases *(34)*.

Many patients suffering from advanced breast cancer had elevated IGFBP-3 proteolytic activity in the circulation. Endocrine therapy with the synthetic progestin, megestrol acetate (MA), led to a significant decrease in IGFBP-3 protease activity *(35)*. However this effect may have been secondary to the effects of MA on metabolism; MA is well known to have stimulatory effects on appetite and increases insulin levels. In a subsequent study we showed that treatment with the anti-estrogen, tamoxifen, had no significant effect on the ratio of fragmented to total IGFBP-3 when the whole group of patients was taken together. In patients responding to treatment with a reduction in tumor burden, however, there was a significant reduction in IGFBP-3 fragmentation. In contrast, there was a trend toward an increase in fragmentation in patients with progressive disease. The reduction in proteolytic activity therefore appeared to be related to tumor burden rather than to the tamoxifen *(36)*.

EXTRAVASCULAR PROTEASES

Physiological: Contrast to the Circulation

LYMPH

That IGFBP-3 proteolysis might not be restricted to the circulation was initially indicated by a study of lymph. Western immunoblotting (WIB) of lymph revealed IGFBP-3 fragments of 30, 21.5, 20, and 16.5 kDa, comparable to those in the serum from pregnant women. There was eightfold greater proteolytic activity in lymph than in serum in an in vitro assay. This activity was again described as a metal ion-dependent serine protease *(37)*. Similarly, we reported that in normal skin interstitial fluid there was a significant level of IGFBP-3 protease activity that was similar to the pregnancy-associated protease by virtue of identical fragmentation of recombinant IGFBP-3 and identical inhibitor profile *(38)*.

Peritoneal Fluid

A WIB of peritoneal fluid from normal women at various stages of the menstrual cycle revealed a decrease in the intensity of the intact IGFBP-3 in samples taken during the late proliferative/early secretory phase when compared with samples taken during either the late secretory or proliferative phases. Although this change did not appear to be accompanied by any alteration in the amount of 29-kDa fragment, there was a marked increase in the ratio of fragmented to intact. This alteration in IGFBP-3 appears to be independent of levels of protease activity, as there was no apparent cycle dependent changes in proteolytic activity when measured in an in vitro assay *(39)*. Alterations in levels of IGFBP-3 in peritoneal fluid may reflect changes in capillary permeability, and therefore influx of proteins from the circulation, at different stages of the cycle.

Synovial Fluid

A WIB of synovial fluid revealed that in the majority of normal subjects almost all the IGFBP-3 present was proteolytically modified. The presence of proteolytic activity was confirmed by an in vitro protease assay *(40)*.

Ovarian Follicular Fluid

In a recent study, we compared IGFBP proteolytic activity in fluid from healthy dominant follicles with fluid from small atretic follicles. We found that despite the presence of a 29-kDa fragment of IGFBP-3 on WIB, the intensity of which was greater in healthy dominant follicular fluid, no IGFBP-3 protease could be detected in the fluid from either follicle type. Fragments of IGFBP-2 were also detectable in follicular fluid, with a 23-kDa fragment being the main form present in dominant follicles. This 23-kDa fragment has not been observed in other fluids, suggesting there may be a novel IGFBP-2 protease in the ovary.

Whilst most of the IGFBP-4 in small atretic follicles was in the intact form, in follicular fluid from dominant healthy follicles almost all of the IGFBP-4 was fragmented. The presence of a metallo/serine IGFBP-4 protease in follicular fluid from human follicles was demonstrated, with a fragmentation pattern identical to that for IGFBP-4 in serum from human third trimester pregnancy. The activity in the follicular fluid of healthy dominant follicles could be greatly reduced by the addition of recombinant IGFBP-1, -2, and -3, with the latter being most effective. Inhibition could be reversed by the presaturation of the recombinant IGFBPs with equimolar IGF-I. From these data we proposed a possible model whereby in atretic follicles where IGF concentrations are low and IGFBP-3 may be unsaturated. The IGFBP-4 protease would be inactive; intact IGFBP-4 would therefore restrict the availability of IGF. In dominant healthy follicles where IGF levels are increased the majority of the IGFBP-3 would be saturated and the IGFBP-4 protease active. Fragmented IGFBP-4 with a reduced affinity for IGF would make more IGF available to push the follicle into final maturation *(41)*.

Urine

In normal urine only very low concentrations of IGFBP-3 were found and a faint doublet of IGFBP-3 was seen by WLB, suggesting there was little detectable protease activity *(42)*.

Seminal Plasma

It has been reported that intact IGFBP-3 was absent from seminal plasma, although low molecular weight fragments were found. Proteolytic activity against IGFBP-3, IGFBP-4, and IGFBP-5 has been demonstrated, although intact IGFBP-4 has been detected in seminal plasma. Prostate-specific antigen (PSA) appeared to be responsible at least partly for the IGFBP-3 and IGFBP-4 proteolytic activity, whereas IGFBP-5 proteolytic activity appeared distinct. No significant differences in binding protein levels or protease activity were found between seminal plasma from normal and azoospermic patients *(43,44).*

Amniotic Fluid

Amniotic fluid has been shown to contain a cation-dependent serine protease very similar to the pregnancy-associated protease, which also degraded IGFBP-3, -4, and -5 *(45).* Amniotic fluid has more recently been shown to also proteolyze IGFBP-1. Highly phosphorylated forms were more resistant to proteolysis. Zymography indicated that the activity consisted of 100-kDa, 60–65-kDa, and 45–50-kDa species and exhibited an inhibitor profile consistent with a metal ion-dependent protease *(46).*

Pathophysiological

Psoriasis

Psoriasis presents as local skin lesions, enabling interstitial fluid from psoriatic lesions to be compared with that from normal skin and with serum, all from within the same subjects. Whereas interstitial fluid from normal skin had high IGFBP-3 protease activity and the majority of the IGFBP-3 was in fragmented forms, corresponding fluid from skin affected by psoriasis had markedly reduced protease activity and much more of the IGFBP-3 was intact. The reduced IGFBP-3 protease activity in the psoriatic lesions appeared to be due to the presence of a local, specific inhibitor *(47).* In normal skin, keratinocyte proliferation is supported by IGFs produced by underlying dermal fibroblasts. Psoriatic lesions are characterized by well-documented overproliferation of keratinocytes. A reduction in proteolysis of IGFBP-3 might be expected to restrict IGF action and the significance of this in relation to the increased keratinocyte proliferation in psoriasis is unclear.

Arthritis

In contrast to that from normal joints, synovial fluid from joints affected by osteoarthritis and rheumatoid arthritis contained more intact IGFBP-3. This was accompanied by a decrease in proteolytic activity measured in an in vitro assay. This decrease in IGFBP-3 proteolysis contrasts with the significant increase in overall extracellular matrix proteolysis that has been widely reported in arthritic joints *(40).* This implies that the regulation of the proteolysis of IGFBP-3 must be very specific. In arthritic joints, if the normal proteolysis of IGFBP-3 is blocked, a consequent restriction in IGF bioavailability may remove a major anabolic signal contributing to the shift toward catabolic joint destruction.

Chronic Renal Failure

Urinary IGFBP-3 (uIGFBP-3) in children with chronic renal failure (CRF) was reported to be markedly decreased or totally absent as assessed by WLB. Significant IGFBP-3 protease activity was found only in the urine and not serum of these children, suggesting that this protease is produced by the kidney and secreted into the urine or that the protease was concentrated by filtration and activated in the kidney *(48)*. A comparison of acute and chronic renal failure in both adults and children revealed three different patterns of fragmentation of uIGFBP-3, with major fragments of either 18 kDa, 14 kDa, or 23 kDa. The first of these was most common and comparable to the pattern of fragmentation seen in normal urine. The protease activity found in normal and renal failure urine was mediated primarily by serine proteases, being completely inhibited by serine protease inhibitors and plasmin inhibitors. Normal urine contains both urinary plasminogen activator (uPA) and, to a lesser extent, tissue plasminogen activator (tPA) and the proteolytic pattern in normal urine and most CRF urine was similar to the pattern generated by plasmin. CRF patients show a markedly diminished platelet pool of PA inhibitors and a low activity of platelet plasminogen activator inhibitor-1 (PAI-1). Proteolysis of uIGFBP-3 may result in increased IGF availability to the renal tubule and may act as a renal growth factor in settings such as compensatory hypertrophy and post-ischemic regeneration of the proximal tubule, and could play a role in the pathogenesis and progression of renal failure *(49)*.

Diabetic Retinopathy

The IGFBP-3 detectable in normal bovine serum by WLB disappeared if this was mixed with bovine vitreous or aqueous humors. This activity was specific to IGFBP-3 and was roughly equal in both humors. The activity appeared to be due to a temperature- and pH-dependent protease that was inhibited by EDTA. In diabetic retinopathy there is dysregulation of blood vessels onto the surface of the retina and into the vitreous; they lack a normal continuous endothelial lining which can lead to an accumulation of serum proteins in the vitreous humor. Increased IGF-1 along with IGFBPs have been found in vitreous humor of diabetic patients, and the proteolysis of IGFBP-3 could release IGF-1 and promote proliferative retinopathy *(50)*.

Cerebrospinal Fluid

IGFBP-3 is only a minor component in cerebrospinal fluid (CSF). However, IGFBP-3 protease activity is elevated in the CSF of children with central nervous system tumors *(51)*.

Asthma

The asthma-associated proinflammatory eicosanoid leukotriene D_4 (LTD_4) has been shown to be comitogenic with IGFs in airway smooth muscle cells. The effect of LTD_4 appeared to be due to the stimulation of MMP-1 production which proteolyzed the IGFBPs and thereby presumably increased IGF availability. Increased levels of MMP-1 were also seen on staining of sections of human airways from asthmatic patients compared to controls. This suggested a role for LTD_4-induced proteolysis of IGFBPs in the proliferation of smooth muscle cells in the airways of asthmatic patients *(52)*.

CHARACTERIZED ENZYMES AS IGFBP PROTEASES

Prostate-Specific Antigen

Produced almost exclusively by prostate epithelial cells, PSA is a 34-kDa androgen-dependent kallikrein-like enzyme. Under normal physiological conditions it would seem that high levels of zinc in the prostate inhibit PSA activity. Elevated levels of PSA in the serum is a widely used marker for prostate carcinoma, although serum PSA is not active as a protease. PSA was the first enzyme to be identified as an in vivo IGFBP-3 protease

(43). Proteolytic activity against IGFBP-3 in seminal plasma generated a pattern of fragmentation identical to that of PSA, and as endogenous levels of PSA increased, so did the IGFBP-3 proteolytic activity. In comparison to the pregnancy-associated protease, PSA generated different-sized IGFBP-3 fragments, indicating different sites of cleavage. Whereas the pregnancy-associated protease is inhibited by EDTA, the metal ion chelator activates PSA, by removing its inhibitor zinc. This activity is also inhibited by phenylmethylsulfonyl fluoride and partially by aprotinin. In in vitro protease assays, PSA has also been shown to cleave IGFBP-4 and IGFBP-5.

As in many other cell culture models, IGFBP-3 reduces the IGF-stimulated growth of prostate epithelial cells. The addition of PSA to the cultures fully restores the stimulatory effect of the IGFs, implying that PSA could functionally alter IGF availability *(53)*. It has also been reported that IGFBP-3 induces apoptosis in the PC-3 prostate cell line, acting via both IGF-dependent and IGF-independent mechanisms, the latter thought to involve binding to putative receptors *(54)*. Proteolysis of IGFBP-3 by PSA prevents both its growth-inhibitory and apoptotic effects.

It is conceivable that during puberty increased production of the androgen-dependent enzyme may contribute to the stimulation of normal prostatic growth. Local foci of metastatic prostate carcinoma cells may also release PSA, thereby increasing available IGFs and facilitating local tumor cell growth.

Nerve Growth Factor

The γ-subunits of nerve growth factor (NGF), a kallikrein-like enzyme, have 64% homology with PSA, and have been shown to have stronger proteolytic activity against IGFBP-3, generating fragments with reduced affinity for IGF *(55)*. In addition to strong proteolytic activity against IGFBP-3, NGF is also a potent IGFBP-4 protease. It has been shown that NGF acts synergistically with IGF to promote neurite growth, and it is possible that some of that effect may be due to its binding protein protease activity. This may have important implications for the regulation of development and repair in brain, prostate, and other tissues.

Plasmin

Plasminogen is readily converted into plasmin in the presence of urinary plasminogen activator (uPA) or tissue plasminogen activator (tPA), with this conversion controlled by plasminogen activator inhibitors (PAIs). Both plasmin and plasminogen (in the presence of uPA or tPA) have been demonstrated to result in dissociation of IGF from IGFBP-3 in osteosarcoma cultures, and this dissociation could be inhibited by PAI-1 and antibodies to uPA *(56)*.

Both IGFBP-3 and IGFBP-4 were proteolyzed when plasminogen was added to cultures of osteosarcoma cells *(57)*. The proteolysis was inhibited by IGFs owing to a direct effect on the regulation of the PA/PAI-1 system. An increase in synthesis of IGFBP-3, IGFBP-4, and IGFBP-2 levels was also stimulated by IGF-I. It seems likely that the effects of IGF on IGFBP production and plasmin activity interact to regulate cell growth. In the presence of high levels of IGF-I, IGFBP-3 production would be favored and proteolytic activity decreased, maintaining levels of intact IGFBP-3 that would sequester IGF and limit its access to cell-surface receptors. When IGF-I levels drop, IGFBP-3 secretion falls and proteolysis increases; therefore more IGF-I is released to the receptor. Osteoblasts produce both PAs and PAIs and, therefore, these interactions may be important for regulating the normal and neoplastic growth of osteoblasts.

Addition of pefabloc (a serine protease inhibitor) to PC-3 adenocarcinoma cells resulted in a dose-dependent inhibition of growth. This was associated with a decrease in fragmentation of IGFBP-3. Conversely, addition of plasminogen resulted in a dose-dependent increase in growth and a reduction in intact IGFBP-3. These cells exhibited a biphasic response to increasing concentrations of IGFBP-3; at low levels proliferation was stimulated, whereas at higher concentrations proliferation was inhibited. However, coincubation of plasminogen with the high concentrations of IGFBP-3 induced the same stimulation as the low concentrations. This implies that at low concentrations of IGFBP-3, most of the IGFBP-3 was proteolyzed to fragments with a reduced affinity for IGF-II and cell growth was stimulated. At higher concentrations, more IGFBP-3 remained intact, binding IGF-II with high affinity and preventing receptor–ligand interaction *(58,59)*.

In skin keratinocyte cultures, IGFBP-3 had both inhibitory and stimulatory effects on IGF actions. Both effects were neutralized by plasminogen. Plasminogen added to cultures of skin keratinocytes resulted in dose-dependent proteolysis of IGFBP-3, presumably following the conversion to plasmin, although the fragmentation pattern and response to inhibitors indicated that proteolysis was not mediated by plasmin directly. IGF added to the cell culture or to a cell-free assay could inhibit the effect of plasminogen. Long R3 IGF, which can bind to the IGF receptor but not to IGFBPs, had no effect, implying that it was interaction with binding protein, not receptor, that was the key to the inhibitory effect in this system. Thus, the protease modulates IGF availability from IGFBP-3 and in turn IGF modulates the activity of this protease *(60)*.

Despite its abundance in serum, plasmin is an unlikely candidate for the circulating and extravascular IGFBP-3 protease, as it is very tightly regulated and active only in the circulation at sites of tissue injury. Although plasminogen activators increase in the circulation in pregnancy, there is a concomitant rise in plasminogen activator inhibitors and therefore no overall increase in activity. A recent study, using monoclonal antibodies raised against specific fragments of IGFBP-3, has shown that the site of cleavage of IGFBP-3 by plasmin is different from the site of cleavage by the pregnancy-associated protease *(61)*.

Acid Cathepsins

Acid cathepsins are secreted from cells as latent precursors, which are autoactivated in extremely acidic conditions that are usually achieved only in lysosomes and endosomes. Acid-activated proteolytic activity against IGFBP-3 has been demonstrated in a number of cell lines, including normal and transformed fibroblasts, osteoblasts, and breast cancer cells. Specific inhibitors, fragmentation patterns, and pH optima were used to identify cathepsin D as the protease responsible *(62)*. All the IGFBPs are sensitive to proteolysis by purified cathepsin D except IGFBP-6 and, in addition, IGF-I and -II are also proteolyzed by this enzyme *(63)*. The activity of cathepsin D as an IGFBP protease may reflect a mechanism of intracellular processing. However, there are extracellular circumstances such as within tumors and the ruffled border of bone-resorbing osteoblasts in which a sufficiently low pH may be generated to allow activation of secreted cathepsin D. Even in these circumstances, a role for cathepsin D in modulating IGF availability is doubtful, as it is active only at a pH at which the IGF would already have dissociated from the IGFBP. The increasing amount of evidence supporting an IGF-independent role for IGFBP-3 and its fragments generated by proteolysis raises other possibilities for acid cathepsins as physiologically relevant IGFBP proteases. In acidified environments, normal extravascular proteases may no longer be active and the activation of acid cathepsins may result in the production of IGFBP-3 fragments with distinct functions; alternatively, these proteases may destroy active IGFBP fragments.

Metalloproteinases

In addition to the observation of IGFBP-3 protease activity in human pregnancy serum, similar activity has also been described in rats and mice *(16,64)*. The activity in serum from pregnant rats has been reported to be inhibited by tissue inhibitor of metalloproteinase-1 (TIMP-1) *(65)*. There are increases in serum levels of metalloproteinase-2 (MMP-2) and MMP-9 during human pregnancy, whereas that of TIMP-1 is reduced. In addition, MMPs have been shown to be produced in placental extracts, and it has been suggested that the placenta is a source of these proteases in serum. Supporting this idea is the fact that animals with only superficial placentas, such as baboons, display little or no pregnancy-associated protease. In contrast, placentas with an invasive trophoblast are present in animals with pregnancy-associated IGFBP-3 protease *(66)*. It is unlikely in humans that the MMPs are solely responsible for the pregnancy-associated protease activity, as inhibitor profiles and more detailed characterization reveal a complex combination/cascade of enzymes.

Metal-dependent proteases have also been described in human adult and fetal fibroblasts *(67)*. The activity of MMPs have been implicated in tumor cell invasion, morphogenesis, trophoblast invasion, cartilage and bone repair and turnover, wound healing, and angiogenesis. The IGFs may play an important role in all of these processes. Matrix-associated IGFBPs bind IGFs that could be subsequently released by secreted MMPs during periods of cell proliferation or differentiation.

Generating Bioactive Fragments of IGFBP-3

That proteolysis may not just result in loss of biological function of IGFBPs but that it could also result in generation of fragments with alternative biological functions has been implied by a number of studies. Incubation of IGFBP-3 with plasmin resulted in the production of two fragments, one of 25/22 kDa that bound IGF weakly on a WLB, and a 16-kDa fragment that was reported to not bind IGF. Stimulation of DNA synthesis in chick embryo fibroblasts by IGF was inhibited 70% by intact IGFBP-3 and 30% by the 22-kDa fragment. The 16-kDa fragment reduced IGF-I stimulation by 70% and also inhibited insulin stimulation by 70%, implying an independent inhibitory effect regardless of its inability to bind IGF *(68)*. Recent studies using IGF receptor-deficient cells showed that this 16-kDa fragment was also a potent inhibitor of FGF-induced mitogenesis *(69)*.

OTHER SPECIFIC IGFBP PROTEASES

IGFBP-4

From the initial observations that the treatment of cell lines with IGF-I and -II leads to a marked decrease in IGFBP-4 levels with little or no effect on mRNA levels, IGF-dependent IGFBP-4 proteolytic activity has been demonstrated in conditioned media from a wide variety of cells *(70)*. These include normal fibroblasts *(71,72)*, human osteoblast-like cells *(73)*, rat neuronal cells *(74)*, and porcine aortic smooth muscle cells *(75)*. The production of distinct fragments of IGFBP-4, as demonstrated by WIB, could be inhibited by EDTA and 1,10-phenanthroline, suggesting a metal-dependent protease *(72)*.

The IGF dependency of the protease did not require receptor binding by the ligand, as antibodies to the receptor did not alter the effect of IGF on IGFBP-4 levels *(76)*. A number of studies demonstrated that IGF-II was more effective than IGF-I in stimulating the protease *(71)*. The effect of IGF may be to directly activate a latent form of protease, or, by binding to IGFBP-4, alter its confirmation so that it becomes susceptible to degradation.

In a fibroblast cell model, we have demonstrated that unsaturated IGFBP-3 could inhibit IGFBP-4 proteolysis. However, the saturation of IGFBP-3 with IGF reversed its ability to inhibit the protease *(77)*. Heparin-binding highly basic regions within the thyroglobulin type 1 repeat of IGFBP-3, -5, and -6 has been shown to inhibit IGFBP-4 degradation by MC3T3-E1 murine osteoblast conditioned media. This inhibitory effect was also reversed by IGF *(78)*.

IGFBP-5 Proteolysis

Proteases degrading IGFBP-5 have been detected in the media from a number of different cell lines, including fibroblasts *(70,79)* and osteosarcoma cells *(80)*. In contrast to the IGFBP-4 protease, IGF-I has been shown to have an inhibitory effect on IGFBP-5 proteolysis *(70)*. Binding of IGFBP-5 to the extracellular matrix (ECM), mediated by heparan sulfate, also protected it from proteolysis *(81)*. In addition, ECM-bound IGFBP-5 has been shown to have a lower affinity for IGF-I, thereby localizing IGF-I and increasing its availability to cell-surface receptors *(82)*. More recently IGFBP-5 has also been shown to bind another ECM component, PAI-I, which also partially protects it from proteolysis *(83)*. By increasing PAI-I synthesis and decreasing IGFBP-5 proteolysis, IGF-I may increase its own availability at specific sites of localization.

Partial purification of IGFBP-5 proteases from fibroblast conditioned medium suggests that the activity is due to a mixture of MMPs and serine proteases *(84)*.

SUMMARY

Despite the numerous descriptions of circulating IGFBP-3 proteolytic activity, the identities of the enzymes involved, the mechanisms of regulation, and their physiological significance remain unclear. Undoubtedly, proteolytic modification of the IGFBPs reduces their affinity for the IGFs and therefore potentially increases the bioavailability of these potent anabolic agents. Identification of the enzymes involved and study of their regulation has proved difficult, as the IGFBPs are extremely sensitive to proteolysis by many common serum enzymes. Once these proteases have been separated from their normal tight regulation in the circulation, their activity then masks that of specific IGFBP proteases that may be present. The precise role played by insulin, GH, or nutritional status in regulating the circulatory IGFBP proteases is unclear, but all these factors appear to be involved in some way.

The potential for IGFBP proteases or their inhibitors in diagnosis or therapy is unclear. Because there is a general increase in extracellular proteolysis around most solid tumors, there were initial hopes that specific inhibition of IGFBP proteolysis may have provided new approaches for restricting tumor growth, as proteases acting on IGFBPs may increase IGF activity. In contrast to the circulation of normal adults, where there is very little proteolysis of the IGFBPs, we have shown in several extravascular fluids that there is considerable IGFBP-3 proteolytic activity. If most IGFBP-3 in tissues is normally cleaved, the significance of increased proteolysis around tumors is then not clear. In view of recent evidence supporting IGF-independent functions of IGFBP-3 and the possible occurrence of fragments, generated by proteolytic cleavage, with distinct actions, the role of IGFBP-3 in extravascular fluids is unclear. The significant decrease in extravascular IGFBP-3 proteolysis in various pathologies appears to occur owing to interactions between a complex system of enzymes and inhibitors.

Our recent studies have indicated that IGFBP-3 proteolysis is due to proteases that are ubiquitously present within the circulation and extravascular fluids. The critical factor in modulating IGFBP-3 proteolysis then appears to be the presence of inhibitors that may protect IGFBP-3 from cleavage. Such inhibitors are normally present in the circulation but not in extravascular fluids, although they appear in tissues where there is local inflammation and increased capillary permeability and in other local pathologies *(85,86)*.

The widespread occurrence of several distinct IGFBP proteases with different IGFBP substrate specificities suggests that proteolytic modification of IGFBPs is a common mechanism for regulating the availability of the IGFs at the tissue level. In addition to the unidentified proteases present in vivo, which are generally serine-type proteases, to date, three classes of enzymes have been identified as IGFBP proteases: kallikreins, cathepsins, and MMPs. By regulating IGF availability, IGFBP proteases may be critical in normal, benign, and malignant cell metabolism and growth.

REFERENCES

1. Jones JI, Clemmons DR. Insulin-like growth factors and their binding proteins: biological actions. Endocr Rev 1995; 16:3–33.
2. Clemmons DR, Underwood LE, Chatelain PG, Van Wyk JJ. Liberation of immunoreactive somatomedin-C from its binding proteins by proteolytic enzymes and heparin. J Clin Endocrinol Metab 1983; 56:384–389.
3. Chatelain PG, Van-Wyk JJ, Copeland KC, Blethen SL, Underwood, LE. Effect of *in vitro* action of serum proteases or exposure to acid on measurable immunoreactive somatomedin-C in serum. J Clin Endocrinol Metab 1983; 56:376–383.
4. Hossenlopp P, Segovia B, Lassarre C, Roghani M, Bredon M, Binoux M. Evidence of enzymatic degradation of insulin-like growth factor-binding proteins in the 150K complex during pregnancy. J Clin Endocrinol Metab 1990; 71:797–805.
5. Giudice LC, Farrell EM, Pham H, Lamson G, Rosenfeld RG. Insulin-like growth factor binding proteins in maternal serum throughout gestation and in the puerperium: effects of a pregnancy-associated serum protease activity. J Clin Endocrinol Metab 1990; 71:806–816.
6. Baxter RC, Martin JL. Radio immunoassay of growth hormone (GH) dependent insulin-like growth factor binding protein (IGFBP) in human plasma. J Clin Invest 1986; 78:1504–1512.
7. Baxter RC. Circulating levels, molecular distribution of the acid labile subunit (ALS) of the high molecular weight insulin-like growth factor binding protein (IGFBP) complex. J Clin Endocrinol Metab 1990; 70:1347–1353.
8. Wilson DM, Bennett A, Adamson GD. Somatomedins in pregnancy: a cross sectional study of insulin-like growth factors (IGFs) I and II and somatomedin peptide content in normal human pregnancy. J Clin Endocrinol Metab 1982; 55:858–861.
9. Liu F, Baxter RC, Hintz RL. Characterisation of the high molecular weight insulin-like growth factor complex in term pregnancy. J Clin Endocrinol Metab 1992; 75:1261–1267.
10. Lamson G, Giudice LC, Rosenfeld RG. A simple assay for the proteolysis of IGFBP-3. J Clin Endocrinol Metab 1991; 72:1391–1394.
11. Lassarre C, Binoux M. Insulin-like growth binding protein-3 is functionally altered in pregnancy plasma. Endocrinology 1994; 134:1254–1262.
12. Davies SC, Holly JMP, Coulson VJ, Cotterill AM, Abdulla AF, Whittaker PG, Chard T, Wass JAH. The presence of cation-dependent proteases for insulin-like growth factor binding proteins does not alter the size distribution of insulin-like growth factors in pregnancy. Clin Endocrinol 1991; 34:501–506.
13. Suikkari A, Baxter RC. Insulin-like growth factor binding protein-3 in pregnancy serum binds native IGF-I but not iodo. J Clin Endocrinol Metab 1992; 73:1377–1379.

14. Holly JMP, Claffey DCP, Cwyfan-Hughes SC, Frost VJ, Yateman ME. Proteases acting on IGFBPs; their occurrence and physiological significance. Growth Regul 1993; 3:88–91.
15. Blat C, Villaudy J, Binoux M. In vivo proteolysis of serum insulin-like growth factor (IGF) binding protein-3 results in increased availability of IGF to target cells. J Clin Invest 1994; 93:2286–2290.
16. Davenport ML, Clemmons DR, Miles MV, Camacho-Hubner C, D'Ercole JA. Regulation of serum insulin-like growth factor-I (IGF-I) and IGF-binding proteins (IGFBPs) during rat pregnancy. Endocrinology 1990; 127:1278–1286.
17. Bang P, Stangenberg M, Westgren M, Rosenfeld RG. Decreased ternary complex formation and predominance of a 29 kDa insulin-like growth factor binding protein-3 (IGFBP-3) fragment in human fetal serum. Growth Regul 1994; 4:68–76.
18. Schwarz AJ, Brasel J, Hintz RL, Mohan S, Cooper DM. Acute effect of brief low- and high-intensity exercise on circulating insulin-like growth factor (IGF) I, II, and IGF-binding protein-3 and its proteolysis in young healthy men. J Clin Endocrinol Metab 1996; 81:3492–3497.
19. Fielder PJ, Guevara-Aguirre J, Rosenbloom AL, Carlsson L, Hintz RL, Rosenfeld RG. Expression of serum insulin-like growth factors, insulin-like growth factor-binding proteins, and the growth hormone binding protein in heterozygote relatives of Ecuadorian growth hormone receptor deficient patients. J Clin Endocrinol Metab 1992; 74:743–750.
20. Cotterill AM, Holly JMP, Taylor AM, Davies SC, Coulson VJ, Preece MA, Wass JAH, Savage MO. The insulin-like growth factor (IGF)-binding proteins and IGF bioactivity in Laron-type dwarfism. J Clin Endocrinol Metab 1992; 74:56–63.
21. Lassarre C, Lalou C, Perin L, Binoux M. Protease-induced alteration of insulin-like growth factor binding protein-3 as detected by radioimmunoassay. Agreement with ligand blotting data. Growth Regul 1994; 4:48–55.
22. Skjaerbaek C, Kaal A, Moller J, Vahl N, Weeke J, Flyvbjerg A. The effect of growth hormone on serum insulin-like growth factor binding protein-3 (IGFBP-3) proteolysis. In: Proceedings of the Fourth International Symposium on Insulin-Like Growth Factors, Tokyo, Oct. 21–24, 1997; A-46.
23. Davies SC, Wass JAH, Ross RJM, Cotterill AM, Buchanan CR, Coulson VJ, Holly JMP. The induction of a specific protease for insulin-like growth factor binding protein-3 in the circulation during severe illness. J Endocrinol 1991; 130:469–473.
24. Cwyfan Hughes SC, Cotterill AM, Molloy AR, Cassell TB, Braude N, Hinds CJ, Wass JAH. The induction of specific proteases for insulin-like growth factor binding proteins following major heart surgery. J Endocrinol 1992; 135:135–145.
25. Timmins AC, Cotterill AM, Hughes SCC, Holly JMP, Ross RJM, Blum W, Hinds CJ. Critical illness is associated with low circulating concentrations of insulin-like growth factor I and growth factor II, alterations in insulin-like growth factor binding-proteins and induction of an insulin-like growth factor binding-protein-3 protease. Crit Care Med 1996; 24:1460–1466.
26. Pucilowska J, Davenport ML, Kabir I, Clemmons DR, Thissen J-P, Butler T, Underwood LE. The effect of dietary protein supplementation on insulin-like growth factors and IGF binding proteins in children with shigellosis. J Clin Endocrinol Metab 1993; 77:1516–1521.
27. Carlsson-Skwirut C, Bhutta Z, Karlsson E, Bang P. Regulation of insulin-like growth factor binding proteins (IGFBPs) and IGFBP-3 proteolysis in children with diarrhea/malnutrition. In: Proceedings of the Fourth International Symposium on Insulin-Like Growth Factors, 1997; A-42.
28. Bang P, Brismark K, Rosenfeld RG, Hall K. Fasting affects serum insulin-like growth factors (IGFs) and IGF binding proteins differently in patients with non-insulin dependent diabetes versus healthy non-obese and obese subjects. J Clin Endocrinol Metab 1994; 78:960–967.
29. Bang P, Brismar K, Rosenfeld RG. Increased proteolysis of insulin-like growth factor binding protein-3 (IGFBP-3) in non-insulin-dependent diabetes serum, with elevation of a 29 kilodalton (kDa) glycosylated IGFBP-3 fragment contained in the approximately 130 to 150 kDa ternary complex. J Clin Endocrinol Metab 1994; 78:1119–1127.
30. Bang P. Serum proteolysis of IGFBP-3. Prog Growth Factor Res 1995; 6:285–292.
31. Bang P, Carlsson-Skwirut C, Nygren J. Insulin regulates an IGFBP-3 proteolytic activity in insulin-resistant subjects. In: Proceedings of the Fourth International Symposium on Insulin-Like Growth Factors, Tokyo, Oct. 21–24, 1997; S2–4.
32. Bereket A, Lang CH, Blethen SL, Fan J, Frost RA, Wilson TA. Insulin-like growth factor binding protein-3 proteolysis in children with insulin-dependent diabetes mellitus: a possible role for insulin in the regulation of IGFBP-3 protease activity. J Clin Endocrinol Metab 1995; 80:2282–2288.
33. Lang CH, Fan J, Frost RA, Gelato MC, Sakurai Y, Herndon DN, Wolfe RR. Regulation of the insulin-like growth factor system by insulin in burn patients. J Clin Endocrinol Metab 1996; 81:2474–2480.

34. Frost VJ, Macaulay VM, Wass JAH, Holly JMP. Proteolytic modification of insulin-like growth factor binding proteins: comparison of conditioned media from human cell lines, circulating proteases and characterized enzymes. J Endocrinol 1993; 138:545–554.
35. Frost VJ, Helle SI, Lonning PE, van der Stappen J, Holly JMP. Effects of treatment with megesterol acetate, aminoglutethimide or formestane on insulin-like growth factor (IGF) I and II and IGF-binding proteins (IGFBPs) and IGFBP-3 protease status in patients with advanced breast cancer. J Clin Endocrinol Metab 1996; 81:2216–2221.
36. Helle SI, Holly JMP, Tally M, Hall K, van der Stappen JWJ, Lonning PE. Influence of treatment with tamoxifen and change in tumour burden on the IGF system in breast cancer patients. Int J Cancer 1996; 69:335–339.
37. Lalou C, Binoux M. Evidence that limited proteolysis of IGFBP-3 occurs in the normal state outside the bloodstream. Regul Peptides 1993; 48:179–188.
38. Xu S, Cwyfan-Hughes SC, van der Stappen JWJ, Sansom J, Burton JL, Donnelly M, Holly JMP. Insulin-like growth factors (IGFs) and IGF binding-proteins in human skin interstitial fluid. J Clin Endocrinol Metab 1995; 80:2940–2945.
39. Cwyfan-Hughes SC, Fernihough JK, Hampton A, Mason HD, Franks S, van der Stappen JWJ, Donnelly M, Holly JMP. Tissue IGFBP-3 protease proteolysis: contrasting pathophysiology to that in the circulation. Prog Growth Factor Res 1995; 6:293–299.
40. Fernihough JK, Billingham MEJ, Cwyfan-Hughes S, Holly JMP. Local disruption of the insulin-like growth factor system in the arthritic joint. Arthritis Rheum 1996; 39:1556–1565.
41. Cwyfan-Hughes SC, Mason HD, Franks S, Holly JMP. Modulation of the IGFBPs by follicle size in the human ovary. J Endocrinol 1997; 154:35–43.
42. Gargosky SE, Pham H, Wilson KE, Liu F, Giudice LC, Rosenfeld RG. Measurement and characterisation of insulin-like growth factor binding protein-3 in human biological fluids: discrepancies between radioimmunoassay and ligand blotting. Endocrinology 1992; 131:3051–3060.
43. Cohen P, Graves HCB, Peehl DM, Kamarei M, Giudice LC, Rosenfeld RG. Prostate-specific antigen (PSA) is an insulin-like growth factor binding protein-3 protease found in seminal plasma. J Clin Endocrinol Metab 1992; 75:1046–1053.
44. Lee K, Oh Y, Giudice LC, Cohen P, Peehl DM, Rosenfeld RG. Identification of insulin-like growth factor-binding protein-3 (IGFBP-3) fragments and IGFBP-5 proteolytic activity in human seminal plasma: a comparison of normal and vasectomised patients. J Clin Endocrinol Metab 1994; 79:1367–1372.
45. Claussen M, Zapf J, Braulke T. Proteolysis of insulin-like growth factor binding protein-5 by pregnancy serum and amniotic fluid. Endocrinology 1994; 134:1964–1966.
46. Gibson JM, Westwood M, Siddal SK, Shaikh T, White A. IGFBP-1 is proteolysed by decidual cell medium, hep G2 medium and amniotic fluid. In: Proceedings of the Fourth International Symposium on Insulin-Like Growth Factors, Tokyo, Oct. 21–24, 1997; C-06.
47. Xu S, Cwyfan-Hughes SC, van der Stappen JWJ, Sansom J, Burton JL, Donnelly M, Holly JMP. Altered insulin-like growth factor II (IGF-II) level and IGF binding protein-3 (IGFBP-3) protease activity in interstitial fluid taken from the skin lesion of psoriasis. J Invest Dermatol 1996; 106:109–112.
48. Lee DY, Cohen P, Krensky AM, Rosenfeld RG, Yorgin PD. Insulin-like growth factor binding protein-3 protease activity in the urine of children with chronic renal failure. Pediatr Nephrol 1993; 7:416–423.
49. Lee DY, Park SK, Yorgin PD, Cohen P, Oh Y, Rosenfeld RG. Alterations in insulin-like growth factor binding-proteins (IGFBPs) and IGFBP-3 protease activity in serum and urine from acute and chronic renal failure. J Clin Endocrinol Metab 1994; 79:1376–1382.
50. Schoen TJ, Waldbillig RJ, Searcy G, Gaudet SJ, Jones BE, Chader GJ, Moshyedi P. Identification and partial characterisation of a proteinase specific for insulin-like growth factor binding protein-3 in aqueous and vitreous humors. Curr Eye Res 1995; 14:127–135.
51. Muller HL, Oh Y, Gargosky SE, Lehrnbecker T, Hintz RL, Rosenfeld RG. Concentrations of insulin-like growth factor (IGF) binding protein-3 (IGFBP-3), IGF and IGFBP-3 protease activity in cerebrospinal fluid of children with leukemia, central nervous system tumor or meningitis. J Clin Endocrinol Metab 1993; 77:1113–1119.
52. Rajah R, Nunn SE, Herrick DJ, Grunstein MM, Cohen P. Leukotriene D4 induces MMP-1, which functions as an IGFBP protease in human airway smooth muscle cells. Am J Physiol 1996; 271:L1014–L1022.
53. Cohen P, Peehl DM, Graves HCB, Rosenfeld RG. Biological effects of prostate specific antigen as an insulin-like growth factor binding protein-3 protease. J Endocrinol 1994; 142:407–415.
54. Rajah R, Valentinis B, Cohen P. Insulin-like growth factor (IGF) binding protein-3 induces apoptosis

and mediates the effects of transforming growth factor on programmed cell death through a p53- and IGF-independent mechanism. J Biol Chem 1997; 272:12181–12188.
55. Rajah R, Bhala A, Nunn SE, Peehl DM, Cohen P. 7S nerve growth factor as an insulin-like growth factor binding protein protease. Endocrinology 1996; 137:2676–2682.
56. Campbell PG, Novak JF, Yanosick TB, McMaster JH. Involvement of the plasmin system in dissociation of the insulin-like growth factor binding protein complex. Endocrinology 1992; 130:1401–1412.
57. Lalou C, Silve C, Rosata R, Segovia B, Binoux M. Interactions between insulin-like growth factor-I (IGF-I) and the system of plasminogen activators and their inhibitors in the control of IGF-binding protein production and proteolysis in human osteosarcoma cells. Endocrinology 1994; 135:2319–2326.
58. Angelloz-Nicoud P, Binoux M. Autocrine regulation of cell proliferation by the insulin-like growth factor (IGF) and IGF-binding protein-3 protease system in a human prostate carcinoma cell line (PC-3). Endocrinology 1996; 136:5485–5492.
59. Angelloz-Nicoud P, Harel L, Binoux M. Recombinant human insulin-like growth factor (IGF) binding protein-3 stimulates prostate carcinoma cell proliferation via an IGF dependent mechanism: role of serine proteases. Growth Regul 1996; 6:130–136.
60. Xu S, Savage P, Burton JL, Sansom J, Holly JMP. Proteolysis of insulin-like growth factor-binding protein-3 by human skin keratinocytes in culture in comparison to that in skin interstitial fluid: the role and regulation of components of the plasmin system. J Clin Endocrinol Metab 1997; 82:1863–1868.
61. Vorwerk P, Oh Y, Lee PDK, Khare A, Rosenfeld RG. Synthesis of IGFBP-3 fragments in a baculovirus system and characterization of monoclonal anti-IGFBP-3 antibodies. J Clin Endocrinol Metab 1997; 82:2368–2370.
62. Conover CA, De Leon DD. Acid activated insulin-like growth factor binding protein-3 proteolysis in normal and transformed cells. Role of cathepsin D. J Biol Chem 1994; 269:7076–7080.
63. Claussen M, Kubler B, Wendland M, Neifer K, Schmidt B, Zapf J, Braulke T. Proteolysis of insulin-like growth factors (IGFs) and IGF binding proteins by cathepsin D. Endocrinology 1997; 138:3797–3803.
64. Davenport ML, Clemmons DR, Miles MV, Camacho-Hubner C, D'Ercole AJ, Underwood LE. Regulation of serum insulin-like growth factor-I (IGF-I) and IGF binding proteins during rat pregnancy. Endocrinology 1990; 127:1278–1286.
65. Fowlkes JL, Suzuki K, Nagase H, Thrailkill KM. Proteolysis of insulin-like growth factor binding protein-3 during rat pregnancy: a role for matrix metalloproteinases. Endocrinology 1994; 135:2810–2813.
66. Giudice LC, Dsupin BA, De Las Fuentes L, Gargosky SE, Rosenfeld RG, Zelinski-Wooten MB, Stouffer RL, Fazleabas AT. Insulin-like growth factor binding proteins in sera of pregnant nonhuman primates. Endocrinology 1993; 132:1514–1526.
67. Fowlkes JL. Degradation of insulin-like growth factor (IGF)-binding protein-3 (IGFBP-3) by a metal-dependent protease produced by human fibroblasts: effects of IGFs on protease activity. Endocr J 1994; 2:63–68.
68. Lalou C, Lassarre C, Binoux M. A proteolytic fragment of insulin-like growth factor (IGF) binding protein-3 that fails to bind IGFs inhibits the mitogenic effects of IGF-I and insulin. Endocrinology 1996; 137:3206–3212.
69. Zadeh MS, Binoux M. The 16kDa proteolytic fragment of insulin-like growth factor binding protein-3 inhibits the mitogenic action of fibroblast growth factor on mouse fibroblasts with a targeted disruption of the type 1 IGF receptor gene. Endocrinology 1997; 138:3069–3072.
70. Camacho-Hubner C, Busby WH, McCusker RH, Wright G, Clemmons DR. Identification of the forms of insulin-like growth factor-binding proteins produced by human fibroblasts and the mechanisms that regulate their secretion. J Biol Chem 1992; 267:11949–11956.
71. Conover CA, Kiefer MC, Zapf J. Posttranslational regulation of insulin-like growth factor binding protein-4 in normal and transformed human fibroblasts. Insulin-like growth factor dependence and biological studies. J Clin Invest 1993; 91:1129–1137.
72. Fowlkes JL, Freemark M. Evidence for a novel insulin-like growth factor (IGF)-dependent protease regulating IGF-binding protein-4 in dermal fibroblasts. Endocrinology 1992; 131:2071–2076.
73. Kanazaki S, Hilliker S, Baylink DJ, Mohan S. Evidence that human bone cells in culture produce insulin-like growth factor binding protein-4 and -5 proteases. Endocrinology 1994; 134:383–392.
74. Cheung PT, Wu J, Banach W, Chernausek SD. Glucocorticoid regulation of a insulin-like growth factor-binding protein-4 protease produced by a rat neuronal cell line. Endocrinology 1994; 135:1328–1334.
75. Parker A, Gockerman A, Busby WH, Clemmons DR. Properties of an insulin-like growth factor binding protein-4 protease that is secreted by smooth muscle cells. Endocrinology 1995; 136:2470–2476.
76. Neely EK, Rosenfeld RG. Insulin-like growth factors (IGFs) reduce IGF-binding protein-4 (IGFBP-4)

concentration and stimulate IGFBP-3 independently of IGF receptors in human fibroblasts and epidermal cells. Endocrinology 1992; 130:985–993.
77. Donnelly M, Holly JMP. The role of IGFBP-3 in the regulation of IGFBP-4 proteolysis. J Endocrinol 1996; 149:R1–R7.
78. Fowlkes JL, Thrailkill KM, George-Nascimento C, Rosenberg CK, Serra DM. Heparin-binding, highly basic regions within the thyroglobulin type-1 repeat of insulin-like growth factor (IGF)-binding proteins (IGFBPs) -3, -5, and -6 inhibit IGFBP-4 degradation. Endocrinology 1997; 138:2280–2285.
79. Nam T, Busby WH, Clemmons DR. Human fibroblasts secrete a serine protease that cleaves insulin-like growth factor binding protein-5. Endocrinology 1994; 135:1385–1391.
80. Thrailkill KM, Quarles LD, Nagase H, Suzuki K, Serra DM, Fowlkes JL. Characterisation of insulin-like growth factor-binding protein 5 degrading proteases produced throughout murine osteoblast differentiation. Endocrinology 1995; 136:3527–3533.
81. Arai T, Arai A, Busby WH, Clemmons DR. Glycosaminoglycans inhibit degradation of insulin-like growth factor-binding protein-5. Endocrinology 1994; 135:2358–2363.
82. Arai T, Parker AJ, Busby WH, Clemmons DR. Heparin, heparan sulfate and dermatan sulfate regulate formation of the insulin-like growth factor-I and insulin-like growth factor-binding protein complexes. J Biol Chem 1994; 269:20388–20393.
83. Nam T, Busby WH, Clemmons DR. Insulin-like growth factor binding protein-5 binds to plasminogen activator inhibitor-1. Endocrinology 1997; 138:2972–2978.
84. Nam T, Busby WH, Clemmons DR. Characterisation and determination of the relative abundance of two types of insulin-like growth factor binding protein-5 proteases that are secreted by human fibroblasts. Endocrinology 1996; 137:5530–5536.
85. Maile LA, Cwyfan-Hughes SC, Xu S, Pell JM, Holly JMP. Characterisation of the components of the IGFBP-3 protease system in adult serum, interstitial fluid and synovial fluid. In: Proceedings of the Fourth International Symposium on Insulin-Like Growth Factors, Tokyo, Oct. 21–24, 1997; A-47.
86. Whellams E, Maile LA, Billingham M, Holly JMP. Are the changes in the IGF system seen in synovial fluid from arthritic joints just related to inflammation? J Endocrinol 1998; 156 (Suppl.):P156.

27 IGF Deficiency

Katie A. Woods, MD, MRCP, *and*
Martin O. Savage, MA, MD, FRCP

Contents

INTRODUCTION

The molecular biology and physiology of the insulin-like growth factor (IGF) system are complex, as detailed in the earlier chapters of this book, and there are consequently many potential mechanisms of IGF deficiency. Broadly, there may be primary IGF deficiency, in which the defect lies in IGF synthesis, or secondary IGF deficiency, in which low levels of IGF arise as a consequence of a deficiency in the pathway leading to IGF production (Table 1).

As IGF synthesis occurs in many tissues, primary IGF deficiency is possible only due to disruption of the IGF gene or its promoter region. A number of elegant murine knock-out studies of IGF-I, IGF-II and their receptors have now been performed and have provided considerable insights into IGF action in the mouse *(1–3)*, as detailed in Chapter 25. The only human example of primary IGF deficiency is our recently described case of a boy with a partial IGF-I gene deletion *(4)*, which is discussed in detail below.

Secondary IGF deficiency arises as a result of a defect in a factor normally required to maintain IGF levels, either by stimulating transcription of the IGF gene or by stabilizing the IGF in the circulation. Here, IGF deficiency may be acquired or genetic, may be specific to certain tissues, or limited to a particular stage of development. Increased IGF clearance, which may arise secondary to a deficiency of IGF binding proteins (particularly IGFBP-3) and/or acid-labile subunit (ALS) (*see* Chapters 6 and 11), may result in IGF deficiency unless increased IGF production can compensate. At present, no disease or syndrome associated with primary IGFBP/ALS deficiency has been described. However, low levels of IGFBP-3 and ALS, in addition to IGF-I deficiency, are found in human growth hormone insensitivity syndrome (GHIS). The abnormal pharmacokinetics of

From: *Contemporary Endocrinology: The IGF System*
Edited by: R. Rosenfeld and C. Roberts Jr © Humana Press, Inc., Totowa, NJ

Table 1
Causes of IGF-I and IGF-II Deficiency

	Primary Deficiency			*Secondary Deficiency*	
	Animal	*Man*		*Genetic*	*Acquired*
IGF-I	Knockout mouse	One case (partial IGF-I gene deletion)	GHD	1. *Pit-1* gene defect 2. GHRH receptor 3. GH gene defect	1. Hypothalamic dysfunction 2. Pituitary defect
			GHI	1. GH receptor defect (Laron syndrome) 2. Post-receptor defect	1. Nutritional deficiency 2. Acute/chronic illness
			Bioinactive GH	GH gene defect	?
IGF-II	Knockout mouse	No examples		Reduced IGF-II levels in GH deficiency and GH receptor defect secondary to low IGFBP-3	As previously; low IGF-II may be seen if IGFBP-3 low

GHD, GH deficiency; GHI, GH insensitivity.

IGF-I observed in GHIS demonstrates the importance of these peptides in stabilizing the IGFs in the circulation and may contribute to the suboptimal growth response to IGF-I therapy observed in GHIS patients *(5,6)*.

IGF resistance may be considered a form of secondary IGF deficiency, as despite high levels of IGF, tissues are functionally IGF deficient. This condition may arise secondary to a defect of the IGF-I receptor or its intracellular signaling pathway, through which both IGF-I and IGF-II are thought to signal to the nucleus.

IGF-I DEFICIENCY

Although IGF-I deficiency is usually the result of acquired rather than genetic causes, the most exciting developments in recent years have been molecular, with a number of new gene defects being described in the human GH–IGF-I axis (Fig. 1). Detailed studies of the phenotypes of these individuals with rare gene defects provide important opportunities to understand the functioning of the GH–IGF-I axis more clearly. This chapter therefore concentrates on the genetic causes of IGF deficiency, and discusses nongenetic IGF deficiency more briefly.

Primary IGF Deficiency

Primary IGF deficiency describes the situation in which low or absent IGF-I arises as a result of a primary inability to produce IGF-I, and not as a consequence of deficiency of another factor. In effect, this is possible only secondary to mutations in the IGF-I gene or its promoter regions.

The IGF-I Knockout Mouse Model

In 1993 two independent groups generated mice with a targeted disruption of the IGF-I gene *(1,2,7)*. Although the homozygous IGF-I knockout mice survived to term in both groups, neonatal survival was low, ranging from < 5% to 68%, dependent on genetic background. The neonatal mortality appeared to result from respiratory failure, possibly

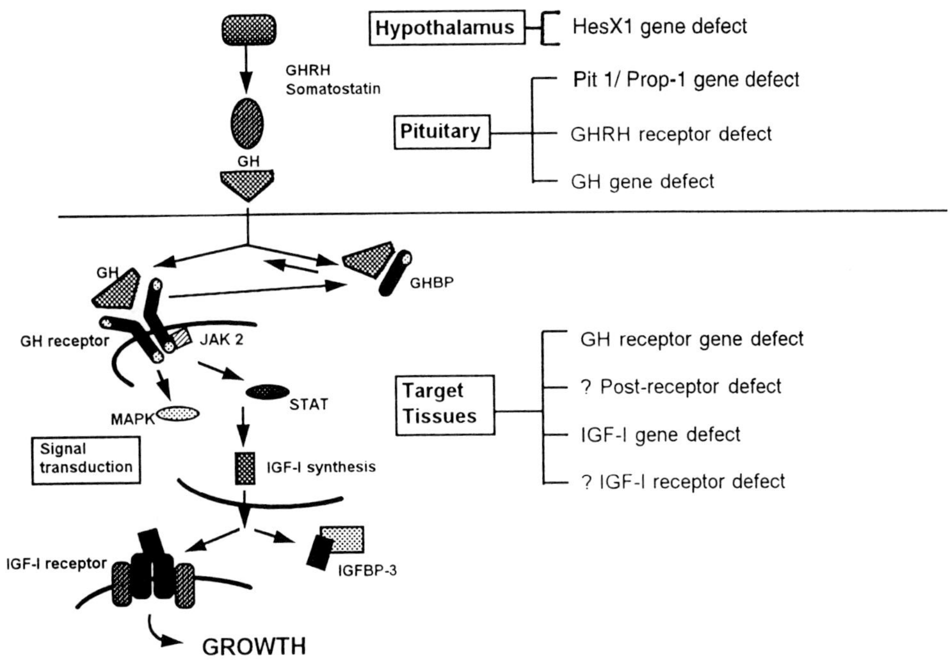

Fig. 1. GH-IGF-I axis, showing the current gene defects described at each level of the axis (? = cases suspected by functional studies but no mutation described, GHRH = GH releasing hormone, JAK2 = Janus associated kinase 2, MAPK = Map kinase, and STAT = signal transducer and activator of transcription.)

secondary to respiratory muscle hypoplasia. The importance of IGF-I for fetal, in addition to postnatal growth, was demonstrated by the severe prenatal growth failure (birth weight 60% of normal) observed. Growth failure continued postnatally, resulting in an adult size reduced by 70% *(2)*. Several other phenotypic abnormalities were observed, most notably reduced brain size with failure of CNS myelination *(8)*, and infertility in both males and females *(9)*, implicating specific roles for IGF-I in the murine neurological and reproductive systems. Placental size was normal, however, suggesting that IGF-I is not involved in placental growth in the mouse *(2)*.

Human IGF-I Gene Deficiency

In 1996 we described the first human case of an IGF-I gene defect *(10)*. The patient, a 15-yr-old boy, had a homozygous partial deletion of the IGF-I gene involving most of the coding region, and had no detectable IGF-I in the circulation. The phenotype of this patient is quite distinct from that observed in secondary IGF-I deficiency, but does share many of the features of the knockout mouse model. He was severely growth retarded at birth, with a birth weight of 1.36 kg (–3.9 SD), birth length 37.8 cm (–5.41 SD), and head circumference 27 cm (–4.94 SD) *(11)*. Despite such a low birth weight, the neonatal period was uneventful, with no respiratory problems. He continued to grow slowly after birth, and by the age of 15.8 yr his height was just 119.1 cm (–6.88 SD). He was mildly dysmorphic, with a small jaw, ptosis, and a low hairline (Fig. 2a). Mirroring the knockout mouse model, abnormalities of central nervous system (CNS) development were apparent, with microcephaly (head circumference 47 cm) and mild mental retardation. He also had severe bilateral sensorineural deafness. The brain was structurally normal as judged

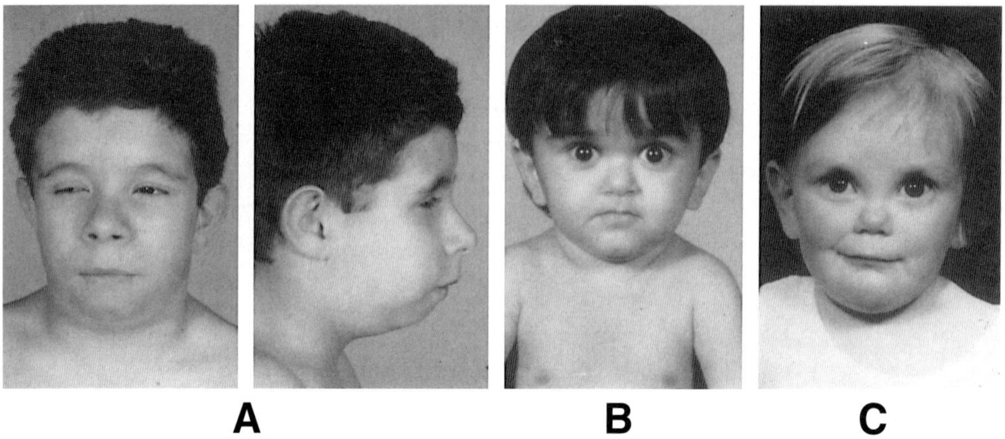

Fig. 2. (**A**) Facial appearance of patient with IGF-I gene deletion, (**B**) facial appearance of patient with genetic GH resistance, (**C**) facial appearance of patient with genetic GH deficiency.

by magnetic resonance imaging, with no evidence of abnormal myelination pattern as seen in the knockout mouse, and nerve conduction studies were normal.

Biochemical evaluation revealed extremely high levels of GH in this patient, both after provocative stimulation to insulin (61 µg/L) and clonidine (94 µg/L) and during an overnight GH profile (peaks up to 175 µg/L). Levels of IGFBP-3 (3.3 mg/L, normal range 2.3–5.2) and ALS were normal, providing supportive evidence that these peptides are primarily under GH control, in contrast to the situation in the rodent *(12–15)*. IGF-II levels (1430 µg/L, pooled serum 1010 µg/L) were elevated, perhaps secondary to the increased capacity of IGFBP-3 and ALS for IGF-II in the absence of IGF-I. Although glucose levels were within the normal range, reduced insulin sensitivity was observed (Bergman modified intravenous glucose and tolbutamide tolerance test: insulin sensitivity 2.06×10^4/min/µU/mL [normal short child 16.9×10^4/min/µU/mL, lean adult 5.1×10^4/min/µU/mL]), consistent with an increase in the "direct" anti-insulin-like effects of GH unopposed by IGF-I. Sex hormone binding globulin was undetectable.

Although the pattern of growth failure in our patient and the IGF-I knockout mice appears to be very similar, their phenotype is not identical, indicating an evolutionary divergence in the role of IGF-I. The placenta of our patient was below average weight (350 g [–1.3 SD]) *(16)*, whereas in the IGF-I knockout mouse placental size was normal. The infertility observed in the male knockout mouse is thought to result primarily from low levels of testosterone secondary to impaired Leydig cell steroidogenesis *(9)*. In contrast, our patient is progressing through puberty normally, although pubertal onset was rather late (15 yr approximately), and has normal basal and human chorionic gonadotropin (hCG)-stimulated levels of testosterone, suggesting that IGF-I does not have a crucial role in testicular steroidogenesis in humans. This is supported by the finding that the IGF-I gene is not expressed in the human testis, whereas it is widely expressed in the rodent testis, and the localization of IGF-I receptor transcripts differs between the rodent and human *(4)*. One of the notable findings in our patient was the relatively mild delay in skeletal maturation bone age compared with chronological age. This is in contrast to the severe delays in bony ossification observed in the knockout mouse *(1,2)*, and to the normally more marked bone age delay seen in severe GH deficiency and GH receptor

Table 2
Growth and Biochemical Features Differentiating Primary and Secondary IGF-I Deficiency

IGF Deficiency		*Growth Failure*	*IGF-I Levels*	*IGFBP-3/ ALS Levels*	*GH Secretion*	*Response to hGH*[b]
Primary		Pre- and postnatal	↓	→	↑	No
Secondary	GHI	Postnatal	↓	↓[a]	↑	No
	GHD	Postnatal	↓	↓	↓	Yes[c]
	Bioinactive GH	Postnatal	↓	↓	↑	Yes[d]

GHI, growth hormone insensitivity; GHD, growth hormone deficiency.
[a] One patient with features of GHIS has been described in whom IGFBP-3 levels were normal. The primary genetic defect in this family has not been established.
[b] Response to exogenously administered hGH on long-term growth and IGF-I levels.
[c] Initial response may not be sustained in GH gene deletion with antibody formation.
[d] In one of the two cases of bioinactive GH syndrome where a GH gene defect has been found response to GH was not sustained, probably owing to the dominant negative effect of the mutant GH.

defects *(17,18)*, perhaps suggesting that in humans, GH can act directly at the growth plate to cause bony maturation independent of GH.

Secondary IGF Deficiency

Although primary IGF-I deficiency is the "purest" form of IGF-I deficiency, low levels of IGF-I in humans usually arise as a consequence of an abnormality in the pathway that leads to transcription of the IGF-I gene (Fig. 1). After birth, as originally proposed by Salmon and Daughaday in 1957 *(19)*, GH is the main stimulus to the production of IGF-I, and thus any condition resulting in GH deficiency (GHD) or GH insensitivity (GHI) will lead to low levels of IGF-I. However, during fetal life, and possibly in certain tissues, IGF-I production is GH independent, and hence, in secondary IGF-I deficiency, IGF-I deficiency is not complete. Furthermore, certain actions of GH, particularly those stimulating metabolic processes, are independent of IGF-I. Thus, patients with a genetic cause of secondary IGF-I deficiency have a distinct phenotype, reflecting the selective loss of GH-dependent IGF-I production and the loss of "direct" GH actions. The clinical and biochemical features, and etiology of, secondary IGF-I deficiency are discussed in the following sections. Table 2 summarizes the basic differences between the clinical features and biochemistry of primary and secondary IGF deficiency.

GH Deficiency

The frequency of GHD is estimated to range from 1 in 4000 to 1 in 10,000 *(20–23)*. In most cases of isolated growth hormone deficiency (IGHD) the disorder is idiopathic, although a high incidence of perinatal problems has been documented *(24)*. However, this chapter focuses on genetic GHD as this condition provides the clearest insight into the biological action of GH in humans.

Genetic GHD. In this condition, GHD arises as a result of a defect in a gene directly involved in the production of GH. Estimates of the proportion of GHD that is familial, suggesting a genetic cause, range from 3% to 30%, and probably depend to a certain extent on the strictness of the criteria used to define GHD *(20–22,25)*.

Clinical Features. Individuals with genetic GHD are usually of normal birth weight, although birth length may be mildly reduced *(26)*. Growth failure usually occurs in the first few months of life in those with severe GHD and is progressive *(27,28)*. Short stature is proportionate, and there is truncal adiposity, thought to reflect lack of the lipolytic effect of GH. Skeletal maturation, as previously noted, is usually delayed by several years *(17)*. In the neonatal period, the infant may have episodes of hypoglycemia, and micropenis sometimes occurs in the male. There is a characteristic facial appearance, which is quite distinct from that in the case of IGF-I gene deletion (Fig. 2b). The facial phenotype includes hypoplasia of the midface and a prominent forehead, resulting in a rather immature appearance, which becomes more apparent with time *(29)*. The prominence of the forehead is due in part to the sparing of growth of the vertex of the skull relative to the bones of the face, in contrast again to the marked microcephaly observed in IGF-I gene deficiency. Motor development may be delayed secondary to poor musculature, but mental development is normal in most individuals, despite infantile hypoglycemia. A high pitched voice is also typical, thought to be due to laryngeal hypoplasia, and hair is sparse and slow growing. Puberty is usually delayed, although in adult life both males and females are fertile.

Biochemical Features. GH levels are usually very low, or undetectable, both on random testing and in response to provocative stimuli. In cases of GH gene deletion, GH antibodies may be present. IGF-I, IGFBP-3, and ALS are all reduced in GHD, reflecting their GH-dependent status *(30–32)*. IGF-II is also low secondary to reduced IGFBP-3/ALS *(33)*. The fasting hypoglycemia that may occur in infancy is presumably secondary to the loss of the anti-insulin-like effects of GH on glycogenolysis and gluconeogenesis, in contrast to the insulin resistance seen in IGF-I gene deletion.

Causes of Genetic GHD. Genetic GHD has been described secondary to defects in the GH gene itself, and also as a result of mutations in four genes involved in GH production, and/or pituitary-hypothalamic development: the growth hormone-releasing hormone (GHRH) receptor gene and the genes encoding the transcription factors *Pit-1, Prop-1,* and *HesX1*. However, in a significant proportion of individuals with an apparently inherited form of GHD the gene defect remains unknown. In those where the GHD is part of a multiple pituitary hormone deficiency (MPHD), a defect in a factor involved in either in development of the pituitary–hypothalamic region, or in a common signaling pathway involved in activating several pituitary hormones, is most likely.

GH Gene Defects. Around half the cases of autosomal isolated GHD (IGHD) have been linked to the GH gene *(GH-1)* locus *(34–36)*. The first defects in *GH-1* were described in 1984 and were homozygous deletions involving *GH-1* and neighboring genes in the human GH gene cluster, and were thought to be secondary to homologous recombination events between sequences flanking this gene *(37–39)*. These patients are at risk of developing GH antibodies during therapy with recombinant hGH due to their lack of exposure to endogenous GH *(40,41)*. Subsequent to this initial report a number of further mutations in the GH gene have been identified, comprising nonsense, frameshift, splice site, and most recently point mutations *(42)*. Phillips has classified familial IGHD into four types depending on the pattern of inheritance and severity of GH deficiency *(42)* (Table 3). It should be noted that heterozygous GH gene defects are being identified in a proportion of cases of sporadic severe GH deficiency, presumably as a result of new mutations, and thus a negative family history does not exclude the possibility of a GH gene defect *(43)*.

GHRH Receptor Defects. In 1993 a mutation in the GHRH receptor gene was identified in the dwarf, GH deficient *lit/lit* mouse *(44)*, and in 1996 human GHRH mutations were described by two independent groups. Wajnrajch and colleagues reported a homozygous nonsense mutation in the human GHRH receptor gene in two members of a consanguineous family, who had extreme short stature and the typical clinical phenotype of severe

Table 3
Classification of GH Gene Defects

Disorder	Mode of Inheritance	GH1 Gene Defect	hGH Protein	GH Ab	Phenotype
IGHD IA	Autosomal recessive	1. Homozygous deletions of 6.7/7.0/7.6kB *(37)*	None	+/–	Some short at birth with hypoglycemia in infancy
		2. Deletion 6.7kB/ frameshift 371del C *(140)*	None	+	Severe dwarfism by 6 mo of age
		3. Homozygous nonsense; W20X *(42)*	None	+	Variable response to hGH depending on development of antibodies
IGHD IB	Autosomal recessive	1. Homozygous G→C donor splice intron IV *(42)* 2. Homozygous G→T donor splice intron IV *(141)* 3. Deletion 6.7kB/ frameshift exon III *(142)*	Reduced	–	Less severe short stature
IGHD II	Autosomal dominant	Heterozygous T→C donor splice intron III *(141)*	Reduced	–	Less severe short stature
IGHD III	X-linked	?	Reduced	–	Short stature +/– hypogammaglobulinaemia

Source: Phillips et al. *(42)*.

GHD *(45)*. The affected patients had undetectable levels of GH, even after provocation with insulin, clonidine, and GHRH, and an extremely good growth response to recombinant human GH. The homozygous Glu72Stop mutation detected in the patients' GHRH receptor gene would be expected to produce a severely truncated receptor with no membrane spanning regions or G-protein binding site. The same mutation was recently reported in patients affected by "dwarfism of Sindh," a cluster of severe dwarfism recently described in the lower Indus valley of Pakistan *(46)*. This new cause of IGHD in man may explain a proportion of those cases of IGHD that are not linked to the *GH-1* locus.

Pit-1 Gene Defects. *Pit-1* is a pituitary-specific transcription factor, involved in somatotroph development and transcription of the GH, prolactin, and thyroid-stimulating hormone (TSH) genes *(47,48)*. *Pit-1* gene defects were originally identified in Jackson and Snell dwarf mice *(49)*, and more recently in humans *(48,50–53)*, and are currently the only described genetic cause of human MPHD. Affected patients have severe GH and prolactin deficiency, with variable TSH deficiency and pituitary hypoplasia. Both dominant and recessive modes of inheritance have been described.

Prop-1 Gene Defects. Recently, homozygous mutations in 'Prophet of Pit-1', or *Prop-1* have been identified in man *(53a)*. This gene, mutated in the Ames dwarf mouse, encodes for a homeodomain transcription factor, which is expressed in the pituitary just before *Pit-1,* and is thought to be important for initial determination of the *Pit-1* lineage (hence the name) *(53b)*. In addition to combined GH, prolactin and TSH deficiency, as seen in *Pit-1* gene defects, the kindred described in this case had gonadotrophin deficiency and partial cortisol deficiency, indicating that the phenotype of *Prop-1* gene defects may be more severe *(53a)*.

HesX1 Gene Defects. Recently, a homozygous missense mutation in the homeobox gene *HesX1* has been identified in one consanguineous kindred with severe familial septo-optic dysplasia (SOD), a heterogeneous condition characterised by any combination of optic nerve hypoplasia, midline abnormalities of the brain, and pituitary hypoplasia. This mutation was shown to severely disrupt the binding of *HesX1* to DNA *(53c)*. The affected sibling pair had severe SOD, with agenesis of the corpus callosum and panhypopituitarism. The authors investigated 18 further patients with sporadic SOD, but found no mutations in *HesX1* in this group, suggesting that other genes may be implicated in this condition.

Acquired GHD.

Clinical Features. The severe phenotype of "classical" GHD described previously is rarely observed in the child with acquired GHD with the exception of infants with sporadic hypopituitarism, who usually have additional problems due to deficiency of other pituitary hormones. The clinical features of the much larger group of patients with acquired GHD/GH insensitivity will depend to a large extent on the severity, timing, and underlying etiology of the problem. At the mildest end on the spectrum, the only physical sign of mild GHD in childhood may be a reduced growth velocity in an otherwise well child, usually of normal or above normal weight for height.

Biochemical Features. The biochemical features of severe acquired GHD may be indistinguishable from those of genetic GHD, and generally do not present diagnostic difficulty. However, the diagnosis of the more commonly encountered partial acquired GHD is more controversial. Owing to the pulsatile nature of GH secretion, investigators have traditionally relied on the peak GH level reached after stimulation with a variety of GH secretagogues for the diagnosis, with an arbitrary cutoff of 10 ng/mL. It has, however, been questioned whether sole reliance on GH provocative testing is appropriate in the diagnosis of partial GHD, as these tests are nonphysiological and poorly reproducible *(54)*.

The situation is complicated further by a number of cases of short stature associated with normal or elevated GH responses to provocative testing, low IGF-I, and good response to exogenous administration of hGH reported in the literature *(55–57)*. In these cases it is likely that there is functional GHD, which may arise either as a result of GH neurosecretory dysfunction, or more rarely bioinactive GH syndrome (described below). In GH neurosecretory dysfunction patients have normal stimulated GH secretion, yet reduced spontaneous GH secretion, as measured by overnight or 24-h GH profiles, presumably secondary to hypothalamic dysfunction *(58–60)*.

On a theoretical basis, the "IGF-I generation test," which involves the measurement of IGF-I and IGFBP-3 before and after administration of exogenous hGH, may be useful in distinguishing children with GHD secondary to neurosecretory dysfunction and GHI. In GHD, a brisk rise in IGF-I, IGFBP-3, and ALS levels would be expected: in four studies using various doses of GH (0.08 IU/kg–0.2 IU/kg per dose for 4–10 d) mean rise in IGF-I in GHD ranged between 175% and 611% *(57,61–63)*. At present, however, the use of this test is limited by the lack of data on the response in normal children.

Causes of Acquired GH Deficiency. Acquired GHD may arise as a consequence of an insult to the pituitary and/or hypothalamic region. The potential causes of such insults are many, and are listed systematically in Table 4. In childhood, CNS tumors (in particular craniopharyngiomas), trauma, or irradiation are the most common causes, usually associated with severe GH deficiency and deficiency of other pituitary hormones. With the advent of high resolution MRI scanning, more subtle pituitary–hypothalamic abnormalities have been identified in patients previously categorized as "idiopathic" GHD, includ-

Table 4
Causes of Acquired GHD

Hypothalamic dysfunction	Congenital malformations	Anenceophaly Holoprosencephaly Septo-optic dysplasia
	Trauma	Traumatic delivery (breech/forceps) Postnatal injury
	Inflammation	Infection (bacterial/viral/fungal) Sarcoidosis
	CNS tumors	Germinomas Meningiomas Craniopharyngioma Histiocytosis X Gliomas Colloid cysts third ventricle Ependymomas Gliomas optic nerve Local extension nasopharyngeal tumour (carcinoma/Hodgkin's disease) Metastatic tumor (rare in childhood)
	CNS irradiation	Dose related
	Psychosocial deprivation	
Pituitary GH deficiency	Congenital malformations	Ectopic neurohypophysis Pituitary aplasia/hypoplasia
	Trauma	See above
	CNS tumors	Pituitary adenoma Any CNS tumor listed above
	Bioinactive GH syndrome	(May be genetic)

Source: Rosenfeld, 1996 *(143)*.

ing ectopic neurohypophysis, absent infundibulum, absent posterior pituitary bright spot, and small anterior pituitary *(64–66)*. These abnormalities are more frequently found in patients with multiple pituitary hormone deficiencies, more severe GHD (peak GH < 10 μg/L after provocation), and a history of perinatal asphyxia, particularly in association with breech delivery.

A nonstructural, reversible cause of GHD has been described in a subgroup of children with psychosocial stress and behavioral disturbance, sometimes known as psychosocial short stature. These children, from inadequate home environments, have growth failure and GHD as demonstrated either by provocative tests or diurnal profiles *(67–69)*. Once removed from the home, GHD reverses and growth improves *(70)*. Characteristic behavioral features have been observed in these children, including hyperphagia and polydipsia *(69)*.

GH Insensitivity

GHI may be genetic (where it may be termed GHI syndrome) or acquired. At a functional level, it may be secondary to a defect in either the GH receptor (GHR), or in an intracellular signaling molecule that mediates the signal of GHR activation to the nucleus. IGF-I gene deficiency could, in one sense, be considered to produce GH insensitivity, but as it causes a distinct syndrome, it has been separated out and was described previously. Again, genetic GHI is discussed first, as it provides a model of the GH-insensitive state.

Genetic GHI (or GHI Syndrome). The first cases of genetic GHI were described in 1966 by Laron and co-workers *(71)*, and this condition is also known as Laron syndrome. There have now been more than 200 cases reported in the world literature *(18)*. As discussed in more detail below, in the majority of cases where molecular studies have been performed a defect in the GHR gene has been described. In recent years, several cohorts of patients with genetic GHI have been investigated in detail, greatly advancing our understanding of this condition. These include 43 patients from Israel *(72,73)*, 56 children from an inbred population in Southern Ecuador *(74,75)*, and 82 genetically heterogeneous European patients *(76,77)*.

Clinical Features. There is striking similarity between the clinical phenotype of patients with inherited GHD and GHI. The growth pattern of genetic GHI has been studied in detail, and growth curves have been constructed from the Israeli GHI patients *(28)*. As in genetic GHD, significant intrauterine growth retardation does not occur, reflecting GH-independent regulation of fetal IGF-I production *(18,76)*. Postnatally, growth failure is progressive and severe, and the final adult stature of patients is usually <-5 SD below the mean *(18,28)*. Truncal adiposity and markedly delayed skeletal maturation are typical *(18,76)*. The facial appearance is indistinguishable from genetic GHD, with a prominent forehead and hypoplasia of the midface, and again cranial size is less abnormal than height *(78)* (Fig. 2b). Mental development is usually normal, and micropenis occurs in some males. Puberty may be mildly delayed, but adults are usually fertile *(18)*.

Biochemical Features. As in GHD, reduced IGF-I *(79–81)*, IGFBP-3 *(75,76)*, ALS *(18,82)*, and IGF-II *(75,76)* are typical. Interestingly, in both the Ecuadorian and European cohorts, IGFBP-3 SDS has been found to be a good marker of severity of disease, correlating with height SDS (Fig. 3). Laron has described one family, however, in which the three affected members had the characteristic phenotype of GHI, elevated GH secretion and low IGF-I, yet normal levels of IGFBP-3 *(83)*.

GH secretion is elevated, reflecting lack of hypothalamic feedback. Overnight studies of GH secretion in GHI patients have demonstrated preservation of GH pulsatility, but the magnitude of the GH peaks is higher with a failure of GH levels to fall to undetectable values between peaks as is usually seen *(84,85)*. Elevated basal GH is frequent but not invariable; for example, in our European survey of 82 GHI patients, individual basal GH values ranged from 0.3 to 319 μg/L *(77)*. Most commonly, GH provocation tests reveal an exaggerated GH rise; however, there is again wide variability: in the European survey of the 23 patients who had provocation tests, the median peak GH was elevated at 74 μg/L, but the range was from 12 to 230 μg/L *(76)*.

The response of IGF-I and IGFBP-3 levels to exogenously administered human GH (the "IGF-I generation test") has proven to be a useful test for the diagnosis of GHI. In the European survey hGH 0.1 U/kg body weight was administered daily for 4 d, with IGF-I and IGFBP-3 measurements on d 1 prior to the first GH injection and d 5 after the last injection *(76)*. Using a cutoff value of an increase in IGF-I from basal of 15 μg/L, and of IGFBP-3 of 400 ng/mL, a clear separation between the accepted and excluded groups was obtained *(86,87)*.

Growth hormone binding protein (GHBP) represents a circulating form of the extra-

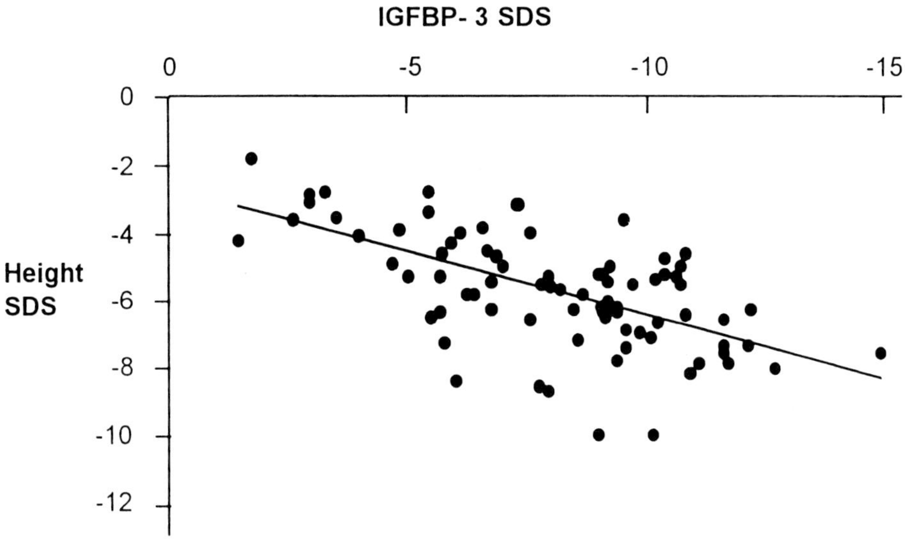

Fig. 3. Correlation of height SDS with IGFBP–3 SDS in European GHIS study (82 patients).

cellular domain of the GHR *(88)* (Fig. 1), and was initially found to be undetectable in patients with GHI *(89,90)*. It is now appreciated, however, that normal GHBP levels do not exclude a diagnosis of GHI, but are simply a marker for a molecular defect affecting the expression or GH binding affinity of the GH receptor *(91)*. In the European GHI cohort, 24% of patients were "GHBP positive," with normal, or even elevated GHBP activity *(76)*, and, in general, these individuals had a milder phenotype with less severe short stature *(77)*.

As with GHD, fasting hypoglycaemia is a frequent feature of GHI in infancy, although as it is often asymptomatic, it may be missed unless specifically looked for *(18,92)*. Serum insulin levels have generally been found to be low and relative insulinopenia is maintained following an oral glucose load *(72)*. IGFBP-1 levels are elevated but show a physiological fall through childhood *(76)*. The hypoglycemia usually improves with age, although may reappear during fasting, and Laron has documented insulin resistance with some patients developing hyperinsulinism in adulthood *(72)*.

Causes of Genetic GHI. GHR MUTATIONS. Mutations of the GH receptor gene, usually affecting both alleles, have been found in many cases of classic genetic GHI. This gene encodes for the GHR, a member of a recently defined family of hormone receptors including the receptors for prolactin, a number of cytokines, and eythropoetin *(93,94)*. These are membrane-bound receptors, possessing an extracellular, hormone binding domain, a transmembrane domain, and an intracellular domain, which interacts with cytoplasmic molecules to allow signaling to the nucleus. As the GHR mediates all of the biological actions of GH, the phenotypic similarity of many cases of genetic GHD and GHI can be understood.

One of the first mutations to be identified was a complex gene deletion *(95)*, which occurs in many of the Jewish-Oriental cases *(96)*. There have now been more than 30 different GHR gene mutations reported in the literature, and all but this first described gene deletion are point mutations, mainly located in the extracellular domain of the GHR *(95,97–102)* (Table 5). As the main function of the extracellular domain of the GHR is

Table 5
GH Receptor Mutations Associated with Genetic GHIs

Type	*Mutation*	*Exon*	*Year Reported*	*Ethnic Origin*
Missense	F96S	4	1989 *(97)*	Mediterranean
	D152H	6	1994 *(106)*	Asian
	R71K	4	1993 *(98)*	Mediterranean
	V125A	5	1993 *(98)*	European
	R161G	6	1993 *(98)*	Middle East
	R211G	7	1993 *(98)*	Mediterranean
	V144D	6	1993 *(98)*	Mediterranean
	C422F	10	1993 *(144)*	United States
	"	10	1997 *(145)*	Asia
	P561T[a]	10	1993 *(144)*	United States
	"	10	1997 *(145)*	Asia
	L526I	10	1997 *(145)*	Asia
	W50R	2	1997 *(146)*	Germany
	C38S	8	1997 *(146)*	Algeria
	S40L	9	1997 *(146)*	Turkey
Nonsense	R43X	4	1991 *(101)*	Mediterranean
	"	4	1993 *(100)*	Ecuador
	R217X	7	1993 *(100)*	US Afrocaribbean
	"	7	1993 *(98)*	States
	C38X	4	1993 *(98)*	Middle East
	"	4	1991 *(101)*	Mediterranean
	Q65X	4	1997 *(146)*	North Europe
	W80X	5	1997 *(146)*	India
	W157X	6	1997 *(146)*	Germany
	E183X	6	1994 *(147)*	Turkey
Splice	189^{-1} G/T	Intron 6	1993 *(100)*	Brazil
	E180splice (594 A/G)	6	1992 *(99)*	Ecuador
	440^{-1} G/T	Intron 5	1993 *(98)*	Mediterranean
	71^{+1} G $\rightarrow$ A	Intron 2	1993 *(100)*	Spain
	70^{+1} G $\rightarrow$ A	Intron 2	1997 *(146)*	Turkey
	723 C $\rightarrow$ T	G332G	1993 *(98)*	Spain
	266^{+1} G/A	Intron 4	1993 *(98)*	N. European
	874 G/C (R274T)	Intron 8	1996 *(105)*	Pakistan
		Intron 8	1997 *(107)*	UK
Frameshift	46 del TT	4	1993 *(100)*	Spain
	230 del AT	7	1993 *(100)*	South Africa
	N118-119 del TT	4	1995 *(134)*	?
	36delC	4	1997 *(146)*	Slovenia

[a]P561T has now been described as a common polymorphism in the Japanese population, and is therefore unlikely to have a pathological role *(148)*. This polymorphism was found on the same allele as C422F: no GHR mutation was found on the other allele. The patient's mother carried the same GHR polymorphisms on one allele and was of normal stature.

hormone binding, most of these mutations are thought to impair receptor action by reducing or abolishing the affinity of the receptor protein for GH, and have been shown, in some cases, to do so *(103,104)*. Most patients with extracellular mutations of the GHR have reduced or absent levels of GHBP, as measured by a functional assay relying on the GH-binding ability of serum, and these low GHBP levels are further evidence for a reduction of binding affinity of the mutant receptors, although reduction in receptor expression is also possible. More recently, we *(105)* and others *(106,107)* have reported GHR mutations in patients with normal or even elevated GHBP, which are thought to affect other functions of the GHR such as dimerization or intracellular signaling *(105,106)*.

Postreceptor Defects. Several patients of genetic GHI have been described in whom the defect appears to be at the postreceptor level, although the specific molecular defect has yet to be defined. These individuals may have severe GHI but have normal levels of GHBP, reflecting an intact GHR. The GHR signals through Janus-associated kinase 2 (JAK2), the signal transducers and activators of transcription (STATs), and mitogen-associated protein kinase (MAPK) (Fig. 1), and the defect in such patients may lie at any point along this signaling pathway.

We identified four GHBP-positive GHI patients from a highly consanguineous family with GHI, in whom segregation studies suggest that the disease is not linked to the GHR *(77)*. Clayton and co-workers have identified two GHBP-positive GHI families, in whom fibroblast studies have indicated abnormalities in intracellular signaling without GHR dysfunction *(108)*. In one family there is reduced activation of MAPK by GH, but normal DNA–STAT DNA binding, indicating the pathology may lie in the MAPK signaling pathway, and in the other family there was both reduced activation of MAPK and reduced DNA–STAT binding, indicating a defect between JAK2 and the GHR *(109)*. Finally, in Laron's family with the classic phenotype of GHI, low levels of IGF-I, and normal IGFBP-3 levels, GHBP was normal, and the authors postulate a defect in the signal transmission pathway to IGF-I synthesis *(83)*.

Partial GHI. In recent years interest has focused on whether "partial" GHI may explain a proportion of cases of short stature with normal or elevated GH secretion. Carlsson and co-workers measured levels of GHBP in a large population of short children and found that children with idiopathic short stature (ISS) had lower average GHBP levels than those with other conditions associated with short stature such as Turner's syndrome or GHD *(110)*. In the >18% of ISS children with GHBP levels > 2 SD below the normal mean for age and sex, mean serum IGF-I was significantly lower, and mean 12-h GH secretion significantly higher than in those with normal GHBP, suggesting a correlation between GHBP levels and the degree of GH sensitivity *(111)*. The authors suggest that the growth failure in this subset of children with ISS may result from partial GHI, the low GHBP levels indicating that this is caused either by an abnormality of GHR expression, or mutations in the extracellular domain resulting in a mild reduction of GH binding. Further support for the concept of "partial GHI" comes from our European survey of GHI patients, in whom we observed a number of less severe GHI cases, some of whom, despite failing our selection criteria, had clear evidence of mild GHI *(77,112)*.

GHR Defects in Partial GHI. Goddard and co-workers studied 14 children with ISS selected for low concentrations of serum GHBP and identified mutations in the extracellular domain of the GHR in four *(104)*. In three patients (height SDS –5.1, –3.3, and –3.2) the defects were heterozygous, and one patient (height SDS –2.9) was a compound heterozygote. This paper raised the possibility that GHR mutations, either in heterozygous form or resulting in mild GHR dysfunction, may explain a subset of ISS. However, in two of the patients in whom mutations were identified, GHBP levels were undetectable, suggesting that these were highly selected patients, not typical of most cases of ISS. Subsequently the same group have studied a less selected group of 86 children with ISS, and identified heterozygous GHR mutations in four children, all of whom had normal or elevated GHBP concentrations *(61)*. However, in studies of the families of the seven cases in total of heterozygous GHR mutations, short stature did not consistently segregate with the heterozygous state, suggesting either that the mutation identified was not the cause of the short stature, or a mutation on the other GHR allele, or even in another gene involved in growth, was also present in the probands.

Study of the first-degree relatives of patients with classic GHI and homozygous GHR mutations is another way of investigating the significance of heterozygous GHR mutations. We observed a reduction in the mean height of the parents of 19 patients with GHR mutations in our European cohort (mean height SDS –1.25, 95% CI for mean –1.6 to –0.9), as compared to Tanner–Whitehouse standards, although the significance of this reduction, given that many of the families were of Asian origin, is unclear *(77)*. Heterozygous first-degree relatives of patients with the Equadorian mutation E180 splice are shorter than their noncarrier sibs (mean height SDS heterozygotes –1.8, noncarriers –1.3), but this difference does not reach significance *(113)*. Thus the relevance of heterozygote GHR defects to the etiology of ISS remains uncertain. The two convincing cases of heterozygous GHR mutations that have been described have both been in patients with severe short stature and a classic severe GHI phenotype, the mutations appearing to have a dominant negative affect (*[107]*, S. Amselem, *personal communication*).

Acquired GHI. Many disease states, both acute and chronic, are associated with the biochemical profile of GHI (Table 6), which may explain the poor growth associated with many chronic illness states in childhood. As prolonged fasting and malnutrition may also induce GHI, it may be that the protein-calorie malnutrition and increased catabolism induced by these disease states may be the underlying causes *(114,115)*. However, in some of these conditions provision of adequate substrate does not completely reverse the GHI, suggesting a role for circulating "cachectic influences" such as cytokines or glucocorticoids *(116)*. In acute illness, the state of GHI may be a secondary adaptive response, whereby there is a shift from the indirect growth promoting actions of GH that are largely mediated by the IGFs to direct GH actions such as lipolysis and insulin antagonism *(117)*. In the setting of catabolism induced by injury or severe illness these changes might provide substrates essential for short-term fuel need but might be counterproductive in the longer term with respect to growth, and the maintenance of body composition.

The mechanism by which GHI is induced in disease states may be secondary to a defect either at the level of the GHR, in the signaling pathway from the GHR to the nucleus, or in reduction in IGF-I bioavailablity and/or clearance secondary to alteration in IGFBP levels. Low levels of GHBP, perhaps indicating reduced GHR expression, have been observed in anorexia nervosa *(118)*, the critically ill *(119)*, and diabetes mellitus *(120,121)*. As the GHR and many cytokine receptors share signaling mechanisms it is possible that cytokine secretion may induce down-regulation of GHR signaling. Interleukin-1-β and tumor necrosis factor-α have recently been shown to inhibit IGF-I messenger ribonucleic acid response to GH in rat hepatocytes *(114)*. Further evidence in support of this theory comes from recent data from a transgenic mouse line overexpressing the cytokine interleukin-6 *(122)*. Transgenic mouse lines expressing high levels of cir-

Table 6
Causes of Acquired GH Insensitivity

Primary nutritional disturbance	Prolonged fasting	
	Malnutrition	
	Anorexia nervosa	
Systemic injury	Critical illness	Trauma
		Sepsis
		Multiorgan failure
	Post major surgery	
	Major burns	
	Wound healing	
"Cachectic states"	Cancer cachexia	
	Chronic infection	Cystic fibrosis
	Chronic inflammation	Inflammatory bowel disease
		Juvenile chronic arthritis
Physiological states	Pregnancy	
	Aging	
Drug therapy	Corticosteriods	
Organ failure	Liver	Fulminant hepatic failure
		Cirrhosis
	Kidney	Acute renal failure
		Nephrotic syndrome
		Uremia
	Pancreas	Insulin-dependent diabetes

Source: Baxter *(116)*, with modifications.

culating IL-6 grew more slowly than control littermates despite having a normal food intake, an effect partly reversed by a monoclonal antibody to the IL-6 receptor. GH secretion was normal and IGF-I levels reduced, suggesting the increased expression of IL-6 had indeed led to a GH-resistant state. Furthermore, alteration of IGFBP levels, and IGF protease activity has been observed in many GH-resistant states, as recently reviewed by Donaghy and Baxter *(116)*. Thus, it appear likely that the etiology of acquired GH resistance is multifactorial, and will vary between disease states.

Bioinactive GH Syndrome

In "bioinactive GH syndrome" the GH molecule is structurally abnormal with reduced biological activity, yet is still measurable by immunoassay. Kowarski and colleagues first reported two cases of this condition in 1978 *(55)*, and there have been several further reports *(56,57,123–125)*. Affected individuals have a clinical phenotype indistinguishable from genetic GHD/GHI. In some respects, the biochemical profile is similar to that seen in GHI, with elevated GH secretion, and low levels of IGF-I. However, unlike GHI, administration of exogenous GH usually results in a rise in IGF-I and improved somatic linear growth. To investigate this condition further, a number of GH bioassays have been developed, including the NB2 bioassay, the IM-9 receptor modulation assay, and a radioreceptor assay *(126,127)*. Studies in children with growth failure using these bioassays indicate that this condition is a rare cause of short stature *(43,128)*.

GH Gene Mutations in Bioinactive GH Syndrome. There have now been two separate cases of GH gene defects, in both cases heterozygous missense mutations, associated with bioinactive GH syndrome. The first, reported in 1996, was identified in a 4.9-yr-old boy with severe short stature (height SDs –6.1) and biochemical evidence of bioinactive GH, with elevated GH secretion, reduced GH bioactivity on NB2 bioassay, and low–normal IGF-I levels *(129)*. The mutant GH had a greatly increased affinity for the GHR over wild-type GH, yet did not stimulate GHR signaling, possibly by inhibiting GHR dimerization. Functional studies indicated the mutant GH had a potent dominant negative action, inhibiting the ability of up to 10 times more wild-type GH to signal through the GHR, potentially explaining the heterozygous effect of the mutation. Interestingly, IGF-I levels did not rise after administration of exogenous hGH in the IGF-I generation test in this patient, and neither IGF-I levels nor rate of linear growth exhibited a sustained rise after long-term hGH therapy. The authors explain these surprising results on the basis of the dominant negative effect of the mutant GH. The second mutation to be described occurred in a 3-yr-old girl with significant short stature (–3.6 SD) *(130)*. Her biochemical findings were typical of bioinactive GH syndrome, and unlike the first case, both rate of growth and IGF-I levels increased substantially after hGH therapy. The mutant GH identified in this case again appeared to inhibit GHR dimerization and signal transduction, but a dominant negative effect could not be demonstrated.

IGF-II DEFICIENCY

No primary syndrome of IGF-II deficiency in humans has been described, although the existing IGF-II knockout mouse model gives us some idea as to the likely phenotype of such a condition. The factors controlling IGF-II levels in humans are unclear, and no defects in other genes have been found to lead to IGF-II deficiency, although a number of overgrowth syndromes have been associated with increased IGF-II expression *(131)*. Low IGF-II levels are seen in defects leading to GH deficiency or resistance, although these are thought to reflect a deficiency in the GH-dependent binding proteins, IGFBP-3 and ALS, as opposed to a reduced IGF-II synthesis.

IGF RESISTANCE

As the growth promoting effects of both IGF-I and IGF-II are thought to be mediated by the IGF-I receptor, any condition associated with IGF-I resistance will automatically be associated with resistance to IGF-II action. This is clinically relevant only in prenatal life, when IGF-II appears to have an additive effect on fetal growth *(1,2)*. Thus IGF-I receptor knockout mice exhibit extreme fetal growth retardation (birth weight 30% expected) and die at birth. No convincing case of a human IGF-I receptor defect has been described, although several cases of heterozygous deletion of the distal arm of chromosome 15, the site of the IGF-I receptor gene in humans, have been identified *(132,133)*. The phenotypic abnormalities in these children include intrauterine growth retardation and postnatal growth failure, which could conceivably result from the loss of one copy of the IGF-I receptor gene, particularly as the IGF-I receptor is a dimer. However, fibroblast studies in two affected individuals failed to provide evidence for impairment of the biological response to IGF-I, despite a reduction in IGF-I receptor expression *(133)*.

It has been suggested that the short stature in the Efe Pygmies of Northern Zaire may be secondary to IGF-I resistance, as lymphocyte studies indicate a combination of decreased IGF-I receptor gene transcription and IGF-I receptor signaling in this population *(134)*. Reduced IGF-I production *(135)* and reduced GHR expression *(136)* have also been suggested to play a part, however, and it is likely that the etiology of the short stature in the African Pygmy is multifactorial. There is a single case report of a child with short stature and normal serum GH and IGF-I levels who had reduced fibroblast responsiveness to IGF-I *(137)*. IGF-I binding was not reduced, suggesting that the IGF-I receptor was not defective, but large quantities of an abnormal IGF binding protein were identified on the cell surface of the fibroblasts, possibly competing with IGF-I binding to the IGF-I receptor *(138)*. Finally, IGF resistance has been observed in acute and chronic renal failure *(139)*, and may have a role in the short stature observed in other chronic conditions.

CONCLUSION

IGF deficiency can present in many ways, and has many potential causes. In recent years there have been great advances in our understanding of this area, mainly as a result of the discovery of new gene defects in children with severe short stature. Thus, the phenotype of complete pre- and postnatal IGF-I deficiency in humans has been illustrated by our recently identified case of human IGF-I gene deletion. The striking differences between the phenotype of this patient with primary IGF deficiency and individuals with IGF deficiency secondary to GHD or GHI demonstrates clearly that the actions of GH and IGF-I action are not always synonymous, particularly during fetal life. As the number of new discoveries at each level of the GH-IGF-I axis increases, the complexities of the endocrine control of human growth become apparent. The major challenge now lying ahead is to apply the knowledge we have gained by these discoveries to improve our understanding of the growth failure in all children with short stature, in order that we might begin to offer therapy appropriate to each individual case.

REFERENCES

1. Liu J, Baker J, Perkins AS, Robertson EJ, Efstratiadis A. Mice carrying null mutations of the genes encoding insulin-like growth factor I (*Igf-1*) and type 1 IGF receptor (*Igf1r*). Cell 1993; 75:59–72.
2. Baker J, Liu J, Robertson EJ, Efstratiadis A. Role of insulin-like growth factors in embryonic and postnatal growth. Cell 1993; 75:73–82.
3. DeChiara TM, Efstratiadis A, Robertson EJ. A growth-deficiency phenotype in heterozygous mice carrying an insulin-like growth factor II gene disrupted by targeting. Nature 1990; 345:78–80.
4. Zhou J, Bondy C. Anatomy of the insulin-like growth factor system in the human testis. Fertil Steril 1993; 60:897–904.
5. Backeljauw PF, Underwood LE, the GHIS Collaborative Group. Prolonged treatment with recombinant insulin-like growth factor-I in children with growth hormone insensitivity syndrome—a clinical research center study. J Clin Endocrinol Metab 1996; 81:3312–3317.
6. Ranke MB. Treatment of growth hormone insensitivity syndrome (GHIS) with insulin-like growth factor (IGF-I). In: Savage MO, Ross RJM, eds. Growth Hormone Resistance. Ballière Tindall, London, 1996, pp. 401–410.
7. Powell-Braxton L, Hollinghead P, Warburton C, Dowd M, Pitts-Meek S, Dalton D, Gillett N, Stewart TA. IGF-I is required for normal embryonic growth in mice. Genes Dev 1993; 7:2609–2617.
8. Beck KD, Powell-Braxton L, Widmer H, Valverde J, Hefti F. Igf1 gene disruption results in reduced brain size, CNS hypomyelination, and loss of hippocampal granule and striatal parvalbumin-containing neurons. Neuron 1995; 14:717–730.
9. Baker J, Hardy MP, Zhou J, Bondy C, Lupu F, Bellve A, Efstratiadis A. Effects of an Igf1 gene null mutation on mouse reproduction. Mol Endocrinol 1996; 10:903–918.
10. Woods KA, Camacho-Hübner C, Savage MO, Clark AJL. Intrauterine growth retardation and postnatal growth failure associated with deletion of the insulin-like growth factor I gene. N Engl J Med 1996; 335:1363–1367.

11. Freeman JV, Cole TJ, Chinn S, Jones PRM, White EM, Preece MA. Cross sectional stature and weight reference curves for the UK. Arch Dis Child 1995; 73:17–24.
12. Gargosky SE, Wilson KF, Fielder PJ, Vaccarello MA, Guevara-Aguirre J, Diamond FB, Baxter RC, Rosenbloom AL. The composition and distribution of insulin-like growth factors (IGFs) and IGF-binding proteins (IGFBPs) in the serum of growth hormone receptor-deficient patients: effects of IGF-I therapy on IGFBP-3. J Clin Endocrinol Metab 1993; 77:1683–1689.
13. Hardouin S, Gourmelen M, Noguiez P, Seurin D, Roghani M, Le Bouc Y, Povoa G, Merimee TJ, Hossenlopp P, Binoux M. Molecular forms of serum insulin-like growth factor (IGF)-binding proteins in man: relationships with growth hormone and IGFs and physiological significance. J Clin Endocrinol Metab 1989; 69:1291–1301.
14. Zapf J, Hauri C, Waldvogel M, Futo E, Hasler H, Binz K, Guler HP, Schmid C, Froesch ER. Recombinant human insulin-like growth factor I induces its own specific carrier protein in hypophysectomized and diabetic rats. Proc Natl Acad Sci USA 1989; 86:3813–3817.
15. Camacho-Hübner C, Clemmons DR, D'Ercole AJ. Regulation of insulin-like growth factor (IGF) binding proteins in transgenic mice with altered expression of growth hormone and IGF-I. Endocrinology 1991; 129:1201–1206.
16. Naeye RL. Do placental weights have clinical significance? Hum Pathol 1987; 18:387–392.
17. Rivarola MA, Phillips JA, 3d, Migeon CJ, Heinrich JJ, Hjelle BJ. Phenotypic heterogeneity in familial isolated growth hormone deficiency type I-A. J Clin Endocrinol Metab 1984; 59:34–40.
18. Rosenfeld RG, Rosenbloom AL, Guevara-Aguirre J. Growth hormone (GH) insensitivity due to primary GH receptor deficiency (Review). Endocr Rev 1994; 15:369–390.
19. Salmon WD, Daughaday WH. A hormonally controlled serum factor which stimulates sulfate incorporation by cartilage in vitro. J Lab Clin Med 1957; 49:825–836.
20. Vimpani GV, Vimpani AF, Lidgard GP, Cameron EHD, Farquhar JW. Prevalence of severe growth hormone deficiency. Br Med J 1977; 2:427–430.
21. Rona RJ, Tanner JM. Aetiology of idiopathic growth hormone deficiency in England and Wales. Arch Dis Child 1977; 52:197–208.
22. Lacey KA, Parkin JM. Causes of short stature: a community study of children in Newcastle-upon-Tyne. Lancet 1974; 1:42–45.
23. Lindsay R, Harris D, Feldkamp M, Robertson J, Rallison M. Utah growth study: growth hormone deficiency prevalence. Pediatr Res 1993; 33:81A.
24. Craft W, Underwood LE, Van Wyk JJ. High incidence of perinatal insult in children with idiopathic hypopituitarism. J Pediatr 1980; 96:397.
25. Phillips JA III, Cogan J. Genetic Basis of Endocrine Disease: Molecular basis of familial human growth hormone deficiency. J Clin Endocrinol Metab 1994; 78:11–16.
26. Gluckman PD, Gunn AJ, Wray A, Cutfield WS, Chatelain PG, Guilbaud O, Ambler GR, Wilton P, Albertsson Wikland K. Congenital idiopathic growth hormone deficiency associated with prenatal and early postnatal growth failure. The International Board of the Kabi Pharmacia International Growth Study. J Pediatr 1992; 121:920–923.
27. Wit JM, van Unen H. Growth of infants with neonatal growth hormone deficiency. Arch Dis Child 1992; 67:920–924.
28. Laron Z, Lilos P, and Klinger B. Growth curves for Laron syndrome. Arch Dis Child 1993; 68: 768–770.
29. Lovinger RD, Kaplan SL, Grumbach MM. Congenital hypopituitarism associated with neonatal hypoglycemia and microphallus: four cases secondary to hypothalamic hormone deficiencies. J Pediatr 1975; 87:1171–1181.
30. Daughaday WH, Laron Z, Pertzelan A, Heins JN. Defective sulfation factor generation: a possible etiological link in dwarfism. Trans Assoc Am Physicians 1969; 82:129–140.
31. Liu F, Hintz RL, Khare A, Diaugustine RP, Powell DR, Lee PD. Immunoblot studies of the IGF-related acid-labile subunit. J Clin Endocrinol Metab 1994; 79:1883–1886.
32. Blum WF, Ranke MB, Kietzmann K, Gauggel E, Zeisel HJ, Bierich JR. A specific radioimmunoassay for the growth hormone (GH)-dependent somatomedin-binding protein: its use for diagnosis of GH deficiency. J Clin Endocrinol Metab 1990; 70:1292–1298.
33. Ranke MB, Blum WF, Bierich JR. Clinical relevance of serum measurements of insulin-like growth factors and somatomedin binding proteins. Acta Paediatr Scand (Suppl) 1991; 347:114–126.
34. Phillips JA III, Parks BL, Hielle JE, Plotnick LP. Genetic analysis of familial isolated growth hormone deficiency. J Clin Invest 1982; 70:489–495.
35. Phillips JA III, Cogan JD, Raskin S, Parker RA. Pedigree analysis of familial isolated growth hormone deficiency. Am J Hum Genet 1992; 51(Suppl):A1459

36. Perez-Jurado LA, Argente J, Raskin S, Garcia F, Phillips JA III. Genetic analysis of familial isolated growth hormone deficiency type II. Horm Res 1992; 37(Suppl 4):A198
37. Phillips JA III, Hjelle BL, Seeburg PH, Zachmann M. Molecular basis for familial isolated growth hormone deficiency. Proc Natl Acad Sci USA 1981; 78:6372–6375.
38. Vnencak-Jones CL, Phillips JA III. Hot spots for growth hormone gene deletions in homologous regions outside of Alu repeats. Science 1990; 250:1745–1748.
39. Vnencak-Jones CL, Phillips JA III, De-fen W. Use of polymerase chain reaction in detection of growth hormone gene deletions. J Clin Endocrinol Metab 1990; 70:1550–1553.
40. Illig R. Growth hormone antibodies in patients treated with different preparations of human growth hormone (HGH). J Clin Endocrinol Metab 1970; 31:679–688.
41. Illig R, Prader A, Ferrandez A, Zachmann M. Hereditary prenatal growth hormone deficiency with increased tendency to growth hormone antibody formation ('a-type' isolated growth hormone deficiency). Acta Paediatr Scand (Suppl) 1971; 60:607.
42. Cogan JD, Phillips JA III, Sakati N, Frisch H, Schober E, Milner RDG. Heterogeneous growth hormone (GH) gene mutations in familial GH deficiency. J Clin Endocrinol Metab 1993; 76:1224–1228.
43. Miyata I, Cogan JD, Prince MA, Kamijo T, Ogawa M, Phillips JA. Detection of growth hormone gene defects by dideoxy fingerprinting (ddF). Endocr J 1997; 44:149–154.
44. Conley ME, Burks AW, Herrod HG, Puck JM. Molecular analysis of X-linked agammaglobulinemia with growth hormone deficiency. J Pediatr 1991; 119:392–397.
45. Wajnrajch MP, Gertner JM, Harbison MD, Chua SC Jr, Leibel RL. Nonsense mutation in the human growth hormone-releasing hormone receptor causes growth failure analogous to the little (lit) mouse. Nature Genet 1996; 12:88–90.
46. Maheshwari HG, Silverman BL, Dupuis J, Baumann G. Phenotype and genetic analysis of a syndrome caused by an inactivating mutation in the growth hormone-releasing receptor: Dwarfism of Sindh. J Clin Endocrinol Metab 1998; 83:4065–74.
47. McCormick A, Brady H, Theill LE, Karin M. Regulation of the pituitary-specific homeobox gene GHF1 by cell-autonomous and environmental cues. Nature 1990, 345:829–832.
48. Ingraham HA, Flynn SE, Voss JW, Albert V, Kapiloff MS, Wilson L, Rosenfeld MG. The POU-specific domain of pit-1 is essential for sequence specific, high affinity DNA binding and DNA-dependant pit-1–pit-1 interactions. Cell 1990; 61:1021–1033.
49. Li S, Crenshaw EB III, Rawson EJ, Simmons DM, Swanson LW, Rosenfeld MG. Dwarf locus mutants lacking three pituitary cell types result from mutations in the POU-domain gene pit-1. Nature 1990; 347:528–533.
50. Tatzumi K, Miyai K, Notomi T, Kaibe K, Amino N, Mizuno Y, Kohno H. Cretinism with combined hormone deficiency caused by a mutation in the pit-1 gene. Nature Genet 1992; 1:56–58.
51. Radovick S, Nations M, Du Y, Berg LA, Weintraub BD. A mutation in the POU-domain of pit-1 is responsible for combined pituitary hormone deficiency. Science 1992; 257:1115–1118.
52. Pfäffle RW, DiMattia GE, Parks JS, Brown MR, Wit JM, Jansen M, Van der Nat H, Van den Brande JL, Rosenfeld MG, Ingraham HA. Mutation of the POU-specific domain of pit-1 and hypopituitarism without pituitary hypoplasia. Science 1992; 257:1118–1121.
53. Woods KA, Weber A, Clark AJL. The molecular pathology of pituitary hormone deficiency and resistance. In: Thakker R, ed. Genetic and Molecular Biological Aspects of Endocrine Disease. Ballière Tindall, London, 1995; pp. 453–487.
53a. Wu W, Cogan JD, Pfäffle RW, Dasen JS, Frisch H, O'Connell SM, Flynn SE, Brown MR, Mullis PE, Parks JS, Phillips JA III, Rosenfeld MG. Mutations in *Prop1* cause familial combined pituitary hormone deficiency. Nature Genet 1998; 18:147–149.
53b. Sornson MW, Wu W, Dasen JS, Flynn SE, Norman DJ, O'Connell SM, Gukovsky I, Carriere C, Ryan AK, Miller AP, Zuo L, Gleiberman AS, Andersen B, Beamer WG, Rosenfeld MG. Pituitary lineage determination by the Prophet of Pit-1 homeodomain factor defective in Ames dwarfism. Nature 1996; 384(6607):327–333.
53c. Dattani MT, Martinez-Barbera J, Thomas PQ, Brickman JM, Gupta R, Mårtensson I, Toresson H, Fox M, Wales JKH, Hindmarsh PC, Krauss S, Beddington RSP, Robinson ICAF. Mutations in the homeobox gene HESX1/Hesx1 associated with septo-optic dysplasia in human and mouse. Nature Genet 1998; 19:125–133.
54. Rosenfeld RG. Editorial: Is growth hormone deficiency a viable diagnosis? J Clin Endocrinol Metab 1997; 82:349–351.
55. Kowarski AA, Schneider J, Ben-Galim E, Weldon VV, Daughaday WH. Growth failure with normal serum RIA-GH and low somatomedin activity: somatomedin restoration and growth acceleration after exogenous GH. J Clin Endocrinol Metab 1978; 47:461–464.

56. Valenta LJ, Sigel MB, Lesniak MA, Elias AN, Lewis UJ, Friesen HG, Kershnar AK. Pituitary dwarfism in a patient with circulating abnormal growth hormone polymers. N Engl J Med 1985; 312:214–217.
57. Rudman D, Kutner MH, Blackston RD, Cushman RA, Bain RP, Patterson JH. Children with normal-variant short stature: treatment with human growth hormone for six months. N Engl J Med 1981; 305:123–131.
58. Spiliotis BE, August GP, Hung W, Sonis W, Mendelson W, Bercu BB. Growth hormone neurosecretory dysfunction. A treatable cause of short stature. JAMA 1984; 251:2223–2230.
59. Shulman DI, Bercu BB. Evaluation of growth hormone secretion: provocative testing vs endogenous 24-hour growth hormone profile. Acta Paediatr Scand Suppl 1987; 337:61–71.
60. Bercu BB, Shulman D, Root AW, Spiliotis BE. Growth hormone (GH) provocative testing frequently does not reflect endogenous GH secretion. J Clin Endocrinol Metab 1986; 63:709–716.
61. Goddard AD, Dowd P, Chernausek S, Geffner M, Gertner J, Hintz R, Hopwood N, Kaplan S, Plotnick L, Rogol A, Rosenfield R, Saenger P, Mauras N, Hershkopf R, Angulo M, Attie K. Partial growth-hormone insensitivity: the role of growth-hormone receptor mutations in idiopathic short stature. J Pediatr 1997; 131:S51–5.
62. Cotterill AM, Camacho-Hbner C, Woods KA, Martinelli C, Duquesnoy P, Savage MO. The insulin-like growth factor generation test in the investigation of short stature. Acta Paediatr Scand Suppl 1994; 399:128–130.
63. Thalange NK, Price DA, Gill MS, Whatmore AJ, Addison GM, Clayton PE. Insulin-like growth factor binding protein-3 generation: an index of growth hormone insensitivity. Pediatr Res 1996; 39:849–855.
64. Fujisawa I, Kikuchi K, Nishimura K, Togashi K, Itoh K, Noma S, Minami S, Sagoh T, Hiraoka T, Momoi T, et al. Transection of the pituitary stalk: development of an ectopic posterior lobe assessed with MR imaging. Radiology 1987; 165:487–489.
65. Abrahams JJ, Trefelner E, Boulware SD. Idiopathic growth hormone deficiency: MR findings in 35 patients. Am J Neuroradiol 1991; 12:155–160.
66. Kuroiwa T, Okabe Y, Hasuo K, Yasumori K, Mizushima A, Masuda K. MR imaging of pituitary dwarfism. Am J Neuroradiol 1991; 12:161–164.
67. Powell GF, Brasel JA, Blizzard RM. Emotional deprivation and growth retardation simulating idiopathic hypopituitarism. I. Clinical evaluation of the syndrome. N Engl J Med 1967; 276:1271–1278.
68. Powell GF, Brasel JA, Raiti S, Blizzard RM. Emotional deprivation and growth retardation simulating idiopathic hypopituitarism. II. Endocrinologic evaluation of the syndrome. N Engl J Med 1967; 276:1279–1283.
69. Skuse D, Albanese A, Stanhope R, Gilmour J, Voss L. A new stress-related syndrome of growth failure and hyperphagia in children, associated with reversibility of growth-hormone insufficiency. Lancet 1996; 348:353–358.
70. Powell GF, Hopwood NJ, Barratt ES. Growth hormone studies before and during catch-up growth in a child with emotional deprivation and short stature. J Clin Endocrinol Metab 1973; 37:674–679.
71. Laron Z, Pertzelan A, Mannheimer S. Genetic pituitary dwarfism with high serum concentration of growth hormone—a new inborn error of metabolism? Isr J Med Sci 1966; 2:152–155.
72. Laron, Z. Laron-type dwarfism (hereditary somatomedin deficiency): a review. In: Frick H, von Harnack G, Kochesiek K, Martini GA, Prader A, eds. Advances in Internal Medicine. Springer-Verlag, Berlin, 1984; 117–150.
73. Laron Z, Klinger B. Laron syndrome: clinical features, molecular pathology and treatment. Horm Res 1994; 42:198–202.
74. Rosenbloom AL, Guevara-Aguirre J, Rosenfeld RG, Fielder PJ. The little women of Loja—growth hormone receptor deficiency in an inbred population of southern Ecuador. N Engl J Med 1990; 323:1367–1374.
75. Guevara-Aguirre J, Rosenbloom AL, Fielder PJ, Diamond FB Jr, Rosenfeld RG. Growth hormone receptor deficiency in Ecuador: clinical and biochemical phenotype in two populations. J Clin Endocrinol Metab 1993; 76:417–423.
76. Savage MO, Blum WF, Ranke MB, Postel-Vinay M, Cotterill AM, Hall K, Chatelain PG, Preece MA, Rosenfeld RG. Clinical features and endocrine status in patients with growth hormone insensitivity (Laron syndrome). J Clin Endocrinol Metab 1993; 77:1465–1471.
77. Woods KA, Dastot F, Preece MA, Clark AJL, Postel-Vinay M, Chatelain PG, Ranke MB, Rosenfeld RG, Amselem S, Savage MO. Phenotype: genotype relationships in growth hormone insensitivity syndrome. J Clin Endocrinol Metab 1997; 82:3529–3535.
78. Schaefer GB, Rosenbloom AL, Guevara-Aguirre J, Campbell EA, Ulrich F, Patil K, Frias JL. Facial

morphometry of Ecuadorian patients with growth hormone receptor deficiency/Laron syndrome. J Med Genet 1994; 31:635–639.
79. Laron Z, Pertzelan A, Karp M, Kowaldo-Silbergeld A, Daughaday WH. Administration of growth hormone to patients with familial dwarfism with high plasma immunoreactive growth hormone: measurement of sulfation factor, metabolic and linear growth responses. J Clin Endocrinol Metab 1971; 33:332–342.
80. New MI, Schwartz E, Parks GA, Landley S, Wiedemann E. Pseudohypopituitary dwarfism with normal plasma growth hormone and low serum sulfation factor. J Pediatr 1972; 80:620–626.
81. Van den Brande JL, Du Caju MVL, Visser HKA, Schopman W, Hackeng WHL, Degenhart HJ. Primary somatomedin deficiency: Case report. Arch Dis Child 1974; 49:297–304.
82. Gargosky SE, Wilson KF, Fielder PJ, Vaccarello MA, Diamond FB, Baxter RC, Rosenbloom AL, Guevara-Aguirre J, Rosenfeld RG. Effects of insulin-like growth factor I treatment on the molecular distribution of insulin-like growth factors among different binding proteins. Acta Paediatr Suppl 1994; 399:159–62.
83. Laron Z, Klinger B, Eshet R, Kaneti H, Karasik A, Silbergeld A. Laron syndrome due to a post-receptor defect: response to IGF-I treatment. Isr J Med Sci 1993; 29:757–763.
84. Cotterill AM, Holly JMP, Taylor AM, Davies SC, Coulson VJ, Preece MA, Wass JAH, Savage MO. The insulin-like growth factor binding proteins and insulin-like growth factor bioactivity in Laron-type dwarfism. J Clin Endocrinol Metab 1992; 74:56–63.
85. Cotterill AM, Holly JMP, Snodgrass GAI, Savage MO. Regulation of growth hormone secretion in patients with growth hormone insensitivity. Acta Paediatr Scand (Suppl) 1991; 377:92–95.
86. Blum WF, Cotterill AM, Postel-Vinay M, Ranke MB, Savage MO, Wilton P. Improvement of diagnostic criteria in growth hormone insensitivity syndrome: solutions and pitfalls. Acta Paediatr Scand (Suppl) 1994; 399:117–124.
87. Blum WF, Ranke MB, Savage MO, Hall K. Insulin-like growth factors and their binding proteins in patients with growth hormone receptor deficiency: suggestions for new diagnostic criteria. Acta Paediatr Suppl 1992; 125:126.
88. Leung DW, Spencer SA, Cachianes G, Hammonds RG, Collins C, Henzel WJ, Barnard R, Waters MJ, Wood WI. Growth hormone receptor and serum binding protein: purification, cloning and expression. Nature 1987; 330:537–543.
89. Daughaday WH, Trivedi B. Absence of serum growth hormone binding protein in patients with growth hormone receptor deficiency (Laron dwarfism). Proc Natl Acad Sci USA 1987; 84:4636–4640.
90. Baumann G, Shaw MA, Winter RJ. Absence of plasma growth hormone-binding protein in Laron-type dwarfism. J Clin Endocrinol Metab 1987; 65:814–816.
91. Woods KA, Savage MO. Laron syndrome: typical and atypical forms. In: Savage MO, Ross RJM, eds. Growth Hormone Resistance. Ballière Tindall, London, 1996; 371–388.
92. Guevara-Aguirre J, Vasconez O, Martinez V, Martinez AL, Rosenbloom AL, Diamond FBJ, Gargosky SE, Nonoshita L, Rosenfeld RG. A randomized, double blind, placebo controlled trial on safety and efficacy of recombinant human insulin-like growth factor-I in children with growth hormone receptor deficiency. J Clin Endocrinol Metab 1995; 80:1393–1398.
93. Boutin JM, Edery M, Shirota M, Jolicoeur C, Leseur L, Ali S, Gould D, Dijane J, Kelly PA. Identification of a cDNA encoding a long form of prolactin receptor in human hepatoma and breast cancer cells. Mol Endocrinol 1989; 3:1455–1461.
94. Bazin JF. Structural design and molecular evolution of a cytokine receptor superfamily. Proc Natl Acad Sci USA 1990; 87:6934–6938.
95. Godowski PJ, Leung DW, Meacham LR, Galgani JP, Helmiss R, Keret R, Rotwein PS, Parks JS, Laron Z, Wood WI. Characterization of the human growth hormone receptor gene and demonstration of a partial gene deletion in two patients with Laron-type dwarfism. Proc Natl Acad Sci USA 1989; 86:8083–8087.
96. Meacham LR, Brown MR, Murphy TL, Keret R, Silbergeld A, Laron Z, Parks JS. Characterization of a noncontiguous gene deletion of the growth hormone receptor in Laron's syndrome. J Clin Endocrinol Metab 1993; 77:1379–1383.
97. Amselem S, Duquesnoy P, Attree O, Novelli G, Bousina S, Postel-Vinay M, Goossens M. Laron dwarfism and mutations of the growth hormone receptor gene. N Engl J Med 1989; 321:989–995.
98. Amselem S, Duquesnoy P, Duriez B, Dastot F, Sobrier M, Valleix S, Goossens M. Spectrum of growth hormone receptor mutations and associated haplotypes in Laron syndrome. Hum Mol Genet 1993; 2:355–359.

99. Berg MA, Guevara-Aguirre J, Rosenbloom AL, Rosenfeld RG, Francke U. Mutation creating a new splice site in the growth hormone receptor genes of 37 Ecuadorian patients with Laron syndrome. Hum Mutat 1992; 1:24–34.
100. Berg MA, Argente J, Chernausek S, Gracia R, Guevara-Aguirre J, Hopp M, Perez-Jurado LA, Rosenbloom AL, Toledo SPA, Francke U. Diverse growth hormone receptor gene mutations in Laron syndrome. Am J Hum Genet 1993; 52:998–1005.
101. Amselem S, Sobrier M, Duquesnoy P, Rappaport R, Postel-Vinay M, Gourmelen M, Dallapiccola B, Goossens M. Recurrent nonsense mutations in the growth hormone receptor from patients with Laron dwarfism. J Clin Invest 1991; 87:1098–1102.
102. Sobrier ML, Dastot F, Duquesnoy P, Kandemir N, Yordam N, Goossens M, Amselem S. Nine novel growth hormone receptor gene mutations in patients with Laron syndrome. J Clin Endocrinol Metab 1997; 82:435–437.
103. Edery M, Rozakis-Adcock M, Goujon L, Finidori J, Levi-Meyrueis C, Paly J, Djiane J, Postel-Vinay M, Kelly PA. Lack of hormone binding in COS-7 cells expressing a mutated growth hormone receptor found in Laron dwarfism. J Clin Invest 1993; 91:838–844.
104. Goddard AD, Covello R, Luoh SM, Clackson T, Attie KM, Gesundheit N, Rundle AC, Wells JA, Carlsson LMS. Mutations of the growth hormone receptor in children with idiopathic short stature. N Engl J Med 1995; 333:1093–1098.
105. Woods KA, Fraser NC, Postel-Vinay M, Savage MO, Clark AJL. A homozygous splice site mutation affecting the intracellular domain of the growth hormone (GH) receptor resulting in Laron Syndrome with elevated GH-binding protein. J Clin Endocrinol Metab 1996; 81:1686–1690.
106. Duquesnoy P, Sobrier M, Duriez B, Dastot F, Buchanan CR, Savage MO, Preece M, Craescu CT, Blouquit Y, Goossens M, Amselem S. A single amino acid substitution in the exoplasmic domain of the human growth hormone (GH) receptor confers familial GH resistance (Laron syndrome) with positive GH-binding activity by abolishing receptor homodimerization. EMBO Journal 1994; 13:1386–1395.
107. Ayling RM, Ross R, Towner P, Von Laue S, Finidori J, Moutoussamy S, Buchanan CR, Clayton PE, Norman MR. A dominant-negative mutation of the growth hormone receptor causes familial short stature (letter). Nature Genet 1997; 16:13–14.
108. Freeth JS, Ayling RM, Whatmore AJ, Towner P, Price DA, Norman MR, Clayton PE. Human skin fibroblasts as a model of growth hormone (GH) action in GH receptor-positive Laron's syndrome. Endocrinology 1997; 138:55–61.
109. Freeth JS, Whatmore AJ, Silva CM, Clayton PE. Growth hormone binding protein positive Laron Syndrome (LS): two models of GH resistance due to abnormal intracellular GH signalling. Horm Res 1997; 48:4 (abstract).
110. Carlsson LM, Attie KM, Compton PG, Vitangcol RV, Merimee TJ. Reduced concentration of serum growth hormone-binding protein in children with idiopathic short stature. National Cooperative Growth Study. J Clin Endocrinol Metab 1994; 78:1325–1330.
111. Attie KM, Carlsson LM, Rundle AC, Sherman BM. Evidence for partial growth hormone insensitivity among patients with idiopathic short stature. The National Cooperative Growth Study. J Pediatr 1995; 127:244–250.
112. Savage MO, Carlsson L, Chatelain PG, Ranke MB, Rosenfeld RG. Growth hormone insensitivity syndromes. Acta Paediatr Suppl 1995; 411:87–90.
113. Rosenbloom AL, Guevara-Aguirre J, Rosenfeld RG, Fielder PJ. Is there heterozygote expression of growth hormone receptor deficiency? Acta Paediatr Suppl 1994; 399:125–127.
114. Thissen JP, Verniers J. Inhibition by interleukin-1 beta and tumor necrosis factor-alpha of the insulin-like growth factor I messenger ribonucleic acid response to growth hormone in rat hepatocyte primary culture. Endocrinology 1997; 138:1078–1084.
115. Ketelslegers JM, Maiter D, Maes M, Underwood LE, Thissen JP. Nutritional regulation of insulin-like growth factor-I. Metabolism 1995; 44:50–57.
116. Donaghy AJ, Baxter RC. Insulin-like growth factor bioactivity and its modification in growth hormone resistant states. Baillieres Clin Endocrinol Metab 1996; 10:421–446.
117. Phillips LS, Unterman TG. Somatomedin activity in disorders of nutrition and metabolism. Clin Endocrinol Metab 1984; 13:145–189.
118. Counts DR, Gwirtsman H, Carlsson LM, Lesem M, Cutler GB Jr. The effect of anorexia nervosa and refeeding on growth hormone-binding protein, the insulin-like growth factors (IGFs), and the IGF-binding proteins. J Clin Endocrinol Metab 1992; 75:762–767.

119. Ross RJ, Miell JP, Holly JM, Maheshwari H, Norman M, Abdulla AF, Buchanan CR. Levels of GH binding activity, IGFBP-1, insulin, blood glucose and cortisol in intensive care patients. Clin Endocrinol Oxf 1991; 35:361–367.
120. Menon RK, Arslanian S, May B, Cutfield WS, Sperling MA. Diminished growth hormone-binding protein in children with insulin-dependent diabetes mellitus. J Clin Endocrinol Metab 1992; 74:934–938.
121. Clayton KL, Holly JM, Carlsson LM, Jones J, Cheetham TD, Taylor AM, Dunger DB. Loss of the normal relationships between growth hormone, growth hormone-binding protein and insulin-like growth factor-I in adolescents with insulin-dependent diabetes mellitus. Clin Endocrinol Oxf 1994; 41:517–524.
122. De Benedetti F, Alonzi T, Moretta A, Lazaro D, Costa P, Poli V, Martini A, Ciliberto G, Fattori E. Interleukin 6 causes growth impairment in transgenic mice through a decrease in insulin-like growth factor–1. J Clin Invest 1997; 99:643–650.
123. Frazer T, Gavin JR, Daughaday WH, Hillman RE, Weldon W. Growth hormone—dependent growth failure. J Pediatr 1982; 101:12–15.
124. Plotnick LP, Van Meter QL, Kowarski AA. Human growth hormone treatment of children with growth failure and normal growth hormone levels by immunoassay: lack of correlation with somatomedin generation. Pediatrics 1983; 71:324–327.
125. Bright GM, Rogol AD, Johanson AJ, Blizzard RM. Short stature associated with normal growth hormone and decreased somatomedin-C concentrations: response to exogenous growth hormone. Pediatrics 1983; 71:576–580.
126. Ilondo MM, Vanderschueren Lodeweyckx M, De Meyts P. Measuring growth hormone activity through receptor and binding protein assays. Horm Res 1991; 36 Suppl, 1:21–26.
127. Tokuhiro E, Dean HJ, Friesen HG, Rudman D. Comparative study of serum human growth hormone measurement with NB2 lymphoma cell bioassay, IM–9 receptor modulation assay, and radio-immunassay in children with disorders of growth. J Clin Endocrinol Metab 1984; 58:549–554.
128. Belgorosky A, Rivarola MA. Progressive increase in non-sex-hormone-binding globulin-bound test-osterone from infancy to late prepuberty in boys. J Clin Endocrinol Metab 1987; 64:482–485.
129. Takahashi Y, Kaji H, Okimura Y, Goji K, Abe H, Chihara K. Brief report: short stature caused by a mutant growth hormone [see comments]. N Engl J Med 1996; 334:432–436.
130. Takahashi Y, Shirono H, Arisaka O, Takahashi K, Yagi T, Koga J, Kaji H, Okimura Y, Abe H, Tanaka T, Chihara K. Biologically inactive growth hormone caused by an amino acid substitution. J Clin Invest 1997; 100:1159–1165.
131. Morison IM, Becroft DM, Taniguchi T, Geoffrey Woods C, Reeve AE. Somatic overgrowth associated with overexpression of insulin-like growth factor II. Nat Med 1996; 3:311–316.
132. Tamura T, Tohma T, Ohta T, Soejima H, Harada H, Abe K, Niikawa N. Ring chromosome 15 involving deletion of the insulin-like growth factor 1 receptor gene in a patient with features of Silver-Russell syndrome. Clinical Dysmorphology 1997; 2:106–113.
133. Siebler T, Lopaczynski W, Terry CL, Casella SJ, Munson P, De Leon DD, Phang L, Blakemore KJ, McEvoy RC, Kelley Rl, et al. Insulin-like growth factor I receptor expression and function in fibroblasts from two patients with deletion of the distal long arm of chromosome 15. J Clin Endocrinol Metab 1995; 80:3447–3457.
134. Counts DR, Cutler GB Jr. Growth hormone insensitivity syndrome due to point deletion and frameshift in the growth hormone receptor. J Clin Endocrinol Metab 1995; 80:1978–1981.
135. Merimee TJ, Zapf J, Hewlett B, Cavalli-Storza LL. Insulin-like growth factors in pygmies. The role of puberty in determining flnal stature. N Engl J Med 1990; 316:1146.
136. Baumann G, Shaw MA, Merimee TJ. Low levels of high affinity growth hormone-binding protein in African Pygmies. N Engl J Med 1989; 320:1705–1709.
137. Heath-Monnig E, Wohltmann HJ, Mills-Dunlap B, Daughaday WH. Measurement of insulin-like growth factor I responsiveness of fibroblasts of children with short stature: identification of a patient with IGF-I resistance. J Clin Endocrinol Metab 1987; 64:501–507.
138. Tollefsen SE, Heath-Monnig E, Cascieri MA, Bayne ML, Daughaday WH. Endogenous insulin-like growth factor binding proteins cause IGF–1 resistance in cultured fibroblasts from a patient with short stature. J Clin Invest 1991; 87:1241–1250.
139. Fouque D. Insulin-like growth factor 1 resistance in chronic renal failure. Miner Electrolyte Metab 1996; 22:133–137.
140. Duquesnoy P, Amselem S, Gourmelen M. Le Bouc Y, Goossens M. A frameshift mutation causing isolated grovvth hormone defidency type 1A. Am J Hum Genet 1990; 47:A110.

141. Cogan JD, Phillips JA III, Sakati N, Schenkman SS, Milner D. Molecular basis of autosomal recessive and autosomal dominant inheritance in familial GH deficiency. Program and Abstracts The Endocrine Society 1993; 376 (abstract).
142. Iharishi Y, Kamijo T, Ogawa M, et al. A new type of inherited growth hormone deficiency: a compound heterozygote of a 6.7kB deletion, including the GH–1 gene, and two base deletion in the third exon of the GH–1 gene. Pediatr Res 1993; 33:S35.
143. Rosenfeld RG. Disorders of growth hormone and insulin-like growth factor secretion and action, in Paediatric Endocrinology (Sperling MA, ed.) 1996; Saunders, Philadelphia, PA, pp. 117–170.
144. Kou K, Lajara R, Rotwein P. Amino acid substitutions in the intracellular part of the growth hormone receptor in a patient with the Laron syndrome. J Clin Endocrinol Metab 1993; 76:54–59.
145. Thanakitcharu, K., Woods KA, Mullis PE, Savage MO, Clark AJL. Severe growth hormone insensitivity associated with homozygous missense mutations in intracellular coding regions of the GH receptor. Horm Res 1997; 48:93 (abstract).
146. Sobrier ML, Dastot F, Duquesnoy P, Kandemir N, Yordam N, Goossens M, Amselem S. Nine novel growth hormone receptor gene mutations in patients with Laron Syndrome. J Clin Endocrinol Metab 1997; 82:435–437.
147. Berg MA, Peoples R, Perez-Jurado L, Guevara-Aguirre J, Rosenbloom AL, Laron Z, Milner RD, Francke U. Receptor mutations and haplotypes in growth hormone receptor deficiency: a global survey and identification of the Ecuadorean E180 splice mutation in an oriental Jewish patient. Acta Paediatr Suppl 1994; 399:112–114.
148. Chujo S, Kaji H, Takahashi Y, Okimura Y, Abe H, Chihara K. No correlation of growth hormone receptor gene mutation P561T with body height. Eur J Endocrinol 1996; 134:560–562.

28 Tumor Hypoglycemia Linked to IGF-II

*Christian A. Koch**, MD, *Kristina I. Rother**, MD, *Jesse Roth*, MD, FACP

CONTENTS

*Contributed equally.

INTRODUCTION

Hypoglycemia is a common medical emergency, most often due to untoward effects of insulin administered to patients with diabetes mellitus. It is also a manifestation of many other medical conditions including tumors. The hypoglycemia associated with tumors occurs within a subset of patients who have "spontaneous" or fasting hypoglycemia, i.e., the fall in blood glucose to subnormal levels occurs apparently unprovoked several hours after the patient has last eaten. Historically, the first and still the most common tumors with hypoglycemia are insulinomas, the insulin-secreting tumors of the endocrine pancreas. All other tumors associated with hypoglycemia are referred to under the broad rubric of non-islet-cell tumors with hypoglycemia (NICTH). The most important group of patients with NICTH are those who harbor tumors that secrete an incompletely processed form of insulin-like growth factor-II (IGF-II), known as big IGF-II. The tumor types included in this group are extremely numerous and are the major focus of our discussion.

In addition to these tumors in which the hypoglycemia is linked to IGF-II production, many other patients with non-islet-cell tumors have hypoglycemia but due to other mechanisms. These include: (1) tumor- or treatment-related dysfunction or destruction of the liver, adrenal cortex, or anterior pituitary; (2) inanition and cachexia; and (3) sepsis

From: *Contemporary Endocrinology: The IGF System*
Edited by: R. Rosenfeld and C. Roberts Jr © Humana Press, Inc., Totowa, NJ

including generation of high levels of cytokines. Another mechanism (*see later* under multiple myeloma) involves production of rare antibodies that bind to and activate the insulin receptor or that reversibly bind endogenous insulin with delayed release.

The history, clinical symptoms, and pathophysiology of both the insulin-secreting and the big IGF-II-secreting tumors are very similar and overlapping. In both, the molecules secreted by the tumor travel through the general circulation, interact with and activate insulin receptors in the full range of insulin-sensitive tissues including muscle, liver, and fat, and thereby provoke hypoglycemia. Another shared feature is that in face of the falling level of glucose, the body's compensatory responses are often blunted to a degree that is inappropriate for the glucose level as described in detail in the next section.

PHYSIOLOGY AND PATHOPHYSIOLOGY

Pathophysiology of Hypoglycemia

The clinical features of patients with IGF-II-linked hypoglycemia closely resemble those of patients with insulinomas. To function, the brain requires a continuous supply of glucose or ketones from blood. Although other tissues such as erythrocytes and kidney medulla are also glucose dependent, it is the brain's need for glucose that dominates clinically. Hypoglycemia also mobilizes the release of catecholamines. Clinicians typically distinguish two sets of hypoglycemic symptoms, those due to brain dysfunction (neuroglycopenia) and those due to catecholamines. Glucose deficiency in the brain is associated with loss of higher cerebral function, disorientation, amnesia, frank psychotic behavior including mania, and, if prolonged and severe, loss of consciousness, generalized seizures, and death. One of the first effects of neuroglycopenia is the loss of good judgment by the patient so that he or she ceases to be a reliable steward of his or her own well-being. The symptoms due to elevated catecholamines include anxiety, sweating, palpitations, tachycardia, and tremor. Although other hormones are typically released in association with acute hypoglycemia, e.g., glucagon, growth hormone, and cortisol, the patient and clinician are typically unaware of the rise in these hormones. With the rapid onset of hypoglycemia, the symptoms associated with epinephrine release often dominate. On the other hand, when the hypoglycemia comes on slowly over a period of hours, as is common with these tumors, or in patients who have had repeated episodes of hypoglycemia, catecholamine release may be blunted or absent so that the symptoms of cerebral dysfunction dominate *(1,2)*.

For a given level of glucose, these patients typically are more impaired than patients with identical blood glucose levels achieved, e.g., by prolonged fasting in a normal person. With the hypoglycemia associated with prolonged fasting, multiple compensatory mechanisms are brought into play. Circulating insulin drops to a very low level, which reduces glucose utilization by muscle to a minimum and permits the mobilization of fatty acids from fat cells to be used as an alternative fuel by nonneural tissues. Simultaneously, incomplete metabolism of fatty acids results in the production of ketones, the only other effective alternative substrate for the brain. In addition, the liver releases glucose at a high rate in the fasting state, provided by breakdown of glycogen and gluconeogenesis augmented by alanine and lactate from skeletal muscle. In contrast, when hypoglycemia is induced by activation of the insulin receptor, as with insulinomas and IGF-II-linked hypoglycemia, these compensatory mechanisms are blunted in muscle, liver, and fat. Not only do fat cells diminish free fatty acid release, but ketogenesis is inhibited as well. This accounts for the phenomenon whereby a blood sugar level of 35–40 mg/dL upon prolonged fasting in normal adults is associated with a clear head, while a similar blood glucose level in a patient with insulin-induced or IGF-II-linked hypoglycemia can cause serious cerebral dysfunction.

Circulating Forms of Hormones

The normal β-cell releases intact insulin with only very small amounts of proinsulin intermediates and proinsulin. (These hormones in the circulation are free and in picomolar concentrations.) Many islet-cell tumors release a relatively high proportion of proinsulin intermediates and proinsulin. However, the dominant defect of the insulinoma is the failure of insulin release to be extinguished when the blood sugar levels fall to subnormal levels; because the levels of big IGF-II (and of anti-receptor antibody) are unaffected by the fall in glucose, a similar scenario occurs. For the patient to become hypoglycemic and symptomatic, the level of insulin receptor activation need not be elevated on an absolute scale, only inappropriate for the level of glucose.

IGF-I and IGF-II are normally synthesized in a wide range of tissues including the liver and circulate at nanomolar concentrations largely bound to one of at least seven IGF binding proteins (IGFBPs) *(3)*. Because the total concentrations of IGF-I and IGF-II in plasma are several orders of magnitude greater than those of insulin, the binding proteins normally play a vital role in tightly restraining the potential insulin-like biological activities of these molecules. A very wide range of normal tissues and many tumors are responsive to the IGFs (with their binding proteins), derived from the circulation and local sites.

Under physiological conditions, IGF-II is produced as a preprohormone containing 180 amino acids. Proteolytic removal of a 24-amino-acid leader sequence at the amino (N)-terminus yields proIGF-II. Further proteolytic cleavage of the 89 amino acids of the carboxy (C)-terminus (E domain) produces mature IGF-II. The circulating 7.5-kDa form of this mature IGF-II molecule is composed of 67 amino acids. Incomplete removal of the E domain yields a series of peptides of varying lengths, known collectively as big IGF-II. Big IGF-II (approximate molecular mass of 15 kDa) accounts for 10–20% of the total IGF-II in the serum of normal adults (about 140 μg/L) *(4–8)*. In IGF-II-linked hypoglycemia, much of serum IGF-II circulates as this partially processed form of proIGF-II *(7)*. ProIGF-II (E1–21), a representative big IGF-II molecule, has been shown to have the same affinity for the IGF-I receptor as normal IGF-II *(4)*.

Why these tumors secrete so much of their IGF-II as partially unprocessed proIGF-II molecules is unknown. Daughaday et al. *(9)* found that normal serum big IGF-II contains O-linked oligosaccharides, whereas big IGF-II of patients with IGF-II-linked hypoglycemia lacks this glycosylation pattern. They speculated that O-glycosylation may be essential for normal peptidase processing of proIGF-II that results in the 67-amino-acid 7.5-kDa form of IGF-II. Other investigators, however, found that big IGF-II is secreted in its glycosylated form *(10,11)*; they hypothesized that a deficiency of the monobasic specific endoprotease that removes the E peptide from mature IGF-II at Arg^{68} caused the secretion of the partially processed IGF-II. Intracellular processing of proIGF-II could also be impaired by a mechanism similar to that which has been described in patients with familial hyperproinsulinemia (due to a mutation of the insulin molecule) where the mutant molecule is misdirected into a constitutive rather than regulated pathway *(12)*.

It is now widely believed that big IGF-II producing tumors cause hypoglycemia almost entirely by an endocrine mechanism, i.e., the IGF-II molecules act systemically on the normal insulin-sensitive tissues. While tumors can be avid consumers of glucose, as established by Warburg, this contribution to hypoglycemia is quantitatively small.

Eastman et al. *(13)* found that the hypoglycemia in a patient with a hepatoma and increased IGF-II production was predominantly due to enhanced glucose uptake by skeletal muscle with little glucose uptake by the hepatoma. Thus, at present it is argued that glucose consumption by the tumor contributes little to the hypoglycemia, although many tumors have been shown to up-regulate their hexokinases and glucose transporters *(14)* as well as to express many receptors, including IGF-I and IGF-II receptors *(15,16)*. The modest glucose uptake of these tumors has been exploited for diagnostic, prognostic, and therapeutic purposes such as 2-deoxyglucose positron emission tomography and transmembrane transport of chemotherapeutic agents *(17,18)*.

Cellular Mechanisms of Hypoglycemia

Insulin- and proinsulin-related molecules at levels encountered in the circulation of patients with insulinomas induce hypoglycemia almost entirely by activating insulin receptors. The activated form of the receptor turns on all the downstream pathways associated with insulin action. Effectively, the insulin receptor *per se* has the full capability of activating all the cellular actions we characteristically associate with insulin, even in the absence of insulin.

We now provide evidence to support the hypothesis that the major pathway through which IGF-II-linked molecules produce hypoglycemia is through the insulin receptor. This represents an example of specificity spillover: one hormone (IGF-II) is present in excess. In addition to possible effects through its own (IGF) receptors, it activates the receptors of another closely related hormone (receptor for insulin) to produce symptoms typically found with excessive amounts of the related hormone (insulin). We will then indicate that in addition some, but probably only a minority, of the effects of IGF-II to produce hypoglycemia may be mediated through the IGF-I receptor and the hybrid IGF-I/insulin receptor but not via the IGF-II (mannose-6-phosphate) receptor. Finally, we will point out that insulin and IGF-II differ most profoundly in that insulin typically does not have any significant binding proteins and therefore leaves the circulation easily and has direct access to its receptors whereas IGF-II interacts with a whole series of binding proteins that affect its egress from the vascular system, its interaction with specific receptors, and its metabolic fate.

The affinity of IGF-II for insulin receptors is 5–10% that of insulin but 5 to 10-fold greater than the affinity of IGF-I or proinsulin. All have roughly equivalent intrinsic activity. Thus, we can expect that a 5 to 10-fold higher concentration of (big) IGF-II will be indistinguishable from insulin in its effects on energy metabolism at the level of the insulin receptor.

IGF-II also interacts with the IGF-I receptor; the affinity of IGF-I is > IGF-II >> insulin. The IGF-I receptor, independent of whether it is activated by one of the IGFs or by insulin, has its characteristic panoply of biological effects, but its effects on glucose metabolism under physiological conditions are either absent or relatively weak. However, in mice where the insulin receptor has been genetically ablated, IGF-I significantly decreases preexisting hyperglycemia *(19)*. Another feature of hypoglycemia with big IGF-II (and insulinomas) is suppression of endogenous insulin secretion from the normal β-cells (via its insulin receptor and IGF-I receptors). This phenomenon explains the clinical presentation of hypoglycemia in the fasting state and hyperglycemia after a glucose load *(20)*. The suppressed levels of circulating glucagon observed with big IGF-II hypoglycemia may be mediated by activation of the IGF-I receptors on the pancreatic α-cells. Insulin-induced suppression of glucagon secretion in patients with insulinomas may also be mediated through this receptor. Likewise the suppressed level of GH observed with big IGF-II hypoglycemia is probably mediated via IGF-I receptors in the hypothalamus.

IGF-II also interacts with IGF-II/mannose-6-phosphate receptors. Under physiological conditions, the IGF-II receptor functions primarily as a scavenger receptor and influences intracellular trafficking of lysosomal enzymes *(21)* but has no known IGF signaling function and no biological effects related to glucose metabolism. This conclusion is supported by studies in which blocking the IGF-II receptor with antibodies had no effect on biological responses to IGF-II *(3)*. In contrast, many IGF-II mediated cellular responses are completely inhibited by α-IR-3, a potent IGF-I receptor blocking antibody *(3)*. The IGF-II receptor's affinity for IGF-I is more than 500-fold lower than for IGF-II *(3)*; insulin does not bind at all to the IGF-II receptor, but by binding to its own receptor, insulin induces a translocation of the IGF-II receptor from endosomes to the plasma membrane. The physiological impact of this phenomenon is currently unknown. It might be anticipated that IGF-II, when it binds to the insulin receptor, might mobilize IGF-II receptors and thereby accelerate degradation of IGF-II. This is highly speculative.

A fourth receptor species are hybrids of the IGF-I and insulin receptors. As noted elsewhere, the IGF-I receptor and the insulin receptor are structurally very similar. Each typically forms homodimers, but is also capable of forming heterodimers in cells that express both receptors. Because skeletal muscle expresses both insulin receptors and IGF-I receptors, it is relatively rich in heterodimers as well as both types of homodimers. Hepatocytes and adipocytes, which are poorly endowed with IGF-I receptors but express high numbers of insulin receptors, presumably have few heterodimers. The affinity of the heterodimers for IGF-I has been reported to be about 20 times that of the affinity for insulin *(22)*. IGF-II binding to the hybrid receptor is not well defined but thought to be intermediate but closer to IGF-I, i.e., IGF-I > IGF-II > insulin. Thus, with tumor hypoglycemia big IGF-II but not insulin would be expected to activate this receptor. The consequences of this activation are unknown. Experimentally, it has been difficult to define precisely the functions of the heterodimers in the presence of the homodimers; the downstream consequences, i.e., effects on glucose metabolism vs other effects, are not yet well defined. Thus, the hybrid receptor is another receptor species through which IGF-II but not insulin might exercise effects on glucose metabolism in patients with tumor hypoglycemia. Before closing, caution should be underscored. With all these receptors, IGF-II has typically been less well studied than insulin and IGF-I. This is especially true for the hybrid receptor which itself is only sketchily known. Further, big IGF-II has been studied in very few systems. The assumption that big IGF-II and normal IGF-II are equivalent in all systems is highly speculative and probably incorrect.

Finally, in early times, it had been postulated that the insulin receptor mediated the metabolic effects of all of the insulin-related hormones while the IGF-I receptor mediated the growth-promoting effects. With the recognition of the overlap in the structure and biochemical function of the two receptors and the post-receptor molecular mediators of the receptors, it is clear that similarities in the functions are inevitable; indeed, their biochemical overlap is so great that so little physiological overlap is quite remarkable.

How similar are insulin and IGF-II-secreting tumors? Because glucose metabolism is dominated by activation of the insulin receptor, the insulin-secreting tumors and IGF-II-producing tumors differ little; IGF-II may have a small extra effect via IGF-I receptors and IGF-I/insulin heterodimer receptors, although this would be manifest largely only in muscle rather than liver or fat.

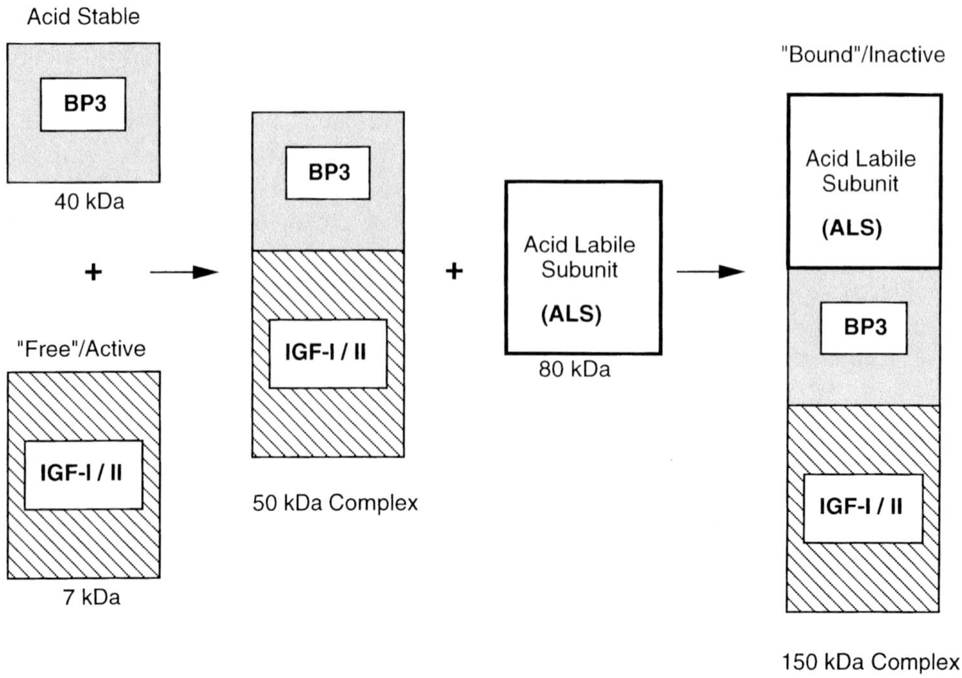

Fig. 1. Schematic representation of the circulating 150-kDa binding protein complex. Normally, IGF-I or IGF-II initially binds to the 40-kDa acid-stable IGFBP-3 subunit, and then the 80-kDa ALS binds to form the 150-kDa complex. In normal subjects, the 150-kDa complex binds virtually all the circulating ligand and thereby renders the IGFs inactive while in the circulation. Modified from ref. *(144)*. That the molecular weight of each component is an "apparent" weight (rather than a true measure of mass) accounts for the lack of additivity.

IGF-Binding Proteins (IGFBPs)

IGFBP-3 normally binds more than 95% of the IGFs in serum by forming a ternary complex of 150 kDa (*see* Figure 1). The remaining 5% or so of the IGFs are found in a 50-kDa IGFBP complex (~3%) or as free IGF (~2%, *[23]*). The bioavailability of the 150-kDa complex is largely restricted by the capillary barrier. The 50-kDa complex can cross the capillary barrier and contains mainly IGFBP-2 and a small amount of IGFBP-3 *(24)*. In contrast to the rather short half-life of the 50-kDa complex (about 30 min), the half-life of the 150-kDa complex is considerably longer (about 16 h; *[23]*). Formation of the 150-kDa complex is dependent on growth hormone (GH), which stimulates production of IGFBP-3, possibly directly or more likely via GH stimulation of IGF-I production. GH increases the serum concentration of the acid-labile subunit (ALS) and IGFBP-3 and inhibits IGFBP-2 secretion *(3)*. Concentrations of IGFBP-2 are increased in patients with tumor hypoglycemia, possibly due to release by the tumor itself, reduced concentrations of GH and of insulin, and increased (free) IGF-II *(25)*.

In IGF-II linked hypoglycemia, about 80% of the circulating IGF-II is carried in the binary 50-kDa complex and hence has access to the interstitial space *(26)*. In addition, free IGF-II may be 20- to 30-fold increased in IGF-II-linked hypoglycemia even when total IGF-II remains at or near normal levels *(27,28)*. The overall reduction in total IGFBPs permits more of the IGF-II to circulate in the free form, becoming available for binding to IGF-I and IGF-II receptors as well as insulin receptors. It has been controversial whether big IGF-II has a decreased affinity for binding to IGFBP and for forming the 150-kDa complex *(7,9)*.

Elevated free IGF-II serum levels inhibit the secretion of GH *(29,30)* at least partially by increasing hypothalamic somatostatin secretion. Low GH levels lead to decreased levels of the ternary complex of IGF-I, IGF-II *(31)*, IGFBP-3, and the ALS, thereby increasing the bioavailability of IGF-II. Elevated levels of big IGF-II appear to suppress normal IGF-II secretion *(7)*.

Responses of serum cortisol, epinephrine, and norepinephrine are reportedly intact in IGF-II hypoglycemia *(13,32,33)*. When hypoglycemia recurs frequently, the epinephrine response may be lost and symptoms of neuroglycopenia may occur without the warning provided by the adrenergic system.

MECHANISMS FOR IGF-II-LINKED HYPOGLYCEMIA

With few exceptions *(20)*, tumor hypoglycemia has been associated with the presence of big IGF-II. In IGF-II-linked hypoglycemia, big IGF-II (10–15 kDa) is more biologically active because of its increased expression and its decreased association with large binding protein containing complexes. As discussed previously, secondary to low concentrations of IGFBP-3, there is also increased bioavailability of 7.5-kDa free IGF-II for binding to the insulin and IGF-I receptors. There is no evidence that big IGF-II has a higher affinity for these receptors than IGF-II. However, increased mitogenic activity of the unglycosylated form of big IGF-II has been reported when compared to completely processed IGF-II or glycosylated big IGF-II *(34)*.

Based on in vitro and animal studies, it appears likely that the hypoglycemic effect of IGF-II is mediated largely via insulin receptors and to a lesser degree via IGF-I receptors. Although it is likely that the observed metabolic effects are mostly due to activation of insulin receptors, it has previously been shown that IGF-I stimulation of IGF-I receptors can lower plasma glucose in knockout mice that completely lack insulin receptors, suggesting that IGF-I receptors have an intrinsic ability to activate pathways leading to glucose uptake *(19)*. IGF-I receptor activation may enhance peripheral glucose uptake in muscle and inhibit hepatic glucose output either directly or indirectly via inhibition of glucagon secretion. The latter hypothesis is supported by our studies in hepatocytes derived from the knockout mice lacking insulin receptors, in which we showed that IGF-I could neither stimulate glycogen synthesis or glucokinase activity nor inhibit glucose production *(35)*.

In the setting of pure IGF-II elevation (such as in patients with Wilms' tumors or in mice lacking IGF-II receptors), hypoglycemia is usually not observed *(36–38)*. This may be due to the increased concentrations of IGFBPs and the formation of the 150-kDa complexes. In contrast, in the presence of big IGF-II, small 50-kDa complexes are preferentially formed and a down-regulation of IGFBPs occurs. There are exceptional cases in humans and animals with increased total IGF-II concentrations associated with hypoglycemia *(39)*. These may be explained by an absolute increase of free IGF-II, although the ratio of IGF-II to free IGF-II remains unchanged and high.

Clinically, several mechanisms have been described leading to hypoglycemia in IGF-II-linked hypoglycemia: decreased hepatic glucose production *(40–42)* due to inhibition of glycogenolysis and gluconeogenesis *(41)*, inhibited lipolysis in adipose tissue *(43)* resulting in low free fatty acid levels, and diminished splanchnic glucose release *(42,44)*. (*See* Figure 2.)

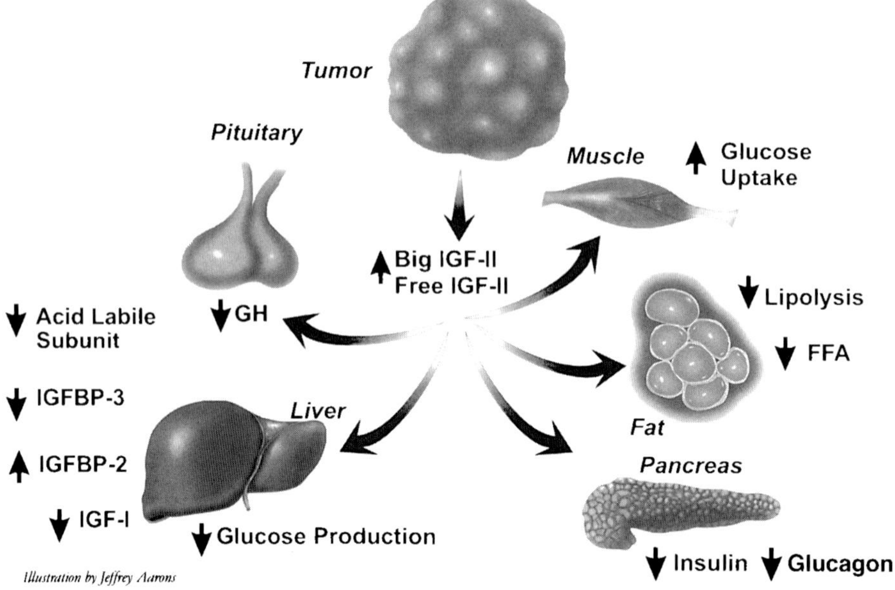

Fig. 2. The role of IGF-II in tumor hypoglycemia. Some tumors secrete large amounts of a prohormone of IGF-II, called big IGF-II. Big IGF-II directly stimulates the uptake of glucose by the tumor and by insulin-responsive tissues such as muscle and fat. Hepatic glucose production and insulin secretion are decreased by hypoglycemia and by the direct inhibitory effects of big IGF-II on the pancreatic β-cells. In addition, big IGF-II inhibits the secretion of pituitary GH, which in turn decreases the synthesis and secretion of IGF-I and IGFBP-3; decreased concentrations of IGFBP-3 enhance the effects of circulating IGF-II. FFA, free fatty acids. Modified from ref. *(145)*.

TUMOR TYPES

Since the initial description of a non-islet-cell tumor associated with hypoglycemia in a patient with primary carcinoma of the liver (also referred to as hepatocellular carcinoma or hepatoma; *45,46*), many other types of neoplasms have been reported in association with hypoglycemia *(27,28,33,47–84)*. These tumors can be of mesenchymal, epithelial, or hematopoetic origin. A representative summary of cases is shown in Table 1.

Unusual tumors were found such as meningioma *(52,70)*, adrenocortical tumors *(49)*, gastric carcinoma *(27,49,54)*, IgA myeloma *(64)*, pancreatic carcinoma *(49,61)*, primary renal carcinoma *(49,50)*, adenocarcinoma of the rectum *(62)*, colon *(27,49,66)*, and prostate *(27,49)*.

The true incidence of IGF-II-linked hypoglycemia is unknown. Marks and Teale *(49)* suggest that the incidence of symptomatic insulinomas is about 0.5–1 per million per year, and that the non-islet-cell tumors are about 25% as common. We will raise an alternative estimate. Among epithelial tumors the most common ones manifesting hypoglycemia are hepatomas. McFadzean and Yeung *(85)* found in Hong Kong that more than 10% of their patients with a hepatoma had a syndrome typical of IGF-II-linked hypoglycemia, confirming earlier studies in Peru. That 10–20% of patients with hepatocellular carcinomas may have IGF-II-linked hypoglycemia is important because these carcinomas are extremely widespread in the developing countries of the world, probably a consequence of the high prevalence of chronic hepatitis B and C. We postulate that on a worldwide basis, the incidence of IGF-II-linked hypoglycemia is substantially greater than that of insulinomas.

Table 1
Tumors Associated with Hypoglycemia

Tumor Type	*Marks (49)*	*Zapf (73)*	*Papaioannou (71)*	*Additional Cases*
Mesenchymal (total):	23	34	90	59
Fibrosarcoma	4	7	48	8
Fibroma	6	3	6	6
Fibroblastoma		1		
Mesothelioma	5	6	13	10
Mesenchymal tumors of the diaphragm, pleura, pancreas		3		7
Pseudomyxoma peritonei			3	
Leiomyosarcoma	2	6	3	7
Hemangiopericytoma	4	3	2	7
Rhabdomyosarcoma			2	2
Myosarcoma of the uterus			1	
Liposarcoma			2	
Lymphosarcoma			2	1
Pelvic clear cell sarcoma				1
Reticulum cell sarcoma			1	
Histiocytoma		1		
Hemangioendothelioma		1		
Hemangiosarcoma				1
Dysgerminoma of the ovary			1	
Seminoma				1
Wilms' tumor			1	1
Mesoblastic nephroma		1		
Neuroblastoma				1
Sarcoma of the kidney	1		1	
Neurofibroma	1		4	1
Leukemia (CML)		1		
Lymphoma		1		1
Meningioma				1
IgA myeloma				1
Neurilemmoma				1
Malignant pheochromocytoma				1
Epithelial:	35	19	55	51
Hepatoma	4	6	31	6
Carcinoma of the			23	1
Stomach	5	8		2
Colon, caecum, sigmoid, rectum				7
Kidney				1
Liver				29
Pancreas	6			1
Esophagus	2			
Breast	1			
Larynx	1			
Ovary	1			
Lung	7	1	1	
Adrenal	2	1		2
Prostate	1	1		1
Medullary thyroid		1		
Undifferentiated	3	1		1
Carcinoid/neuroendocrine tumor	1			2
Unknown histology	9	8		
Total	68	61	145	112

In large series, mesenchymal tumors represent the majority of tumors associated with IGF-II-linked hypoglycemia *(7,27,49)*, with fibrosarcomas and fibromas being most common. However, in patients with fibromatous tumors, hypoglycemia was detected in only 4% of patients with benign *(86)* and 11% with malignant tumors *(87)*. The IGF-II-linked hypoglycemia is not necessarily an indicator of size or invasiveness of the tumor, although fibromas and fibrosarcomas tend to be large. Rarely, such tumors weigh less than 500 g. The largest fibrous tumor recorded weighed more than 20 kg *(88)*. Despite their large size, many of these tumors have been clinically silent and localized in the thorax or retroperitoneum.

Interestingly, Hodzic *(56)* analyzed the parental imprinting status and the promoter usage of the IGF-II gene of a patient with mesothelioma and IGF-II-linked hypoglycemia showing loss of imprinting in the mesothelioma. Other examples of tumors known to be associated with loss of imprinting (or uniparental disomy) leading to overexpression of IGF-II are neuroblastoma, rhabdomyosarcoma, and Wilms' tumor *(10,36,37)*. However, hypoglycemia is rarely observed with these tumors. Haselbacher *(89)* found increased concentrations of IGF-II in tumors of two patients with Wilms' tumors of the kidney (220 and 380 ng/g) and much higher concentrations of IGF-II in tumors of three patients with pheochromocytomas (53,000–70,000 ng/g). None of these patients had hypoglycemia. On the other hand, exceptional cases are a neonate and a 2-yr-old girl with neuroblastoma *(51,90)* as well as two 5-yr-old children with IGF-II-linked hypoglycemia, one with an embryonic Wilms' tumor *(91)*, and one with a primary renal tumor *(50)*.

CLINICAL PRESENTATION AND DIAGNOSIS

Patients with IGF-II-linked hypoglycemia present with similar clinical findings compared to patients with insulinomas except that the presence of a neoplasm may have already been established by the time hypoglycemia occurs. Patients with insulinomas typically experience hypoglycemia as the presenting symptom before the tumor is discovered. An unknown number of cases with IGF-II-linked hypoglycemia, however, may remain unrecognized, in particular patients with advanced tumors receiving palliative care, often including narcotics. Depending on the location of the tumor, especially when abdominal, retroperitoneal, or thoracic, it can grow to a large size which allows big IGF-II production to increase sufficiently to induce hypoglycemia.

In addition to its effects on glucose metabolism, on circulating levels of other hormones, and IGFBPs, big IGF-II in occasional patients can manifest growth-promoting effects. Trivedi *(57)* reported a woman with a pelvic clear cell sarcoma and acromegaloid features including skin tags and coarse facial appearance who experienced postoperatively not only a decrease of serum big IGF-II, but also disappearance of the acromegaloid features. Acromegaloid patients with IGF-II-linked hypoglycemia have also been described by Papaioannou *(71)*. In his literature review of 145 patients with IGF-II-linked hypoglycemia, he mentioned a 59-yr-old woman with a large diaphragmatic mesothelioma and skin papillomas who experienced disappearance of hypoglycemia and skin lesions after the removal of the tumor. However, the combination of tumor-associated hypoglycemia in patients with acromegalic features is not unique to IGF-II-producing tumors. The same combination may occur secondary to coexistence of insulin- and GH-secreting tumors in the context of multiple endocrine neoplasia.

A rational approach to the clinical workup of fasting hypoglycemia is suggested in Figure 3 *(also see ref. 92)*. In addition, one should exclude artifactual hypoglycemia

Work-up of a Patient with Hypoglycemia

A. Confirm fasting hypoglycemia before embarking on extensive work-up. (Glucose ≤40, preferably with symptoms)

B. Exclude non-tumor causes before proceeding

1. Be certain to exclude adrenal insufficiency (for certainty, consider ACTH-stimulation test)
2. Exclude severe renal or hepatic failure
3. Exclude serious degree of starvation
4. Exclude side-effects of medications, especially those not related to treatment of diabetes

C. Focus on tumor hypoglycemia: First consider insulinomas and factitious hypoglycemia

1. Measure anti-insulin antibodies
   - †a. Positive:
     - (i) Surreptitious insulin administration
     - (ii) Auutoimmune disorder } Typically, hypoglycemia is post-prandial
     - (iii) Monoclonal gammopathy }
     - (iv) Unrelated finding:
       - (a) Patient with diabetes who is or has been on insulin treatment
       - (b) Patient who has just come down with Type I diabetes
       - (c) Patient who is euglycemic but has a high propensity to develop Type I diabetes
   - b. Negative:
     - (i) Does not exclude surreptitious insulin administration
2. Measure plasma insulin and proinsulin-related elements (C-peptide and percent proinsulin related components) in blood sample obtained during fasting hypoglycemia.
   - a. All three elevated: Indicates insulinoma
   - b. Insulin and C-peptide elevated (but not proinsulin): Supportive of an insulinoma, but also could occur with surreptitious administration of sulfonylurea
   - c. Insulin-elevated but not C-peptide or proinsulin: Strongly suggests surreptitious insulin administration or administration of sulfonylurea
   - d. Insulin and other elements not elevated: Suggests non-islet cell tumor, but surreptitious drug use not excluded
3. Measure blood levels of sulfonylureas or other anti-diabetic drugs in cases of suspected surreptitious administration.
4. Measure blood levels of antibodies against the insulin receptor in patients who have features to suggest Type B extreme insulin resistance:
   - a. Well-defined autoimmune disorder (e.g., lupus) or clinical features of autoimmunity (e.g., hemolytic anemia)
   - b. Significant acanthosis nigricans, now or in the past
   - c. Diabetes with severe insulin resistance, now or in the past

D. Focus on non-islet cell hypoglycemia in patients with clear cut fasting hypoglycemia but plasma insulin levels that are low

1. Measure levels of big-IGF-II: Elevated level expected
2. Measure total IGF-II: Elevated level helpful; normal level is not exclusionary
3. Search for large tumors, especially in the thorax or retroperitoneum: Large tumors supportive of diagnosis; small tumors may be unrelated
4. Look for depressed levels of IGF-I, GH, and IGF BP 3. Any of these would support the diagnosis

† ***Caution***: Anti-insulin antibodies can interfere with many methods of measurement of plasma insulin. Consult with experts to help you decide how antibodies to insulin in patient's blood can be circumvented in the measurement of endogenous circulating insulin.

Fig. 3. Differential diagnosis of fasting hypoglycemia. Modified from ref. *(92)*.

which can occur in conditions such as leukemia when significantly elevated leukocyte concentrations lead to excessive in vitro consumption of glucose *(93,94)*. Other rare causes of tumor-associated hypoglycemia have been described in patients with multiple myeloma and Hodgkin's disease that produced antibodies directed against the insulin receptor; presumably, the antibody binds to the insulin receptor and activates it in a fashion similar to insulin and big IGF-II and produces a very similar clinical picture *(95,96)*. Hypoglycemia in patients with multiple myeloma or a related condition, benign gammopathy, who have monoclonal antibodies that bind insulin, may be another example *(95)*.

Most current assays fail to distinguish between regular IGF-II and big IGF-II, the larger forms that retain some part of the E domain. However, several investigators developed radioimmunoassays (RIAs) using antibodies against the E domain *(4,97–99)*. Daughaday *(4)* described an RIA that selectively detected the first 21 residues of the E domain (E1–21) of proIGF-II [68–88]. Similarly, Liu et al. *(98)* developed an assay against proIGF-II [69–84]. Tally et al. *(99)* reported an RIA for a 15-amino-acid part of the E domain that was unaffected by the presence of IGFBP in the samples.

The IGF-II:IGF-I ratio may also be helpful in the differential diagnosis of IGF-II-linked hypoglycemia. The normal molar ratio of IGF-II:IGF-I in plasma is 3:1. Especially when IGF-II levels are normal, the IGF-II:IGF-I ratio is useful diagnostically *(100)*. A ratio of >10:1 is virtually pathognomonic for IGF-II-linked hypoglycemia. Fukuda et al. *(101)* found IGF-II:IGF-I ratios as high as 64:1. Abnormally high IGF-II:I ratios can also occur in sepsis and severe cachexia. However, in those cases both IGF-II and IGF-I levels are subnormal. Another caveat rests on the markedly increased accumulation of low molecular weight fragments of the E domain in plasma of uremic patients; measurements of big IGF-II by RIA become unreliable.

In spite of earlier optimism, measurements of IGFBPs are not yet useful. Measurements of free IGF-II and IGF-I, however, are helpful *(27,28,102)*. Other tests useful for the differential diagnosis of hypoglycemia consistent with IGF-II-linked hypoglycemia are the measurements of plasma insulin (low), proinsulin (low), C peptide (low), growth hormone (low), glucagon (low), ketones (normal or low), and free fatty acids (low).

TREATMENT

The preferential therapeutic management of patients with IGF-II-linked hypoglycemia is surgery, but conservative measures or radiation can also be helpful.

Curative Surgery

Because big IGF-II and subsequent hormonal changes (such as increased free IGF-II) appear to be responsible for the hypoglycemia, therapy is directed at controlling big IGF-II secretion. Ideally, complete surgical resection of the tumor is performed because this rapidly corrects the hypoglycemia *(71,103)*. After successful surgery, the hormonal and metabolic abnormalities disappear; big IGF-II concentrations fall to very low levels; IGF-II levels normalize, IGF-I levels rise, and the IGF-II:IGF-I ratio declines. IGFBP-3 concentrations rise. Responses of GH, glucagon, and insulin are restored *(48,50,57,58,63,65,72)*.

Palliative Surgery

Even when distant metastases exist or the primary tumor is not completely resectable, debulking or partial removal often results in a prolonged remission of the hypoglycemia *(71)*. With hepatomas, hepatic lobar resection or hepatectomy with liver transplantation may be an option *(47)*. Hypoglycemia may be the presenting feature of tumor recurrence. Sturrock et al. *(61)* reported a patient with a pancreatic carcinoma who underwent surgery without any previous episodes of hypoglycemia. Five years later, she developed sudden hypoglycemic attacks associated with low insulin, high IGF-II, and low IGF-I concentrations. At that time, liver metastases were detected. Foger et al. *(66)* reported a similar case; a patient with adenocarcinoma of the colon underwent hemicolectomy and developed acute hypoglycemic attacks 2 mo postoperatively, when liver metastases became evident. Wasada et al. *(20)* reported hypoglycemia secondary to histiocytoma 8 yr after initial presentation.

Radiation Therapy

This treatment modality is rarely used in patients with IGF-II-linked hypoglycemia, but has been shown to be of value in some patients *(71)*. Phuphanich et al. *(52)* reported a patient with spinal cord meningioma who became euglycemic after treatment with 3770 Gy. Gorden et al. *(76)* described a marked amelioration of hypoglycemia in a patient with a hemangiopericytoma involving the liver. After treatment with radiation (3000 rad locally), the level of IGF-like material was reduced to less than one third of the preradiation value. Another patient with neurofibromatosis underwent radiotherapy to the liver for a hepatic fibrosarcoma and after radiation experienced improvement of the hypoglycemia and of the associated skin lesions *(104)*.

Medical Treatment

Non-islet-cell tumors with hypoglycemia due to IGF-II rarely respond to chemotherapy. An exceptional case was reported by Hunter et al. *(67)*; a patient with an unresectable hepatoma had remission of hypoglycemia after hepatic artery chemoembolization with adriamycin.

Some cases of IGF-II-linked hypoglycemia can be controlled adequately by frequent feedings and/or intravenous glucose administration. Commonly, glucose needs to be ingested every 2–3 h around the clock. Continuous glucagon administration has also been described *(105)*. A patient with metastatic neurofibrosarcoma received intravenous glucagon at a rate of 0.31 mg/h over 16 wk to ameliorate hypoglycemia. Hunter *(67)* administered octreotide unsuccessfully in a patient with hepatoma as did Perros et al. *(68)* in a patient with pleural fibroma. In general, agents that suppress insulin secretion in some insulin-producing islet cell tumors (such as octreotide, diazoxide, and calcium channel blockers) are ineffective in IGF-II-producing tumors.

GH and glucocorticoids, each of which interact with multiple elements in the metabolism of glucose, are broad-spectrum agents that have been used with some success in patients with hypoglycemia of many etiologies. In patients with IGF-II-producing tumors, these agents may have additional benefits; down-regulation of the production of IGF-II on the transcriptional level (glucocorticoids) and modification of the expression of IGFBPs (both agents) *(106–108)*.

Wing et al. *(109)* and Zapf *(73)* reported improvement of hypoglycemia with GH doses that produce hormone levels comparable to those seen in acromegalic patients. Teale et al. *(74)* was able to prevent hypoglycemia in three patients (pleural fibroma, bladder fibrosarcoma, retroperitoneal tumor) with 12 U GH/d over several days. This reflects a dose that is 10 to 20 × higher than a replacement dose and was associated with a rise in serum IGF-I and IGFBP-3 during the therapy. Hunter *(67)* reported beneficial effects of 2 U GH/d in a patient with an inoperable hepatoma. Samaan *(105)* observed less frequent and less intense hypoglycemic episodes in a patient treated with 0.1 mg/kg GH twice a day for 4 wk. The lack of response to GH in some other cases may be explained by lower doses used or, with prolonged treatment, by the possible development of antibodies against the applied hGH *(51,75,110)*.

Control of IGF-II-linked hypoglycemia has been occasionally achieved with glucocorticoids alone or in combination with GH *(71,111)*. Foger et al. *(66)* described an 80-yr-old man with adenocarcinoma of the colon who developed severe hypoglycemia 2 mo after hemicolectomy when liver metastases and increased serum big IGF-II were discovered. Initial therapy consisted of intravenous glucagon without much success. The patient also received 350 g of glucose per day intravenously and octreotide (100 μg three times a day) for 4 d without significant improvement. Prednisolone (65 mg/d) abolished the IGF-II-linked hypoglycemia for a short period before the patient expired. Teale and Marks *(112)* compared hGH doses (4 U and 36 U/d) with prednisolone (30 mg/d). They found a 29% increase of big IGF-II in hGH-treated cases compared to a 75% decrease of big IGF-II in glucocorticoid-treated patients. The latter also had a threefold increase in serum ALS. Perros et al. *(68)* noted a decline of big IGF-II and a rise of ALS and IGFBP-

3 after 7 d of treatment with 2 U/d GH plus prednisolone 30 mg/d. Baxter et al. *(69)* also observed a decrease of big IGF-II associated with rising IGF-I, IGFBP-3, and ALS levels after treatment with prednisolone and GH doses between 1 and 8 U/d for 97 d in a patient with pleural fibroma.

POTENTIALLY APPLICABLE FUTURE THERAPIES

Theoretically, two approaches to treat or ameliorate IGF-II-linked hypoglycemia appear promising: to decrease circulating levels of big and total IGF-II or to block the effect of big IGF-II.

With the global function of IGF-II in healthy adults not being clear, it appears most suitable to target tumor production of IGF-II. The administration of antisense RNA via an arterial catheter placed in vessels feeding the tumor is conceivable, following the example of successful inhibition of translation of proteins with c-*myc* antisense oligonucleotides in the treatment of leukemia. On the other hand, the effect of (big) IGF-II could be diminished specifically by administration of antibodies directed solely against epitopes of the E domain. Treatment with recombinant IGF binding proteins parenterally might also decrease the bioavailability of IGF-II.

HISTORICAL BACKGROUND

We describe the history of hypoglycemia associated with tumors producing IGF-II-related molecules in parallel with the scientific advances of the times. Surprisingly, tumor hypoglycemia as a clinical entity did not become recognized until after the isolation and purification of insulin in 1921 *(113,114)*.

In the 19th century, experimental surgery involving brain, liver, and pancreas by Bernard *(115)* and by von Mehring and Minkowski *(116)* implicated these organs in glucose homeostasis. Measurements of glucose in blood were developed, yet recognition of hypoglycemia in patients waited until the second decade of the 20th century. The first patients suffered from Addison's disease *(117)*, chromophobe adenoma *(118)*, and terminal starvation as part of the treatment for diabetes in the pre-insulin era *(119)*.

It was the introduction of insulin into clinical practice in the early 1920s that brought hypoglycemia into clear focus for the medical community. The hormone, discovered by F. G. Banting and J. J. R. Macleod *(120)*, in collaboration with Collip and Best, was brought to clinical use quickly through close cooperation with pharmaceutical companies, Eli Lilly and Connaught. The early use of insulin was associated with high rates of hypoglycemic reactions, not only because of the inexperience in using this new drug but because the patients were hyperglycemic youngsters with type 1 (insulin-dependent) diabetes receiving regular insulin subcutaneously without the availability of frequent blood glucose measurements. Compounding the problem was that the early preparations of insulin were highly variable in their potency from batch to batch and often were pyrogenic as well. Thus, excessive insulin in patients with diabetes (still the major cause of hypoglycemia) brought hypoglycemia to the forefront of clinical medicine.

With the insulin reaction and its symptoms in mind, Harris *(121)* proposed the idea of hyperinsulinism with hypoglycemia as a possible spontaneous condition in otherwise normal nondiabetic subjects. He used the analogy of hyperthyroidism and hypothyroidism as the basis for his postulate. Recall that at that time the medical community viewed diabetes exclusively as a disorder of insulin deficiency so that Harris *(121)* comfortably postulated the existence of hyperinsulinism with hypoglycemia as a mirror image of diabetes. In his published description of five patients, only two had blood sugar concentrations below 50 mg/dL and none were below 40 mg/dL; it is hard to be sure in retrospect that any of his five patients actually had an insulinoma or one of the serious forms of hypoglycemia. His theoretical concept remained valid although he may not have reported the first patients with this disorder.

The first publication of tumor hypoglycemia recorded a patient with a classic history and a carcinoma of the pancreatic islets with metastases to liver and lymph nodes at surgery and autopsy. Based on the authors' careful review of the literature, this does appear to be the first publication of a functioning islet-cell tumor, presumably associated with inappropriate insulin secretion. The concept was further strengthened a couple years later when an islet-cell tumor of the pancreas associated with severe hypoglycemia was cured by surgery.

The first extrapancreatic tumors associated with hypoglycemia were in liver; Nadler and Wolfer *(45)* as well as Elliot *(46)* reported hypoglycemia associated with a hepatocellular carcinoma in 1929 followed by several reports thereafter. In 1930, Doege *(77)* described a patient with a large fibrocarcinoma in the mediastinum and symptoms suggestive of hypoglycemia that were relieved by removal of the tumor. Later recurrence of the tumor and of the hypoglycemia confirmed the relationship. In the same year, a patient was reported with fatal hypoglycemia associated with an adrenal tumor *(122)*.

During the next several decades, reports of islet-cell tumors burgeoned while those of non-islet cells accumulated slowly. Whipple *(123)*, enunciator of the classic triad of (1) characteristic symptoms coming on with fasting, (2) blood glucose less than 50 mg/dL, and (3) relief of symptoms by glucose, in reviewing more than 100 patients operated on for islet-cell tumors, warned about confusion of these patients with those who have dysfunction of liver, adrenal, pituitary, and other sites but failed to include the types of tumors we associate now with IGF-II production. In 1949, Crain and Thorn *(1)* reviewed 258 cases of islet-cell tumors; their differential diagnosis did not include extrapancreatic tumors. Diseases of the liver were included but not hepatocellular carcinomas. Adrenal hypofunction was mentioned but adrenal tumors were discussed only in the context of destruction of the gland. Zimmerman et al. *(124)*, in their review of hepatic diseases, noted the rarity of hypoglycemia and did not include hepatocellular carcinoma. Conn and Seltzer *(125)*, in their classic review in 1955, covered extensively spontaneous hypoglycemia and listed hyperinsulinism as the major cause. Hepatic disease, with eight subcategories, was the second major group. Hepatocellular carcinoma was not listed; diffuse carcinomatosis was the subcategory to which they linked literature citations of hepatoma with hypoglycemia. Apparently they failed to make the distinction between hepatoma, i.e., hepatocellular carcinoma, and other disorders of the liver. Their review of the literature revealed only 12 cases of large sarcomas and fibromas associated with severe fasting hypoglycemia. Eight were excised with prompt disappearance of the hypoglycemia; recurrence of the tumor occurred in five with the return of the hypoglycemia. They postulated that these tumors might secrete insulin or a similar substance, although the two tumors that were assayed for insulin were negative. However, they preferred the idea that these large neoplasms selectively metabolized large amounts of glucose.

In 1956, Porter and Frantz *(126)* presented an extensive review of tumor hypoglycemia, but noted only a handful of cases of the type associated with IGF-II hypoglycemia while they did recount hundreds of patients with islet-cell tumor hypoglycemia. Williams in 1960, in summarizing the etiology of hypoglycemia, noted the disorders of the liver but not hepatocellular carcinoma. Under disorders of the adrenal, adrenal tumors were not listed except in possible relationship to Addison's disease. Other extrapancreatic tumors are not mentioned.

In summary, by 1960, hundreds of cases of insulinoma were reported and well defined. On the other hand, published descriptions of extrapancreatic tumors with hypoglycemia were sparse, possibly only a dozen and limited to the large sarcomas and fibromas. Hepatocellular carcinoma was lost within other disorders of the liver, and generally liver disease was thought to be associated with hypoglycemia only uncommonly. In the differential diagnosis of hypoglycemia, adrenal insufficiency was a dominant theme and adrenal tumors were considered exclusively as destructive lesions producing hypoadrenalism, not as the cause of hypoglycemia on their own. McFadzean and Yeung's study *(85)* of more than 100 patients with hepatomas distinguished clinically a small subset corresponding to what we now consider to be the IGF-II secretors from those with hypoglycemia due to more broad-spectrum consequences of harboring a malignant tumor in the liver. A similar clarification of the adrenal tumors awaited the radioreceptor assay of IGF-II in the 1970s *(76,127)*.

In 1960, with the introduction by S. A. Berson and R. S. Yalow *(128)* of the RIA for insulin (which detected insulin but not IGF). It became clear that insulinomas did produce and release insulin but non-islet-cell tumors did not. For the first decade or more after the introduction of the RIA, rare reports appeared of immunoassayable insulin in blood and tumors of patients with NICTH *(129–131)*. At present, it is widely believed that NICTH is not due to ectopic production of insulin, i.e., overproduction of insulin by a non-islet-cell tumor, despite increasing evidence that many normal tissues outside of the pancreas do produce small amounts of insulin.

Prior to 1960, insulin was measured solely by bioassay, in vivo and later in vitro. The reactivity of plasma or serum in bioassays in vitro was referred to as insulin-like activity (ILA). With rat diaphragm muscle and epididymal fat pad (isolated adipocytes were introduced later), good results were obtained with pure insulin and pancreatic extracts but plasma and plasma extracts gave results that were difficult to interpret. In contrast to pure insulin, whose bioactivity is "suppressed" by anti-insulin antibodies, most of the serum ILA was unaffected (i.e., nonsuppressible) by anti-insulin antibody, hence the designation NSILA, or nonsuppressible insulin-like activity for this part of ILA. The total level of plasma insulin bioactivity or ILA was largely unchanged by fasting, by feeding glucose, or by pancreatectomy with diabetes and was not elevated in patients with insulinoma. Much of the bioactivity was of high molecular weight, releasable at low pH or with other manipulations. Other designations for this activity were "bound" insulin and "atypical" insulin. Only with the heroic isolation and purification from plasma and sequencing of IGF-I and II in 1976 *(132,133)* and the recognition of the role of IGFBPs that the nature of insulin-like bioactivities of plasma was made clear.

With regard to massive neoplasmas and hypoglycemia, there have been many speculations *(41,44,130,134–137)*. Unger *(134)* clearly summarized the scientific community's collected wisdom in 1966 just when the general concepts of ectopic hormone production by other tumors were developing and after the RIA emerged with accurate measures of circulating insulin but before any clear notions of the nature or significance or reality of other ILA from blood. "The most plausible of the theories proposed to explain tumor hypoglycemia are (1) release of a suppressor of counterregulatory hormone(s) from the tumor; (2) release of a betacytotropin from the tumor; (3) release of insulin from tumor; (4) release of an insulinoid substance from tumor; (5) release of an insulin potentiator from tumor; and (6) overutilization of glucose by tumor."

Overutilization of glucose by the tumor was a widely held theory, based on well-established observations that tumors metabolize large amounts of glucose, without the need for insulin. A corollary was that the tumors outstrip the ability of the liver to compensate. Unger *(134)* marshaled data to counter this notion and postulated that the liver is very likely able to meet the demands of even the most voracious tumors. Unger *(134)* pointed instead to evidence of deficient counterregulation: "deficient exclusion of glucose from peripheral tissues" and "deficient compensatory glucose production by liver" at the same time that the liver has adequate stores of glycogen (that can be mobilized by exogenous administration of glycogenolytic hormones). This led him to favor humoral mechanisms linked to an insulin action but the molecules responsible were as yet undefined. Eastman et al. *(13)* in 1992 verified that glucose consumption by the tumor was modest whereas peripheral (muscle) utilization of glucose occurred at a high rate.

The year 1974 marked a major advance in studies of the ILA when radioreceptor assays were developed that could distinguish the insulin-like molecules from insulin *(127)*. From 1974 to 1981, Megyesi *(127)* in our group at the NIH published a series of studies that linked IGF-II to the hypoglycemia in patients with non-islet-cell tumors. Using partially purified peptides prepared by Hummel and Froesch, Megyesi *(127)* developed a reasonably sensitive and specific radioreceptor assay for NSILA-s (the designation for plasma-derived mixtures highly enriched in IGFs). Later the results were extensively verified with more highly purified IGF preparations, including purified IGF-II. We demonstrated elevated levels of this material in blood from some patients with NICTH (but not elevated in patients with insulinomas or non-islet-cell tumors that were without hypoglycemia) and plasma levels fell dramatically with successful treatment. These studies also showed a higher molecular weight form of this material in blood from patients with NICTH (but not in others), probably corresponding in retrospect to big IGF-II. We also provided evidence that free IGF-II components are important; we gel filtered plasma at pH 7.4 and demonstrated a high level of insulin bioactivity in the fractions corresponding to the region of insulin and free IGFs that was unaffected by anti-insulin antibody. High levels of these materials were found in extracts of a tumor of a patient with NICTH; the failure to detect these materials in prior studies was due to inappropriate choice of solvents *(146)*. That the findings in the 19 positive patients of the first 52 studied included large mesenchymal tumors as well as an adrenal corticol tumor, several hepatomas, and pheochromoytomas provided a previously unrecognized basis for unifying several groups of tumor hypoglycemia entities *(76)*.

When IGF-I could be distinguished from IGF-II by more specific immunoassays and receptor assays in the late 1970s *(138)*, the possible role of IGF-II in NICTH became a subject of debate. Whereas Daughaday et al. *(139)* reported that IGF-II concentrations were increased in 10 of 14 serum samples from patients with non-islet-cell tumors and hypoglycemia, Widmer et al. *(140)* found no increases in IGF-II in sera from 22 patients with NICTH. Merimee *(141)* provided a possible explanation for the controversy. He found large decrements in receptor active IGF-II in blood samples analyzed 8–15 wk after collection compared to the same samples analyzed by the same receptor assay on day 1. Subsequently, in a woman with leiomyosarcoma and hypoglycemia, Daughaday et al. *(33)* discovered that IGF-II immunoactivity was predominantly present as a molecular species larger than IGF-II. He identified a similar substance in both tumor and serum by acidic gel-permeation chromatography. This higher molecular weight substance (big IGF-II) was shown to be a derivative of proIGF-II. One of the actions was to suppress GH secretion and thereby also IGF-I production. With removal of the leiomyosarcoma, serum big IGF-II dropped to nearly zero. The larger size IGF-II-like material described by Gorden et al. *(76)* by radioreceptor assay has not been compared to the big IGF-II of Daughaday.

Let us now use the historical background as a guide to the future. At present, the etiologies of non-islet-cell tumor hypoglycemia are most often big IGF-II, or less often, large amounts of free IGF-II, or rarely antibodies to insulin receptors or insulin. In patients in whom IGF-II or antibodies cannot be implicated at this time, the astute physician and investigator will consider the possibility of ectopic secretion of insulin *(129,130,131)*; secretion of insulinotropic agents *(134,142)*; proliferation of insulin receptors *(143)*; insulin-like material that reacts with the insulin receptor but is not detected in the RIAs for insulin or IGF-II, or materials that activate the insulin pathway at sites beyond the receptor.

SUMMARY

Non-islet-cell tumors associated with hypoglycemia do not produce insulin; most, but not all, secrete partially processed forms of proIGF-II, known collectively as big IGF-II. These variant peptides usually circulate in small amounts in healthy subjects, but are the predominant forms in patients with IGF-II-linked hypoglycemia. Big IGF-II has the same capacity as regular IGF-II to bind and activate insulin receptors and IGF-I receptors but its binding characteristics in the circulation are different. Regular (7.5-kDa) IGF-II is transported in large (150-kDa) complexes and is biologically inactive in this bound form. Only a small percentage of IGF-II exists in its free and active form. Big IGF-II, however, circulates predominantly as part of small (50-kDa) complexes that can cross the capillary barrier and therefore has increased bioavailability. Our current understanding is that multiple factors allow big IGF-II to interact with receptors for both insulin and IGF-I, leading to the observed glucose lowering effects (and only occasionally to excessive growth). Big IGF-II has a high bioavailability and also suppresses production of GH, IGF-I, and subsequently of IGFBP-3 which cause an increase in the amount of free IGF-II. Why certain tumors produce this partially processed form of IGF-II remains unknown. Big IGF-II-producing tumors include large tumors of mesenchymal origin, hepatocellular carcinomas, and a wide range of other tumors. New diagnostic methods (RIAs for big IGF-II) and therapeutic techniques (GH combined with glucocorticoids) have substantially improved the management of IGF-II-linked hypoglycemia.

ACKNOWLEDGMENTS

We thank Drs. Domenico Accili and Derek LeRoith for helpful comments on our manuscript, and Drs. Richard A. Mains, Vincent Marks, and J. Derrick Teale for their advice and sharing of unpublished data and manuscripts.

REFERENCES

1. Crain EL, Thorn G. Functioning pancreatic islet-cell adenomas. A review of the literature and presentation of two new differential tests. Medicine 1949; 28:427–447.
2. Cryer PE. Hypoglycemia. Pathophysiology, diagnosis, and treatment. Oxford University Press, New York; 1997.
3. Jones JI, Clemmons D. Insulin-like growth factors and their binding proteins: Biological actions. Endocr Rev 1995; 16:3–29.

4. Daughaday WH, Trivedi B. Measurement of derivates of pro-IGF II in serum by a radioimmunoassay directed against the E domain in normal subjects and patients with non-islet cell tumor hypoglycemia. J Clin Endocrinol Metab 1992; 75:110–115.
5. Haselbacher G, Humbel R. Evidence for two species of insulin-like growth factor II (IGF II and big IGF II) in human spinal fluid. Endocrinology 1982; 110:1822–1824.
6. Zumstein PP, Luthi C, Humbel R. Amino acid sequence of a variant proform of insulin-like growth factor II. Proc Natl Acad Sci USA 1985; 82:3169–3172.
7. Zapf J, Futo E, Peter M, Froesch E. Can big insulin-like growth factor II in serum of tumor patients account for the development of extrapancreatic tumor hypoglycemia? J Clin Invest 1992; 90:2574–2584.
8. Perdue JF, Gowan LK, Hudgins WR, Scheuermann J, Foster B, Northcutt-Brown S. Development of a specific radioimmunoassay for E domain containing forms of insulin-like growth factor II. In: Raizada MK, LeRoith D, eds. Molecular Biology and Physiology of Insulin, IGFs, and their Receptors. Plenum Press, New York, 1991, pp. 45–56.
9. Daughaday WH, Trivedi B, Baxter R. Serum big insulin-like growth factor II from patients with tumor hypoglycemia lacks normal E-domain O-linked glycosylation, a possible determinant of normal propeptide processing. Proc Natl Acad Sci USA 1993; 90:5823–5827.
10. El-Badry OM, Minniti C, Kohn EC, Houghton PJ, Daughaday WH, Helman L. Insulin-like growth factor II acts as an autocrine growth and motility factor in human rhabdomyosarcoma tumors. Cell Growth Differ 1990; 1:325–331.
11. Rechler M. Insulin-like growth factor-II: gene structure and expression into messenger RNA and protein. In: LeRoith D, ed. Insulin-Like Growth Factors: Molecular and Cellular Aspects. CRC Press, Boca Raton, FL, 1991, pp. 87–110.
12. Carroll RJ, Hammer RE, Chan SJ, Swift HH, Rubenstein AH, Steiner D. A mutant human proinsulin is secreted from islets of Langerhans in increased amounts via an unregulated pathway. Proc Natl Acad Sci USA 1988; 85:8943–8947.
13. Eastman RC, Carson RE, Orloff DG, Cochran CS, Perdue JF, Rechler MM. Lanau F, Roberts CT Jr, Shapiro J, Roth J, LeRoith D. Glucose utilization in a patient with hepatoma and hypoglycemia, assessment by a positron emission tomography. J Clin Invest 1992; 89:1958–1963.
14. Yamamoto T, Seino Y, Fukumoto H, Koh G, Yano H, Inagaki N. Overexpression of facilitative glucose transporter genes in human cancer. Biochem Biophys Res Commun 1990; 170:223–230.
15. Werner H, Re GG, Drummond IA, Sukhatme VP, Rauscher FJ, Sens DA, Garvin AJ, LeRoith D, Roberts CT Jr. Increased expression of the insulin-like growth factor I receptor gene, IGF1R, in Wilms' tumor is correlated with modulation of IGF1R promoter activity by the WT1 Wilms' tumor gene product. Proc Natl Acad Sci USA 1993; 90:5828–5832.
16. Daughaday WH, Deuel T. Tumor secretion of growth factors. Endocrinol Metab Clin North Am 1991; 20:539–563.
17. Haberkorn U, Bellemann ME, Altmann A, Gerlach L, Morr I, Oberdorfer F, Brix G, Doll J, Blatter J, van Kaick G. PET 2-fluoro-2-deoxyglucose uptake in rat prostate adenocarcinoma during chemotherapy with gemcitabine. J Nucl Med 1997; 38:1215–1221.
18. Pohl J, Bertram B, Hilgard P, Nowrousian MR, Stuben J, Wiessler M. D-19575—a sugar-linked isophosphoramide mustard derivative exploiting transmembrane glucose transport. Cancer Chemother Pharmacol 1995; 35:364–370.
19. Di Cola G, Cool MH, Accili D. Hypoglycemic effect of insulin-like growth factor I in mice lacking insulin receptors. J Clin Invest 1997; 99:2538–2544.
20. Wasada T, Hizuka N, Yamamoto M, et al. An insulin-like growth factor II producing histiocytoma associated with hypoglycemia: analysis of the peptide, its gene expression, and glucose transporter isoform. Metabolism 1992; 41:310–316.
21. Kornfeld S. Structure and function of the mannose 6-phosphate/insulin-like growth factor II receptors. Annu Rev Biochem 1992; 61:307–330.
22. Soos MA, Field CE, Siddle K. Purified hybrid insulin/insulin-like growth factor-I receptors bind insulin-like growth factor-I, but not insulin, with high affinity. Biochem J 1993; 290:419–426.
23. Guler HP, Zapf J, Schmid C, Froesch E. Insulin-like growth factors I and II in healthy man. Estimation of half-lives and production rates. Acta Endocrinol 1989; 121:753–758.
24. Harduoin S, Fourmelon M, Noguiez P, Seurin D, Roghani M, LeBouc KY, Povoa G, Merimee TJ, Hossenlapp P, Binoux M. Molecular forms of serum insulin-like growth factor (IGF) binding proteins in man: relationships with growth hormone and IGFs and physiological significance. J Clin Endocrinol Metab 1989; 69:1291–1301.

25. Blum WF, Horn N, Kratsch J, Jorgensen JO, Juul A, Teale JD, Mohnike K, Ranke MB. Clinical studies of IGFBP-2 by radioimmunoassay. Growth Regul 1993; 3:100–104.
26. Binoux M, Hossenlopp P. Insulin-like growth factor (IGF) and IGF binding proteins: comparison of human serum and lymph. J Clin Endocrinol Metab 1988; 67:509–514.
27. Frystyk J, Skjaerbaek C, Zapf J, Orskov H. Increased levels of circulating free insulin-like growth factors in patients with non-islet cell tumor hypoglycemia. Diabetologia 1998; 41:589–594.
28. Moller N, Frystyk J, Skjaerbaek C, Pedersen HB, Christensen C, Schmitz O, Gee A, Gjedde A, Zapf J, Orskov H. Systemic and regional tumour metabolism in a patient with non-islet cell tumour hypoglycemia: role of increased levels of free insulin-like growth factors. Diabetologia 1996; 39:1534–1535.
29. Frohman LA, Jansson J. Growth hormone-releasing hormone. Endocr Rev 1986; 7:223–253.
30. Frohman LA, Downs TR, Chomczynski P. Regulation of growth hormone secretion. Front Neuroendocrinol 1992; 13:344–405.
31. Zapf J, Walter H, Froesch E. Radioimmunological determination of insulin-like growth factors I and II in normal subjects and in patients with growth disorders and extrapancreatic tumor hypoglycemia. J Clin Invest 1981; 68:1321–1330.
32. Zenobi PD, Graf S, Ursprung H, Froesch E. Effects of insulin-like growth factor I on glucose tolerance, insulin levels and insulin secretion. J Clin Invest 1992; 89:1908–1913.
33. Daughaday WH, Emanuele MA, Brooks MH, Barbato AL, Kapadia M, Rotwein P. Synthesis and secretion of insulin-like growth factor II by a leiomyosarcoma with associated hypoglycemia. N Engl J Med 1988; 31:1434–1440.
34. Yang C, Zhan X, Hu X, Kondepudi A, Perdue J. The expression and characterization of human recombinant proinsulin-like growth factor-II and a mutant that is defective in the O-glycosylation of its E domain. Endocrinology 1996; 137:2766–2773.
35. Rother KI, Imai Y, Caruso M, Beguinot F, Formisano P, Accili D. Evidence that IRS-2 phosphorylation is required for insulin action in hepatocytes. J Biol Chem 1998; 273:17491–17497.
36. Scott J, Cowell J, Robertson ME, Priestly LM, Wadey R, Hopkins B, Pritchard J, Bell GI, Rall LB, Graham CF, Knott TJ. Insulin-like growth factor II gene expression in Wilms' tumor embryonic tissues. Nature 1985; 317:260–266.
37. Reeve AE, Eccles MR, Wilkins RJ, Bell GI, Millow L. Expressions of insulin-like growth factor-II transcripts in Wilms' tumor. Nature 1985; 317:258–259.
38. Lau MMH, Stewart CEH, Liu Z, Bhatt H, Rotwein P, Stewart C. Loss of the imprinted IGF 2/cation-independent mannose 6-phosphate receptor results in fetal overgrowth and perinatal lethality. Genes Dev 1994; 8:2953–2963.
39. Rogler CE, Yang D, Rossetti L, Donohoe J, Alt E, Chang CJ, Rosenfeld R, Neely K, Hintz R. Altered body composition and increased frequency of diverse malignancies in insulin-like growth factor II transgenic mice. J Biol Chem 1994; 269:13779–13784.
40. Benn JJ, Firth RGR, Sonksen P. Metabolic effects of an insulin-like factor causing hypoglycemia in a patient with a hemangiopericytoma. Clin Endocrinol 1990; 32:769–780.
41. Jakob A, Meyer UK, Flury R, Ziegler WH, Labhart A, Froesch E. The pathogenesis of tumor hypoglycemia: blocks of hepatic glucose release and of adipose tissue lipolysis. Diabetologia 1967; 3:506–514.
42. Moller N, Blum WF, Mengel A, Hansen LB, Alberti KGM, Schmitz O. Basal and insulin stimulated substrate metabolism in tumor induced hypoglycemia. Evidence for increased muscle glucose uptake. Diabetologia 1991; 34:17–20.
43. Yki-Jarvinen H, Bogardus C, Howard B. Hypoglycemia stimulates carbohydrate oxidation in humans. Am J Physiol 1987; 253:E376–E382.
44. Nissan S, Bar-Maoar A, Shafrir E. Hypoglycemia associated with extrapancreatic tumors. N Engl J Med 1968; 278:177–183.
45. Nadler WH, Wolfer J. Hepatogenic hypoglycemia associated with primary liver cell carcinoma. Arch Intern Med 1929; 44:700–705.
46. Elliot C. Hepatic hypoglycemia associated with primary liver cell carcinoma. Trans Assoc Am Physicians 1929; 44:121–126.
47. Kotani K, Tsuji M, Oki A. IGF-II producing hepatic fibrosarcoma associated with hypoglycemia. Intern Med 1993; 32:897–901.
48. Ron D, Powers AC, Pandian MR, Godine JE, Axelrod L. Increased insulin-like growth factor II production and consequent suppression of growth hormone secretion: a dual mechanism for tumor-induced hypoglycemia. J Clin Endocrinol Metab 1989; 68:701–706.
49. Marks V, Teale JD. Tumors producing hypoglycemia. In: James V, ed. Hypoglycemia. Endocr Rel Cancer 1998; 5:111–129.

50. Korn E, Van Hoff J, Buckley P, Daughaday WH, Carpenter T. Secretion of a large molecular weight form of insulin-like growth factor by a primary renal tumor. Med Pediatr Oncol 1995; 24:392–396.
51. Agus MSD, Levit Katz LE, Satin-Smith M, Meadows AT, Hintz RL, Cohen P. Non-islet cell tumor associated with hypoglycemia in a child: successful long-term therapy with growth hormone. J Pediatr 1995;127:403–407.
52. Phuphanich S, Jacobs L, Poulos JE, Vesely D. Case report: Hypoglycemia secondary to a meningioma. Am J Med Sci 1995; 309:317–321.
53. Hoog A, Sandberg Nordqvist AC, Hulting AL, Falkmer U. High-molecular weight IGF-2 expression in a haemangiopericytoma associated with hypoglycemia. APMIS 1997; 105:469–482.
54. Horiuchi T, Shinohara Y, Sakamoto Y, Hizuka N, Watabe T, Mokuda O, Tanaka K, Shimizu N, Sugano I, Nagao K. Expression of insulin-like growth factor II by a gastric carcinoma associated with hypoglycemia. Virchows Arch Pathol 1994; 424:449–452.
55. Sakamoto T, Kaneshige H, Takeshi A, Tsushima T, Hasegawa S. Localized pleural mesothelioma with elevation of high molecular weight insulin-like growth factor II and hypoglycemia. Chest 1994; 106:965–967.
56. Hodzic D, Delacroix L, Willemsen P, Bensbaho K, Collette J, Broux R, Lefebvre P, Legros JJ, Grooteclaes M, Winkler R. Characterization of the IGF system and analysis of the possible molecular mechanisms leading to IGF-II overexpression in a mesothelioma. Horm Metab Res 1997; 29:549–555.
57. Trivedi N, Mithal A, Sharma AK, Mishra SK, Pandey R, Trivedi B, Daughaday WH. Non-islet cell tumour induced hypoglycemia with acromegaloid facial and acral swelling. Clin Endocrinol 1995; 42:433–435.
58. Yokosaki Y, Kido M, Nagata N, Nikaido Y, Tsuda T, Miyake J, Manabe H. Hypoglycemia associated with localized fibrous mesothelioma of the pleura. Sangyo Ika Daigaku Zasshi 1995; 17:191–197.
59. Ishida S, Noda M, Kuzuya N, Kubo F, Yamada S, Yamanaka T, Isozaki O, Hizuka N, Kanazawa Y. Big insulin-like growth factor II-producing hepatocellular carcinoma associated with hypoglycemia. Intern Med 1995; 34:1201–1206.
60. Sodha T, Yun K. Insulin-like growth factor II expression in primary meningeal hemangiopericytoma and its metastasis to the liver accompanied by hypoglycemia. Hum Pathol 1996; 27:858–861.
61. Sturrock ND, Selby C, Hosking D. Spontaneous hypoglycemia in a noninsulin-dependent diabetes mellitus patient with disseminated pancreatic carcinoma. Diabet Med 1997; 14:324–326.
62. Kuenen BC, van Doorn J, Slee P. Non-islet-cell tumour induced hypoglycemia: a case report and review of the literature. Neth J Med 1996; 48:175–179.
63. Schweichler M, Hennessey JV, Cole P, Perdue JF, LeRoith D. Hypoglycemia in pregnancy secondary to a non-islet cell tumor of the pleura and ectopic insulin-like growth factor II hormone production. Obstet Gynecol 1995; 85:810–813.
64. Snowden JA, Greaves M, Page K. Reversal of diabetes associated with escape of myeloma: evidence for inappropriate IGF-II secretion. Br J Haematol 1994; 87:202–204.
65. Strauss G, Christensen L, Zapf J. Tumour-induced hypoglycemia due to "big" IGF-II. J Intern Med 1994; 236:97–99.
66. Foger B, Zapf J, Lechleitner M, Konwalinka G, Patsch J. Prevention with glucocorticoids of extrapancreatic tumor hypoglycemia as a result of increased big insulin-like growth factor II. J Intern Med 1994; 236:692–693.
67. Hunter SJ, Daughaday WH, Callender ME, McKnight JA, McIlrath EM, Teale JD, Atkinson AB. A case of hepatoma associated hypoglycemia and overproduction of IGF II (E21): beneficial effects of treatment with growth hormone and intrahepatic adriamycin. Clin Endocrinol 1994; 41:397–401.
68. Perros P, Simpson J, Innes JA, Teale JD, McKnight J. Non-islet cell tumor associated hypoglycemia: ^{111}In-octreotide imaging and efficacy of octreotide, growth hormone and glucocorticosteroids. Clin Endocrinol 1996; 44:727–731.
69. Baxter RC, Holman SR, Corbould A, Stranks S, Ho PJ, Braund W. Regulation of the insulin-like growth factors and their binding proteins by glucocorticoid and growth hormone in nonislet cell tumor hypoglycemia. J Clin Endocrinol Metab 1995; 80:2700–2708.
70. Ferguson JM, Flinn J. Intracranial meningioma with hepatic metastases and hypoglycemia treated by selective hepatic arterial chemoembolization. Austr Radiol 1995; 39:97–99.
71. Papaioannou AN. Tumors other than insulinoma associated with hypoglycemia. Surg Gyn Obstet 1966; 123:1093–1109.
72. Cole FH, Ellis RA, Goodman RC, Weber BC, Courington D. Benign fibrous pleural tumor with elevation of insulin-like growth factor and hypoglycemia. South Med J 1990; 83:690–694.

73. Zapf J. IGFs: function and clinical importance. Role of insulin-like growth factor II and IGF binding proteins in extrapancreatic tumor hypoglycemia. J Intern Med 1993; 234:543–552.
74. Teale JD, Blum WF, Marks V. Alleviation of non-islet cell tumor hypoglycemia by growth hormone therapy is associated with changes in IGF binding protein 3. Ann Clin Biochem 1992; 29:314–323.
75. Khaleeli A, Perumanar M, Spedding AV, Teale JD, Marks V. Treatment of tumor induced hypoglycemia with human growth hormone. J R Soc Med 1991; 85:303–306.
76. Gorden P, Hendricks CM, Kahn CR, Megyesi K, Roth J. Hypoglycemia associated with non-islet cell tumor and insulin-like growth factors. A study of the tumor types. N Engl J Med 1981; 305:1452–1455.
77. Doege K. Fibrosarcoma of the mediastinum. Ann Surg 1930; 92:955–960.
78. McPeak CJ, Papaioannou A. Nonpancreatic tumors associated with hypoglycemia. Arch Surg 1966; 93:1019–1024.
79. Gaspary F, Guimpel A. Severe hypoglycemia associated with an intrauricular rhabdomyosarcoma. Mal Cardiovasc 1969; 10:605–615.
80. Wu JC, Daughaday WH, Lee SD, Hsiao TS, Chou CK, Lin HD, Tsai YT, Chiang BN. Radioimmunoassay of serum IGF-I and IGF-II in patients with chronic liver diseases and hepatocellular carcinoma with or without hypoglycemia. J Lab Clin Med 1988; 112:589–594.
81. Yonei Y, Tanaka M, Ozawa Y, Miyazaki K, Tsukada N, Inada S, Inagaki Y, Miyamoto K, Suzuki O, Okawa H. Primary hepatocellular carcinoma with severe hypoglycemia: involvement of insulin-like growth factors. Liver 1992; 12:90–93.
82. Daughaday WH, Wu JC, Lee SD, Kapadia M. Abnormal processing of pro-IGF-II in patients with hepatoma and in some hepatitis B virus antibody positive asymptomatic individuals. J Lab Clin Med 1990; 116:555–562.
83. Barzilai N, Cohen P, Bar-Illan R, McIntyre N, Karnieli E. Case report:increased insulin sensitivity in tumour hypoglycemia in a diabetic patient:glucose metabolism in tumor hypoglycemia. Am J Med Sci 1991; 302:229–234.
84. Lowe WL, Roberts CT, LeRoith D, Merimee TJ, Fui ST, Keen H, Arnold D, Mersey J, Gluzman S, Spratt D, Eastman R, Roth J. Insulin-like growth factor-II in non-islet cell tumors associated with hypoglycemia: increased levels of messenger ribonucleid acid. J Clin Endocrinol Metab 1989; 69: 1153–1159.
85. McFadzean AJS, Yeung R. Further observations on hypoglycemia in hepatocellular carcinoma. Am J Med 1969; 47:220–235.
86. Briselli M, Mark EJ, Dickerson G. Solitary fibrous tumors of the pleura. Eight new cases and review of 360 cases in the literature. Cancer 1981; 47:2678–2689.
87. England DM, Hochholzer L, McCarthy M. Localized benign and malignant fibrous tumors of the pleura. A clinicopathologic review of 223 cases. Am J Surg Pathol 1989; 13:640–658.
88. Marks V, Rose F. Hypoglycemia. Blackwell, Oxford, 1981.
89. Haselbacher GK, Irminger JC, Zapf J, Ziegler WH, Humbel R. Insulin-like growth factor II in human adrenal pheochromocytomas and Wilms' tumors: expression at the mRNA and protein level. Proc Natl Acad Sci USA 1987; 84:1104–1106.
90. Shapiro M, Simcha A, Rosenmann E, Shafrir E. Hypoglycemia associated with neonatal neuroblastoma and abnormal responses of serum glucose and free fatty acids to epinephrine injection. Isr J Med Sci 1966; 2:705–714.
91. Loutfi AH, Mehrez I, Shabender S, Abdine F. Hypoglycemia with Wilms' tumor. Arch Dis Child 1964; 39:197–203.
92. Lorenzi M. Hypoglycemia. In: Fitzgerald R, ed. Handbook of Clinical Endocrinology, 2nd ed. Appleton & Lange, San Mateo, CA, 1992; pp. 561–581.
93. Astles JR, Petros WP, Peters WP, Sedor F. Artifactual hypoglycemia associated with hematopoetic cytokines. Arch Pathol Lab Med 1995; 119:713–716.
94. Goodenow TJ, Malarkey W. Leukocytosis and artifactual hypoglycemia. JAMA 1977; 237:1961–1962.
95. Walters EG, Tavare JM, Denton RM, Walters G. Hypoglycemia due to an insulin-receptor antibody in Hodgkin's disease. Lancet 1987; 1:241–243.
96. Arnqvist HJ, Halban PA, Mathiesen UL, Zahnd G, von Schenck H. Hypoglycemia caused by atypical insulin antibodies in a patient with benign monoclonal gammopathy. J Intern Med 1993; 234:421–427.
97. Mason BH, Tatnell MA, Holdaway I. Measurement of human IGF II using Sephacryl extraction: a rapid and reliable assay method. Ann Clin Biochem 1996; 33:201–208.
98. Liu F, Baker BK, Powell DR, Hintz R. Characterization of proinsulin-like growth factor II E-region immunoreactivity in serum and other biological fluids. J Clin Endocrinol Metab 1993; 76:1095–1100.

99. Tally M, Eriksson U, Thoren M, Brismar K, Hall K. Immunoreactive proinsulin-like growth factor II levels in healthy subjects, patients with growth hormone deficiency, and patients with type I diabetes: effects of insulin-like growth factor I and insulin. J Clin Endocrinol Metab 1994; 79:1576–1581.
100. Teale JD, Marks V. Inappropriately elevated plasma insulin-like growth factor II in relation to suppressed insulin-like growth factor I in the diagnosis of non-islet cell tumor hypoglycemia. Clin Endocrinol 1990; 33:87–98.
101. Fukuda I, Hizuka N, Takano K, Asakawa-Yasumoto K, Shizume K, Demura H. Characterization of insulin-like growth factor II and IGF binding proteins in patients with non-islet cell tumor hypoglycemia. Endocr J 1993; 40:111–119.
102. Chung J, Henry R. Mechanisms of tumor induced hypoglycemia with intraabdominal hemangiopericytoma. J Clin Endocrinol Metab 1996; 81:879–885.
103. Howard JW, Davis P. Retroperitoneal hemangiopericytoma associated with hypoglycemia and masculinization. Del Med J 1959; 31:29–33.
104. Snapper I, Schraft WC, Ginsberg D. Severe hypoglycemia due to fibrosarcoma of the liver. Maandschr Kindergeneesk 1964; 32:337–340.
105. Samaan NA, Pham FK, Sellin RV, Fernandez JF, Benjamin R. Successful treatment of hypoglycemia using glucagon in a patient with an extrapancreatic tumor. Ann Intern Med 1990; 113:404–406.
106. Li J, Saunders JC, Fowden A, Dauncey M, Gilmour R. Transcriptional regulation of insulin-like growth factor-II gene expression by cortisol in fetal sheep during late gestation. J Biol Chem 1998; 273:10586–10593.
107. Bach LA, Leeding KS, Leng S. Regulation of IGF-binding protein-6 by dexamethasone and IGFs in PC12 rat phaeochromocytoma cells. J Endocrinol 1997; 155:225–232.
108. Conover CA, Lee PD, Riggs BL, Powell D. Insulin-like growth factor-binding protein-1 expression in cultured human bone cells: regulation by insulin and glucocorticoid. Endocrinology 1996; 137:3295–3301.
109. Wing JR, Panz VR, Joffe B. Hypoglycemia in hepatocellular carcinoma: Failure of short term growth hormone administration to reduce enhanced glucose requirements. Metabolism 1991; 40:508–512.
110. Mitchell ML, Ernesti M, Raben MS, Scaliter H. Control of hypoglycemia with diazoxide and hormones. Ann NY Acad Sci 1968; 159:404–406.
111. Nesbitt KA, Boswell JT, Gonzales MA, Sarkisan S. Malignant mesothelioma associated with hypoglycemia. Am J Clin Pathol 1958; 30:148–157.
112. Teale JD, Marks V. Glucocorticoid therapy suppresses abnormal secretion of big IGF II by non-islet cell tumors inducing hypoglycemia (NICTH). Clin Endocrinol 1998; 49:491–498.
113. Bliss M. Banting's, Best's, and Collip's accounts of the discovery of insulin. Bull Hist Med 1982; 56:554–568.
114. Bliss M. J.J.R. Macleod and the discovery of insulin. Q J Exp Physiol 1989; 74:87–96.
115. Bernard C. An introduction to the study of experimental medicine. 1865.
116. von Mehring J, Minkowski O. Diabetes mellitus nach Pankreasexstirpation. Arch Exp Pathol Pharmakol 1890; 26:371–387.
117. Porges O. Ueber den Einfluss der Nebennieren auf den Kohlenhydratstoffwechsel. Verhandl des deutschen Kongr fuer Innere Med 1910; 27:591–595.
118. Kauvar A, Goldener M. Hypoglycemia and the hypoglycemic syndromes. Charles C Thomas, Springfield, IL, 1954, p. 3.
119. Joslin E. Practical lesson in diabetes. Med Clin North Am 1921; 4:1723.
120. Banting FG, Best CH, Macleod JJR. The internal secretion of the pancreas. Am J Physiol 1922; 59:479.
121. Harris S. Hyperinsulinism and dysinsulinism. JAMA 1924; 83:729–733.
122. Anderson H. A tumor of the adrenal gland with fatal hypoglycemia. Am J Med Sci 1930; 180:71–79.
123. Whipple A. Present day surgery of the pancreas. N Engl J Med 1942; 226:515–526.
124. Zimmerman HJ, Thomas LJ, Scherr E. Fasting blood sugar in hepatic disease with reference to infrequency of hypoglycemia. Am Arch Intern Med 1953; 91:577–584.
125. Conn JW, Seltzer H. Spontaneous hypoglycemia. Am J Med 1955; 19:460–478.
126. Porter MR, Frantz V. Tumors associated with hypoglycemia—pancreatic and extrapancreatic. Am J Med 1956; 21:944–961.
127. Megyesi K, Kahn CR, Roth J, Gorden P. Hypoglycemia in association with extrapancreatic tumors: demonstration of elevated plasma NSILAs by a new radioreceptor assay. J Clin Endocrinol Metab 1974; 38:931–934.
128. Yalow R, Berson S. Immunoassay of endogenous plasma insulin in man. J Clin Invest 1960; 39:1157–1175.

129. Pavelic K, Popovic M. Insulin and glucagon secretion by renal adenocarcinoma. Cancer 1981; 48:98–100.
130. Kahn C. The riddle of tumor hypoglycemia revisited. Clin Endocrinol Metab 1980; 9:335–360.
131. Baltic V, Levanat S, Petek M. Elevated levels of substances immunologically cross-reactive with insulin in blood of patients with malignant and nonmalignant pulmonary tissue proliferation. Oncology 1985; 42:174–178.
132. Rinderknecht E, Humbel P. Polypeptides with nonsuppressible insulin-like and cell growth promoting activities in human serum: isolation, chemical characterization, and some biological properties of forms I and II. Proc Natl Acad Sci USA 1976; 73:2365–2369.
133. Rinderknecht E, Humbel P. Primary structure of human insulin-like growth factor II. FEBS Lett 1978; 89:283–286.
134. Unger R. The riddle of tumor hypoglycemia. Am J Med 1966; 40:325–330.
135. Kreisberg RA, Hershman JM, Spenney JG, Boshell BR, Pennington L. Biochemistry of extrapancreatic tumor hypoglycemia. Diabetes 1970; 19:248–258.
136. Chandalia HB, Boshell B. Hypoglycemia associated with extrapancreatic tumors. Arch Intern Med 1972; 129:447–456.
137. Silbert CK, Rossini AA, Ghazvinian S, Widrich WC, Marks LJ, Sawin C. Tumor hypoglycemia: deficient splanchnic glucose output and deficient glucagon secretion. Diabetes 1976; 25:202–206.
138. Furlanetto RW, Underwood LE, Van Wyk J. Estimation of somatomedin C levels in normals and patients with pituitary disease by radioimmunoassay. J Clin Invest 1977; 60:648–659.
139. Daughaday WH, Trivedi B, Kapadia M. Measurement of insulin-like growth factor II by a specific radioreceptor assay in serum of normal individuals, patients with abnormal growth hormone secretion, and patients with tumor associated hypoglycemia. J Clin Endocrinol Metab 1981; 53:289–294.
140. Widmer U, Zapf J, Froesch E. Is extrapancreatic tumor hypoglycemia associated with elevated levels of insulin-like growth factor II? J Clin Endocrinol Metab 1982; 55:833–839.
141. Merimee T. Insulin-like growth factors in patients with nonislet cell tumors and hypoglycemia. Metabolism 1986; 35:360–363.
142. Lyall SS, Marieb NJ, Wise JK, Cornog J. Hyperinsulinemic hypoglycemia associated with a neurofibrosarcoma. Arch Intern Med 1975; 135:865–867.
143. Stuart CA, Prince MJ, Peters EJ, Smith FE, Townsend CM, Poffenbarger P. Insulin receptor proliferation: a mechanism for tumor associated hypoglycemia. J Clin Endocrinol Metab 1986; 63:879–885.
144. LeRoith D, Clemmons D, Nissley P, Rechler M. Insulin-like growth factors in health and disease. Ann Intern Med 1992; 116:854–862.
145. LeRoith D. 1997. Insulin-like growth factors. N Engl J Med 336:633–640.
146. Hyodo T, Megyesi K, Kahn CR, McLean JP, Freissen HE. Adrenocortical carcinoma and hypoglycemia: evidence for production of nonsuppressible insulin-like activity of the tumor. J Clin Endocrinol Metab 1977; 44:1175–1184.

29 IGFs in the Evaluation of Acromegaly

Marlys R. Drange, MD, PHD,
and Shlomo Melmed, MD

Contents

INTRODUCTION

Acromegaly is a condition of dysregulated growth hormone (GH) secretion characterized by excessive GH levels. Generally, acromegaly is the result of autonomous GH secretion, most often from a benign pituitary somatotropinoma, although rarely from GH-secreting pituitary carcinoma, ectopic pituitary tumors, or extrapituitary GH-producing tumors. Less frequently, acromegaly may be the result of excess GH-releasing hormone (GHRH) secretion from GHRH-producing tumors such as carcinoid or pancreatic islet cell tumors *(1,2)*. Annual incidence estimates for acromegaly range from three to four cases per million persons, with a prevalence of 60–70 cases per million *(2)*. Onset is insidious, typically resulting in a 10- to 15-yr delay after onset of GH hypersecretion before diagnosis is made. Acromegaly is associated with disfigurement, disability, and a two- to threefold increased mortality rate due to deaths from cardiac, respiratory, metabolic, or malignant complications *(2)*. Early diagnosis and effective therapy are crucial to improve quality of life and normalize life expectancy.

This chapter explores the role of the somatomedins in evaluation of acromegaly. Stimulation of insulin-like growth factor (IGF) production is one of the principal actions of GH. Thus, the common lesion in acromegaly is increased GH, but the clinical and biochemical sequelae characteristic of the condition are largely the result of excessive GH-dependent production of IGF-I. Indeed, IGF-I concentrations are markedly elevated in acromegaly and reduced in states of GH deficiency. In contrast, IGF-II levels are unchanged in acromegaly and modestly reduced in GH deficiency *(3)*. These observations suggest that IGF-I is the primary mediator of the sequelae of GH action in acromegaly. Moreover, IGF-I exhibits greater GH dependency, is more potent in growth promotion than IGF-II, and participates in the negative feedback of GH synthesis and

From: *Contemporary Endocrinology: The IGF System*
Edited by: R. Rosenfeld and C. Roberts Jr © Humana Press, Inc., Totowa, NJ

secretion. Conversely, IGF-II has more insulin-like activity compared to IGF-I *(4)* and does not appear to be a useful marker for diagnosis or monitoring of acromegaly.

Review of GHRH–GH–IGF-I Axis

Physiology of GH Secretion

GH is required for normal linear growth, for maintaining normal carbohydrate and lipid metabolism, and for normal body composition. The GH gene consists of a cluster of five highly conserved genes spanning approx 66 kb on the long arm of chromosome 17 *(5)*. The most 5′ of these genes is designated *hGH-N*, which is expressed exclusively in somatotrophs of the anterior pituitary, and encodes a 22-kDa protein consisting of 191 amino acids *(6,7)*. Absolute concentration of GH secreted is age dependent. Circulating levels are elevated at puberty, decline after adolescent growth, remain stable until mid-adulthood, and then progressively decline with age. Moreover, levels are also gender dependent with higher integrated concentrations in premenopausal women than in men *(8)*. GH levels in the normal individual are modulated by a variety of physiological and pharmacological stimuli necessitating coordinated, multifaceted regulation. The main regulatory loop for the integrated control of GH secretion is composed of GH, IGF-I, GHRH, and somatostatin.

GHRH mediates enhanced GH secretion and gene transcription. In contrast to GHRH, somatostatin attenuates GH secretion and has little effect on GH synthesis. GHRH circulates as 37-, 40-, or 44-amino-acid peptides that derive from a single gene encoding a preprohormone for GHRH-44 *(9)*. Somatostatin molecules are produced from a prohormone and are of varying amino acid length *(10)*. Both peptides are produced in the hypothalamus and are transported via axons to the hypophyseal portal system. Five somatostatin receptors have been characterized to date *(11)*. Variable tissue-specific receptor subtype expression likely determines the distinct biologic effects of somatostatin *(12,13)*.

GH secretion from the anterior pituitary is pulsatile. Normally intermittent secretory bursts occur on a background interpulse level of secretion that falls below the limit of detection of standard assays. More than 95% of total GH is secreted in the pulses, while tonic interpulse secretion accounts for the remainder *(14,15)*. The pattern of GH secretion may determine tissue responses and is dependent upon a number of factors including nutritional status, sex steroids, sleep, stress, body composition, and physical activity *(16,17)*. Moreover, there is a distinct diurnal rhythm to GH secretion, with the majority of GH secreted near the onset of deep sleep when secretory peaks occur with greater amplitude and increased frequency *(18,19)*.

Pulsatile secretion of GH is elicited by the opposing actions of GHRH and somatostatin, which are likewise secreted in asynchronous periodic pulses from the hypothalamus. Human and animal studies suggest that hypothalamic GHRH secretion is primarily responsible for generating GH pulses *(20–22)*, whereas interpulse GH secretion is dependent upon tonic somatostatin inhibition *(21,22)*. Interestingly, pulsatility continues despite persistently elevated GHRH, as seen in patients with GHRH-producing tumors or receiving GHRH continuous infusion *(6,23)*. These observations suggest that intermittent somatostatin decrements may also contribute to pulse generation. Chronic GHRH stimulation in normal individuals results in somatotroph desensitization and subsequent down-regulation of GH release.

In acromegaly, somatotrophs fail to desensitize to administered GHRH *(24,25)*. Indeed, GHRH receptor gene expression is up-regulated specifically in GH-secreting adenomas

(26). The secretory pattern is also abnormal. Pulse frequency is increased, basal GH levels are elevated, failing to reach a nadir in the undetectable range, and peak amplitude is blunted *(18,27)*. Thus, the mode of secretion is shifted from pulsatile to tonic. Indeed, only 50% of total GH is secreted in the pulsatile bursts, compared to greater than 95% in normal persons *(15)*.

GH Regulation

GH is the major transcriptional inducer of IGF-I gene expression in the liver and several extrahepatic tissues. IGF-I acts as an effector of GH action and also participates in the regulation of GH at both the pituitary and hypothalamic levels *(6,17)*. In the pituitary, it inhibits GH secretion and suppresses GH messenger RNA transcription *(28)*. At the hypothalamic level, IGF-I stimulates somatostatin release *(29)* and gene expression, while down-regulating GHRH gene expression *(30,31)*. IGF-I also suppresses GH responsiveness to GHRH, likely through increased somatostatin secretion *(32)*. Somatostatin inhibits GHRH release centrally through direct synaptic connections between somatostatin- and GHRH-containing axons *(33)*. In contrast, GHRH stimulates somatostatin secretion creating a negative feedback loop. GHRH and somatostatin also inhibit their own secretion. GH directly regulates synthesis and secretion of GHRH, and also stimulates somatostatin release *(30,31)*. Moreover, several neuropeptides, neurotransmitters, and endogenous opiates modulate release of these hypothalamic hormones *(16)*.

In addition to regulation of GH by hypothalamic and end-product feedback, a number of other hormones serve to influence GH synthesis and secretion *(16)*. For instance, gastrointestinal neuropeptides, glucagon, and sex steroids are stimulators of GH secretion. Glucocorticoids induce both stimulation and inhibition of GH release depending on the timing of administration. Thyroid hormone suppresses GH secretion, while it augments basal and stimulated GH synthesis.

In summary, complex and redundant feedback mechanisms exist for the regulation of GH secretion from the anterior pituitary. This complexity is compounded further by the influence of a number of physiological and pharmacological factors.

Role of IGF-I in GH Action

GH exerts its effects on peripheral target tissues by direct action, through its own receptor, and indirectly, via hepatic or local production of IGF-I. IGFs role in the action of GH is then mediated both by a classic endocrine mechanism and also by local tissue production of IGF that acts as an autocrine/paracrine factor. Four decades ago, Salmon and Daughaday's somatomedin hypothesis proposed that the actions of GH were mediated in a classic endocrine fashion by plasma GH-dependent factors produced in the liver *(34)*. Indeed, IGF-I and IGF-II were isolated on the basis of GH-like activities *(4,35)*. That IGF-I and IGF-II could mimic the growth-promoting action of GH in hypophysectomized rats *(36)* provided strong evidence to support the somatomedin hypothesis. However, when the growth-promoting effects of systemic administration of GH and IGF-I were compared, significant differences were noted. Although both IGF-I and GH stimulate growth in GH-deficient animals, the quantitative effect of GH is greater on total weight gain and linear growth, and the effect of IGF-I is selectively greater on kidney, spleen, and thymus *(36,37)*. These differences were reconciled by the recognition that IGF-I is produced locally in most tissues in response to GH or other trophic hormone stimulation and acts in an autocrine/paracrine manner. These observations led to proposal of the "dual effector model of growth hormone action" by Green et al. *(38)*. This model proposes that GH and the IGFs perform synergistically in target tissues. GH acts both directly, to stimulate precursor cells to differentiate, and indirectly, to stimulate local IGF production. Somatomedins then stimulate clonal expansion of the differentiated cells.

GH and IGF-I in the Circulation

Circulating IGF-I levels primarily reflect hepatic production, where GH-activated GH receptors exquisitely regulate IGF-I gene expression, synthesis, and release. IGF-I is bound to a family of six specific IGF-binding proteins (IGFBPs) in the circulation *(39)*. IGFBP-3, the most abundant binding protein, is positively regulated by GH, and becomes elevated in acromegaly and reduced in GH-deficient states. These proteins function to prolong IGF-I plasma half-life and to modulate IGF actions in both a positive and a negative manner by regulating their availability to target tissues *(40)*. The IGFBPs provide a reservoir for IGF-I that allows for relatively stable serum concentrations. In the circulation, more than 99% of IGF-I is complexed with the various binding proteins *(40)*. When injected into normal rats, half-life of IGF-I and IGF-II is extended owing to binding to the IGFBP-3/acid-labile subunit complex to about 4 h, compared to a half-life of 20–30 min in hypophysectomized rats *(41)*.

GH molecules in the circulation are heterogeneous, consisting of at least three monomeric and several oligomeric forms *(42)*. The monomeric moieties include a 22-, a 20-kDa, and an incompletely characterized acidic form. The 22-kDa peptide is the major physiologic class. The less abundant 20-kDa monomer is an alternative splice product of the *hGH-N* gene that results in deletion of amino acids 32–46 *(43)*. The oligomeric variants are aggregates of as many as five monomers linked primarily by noncovalent, although infrequently by intermolecular disulfide bonds. The 22-kDa form possesses growth-promoting activity, but lacks the diabetogenic effects that are more pronounced in the 20-kDa peptide.

GH, like IGF-I, is also complexed with specific binding proteins (GHBPs) in plasma. The principal binding protein, GHBP-1, is a 60-kDa soluble receptor fragment identical to the extracellular domain of the GH receptor *(44)*. GHBP-1 binds 50% of circulating GH. It decreases metabolic clearance and acts as a competitive inhibitor of binding to the cell surface receptor *(45)*. The circulating half-life of human GH is short, about 20 min, and plasma concentrations vary widely throughout the day owing to the pulsatile nature of release.

USE OF IGF-I IN DIAGNOSIS OF ACROMEGALY

Correlation of IGF-I Levels with Clinical and Biochemical Disease Activity

The biochemical hallmark of acromegaly is elevated plasma GH levels. Circulating IGF-I concentrations provide an alternative marker for the disease. IGF-I has the advantage of a long, relatively stable serum half-life in contrast to the wide spontaneous fluctuations of GH levels. Serum IGF-I levels are consistently elevated in active acromegaly *(3,46,47)*. In normal individuals, plasma IGF-I concentrations are proportional to the mean integrated GH concentration determined by frequent sampling during a 24-h period *(8)*. However, in acromegalic patients, early reports failed to demonstrate a correlation between circulating IGF-I and GH concentrations *(3,48–50)*. This correlation was confirmed only when it was recognized that the relationship between GH and IGF-I concentrations is nonlinear, until converted to a log dose–response plot *(51–55)*. Application of this type of dose–response relationship reveals that IGF-I concentrations plateau at high

Table 1
Factors Affecting Circulating IGF-I Measurements

Factor	*Effect on IGF-I Level*
Age	Increased at puberty
	Declines with increasing age in adults
Gender	Higher in females
Nutritional status	Decreased by energy deprivation
Acute and chronic illness	Decreased
Hypoproteinemia	Decreased
Renal failure	Increased binding capacity
Estrogen	Dose-dependent increase
Hypoinsulinemia	Decreased
Hypothyroidism	Decreased
Pregnancy	Increased
Hyperprolactinemia	No effect in normal subjects or acromegaly
Physical activity	Inconsistent data
Light activity	No effect

GH concentrations. In vitro and in vivo data support this observation and suggest that IGF-I production may be maximally stimulated at a GH concentration of approx 20–40 µg/L *(51–53,55)*. Additional increases in GH concentration above this level, as may occur in patients with acromegaly, are unlikely to raise IGF-I levels further. Thus, absolute IGF-I values may be unreliable in the circumstances of high GH levels. Conversely, the nature of the relationship between IGF-I and GH levels implies that IGF-I is a very sensitive indicator of minimally abnormal GH elevations.

IGF-I levels correlate with several clinical manifestations of hypersomatotropism, including heel pad thickness, glucose intolerance, ring size, and subjective improvement *(3,48,49,54)*. Several investigators have suggested that IGF-I levels not only correlate with disease activity, but that IGF-I correlates better than GH levels, especially at marginally elevated GH levels *(3,53–55)*. Indeed, a wide range of plasma GH levels is found in acromegaly patients with similar clinical severity and reduction of GH is not always accompanied by a corresponding improvement of symptoms. The lower correlation of GH levels to clinical activity, however, may in part be attributed to the method by which GH levels were measured. Many reports utilized fasting, postprandial, or glucose-suppressed GH levels to make these comparisons. These are approximations of GH concentration that do not always accurately reflect the 24-h integrated levels, particularly in acromegaly, because GH secretion is known to be aberrant, with increased pulse frequency, elevated basal levels, and paradoxical responses to glucose *(27,56)*. The finding of elevated IGF-I levels in patients with supposedly normal GH levels or absent subjective disease activity has led to reevaluation of the definition of "normal"plasma GH concentrations and the criteria for biochemical "cure."

Factors Affecting IGF-I Levels—Potential Confounders (Table 1)

Because the majority of potential confounding factors decrease IGF-I levels, patients may present with falsely low IGF-I measurements. Indeed, there are examples in the literature of patients with acromegaly who had normal IGF-I values *(57,58)*. Circulating IGF-I levels can be affected by a number of physiological factors. Some of these factors may override the influence of GH. These must be taken into consideration to accurately interpret IGF-I results, if the IGF-I level is to be employed as a diagnostic test for acromegaly or for monitoring disease activity.

Age and Gender

IGF-I concentrations are very low during fetal life and rise during childhood to reach peak values during the second stage of puberty *(46,59)*. Levels then decline with increasing age, so that values in the seventh decade are half of those in the second and third decades *(47,59)*. The peak, which is 2.5- to 3-fold adult levels, occurs 1–2 yr earlier in girls compared to boys, suggesting a link to pubertal development *(60)*. Moreover, values are generally 10–20% higher in girls. As with children, adult females tend to have approx 20% higher levels than men. Data from Yamamoto et al. demonstrated a steeper rate of decline in young (second/third decades) adult females compared to males, such that plasma IGF-I levels equalized in age-matched elderly subjects *(61)*. Tables of age- and sex-adjusted normal values based on cross-sectional studies are available for use in interpreting IGF-I values. It is important to note that in children, IGF-I concentrations should be interpreted based upon developmental age rather than chronological age owing to the influence of pubertal status as discussed previously.

Several findings suggest GH plays a causal role in the age and gender dependence of IGF-I production. Indeed, GH secretion exhibits corresponding age and sex differences in the adult *(8)*. Furthermore, administration of GH to older adults causes a rise in IGF-I, which suggests that the age-related decline of IGF-I is due to attenuated GH secretion *(62)*. Gender differences may be explained by the higher estrogen levels found in postpubertal girls and young women, as estrogen is known to increase GH secretion *(8)*. Recent data propose that the IGFBP system also controls circulating IGF-I levels by development- and age-related changes in concentrations of the various IGFBPs and their ratio to IGF-I *(63)*.

Nutritional Status

Nutritional status is nearly as significant in the regulation of IGF-I levels as GH. Chronic protein-calorie malnutrition (kwashiorkor) is associated with poor growth in children and low levels of IGF-I in adults and children *(64)*. Fasting is accompanied by a dramatic reduction in circulating IGF-I levels that is reversible with refeeding *(65,66)*. Regulation of IGF-I production is dependent upon both caloric content and dietary composition *(66)*. An adequate supply of energy appears to be essential for the rise in IGF-I after fasting. Conversely, protein deficiency results in a delayed response to refeeding, although eventually IGF-I levels normalize.

Paradoxically, obesity is also negatively correlated with IGF-I activity and plasma concentrations despite the energy replete status *(67,68)*. These observations may reflect impaired GH secretion in obesity. Indeed, in overfed or obese individuals GH responses to provocative stimuli are blunted and normalize with weight reduction *(69)*. Interestingly, increased concentrations of serum free fatty acids blunt the effect of various stimuli of GH secretion *(70,71)*, suggesting that altered lipid metabolism in obese individuals may play a role in decreasing IGF-I generation.

Acute and Chronic Illness

Critical Illness. Serum IGF-I levels have been consistently reported as low in acutely ill individuals despite raised GH levels. Although it may be difficult to separate the effects of critical illness from those of acute undernourishment, a study of critically ill patients and age-matched controls in the fasted and fed state showed attenuated oscillatory activity of GH secretion in patients compared to controls *(72)*. Aberrant GH secretory patterns persisted despite parenteral nutritional support in the critically ill patients, suggesting that this effect may be independent of nutritional status.

Liver Disease. Chronic liver disease is associated with reduced levels of IGF-I *(73)*. In general, any etiology of diminished hepatic synthetic function results in decreased plasma IGF-I levels. Indeed, concentrations of somatomedins correlate directly with serum albumin. Furthermore, hypoproteinemia due to protein loss (i.e., nephrotic syndrome or protein-losing enteropathy) rather than decreased production has also been associated with inappropriately low IGF-I values *(57,74)*.

Kidney Disease. Children with renal insufficiency exhibit poor growth despite supranormal GH levels. Renal failure is associated with reduced free IGF-I levels although the total IGF-I is normal or elevated. Investigators employing an experimental model of uremia have demonstrated diminished hepatic IGF-I gene expression, while IGFBP-1 and IGFBP-2 production are augmented *(75)*. Thus, binding capacity for IGF-I is increased owing to a marked elevation of several IGFBPs *(40)*.

Other Hormones

Estrogen. The relationship between estrogens and IGF-I is complex and dose dependent. In high doses, estrogen is known to inhibit growth and decrease bioassayable and radioimmunoassayable IGF-I *(76,77)*. Indeed, estrogen has been used to limit linear growth in girls with excessive predicted height *(78,79)*, and in acromegalic patients to block disease activity *(76)*. There is no evidence that estrogen affects IGF action. Thus, IGF-I concentrations decrease because GH-stimulated IGF-I production is diminished.

In contrast, lower doses of estrogen may enhance IGF-I generation *(80,81)*. Administration of estrogen in low doses stimulated ulnar growth rate in girls with gonadal dysgenesis without increasing IGF-I, whereas intermediate doses increased IGF-I without resulting in growth augmentation *(80)*. These data suggest that tissue response to estrogen is subject to dose-dependent regulation. Recent data indicate that circulating levels of IGF-I are not affected by either estrogen or progesterone, whereas progesterone increased IGF-I gene transcription locally in bone, while estrogen had the opposite effect *(82)*. These observations suggest that tissue response may also be the net result of local progesterone and estrogen effects.

Insulin. IGF-I concentrations appear to be regulated by insulin. Poor growth with low insulin and IGF-I levels, despite high GH levels, has been observed in patients with poorly controlled type 1 (insulin-dependent) diabetes *(83)*. Animal studies indicate that a state of insulin insufficiency may represent a cellular energy deficit analogous to malnutrition. Indeed, streptozocin-induced diabetic rats demonstrate a marked fall in somatomedin activity similar to that observed in fasting. Administration of GH will not prevent this decrease, whereas insulin administration normalizes IGF activity *(84)*. Insulin therapy has also been shown to progressively restore circulating free and total IGF-I concentrations in diabetic children *(85)*.

Thyroxine. Several studies have reported that IGF-I levels are reduced in hypothyroidism *(86,87)*. This effect is reversible upon thyroxine replacement. The reduction of IGF-I concentrations may reflect altered GH secretory patterns with diminished absolute GH concentrations *(87)*.

Glucocorticoids. Glucocorticoid excess, in doses as low as two- to threefold daily replacement, attenuates growth *(88)*. The role of IGF-I in glucocorticoid-induced growth retardation is not entirely clear. It appears that IGF-I levels are either normal *(89)* or slightly elevated *(46,90)*. However, there is evidence that IGF-I bioactivity is inhibited *(90)*.

Prolactin. The effect of prolactin on IGF-I plasma concentrations in humans is unclear. Prolactin had a weakly stimulatory effect on IGF-I production in GH-deficient patients with prolactin-secreting tumors *(91)*. Conversely, other investigators report either no effect of hyperprolactinemia on IGF-I levels *(92,93)*, or IGF reduction *(94)* in patients with acromegaly. Perhaps the effect on IGF-I is dependent on GH status, with high levels of prolactin capable of exerting some cross-reactivity with GH only in the setting of GH deficiency.

Pregnancy

IGF-I levels are elevated in the third trimester of pregnancy and decline promptly at delivery *(95)*. A possible role for placental lactogen has been implicated *(59,95)*. Alternatively, it has been suggested that a pregnancy-induced protease cleaves IGFBP-3 and results in a greater proportion of free IGF-I *(96)*.

Physical Activity

Several investigators report that there is no significant correlation between general fitness, resistance exercise, endurance exercise, or acute moderate activity and IGF-I *(97–100)*. Conversely, Poehlman and Copeland suggest that maximal aerobic capacity and leisure time physical activity positively correlated with IGF-I levels after controlling for the effects of age *(101)*. It has been postulated that the level of exercise intensity is important in discerning the augmentation of IGF-I (i.e., higher intensity or maximal exercise results in greater IGF-I rise) *(101,102)*. The positive association of exercise with IGF-I levels may merely reflect GH status, as exercise is known to stimulate GH secretion. However, Schwarz et al. report that the IGF-I response appears to be independent of GH *(102)*. It is generally agreed that light exercise does not significantly alter IGF-I levels.

Influence of IGF-Binding Protein Levels

The specific composition of the IGFBP milieu results in modest variability of IGF-I concentrations and bioavailability *(40)*. Moreover, IGFBPs can interfere with IGF-I measurements in direct assays of unextracted serum samples *(47)*. Assays of unextracted sera are also subject to false readings if improperly collected. Thus, it has been recommended that serum samples be extracted to remove the binding proteins prior to assaying IGF-I concentrations *(92,103–105)*. Although some investigators feel extraction is not mandatory providing the tracer utilized in radioimmunoassay has sufficiently high affinity *(59)*.

IGF-I is a Cost-Effective Biochemical Disease Screen for Acromegaly

Prolonged exposure to high levels of GH and its target hormone, IGF-I, leads to chronic pain, disfigurement, and shortened life expectancy. Large pituitary tumors may exert mass effects by compressing or destroying adjacent structures that can result in visual impairment, pituitary hormone deficiencies, or even death. Thus, there are distinct advantages to earlier diagnosis and medical intervention for acromegaly. A diagnostic test for acromegaly should ideally be sensitive and specific, require minimal time and effort from patient and staff, be applicable to the outpatient setting, and not be technically difficult or labor intensive. This test should permit identification of patients with mild hypersomatotropism and minimal clinical signs or symptoms.

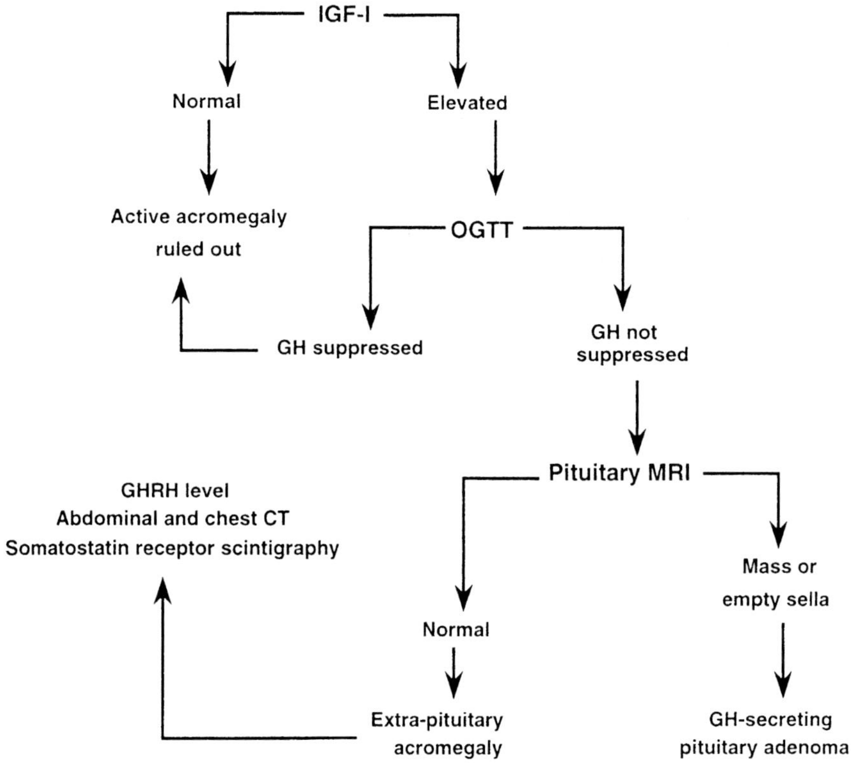

Fig. 1. Schematic for diagnosis of acromegaly. Abbrev.: CT = computed tomography; GH = growth hormone; GHRH = growth hormone-releasing hormone; IGF-I = insulin-like growth factor-I; MRI = magnetic resonance imaging; OGTT = oral glucose suppression test. [Adapted from Melmed, S. Anterior pituitary. In Current Practice of Medicine, Vol. 1. (Bone, CC, ed) Churchill Livingstone, New York, 1996, pp. 2.1–2.12.]

A single random IGF-I level has been advocated as a cost-effective biochemical screening test for initial diagnosis of acromegaly *(18,53,105–107)* (Fig. 1). Serum IGF-I levels are universally elevated in active acromegaly with rare exception. There is clear discrimination between normal controls and patients with acromegaly. Indeed, in one study of 57 patients with active acromegaly, mean IGF-I values were 10-fold higher than in controls *(3)*. IGF-I is proportional to the logarithm of mean integrated 24-h GH concentration, and has the advantage of a long, relatively stable, serum half-life. These attributes obviate the need for multiple assessments or concern regarding temporal relationship of specimen collection to factors that may transiently alter serum GH levels such as sleep, meals, or moderate physical activity. Although IGF-I levels are unreliable at high GH concentrations owing to the curvilinear nature of the dose–response relationship between these variables, IGF-I is a very sensitive indicator of marginally elevated GH levels. This observation implies that IGF-I assessment has a very high positive predictive value. Potential confounding factors affecting sensitivity were discussed previously. Elevated IGF-I concentrations are also very specific for acromegaly, with the caveat that levels are known to be modestly elevated in pregnancy and at puberty in normal individuals. Data regarding true sensitivity and specificity are lacking, primarily because calculation of those percentages requires a "gold standard" for determining those patients with disease and those without.

Such a standard is, as yet, undefined. It is clear that in a generally healthy, outpatient population, IGF-I fulfills the criteria for an ideal diagnostic test for acromegaly.

Oral Glucose Suppression Test (OGTT)

An OGTT should be employed to confirm abnormal regulation of GH secretion. It is recommended in conjunction with a plasma IGF-I level for diagnosis of acromegaly by the Acromegaly Therapy Consensus Development Panel *(106)*. Oral ingestion of 50–100 g of glucose after an overnight fast will cause suppression of GH release to less than 1 µg/L over the subsequent 30–90 min in normal individuals. More than 80% of patients with acromegaly fail to suppress GH secretion following an oral glucose load and approx 30% demonstrate a paradoxical rise in GH concentrations *(56)*. Advantages of the OGTT are: it more accurately reflects disease activity than a day series *(108)*; it is relatively easy and inexpensive to perform; and it imparts information about aberrant GH secretion. The major disadvantage of this test has been some overlap with normals as a number of investigators have reported "normal" responses in patients with active acromegaly *(3,109,110)*. This issue is likely to be obviated soon by the development of a chemiluminescent GH assay with markedly enhanced sensitivity *(111)*. Results generated by this assay suggest that normal GH response post-glucose is much lower than previously accepted (i.e., 2 µg/L) and is gender dependent. A normal suppression to a 100-g OGTT would be less than 0.057 µg/L in men and less than 0.71 µg/L in women. Increased sensitivity of the OGTT for diagnosis and monitoring of acromegaly could be achieved by adoption of lower criteria for a normal suppressive response. Utilization of the enhanced chemiluminescence assay and stricter criteria may lead to earlier detection of acromegaly and better recognition of persistent disease activity following treatment.

Direct Measurement of Serum GH

Since excess GH is the primary biochemical aberration of acromegaly and most often the secretory product of the underlying tumor, it has been argued that direct evaluation of GH status is best for diagnosis. Optimally, this would entail frequent sampling over 24 h to assess integrated 24-h GH concentrations and secretory patterns. Measurement of the 24-h integrated GH concentration involves continuous blood collection via a nonthrombogenic withdrawal pump or indwelling catheter in patients whose food and exercise are not restricted. Sampling intervals of, at maximum, 20 min are most often used *(112)*, but shorter intervals (i.e., 5 min or even 30 s) detect significantly more pulses *(113)*. This method is impractical and too costly for routine use as it requires laboratory services to process more than 70 samples, clinical facilities for 24-h inpatient stay, and is time consuming for patient and staff.

Single random or basal morning GH levels are not useful for accurately assessing GH status given the pulsatile nature of GH secretion and short serum half-life. Serum concentrations of GH vary from undetectable (in standard assays) to 30 µg/L and are below the limit of detection about half the time in normal individuals *(8,114)*. In contrast, despite retention of pulsatile secretion, acromegalic patients have blunted secretory excursions, elevated basal GH levels, and failure to reach nadir values in the undetectable range *(18)*. The probability of a diagnosis of acromegaly is very high in a patient with a fasting morning GH level above 5 µg/L. Yet, there is an unacceptable overlap with the fluctuating values of normal individuals. Indeed, random GH measurements missed nearly 40% of patients with positive IGF-I elevations when a threshold of approx 5 µg/L was used *(18)*. Moreover, factors that transiently alter GH levels, such as meals, physical activity, or stress, can easily result in misinterpretation of single GH values.

The average of a series of four to eight random GH measurements throughout the day, known as a day series, has also been advocated. Compared to the single morning GH, a day series is slightly more sensitive and definitely more specific *(105)*. Intuitively, this method has little advantage over a single random GH level, as each sample is subject to the same unpredictable influences of pulsatile GH secretion and various confounding factors. Furthermore, the day series, like a 24-h profile, is expensive; requires day care facilities; and expends considerable laboratory, patient, and staff effort. In contrast to the 24-h profile, this method is unable to provide any information about GH secretory patterns. Thus, none of the direct measurements of serum GH provide a convenient, cost-effective, reliable diagnostic indicator for acromegaly.

Urinary GH

As an alternative to serial measurements of serum GH, measurement of urinary GH levels is of potential use in estimating GH production. Urinary GH levels reflect integrated serum GH concentrations over the course of the study period in patients with normal renal function *(115,116)*, and are a more sensitive test in reflecting increased GH secretion than a day series *(117)*. Urinary GH has been demonstrated to be elevated in patients with acromegaly *(116–123)* and strongly correlated with serum GH *(120–122)*, GH following oral glucose *(122,123)*, and IGF-I values *(122)*. Assessment of GH status by urinary GH assessment has several advantages. This test is noninvasive. Clinical and laboratory requirements are minimal, although some sample preparation is required. The specimen can be collected at the patient's convenience at home and only a single sample from the pooled specimen need be assayed. Overnight collections measure the physiological sleep-induced GH peaks without potential interference by meals or physical activity. Moreover, there is some evidence that even random urine samples may be adequate for diagnosis of acromegaly *(123)*. However, there is disagreement in the literature as to whether urinary GH levels exhibit age and sex differences in normal controls comparable to those observed in serum GH levels *(116,120,123)*. These discrepancies may be explained by differences in sensitivity of the assays employed. Alternatively, minimal alterations in renal function in the subjects studied may account for the failure to detect appropriate age- or gender-related differences, as reduced creatinine clearance may increase urinary GH output 100- to 1000-fold *(124)*. Several investigators using highly sensitive assays report complete discrimination of patients with acromegaly from normal individuals *(118,123)*. Pholsena and co-workers also demonstrated no overlap between normal controls and untreated acromegalic patients, whereas only 83% of partially treated patients were distinguished from controls *(122)*. Likewise, Main et al. observed some overlap between patients with low serum GH concentrations and normal controls *(117)*. Another problematic aspect of urinary GH measurement is the intrapatient variability; estimates range from 16% to 49% depending upon the study *(116,123,125)*. These findings suggest that assessment of urinary GH may be more useful as a screening tool than as a method to monitor therapy. Furthermore, technical aspects of the assay and normal ranges may need improvement prior to widespread acceptance of urinary GH determinations for monitoring GH status.

IGFBP-3

Another potentially useful diagnostic assessment for acromegaly is measurement of serum IGFBP-3. IGFBP-3 is GH dependent and parallels changes in GH concentrations. Age-adjusted reference ranges are necessary. Levels are uniformly elevated in patients with acromegaly *(126)*. Like IGF-I, IGFBP-3 is thought to function as an integrated marker of GH activity with a prolonged half-life and stable serum concentrations. It correlates well with IGF-I and glucose-suppressed GH results in treated and untreated patients. In a study of 44 acromegaly patients with pathologically confirmed somatotroph adenomas, the IGFBP-3 assay identified an additional 11% of patients in whom GH levels following oral glucose were less than 2 μg/L *(126)*. However, there is disagreement as to whether IGFBP-3 measurements effectively discriminate acromegaly patients from normal controls. Several studies demonstrate significant overlap between the two groups *(127–129)*, whereas others show clear distinction *(126,130)*. There is no indication that this test is superior to IGF-I assessments. Indeed, IGFBP-3 production is mediated at least in part by IGF-I *(40,131)*. These data support the contention that IGF-I is a more direct measure of GH status. IGFBP-3 assessment may be useful as an adjunctive indicator of GH status in cases where glucose-suppressed GH levels and IGF-I results are discordant, but cannot be recommended for routine diagnostic use.

Paradoxical Responses

Determinations of abnormal regulation of GH secretion, such as paradoxical responses to thyrotropin-releasing hormone (TRH), gonadotropin-releasing hormone (GnRH), or L-dopa, can be found in 60–80% of patients with acromegaly *(108,132,133)*. In responsive patients, intravenous administration of TRH or GnRH stimulates increased GH secretion, unlike in normal subjects, who rarely (<10%) exhibit a GH response to these hormones. Likewise, L-dopa or exercise, which normally induce GH secretion, have the opposite effect in some, but not all patients. These tests lack the sensitivity and specificity necessary to play a consequential role in diagnosis. Indeed, TRH may also stimulate GH release in renal failure, liver disease, poorly controlled diabetes, anorexia nervosa, or depression. These paradoxical responses are found in acromegaly caused by ectopically produced GHRH, as well as by pituitary adenomas, so render no etiological information. However, these tests provide useful markers of persistent disease activity following medical intervention in those patients in whom the abnormal responses are present prior to treatment.

USE OF IGF-I IN MONITORING THERAPEUTIC RESPONSE

Three therapeutic modes are currently available for management of acromegaly; surgery, medical therapy with either bromocriptine or octreotide, and external beam irradiation. Transsphenoidal surgical excision of GH-secreting pituitary adenomas is the therapy of choice *(106)*. Medical management may be used as the primary therapeutic option in selected patients. The basic management strategy for acromegaly has the underlying objective of amelioration of prolonged exposure to excess GH. If GH and IGF-I levels have normalized after surgery no additional therapy is indicated. However, if surgery fails to achieve biochemical remission, medical management should be initiated. Irradiation is considered an adjuvant therapy for acromegaly when surgical and medical treatment have failed to control the disease.

A significant percentage of patients have evidence of persistent disease or recurrence of disease following the initial therapeutic intervention. Precise interpretation of surgical results in acromegaly is difficult as most published series report only short-term data and do not report glucose-suppressed GH or IGF-I levels. In most large surgical series, 60% of acromegalic patients had postoperative random GH levels less than 5 μg/L; however, many of these had elevated IGF-I levels or increased GH concentrations when retested a year or more postoperatively *(2)*. Indeed, acromegaly may recur several years after surgery in 5–10% of operated patients despite normal dynamic GH responses shortly after tumor resection. Bromocriptine therapy, which requires up to 30 mg/d divided into three to four oral doses, decreases random GH levels to less than 10 μg/L in approx 50%, less than 5 μg/L in 20%, and normalizes IGF-I in only 10% of patients *(134)*. Alternatively, the somatostatin analog octreotide is administered in three daily subcutaneous injections in dosages ranging from 300 to 1500 μg/day. This drug significantly suppresses GH levels in more than 90% of acromegalic patients *(135)*, and normalizes IGF-I in approx 50% *(2,136)*. These data demonstrating incomplete cure in a significant percentage of patients regardless of treatment modality confirm the importance of adequate therapeutic surveillance in treated acromegaly.

Attributes of the various biochemical determinations for monitoring therapeutic response of acromegaly are essentially equivalent to those for diagnosis, with one qualitative difference. Diagnosis merely requires an indication of GH hypersecretion that can be reliably distinguished from normal, without specific regard for the degree of excess. For effective monitoring of treatment, however, serial determinations are obtained and evaluated to determine trends in disease activity or adequacy of a specific therapy. It is crucial that subtle changes in disease activity suggestive of a recurrence or inadequate therapy be detected early. Technical precision within the assay employed is required. Once again, the stability of IGF-I levels and the capacity of this parameter to reflect cumulative GH secretion over the previous day or days is advantageous. This may be particularly true in the case of medical therapy with the somatostatin analog octreotide. This somatostatin analog has a half-life of 1.7 h. Following an octreotide injection, GH levels are suppressed for 2–5 h, then increase toward the next injection, creating an artificial oscillatory GH secretory pattern *(137)*. It is sufficient to measure a IGF-I plasma level to optimize the number of daily injections and dosage of the drug, precluding the need for multiple assessments. Furthermore, IGF-I, unlike direct measurements of GH concentration, may also give information regarding the compliance of the patient with the drug regimen.

Distinguishing abnormal from normal becomes increasingly difficult as values approach the normal range. In this regard, the variability intrinsic to serum or urinary GH measurements becomes increasingly problematic. Conversely, IGF-I determinations are a remarkably sensitive indicator of GH hypersecretion. Indeed, IGF-I plasma levels remain elevated in response to the cumulative effect of total GH secretion even when GH levels appear to be in the accepted normal range most of the time. The recommended strategy for monitoring of treatment as well as diagnosis is to measure IGF-I in conjunction with an OGTT *(106)*. Initially, these determinations should be performed every 3 mo following treatment. Semiannual residual anterior pituitary hormone evaluation and annual MRI should also be obtained in the first years following successful treatment. Ancillary tests for long-term follow-up include colonoscopic, cardiac, and dental evaluations. Patients with residual tumor require regular visual field assessment by perimetry.

Controversy Regarding Definition of Biochemical Cure of Acromegaly

Once acromegaly has been diagnosed and the source of excess GH secretion identified, the goal of management is to achieve a biochemical and clinical cure. Criteria for cure include: (1) elimination of the causative tumor, ideally with preservation of residual anterior pituitary function; (2) resolution of clinical signs and symptoms; (3) normalization of GH and IGF-I concentrations with restoration of normal GH physiological responses. Appropriate GH secretory patterns, glucose-suppressability, and loss of

Table 2
Factors Resulting in Discordant Circulating IGF-I and GH Values

Unreliable definition of "normal"
Delayed normalization of IGF-I levels following surgical intervention
GH secretory pattern that more effectively stimulates IGF-I production
Persistently elevated and erratic GH pulse frequency following treatment
Contribution of local IGF-I production
Variable sensitivity of assays employed

paradoxical responses to provocative stimuli such as glucose, TRH, or dopamine should be achieved to ensure that acromegaly is cured.

It has been difficult to define biochemical "cure" of acromegaly because there is no established consensus regarding the post-therapy value for GH necessary to eliminate long-term pathophysiological sequelae of the disease. Various limits between 2 and 10 μg/L have been used to denote a successful treatment outcome in most published studies. These limits are arbitrary and have largely reflected symptomatic response. Unfortunately, by the stringent criteria enumerated earlier, treatment of acromegaly in many patients results in remission, but not cure. Through the development of more sensitive GH assays and a greater understanding of the dynamics of normal GH secretion, the definition of biochemical cure has been modified to include normalization of age- and sex-adjusted IGF-I values and lowered serum GH concentrations *(106)*. The currently acceptable limit for serum GH levels is a value of less than 1 μg/L after glucose ingestion. The test of time is required to prove that the treated acromegalic patient is in fact "cured" before biochemical cure can be accurately defined.

Factors Resulting in Discrepancy Between IGF-I and GH Determinations After Therapeutic Intervention (Table 2)

Erroneous acceptance of patients as being cured based on arbitrary limits has led to inconsistencies in the literature. Indeed, the utility of IGF-I as an indicator of active acromegaly was questioned in the past because IGF-I was often elevated in the face of supposedly normal GH levels. Although the majority of these discrepancies can be explained by lack of agreement over the definition of "normal," many other discrepancies between IGF-I and GH determinations after treatment can be reconciled by one or a combination of the following factors.

Time Elapsed Since Therapeutic Intervention

GH levels usually return to normal within hours after successful GH-secreting tumor resection. Conversely, IGF-I levels do not normalize immediately after surgery. In a retrospective study conducted at the Mayo Clinic, no IGF-I levels were normalized on the first postoperative day, three of four had normalized by the fourth postoperative day, and all had normalized 2 mo following surgery *(138)*. Thus, IGF-I levels generally normalize over a course of days, but may remain elevated for several months even when GH levels are in remission. The IGFBPs, which provide a reservoir for IGF-I, most likely account for the delay in normalization. The "stored" IGF-I must turnover before a new steady state reflecting the lowered GH levels is achieved.

GH Secretory Pattern

The pattern of GH secretion itself can significantly modulate IGF-I. Ho and Weissberger have shown that GH secretion in acromegalic patients has a disturbed diurnal pattern, elevated basal level and attenuated pulsatile excursions, although pulse frequency was increased *(18)*. These authors suggest that the blunted pattern of secretion is more effective in elevating IGF-I concentrations. Indeed, Jorgensen and colleagues observed that continuous infusion of GH results in greater IGF-I response than the same dosage administered as two boluses *(139)*. Thus, IGF-I values may be elevated in acromegalic patients with apparently normal mean 24-h GH concentrations.

Interestingly, recent reports demonstrated that the increased GH pulse frequency observed in active acromegaly persists in apparently "cured" patients (normal IGF-I, basal and dynamic GH levels) *(140,141)*. This finding suggests an inherent defect at the hypothalamic level in acromegaly. Conversely, van den Berg et al. observed no differences in pulse characteristics between controls and surgically treated patients *(15)*. Resolution of these discordant findings has implications for both understanding the pathology underlying somatotropinoma formation and for determining long-term prognosis of treated acromegaly. Furthermore, persistently accelerated pulse generation may explain some cases where GH assessments indicate elevated concentrations, while IGF-I levels are normal.

Contribution of Local IGF-I Production

Plasma IGF-I determinations primarily measure GH-mediated IGF-I production by the liver. However, IGF-I is also produced locally in most tissues. Local production is stimulated not only by GH, but also by adrenocorticotropin, thyrotropin, lutenizing hormone, and follicle-stimulating hormone. The relative role of circulating and locally produced IGF-I in the control of cellular activity, acromegalic features, and complications of the disease is unknown *(142)*. The negative feedback of IGF-I upon GH gene expression could be exerted via endocrine actions of the circulating peptide or possibly through paracrine/autocrine action of IGF-I produced locally in the pituitary *(143)*.

Variability of GH Assays

The various assays differ in many of their components, including tracer specificity and purity, standards used, methods of detection (radiolabel vs colorimetric vs chemiluminescence), methods of measurement (sandwich vs competitive binding), and data reduction technique. Differences in the ability of commercial immunoassays to detect the various monomeric and dimeric molecular forms of GH have been described *(114,144,145)*. This differential cross-reactivity may in part account for the wide range of interassay variability. Indeed, a two- to fourfold difference in serum GH measurements has been demonstrated using several available GH radioimmunoassays *(144–147)*. Immunoradiometric assay (IRMA) employing a double-antibody detection system is widely used because of its sensitivity, accuracy, and rapid quantification of serum samples. It may be superseded by a modified chemiluminescence assay which has a significantly enhanced sensitivity, 0.003 μg/L compared to 0.2 μg/L for IRMA *(114)*. Normal ranges vary significantly between assays and between laboratories for that matter. Results must be interpreted within the context of local normative data.

Chapman et al. observed that recombinant GHBP-1, the predominant binding protein, competes with immunoassay antisera for GH binding *(114)*. Conversely, Jan and co-workers reported that addition of native GHBP did not affect GH measurements *(148)*. These discordant findings may be the result of the different GHBP preparations employed. Further studies are necessary to determine if endogenous GHBPs interfere with accurate measurement of serum GH.

CONCLUSIONS

Early diagnosis and successful treatment are essential to obviate the deleterious effects of acromegaly. Prompt recognition of those patients not truly cured in whom mild hypersomatotropism persists or recurrence occurs can minimize excess morbidity and mortality associated with prolonged exposure to elevated GH and IGF-I concentrations. Measurement of IGF-I plays an integral role in these endeavors, both as a cost-effective biochemical screen and in monitoring therapeutic response. As our understanding of the molecular and genetic basis for the dysregulation of GH secretion underlying acromegaly advances, our methods for assessing GH status will become increasingly refined. This knowledge may also resolve the controversy regarding the definition of biochemical cure.

REFERENCES

1. Melmed S. Acromegaly. N Engl J Med 1990; 322:966–77.
2. Shimon I, and Melmed S. Acromegaly: differential diagnosis and treatment. In Wierman ME, ed. Contemporary Endocrinology, Vol 3: Diseases of the Pituitary: Diagnosis and Treatment. Humana Press, Totowa, NJ, 1997, pp. 135–151.
3. Clemmons DR, Van Wyk JJ, Ridgeway E, Kliman B, Kjellberg RN, Underwood LE. Evaluation of acromegaly by radioimmunoassay of somatomedin C. N Engl J Med 1979; 301:1138–1142.
4. Rinderknecht E, Humbel RE. Polypeptides with nonsuppressible insulin-like and cell-growth promoting activities in human serum; isolation, chemical characterization, and some biological properties of forms I and II. Proc Natl Acad Sci USA 1976; 73:2365–2369.
5. Miller WL, Eberhardt NL. Structure and evaluation of growth hormone gene family. Endocrinol Rev 1983; 4:97–130.
6. Berelowitz M, Szabo M, Frohman LA, Fireston S, Chu L, Hintz RL. Somatomedin-C mediates growth hormone negative feedback by effects on both hypothalamus and the pituitary. Science 1982; 212:1279–1281.
7. Chen EY, Yu-Cheng L, Smith DH, Berrera-Saldana HA, Gelinas RE, Seeburg PH. The human growth hormone locus: nucleotide sequence, biology, and evolution. Genomics 1989; 4:479–497.
8. Ho KY, Evans WS, Blizzard RM, Veldhuis JD, Merriam GR, Samojlik E, Furlanetto R, Rogol AD, Kaiser DL, Thorner MO. Effects of sex and age on the 24-hour profile of growth hormone secretion in man: importance of endogenous estradiol concentrations. J Clin Endocrinol Metab 1987; 64:51–58.
9. Mayo KE, Vale W, Rivier J, Rosenfeld MG, Evans RM. Expression, cloning, and sequencing of a cDNA encoding human growth hormone-releasing factor. Nature 1983; 306:86–88.
10. Shen LP, Picket RL, Rutter WJ. Human somatostatin I: sequence of the cDNA. Proc Nat Acad Sci USA 1982; 79:4575–4579.
11. Yamada Y, Post SR, Wang K, Tager H, Bell GI, Seino S. Cloning and functional characterization of a family of human and mouse somatostatin receptors expressed in brain, gastrointestinal tract, and kidney. Proc Nat Acad Sci USA 1992; 89:251–255.
12. Schonbrunn A, Gu YZ, Dournard P, Beaudet A, Tannenbaum GS, Brown PJ. Somatostatin receptor subtypes: specific expression and signaling properties. Metab Clin Exp 1996; 45:8–11.
13. Shimon I, Yan X, Taylor JE, Weiss ML, Culler MD, Melmed S. Somatostatin receptor (SSTR) subtype-selective analogues differentially suppress in vitro growth hormone and prolactin in human pituitary adenomas. J Clin Invest 1997; 100:2386–2392.
14. Reutens AT, Veldhuis JD, Hoffman DM, Leung KC, Ho KK. A highly sensitive growth hormone (GH) enzymed-linked immunosorbent assay uncovers increased contribution of a tonic mode of GH secretion in adults with organic GH deficiency. J Clin Endocrinol Metab 1996; 81:1591–1597.

15. van den Berg G, Frolich M, Veldhuis JD, Roelfsema F. Growth hormone secretion in recently operated acromegalic patients. J Clin Endocrinol Metab 1994; 79:1706–1715.
16. Herman-Bonert VS, Prager D, Melmed S. Growth hormone. In Melmed S, ed. The Pituitary. Blackwell Science, Inc. Cambridge, MA, 1995, pp. 98–135.
17. Hartman ML, Veldhuis JD, Thorner MO. Normal control of growth hormone secretion. Hormone Res 1993; 40:37–47.
18. Ho K, Weissberger AJ. Characterization of 24-hour growth hormone secretion in acromegaly: implications for diagnosis and therapy. Clin Endocrinol 1994; 41:75–83.
19. Veldhuis JD, Iranmanesh A. Physiological regulation of the human growth hormone (GH)-insulin-like growth factor type I (IGF-I) axis: predominant impact of age, obesity, gonadal function, and sleep. Sleep 1996; 19:S221–S224.
20. Jaffe CA, DeMott-Friberg R, Barkan AL. Suppression of growth hormone (GH) secretion by a selective GH-releasing hormone (GHRH) antagonist: direct evidence for involvement of endogenous GHRH in the generation of GH pulses. J Clin Invest 1993; 92:695–701.
21. Plotsky PM, Vale W. Patterns of growth hormone-releasing factor and somatostatin secretion into the hypophysial-portal circulation of the rat. Science 1985; 230:461–463.
22. Frohman LA, Downs TR, Clarke IJ, Thomas GB. Measurement of growth hormone-releasing hormone and somatostatin in hypothalamic-portal plasma of unanesthetized sheep. Spontaneous secretion and response to insulin-induced hypoglycemia. J Clin Invest 1990; 86:17–24.
23. Thorner MO, Perryman RL, Cronin MJ, Rogol AD, Draznin M, Johanson A, Vale W, Horvath E, Kovacs K. Somatotroph hyperplasia: successful treatment by removal of a pancreatic islet tumor secreting a growth hormone-releasing factor (GRF). J Clin Invest 1982; 70:965–977.
24. Losa M, Chiodini PG, Liuzzi A, Konig A, Muller OA, Schopohl J, von Werder K. Growth hormone releasing-hormone infusion in patients with active acromegaly. J Clin Endocrinol Metab 1986; 63:88–93.
25. Wood SM, Ch'ng JL, Adams EF, Webster JD, Joplin GF, Mashiter K, Bloom SR. Abnormalities of growth hormone release in response to human pancreatic growth hormone releasing factor (GRF (1-44)) in acromegaly and hypopituitarism. Br Med J 1983; 286:1687–1691.
26. Lopes MB, Gaylinn BD, Thorner MO, Stoler MH. Growth hormone-releasing hormone receptor mRNA in acromegaly pituitary tumors. Am J Pathol 1997; 150:1885–1891.
27. Hartman ML, Veldhuis JD, Vance ML, Faria AC, Furlanetto RW, Thorner MO. Somatotropin pulse frequency and basal concentrations are increased in acromegaly and are reduced by successful therapy. J Clin Endocrinol Metab 1990; 70:1375–1384.
28. Yamashita S, Weiss M, Melmed S. Insulin-like growth factor I regulates growth hormone secretion and messenger ribonucleic acid levels in human pituitary tumor cells. J Clin Endocrinol Metab 1986; 63:730–735.
29. Gil Ad I, Weizman A, Silbergeld A, Dickerman Z, Kaplan B, Laron Z, Koch Y. Differential effects of insulin-like growth factor-I and growth hormone on hypothalamic regulation of growth hormone secretion in the rat. J Endocrinol Invest 1996; 19:542–547.
30. Sato M, Frohman LA. Differential effects of central and peripheral administration of growth hormone (GH) and insulin-like growth factor on hypothalamic GH-releasing hormone and somatostatin gene expression in GH-deficient dwarf rats. Endocrinology 1993; 133:793–799.
31. Uchiyama T, Kaji H, Abe H, Chihara K. Negative regulation of hypothalamic growth hormone-releasing factor messenger ribonucleic acid by growth hormone and insulin-like growth factor I. Neuroendocrinology 1994; 59:411–450.
32. Bermann M, Jaffe CA, Tsai W, DeMott-Frieberg R, Barkan AL. Negative feedback regulation of pulsatile growth hormone secretion by insulin-like growth factor I. Involvement of hypothalamic somatostatin. J Clin Invest 1994; 94:138–145.
33. Epelbaum J, Moyse E, Tannenbaum GS. Combined autoradiographic and immunohistochemical evidence for an association of somatostatin binding sites with growth hormone-releasing factor-containing nerve cell bodies in the rat arcuate nucleus. J Neuroendocrinol 1989; 1:109–116.
34. Salmon WD Jr, Daughaday WH. A hormonally-controlled serum factor which stimulates sulfate incorporation by cartilage *in vitro*. J Lab Clin Med 1957; 49:825–836.
35. Bell GI, Gerhard DS, Fong NM, Sanchez-Pescador K, Rall LB. Isolation of the human insulin-like growth factor genes: insulin-like growth factor II and insulin genes are contiguous. Proc Natl Acad Sci USA 1985; 82:6450–6454.
36. Schoenle E, Zapf J, Humbel RE, Froesch ER. Insulin-like growth factor I stimulates growth in hypophysectomized rats. Nature 1982; 296:252–253.

37. Pell JM, Bates PC. Differential actions of growth hormone and insulin-like growth factor I on tissue protein metabolism in dwarf mice. Endocrinology 1992; 130:1942–1950.
38. Green H, Morikawa M, Nixon T. A dual effector theory of growth hormone action. Differentiation 1985; 29:195–198.
39. Clemmons DR. Insulin-like growth factor binding proteins. TEM 1990; 1:412–417.
40. Rajaram S, Baylink DL, Mohan S. Insulin-like growth factor binding proteins in serum and other biological fluids: regulation and functions. Endocr Rev 1997; 18:801–831.
41. Zapf J, Waldvogel M, Froesch ER. Acute metabolic effects and half-lives of intravenously administered insulin-like growth factors in healthy adults. J Clin Invest 1986; 77:1768–1775.
42. Baumann G, Winter RJ, Shaw M. Circulating molecular variants of growth hormone in childhood. Pediatr Res 1987; 21:21–22.
43. Cooke NE, Ray J, Watson MA, Estes PA, Kuo BA, Liebhaber SA. Human growth hormone gene and the highly homologous growth hormone variant gene display different splicing patterns. J Clin Invest 1988; 82:270–275.
44. Leung DW, Spencer SA, Cachianes G, Hammonds RG, Collins C, Henzel WJ, Barnard R, Waters MJ, Wood WI. Growth hormone receptor and serum binding protein: purification, cloning, and expression. Nature 1987; 330:537–543.
45. Manner DA, Winer LM, Shaw MA, Baumann G. Plasma growth hormone (GH)-binding proteins: effect of GH binding to receptors and GH action. J Clin Endocrinol Metab 1991; 73:30–34.
46. Furlanetto RW, Underwood LE, Van Wyk JJ, D'Ercole AJ. Estimation of somatomedin-C levels in normals and patients with pituitary disease by radioimmunoassay. J Clin Invest 1977; 60:648–657.
47. Zapf J, Walther H, Froesch ER. Radioimmunological determination of insulin-like growth factors I and II in normal subjects and in patients with growth disorders and extrapancreatic tumor hypoglycemia. J Clin Invest 1981; 68:1321–1330.
48. Wass JAH, Clemmons DR, Underwood LE, Barrow I, Besser GM, Van Wyk JJ. Changes in circulating somatostatin-C levels in bromocriptine-treated acromegaly. Clin Endocrinol (Oxf) 1982; 17:369.
49. Stonesifer LD, Jordan RM, Kohler PO. Somatomedin C in treated acromegaly: poor correlation with growth hormone and clinical response. J Clin Endocrinol Metab 1981; 53:931–934.
50. Nortier JWR, Croughs RJM, Thijssen JHH, Schwartz F. Bromocriptine therapy in acromegaly: effects on plasma levels, somatomedin C levels and clinical activity. Clin Endocrinol (Oxf) 1985; 22:209.
51. Clemmons DR, Underwood LE, Van Wyk JJ. Hormonal control of immunoreactive somatomedin production by cultured human fibroblasts. J Clin Invest 1981; 67:10.
52. Scott CD, Martin JL, Baxter RC. Rat hepatocyte insulin-like growth factor I and binding protein: effect of growth hormone in vitro and in vivo. Endocrinology 1985; 116:1102.
53. Barkan AL, Beitins IZ, Kelch RP. Plasma insulin-like growth factor-I/somatomedin-C in acromegaly: correlation with the degree of growth hormone hypersecretion. J Clin Endocrinol Metab 1988; 67:69–73.
54. Rieu M, Girard G, Bricaire H, Binoux M. The importance of insulin-like growth factor (somatomedin) measurements in the diagnosis and surveillance of acromegaly. J Clin Endocrinol Metab 1982; 55:147–153.
55. Lamberts SWJ, Uitterlinden P, Verleum T. Relationship between growth hormone and somatomedin-C levels in untreated acromegaly, after surgery and radiotherapy and during medical therapy with Sandostatin (SMS 201-995). Eur J Clin Invest 1987; 17:354–359.
56. Beck P, Parker ML, Daughaday WH. Paradoxical hypersecretion of growth hormone in response to glucose. J Clin Endocrinol Metab 1966; 26:463–469.
57. Gama R, Labib M, Teale JD, Marks V. Acromegaly with a misleading normal plasma insulin-like growth factor 1. Ann Clin Biochem 1989; 26:102–103.
58. Jaffe LS, Rosen HN. Normal insulin-like growth factor-I in an elderly woman with acromegaly and medical illness. Arch Intern Med 1994; 154:1533–1535.
59. Clemmons DR, Van Wyk JJ. Factors controlling blood concentration of somatomedin C. Clin Endocrinol Metab 1984; 13:113–143.
60. Luna AM, Wilson DM, Wibbelsman CJ, Brown RC, Nagashima RJ, Hintz RL, Rosenfeld RG. Somatomedins in adolescence: a cross-sectional study of the effect of puberty on plasma insulin-like growth factor I and II levels. J Clin Endocrinol Metab 1983; 57:268–271.
61. Yamamoto H, Sohmiya M, Oka N, Kato Y. Effects of aging and sex on plasma insulin-like growth factor I (IGF-I) levels in normal adults. Acta Endocrinol 1991; 124:497–500.
62. Johanson AJ, Blizzard RM. Low somatomedin-C levels in older men rise in response to growth hormone administration. Johns Hopkins Med J 1981; 149:115–117.
63. Juul A, Dalgaard P, Blum WF, Bang P, Hall K, Michaelsen KF, Muller J, Skakkebaek NE. Serum levels

of insulin-like growth factor (IGF)-binding protein-3 (IGFBP-3) in healthy infants, children, and adolescents: the relation to IGF-I, IGF-II, IGFBP-1, IGFBP-2, age, sex, body mass index, and pubertal maturation. J Clin Endocrinol Metab 1995; 80:2534–2542.
64. Hintz RL, Suskind R, Amatayakul K, Thanangkul O, Olson R. Plasma somatomedin and growth hormone values in children with protein-calorie malnutrition. J Pediatr 1978; 92:153–156.
65. Clemmons DR, Klibanski A, Underwood E, McArthur JW, Ridgeway EC, Beitins IZ, Van Wyk JJ. Reduction in plasma immunoreactive somatomedin-C during fasting in humans. J Clin Endocrinol Metab 1981; 53:1247–1250.
66. Isley WL, Underwood LE, Clemmons DR. Dietary components that regulate serum somatomedin-C concentrations in humans. J Clin Invest 1983; 71:175–182.
67. Grzywa M. Serum somatomedin activity and growth hormone level in obese men: dependency on degree of obesity and hyperlipidemia. Exp Clin Endocrinol 1986; 88:325–333.
68. Copeland KC, Colletti RB, Devlin JT, McAuliffe TL. The relationship between insulin-like growth factor-I, adiposity, and aging. Metabolism 1990; 39:584–587.
69. Williams T, Berelowitz M, Joffe SN, Thorner MO, Rivier J, Vale W, Frohman LA. Impaired growth hormone responses to growth hormone-releasing factor in obesity. N Engl J Med 1984; 311:1403–1407.
70. Blackard WG, Hull EW, Lopez A. Effect of lipids on growth hormone secretion in humans. J Clin Invest 1971; 50:1439–1441.
71. Fineberg SE, Horland AA, Merimee TJ. Free fatty acid concentrations and growth hormone secretion in man. Metabolism 1972; 21:491–493.
72. Ross R, Miell J, Freeman E, Jones J, Matthews D, Preece M, Buchanan C. Critically ill patients have high basal growth hormone levels with attenuated oscillatory activity associated with low levels of insulin-like growth factor-I. Clin Endocrinol 1991; 35:47–54.
73. Wu A, Grant DB, Hambley J, Levi AJ. Reduced serum somatomedin activity in patients with chronic liver disease. Clin Sci Mol Med 1974; 47:359–366.
74. Hirschberg R, Kaysen GA. Insulin-like growth factor I and its binding proteins in the experimental nephrotic syndrome. Endocrinology 1995; 136:1565–1571.
75. Tonshoff B, Powell DR, Zhao D, Durham SK, Coleman ME, Domene HM, Blum WF, Baxter RC, Moore LJ, Kaskel FJ. Decreased hepatic insulin-like growth factor (IGF)-I and increased IGF binding protein-1 and -2 gene expression in experimental uremia. Endocrinology 1997; 138:938–946.
76. Weidemann E, Schwartz E. Suppression of growth hormone-dependent human sulfation factor by estrogen. J Clin Endocrinol Metab 1972; 34:51–58.
77. Clemmons DR, Underwood LE, Ridgeway EC, Kliman B, Kjellberg RN, Van Wyk JJ. Estradiol treatment of acromegaly. Reduction of immunoreactive somatomedin-C and improvement of metabolic status. Am J Med 1980; 69:571–575.
78. Whitelaw MJ. Experiences in treating excessive height in girls with cyclic oestradiol valerate; a ten year survey. Acta Endocrinol 1967; 54:473–484.
79. Wettenhall HNB, Cahill C, Roche AF. Tall girls: a survey of 15 years of management and treatment. J Pediatr 1975; 86:602–610.
80. Ross JL, Cassorla FG, Skerda MC, Valk IM, Loriaux DL, Cutler GB. A preliminary study on the effect of estrogen dose on growth in Turner's syndrome. N Engl J Med 1983; 309:1104.
81. Cuttler L, Van Vliet G, Conte FA, Kaplan SL, Grumbach MM. Somatomedin-C levels in children and adolescents with gonadal dysgenesis: differences from age-matched normal females and effect of chronic replacement therapy. J Clin Endocrinol Metab 1985; 60:1087.
82. Barengolts EI, Kouznetsova T, Segalene A, Lathon P, Odvina C, Kukreja SC, Unterman TG. Effects of progesterone on serum levels of IGF-1 and on femur IGF-1 mRNA in ovariectomized rats. J Bone Min Res 1996; 11:1406–1412.
83. Maes M, Underwood LE, Ketelslegers J-M. Low serum somatomedin-C in insulin-dependent diabetes: evidence for a post receptor mechanism. Endocrinology 1986; 118:377–382.
84. Phillips LS, Orawski AT. Nutrition and somatomedin. III. Diabetic control, somatomedin, and growth in rats. Diabetes 1977; 26:864–869.
85. Bereket A, Lang CH, Blethen SL, Ng LC, Wilson TA. Insulin treatment normalizes reduced free insulin-like growth factor-I concentrations in diabetic children. Clin Endocrinol 1996; 45:321–326.
86. Baxter RC. Radioimmunoassay of somatomedin-C: components with radioreceptor assay in patients with growth hormone disorders, hypothyroidism, and renal failure. Clin Chem 1982; 28:488–495.
87. Chernausek SD, Underwood LE, Utiger RD, Van Wyk JJ. Growth hormone secretion and plasma somatomedin-C in primary hypothyroidism. Clin Endocrinol 1983; 19:337–344.

88. Dieguez C, Mallo F, Senaris R, Pineda J, Martul P, Leal-Cerro A, Pombo M, Casanueva FF. Role of glucocorticoids in the neuroregulation of growth hormone secretion. J Pediatr Endocrinol Metab 1996; 9:S255–S260.
89. Gournielen M, Girard F, Binoux M. Serum somatomedin/insulin-like growth factor (IGF) and IGF carrier levels in patients with Cushing's syndrome or receiving glucocorticoid therapy. J Clin Endocrinol Metab 1982; 54:884–891.
90. Miell JP, Taylor AM, Jones J, Holly JM, Gaillard RC, Pralong FP, Ross RJ, Blum WF. The effects of dexamethasone treatment on immunoreactive and bioactive insulin-like growth factors (IGFs) and IGF binding protein in normal male volunteers. J Endocrinol 1993; 136:525–533.
91. Clemmons DR, Underwood LE, Ridgeway EC, Liman B, Van Wyk JJ. Hyperprolactinemia is associated with increased immunoreactive somatomedin-C in hypopituitarism. J Clin Endocrinol Metab 1981; 52:731–735.
92. Teale JD, Marks V. The measurement or insulin-like growth factor I: clinical applications and significance. Ann Clin Biochem 1986; 23:413–424.
93. Roelfsema F, Frolich M, Van Dulken H. Somatomedin-c levels in treated and untreated patients with acromegaly. Clin Endocrinol 1987; 26:137–144.
94. Hulting A-L, Hall K, Werner S. Somatomedin A increments are diminished in acromegaly with concomitant hyper prolactinemia. Acta Endocrinol 1984; 106:305–310.
95. Furlanetto RW, Underwood LE, Van Wyk JJ, Handwerger S. Serum immunoreactive somatomedin-C is elevated in late pregnancy. J Clin Endocrinol Metab 1979; 47:695–697.
96. Hasegawa Y, Hasegawa T, Takada M, Tsuchiya Y. The free form of insulin-like growth factor I increases in circulation during normal human pregnancy. J Clin Endocrinol Metab 1995; 80:3284–3286.
97. Rudman D, Mattson DE. Serum insulin-like growth factor I in healthy older men in relation to physical activity. J Am Geriatr Soc 1994; 42:71–76.
98. Kreamer WJ, Aguilera BA, Terada M, Newton RU, Lynch JM, Rosandaal G, McBride JM, Gordon SE, Hakkinen K. Responses of IGF-I to endogenous increases in growth hormone after heavy-resistance exercise. J Appl Phys 1995; 79:1310–1315.
99. Vitiello MV, Wilkinson CW, Merriam GR, Moe KE, Prinz PN, Ralph DD, Colasurdo EA, Schwartz RJ. Successful 6-month endurance training does not alter insulin-like growth factor-I in healthy older men and women. J Gerontol 1997; 52:M149–M154.
100. Di Luigi I, Conti FG, Casini A, Guidetti L, Zezza G, Pigozzi f, Spera G, Fortunio G, Romanelli F. Growth hormone and insulin-like growth factor I responses to moderate submaximal acute physical exercise in man: effects of octreotide, a somatostatin analogue, administration. Int J Sports Med 1997; 18:257–263.
101. Poehlman ET, Copeland KC. Influence of physical activity on insulin-like growth factor-I in healthy younger and older men. J Clin Endocrinol Metab 1990; 71:1468–1473.
102. Schwarz AJ, Brasel JA, Hintz RL, Mohan S, Cooper DM. Acute effect of brief low- and high-intensity exercise on circulating insulin-like growth factor (IGF) I, II and IGF-binding protein-3 and its proteolysis in young healthy men. J Clin Endocrinol Metab 1996; 81:3492–3497.
103. Horner JM, Liu F, Hintz RL. Comparison of ^{125}I-somatomedin-A and ^{125}I-somatomedin-C radioreceptor assays for somatomedin peptide content in whole and acid-chromatographed plasma. J Clin Endocrinol Metab 1978; 47:1287–1295.
104. Daughaday WH, Mariz IK, Blethin SL. Inhibition of access of bound somatomedin to membrane receptor and immunobinding sites: a comparison of radioreceptor and radioimmunoassay of somatomedin in native and acid-ethanol extracted serum. J Clin Endocrinol Metab 1980; 51:781–788.
105. Barreca A, Ciccarelli E, Minuto F, Bruzzi P, Giordano G, Camanni F. Insulin-like growth factor I and daily growth hormone profile in the assessment of active acromegaly. Acta Endocrinol 1989; 120:629–635.
106. Acromegaly therapy consensus development panel. Consensus statement: benefits versus risks of medical therapy for acromegaly. Am J Med 1994; 97:468–473.
107. Barkan A. Controversies in the diagnosis and therapy of acromegaly. Endocrinologist 1997; 7: 300–307.
108. Dobrashian RD, O'Halloran DJ, Hunt A, Beardwell CG, Shalet SM. Relationships between insulin-like growth factor-I levels and growth hormone concentrations during diurnal profiles and following oral glucose in acromegaly. Clin Endocrinol 1993; 38:589–593.
109. Earll JM, Sparks LL, Forsham PH. Glucose suppression of serum growth hormone in the diagnosis of acromegaly. JAMA 1967; 201:134.

110. Jaquet P, Guibout M, Jaquet C, Grisoli F, Devolx BC, Dumas D, Bert J. Circadian regulation of growth hormone secretion after treatment in acromegaly. J Clin Endocrinol Metab 1980; 50:322–328.
111. Chapman IM, Hartman ML, Straume M, Johnson ML, Veldhuis JD, Thorner MO. Enhanced sensitivity growth hormone (GH) chemiluminescence assay reveals lower postglucose nadir GH concentrations in men than in women. J Clin Endocrinol Metab 1994; 78:1312–1319.
112. Thompson RG, Rodriguez A, Kowarski AA, Blizzard RM. Growth hormone: metabolic clearance rates, integrated concentrations and production rates in normal adults and the effect of prednisone. J Clin Invest 1972; 51:3193–3199.
113. Evans WS, Faria ACS, Christiansen E, Ho KY, Weiss J, Rogol AD, Johnson ML, Blizzard RM, Veldhuis JD, Thorner MO. Impact of intensive venous sampling on characterization of pulsatile GH release. Am J Physiol 1987; 252:E549–E556.
114. Chapman IM, Hartman ML, Straume M, Johnson ML, Veldhuis JD, Thorner MO. Enhanced sensitivity growth hormone (GH) chemiluminescence assay reveals lower postglucose nadir GH concentrations in men than in women. J Clin Endocrinol Metab 1994; 78:1312–1319.
115. Okuno A, Yano K, Itoh Y, Hashida S, Ishikawa E, Mohri Z, Murakami Y. Urine growth hormone determinations compared with other methods in the assessment of growth hormone secretion. Acta Paediatr Scand 1987; 337:Suppl. 74–81.
116. Lunt H, Tucker AJ, Bullen H, Gibbs C, Wilkin TJ. Overnight urinary growth hormone measurement in the diagnosis of acromegaly. Clin Endocrinol 1990; 33:205–212.
117. Main KM, Lindholm J, Vandeweghe M, Skakkebaek NE. Urinary growth hormone excretion in acromegaly: diagnostic value in mild disease activity. Acta Endocrinol 1993; 129:409–413.
118. Sukegawa I, Hizuka N, Tanako K, Asakawa K, Horikawa R, Hashida S, Ishikawa E, Mohri Z, Murakami Y, Shizuma K. Urinary growth hormone (GH) measurements are useful for evaluating endogenous GH secretion. J Clin Endocrinol Metab 1988; 66:1117–1123.
119. Girard J, Celniker A, Price A, Tanaka T, Walker J, Welling K, Albertsson-Wikland K. Urinary measurement of growth hormone secretion. Acta Paediatr Scand Suppl 1989; 366:149–154.
120. Winer LM, Shaw MA, Baumann G. Urinary growth hormone excretion rates in normal and acromegalic man: a critical appraisal of its potential clinical utility. J Endocrinol Invest 1989; 12:461–467.
121. Hanssen KF, Rasmussen P. Urinary growth hormone in acromegaly. Acta Med Scand 1974; 195:381–384.
122. Pholsena M, Le Bouc Y, Rousseau E, Christol R, Birman P, Perin L, Girard F. Evaluation of acromegaly by measurement of 24-hourly urinary growth hormone excretion. Acta Endocrinol 1993; 128:9–14.
123. Evans AJ, Willis DS, Wood PJ. The assay of urinary growth hormone in normal and acromegalic adults. Clin Endocrinol 1991; 35:413–418.
124. Hattori N, Kato Y, Murakami Y, Hashida S, Ishikawa E, Mohri Z, Imura H. Urinary growth hormone levels measured by ultrasensitive enzyme immunoassay in patients with renal insufficiency. J Clin Endocrinol Metab 1988; 66:727–732.
125. Hattori N, Shimatsu A, Kato Y, Koshiyama H, Ishikawa Y, Tanoh T, Assadian H, Imura H. Urinary excretion of human growth hormone: daily variation and relationship with albumin and a1-microglobulin in urine. Acta Endocrinol 1989; 121:533–537.
126. Grinspoon S, Clemmons D, Swearingen B, Klibanski A. Serum insulin-like growth factor-binding protein-3 levels in the diagnosis of acromegaly. J Clin Endocrinol Metab 1995; 80:927–932.
127. Rasat R, Livesey JL, Espiner EA, Abbott GD, Donald RA. IGF-I and IGFBP-3 screening for disorders of growth hormone secretion. NZ Med J 1996; 109:156–159.
128. de Herder WW, van der Lely AJ, Janssen JA, Uitterlinden P, Hofland LJ, Lamberts SW. IGFBP-3 is a poor parameter for assessment of clinical activity in acromegaly. Clin Endocrinol 1995; 43:501–505.
129. Paramo C, Andrade OM, Fluiters E, Luna R, de la Fuente J, Garcia-Mayor RV. Comparative study of insulin-like growth factor-I (IGF-I) and IGF-binding protein-3 (IGFBP-3) level and IGF-I/IGFBP-3 ratio measurements and their relationship with an index of clinical activity in the management of patients with acromegaly. Metab Clin Exp 1997; 46:494–498.
130. Thissen JP, Ketelslegers JM, Maiter D. Use of insulin-like growth factor-I (IGF-I) and IGF-binding protein-3 in the diagnosis of acromegaly and growth hormone deficiency in adults. Growth Regul 1996; 6:222–229.
131. Clemmons DR, Thissen JP, Maes M, Ketelslegers JM, Underwood LE. Insulin-like growth factor-I (IGF-I) infusion into hypophysectomized or protein-deprived rats induces specific IGF-binding proteins in serum. J Clin Endocrinol Metab 1989; 125:2967–2972.
132. Irie M, Tsushima T. Increase in serum growth hormone concentration following thyrotropin-releasing hormone injection in patients with acromegaly or gigantism. J Clin Endocrinol Metab 1972; 35:97–100.

133. Shibasaki T, Hotta M, Masuda A, Imaki T, Obara N, Hizuka N, Takanok, Wakabayashi I, Demura H, Ling N. Studies on the response of growth hormone (GH) secretion to GH-releasing hormone, thyrotropin-releasing hormone, gonadotropin-releasing hormone, and somatostatin in acromegaly. J Clin Endocrinol Metab 1986; 63:167–173.
134. Jaffe CA, Barkan AL. Treatment of acromegaly with dopamine agonists. Endocrinol Metab Clin North Am 1992; 21:713–735.
135. Ezzat S, Snyder PJ, Young WF, Boyajy LD, Newman C, Klibanski A, Molitch ME, Boyd AE, Sheeler L, Cooke DM. Octreotide treatment of acromegaly: a randomized, multicenter study. Ann Int Med 1992; 117:711–718.
136. Newman CB, Melmed S, Snyder PJ, Young WF, Boyajy LD, Levy R, Stewart WN, Klibanski A, Molitch ME, Gagel RF. Safety and efficacy of long-term octreotide therapy of acromegaly: results of a multicenter trial in 103 patients—a clinical research center study. J Clin Endocrinol Metab 1995; 80:2768–2775.
137. Lamberts SWJ, Uitterlinden P, Schuijff PC, Klijn JGM. Therapy of acromegaly with sandostatin: the predictive value of an acute test, the value of serum somatomedin-C measurements in dose adjustment and the definition of a biochemical "cure." Clin Endocrinol 1988; 29:411–420.
138. Kao PC, Laws ER Jr, Zimmerman D. Somatomedin C/insulin-like growth factor I levels after treatment of acromegaly. Ann Clin Lab Sci 1992; 22:95–99.
139. Jorgensen JOL, Moller N, Lauritzen T, Christiansen JS. Pulsatile versus continuous intravenous administration of growth hormone (GH) in GH-deficient patients: effects on circulating insulin-like growth factor-I and metabolic indices. J Clin Endocrinol Metab 1990; 70:1616–1623.
140. Ho PJ, Jaffe CA, Friberg RD, Chandler WF, Barkin AL. Persistence of rapid growth hormone (GH) pulsatility after successful removal of GH-producing pituitary tumors. J Clin Endocrinol Metab 1994; 78:1403–1410.
141. Semer M, Faria AC, Nery M, Salgado LR, Knoepfelmacher M, Wajchenberg BL, Liberman B. Growth hormone pulsatility in active and cured acromegalic subjects. J Clin Endocrinol Metab 1995; 80:3767–3770.
142. Daughaday WH. A personal history of the origin the somatomedin hypothesis and recent challenges to its validity. Perspect Biol Med 1989; 39:194–211.
143. Melmed S, Yamashita S, Yamasaki H, Fagin J, Namba H, Yamamoto H, Weber M, Morita S, Webster J, Prager D. IGF-I receptor signalling: lessons from the somatotroph. Recent Prog Horm Res 1996; 51:189–216.
144. Celniker AC, Chen AB, Wert RM, Sherman BM. Variability in the quantitation of circulating growth hormone using commercial immunoassays. J Clin Endocrinol Metab 1989; 68:469–476.
145. Bowsher RR, Apathy JM, Ferguson AL, Riggin RM, Henry DP. Cross-reactivity of monomeric and dimeric biosynthetic human growth hormone in commercial immunoassays. Clin Chem 1990; 36:362–366.
146. Blethen SL, Chasalow FI. Use of a two-site immunoradiometric assay for growth hormone (GH). J Clin Endocrinol Metab 1983; 57:1031–1035.
147. Reiter EO, Morris AH, MacGillivray MH, Weber D. Variable estimates of serum growth hormone concentrations by different radioassay systems. J Clin Endocrinol Metab 1988; 66:68–71.
148. Jan T, Shaw MA, Baumann G. Effects of growth hormone-binding proteins on serum growth hormone measurements. J Clin Endocrinol Metab 1991; 71-2:387–391.

30 IGF in the Treatment of Diabetes

Robert A. Gabbay, MD, *and Alan C. Moses,* MD

CONTENTS

Since the identification of insulin-like growth factor-I (IGF-I) as one of the proteins responsible for the "nonsuppressible insulin-like activity" described by Froesch et al. *(1)*, IGF-I has been an attractive candidate for use in the treatment of diabetes mellitus. IGF-I shares not only structural homology with insulin, but cognate receptor and post-receptor signaling similarities as well. The large-scale production of recombinant hIGF-I has led to an explosion in research characterizing the metabolic effects and therapeutic actions of IGF-I in different forms of diabetes. Initial studies focused on the use of IGF-I in patients with severe insulin resistance, a group in whom effective therapeutic options are quite limited. Early reports with this patient population provided the rationale for using IGF-I in both type 2 (non-insulin-dependent) and type 1 (insulin-dependent) diabetes. This chapter reviews the rationale for the use of exogenous IGF-I in different types of diabetes mellitus and summarizes the therapeutic trials that have been published based on treatment with IGF-I.

RATIONALE FOR THE USE OF IGF-I IN DIABETES MELLITUS

IGF-I shares sufficient structural and functional homologies with insulin, suggesting the possibility of similar biologic actions. Insulin and IGF-I have approx 50% sequence homology and both peptides appear to derive from the same ancestral gene. The cognate receptors for the two hormones also are similar. Both are high-affinity heterotetrameric tyrosine kinase receptors. In fact, it appears that following receptor binding, both receptors utilize many of the same signaling cascades (including insulin-receptor substrate-1 [IRS-1] and IRS-2 phosphorylation and *ras*, mitogen-activated protein [MAP] kinase,

From: *Contemporary Endocrinology: The IGF System*
Edited by: R. Rosenfeld and C. Roberts Jr

and phosphatidylinositol 3-kinase activation) (*see* Chapter 8). Initially, many of the insulin-like actions of IGF-I were believed secondary to binding of IGF to the insulin receptor; however, it appears that IGF-I can exert some insulin-like effects through high-affinity binding to its own receptor. The relative specificity of insulin and IGF-I appears to derive in large part from the tissue distribution of their respective receptors. In contrast to the presence of insulin receptors on liver, muscle, and adipose tissue, IGF receptors do not appear to be present in appreciable amounts on human hepatocytes or adult adipocytes. This discrepancy in tissue distribution has important implications for the potential mechanisms through which IGF mediates insulin-like effects.

One important contrast between insulin and IGF-I is the plasma half-lives of these two homologous polypeptides. IGF-I circulates as a protein-bound complex with several distinct specific, high-affinity carrier proteins. The principal IGF binding protein is IGF binding protein 3 (IGFBP-3), which together with its acid-labile subunit (ALS) creates a macromolecular complex of molecular mass > 150 kDa. IGFBP-3 and the other circulating IGF binding proteins produce relatively stable IGF-I concentrations over a 24-h period. This is in sharp contrast to the pulsatile release of insulin into the portal circulation in response to various secretagouges and the rapid disappearance of insulin from the plasma compartment with a half-life measured in minutes. In all there are at least six specific high-affinity IGF binding proteins that, through complex mechanisms, control the circulating half-life, tissue distribution, and biologic activity of IGF-I. This regulation of IGF-I is complicated further by a set of binding protein-specific proteases that cleave the binding proteins and thereby alter the half-life and local tissue availability of IGF-I (*see* Chapter 11). It is possible that the prolonged activation of mitogenic signaling pathways through the IGF-I receptor (i.e., *ras* and MAP kinase) by levels of IGF-I that are sustained in both the plasma and extracellular compartments may help to explain the greater mitogenic potential of IGF-I compared to insulin.

Early studies examining the in vitro effects of IGF-I supported the hypothesis that IGF-I might have therapeutic potential as a hypoglycemic agent. Whereas early studies concentrated on the ability of IGF-I to stimulate DNA synthesis and cell replication, IGF-I also was found to stimulate glycogen synthesis and glucose and amino acid uptake in vitro. IGF-I binds to and activates the insulin receptor with a potency less than 1% the potency of insulin itself. The concentration of total IGF-I in healthy adults is 30–40 n*M* compared with mean insulin concentrations in the range of 50–100 p*M*. Therefore, from a mass basis, the insulin-like potential of IGF-I dwarfs that of insulin itself. However, the presence of the multiple IGFBPs described previously results in much lower local concentrations of IGF-I that have been difficult to estimate accurately. It has therefore been difficult to determine if the insulin-like effects of IGF-I in vivo are mediated through local interactions with the insulin receptor or directly through the IGF-I receptor. Despite the inability to resolve this issue unequivocally, the early in vitro observations with IGF-I served as the impetus for examining the in vivo actions of IGF-I.

PHYSIOLOGIC EFFECTS OF IGF-I IN HEALTHY HUMANS

The availability of large quantities of recombinant human IGF-I (rhIGF-I) provided the opportunity to examine the metabolic effects of rhIGF-I in humans. Early studies focused on the effects of a single intravenous bolus of IGF-I on healthy adults.Guler et al. *(2)* demonstrated that an intravenous bolus of 100 μg/kg lowered blood glucose with a temporal pattern similar to that of insulin. Subsequently, IGF-I was calculated to have

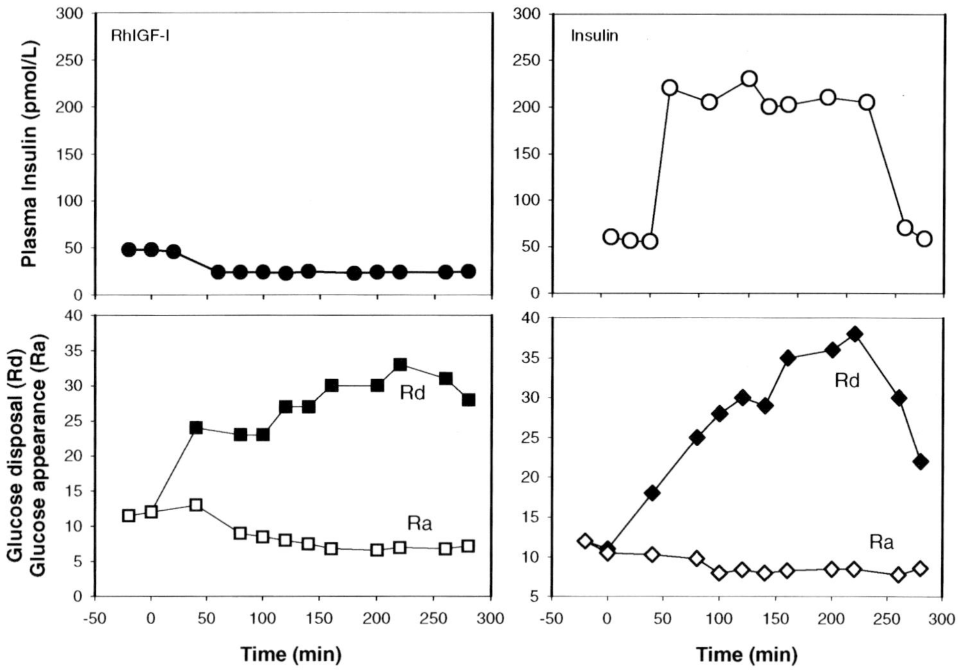

Fig. 1. The effects of continuously infused rhIGF-I (49 pmol/kg/min) and insulin (2.25 pmol/kg/min) on overall glucose disposal rate (R_d), hepatic glucose production (or glucose appearance rate, R_a), and plasma insulin levels in healthy adults. (From Elahi et al. *[3]*)

a hypoglycemic potency approx 5–7% that of insulin on a mass basis *(3,4)*. This suggests that the hypoglycemic effect of IGF-I probably cannot be accounted for strictly through binding to the insulin receptor, for which IGF-I has only 1% the affinity of insulin. In fact, the hypoglycemic effects of IGF-I were identical in transgenic mice lacking the insulin receptor when compared to normal mice *(5)*. In addition, older individuals appear to retain their responses to IGF-I while developing resistance to those of insulin *(4)*, providing further evidence that insulin receptor interactions are unlikely to provide the sole mechanism for IGF-I effects on carbohydrate metabolism.

Studies employing euglycemic clamp procedures in humans using doses of IGF-I adjusted to yield a hypoglycemic effect similar to that of insulin (i.e., a 20-fold increased mass of IGF-I compared to insulin) examined the relative metabolic actions of these two hormones *(3,4)* (Fig. 1). Studies from two groups of investigators indicated that insulin and IGF-I stimulated glucose and amino acid uptake, suppressed hepatic glucose production, inhibited protein degradation, lowered free fatty acid levels, and decreased glucagon concentrations. There were, however, some conflicting results regarding the relative potencies of the two agents that may be explained by differences in methodology and in the patient populations of the two studies. In particular, Elahi et al. *(3)* found that IGF was far less potent that insulin in inhibiting either hepatic glucose production or lipolysis whereas the two hormones were found to be equipotent in the study by Boulware et al. *(4)*. The actions of IGF-I on hepatic glucose production and lipid oxidation are intriguing given the lack of (or paucity of) IGF-I receptors on human hepatocytes and adult adipocytes *(6)*. Whether these effects of IGF-I, generally reported at somewhat higher doses, occur through direct actions or through changes in other mediators (i.e., changes in substrate concentrations, insulin/glucagon ratio, or cross-reactivity with the insulin receptor) await further investigation. Although euglycemic clamp studies have found that IGF-I lowers glucagon levels *(3,4)*, other infusion protocols have failed to observe significant changes in glucagon despite effects on insulin sensitivity *(5)*. IGF-I is more potent than insulin at inhibiting growth hormone release and at inhibiting insulin secretion (as reflected by absolute insulin and C peptide levels) *(3,4)*. It is unclear how the acute effects of IGF-I during these intravenous infusions, where high levels of free IGF-I are likely, correlate with the chronic effects of subcutaneous IGF-I administered over several doses where free IGF-I levels are not as high and probably are not sufficient to interact extensively with insulin receptors.

IGF-I alters the response to hypoglycemia. Healthy volunteers treated with a continuous IGF-I infusion demonstrated a delayed recovery from hypoglycemia that was characterized by decreased glucagon and growth hormone (GH) responses with unchanged epinephrine secretion *(7,8)*. Interestingly, there was an augmentation of the norepinephrine response *(7,8)* leading to increased sympathetic nervous system-mediated symptoms with consequent heightened hypoglycemia awareness *(8)*. This increased hypoglycemia awareness may be one reason there have not been reports of increased frequency of severe hypoglycemic events in patients treated with rhIGF-I.

ABNORMALITIES OF THE IGF AXIS IN DIABETES MELLITUS

It has long been recognized that the GH–IGF-I axis is deranged in type 1 diabetes mellitus *(9)*. Tamborlane et al. *(10)* demonstrated that IGF-I levels in children with type 1 diabetes are inversely related to their glycemic control. Improvement of glycemic control with aggressive insulin treatment led to an increase in IGF-I levels and an acceleration in height velocity. Patients with poorly controlled type 1 diabetes are functionally resistant to GH as characterized by hypersecretion of GH, deceased GH binding protein (an indicator of decreased GH binding and action), and decreased IGF-I levels *(11)*. In fact levels of free IGF-I also are diminished in type 1 diabetes *(14)*. Abnormalities in the level and regulation of IGFBP-1 and -3 also have been described. In addition, it has been suggested that IGFBP-1 levels in patients with type 1 diabetes may be more important than nocturnal GH in potentiating fasting hyperglycemia, the “dawn phenomenon” *(12,13)*.

Insulin administration profoundly affects the circulating levels of IGF-I and its binding proteins in patients with type 1 diabetes. Bereket et al. *(15)* prospectively followed 14 newly diagnosed patients with type 1 diabetes after the initiation of insulin therapy. Initially, IGFBP-1 levels dropped from their elevated baseline with the administration of subcutaneous insulin. This effect was followed by IGF-I rising to normal levels after 1 wk of treatment with a slower rise in IGFBP-3. In contrast ALS levels were suppressed only moderately in newly diagnosed type 1 diabetes and rose more slowly than did either IGF-I or IGFBP-3 and did not normalize with insulin treatment. These findings suggest that IGFBP-1 (which responds rapidly to insulin therapy), and not ALS, is responsible for determining the bioavailability of IGF-I in type 1 diabetes.

The route of insulin delivery also appears to influence its actions on the IGF system. Shishko et al. *(16)* compared the effects of continuous portal infusion of insulin (via the umbilical vein) to insulin administered as a continuous subcutaneous infusion (CSII) and to “standard” intermittent subcutaneous insulin therapy in 36 patients with newly diagnosed type 1 diabetes (Fig. 2). Intraportal delivery of insulin was more effective than CSII

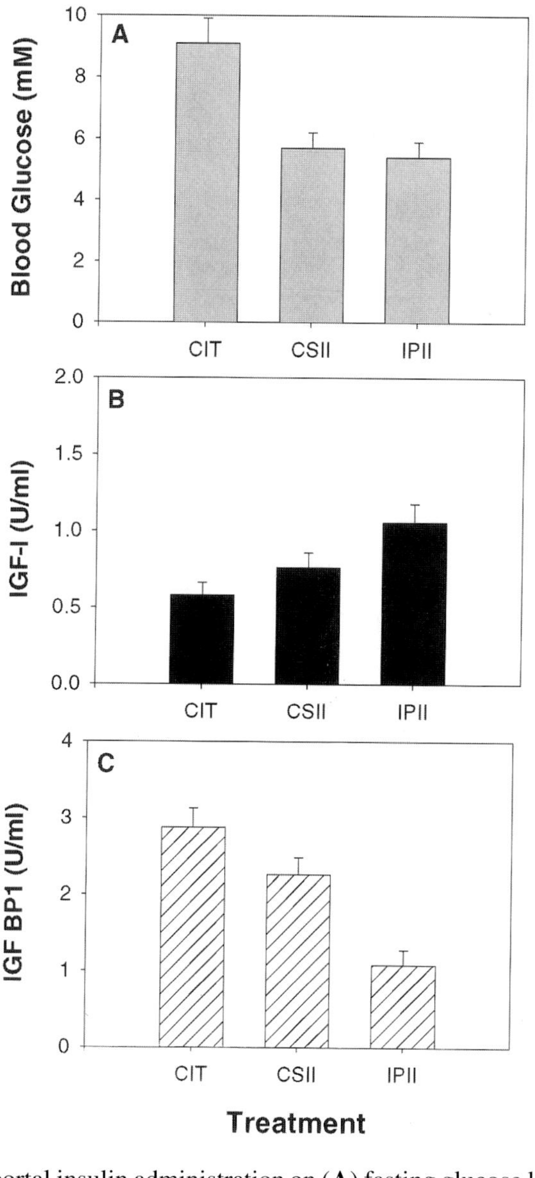

Fig. 2. Effects of intraportal insulin administration on (**A**) fasting glucose level, (**B**) serum IGF-I levels, and (**C**) IGFBP-1 levels. Data represent measurements obtained after 1 wk of therapy with conventional subcutaneous insulin therapy (CIT), with continuous subcutaneous insulin infusion (CSII), or with intraportal insulin (IPII). (Data from Shishko et al. *[15]* as modified from *[16]*.)

in elevating IGF-I levels, despite equivalent glycemic control in the two treatment groups. IGF-I levels reached a steady state within 1 wk and remained stable for the entire 6 mo of the study. Conventional intermittent subcutaneous insulin therapy (CIT) resulted in poorer glycemic control and less improvement in IGF-I levels. Intraportal insulin therapy also was more potent than the other two treatments in raising IGFBP-1 levels. Further substantiation for the important role of hepatic insulinization was obtained using an intraperitoneal insulin infusion protocol *(17)*. Intraportal insulin again appeared to be an important regulator of IGF-I and IGFBP-3. Thus, the relative hepatic insulinopenia observed even in the presence of conventional insulin therapy in type 1 diabetes may explain why serum IGF-I and IGFBP-3 tend to remain suppressed under these conditions.

Alterations in the GH–IGF-I axis in type 2 diabetes are somewhat less clear. Some studies have demonstrated a relationship between poor glycemic control in type 2 diabetes and IGF-I levels similar to that seen in type 1 diabetes. Tan and Baxter *(18)* have reported that IGF-I levels are suppressed in older type 2 diabetic patients even after the effects of age are taken into account. It is interesting to speculate that the decrease in IGF-I that occurs with aging may play a role in the increased propensity toward both insulin resistance and type 2 diabetes that occurs with aging. Collectively, the reduced IGF-I levels in type 1 diabetes and possibly type 2 diabetes provide a further rationale for the possible use of IGF-I as "replacement therapy" in these two populations.

USE OF IGF IN SEVERE INSULIN RESISTANCE

The ability of IGF-I to stimulate glucose uptake and oxidation, suppress fatty acid oxidation, and inhibit hepatic glucose production in healthy subjects, all in the presence of lowered insulin concentrations, gave rise to protocols to examine IGF-I use in diabetes. Patients with severe insulin resistance represented a particularly attractive subset of diabetes to examine the effects of IGF-I as there are no effective treatment options for this patient population. Clearly, an agent that has insulin-like effects while lowering the extremely high endogenous insulin levels found in these patients could be a powerful tool in the treatment of these disorders. Unfortunately, many of the reports of IGF-I use in severe insulin resistance have been limited by the small number of patients with differing pathophysiologic causes of insulin resistance to whom varying dosing regimens of rhIGF-I have been administered by varying routes and for varying periods of time. However, even with these methodologic limitations, the demonstrated ability of IGF-I to ameliorate the poor glycemic control in at least some patients with severe insulin resistance helped to encourage studies to extend these results to the larger number of patients with type 1 or 2 diabetes mellitus.

Shoenle et al. *(19)* were the first to demonstrate that IGF-I decreased blood glucose and lowered insulin levels in three patients with the type A insulin resistance syndrome. These results also were confirmed in a patient with Rabson–Mendenhall syndrome *(20)*. Morrow et al. *(21)* extended these results in subjects with type A insulin resistance by administering rhIGF-I (100 μg/kg of body weight twice daily) for 1 mo and followed measures of insulin sensitivity. Glycemic control improved with a concomitant decrease in endogenous insulin levels and an increase in first-phase insulin release in response to an intravenous glucose challenge. Insulin sensitivity (as measured in a steady-state plasma glucose test) also improved. Interestingly, the effects of IGF-I were gradual in onset and in one patient the effects persisted several weeks after IGF-I was discontinued. Zenobi et al. *(22)* found that two patients with type A insulin resistance secondary to documented insulin receptor mutations responded to IGF-I treatment (150 μg/kg body weight) for 5 d with a normalization of fasting blood glucose in the setting of decreased serum insulin and C peptide levels. In an attempt to use lower doses of rhIGF-I to potentially minimize side effects (80 μg/kg twice a day for 2 wk) similar decreases in fasting glucose and insulin levels were observed in four patients with the type A syndrome *(23)*. A reduction of the dose to only 40 μg/kg/d during an additional 10 wk retained the effect on fasting glucose levels but insulin levels returned to baseline, suggesting a waning of the effect on insulin sensitivity that may be dose dependent. Kajimoto et al. *(24)* recently described the treatment of an insulin-resistant patient with anti-insulin antibodies with rhIGF-I. Blood glucose levels improved but more importantly the patient's frequent hypoglycemia, presumably mediated by release of insulin from the antibody complex, improved markedly.

Several other case reports have documented a response of patients with type A insulin resistance to IGF-I. Nakashima et al. *(25)* demonstrated that hyperglycemia was alleviated during 9 mo of treatment with IGF-I in a type A insulin-resistant patient with a homozygous mutation in the α-subunit of the insulin receptor. Ishihama et al. *(26)* demonstrate prolonged effects of 2 yr of treatment with IGF-I in a child with type A insulin resistance syndrome. Glycemic control improved in this patient, but there was an exacerbation of the patient's hyperandrogenism, a potential side effect of rhIGF-I therapy in these patients who often are hyperandrogenemic at the time of presentation. Theoretically, IGF-I might increase ovarian androgen production in vivo as it has been shown to do in vitro *(27)*. Another report by Dozio et al. *(28)*, however, found that IGF treatment was unable to improve glycemic control in a patient with a mutation in the α-subunit of the insulin receptor. Interestingly, IGF-I was effective in decreasing free fatty acid levels in this patient, demonstrating the further complexity in the metabolic actions of IGF-I.

The effects of IGF-I in patients with the type A syndrome of insulin resistance also provide clues to the mechanism of IGF-I action. The pathophysiologic mechanisms underlying both type A insulin resistance and Rabson–Mendenhall syndrome are believed to be mutations in either the insulin receptor or its downstream signaling molecules. The efficacy of IGF-I in these patients, including several patients with documented mutations in the insulin receptor *(22,23,25,26,28)*, suggests that IGF-I is acting through a non-insulin-receptor-mediated pathway. The gradual onset of action of IGF-I in these syndromes suggests that it may be generating some activated intermediary signaling molecule that can supplant the actions of insulin or that it is altering substrate metabolism in such a way as to improve insulin resistance. Clearly, further investigation will be required to delineate the molecular mechanism responsible for these metabolic actions of IGF-I.

Two somewhat larger studies have explored the effects of IGF-I in various phenotypes of severe insulin resistance. Kuzuya et al. *(29)* demonstrated both short-term and longer term responses to rhIGF-I in nine patients with several distinct phenotypes of severe insulin resistance. rhIGF-I lowered blood glucose and decreased serum insulin levels; however, in this study no direct measure of insulin sensitivity was performed. Interestingly, there was improvement in the skin lesion, acanthosis nigricans, a marker of severe insulin resistance. Moses et al. *(30)* provided detailed information on the effects of IGF-I in six subjects with different phenotypes of severe insulin resistance including two subjects with the type A syndrome, three subjects with lipodystrophy, and one subject with severe insulin resistance in the setting of combined luteinizing hormone-releasing hormone (LHRH) and gonadotropin-releasing hormone (GRH) deficiency. The responses of IGF-I were gradual in onset and included: (1) a 60–80% decrease in both fasting and mean 24-h serum insulin levels; (2) improvement in glucose tolerance in patients with impaired glucose tolerance or overt diabetes mellitus; (3) decreased fasting and postprandial glucose levels in patient with baseline hyperglycemia; (4) increased insulin sensitivity as measured by the steady-state plasma glucose test; and (5) statistically and clinically significant improvement in triglyceride levels (Fig. 3). It also is important to note that not all patients with severe insulin resistance respond to rhIGF-I with an improvement in glycemic control.

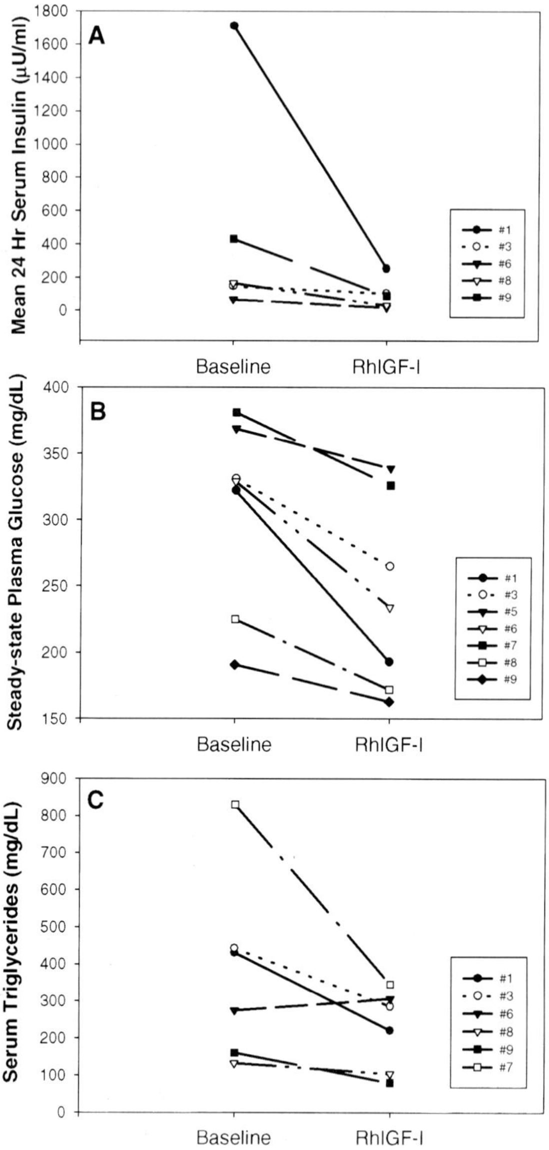

Fig. 3. rhIGF-I treatment of patients with severe insulin resistance. Administration of rhIGF-I for 4 wk to several patients with different phenotypes of severe insulin resistance (**A**) decreased mean 24-h serum insulin levels, (**B**) improved insulin sensitivity as measured by the steady-state glucose level, and (**C**) lowered serum triglyceride levels. (Data are derived from Moses et al. *[30]*.)

The ability of IGF-I treatment to improve insulin sensitivity and lower endogenous insulin levels in these different types of severe insulin resistance provided the rationale for the use of rhIGF-I in the treatment of the more prevalent type 2 diabetes, a disease whose hallmarks are insulin resistance and, in the early phase of the disease, hyperinsulinemia. A strong rationale for using an agent that increases insulin sensitivity and lowers endogenous insulin or the need for exogenous insulin also comes from epidemiologic data that support hyperinsulinemia as a direct mediator of cardiovascular risk and mortality in type 2 diabetes and in insulin resistance (syndrome X) not associated with frank hyperglycemia *(31)*.

USE OF IGF IN TYPE 2 DIABETES

Many of the studies that have examined the effects of rhIGF-I administration in type 2 diabetes have used relatively high doses based on the preliminary observations of in vivo effects of IGF-I in normal volunteers. Indeed, classic dose-ranging studies were not performed until relatively recently. Initially, Zenobi et al. *(32)* treated eight patients with type 2 diabetes with 120 μg/kg twice a day for 5 d, resulting in an elevation of free IGF-I levels 2.5-fold above basal levels. Fasting blood glucose decreased by 30% while insulin and C peptide levels were reduced by 50%, suggesting that rhIGF-I decreased insulin resistance. These effects were short lived, returning to baseline within 48–72 h after discontinuation of therapy. Schalch et al. *(33)* subsequently used doses of rhIGF-I as high as 160 μg/kg in 12 patients and found improvements in fasting glucose and insulin and C peptide levels similar to those seen by Zenobi et al. *(32)*. However, significant side effects were noted in many subjects in this study, calling into question the tolerability of such high doses of IGF-I. An attempt to treat a subset of these patients for a longer time period (8 wk) with either 120 or 160 μg/kg led to a withdrawal of therapy secondary to side effects in all seven patients *(34)*. Based on these studies, it became clear that if rhIGF-I was to become an effective and tolerated therapy for type 2 diabetes, lower doses with a better side effect profile would have to be employed.

Moses et al. *(35)* treated 12 patients with obesity and poorly controlled type 2 diabetes with 100 μg/kg of rhIGF-I as monotherapy for 4–6 wk following 2 wk of observation off conventional therapy (Fig. 4). In these subjects, fasting blood glucose decreased dramatically as did mean 24-h glucose. In addition, fasting and mean 24-h insulin levels also declined, suggesting an improvement in insulin sensitivity. Direct measurement of insulin sensitivity by the frequent sampled intravenous glucose tolerance test (FSIVGTT) revealed a 3.4-fold increase in insulin sensitivity. Of the initial 12 patients enrolled in this study, nine completed 4.5 wk of the protocol and half completed the entire 6 wk of the study. This dose of rhIGF-I (100 μg/kg bid) also appears to be above the range tolerated by the majority of individuals. Subjects withdrew secondary to a series of side effects reported in other studies with rhIGF-I including edema, jaw pain, and arthralgias. In all patients the side effects resolved after discontinuation of therapy. In an attempt to determine the minimum effective dose of rhIGF-I in type 2 diabetes, a large multicenter, randomized, double-blind, placebo-controlled trial with 212 patients was undertaken. Preliminary results of 12 wk of treatment indicate that a dose of rhIGF-I as low as 40 μg/kg twice daily significantly improved glycemic control (measured by a decrease in fasting glucose levels and HbA1C) (Fig. 4) *(36)*. These studies also clearly demonstrated a dose–response relationship between both efficacy with rhIGF-I and toxicity from rhIGF-I *(37)*. Longer term studies with these better tolerated low doses of rhIGF-I have been initiated.

The mechanisms through which IGF-I improves glycemic control and insulin sensitivity remain unresolved. The decrease in fasting blood glucose that has been observed suggests an effect of rhIGF-I on hepatic glucose production (the primary determinant of fasting glucose levels). However, adult hepatocytes have low numbers or absent IGF-I receptors *(38)*. Whether decreases in the counterregulatory hormones glucagon and

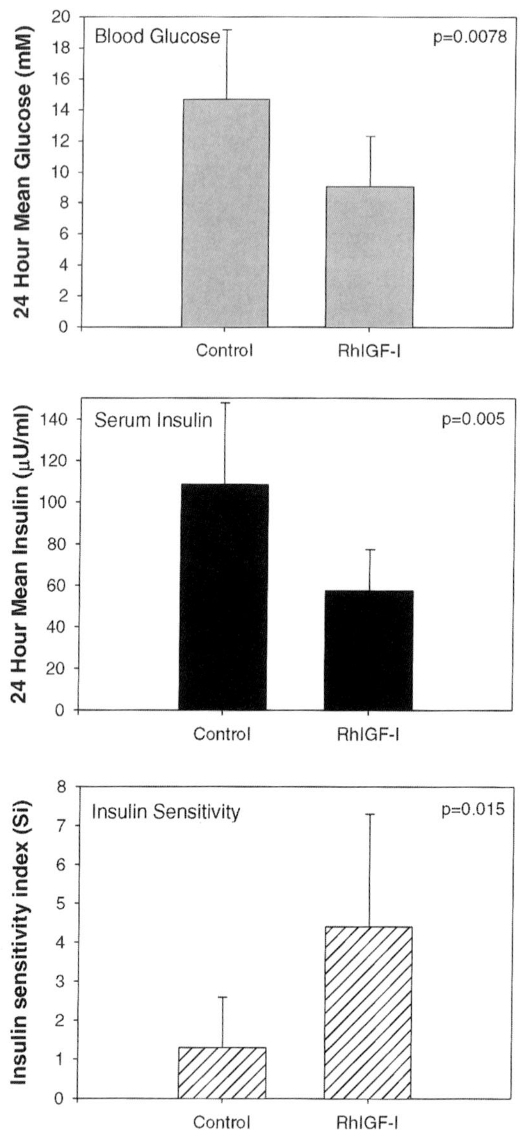

Fig. 4. rhIGF-I improves insulin sensitivity and glycemic control in patients with type 2 diabetes mellitus. (*top*) 24-h mean glucose and (*middle*) 24-h mean insulin levels were measured in nine subjects with type 2 diabetes before and after 6 wk of rhIGF-I therapy. The insulin sensitivity index (S_i), as calculated from a frequently sampled intravenous glucose tolerance test, was also markedly improved. The data are means ± SD and are from Moses et al. *(35)*.

growth hormone may account for the improvement in insulin sensitivity remains uncertain given the lack of data on these hormones in patients with type 2 diabetes mellitus treated with rhIGF-I. In contrast to the situation in patients with type 1 diabetes, growth hormone excess is not considered to have a significant impact on glycemic control in type 2 diabetes. Recently, Ciaraldi et al. *(39)* studied the in vitro effects of IGF-I on skeletal muscle removed from patients with type 2 diabetes. Glucose transport and glycogen synthesis both were more sensitive to IGF-I than to insulin when compared to muscle

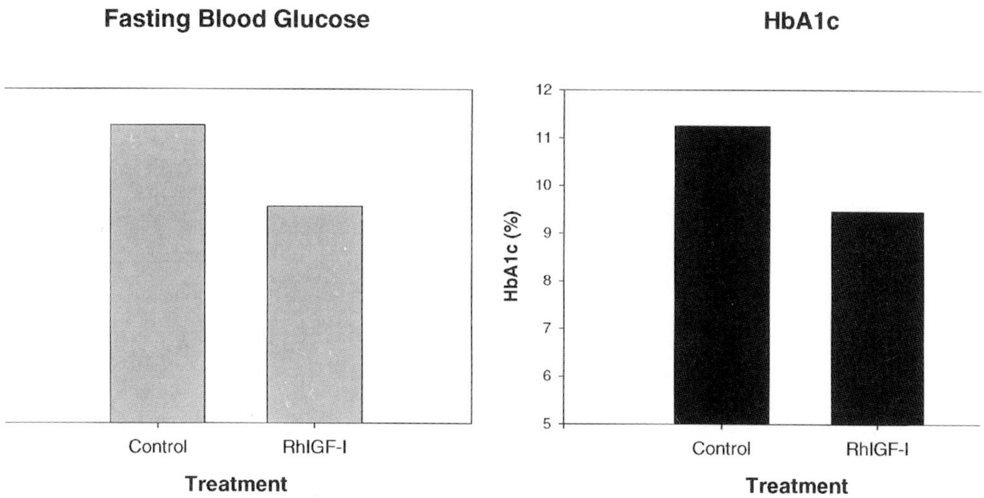

Fig. 5. rhIGF-I improves glycemic control in a large cohort of patients with type 2 diabetes. Twelve weeks of treatment with low-dose rhIGF-I (40 µg/kg twice daily) significantly improved fasting glucose and HbA1C ($p < 0.05$ compared with matched untreated subjects). Data are derived from rhIGF-I in NIDDM (RINDS) Study Group *(36)*.

from normal control patients. This suggests that the primary target of IGF-I action may be muscle (a tissue with a plentiful supply of IGF-I receptors). Unfortunately, no glucose clamp studies have been performed on patients with type 2 diabetes who have been treated for several days or longer with subcutaneous rhIGF-I to ascertain the relative contribution of hepatic glucose production and muscle metabolism to the effects of this treatment.

The possibility that rhIGF-I treatment may ameliorate some of the parameters of the so-called syndrome X (hyperinsulinemia, insulin resistance, obesity, hypertension, hyperlipidemia, and increased cardiovascular disease) is provocative. RhIGF-I treatment of type 2 diabetes lowers both total and low-density lipoprotein (LDL) cholesterol and triglyceride levels *(33,40,41)*. Obesity and body composition also are improved in type 2 diabetic patients treated with rhIGF-I *(36,41)*. In addition, lipoprotein (a), an important risk factor for coronary artery disease, also declined following short-term administration of rhIGF-I *(40)*. Finally, it appears that the risk of coronary thrombosis (as measured by plasminogen activator inhibitor-1 (PAI-1) and fibrinogen levels) also is improved with IGF-I treatment *(41,42)*. Of potential concern, however, is the observation in several studies that rhIGF-I leads to an asymptomatic increase in heart rate. It will be of great interest to determine if these changes in cardiovascular risk factors ultimately will translate into a decrease in cardiovascular disease.

USE OF IGF IN TYPE 1 DIABETES

The dysregulation of the GH–IGF axis in type 1 diabetes is characterized by a state of functional GH resistance in which GH is elevated and IGF-I is decreased *(9)*. It has been suggested that this elevation in GH may underlie the worsening of glycemic control seen in many adolescents with type 1 diabetes. There are several potential mechanisms that may explain these disturbances. Undernutrition can produce a similar pattern of GH and IGF-I secretion and this may help to explain why disturbances in the GH–IGF system are

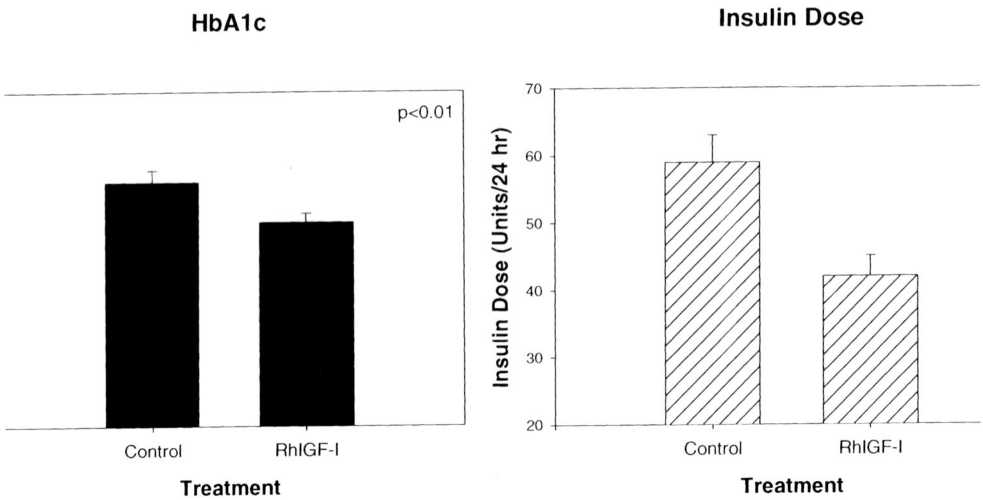

Fig. 6. rhIGF-I improves glycemic control and insulin requirements in type 1 diabetes mellitus. rhIGF-I (40 μg/kg twice daily) administered to 54 type 1 diabetic patients for 12 wk improved (**A**) glycemic control (% HbA1C) and (**B**) mean 24-h insulin requirements. Results are the means ± SEM (with $p < 0.01$ for IGF-I effects compared with placebo control. (From Thrailkill et al. *[50]*.)

observed more readily in type 1 than in type 2 diabetes. In animal studies, undernutrition is associated with down-regulation of the number of GH receptors in liver, possibly mediating a functional GH resistance *(43)*. Relative hepatic insulinopenia also appears to be important for normal IGF-I secretion in type 1 diabetes *(16,17)*. Diminished IGF-I levels would consequently lead to reciprocal feedback regulation of pituitary GH secretion. Finally, elevated levels of IGFBP-1 found in patients with type 1 diabetes *(11–13)* may directly cause hyperglycemia and could lower the free IGF-1 available at the tissue level *(44)*, further enhancing GH release at the pituitary level.

Several studies have examined the efficacy of rhIGF-I in type 1 diabetic patients. Initially, a continuous 10-h infusion of rhIGF-I was shown to decrease overnight arginine-stimulated GH levels and to lower nocturnal insulin requirements *(45)*. Cheetham et al. *(46)* then proceeded to demonstrate that a subcutaneous dose of 40 μg/kg at bedtime similarly decreased GH secretion. Extending this dosing regimen for 4 wk, these authors subsequently demonstrated that not only was GH decreased, but there was also an improvement in glycemic control (as measured by a 1% decrease in HbA1c) in the six adolescent type 1 diabetic patients treated with rhIGF-I *(47)*. Quattrin et al. *(48)* randomly assigned 43 patients to receive rhIGF-I (80 μg/kg/d) for 4 wk and found a smaller improvement in glycemic control (HbA1C decreased by 0.5%). These investigators also found a statistically significant decrease in insulin requirements. They subsequently proceeded to demonstrate a suppression of baseline elevated IGFBP-1 in response to rhIGF-I *(49)*. Recently, preliminary results of a 12-wk, double-blind trial of different doses of rhIGF-I in 224 patients with type 1 diabetes suggested that, at a well-tolerated dose (40 μg/kg twice a day), there was a statistically significant improvement in glycemic control in the setting of reduced insulin requirements *(50)* (Fig. 6).

A question has been raised as to whether these promising results will continue to be seen during longer term treatment. Recently, Acerini et al. *(51)* reported the preliminary results of a 6-mo multicenter, placebo-control, double-blind study in 53 type 1 diabetic patients treated with either 20 or 40 μg/kg rhIGF-I. There was a dose-dependent reduction in HbA1C at 3 mo; however, this effect was lost at 6 mo. In addition, there was no change in insulin requirements in the patients studied. Whether there exists some degree of tachyphylaxis to the glucose lowering effects of IGF-I remains to be determined. It appears that age also may impact the efficacy of rhIGF-I. In a small trial of six adult patients with type 1 diabetes, 19 d of treatment with relatively higher doses of rhIGF-I (50 μg/kg twice a day) led to a decrease in GH and insulin requirements but did not improve glycemic control *(52)*. One might speculate that the contribution of a dysregulated GH–IGF axis to glycemic control in adults may not be as significant as it appears to be in adolescents. Further analysis of these trials with respect to the influence of age on treatment success will be of interest.

It appears that the primary mechanism through which IGF-I improves glycemic control in type 1 diabetes is through antagonizing the effects of elevated GH. In an elegant study, Crowne et al. *(53)* performed overnight euglycemic clamps on adolescents treated with rhIGF-I. Endogenous GH secretion was suppressed with octreotide and three exogenous GH pulses were reintroduced. In a double-blind study, rhIGF-I blocked the increased insulin requirements that occurred secondary to GH pulses. Therefore, it appears that rhIGF-I is capable of directly preventing the consequences of increased GH secretion in the absence of any change in GH level. Interestingly, there was no change in free fatty acids, ketone bodies, or glucagon under these conditions, suggesting that some of the changes in these parameters reported in other studies may be secondary to IGF-induced decreases in GH secretion. Ultimately, IGF-I likely is acting to inhibit both the secretion and metabolic actions of GH.

SIDE EFFECTS AND POTENTIAL LONG-TERM TOXICITY

The development of rhIGF-I as a therapy for diabetes, and indeed, for the therapy of other conditions, has been a bit atypical in the sense that the doses of drug initially chosen for clinical trials were based on the assumption that they had to be equipotent to insulin. These data were derived from acute administration of IGF-I to normal human volunteers and did not take into account cumulative or time-dependent actions. Unfortunately, when these doses (100 μg/kg bid sc) were administered over days or weeks, a series of troubling side effects became apparent. Many of these side effects also have been seen in patients with GH resistance syndrome (Laron dwarfism) and include Bell's palsy, headaches, muscle aches, peripheral edema, tachycardia, and rarely, evidence of increased intracranial pressure. The mechanisms of most of these side effects remain obscure except for the ability of IGF-I to increase renal tubule sodium reabsorption, which probably contributes to the formation of edema *(54)*. Unfortunately, the high frequency of side effects with "high" doses of rhIGF-I have discouraged some investigators from pursuing this agent as a new therapeutic entity. More recently, dose–response studies in both type 1 and type 2 diabetes demonstrated that lower doses of rhIGF-I (20–40 μg/kg bid sc) have both clinically significant biological effects and have a much reduced potential to produce intolerable side effects. In a recent preliminary study, there was a direct relationship between side effects and dose of rhIGF-I. There also was a direct relationship between the dose of rhIGF-I and beneficial therapeutic effects. Fortunately, benefit preceded toxicity in the dose–response curve and it appears that doses of <40 μg/kg bid sc are well tolerated and efficacious.

The clinical trials reported to date have been of relatively short duration. There are no data in patients with diabetes for long-term IGF-I administration. Several patients with severe insulin resistance have been treated with rhIGF-I for more than 1 yr but their tolerance of this agent has not been reported and may differ from patients with more classic forms of diabetes. Concerns have been raised about the potential adverse effects of the mitogenic potential of exogenous IGF-I although there are no prospective data in normal animals or humans to support this problem. Concerns also have been raised about the potential role of GH and IGF-I in the genesis or progression of proliferative retinopathy. Current clinical trials are looking at this issue in detail but are hampered by the difficulty of separating effects of IGF-I itself from effects induced by rapidly improving glycemic control in patients who have been in notoriously poor control prior to entry into the study. Animal studies do suggest that IGF-I may be playing a role in the development of proliferative retinopathy although the data support the notion that the presence of sufficient IGF-I is necessary for retinopathy to develop and that inhibition of IGF-I levels may protect against the development of retinopathy *(55)*. These issues must be worked out during larger scale clinical trials and with careful attention to the dose–response curve of efficacy and of toxicity from IGF-I. Overall, the risk–benefit ratio of a therapy such as rhIGF-I will lie in the population exposed and the ability to add value to their treatment without undue risk.

SUMMARY

IGF-I shares many structural and functional similarities to its close homolog, insulin. Dysregulation of the GH–IGF-I axis in type 1 diabetes mellitus is clear and data support a degree of dysregulation in type 2 diabetes as well. Compelling data support that the deficiency of IGF-I in poorly controlled type 1 diabetes results from inadequate hepatic insulinization although a direct relationship between low IGF-I levels and either insulin resistance or poor glycemic control has been presumed, but not proven. The pharmacokinetics of IGF-I differ dramatically from those of insulin, and as such make understanding the pharmacodynamics of the drug somewhat obscure. The effects of rhIGF-I on glycemic control are gradual in onset and are associated with a lowering of endogenous insulin despite improvement in blood glucose concentration. Side effects of rhIGF-I treatment are dose dependent and tend to occur at doses higher than those required for efficacy. Clinical trials of IGF-I in both type 1 and type 2 diabetes and in patients with severe insulin resistance of different phenotypes support the concept that this agent is of therapeutic value in at least some patients. The long-term efficacy and the long-term safety of rhIGF-I remain to be elucidated through additional clinical trials.

REFERENCES

1. Froesch ER, Burgi H, Ramsier EB, Bally P, Labhart A. Antibody suppressible and nonsuppressible insulin-like activities in human sera and their physiologic significance. An insulin assay with adipose tissue of increased precision and specificity. J Clin Invest 1963; 42:1816–1834.
2. Guler HP, Zapf J, Froesch ER. Short-term metabolic effects of recombinant human insulin-like growth factor I in healthy adults. N Engl J Med 1987; 317:137–140.
3. Elahi D, McAloon-Dyke M, Fukagawa NK, Sclater AL, Wong GA, Shannon RP, Minaker KL, Miles JM, Rubenstein AH, Vanderpol CJ, Guler H-P, Good WR, Seaman JJ, Wolfe RA. Effects of recombinant human IGF-I on glucose and leucine kinetics in men. Am J Physiol 1993; 265:E831–838.
4. Boulware SD, Tamborlane WV, Rennert NJ, Gesundheit N, Sherwin RS. Comparison of the metabolic effects of recombinant human insulin-like growth factor-I and insulin. Dose–response relationships in healthy young and middle-aged adults. J Clin Invest 1994; 93:1131–1139.

5. Dicola GD, Cook MH, Accili D. Hypoglycemic effect of insulin-like growth factor-1 in mice lacking insulin receptors. J Clin Invest 1997; 99:2538–2544.
6. Hussain MA, Schmitz O, Mengel A, Keller A, Christiansen JS, Zapf J, Froesch ER. Insulin-like growth factor I stimulates lipid oxidation, reduces protein oxidation, and enhances insulin sensitivity in humans. J Clin Invest 1994; 92:2249–2256.
7. Kerr D, Tamborlane WV, Rife F, Sherwin RS. Effect of insulin-like growth factor-1 on the responses to and recognition of hypoglycemia in humans. A comparison with insulin. J Clin Invest 1993; 91:141–147.
8. Hussain MA, Schmitz O, Christiansen JS, Christensen NJ, Alberti KGMM, Froesch ER. IGF-I alters skeletal muscle substrate metabolism and blunts recovery from insulin-induced hypoglycemia. Am J Physiol 1996; 270:E1–E7.
9. Dunger DB, Cheetham TD, Crowne EC. Insulin-like growth factors (IGFs) and IGF-1 treatment in the adolescent with insulin-dependent diabetes mellitus. Metabolism 1995; 44 (Suppl 4): 119–123.
10. Tamborlane WV, Hintz RL, Bergman M, Genel M, Felig P, Sherwin R. Insulin infusion pump treatment of diabetes: influence of improved metabolic control on plasma somatomedin levels. N Engl J Med 1981; 305:303–307.
11. Clayton KL, Holly JM, Carlsson LM, Jones J, Cheetham TD, Taylor AM, Dunger DB. Loss of the normal relationships between growth hormone, growth hormone-binding protein and insulin-like growth factor-I in adolescents with insulin-dependent diabetes mellitus. Clin Endocrinol (Oxf) 1994; 41:517–524.
12. Cotterill AM, Daly F, Holly JMP, Hughes SC, Camacho-Hubner C, Abdullah AF, Gale EAM, Savage MO. The 'dawn phenomenon' in adolescents with insulin-dependent diabetes mellitus: possible contribution of insulin-like growth factor binding protein-1. Clin Endocrinol 1995; 43:567–574.
13. Kobayashi K, Amemiya S, Sawabibiru E, Higashida K, Ishihara T, Kobayashi K, Kato K, Nakazawa S. Role of IGF binding protein-1 in the dawn phenomenon and glycemic control in children and adolescents with IDDM. Diabetes Care 1997; 20:1442–1447.
14. Quattrin T, Thrailkill KM, Baker L, Quarmby V, Compton P, Pappenheimer M, Martha PM. Low plasma free IGF-I levels in IDDM: additional evidence for a bihormonal defect in diabetes. Diabetes 1996; 45 (Suppl 2):17A.
15. Bereket A, Lang CH, Blethen SL, Gelato MC, Fan J, Frost RA, Wilson TA. Effect of insulin on the insulin-like growth factor system in children with new-onset insulin-dependent diabetes mellitus. J Clin Endocrinol Metab 1995; 80:1312–1317.
16. Shishko PI, Dreval AV, Abugova IA, Zajamy IU, Goncharov VC. Insulin-like growth factors and binding proteins in patients with recent onset type I diabetes mellitus: influence of diabetes control and intraportal insulin infusion. Diabetes Res Clin Pract 1995; 25:1–12.
17. Hanaire-Boutin H, Sallari-Caute B, Poncet MF, Tauber M, Bastide R, Rosenfeld R, Tauber JP. Insulin therapy and GH–IGF-1 axis disorders in diabetes: impact of glycemic control and hepatic insulinization. Diab Metab 1996; 22:245–250.
18. Tan K, Baxter RC. Serum insulin-like growth factor I levels in adult diabetic patients: the effect of age. J Clin Endocrinol Metab 1986; 63:651–655.
19. Schoenle EJ, Zenobi PD, Torresani T, Werder EA, Zachmann M, Froesch ER. Recombinant human insulin-like growth factor I (rhIGF I) reduces hyperglycemia in patients with extreme insulin resistance. Diabetologia 1991; 34:675–691.
20. Quin JD, Fisher BM, Paterson KR, Inoue A, Beastall GH, MacCuish AC. Acute response to recombinant insulin-like growth factor I in a patient with Mendenhall's syndrome. N Engl J Med 1990; 323:1425–1426.
21. Morrow LA, O'Brien MB, Moller DE, Flier JS, Moses AC. Recombinant human insulin-like growth factor-I therapy improves glycemic control and insulin action in the type A syndrome of severe insulin resistance. J Clin Endocrinol Metab 1994; 79:205–210.
22. Zenobi PD, Glatz Y, Keller A, Graf S, Jaeggi-Groisman SE, Riesen WF, Schoenle EJ, Froesch ER. Beneficial metabolic effects of insulin-like growth factor I in patients with severe insulin-resistant diabetes type A. Eur J Endocrinol 1994; 131:251–257.
23. Vestergaard H, Rossen M, Urhammer SA, Muller J, Pedersen O. Short- and long-term metabolic effects of recombinant human IGF-1 treatment in patients with severe insulin resistance and diabetes mellitus. Eur J Endocrinol 1997; 136:475–482.
24. Kajimoto Kajimoto Y, Kawamori R, Fujitani Y, Kishimoto M, Kubota M, Yamasaki Y, Morishima T, Kamada T. A case of non-insulin dependent diabetes mellitus with antiinsulin antibody: effect of subcutaneous injection of human recombinant insulin-like growth factor-I. Endocr J 1995; 42:101–105.

25. Nakashima N, Umeda F, Yanase T, Nawata H. Insulin resistance associated with substitution of histidine for arginine 252 in the alpha-subunit of the human insulin receptor: trial of insulin-like growth factor I injection therapy to enhance insulin sensitivity. J Clin Endocrinol Metab 1995; 80:3662–3667.
26. Ishihama H, Suzuki Y, Muramatsu K, Nagai M, Kokubo M, Shiraya H, Kawakita A, Nishimura Y, Imamura T, Koboyashi M. Long-term follow-up in type A insulin resistant syndrome treated by insulin-like growth factor-I. Arch Dis Child 1994; 71:144–146.
27. Hernandez ER, Resnick CE, Svoboda ME, Van Wyck JJ, Payne DW, Adashi EY. Somatomedin C/insulin-like growth factor-1 as an enhancer of androgen biosynthesis by cultured rat ovarian cell. Endocrinology 1988; 122:1603–1612.
28. Dozio N, Scavini M, Beretta A, Sartori S, Meschi F, Sarugeri E, Pozza G. In vivo metabolic effects of insulin-like growth factor-I not mediated through the insulin receptor. J Clin Endocrinol Metab 1995; 80:1325–1328.
29. Kuzuya H, Matsuura N, Sakamoto M, Makino H, Sakamoto Y, Kadowaki T, Suzuki Y, Kobayashi M, Akazawa Y, Nomura M, Yoshimasi Y, Kasuga M, Goji K, Nagataki S, Oyasu H, Imura H. Trial of insulinlike growth factor I therapy for patients with extreme insulin resistance syndromes. Diabetes 1993; 42:696–705.
30. Moses AC, Morrow LA, O'Brien M, Moller DE, Flier JS. Insulin-like growth factor I (rhIGF-I) as a therapeutic agent for hyperinsulinemic insulin-resistant diabetes mellitus. Diabetes Res Clin Pract 1995; 28 (Suppl):S185–194.
31. Reaven GM. Pathophysiology of insulin resistance in human disease. Physiol Rev 1995; 75:473–486.
32. Zenobi PD, Jaeggi-Groisman SE, Riesen WF, Roder ME, Froesch ER. Insulin-like growth factor-I improves glucose and lipid metabolism in type 2 diabetes mellitus. J Clin Invest 1992; 90:2234–2241.
33. Schalch DS, Turman NJ, Marcsisin VS, Heffernan M, Guler HP. Short-term effects of recombinant human insulin-like growth factor I on metabolic control of patients with type II diabetes mellitus. J Clin Endocrinol Metab 1993; 77:1563–1568.
34. Jabri N, Schalch DS, Schwartz SL, Fischer JS, Kipnes MS, Radnik BJ, Turman NJ, Marcsisin VS, Guler HP. Adverse effects of recombinant human insulin-like growth factor I in obese insulin-resistant type II diabetic patients. Diabetes 1994; 43:369–374.
35. Moses AC, Young SCJ, Morrow LA, O'Brien M, Clemmons DR. Recombinant human insulin-like growth factor-1 increases insulin sensitivity and improves glycemic control in type II diabetes. Diabetes 1996; 45:91–100.
36. RHIGF-1 in NIDDM Study Group. Evidence from a dose-ranging study that recombinant insulin-like growth factor-1 (RhIGF-1) effectively and safely improves glycemic control in non-insulin dependent diabetes mellitus. Diabetes 1996; 45 (Suppl 2):27A
37. RHIGF-1 in NIDDM Study Group. Safety profile of rhIGF-1 therapy in patients with NIDDM: a dose ranging, placebo control trial. Diabetes 1996; 45 (Suppl 2):71A
38. Caro JF, Poulos J, Ittoop O, Pories WJ, Flickinger EG, Sinha MK. Insulin-like growth factor 1 binding in hepatocytes from human liver, human hepatoma, and normal, regenerating, and fetal rat liver. J Clin Invest 1988; 81:976–981.
39. Ciaraldi TP, Abrams-Carter L, Park KS, Mudaliar S, Henry RR. IGF-1 compensates for impaired insulin action in skeletal muscle of NIDDM. Diabetes 1997; 46(Suppl 1):352A.
40. Zenobi PD, Holzman P, Glatz Y, Riesen WF, Froesch ER. Improvement of lipid profile in type 2 (non-insulin dependent) diabetes by insulin-like growth factor I. Diabetologia 1993; 16:465–469.
41. Moses A, Philips L, Martha P, Compton P, Zysow B for the RINDS and RICS Groups. RhIGF-1 improves coronary risk factors in type II diabetes (abstract). Diabetologia 1997; 40 (Suppl 1):1484.
42. Padayatty SJ, Orme S, Zenobi PD, Strickland MH, Belchetz PE, Grant PJ. The effects of insulin-like growth factor 1 on plasminogen activator inhibitor-1 synthesis and secretion: results from in vitro and in vivo studies. Thromb Heaemost 1993; 70:1009–1013.
43. Ketelslegers JM, Maiter D, Maes M, Underwood LE, Thissen JP. Nutritional regulation of insulin-like growth factor-1. Metabolism 1995; 44 (Suppl 1):50–57.
44. Lewitt MS, Denyer GS, Cooney GJ, Baxter RC. Insulin-like growth factor binding protein-1 modulates blood glucose levels. Endocrinology 1991; 129:2254–2256.
45. Bach MA, Chin E, Bondy CA. The effects of subcutaneous insulin-like growth factor-1 infusion in insulin-dependent diabetes mellitus. J Clin Endocrinol Metab 1994; 79:1040–1045.
46. Cheetham TD, Clayton KL, Taylor AM, Holly J, Matyhews DR, Dungar DB. The effects of recombinant insulin-like growth factor-1 on growth hormone secretion in adolescents with insulin dependent diabetes mellitus. Clin Endocrinol 1994; 40:515–522.

47. Cheetham TD, Holly JM, Clayton K, Cwyfan-Hughes S, Dungar DB. The effects of repeated daily recombinant human insulin-like growth factor 1 administration in adolescents with type 1 diabetes. Diabetic Med 1995; 12:885–892.
48. Quattrin T, Thrailkill K, Baker L, Litton J, Dwigun K, Rearson M, Poppenheimer M, Giltinan D, Gesundheit N, Martha P. Dual hormonal replacement with insulin and insulin-like growth factor 1 in IDDM. Effects on glycemic control, IGF-1 levels and safety profile. Diabetes Care 1997; 20:374–380.
49. Thrailkill K, Quattrin T, Baker L, Litton J, Dwigun K, Rearson M, Poppenheimer M, Kotlovker D, Giltinan D, Gesundheit N, Martha P. Dual hormonal replacement therapy with insulin and insulin-like growth factor 1 in insulin-dependent diabetes mellitus. Effects on the growth hormone/IGF/IGF-binding protein system. J Clin Endocrinol Metab 1997; 82:1181–1187.
50. Thrailkill K, Quattrin T, Baker L, Kuntze J, Bajamonde A, Martha P, for the rhIGF-1 in IDDM Study Group. Co-therapy with recombinant human insulin-like growth factor-1 and insulin improves glycemic control in IDDM. Diabetes 1997; 46(Suppl 1):43A.
51. Acerini CL, Patton CM, Kernell A, Savvage MO, Westphal O, Dungar DB. A trial of insulin-like growth factor-1 as adjunct to multiple injection therapy in insulin dependent diabetes. Diabetologia 1997; 40 (Suppl 1):1299.
52. Carroll PV, Umplegy AM, Ward GS, Imuere S, Alexander E, Dungar D, Sonksen PH, Russel-Jones DL. RhIGF-1 reduces growth hormone secretion, lipid levels, and insulin requirement in adults with IDDM. Diabetologia 1997; 40 (Suppl 1):1487.
53. Crowne EC, Samra JS, Cheetham T, Watts A, Holly JMP, Dunger DB. Low dose recombinant human IGF-1 abolishes changes in insulin sensitivity consequent upon GH pulsatility in young adults with type 1 diabetes mellitus. Metabolism, in press.
54. Kopple JD, Ding H, Hirschberg R. Effects of recombinant human insulin-like growth factor-1 on renal handling of phosphorus, calcium, and sodium in normal humans. Am J Kidney Dis 1995; 26:818–824.
55. Smith Le, Kopchick JJ, Chen W, Knapp J, Kinose F, Daley D, Foley E, Smith RG, Schaeffer JM. Essential role of growth hormone in ischemia-induced retinal neovascularization. Science 1997; 276:1706–1709.

31 IGF-I Treatment of Growth Hormone Insensitivity

Arlan L. Rosenbloom, MD

Contents

DEFINITION AND CLASSIFICATION OF GROWTH HORMONE INSENSITIVITY

Need for a New Classification of GHI

The classification proposed in 1993 divided growth hormone insensitivity (GHI) into primary and secondary groups, with primary GHI referred to as Laron syndrome *(1)*, recognizing the initial description of the apparent paradoxical clinical appearance of severe GH deficiency (GHD) despite elevated levels of GH by Laron et al. *(2)* (Table 1). This classification now requires modification. Since its promulgation, defective insulin-like growth factor I (IGF-I) synthesis has been described due to deletion of the IGF-I gene *(3)*, without the clinical phenotype of GHD, and dominant negative heterozygous mutations of the cytoplasmic domain of the GH receptor (GHR) have been found that result in less severe short stature and absence of or mild GHD clinical phenotype *(5)*. Three Arab siblings with the clinical picture of Laron syndrome but apparently normal binding to the extracellular domain of the GHR, indicated by normal GH binding protein (GHBP) and IGFBP-3 levels, were thought to have a post-receptor defect *(6)*, as were GHBP-positive children with the growth and biochemical characteristics of GHI *(7,8)*. It is logical to consider primary GHI (Laron syndrome) to be more tightly defined as is done with other hormone resistance conditions, restricted to GHR deficiency or GH-GHR signal transduction abnormality and usually, but not always, associated with the clinical phenotype of severe GHD. IGF-I deficiency would be separately classified as either primary, due to intrinsically defective IGF-I synthesis; secondary, due to GH receptor dysfunction; or tertiary, as the result of GHD (Table 2).

The concept of secondary GHI (Table 1) is problematic and should probably be abandoned, as these conditions are not consistently associated with elevated GH or low levels

From: *Contemporary Endocrinology: The IGF System*
Edited by: R. Rosenfeld and C. Roberts Jr © Humana Press, Inc., Totowa, NJ

Table 1
Current Classification of GHI

1. Primary GH insensitivity syndromes (Laron syndrome; hereditary congenital defects)
   A. GHRD (encompassing quantitative and qualitative defects in the GHR)
   B. Abnormalities of GH signal transduction (post-receptor defects)
   C. Primary defects of synthesis of IGF-1

2. Secondary GHI syndromes (acquired conditions; sometimes transitory)
   A. Circulating antibodies to GH that inhibit GH action
   B. Antibodies to the GHR
   C. GHI caused by malnutrition
   D. GHI caused by liver disease
   E. Other conditions that cause GHI

See ref. *(1)*.

Table 2
Classification of IGF-I Deficiency and IGF-I Resistance with Growth and Biochemical Features

		Biochemistry			
Condition	*Ht SDS*	*GH*	*GHBP*	*IGF-I*	*IGFBP-3*
Primary IGF-I deficiency					
Congenital					
Defect in IGF-I synthesis	–6.9	High	Normal	Very low	Normal
Acquired					
Alagille syndrome *(12)*	Varies	High	High	Low	Normal
Secondary IGF-I deficiency					
Congenital					
GH receptor deficiency	–4 to –12	High	Low/normal/high	Very low	Low
GH-GHR signal transduction defect *(6,7)*	–3.4 to –6	High	Normal	Very low	Normal/low
Acquired *(48–56)*					
(GH receptor antibodies)[a]					
Catabolic states/chronic illness	Normal–low	High	Low/normal	Low	Normal/low
Tertiary IGF-I deficiency					
Congenital					
GH deficiency (GHD)	< –3	Low	Normal	Low	Low
Acquired					
GH inhibiting antibodies	–3 to –8.5	Low	Normal	Low	Low
GHD	Varies	Low	Normal	Low	Low
IGF-I insensitivity					
Congenital					
(IGF receptor defects)[a]					
IGF-IGFR signal transduction defects	–2 to –4.6	Normal	?	High	?
Acquired					
(IGF antibodies)[a]					
(IGF receptor antibodies)[a]					

Ht SDS, standard deviation score for height.
[a]Conditions in parentheses have not been reported.

of IGF-I. Abnormally low levels of GH are found in the GH gene deletion syndrome with development of GH inhibiting antibodies following treatment with GH *(9)* and in renal disease GH may be high or normal and IGF-I concentrations are usually normal.

The revised classification proposed in Table 2 is based on the fundamental problem of IGF-I deficiency rather than of one step in the GH–IGF pathway, GHR abnormality, although almost all instances of what has been called primary GHI have been due to such an abnormality. This classification is consistent with those used for other target hormone deficiencies. In this proposed classification, most of the circumstances classified as secondary GHI in the earlier classification are considered acquired primary or acquired secondary IGF-I deficiency, reflecting inability of the IGF-I producing tissues, primarily liver, to respond appropriately in the chronically ill, abnormal metabolic, or undernourished state. This disruption of the GH–IGF-I axis can be compared to the sick euthyroid syndrome, which is not perceived as a thyroid-stimulating hormone (TSH) resistant state.

Primary IGF-I Deficiency

A single case of defective IGF-I synthesis due to a gene deletion has been described in a patient with a homozygous partial deletion of the IGF-I gene. His profound intrauterine growth retardation (IUGR) persisted into adolescence and he had sensorineural deafness with mental retardation *(3)*. The absence of the craniofacial phenotype of severe GHD and the presence of normal IGFBP-3 in this patient, despite unmeasurable levels of IGF-I, indicates that the craniofacial features and low IGFBP-3 of GHD and GHR deficiency are related to an absence of the direct effects of GH that do not act through the medium of IGF-I synthesis. It is also noteworthy that profound IUGR and mental retardation are not characteristic of GHD or GHRD *(10)*, but IGF-I knockout mice have defective neurological development as well as growth failure *(11)*. Thus, IGF-I production in utero does not appear to be GH–GHR-dependent. Elevated GH levels would be expected in this circumstance as a result of the absence of IGF-I suppression of somatostatin; diurnal levels of GH were high, with an overnight GH peak of 171 ng/mL *(3)*.

Alagille syndrome (chronic intrahepatic cholestasis) is associated with moderate growth failure in 50% of instances along with elevated GH and GHBP levels, and normal IGFBP-3 levels. Thus, this does not appear to be a GH-resistant state but one involving failure of IGF-I generation due to hepatic dysfunction *(12)*.

Secondary IGF-1 Deficiency

Congenital (Laron Syndrome)

GHR Deficiency. Following the initial report *(2)* of three Yemenite Jewish siblings "with hypoglycemia and other clinical and laboratory signs of growth hormone deficiency, but with abnormally high concentrations of immunoreactive serum growth hormone," 22 patients were reported from Israel, all Oriental Jews, with an apparent autosomal recessive mode of transmission in consanguineous families *(13)*. These reports preceded the recognition of the critical role of cell surface receptors in hormone action and it was postulated that the defect was in the GH molecule that these patients were producing. This impression was substantiated by the observation of free fatty acid mobilization, nitrogen retention, and growth in patients being administered exogenous GH *(2,13)*. These effects may have been due to other pituitary hormones in the crude extracts administered or to nutritional changes in the investigative setting. In the first patient reported outside of Israel, in 1968, there was no response to exogenous GH, leading to the hypothesis that the defect was in the GH receptor *(14)*. This hypothesis was substantiated by the failure to demonstrate sulfation factor activation with exogenous GH administration, reported in 1969 *(15)* and reports in 1973 and 1976 that the GH of the patients was normal on fractionation, in its binding to antibodies, and in its binding to hepatic cell membranes from normal individuals *(16–19)*. In vitro demonstration of cellular unresponsiveness to GH was demonstrated by the failure of erythroid progenitor cells from the peripheral blood of two patients to respond to exogenous GH *(20)*. The failure of radioiodine-labeled GH to bind to liver cell microsomes obtained from biopsy of two patients with Laron syndrome confirmed that the defect was in the GHR *(21)*.

The recognition that circulating GHBP in rabbit serum corresponded to liver cytosolic GHBP *(22)* was followed by reports of absence of circulating GHBP in patients with Laron syndrome *(23,24)*. At the same time, human GHBP was purified, cloned, and sequenced, and found to be structurally identical to the extracellular hormone binding domain of the membrane bound GHR *(25)*. The entire human GHR gene was characterized by Godosky et al. *(26)*, who demonstrated that the coding and 3′-untranslated region of the receptor are encoded by nine exons, numbered 2–10. Exon 2 encodes a secretion signal peptide, exons 3–7 the extracellular domain, exon 8 the transmembrane domain, and exons 9–10 the cytoplasmic domain and 3′-untranslated region.

This report of the characterization of the gene included the first description of a genetic defect of the GHR, a noncontiguous deletion of exons 3, 5, and 6 in two Israeli patients with Laron syndrome. The deletion of exon 3 is an alternatively spliced variant without functional significance, rather than a component of the defect in these patients *(27)*. Only 5 Israeli patients have been found to have this defect, out of more than 30 of Oriental Jewish origin in Israel, emphasizing that, even in this ethnically homogeneous population, there is heterogeneity for the genetic defects in the GHR *(28)*. In fact, 37 additional defects have been described in subjects with Laron syndrome, including 14 missense mutations, 8 nonsense mutations, 10 splice mutations, and 5 frameshift mutations *(4,5,29–45)*. No other exon deletions have been described.

In Table 3, the deletion defect and 35 point mutations associated with typical autosomal recessive GHRD when present as homozygotes or compound heterozygotes are listed, along with two heterozygous mutations demonstrated to have a dominant negative effect *(4,5)*. The 35 point mutations that result in severe GHI, when present in the homozygous state or as a compound heterozygote, are all associated with the typical phenotype of severe GHD. All but three of the defects result in absent or extremely low levels of GHBP. Noteworthy is the D152H missense mutation which affects the dimerization site, thus permitting the production of the extracellular domain in normal quantities but failure of dimerization at the cell surface, which is necessary for signal transduction and IGF-I production *(40)*. The two defects that are close to (G223G) or within (R274T) the transmembrane domain result in extremely high levels of GHBP *(32,43,45)*. These defects interfere with the normal splicing of exon 8 which encodes the transmembrane domain with the mature GHR transcript being translated into a truncated protein that retains GH binding activity but cannot be anchored to the cell surface.

As noted, all these homozygous defects and the four compound heterozygotes, whether involving the extracellular domain or the transmembrane domain and whether associated with very low or unmeasurable GHBP, result in a typical phenotype of severe GHD. In contrast, the intronic mutation present in the heterozygous state in a mother and daughter with relatively mild growth failure (both SDS for height –3.6), and resulting in a dominant negative effect on GHR formation, is not associated with other phenotypic features of GHD. This splice mutation preceding exon 9 results in an extensively attenuated,

Table 3
Mutations of the GHR Associated with Growth Failure

ype	*Molecular Defect*	*Exon*	*Origin*	*Number of Pts/Families*	*Reference*
eletion	Exon (3) 5,6	5,6	Oriental Jewish	5[a]/3	*(26,28)*
onsense mutation	C38X	4	Northern European/ Mediterranean	2/2	*(29)*
	R43X	4	Mediterranean/ Ecuador/Russia	6/5	*(29–31,35)*
	Q65X	4	India	1	*(32)*
	W80X	5	Germany	1	*(32)*
	W157X	6	Turkey	1	*(32)*
	E183X	6	Mexico	1	*(33)*
	R217X	7	African-American/ Saudi Arabia	3/3	*(33–35)*
	G224X[b]	7	Japanese	1	*(37)*
issense mutation	C38S	4	Algeria	1	*(32)*
	S40L	4	Turkey	1	*(32)*
	W50R	4	Germany	1	*(32)*
	R71K	4	Mediterranean	1	(35)
	F96S	5	Tunisia	4/1	*(38)*
	V125A	5	European	1	(35)
	P131Q	6	Vietnamese	1	*(39)*
	V144D[c]	6	Mediterranean	1	*(35)*
	D152H	6	Pakistani	4/2	*(40)*
	I153T[d]	6	European	1	*(41)*
	Q154P	6	Hispanic	1	*(41)*
	V155G	6	Saudi Arabian	1	*(41)*
	R161C	6	Middle East	1	*(35)*
	R211G[c]	7	Mediterranean	1	*(35)*
rameshift Mutation	36delC	4	Slovenia	1	*(32)*
	46delTT[e]	4	Spain	2/2	*(34)*
	230delTA	7	African	2/1	*(34)*
	282del[d]→	9	European	1	*(41)*
	309delC[b]	10	Japanese	1	*(37)*
plice mutation	intron 2		Turkey	2/1	*(32)*
	intron 4[e]		N. European/Spain/ Japanese	3/3	*(34–36)*
	intron 5		Mediterranean	1	*(35)*
	intron 6		Brazil	1	*(34)*
	E180 splice	6	Ecuador Converso/ Oriental Jewish	71/52	*(42)*
	G223G	7	Druse	1	*(32,43)*
	G236 splice	7	Caucasian (Bahamas)	8/4	*(44)*
	R274T	8	Pakistan	2/1	*(45)*
	intron 8[f]		Caucasian (England)	2/1	*(4)*
	intron 9[g]		Japanese	3/1	*(5)*

[a]One of these patients is heterozygous for this deletion, with unidentified mutation of the other allele.
[b]Compound heterozygote.
[c]Compound heterozygote.
[d]Compound heterozygote.
[e]Compound heterozygote *(34)*. These mutations also occur as homozygous defects *(35,36)*.
[f]Heterozygous mutation with dominant negative effect in mother and daughter.
[g]Heterozygous mutation with dominant negative effect in mother and two children.

virtually absent intracellular domain *(4)*. Japanese siblings and their mother have a similar heterozygous point mutation of the donor splice site in intron 9, also resulting in mild growth failure compared to GHRD, but with definite, although mild, phenotypic features of GHD *(5)*.

Of 250 reported cases of typical GHRD, ethnic origin is known for 232 *(39,41,46)*. Of these, nearly 50% are Oriental Jews as described in the original report, or known descendants of Iberian Jews who converted to Catholicism during the Spanish Inquisition. The latter comprise the largest cohort (*n* = 70) and the only genetically homogenous group, all but one subject having the E180 splice site mutation which is shared with at least one Israeli patient of Moroccan heritage. Most of the other defects appear to be highly family specific, with the R43X mutation that is seen in a single Ecuadorian patient, two other nonsense mutations (C38X, R217X), and the intron 4 splice mutation being the only ones thus far described that appear in disparate populations, on different genetic backgrounds, indicating mutational hotspots *(29,34,35)*. Because the molecular defect in the GHR has been identified in about one third of the patients with GHRD outside of Ecuador, it is likely that this list will continue to grow and provide further insight into the function of the GHR.

Of particular interest is the opportunity for genetic anthropologic study. The finding of a Jewish patient of Moroccan origin with the same mutation as the Ecuadorian patients supports the hypothesis that this mutation was brought to the New World by Spanish conversos (new Christians) fleeing the Inquisition *(33)*. Thus far, there is no explanation for the Middle Eastern predominance of this condition, although the high frequency of consanguinity in Arab and traditional Jewish populations is certainly a factor. Nearly 90% of patients are either Oriental Jews, Arabs, or other Middle Easterners, from elsewhere in the Mediterranean area, or from the Indian subcontinent. Many of those from other parts of the world may have Middle Eastern Semitic origins. The small numbers of non-Mediterranean, non-Indian patients include a genetic isolate of Anglo Saxons from a Bahamian island *(44)*, five Africans, five Japanese, two siblings from Cambodia *(47)*, a Vietnamese *(39)*, and several from northern Europe *(46)*.

GH–GHR Signal Transduction Defect. Evidence of a post GHR defect in three siblings of Palestinian Arab origin has been presented by Laron et al. *(6)*. The children had typical features of severe GHD, but normal GHBP and IGFBP-3 levels with severe IGF-I deficiency. The molecular defect has yet to be identified. In contrast to these patients, four GHBP-positive children from two unrelated Pakistani families, who had high GH, low IGF-I, and low IGFBP-3 serum concentrations, had no phenotypic features of Laron syndrome except for short stature which was less severe than in typical GHRD *(7)*. Fibroblasts from the children in one of the families demonstrated failure to activate part of the GH signaling pathway *(8)*.

Acquired Secondary IGF-I Deficiency

The possibility of GHR antibodies resulting in GHI and IGF-I deficiency remains theoretical. Malnutrition and diabetes mellitus, however, would appear to fall under the category of acquired GHI producing IGF-I deficiency, although growth failure is a variable result. In those conditions, GHBP is diminished for reasons that remain unexplained *(48–53)*. In malnutrition states this is associated with elevated GH levels, decreased IGF-I, and normal to decreased IGFBP-3 levels *(48–52,54)*. Children with diabetes mellitus also have elevated GH levels and IGF-I deficiency, but normal levels of IGFBP-3; IGFBP-1 levels are elevated, presumably reflecting relative insulin deficiency for a portion of the day *(51)*. Children with end-stage renal disease may have, in addition to decreased GHBP, elevated GH levels and normal IGF-I levels; high levels of IGFBPs decreasing IGF-I bioavailability are thought to contribute to growth failure in these children *(53,56)*. In the proposed classification , renal disease is not included because there is neither IGF-I deficiency nor resistance. It is unclear why the increased binding proteins would not be compensated for, as occurs in thyroid binding protein excess, unless the problem were simply one of chronic disease/cachexia *(48–51)*.

Tertiary IGF-I Deficiency

Congenital Abnormalities Causing Tertiary IGF-I Deficiency

In this category are included all causes of GHD, whether of hypothalamic or pituitary origin, which are congenital. Defects may be in the GHRH gene or its pituitary somatotroph receptor, in pituitary differentiating factors, or in the GH gene. In addition, a wide variety of malformations may result in congenital pituitary deficiency or isolated GHD.

Acquired Conditions Resulting in Tertiary IGF-I Deficiency

GH inhibiting antibodies can develop after several months of therapy with recombinant human GH in subjects lacking the GH gene. Because these patients have never produced their own GH, the exogenous hormone is perceived as a foreign protein, generating an immune response. Instances of spontaneously occurring GH inhibiting antibodies have not been recorded. There are numerous causes of acquired GHD, including trauma, infection, neurosurgery, radiation, and hypophysitis. The degree of growth failure and clinical appearance will depend n the age of onset, etiology, and severity of IGF-I deficiency.

IGF-I Insensitivity

Congenital IGF-I Insensitivity

Defects in the IGF type I receptor or in the IGF receptor signal transduction pathway have been hypothesized in several patients. IGF-I resistance has been invoked as a cause of short stature in the Pygmy, demonstrated by failure of response of T cells from such patients to IGF-I in vitro; the absence of IGF-I elevation that should accompany such resistance, however, requires explanation *(57)*. Loss of a single allele for the IGF receptor caused by distal long arm deletion of chromosome 15 fails to abolish in vitro effects of IGF-I and, therefore, would not appear to explain short stature in affected patients *(58)*. Homozygous IGF-I receptor deletion is probably lethal *(59)*.

Three patients have been reported with normal or high GH levels and elevated IGF-I levels. In two of these patients, normal bioactivity was demonstrated for the circulating IGF-I and fibroblasts from skin biopsies of these patients responded normally to recombinant IGF-I *(60–62)*. It was hypothesized that these patients might have a tissue-specific (i.e., skeletal) IGF-I unresponsiveness. Such an explanation could also be invoked to explain the findings of normal IGF-I levels, lack of response to exogenous GH, and normal fibroblast response to IGF-I in those patients with heterozygous deletion of the IGF-I receptor gene *(58)*.

Acquired IGF-I Insensitivity

No circumstances have been described that would implicate IGF-I or IGF-I receptor antibody formation. It might be expected that patients such as the boy recently described with IGF-I gene deletion and undetectable circulating IGF-I *(3)* might develop secondary

Table 4
Growth in IGF-I Deficiency Due to GHRD

Birth weight—normal
Birth length—usually normal
Growth failure, from birth, with velocity 1/2 normal
Height deviation correlates with (low) serum levels of IGF-I, -II, and IGFBP-3; no correlation with first-degree relatives for height age
Delayed bone age, but advanced for height age
Segmental ratios and arm span: normal for bone age (Ecuador, Europe), abnormal for chronologic age (Israel) in children; abnormal (reduced upper/lower, arm span) in adults.
Small hands or feet (< 10th %ile for height) in 70%

resistance with IGF-I antibodies as a result of interpreting exogenous recombinant human IGF-I as a foreign protein, similar to what occurs with exogenous GH administration in some patients with GH gene deletion.

EFFECTS OF IGF-I DEFICIENCY DUE TO GHR DEFICIENCY

Clinical Effects

Mortality from GHRD

The only available report of the effect of GHRD on mortality comes from the Ecuadorian population *(10)*. Because families in the relatively small area from which the Ecuadorian patients originate have intensive experience with this condition, lay diagnosis was considered reliable; of 79 affected individuals for whom information could be obtained, 15 (19%) died under 7 yr of age, as opposed to 21 out of 216 of their unaffected siblings (9.7%, $p < 0.05$). The kinds of illnesses resulting in death, such as pneumonia, diarrhea, and meningitis, were no different for affected vs unaffected siblings *(63)*.

The complete lifespan represented in the Ecuadorian cohort provided an opportunity to look at adult mortality risk factors. This is of interest because GHD in adults is associated with premature atherosclerosis and increased cardiovascular mortality, with GH replacement therapy improving the risk factors of hyperlipidemia and obesity *(64)*. Twenty-three adult patients with GHD studied by Guevara-Aguirre et al. *(65)* had, compared to related and unrelated community controls, elevated cholesterol levels, normal high-density lipoprotein (HDL)-cholesterol levels, elevated low-density lipoprotein (LDL)-cholesterol levels, and normal triglycerides. It was postulated that the effect of IGF-I deficiency due to GHRD was to decrease hepatic clearance of LDL-C, as the triglyceride and HDL-C levels were unaffected. This effect was independent of obesity or of IGFBP-1 levels, used as a surrogate for insulinemia. The key pathogenic factor was thought to be the absence of GH induction of LDL receptors in the liver *(66)*.

Of eight Ecuadorian patients over 50 yr of age followed for greater than 7 yr, two died of heart disease, thought to be an uncommon problem in the Andean setting *(63)*. This might suggest increased cardiovascular risk comparable to that seen with GHD in adults.

Growth in GHRD

The growth characteristics of GHRD are given in Table 4. The role of IGF-I in normal intrauterine growth is emphasized by the patient with severe IUGR with a proven IGF-I gene defect *(3)*, but this IGF-I synthesis does not appear to be GH dependent, as many,

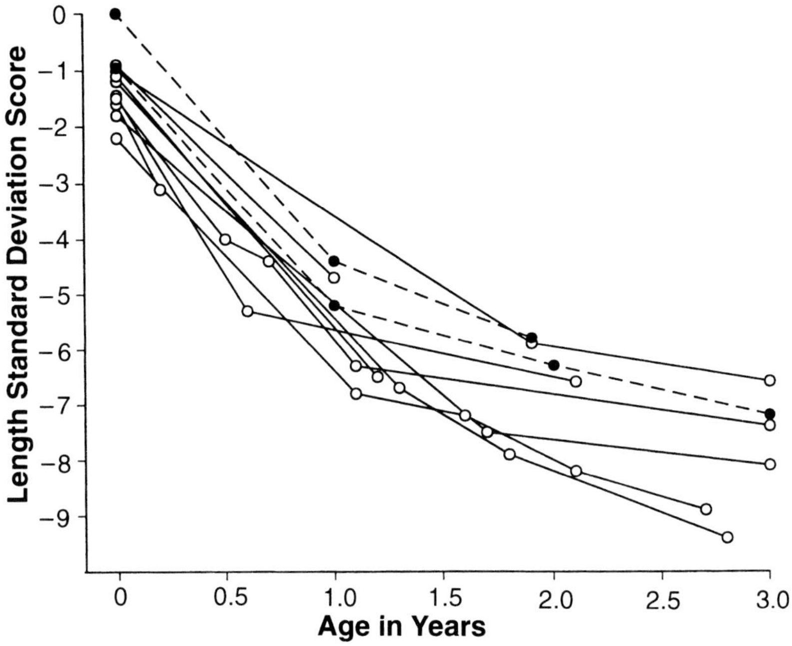

Fig. 1. Length standard deviation scores in nine female patients from Ecuador (*open circles and solid lines*) and two brothers from Southern Russia (*solid circles and dashed lines*) with known birth lengths, followed during infancy. (Reproduced with permission from ref. *[67]*).

if not most, GHRD patients have normal intrauterine growth *(67)*. Children with GH gene deletion also have normal intrauterine growth despite total absence of endogenous GH *(9)*. There is rapid postnatal decline in SDS for length (Fig. 1), emphasizing the GH dependency of postnatal growth; earlier anecdotal observations of GH-deficient patients had suggested that the first 6 mo of growth were not GH dependent. The growth velocity with severe GHD or GHI is approximately half normal (Fig. 2). Occasional periods of normal growth velocity may be related to improved nutrition. There is generally an absence of a pubertal growth spurt despite normal sexual maturation in the most extensive available data from Israel and Ecuador, emphasizing the importance of IGF-I for this phenomenon *(67,68)*. Among 24 Israeli patients followed from infancy to adulthood, persistent growth beyond the normal time of adolescence was seen only in boys. In the Ecuadorian population girls also showed this phenomenon (Fig. 2). Adult stature in GHRD varies from –5.3 to –12 SDS in Ecuadorian patients and –3.8 to –9 SDS in others in the literature, using the US standards *(67)*. This is a height range of 95 cm to 124 cm for women and 106 cm to 141 cm for men in the Ecuadorian population. This wide variation in the effect of GHRD on stature was seen not only within the population but also within affected families, and this has also been described with severe GHD due to GH gene deletion *(9)*.

Some patients with GHRD, as with GHD, will have an appetite problem in addition to their IGF-I deficiency. Crosnier et al. *(69)* studied a child aged 3 yr 7 mo with GHRD as a result of a nonsense mutation of the GHR, who had severe anorexia. With his usual intake of approx 500 cal/d he grew at a rate of 2 cm/yr. With moderate hyperalimentation to approx 1300 cal/d, growth rate increased to 9 cm/yr without significant change in

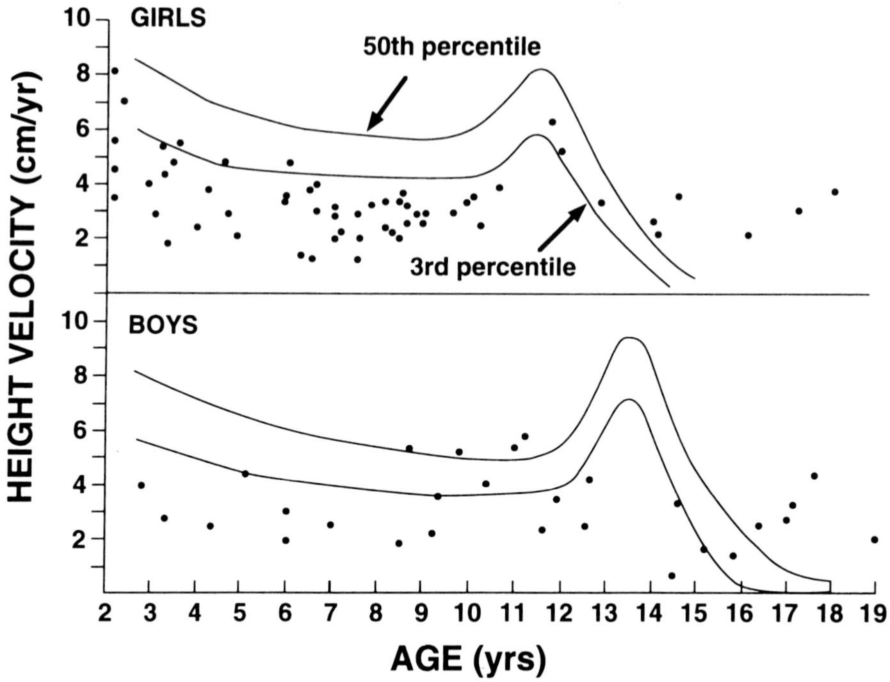

Fig. 2. Growth velocities of 30 Ecuadorian patients (10 males) with GH receptor deficiency: repeated measures are at least 6 mo apart. Reference standards are for North American children. (Reproduced with permission from ref. *[67]*).

plasma IGF-I levels. The hyperalimentation period was associated with an increase in the IGFBP-3 bands on Western ligand blots to levels comparable to those seen in GHD from total absence in the anorexic period. The catchup growth noted could not be explained by hyperinsulinism, which has provided the explanation for accelerated or normal growth in children with GHD and obesity following removal of a craniopharyngioma. There was no appreciable increase in circulating basal or stimulated insulin during the hyperalimentation. In this patient there was speculation that a nutrition-dependent autocrine/paracrine increase in IGF-I concentration of the cartilage growth plate might have occurred, independent of the GHR. Not considered at the time was the possibility that IGFBP-3 itself might have growth promoting effects. In any case, the importance of adequate nutrition for catchup growth was emphasized and strengthened the notion that normal periods of growth documented in patients with GHRD without IGF-I replacement therapy might be explained by periods of improved nutrition alone.

Craniofacial Characteristics of GHRD *(10)*

Craniofacial characteristics associated with severe GHD or GHRD are listed in Table 5 and shown in Fig. 3. Affected children are recognized by knowledgeable family members at birth because of frontal prominence, depressed nasal bridge, sparse hair, small hands or feet, and hypoplastic fingernails. Decreased vertical dimension of the face is demonstrable by computer analysis of the relationships between facial landmarks and is present in all patients when compared with their relatives, including those with reasonably normal facial appearance, such as the 21-yr-old man (number 10) in Fig. 3. Blue scleras,

Table 5
Craniofacial Characteristics of IGF-I Deficiency Due to GHRD

Sparse hair before age 7 in 70%; frontotemporal hairline recession in all
Prominent forehead
Head size more normal than stature with impression of large head
Hypoplastic nasal bridge, shallow orbits
Decreased vertical dimension of face
A setting sun sign in 25% less than age 10 yr
Blue scleras
Prolonged retention of primary dentition with decay; normal permanent teeth, may be crowded; absent third molars in 90%
Sculpted chin
Unilateral ptosis, facial asymmetry in 15%

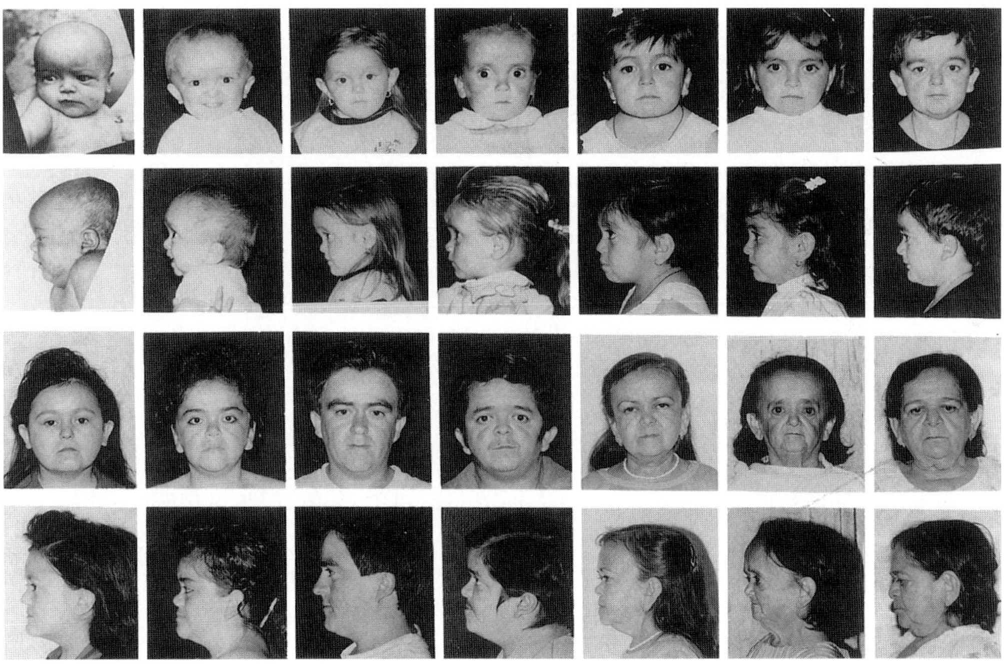

Fig. 3. Front and side views of 14 Ecuadorian patients with GHRD, ranging in age from 1 mo to 66 yr, demonstrating craniofacial abnormalities and phenotypic heterogeneity despite genetic homogeneity for the mutation in the GHR. From *left to right*: (1) 1-mo-old girl of normal length and weight for family; (2) her 26-mo-old brother (height, –8.2 SD); (3) 4.8-yr-old patient (height, –7.4 SD); (4) 6.5-yr-old patient (height, –9.4 SD); (5) 8.2-yr-old patient (height, –8.3 SD); (6) her 10-yr-old sister (height, –7.5 SD); (7) 14.8-yr-old patient (height, –8.3 SD; bone age 7 yr); (8) 17-yr-old woman who had menarche at 15 yr of age and stopped growing at 16 yr (height, –7.8 SD); (9) 19-yr-old woman (tallest female [height, 123.9 cm; –6.5 SD]) with menarche at 15 yr of age; (10) 21-yr-old man with normal timing of puberty and cessation of growth at 16 yr (height, 126.6 cm; –7.6 SD); (11) 28.5-yr-old man (height, –9.2 SD); (12) 40-yr-old woman (height, –8.6 SD) with advanced aging of skin; (13) 52-yr-old woman (height, 94.8 cm; –11.5 SD); (14) 66-yr-old woman (height, –9.5 SD). (Reproduced with permission from ref. *[79]*).

Table 6
Musculoskeletal/Body Composition Characteristics of IGF-I Deficiency Due to GHRD

Hypomuscularity in infancy with delayed walking in 70%
Hip dysplasia: avascular necrosis of femoral head in 25%; congenital hip dislocation in 5%
High pitched voices all children, 90% of adults
Thin, prematurely aged skin
Limited elbow extensibility in 85% of those over 5 yr of age
Children underweight to normal for height, most adults overweight for height with markedly decreased lean/fat mass.
Osteopenia indicated by DPA, DEXA; normal bone histomorphometry

originally described in the Ecuadorian population, have now been recognized in other populations with GHRD, as well as in GHD. This feature is due to decreased thickness of the scleral connective tissue, permitting visualization of the underlying choroid, and is more prominent in younger patients. Unilateral ptosis and facial asymmetry likely reflects positional deformity due to decreased muscular activity *in utero*, although mothers do not recognize decreased fetal movement in pregnancies with affected infants.

Musculoskeletal and Body Composition Characteristics *(10)*

The musculoskeletal and body composition features of GHRD are listed in Table 6. Hypomuscularity, apparent in roentgenograms, is thought to be responsible for delayed walking despite normal intelligence and time of speech onset. Although radiographs of the children suggest osteopenia and dual-photon absorptiometry and dual-energy X-ray absorptiometry in children and adults confirms this, dynamic bone histomorphometry in adults has been found to be substantially normal, suggesting that some of these findings are artifactual, based on small bone size *(70)*.

Limited elbow extensibility seen in most patients over 5 yr of age in the Ecuadorian population is an acquired characteristic, absent in younger children and increasing in severity with age. That this feature is not peculiar to the Ecuadorian population or to IGF-I deficiency due to GHRD has recently been confirmed by finding a Brazilian patient with GHRD with limited elbow extension *(71)* and observing this finding in all but the youngest patient in a family with eight individuals affected by multiple pituitary deficiencies *(72)*. The basis for this finding is unclear.

Although children appear overweight, they are actually underweight to normal weight for height, whereas most adults, especially females, are overweight with markedly decreased lean to fat ratios.

Reproduction *(10)*

It has long been recognized that severe GHD is associated with small penis size but normal penile growth at adolescence or with testosterone treatment in childhood. This is also true of GHRD. Although puberty may be delayed 3–7 yr in some 50% of individuals, there is normal adult sexual function with documented reproduction by males and females. Females require cesarean section delivery.

Intellectual and Social Development

Concerns about the effect of severe IGF-I deficiency due to GHRD on intellectual development arose early with the report that affected Israeli patients had intellectual impairment when compared to children with GHD *(73)*. Among 18 children and adolescents with GHRD administered the Wechsler Intelligence Scale for Children, only 3 had IQs within the average range (90–110); of the remaining 15 subjects, 3 were in the low average range (80–89), 3 in the borderline range (70–79), and 9 in the intellectually disabled range (< 70). These studies were done without family controls, so that the possibility of other factors related to consanguinity that might affect intellectual development could not be addressed. In a followup study 25 yr later, the investigators reexamined 8 of the original 18 patients and 4 new patients with GHRD, excluding five patients with mental disabilities who were in the original study *(74)*. This group had mean verbal and performance IQs of 86 and 92 on the Wechsler scale without evidence of visual–motor integration difficulties that had been noted in the earlier group, but there was a suggestion of deficient short-term memory and attention. The investigators hypothesized that early and prolonged IGF deficiency might impair normal development of the central nervous system, or that hypoglycemia common in younger patients may have had a deleterious effect. The recent description of intellectual impairment with severe IGF-I deficiency due to IGF-I gene deletion has added to the concern about potential effects of severe IGF-I deficiency in utero *(3)*. Nonetheless, patients with GH gene deletion and severe IGF-I deficiency have not been intellectually impaired *(9)*. Sporadic anecdotal reports of patients with GHRD suggested a normal range of intelligence, with retardation due to hypoglycemia being unusual. The collective data from the European IGF-I treatment study group, which includes a wider range of clinical abnormality than either the Ecuadorian or Israeli population, notes a mental retardation rate of 13.5% among 82 patients, but formal testing was not carried out *(75)*. Here again, the high rate of consanguinity was proposed as a possible explanation; hypoglycemia could not be correlated with these findings.

In the Ecuadorian population, investigators documented exceptional school performance among 51 affected individuals of school age or older who had attended school, with 44 typically in the top three places in their classes and most thought to be as bright or brighter than the smartest of their unaffected siblings within the family *(76)*.

The first controlled documentation of intellectual function in a population with GHRD was carried out by Kranzler et al. *(77)* in the Ecuadorian patients, focusing on schoolage individuals and comparing them to their close relatives and community controls. No significant differences in intellectual ability could be detected among these groups, with the use of nonverbal tests with minimal cultural limitations. It was hypothesized that the exceptional school performance in this population might be related to the lack of social opportunities due to extreme short stature, permitting greater devotion to studies and superior achievement in school for IQ level.

The clinical findings of intellectual impairment with IGF-I gene deletion *(3)* and intellectual normality with GHRD corresponds with gene disruption studies in mice. The IGF-I deleted mouse is neurologically impaired, whereas the GHRD mouse is behaviorally normal *(11,78)*. Thus, GH-dependent IGF-I production does not appear to be necessary for normal brain development and function.

GH–IGF Axis in IGF-I Deficiency Due to GHRD

The various biochemical characteristics of IGF-I deficiency states are noted in Table 2. Following are the distinctive features of IGF-I deficiency due to GHRD *(10,79)*.

Growth Hormone

Affected children have random GH levels that are >10 ng/mL and may be as high as 200 ng/mL with enhanced responsiveness to stimulation and paradoxical elevations following oral and intravenous glucose, as is seen in acromegaly. The GH levels show normal diurnal fluctuation. Twenty-four-hour profiles demonstrate marked variability among adult patients with some showing normal dynamics and suppression with exogenous recombinant human IGF-I. Thus, the normal sensitivity of the GH secretion is preserved, despite elevated levels and lack of feedback suppression from IGF-I.

Postpubertal patients may have normal or elevated basal levels but invariably demonstrate hyperresponsiveness to stimulation, which is all the more impressive with their obesity, which typically suppresses GH responses in normal individuals. In the Ecuadorian population, mean basal GH level in adults was significantly lower than that in children (11 ± 11 ng/mL vs 32 ± 22 ng/mL; $p < 0.0001$). This is thought to be related to the greater, although still markedly abnormal, IGF-I levels in the adults, resulting in some feedback inhibition of GH secretion.

Growth Hormone Binding Protein

It was initially thought that absence of GHBP in the circulation was a requirement for the diagnosis of GHRD, together with the clinical phenotype, very low IGF-I levels, and elevated (in children) or normal to elevated (in adults) GH levels. Early chromatographic analysis, however, showed measurable although reduced levels in a number of patients. The ligand-mediated immunofunction assay (LIFA), used to measure GHBP serum levels since 1990, uses an anti-GH monoclonal antibody to measure the amount of GH bound to GHBP. As a largely functional assay, this should not detect structurally abnormal GHBP.

As noted previously, certain genetic defects in the GHR, affecting dimerization or anchoring of the GHBP to the cell membrane, can result in normal or markedly elevated GHBP levels. In the Ecuadorian population, despite in vitro evidence for failure of production of normally spliced receptor, four children and four adults out of 49 patients had serum GHBP levels higher than 40% of the sex specific lower limit for controls and one adult male had a level in the lower portion of the normal adult male range. The presence or amount of GHBP measured did not relate to stature. There were no age-dependent changes, indicating that the difference in IGF values between children and adults was not related to the GHBP levels and the GHBP levels did not correlate with stature or with serum IGF-I levels. It would appear that extremely low or undetectable levels of GHBP constitute an important finding but not a *sine qua non* for the diagnosis of GHRD.

Insulin-Like Growth Factors

The severity of IGF-I deficiency is related to the cause, with the lowest levels of IGF-I reported in severe congenital defects in GH synthesis, deletion of the IGF-I gene, and GHRD. IGF-II is comparably suppressed. Among 50 Ecuadorian patients homozygous for the E180 splice-site mutation, IGF-I levels were significantly greater in adults 16–67 yr of age ($n = 31$, 25 ± 19 mcg/L) than in the 19 subjects under 16 yr of age (3 ± 2 mcg/L, $p < 0.0001$), although still markedly below the normal range of 96–270 mcg/L. The children's levels were too low to correlate with stature, but, in the adults, IGF-I levels correlated inversely with statural SDS with a coefficient of 0.64, $p < 0.001$. IGF-II levels in adults were also significantly greater than in children (151 ± 75 mcg/L vs 70 ± 42 mcg/L, normal 388 to 792 mcg/L, $p < 0.0001$). The correlation between serum IGF-I and IGF-II levels was highly significant, $r = 0.53$, $p < 0.001$. As noted previously, there was no indication of age difference in GHBP levels. Therefore, the increased levels of IGF-I and II suggest effects on syntheses of these growth factors that are not mediated through the GHR and are presumably under the influence of sex steroids. The correlation of IGF-I levels with stature in the adults indicates that, despite the markedly low levels, the influence of IGF-I on stature remains important.

IGF Binding Proteins

The principal binding protein for circulating IGFs, IGFBP-3, is considered to serve as a reservoir and delivery mechanism for IGF to tissues. In IGF deficiency states that are the result of GHD or GHI, IGFBP-3 is reduced and in children and adults with GHRD, this reduction correlates with statural impairment. In renal disease, elevated IGFBP-3, as well as IGFBP-1 IGFBP-2, are thought to prevent the effect of normal levels of IGF-I.

Short-term and extended treatment of GHI with IGF-I has failed to result in increases in IGFBP-3, whereas treatment of GHD with recombinant human GH restores levels to normal. This indicates that IGFBP-3 is under the direct influence of GH.

IGFBP-I is elevated in GHD and GHRD; in GHRD it is the most abundant IGFBP and is strongly inversely related to insulinemia. IGFBP-2 is present at a mean 300% of control concentrations in children with GHRD and 175% of control in affected adults. This is a significant difference, and the IGFBP-3 levels in adults with GHRD are significantly greater than in affected children, as well.

TREATMENT OF IGF-I DEFICIENCY WITH RECOMBINANT HUMAN IGF-I

Tertiary IGF-I deficiency, i.e., due to congenital or acquired GHD, and not associated with the development of GH autoantibodies, is effectively treated with recombinant human GH in daily dosages. The advantage of GH therapy is that it can be given once daily, the hormone is commercially available, and its administration in deficient states corrects both IGF-I and IGFBP-3 deficiencies. IGF-I remains an experimental drug with limited specific indication (GHRD and GHI due to GH inhibiting antibodies), with other possible applications (e.g., diabetes mellitus, osteoporosis, renal failure) under study.

IGF-I Administration to Normal Subjects

Soon after the cloning of the human IGF-I cDNA, human IGF-I was synthesized by recombinant DNA techniques *(80)*. The first report of its administration to humans was by Guler et al. in 1987 *(81)*. The initial preparations were available only for intravenous injection. Eight healthy adults developed hypoglycemia 15 min post-injection and a nadir of 50% of baseline glycemia was reached at 30 min, with a return to normal glycemia by 120 min. Similar curves were produced by 100 mcg/kg of IGF-I or 0.15 IU/kg of insulin, resulting in an estimate that IGF-I had 6% of the potency of insulin on a molar basis. The free IGF-I half-life after bolus injection was estimated to be 20 min, compared to 10 min for insulin injection. The insulin-like effect of IGF-I was reflected further in a comparable decrease in free fatty acids and increase in counterregulatory hormones in response to IGF-I and insulin. The authors considered the insulin-like effects of IGF-I to reflect the marked rise in free IGF-I in the serum, from 19% of total circulating IGF-I at baseline to 80% at 15 min.

Two groups of investigators reported results of euglycemic glucose clamp studies with intravenous infusion of IGF-I *(82,83)*. Both studies demonstrated potent metabolic effects resembling those of insulin, with, however, reduction in insulin secretion. These effects included increased glucose consumption and decreased hepatic production and decreased free fatty acids and branched chain amino acids with decreased whole body leucine flux. In another study designed to achieve levels of IGF-I in the circulation comparable to those achieved with 7 d of GH injection, eight normal individuals were administered a 28-h infusion of IGF-I that resulted in suppression of plasma insulin, C peptide, and glucagon concentrations without affecting plasma free fatty acid levels. The rates of leucine oxidation and of protein synthesis were unaffected and no increase in IGFBP-3 levels was noted, in contrast to what had been seen in a previous study with comparable levels of IGF-I in response to GH infusion, where IGFBP-3 increased 50% and there was a potent protein anabolic effect. These studies led to concern that IGF-I treatment in dosages sufficient to achieve levels of IGF-I in the circulation comparable to those induced by GH treatment of GHD would not have a substantial anabolic effect *(84)*.

Studies of intravenous administration of IGF-I were curtailed when cardiovascular side effects were noted, including asystole and hypotension, thought to be a result of acute hypophosphatemia *(85)*. Fortunately, subcutaneous preparations became available at the same time as these side effects were noted.

Guler and colleagues *(86)* went on to administer IGF-I subcutaneously at a dose of 20 mcg/kg body weight per hour over 6 d to two healthy adult males resulting in serum IGF-I concentrations increasing from 150 to 700 ng/mL, without hypoglycemia. The increased IGF-I level was accompanied by a decrease in serum IGF-II levels while insulin levels remained normal; C peptide levels were decreased, as was GH secretion. Glomerular filtration rate, estimated by creatinine clearance, increased to 130% of preinfusion values during the IGF-I infusion with no change in total creatinine and urea excretion, indicative of increased protein synthesis. The authors noted that a major difference between the effects of GH and IGF-I was that GH causes hyperinsulinism whereas IGF-I decreases insulin secretion. Somewhat different results were obtained in a study of daily subcutaneous injection of IGF-I in a considerably smaller dosage of 100 mcg/kg body weight for 7 consecutive days *(87)*. Six normal subjects had a 2.5-fold increase in their serum IGF-I levels with significant suppression of blood glucose levels 4 h after injection. Free IGF-I rose from 2.3 ± 0.3 ng/mL to a maximum of 43.5 ± 5.1 ng/mL 2 h after the initial IGF-I injection and did not change appreciably after the seventh injection. No changes were noted in the excretion of urea nitrogen, creatine, creatinine, sodium, potassium, chloride, calcium, or C peptide in the urine during the 7 d of IGF-I administration.

The effect of IGF-I on capillary permeability in the skin and retina was studied by Hussain et al. *(88)* because of the mild generalized edema that develops with IGF-I injections and experimental evidence for a role of IGF-I in diabetic retinopathy. Eight healthy volunteers were given saline or IGF-I subcutaneous injections in the dose of 10 mcg/kg/h in a randomized crossover study. They looked at transcapillary diffusion of sodium fluoride by videodensitometry in the skin on the fourth treatment day and plasma penetration of NaF at the blood retinal barrier using vitreous fluorometry on the third and fifth days of treatment. IGF-I treatment was accompanied by increased transcapillary diffusion of NaF in skin and in the retinal vessels.

Administration of IGF-I in Catabolic States

IGF-I has been proposed as an alternative to GH for treatment of the physiologic decline with age, and with cachexia from glucocorticoid therapy, renal failure, gastrointestinal surgery, burns, and AIDS. GH treatment in these conditions has been associated with problems of edema, arthralgas, hyperglycemia, and recently with increased mortality in acute catabolism (64, drug warning letter November 10, 1997). In a study of six normal young adult volunteers who had restriction of their dietary intake by 50% to produce moderate catabolism for 2 wk, either IGF-I (12 mcg/kg/h intravenously over 16 h) or GH (0.05 mg/kg/d subcutaneously) was administered during the last 6 d of diet restriction. IGF-I and GH produced comparable improvement in nitrogen balance despite the catabolic diet; while IGF-I decreased fasting blood glucose levels, GH raised them, concomitant with decreased serum insulin levels with IGF-I and elevated insulin levels with GH. Diet restriction alone affected neither IGFBP-1 nor IGFBP-2 levels significantly *(89)*. The IGF-I treatment increased IGFBP-1 and IGFBP-2 approximately two-fold while GH injections resulted in a modest decline in IGFBP-1 concentrations and no effect on IGFBP-2. Both IGF-I and GH caused a modest increase in IGFBP-3 serum levels, which were assessed by ligand blotting. Free IGF-I levels were increased disproportionately during the IGF-I infusions with hypoglycemia occurring despite insulin suppression, presumably as a result of the failure of concomitant increase in IGF binding capacity in the serum *(90)*.

Kupfer et al. *(91)* postulated that combined therapy with GH and IGF-I might be preferable in catabolic states to avoid, on the one hand, the worsening of glucose intolerance that accompanies glucocorticoid therapy or severe stress in these situations and, on the other hand, the IGF-I-induced hypoglycemia. It was also considered that GH might enhance recovery by stimulating precursor cells to differentiate and become responsive to IGF-I. Caloric intake of seven normal volunteers was restricted as in this group's earlier studies *(89,90)* and after 7 d either IGF infusion (12 mcg/kg/h for 16 h per day intravenously) for 6 d or IGF-I combined with GH (0.05 mg/kg/day subcutaneously) for 6 d was administered. There was a marked anabolic (nitrogen retention) response to the combination from –170 mmol/d to +46 mmol/d. It was noteworthy that the peak IGF value with IGF alone occurred after 24 h (1100 ng/mL), but declined to 750 ng/mL by the last day of infusion; this decrease was thought to be due to suppression of GH secretion that is needed for IGFBP-3 production. Combination therapy resulted in an increase in IGFBP-3 and acid-labile subunit (ALS) while IGF-I alone resulted in a decrease in these two proteins. With combined therapy, the serum IGF-I value did not reach its peak until the fourth day (1800 ng/mL) and remained at this level throughout treatment. With combination therapy, serum insulin and blood glucose levels remained in the normal range.

Single injections of IGF-I (40 mcg/kg subcutaneously) were given to 20 patients following major gastrointestinal surgery in a double-blind placebo-controlled study resulting in normalization of circulating IGF-I levels from 4 h after injection to 15 h post-injection; brief insulin suppression; and decrease in serum cholesterol, potassium, and creatinine levels with an increase in creatinine clearance *(92)*. Long-term studies to determine the anabolic effects of IGF-I in this patient group have not been reported.

In 21 patients with AIDS, a single injection of GH produced a smaller increase in circulating IGF-I than in 23 age-matched controls, indicating partial GH resistance. Ten subjects with AIDS-associated cachexia were given either low or high dose intravenous IGF-I for 12 h daily for 10 d. Although cumulative nitrogen retention was positive for both dosage groups, a significant increase in daily nitrogen retention was seen only in the low-dose group on half of the days. The effect was a waning one, with nitrogen balance and protein turnover, as estimated by stable isotope studies, unchanged during the last 3 d of treatment when compared to baseline values. Further, the IGFBP-3 levels decreased *(93)*. This would appear to be a circumstance in which the combination of IGF-I and GH administration might be effective.

Another model of malnutrition and GH insensitivity, anorexia nervosa, is associated with low bone density. Twenty-three young women had IGF-I levels in the low normal range with IGFBP-3 levels normal to elevated. The group included 20% with bulimia, and an overall relatively high caloric intake of 1500 per day. Six-day treatment periods with subcutaneous injections of IGF-I at 100 or 30 mcg/kg twice daily were compared with placebo. Markers of bone formation, type I pro-collagen C-terminal propeptide (PICP) and osteocalcin increased significantly, as did markers of bone resorption, urine pyridinoline, and deoxypridinoline at the higher dose of IGF-I, whereas the lower dose produced a significant increase only in PICP *(94)*. Thus, in this form of malnutrition with mild GH resistance, there is a short-term effect of IGF-I on bone remodeling. Whether such an effect is sustainable or not remains to be demonstrated.

Thompson et al. *(95)* and Ghiron et al. *(96)* compared low-dose (15 mcg/kg/d) and high-dose (60 mcg/kg/d) IGF-I with GH (0.025 mg/kg/d) administered over 28 d to 14 healthy elderly women. There were numerous side effects with both GH and high-dose IGF-I, including headaches, lethargy, joint swelling and pain, and bloating *(96)*. Fat mass decreased in all groups, with significant increases in lean mass and nitrogen retention only in the GH and high-dose IGF-I groups *(95)*. The markers of resorption and formation of bone were both increased by GH and high-dose IGF-I; formation indices were increased by low-dose IGF-I, but there was no effect on resorption markers *(95)*. Longer term studies are necessary and pending.

Administration of IGF-I to Patients with GHD

Two young adult patients were studied 3 yr after completing courses of GH administration for growth, with 7 d of administration of IGF-I in a dose of 0.1 mg/kg/d. This resulted in the attainment of normal levels of IGF-I within an hour of injection that were sustained for at least 12 h. Peak levels were reached 4 h after injection and there was no hypoglycemia with the injection being given after breakfast. Maximal levels were significantly increased on the seventh day of administration compared to the first day, suggesting induction of IGFBP-3 *(87)*. In another study, six adults with GHD were given IGF-I in a dose of 40 mcg/kg subcutaneously as a single injection. Mean maximal IGF-I concentrations were attained at 6 h after injection but remained above baseline at 24 h. As had been found in normal individuals, there was a concomitant decrease in IGF-II concentrations. These studies were thought to indicate the need for the administration of IGF-I more frequently than once a day *(97)*. These same investigators subsequently reported the effect of repeated subcutaneous injections over 7 d in these patients, now noting that maximal IGF-I concentration was achieved 3–4 h after the injection and that higher peak and basal values were noted on d 7 compared to those observed on d 1. The IGF-I area under the curve on d 1 and 7 correlated with IGFBP-3 levels. The patient with the highest IGFBP-3 level had the only normal concentration of IGF-I 24 h after injection. The conclusion was again that normal diurnal IGF-I levels required administration of the recombinant hormone two or three times per 24 h in subjects with subnormal IGFBP-3 levels *(98)*.

Administration of IGF-I to Subjects with GHI

Short-Term Administration of IGF-I in Children and Adults with GHI due to GHRD

The hypoglycemic effect of an intravenous bolus of IGF-I (75 mcg/kg) following an overnight fast was demonstrated in nine patients with GHRD aged 11–33 yr. The hypoglycemia was symptomatic and associated with a fall in plasma insulin level with recovery only after food was taken at 2 h post-injection. In the seven older individuals, there was a marked increase in serum GH concentration, reflecting the overriding effect of hypoglycemia despite the expected somatostatin response to the increased levels of free IGF-I *(99)*. Indirect evidence of somatostatin stimulation was seen in suppression of TSH levels *(100)*. The elimination time for intravenously injected IGF-I was found to be markedly reduced in three children and seven adults with GHRD compared to three healthy children and three healthy adults. This was attributed to the absence of IGFBP-3 in the patients *(101)*. When IGF-I was administered subcutaneously (120–150 mcg/kg/d) for 7 d, followed by breakfast, no significant hypoglycemia occurred, plasma GH was markedly reduced, and type 3 procollagen markedly increased *(102)*. A 9-yr-old child given intravenous IGF-I for 11 d demonstrated substantial anabolic effects, including decreased serum urea nitrogen, increased urinary calcium excretion, and decreased urinary phosphate and sodium excretion. As previously noted with intravenous IGF-I in normals, IGF-II levels were suppressed and asymptomatic hypoglycemia documented *(103)*.

Preparatory to treatment of the large number of children in the Ecuadorian population with GHRD, Vaccarello et al. *(104)* administered IGF-I subcutaneously at a dose of 40 mcg/kg every 12 h over 7 d to six Ecuadorian adults with GHRD. Although insulin levels were suppressed, no hypoglycemia occurred. There was a twofold increase in urinary calcium excretion without a change in serum calcium levels. Mean integrated 24-h GH levels were suppressed, as were the number of peaks, the area under the curve, and clonidine-stimulated GH release. The mean peak serum IGF-I levels were 253 ± 11 ng/mL reached between 2 and 6 h after injection and mean trough levels were 137 ± 8 ng/mL before the next injection, values not significantly different from those of normal control Ecuadorian adults. As previously noted, IGF-II levels decreased and correlated inversely with IGF-I levels. There were no significant changes in the half-life or metabolic clearance of IGF-I between d 1 and 7, although the distribution volume did increase over this time. Although IGFBP-3 levels did not increase, elevated baseline IGFBP-2 levels (153% of control) increased 45% ($p < 0.01$).

The short-term studies demonstrated that there was an insignificant risk of hypoglycemia despite low levels of IGFBP-3. There remained, however, concern whether the low IGFBP-3 levels would result in more rapid clearance of IGF-I as demonstrated by Laron et al. *(101)*, with blunting of the therapeutic effect. Four children and four adults with GHRD treated for 7 d with a single daily subcutaneous injection of IGF-I in a dose of 120–150 mcg/d experienced a significant decrease in serum IGFBP-3 levels measured by specific RIA *(105)*. As noted previously, Vaccarello et al. *(104)* did not find a significant increase in IGFBP-3 levels following 7 d of twice daily subcutaneous injection of doses of IGF-I sufficient to raise serum levels to normal in adults with GHRD. Further studies were carried out on composition and distribution of IGFs and IGF binding proteins in patients from this study. It was found that the two forms of IGFBP-3 associated with IGF and ALS that are able to form the ternary 150-kDa complex are abnormally distributed in GHRD patients. This distribution was unchanged by IGF-I treatment which, in addition to not increasing the IGFBP-3, did not increase ALS levels *(106)*. The failure of short-term IGF-I administration to increase IGFBP-3 levels was, as noted, of concern for long-term therapy. Although it was hoped that chronic treatment would result in stimulation of ALS and IGFBP-3, this did not seem likely in view of the experience with GH treatment of GHD. In this situation, where serum IGFBP-3 levels are also reduced, exogenous GH rapidly increases IGFBP-3 levels *(107)*.

Long-Term (6 Mo Plus) Treatment with IGF-I in GHI

In addition to concerns regarding the failure of short-term therapy to increase IGFBP-3 levels, there was concern whether catchup growth in children with GHI due to GHRD or to GH antibodies would be as substantial as occurs with GH replacement therapy in patients with GHD in the absence of a direct effect of GH on bone. This direct effect is considered to be the differentiation of prechondrocytes into early chondrocytes which then secrete IGF-I that in turn stimulates clonal expansion and maturation of the chondrocytes, or growth *(108)*. It is postulated that 20% of GH induced growth is a result of such direct effects *(108)* which would, of course, not occur in the absence of a GH receptor or in the presence of GH inhibiting antibodies.

Growth acceleration comparable to that seen with GH treatment of GHD was reported in five children aged 3.3 yr to 14.5 yr with GHRD treated for 3–10 mo with single daily doses of 150 mcg/kg. Baseline growth velocities of 2.8 to 5.8 cm/yr increased by approx 1.5- to 4-fold to 8.8–13.6 cm/yr, with the youngest patients responding most impressively. A reduction of subcutaneous fat, measured by skinfold thickness, was noted in the four patients who were considered obese *(109)*. No long-term effects were noted on prolactin levels, which were mildly elevated in some subjects at baseline *(110)*. Consistent with the studies of single injections of IGF-I intravenously, there was a transitory decrease of both serum TSH and free thyroxine (T_4) followed by a spontaneous rise to pretreatment or even higher values without any change in serum triiodothyronine (T_3) levels. These effects were explained as an early stimulation of somatostatin release causing a decrease in TSH, followed by development of compensatory mechanisms *(111)*. In contrast to the single daily dosage of 150 mcg/kg administered by Laron et al. *(109)*, Walker et al. *(112)* gave 120 mcg/kg twice daily to a 9.7-yr-old child with GHRD with an increase in growth velocity over a 9-mo period from 6.5 to 11.4 cm/yr. Although mean pretreatment IGF-I levels of 9 ± 2 mcg/L increased to 347 ± 26 mcg/L after 2 h, serum concentrations of IGF-II were unchanged, contrasting with earlier short-term studies. Serum and urinary urea nitrogen decreased whereas creatinine clearance, urine volume, and urinary calcium increased. There were no abnormalities in glucose metabolism.

The first report of treatment for longer than 10 mo was in two children with GHRD with pretreatment height velocities of 4.3 and 3.8 cm/yr at 8.4 and 6.8 yr of age, respectively. They had suppression of greatly elevated serum GH levels and increase in procollagen I levels shortly after starting treatment and 6-mo height velocities increased to 7.8 and 8.4 cm/yr without side effects. In the second 6 mo of treatment the velocities decreased to 6.6 and 6.3 cm/yr and in the subsequent 5 mo growth rate returned to pretreatment values. Improvement in bone density was documented by photon absorptiometry. These patients were treated with a dose of 40 mcg/kg subcutaneously twice daily and the waning of their growth response after a year indicated that this dosage was not adequate for sustained effect *(113)*.

Since 1995, data have become available on 70 patients with GHI treated with subcutaneous injections of IGF-I for 12 mo or longer. Seven of these patients had tertiary IGF-I deficiency with GH inhibiting antibodies developing after treatment with GH injections for GHD due to GH gene deletion. The rest of the patients had secondary IGF-I deficiency due to congenital abnormalities in GH receptor function (Laron syndrome), including Pakistani and Arab patients thought to have post-receptor defects *(6–8)*. The data for 69 of these patients are summarized in Table 7. The largest group of patients was recruited for the European multicenter study which included 33 patients age 3.7 to 23 yr, 2 with GH

Table 7
IGF-I Treatment of Children with GHI for 1 yr or Longer

mber (ales)	*Age (yr)*	*IGF-I dose (µg/kg)*	*Growth velocity[a] (cm/yr)* Pre Rx	1 yr	2 yr	*Height SDS[a]* Pretreatment	1 yr	2 yr
ropean Study Group *(114)*								
(14)	3.7–19.6	40–120 bid	NA	8.5 (2.1)	X	–6.8 (1.6)	–6.1 (1.5)	X
(10)[b]	3.7–16.7	40–120 bid	NA	8.6 (1.7)	6.4 (2.2)	–6.4 (1.7)	–5.6 (1.6)	–5.2 (1.9)
uador *(116,123)*								
1)[c]	18/17.5	120 bid	1.9/2.7	6.1/8.0	X	–8.0/–9.1	–7.1/ –7.9	X
(5)	4.7–17.1	120 bid	3.4 (1.4)	8.8 (1.1)	6.4 (1.1)	–8.5 (1.3)	–7.5 (1.1)	–7.0 (1.2)
2)	3.1–15.2	80 bid	3.0 (1.8)	9.1 (2.2)	5.6 (2.1)	–8.0 (1.8)	–7.2 (1.8)	–6.7 (1.8)
ael *(124)*								
6)	0.5–14.6	150–200/d	4.6 (1.3)[d]	8.2 (0.8)	X	–5.6 (1.5)	–5.2 (1.7)	X
4)[e]	0.5–14.6	150–200/d	5.0 (1.2)[d]	8.3 (1.0)	6.0 (1.3)	–6.2 (1.5)	–6.0 (1.6)	–5.8 (1.2)
rth Carolina Study *(125)*								
6)	2.3–11	80–120 bid	4.0 (1.0)	9.3	6.2	–5.6 (1.1)	NA	–4.5 (1.3)
azil/Vanderbilt *(128)*								
1)	10.4/8.8	40–120 bid	4.5/1.8	7.5/6.7	7.0/6.3	–6.7/–5.7	NA	NA

NA, = not available.
[a]SD in parentheses.
[b]Same cohort as all 26.
[c]Also treated with GnLRH analog *(106)*.
[d]Patient age 0.5 excluded because of rapid GV first 6 mo of life before Rx (18 cm/yr).
[e]Same cohort as all 9.

gene deletion and the rest with GHRD. One year growth data were reported for 26 of these patients and 2-yr data for 18. The 7 patients not reported included 2 who stopped treatment before 12 mo because of insufficient growth response and 5 because of adverse events. Three did not continue for a full 24 mo because of poor compliance, headache, and weight gain. The adverse events noted were headache, hypoglycemia, papilledema (one instance), Bell's palsy (one instance), lipohypertrophy, and tonsillectomy/ adenoidectomy (three instances). The IGF-II levels decreased and IGFBP-3 remained constant during this long-term therapy *(114)*.

The longest term treatment report is of one of the patients in the European study cohort, results from age 9.3 to 14.5 yr while receiving IGF-I 120 mcg/kg twice daily. During this time he grew 40 cm and improved his height SDS from –4.5 to –2.1 while advancing his bone age concomitantly with chronologic age. His response was considered to be better than that usually seen in patients with Laron syndrome, but was accompanied by remarkable weight gain, thought to be the result of insulin-like activity of the IGF-I acting through the insulin receptor *(115)*.

The first IGF-I treatment report from the large Ecuadorian cohort was of growth and body composition changes in two adolescent patients treated with a combination of IGF-I (120 mcg/kg bid) and LRH analog to forestall puberty. A girl aged 18 and boy age 17.5 yr, with bone ages of 13 1/2 and 3 yr experienced an approximate tripling of growth velocity, increased bone mineral density, and maturation of facial features on IGF-I for

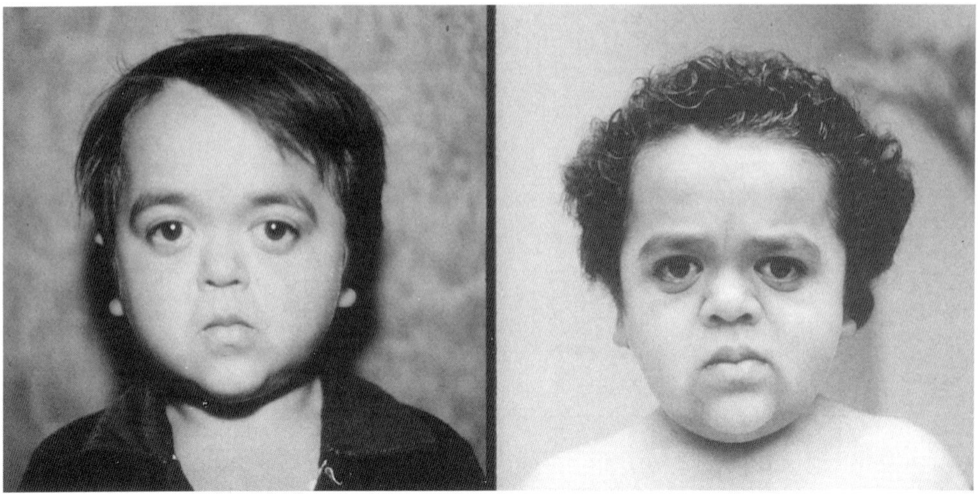

Fig. 4. Face and hair changes in 17.5-yr-old patient (bone age 13 yr) with GHRD during 6 mo treatment with IGF-I, 120 mcg/kg bid and depot GnRH agonist begun at age 16.5 yr *(116)*.

1 yr. There was initial hair loss followed by recovery of denser and curly hair with filling of the frontotemporal baldness, the appearance of axillary sweating, loss of deciduous teeth, and appearance of permanent dentition. The boy had coarsening of his facial features (Fig. 4), Submaxillary gland enlargement was also noticed in one patient and fading of premature facial wrinkles, commonly associated with GHD, noted in the other patient. Serum IGF-I levels were seen to increase into the normal range during the first 2–8 h following IGF-I injection *(116)*. Studies were done at doses of 40, 80, and 120 mcg/kg with pharmacokinetic profiles suggesting a plateau effect between 80 and 120 mcg/kg per dose. It was considered that the carrying capacity of the IGFBPs was saturated at this level. As previously noted, mean serum IGF-II levels decreased concurrently with the increase in IGF-I and serum IGFBP-3 levels did not respond to prolonged IGF-I treatment. There was no apparent change in the half-life of IGF-I during the treatment period, indicating no alteration of IGF-I pharmacokinetics induced by prolonged treatment *(117)*.

Tachycardia was noted in these adolescent patients and further studied in 16 prepubertal Ecuadorian patients during induction of treatment at progressive dosages of 40, 80, and 120 mcg/kg bid. Both repeated palpation of the radial artery and continuous portable Holter monitoring were used. There was a direct dose-related increase to approx 125% of baseline rate at the highest dosage. This was unrelated to glucose or electrolyte changes which were not significant *(118)*. There is considerable experimental evidence for a chronotropic effect of IGF-I *(119,120)*.

Two prepubertal girls and two adult women with GHRD in Israel developed clinical features of hyperandrogenism and elevated androgen levels during treatment with IGF-I *(121)*.

Seventeen prepubertal Ecuadorian patients were entered into a randomized double-blind, placebo-controlled trial of IGF-I at 120 mcg/kg subcutaneously bid for 6 mo, following which all subjects received IGF-I. Such a study was considered necessary because of the observation of spontaneous periods of normal growth in these youngsters, the suggestion that nutritional changes that might accompany intervention would be an independent variable, and the need to control for side effects, particularly hypoglycemia, that occur in the untreated state. The nine placebo-treated patients had a modest but not significant increase in height velocity from 2.8 ± 0.3 to 4.4 ± 0.7 cm/yr, entirely attributable to three individuals with 6-mo velocities of 6.6–8 cm/yr. Although this response was attributed to improved nutritional status there was no accompanying increase in IGFBP-3 as noted with nutrition induced catch up growth by Crosnier et al. *(65)* in their GHRD patient with anorexia. For those receiving IGF-I, the height velocity increased from 2.9 ± 0.6 to 8.8 ± 0.6 cm/yr and all 16 patients had accelerated velocities during the second 6-mo period when all were receiving IGF-I. No changes or differences in IGFBP-3 were noted. There was no difference in the rate of hypoglycemia events, nausea or vomiting, headaches, or pain at the injection site between the placebo and IGF-I treated groups. Six of the seven IGF-I treated patients experienced hair loss *(122)*.

Two-year treatment results in the Ecuadorian group have been reported, comparing the 120 mg/kg bid dosage to 80 mg/kg bid and also comparing responses to those of GH-treated GHD in the Ecuadorian population. There were no baseline differences between the low- and high-dose groups for growth velocity, bone age, SDS for height, or mean percent body weight for height (MBWH). No differences in growth velocity or changes in height SDS, height age, or bone age between the two dosage groups were seen and a group of six subjects receiving the higher dose followed for a third year continued to maintain second year growth velocities. The annual changes in height age in both the first and the second year of treatment correlated with IGF-I trough levels which tended to be in the low normal range despite a failure of serum IGFBP-3 levels to increase. The comparable growth responses to the two dosage levels and the similar IGF-I trough levels were thought to indicate a plateau effect at or below 80 mcg/kg body weight twice daily *(123)*.

When all 22 of these patients were compared to 11 GH-treated GHD subjects, the GHD group was found to have a greater change in SDS for height and height age but did not differ in the change in bone age or in the ratio of height age to bone age changes over the 2-yr period. There was, however, a greater change in percent MBWH in the GHRD group treated with IGF-I, thought to indicate comparable effects on lean body mass of the two treatment programs, without the lipolytic effects of GH in the IGF-I-treated GHRD subjects. The difference in growth response between IGF-I-treated GHRD and GH-treated GHD was consistent with the hypothesis that 20% or more of GH-influenced growth is due to the direct effects of GH on bone *(108)*. Nonetheless, the comparable ratios of height age change to bone age change suggested similar long-term effects for replacement therapy in these two conditions *(123)*.

The Israeli report of 3 yr treatment of nine patients, including the three with a presumed post-receptor defect, is the only one in which patients were given IGF-I as a single daily dosage (150–200 mcg/kg). While the European study group noted height SDS improvement of 0.7 over 1 yr and 1.2 over 2 yr, and the Ecuadorian patients had an improvement of 1 SDS over 1 yr and 1.5 over 2 yr at the higher dose and 0.8 SDS over 1 yr and 1.3 SDS over 2 yr at the lower dosage, the Israeli patients had an improvement of only 0.4 over 1 yr and for the six with 2-yr data, 0.2 over 1 yr and 0.4 over 2 yr *(124)*. The kinetic studies that originally formed the rationale for twice daily administration are supported by these observations. Of the five patients completing 3 yr of IGF-I treatment in the Israeli report, three were growing at slower rates than before treatment while the other two remained at comparable velocities to the second year of treatment. The subgroup with normal IGFBP-3 levels is particularly interesting, as the growth responses do not differ from those of the rest of the group, suggesting that the IGFBP-3 deficiency is not the explanation for the relatively deficient growth response to replacement therapy with IGF-I when compared to GH treatment of GHD *(6)*.

Five children with GHRD and three with growth attenuating antibodies to GH have been treated for 4 yr by Backeljauw and colleagues *(125)*. The 2-yr data are given in Table 7 and are comparable to data in the European and Ecuadorian populations, with an improvement in height SDS over the 2-yr period of 1.1. Bone age was noted to advance consistent with chronologic age. There was rapid growth of the spleen and kidney, as determined by ultrasound during the first year with a slowing of spleen growth but a continued rapid kidney growth during the second year. In a preliminary report, the third year and fourth year results of treatment in this group were reported; a further decrease in height velocity to a mean 5.1 cm/yr occurred during the third year, but the fourth year mean height velocity was 5.7 cm/yr. Kidney growth continued in five of the eight patients who had renal length for height at or above the 95th percentile. There were no functional or structural abnormalities of the kidneys, however *(126)*. In five of these patients, craniofacial growth has been monitored with the suggestion of mandibular overgrowth, most noticeable in the last 2 yr of treatment *(127)*.

Two unrelated Brazilian patients with GH gene deletions who had responded briefly to GH were treated with increasing doses of IGF-I from 40 to 120 mcg/kg bid subcutaneously for 2 yr. Their data are summarized in Table 7. They experienced no side effects and IGFBP-3 levels remained low. As noted in the GHRD patients, there was not a decrease in body fat but an increase in lean mass *(128)*. The Vietnamese girl treated by Walker et al. *(39)* with IGF-I and LHRH analog, not included in Table 3, had modest improvement of growth over the first 3 yr of treatment, a height SDS increase of 1.2 with dramatic slowing of growth response thereafter. As with the patient in Fig. 4, facial coarsening was noted.

CONCLUSIONS

Recombinant human IGF-I has been proven effective in growth promotion for children with IGF-I deficiency due to GHI as the result of GHRD, presumed defective signal transduction, or GH inactivating antibodies. Twice daily injection in a dose of 80–120 mcg/kg appears optimal. Whether normal stature is attainable remains uncertain. Less growth acceleration than is seen with GH treatment of GHD and more rapid and greater waning of effect have been noted. This could be attributable to failure to increase IGFBP-3 levels, a direct GH effect, but three children who had defective IGF-I synthesis because of a presumed post-receptor defect did not grow better while receiving IGF-I than did subjects with GHRD, despite normal IGFBP-3 levels *(6,124)*. Furthermore, a profound effect of treatment, acromegaloid facial changes (*[127,129]*; J. Guevara-Aguirre, *personal communication*) indicates that the amount of IGF-I reaching tissues is supraphysiologic.

If the tissue dose of IGF-I in patients with GHI treated with IGF-I is supraphysiologic, why then do we not see sustained growth acceleration as in GH-treated GHD? The dual effector hypothesis remains the preeminent explanation *(108)*. With diminished ability to stimulate prechondrocyte differentiation and local IGF-I production, children with GHI can expect only partial recovery of normal growth with IGF-I replacement. The comparable ratios of change in height age to change in bone age over 2 yr of treatment in GHI children treated with IGF-I and GHD children treated with GH, however, suggests that the absence of a direct GH effect in GHI may have a temporal, rather than absolute, effect on long-term response to treatment *(123)*. Thus, IGF-I replacement therapy of GI may need to continue longer than GH treatment of GHD to achieve normal height. This goal will likely require suppressing adolescence in most children with GHI, using GnRH analogues *(116)*.

In addition to statural attainment, goals of replacement therapy with IGF-I in GHI include improvement in body composition, normalization of facial appearance, and possible reduction of risk factors for childhood and adult mortality. All studies that have monitored body composition have verified lean mass increases, including increased bone density. Normalization of craniofacial features has also been apparent *(129)*. Voice change has not been remarked on, but can be expected. It is likely that maintenance of body composition changes will require continued treatment in adulthood.

The reduction of risk factors for the higher mortality in infancy and childhood with GHRD is to be expected with IGF-I therapy, but the reason for this increased risk is unknown. Leukocytes share in the general up-regulation of IGF-I receptors in GHRD, and appear to function normally in this condition *(130)*. In a study of one affected infant (who died at 7 mo with bronchitis) and five adults with GHRD from Ecuador, Diamond et al. *(131)* demonstrated a variety of immune disturbances in the infant and three of the adults. The pathologic significance of these findings remains uncertain *(132)*.

If there is increased cardiovascular risk in adults with GHI, as there is with GHD, and as suggested by the anecdotal experience from Ecuador, IGF-I replacement therapy of adults with GHI would be justified. Five adult patients from Israel with GHRD given IGF-I injections for 9 mo experienced reductions in the risk factors of circulating lipoprotein (a) and cholesterol (total and LDL). Their skinfold thickness measurements also decreased *(133,134)*. Data from a 2-yr IGF-I treatment study of eight Ecuadorian adults with GHRD are currently being analyzed.

NOTE ADDED IN PROOF

The recent description of the clinical and biochemical phenotype in 18 patients with severe tertiary IGF-I deficiency resulting from an inactivating mutation of the growth hormone releasing hormone receptor (GHRH-R) is of considerable interest because of striking differences from the comparably IGF-I deficient states characterized by the classic GHD phenotype (severe GHD and GHRD). The GHRH-R–deficient patients have normal facies, asymptomatic hypotension, relatively small heads, and do not have micropenis, truncal obesity, body segment disproportion, or hypoglycemia. The hypothesis that nonGH mediated IGF-I and IGFBP-3 production occurs as a result of sex steroids in adults with GHRD, explaining their higher levels of these proteins and lower GH levels compared to affected children, is not supported by findings in the GHRH-R deficient subjects. The adults had identically low concentrations of IGF-I and IGFBP-3 as the children *(135)*.

REFERENCES

1. Laron Z, Blum W, Chatelain P, Ranke M, Rosenfeld R, Savage M, Underwood L. Classification of growth hormone insensitivity syndrome. J Pediatr 1993; 122:241.
2. Laron Z, Pertzelan A, Mannheimer S. Genetic pituitary dwarfism with high serum concentration of growth hormone—a new inborn error of metabolism? Isr J Med Sci 1996; 2:152–155.
3. Woods KA, Camacho-Hübner C, Savage MO, Clark AJL. Intrauterine growth retardation and postnatal growth failure associated with deletion of the insulin-like growth factor I gene. N Engl J Med 1996; 335:1363–1367.

4. Ayling RM, Ross R, Towner P, Von Laue S, Finidori J, Moutoussamy S, Buchanan CR, Clayton PE, Norman MR. A dominant-negative mutation of the growth hormone receptor causes familial short stature. Nature Genet 1997; 16:13–14.
5. Iida K, Takahashi Y, Kaji H, Nose N, Okimura Y, Abe H, Chihara K. Growth hormone (GH) insensitivity syndrome with high serum GH binding protein levels caused by a heterozygous splice site mutation of the GH receptor gene producing a lack of intracellular domain. J Clin Endocrinol Metab 1998; 83:531–537.
6. Laron Z, Klinger B, Eshet R, Kaneti, Karasik A, Silbergeld A. Laron syndrome due to a post-receptor defect: response to IGF-I treatment. Isr J Med Sci 1993; 29:757–763.
7. Freeth JS, Ayling RM, Whatmore AJ, Towner P, Price DA, Norman MR, Clayton PE. Human skin fibroblasts as a model of growth hormone (GH) action in GH receptor-positive Laron's syndrome. Endocrinology 1997; 138:55–61.
8. Freeth JS, Silva CM, Whatmore AJ, Clayton PE. Activation of the signal transducers and activators of transcription signaling pathway by growth hormone (GH) in skin fibroblasts from normal and GH binding protein-positive Laron syndrome children. Endocrinology 1998; 139:20–28.
9. Rivarola MA, Phillips JA III, Migeon CJ, Heinrich JJ, Hjelle BJ. Phenotypic heterogeneity in familial isolated growth hormone deficiency type I-A. J Clin Endocrinol Metab 1984; 59:34–40.
10. Rosenbloom AL, Rosenfeld RG, Guevara-Aguirre J. Growth hormone insensitivity. Pediatr Clin North Am 1997; 44:423–442.
11. Beck KD, Powell-Braxton L, Widmer H-R, Valverde J, Hofti F. IGF-I gene disruption results in reduced brain size, CNS hypomyelination, and loss of hippocampal granule and striatal parvalbumin-containing neurons. Neuron 1995; 14:717–730.
12. Bucuvalas JC, Horn JA, Carlsson L, Balistereri WF, Chernausek SD. Growth hormone insensitivity associated with elevated circulating growth hormone-binding protein in children with Alagille syndrome and short stature. J Clin Endocrinol Metab 1993; 76:1477–1482.
13. Laron Z, Pertzelan A, Karp M. Pituitary dwarfism with high serum levels of growth hormone. Isr J Med Sci 1968; 4:883–894.
14. Merimee TJ, Hall J, Rabinovitz D, McKusick VA, Rimoin DI. An usual variety of endocrine dwarfism: subresponsiveness to growth hormone in a sexually mature dwarf. Lancet 1968; ii:191–193.
15. Daughaday WH, Laron Z, Pertzelan A, Heins JN. Defective sulfation factor generation: a possible etiological link in dwarfism. Trans Assoc Am Physicians 1969; 82:129–138.
16. Bala RM, Beck JC. Fractionation studies on plasma of normals and patients with Laron dwarfism and hypopituitary gigantism. Can J Physiol Pharmacol 1973; 91:845–852.
17. Eshet R, Laron Z, Brown M, Arnon R. Immunoreactive properties of the plasma hGH from patients with the syndrome of familial dwarfism and high plasma IR-hGH. J Clin Endocrinol Metab 1973; 37:819–821.
18. Tsushima T, Shiu RPC, Kelly PA, Friesen HG. Radioreceptor assay for human growth hormone and lactogens: structure–function studies and clinical applications. In: Raiti S, ed. Advances in Human Growth Hormone Research. USPHS-DHEW Publication 74-612, 1973, pp. 372–387.
19. Jacobs LS, Sneid DS, Garland JT, Laron Z, Daughaday WH. Receptor-active growth hormone in Laron dwarfism. J Clin Endocrinol Metab 1976; 42:403–406.
20. Golde DW, Bersch N, Kaplan SA, Rimoin DL, Li CH. Peripheral unresponsiveness to human growth hormone in Laron dwarfism. N Engl J Med 1980; 303:1156–1159.
21. Eshet R, Laron Z, Pertzelan A, Arnon R, Dintzman M. Defect of human growth hormone receptors in the liver of two patients with Laron-type dwarfism. Isr J Med Sci 1984; 20:8–11.
22. Ymer SI, Herington AC. Evidence for the specific binding of growth hormone to a receptor-like protein in rabbit serum. Mol Cell Endocrinol 1985; 41:153–161.
23. Daughaday WH, Trivedi B. Absence of serum growth hormone binding protein in patients with growth hormone receptor deficiency (Laron dwarfism). Proc Natl Acad Sci USA 1987; 84:4636–4640.
24. Baumann G, Shaw MA, Winter RJ. Absence of plasma growth hormone-binding protein in Laron-type dwarfism. J Clin Endocrinol Metab 1987; 65:814–816.
25. Leung DW, Spencer SA, Cachianes G, Hammonds RG, Collins C, Henzel WJ, Barnard R, Waters MJ, Wood WI. Growth hormone receptor and serum binding protein: purification, cloning and expression. Nature 1987; 330:537–543.
26. Godowski PJ, Leung DW, Meacham LR, Galgani JP, Hellmiss R, Keret R, Rotwein PS, Parks JS, Laron Z, Wood WI. Characterization of the human growth hormone receptor gene and demonstration of a partial gene deletion in two patients with Laron-type dwarfism. Proc Natl Acad Sci USA 1989; 86:8083–8087.
27. Stallings-Mann ML, Ludwiczak RL, Klinger KW, Rottman F. Alternative splicing of exon 3 of the

human growth hormone receptor is the result of an unusual genetic polymorphism. Proc Natl Acad Sci USA 1996; 93:12394–12399.
28. Meacham LR, Brown MR, Murphy TL, Keret R, Silbergeld A, Laron Z, Parks JS. Characterization of a noncontiguous gene deletion of the growth hormone receptor in Laron's syndrome. J Clin Endocrinol Metab 1993; 77:1379–1383.
29. Amselem S, Sobrier M-L, Duquesnoy P, Rappaport R, Postel-Vinay M-C, Gourmelen M, Dallapiccola B, Goossens M. Recurrent nonsense mutations in the growth hormone receptor from patients with Laron dwarfism. J Clin Invest 1991; 87:1098–1102.
30. Berg MA, Argente J, Chernausek S, Gracia R, Guevara-Aguirre J, Hopp M, Perez-Jurado L, Rosenbloom AL, Toledo SPA, Francke U. Diverse growth hormone receptor gene mutations in Laron syndrome. Am J Hum Genet 1993; 52:998–1005.
31. Rosenbloom AL, Berg MA, Kasatkina EP, Volkova TN, Ckorobogatova VF, Sokolovskaya VN, Francke U. Severe growth hormone insensitivity (Laron syndrome) due to nonsense mutation of the GH receptor in brothers from Russia. J Pediatr Endocrinol Metab 1995; 8:159–165.
32. Sobrier M-L, Dastot F, Duquesnoy P, Kandemir N, Yordam N, Goossens M, Amselem S. Nine novel growth hormone receptor gene mutations in patients with Laron syndrome. J Clin Endocrinol Metab 1997; 82:435–437.
33. Berg MA, Peoples R, Perez-Juardo L, Guevara-Aguirre J, Rosenbloom AL, Laron Z, Milner RDG, Francke U. Receptor mutations and haplotypes in growth hormone receptor deficiency: a global survey and identification of the Ecuadorean E180 splice mutation in an Oriental Jewish patient. Acta Paediatr Suppl 1994; 399:112–114.
34. Berg MA, Argente J, Chernausek S, Gracia R, Guevara-Aguirre J, Hopp M, Perez-Jurado L, Rosenbloom A, Toledo SPA, Francke U. Diverse growth hormone receptor gene mutations in Laron syndrome. Am J Genet 1993; 52:998–1005.
35. Amselem S, Duquesnoy P, Duriez B, Dastot F, Sobrier M-L, Valleix S. Spectrum of growth hormone receptor mutations and associated haplotypes in Laron syndrome. Hum Mol Genet 1993; 4:355–359.
36. Otsuko T, Iwatani N, Kodama M, Sakakida M. Chichiri M, Jinno Y, Niikawa N, Miike T. The growth hormone receptor gene mutation of a Japanese patient with Laron syndrome. Jpn J Hum Genet 1997; 42:323–329.
37. Kaji H, Nose O, Tajiri H, Takahashi Y, IIDA K, Takahashi T, Okimura Y, Abe H, Chihara K. Novel compound heterozygous mutations of growth hormone (GH) receptor gene in a patient with GH insensitivity syndrome. J Clin Endocrinol Metab 1997; 82:3705–3709.
38. Amselem S, Duquesnoy P, Attree O, Novelli G, Bousnina S, Postel-Vinay M-C, Goossens MJ. Laron dwarfism and mutations of the growth hormone-receptor gene. N Engl J Med 1989; 321:989–995.
39. Walker JL, Crock PA, Behncken SN, Rowlinson SW, Nivholson LM, Boulton TJC, Walters M. A novel mutation affecting the interdomain link region of the growth hormone receptor in a Vietnamese girl, and response to long-term treatment with recombinant human insulin-like growth factor-I and luteinizing hormone-releasing hormone analogue. J Clin Endocrinol Metab 1998; 83:2554–2561.
40. Duquesnoy P, Sobrier M-L, Duriez B, Dastot F, Buchanan CR, Savage MO, Preece MA, Craescu CT, Blouquit Y, Goossens M, Amselem S. A single amino acid substitution in the exoplasmic domain of the human growth hormone (GH) receptor confers familial GH resistance (Laron syndrome) with positive GH-binding activity by abolishing receptor homodimerization. EMBO J 1994; 13:1386–1395.
41. Wojcik J, Berg MA, Esposito N, Geffner ME, Sakati N, Reiter EO, Dower S, Francke U, Postel-Vinay M-C, Finidori J. Four contiguous amino acid substitutions, identified in patients with Laron syndrome, differently affect the binding affinity and intracellular trafficking of the growth hormone receptor. J Clin Endocrinol Metab 1998; 83:4481–4489.
42. Berg MA, Guevara-Aguirre J, Rosenbloom AL, Rosenfeld RG, Francke U. Mutation creating a new splice site in the growth hormone receptor genes of 37 Ecuadorean patients with Laron syndrome. Hum Mutat 1992; 1:24–34.
43. Silbergeld A, Dastot F, Klinger B, Kanety H, Eshet R, Amselem S, Laron Z. Intronic mutation in the growth hormone (GH) receptor gene from a girl with Laron syndrome and extremely high serum GH binding protein: extended phenotypic study in a very large pedigree. J Pediatr Endocrinol Metab 1997; 10:265–274.
44. Baumbach L, Schiavi A, Bartlett R, Perera E, Day J, Brown MR, Stein S, Eidson M, Parks JS, Cleveland W. Clinical, biochemical, and molecular investigations of a genetic isolate of growth hormone insensitivity (Laron's syndrome). J Clin Endocrinol Metab 1997; 82:444–451.
45. Woods KA, Fraser NC, Postel-Vinay M-C, Savage MO, Clark AJL. A homozygous splice site mutation

affecting the intracellular domain of the growth hormone (GH) receptor resulting in Laron syndrome with elevated GH-binding protein. J Clin Endocrinol Metab 1996; 81:1686–1690.
46. Rosenbloom AL, Guevara-Aguirre J. Lessons from the genetics of Laron syndrome. Trends Endocrinol Metab 1998; 9:276–283.
47. Vesterhus P. Veksthormoninsensitivitet: Genetiske defekter og terapi ved Larons syndrom og liknende tilstander. Tidsskr Nor Loegeforen 1997; 117:948–951.
48. Maes M, Maiter M, Thissen JP, Underwood LE, Ketelslegers JM. Contributions of growth hormone receptor and postreceptor defects to growth hormone resistance in malnutrition. Trends Endocrinol Metab 1991; 2:92–97.
49. Hochberg Z, Phillip M, Youdim MBH, Amit T. Regulation of the growth hormone (GH) receptor and GH-binding protein by GH pulsatility. Metabolism 1993; 42:1617–1623.
50. Laursen EM, Juul A, Lanng S, Holby N, Koch C, Müller J, Skakkebaek NE. Diminished concentrations of insulin-like growth factor I in cystic fibrosis. Arch Dis Child 1995; 72:494–497.
51. Zamboni G, Dufillot D, Antonizaai F, Valentini R, Gendrel D, Tato L. Growth hormone-binding proteins and insulin-like growth factor-binding proteins in protein-energy malnutrition, before and after nutritional rehabilitation. Pediatr Res 1996; 39:410–414.
52. Menon RK, Arslanian S, May B, Cutfield WS, Sperling MA. Diminished growth hormone-binding protein in children with insulin-dependent diabetes mellitus. J Clin Endocrinol Metab 1992; 74:934–938.
53. Tönshoff B, Cronin MJ, Reichert M, Haffner D, Wingen A-M, Blum WF, Mehls O, the European Study Group for Nutritional Treatment of Chronic Renal Failure in Childhood, and members of the German Study Group for Growth Hormone Treatment in Chronic Renal Failure. Reduced concentration of serum growth hormone (GH)-binding protein in children with chronic renal failure: correlations with GH insensitivity. J Clin Endocrinol Metab 1997; 82:1007–1013.
54. Scacchi M, Pincelli AI, Caumo A, Tomasi P, Delitala G, Baldi G, Cavagnini F. Spontaneous nocturnal growth hormone secretion in anorexia nervosa. J Clin Endocrinol Metab 1997; 82:3225–3229.
55. Munoz MT, Barrios V, Pozo J, Argente J. Insulin-like growth factor I, its binding proteins 1 and 3, and growth hormone-binding protein in children and adolescents with insulin-dependent diabetes mellitus: clinical implications. Pediatr Res 1996; 39:992–998.
56. Hodson EM, Brown AS, Roy LP, Rosenberg AR. Insulin-like growth factor I, growth hormone-dependent insulin-like growth factor-binding protein and growth in children with chronic renal failure. Pediatr Nephrol 1992; 6:433–438.
57. Geffner ME, Bersch N, Bailey RC, Golde DW. Insulin-like growth factor I resistance in immortalized T cell lines from African Pygmies. J Clin Endocrinol Metab 1995; 80:3732–3738.
58. Siebler T, Lopaczynski W, Terry CL, Casella SJ, Munson P, DeLeon DD, Phang L, Blakemore KJ, McEvoy RC, Kelley RI, Nissley P. Insulin-like growth factor I receptor expression and function in fibroblasts from two patients with deletion of the distal long arm of chromosome 15. J Clin Endocrinol Metab 1995; 80:3447–3457.
59. Jain S, Golde DW, Bailey R, Geffner ME. Insulin-like growth factor-I resistance. Endocr Rev 1998; 19:625–646.
60. Lanes R, Plotnick LP, Spencer EM, Daughaday WH, Kowarski AA. Dwarfism associated with normal serum growth hormone and increased bioassayable, receptor-assayable and immunoassayable somatomedin. J Clin Endocrinol Metab 1980; 50:485–488.
61. Heath-Monnig E, Wohltmann HJ, Mills-Dunlap B, Daughaday WH. Measurement of insulin-like growth factor I (IGF-I) responsiveness of fibroblasts of children with short stature: identification of a patient with IGF-I resistance. J Clin Endocrinol Metab 1987; 64:501–507.
62. Momoi T, Yamanaka C, Kobayashi M, Haruta T, Sasaki H, Yorifuji T, Kaji M, Mikawa H. Short stature with normal growth hormone and elevated IGF-I. Eur J Pediatr 1992; 151:321–325.
63. Rosenbloom AL, Guevara-Aguirre J, Fielder PJ, Gargosky S, Rosenfeld RG, Berg MA, Francke U, Diamond Jr FB, Vaccarello MA. Laron syndrome (growth hormone receptor deficiency) in Ecuador: clinical and biochemical characteristics. Pediatr Adolesc Endocrinol 1992; 24:34–51.
64. Jorgensen JOL, Müller J, Moller J, Wolthers T, Vahl N, Juul A, Skakkebaek NE, Christiansen JS. Adult growth hormone deficiency. Horm Res 1994; 42:235–241.
65. Rosenbloom AL, Martinez V, Kranzler J, Bachrach LK, Rosenfeld RG, Guevara-Aguirre J. Natural history of growth hormone receptor deficiency. Acta Paediatr Suppl 1999; 428:153–156.
66. Rudling M, Norstedt G, Olivecrona H, Reihnér E, Gustafsson J-A, Angelin B. Importance of growth hormone for the induction of hepatic low density lipoprotein receptors. Proc Natl Acad Sci USA 1992; 89:6983–6987.

67. Rosenbloom AL, Guevara-Aguirre J, Rosenfeld RG, Pollock BH. Growth in growth hormone insensitivity. Trends Endocrinol Metab 1994; 5:296–303.
68. Laron Z, Lilos P, Klinger B. Growth curves for Laron syndrome. Arch Dis Child 1993; 68:768–770.
69. Crosnier H, Gourmelen M, Prëvot C, Rappaport R. Effects of nutrient intake on growth, insulin-like growth factors, and their binding proteins in a Laron-type dwarf. J Clin Endocrinol Metab 1993; 76:248–250.
70. Bachrach LK, Marcus R, Ott SM, Rosenbloom AL, Vasconez O, Martinez V, Martinez AL, Rosenfeld RG, Guevara-Aguirre J. Bone mineral, histomorphometry, and body composition in adults with growth hormone receptor deficiency. J Bone Miner Res 1998; 13:415–421.
71. Bandeira F, Camargo K, Caldas G, Rosenbloom AL, Stabler B, Underwood LE. Primary growth hormone insensitivity: case report. Arquiv Brasil Endocrinol Metabolismo 1997; 41:155–162.
72. Rosenbloom AL, Selman-Almonte A, Brown MR, Fisher DA, Baumbach L, Parks JS. Clinical and biochemical phenotype of familial anterior hypopituitarism from mutation of PROP1 gene. J Clin Endocrinol Metab 1999; 84:50–57.
73. Frankel JJ, Laron Z. Psychological aspects of pituitary insufficiency in children and adolescents with special reference to growth hormone. Isr J Med Sci 1968; 4:953–961.
74. Galatzer A, Aran O, Nagelberg N, Rubitzek J, Laron Z. Cognitive and psychosocial functioning of young adults with Laron syndrome. Pediatr Adolesc Endocrinol 1993; 24:53–60.
75. Woods KA, Dastot F, Preece MA, Clark AJL, Postel-Vinay M-C, Chatelain PG, Ranke MB, Rosenfeld RG, Amselem S, Savage MO. Extensive personal experience. Phenotype: genotype relationships in growth hormone insensitivity syndrome. J Clin Endocrinol Metab 1997; 82:3529–3535.
76. Guevara-Aguirre J, Rosenbloom AL. Psychosocial adaptation of Ecuadorian patients with growth hormone receptor deficiency/Laron syndrome. Pediatr Adolesc Endocrinol 1993; 24:61–64.
77. Kranzler J, Rosenbloom AL, Martinez V, Guevara-Aguirre J. Normal intelligence with severe IGF-I deficiency due to growth hormone receptor deficiency: a controlled study in a genetically homogenous population. J Clin Endocrinol Metab 1998; 83:1953–1958.
78. Zhou Y, Xu BC, Maheshwari HG, He H, Reed M, Lozkowski M, Okada S, Cataldo L, Coschigamo K, Wagner TE, Baumann G, Kopchick JJ. A mammalian model for the Laron syndrome produced by targeted disruption of the mouse growth hormone receptor/binding protein gene (the Laron mouse). Proc Natl Acad Sci USA 1997; 94:13215–13220.
79. Rosenfeld RG, Rosenbloom AL, Guevara-Aguirre J. Growth hormone (GH) insensitivity due to primary GH receptor deficiency. Endocr Rev 1994; 15:369.
80. Jansen M, van Schaik FMA, Ricker AT, Bullock B, Woods DE, Gabbay KH, Nussbaum AL, Sussenbach JS, Van den Brande JL. Sequence of cDNA encoding human insulin-like growth factor I precursor. Nature 1983; 255:306–312.
81. Guler HP, Zapf J, Froesch ER. Short-term metabolic effects and half-lives of intravenously administered insulin-like growth factors I in healthy adults. N Engl J Med 1987; 317:137–140.
82. Boulware SD, Tamborlane WV, Matthews LS, Sherwin RS. Diverse effects of insulin-like growth factor I on glucose, lipid, and amino acid metabolism. Am J Physiol 1992; 262:E130–E133.
83. Turkal JI, Keller U, Ninnis R, Vosmeer S, Stauffacher W. Effect of increasing doses of recombinant human insulin-like growth factor-I on glucose, lipid, and leucine metabolism in man. J Clin Endocrinol Metab 1992; 75:1186–1191.
84. Mauras N, Horber FF, Haymond MW. Low dose recombinant human insulin-like growth factor-I fails to affect protein anabolism but inhibits islet cell secretion in humans. J Clin Endocrinol Metab 1992; 75:1192–1197.
85. Usala A-L, Madigan T, Burguera B, Sinha MK, Caro FJ, Cunningham P, Powell JG, Butler PC. Brief report: Treatment of insulin-resistant diabetic ketoacidosis with insulin-like growth factor I in an adolescent with insulin-dependent diabetes. N Engl J Med 1992; 327:853–857.
86. Guler HP, Schmid C, Zapp J, Froesch ER. Effects of recombinant insulin-like growth factor I on insulin secretion and renal function in normal human subjects. Proc Natl Acad Sci USA 1989; 86:2868–2872.
87. Takano K, Hitzuka N, Shizume K, Asakawa K, Fukuda I, Demura H. Repeated sc administration of recombinant human insulin-like growth factor I (IGF-I) to human subjects for 7 days. Growth Regul 1991; 1:23–28.
88. Hussain MA, Studer K, Messmer EP, Froesch R. Treatment with insulin-like growth factor I alters capillary permeability in skin and retina. Diabetes 1995; 44:1209–1212.
89. Clemmons DR, Smith-Banks A, Underwood LE. Reversal of diet-induced catabolism by infusion of recombinant insulin-like growth factor-I in humans. J Clin Endocrinol Metab 1992; 75:234–238.

90. Young SC, Underwood LE, Celniker A, Clemmons DR. Effects of recombinant insulin-like growth factor-I (IGF-I) and growth hormone on serum IGF-binding proteins in calorically restricted adults. J Clin Endocrinol Metab 1992; 75:603–608.
91. Kupfer SR, Underwood LE, Baxter RC, Clemmons DR. Enhancement of the anabolic effects of growth hormone and insulin-like growth factor I by use of both agents simultaneously. J Clin Invest 1993; 91:391–396.
92. Miell JP, Taylor AM, Jones J, Buchanan CR, Rennie J, Sherwood R, Leicester R, Ross RJ. Administration of human recombinant insulin-like growth factor-I to patients following major gastrointestinal surgery. Clin Endocrinol 1992; 37:542–551.
93. Lieberman SA, Butterfield GE, Harrison D, Hoffman AR. Anabolic effects of recombinant insulin-like growth factor-I in cachectic patients with the acquired immunodeficiency syndrome. J Clin Endocrinol Metab 1994; 78:404–410.
94. Grinspoon S, Baum H, Lee K, Anderson E, Herzog D, Kilbanski A. Effects of short-term recombinant human insulin-like growth factor-I administration on bone turnover in osteopenic women with anorexia nervosa. J Clin Endocrinol Metab 1996; 81:3864–3870.
95. Thompson JL, Butterfield GE, Marcus R, Hintz RL, Van Loan M, Ghiron L, Hoffman AR. The effects of recombinant human insulin-like growth factor-I and growth hormone on body composition in elderly women. J Clin Endocrinol Metab 1995; 80:1845–1852.
96. Ghiron LJ, Thompson JL, Holloway L, Hintz RL, Butterfield GE, Hoffman AR, Marcus R. Effects of recombinant insulin-like growth factor-I and growth hormone on bone turnover in elderly women. J Bone Miner Res 1995; 10:1844–1852.
97. Thorén M, Wivall-Helleryd I-L, Hall K. Effects of recombinant human insulin-like growth factor I administration in adults with growth hormone deficiency. Acta Paediatr Suppl 1993; 388:45–48.
98. Thorén MC, Wivall-Helleryd I-L, Blum WF, Hall KE. Effects of repeated subcutaneous administration of recombinant human insulin-like growth factor I in adults with growth hormone deficiency. Eur J Endocrinol 1994; 131:33–40.
99. Laron Z, Erster B, Klinger B, Anin S. Effect of acute administration of insulin-like growth factor I in patients with Laron-type dwarfism. Lancet 1988; ii:1170–1172.
100. Laron Z, Klinger B, Silbergeld A, Lewin R, Erster B, Gil-Ad I. Intravenous administration of recombinant IGF-I lowers serum GHRH and TSH. Acta Endocrinol 1990; 123:378–382.
101. Klinger B, Garty M, Laron Z. Elimination characteristics of intravenously administered rIGF-I in Laron-type dwarfs. Div Pharmacol Ther 1990; 15:196–199.
102. Laron Z, Klinger B, Jensen LT, Erster B. Biochemical and hormonal changes induced by one week of administration of rIGF-I to patients with Laron type dwarfism. Clin Endocrinol 1991; 35:145–160.
103. Walker JL, Ginalska-Malinowska M, Romer TE, Pucilowska JB, Underwood LE. Effects of the infusion of insulin-like growth factor I in a child with growth hormone in sensitivity syndrome (Laron dwarfism). N Engl J Med 1991; 324:1483–1488.
104. Vaccarello MA, Diamond FB Jr, Guevara-Aguirre J, Rosenbloom AL, Fielder PJ, Gargosky S, Cohen P, Wilson K, Rosenfeld RG. Hormonal and metabolic effects and pharmacokinetics of recombinant insulin-like growth factor-I in growth hormone receptor deficiency/Laron syndrome. J Clin Endocrinol Metab 1993; 77:273–280.
105. Laron Z, Klinger B, Blum WF, Silbergeld A, Ranke MB. IGF binding protein 3 in patients with Laron type dwarfism: effect of exogenous rIGF-I. Clin Endocrinol 1992; 36:301–304.
106. Gargosky SE, Wilson KF, Fielder PJ, Vaccarello MA, Guevara-Aguirre J, Diamond FB Jr, Baxter RC, Rosenbloom AL, Rosenfeld RG. The composition and distribution of insulin-like growth factors (IGFs) and IGF-binding proteins (IGFBPs) in the serum of growth hormone receptor-deficient patients: effects of IGF-I therapy on IGFBP-3. J Clin Endocrinol Metab 1993; 77:1683–1689.
107. Lee PDK, Durham SK, Martinez V, Vasconez O, Powell DR, Guevara-Aguirre J. Kinetics of insulin-like growth factor (IGF) and IGF-binding protein responses to a single dose of growth hormone. J Clin Endocrinol Metab 1997; 82:2266–2274.
108. Daughaday WH, Rotwein P. Insulin-like growth factors I and II: peptide, messenger ribonucleic acid and gene structures, serum and tissue concentrations. Endocr Rev 1989; 10:68–91.
109. Laron Z, Anin S, Klipper-Aurbach Y, Klinger B. Effects of insulin-like growth factor on linear growth, head circumference, and body fat in patients with Laron-type dwarfism. Lancet 1992; 339:1258–1261.
110. Silbergeld A, Klinger B, Schwartz H, Laron Z. Serum prolactin in patients with Laron-type dwarfism: effect on insulin-like growth factor I. Horm Res 1992; 37:160–164.
111. Klinger B, Tonesco A, Anin S, Laron Z. Effect of insulin-like growth factor I on the thyroid axis in patients with Laron-type dwarfism and healthy subjects. Acta Endocrinol 1992; 127:515–519.

112. Walker JL, Van Wyk JJ, Underwood LE. Stimulation of statural growth by recombinant insulin-like growth factor I in a child with growth hormone insensitivity syndrome (Laron type). J Pediatr 1992; 121:641–646.
113. Heinrichs C, Vis HL, Bergmann P, Wilton P, Bourguignon JP. Effects of 17 months treatment using recombinant insulin-like growth factor-I in two children with growth hormone insensitivity (Laron) syndrome. Clin Endocrinol 1993; 38:647–651.
114. Ranke MB, Savage MO, Chatelain PG, Preece MA, Rosenfeld RG, Blum WF, Wilton P. Insulin-like growth factor I improves height in growth hormone insensitivity: two years' results. Horm Res 1995; 44:253–264.
115. Kržišnik C, Battelino T. Five year treatment with IGF-I of a patient with Laron syndrome in Slovenia (a follow-up report). J Pediatr Endocrinol Metab 1997; 10:443–447.
116. Martinez V, Vasconez O, Martinez AL, Moreno Z, Davila N, Rosenbloom AL, Diamond FB, Bachrach L, Rosenfeld RG, Guevara-Aguirre J. Body changes in adolescent patients with growth hormone receptor deficiency receiving recombinant human insulin-like growth factor I and luteinizing hormone-releasing hormone analogue: preliminary results. Acta Pediatr Suppl 1994; 399:133–136.
117. Wilson KF, Fielder PJ, Guevara-Aguirre J, Cohen P, Vasconez O, Martinez V, Martinez A, Rosenbloom AL, Rosenfeld RG. Long-term effects of insulin-like growth factor (IGF)-I treatment on serum IGFs and IGF binding proteins in adolescent patients with growth hormone receptor deficiency. Clin Endocrinol 1995; 42:399–407.
118. Vasconez O, Martinez V, Martinez AL, Hidalgo F, Diamond FB, Rosenbloom AL, Rosenfeld RG, Guevara-Aguirre J. Heart rate increases in patients with growth hormone receptor deficiency treated with insulin-like growth factor I. Acta Paediatr Suppl 1994; 399:137–139.
119. Duerr RL, McKirnan MD, Gim RD, Clark RG, Chien KR, Ross J Jr. Cardiovascular effects of insulin-like growth factor-I and growth hormone in chronic left ventricular failure in the rat. Circulation 1996; 93:2188–2196.
120. Donath MY, Sütsch G, Yan X-W, Piva B, Brunner HP, Glatz Y, Zapf J, Follath F, Froesch ER, Kiowski W. Acute cardiovascular effects of insulin-like growth factor I in patients with chronic heart failure. J Clin Endocrinol Metab 1998; 83:3177–3183.
121. Klinger B, Anin S, Sibergeld A, Eshet R, Laron Z. Development of hyperandrogenism during treatment with insulin-like growth factor-I (IGF-I) in female patients with Laron syndrome. Clin Endocrinol 1998; 48:81–87.
122. Guevara-Aguirre J, Vasconez O, Martinez V, Martinez AL, Rosenbloom AL, Diamond FB Jr, Gargosky SE, Nonoshita L, Rosenfeld RG. A randomized, double-blind, placebo controlled trial on safety and efficacy of recombinant human insulin-like growth factor-I in children with growth hormone receptor deficiency. J Clin Endocrinol Metab 1995; 80:1393–1398.
123. Guevara-Aguirre J, Rosenbloom AL, Vasconez O, Martinez V, Gargosky SE, Rosenfeld RG. Two year treatment of growth hormone receptor deficiency (GHRD) with recombinant insulin-like growth factor-I in 22 children: comparison of two dosage levels and to GH treated GH deficiency. J Clin Endocrinol Metab 1997; 82:629–633.
124. Klinger B, Laron Z. Three year IGF-I treatment of children with Laron syndrome. J Pediatr Endocrinol Metab 1995; 8:149–158.
125. Backeljauw PF, Underwood LE, The GHIS Collaborative Group. Prolonged treatment with recombinant insulin-like growth factor-I in children with growth hormone insensitivity syndrome—a clinical research center study. J Clin Endocrinol Metab 1996; 81:3312–3317.
126. Backeljauw PF, Kissoondial A, Underwood LE, Simmons KE. Effects of 4 years treatment with recombinant human insulin-like growth factor-I (rh IGF-I) on craniofacial growth in children with growth hormone insensitivity syndrome (GHIS). Horm Res 1997; 48 (Suppl 2):40, Abstract 243.
127. Backeljauw PF, Underwood LE, The GHIS Collaborative Group. Prolonged treatment with recombinant insulin-like growth factor-I (rh IGF-I) in children with growth hormone insensitivity syndrome (GHIS): 4 year update. Horm Res 1997; 48 (Suppl 2):15, Abstract 58.
128. Oliveira SB, Arnhold IJP, Phillips JA III, Mendonca BB. Treatment with recombinant IGF-I for patients with growth hormone (GH) gene deletions. In: Proceedings of the 10th International Congress on Endocrinology (ICE '96) June 12–15, 1996, San Francisco, CA, Abstract 3-26.
129. Leonard J, Samuels M, Cotterill AM, Savage MO. Effects of recombinant insulin-like growth factor I on craniofacial morphology in growth hormone insensitivity. Acta Paediatr Suppl 1994; 399:140–141.
130. Bjerknes R, Vesterhus P, Aarskog D. Increased neutrophil insulin-like growth factor-I (IGF-I) receptor expression and IGF-I-induced functional capacity in patients with untreated Laron syndrome. Eur J Endocrinol 1997; 136:92–95.

131. Diamond F, Martinez V, Guevara-Aguirre J, Vasconez O, Martinez A, Rosenbloom A, Good R, Day NB. Immune function in patients with growth hormone receptor deficiency (GHRD). In: Proceedings of the Endocrine Society 78th Annual Meeting, San Francisco, CA, Vol. 1, 276.
132. Le Roith D, Blakesley VA. The yin and the yang of the IGF system: immunological manifestations of GH resistance. Soc Eur J Endocrinol 1997; 136:33–34.
133. Laron Z, Klinger B. IGF-I treatment of adult patients with Laron syndrome: preliminary results. Clin Endocrinol 1994; 44:631–638.
134. Laron Z, Wsang XL, Klinger B, Sittergeld A, Wilcken DE. Insulin-like growth factor-I decreases serum lipoprotein (a) during long-term treatment of patients with Laron syndrome. Metabolism 1996; 45:1265–1266.
135. Maheshwari HG, Silverman BL, Dupuis J, Baumann G. Phenotype and genetic analysis of a syndrome caused by an inactivating mutation in the growth hormone-releasing hormone receptor: dwarfism of Sindh. J Clin Endocrinol Metab 1998; 83:4065–4074.

Index

A

B

C

D

E

F

G

K

U, V, W

ABOUT THE EDITORS

Ron G. Rosenfeld, MD is Professor and Chairman of the Department of Pediatrics and Professor of Cell and Developmental Biology at Oregon Health Sciences University, and Physician-in-Chief of Doernbecher Children's Hospital. He is a graduate of Stanford University School of Medicine, where he also completed a postdoctoral fellowship in pediatric endocrinology. In 1980, he joined the faculty of the Department of Pediatrics at Stanford University, where he was appointed Associate Professor in 1985 and Professor of Pediatrics in 1989.

Dr. Rosenfeld's research has resulted in more than 450 published papers and edited books. He is on the editorial boards of *Growth Hormone and IGF Research, Molecular Genetics and Metabolism, Journal of Pediatric Endocrinology and Metabolism, Clinical Pediatric Endocrinology, Growth and Growth Factors,* and *The Endocrinologist.* He has chaired the Endocrinology Study Section at NIH and chairs the Research Advisory Board for the March of Dimes Birth Defects Foundation. Dr. Rosenfeld's extensive experience in IGF research, growth hormone, and sexual development has led to a listing in the first and second editions of Best Doctors in America. He has recently completed a term as President of Lawson Wilkins Pediatric Endocrine Society.

Charles T. Roberts, Jr., PhD is Professor and Associate Chair for Research in the Department of Pediatrics and Professor of Cell and Developmental Biology at Oregon Health Sciences University, and director of the Doernbecher Pediatrics Research Laboratories. Dr. Roberts received a Ph.D. in biochemistry and molecular biology from the Department of Biological Sciences at the University of California at Santa Barbara (UCSB) in 1975. From 1975 to 1977, he was an American Cancer Society postdoctoral fellow at UCSB and, from 1977 to 1984, was an Assistant Research Professor. In 1984, Dr. Roberts moved to the Diabetes Branch of the NIDDK in Bethesda, MD, as a Special Expert. At that time, he initiated his studies on the insulin-like growth factor (IGF) system. Dr. Roberts was awarded tenure as a Research Biologist in 1989 and also served as Associate Chief of the Section on Molecular and Cellular Physiology in the Diabetes Branch from 1989-1994. In 1994, he moved to his present position at Oregon Health Sciences University. Dr. Roberts' research has encompassed the areas of IGF-1 and IGF-1 receptor gene expression, and structure-function analysis of the IGF-1 and insulin receptors, and has resulted in more than 200 publications. His current research focuses on control of IGF action in malignancy.